Physical Constants

Quantity	Symbol	
Atomic mass unit	u	1.6[illegible]
Avogadro's number	N_a	6.023×10^{23} molecules/mole
Boltzmann's constant	k	1.38×10^{-23} J/K, or 8.61×10^{-5} eV/K, or 8.31 J/(mole · K)
Coulomb's constant	k, or $\frac{1}{4\pi\epsilon_0}$	8.99×10^{9} N · m²/C²
Electron		
charge	e	1.60×10^{-19} C
mass	m_e	9.11×10^{-31} kg
Gravitational constant	G	6.67×10^{-11} N m^2/kg^2, or 3.44×10^{-8} lb $ft^2/slug^2$
Neutron mass	m_n	1.674×10^{-27} kg
Permeability of free space	μ_0	$4\pi \times 10^{-7}$ N/A^2 or Wb/A
Permittivity of free space	ϵ_0	8.85×10^{-12} $C^2/(N \cdot m^2)$
Planck's constant	h	6.63×10^{-34} J · s
Proton mass	m_p	1.673×10^{-27} kg
Speed of light	c	3.00×10^{8} m/s, or 1.86×10^{5} mi/s
Stefan-Boltzmann constant	σ	5.67×10^{-8} $J/(m^2 \cdot s \cdot K^4)$
Universal gas constant	R	8.13 J/(mole · K), or 0.0821 atm · 1/(mole · K), or 3.40 ft · lb/(mole · R)

Commonly Used Data

Quantity	Value
Absolute zero	0 K = −273° C, or 0 R = −460°F
Acceleration of gravity on earth's surface	**g** = 32.2 ft/s² = 9.81 m/s²
Atmosphere of pressure (STP)*	1 atm = 14.7 lb/in² = 1.013×10^{5} Pa = 1013 millibar = 76.0 in · Hg
Density of air (STP)*	$\rho_{air} = 2.50 \times 10^{-3}$ slug/ft³ = 1.3 kg/m³
Density of fresh water	ρ_{water} = 62.4 lb/ft³ = 1.94 slug/ft³ = 1000 kg/m³ = 1.00 g/cm³
Earth	
mass	$M_e = 5.98 \times 10^{24}$ kg
orbit (average)	$R_0 = 1.49 \times 10^{11}$ m $= 9.25 \times 10^{7}$ mi
radius	$R_e = 6.38 \times 10^{6}$ m $= 3.98 \times 10^{3}$ mi
One molar volume	V_0 = 22.4 l = 1 gram-equivalent mass (GEM) = 1 mole = 6.023×10^{23} molecules

* Standard Temperature and Pressure.

TECHNICAL PHYSICS

ERWIN
SELLECK

DELMAR PUBLISHERS INC.®

NOTICE TO THE READER

Publisher does not warrant or guarantee any of the products described herein or perform any independent analysis in connection with any of the product information contained herein. Publisher does not assume, and expressly disclaims, any obligation to obtain and include information other than that provided to it by the manufacturer.

The reader is expressly warned to consider and adopt all safety precautions that might be indicated by the activities described herein and to avoid all potential hazards. By following the instructions contained herein, the reader willingly assumes all risks in connection with such instructions.

The publisher makes no representations or warranties of any kind, including but not limited to, the warranties of fitness for particular purpose or merchantability, nor are any such representations implied with respect to the material set forth herein, and the publisher takes no responsibility with respect to such material. The publisher shall not be liable for any special, consequential or exemplary damages resulting, in whole or in part, from the readers' use of, or reliance upon, this material.

Delmar Staff

Executive Editor: David C. Gordon
Developmental Editor: Mary Ormsbee
Project Supervisor: Marlene McHugh Pratt
Project Editor: Mary Robinson
Production Supervisor: Karen Seebald
Art Supervisor: John Lent
Design Supervisor: Susan C. Mathews

For information, address Delmar Publishers Inc.
3 Columbia Circle Drive, PO Box 15-015
Albany, NY 12212-5015

Printed in the United States of America
Published simultaneously in Canada
by Nelson Canada,
a division of The Thomson Corporation

10 9 8 7 6 5 4 3 2

Selleck, Erwin.
Technical physics / Erwin Selleck.
p. cm.
Includes index.
ISBN 0-8273-4607-7
1. Physics. I. Title.
QC23.S445 1991
530—dc20

90-43240
CIP

CONTENTS

PREFACE

This introductory textbook is written for students majoring in technical fields. The material is presented in simple language for clarity and easy comprehension. Students learning from this text should be able to use intermediate algebra and should have a very basic working relationship with trigonometry. The level of math needed to master the material in this text is reviewed in Appendix A.

OBJECTIVES

The broad aim of this text is threefold. All three objectives are equally important.

1. **Principles of Physics:** Physics builds its structure from a small foundation of basic ideas. These ideas are presented in a simple, understandable fashion.
2. **Patterns:** There are many recurrent patterns in the world around us. This textbook points out some of these patterns and encourages the student to seek out additional patterns.
3. **Problem solving:** One of the great uses of physics is the ability to predict events. Each chapter contains several example problems. A wide selection of problems varying in type and difficulty is presented at the end of each chapter. Problem-solving procedures are emphasized rather than memorization.

TEXTBOOK STRUCTURE

Main sequence ideas are presented at the beginning of each chapter; applications are placed toward the end of the chapter. This separation reduces the confusion between mainstream physics and its applications. The instructor can easily choose which applications to include. A summary of important ideas and equations is located at the end of each chapter. A section of exercises is followed by an assortment of problems. More difficult problems are noted with asterisks.

The order of presentation of the material is a bit different from most texts. Rotational motion is introduced early. Rotational dynamics is integrated in the linear dynamics chapters rather than isolating it in separate chapters in the middle of the text. This organization lets the student become familiar with rotation throughout the semester rather than in a brief, concentrated topic toward the end of the semester. This early introduction will benefit technical students who must work with power equipment and electric motors throughout the semester.

Waves are related to vibrational motion and follow the chapter on harmonic motion. In many texts, optics follow electricity and magnetism. Here, we have placed optics in a position to emphasize the wave properties of light.

The chapters can be broken into the following broad topics.

Chapters 1–3: Measurement and mathematics skills
Chapters 4 and 5: Kinematics
Chapters 6–14: Classical dynamics
Chapters 15, 16, 18, and 19: Properties of materials
Chapters 17, 20, 21, and 22: Heat and thermodynamics
Chapters 23–27: Vibrational motion, waves, and optics
Chapters 29–35: Electricity and magnetism
Chapter 36: Nuclear physics

SOME ADDITIONAL FEATURES

Graphics

Because a successful technician must be able to read graphical information, an emphasis has been placed on graphics. In Chapter 3, graphs are related to proportion and power laws. There are several opportunities to use graphical information throughout the text.

Inductive Reasoning

For the most part we learn by drawing generalizing rules from the things we see in the world around us. This book emphasizes pattern recognition. Whenever it is reasonably appropriate in this text, physical laws are developed from specific examples using induction. The name of a law or principle is given only after the underlying pattern is developed. This process is most frequently used in the classical mechanics chapters of the text.

The Real World

Sometimes a student leaves a physics course convinced the world is filled with frictionless inclined planes and weightless beams. It's difficult to transfer physics to a technical course if physics remains abstract. There are many real-world problems in this text, either as example problems or assigned problems. This book also includes many more up-to-date applications than the average introductory text. You'll find topics such as smart buildings, sedimentation of atmospheric pollutants, optical fibers, body mechanics, wind turbines, power trains, R values, the centrifugal clutch, ultrasonics, the structure of electric cells, the greenhouse effect, how to get more miles per gallon from your car, and much more.

Problems and Exercises

The end-of-the-chapter exercises and problems are listed according to increasing difficulty. Asterisks indicate challenge problems.

Calculators

Some example problems contain instructions on using some of the less familiar keys found on the more popular scientific pocket calculators. A minicourse in using your calculator follows the preface material.

Vector Notation

Boldface is used to indicate vectors with more than one dimension. Usually boldface is dropped for one dimensional vectors and components; + and − signs are used to indicate direction in these situations. This is consistent with most introductory physics texts.

Applications Features

Each chapter has a special applications feature. These features describe non-mathematical applications for fun. How does a cat land on its feet? Why do soda cans fizz? How does fire walking work? What would King Kong really look like? These are samples of the questions answered in the application boxes.

Historical Features

Each chapter has its own historical time line. These historical notes are keyed to the ideas in the chapter. They give a flavor of what was going on when specific discoveries were made.

Scattered throughout the book you will also find some boxed historical articles. Discoveries in science usually are based on a sequence of ideas. The history articles explore such concepts as the development of time measurement, the growth and demise of the caloric theory, impetus and inertia, and the atom.

ACKNOWLEDGMENTS

I wish to thank all the people who have reviewed the manuscript and made many helpful and useful criticisms. I particularly would like to thank Paul Boettcher for his assistance and for preparing some of the solutions to the assigned problems.

Cornelius J. Noonan
Monroe Community College

Gerald B. Ploen
Northeast Wisconsin Technical College

Dr. Albert Stwertka
U.S. Merchant Marine Academy

Carolyn R. Mallory
La Pierce College

John S. Splett
Erie Community College

Bill Harris
Mountain Empire Community College

Robert W. Eshelman
Henry Ford Community College

Armen S. Casparian
Wentworth Institute of Technology

Paul Feldker
St. Louis Community College

James H. Abbott
Thornton Community College

Much time is taken from an author's everyday life when a book like this is written. The drippy upstairs faucet goes unrepaired, ball games are missed, meals are late, conversations are lost. For these reasons I'd like to thank my family for their forbearance—Carolyn, Sonya, Clay, Charlene, Inky Pooh, and Fella.

USING SCIENTIFIC CALCULATORS

If you don't already have a calculator, you'll want to get one for this physics course. You will want a scientific calculator, one with trigonometric functions and logarithmic functions. When you buy your calculator, a manual will come with it.

Keep the manual.

We'll discuss some of the functions on the calculator. Please, realize different brands and models of calculators have different ways of labeling the keys and different ways of operating. Keep the manual handy as you read this.

Turn your calculator on.

Examine the keyboard. Most calculators have an [=] key. A few, such as a Hewlett-Packard [HP] calculator, have an [ENTER] key instead. A Hewlett-Packard uses inverse Polish notation. The way numbers are entered is a bit different from other calculators. We'll show the HP operations to the right on the following exercises.

Keep your manual handy.

First let's do some arithmetic calculations. Assume we want to multiply 7 by 12.

ARITHMETIC CALCULATIONS

Most calculators	HP
Enter: 7	Enter: 7
Press: [×]	Press: [ENTER]
Enter: 12	Enter: 12
Press: [=]	Press: [×]

Your calculator should display: 84

CHAINING OPERATIONS

Often we want to do several operations one after another. Let's do this calculation:

$$(7 \times 12)/3$$

Most calculators	HP
Enter: 7	Enter: 7
Press: [×]	Press: [ENTER]
Enter: 12	Enter: 12
Press: [=]	Press: [×]
Press: [÷]	Press: [ENTER]
Enter: 3	Enter: 3
Press: [=]	Press: [÷]

Your calculator should display: 28

SECOND FUNCTIONS—LOG AND 10^x

Your calculator probably has two sets of symbols. One set of symbols is on the keys. The second set of symbols is on the board itself, just above the keys. Look for a key labeled [2ndF] or [INV] or [ARC]. Different brands of calculators use different notations. This key gives you access to the operations listed on the board above the keys. If you press [2ndF], the next key you punch will tell the calculator to use the operation above the key.

We'll take the logarithm of 1000. (See Section 1.5 for a discussion of logarithms.) When we take the logarithm we're asking, "1000 is equal to 10 to what power?"

Most calculators	HP
Enter: 1000	Enter: 1000
Press: [log]	Press: [ENTER]
	Press: [log]

Your calculator should display: 3

Now let's take the antilogarithm or inverse function. When we take the antilogarithm we're asking the question, "What number is 10 raised to the 3 power?" Notice the function [10^x] is above the [log] key. Make sure the 3 is still displayed on the calculator from the last calculation.

Most calculators	HP
Press: [2ndF] or [INV] or [ARC]	Press: [ENTER]
Press: [log]	Press: [2ndF] or [INV] or [ARC]
	Press: [log]

Your computer should display: 1000

RECIPROCAL [1/X]

The product of a number by its reciprocal is 1. x (1/x) = 1. When we multiply by the reciprocal of a number this gives us the same answer as if we had divided by the number. For example, the reciprocal of 2 is 1/2. If we multiply 4 by 1/2 the answer is 2. Suppose we plan to divide 27.6 by 12.3. We can do it this way. [NOTE: The function [1/x] is a second function on some calculators. You may have to use a [2ndF] key to get to this operation.]

Most calculators	HP
Enter: 12.3	Enter: 12.3
Press: [1/x]	Press: [ENTER]
Press: [×]	Press: [1/x]
Enter: 27.6	Enter: 27.6
Press: [=]	Press: [×]

The computer display should be: 2.243902439

You can use the key [y^x] to find the power or root of any number. Find 12^3. Find the fourth root of 81.

Most calculators	12^3	HP
Enter: 12		Enter: 12
Press: [y^x]		Press: [ENTER]
Enter: 3		Enter: 3
Press: [=]		Press: [y^x]

The display should be: 1728

	$(81)^{1/4}$	
Enter: 81		Enter: 81
Press: [y^x]		Press: [ENTER]
Enter: 0.25		Enter: 0.25
Press: [=]		Press: [y^x]

The display should be: 3

USING MEMORY

Most calculators have three memory keys. Memory is just a place we can store a number temporarily. You may have to check your manual to see how these keys are labeled. Here are the most common labels.

[×→M] or [M] This stores the number on display in memory. It erases anything already in memory.

[×←M] or [RM] Returns the number stored in memory.

[M+] This adds the number on display to what is already stored in memory.

Here's an example of using memory.
Calculate $(23 \times 3.2) - (4.9 \times 3.2^2)$

Most calculators	HP
Enter: 23	Enter: 23
Press: [×]	Press: [ENTER]
Enter: 3.2	Enter: 3.2
Press: [=]	Press: [=]

The display is: 73.6. We'll store this in memory while we calculate the other term.

Most calculators	HP
Press: [×→] or [M]	Press: [×→M] or [M]
Enter: 4.9	Enter: 4.9
Press: [×]	Press: [ENTER]
Enter: 3.2	Enter: 3.2
Press: [$\times^2$] This may be a [2ndF]	Press: [$\times^2$]
Press: [=]	Press: [×]
Press: [×/−] Changes sign	Press: [+/−]

The display is: −50.176. We have two ways to proceed. We can add the number in memory to the display or we can add the display to memory.

Most calculators	HP
Press: [+]	Press: [ENTER]
Press: [RM] or [×←M]	Press: [RM] or [×←M]
Press: [=]	Press: [×]

or

Most calculators	HP
Press: [M+]	Press: [M+]
Press: [RM] or [×←M]	Press: [RM] or [×←M]

The display is: 23.424

FIXED POINT AND EXPONENTIAL (SCIENTIFIC) NOTATION

By fixed point we mean a number is in common decimal notation. The number 2302.4 is in fixed point notation. We can write the same number as a number between 1 and 10 multiplied by a power of 10. (Scientific notation is developed more thoroughly in Section 1.4.) In exponential or scientific notation the number is 2.3024×10^3.

Most calculators have a key to convert the display on the calculator from fixed point to scientific notation. The key is labeled [F↔E] or [F↔S]. Some calculators have a key labeled [FES]. This key converts between fixed point, engineering and scientific notation. Convert 230000 to scientific notation.

Enter: 230000
Press: [F↔S] or [F↔E]

Enter: 230000
Press: [ENTER]
Press: [F↔S] or [F↔E]

The display is 2.3 05. The last three places on the display indicate the power of 10. We read this as 2.3×10^5.

If you have a [FES] key, the key has to be pressed twice to get the scientific notation.

Enter: 230000
Press: [FES]
Press: [FES]

The display is 2.3 05.

We can enter numbers into the computer using scientific notation. Your calculator has a key labeled [EX] or [EE] or [EXP]. This is used to enter the power of 10.

$$2.3 \times 10^5$$

Enter: 2.3
Press: [EE] or [EX] or [EXP]
Enter: 5

Enter: 2.3
Press: [EE] or etc.
Enter: 5
Press: [ENTER]

The display is: 2.3 05

Numbers less than 1 have a negative exponent. $0.0014 = 1.4 \times 10^{-3}$. Enter 1.6×10^{-19}.

Enter: 1.6
Press: [EE] or [EX] or [EXP]
Enter: 19
Press: [+/−]

Enter: 1.6
Press: [EE] or etc.
Enter: 19
Press: [+/−]
Press: [ENTER]

The display is: 1.6 − 19.

TRIGONOMETRIC FUNCTIONS: SEE APPENDIX A

This minicourse gives you an idea of some of the things you can do with your calculator. Trigonometric functions are discussed in Appendix A, Section 2. There are other functions in your calculator that are described in your manual.

In the early chapters of this book some of the example problems remind you how to use your calculator.

CAUTION: Your calculator is not perfect. Different calculators use different methods of calculating values and use different round-off routines. Try this calculation. Enter the number 1.0000001. There are 6 zeros in the number. Now press [$\times^2$] 27 times. The correct result is 674,530.47. What do you get on your calculator?

SUMMARY OF SOME SELECT KEYS

Key	Function
[+]	Addition
[−]	Subtraction
[×]	Multiplication
[÷]	Division
[+/−]	Change sign
[log]	Gives the base 10 logarithm of number on display.
[10^x]	Finds the antilogarithm of the number on display.
[1/x]	Finds the reciprocal of the number on display.
[x^2]	Squares number on display.
[y^x]	Raises y to the power x. Gives powers and roots.
[EXP] [EE]	Enters exponential part of scientific notation.
[ENTER]	Enters first number on Hewlett-Packard calculators.
[FS] [FE] [FES]	Converts between fixed point and exponential notation.
[DRG]	Converts angles among degrees, radians, and grads.
[SIN]	Gives the sine of an angle.
[SIN^{-1}]	Gives the angle corresponding to the sine function.
[M] [x→M]	Moves number on display to memory.
[RM] [x←M]	Returns number from memory.
[M+]	Adds number on display to memory.

Chapter 1

AN INTRODUCTION

In 1623, two scientists invented the slide rule. Technicians used slide rules to make calculations until electronic calculators replaced them in the 1960s. As you read this text, you will see how scientists observe patterns in the physical world and help shape our understanding of physics. (Photograph courtesy of Hewlett-Packard Company.)

OBJECTIVES

In this chapter you will learn to:

- use the scientific process
- answer the question, "What is physics?"
- learn to recognize patterns in order to reduce the material you need to memorize
- write numbers in scientific notation and convert between scientific notation and decimal notation
- find base 10 logarithms and find natural logarithms using a hand calculator
- solve logarithmic equations

Time Line for Chapter 1

2900 BC	Sumerians develop a set of symbols to represent word syllables. First known school room established by Sumerians.
2400 BC	Sumerians develop positional notation in writing numbers.
2000 BC	Mesopotamian culture solves the quadratic equation.
1900 BC	Mesopotamian mathematicians discover the Pythagorean Theorem.
1750 BC	Egyptian knowledge of geometry summarized on papyrus.
1350 BC	Decimal numerals used in China.
876 BC	First known use of zero in number notation.
190	Chinese mathematicians use powers of 10 to express numbers.
1253	The decimal system is introduced to England.
1614	John Napier publishes common (base 10) logarithm tables along with instructions on their use.
1623	Wilhelm Schickardt invents a slide rule.
1668	Nicklaus Kauffman publishes calculations using natural (base e) logarithms.

In this chapter we will try to define what we mean by science and what we mean by physics. We also will begin to look for patterns. When we see patterns and connections tying ideas together there is less for us to memorize.

Before we get very far into this course, you may want to look at Appendix D. It contains some hints concerning study skills.

1.1 THE SCIENTIFIC PROCESS

Many people think science is a group of facts. Other people believe it is a collection of ideas. Both views are wrong, although the second view may be a bit closer to the truth. It would be better to describe science as a process that helps us look for patterns that we can use to predict the way things in our world behave. Here is the process.

Step 1: **Look for patterns.** You may have wads of information on computer printout paper or you may have only a few observations of naturally occurring events. Whatever your source of data, look for a cause-and-effect relationship.

Example: You are at a basketball game. You notice the ball bounces higher when it falls from a greater height.

Step 2: **Explain what causes the pattern.** This is known as making a **hypothesis** or theory that may explain why something is happening.

Example: You have the idea that the height of the bounce of a ball is directly proportional to its initial height.

Step 3: **Check the idea with the real world.**

A. Does the hypothesis fit all the data you have collected? If there are exceptions to the rule, you will have to change the idea.

B. Experiment! A scientific hypothesis is useful only if it can predict what is going to happen.

C. Share your hypothesis with other scientists. If there is something wrong with it, they will let you know.

Example: At another basketball game you notice that the height of the bounce can vary depending on whether a player throws the ball up or down. You may decide to change the hypothesis to apply only to *dropped* balls.

Step 4: **Perform a careful experiment.**

Example: You measure the effect caused by dropping a ball from different heights. You have noticed that hard basketballs bounce differently from soft basketballs. You decide to drop a single basketball repeatedly from different heights. You also suspect that a basketball would bounce differently on a sheet of jello or on a shag rug than it would on a hardwood floor. You decide to use the same area of a gym floor for your experiment. You do this to make sure that changes of the bounce are caused only by a change in the ball's initial height. This is called **controlling variables.**

Measurements show you that there is a mathematical relationship between the initial height of the basketball and the height of its bounce. When the initial height is doubled, the bounce doubles in height. When the initial height is cut in half, the ball bounces only half as high as it did before. This shows us that the bounce height is directly proportional to the initial height. This can be written mathematically.

Let the symbol B equal the bounce, H equal the initial height, and R equal the constant ratio of bounce divided by initial height. In the language of algebra, your idea looks like this:

$$B = R\,H$$

Step 5: **Share your idea with the rest of the scientific community.** Skepticism and criticism are part of the scientific process.

Example: Someone questions whether your result with basketballs is also true for baseballs, snowballs, crystal balls, footballs, or other balls. More experiments are needed to find out how well your theory applies to other kinds of balls. After a period of time, people find the kinds of systems for which your theory gives good predictions.

Other scientists become interested in the other variables of your bouncing ball problem. One group becomes interested in the bouncing surface. They bounce balls on concrete floors, wrestling mats, and icy highways. Another group measures the bounce of a basketball as a function of the amount of air in the ball.

Still other scientists think they can improve your theory by measuring the speed of the ball before and after collision with photocells. They think speed can be used to get the height. They experimentally verify their theory. Their improved theory can predict the bounce when a ball is thrown as well as dropped.

A scientist who plays billiards reads the reports of all these different activities. The scientist applies all the ideas to two-dimensional collisions of balls rolling on a felt surface. He adds and tests some additional ideas. After many years of research, the billiard player publishes a book entitled *The Dynamic Collision Theory of Eight Ball.*

Science *is a process of looking for new patterns in the world around us and creating hypotheses* (*theories*), *supported by experience, that explain those patterns.*

A Classy People

By 5000 BC, the delta region of the Tigris and Euphrates rivers changed from a marsh to solid, fertile ground. A people known as the Sumerians settled this delta land. The Sumerians were highly intelligent. They developed new ideas in mathematics, technologies, and law that eventually diffused westward to the countries bordering the Mediterranean Sea and eastward into India. Some of the Sumerian achievements in the field of mathematics are shown in the time line at the beginning of this chapter.

Eventually Sumer merged with the country Akkadia further up the river valley to form the core of what is called the Mesopotamian culture.

Early civilizations used pictures as their written language. A picture represented one idea or one word. Thousands of pictures were needed to make a language. Only a small group of people called scribes memorized enough of the pictures to write and read this language effectively.

The Sumerian genius overcame this symbolic stumbling block. Around 3000 BC, they invented a language with symbols representing syllables instead of whole words. This reduced the number of symbols needed to write the language from thousands to only 700. This less cumbersome manner of writing made literacy a goal that could be achieved by many rather than a select few.

This brought on yet another invention, the classroom. What is thought to be the earliest classroom has been unearthed in Mesopotamia.

When we find something that we cannot explain with old ideas, we use ingenuity and the ideas of others to make up a new explanation. We test this new theory with experiments. Careful measurement, observation, and the use of mathematical tools guide the experiments. The ideas are then presented to the scientific community for evaluation and application. If the idea is good, it becomes a building block for newer ideas in science and technology.

1.2 PHYSICS

Physics *is a science that studies the basic rules governing our universe.*

Physics is the most fundamental of sciences. The branch of physics that explains why things move is called dynamics. Atomic and nuclear physics explain the composition of matter. Thermodynamics explains the behavior of heat, temperature, and groups of many small particles such as gas molecules. The nature of waves is explained in optics. The study of electricity, magnetism, and gravity explores the behavior of things that can interact without touching.

Physics is used by all the other physical sciences; chemistry, geology, meteorology, and astronomy all apply physics. Life sciences use physical principles as well. Environmental science uses ideas of energy flow through the food chain. The understanding of the circulation system requires the understanding of fluid dynamics.

Many technologies are direct applications of physics. People in heating, ventilation, and refrigeration technologies must understand thermodynamics and the behavior of fluids. Civil engineers, who design roads and dams, require an understanding of the equilibrium of vector forces. Electrical technicians need to understand the concepts involved with electricity and magnetism.

As we proceed through this book, we will see many real-world applications of physics.

1.3 RECOGNIZING PATTERNS

Physics is a science. A science involves recognizing and explaining patterns. By **pattern** we mean some systematic relationship among different things. Figure 1–1 shows a collection of symbols with several patterns relating them. See how many of the patterns you can find before you read further.

A X H I M L I F E

Figure 1–1: Find patterns that relate the symbols.

Here are most of the patterns.

- The symbols are all the same color—black.
- They lie in a straight line.
- They are all found in the alphabet.
- They are all capital letters.
- They are all formed by straight line strokes; there are no curved lines.
- Their heights are proportional to the number of line segments that form the symbols.
- The letters can be grouped into English words.
- Each successive word has one more letter.

Much of physics involves looking for patterns and converting them into useful mathematical forms or unifying ideas. Much of this book involves looking for patterns and applying them to technical situations. The following is an example of developing a mathematical pattern.

Figure 1–2 shows some geometric solids with the formulas for calculating lateral surface. The **lateral surface** is the area on the side of a geometric solid. It includes practically everything except the top and bottom of the object. The lateral surfaces are the shaded areas on the different shapes. Let us look for a pattern.

The solids shown in the diagram have one thing in common. Each has a constant cross section. If we cut perpendicular to the cylinder's

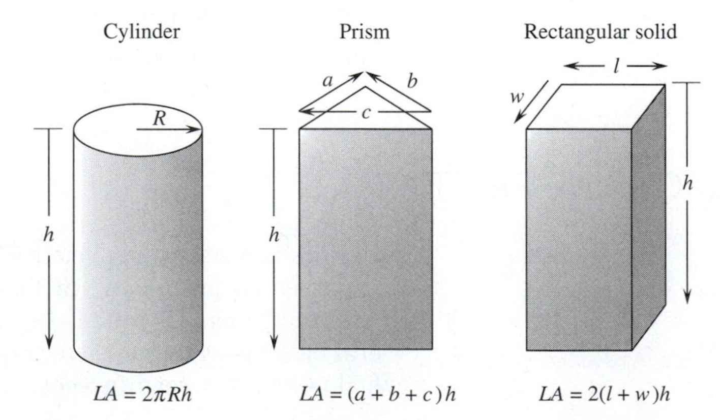

Figure 1–2: The shaded areas are the lateral surface areas (LA) of the objects. Each of the objects has a constant cross section. h, *height;* R, *radius;* w, *width;* l, *length.*

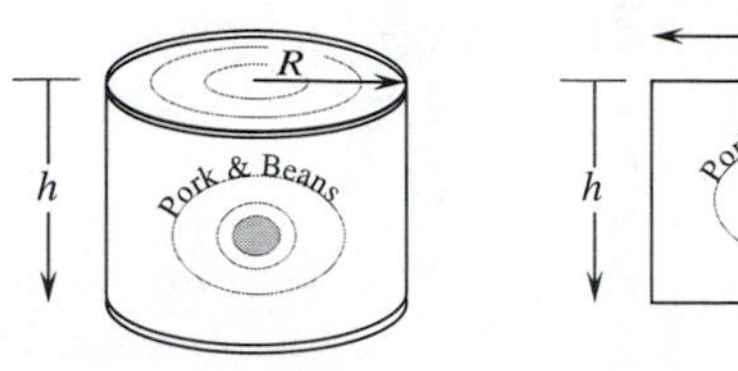

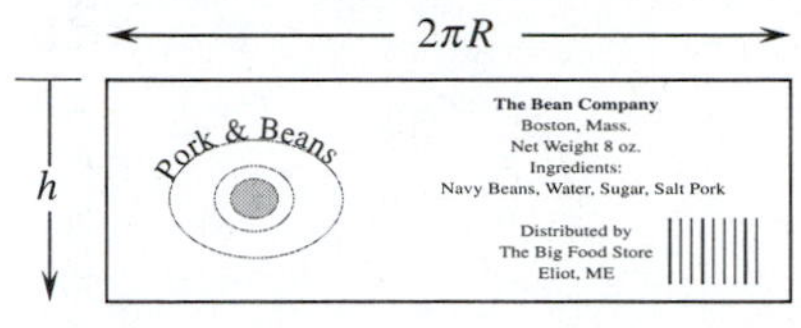

Figure 1–3: The label on a can of pork and beans is the lateral surface of a cylinder. When it is peeled off and laid flat, it is a rectangle.

long axis, the cut surface (the **cross section**) is a circle. No matter where we make a cut at right angles to the long axis, the area of the circle is the same. The cross section of the prism is a triangle. The cross section of the rectangular solid is a rectangle. In each case the cut surface has the same area no matter where we make the cut at right angles to the height (h).

The label on a can of pork and beans is the lateral surface. If we peel off the label, we get a rectangle (Figure 1–3). One side of the label is the height (h) of the can. The other side is the distance around the cylinder, its **perimeter.** The perimeter of a circle is called the circumference, $C = 2\pi R$, where R is the radius of the circle. The lateral surface of the cylinder is

$$LA(\text{cylinder}) = 2\pi R \times h.$$

If we peeled a label off from the rectangular solid and laid the label flat, we would get another rectangle. One side would be the height of the rectangular solid; the other side would be its perimeter, $P = 2\,(1 + \text{w})$, where l is the length and w is the width. This gives us a lateral surface of

$$LA(\text{rectangular solid}) = 2(l + w) \times h.$$

You should be able to see the pattern. For any solid with a constant cross section the lateral surface is simply the height times the perimeter. This relationship is true for each of the objects in Figure 1–2. We do not need to learn a lot of new formulas. All we need to know is how to find the perimeter, the distance around the object.

$$LA(\text{solid with constant cross section}) = \text{perimeter} \times \text{height}$$
$$LA = P \times h$$

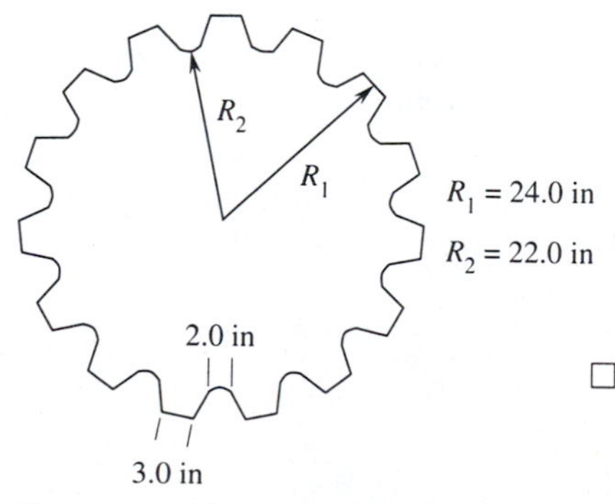

Figure 1–4: The cross section of a fluted column.

□ **EXAMPLE PROBLEM 1–1: THE PAINT JOB**

Penny earned a living as a house painter. She was called on a job to paint 30 deeply fluted columns that were 20 ft tall. Figure 1–4 shows the cross section of the columns. How did she figure out the total surface to be painted?

■ *SOLUTION*

The total surface (A) is 30 times the lateral surface ($P \times h$) of one column.

$$\begin{aligned} A(\text{total}) &= 30(P \times h) \\ &= 30(P \times 20 \text{ ft}) \end{aligned}$$

To calculate the perimeter Penny did not need to use the measurements shown in Figure 1–4. She wrapped a piece of masking tape around the pillar, fitting it tightly to the flutes. She then peeled off the tape and measured its length. This gave her the perimeter to finish the calculation.

1.4 SCIENTIFIC NOTATION

Here are some patterns in the way we write numbers. When we write the number 3, we mean three individual things. When we write the number 30, we mean three groupings of 10. When we write the number 300, we mean three groupings of 100.

We can extend this pattern.

$$3 = 3 \times 1$$

$$30 = 3 \times 10$$

$$300 = 3 \times 100$$

$$3000 = 3 \times 1000$$

$$30{,}000 = 3 \times 10{,}000$$

The zeros act as placeholders to tell us the size of the grouping, and the digit tells us how many groupings of this size we have. If we have more than one group that is not empty, we add the two groups together. For instance:

$$32{,}000 = (3 \times 10{,}000) + (2 \times 1000)$$

Notice that the small grouping is 1/10 the size of the larger grouping. We can use this **decimal notation** to express the whole number in terms of the larger grouping size. Thus

$$2 \times 1000 = 0.2 \times 10{,}000$$

and

$$32{,}000 = 3.2 \times 10{,}000$$

All the grouping sizes are multiples of 10 because our ancestors counted on their fingers. (That is why the word *digit* can refer to either a numeral or a finger.) To get the grouping size 10,000, we multiply 10 four times.

$$10{,}000 = 10 \times 10 \times 10 \times 10$$

We can abbreviate this by using a superscript to indicate how many times 10 must be multiplied. The superscript is called an **exponent** or power of 10. The number we are multiplying is called the **base.**

$$10{,}000 = 10^4$$
$$1000 = 10^3$$
$$100 = 10^2$$

For each succeeding smaller group the exponent is reduced by one. We can write 10 and 1 in this notation by continuing the pattern.

$$100 = 10^2$$
$$10 = 10^1$$
$$1 = 10^0$$

Using this notation to write the number 32,000, we get

$$32{,}000 = 3.2 \times 10^4$$

This is called **scientific notation.**

In scientific notation *a number is written as a number between 1 and 10 multiplied by a power of 10.*

Scientific notation is more useful than decimal notation when we want to write large numbers. For example, the mass of the earth (M_e) is

$$M_e = 5{,}990{,}000{,}000{,}000{,}000{,}000{,}000 \text{ metric tons}$$

or

$$M_e = 5.99 \times 10^{21} \text{ metric tons}$$

It is much easier to compare large numbers when they are written in scientific notation. We simply look at the exponent to find the grouping size. We can easily see that 3.4×10^{19} is larger than

7.8×10^{18}. In decimal notation these same two numbers are 34,000,000,000,000,000,000 and 7,800,000,000,000,000,000.

To convert back and forth between decimal notation and scientific notation simply move the decimal point. When the decimal point moves to the left, the exponent increases by one. When the decimal point moves to the right, the exponent decreases by one.

$$743{,}000 = 743{,}000. \times 1$$
$$= 743{,}000 \times 10^0$$
$$743{,}000 = 74{,}300.0 \times 10^1$$
$$743{,}000 = 7430.00 \times 10^2$$
$$743{,}000 = 743.000 \times 10^3$$
$$743{,}000 = 74.3 \times 10^4$$
$$743{,}000 = 7.43 \times 10^5$$

The decimal point indicates the end of a whole number. Digits to the left of the decimal are whole numbers. Digits to the right of it are fractions. Zeros are used as placeholders to show the size of the fraction. For example,

$$0.3 = 3 \times \frac{1}{10} = 3 \times \frac{1}{10^1}$$
$$0.03 = 3 \times \frac{1}{100} = 3 \times \frac{1}{10^2}$$
$$0.003 = 3 \times \frac{1}{1000} = 3 \times \frac{1}{10^3}$$
$$0.0003 = 3 \times \frac{1}{10{,}000} = 3 \times \frac{1}{10^4}$$

We can maintain the pattern of reducing the exponent by one as we move the decimal point to the right, by expressing fractions as negative exponents. Let n be the exponent, then:

$$\frac{1}{10^n} = 10^{-n}$$

A negative exponent means the number is less than one.

$$0.3 = 3 \times 10^{-1}$$
$$0.03 = 3 \times 10^{-2}$$
$$0.003 = 3 \times 10^{-3}$$
$$0.0003 = 3 \times 10^{-4}$$

In scientific notation, the number 0.0082 is

$$0.0082 = 0.0082 \times 1$$
$$= 0.0082 \times 10^{0}$$
$$0.0082 = 0.082 \times 10^{-1}$$
$$0.0082 = 0.82 \times 10^{-2}$$
$$0.0082 = 8.2 \times 10^{-3}$$

Your scientific calculator uses scientific notation to express answers. Let us do some calculations on a calculator.

Enter the number 23000 on your calculator

Press [×] (multiplication)

Enter the number 47300

Press [=]

Your answer is 1.0879 09

The calculator saves space by using a gap between the multiplier and the power of 10. We read this answer as 1.0879×10^{9}.

You can also enter numbers in scientific notation. Look for a key labeled either [EXP] or [EE]. This key lets the calculator know you are entering an exponent. To enter the number 1.602×10^{-19} do the following.

Enter: 1.602

Press: [EXP] or [EE]

Enter: 19

Press: [+/−] This changes the sign of the entry. Do not confuse it with the subtraction operation labeled [−].

Your display should be 1.602 −19

Your calculator will not always give your answer in scientific notation. When it does not, you will have to use the rules of moving the decimal to convert the answer to scientific notation. Here is an example.

☐ **EXAMPLE PROBLEM 1–2: SCIENTIFIC NOTATION**

Use your calculator to multiply 3.1×10^{3} by 5.2×10^{2}. Give your answer in scientific notation.

■ *SOLUTION*

Enter: 3.1

Press: [EXP] or [EE]

Enter: 3

Press: [×]

Enter: 5.2

Press: [EXP] or [EE]

Enter: 2

Press: [=]

Your answer is 1612000. Move the decimal point six places to the left to get the answer in scientific notation.

$$1.612 \times 10^6$$

(NOTE: Some calculators have a key labeled [F<−>E], which converts between scientific notation and decimal notation. If you have this key, press it instead of counting decimal movements.)

1.5 LOGARITHMS

We can extend the idea of scientific notation one step further. We can write the entire number as 10 raised to some power. Look at the number 1600. The number is larger than 10^3, but smaller than 10^4. It can be written with an exponent that lies between 3 and 4 if we do not restrict ourselves to whole-number exponents. Another name for such fractional exponents is **logarithms.**

You can find logarithms using your calculator. Look for a key labeled [LOG]. To find the logarithm for 1600 do the following.

Enter: 1600

Press: [LOG]

Your answer is 3.20412.

$$1600 = 10^{3.20412}$$

This way of writing a number is much like scientific notation. It uses a power of 10, but doesn't need a multiplier.

Logarithmic Distances

In 1594, the Scottish mathematician John Napier got the idea for logarithms. It took him twenty years to fully develop this idea. He was 64 years old when he published a book with common (base 10) logarithms in 1614. Just before his death three years later, he described a device for multiplication.

Wilhelm Schickardt developed Napier's idea and produced a wooden calculating device in 1623 that could multiply and divide. The idea is based on logarithms and number lines.

A number line is simply a line with numbers marking off distances along the line. The x axis of a graph is a number line.

We can use two number lines to add. If we want to add 2 + 2, we'll place the beginning (0) of the second number line above the "2" on the first number line. The "2" on the second number line will lie over the "4" of the first line. The second number line simply tells us how far we should move along the first line to travel 2 units of distance. We get 2 + 2 = 4.

Napier and Schickardt realized special number lines could be used to multiply and divide. Instead of spacing numbers evenly along the number line, a number line was developed with the distances proportional to the logarithms of the numbers. Number lines of this sort can be seen in Appendix C along the axes of log-log graphs.

Multiplication occurs when the logarithms of numbers are added. When distances are proportional to logarithms along a number line, we can use number line addition to multiply!

(Photograph courtesy of Hewlett-Packard Company.)

Wilhelm Schickardt invented the slide rule. Technicians and engineers carried slide rules everywhere they went through the 1960s, until the electronic pocket calculator replaced it.

If you have a logarithm, you can use your calculator to find the number in decimal notation. Look for a key labeled either [2ndF] or [INV].

Enter: 3.20412

Press: [2ndF] or [INV]

Press: [LOG]

Your answer should be very close to 1600. There may be some round-off error in the way the calculator does its conversions.

In our number system we use a base of 10. But there is a naturally occurring base that emerges from probability. The base number is symbolized by the letter e. It is an irrational number, like π, for example. The value of e is approximately 2.71828.

The number e comes up in studies of radioactivity, the speed of an object slowing down in a fluid, the forgetting curves discussed in Appendix D of this book, the current flow through a semiconductor diode, the evaporation of liquids, and many others.

The exponent for a number written as a power of e is called a **natural logarithm.** The key labeled [ln] on your calculator will give you natural logarithms. It is used in the same way as the [LOG] key, but [ln] calculates with the base e = 2.71828.

☐ **EXAMPLE PROBLEM 1–3: NATURAL LOGARITHMS**

Answer the following questions.

A. What number has a natural logarithm of 2.3025851?

B. What is the natural logarithm of 230?

C. What is the value of e raised to the -4.230 power?

■ ***SOLUTION***

A.

Enter: 2.3025851

Press: [2ndF] or [INV]

Press: [ln]

Display: 10.

B.

Enter: 230

Press: [ln]

Display: 5.4380793

C. This is a lot like Part A. -4.230 is the natural logarithm of the number.

Enter: 4.230

Press: [+/−]

Press: [2ndF] or [INV]

Press: [ln]

Display: 0.0145523

SUMMARY

Science is a process of examining things around us and explaining why they occur. Any new scientific idea must be subjected to real-world tests and to the criticism of the scientific community.

Scientific notation helps us to express our observations and ideas in mathematical terms. In scientific notation a number is written with a multiplier between 1 and 10 times a power of 10. When we write a number using 10 to the *n*th power (10^n), n is called the exponent. When the decimal point is moved one place to the right, the exponent decreases by one. When the decimal point is moved to the left, the exponent increases by one for each place moved.

Fractional exponents called logarithms can be used in place of the multiplier to write a number in powers of 10. To find logarithms to the base 10 use the [LOG] key on your calculator.

Sometimes numbers are written as a power of e, where e is an irrational number approximately equal to 2.71828. These are called natural logarithms. To find natural logarithms use the [ln] key on your calculator.

KEY TERMS

If you can explain the following terms to a friend or classmate, you understand their meaning. If you cannot explain the terms, you should reread the sections in which they are discussed.

base	**logarithm**
controlling variables	**natural logarithm**
cross section	**pattern**
decimal notation	**perimeter**
exponent	**physics**
hypothesis	**scientific notation**
lateral surface	**scientific process**

EXERCISES

Sections 1.1–1.3:

1. Many introductory science books describe the scientific process using the key words, observation, inference, hypothesis, experimentation, and verification. Describe these five concepts in your own words. How is each related to the scientific process?

2. Find as many patterns as you can for the following display.

   NOPE
   FIRST
   DEFINE
   HIJACKS

3. Write the number 20 using six 9s. Give at least two possible solutions.

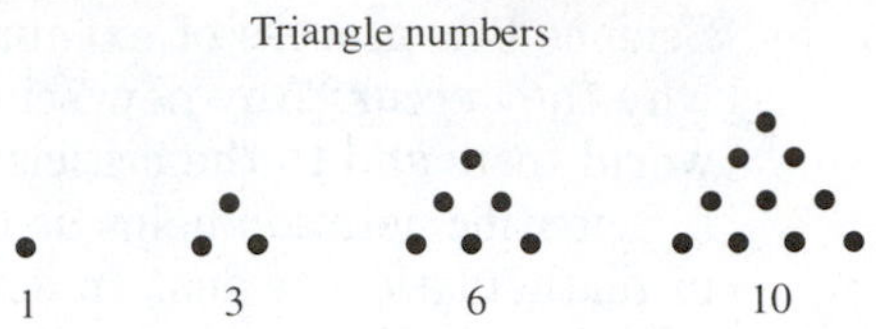

Triangle numbers

1 3 6 10

Figure 1–5: Diagram for Exercise 4. The first four triangle numbers.

4. Figure 1–5 shows arrays of triangle numbers. Look for a numerical pattern. Write down the next four triangle numbers in the series without resorting to drawing dot patterns.

Section 1.4:

5. Use your calculator to perform the following multiplications. From the answers you get, find a rule or pattern for combining exponents when multiplying numbers in scientific notation.

   a. $(2 \times 10^{7}) \times (3 \times 10^{8})$ **b.** $(4.1 \times 10^{5}) \times (1.2 \times 10^{6})$
   c. $(4.0 \times 10^{-6}) \times (1.1 \times 10^{22})$ **d.** $(2 \times 10^{-5}) \times (2.1 \times 10^{-6})$

6. Use your calculator to perform the following divisions. From the answers you get, find a rule or pattern for combining exponents when dividing numbers in scientific notation.

   a. $\dfrac{4.6 \times 10^{18}}{2.3 \times 10^{6}}$ **b.** $\dfrac{7.9 \times 10^{12}}{4.6 \times 10^{3}}$
   c. $\dfrac{8.9 \times 10^{-5}}{3.2 \times 10^{14}}$ **d.** $\dfrac{5.6 \times 10^{6}}{2.4 \times 10^{-3}}$

Section 1.5:

7. Take the logarithms of the following numbers using your calculator. Record your answers. From this information, predict what the logarithm of zero must be.

   log 100 = ? **log 10 = ?** **log 1 = ?** **log 0.1 = ?**
   log 0.01 = ? **log 0.001 = ?** **log 0.0001 = ?** **log 0.00001 = ?**

8. Find the inverse logarithms of the following logarithms using your calculator.

   a. $10^{2.3032}$ **b.** $10^{1.00}$ **c.** $10^{0.00}$ **d.** $10^{-4.56}$
   e. $e^{2.3032}$ **f.** $e^{1.00}$ **g.** $e^{0.00}$ **h.** $e^{-4.56}$

PROBLEMS

Section 1.3:

1. Find the lateral surface of a circular cylinder with a height of 12.0 in and a radius of 4.2 in.
2. Figure 1–6 shows the cross section of an I beam. If the beam is 24.3 ft long, what is its lateral surface?
3. Find a general pattern that relates the volumes of the objects shown in Figure 1–7. Each of the objects has a constant cross section.

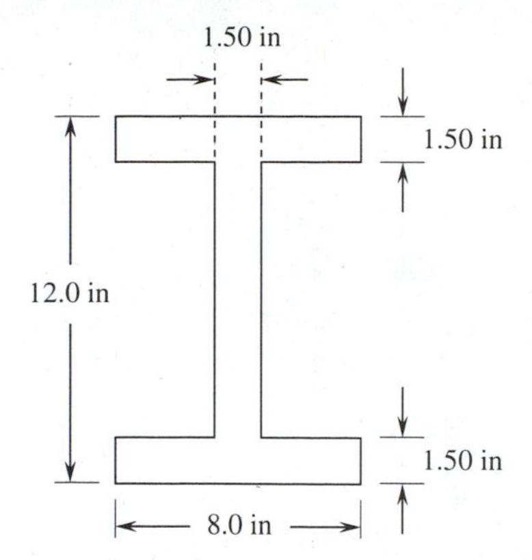

Figure 1–6: Diagram for Problem 2. Find the lateral surface of the beam with this cross section.

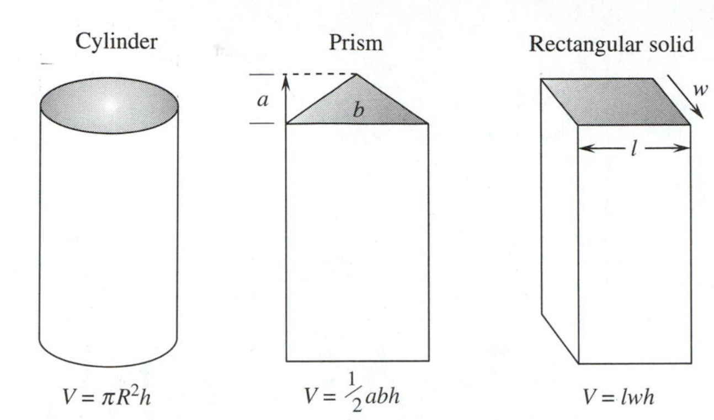

Figure 1–7: Diagram for Problem 3. Find a pattern that relates the volume (V) of each object with a constant cross section. R, radius; h, height; l, length; w, width.

4. The area of a circle is πR^2. The area of a semicircle is (1/2) πR^2.
   A. What is the area of a quarter of a circle?
   B. Write an equation to find the area of any sector with an angle θ. A sector is the pie-shaped area shown in Figure 1–8.
   C. Use your answer to Part B to find the area of a sector with a radius of 23.4 cm and angle $\theta = 17.2°$.

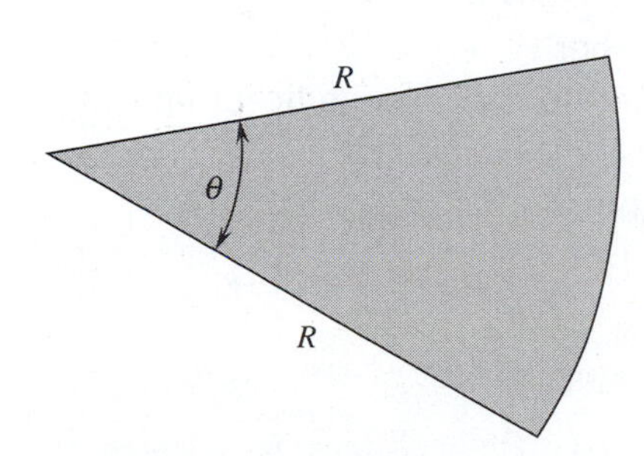

Figure 1–8: Diagram for Problem 4. A sector from a circle.

Section 1.4:

5. Convert the following decimal numbers into scientific notation.

   a. 0.0032 **b.** 130 **c.** 760000 **d.** 0.000065
   e. 23000000 **f.** 14 **g.** 1/37 **h.** 0.76

6. Write the following numbers in decimal notation.

   a. 3.2×10^{5} **b.** 3.2×10^{-5} **c.** 7.40×10^{3}
   d. 1.8×10^{-4} **e.** 5.4×10^{9} **f.** 9.1×10^{0}

7. ***A.*** Which of the numbers listed below are *NOT* in scientific notation?
   B. Convert those numbers to scientific notation.

   a. 2.2×10^{4} **b.** 23 **c.** 11.6×10^{-4}
   d. 0.034 **e.** 6.5×10^{7} **f.** 0.93×10^{14}

Section 1.5:

8. Use your calculator to evaluate the value of y in each of the following expressions. Put your answer in scientific notation.

   a. $y = 5.3 \times \log(57.8)$ **b.** $b \cdot y = 11.6 \times \log(0.238)$
   c. $y = \dfrac{78.2}{\ln(3.72)}$ **d.** $5.0 = \ln(y)$
   e. $2.3 \log(y) = 7.2$ **f.** $\ln(y - 3) = -0.126$

9. Find the value of y in each of the following equations.

   a. $y = 12.5 \times 10^{2.870}$ **b.** $y = 45 \times e^{-4.622}$
   c. $12 = y \times 10^{2.111}$ **d.** $6 = 3.2 \times e^{y}$

Chapter 2

MEASUREMENT

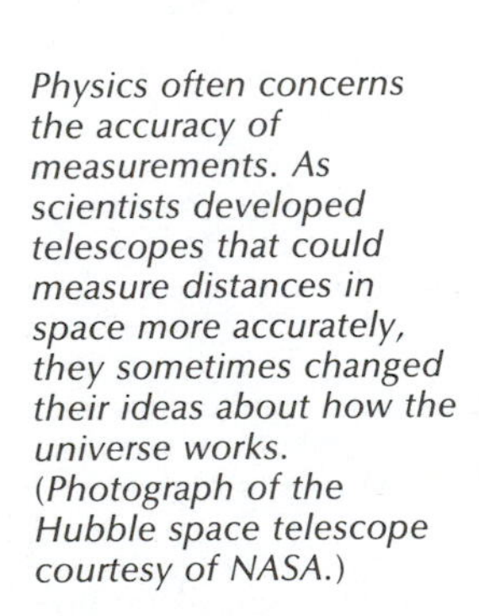

Physics often concerns the accuracy of measurements. As scientists developed telescopes that could measure distances in space more accurately, they sometimes changed their ideas about how the universe works. (Photograph of the Hubble space telescope courtesy of NASA.)

OBJECTIVES

In this chapter you will learn:

- the standard units for the British system of measurement and the metric system of measurement
- the prefixes used to express magnitude in standard units
- how to convert from one set of units to another
- how to record scale readings with the proper number of significant figures
- to determine the precision and accuracy of a measurement
- how to calculate the correct number of significant figures in an answer obtained from measured quantities

Time Line for Chapter 2

2600 BC	Sundials are used in China.
2500 BC	Standards of length, weight, and volume are set by Mesopotamians.
700 BC	Water clocks are used in Assyria.
1215	King John signs the Magna Carta. Part of the document sets a unified system of measure for England.
1310	First mechanical clock appears in Europe, powered by a falling weight.
1631	Pierre Vernier invents the vernier scale.
1639	William Gascoigne invents a micrometer that can measure angular distance between stars.
1641	Galileo's son builds a pendulum clock using his father's ideas.
1758	English commission sets standards for the "Imperial" standard.
1791	The metric system of measurement is proposed in France.
1829	The first electric clock is invented.
1875	The standard kilogram, a bar of platinum-iridium, is stored in Paris.
1928	Two Americans develop the first quartz clock.
1948	An atomic clock operating on cesium is invented.
1983	The meter is redefined as the distance light travels in 1/299,792,458th of a second.

Three athletic students measure the length of their dorm room by pacing out the distance. The basketball player finds the room is 15 ft long, more or less. The gymnast measures the room to be a little more than 20 ft long. The third student, a soccer player, finds the room is 17 ft long. The problem is, of course, they all have different size shoes. None of them is using a *standard* unit of measurement.

If they all had used a standard foot-long ruler, their measurements would have been closer. There would still have been disagreement of an inch or two because of random errors made in sliding the ruler end to end. If they each had used a steel tape measure, the random error would have been even less. They would probably have agreed on the length of the room within an inch. There is always a bit of uncertainty in any measurement.

Measurement is a process of comparing the size of something with a standard unit. The size of the thing measured is given in multiples of the standard unit. The thing measured is so many times as large or as small as the standard. A sack of potatoes weighs 10 lbs. This means that it is 10 times as heavy as a standard 1-lb weight. A room that is 3.56 m long is 3.56 times as long as a standard meter. When we write a measured quantity, we need to show the kind of unit used as well as the number of units.

In this chapter we will look at different kinds of standard units and how to convert from one sized unit to another. We will also look at the errors that occur when we make a measurement.

2.1 STANDARD UNITS

Many kinds of units of measurement are used in industry and technology. Some are special to a particular technology. Lumber is measured in board feet. A board foot is a volume of wood with an area of 1 ft^2 and a thickness of 1 in. The water held in a reservoir is measured by the acre-foot, a volume of water covering 1 acre to a depth of 1 ft. People who work with heating, venting, and air conditioning measure pressure in inches of water (Figure 2–1). This is the pressure that a column of water 1 in high would exert on the bottom of its container. It is a rather small unit of pressure, but it is a practical unit to use for measuring airflow in ducts.

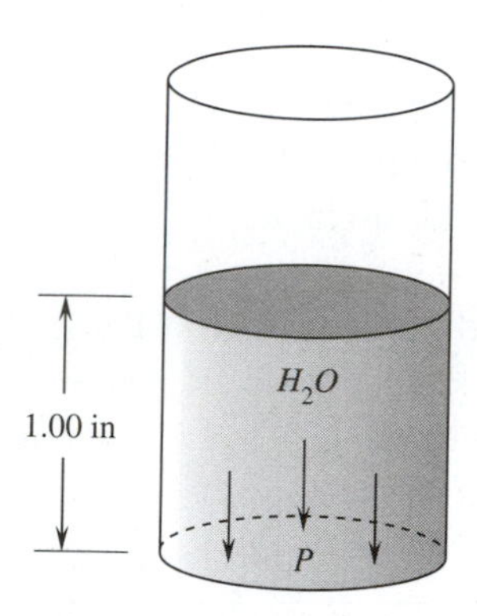

Figure 2–1: A unit of pressure used in heating and air conditioning is 1 in of water. This is the pressure a 1-in high column of water would exert on the bottom of the container.

In this text we will not use many specialized units. Instead, we will restrict ourselves mostly to two standard sets of units, the British engineering system and the metric system.

All mechanical measurements can be broken down into three basic quantities. They are distance, time, and mass (or weight). All other units can be expressed as a combination of these three basic units. For example, speed is distance divided by time, and volume is the product of three distance measurements.

British Engineering Units

The standard unit of **distance** is the foot, abbreviated ft. Larger units of distance, or length, may be defined as multiples of this basic standard; smaller units, as fractions.

$$1.00 \text{ yard (yd)} = 3.00 \text{ ft}$$

$$1.00 \text{ mile (mi)} = 5280 \text{ ft}$$

$$1.00 \text{ inch (in)} = \frac{1}{12} \text{ ft}$$

Time is measured in seconds, abbreviated s. Larger units are expressed as multiples of the basic unit.

$$1.00 \text{ minute (min)} = 60 \text{ s}$$

$$1.00 \text{ hour (h)} = 3600 \text{ s}$$

$$1 \text{ day} = 8.64 \times 10^4 \text{ s}$$

In the British system, a unit of weight is used as the third unit. **Weight** is the amount of pull the earth exerts on an object as it tries to draw the object to its center. The unit of weight used is the pound, abbreviated lb.

$$1.00 \text{ ton} = 2000 \text{ lb}$$

$$1.00 \text{ ounce (oz)} = \frac{1}{16} \text{ lb}$$

The Metric System

The metric system is also known as the Système International d'Unites or SI. The basic metric unit of distance is the meter (m); of time, the second (s); and of mass, the kilogram (kg).

In SI units, mass is used instead of weight. **Mass** is independent of the gravitational pull of the earth; it is a measure of an object's resistance to change of motion. This resistance is sometimes called **inertia.** We will find in Chapter 6 that weight is proportional to mass.

In SI, units larger and smaller than the three basic units are expressed as multiples of 10. We do not need to memorize a variety of conversion factors as we do with the British system. Table 2–1 shows a set of **prefixes** that are used to indicate the multiples of 10. For instance, 2.3 *centi*meters is 2.3 times 1/100 of a meter. Some prefixes can be used with British units. A *mega*ton is 1,000,000 tons. You should become familiar with these prefixes; you will encounter them often.

Table 2–1: Prefixes indicating a power of 10, used with units of measurement.

POWER OF 10	DECIMAL NOTATION	PREFIX	SYMBOL	PRONUNCIATION
10^{12}	1,000,000,000,000	tera	T	ter'a
10^{9}	1,000,000,000	giga	G	jig'a
10^{6}	1,000,000	mega	M	meg'a
10^{3}	1,000	kilo	k	kil'o
10^{2}	100	hecto	h	hek'to
10^{1}	10	deka	da	dek'a
10^{-1}	0.1	deci	d	des'i
10^{-2}	0.01	centi	c	sent'i
10^{-3}	0.001	milli	m	mil'i
10^{-6}	0.000001	micro	μ	mi'kro
10^{-9}	0.000000001	nano	n	nan'o
10^{-12}	0.000000000001	pico	p	pe'ko

□ **EXAMPLE PROBLEM 2–1: PREFIXES**

A. How many grams are there in 2.3 kg?

B. Which is smaller: 2.3 μs or 1.2×10^3 ns?

■ ***SOLUTION***

A. The prefix *kilo* means 10^3, or 1000.

$$2.3\ \textit{kilo}\text{grams} = 2.3 \times 10^3\ \text{g}$$

B. The prefix *micro* means 10^{-6}.

$$2.3\ \textit{micro}\text{seconds} = 2.3 \times 10^{-6}\ \text{s.}$$

The prefix *nano* means 10^{-9}.

$$1.2 \times 10^3\ \textit{nano}\text{seconds} = 1.2 \times 10^3 \times 10^{-9}\ \text{s}$$

or

$$1.2 \times 10^{-6}\ \text{s}$$

Therefore the second number is the smaller interval of time.

2.2 UNITS CONVERSIONS

We can convert a measurement from one set of units to another set by using a **conversion equation.** If we carefully measure the length of a meter stick using a foot ruler, we find that the meter stick is 3.28 ft long. This is our conversion equation between metric and British units of length.

$$1.00\ \text{m} = 3.28\ \text{ft}$$

The ratio of these two measurements, 1.00 m and 3.28 ft, is one. The two measurements represent the same quantity or distance.

$$\frac{1.00\ \text{m}}{3.28\ \text{ft}} = 1 \quad \text{or} \quad \frac{3.28\ \text{ft}}{1.00\ \text{m}} = 1$$

We know we can multiply a quantity by one without changing its value. If we want to express 2.80 m in terms of feet, we can multiply by the ratio of feet to meters. The units behave algebraically like

Measuring the Earth's Orbit with a Micrometer

When we measure something, we often find that nature has a surprise, a new puzzle to explain.

In 1639, William Gascoigne invented a micrometer to be placed in the focus of a telescope. With this device, the angular distance between stars could be measured. Over the next few years, there was a boom in the development of devices to measure the relative position of stars.

In 1667, an astronomer named Jean Picard invented a rather good micrometer. He was surprised to find a seasonal change of the positions of stars. Stars along the axis of the earth's orbit moved in an ellipse that took exactly one year to complete. Stars along the plane of the earth moved back and forth in a straight line, also taking one year to complete one cycle. Stars at an angle to the plane moved in ellipses, with the narrower ellipses near the earth's plane. This odd behavior is called *stellar aberration*.

It took 60 years for someone to unravel this problem. In 1728, James Bradley came up with the solution.

Imagine you're walking in the rain holding a piece of stovepipe in a vertical position. If you were standing still, rain would fall straight down through the pipe. Things are different when you walk. As a raindrop falls through the pipe, the back of the pipe moves forward and strikes the raindrop.

In a moving car, you may have noticed raindrops slant toward the windshield because of the vehicle's motion. Those same raindrops fall straight down when the car stops.

(Photograph of the Hubble space telescope courtesy of NASA.)

The raindrop can get through the pipe if you tip the pipe forward. The amount of tip depends on your speed. The faster you move, the more you'll need to tip the pipe forward. The pipe needs to match the slant of the raindrops.

Replace the stovepipe with a telescope and the raindrop with light. We have the puzzle discovered by Jean Picard. Astronomers must tip their telescopes in the direction of the earth's orbit for the light to get through the center of the telescope.

Because the earth's orbit is an ellipse, the direction of the tilt changes slightly each day. Picard had discovered an experimental proof that the earth orbits the sun.

numbers. If we divide the unit meters by a unit of meters, the units cancel out. Their ratio is the unitless number one.

$$2.80 \text{ m} = 2.80 \text{ m} \times \frac{3.28 \text{ ft}}{1.00 \text{ m}} = \frac{2.80 \cancel{\text{m}}}{1.00 \cancel{\text{m}}} \times 3.28 \text{ ft}$$

$$2.80 \times 3.28 \text{ ft} = 9.18 \text{ ft}$$

To convert feet to meters we can use the same ratio upside down to put the unit of foot in the denominator because we want that unit to cancel out.

$$6.93 \text{ ft} = 6.93 \cancel{\text{ft}} \times \frac{1.00 \text{ m}}{3.28 \cancel{\text{ft}}} = 2.11 \text{ m}$$

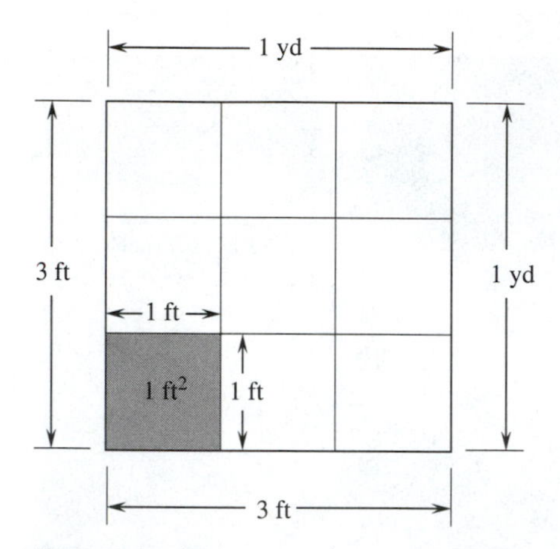

Figure 2–2: A square yard is a square 3 ft to a side.

When we calculate area we multiply two distance measurements together. Figure 2–2 shows a square area 1 yd on a side. It takes nine tiles, each with an area of 1 ft^2, to make 1 yd^2.

$$\textit{area} \text{ (square)} = \textit{length} \times \textit{width} = 1 \text{ yd} \times 1 \text{ yd} = 1 \text{ yd}^2$$

$$\textit{area} = 3 \text{ ft} \times 3 \text{ ft} = 9 \text{ ft}^2$$

The conversion equation is $1 \text{ yd}^2 = 9 \text{ ft}^2$.
Notice that we can get the area conversion equation from the linear conversion equation. All we need to do is square both sides of the equation.

$$1 \text{ yd} = 3 \text{ ft}$$
$$(1 \text{ yd})^2 = (3 \text{ ft})^2$$
$$1^2 \text{ yd}^2 = 3^2 \text{ ft}^2$$
$$1 \text{ yd}^2 = 9 \text{ ft}^2$$

To convert volume we would cube both sides of the conversion equation.

$$(1 \text{ yd})^3 = (3 \text{ ft})^3$$
$$1^3 \text{ yd}^3 = 3^3 \text{ ft}^3 = 27 \text{ ft}^3$$

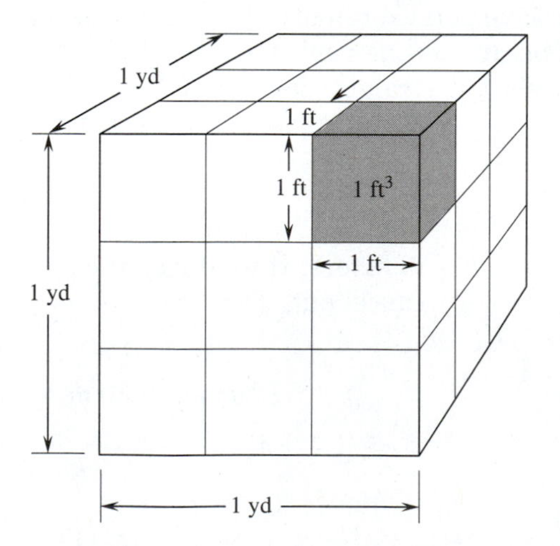

Figure 2–3: One cubic yard contains 27 cubes with an edge length of 1 ft.

Twenty-seven 1-ft cubes are needed to build a cube 1 yd to a side (see Figure 2–3).

□ **EXAMPLE PROBLEM 2–2: THE WASHER**

A $\frac{1}{8}$-in thick washer has the cross-sectional area shown in Figure 2–4. Calculate the volume in inches, and convert the answer to centimeters given 1.00 in = 2.54 cm.

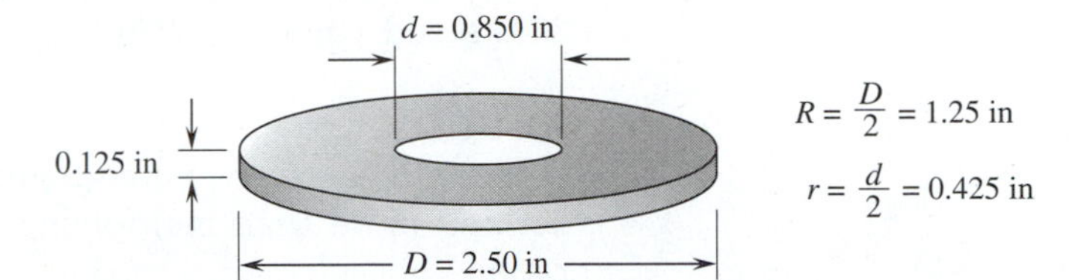

Figure 2–4: *Diagram for Example Problem 2–2.*

■ ***SOLUTION***

The net cross-sectional area (A) is the area of a large circle with a smaller circle bored out of it. Letting R = the radius of the large circle and r = the radius of the smaller circle,

$$A = \pi R^2 - \pi r^2 = \pi(R^2 - r^2)$$

Since the cross-sectional area is constant, we can multiply this area by the thickness (t) to get the volume (V).

$$V = A\,t = \pi(R^2 - r^2)\,t$$

$$V = 3.14 \times [(1.25 \text{ in})^2 - (0.425 \text{ in})^2] \times 0.125 \text{ in}$$

$$V = 0.542 \text{ in}^3$$

We can find the volume conversion between inches and centimeters by cubing the linear conversion equation.

$$(1.00 \text{ in})^3 = (2.54 \text{ cm})^3$$

$$1.00 \text{ in}^3 = 16.4 \text{ cm}^3$$

NOTE: To find the cube of 2.54 on your calculator

Enter: 2.54

Press: [y^x]

Enter: 3

Press: [=]

Now that we have the volume conversion equation, we can convert the volume units.

$$V = 0.542 \cancel{\text{in}^3} \times \frac{16.4 \text{ cm}^3}{1.00 \cancel{\text{in}^3}}$$

$$V = 8.89 \text{ cm}^3$$

□ **EXAMPLE PROBLEM 2–3: SPEED**

Convert 45 mph to feet per second. Use the conversions 1 h = 3600 s and 1 mile = 5280 ft.

■ ***SOLUTION***

We can chain conversions together in one calculation. We are doing nothing more than multiplying by one when we use a conversion ratio.

$$45 \frac{\text{mi}}{\text{h}} = 45 \frac{\cancel{\text{mi}}}{\cancel{\text{h}}} \times \frac{1 \cancel{\text{h}}}{3600 \text{ s}} \times \frac{5280 \text{ ft}}{1 \cancel{\text{mi}}}$$

$$45 \frac{\text{mi}}{\text{h}} = 66 \frac{\text{ft}}{\text{s}}$$

2.3 READING SCALES

Anytime we read a scale we need to estimate part of the measurement. The coarser the scale, the rougher the estimate. Figure

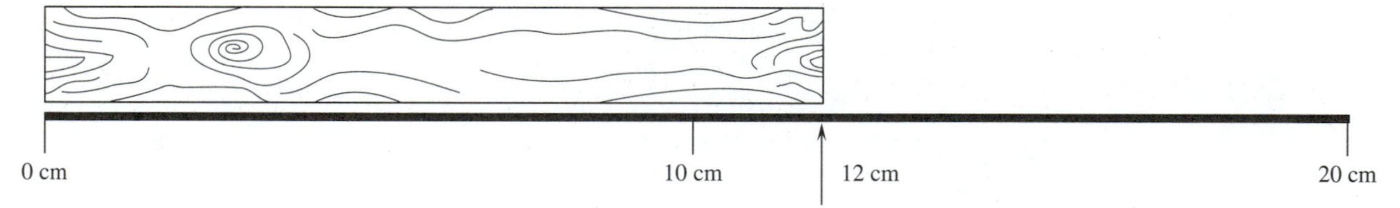

Figure 2–5: A piece of wood is measured by a scale marked in decimeters. The scale is read to two-figure accuracy.

2–5 shows a scale in which the smallest division is 10 cm, or a decimeter. The length of the object is more than 10 cm, but less than 20 cm. We will estimate the object to be 12 cm long. We are a bit uncertain about the 2. The object might be a little larger or smaller.

There are two significant figures in the measurement. The last digit, the one with the uncertainty in it, is at the 1 cm spot. This is called the **least significant digit** (LSD).

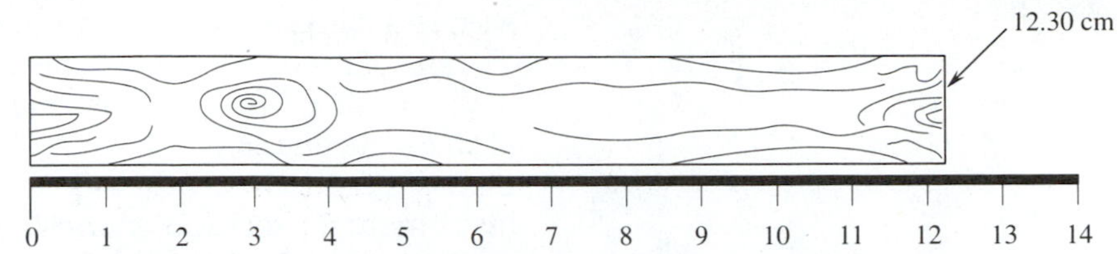

***Figure 2–6:** The same piece of wood is measured by a scale marked in centimeters. We can make a more precise estimate of its length.*

Figure 2–6 shows the same object measured with a scale with centimeter markings. Now we see that the object is a little more than 12 cm long. We can estimate a third digit. We will guess 12.3 cm. We now have a measurement with three meaningful digits, and the least significant digit is in the 0.1-cm grouping.

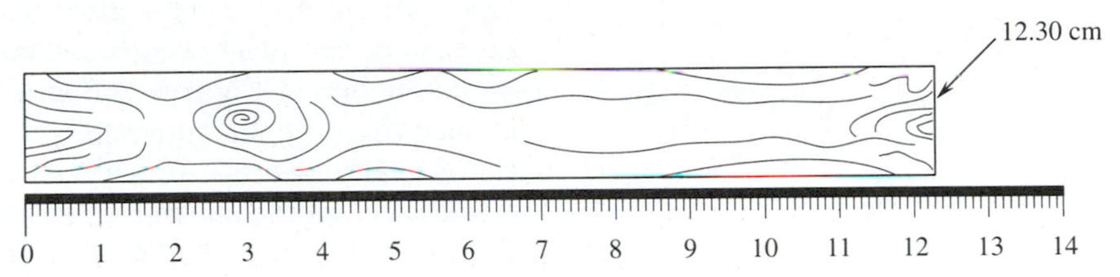

***Figure 2–7:** The piece of wood is measured with a millimeter scale.*

Figure 2–7 shows the same measurement using a scale with millimeter markings. The end of the object lines up with the 12.3 mm marking. There is no longer a need to guess about the 3. We should indicate that this measurement has more precision than the measurement made using the centimeter rule shown in Figure 2–6. We can do this by adding a zero to the end of the reading. This shows that there is no uncertainty about the 3 and that we can guess the measurement to a precision of 0.005 cm. We are fairly sure that the scale reading is larger than 12.295 and smaller than 12.305.

- The measurement for Figure 2–7 is 12.30 cm.
- The LSD is 0.01 cm.
- The number is written with four significant figures.

When we read a scale, the LSD is an estimated number. The size of this digit depends on the fineness of the measuring scale. We call this precision.

Precision *determines the exactness with which a quantity can be measured.*

There is more than one way to express precision mathematically. When we take a single measurement from a measuring instrument such as a meter stick or a stopwatch, we use instrumental precision. The size of the divisions of the measuring scale controls the exactness of the reading. We define instrumental precision in terms of the size of the scale divisions.

Instrumental precision *is one-half the smallest count or division on the measuring scale.*

In Figure 2–5 the smallest division of the scale is 10 cm; the instrumental precision is 5 cm. In Figure 2–6 the smallest division of the scale is 1 cm; the instrumental precision is 0.5 cm. In Figure 2–7 the smallest division is 0.1 cm; the instrumental precision is 0.05 cm.

Notice that the LSD is closely related to the instrumental precision. As the scales in the illustrations get finer (that is, have higher precision), the LSD becomes smaller.

Also notice in Figures 2–5 and 2–6 the **readability** of the scales is finer than the instrumental precision. With the coarse scales we can estimate a tenth of a division fairly well. As we get to finer scales, such as the one in Figure 2–7, the instrumental precision and the readability are pretty much the same.

Precision or an LSD give us some idea of how well we have read a scale. Precision or LSD will *not* tell us how well the measurement agrees with what it should be. For example, an experimenter uses very precise measuring instruments to find the ratio of the circumference of a circle to the diameter (π). The result is 3.13924. The LSD is 0.0001. According to the precision of the instruments used in this experiment, 0.0001 is the correct LSD for the result. However, the accepted value of (π) to six significant figures is 3.14159. The experimental result has a great deal of precision, but it disagrees with the accepted result in the third figure. We say the result has three-figure accuracy.

Accuracy *as applied to measuring systems is the agreement between a measurement and a corresponding accepted or true value.*

There is more than one way to express accuracy mathematically. We will express accuracy in terms of **figures of accuracy.** In the experimental determination of π the result had only three-figure accuracy; there was error in the third figure. The accuracy may not have been as good as expected for a variety of reasons. The circle may have been out of round. The measuring instruments may not have been properly zeroed. The experimenter may not have read the scales carefully. Instrumental precision is only one factor in determining the accuracy of a reading.

The fineness of the instrumental precision we need at any time depends on the kind of measurement we are taking. At the turn of the century, astronomers could measure the earth-moon distance with a precision of 10 km. After World War II, the new technology of radar made it possible to get an instrumental precision of less than 1 km. Recent use of lasers has given astronomers a precision of less than 1 m. Because the moon revolves in an elliptical orbit around the earth, it is not always the same distance away. A typical measurement at some instant of time might be 384,412.238 km using laser-ranging techniques. An instrumental precision of 0.5 m gives us nine-figure accuracy for this case. Planetary astronomers are delighted with this precision. On the other hand, a carpenter who measures and cuts boards to the nearest meter would find it difficult to get employment. In most cases, the accuracy of the carpenter's work is to only one significant figure. A carpenter needs a precision a thousand times smaller than the precision that delights an astronomer.

Look again at the measurement for Figure 2–7. It is 12.30 cm. The LSD is 0.01 cm. The instrumental precision is 0.05 cm. The measurement has four figures of accuracy.

□ **EXAMPLE PROBLEM 2–4: READING SCALES**

A. Write down the measurements (circled numbers 1–6) in Figure 2–8.

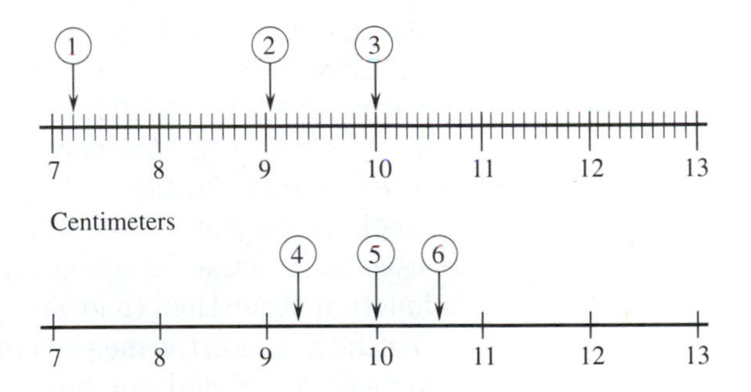

Figure 2–8: *Diagram for Example Problem 2–4.*

B. What is the least significant digit (LSD) of each measurement?
C. How many figures of accuracy does each measurement have?

■ ***SOLUTION***

1. 7.20 cm, LSD 0.01 cm, accuracy 3 figures.
2. 9.05 cm, LSD 0.01 cm, accuracy 3 figures.
3. 10.00 cm, LSD 0.01 cm, accuracy 4 figures.
4. 9.3 cm, LSD 0.1 cm, accuracy 2 figures.

5. 10.0 cm, LSD 0.1 cm, accuracy 3 figures.

6. 10.8 cm, LSD 0.1 cm, accuracy 3 figures.

2.4 SIGNIFICANT ZEROS

We estimate the accuracy of a measurement simply by counting significant digits. It is easy to do this for numerals 1 through 9. But zero causes a problem. It is sometimes used for no purpose other than to locate the decimal point.

A zero that has no function other than to locate a decimal point is not significant.

Key the number .13 on your calculator. (Do not forget the decimal point.) The display will be 0.13. The **leading zero** appears even though you did not key it into the calculator. The leading zero helps to locate the decimal point. It is always good form to use a leading zero with numbers less than one.

The leading zero (the zero to the left of the decimal point) in a number less than one is not significant.

In scientific notation zeros are not used to locate the decimal point. We use a power of 10 instead. One way to check to see if a zero is significant or not is to convert the number into scientific notation. Look at the measurement 0.0074 mm. In scientific notation it is 7.4×10^{-3} mm. All the zeros were used to locate the decimal point in the decimal notation.

Look at the number 0.00740 mm. The zero to the extreme right is used to indicate the precision of the measurement. Because it has a function other than to locate the decimal point, it is significant. In scientific notation the measurement is 7.40×10^{-3} mm. Incidentally, your calculator will not help you with right-hand zeros. When it converts to scientific notation, it drops *all* right-hand zeros whether they are significant or not.

Here are two more rules:

All zeros to the immediate right of the decimal point in numbers less than one are not significant.

Any zero used to indicate precision is significant.

Let us look at some numbers larger than 1. In the measurement 3007 ft the two zeros do not locate the decimal place. They are used

Why Is a Pound a lb?

In the era of the Caesars, grain in the Roman marketplace was measured by an equal-armed balance. This is the same kind of balance you see in the Statue of Justice. A standard weight called a *libra pondo* was placed in one pan of the scale. Grain was added to the other pan until it was equal in weight to the standard weight. When the scales balanced, there was equal libra pondo in each balance pan. From this we get the word equilibrium, which means a system is in balance.

Julius Caesar invaded the British Islands with his cohorts and legions to subdue the barbarians living there. As a conquered people, the Britons adopted some of the Roman language. The libra pondo became a standard measure, abbreviated lb, for *libra*. After a period of time, the name was shortened by dropping the first word, libra. The weight became known as a *pound*, while its abbreviation remained lb, after the missing part of the name.

---

as placeholders to show that a grouping size is empty. In scientific notation the measurement is 3.007×10^3 ft. For the measurement 200.0 m the right-hand zero is used to indicate the precision of the measurement. When we convert to scientific notation we should retain that zero to show the precision of 0.1 m. The measurement is 2.000×10^2 m in scientific notation. These two examples give us another rule.

A zero located between two significant figures is significant.

Look at the measurement 1400 mi. The precision is not clear. Maybe the zeros were used only to place the decimal. If so, the distance was estimated to the nearest hundred miles, and in scientific notation would be written 1.4×10^3 mi. If the distance was measured to the nearest 10 mi, the left-hand zero is significant. To indicate a 10-mi precision in scientific notation the measurement would be written 1.40×10^3 mi. If the distance was measured to the nearest mile, both zeros are significant. In scientific notation we get 1.400×10^3 mi. Obviously, we need a scheme to clearly indicate the least significant zero. We can do this by placing an indicator over the zero corresponding to the precision of the measurement. A bar is commonly used for this purpose.

$$\text{Precision} = 1 \text{ mi: } 140\bar{0} = 1.400 \times 10^3 \text{ mi}$$

$$\text{Precision} = 10 \text{ mi: } 14\bar{0}0 = 1.40 \times 10^3 \text{ mi}$$

$$\text{Precision} = 100 \text{ mi: } 1400 = 1.4 \times 10^3 \text{ mi}$$

☐ **EXAMPLE PROBLEM 2–5: SIGNIFICANT ZEROS**

Indicate the least significant digit (LSD) and the accuracy of each of the following measurements. Write each measurement in scientific notation.

a. 10.0 ft **b.** 210 volt **c.** 0.0032 cm **d.** 0.0400 in^3

e. $23\bar{0}0$ ft/s **f.** 7200 m/s **g.** 23.0 kg **h.** $120\bar{0}$ lb

■ *SOLUTION*

MEASUREMENT	LSD	FIGURE OF ACCURACY	SCIENTIFIC NOTATION
a. 10.0 ft	0.1 ft	3	1.00×10^1 ft
b. 210 volt	10 volt	2	2.1×10^2 volt
c. 0.0032 cm	0.0001 cm	2	3.2×10^{-3} cm
d. 0.0400 in^3	0.0001 in^3	3	4.00×10^{-2} in^3
e. $23\bar{0}0$ ft/s	10 ft/s	3	2.30×10^3 ft/s
f. 7200 m/s	100 m/s	2	7.2×10^3 m/s
g. 23.0 kg	0.1 kg	3	2.30×10^1 kg
h. $120\bar{0}$ lb	1 lb	4	1.200×10^3 lb

2.5 CALCULATED VALUES

Measured quantities are often used in calculations. We would like to know how the precision and accuracy of a measurement affect the results of a calculation.

A room is measured with a 16-ft steel tape. The width is 15.26 ft. The length is 26.2 ft. We have less accuracy for the longer length because of the way we needed to use the tape. We can plug these values into a calculator to get the area of the room.

$$A = l \times w = 26.2\text{ ft} \times 15.26\text{ ft}$$

The calculator reads 399.812. We need to know how many digits to write down. The last digit in each measurement is uncertain; it is an estimated value. Each uncertain digit in the calculation below is indicated by an overbar. A digit multiplied by an uncertain digit will also be uncertain. Let us calculate the area, keeping track of the uncertain digits.

$$\begin{array}{rr} & 15.2\bar{6}\text{ ft} \\ \times & 26.\bar{2}\text{ ft} \\ \hline & \overline{3052} \\ & 915\bar{6} \\ & 305\bar{2} \\ \hline & 39\bar{9}.\overline{812}\text{ ft}^2 \end{array}$$

There is no sense in keeping all the uncertain digits. We will round off the result to the largest uncertain digit. This will give an area of $40\bar{0}$ ft^2.

Mathematical operations generate uncertainties. Notice that the accuracy of the answer is 3. The least accurate measurement (input) put into the calculation is 3. The least accurate input determines the accuracy of the result in multiplication. This is also true for the inverse operation of division. We can make a general rule.

In multiplication and division the result should have only as many figures of accuracy as the least accurate input.

□ **EXAMPLE PROBLEM 2–6: THE CYLINDER**

The diameter (D) of a cylinder is measured by a micrometer (Figure 2–9). It is found to be 11.254 mm. Vernier calipers are used to measure the height because the cylinder is too long to fit into the jaws of the micrometer. The height (h) is 6.37 cm. Calculate the volume (V) of the cylinder, and write the result with the correct figures of accuracy.

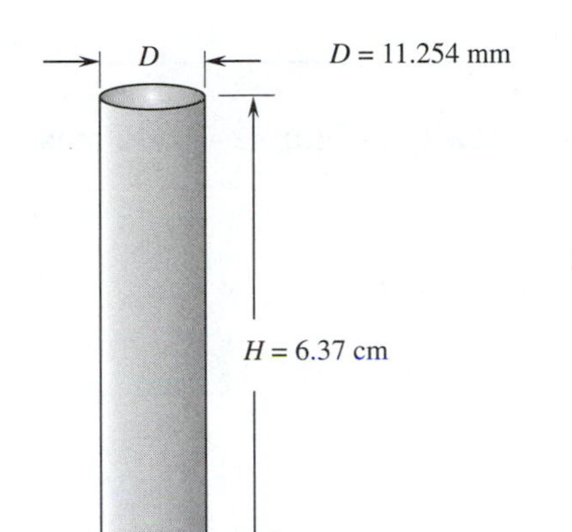

Figure 2–9: *Diagram for Example Problem 2–6.*

■ ***SOLUTION***

$$V = \pi \frac{D^2}{4} H$$

The calculator will give us π to several significant figures. The 4 is a geometric factor. We can assume it is *exact*. We can treat it as though it had many significant figures even though we have not written out the significant zeros. The height measurement is the least accurate input, with an accuracy of 3. The result should have an accuracy of 3.

$$V = 3.141593 \times \frac{(1.1254 \text{ cm})^2}{4} \times 6.37 \text{ cm}$$

The calculator reads 6.336408025.

The correct result is:

$$V = 6.34 \text{ cm}^3$$

□ **EXAMPLE PROBLEM 2–7: DENSITY**

The cylinder in Example Problem 2–6 is placed on a triple-beam balance. Its mass is measured to be 17 g. Calculate its density.

■ *SOLUTION*

Density (D) is the mass of something per unit volume. We can find this by dividing mass (m) by volume (V).

$$D = \frac{m}{V}$$

$$D = \frac{17\ \text{g}}{6.34\ \text{cm}^3}$$

The calculator reads 2.681388013. The least accurate input is the mass, with an accuracy of 2. Round off to two significant figures.

$$D = 2.7\ \text{g/cm}^3$$

The sides of the four-sided figure shown in Figure 2–10 are measured to be:

$$\text{A} = 3.50\ \text{in}$$

$$\text{B} = 12\ \text{in}$$

$$\text{C} = 4.352\ \text{in}$$

$$\text{D} = 11.6\ \text{in}$$

We can find the perimeter by adding the measurements. Let us keep track of uncertain digits.

$$\begin{array}{r} 3.5\bar{0}\ \text{in} \\ 1\bar{2}\ \text{in} \\ 4.35\bar{2}\ \text{in} \\ +\ 11.\bar{6}\ \text{in} \\ \hline 3\bar{1}.\overline{452}\ \text{in} \end{array}$$

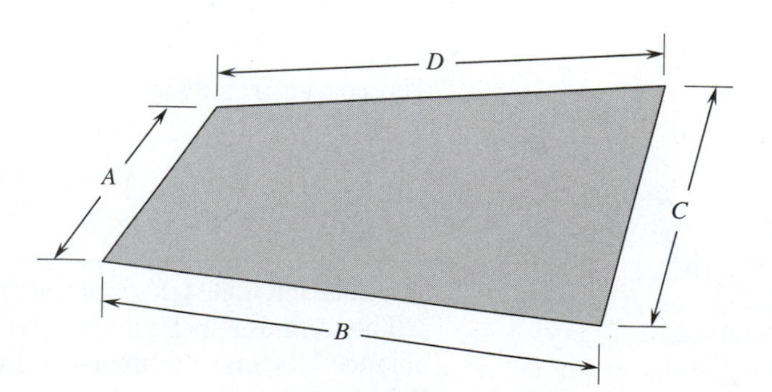

Figure 2–10: *Find the perimeter of an irregular four-sided figure by adding up the lengths of its sides.*

There is no reason to keep so many uncertain digits in the result. Rounding off to the largest uncertain digit gives a perimeter of 31 in. Notice that the least precise input determines the precision of the result in addition. This is also true for the inverse operation of subtraction.

In addition and subtraction the result is only as good as the least precise input.

☐ **EXAMPLE PROBLEM 2–8: ADDITION**

Add the following measurements: 2.31 m, 45 m, 7.832 m, 17.34 m.

■ ***SOLUTION***

$$\begin{array}{r} 2.31\ \ \ \text{m} \\ 45\ \ \ \ \ \ \ \ \text{m} \\ 7.832\ \text{m} \\ +\ 17.34\ \ \ \text{m} \\ \hline 72.482\ \text{m} \end{array}$$

The least precise input has an LSD of 1 m. Our result is good only to 1 m.

Answer: 72 m

☐ **EXAMPLE PROBLEM 2–9: THE RING**

Find the net area of the ring shown in Figure 2–11.

■ ***SOLUTION***

We can find the net area by subtracting the area of the hole (A_1) from the area of the disk (A_2).

$$A(\text{net}) = A_2 - A_1$$

$$A(\text{net}) = \pi(R_2^2 - R_1^2)$$

$$A(\text{net}) = 3.14159 \times [(1.340\ \text{in})^2 - (0.71\ \text{in})^2]$$

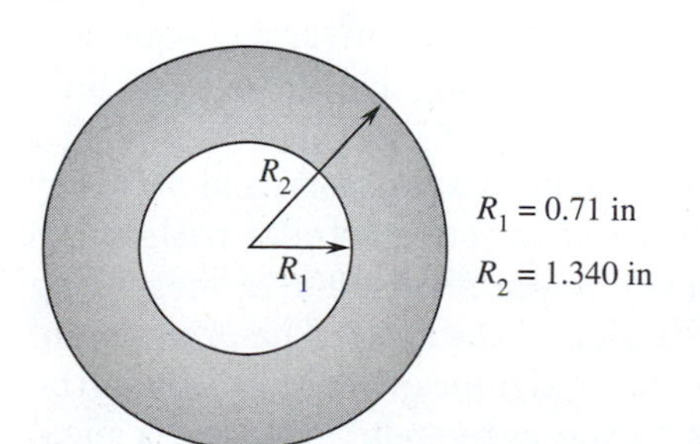

Figure 2–11: *Diagram for Example Problem 2–9. Find the area of the ring.*

When we square the terms (multiply a number by itself), we can retain four significant figures in the first term but only two in the second term.

$$A(\text{net}) = 3.14159 \times (1.796\ \text{in}^2 - 0.50\ \text{in}^2)$$

We can retain an LSD of 0.01 cm in the subtraction.

$$A(\text{net}) = 3.14159 \times (1.30 \text{ in}^2)$$

The final result has an accuracy of 3.

$$A(\text{net}) = 4.08 \text{ in}^2$$

In Example Problem 2–9 we rounded off at each step of the calculation. If we simply had keyed values into the calculator without rounding off at each step we would have gotten a calculator display of 4.057363485. This final result rounds off to give us a slightly different answer for the area.

$$A = 4.06 \text{ in}^2$$

Most people prefer not to round off numbers until the final result of a calculation has been obtained. Rounding off changes the values of the original inputs slightly and introduces **round-off error** into the calculation.

SUMMARY

Measurement is a process of comparing the size of something with a standard unit of measurement. The three basic units of measurement are distance, weight (or mass), and time. This book uses two systems of standard measurement: the British system and the metric system. In the British system, distance, weight, and time are measured in feet (ft), pounds (lb), and seconds (s). The metric system uses SI units: meters (m), kilograms (kg) for mass, and seconds (s). Prefixes (see Table 2–1) are often used with basic units to indicate multiples or factors of the base unit.

A conversion equation can be obtained by measuring one unit against another. The ratio of the two sides of the conversion is one, since they represent the same quantity. The ratio can be used as a multiplier to convert units from one system to another. Units of area may be obtained by squaring linear units in the conversion equation. Units of volume may be obtained by cubing the linear units in a conversion equation.

It is important to be able to determine the precision and accuracy of measurements. Precision is related to the exactness with which a quantity is measured. The instrumental precision of a measuring device is one-half the smallest division of the scale. The precision of a measurement is related to the least significant digit (LSD) in the measurement. Accuracy refers to the agreement between a measurement and an accepted or true value. The number of significant digits in a measurement estimates the accuracy of the measurement.

In decimal notation zeros that have no function other than to locate the decimal point are *not* significant. In scientific notation all zeros are significant.

Precision and accuracy of measurements are also important in calculations. For the operations of multiplication and division the result is only as accurate as the least accurate input. For the operations of addition and subtraction the result is only as precise as the least precise input. It is good practice not to round off numbers in a calculation until the final answer is obtained. This reduces round-off error.

KEY TERMS

If you can explain the following terms to a friend or classmate, you understand their meaning. If you cannot explain the terms, you should reread the sections in which they are discussed.

accuracy
conversion equation
distance
figures of accuracy
inertia
instrumental precision
leading zero
least accurate input
least precise input
least significant digit (LSD)
mass
measurement
precision
prefixes
readability
round-off error
time
weight

EXERCISES

Section 2.1:

1. Write the following measurements in base units. Use scientific notation. Check Table 2–1 for abbreviations. (Example: 11.2 ms = 1.12×10^{-2} s.)

   a. 64.0 μs **b.** 4.5 pm **c.** 23 km **d.** 34 kiloton
   e. 37 Mw **f.** 58.2 cm **g.** 1.4 mm **h.** 7.4 mw

2. A major country has a national debt of approximately 2.3 terabucks. Express this number in dollars (1 buck = \$1).

Section 2.2:

3. Convert the following lengths to meters (1.00 m = 3.28 ft).

   a. 42.0 ft **b.** 17.0 in **c.** 1.60 mi **d.** 18.7 yd **e.** 143 cm

4. Convert the following lengths to feet (1.00 m = 3.28 ft).

   a. 2.78 m **b.** 243 cm **c.** 1.20 km **d.** 2.1 mi
   e. 2.33×10^4 mm

5. Make the indicated area conversions.

   a. 7.20 yd^2 = ? ft^2 b. 32.0 ft^2 = ? in^2
   c. 23.8 ft^2 = ? m^2 d. 20.6 cm^2 = ? in^2
   e. 1.80 m^2 = ? cm^2 f. 1.20 mi^2 = ? km^2

6. Make the indicated volume conversions.

   a. 7.20 yd^3 = ? ft^3 b. 52.0 ft^3 = ? in^3
   c. 19.3 m^3 = ? ft^3 d. 11.8 cm^3 = ? in^3
   e. 2.00 m^3 = ? cm^3 f. 5.70 km^3 = ? cm^3

7. A building is 43.2 m high. If the height were expressed in feet rather than meters would there be more foot units than meter units?

Section 2.3:

8. Write down the measurements shown in Figure 2–12.

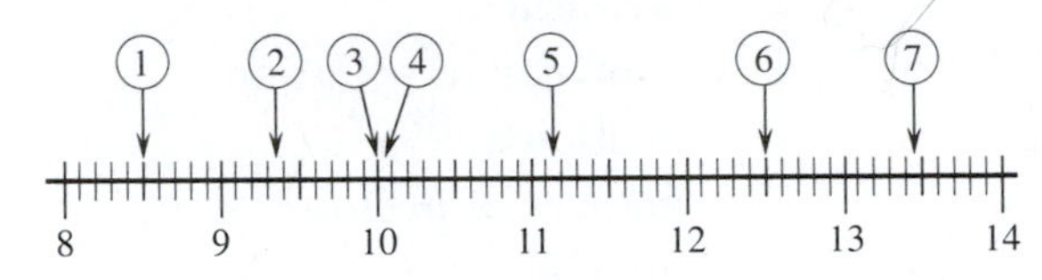

Figure 2–12: *Diagram for Exercise 8. Read the values from the scale.*

9. Write down the measurements shown in Figure 2–13.

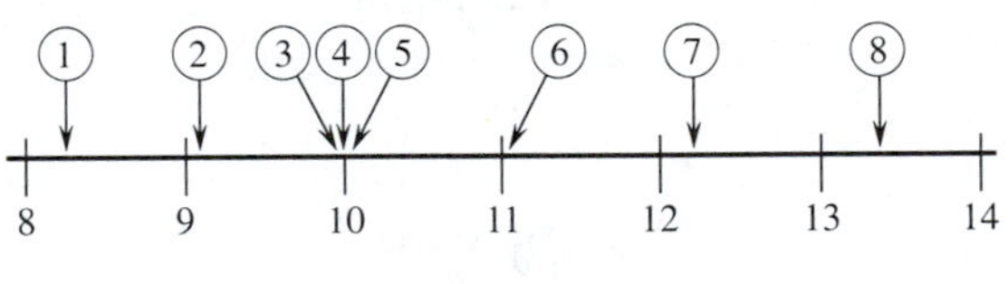

Figure 2–13: *Diagram for Exercise 9. Read the values from the scale.*

Section 2.4:

10. Underline the significant digits in each of the following numbers.

    a. 3400 ft b. 24,0$\bar{0}$0 in c. 0.00230 mm d. 40.00 ft^2
    e. 0.0200 lb f. 19.0 kg g. 23,000 mg h. 2003.0 cm^3

11. Convert the following measurements to scientific notation. Retain all significant zeros.

    a. 320 lb b. 0.000560 mm c. 23,0$\bar{0}$0 mg d. 73$\bar{0}$0 lb
    e. 803$\bar{0}$ ft f. 90 ton g. 0.0730 m h. 530.0 kg

12. Indicate the least significant digit (LSD) and the accuracy of each of the following measurements.

    a. 12.02 kg b. 12$\bar{0}$0 mi c. 0.0034 mm d. 72.00 w
    e. 10,$\bar{0}$00 ft f. 12.030 mm g. 10.0 ft/s h. 109.0 km/h

Section 2.5:

13. Perform the indicated operations. Give your answer with the proper number of significant figures.

    a. $12.3\text{ m} \times 0.72\text{ m} = ?$ b. $20.030\text{ lb} \times 0.720\text{ ft} = ?$

    c. $\dfrac{123.66\text{ g}}{18.3\text{ cm}^3} = ?$ d. $\dfrac{1200\text{ mi}}{40.0\text{ h}} = ?$

    e.
    $$\begin{array}{r} 12.023\text{ cm} \\ 105.0\ \ \text{ cm} \\ +\ 11.96\ \ \text{cm} \\ \hline \end{array}$$

    f. $43.19\text{ ft} - 11.2\text{ ft} = ?$

14. Perform the indicated operations. Give your answer with the appropriate number of significant figures.

    a. $3.14159 \times (2.30\text{ cm})^2$ b. $1450\text{ kg} \times 9.80\text{ m/s}^2 \times 2.0\text{ m}$

    c. $\dfrac{4000\text{ lb}}{21\text{ ft}}$ d. $\dfrac{(23.8\text{ ft}^2 - 19\text{ ft}^2)}{4.056\text{ ft}^2}$

    e.
    $$\begin{array}{r} 23.0\ \ \text{g} \\ 528.0\ \ \text{g} \\ 11.08\ \text{g} \\ +\ \ 9.276\ \text{g} \\ \hline \end{array}$$

    f. $(240\text{ slug} - 34\text{ slug}) \times 32.0\text{ ft/s}$

PROBLEMS

Section 2.2:

1. The average distance between the earth and the sun is 1.5×10^8 km. This distance is called an astronomical unit (AU). Neptune is 4.5×10^9 km from the sun. What is this distance in astronomical units?

2. A plumbing nipple is a short pipe threaded at both ends. How many 2.0-in nipples can be cut from a 2.0-m length of pipe? Allow 1/8 in (0.125 in) loss for each hacksaw cut.

3. In Europe, clothing sizes are in centimeters.
   A. What would a 30-in waist measurement be in centimeters?
   B. What would be the size in centimeters of a 34-in inseam?

4. In the Canadian Football League the playing field is 110.0 yd long. How close in length is this to a 100.0 m soccer field? (In Europe, soccer has gone metric.)

5. A crescent wrench opens to a maximum width of 2.0 in. Would it fit over a 5.0-cm handlebar adjustment?

6. What is the cross-sectional area of the I beam shown in Figure 2–14?

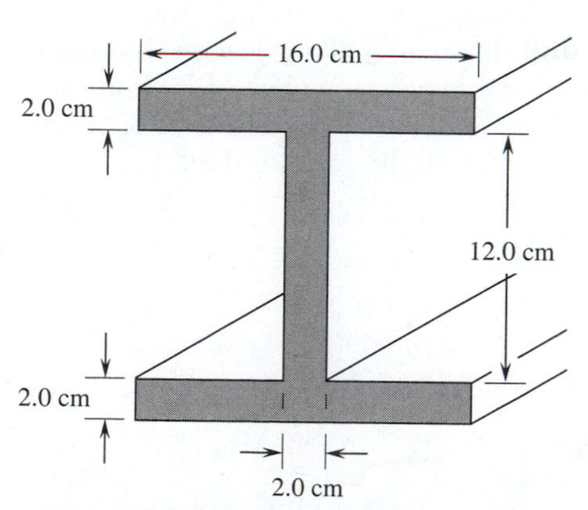

Figure 2–14: *Diagram for Problem 6. Find the cross-sectional area.*

7. Concrete is ordered to pour a patio that is 25.0 ft long and 17.8 ft wide. If the concrete is 6.0 in thick, how much concrete should be ordered? Express your answer in cubic yards.

8. Estimate the number of square tiles 10.0 in on a side needed to cover a kitchen floor 15.0 ft long and 10.0 ft wide.

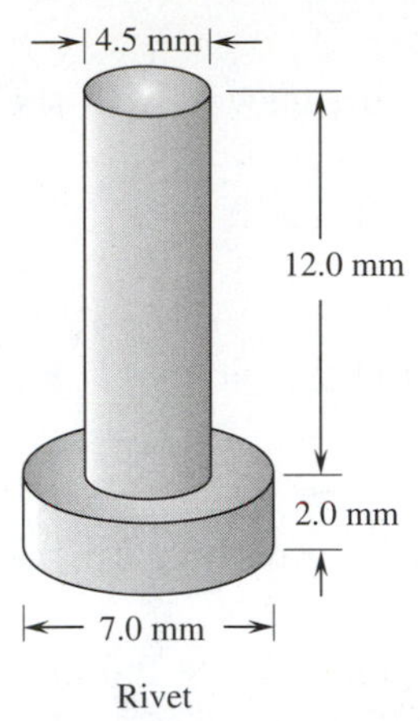

Figure 2–15: *Diagram for Problem 10.*

Figure 2–16: *Diagram for Problem 12. Does the snail have enough time to get across the road?*

9. An acre-foot is a unit of volume used in irrigation. An acre-foot is the volume of 1 acre of water 1 ft deep (1 acre = 43,560 ft^2). How many acre-feet are in a reservoir that covers 2.30 acres and has an average depth of 8.20 ft? How many cubic feet of water would this be?

***10.** Aluminum rivets are formed with the specifications shown in Figure 2–15. How many cubic feet of aluminum would be used to make 100,000 rivets?

***11.** Each cubic centimeter of aluminum has a mass of 2.70 g. What would be the total mass in kilograms of 100,000 rivets of the kind shown in Figure 2–15?

***12.** The operator of a power roller stops for a 1.00-h lunch break. At that moment, a snail races across the road in front of the roller at a speed of 12.0 furlongs per fortnight (Figure 2–16). After the lunch break the operator intends to roll the entire 20.0-ft width of the road. Will the snail make it across the road in time, or will it become an escargot pancake? (A furlong is used in horse racing. It is 1/8 mi. A fortnight is a unit of time found in old English novels. 1 fortnight = 14.0 days.)

Chapter 3

TOOLS OF THE TRADE

Users of the Chinese abacus can perform complex calculations with remarkable speed and accuracy. This chapter will introduce you to several ways in which physics manipulates quantities. (Photograph courtesy of International Business Machines Corporation.)

OBJECTIVES

In this chapter you will learn:

- to recognize a direct proportion
- how to relate direct proportion with a straight line
- how to find the equation of a straight line from a graph
- how to plot a power law equation to get a straight line and find the proportionality constant graphically
- to distinguish between scalars and vectors
- to add vectors using the graphical method
- to find the components of a vector
- to add vectors using the component method
- to multiply vectors in two ways

Time Line for Chapter 3

1000	Mathematician and poet Omar Khayyam unifies algebra and geometry.
1328	Thomas Brandwardine develops a theory of proportion.
1837	Rene Descartes publishes a work on analytical geometry.
1881	J. Willard Gibbs develops a system of three-dimensional vectors.

One of the useful things about physics is that it helps us find cause and effect relationships so that we can predict events. For instance, we can predict whether or not a brick will slide down a pitched roof if we know something about gravity, force, and friction. We can predict the path of a communications satellite if we know something about the pull of the earth's gravitation and the velocity of the satellite at some point in its orbit. The front wheels of a car exert a force on the pavement. If the car is on a hill, it may or may not move forward. It might slip backward. Given some information about the car and the road we could predict the direction of motion and we could calculate the car's speed.

In each of these cases we want to know two things: the direction of the effect and the size of the change that will occur. Two mathematical tools help us to find this information: proportion and vectors.

In this chapter we will look at different kinds of proportion. We will then develop the idea of a vector, a quantity that has both a size and a direction.

3.1 DIRECT PROPORTION

Two students want to predict the weight of different-sized sacks of beach sand. By checking the weight of different volumes of sand they hope to find a mathematical relationship that will help them in their predictions. In the laboratory they measure the weight of different quantities of sand in a graduated cylinder (Figure 3–1). Table 3–1 shows the data found by the first student. The column on the right shows the ratio of weight to volume for each pair of measurements. Because this is an experimental result, we expect a little random error in the measurements, but the ratio is fairly constant. We say we have a **direct proportion** when two quantities are related by a constant ratio. The ratio is called the **proportionality constant** **(*k*)**. For this experiment the proportionality constant is the average of the experimental ratios, or 0.080 lb/in^3.

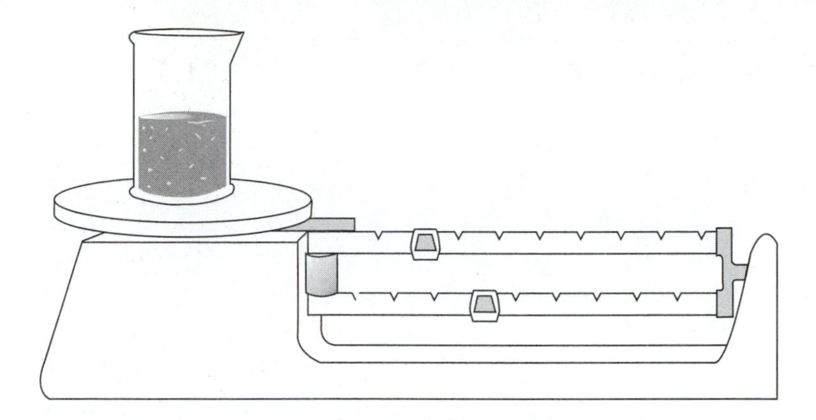

Figure 3–1: The weight of a volume of sand is measured on a balance.

One way to find the average proportionality constant is to graph the data. Figure 3–2 shows a graph of the information found by the first student. The direct proportion plots into a straight line through the origin. To make sure that the line has the best fit for all of the data points choose two points on the line and find the ratio of the change of y to the change of x between the two points. The symbol Δ means "the change of." The ratio of $\Delta y/\Delta x$ is called the **slope** of the line. The slope of the line ($\Delta y/\Delta x$) has units of pounds per cubic inch. We have found the average proportionality constant graphically. We get a slope of 0.080 lb/in^3.

A direct proportion *plots into a straight line through the origin. The slope of the line is the* proportionality constant.

$$\frac{\text{change of } y}{\text{change of } x} = \frac{\Delta y}{\Delta x} = \text{slope}$$

Table 3–1: Weights and volumes of sand, first experiment.

WEIGHT (lb)	VOLUME (in^3)	$\frac{\text{WEIGHT}}{\text{VOLUME}}$ $\left(\frac{\text{lb}}{\text{in}^3}\right)$
0.80	10.0	0.080
1.62	20.0	0.081
2.43	30.0	0.081
3.17	40.0	0.079
4.00	50.0	0.080
4.72	60.0	0.079
5.60	70.0	0.080
6.41	80.0	0.080

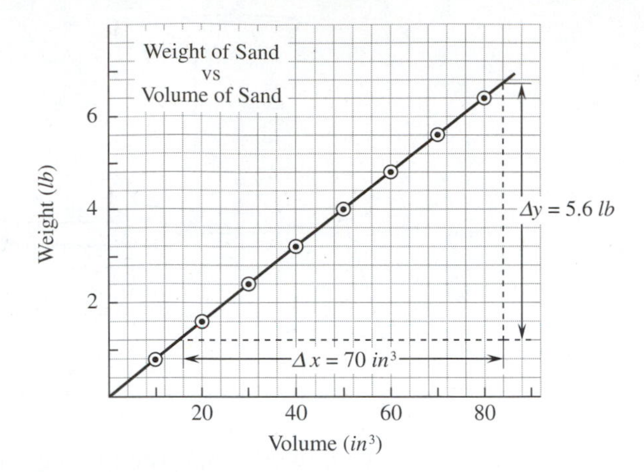

Figure 3–2: A graph of the data in Table 3–1. The ratio of weight to volume is the slope of the line.

Once we have the proportionality constant, we can predict the weight for any particular volume.

☐ **EXAMPLE PROBLEM 3–1: THE SACK OF SAND**

What would be the weight of a 1.00-ft^3 sack of sand using the information found by the first student?

■ ***SOLUTION***

We can use the direct proportion to find the weight. Weight (W) = proportionality constant (k) × volume (V).

$$W = k\,V$$

$$W = 0.080\,\frac{\text{lb}}{\text{in}^3} \times 1.00\ \text{ft}^3 \times \left(\frac{12\ \text{in}}{1\ \text{ft}}\right)^3$$

$$W = 138\ \text{lb}$$

☐ **EXAMPLE PROBLEM 3–2: A SMALLER SACK**

We would like to package the sand for sale in supermarkets for sandboxes. Sacks weighing 140 lb are too heavy. We will package the sand in 75-lb plastic sacks instead. Calculate the volume of sand needed for a 75-lb sack of sand. Express the answer in cubic feet.

■ ***SOLUTION***

We can use the proportionality backwards.

$$V = \frac{W}{k}$$

$$V = \frac{75 \text{ lb}}{0.080 \text{ lb/in}^3}$$

$$V = 937.5 \text{ in}^3 \times \frac{1 \text{ ft}^3}{1728 \text{ in}^3}$$

$$V = 0.54 \text{ ft}^3$$

If the student had taken data for weight versus volume for a different material, such as water, cooking oil, or sugar, the slope would have been different (see Figure 3–3). Each slope represents a different proportionality constant. In each case the relationship is the same, a direct proportion between weight and volume. Only the proportionality constant is different. Because the constants depend only on the type of material used in the experiments, they are called **material constants.** The ratio of weight to volume for different materials can be tabulated in handbooks for future use. Material constants are usually given a name. The ratio of weight to volume is called **weight density.** We will use the symbol d to represent this ratio. For a wide range of materials we can write a direct proportional equation to predict weights or volumes.

$$W = d\,V \qquad \textbf{(Eq. 3–1)}$$

In the meantime, the second student has also gotten a set of data for the same experiment. These data are shown in Table 3–2. Notice that the ratio of weight to volume is not very constant. This may not be a direct proportion. A plot of the data is shown in Figure 3–4. The first graph is shown for comparison.

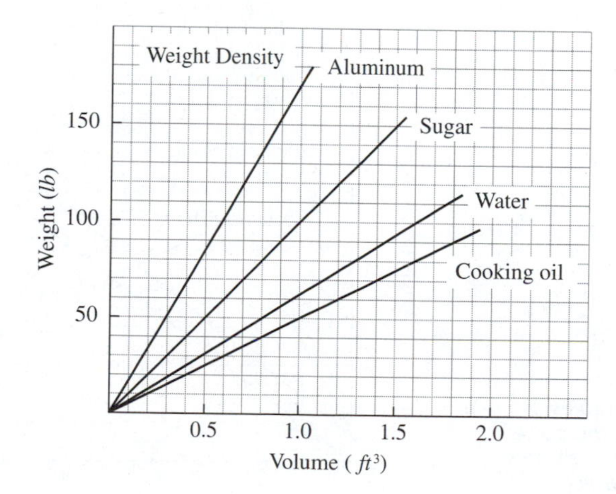

Figure 3–3: The ratio of weight to volume is different for different materials. The proportionality constant is called weight density.

Table 3–2: Weights and volumes of sand, second experiment.

WEIGHT (lb)	VOLUME (in^3)	$\frac{\text{WEIGHT}}{\text{VOLUME}}$ $\left(\frac{lb}{in^3}\right)$
1.23	10.0	0.123
2.02	20.0	0.101
2.90	30.0	0.097
3.61	40.0	0.090
4.41	50.0	0.088
5.15	60.0	0.086
6.02	70.0	0.086
6.83	80.0	0.085

The data in Table 3–2 plot into a straight line. However, the line does not go through the origin, even though the slope, 0.080 lb/in^3, agrees with the data of the first experimenter. The second student reexamines his data. He finds that he did not subtract the weight of the beaker from the total reading of container plus sand. The data and the graph show a direct proportion, but a constant value has been added. The easiest way to find the added constant is to look at the point where the volume is zero. This is where the line crosses the vertical (y) axis of the graph. This point is called the **y intercept.** The weight is 0.42 lb at a volume of zero sand. This is the weight of the graduated cylinder that was added to each of the measurements. We can write the second experimenter's result as a direct proportion with an added constant (b).

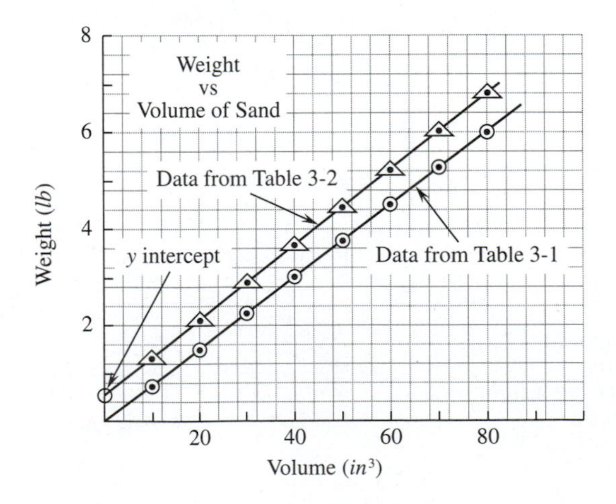

Figure 3–4: The data from Tables 3–1 (⊙) and 3–2 (△) are plotted on the same graph. The upper line has the same slope as the bottom line. It is a direct proportion with an added constant.

$$W = (d\,V) + b$$

$$W = 0.080 \text{ lb/in}^3 + 0.42 \text{ lb}$$

In general we can view any straight-line graph as a direct proportion with an added constant. We will use the symbol m to mean the slope and b to mean the y intercept; y and x are the variables. The equation for any straight line is

$$y = (m\,x) + b.$$

☐ **EXAMPLE PROBLEM 3–3: AVERAGE SPEED**

The proportionality ratio between total distance traveled (s) and the elapsed time (t) is called **average speed** (v_{avg}). Figure 3–5 is a plot of the position versus time of a basketball rolling across a gymnasium floor.

A. Write the equation for the graph of distance versus time shown in Figure 3–5.

B. What does the slope represent?

C. What does the y intercept mean?

■ ***SOLUTION***

A. The slope of the curve is:

$$m = \frac{\Delta s}{\Delta t} = \frac{(40.0 - 8.0) \text{ ft}}{(8.0 - 1.0) \text{ s}}$$

$$m = 4.6 \text{ ft/s}$$

The y intercept (b) is 3.4 ft.
The equation is

$$s = (4.6 \text{ ft/s})\,t + 3.4 \text{ ft}$$

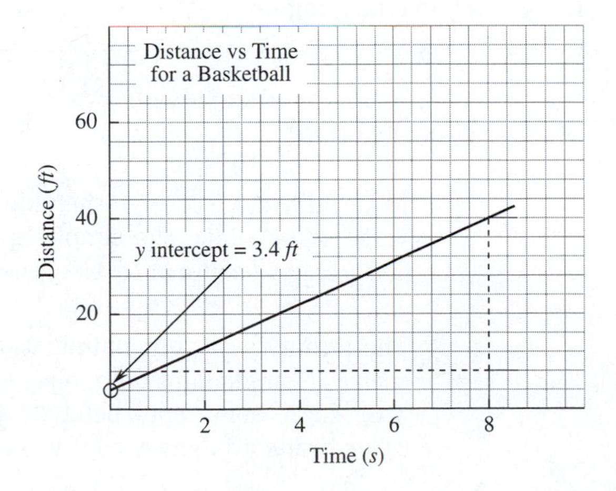

Figure 3–5: *Graph for distance versus time for a basketball rolling across a gym floor.*

B. The slope is the average speed since it is the proportionality constant between distance and time.

C. When we started our stopwatch, the basketball was already at the distance 3.4 ft.

Language, Symbol, and Thought

For you and I to understand each other, we need some sort of code, a set of symbols, a language. The better the code, the better will be the information exchanged. The amount and kind of information we exchange depend on the limits of the code we use.

The earliest written languages were mathematical in nature. Tally bones recording numbers reach back to 20,000 BC. About the same time cave paintings began to appear. Eventually, written picture languages developed. Simplified, standard pictures began to represent ideas. Egyptian hieroglyphics is an example. These pictographic languages needed one complete picture to represent one idea. One needed to memorize a catalog of thousands of pictures to make up all the ideas found in a complex culture. If someone wanted to express a new idea, a new symbol needed to be invented.

Sumerians developed a better idea. They keyed their written language based on their oral language. Each syllable had a symbol. This reduced the number of symbols down to about 700 that could be learned by many people in the culture. New ideas could be expressed by scrambling the old symbols in a new way. This made for a flexible language.

The Phoenicians extended the idea a bit farther. Instead of having a symbol for each syllable in the oral language, they had a symbol for each consonant sound. This reduced the symbols to 22. There were no vowel symbols. You were expected to figure them out by context. This was particularly adaptable to different languages. This language shorthand quickly spread throughout Europe and Eastern Asia.

Special mathematical languages have been developed over the ages. Here are a few of them:

- *Arithmetic* A counting language to keep track of the number of things you have.
- *Geometry* A language used to describe the shape of things.
- *Algebra* A language used to compare the relative size of different things.
- *Calculus* A language used to measure how things change.
- *Fortran, Basic, Pascal, C, Cobol* Some of the many languages used to communicate with a computer.

3.2 OTHER KINDS OF PROPORTION

Physical systems obey other kinds of proportion. They are the inverse square law, the simple inverse law, y is proportional to x^2, and y is proportional to $x^{1/2}$. These are called the **power laws.**

Inverse Square Law

The intensity of light illumination (I) from a point source of light, such as a single light bulb, obeys an **inverse square law** (see Figure 3–6). Light sometimes behaves as if it were made up of waves. At other times it behaves as if it were made up of tiny particles. Here

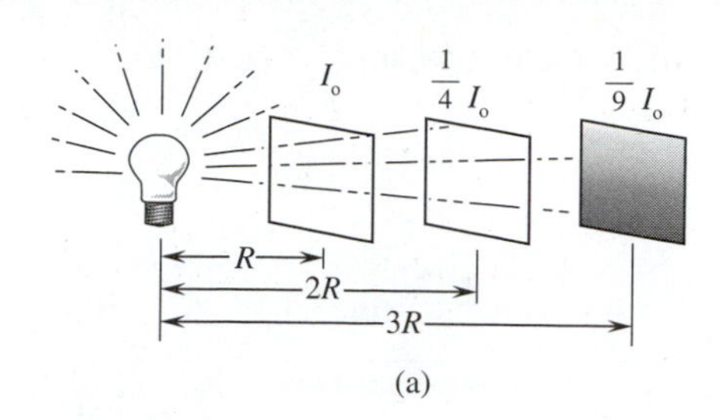

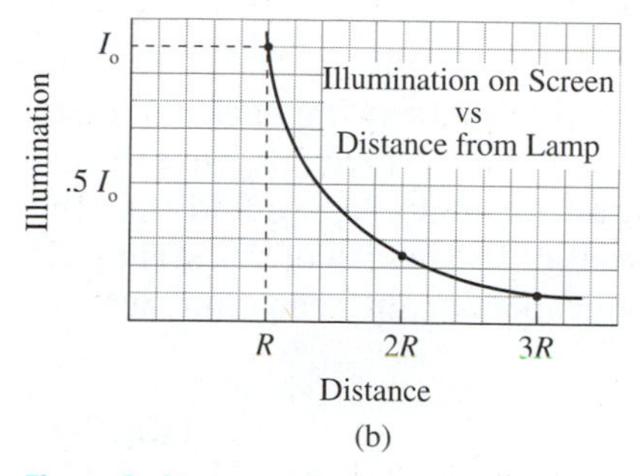

Figure 3–6: *An example of the inverse square law.* (a) *The illumination on a screen falls off as an inverse square as the screen is moved away from the lamp.* (b) *A plot of illumination versus distance.*

we will assume that light is made up of many small particles called **photons.** The number of light photons striking a square foot of area at right angles to the path of the light decreases as distance increases. A screen is placed at a distance R from the bulb. When the distance is doubled, the illumination on the screen falls to one-fourth of the original value. If the distance is tripled, the illumination is only one-ninth of the original value. We can write this as a proportion. The symbol $\propto$ means "proportional to."

$$I \propto \frac{1}{R^2}$$

Let I_0 be the illumination when $R = 1$ unit of distance. I_0 is our proportionality constant. We can write the equation

$$I = \frac{I_0}{R^2}$$

or

$$I = I_0 R^{-2} \qquad \textbf{(Eq. 3–2)}$$

Simple Inverse Law

Most waves, such as sound waves and light waves, obey a simple **inverse law.** The frequency (f) is inversely proportional to the wavelength (λ). If the frequency is doubled, the wavelength is half its original value. If the frequency is increased by a factor of four, the wavelength falls to one-fourth of its original value (Figure 3–7).

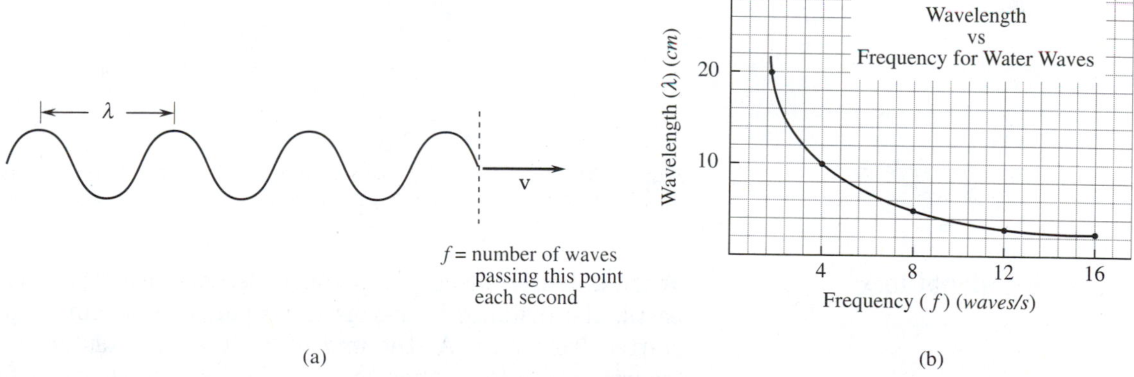

Figure 3–7: *An example of simple inverse proportion.* (a) *Wavelength is the distance between wave peaks. Frequency is found by counting the number of waves passing a reference point* (f) *per second.* λ, *wavelength;* v, *proportionality constant.* (b) *A plot of wavelength versus frequency for water waves.*

If v is the proportionality constant, we can write

$$\lambda = \frac{v}{f}$$

or

$$\lambda = v\,f^{-1} \qquad \textbf{(Eq. 3–3)}$$

y Is Proportional to $x^{1/2}$

The amount of time required for a pendulum to swing back and forth once is called the **period** (P) of the pendulum. If we put a heavy bob on a string and let it swing through a small angle, we find that the period increases with the square root of the length of the string (L) (see Figure 3–8). If we let k be the proportionality constant, we can write

$$P = k\,L^{1/2} \qquad \textbf{(Eq. 3–4)}$$

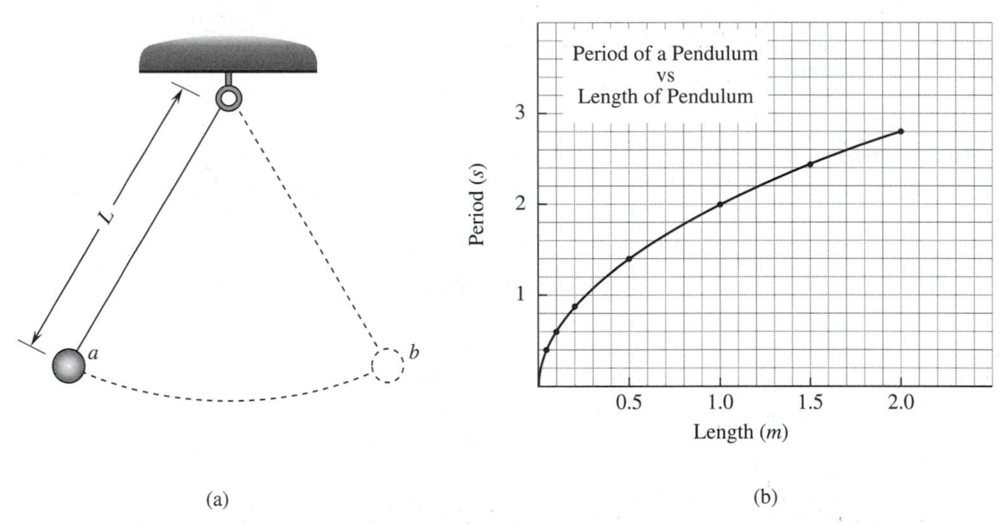

Figure 3–8: (a) *The period is the amount of time for the bob to swing from a to b and back again.* (b) *A plot of period versus length of a simple pendulum.*

y Is Proportional to x^2

When a dense object falls a short distance near the surface of the earth, the distance it travels is proportional to time squared if it started from rest. At the end of 2 s the total distance the object travels will be four times as far as in the first second. After 3 s the total distance traveled will be nine times the distance traveled during the first second. If k is the proportionality constant we can write

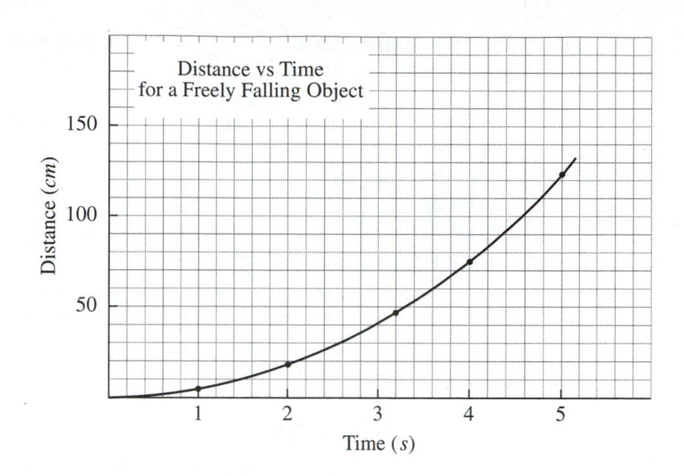

Figure 3–9: The distance traveled by a freely falling object plotted against time.

$$s = k\,t^2 \qquad \textbf{(Eq. 3–5)}$$

Equations 3–1 through 3–5 are a few of the many proportions that will be developed in more detail later on in this text. They all have one thing in common. One variable is proportional to a power of the other variable. They can be written as one general equation known as the power law. If m is the proportionality constant:

$$y = m\,x^n \qquad \textbf{(Eq. 3–6)}$$

If $n = 1$, the equation is a direct proportion. If $n = -1$, the equation is a simple inverse law. If $n = -2$, the equation is an inverse square law. Values of $n = 1/2$ and $n = 2$ produce the other two power laws we have discussed.

There are two graphical methods for analyzing data for a power law. The first method is to plot the data on logarithm paper. This method is shown in Appendix B. The second method is to recognize that y is not proportional to x, but is proportional to a power of x. If we plot y versus x^n, rather than y versus x, we get a straight line. The slope of the straight line is the proportionality constant. Here are a couple of examples.

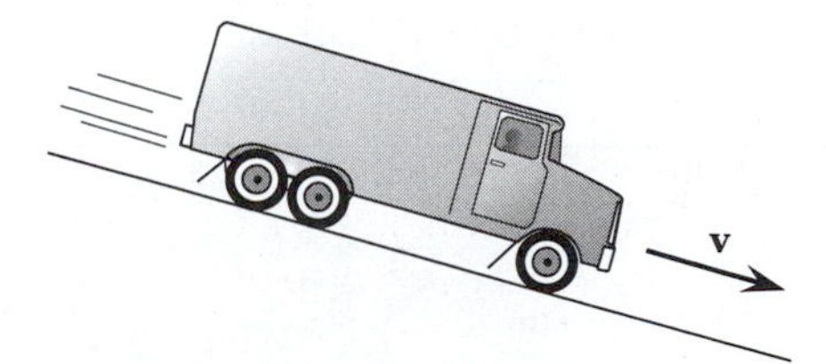

Figure 3–10: A truck rolls down an incline at velocity, **v**.

EXAMPLE PROBLEM 3–4: THE ROLLING TRUCK

Table 3–3 shows the total distance (s) traveled by a truck rolling down a hill as a function of time (t). Find the proportionality constant between distance and time, and write the equation.

Table 3–3: The rolling truck.

DISTANCE TRAVELED (m)	TIME (s)
2.8	1.0
10.7	2.0
24.2	3.0
43.3	4.0
67.7	5.0

Table 3–4: Rolling truck with time squared.

DISTANCE (m)	TIME (s)	TIME SQUARED (s^2)
2.8	1.0	1.0
10.7	2.0	4.0
24.2	3.0	9.0
43.3	4.0	16.0
67.7	5.0	25.0

■ ***SOLUTION***

First let us plot the data (see Figure 3–11a). Notice that y increases more rapidly than x. This suggests that n is larger than one. We will try $n = 2$. In Table 3–4 we have added the values of x^2 to Table 3–3. We now plot y versus x^2 to get the graph shown in Figure 3–11b. This is a straight line. We have made the right guess for the power of n. The slope of the line in Figure 3–11b is $m = 2.7\ \text{m/s}^2$. The equation is

$$y = m\,x^{n}$$

$$D = \left(2.7\,\frac{\text{m}}{\text{s}^2}\right)t^2$$

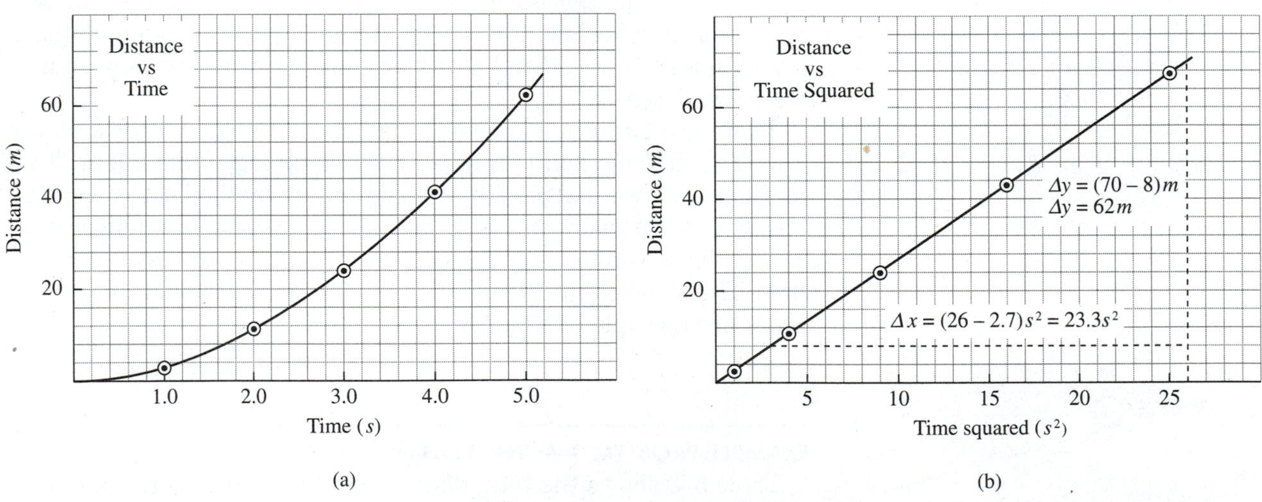

Figure 3–11: (a) *A plot of distance* (D) *versus time* (t) *for the rolling truck.* (b) *The straight line indicates that* D *is proportional to* t^2. *The slope of the line is the proportionality constant.*

☐ EXAMPLE PROBLEM 3–5: FREQUENCY AND WAVELENGTH

A sine wave is generated by a signal generator. The wave is shown on an oscilloscope. As the frequency (f) is increased, the wavelength (λ) of the pattern on the oscilloscope is recorded. The data are shown in Table 3–5. Find the proportionality constant and equation for this situation.

Table 3–5: Wavelengths and frequencies.

WAVELENGTH (cm)	FREQUENCY [(waves/s) × 10^4]
4.0	0.30
2.4	0.50
1.2	1.0
0.6	2.0
0.4	3.0
0.3	4.0

■ *SOLUTION*

Figure 3–12a shows a plot of wavelength versus frequency. We suspect that this is an inverse law. Values of $1/f$ are shown in Table 3–6. Figure 3–12b is the plot of wavelength vs. $1/f$. Here we get a straight line. The slope is $m = 1.2 \times 10^4$ cm/s. The equation is

$$y = m\,x^n$$

$$\lambda = (1.2 \times 10^4 \text{ cm/s})f^{-1}$$

or

$$\lambda = \frac{1.2 \times 10^4 \text{ cm/s}}{f}$$

Table 3–6: Wavelengths, frequencies, and reciprocal frequencies.

WAVELENGTH (cm)	FREQUENCY [(waves/s) × 10^4]	1/FREQUENCY [(s/wave) × 10^{-4}]
4.0	0.30	3.3
2.4	0.50	2.0
1.2	1.0	1.0
0.6	2.0	0.50
0.4	3.0	0.33
0.3	4.0	0.25

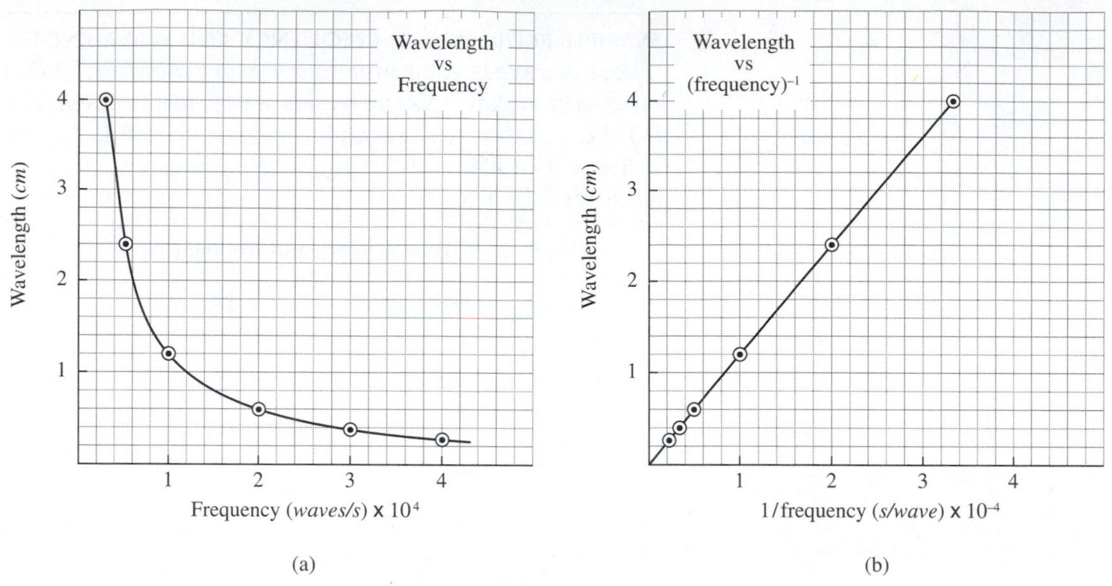

Figure 3–12: (a) *A plot of wavelength versus frequency using the data from Table 3–5.* (b) *A plot of wavelength versus reciprocal frequency* (1/f) *using the data from Table 3–6.*

SECTION 3.3: SCALARS AND VECTORS

Two kinds of quantities are used to solve physics problems: scalars and vectors. Here are a couple of problems to help identify them.

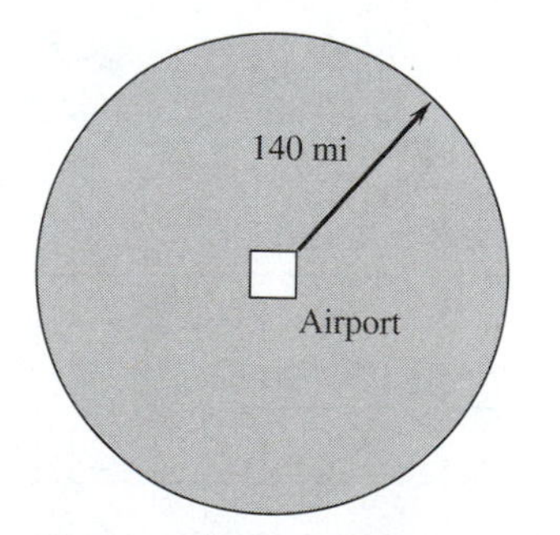

Figure 3–13: *The airplane can be anywhere inside a circle with a radius of 140 mi.*

☐ **EXAMPLE PROBLEM 3–6: THE AIRPLANE**

A light airplane leaves an airport. It travels at a speed of 140 mph for 1 h. Where is the airplane at the end of 1 h?

■ ***SOLUTION***

All we know is that the airplane traveled at a constant speed. If it traveled in a straight line it could be as far as 140 mi away in any direction. But if the airplane circled the airport for an hour it would be back at its starting point. Figure 3–13 shows the solution. The airplane is somewhere inside a circle 140 mi in radius with the airport at the center.

EXAMPLE PROBLEM 3–7: THE AIRPLANE AGAIN

A light airplane leaves an airport. It travels at a constant velocity of 140 mph in a heading of due east. Where is the airplane at the end of 1 h?

SOLUTION

Here we have the added information of direction. We now know that the airplane will be 140 mi east of the airport at the end of the hour.

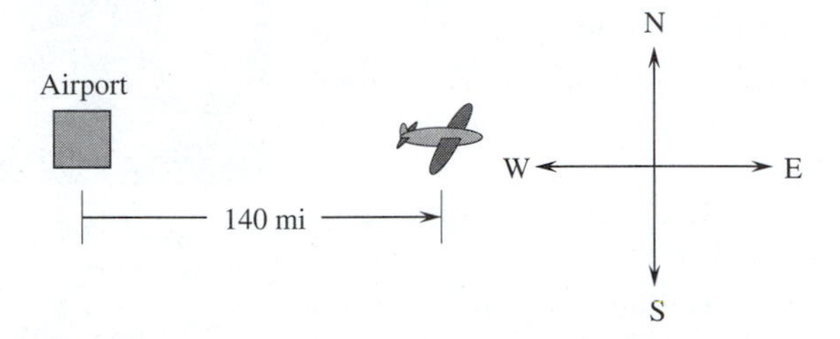

Figure 3–14: *The airplane is pinpointed 140 mi east of its starting point.*

In Example Problem 3–6 we knew only how fast the plane was traveling. We did not know its direction. This is called a **scalar** quantity. A scalar has only a magnitude. There will also be a set of measuring units associated with the measured quantity. Counting numbers are scalars. The number of bolts in a cardboard box is a scalar quantity. The speedometer in a car gives the scalar quantity **speed.** The speedometer tells you how fast you are moving, but not where you are going. Most speedometers also have an odometer, which tells how far the car has traveled, but it does not tell in what direction the car has traveled. This is a scalar quantity called **distance.**

A measured scalar *quantity has a magnitude and units.*

In Example Problem 3–7 we had the added information of direction. This made it possible to pinpoint the position of the plane exactly. We were given the **vector** quantity called velocity. A vector has both a magnitude and a direction. The measured quantity has a set of units associated with the vector.

When you push on something, you exert a force. The force or push is in a particular direction. It is a vector quantity. The distance walked in a particular direction is also a vector quantity because a direction is given. This vector quantity is distinguished from distance by calling it **displacement.**

A measured vector *quantity has a magnitude, a direction, and a set of units.*

2 + 2 = ?

One of the first number "facts" you learned in grade school was 2 + 2 = 4.

You probably learned another number fact at about the same time: You can't add apples and oranges. Actually, you *can* add apples and oranges—as long as you call them all objects or pieces of fruit. The idea here is that you need to add the same kind of things. In this scalar addition, we place all of the objects in one pile and then count the total number of objects in this larger pile. We get an answer of 4.

When we add 2 and 2, can the answer be something other than 4? Let's count the ways.

(Photograph courtesy of International Business Machines Corporation.)

First let's try this way. Let's write our numbers using a base 3 system rather than the usual base 10 system. In a base 3 system, there are only three symbols for numbers: 0, 1, and 2. When a grouping size reaches 3, we put a marker to the left of the unit's position. In this notation, one grouping of 3 (in base ten) is written 10 (in base three). 2 apples + 2 oranges gives us one grouping of three pieces of fruit with one left over. In base three notation, 2 + 2 = 11.

Skeptical? We still have four pieces of fruit. The only thing that's changed is the way we write it down on this page. The concepts—the numbers—haven't changed. Only the symbols, or numerals, we've chosen to represent the numbers have changed; 2 + 2 still equals 4.

Let's try something else. We'll fill a two-pint container with pebbles. Another two-pint container is filled with fine sand. We add the sand and the pebbles by mixing them in a four-pint container. Whoops! The four-pint container is only partly filled. $2 + 2 \neq 4$.

What went wrong?

We tricked ourselves. We were really counting the pebbles, sand, *and the empty space between them*. When we mixed the solid particles, sand replaced some of the empty space between the pebbles and forced the empty space to the top of the container. The four-pint container has the same total of pebbles, sand, and empty space as the two separate containers: 2 pints + 2 pints = 4 pints. The same thing happens if we mix sugar and water or alcohol and water.

We were too lazy to count all the pebbles and all the grains of sand. Instead we used volume and counted the space between the particles as well. If we only wanted to consider the solid particles, we should have used some property that depended only on the particles, such as weight.

Let's try one final addition.

Add a displacement of 2 feet to the east to a displacement of 2 feet to the north. Graphically this is done by drawing arrows, the second arrow starting from the head of the first. We then measure the distance from where we started to where we finished. In this case, we're only 2.83 feet from where we started. If the arrows pointed in the same direction, the answer is 4. If the arrows point in opposite directions, the answer is 0.

In vector addition, we mean something very different from scalar addition. They're not the same kind of operation.

So, does 2 + 2 = 4? In careful scalar addition, the answer is yes. In vector addition, all bets are off.

EXAMPLE PROBLEM 3–8: VECTORS AND SCALARS

Identify the following quantities as vectors or scalars.

A. Your shoe size.
B. A $\frac{1}{8}$-in diameter machine bolt.
C. A car speeding northward at 68 mph.
D. The density of water.
E. A 120-lb weight pressing down on a table.
F. A bike hike of 5.7 km to the east.

SOLUTION

A. Scalar quantity.
B. Scalar quantity.
C. Vector quantity. This is a velocity in the direction north.
D. Scalar quantity. This is a direct proportion with no direction.
E. Vector quantity. The weight acts downward.
F. Vector quantity. This is a displacement to the east.

3.4 GRAPHICAL METHOD OF VECTOR ADDITION

Figure 3–15 shows a map of a camping area. Like most maps it shows some land features drawn to proportion. The distances between the features are proportional to real, life-size distances. A legend shows the proportionality constant or **scale factor.** Directions are measured

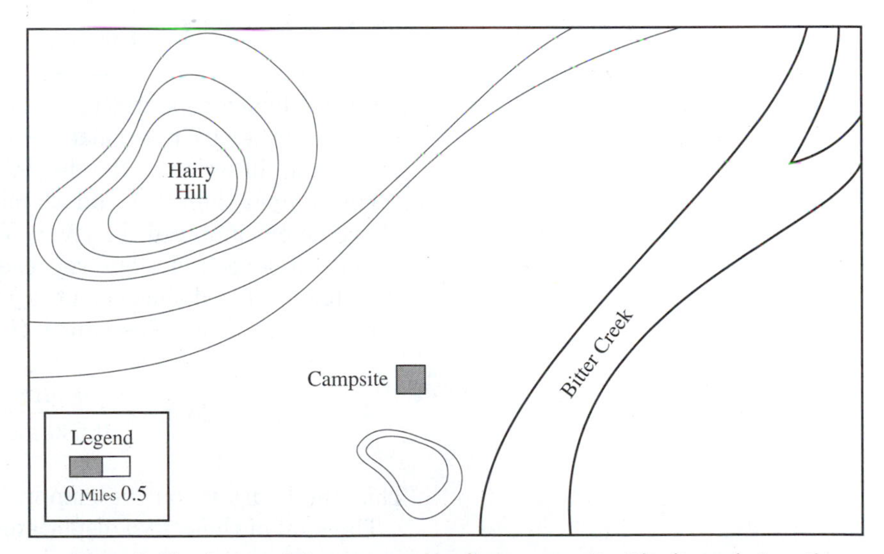

Figure 3–15: *A map of the area surrounding a campsite. The legend gives the scale factor.*

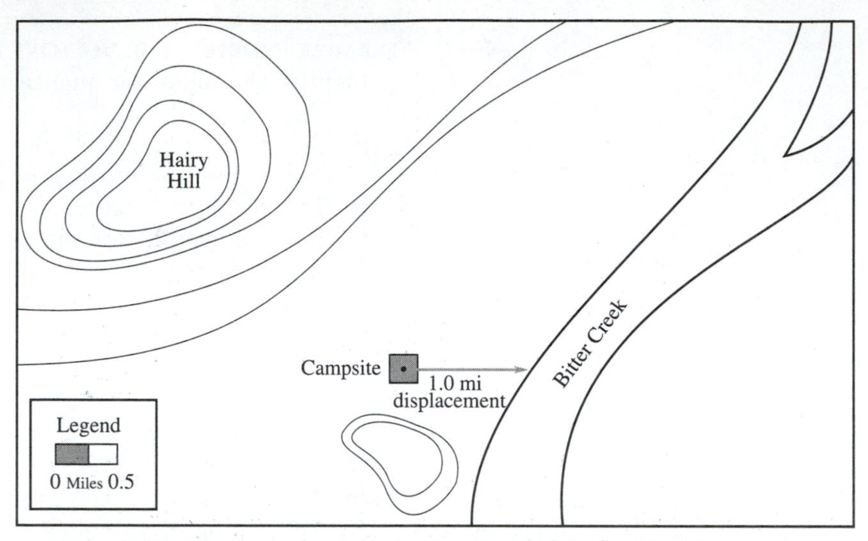

Figure 3–16: *The arrow shows the camper traveled 1 mi eastward.*

by the points of a compass. A camper leaves the campsite traveling eastward until he encounters the river (see Figure 3–16). We can measure the length of the line and compare it with the legend to find the magnitude of the camper's eastward displacement. The line is 2.00 units long. The scale factor is 1.00 unit = 0.500 miles. The displacement ($\mathbf{s}_1$) of the camper is:

$$\mathbf{s}_1 = 2.00 \text{ units} \times \frac{0.500 \text{ mi}}{1.00 \text{ unit}} = 1.00 \text{ mi eastward}$$

Note: (Boldface type in the equation indicates that the quantity is a vector rather than a scalar.) We have determined both the magnitude and direction of the displacement vector from the map.

The camper then travels 1.25 mi toward the north from the point where he encountered the river. To draw this on the map, we need to find the length of a line representing 1.25 mi. Again we use the scale factor. The displacement ($\mathbf{s}_2$) is 1.25 mi. The scale factor is 1.00 unit = 0.500 miles. The length (L) of the line is:

$$L = 1.25 \text{ mi} \times \frac{1.00 \text{ unit}}{0.500 \text{ mi}} = 2.50 \text{ units}$$

This line is drawn on the map in Figure 3–17.

The sum of these two displacements tells us where the camper is relative to the campsite. Look at Figure 3–18. The displacement vector labeled **R** is the sum of the other two vectors. It is called the **resultant.** We can write a vector equation.

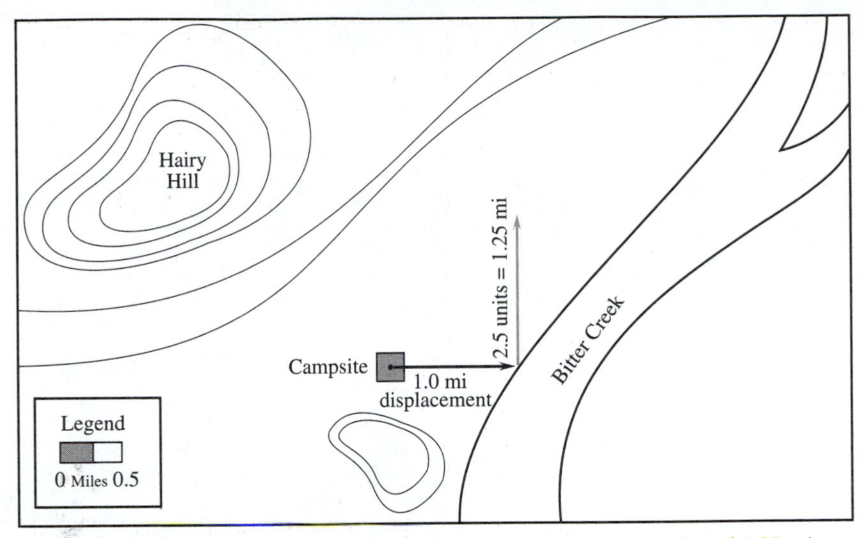

Figure 3–17: The vertical arrow shows that the camper was displaced 1.25 mi north from his location at the riverbank.

$$\mathbf{R} = \mathbf{s}_1 + \mathbf{s}_2$$

R measures 3.2 units long. The magnitude of the resultant is

$$|\mathbf{R}| = 3.20 \text{ units} \times \frac{0.500 \text{ mi}}{1.00 \text{ unit}} = 1.60 \text{ mi}$$

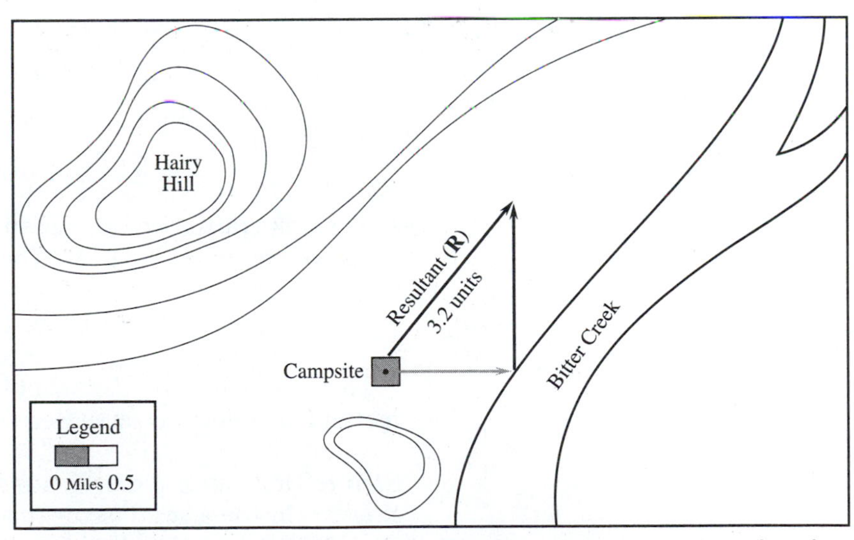

Figure 3–18: The resultant **(R)** is the net displacement vector. It shows how far the camper is from the campsite and in which direction.

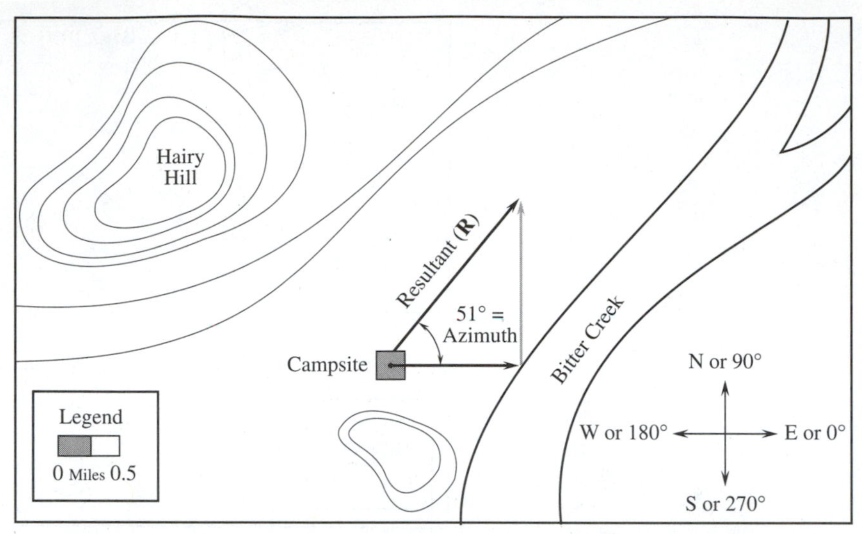

Figure 3–19: *The direction of the resultant displacement is found using an azimuth angle.*

The direction of **R** is a little north of northeast. We might get a better estimate of the direction if we use an angle rather than a point of the compass. We will call the direction east ($+x$ axis), the angle 0°. We can measure angles counterclockwise through a whole circle of 360°. The angle measured counterclockwise from the $+x$ axis is called the **azimuth angle.** Each direction of the compass is uniquely defined as an angle (see Figure 3–19).

$$\text{East } (+x \text{ axis}) = 0^\circ$$

$$\text{North } (+y \text{ axis}) = 90^\circ$$

$$\text{West } (-x \text{ axis}) = 180^\circ$$

$$\text{South } (-y \text{ axis}) = 270^\circ$$

Using a protractor we can measure the azimuth angle of the resultant. We get an angle of 51°. The resultant is

$$\mathbf{R} = 1.60 \text{ mi } \angle 51^\circ$$

There is a method for mapping vectors called the **graphical method** of adding vectors. Here are the steps for this process.

Step 1: Draw an x–y coordinate system (map directions).
Step 2: Choose a scale factor (map legend).
Step 3: Draw the first line from the origin (starting point).

Step 4: The length of the line is proportional to the magnitude of the vector. Its direction is given by an azimuth angle (point of the compass).

Step 5: Draw the next vector from the head of the first vector.

Step 6: Additional vectors can be added by drawing them to head to tail to form a polygon.

Step 7: The resultant is the vector drawn from the origin to the head of the last vector.

□ **EXAMPLE PROBLEM 3–9: DISPLACEMENT**

Add the displacement vectors of $\mathbf{A} = 200 \text{ m} \angle 30°$ and $\mathbf{B} = 150 \text{ m} \angle 120°$.

■ *SOLUTION*

You will need a pencil, a centimeter ruler, and a protractor.

1. First draw coordinate axes like those shown in Figure 3–20a.
2. Next choose a scale factor. We will try a scale factor of 1.00 cm = 25.0 m.
3. Convert the vectors into ruler lengths.

$$|\mathbf{A}| = 200 \text{ m} \times \frac{1.00 \text{ cm}}{25 \text{ m}} = 8.0 \text{ cm}$$

$$|\mathbf{B}| = 150 \text{ m} \times \frac{1.00 \text{ cm}}{25 \text{ m}} = 6.0 \text{ cm}$$

4. Use the protractor to find a heading of 30°. Along this direction draw a line 8.0 cm long (Figure 3–20b).
5. Next draw a horizontal reference line at the head of vector **A**. Remember the azimuth angles are points of the compass. We always have to measure angles from a line relative to a horizontal line directed parallel to the $+x$ axis.
6. Measure an angle of 120° relative to the reference line.
7. Draw a 6.0-cm line along the 120° heading (Figure 3–20c).
8. Draw the resultant vector (Figure 3–20d, **R**). Measure its length and angle. Here we get a length of 10.0 cm and an angle of 67°.
9. Use the scale factor to convert centimeters to meters.

$$|\mathbf{R}| = 1\bar{0} \text{ cm} \times \frac{25 \text{ m}}{1.00 \text{ cm}} = 250 \text{ m}$$

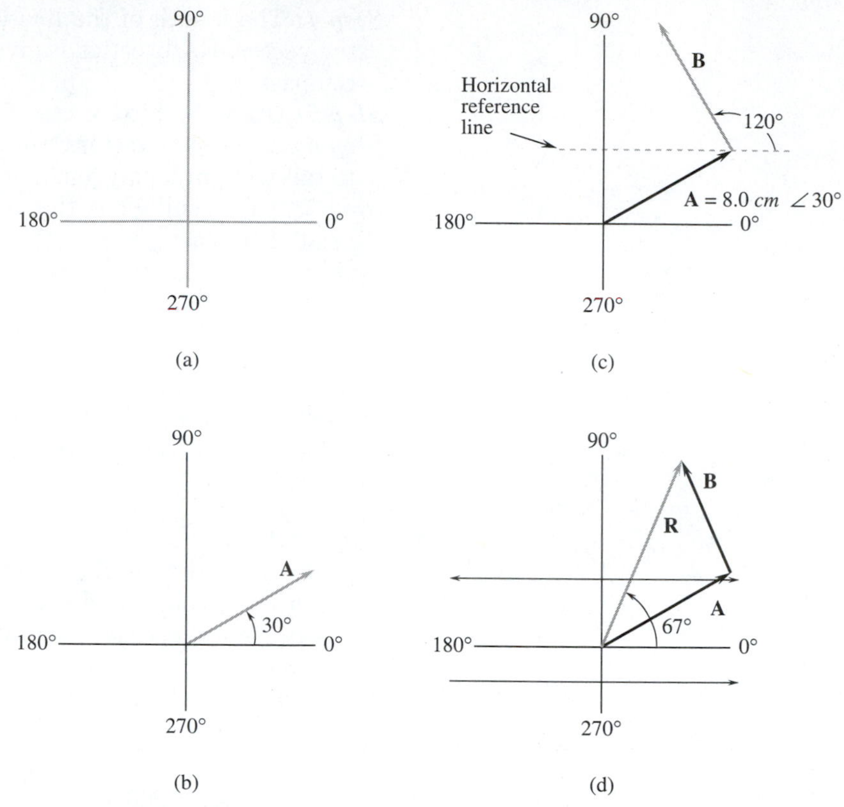

Figure 3–20: *(Diagrams in this figure have been reduced in size.) The scale in each of the four diagrams is 1.00 cm = 25.0 m. (a) Draw a coordinate system. Label the azimuth angles of the axes. (b) Vector* **A** *is drawn at an angle of 30°. (c) Vector* **B** *is drawn from the head of vector* **A.** *A horizontal reference line is used to measure the angle. (d) The resultant is drawn from the starting point to the head of the second vector.*

The resultant displacement vector is:

$$\mathbf{R} = 250 \text{ m } \angle 67°$$

The graphical method can be used for any kind of vector, not just displacement vectors. Example Problem 3–10 shows an example of adding velocity vectors. A pilot wants to know the direction and speed of his plane relative to the ground. This is the sum of the velocity of the plane relative to the air and the amount of drift caused by the wind.

☐ **EXAMPLE PROBLEM 3–10: GROUND SPEED**

A single-engine airplane travels at a speed of 160 mph with a heading of 143° according to an on-board compass. If the wind velocity

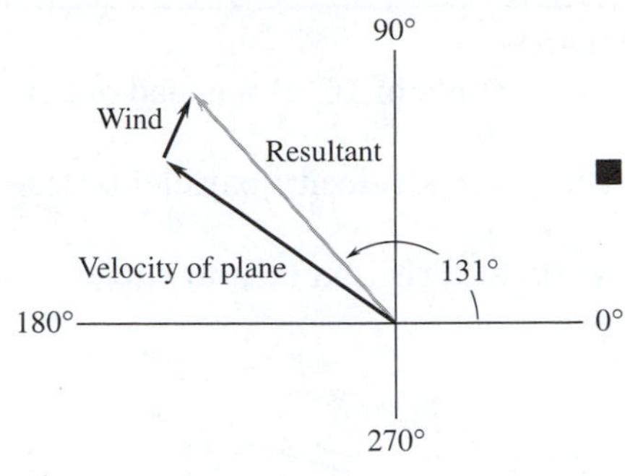

Figure 3–21: Graphical solution for Example Problem 3–10.

is 35 mph ∠80°, find the velocity of the airplane relative to the ground (known as the **ground speed**).

■ *SOLUTION*

Figure 3–21 shows the solution. The scale factor is 1.00 cm = 25 mph. The resultant has a length of 8.0 cm at an azimuth angle of 131°.

$$|\mathbf{R}| = 7.0 \text{ cm} \times \frac{25 \text{ mph}}{1.00 \text{ cm}} = 175 \text{ mph or } 180 \text{ mph}$$

$$\mathbf{R} = 180 \text{ mph } \angle 131°$$

3.5 VECTOR COMPONENTS

A scalar quantity, such as the distance 7.2 in, can be written as the sum of two other numbers. Here are some examples:

$$5.2 \text{ in} + 2.0 \text{ in} = 7.2 \text{ in}$$

$$1.0 \text{ in} + 6.2 \text{ in} = 7.2 \text{ in}$$

$$1.23 \text{ in} + 5.97 \text{ in} = 7.2 \text{ in}$$

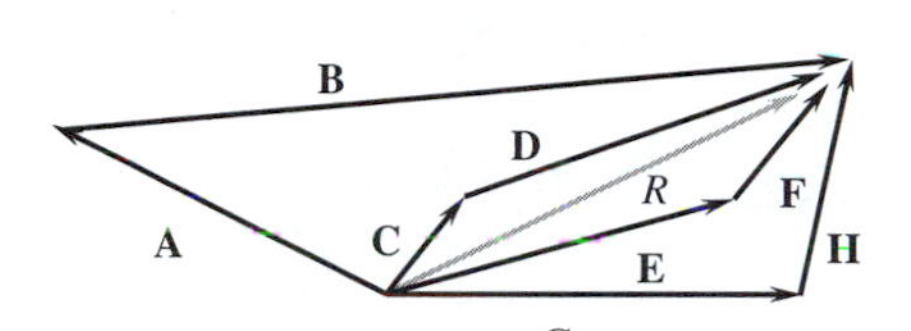

Figure 3–22: A vector can be the resultant of several different pairs of vectors.

The same thing is true of a vector. Any vector can be written as the resultant of two other vectors. Figure 3–22 shows some examples.

We can write vectors as the sum of vectors parallel to the coordinate axes. Look at Figure 3–23. The two vectors parallel to the coordinate axes form the two legs of a right triangle. The resultant is the hypotenuse. The vectors parallel to the axes are called the **components of the vector.**

The x component (V_x) is parallel to the x axis. Since it is the adjacent side of a right triangle, we can calculate its value using the cosine function.

$$V_x = |V| \cos \theta$$

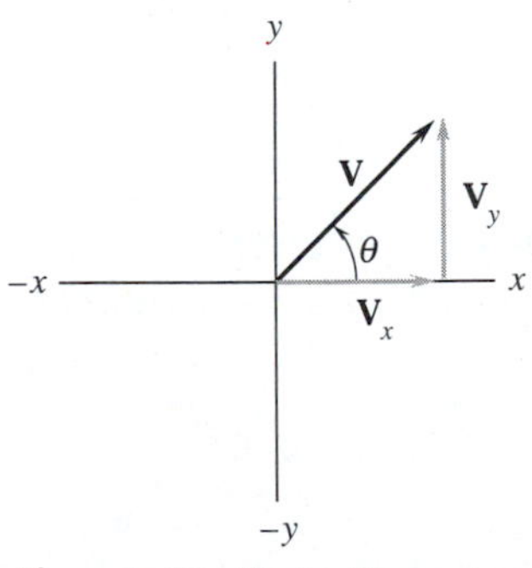

Figure 3–23: Vector **V** *can be written as the resultant of a vector parallel to the x axis* ($\mathbf{V_x}$) *and a vector parallel to the y axis* ($\mathbf{V_y}$)*. These two vectors are the x and y components of* **V**.

The y component ($\mathbf{V}_y$) is parallel to the y axis. Since it is the opposite side of the triangle, we can use the sine function to calculate its value.

$$V_y = |V| \sin \theta$$

EXAMPLE PROBLEM 3–11: THE AIRPLANE

An airplane takes off climbing at an angle of 16° at a speed of $21\bar{0}$ mph.

A. What is the component of the plane's velocity parallel to the ground?

B. What is its velocity in the vertical direction (**climb rate**)?

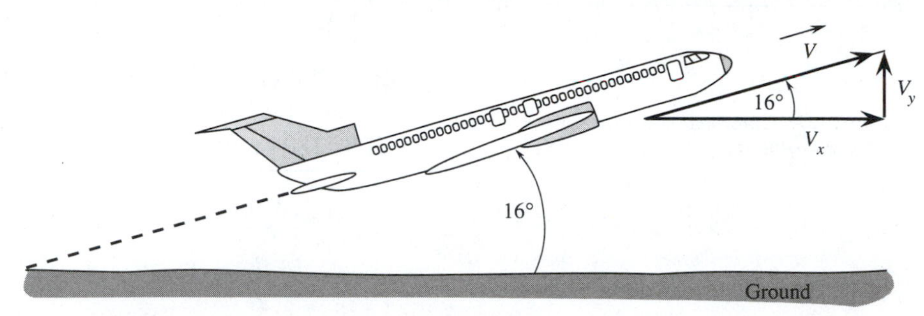

Figure 3–24: *Diagram for Example Problem 3–11. The airplane's velocity can be broken into horizontal and vertical motion.*

SOLUTION

A. The horizontal component (ground speed) is:

$$\mathbf{V}_x = |\mathbf{V}| \cos \theta$$

$$\mathbf{V}_x = 21\bar{0} \text{ mph} \cos 16°$$

$$\mathbf{V}_x = 202 \text{ mph}$$

B. The vertical component (climb rate) is:

$$\mathbf{V}_y = |\mathbf{V}| \sin \theta$$

$$\mathbf{V}_y = 21\bar{0} \text{ mph} \sin 16°$$

$$\mathbf{V}_y = 58 \text{ mph}$$

EXAMPLE PROBLEM 3–12: COMPONENTS OF DISPLACEMENT

Find the components of a displacement of $72\bar{0}$ m $\angle 245°$. Interpret the results.

SOLUTION

$$\mathbf{S}_x = |\mathbf{V}| \cos \theta \qquad \mathbf{S}_y = |\mathbf{V}| \sin \theta$$

$$\mathbf{S}_x = 72\bar{0} \text{ m} \cos 245° \qquad \mathbf{S}_y = 72\bar{0} \text{ m} \sin 245°$$

$$\mathbf{S}_x = -304 \text{ m} \qquad \mathbf{S}_y = -653 \text{ m}$$

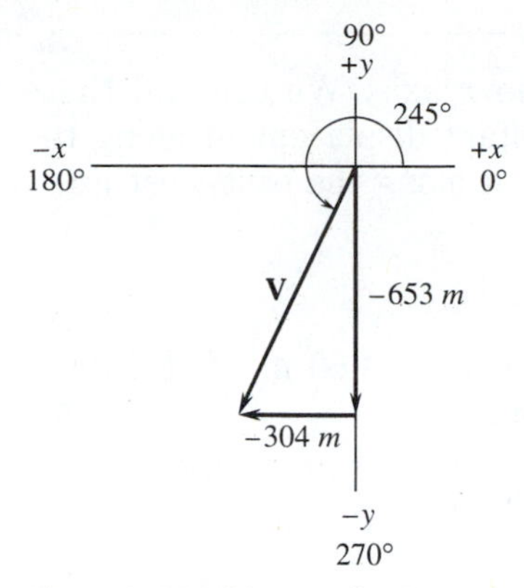

Figure 3–25: Diagram for Example Problem 3–12. Plus (+) and minus (−) signs indicate the direction of the components.

Look at Figure 3–25. The negative signs indicate that the components are pointing in the negative directions. $\mathbf{S}_x$ points in the minus x direction; $\mathbf{S}_y$ points in the minus y direction.

A vector component is parallel to a coordinate axis. It can have only two directions. Plus and minus signs indicate the direction of the component.

☐ **EXAMPLE PROBLEM 3–13: EAST-WEST DISPLACEMENTS**

A pickup truck travels east 3.0 km. The driver discovers a crate has fallen off the back of the truck. The driver turns around and travels west 0.8 km to find the crate. What is the net displacement?

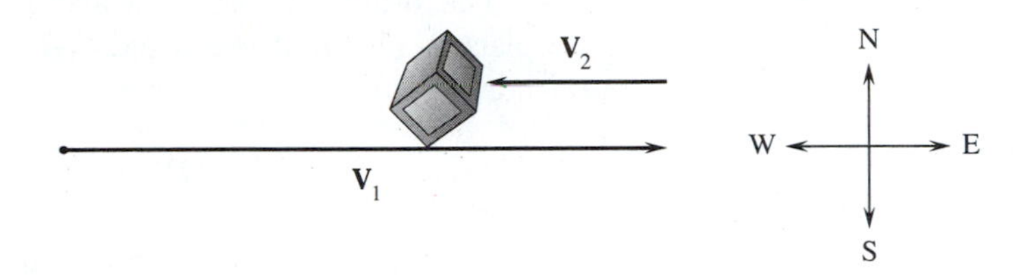

Figure 3–26: Diagram for Example Problem 3–13.

■ ***SOLUTION***

The first displacement is

$$\mathbf{V}_{1x} = 3.0 \text{ km} \cos 0° = +3.0 \text{ km}$$

The second displacement is

$$\mathbf{V}_{2x} = 0.8 \text{ km} \cos 180° = -0.8 \text{ km}$$

There is no vertical component of displacement in this case. We can use **algebraic addition** of the two components to get the resultant.

$$\mathbf{R} = \mathbf{R}_x = +3.0 \text{ km} + (-0.8 \text{ km}) = +2.2 \text{ km}$$

The truck is 2.2 km east of the starting point.

☐ **EXAMPLE PROBLEM 3–14: TWO-DIMENSIONAL DISPLACEMENT WITH COMPONENTS**

Find the resultant of the following displacements:

$\mathbf{A} = 60.0 \text{ m} \angle 0°$ (east) $\quad \mathbf{B} = 50.0 \text{ m} \angle 9\overline{0}°$ (north)
$\mathbf{C} = 25.0 \text{ m} \angle 18\overline{0}°$ (west) $\quad \mathbf{D} = 25.0 \text{ m} \angle 9\overline{0}°$ (north)

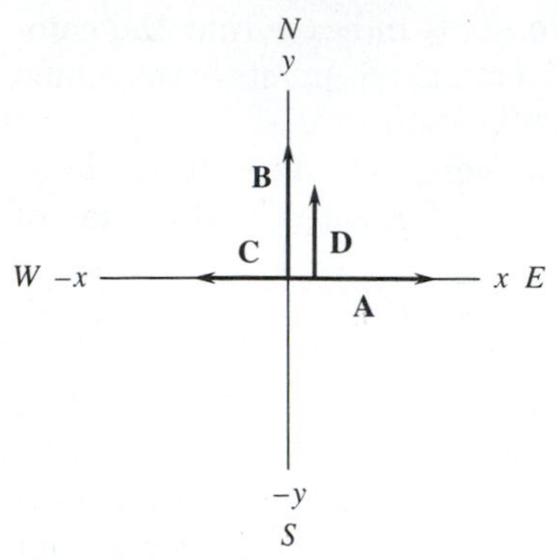

Figure 3–27: Diagram for Example Problem 3–14. Four displacement vectors.

■ *SOLUTION*

We have two vectors parallel to the x axis. We can add these together algebraically to get the resultant displacement along the x axis ($\mathbf{R}_x$). This is the net displacement along the east-west line.

$$\mathbf{R}_x = \mathbf{A} + \mathbf{C}$$

A points in the positive x direction ($|\mathbf{A}| = +60$ m). **C** points in the negative x direction ($|\mathbf{C}| = -25$ m).

$$\mathbf{R}_x = +60.0\text{ m} + (-25.0\text{ m})$$

$$\mathbf{R}_x = +35.0\text{ m}$$

The other two vectors are along the y axis. The resultant displacement along the y axis ($\mathbf{R}_y$) is the net displacement along the north-south line.

$$\mathbf{R}_y = \mathbf{B} + \mathbf{D}$$

B points in the positive y direction ($|\mathbf{B}| = +50$ m). **D** points in the positive y direction ($|\mathbf{D}| = +25$ m).

$$\mathbf{R}_y = +50.0\text{ m} + 25.0\text{ m}$$

$$\mathbf{R}_y = 75.0\text{ m}$$

$\mathbf{R}_x$ and $\mathbf{R}_y$ are shown in Figure 3–28. They form the legs of a right triangle. The magnitude of their resultant (**R**) is the hypotenuse of the triangle. The ratio of $\mathbf{R}_y/\mathbf{R}_x$ is the tangent of the angle.

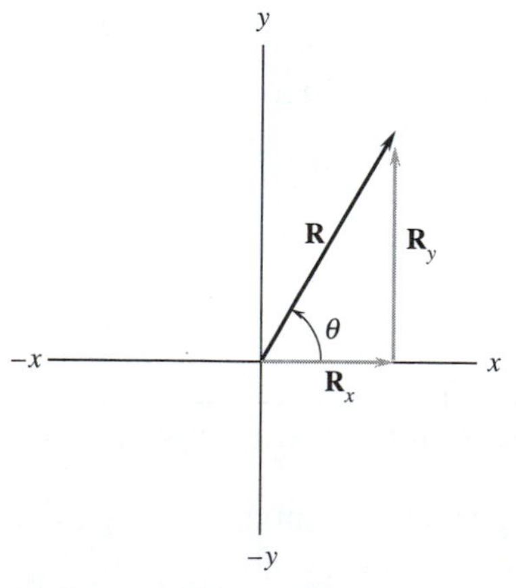

Figure 3–28: The resultant of the four displacements for Example Problem 3–14. $\mathbf{R}_x$ and $\mathbf{R}_y$ form the legs of a right triangle.

To get the magnitude of **R** use the Pythagorean theorem.

$$|\mathbf{R}| = (\mathbf{R}_x^2 + \mathbf{R}_y^2)^{1/2}$$

$$|\mathbf{R}| = [(35.0\text{ m})^2 + (75.0\text{ m})^2]^{1/2}$$

$$|\mathbf{R}| = 82.8\text{ m}$$

We can find the angle using the tangent function.

$$\tan\theta = \frac{\mathbf{R}_y}{\mathbf{R}_x}$$

$$\tan\theta = \frac{75.0\text{ m}}{35.0\text{ m}} = 2.1428$$

$$\theta = \tan^{-1}(2.1428)$$

$$\theta = 65.0°$$

$$\mathbf{R} = 82.8\text{ m} \angle 65.0°$$

NOTE: To find the inverse tangent ($\tan^{-1}$) on your calculator do the following steps.

Enter: 75

Press: [÷]

Enter: 35

Press: [=]

Your display should be [2.142857143].

Press: [2ndF] or [INV], depending on the make of your calculator.

Press: [tan]

If your calculator is set for degrees, the display is

[64.983106]

3.6 COMPONENT METHOD OF VECTOR ADDITION

Example Problem 3–14 used a method of adding vectors without mapping them out with a ruler and protractor. This procedure is called the **component method.** If we break vectors down into components parallel to the x and y axes, we can use algebraic addition of the components to find the resultant components along the two axes. The resultant is then found using right angle trigonometry. Here is the component method strategy:

1. Find the x and y components of the vectors ($\mathbf{V}_i$).

$$\mathbf{V}_{xi} = |\mathbf{V}_i| \cos \theta_i \qquad \mathbf{V}_{yi} = |\mathbf{V}_i| \sin \theta_i$$

2. Add the components algebraically to find the resultant $\mathbf{R}_x$ and $\mathbf{R}_y$ components.

$$\mathbf{R}_x = \sum_i \mathbf{V}_{xi} \qquad \mathbf{R}_y = \sum_i \mathbf{V}_{yi}$$

3. Use the Pythagorean theorem to find the magnitude of the resultant ($\mathbf{R}$).

$$|\mathbf{R}| = (\mathbf{R}_x^2 + \mathbf{R}_y^2)^{1/2}$$

4. The angle relative to the $+x$ axis (ϕ) is:

$$\phi = \tan^{-1}\left(\frac{\mathbf{R}_y}{\mathbf{R}_x}\right)$$

☐ **EXAMPLE PROBLEM 3–15: COMPONENT ADDITION**

Add the following two vectors using the component method.

$$\mathbf{V}_1 = 23\bar{0}\ \text{mph}\ \angle 35° \qquad \mathbf{V}_2 = 178\ \text{mph}\ \angle 137°$$

■ *SOLUTION*

First find the components of the two vectors.

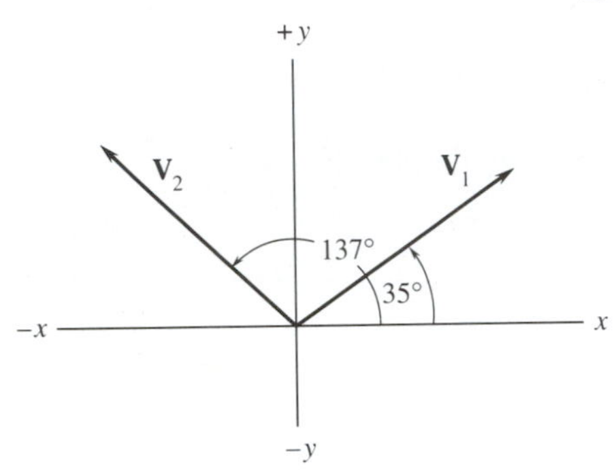

Figure 3–29: *Diagram for Example Problem 3–15.*

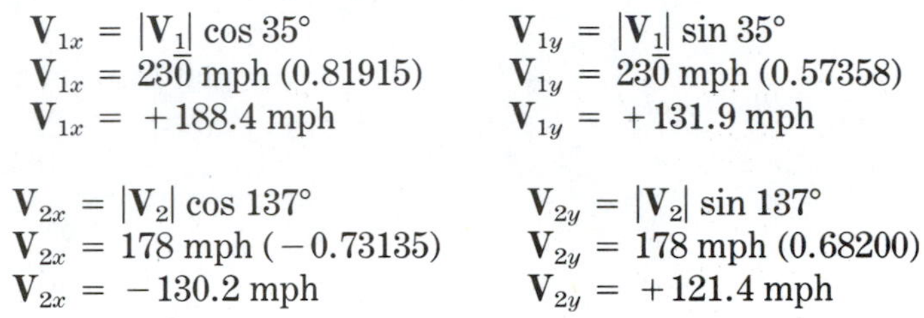

$$\mathbf{V}_{1x} = |\mathbf{V}_1| \cos 35° \qquad \mathbf{V}_{1y} = |\mathbf{V}_1| \sin 35°$$
$$\mathbf{V}_{1x} = 23\bar{0}\ \text{mph}\ (0.81915) \qquad \mathbf{V}_{1y} = 23\bar{0}\ \text{mph}\ (0.57358)$$
$$\mathbf{V}_{1x} = +188.4\ \text{mph} \qquad \mathbf{V}_{1y} = +131.9\ \text{mph}$$

$$\mathbf{V}_{2x} = |\mathbf{V}_2| \cos 137° \qquad \mathbf{V}_{2y} = |\mathbf{V}_2| \sin 137°$$
$$\mathbf{V}_{2x} = 178\ \text{mph}\ (-0.73135) \qquad \mathbf{V}_{2y} = 178\ \text{mph}\ (0.68200)$$
$$\mathbf{V}_{2x} = -130.2\ \text{mph} \qquad \mathbf{V}_{2y} = +121.4\ \text{mph}$$

Next add up the x and y components to find the components of the resultant.

$$\mathbf{R}_x = \mathbf{V}_{1x} + \mathbf{V}_{2x} = +188.4\ \text{mph} + (-130.2\ \text{mph}) = 58.2\ \text{mph}$$

$$\mathbf{R}_y = \mathbf{V}_{1y} + \mathbf{V}_{2y} = +131.9\ \text{mph} + 121.4\ \text{mph} = 253.3\ \text{mph}$$

The Pythagorean theorem gives us the magnitude of the resultant.

$$|\mathbf{R}| = (\mathbf{R}_x^2 + \mathbf{R}_y^2)^{1/2}$$
$$|\mathbf{R}| = [(58.2\ \text{mph})^2 + (253.3\ \text{mph})^2]^{1/2}$$
$$|\mathbf{R}| = 26\bar{0}\ \text{mph}$$

The inverse tangent gives us the angle.

$$\tan\phi = \frac{\mathbf{R}_y}{\text{R}_x}$$

$$\tan\phi = \frac{253.3\ \text{mph}}{58.2\ \text{mph}} = 4.35223$$

$$= \tan^{-1}(4.35223) = 77.1°$$

$$\mathbf{R} = 26\bar{0}\ \text{mph}\ \angle 77.1°$$

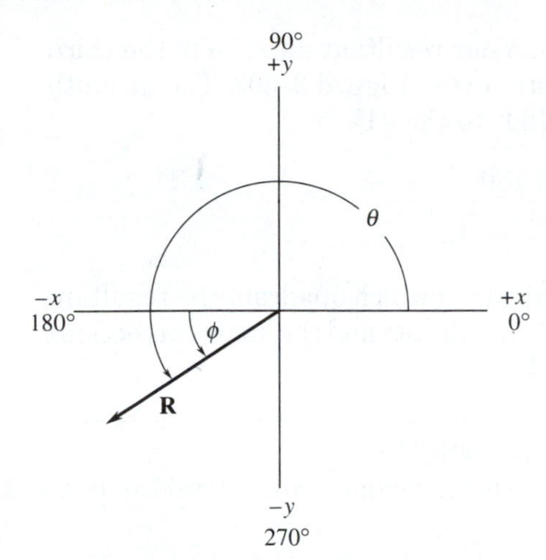

Figure 3–30: θ *is the azimuth angle.* ϕ *is the angle to the nearest x axis.* θ = *180° +* ϕ.

Your calculator does not give you the azimuth angle of the resultant. The best it can do is give an angle to the nearest x axis. Suppose you have performed the first two steps of the component method of vector addition. You have calculated the components of the resultant. Your values are $\mathbf{R}_x = -23\bar{0}$ km and $\mathbf{R}_y = -20\bar{0}$ km. Your calculator gives you an angle.

$$\phi = \tan^{-1}\left(\frac{-200 \text{ km}}{-230 \text{ km}}\right)$$

$$\phi = 41.0°$$

This angle is to the nearest x axis. Look at the components of the resultant. $\mathbf{R}_x$ points to the left along the negative x axis. $\mathbf{R}_y$ points

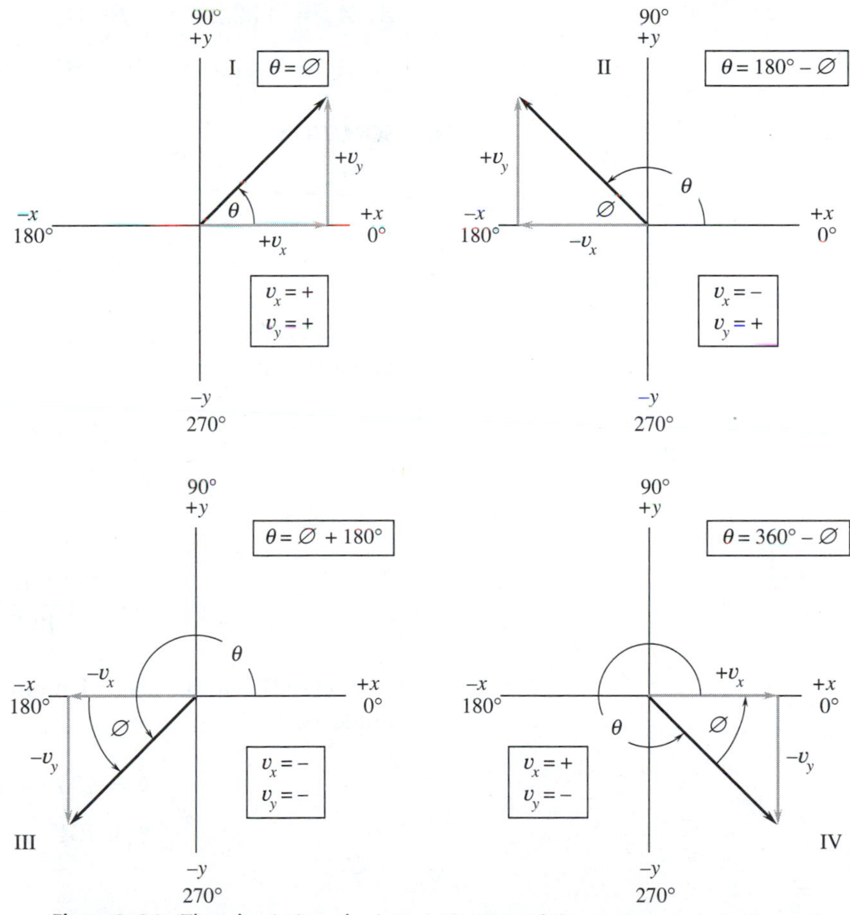

Figure 3–31: The plus (+) and minus (−) signs of the components indicate the quadrant in which the resultant is located. The four diagrams show how to convert from the angle to the nearest x axis to the azimuth θ.

down along the negative y axis. Your resultant must be in the third quadrant of the coordinate system (see Figure 3–30). The azimuth angle is calculated by adding 180° to the 41°.

$$\phi = 41° + 180°$$

$$\phi = 221°$$

The signs of the components indicate in which quadrant the resultant is located. Figure 3–31 shows the quadrants and the angle corrections needed to get the azimuth angle.

☐ **EXAMPLE PROBLEM 3–16: AZIMUTH ANGLES**

Find the azimuth angles for the following pairs of vector components.

A. $A_x = +32.2\text{ lb}$ **B.** $B_x = -12.2\text{ ft}$ **C.** $C_x = +12\text{ m/s}$

$A_y = -47.7\text{ lb}$ $B_y = +3.6\text{ ft}$ $C_y = +12\text{ m/s}$

■ ***SOLUTION***

A.

$$\phi = \tan^{-1}\left(\frac{\mathbf{A}_y}{\mathbf{A}_x}\right) = \tan^{-1}\left(\frac{47.7}{32.2}\right) = 56.0°$$

The vector **A** is in the fourth quadrant (see Figure 3–31). The azimuth angle is:

$$\theta = 360° - 56°$$

$$\theta = 304°$$

B.

$$\phi = \tan^{-1}\left(\frac{3.6}{12.2}\right) = 16°$$

Vector **B** is in the second quadrant (see Figure 3–31). The azimuth angle is:

$$\theta = 180° - 16°$$

$$\theta = 164°$$

C.

$$\phi = \tan^{-1}\left(\frac{12}{12}\right) = 45°$$

Both components are positive. In the first quadrant, the angle with respect to the nearest x axis *is* the azimuth angle.

$$\theta = 45°$$

3.7 VECTOR MULTIPLICATION

Sometimes we will want to multiply two vectors. In problems involving work and energy, the dot product method is used. In problems involving torque and rotational motion, the cross product method is used.

Dot Product

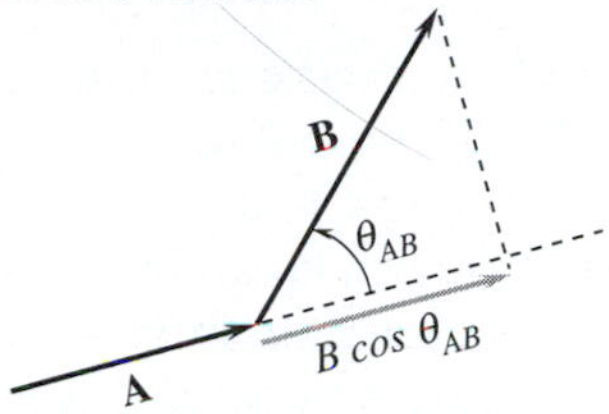

Figure 3–32: In the dot product, the parallel component of one vector (in this case **B***) is multiplied by the magnitude of the other vector* **(A)***.*

Figure 3–32 shows two vectors, **A** and **B.** One way to find a product is to multiply **A** by the component of vector **B** that is parallel to **A.** We can find this by multiplying **B** by the cosine of the angle between the two vectors. We will use a dot (·) to indicate this kind of multiplication, known as the **dot product.** When we multiply two vectors with the dot product the answer is a scalar. Thus the **dot product** is sometimes called the **scalar product.**

$$\mathbf{A} \cdot \mathbf{B} = |\mathbf{A}||\mathbf{B}| \cos \theta_{\mathbf{AB}}$$

Cross Product

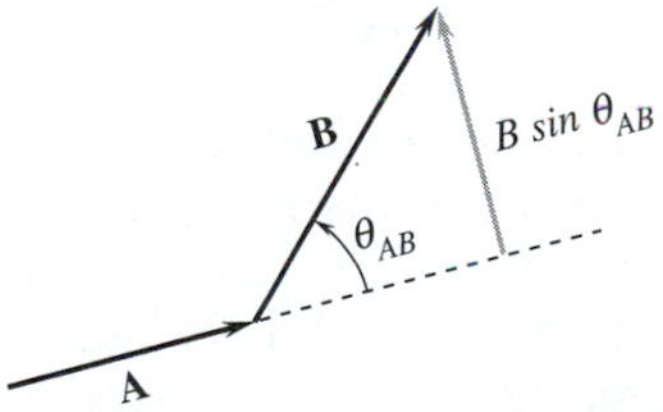

Figure 3–33: In the cross product, the vertical component of one vector **(B)** *is multiplied by the magnitude of the other vector* **(A)***.*

Figure 3–33 shows the same vectors, **A** and **B.** Instead of multiplying vector **A** by the parallel component of **B,** we will multiply **A** by the perpendicular component of **B.** This method of multiplying vectors is known as the **cross product.**

We find the component of **B** that is perpendicular to **A** by using the sine of the angle between the two vectors. We will denote this kind of multiplication by using an ×. The cross product is a vector. The direction will point along the axis of rotation as **A** rotates into **B.** For the moment we will worry only about the magnitude of the cross product. The cross product is also called the **vector product.**

$$|\mathbf{A} \times \mathbf{B}| = |\mathbf{A}||\mathbf{B}| \sin \theta_{\mathbf{AB}}$$

☐ **EXAMPLE PROBLEM 3–17: VECTOR MULTIPLICATION**

$$\mathbf{A} = 46\overline{0} \text{ ft} \angle 35° \text{ and } \mathbf{B} = 342 \text{ lb} \angle 75°$$

A. Find **A · B.**

B. Calculate the magnitude of **A × B.**

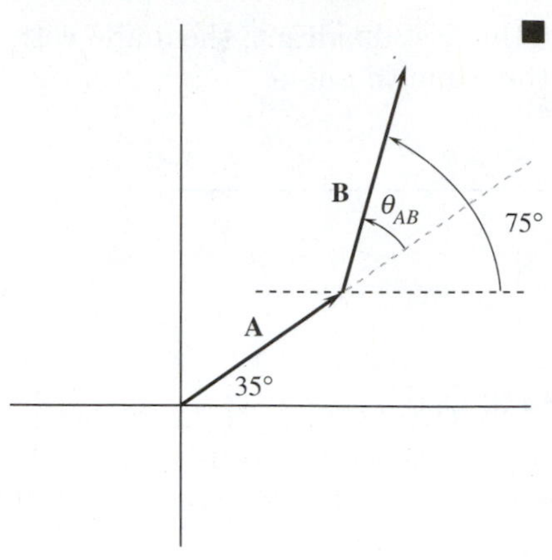

Figure 3–34: Diagram for Example Problem 3–17.

■ *SOLUTION*

The angle between **A** and **B** is 75° − 35° = 40°.

A.

$$\mathbf{A} \cdot \mathbf{B} = |\mathbf{A}||\mathbf{B}| \cos 40°$$

$$\mathbf{A} \cdot \mathbf{B} = (460 \text{ ft})\,(342 \text{ lb})\,(0.7660)$$

$$\mathbf{A} \cdot \mathbf{B} = 1.21 \times 10^5 \text{ ft} \cdot \text{lb}$$

B.

$$\mathbf{A} \times \mathbf{B} = |\mathbf{A}||\mathbf{B}| \sin 40°$$

$$\mathbf{A} \times \mathbf{B} = (460 \text{ ft})\,(342 \text{ lb})\,(0.6428)$$

$$\mathbf{A} \times \mathbf{B} = 1.01 \times 10^5 \text{ ft} \cdot \text{lb}$$

SUMMARY

A direct proportion ($y = k\,x$) plots into a straight line passing through the origin of a standard y versus x graph. The slope (m) of the line is the proportionality constant (k). In general, the equation of a straight line has the form of a direct proportion with an added constant (b), $y = (\text{m x}) + b$, where m is the slope and b is the y intercept.

Power laws, which apply to other kinds of proportion, have the form of $y = m\,x^n$, where n is the power and m is the proportionality constant. In these cases, where the proportion is not direct, y is plotted against x^n rather than against x, and a straight line results. The proportionality (m) is the slope of this straight line.

The two kinds of quantities used to solve problems in physics are scalars—measured quantities that have a magnitude and units—and vectors—measured quantities that have magnitude, units, and a direction.

Vectors may be added graphically. In the graphical method a protractor and a ruler are used to draw line segments proportional to the magnitudes of the vectors and pointing in the appropriate directions. The sum of two or more vectors is called a resultant. Any vector may be written as the resultant of two vectors parallel to the coordinate axes. These vectors are called the components of the vector. In the component method of vector addition vectors are broken into components. If the azimuth angles of the vectors are used, then:

$$\mathbf{V}_{xi} = |\mathbf{V}_i| \cos \theta_i \text{ and } \mathbf{V}_{yi} = |\mathbf{V}_i| \sin \theta_i$$

The x and y components are added algebraically to find the components of the resultant.

$$\mathbf{R}_x = \sum_i \mathbf{V}_{xi} \qquad \mathbf{R}_y = \sum_i \mathbf{V}_{yi}$$

The magnitude and angle of the resultant are found using right angle trigonometry.

$$|\mathbf{R}| = (\mathbf{R}_x{}^2 + \mathbf{R}_y{}^2)^{1/2}$$

$$\phi = \tan^{-1}\left(\frac{R_y}{R_x}\right)$$

Vectors may be multiplied in two ways. In the dot product method, the magnitude of one vector is multiplied by the parallel component of the second vector. The dot product is a scalar. If θ_{AB} is the angle between the two vectors, then:

$$\mathbf{A} \cdot \mathbf{B} = |\mathbf{A}||\mathbf{B}| \cos \theta_{AB}$$

In the cross product method, the magnitude of one vector is multiplied by the perpendicular component of the other vector. The cross product is a vector. If θ_{AB} is the angle between the two vectors, then:

$$|\mathbf{A} \times \mathbf{B}| = |\mathbf{A}||\mathbf{B}| \sin \theta_{AB}$$

KEY TERMS

If you can explain the following terms to a friend or classmate, you understand their meaning. If you cannot explain the terms, you should reread the sections in which they are discussed.

algebraic addition
average speed
azimuth angle
climb rate
component method
components of the vector
cross product
direct proportion
displacement
distance
dot product
graphical method
ground speed
inverse law
inverse square law
material constants
period
photons
power law
proportionality constant (k)
resultant
scalar
scalar product
scale factor
slope
speed
vector
vector product
weight density
y intercept

EXERCISES

Section 3.1:

1. The ratio of voltage (V) to current (I) through an electric circuit is found to be constant. Are voltage and current related by a direct proportion?
2. The pressure (P) of a gas is directly proportional to the temperature (T). If the temperature is doubled, what happens to the pressure?
3. As the volume of a liquid in a beaker increases, what happens to the liquid's weight? To its density?

Section 3.2:

4. Frequency (f) and wavelength (λ) of sound waves are inversely proportional ($\lambda f = c$, where c is constant). If the frequency is doubled, how does the wavelength change?
5. The period (P) of a pendulum is proportional to the square root of its length (L). By what factor must the length of the pendulum be changed to double the period?

Section 3.3:

6. In Figure 3–35, which are pairs of equal vectors?
7. Some of the pairs of vectors in Figure 3–35 are not equal. Explain why they are not.
8. Identify each of the following as a vector or a scalar.
   A. Your age.
   B. A hovercraft traveling 83 mph northward.
   C. The length of a football field.
   D. The density of water.
   E. A rock falling downward at 67 ft/s.

Section 3.4:

9. In a graphical solution to vector addition the resultant is found to be 3.4 in long. If the scale factor is 1.00 in = 25 lb, what is the magnitude of the vector?
10. A vector is to be plotted with a scale factor of 1.00 cm = 4.0 km. How long a line on the plot is needed to represent a displacement vector of 25 km?

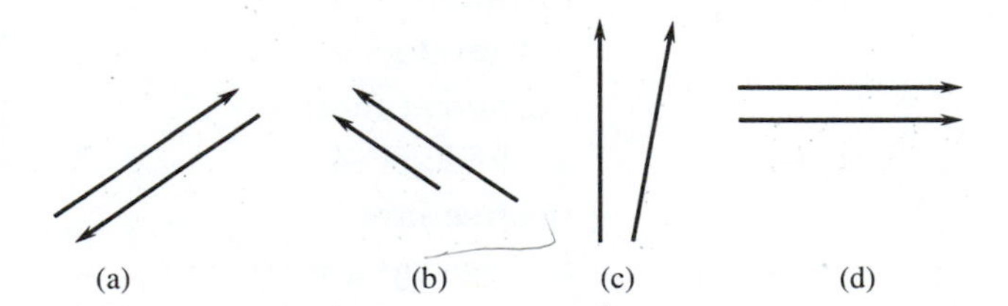

Figure 3–35: Diagram for Exercise 6.

Section 3.5:

11. Which of the vectors sketched in Figure 3–36 have:

A. negative x components?

B. negative y components?

C. both components negative?

D. both components positive?

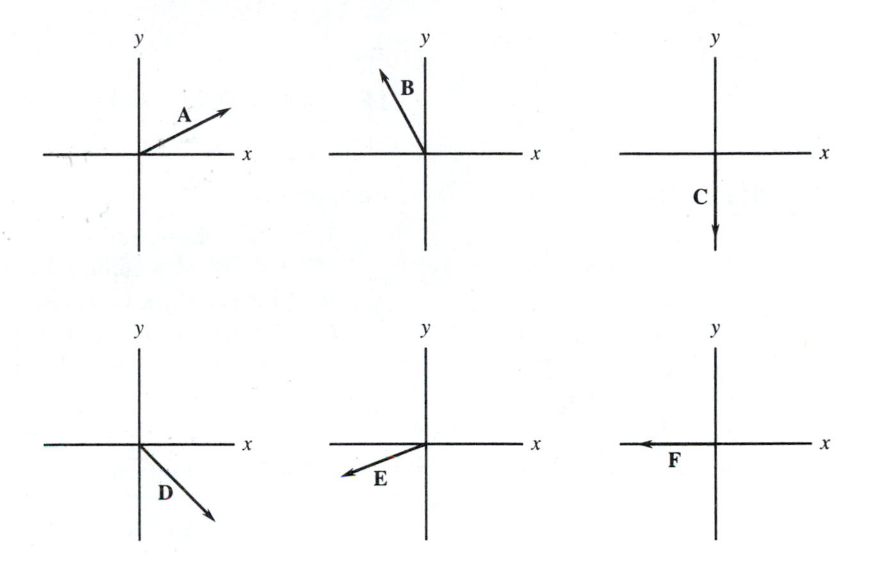

Figure 3–36: *Diagram for Exercise 11.*

Section 3.6:

12. Find the azimuth angles for the resultant vectors shown in Figure 3–37.

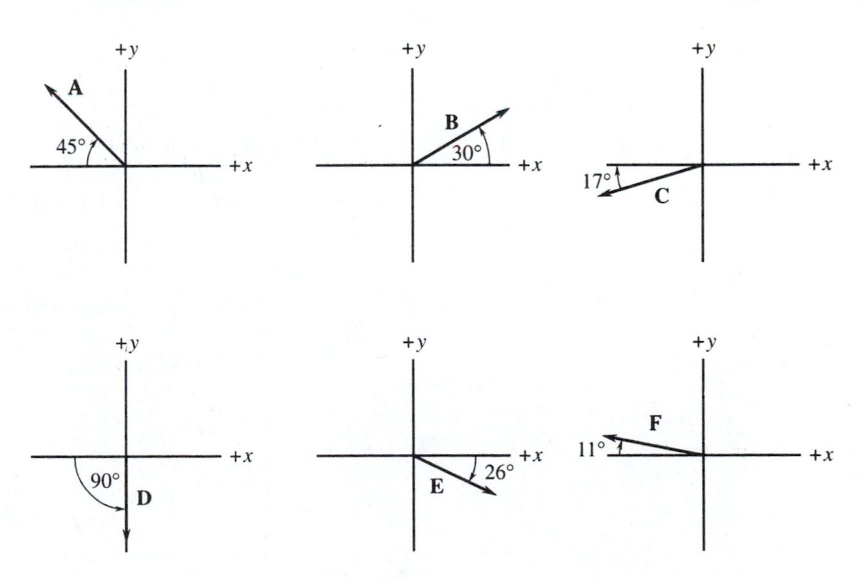

Figure 3–37: *Diagram for Exercise 12.*

13. Find the components of the resultant for the addition of the following three vectors.

Vector	Component V_x	V_y
V_1	−34.0 m	+48.3 m
V_2	+53.0 m	+9.2 m
V_3	+12.5 m	−19.6 m

14. What are the x and y components of $V = 56 \text{ mph } \angle 27\bar{0}°$?

PROBLEMS

Section 3.1:

1. Table 3–7 shows data for an experiment measuring the change of pressure of a confined gas with temperature.
   A. Plot the data in Table 3–7.
   B. Find the proportionality constant.
   C. Calculate the pressure at a temperature of 700 R.

Table 3–7: Pressure and temperature of a confined gas.

PRESSURE (lb/in²)	TEMPERATURE (Rankine [R])
3.2	$10\bar{0}$
6.5	$20\bar{0}$
9.7	$30\bar{0}$
13.0	$40\bar{0}$
16.2	500

2. The voltage and current for an electric circuit are shown in Table 3–8. Plot the data and find the proportionality constant between voltage and current from the graph.

Table 3–8: Voltage and current of an electric circuit.

VOLTAGE (volts)	CURRENT (amperes)
3.2	0.2
6.5	0.4
9.7	0.6
13.0	0.8
16.2	1.0

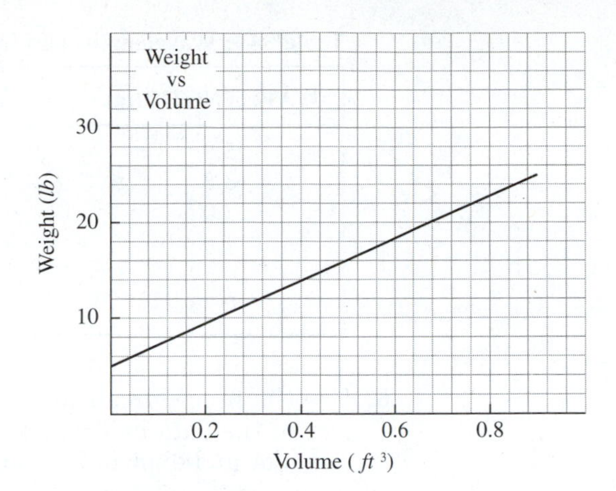

Figure 3–38: *Diagram for Problem 3.*

3. Find the equation of the straight line that fits the graph shown in Figure 3–38.

Section 3.2:

4. What are the proportionality constant and the exponent (n) of the power law plotted in Figure 3–39?

5. Table 3–9 shows wavelengths (λ) and frequencies (f) of microwaves. The variables are related by the equation $f\lambda = c$, where c is the speed of light.

A. Find and plot the power of λ that produces a straight line.

B. Show that the slope of the line is the speed of light.

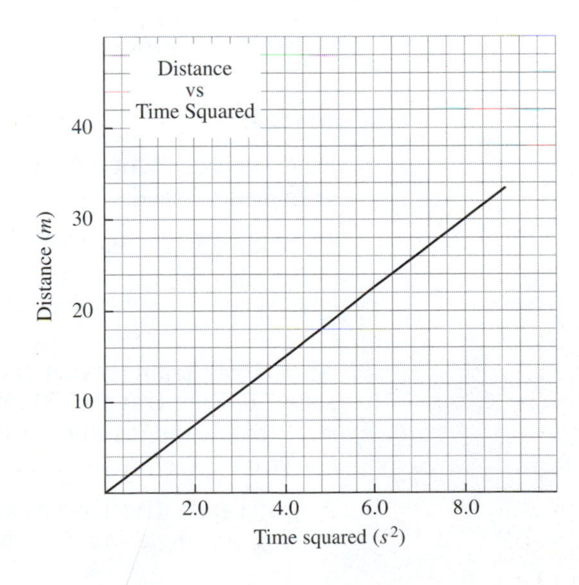

Figure 3–39: *Diagram for Problem 4.*

Table 3–9: Wavelengths and frequencies of microwaves.

WAVELENGTH (m)	FREQUENCY [(cycles/s) × 10^7]
5.9	5.0
3.8	8.0
3.0	10.0
1.6	20.0
0.59	50.0
0.38	80.0

6. Use the procedures shown in Appendix B to produce a log-log plot of the data in Table 3–9. Calculate the slope of the line in this plot and explain its meaning.
7. Table 3–10 contains the periods of revolution and average distance from the sun of the planets. Use the log-log techniques shown in Appendix B to find the exponent in the power law relating planetary period of revolution with distance from the sun.

Table 3–10: Distance from sun and orbital periods of planets.

PLANET	DISTANCE (AU)*	ORBITAL PERIOD (years)
Mercury	0.34	0.24
Venus	0.72	0.62
Earth	1.00	1.00
Mars	1.52	1.88
Jupiter	5.20	11.86
Saturn	9.54	29.50
Uranus	19.20	84.00

* One astronomical unit (AU) is the earth's average distance from the sun (93,000,000 mi).

Section 3.4:

8. Find the resultant of the following sets of vectors using the graphical method.

   a. 12 km $\angle 0°$
 15 km $\angle 90°$

   b. 15 ft $\angle 3\bar{0}°$
 33 ft $\angle 14\bar{0}°$

   c. 4.5 mi $\angle 21\bar{0}°$
 5.0 mi $\angle 32\bar{0}°$
 3.0 mi $\angle 11\bar{0}°$

9. An Amtrak train travels due west from city A to city B 255 mi away. It then travels 150 mi southwest to city C.
   A. What is the distance traveled by the train?
   B. What is its total displacement? Use the graphical method.
10. A boat heads directly across a river with a speed of 7.8 knots making an angle of 90° with the riverbank. The river has a

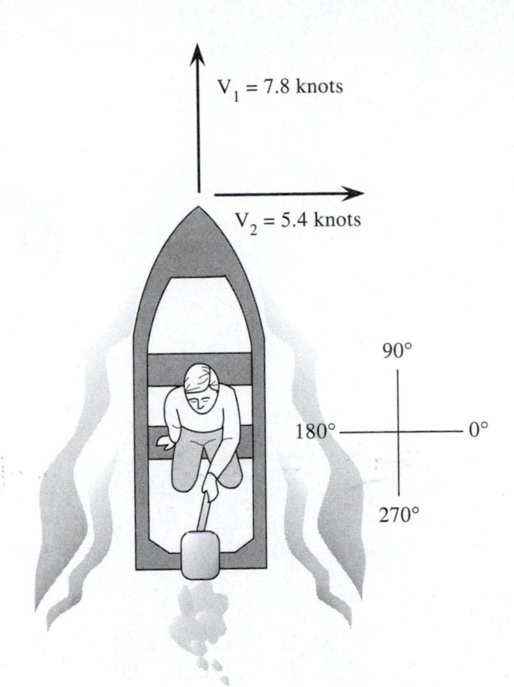

Figure 3–40: Diagram for Problem 10.

current of 5.4 knots that pushes the boat downstream as it crosses. Find the angle the boat's motion makes with the bank (see Figure 3–40). Use a graphical solution.

Section 3.5:

11. Find the x and y components of the following vectors.

a. $12\bar{0}$ mph $\angle 67°$ **b.** 23.7 km $\angle 152°$
c. 74.1 lb $\angle 223°$ **d.** $46\bar{0}$ m/s² $\angle 27\bar{0}°$

12. A power boat travels at a speed of 11.3 knots at an angle of 55° relative to a riverbank (see Figure 3–41). What are its components of velocity parallel and perpendicular to the riverbank?

Section 3.6:

13. Find the resultant magnitude and azimuth angles of the following pairs of resultant components.

a. $\mathbf{R}_x = 35$ ft, $\mathbf{R}_y = 42$ ft
b. $\mathbf{R}_x = -54$ m, $\mathbf{R}_y = 14$ m
c. $\mathbf{R}_x = -23$ m/s, $\mathbf{R}_y = 0.0$ m/s
d. $\mathbf{R}_x = 12$ mph, $\mathbf{R}_y = 32$ mph
e. $\mathbf{R}_x = -234$ ft/s, $\mathbf{R}_y = -125$ ft/s

14. Find the resultants of the following sets of vectors using the component method.

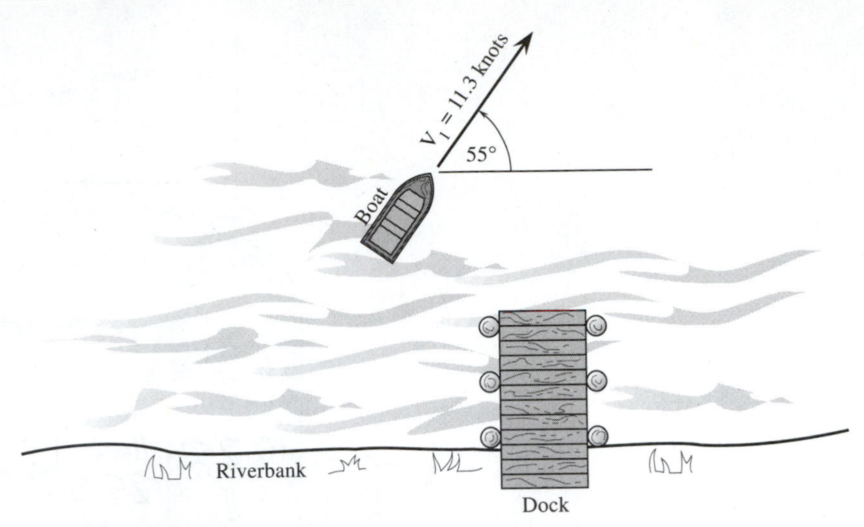

Figure 3–41: *Diagram for Problem 12.*

a. $\mathbf{A} = 12\bar{0}$ lb $\angle 3\bar{0}°$
$\mathbf{B} = 12\bar{0}$ lb $\angle 6\bar{0}°$

b. $\mathbf{A} = 48.2$ ft $\angle 11\bar{0}°$
$\mathbf{B} = 63.0$ ft $\angle 243°$

c. $\mathbf{A} = 26$ m/s $\angle 72°$
$\mathbf{B} = 33$ m/s $\angle 112°$
$\mathbf{C} = 37$ m/s $\angle 34\bar{0}°$

d. 100 m $\angle 32°$
100 m $\angle 122°$
141 m $\angle 257°$

15. An airplane has an airspeed of 210 mph $\angle 127°$. The wind speed is 46 mph $\angle 27\bar{0}°$. Find the plane's velocity relative to the ground using the component method.

16. Calculate the resultant of displacements of $12\bar{0}$ m eastward, $23\bar{0}$ m southeast, and 180 m southwest. Use the component method.

17. A guy wire is connected to a power pole at a height of 23 ft above the ground. The bottom end is anchored 14 ft from the base of the pole. How long is the guy wire?

Section 3.7:

18. What is the angle between the following pairs of vectors?

a. 24 ft $\angle \bar{0}°$
32 ft $\angle 45°$

b. 32 m $\angle 18°$
18 m $\angle 122°$

c. 35 mph $\angle 53°$
55 mph $\angle 286°$

19. Find the magnitudes of the cross products ($|\mathbf{A} \times \mathbf{B}|$) for the following pairs of vectors.

a. $\mathbf{A} = 122$ ft $\angle 0°$
$\mathbf{B} = 23.0$ lb $\angle 40°$

b. $\mathbf{A} = 45$ N $\angle 33°$
$\mathbf{B} = 0.87$ m $\angle 102°$

c. $\mathbf{A} = 44$ m/s $\angle 46°$
$\mathbf{B} = 1.2$ m $\angle 136°$

d. $\mathbf{A} = 23$ ft $\angle 112°$
$\mathbf{B} = 3.4$ lb $\angle 292°$

20. Find the dot products ($\mathbf{A} \cdot \mathbf{B}$) for the vector pairs given in Problem 19.

Chapter 4

LINEAR MOTION

How fast is this aircraft traveling? That depends on where in the universe you are standing. In this chapter, you will see how all motion can be relative to your vantage point. (Photograph courtesy of NASA.)

OBJECTIVES

In this chapter you will learn:

- how to tell the difference between speed and velocity
- to define average acceleration
- to combine defining equations of motion to create derived equations fitting special conditions
- to solve motion problems in one dimension
- to solve problems involving gravitational acceleration
- to solve motion problems in two dimensions

Time Line for Chapter 4

370 BC	Aristotle states free fall is accelerated motion.
340 BC	Strato performs experiments that seem to support Aristotle's ideas of motion.
1590	Galileo refutes Aristotelean physics.
1686	Isaac Newton publishes the *Principia, De Motu Corporu,* the motion of bodies.
1784	George Atwood finds value of the acceleration of gravity accurately.

In Chapter 3 we encountered the ratio between distance and time, which we called speed. The ratio between displacement and time is the vector quantity called **velocity.** This proportion helps us describe motion—the change of position of something with time. It works quite well as long as the object does not speed up or slow down. We will need to introduce some new ideas to handle these more complicated situations.

In this chapter we will develop the techniques to describe the linear motion of an object. **Linear motion** is motion in a straight line. Spinning objects have **rotational motion.** Rotational motion will be covered in the next chapter.

4.1 AVERAGE SPEED AND AVERAGE VELOCITY

Average speed is defined as the total distance traveled divided by the elapsed time. This is the *constant* speed one could travel for the same interval of time to cover the same distance. If s_f is the final position and s_i is the initial position, we can write:

$$v_{\text{avg}} = \frac{s_f - s_i}{t} \qquad \textbf{(Eq. 4–1)}$$

Figure 4–1 shows a plot of distance versus time for a bicycle trip. The average speed for the entire trip is represented by the dashed line. It is the total distance divided by the total time. The average speed *is not* calculated by adding up the three speeds and dividing by three.

□ **EXAMPLE PROBLEM 4–1: A BICYCLE TRIP**

For the bicycle trip shown in Figure 4–1:

A. Find the average speed between A and B.

B. Find the average speed between B and C.

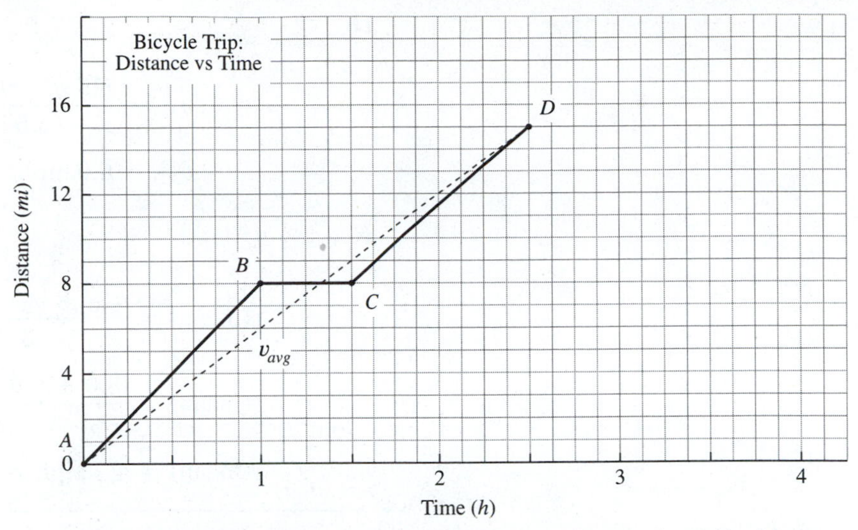

Figure 4–1: Graph of distance versus time for a bicycle trip. The dashed line represents the average speed for the entire trip.

C. Find the average speed between C and D.
D. Find the average speed for the total trip.
E. Show that the mean average of the three speeds is *not* the average speed for the trip defined in Equation 4–1.

■ ***SOLUTION***

A.

$$v_{AB} = \frac{s_f - s_i}{t}$$

$$v_{AB} = \frac{(8.0 - 0.0)\text{ mi}}{1.0\text{ h}}$$

$$v_{AB} = 8.0\text{ mph}$$

B.

$$v_{BC} = \frac{(8.0 - 8.0)\text{ mi}}{0.5\text{ h}}$$

$$v_{BC} = 0.0\text{ mph}$$

C.

$$v_{CD} = \frac{(15.0 - 8.0)\text{ mi}}{1.0\text{ h}}$$

$$v_{CD} = 7.0\text{ mph}$$

D.

$$v_{AD} = \frac{(15.0 - 0.0) \text{ mi}}{2.5 \text{ h}}$$

$$v_{AD} = 6.0 \text{ mph}$$

E.

$$v(\text{mean}) = \frac{v_{AB} + v_{BC} + v_{CD}}{3}$$

$$v(\text{mean}) = \frac{(8.0 + 0.0 + 6.0) \text{ mph}}{3}$$

$$v(\text{mean}) = 4.7 \text{ mph} \neq v_{AD}$$

Quite often we want to know the average velocity rather than the average speed. The **average velocity** is the total *displacement* divided by the elapsed time. The direction of the motion becomes important. To find the velocity use Equation 4–1, but use the initial and final displacements rather than distances.

□ **EXAMPLE PROBLEM 4–2: THE POWER ROLLER**

At a road construction job a power roller has the motion shown in Figure 4–2. Find

A. The average velocity between A and B.
B. The average velocity between B and C.
C. The average velocity between A and C.

■ ***SOLUTION***

A.

$$v_{AB} = \frac{s_f - s_i}{t}$$

$$v_{AB} = \frac{(160 - 0) \text{ m}}{120 \text{ s}}$$

$$v_{AB} = 1.33 \text{ m/s}$$

B.

$$v_{BC} = \frac{[(-40) - 160)] \text{ m}}{300 \text{ s}}$$

$$v_{BC} = -0.67 \text{ m/s}$$

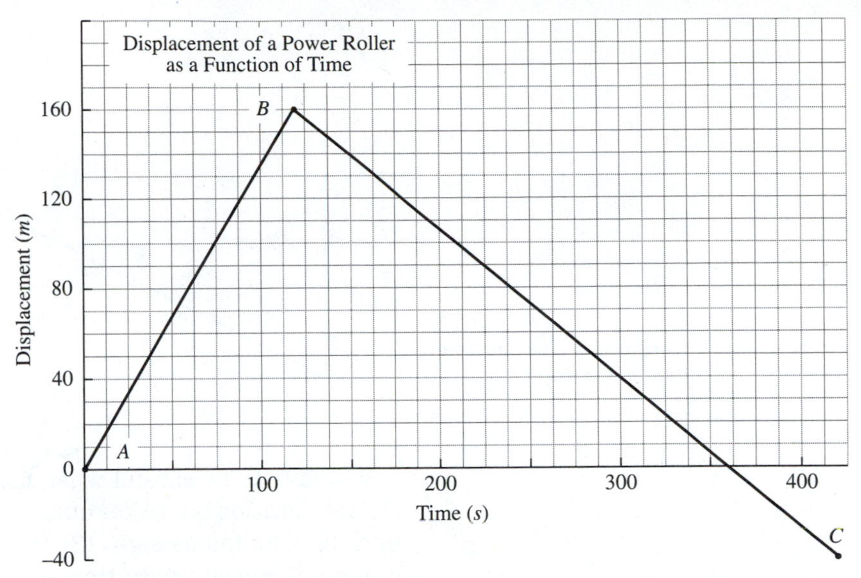

Figure 4–2: Plot of displacement versus time for a power roller. Between points A and B the motion is forward. Between points B and C the roller travels backward.

C.

$$v_{AC} = \frac{[(-40.0) - 0.0]\text{ m}}{420\text{ s}}$$

$$v_{AC} = -0.1\text{ m/s}$$

If the roller had returned to its original position the overall displacement would have been zero. The average velocity would have been zero.

4.2 AVERAGE ACCELERATION

A rock is thrown into the air. Its velocity decreases as it rises. It stops at a maximum height and plummets downward with an increasing speed. An automobile traveling at 55 mph comes to a stop at an intersection. A piston moves rapidly up and down inside a cylinder block. These are all examples of situations in which velocity is changing with time. We need a relationship to describe this change. The simplest solution would be to use an equation much like Equation 4–1. An average change of velocity with time is called **average acceleration** (a).

$$a_{\text{avg}} = \frac{v_f - v_i}{t} \qquad \textbf{(Eq. 4–2)}$$

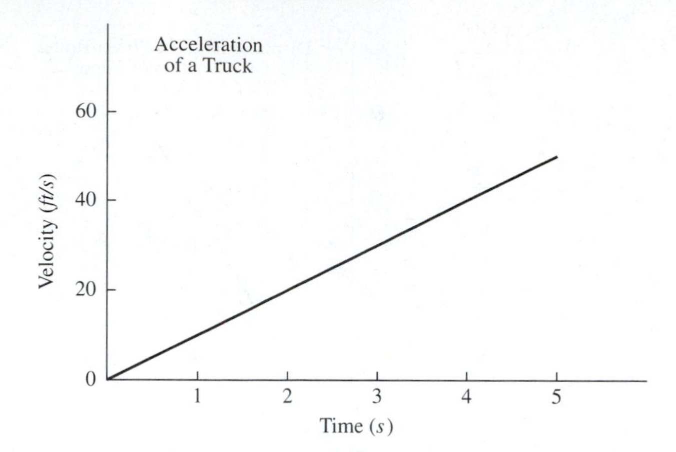

Figure 4–3: Graph of velocity versus time for a truck. A uniform acceleration is indicated by the constant slope of the plot. Equation 4–2 can be used.

We have to be careful when using this equation for two reasons: (1) Acceleration is a *vector* quantity. Its direction should be indicated by a plus or minus sign. (2) In this equation we assume that the change of velocity with time is a constant ratio. That is to say the acceleration is constant. Figure 4–3 shows a case where Equation 4–2 does a good job of describing the motion. Figure 4–4 shows a situation in which Equation 4–2 does not work very well. We can use Equation 4–2 to find the average acceleration in going from A to B. It would be the slope of the dashed line connecting points A and B. It would do a poor job of predicting the velocity at points between A and B. It would do an even poorer job of estimating distances.

In Figure 4–4 the average acceleration for the change of velocity during a very short interval of time fits fairly well to the curve. Look at points C and D. A short segment of a curved line appears to be straight. We can define **instantaneous velocity** using this idea.

$$a = \lim_{t \to 0} \frac{\Delta v}{\Delta t}$$

Instantaneous acceleration is an exact fit of the slope of the curve at any point along the velocity-time curve. In Figure 4–4 the instantaneous acceleration is quite large near point A and nearly zero at point B.

Acceleration has units of velocity divided by time. In British units it is feet per second per second, or feet per second squared. If an automobile has a uniform acceleration of 7.2 ft/s^2, its velocity will increase by 7.2 ft/s for each second it travels. If it starts from rest, it will be traveling at 7.2 ft/s at the end of the first second. At the end of 2 s its velocity will be 14.4 ft/s. At the end of 3 s its velocity will be 21.6 ft/s. The acceleration is the amount of change of velocity the car undergoes during each unit of time.

Figure 4–4: Velocity plotted against time for a motorboat. The velocity increases, but not at a constant rate. Equation 4–2 cannot be used to describe the boat's motion.

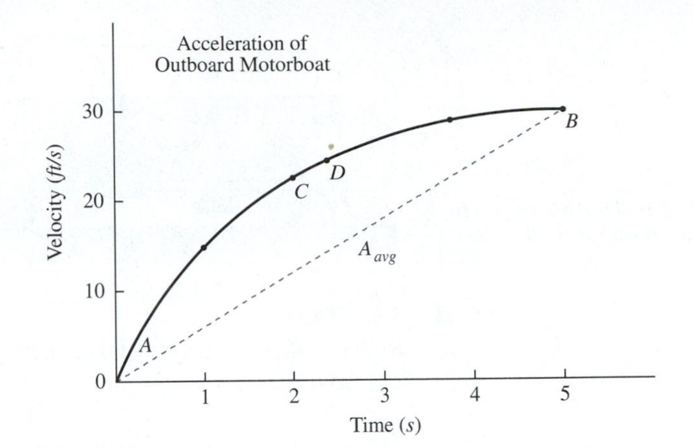

☐ **EXAMPLE PROBLEM 4–3: THE ELECTRON BEAM**

A beam of electrons initially at rest experiences an average acceleration of 8.80×10^{14} m/s^2 as the electrons pass between a set of electric plates. The electrons reach a final velocity of 2.14×10^7 m/s. How much time elapses while the electrons are accelerated?

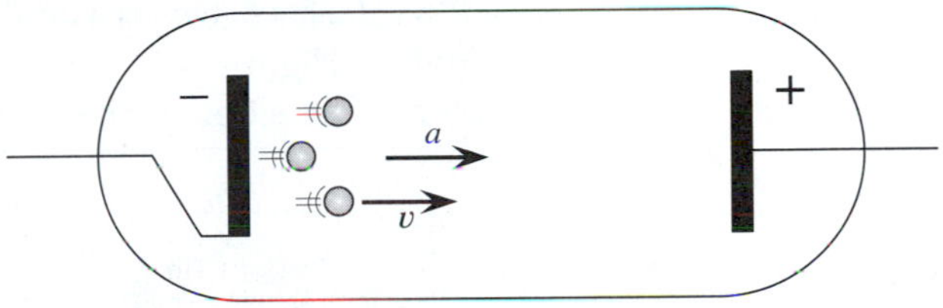

Figure 4–5: Electrons are accelerated between a pair of electric plates. Acceleration and velocity are in the same direction.

■ ***SOLUTION***

Solve Equation 4–2 for time.

$$t = \frac{v_f - v_i}{a}$$

$$t = \frac{2.14 \times 10^7 \text{ m/s}}{8.80 \times 10^{14} \text{ m/s}^2}$$

$$t = 2.43 \times 10^{-8} \text{ s}$$

☐ **EXAMPLE PROBLEM 4–4: THE BRAKING TRUCK**

A truck traveling at 83 ft/s brakes to a stop in 6.3 s. What is its average acceleration?

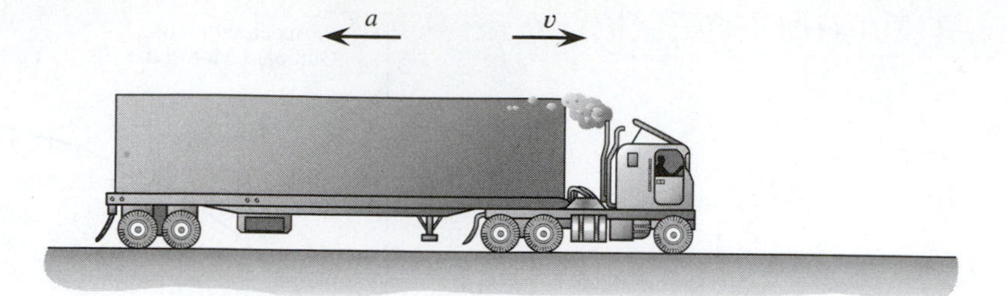

Figure 4–6: A truck brakes to a stop. The acceleration is in the opposite direction to the velocity.

■ *SOLUTION*

We will take the initial direction of the truck's motion to be the positive direction.

$$a_{\text{avg}} = \frac{v_f - v_i}{t}$$

$$a_{\text{avg}} = \frac{(0.0 - 83)\ \text{ft/s}}{6.3\ \text{s}}$$

$$a_{\text{avg}} = -13.2\ \text{ft/s}^2$$

The negative sign means that the acceleration is in the opposite direction of the truck's initial velocity. In this case it means the truck's velocity decreases on the average by 13.2 ft/s for each second it travels. Figure 4–7 is a graph of velocity versus time for this problem. The slope of the line (acceleration) is negative.

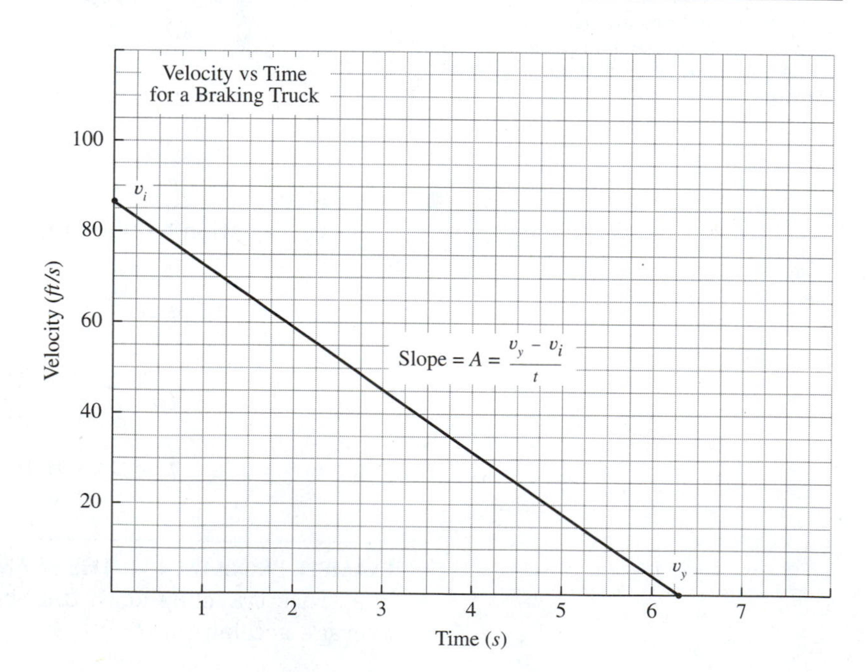

Figure 4–7: Graph of velocity versus time for Example Problem 4-4. The negative slope of the plot is the truck's acceleration.

4.3 AVERAGE VELOCITY REVISITED

If something has a uniform acceleration as depicted in Figures 4–3 and 4–8, there is another way to define average velocity. Notice that velocity versus time plots into a straight line. The slope of the graph, the acceleration ($\Delta v/\Delta t$), is constant. The shaded area under the line is calculated by multiplying velocity by time. The area represents the distance traveled.

We can break the shaded area into two parts: a rectangle with a height of v_0 and a triangle (darker area) with a height of $v_f - v_0$. The total distance traveled is:

$$s = \textit{area of rectangle} + \textit{area of triangle}$$

$$s = (\text{base}_1 \times \text{height}_1) + \left(\frac{\text{base}_2 \times \text{height}_2}{2}\right)$$

$$s = (v_i\,t) + \frac{(v_f - v_i)t}{2}$$

or

$$s = \frac{(v_f + v_i)}{2}\,t$$

Compare this with Equation 4–1, which represents the same distance (s) traveled during the time interval (t). The things in parentheses represent the same thing, average velocity.

$$s = v_{\text{avg}}\,t = \left(\frac{v_i + v_f}{2}\right)t \qquad \text{or}$$

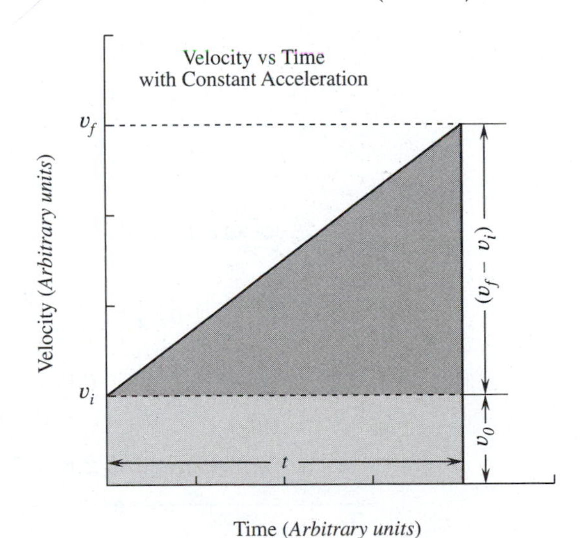

Figure 4–8: The area under the velocity versus time curve is velocity multiplied by time or the distance traveled. If the acceleration is constant, then $(v_f + v_i)/2 = v_{\text{avg}}$.

$$v_{avg} = \frac{v_i + v_f}{2} \qquad \textbf{(Eq. 4–3)}$$

Equation 4–3 is more specific than Equation 4–1. It can be used to determine average velocity only if acceleration is uniform.

□ **EXAMPLE PROBLEM 4–5: THE PISTON**

A piston in a gasoline engine has an initial velocity of zero at the beginning of its power stroke. It is accelerated to a final velocity of 20.6 m/s in 0.0073 s.

A. What is its average velocity?

B. What distance does the piston move during the stroke?

C. What is the piston's average acceleration?

■ ***SOLUTION***

A. Average Velocity:

$$v_{avg} = \frac{v_i + v_f}{2}$$

$$v_{avg} = \frac{(0.0 + 20.6)}{2} \text{ m/s}$$

$$v_{avg} = 10.3 \text{ m/s}$$

B. Distance:

$$s = v_{avg}\, t$$

$$s = 10.3 \text{ m/s} \times 0.0073 \text{ s}$$

$$s = 0.075 \text{ m or } 7.5 \text{ cm}$$

C. Acceleration:

$$a_{avg} = \frac{v_f - v_i}{t}$$

$$a_{avg} = \frac{(20.6 - 0.0) \text{ m/s}}{0.0073 \text{ s}}$$

$$a_{avg} = 2.8 \times 10^3 \text{ m/s}^2$$

4.4 DERIVED EQUATIONS

We have developed three equations that explain what we mean by average velocity and average acceleration. These are called **defining equations.** With only these three equations we can solve any problem involving uniform acceleration. Unfortunately all three equations may be needed to solve some kinds of problems. It would be helpful to combine the equations to make a single equation that would better suit a particular situation. Equations made by combining defining equations are called **derived equations.**

All three of the defining equations include the final velocity, v_f. In many kinds of problems we would like to find the displacement of something given the initial conditions and time. Let us find a derived equation that does not contain v_f.

Substitute Equation 4–2 into Equation 4–1.

$$s = v_{avg}\,t = \frac{(v_i + v_f)}{2}\,t$$

or

$$s = \left(\frac{v_i\,t}{2}\right) + \left(\frac{v_f\,t}{2}\right) \qquad \textbf{(Eq. 4–4A)}$$

Solve Equation 4–2 for final velocity.

$$v_f = v_i + (a_{avg}\,t) \qquad \textbf{(Eq. 4–4B)}$$

Substitute Equation 4–4B into Equation 4–4A.

$$s = \left(\frac{v_i\,t}{2}\right) + \left[\frac{v_i + (a_{avg}\,t)}{2}\right]t$$

Rearranging the terms, we get:

$$s = v_i\,t + \frac{a_{avg}}{2}\,t^2 \qquad \textbf{(Eq. 4–5)}$$

☐ **EXAMPLE PROBLEM 4–6: THE PICKUP TRUCK**

A pickup truck initially traveling at 86.4 ft/s brakes with a constant deceleration of 9.72 ft/s^2. What distance (s) does the pickup travel while braking in 3.3 s?

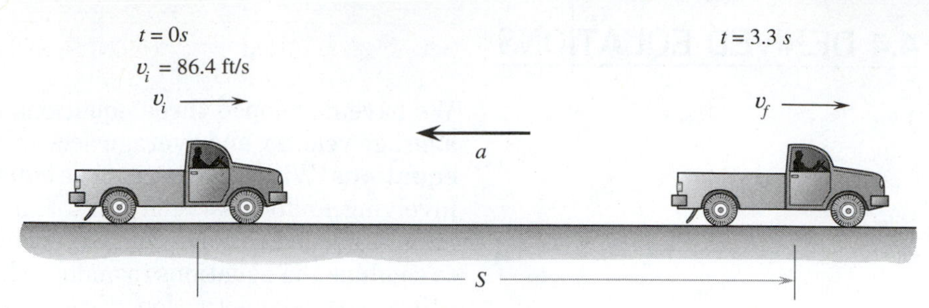

Figure 4–9: *A pickup truck travels a distance s while it slows down.*

SOLUTION

The acceleration is in the opposite direction of the initial velocity. If the initial velocity is positive, the acceleration is negative. Let us list the information we have.

$$v_i = 86.4 \text{ ft/s}$$

$$a = -9.72 \text{ ft/s}^2$$

$$t = 3.3 \text{ s}$$

$$s = ?$$

$$s = (v_i\, t) + \left(\frac{a\, t^2}{2}\right)$$

$$s = (86.4 \text{ ft/s} \times 3.3 \text{ s}) + \frac{(-9.72 \text{ ft/s}^2)(3.3 \text{ s})^2}{2}$$

$$s = 232 \text{ ft}$$

Let us now derive an equation that will find the distance traveled in an accelerated system when we do not know the elapsed time. Solve Equations 4–2 and 4–4A for time.

$$t = \frac{v_f - v_i}{a}$$

$$t = \frac{2s}{v_i + v_f}$$

These two equations can be set equal to each other to eliminate t.

$$\frac{v_f - v_i}{a} = \frac{2s}{v_i + v_f}$$

$$2(a\, s) = v_f^2 - v_i^2 \qquad \textbf{(Eq. 4–6)}$$

EXAMPLE PROBLEM 4–7: THE RIFLE BULLET

A bullet is accelerated to a speed of 380 m/s as it travels down a 76.3-cm rifle barrel. What is the average acceleration of the bullet?

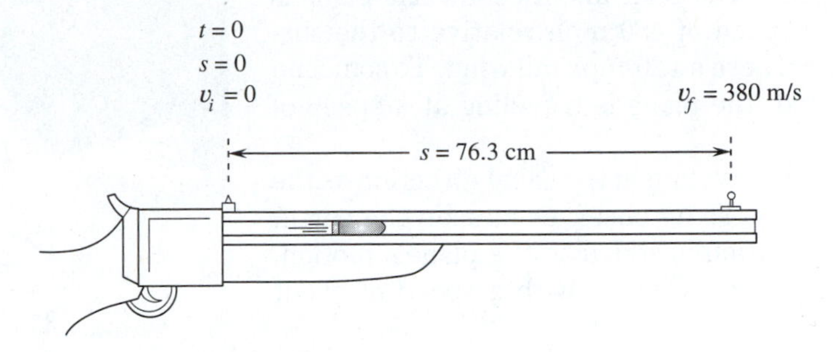

Figure 4–10: A bullet is accelerated by expanding gases as it speeds the length of a rifle barrel.

SOLUTION

$$s = 76.3 \text{ cm} = 0.763 \text{ m}$$

$$v_i = 0.00$$

$$v_f = 38\bar{0} \text{ m/s}$$

$$a = ?$$

$$a = \frac{v_f^2 - v_i^2}{2s}$$

$$a = \frac{(380 \text{ m/s})^2 - (0.00 \text{ m/s})^2}{2 \times 0.763 \text{ m}}$$

$$a = 9.46 \times 10^4 \text{ m/s}^2$$

4.5: GRAVITATIONAL ACCELERATION

Toss a coin straight up in the air. It moves upward with a decreasing velocity, comes to a stop, and plummets downward with an increasing velocity. The motion of the coin can be described as having a downward acceleration. In the upward motion, the acceleration—down toward the center of the earth—is in the opposite direction of the velocity. The downward acceleration brings the coin momentarily to a halt at the top of its path. As the coin begins to fall, the velocity and the acceleration are both in the same direction, downward. The coin speeds up. The downward acceleration of the tossed coin is caused by the planet's gravitational field. It is always directed toward the center of the planet. With very small variations the acceleration of a freely falling body has the same value near the surface of the

Motion Is Relative

A large jet airplane is cruising on the level with a constant speed. The instruments show the plane is moving at a speed of 480 mph relative to the surrounding air. There's a 20-mph tail wind. To someone on the ground, the plane is traveling at a speed of 500 mph.

The plane is traveling in the same direction as the earth's rotation. To an observer outside the earth, the earth's rotation is added to the plane's motion. The plane appears to move with a speed of about 1,500 mph.

Someone stationed near the sun sees the motion of the earth. This adds a speed of 18.5 miles a second to the other motions.

(Photograph courtesy of NASA.)

The sun and planet earth are located on a spiral arm about two thirds of the way to the outer edge of the Milky Way galaxy. We're rotating around the galactic center at a speed of 155 miles per second. At the moment, we're moving toward the constellation Cygnus.

We're part of a small cluster of galaxies called "The Local Group." The galaxies mill around each other. An observer on a planet in one of these neighboring galaxies would see the motion of our galaxy relative to his/her/its galaxy added onto the other motions.

It doesn't stop here. All of the galaxies in our neighborhood of the universe seem to be moving in the same direction. It's as though we were being pulled by an immense black hole dubbed "The Big Attracter."

You're probably sitting in a chair as you read this. You probably think you're at rest. That's your opinion. All motion is relative.

---

earth. It can be treated as constant acceleration. The acceleration caused by a gravitational field is called **gravitational acceleration** (g). On the surface of the earth

$$g = 32 \text{ ft/s}^2 = 9.8 \text{ m/s}^2$$

Gravitational acceleration is not unique to the earth. All planets and their moons have a surface gravitational acceleration. The only difference is that the value of g will not be the same as it is here on earth. For example, on Mars $g = 3.7 \text{ m/s}^2$ or 12 ft/s^2. On the moon $g = 1.6 \text{ m/s}^2$ or 5.3 ft/s^2.

Let us return to the surface of the earth. We will perform an experiment in which we can ignore air friction. If a compact, dense object is dropped near the earth's surface, it will be falling downward at 32 ft/s at the end of 1 s. At the end of 2 s its velocity will be 64 ft/s. At the end of 3 s its final velocity is 96 ft/s downward. The change of velocity is 32 ft/s downward for every second of motion.

The five equations of motion can be used to describe an object's motion under the influence of gravity when friction is ignored. The only adjustment we need to make is to recognize that the acceleration has the value of g, and g is always directed down. The three equations of motion containing g are:

$$s = (v_i\, t) + \left(\frac{g\, t^2}{2}\right)$$

$$v_f = v_i + (g\, t)$$

$$2\,(g\, s) = v_f^2 - v_i^2$$

□ **EXAMPLE PROBLEM 4–8: THE FALLING BOLT**

At a building site, a rigger drops a large bolt from a height of 12.7 m.

A. How long will it take the bolt to hit the ground?

B. What velocity will the bolt have just before it hits the ground?

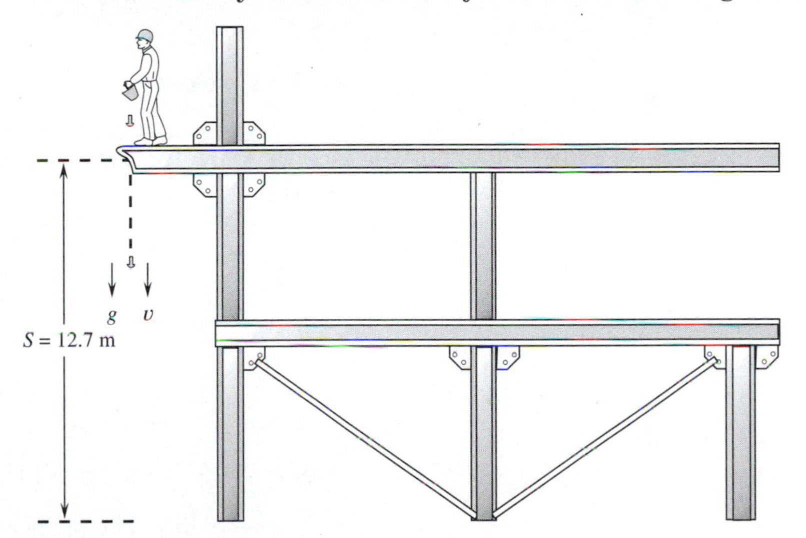

Figure 4–11: *A bolt is dropped. It accelerates downward under the influence of gravity.*

■ ***SOLUTION***

First list the information. This will help us choose the easiest equation of motion for the problem.

v_i = 0.0 (The bolt was dropped, not thrown.)

$a = g = -9.8\ \text{m/s}^2$ [Downward is the negative (–) direction.]

$s = -12.7$ m (The displacement is down.)

t = ?

v_f = ?

A. Time of flight:

Equation 4–5 best fits the data we have listed.

$$s = (v_i\, t) + \left(\frac{g\, t^2}{2}\right)$$

The first term on the right-hand side of the equation is zero, because $v_i = 0.0$. Solve the equation for t.

$$t = \left(\frac{2s}{g}\right)^{1/2}$$

$$t = \left[\frac{2(-12.7\text{ m})}{-9.8\text{ m/s}^2}\right]^{1/2}$$

$$t = 1.6\text{ s}$$

B. Final velocity:

Since we know the time, we can use Equation 4–2.

$$a = g = \frac{v_f - v_i}{t}$$

or

$$v_f = v_i + (g\, t)$$

$$v_f = 0.0 + [(-9.8\text{ m/s}^2)\ 1.6\text{ s}]$$

$v_f = -16$ m/s (The negative sign means motion is downward.)

☐ **EXAMPLE PROBLEM 4–9: THE LUNAR FEATHER**

An astronaut stands on the edge of a small lunar crater. The astronaut tosses a feather upward with a speed of 22 ft/s.

A. Why can we use the equations of motion in this chapter for a feather tossed on the moon, but not on earth?

B. How high does the feather travel from the point of release?

C. How long is the feather in flight before it strikes the surface of the moon 12.3 ft below the point of release?

Figure 4–12: *An astronaut tosses a feather upward at the edge of a lunar crater.*

■ *SOLUTION*

A. The friction from the earth's atmosphere quickly slows down a feather dropped near the planet's surface. But the moon has no atmosphere, so there is no air friction to slow down the feather. It moves with the same motion as a heavy, dense object such as a hammer or a rock.

B. Maximum height:

First choose the positive and negative directions. Let us take upward as positive.

$$v_i = 22 \text{ ft/s}$$

$$g = -g_{\text{moon}} = -5.3 \text{ ft/s}^2$$

When the feather reaches its highest point it comes to a stop momentarily before traveling downward. To find the highest point we can set the final velocity at zero.

$$v_f = 0.0 \text{ (at the peak of the flight)}$$

$$2g\, s_{\text{max}} = v_f^2 - v_i^2$$

or

$$s_{\text{max}} = \frac{v_f^2 - v_i^2}{2g}$$

$$s_{\text{max}} = \frac{(0.0 \text{ ft/s})^2 - (22 \text{ ft/s})^2}{2(-5.3 \text{ ft/s}^2)}$$

$$s_{\text{max}} = 46 \text{ ft}$$

C. Time of flight:

At the end of the flight the feather is below the point of release. Since we chose upward to be the positive direction, the final displacement is negative.

$$s = -12.3 \text{ ft}$$

$$s = (v_1 t) + \left(\frac{g t^2}{2}\right)$$

or

$$\left(\frac{g t^2}{2}\right) + (v_i t) - s = 0$$

$$\left[\frac{(-5.3 \text{ ft/s}^2)}{2} t^2\right] + \left[(22 \text{ ft/s})\, t\right] - (-12.3 \text{ ft}) = 0$$

Divide the equation by the leading coefficient. We get:

$$t^2 - [(8.3 \text{ s})\, t] - 4.64 \text{ s}^2 = 0$$

We can now use the quadratic equation to find time.

$$t = \frac{-(-8.3 \text{ s}) \pm [(-8.3 \text{ s})^2 - (4 \times 1) \times (-4.64 \text{ s}^2)]^{1/2}}{2 \times 1}$$

$$t = (4.15 \pm 4.67) \text{ s}$$

$$t = 8.8 \text{ s or } -0.52 \text{ s}$$

Only the positive solution fits the conditions of the problem. We started our clock at $t = 0$.

$$t = 8.8 \text{ s}$$

☐ **EXAMPLE PROBLEM 4–10: THE LUNAR FEATHER AGAIN**

Here is another way to solve the feather problem if you don't like using the quadratic formula. Break the flight into two parts: the flight upward and the flight downward (see Figure 4–13).

$$t = t_{up} + t_{down}$$

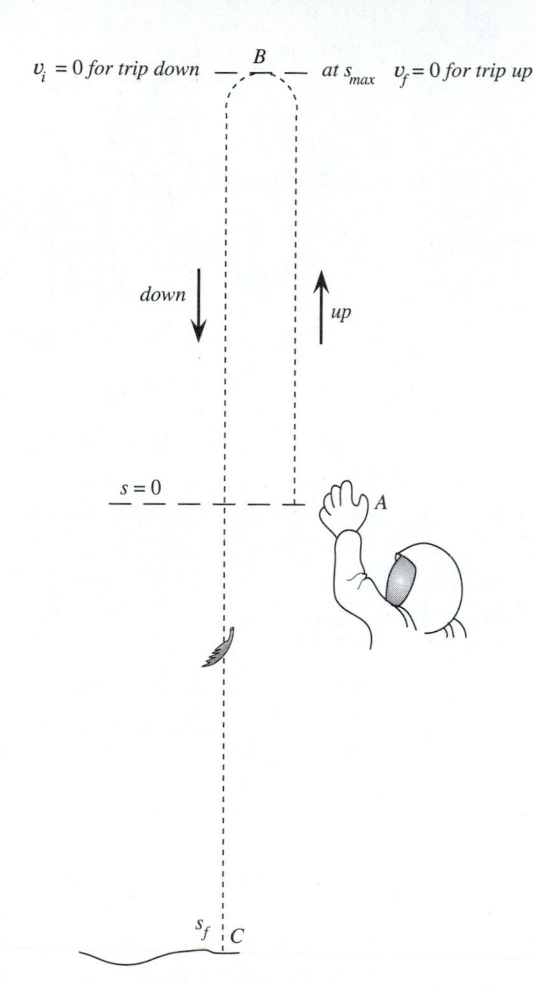

Figure 4–13: *The motion of the feather can be broken down into two parts.* ***A to B.*** *The feather moves upward reaching a velocity of zero at its maximum height.* ***B to C.*** *The feather falls from the maximum height with an initial speed of zero.*

t_{up} can be found using Equation 4–2. The speed of the feather is zero at the top of the flight.

$$t_{up} = \frac{v_f - v_i}{-g}$$

$$t_{up} = \frac{0.0 - 22 \text{ ft/s}}{-5.3 \text{ ft/s}^2}$$

$$t_{up} = 4.15 \text{ s}$$

From the top of the flight the feather falls 58 ft: 45.7 ft to get back to its original height plus the extra 12.3 ft. Since the velocity is zero at s_{max}, we can take the initial velocity as zero for the second part of the flight.

$$s = \frac{a\, t_{\text{down}}^2}{2} \quad \text{for } v_i = 0.0$$

$$t_{\text{down}} = \left(\frac{2\,s}{a}\right)^{1/2}$$

$$t_{\text{down}} = \left[\frac{2\,(-58\text{ ft})}{-5.3\text{ ft/s}^2}\right]^{1/2}$$

$$t_{\text{down}} = 4.68\text{ s (Use only the positive root.)}$$

$$t = t_{\text{up}} + t_{\text{down}}$$

$$t = 4.15\text{ s} + 4.68\text{ s}$$

$$t = 8.8\text{ s}$$

4.6 TWO-DIMENSIONAL MOTION

A basketball player races across the court, jumps into the air to intercept a pass, and lands 8 ft in front of his take-off point. A mechanic tosses a socket wrench to her partner. A small piece of wood catches on the blade of a table saw. It is kicked back and flies across the room with its path forming an arc. These are examples of two-dimensional motion. It is called **projectile motion** when the only acceleration is caused by gravity.

The analysis of projectile motion is not difficult when we remember that displacement, velocity, and acceleration are all vector quantities. Because vectors can be broken into components, motion can

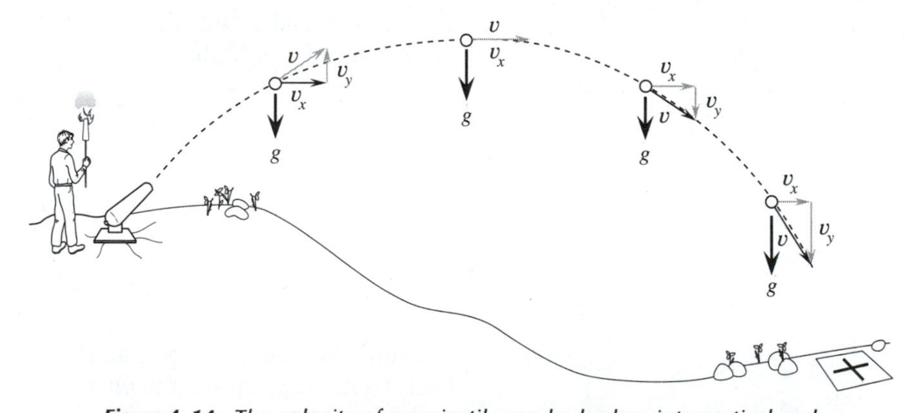

Figure 4–14: *The velocity of a projectile can be broken into vertical and horizontal components. If friction is neglected, the horizontal component is constant. The vertical component of velocity is accelerated downward at a constant gravitational acceleration.*

be broken into components. Figure 4–14 shows how the velocity components of projectile motion may be broken into vertical and horizontal components.

Let us look at the horizontal motion. When something is thrown in the air, it accelerates downward. There is no horizontal component of gravitational acceleration. Gravity does not pull things sideways. If we restrict ourselves to situations in which air friction can be ignored, there is no acceleration on the x plane. This means the horizontal component of velocity cannot change. The horizontal initial velocity (v_{ix}), final velocity, and average velocity are all the same. For the horizontal components of motion we can write:

$$v_{ix} = v_{x\text{avg}} = v_{xf}$$

The horizontal distance the object travels is often called the **range.** The range can be found easily if we know how long the object is in the air.

$$range = x = v_x t \qquad \textbf{(Eq. 4–7)}$$

In Equation 4–7 we have not specified whether we mean initial, final, or average horizontal velocity, because in projectile motion they are all the same.

Now let us look at the vertical component of motion. In projectile motion there is always an acceleration in the vertical direction acting downward. Its magnitude is g, the acceleration of gravity. The vertical component of velocity will continually change. Look again at Figure 4–14. The vertical component of velocity changes during the flight. It decreases in magnitude as the object rises, has a value of zero at the top of the flight, and then increases as the object falls downward. This is the same behavior described for a coin in Section 4.5, vertical motion under the influence of gravity.

The two components of motion are independent. What they have in common is the time of flight. This leads us to the following strategy for solving two-dimensional, or projectile motion, problems.

Step 1: **Read the problem.**

Step 2: **Find the horizontal and vertical components of the initial velocity.**

Step 3: **Write down the available data for horizontal motion and vertical motion separately.** Treat the horizontal and vertical motion as separate problems.

Step 4: **Find the time of flight (t) from the equation of motion of one component of motion** (usually the vertical component).

Step 5: **Plug the time of flight (t) into the other component equation of motion and solve for the unknown.**

EXAMPLE PROBLEM 4–11: THE HAYBALER

A haybaler has a kicker attached to it. The kicker throws completed bales at an angle of 52° above the horizontal and at a speed of 26.0 ft/s onto a trailer attached to the baler. What are the horizontal and vertical components of a bale's initial velocity?

Figure 4–15: A kicker at the end of a baler tosses bales into a following wagon.

SOLUTION

The initial velocity is v_i. The horizontal component is:

$$v_{ix} = v_i \cos \theta$$

$$v_{ix} = 26.0 \text{ ft/s} \times \cos 52^\circ$$

$$v_{ix} = 16.0 \text{ ft/s}$$

The vertical component is:

$$v_{iy} = v_i \sin \theta$$

$$v_{iy} = 26.0 \text{ ft/s} \times \sin 52^\circ$$

$$v_{iy} = 20.5 \text{ ft/s}$$

EXAMPLE PROBLEM 4–12: THE FIRE HOSE

A nozzle of a fire hose ejects water at a speed of 78.6 ft/s at an angle of 32° above the horizontal. The water strikes its target at the same elevation as the nozzle.

A. What is the time of flight of the water?

B. What is the horizontal distance between the nozzle and the target?

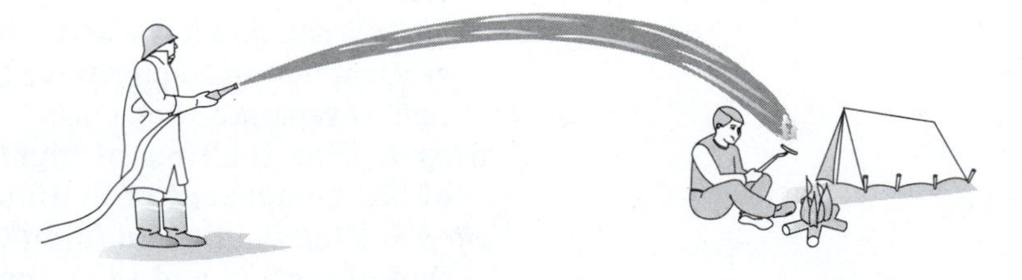

Figure 4–16: A stream of water has projectile motion.

■ *SOLUTION*

First find the components of the initial velocity.

$$v_{ix} = v_i \cos \theta$$

$$v_{ix} = 78.6 \text{ ft/s} \times \cos 32°$$

$$v_{ix} = 66.7 \text{ ft/s}$$

$$v_{iy} = v_i \sin \theta$$

$$v_{iy} = 78.6 \text{ ft/s} \times \sin 32°$$

$$v_{iy} = 41.7 \text{ ft/s}$$

Next list the data for each component of motion. We will take upward as the positive direction.

Vertical Motion	**Horizontal Motion**
$y = 0.0$ ft	$x = \ ? = v_{avg}\, t$
$v_{iy} = 41.7$ ft/s	$v_{avg} = v_{ix} = 66.7$ ft/s
$a_y = g = -32$ ft/s^2	$a_x = 0.0$ ft/s^2
$t = \ ?$	$t = \ ?$

A. Time of flight:

We can find the time of flight from the vertical motion.

$$y = (v_{iy}\, t) + \left(\frac{g\, t^2}{2}\right)$$

$$0.0 = [(41.7 \text{ ft/s})\, t] + \left[\frac{(-32 \text{ ft/s}^2)}{2}\, t^2\right]$$

$$t[(41.7 \text{ ft/s}) - (16 \text{ ft/s}^2\, t)] = 0.0$$

$$t = 0.0 \text{ s or } 2.61 \text{ s}$$

$t = 0.0$ corresponds to the time of release of the water. This is the first time the water passes through a height of 0.0 ft. What we want is the other solution, a time when the water again falls to an elevation of 0.0 ft.

$$t = 2.61 \text{ s}$$

B. Range:

Use the time of flight from Part A of the problem.

$$x = v_{ix}\,t$$

$$x = 66.7 \text{ ft/s } (2.61 \text{ s})$$

$$x = 174 \text{ ft}$$

☐ **EXAMPLE PROBLEM 4–13: THE BASKETBALL PLAYER**

A basketball player is racing at 6.44 m/s across the court toward the out-of-bounds line. She leaps 0.95 m into the air to intercept a pass.

A. What is her time of flight?

B. If she jumps into the air 2.5 m in front of the out-of-bounds line, will she land inside or outside the playing court?

C. At what angle does she leave the floor?

Figure 4–17: A basketball player leaps to intercept the ball.

■ ***SOLUTION***

List the information we have. We will take upward as positive.

Vertical Components	Horizontal Components
$y = 0.0$ ft	$x = ?$
$v_{iy} = ?$	$v_{ix} = 6.44$ ft/s
$a_{iy} = g = -9.8 \text{ m/s}^2$	$t = ?$
$t = ?$	

A. Time of flight:

We do not have the initial velocity in the vertical motion. There are two approaches to finding the time. Both approaches take advantage of the fact that we know how high the player leaps. Let us see how they work.

1. Find the initial velocity using y_{max}. Then solve the problem in the same way as Problem 4–12.
2. The flight is symmetric; time up equals time down. The time of flight is twice the time to fall from the maximum height with an initial speed of 0.0.

Method 1: At the peak of the path $v_{fy} = 0$; $y_{\text{max}} = 0.95$ m.

$$2\,y_{max}\,g = v_{fy}^{\,2} - v_{iy}^{\,2}$$
$$v_{iy}^{\,2} = -2\,(y_{max}\,g) + v_{fy}^{\,2}$$
$$v_{iy}^{\,2} = -2\,[(0.95\text{ m})(-9.8\text{ m/s}^2)] + 0.0$$
$$v_{iy} = +4.3\text{ m/s (Upward is positive.)}$$
$$y = (v_{iy}\,t) + \left(\frac{g}{2}\,t^2\right)$$
$$0.0 = [(4.3\text{ m/s})\,t] + \left[\frac{(-9.8\text{ m/s}^2)}{2}\,t^2\right]$$
$$0.0 = t\,[(4.3\text{ m/s}) - (4.9\text{ m/s}^2\,t]$$
$$t = 0.88\text{ s}$$

Method 2: The player falls 0.95 m from an initial speed of 0.0.

$$y = (v_{iy}\,t) + \left[\frac{g}{-2}\,(t_{down})^2\right]$$
$$-0.95 = 0.0 + \left[\frac{(-9.8\text{ m/s}^2)}{2}\,(t_{down})^2\right]$$
$$t_{down} = 0.44\text{ s}$$
$$t = 2 \times t_{down} = 0.88\text{ s}$$

B. Range:

$$x = v_{ix}\,t$$
$$x = 6.44\text{ m/s} \times 0.88\text{ s}$$
$$x = 5.7\text{ m}$$

If she gets the ball, she had better pass off to a teammate before she lands. She will be way out of bounds.

C. We can take the ratio of initial velocities to find the angle. We have v_{iy} from Part A, Method 1.

$$\tan\theta = \frac{v_{iy}}{v_{ix}} = \frac{4.3\text{ m/s}}{6.44\text{ m/s}}$$
$$\theta = 34°$$

SUMMARY

Three defining equations can be used to describe linear motion. The equation for **average speed** is

$$v_{avg} = \frac{s_f - s_i}{t}$$

The equation for **uniform** or **average acceleration** is

$$a = \frac{v_f - v_i}{t}$$

The equation for **average velocity when an object has uniform acceleration** is

$$v_{avg} = \frac{v_i + v_f}{2}$$

Two additional equations can be derived from the defining equations.

$$2\,(a\,s) = v_f^2 - v_i^2$$

$$s = (v_i\,t) + a\,\frac{t^2}{2}$$

When an object is near a planet, gravity accelerates the object downward toward the center of the planet. The magnitude of the acceleration (g) is 32 ft/s^2, or 9.8 m/s^2, near the surface of earth.

Projectile motion is two-dimensional motion with gravitational acceleration. The vectors of displacement, velocity, and acceleration can be broken into horizontal and vertical components, which are independent of each other. They have the same time of flight. If there is no air friction, the horizontal component of velocity is constant ($v_{ix} = v_{fx} = v_{avgx}$). In component form, the equations for projectile motion are:

$$v_{ix} = v_i \cos\theta \qquad v_{iy} = v_i \cos\theta$$

$$x = v_{ix}\,t \qquad y = (v_{iy}\,t) + (g\,t^2)$$

KEY TERMS

If you can explain the following terms to a friend or classmate, you understand their meaning. If you cannot explain the terms, you should reread the sections in which they are discussed.

acceleration
average acceleration
average speed
average velocity
defining equations
derived equations
gravitational acceleration

instantaneous velocity
linear motion
projectile motion
range
rotational motion
velocity

EXERCISES

Section 4.1:

1. An automobile speedometer tells you how fast you are traveling. Is it possible to have a constant speed (scalar) and yet have a change of velocity (vector)? If your answer is yes, give an example.

2. A runner runs around a quarter-mile track four times in exactly 4 min. What is the average speed in miles per hour? What is the average velocity in miles per hour?

Section 4.2:

3. Can a negative acceleration cause something to increase its speed? If your answer is yes, give an example.

4. A motor scooter accelerates at 12 m/s^2. Which of the following statements *must* be true?
   A. The motor scooter travels a distance of 12 m in 1 s.
   B. The motor scooter will have a speed of 12 m/s at the end of 1 s.
   C. The motor scooter will have a change of speed of 12 m/s during a 1-s interval.
   D. The motor scooter's change of speed in 1 s is 9.8 m/s^2.

Section 4.3:

5. For which of the situations shown in Figure 4–18 (page 108) can the Equation 4–3 [$v_{avg} = (v_i + v_f)/2$] be used to find the average velocity?

6. Is it possible for something to undergo a constant acceleration and have an average velocity of zero? Explain why or why not, using Equation 4–3.

7. Is it possible for something to undergo a uniform acceleration and have an average speed of zero? Explain why or why not, using Equation 4–3.

Section 4.4:

8. A car has a constant acceleration in a straight line. Which of the following statements *must* be true?
   A. The car goes faster and faster.
   B. The car has a constant speed.
   C. The car's velocity must be continually changing.
   D. A plot of velocity versus time is a curved line.

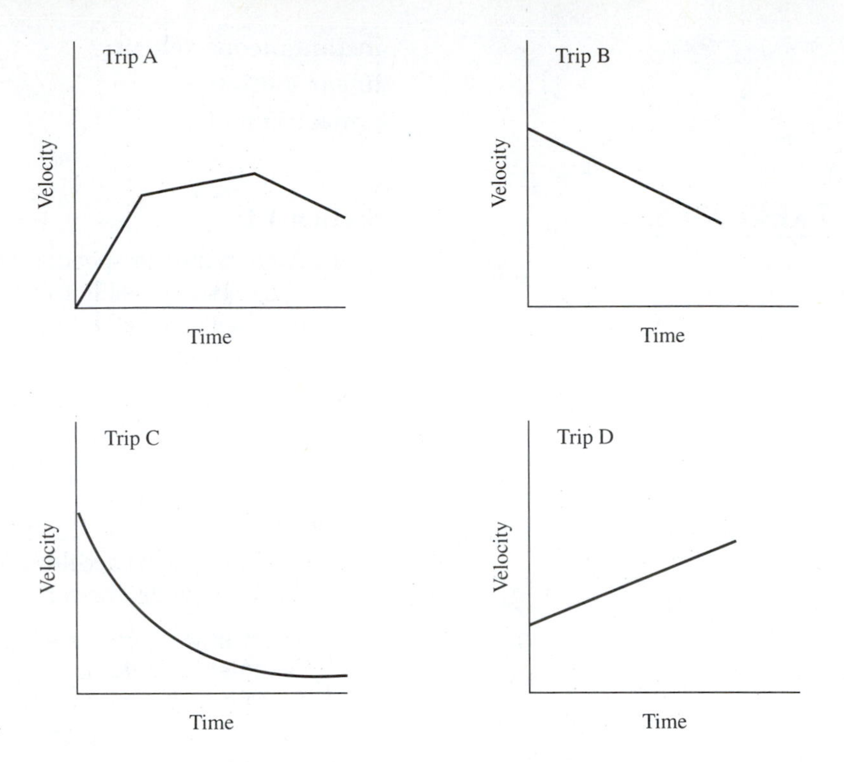

Figure 4–18: *Diagrams for Exercise 5.*

9. Figure 4–19 shows the velocity of some objects plotted against time.

A. Which object(s) is (are) slowing down?

B. Which object(s) has (have) constant acceleration?

C. Which object(s) has (have) an acceleration of zero?

D. Which object(s) has (have) an initial speed of other than zero?

10. Two cars have the same initial velocity. When they brake, car A decelerates at a rate of $2a$ and car B decelerates at a rate of a. What is the ratio of the stopping distances?

11. Two cars have the same rate of deceleration. Car A has twice the initial speed of car B. What is the ratio of the stopping distances?

****12.** A Ford and a Chevrolet race along a straight course from a standing start. The Ford accelerates at a rate of a for the first half of the distance. For the second half of the course it accelerates at $2a$. The Chevrolet accelerates at $2a$ for the first half of the distance and at a for the second half. Which car arrives at the finish line first, or do they arrive at the same time? (Hint: Look at Equation 4–5.)

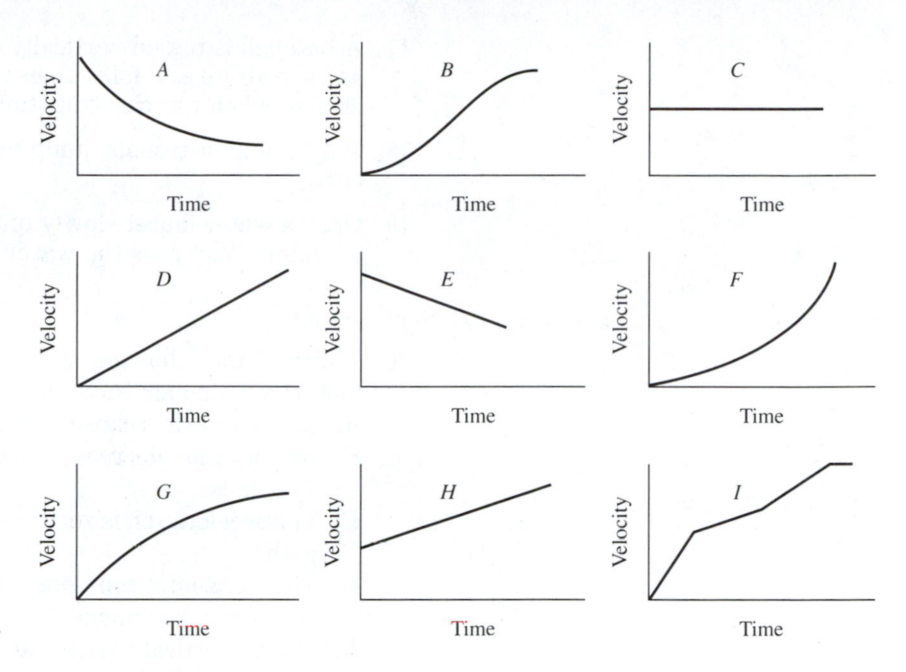

Figure 4–19: *Diagrams for Exercise 9.*

Section 4.5:

13. A hammer is dropped from a tall building. How far does it travel in 1 s? How far has it traveled after 2 s? After 3 s? Show that the ratio of total distances traveled is 1:4:9.

14. A rock falls down a mine shaft. What distance does it travel *during* the first second? What distance does it travel *during* the second second? (This is not total distance, but only the distance traveled during that 1-s interval.) During the third second? Show that the ratio of distances is 1:3:5.

15. Two rocks are dropped from a cliff. Rock B is dropped 1.0 s later than rock A. Which of the following statements is (are) true if we neglect friction?

A. The vertical distance between the rocks is a constant distance of 4.9 m as the two rocks fall.

B. The vertical distance between the rocks increases as they fall.

C. The vertical distance between the rocks decreases as they fall.

D. The two rocks hit the ground at the same time.

E. Rock B hits the ground 1 s after rock A.

F. Rock B hits the ground after rock A, but more than a second later.

16. A baseball is tossed vertically into the air. At its highest point its velocity is zero. Is the acceleration zero at this point? Explain.

17. A baseball is tossed vertically upward. It slows up as it rises and speeds up as it falls. Does the direction of the acceleration change when the baseball starts to fall? Explain.

18. Why can an astronaut jump higher on the moon than on the earth?

19. Open a water faucet slowly until there is a thin, even stream of water. Why does the water break into droplets as it falls?

Section 4.6:

20. Which of the following statements is (are) true of projectile motion with no air friction?
    - ***A.*** Gravity always causes the speed of the object to increase.
    - ***B.*** Acceleration decreases as the object rises, and decreases as it falls.
    - ***C.*** The acceleration is zero at the highest point of the object's path.
    - ***D.*** The horizontal components of motion are independent of the vertical components.
    - ***E.*** As the vertical velocity decreases, the horizontal velocity increases.
    - ***F.*** The vertical component of velocity is zero at the object's maximum height.

21. Figure 4–20 shows the motion of some compact, dense objects thrown into the air. (Air resistance is not a factor.) Which of the objects will stay in the air the longest? Which the shortest?

22. Two arrows are fired simultaneously and horizontally from a height of 1.00 m. Arrow A has twice the speed of arrow B. Neglecting air friction, compare their times of flight.

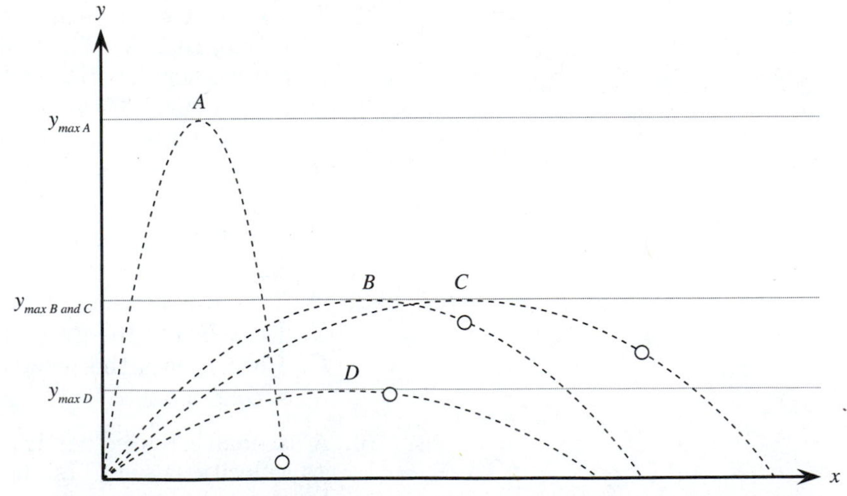

Figure 4–20: *Diagram for Exercise 21. Objects are projected at different initial speeds and angles of elevation.*

23. When a projectile is fired, what effect does the angle of elevation have on each of the following?
    A. The time of flight?
    B. The range?
    C. The maximum height of the path?

PROBLEMS

Section 4.1:

1. Find the average speed for the graph of distance versus time shown in Figure 4–21.
2. Find the indicated quantity.
   A. $s = 23.5$ mi; $t = 0.65$ h: $v_{avg} = ?$ mph
   B. $s = 23{,}400$ m; $t = 24$ min: $v_{avg} = ?$ km/h
   C. $v_{avg} = 126$ m/s; $t = 12.3$ s: $s = ?$ m
   D. $v_{avg} = 66.2$ ft/s; $s = 820$ ft: $t = ?$ s
   E. $t = 2.3$ min; $s = 1.24$ mi: $v_{avg} = ?$ ft/s
3. A participant in a track meet races around a quarter-mile oval track covering 1 mi in 3:58:2 min. What is the average speed? What is the average velocity?
4. The average speed of electrons in an electrical conductor is called the drift velocity. If the drift velocity of electrons in a copper wire is 0.32 mm/s, how long will it take electrons to travel through a length of wire 4.55 m long?
5. A cyclist travels at a speed of 12.4 mph for 40 min. The cyclist increases the speed to 17.3 mph for the next 20 min.
   A. What distance is traveled in the hour?
   B. What is the average speed?
6. The linear flow rate of water through a ½-in pipe is 14.0 ft/s. How many US gallons of water are delivered in 1 min? (One US gallon = 231 in^3.) See Figure 4–22 (page 112).

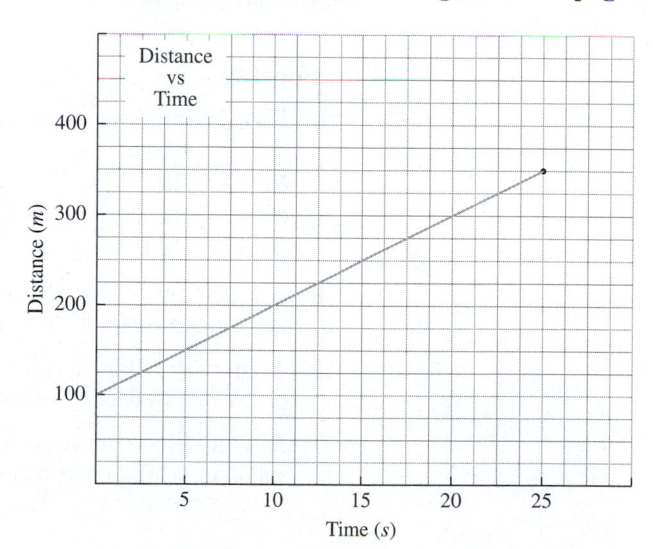

Figure 4–21: *Diagram for Problem 1.*

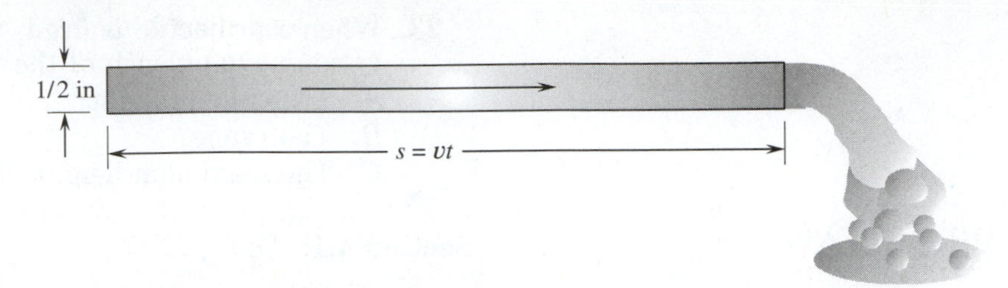

Figure 4–22: Diagram for Problem 6. The linear velocity of water passing through a pipe determines its volumetric flow rate.

**7. Two joggers are 2.0 mi apart. Each jogs at 4.0 mph toward the other runner. A dragonfly initially flies from runner A to runner B. It then turns around and flies back to runner A. The dragonfly continues to move back and forth between the two runners until they meet. The dragonfly moves at a constant rate, in which it can cover 2.0 mi in 10 min.

   A. What is the dragonfly's average speed?
   B. What is the dragonfly's average velocity?

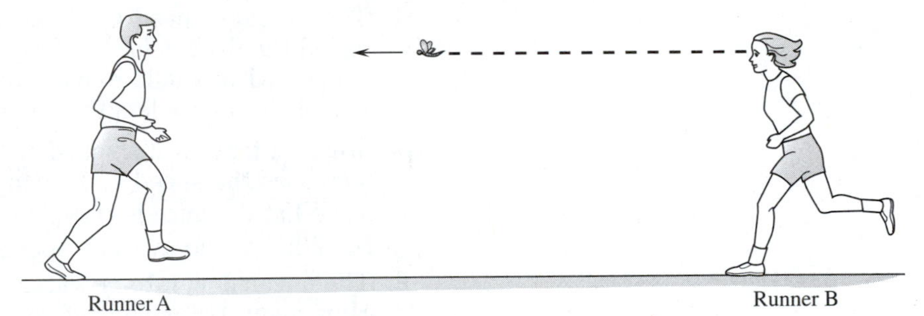

Figure 4–23: Diagram for Problem 7. A dragonfly races back and forth at a constant speed between two joggers.

Section 4.2:

8. Find the indicated quantity.
   A. $v_i = 0.0$ m/s; $v_f = 54.6$ m/s; $t = 12.6$ s: $|a| = ?$ m/s^2
   B. $v_i = 12.0$ ft/s; $v_f = 48.0$ ft/s; $t = 8.0$ s: $|a| = ?$ ft/s^2
   C. $v_i = 14.8$ ft/s; $t = 2.6$ s; $a = 2.70$ ft/s^2: $v_f = ?$ ft/s
   D. $v_i = 12.0$ m/s; $v_f = 32.6$ m/s; $a = 3.2$ m/s^2: $t = ?$ s

9. A truck traveling at 55 mph comes to a stop in 5.2 s. What is the magnitude of its acceleration in feet per seconds squared?

10. A conveyor belt accelerates from rest to a final linear speed of 1.7 m/s in 0.72 s. What is the magnitude of its acceleration?

11. A small piece of wood catches on the blade of a table saw. It undergoes an acceleration of 8.15×10^2 m/s^2 for 0.034 s. What is its kickback speed?

12. How long does it take a car to accelerate from 35 mph to 55 mph at a uniform acceleration of 8.70 ft/s²?

13. A rubber ball with an initial velocity of +87.0 ft/s strikes a wall and rebounds at −74.0 ft/s. If the change of velocity occurs in a period of 0.017 s, what is the acceleration of the ball?

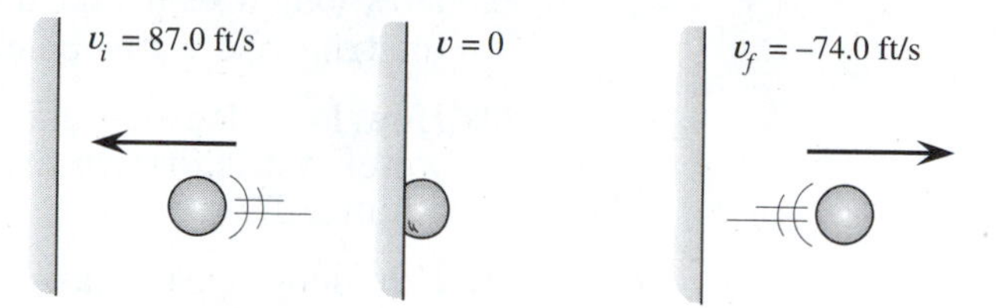

Figure 4–24: *Diagram for Problem 13. A rubber ball striking a wall is accelerated during the period of contact.*

Section 4.3:

14. A frozen pizza rolls down a hill at a uniform acceleration. What is its average speed if it started at 2.0 ft/s and attains a final speed of 8.2 ft/s?

15. A vise grip plummets from a rooftop and attains a final velocity of −23.0 m/s. If its average velocity for the flight is −16.5 m/s, what was its initial velocity? (Assume downward is negative.)

16. What are the average speed and average velocity of the rubber ball in Problem 13?

Section 4.4:

17. A car moving at +55 mph brakes, coming to rest in 4.6 s.
    A. What is its acceleration?
    B. What distance does it travel while coming to a stop?

18. A freight train slows from 34.0 m/s to 12.0 m/s at an acceleration of −2.76 m/s². What distance does it travel?

19. A hydraulic punch is used to fabricate sheet-metal parts. The punch is accelerated from rest to a speed of 93 ft/s in a distance of 8.60 in. What is the magnitude of its average acceleration?

20. A rifle with a 32-in barrel fires a bullet with a muzzle velocity of 1230 ft/s. What is the average acceleration of the bullet while it is in the barrel?

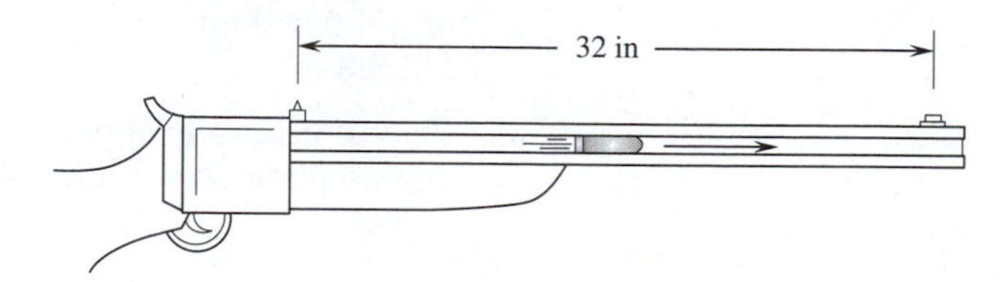

Figure 4–25: *Diagram for Problem 20. A bullet is accelerated in a rifle barrel.*

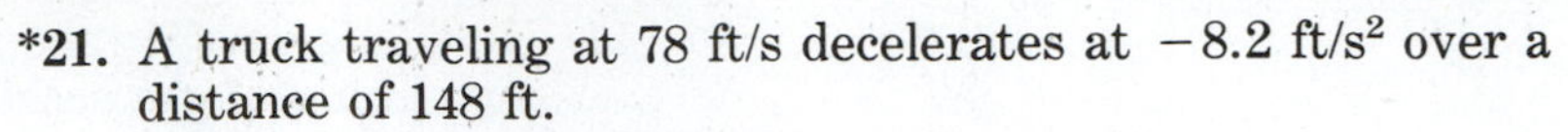

*21. A truck traveling at 78 ft/s decelerates at -8.2 ft/s^2 over a distance of 148 ft.

A. How much time elapses?

B. What is the final velocity?

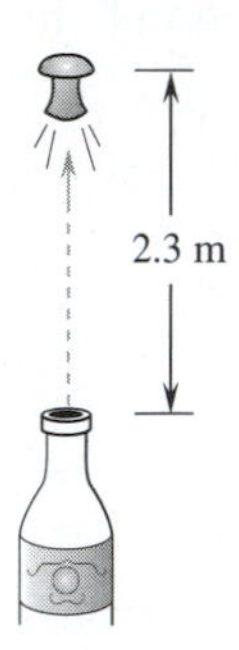

Figure 4–26: Diagram for Problem 25. A champagne cork pops upward.

Section 4.5:

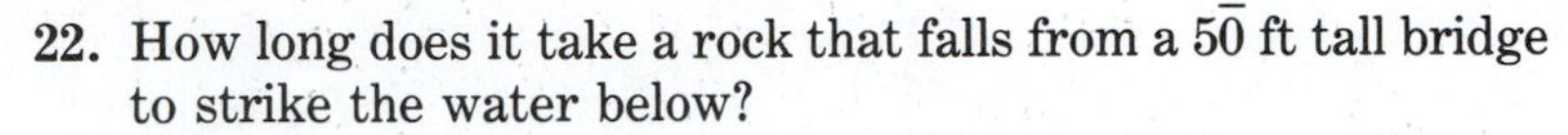

22. How long does it take a rock that falls from a $5\bar{0}$ ft tall bridge to strike the water below?

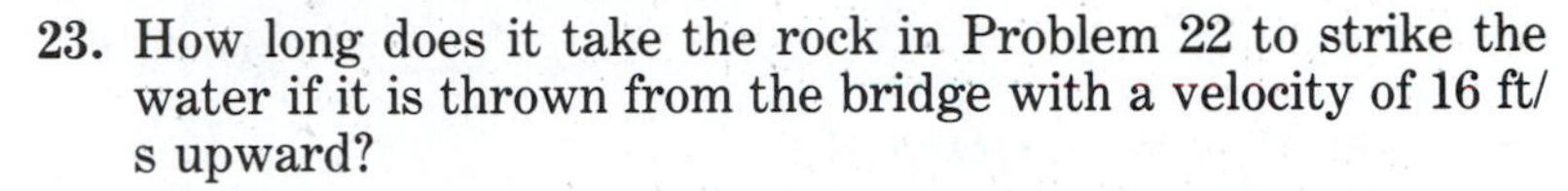

23. How long does it take the rock in Problem 22 to strike the water if it is thrown from the bridge with a velocity of 16 ft/s upward?

24. How long does it take the rock in Problem 22 to strike the water if it is thrown downward at a speed of 16 ft/s?

25. A champagne cork pops straight up, reaching a height of 2.3 m. What was its initial velocity?

26. An astronaut jumps up in the air with an initial speed of 4.6 ft/s.

A. How high does the astronaut jump if he is on the earth, where $g = -32$ ft/s^2?

B. How high does the astronaut jump if he is on the moon, where $g = -5.3$ ft/s^2?

Section 4.6:

27. Which of the objects described below will be in the air the longest? (Ignore friction. Assume level ground.)

A. Initial velocity of 45 m/s at 30°.

B. Initial velocity of $3\bar{0}$ m/s at 45°.

C. Initial velocity of 23 m/s at 75°.

28. A discus is thrown at an angle of 47° above the horizontal with an initial speed of 56 ft/s.

A. How long is it in the air?

B. What is its range?

29. In a baseball game, a first-base player leaps upward 2.5 ft and to his right to rob the batter of a single. How long is the first-base player in the air?

30. In a manufacturing process, metal parts roll off the end of a conveyor belt and fall into a bin. If the conveyor has a speed of 4.6 ft/s, how far horizontally will the parts travel as they fall 2.3 ft?

31. Provisions are dropped to a lighthouse keeper from a single-engine plane (see Figure 4–27). The plane is moving horizon-

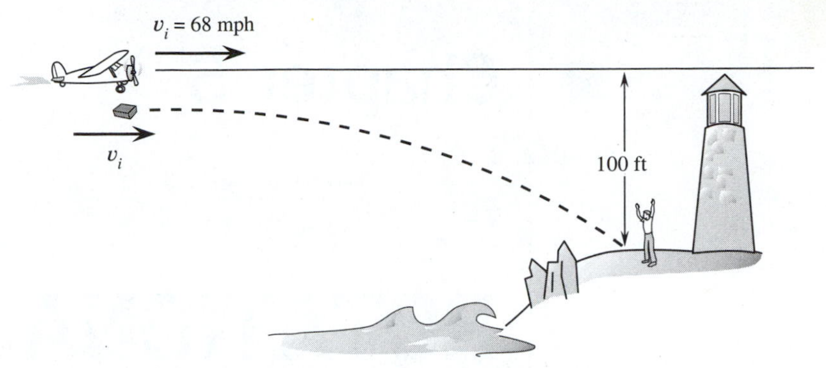

Figure 4–27: *Diagram for Problem 31. Supplies are dropped to a lighthouse keeper. The package has the same initial horizontal velocity as the airplane.*

tally with a speed of 68 mph at an elevation of $10\overline{0}$ ft when a bundle is dropped. What horizontal distance will the bundle travel as it falls? (Neglect friction.)

32. A bullet is fired horizontally with a muzzle velocity of 365 m/s. How far will the bullet drop as it travels $20\overline{0}$ yds?

***33.** A slingshot fires a pellet at an angle of 35° above the horizontal with an initial speed of 42.0 m/s. The pellet reaches its target 8.4 m above the starting point. How far did it travel horizontally?

Chapter 5

ROTATIONAL MOTION

How does the speedometer know how fast this car is traveling? As you put together the concepts you are learning about speed and motion, you will understand how speedometers work. (Photograph courtesy of General Motors.)

OBJECTIVES
In this chapter you will learn to:

- calculate angular displacement in radians
- find the angular speed of a spinning object
- find the angular velocity of a spinning object
- find the tangential velocity for some point on something that is rotating
- find the angular acceleration of an object that is increasing or decreasing its angular velocity
- calculate the tangential acceleration related to angular acceleration
- solve angular motion problems using equations similar to linear motion equations

Time Line for Chapter 5

180 BC	360-degree circle is introduced to Greek mathematicians.
1543	Copernicus advances the idea that planets orbit the sun.
1609	Johannes Kepler publishes rules governing planetary motion.
1765	Leonard Euler finds equations to describe complex rotational motion and the details of the earth's motion.
1783	Pierre-Simon Laplace develops a new method for describing the motion of celestial bodies.

If you work with machinery, you will encounter rotational motion. Drill presses, lathes, electric motors, drive shafts, pulleys, fans, and many other machines operate with rotational motion. Most diesel and gasoline engines convert the linear motion of a piston into rotational motion. Other machines and devices convert rotational motion into linear motion. Wheels or drums are often used to do this. Bicycles, trucks, winches, and cable elevators are other examples. A wheel with a hinged crank is another device used to convert motion. Augers are used to transport bulk feeds in mixing plants and to deliver a bag of potato chips from a vending machine. Rack and pinion gears are sometimes used to convert between rotational and linear motion.

If we use machines, we need to know two things: how to describe rotational motion, and how to convert between linear and rotational motion. In this chapter we will examine these two problems.

5.1 ANGULAR DISPLACEMENT

An angle measures the separation of two intersecting straight lines. Three units used to represent the divergence of the intersecting lines are degrees, grads, and mils. The degree is a very old unit that originated when calendars were not very accurate. The year was thought to be about 360 days long, and the sun was thought to orbit the earth. The degree was approximately the angle the sun travels relative to the stars in one day. There are 360 degrees in a circle. The grad represents an attempt to fit angles into a decimal system. A right angle (90°) is broken into 100 grads. This means that there are 400 grads in a circle. Some pocket calculators will calculate angles in grads. The *G* in the [DGR] button is the grad scale. Mils are used in two ways. One is to measure angles. Traditionally angles were expressed in degrees, minutes, and seconds. Conversion to a decimal fraction was cumbersome. For some uses,

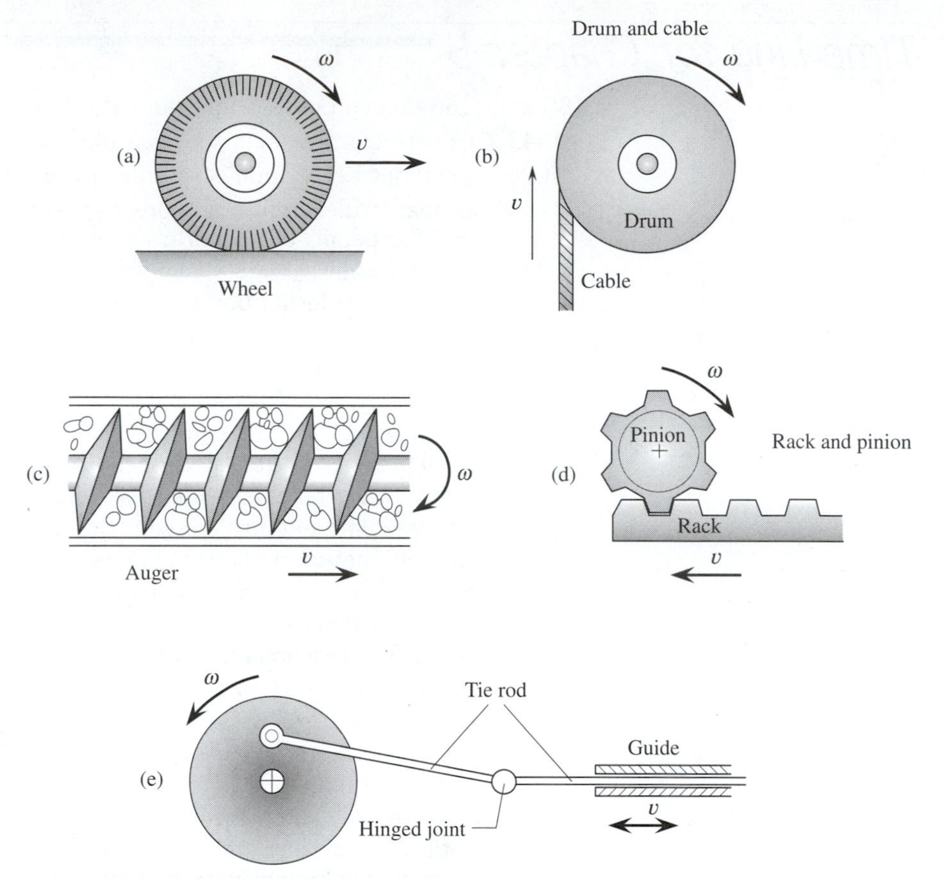

Figure 5–1: *Some different ways rotary motion can be converted to linear motion.* (a) *A rolling wheel.* (b) *Drum and cable system used on a winch.* (c) *Auger or helix. As the auger turns, material is carried along the threads.* (d) *A rotating pinion moves a rack linearly.* (e) *An eccentric rod on a wheel causes linear oscillatory motion.*

such as artillery, a small unit was devised for rapid adjustment of guns. There are 6400 mils in a circle (1 mil = 0.05625 degrees). Mils are also used to indicate wire size. A 1-mil wire has a diameter of 0.001 in.

All three of these units of angle are arbitrary. One of our objectives is to convert between rotational and linear motion. Let us define a new angular unit that is a simple unitless ratio.

If a wheel rotates through one revolution, or 360°, it will travel a length equal to its circumference (C) (see Figure 5–2). If R is the radius of the wheel, then:

$$s = C = 2\pi R$$

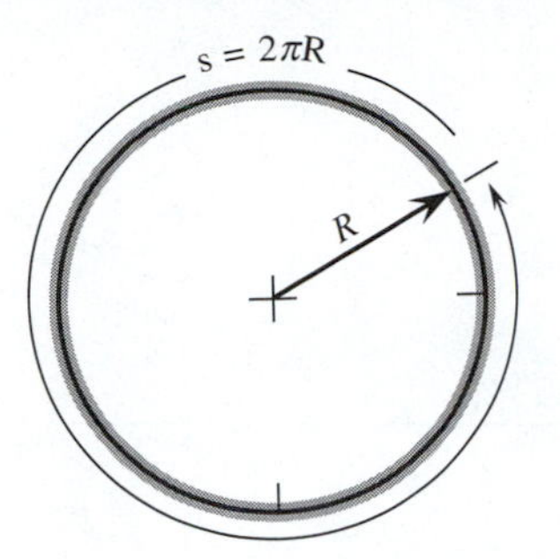

Figure 5–2: The distance around a circle is 2πR.

For one rotation we always get the same ratio of circumference to the radius of a circle. This is an angle measurement equal to 360°.

$$\theta = \frac{s}{R} = \frac{C}{R} = \frac{2\pi R}{R} = 2\pi = 360°$$

The angle unit based on the ratio of a circle's circumference to its radius is called a **radian,** abbreviated *rad*. There are 2π radians in a circle. We can find the size of one radian in degrees by dividing 360° by 2π.

$$1 \text{ radian} = \frac{360°}{2\pi} = 57.2958°$$

The radian is a rather large unit of angle, but it has a couple of convenient properties. (1) The radian is a unitless quantity. It is the ratio of the length of an arc of a circle to the radius—length divided by length. The units of length cancel. (2) The radian relates an angle with linear distance. To find the distance a wheel rolls, all we need to do is to multiply its angle of rotation, called the **angular displacement,** by its radius.

$$s = \theta R \qquad \textbf{(Eq. 5–1)}$$

EXAMPLE PROBLEM 5–1: A PULLEY

A V-belt pulley with a diameter of 14.0 cm rotates through an angle of 16.7 radians.

A. What is the angle in degrees?
B. How many rotations does the pulley make?
C. What length of belt passes over the pulley?

SOLUTION

A. Degrees:

$$\theta = 16.7 \text{ rad} \times 57.3°/\text{rad}$$

$$\theta = 957°$$

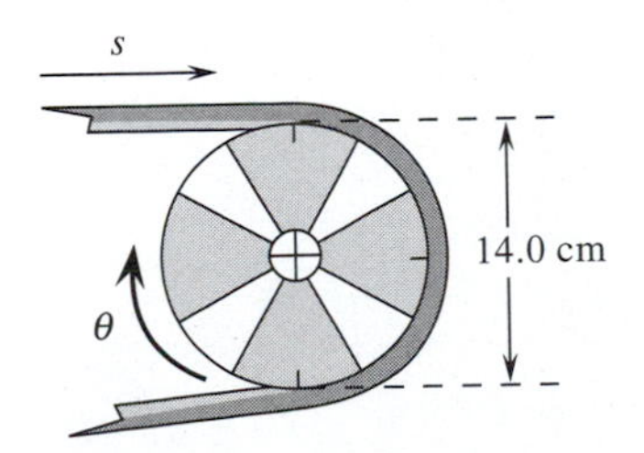

Figure 5–3: A spinning pulley pulls a length of V-belt along its rim.

B. Revolutions:

$$\theta = \left(\frac{16.7 \text{ rad}}{(16.7 \text{ rad})}\right)\left(\frac{1 \text{ rev}}{2\pi \text{ rad}}\right)$$

$$\theta = 2.66 \text{ rev}$$

C. Distance:

$$s = \theta R$$

$$s = 16.7 \text{ rad} \times \frac{0.140}{2} \text{ m}$$

$$s = 1.17 \text{ m}$$

5.2 ANGULAR VELOCITY

We need a quantity to show how fast something is spinning. Let us develop some relationships that are a lot like the equations for linear motion.

Average angular velocity is the change in angular position divided by the elapsed time. We will use the symbol ω to indicate angular velocity or angular speed. We will also use the symbol Δ to indicate the change of something.

$$\omega = \frac{\theta_f - \theta_i}{t} = \frac{\Delta\theta}{t} \qquad \textbf{(Eq. 5–2)}$$

When a pulley turns on a belt drive, the belt has a velocity related to the rotational speed of the pulley. This is the **tangential velocity.** A stone is whirled at the end of a string. When the string breaks, the stone no longer whirls in a circle. Instead, it travels in a straight line *tangent* to the radius connecting the stone to the spin axis. By tangent we mean "at right angles to the radius." The tangential velocity has a direction at right angles to the string as the stone whirls around in a circle. The tangent direction changes as the string changes position. The tangential velocity, v_t, can be found by substituting Equation 5–1 into Equation 5–2.

$$\omega = \frac{s_f/R - s_i/R}{t}$$

or

$$\omega = \frac{(s_f - s_i)/t}{R}$$

$$\omega = \frac{v_t}{R}$$

The relationship between tangential velocity and rotational velocity is:

$$v_t = \omega R \qquad \text{(Eq. 5–3)}$$

where v_t is perpendicular to R.

☐ **EXAMPLE PROBLEM 5–2: THE LATHE**

A lathe turns 96.6 revolutions in 1 min. What is its angular velocity in revolutions per minute? In revolutions per second? In radians per second?

■ ***SOLUTION***

A. Revolutions per minute:

$$\omega = \frac{\Delta\theta}{t}$$

$$\omega = \frac{96.6 \text{ rev}}{1.00 \text{ min}}$$

$$\omega = 96.6 \text{ rpm}$$

B. Revolutions per second:

$$\omega = \frac{\Delta\theta}{t}$$

$$\omega = \frac{96.6 \text{ rev}}{60.0 \text{ s}}$$

$$\omega = 1.61 \text{ rev/s}$$

C. Radians per second:

$$\omega = 1.61 \frac{\text{rev}}{\text{s}} \times \frac{2\pi \text{ rad}}{1 \text{ rev}}$$

$$\omega = 10.1 \text{ rad/s}$$

☐ **EXAMPLE PROBLEM 5–3: THE WIND TURBINE**

A wind turbine with 2.60 ft long vanes rotates with an angular velocity of 168 rpm. What is the tip speed (the tangential speed of the blade tips) of one of the vanes?

■ *SOLUTION*

First convert the angular velocity into radians per second.

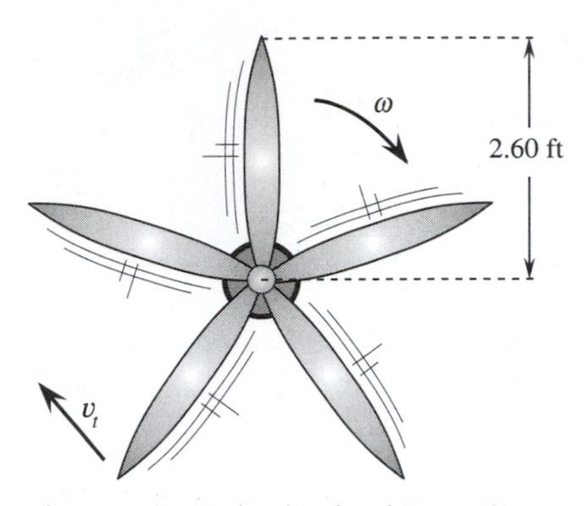

Figure 5–4: *As the wind turbine rotates, the tips of the vanes move in a circle with a tangential velocity* v_t*.*

$$\omega = 168\,\frac{\text{rev}}{\text{min}} \times \frac{1\text{ min}}{60\text{ s}} \times \frac{2\pi\text{ rad}}{1\text{ rev}}$$

$$\omega = 17.6\text{ rad/s}$$

$$v_t = \omega R$$

$$v_t = 17.6\text{ rad/s} \times 2.60\text{ ft}$$

$$v_t = 45.8\text{ ft/s}$$

We did not need to indicate units of radians in the answer. Remember the radian is a unitless ratio.

5.3 AVERAGE ANGULAR ACCELERATION

We can define the average angular acceleration in a way that is similar to average linear acceleration. Use the initial and final angular velocities to find the time rate of change. We will use the symbol α to indicate angular acceleration.

$$\alpha = \frac{\Delta\omega}{t} = \frac{\omega_f - \omega_i}{t} \qquad \textbf{(Eq. 5–4)}$$

Here we have the same restrictions that we had on average linear acceleration. The acceleration must be constant. A plot of angular velocity versus time would be a straight line.

When a disk is given an angular acceleration, a point on the rim will undergo a corresponding change of tangential velocity. This is called the **tangential acceleration.** It exists only if the disk's spin is increasing or decreasing.

The relationship between tangential acceleration (a_t) and angular acceleration (α) can be found by substituting Equation 5–3 into Equation 5–4.

$$\alpha = \frac{\omega_f - \omega_i}{t}$$

$$\alpha = \frac{v_f/R - v_i/R}{t}$$

$$\alpha = \frac{(v_f - v_i)/t}{R}$$

$$\alpha = \frac{a_t}{R}$$

or

$$a_t = \alpha R \qquad \textbf{(Eq. 5–5)}$$

□ **EXAMPLE PROBLEM 5–4: THE TIRE**

A truck accelerates from 18.5 m/s to 37.9 m/s in 4.70 s. If its tires have a diameter of 73.2 cm, find:

A. The tangential acceleration of the tires.

B. The angular acceleration of the tires.

■ ***SOLUTION***

Use the equations for linear motion.

A. Tangential acceleration:

$$a_t = \frac{v_f - v_i}{t}$$

$$a_t = \frac{37.9 \text{ m/s} - 18.5 \text{ m/s}}{4.70 \text{ s}}$$

$$a_t = 4.13 \text{ m/s}^2$$

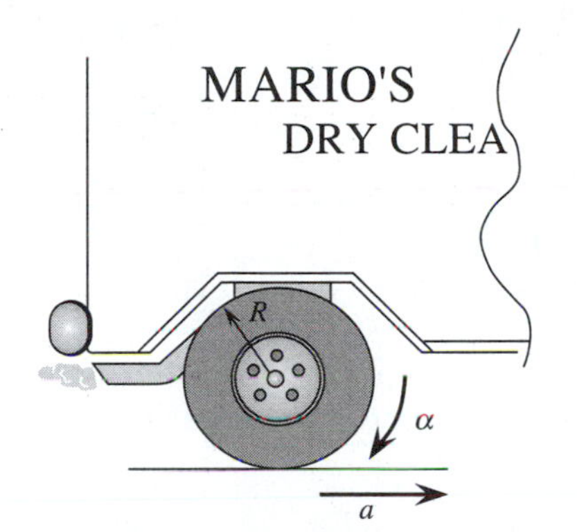

Figure 5–5: *As the truck accelerates forward, the tires undergo an angular acceleration.*

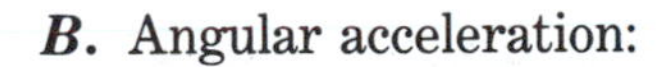

B. Angular acceleration:

$$\alpha = \frac{a_t}{R}$$

$$\alpha = \frac{4.13 \text{ m/s}^2}{0.366 \text{ m}}$$

$$\alpha = 11.3 \text{ rad/s}^2$$

5.4 MORE EQUATIONS OF ANGULAR MOTION

We have established a pattern between linear and rotational motion. We have three equations that convert between rotational motion and tangential (linear) motion. They are: $a_t = \alpha$ R; $v_t = \omega R$; and $s = \theta R$.

Let us find the analogous equation to average velocity given the initial and final velocities. The linear equation is: $v_{avg} = \frac{v_f + v_i}{2}$.

Use the conversion between angular velocity and linear velocity.

$$\omega_{avg} R = \frac{(\omega_f R) + (\omega_i R)}{2}$$

Dividing the equation by R, we get:

$$\omega_{avg} = \frac{\omega_f + \omega_i}{2} \quad \textbf{(Eq. 5–6)}$$

We convert the derived equations for linear motion into rotational motion in a similar fashion using the conversion equations. This is left as an exercise for you to do. The derived equations are:

$$\theta = (\omega_i t) + \left(\frac{\alpha t^2}{2}\right) \quad \textbf{(Eq. 5–7)}$$

$$2\theta\alpha = \omega_f^2 - \omega_i^2 \quad \textbf{(Eq. 5–8)}$$

☐ **EXAMPLE PROBLEM 5–5: ELECTRIC MOTOR**

An electric motor initially at rest accelerates to $23\bar{0}0$ rpm in 2.75 s.

A. What is its average angular velocity for the 2.75 s?

B. Through how many revolutions does the drive shaft turn?

■ ***SOLUTION***

A. Average angular velocity:

$$\omega_{avg} = \frac{\omega_f + \omega_i}{2}$$

$$\omega_{avg} = \frac{2300 \text{ rpm} + 0.00 \text{ rpm}}{2}$$

$$\omega_{avg} = 1150 \text{ rpm}$$

B. Angle:

$$\theta = \omega_{avg}\, t$$

$$\theta = 1150 \text{ rpm} \times 2.75 \text{ s} \times \frac{1 \text{ min}}{60 \text{ s}}$$

$$\theta = 52.7 \text{ rev}$$

How Speedometers Measure Speed

Did you ever wonder how your car knows how fast it's going?

The angular speed of an axle is monitored. The rotational motion is transmitted by a speedometer cable and gearing system to the head of an analog speedometer, or the information is fed to a computer. In either case, the car infers its linear speed from the angular speed by assuming it is equipped with standard-size tires. The linear speed is the angular speed multiplied by the tire radius.

If the car has worn, underinflated tires, the speedometer will overestimate the car's speed. If the car is equipped with oversized tires, the speedometer reading will be too low.

Using the wrong size tire can cause more problems than a poor speedometer reading. Many recent models have an on-board computer. The computer makes continual adjustments to the motor, based on the car's speed, to improve the car's efficiency. Wrong-sized tires can completely mess up the timing.

(Photograph courtesy of General Motors.)

☐ **EXAMPLE PROBLEM 5–6: THE TURNTABLE**

A 33⅓-rpm turntable comes to rest in 10.3 s.

A. What is its angular acceleration?

B. Through how many radians does the turntable rotate while stopping?

C. What tangential acceleration is experienced by a cockroach sitting on the rim of a 12-in record?

D. Through what distance does the bug travel?

■ ***SOLUTION***

First convert the rotational speed into radians per second.

$$\omega = 33.33\,\frac{\text{rev}}{\text{min}} \times 2\,\pi\,\frac{\text{rad}}{\text{rev}} \times \frac{1.00\ \text{min}}{60.0\ \text{s}}$$

$$\omega = 3.49\ \text{rad/s}$$

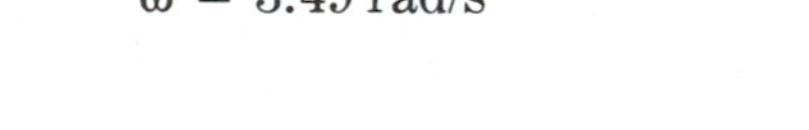

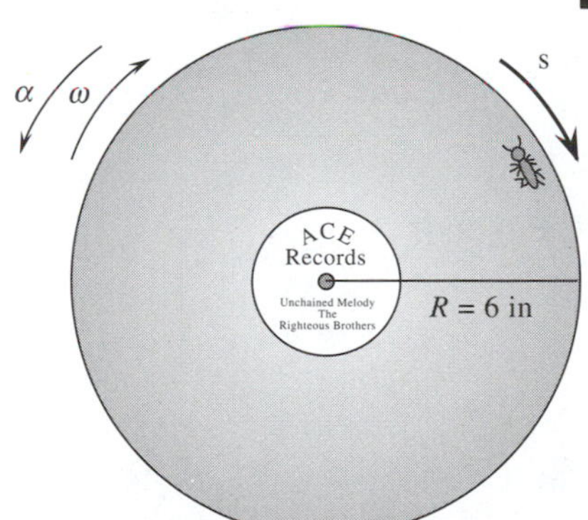

Figure 5–6: *The bug on the rim of a 12-in record rides along the circumference of a circle. It travels a distance s as the record comes to a stop.*

The information we have is:

$$\theta = ?\ \text{radians}$$

$$\omega_i = 3.49\ \text{rad/s}$$

$$\omega_f = 0.00$$

$$\alpha = ?\ \text{rad/s}^2$$
$$t = 10.3\ \text{s}$$
$$a_t = ?\ \text{ft/s}^2$$
$$R = 0.50\ \text{ft}$$

A. Angular acceleration:

$$\alpha = \frac{\omega_f - \omega_i}{t}$$

$$\alpha = \frac{(0.00 - 3.49)\ \text{rad/s}}{10.3\ \text{s}}$$

$$\alpha = -0.339\ \text{rad/s}^2$$

B. Angular displacement:

$$2\alpha\theta = \omega_f^2 - \omega_i^2$$

$$\theta = \frac{\omega_f^2 - \omega_i^2}{2\alpha}$$

or

$$\theta = \frac{(0.00\ \text{rad/s})^2 - (3.49\ \text{rad/s})^2}{2 \times (-0.339\ \text{rad/s}^2)}$$

$$\theta = 18.0\ \text{radians}$$

C. Tangential acceleration:

$$a_t = \alpha R$$

$$a_t = -0.339\ \text{rad/s}^2 \times 0.50\ \text{ft}$$

$$a_t = -0.17\ \text{ft/s}^2$$

D. Linear distance:

$$s = \theta R$$

$$s = 18.0\ \text{rad} \times 0.50\ \text{ft}$$

$$s = 9.0\ \text{ft}$$

SUMMARY

There are several ways to describe rotational motion and to convert between rotational and linear motion.

Angular displacement is the angle through which something turns. The unit of angle, called the radian, is the ratio of arc length (s) to radius (R) of the circle.

$$\theta = \frac{s}{R}$$

The angular velocity is the time rate of change of angular displacement.

$$\omega = \frac{\theta_f - \theta_i}{t}$$

The angular acceleration is the time rate of change of the angular velocity.

$$\alpha = \frac{\omega_f - \omega_i}{t}$$

The tangential velocity of some point on a rotating body is the linear speed the point has a distance R from the spin axis. The direction is at right angles to the radius connecting the point to the spin axis.

$$v_t = \omega R$$

The tangential acceleration is the linear acceleration a point has at a distance R from the center of rotation.

$$a_t = \alpha R$$

The equations of rotational motion are analogous to the equations for linear motion. They have the same form. Linear equations may be transformed into rotational equations by replacing the linear symbols by the appropriate rotational symbols as shown in Table 5–1.

Table 5–1: Transformations between linear equations of motion and rotational equations of motion.

TRANSFORMATION	TRANSFORMATION EQUATION	LINEAR EQUATION	ANGULAR EQUATION
acceleration: $a \rightarrow \alpha$	$a = \alpha R$	$s = v_{avg}t$	$\theta = \omega_{avg}t$
velocity: $v \rightarrow \omega$	$v = \omega R$		
distance: $s \rightarrow \theta$	$s = \theta R$	$v_f = v_i + (a\ t)$	$\omega_f = \omega_i + (\alpha\ t)$
		$v_{avg} = \frac{v_f + v_i}{2}$	$\omega_{avg} = \frac{\omega_f + \omega_i}{2}$
		$s = v_i t + \left(\frac{a\ t^2}{2}\right)$	$\theta = \omega_i t + \left(\frac{\alpha\ t^2}{2}\right)$
		$2(a\ s) = v_f^2 - v_i^2$	$2(\alpha\ \theta) = \omega_f^2 - \omega_i^2$

KEY TERMS

If you can explain the following terms to a friend or classmate, you understand their meaning. If you cannot explain the terms, you should reread the sections in which they are discussed.

angular displacement
average angular velocity
radian
tangential acceleration
tangential velocity

EXERCISES

Section 5.1:

1. Equation 5–1 can be used to convert between the linear distance traveled by an object at a distance R from the center of rotation and the angle through which it rotates. Will this work if we use angle units of degrees rather than radians? Explain why or why not.

2. Which of the following angles is the largest? Which is the smallest?
   A. $11\bar{0}$°
   B. ⅓ rev
   C. 2.0 rad
   D. $20\bar{0}0$ mils
   E. $10\bar{0}$ grads

Section 5.2:

3. Toss a hammer into the air giving it a spin. Does the head of the hammer have the same angular velocity as the end of the handle? Does the head of the hammer have the same tangential velocity as the end of the handle?

4. A bug crawls outward on a rotating 45-rpm phonograph record. As it moves what can you say about its angular velocity? About its tangential velocity?

5. Is a revolution per minute larger or smaller than a radian per second?

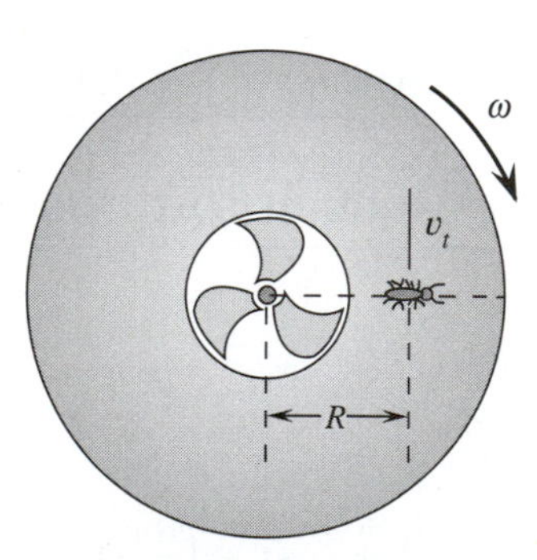

Figure 5–7: Diagram for Exercise 4. A bug crawls radially outward on a rotating 45-rpm record.

Section 5.3:

6. The shaft of an electric motor with a changing spin rate has positive values for both the initial and final angular velocities. Does this mean the angular acceleration must also be positive? Explain.

7. A wind turbine has a negative angular acceleration and a positive velocity. Is the turbine speeding up or slowing down?

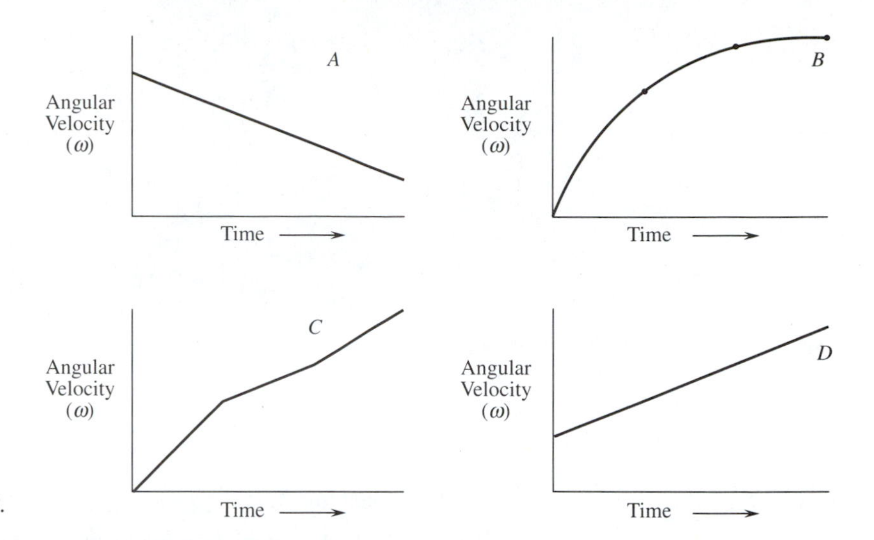

Figure 5–8: Diagram for Exercise 8.

8. Which of the objects shown in Figure 5–8 has (have) uniform angular acceleration? Decreasing angular acceleration? Negative angular acceleration?

9. Is it possible for something to have a negative angular acceleration and yet have an increasing spin rate?

Section 5.4:

10. Is it possible for something to have an average angular velocity of zero when the initial and final angular velocities are *not* zero?

11. Three of the equations in Table 5–1 are true only when the angular displacement is expressed in radians. Which three equations are they?

12. A weight is hung from a cord wrapped around a pulley (see Figure 5–9). When the weight is released from rest it has a uniform acceleration downward, causing the pulley to spin. Which of the following statements is (are) *not* correct? (Ignore friction.)
    - ***A.*** The angular displacement of the pulley is proportional to the time squared.
    - ***B.*** The angular velocity of the pulley is proportional to the time squared.
    - ***C.*** The angular velocity of the pulley is proportional to the time.
    - ***D.*** The speed of the weight increases proportionally to the angular velocity of the pulley.
    - ***E.*** The linear distance the weight travels is proportional to the angular displacement of the pulley.
    - ***F.*** As the weight falls the angular acceleration of the pulley increases.

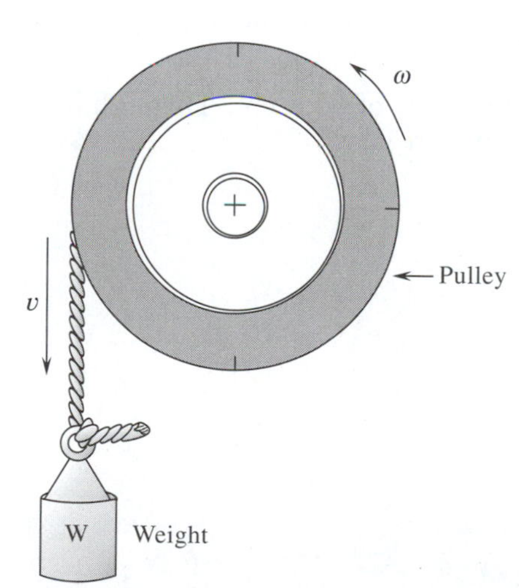

Figure 5–9: Diagram for Exercise 12. A cord attached to a weight is wrapped around a pulley. As the weight falls, the pulley rotates.

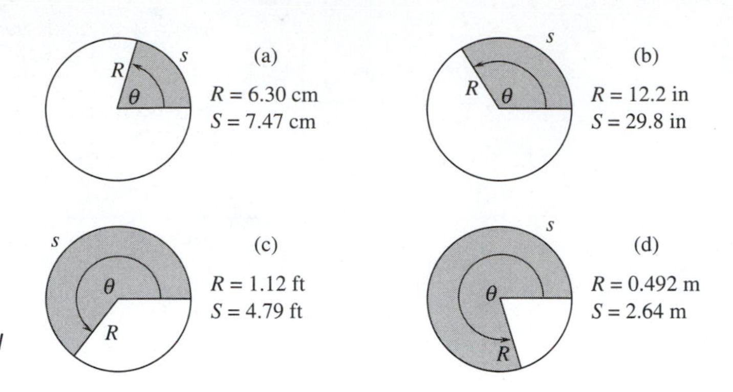

Figure 5–10: Diagram for Problem 2. Find the angles.

PROBLEMS

Section 5.1:

1. Convert the following quantities to the indicated units.

   a. 2.73 rad = ? degrees **b.** 3.75 rev = ? rad
   c. 1230° = ? rev **d.** 37.2 rev = ? rad

2. Find the angles for the circles in Figure 5–10.
3. From the information given in Figure 5–11 find the radii of the circles.
4. Which of the following angles is the smallest?

   a. 39° **b.** 0.70 rad **c.** 0.12 rev **d.** 43 grad

5. The tire on an automobile usually has a radius of 12.5 in. Through how many revolutions does an automobile tire turn in 1 mi?
6. A bicycle has tires with a diameter of 66.0 cm. Through how many radians does the tire turn in 1 km?

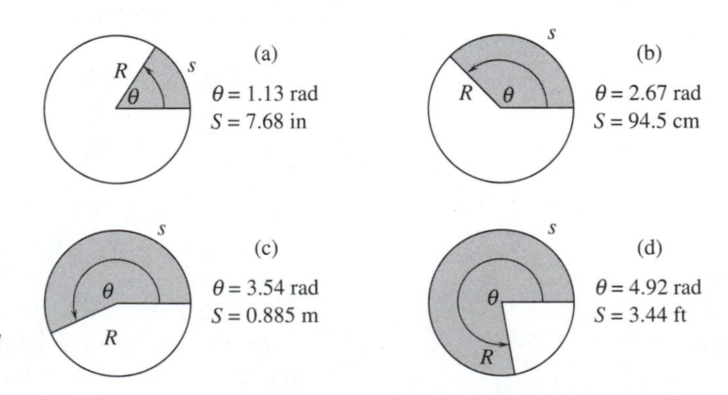

Figure 5–11: Diagram for Problem 3. Find the radii of the circles.

Table 5–2: Angle measurements and time for Problem 7.

ANGLE (rad)	TIME (s)
0.0	0.0
6.1	1.0
11.9	2.0
36.0	3.0
39.8	4.0
36.1	5.0
42.2	6.0
48.1	7.0

Section 5.2:

7. Plot the information in Table 5–2. Using slopes from your graph find:
   A. the initial angular velocity.
   B. the largest angular velocity.
   C. the average angular velocity for the entire graph.

8. Find the average angular speed for the graph shown in Figure 5–12.

9. Calculate the conversion factor between revolutions per minute and radians per second.

10. Perform the indicated units conversions.

    a. 3.42 rad/s = ? rpm **b.** 3.68 rpm = ? rad/s
    c. 6.47 rev/s = ? rad/s **d.** 392°/s = rad/s

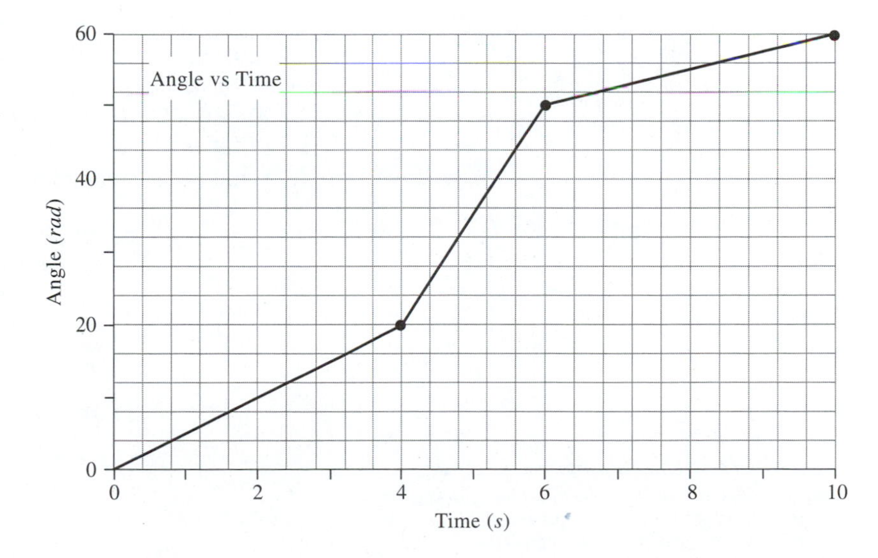

Figure 5–12: Diagram for Problem 8. Angle versus time. Find the average angular velocity.

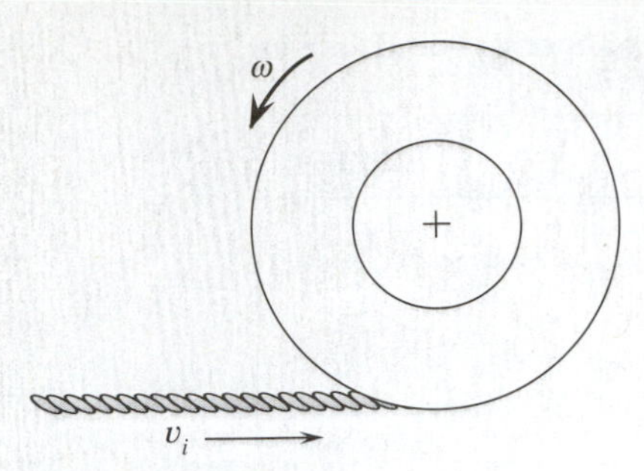

Figure 5–13: Diagram for Problem 12. As the barrel rotates, the cable is wrapped around it.

11. One metal casting process is called centrifugal casting. A mold is spun at a very high speed as molten metal is poured into it. As the metal cools it is pressed tightly against the face of the mold, giving a high-quality imprint. If a mold is 76.4 cm in radius and has a spin rate of 18,000 rpm, what is:

A. the angular speed in radians/s?

B. the tangential velocity of the outer edge of the mold?

12. A 27.2 in diameter barrel on a winch rotates with an angular speed of 14.5 rpm (see Figure 5–13). What is the linear speed of the cable wrapped around the barrel?

***13.** One type of variable clutch consists of a rotating large rubber disk connected to the input power. A smaller disk rides on the face of the driver disk (see Figure 5–14). If R = 5.2 cm, r = 0.88 cm, and the large disk rotates at 1450 rpm, what is the angular velocity of the smaller disk?

Section 5.3:

14. Perform the indicated units conversions.

a. 4.72 rev/s^2 = ? rad/s^2 **b.** 8.62 rad/s^2 = ? rev/s^2

c. 0.453 rpm/s = ? rad/s^2 **d.** 93.2 rad/s^2 = ? rpm/s

15. Find the angular acceleration for the graph in Figure 5–15.

16. A truck with 16 in radius tires accelerates from 34.7 ft/s to 89.2 ft/s in 12.6 s.

A. What is the linear acceleration of the truck?

B. What is the angular acceleration of the tires?

17. A drive shaft reduces its angular speed from 93.4 rad/s to 46.3 rad/s in 7.1 s. Find the angular acceleration.

18. A laboratory centrifuge initially at rest reaches an angular speed of 9220 rpm in 4.67 s. The bottom of the test tubes in the apparatus are 27.0 cm from the center of rotation.

A. What is the angular acceleration?

B. What is the tangential acceleration of the test tube bottoms?

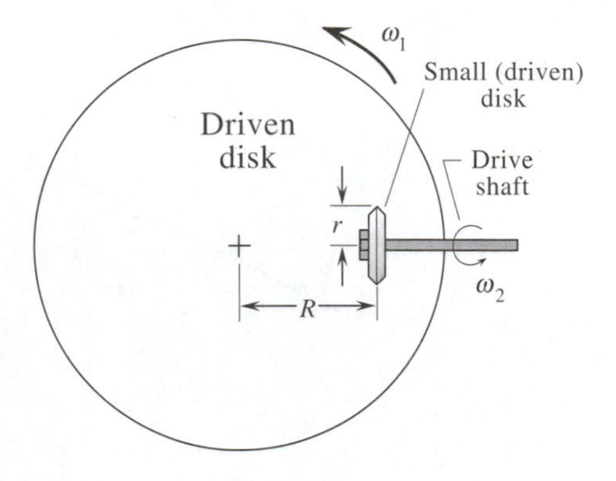

Figure 5–14: Diagram for Problem 13. A variable-speed friction clutch. The speed of the small disk depends on its distance from the center of the larger disk (R). r, radius of the small disk.

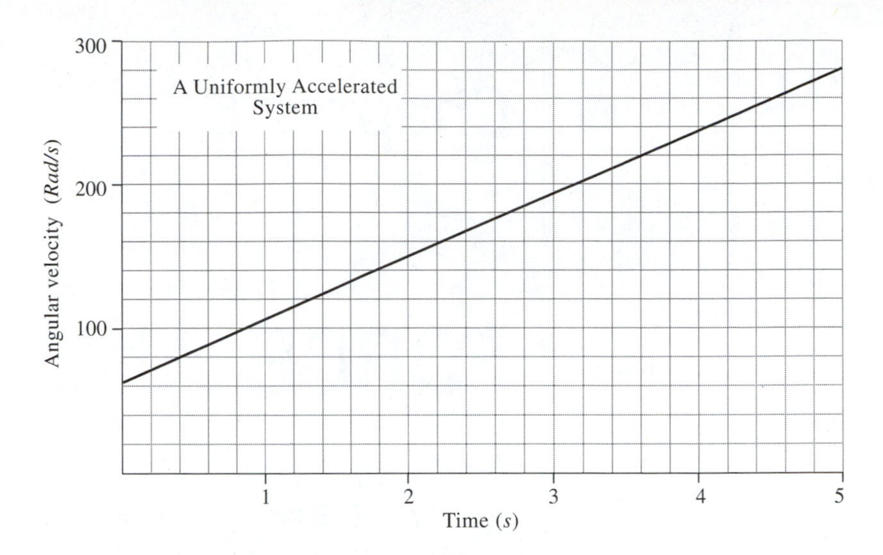

Figure 5–15: Diagram for Problem 15. A uniformly accelerated system.

Section 5.4:

19. A 25.0 cm diameter ball rolls up an incline with an initial velocity of 12.0 m/s. The ball reaches its maximum height and rolls back past its starting position attaining a final velocity of −17.3 m/s. What is the ball's average angular velocity?

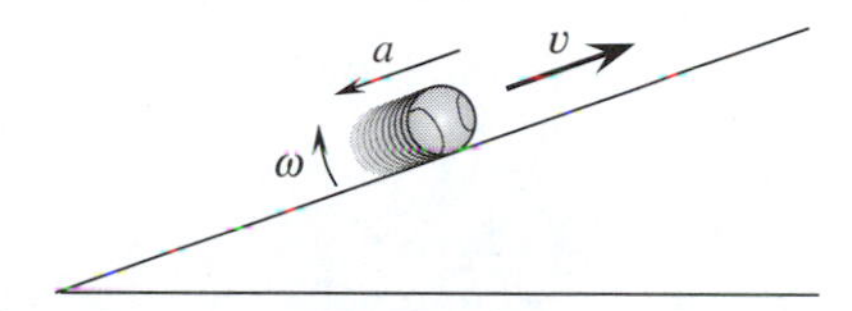

Figure 5–16: Diagram for Problem 19. A ball rolls up an incline. The acceleration (a) is down the incline.

20. A flywheel initially rotating at 23.4 rad/s undergoes 6.7 revolutions as it accelerates to 46.6 rad/s. What is the angular acceleration?

21. A turntable initially spinning at 78.0 rpm slows to a stop in 21.2 s.
    A. What is the angular acceleration?
    B. Through how many revolutions does the turntable rotate as it stops?

22. When the power is turned off, a metal lathe initially turning at 12.9 rad/s makes 0.78 rev as it comes to rest.
    A. How long does it take the lathe to stop turning?
    B. What is the acceleration?

23. The rollers in a rotary printing press slow from 9.2 rad/s at a rate of 0.23 rad/s^2.
    A. Through what angle do they turn in 2.4 s?
    B. What is their velocity at the end of 2.4 s?

Chapter 6

FORCE AND MOTION

This painting of the USS Constitution *and the* Guerriere *shows the gunports and cannons used in battle. As you read about force and motion in this chapter, you will see how recoil affected the use of these cannons. (Photograph courtesy of the Smithsonian Institution, photo 84–1511.)*

OBJECTIVES

In this chapter you will learn:

- to identify whether or not an unbalanced force is acting on something
- to distinguish between internal forces and external forces
- to distinguish between weight and mass
- to calculate the motion for a constant mass when an unbalanced force acts on it
- that forces occur in acting and reacting pairs
- to calculate the center of mass of a system
- to calculate the thrust created by the ejection of a mass from a system at constant velocity
- to calculate the momentum of a moving mass

Time Line for Chapter 6

370 BC	Aristotle states free fall is accelerated motion.
1586	Simon Stevinus drops two objects of different weights and notes that they hit the ground at the same time.
1590	Galileo refutes Aristotelean physics.
1632	Galileo publishes *Dialogue Concerning Two World Systems* and describes relative motion.
1686	Isaac Newton publishes the Principia, De Motu, the motion of bodies.

We have looked at ways to describe motion, but we have not answered the question, "What causes motion?" The fuel-air mixture in a cylinder ignites, driving the piston of a gasoline engine downward. The thrust of a rocket engine pushes a space shuttle up and away from the earth. A rock dropped from a bridge is pulled downward at an increasing speed by the earth's gravity. A gust of wind presses against a wind turbine, causing it to spin.

In each of these examples there is a push or a pull in a particular direction. We call this vector quantity **force.** In this chapter we will explore how forces affect motion. When do we know that a force is acting on an object? How does force alter an object's motion? How does the object interact with the external agent applying the force?

6.1 DETECTING FORCES

Let us look at some systems of forces. You are probably sitting in a chair as you read this. The gravitational pull of the earth is exerting a force downward, and yet you do not fall. The chair is pushing up on your body to balance your weight. You probably did not notice this force until you were reminded of it. In a tug of war, the flag at the center of the rope does not move if both teams are pulling with the same amount of force. When one team pulls harder, the flag moves in the direction of the larger force. An object at rest will start to move only if the **vector sum** of the forces acting on it is different from zero ($\mathbf{F} \neq 0$). The sum of forces acting on an object is called the **net force.** If the net force is not zero, an **unbalanced force** exists. Let us look at an example.

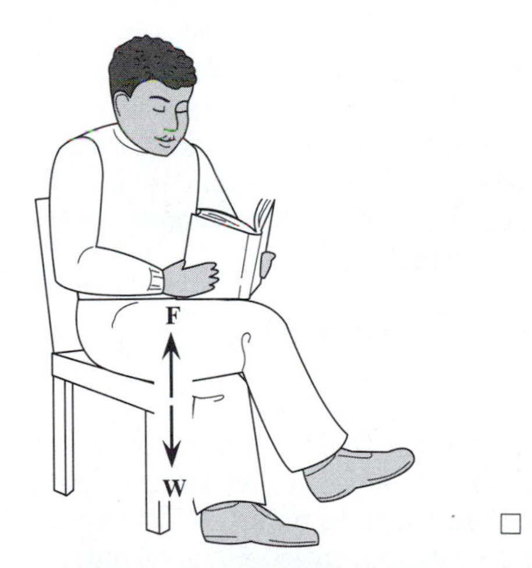

Figure 6–1: *The forces on a reader.* **F,** *the upward force of the chair;* **W,** *the downward force of gravity (weight).*

EXAMPLE PROBLEM 6–1: A TUG OF WAR

Figure 6–2 depicts a tug of war. The three team members on the left pull with forces of 152 lb, 198 lb, and 195 lb. The team on the right pulls with forces of 178 lb, 155 lb, and 192 lb. What is the unbalanced force on the rope?

Figure 6–2: *Two teams in a tug of war.*

■ *SOLUTION*

Define the forces acting on the right as positive, those on the left as negative.

$$\mathbf{F} = (-152\ \text{lb}) + (-198\ \text{lb}) + (-195\ \text{lb}) + 178\ \text{lb} + 155\ \text{lb} + 192\ \text{lb}$$

$$\mathbf{F} = -2\bar{0}\ \text{lb}$$

The unbalanced force is $2\bar{0}$ lb to the left.

When a balloon is gently heated it will expand. If the balloon is long and skinny, its length will increase. The forces inside the balloon caused by the heated air can change the size and shape of the balloon, but they cannot move the balloon to a new location. They do not change the balloon's motion.

Forces that act between different parts of an object are called internal forces.

The pressure of the gas inside the balloon is called an **internal force.** It pushes against the rubber membrane. Thermal forces involved with expansion, molecular binding forces, and any other forces originating from inside an object and acting on the object itself (internal forces), cannot change the object's position. Only forces from outside the object (**external forces**) can cause it to move from one place to another. Experience tells us that this book will not move unless pushed, pulled, lifted, carried, or dropped.

An object at rest will remain at rest unless acted upon by an unbalanced external force.

The rule above can be applied as a test to determine whether or not a net external force other than zero is acting on something. If the object is initially at rest and then starts to move, we know that there is an unbalanced force acting on it from outside. Not all moving things have an unbalanced external force acting on them. For instance, early in 1972 an Atlas-Centaur rocket lifted Pioneer 10 off the face of the earth and hurled it toward Jupiter. As the unmanned craft skimmed past the planet, Jupiter's gravitational field pulled the probe into a curved path. After photographing the planet, Pioneer 10 moved out of the solar system. All these activities involved unbalanced forces, but now Pioneer 10 drifts through space, moving at a constant speed in a straight line toward the star Aldebaran. It is far away from the gravitational forces of the sun and other stars. The frictional drag of the nearly perfect vacuum of space is negligible. Yet, Pioneer 10 moves steadily on its voyage through space.

An object moving at a constant speed in a straight line will continue to do so unless acted upon by an unbalanced external force.

This rule can also be used as a test to determine if an unbalanced force is acting on an object. If something changes speed or direction (i.e., if its velocity changes) we know that an external force is acting on it. Because it is a vector quantity, for velocity to be constant, both its magnitude and direction must remain unchanged. A velocity of zero is a perfectly acceptable velocity. We can summarize these two tests for detecting an unbalanced force in one simple statement, known as **Newton's first law of dynamics.**

An object's velocity will be constant unless acted upon by an unbalanced external force.

□ **EXAMPLE PROBLEM 6–2: UNBALANCED FORCES**

For each of the following situations determine whether or not an unbalanced external force is acting on the object.

A. The piston in a compressor comes to a stop as it completes a stroke.

B. A car accelerates from rest to 35 mph.

C. In an oil-drilling rig, a winch lifts the drilling bit at a constant speed of 8.9 ft/s.

D. A crane moves a heavy load horizontally through an arc of a circle at a constant speed.

■ ***SOLUTION***

A. An unbalanced external force acts on the piston. Its speed is decreasing.

B. An unbalanced external force acts on the car. Its speed is increasing.

C. There is no unbalanced external force. Velocity is constant.

D. There is an unbalanced external force on the load. Although the speed is constant, the load is moving through a curved path. The direction is changing.

6.2 FORCES ON A CONSTANT MASS

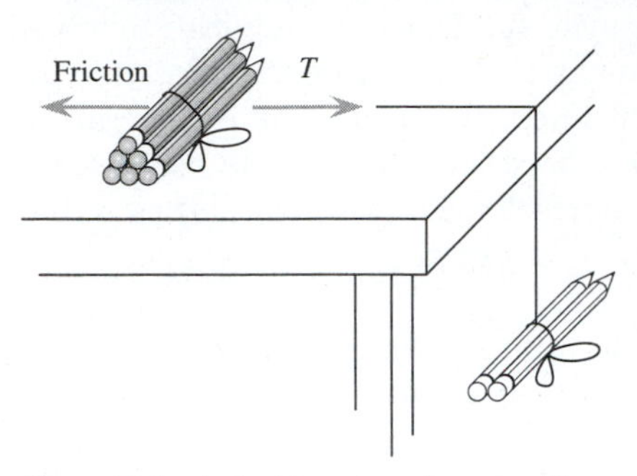

Figure 6–3: A desktop experiment with pencils. T is the tension on the string.

From Newton's first law of dynamics we know when an unbalanced external force is acting on an object. Here is a desktop experiment we can perform with some pencils and a piece of string to illustrate some of the forces that affect motion. Tie a bundle of six pencils at one end of the string and two pencils at the other end. Hold the bundle of six pencils on the desktop and hang the end with two pencils over the side (see Figure 6-3). Release the bundle of pencils. If there are enough pencils hung over the side of the desk, all the pencils will move. The net external force is the difference between the weight of the hung pencils and the frictional drag of the bundle on the desktop. The **system** is all eight pencils on the string. The pencils on the desktop move faster as they approach the edge of the desk. Apparently forces cause acceleration.

Now take one pencil from the bundle of six and add it to the two hung over the edge. We still have a system of eight pencils, but now the net gravitational force on it is three pencil weights instead of two and the frictional drag is on five pencils instead of six. We find the acceleration is larger than before. Larger accelerations can be obtained by moving more pencils. This leads us to suspect that acceleration is directly proportional to the net force acting on an object. More sophisticated laboratory experiments support this idea. The direct proportion may be written as an equation. This is **Newton's second law** for constant mass.

$$\mathbf{F} = m\,a \qquad \textbf{(Eq. 6–1)}$$

where m is mass, the proportionality constant.

Experience tells us that mass is a property of the object. Only a little force is required to accelerate a pencil. To give a textbook the same acceleration requires a larger force; a much larger force is needed to accelerate a stalled automobile. The resistance to change of motion of an object is called **inertia. Mass is a measure of an object's inertia.** The greater the mass of an object, the more resistance there will be to a change of velocity. Remember when using Equation 6–1 that it applies only to situations in which the mass of the system is not changing. It is a special case of Newton's second law of dynamics.

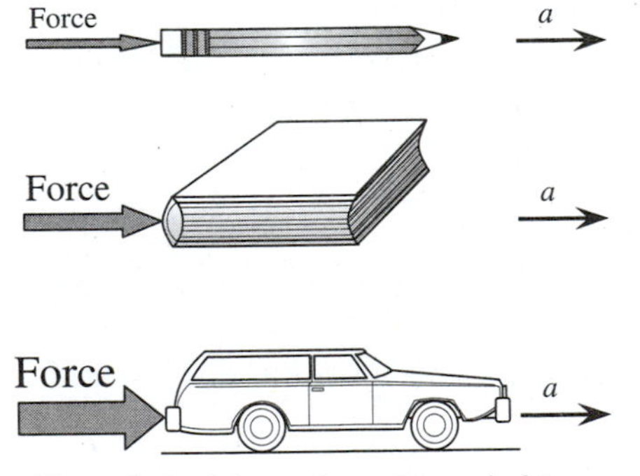

Figure 6–4: A larger force is needed to give the same acceleration to an object with a larger inertia. a, acceleration.

6.3 WEIGHT AND MASS

The British system of measurement is commonly used in the United States. Food, freight, and other items are usually expressed in pounds, a unit of weight. Most other countries use metric (SI) units,

Motion and Impetus

As a general rule, great scientists invent new ideas from the thoughts of those who have gone before them. There's a string of ideas and arguments connecting the past with the present.

Take the example of motion.

Around 370 BC, Aristotle promoted the idea that freely falling bodies are accelerated and heavier bodies fall faster than lighter bodies. He further asserted that a force must be exerted upon a body to keep it in motion in a horizontal direction.

A few years later another Greek scientist named Strato performed some experiments that seemed to support Aristotle's ideas. (Galileo was not the first experimenter, but he did popularize the idea that any theory *must* fit real-world observations.)

For quite a long time, Aristotle's views on motion were accepted as truth. There were some detractors. In the sixth century, Johannas Philoponus advanced the idea of *impetus*. An object set in motion had an impetus and did not need to be continually pushed to continue its motion. The sphere of stars overhead could continue to rotate on its own accord because nothing resisted its motion.

In the thirteenth century, Thomas Aquinas believed in Aristotle's rule of motion. He thought God supplied the force to rotate the spheres of stars and planets that wheel overhead in the night sky. This was a proof of the existence of God. This view was made an official position of the Christian church in 1277.

In the mid-fourteenth century, Jean Buridan, a French scientist, further developed the impetus theory. He rejected the necessity of God as a prime mover or the efforts of angels to keep the starry sky rotating. Initial impetus could explain the motion.

In 1586, Simon Stevinus dropped two objects of different weights. He noted they struck the ground at the same time. This refuted Aristotle's idea that bodies fell with speeds proportional to their weight.

This set the stage for Galileo. He performed more experiments on motion and published his first paper on motion in 1590, debunking Aristotelean physics. By 1604, Galileo had deduced the distance a body falls was proportional to time squared. In 1632, he published *Dialogue Concerning Two World Systems*. The book supported the Copernican idea that the sun was at the center of the solar system. He also described relative motion. This view of relativity was finally modified by Albert Einstein centuries later.

---

and measure food and freight in kilograms, which are units of mass. This leads to confusion between the different kinds of quantities.

Placing the bathroom scales on your head is not a good way to find your weight. Weight has a direction—down toward the center of the earth; it is a vector quantity. The scale needs to be placed between you and the earth to get a reading. **Weight** is the gravitational force acting on you or any other object. On the surface of the earth, weight is fairly constant. If we move up away from the earth's surface, the gravitational pull is less. The weight decreases. An object on another planet would experience a different gravitational pull to that planet's center than here on earth. Its weight would be different.

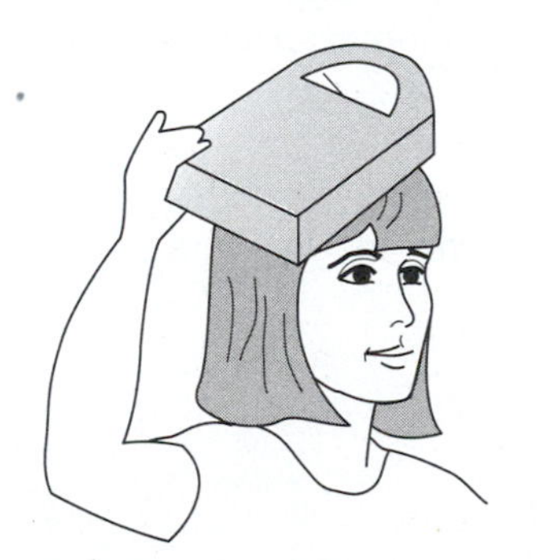

Figure 6–5: *An improper way to measure one's weight.*

Mass is something else. There is no direction associated with it. It is a scalar, not a vector. According to Newton's second law of dynamics, **mass** is the ratio between the net force acting on an object and its acceleration. When you push harder on the pedals of a bicycle to speed up, or try to get a stalled car moving, the resistance you encounter is caused partly by inertia. It does not matter whether the bike is moving north or west. It does not matter in which direction you push the car. It does not matter whether you are on the earth or the moon. The inertia associated with the mass of the object is the same. Anything will have a mass that resists the change of motion no matter where it is or where it is going.

Weight and force are related by Newton's second law of dynamics. Weight is the force caused by gravity that gives a mass an acceleration of 32 ft/s^2 or 9.80 m/s^2 on the surface of the earth. These quantities placed into Newton's second law give us:

$$\mathbf{W} = m\,\mathbf{g}, \qquad \textbf{(Eq. 6–2)}$$

where **W** is the gravitational force on the mass (m) and **g** is the acceleration caused by gravity.

In the British system of measurement, mass is measured in units of **slugs.**

A pound (lb) *is the force required to give a one-slug mass an acceleration of 1.00 ft/s*^{2}.

In SI units, mass is measured in kilograms.

A Newton (N) *is the force required to give a one-kilogram mass an acceleration of 1.00 m/s*^{2}.

The conversion equation between SI and British units of force is:

$$1.00 \text{ N} = 0.225 \text{ lb} \qquad \textbf{(Eq. 6–3)}$$

□ **EXAMPLE PROBLEM 6–3: AN ACCELERATED AUTOMOBILE**

An unbalanced force of 1320 lb gives an automobile an acceleration of 12.0 ft/s^2. What is the weight of the car?

■ ***SOLUTION***

Newton's second law for constant mass gives us the car's mass.

$$m = \frac{|\mathbf{F}|}{|\boldsymbol{a}|}$$

$$m = \frac{1320 \text{ lb}}{12.0 \text{ ft/s}^2}$$

$$m = 110 \text{ slug}$$

The weight is:

$$|\mathbf{W}| = m\,|\mathbf{g}|$$
$$|\mathbf{W}| = (110 \text{ slug})(32 \text{ ft/s}^2)$$
$$|\mathbf{W}| = 3520 \text{ lb}$$

□ **EXAMPLE PROBLEM 6–4: AN ELEVATOR**

A cable winch is used to lift an elevator car out of a zinc mine. As the car starts up the shaft, an unbalanced force of 16,800 N acts on the 5440-lb elevator car. What is the acceleration in SI units?

■ ***SOLUTION***

Use Equation 6–3 to convert the weight of the car to metric units.

$$\mathbf{W} = m\,\mathbf{g} = 5440 \text{ lb} \times \frac{1.00 \text{ N}}{0.225 \text{ lb}}$$
$$\mathbf{W} = 2.42 \times 10^4 \text{ N}$$

Next, find the mass of the elevator car.

$$m = \frac{|\mathbf{W}|}{|\mathbf{g}|} = \frac{2.42 \times 10^4 \text{ N}}{9.80 \text{ m/s}^2}$$
$$m = 2470 \text{ kg}$$

The acceleration can now be found using Newton's second law.

$$|\boldsymbol{a}| = \frac{|\mathbf{F}|}{m} = \frac{16{,}800 \text{ N}}{2470 \text{ kg}}$$
$$|\boldsymbol{a}| = 6.80 \text{ m/s}^2$$

In a problem like this be careful to sort out the forces. Some forces will cause acceleration. Other forces give you the weight of the object, from which the mass of the system can be calculated.

□ **EXAMPLE PROBLEM 6–5: ON THE MOON**

What would be the weight of a 16-lb ball on the surface of the moon? (The acceleration of gravity on the moon's surface is 5.28 ft/s^2.)

■ *SOLUTION*

The magnitude of the weight of the ball on earth is:

$$|\mathbf{W}_e| = m\,|\mathbf{g}_e|$$

On the moon, the pull of gravity is less, causing a smaller acceleration for the same mass. The weight of the ball on the moon is:

$$|\mathbf{W}_m| = m\,|\mathbf{g}_m|$$

Divide one weight equation by the other. The mass cancels because the mass of an object does not depend on its location.

$$\frac{|\mathbf{W}_m|}{|\mathbf{W}_e|} = \frac{m\,|\mathbf{g}_m|}{m\,|\mathbf{g}_e|}$$

or

$$|\mathbf{W}_m| = \frac{|\mathbf{g}_m|\,|\mathbf{W}_e|}{|\mathbf{g}_e|}$$

$$|\mathbf{W}_m| = \frac{5.28\text{ ft/s}^2 \times 16.0\text{ lb}}{32.0\text{ ft/s}^2}$$

$$|\mathbf{W}_m| = 2.64\text{ lb}$$

6.4 ACTION AND REACTION

Push against a wall. You will feel it push back. Push harder. The wall reacts by pushing against your hand with a larger force. A shotgun is fired. A force propels the shot forward; another force causes the gun to recoil in the opposite direction. At a construction site, a bulldozer pushes against an embankment of gravel; the gravel resists with a backward force. A camper steps out of a canoe, accelerating it backward. At the same time, the canoe pushes against her foot to move her forward on to land.

These examples demonstrate the third nature of forces. Forces occur in pairs; whenever a force is applied, a second force will react in the opposite direction. This is **Newton's third law.**

For every applied force there exists an equal but oppositely directed reacting force.

The action-reaction principle explains why internal forces cannot cause an object to change its position. For every internal force there

Loose Cannons

Old swashbuckler movies popular in the 1940s and 1950s always had a scene with a large warship in full sail firing a broadside salvo of cannon shot on another ship. By *broadside,* we mean all the cannons are fired at once. This is not a realistic event.

Typically a cannonball weighed 4 to 6 pounds. The heavy brass cannon that fired the ball was on a wheeled carriage so the gun could be aimed, fired, and then moved so it could be recharged through the muzzle. Chains connected the cannon to the ship's gunwales. When a cannonball was fired, equal and opposite forces were exerted on the cannon and projectile. The backward force on the cannon is called *recoil.* Even a 4-pounder had a large recoil when it fired. If several guns were fired at the same time, the ship would roll violently. In rough water, such a salvo could roll the ship on its side.

(Courtesy of the Smithsonian Institution, photo 84–1511.)

If a chain were loose, or worse, broken, the cannon's recoil could drive the carriage backward during battle, causing serious damage to anything in its way.

From this we get the expression "loose cannon" for someone out of control or acting in a manner that could cause damage.

exists another force inside the object with the same magnitude, but pointing in the opposite direction. The sum of each pair is zero.

The principle also explains how an object interacts with an outside agent. When something from the outside pushes on an object, the object reacts by pushing back. Only the applied force is acting on the object. The reacting force is pushing against something outside it.

When we use Newton's second law to describe the motion of an object we ignore the reacting force. It is a force acting on the outside agent, not on the object. If we want to calculate the acceleration of a bullet down a gun barrel, we are interested only in the force the gun exerts on the bullet. The reacting force of the bullet on the gun is a force on the gun, not on the object we want to analyze.

Because we are looking at only one of the forces in the action-reaction pair, the sum of the forces *acting on the object* may turn out to be something other than zero.

6.5 CENTER OF MASS: AN IMPORTANT POINT

There is a problem with the equations used thus far to describe motion. They do not describe where different parts of an object are located. In part (c) of Figure 6–6 assume the bulldozer is 11.2 m

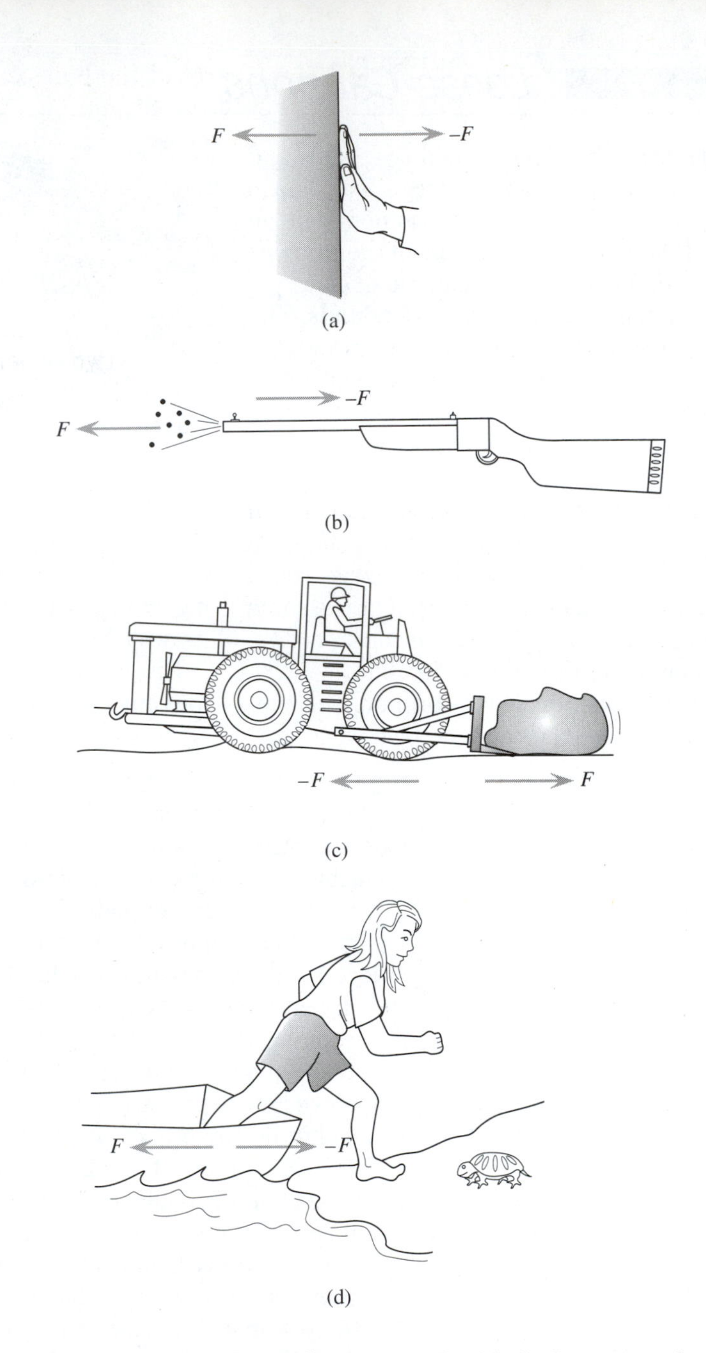

Figure 6–6: Some examples of action-reaction pairs of forces (F).

from the tree. Does that mean the blade is 11.2 m from the tree? Or is the back end of the bulldozer 11.2 m from the tree? Or the steering lever?

In the process of generating useful equations of motion we have reduced the bulldozer to a single point in space. The equations of motion explain what is happening to the mathematical point, not to

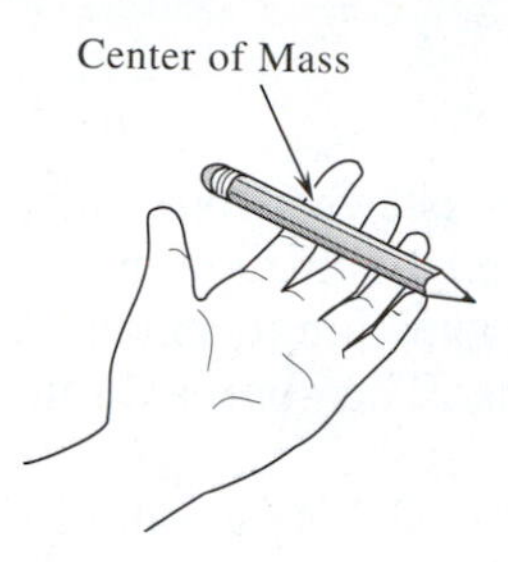

Figure 6–7: A pencil balances at its center of mass.

the different parts of the bulldozer. We need to find where on an object that point is located.

Balance a pencil on your finger. It will balance a bit off center near the eraser end. This is called the **center of mass** of the pencil.

Hold the pencil by the point and toss it straight up into the air. The pencil will rotate around the center of mass. When we use Newton's second law, the equation describes the motion of the center of mass.

Here is another experiment. Balance a stiff plastic ruler on a pencil laid on a desktop. Place two paper clips on the 4-in mark. You will find a new balance point; the center of mass has moved from the 6-in mark. It is a bit closer to the two paper clips. Apparently the center of mass depends on the way mass is distributed in a system.

Take an additional paper clip and place it on the 10-in mark. The center of mass moves back to the original position. One paper clip 4 in from the center of mass balances off two paper clips at only 2 in. Try different numbers of paper clips in different positions. The product of the mass of paper clips to the right of the center of mass times their distance from the center of mass balances an equal product to the left.

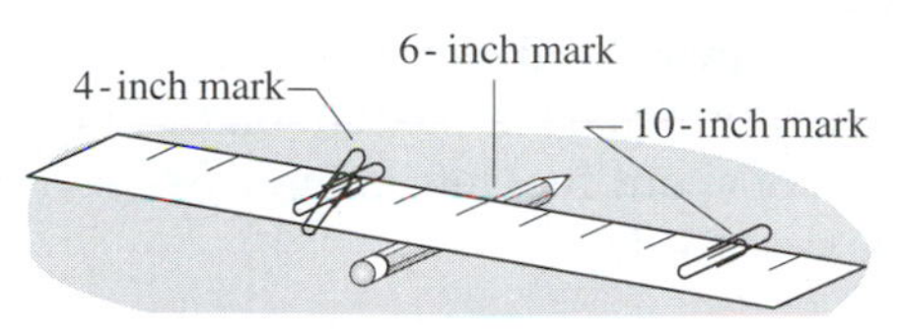

Figure 6–8: Two masses at one distance from the center of mass are balanced by one mass at twice the distance.

$$M_r\, x_r = m_l\, x_l \qquad \textbf{(Eq. 6–4)}$$

where M_r is the mass to the right of the center of mass, m_l is the mass to the left of the center of mass, and x_r and x_l are the corresponding displacements.

□ **EXAMPLE PROBLEM 6–6: CENTER OF MASS OF A CAR**

A car has a 102-in wheelbase. The front axle supports 3200 lb of the car's weight; the rear axle 2240 lb. Where is the center of mass of the car relative to the front axle?

■ ***SOLUTION***

Let X be the distance between the front axle and the center of mass. The distance between the center of mass and the rear axle is then 102 in $-X$. Because the mass is proportional to the weight (see Equation 6–2), we can use the weight rather than the mass in Equation 6–4.

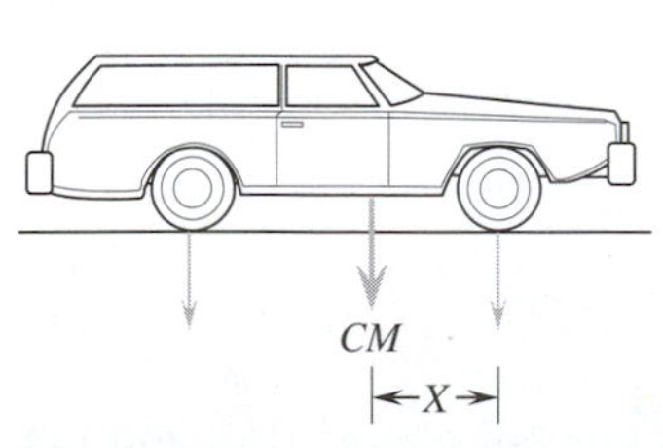

Figure 6–9: The balance point (center of mass, CM) of a car is between the front and rear axles.

$$m_{\text{rear}}\, \mathbf{g}\, (102 \text{ in} - X) = m_{\text{front}}\, \mathbf{g}\, X$$

$$2240 \text{ lb } (102 \text{ in} - X) = 3200 \text{ lb } (X)$$

$$(5.44 \times 10^3 \text{ lb})\, X = 2.28 \times 10^5 \text{ lb} \cdot \text{in}$$

$$X = 42 \text{ in}$$

6.6 NEWTON'S SECOND LAW REVISITED

A rocketship of mass M drifts through space between our solar system and the star Alpha Centauri. There is no air friction in this near-vacuum of space. Gravitational forces are vanishingly small. A small amount of fuel is burned for a short time. The spaceship increases its velocity.

There are no external forces in this case. The rocketship and its fuel, burned and unburned, is a single system. The forces generated by the combustion of rocket fuel are internal forces, which can change the shape and size of something, but cannot cause acceleration. How does the rocketship change velocity?

One answer is to define the system as only the ship and its unspent fuel. The forces from the burned fuel would then be the external forces that push the ship and the remaining fuel. This is a pretty good idea. However, we can develop some new ideas if we use the original definition of the system—the system is the spaceship with all of its fuel, burned and unburned. Under these conditions there is no unbalanced external force, and the system cannot accelerate. This seems absurd, but it is true.

Remember that Newton's second law describes *the motion of the center of mass* of the system. The center of mass is located somewhere between the ship and the long trail of spent fuel left behind. We can write an equation similar to Equation 6–4.

$$m\ x_f = M\ X_r$$

where x_f is the average location of the mass of the spent rocket fuel relative to the center of mass of the system, X_r is the position of the rocket fuel relative to the center of mass, m is the mass of burned fuel, and M is the mass of the rocket with the unspent fuel.

The mass of spent fuel (Δm) and that of the rocket (M) are moving in opposite directions away from the center of mass of the system. We can show this change in the equation.

$$\frac{\Delta m\ x_f}{t} = \frac{M\ X_r}{t}$$

or

$$\Delta m\ v_f = M\ V_r$$

where the center of mass of the system has not changed its motion. The system is simply becoming very long as the two parts of the system spread out. This is a change of shape consistent with internal forces. The center of mass is still moving along at a constant velocity toward Alpha Centauri. During the short time interval the fuel is burned, the velocity of the rocketship changes. (Remember, the rocketship is only part of the system.) This can be shown by dividing the equation above by the interval of time during which the fuel is burning.

$$\frac{\Delta m\, v_f}{\Delta t} = \frac{M \Delta V_r}{\Delta t}$$

The right-hand side of the equation is nothing more than the acceleration of the spaceship (A_r) multiplied by the spaceship's mass.

The left-hand side of the equation needs to be handled differently. The fuel burned in the rocket exits at a *constant* speed (V_c) relative to the rocket. It is like water ejected from the nozzle of a garden hose. The speed is determined by the size and shape of the rocket's tail and the temperature of the burning fuel. It is the mass of ejected fuel (Δm) that is dependent on the elapsed time (Δt). This can be shown in the equation

$$v_f \left(\frac{\Delta m}{\Delta t}\right) = M A_r$$

where the left-hand side of the equation must be force because it accelerates a mass. As fuel is pushed out the rear of the rocket, it creates a reacting force pushing back on the rocket. This force is called **thrust.**

$$\text{thrust} = -v \left(\frac{\Delta m}{\Delta t}\right)$$

This leads to a new idea. Force is related to mass and velocity in two ways: either the velocity of an object may change or its mass may change. A general form of Newton's second law may be written:

$$\mathbf{F}_{\text{net}} = \left(\frac{\mathbf{v}\, \Delta m}{\Delta t}\right) + \left(\frac{m\, \Delta \mathbf{v}}{\Delta t}\right)$$

We can write this in a simpler form. Let

$$\frac{\Delta(m\, \mathbf{v})}{\Delta t} = \left(\frac{\mathbf{v}\, \Delta m}{\Delta t}\right) + \left(\frac{m\, \Delta \mathbf{v}}{\Delta t}\right)$$

then

$$\mathbf{F}_{\text{net}} = \frac{\Delta(m\mathbf{v})}{\Delta t}$$

The product of mass and velocity, $m\mathbf{v}$, seems to have some special importance. It is called **momentum**, represented by the symbol **p**.

$$\mathbf{p} = m\,\mathbf{v} \qquad \textbf{(Eq. 6–6)}$$

Newton's second law can be written in terms of momentum.

$$\Sigma\,\mathbf{F} = \frac{\Delta\mathbf{p}}{\Delta t}$$

☐ **EXAMPLE PROBLEM 6–7: MOMENTUM**

Find the magnitude of the momentum of each of the following objects.

A. An electron ($m_e = 9.11 \times 10^{-31}$ kg) in a cathode ray tube (CRT) used in a computer monitor has a speed of 5.0×10^7 m/s.

B. A $12\bar{0}0$-lb wrecking ball swings with a speed of $4\bar{0}$ ft/s.

■ ***SOLUTION***

A. Momentum of electron:

$$p = m\,v$$

$$p = 9.11 \times 10^{-31}\ \text{kg} \times 5.0 \times 10^{7}\ \text{m/s}$$

$$p = 4.6 \times 10^{-23}\,\frac{\text{kg} \cdot \text{m}}{\text{s}}$$

B. Momentum of wrecking ball: First find the mass of the ball.

$$m = \frac{|\mathbf{W}|}{|\mathbf{g}|}$$

$$m = \frac{12\bar{0}0\ \text{lb}}{32\ \text{ft/s}^2}$$

$$m = 37.5\ \text{slug}$$

$$p = m\,v$$

$$p = 37.5\ \text{slug} \times 4\bar{0}\ \text{ft/s}$$

$$p = 1.5 \times 10^{3}\ \text{slug} \cdot \text{ft/s}$$

☐ **EXAMPLE PROBLEM 6–8: FORCE AND MOMENTUM**

At a large paper mill a blower is used to move wood chips along a large pipe from the chipper in the log yard to the paper-making plant. Neglecting friction, how large a force must the blower develop to transport $50\bar{0}$ lb of chips through a 500-ft pipe in 12 s?

■ *SOLUTION*

First find the mass of the chips.

$$m = \frac{|\mathbf{W}|}{|\mathbf{g}|}$$

$$m = \frac{50\bar{0}\ \text{lb}}{32\ \text{ft/s}^2}$$

$$m = 15.6\ \text{slug}$$

Using s for distance and t for time, the speed of the chips is:

$$\mathbf{v} = \frac{s}{t}$$

$$\mathbf{v} = \frac{50\bar{0}\ \text{ft}}{12\ \text{s}}$$

$$\mathbf{v} = 41.7\ \text{ft/s}$$

The general form of Newton's second law is:

$$\mathbf{F} = \frac{\Delta \mathbf{p}}{\Delta t} = \left(\frac{\Delta m\ \mathbf{v}}{\Delta t}\right) + \left(\frac{m\ \Delta \mathbf{v}}{\Delta t}\right)$$

Because the chips move at a constant speed, $a = 0$, or $\Delta \mathbf{v}/\Delta t = 0$.

$$|\mathbf{F}_{\text{net}}| = \frac{\Delta m\ \mathbf{v}}{\Delta t}$$

$$|\mathbf{F}_{\text{net}}| = \frac{15.6\ \text{slug} \times 41.7\ \text{ft/s}}{12\ \text{s}}$$

$$|\mathbf{F}_{\text{net}}| = 54\ \text{lb}$$

Because we have neglected both friction and the mass of the air blown through the pipe, this estimate is low.

□ **EXAMPLE PROBLEM 6–9: FIRE HOSE**

A fire hose delivers 1.3 ft^3 of water with a nozzle velocity of 50 ft/s. With what force does the nozzle recoil? The density of water is 1.96 slug/ft^3.

■ *SOLUTION*

Since the water has a constant velocity, $a = 0$,

$$|\mathbf{F}| = \frac{\Delta m\, \mathbf{v}}{\Delta t}$$

where the mass is volume of the water times density.

$$|\mathbf{F}| = 1.3 \text{ ft}^3 \times 1.96 \frac{\text{slug}}{\text{ft}^3} \times 50 \frac{\text{ft}}{\text{s}}$$

$$|\mathbf{F}| = 130 \text{ lb backward}$$

SUMMARY

A system is one or more objects connected by internal forces. Internal forces can cause a system to change size or shape, but only external forces can cause a system to change its velocity.

Equations of motion describe the behavior of the center of mass of a system. Mass is a measure of an object's resistance to change of velocity, its inertia. Mass is a scalar quantity. The weight of an object is the gravitational force acting on a mass; it is a vector quantity. Mass and weight are related by Newton's second law: $\mathbf{W} = m\, \mathbf{g}$, where $\mathbf{g}$ is the acceleration caused by gravitational forces.

Momentum ($\mathbf{p}$) is the product of mass and velocity ($\mathbf{p} = m\, \mathbf{v}$). Momentum is a vector quantity in the direction of the velocity.

Newton's laws of dynamics suggest these principles:

1. **Inertia:** A system will not change its velocity unless acted upon by an unbalanced force.
2. **Force and motion:**

$$\mathbf{F} = \frac{\Delta \mathbf{p}}{\Delta t} = \left(\frac{\Delta m\, \mathbf{v}}{\Delta t}\right) + \left(\frac{m\, \Delta \mathbf{v}}{\Delta t}\right)$$

   When the mass of a system is constant ($\Delta m = 0$), $\mathbf{F} = m\, a$, where a is acceleration. When the velocity of a system is constant ($a = 0$), $\mathbf{F} = \mathbf{v}\, \Delta m / \Delta t$. When both velocity and mass are constant, then $\mathbf{F} = 0$.
3. **Action and reaction:** Forces occur in pairs. For every force there is another force that is equal in magnitude, but acting in the opposite direction.

KEY TERMS

If you can explain the following terms to a friend or classmate, you understand their meaning. If you cannot explain the terms, you should reread the sections in which they are discussed.

center of mass
external forces
inertia
internal forces
force
mass

momentum
net force
Newton
Newton's laws of dynamics
slug
system
thrust
unbalanced force
vector sum
weight

Section 6.1:

1. Explain each of the following terms to a classmate in your own words: internal force, external force, system, inertia, net force.
2. Explain the difference between internal force and external force. Between speed and velocity.
3. In each of the following systems, is an unbalanced external force acting on the object? Explain why or why not.
   A. At a constant speed a truck climbs a hill, goes over the top, and comes down the opposite side.

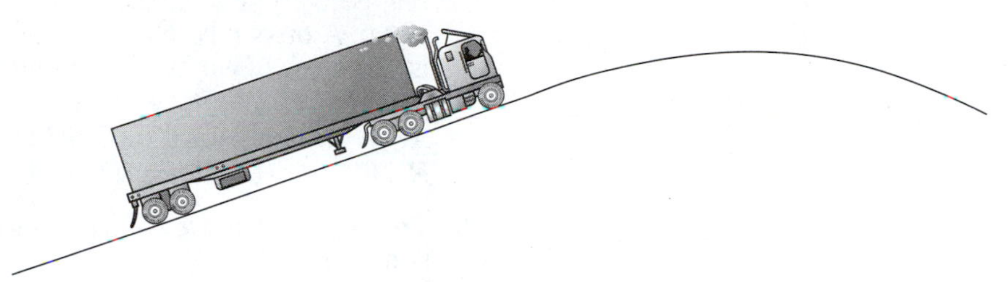

Figure 6–10: *A truck travels over the top of a hill.*

   B. A parachutist falls at a constant speed of 10 mph.
   C. A rock dropped from a bridge accelerates downward at 32 ft/s^2.
   D. Earth orbits the sun at a constant speed.
   E. A crane lifts a steel girder at a constant speed.
4. Determine whether or not an unbalanced force is acting on each of the following systems. Explain why or why not.

Figure 6–11: *A train moves up a constant grade at a constant speed.*

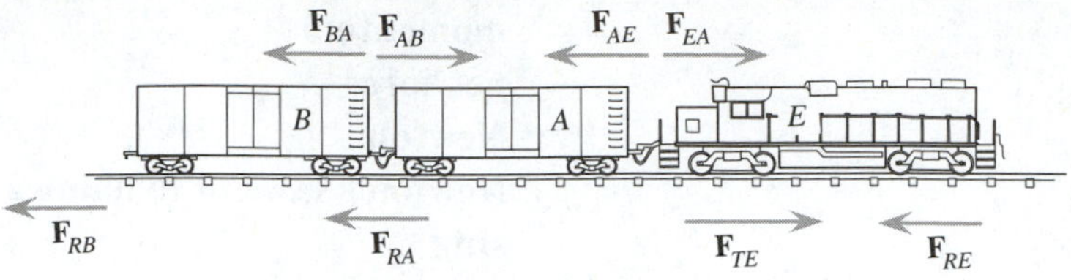

Figure 6–12: Some of the forces acting on an engine (E) with two railroad cars.

A. At a steady speed a freight train climbs a hill of constant grade.
B. An airplane makes a loop-the-loop.
C. A motorcycle makes a left-hand turn.
D. A basketball rolls at a constant speed along a gym floor parallel to the out-of-bounds line.
E. A baseball flies to centerfield.

5. In Figure 6–12 an engine (E) pulls two railroad cars at a constant speed. The system is the engine and both cars. $\mathbf{F}_{RE}$, $\mathbf{F}_{RA}$, and $\mathbf{F}_{RB}$ are the frictional drags on the engine, car A, and car B, respectively. $\mathbf{F}_{BA}$ is the force of car B on car A. $\mathbf{F}_{AB}$ is the force of car A on car B. $\mathbf{F}_{EA}$ is the force of the engine on car A. $\mathbf{F}_{AE}$ is the force of car A on the engine. $\mathbf{F}_{TE}$ is the force of the track pushing on the drive wheels of the engine.
   A. What forces are internal to the system?
   B. $\mathbf{F}_{TE}$ must balance what other forces?

6. The train in Figure 6–12 moves at a constant speed. The system is only car A.
   A. What external forces act on car A?
   B. $\mathbf{F}_{EA}$ must balance what other forces?
   C. Is $\mathbf{F}_{EA}$ the same size as $\mathbf{F}_{AB}$? Why or why not?

7. What is the sum of all of the forces in the universe?

8. Farmer Brown rides a cart pulled by an educated mule. One day they were crossing a field when the mule stopped and turned his head toward Farmer Brown. The animal said, "I'm an educated mule and former physics student. I happen to know that no matter how hard I pull on this cart it will pull back with an equal but opposite force. Therefore, I couldn't possibly move this cart. So I won't try." He then put his head down and sedately munched on grass. Is the animal correct? Why or why not?

9. Figure 6–13 is a simplified version of the power train of an automobile.
   Expanding gases in the cylinder push on the piston (F_{12}).
   The piston pushes on the piston rod (F_{23}).
   The rod pushes on the crankshaft (F_{34}).
   The crankshaft exerts a force on the flywheel (F_{56}).
   The flywheel exerts a force on the clutch, which transmits the force to the drive shaft (F_{78}).

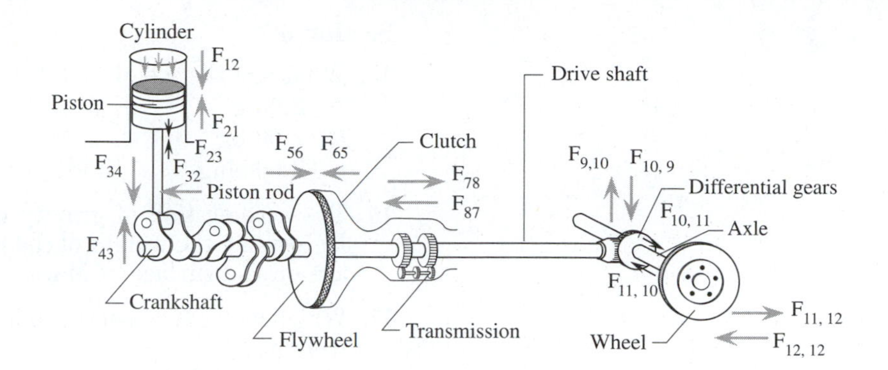

Figure 6–13: *The power train for a car with rear-wheel drive.*

The drive shaft exerts a force on the differential gears ($F_{9,10}$).
The differential gears exert a force on the wheels through the axle ($F_{10,11}$).
The wheels push against the highway ($F_{11,12}$).
The highway pushes back with a reacting force ($F_{12,11}$).

A. The system is the automobile. What forces are internal forces and what forces are external?

B. The system is the engine (everything in front of the clutch). What forces are internal forces and what forces are external?

C. The system is the piston. What forces act on the piston?

Section 6.2:

10. The gravitational force on a 2-lb rock is twice as large as the force on a 1-lb rock. Why does the heavier rock not fall faster?

11. How large an acceleration would result from a 1.00-N force acting on each of the following objects?

A. 200-g mass
B. 1.00-kg mass
C. 1.00-N weight
D. 200-kg mass
E. 1.00-g mass
F. 1.00-slug mass
G. 1.00-lb weight

12. A pole-vaulter would probably be hurt if he landed on an asphalt surface rather than on a thick spongy pad. Use Newton's second law to explain why.

13. A car stops abruptly. Why is a passenger who is not wearing a seat belt thrown forward?

14. A passenger is eating dinner on an airplane. A pea on his plate suddenly rolls toward him. Give at least two possible explanations for the pea's behavior.

Section 6.3:

15. What are the weights of the following masses?
    A. 2.00 g
    B. 2.00 kg
    C. 2.00 slug

16. The acceleration of gravity on Mars is 3.72 m/s^2 or 12.2 ft/s^2. Determine the weight of the masses given in Exercise 15 if they are on the surface of Mars.

17. Which item in each of the following pairs of objects has the larger weight?
    A. A 2.0-kg box of patching plaster or a 4.6-lb pail of latex paint.
    B. A 15-ft chain with a mass of 210 g/ft or a 45-ft rope with a weight of 0.15 lb/ft.
    C. A 44-kg bag of Portland cement or a 100-lb box of nails.

18. In the year A.D.2031 a moon-based colony orders 100 kg of canned tomatoes from a supplier on earth. Does it matter whether the tomatoes are measured on the earth or the moon? If the order had been 220 lb of tomatoes, would it matter whether they were measured in earth pounds or moon pounds?

19. Two kinds of scales are commonly used to measure mass or weight (see Figure 6–14). On a spring scale, the stretch of the spring is proportional to the weight hung on it. A balance scale balances known weights or masses against an unknown quantity. If a spring scale and a balance scale both read 2.3 kg for a particular rock here on earth, would they read the same for the same rock on the moon?

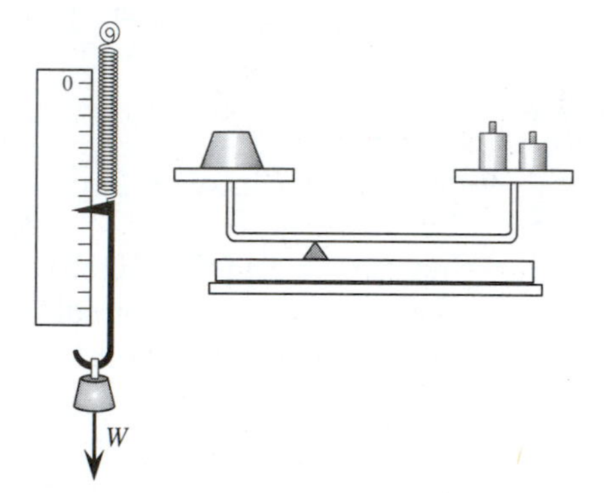

Figure 6–14: A spring scale (left) and a balance scale (right).

Section 6.4:

20. A pickup truck crashes into the rear end of a stationary car. The kinds of injuries sustained by the two drivers will be different. Explain the causes of the injuries.

21. A basketball player jumps into the air. Which of the following statements best describe(s) the forces involved?
    A. The force that pushes the player into the air equals the force with which she pushes against the floor.
    B. The floor pushes up on the player with a force greater than her weight.
    C. The force of the floor pushing on the player cannot be larger than her weight.
    D. There is no external force acting on the basketball player.

22. Describe the effect of these action-reaction forces.
    A. A sledgehammer hits a rock without breaking it.
    B. A stack of construction blocks to be loaded on a flatbed truck is lifted by a cable winch.

C. The blade of an outboard motor pushes against water.
D. A cement pier holds up a section of a bridge.

Section 6.5:

23. A $30\bar{0}$-g mass and a $10\bar{0}$-g mass are connected by a slender rod of negligible mass. The point of balance is at a point
   A. in the middle of the rod.
   B. three-fourths the length of the rod nearest the $10\bar{0}$-g mass.
   C. three-fourths the length of the rod nearest the $30\bar{0}$-g mass.
   D. two-thirds the length of the rod nearest the $30\bar{0}$-g mass.
24. The moon does not orbit around the earth. The earth-moon system rotates around a common center of gravity. Why does the moon appear to be orbiting around the earth?
25. When a carpenter's hammer is tossed in the air, the head rotates through a smaller circle than the end of the handle. Why?
26. A bomb falling straight downward explodes. Describe the motion of the center of mass.

Section 6.6:

27. Blow up a balloon. Release it without tying the neck. Use the general form of Newton's second law to explain the motion.
28. What causes a lawn sprinkler to rotate?

PROBLEMS

Section 6.1:

1. A crane lifts an $80\bar{0}$-lb girder vertically upward at a speed of 8.0 ft/s. What net force acts on the girder?
2. College students are playing push ball with a 5.0 ft diameter ball. The following forces act on the ball: 47 lb north, 56 lb east, 27 lb west, 14 lb north, 46 lb south, 29 lb west, and 25 lb south. What is the net force? In what direction will the ball move?
3. Which of the following systems has the largest unbalanced force acting on it?
   A. A winch lifts an $80\bar{0}$-lb bucket of concrete. The tension on the cable attached to the bucket is $87\bar{0}$ lb.
   B. The propeller of an outboard motor exerts a force of 195 lb against the water. The drag on the hull of the boat is 130 lb.
   C. Wind exerts a force of $70\bar{0}$ lb on a wind turbine. The frictional drag of the system is $7\bar{0}$ lb. The electrical generator resists with a force of $59\bar{0}$ lb.

4. Find the net force of each of the following sets of forces.
   A. $3\bar{0}$ N northward, $2\bar{0}$ N southward, $1\bar{0}$ N eastward, $1\bar{0}$ lb westward
   B. $4\bar{0}$ N at $3\bar{0}°$, $5\bar{0}$ N at $12\bar{0}°$, $4\bar{0}$ N at $27\bar{0}°$

Section 6.2:

5. Haybalers often have an attached mechanical device called a kicker. The kicker throws the bale onto a trailing wagon. How large a force must a kicker exert on a $6\bar{0}$-lb bale to give it an acceleration of 25 ft/s^2?

6. A bullet accelerates down the barrel of a rifle, reaching a muzzle velocity of 400 m/s in 0.0020 s. What is the acceleration of the bullet and what net force acts on it?

7. Which of the following systems has the largest unbalanced force acting on it?
   A. A 2.0-kg rock falling with an acceleration of 9.8 ft/s^2.
   B. An 85.0-lb sky diver falling at 67.0 m/s.
   C. A 70-kg cyclist on a 7.0-kg bicycle accelerating at 1.3 m/s^2.

8. Find the acceleration of each of the following systems.
   A. A 1.8-lb piston is accelerated downward by a net force of 65 lb.
   B. A 3800-lb car is accelerated by a net force of 970 lb.
   C. A 545-kg elevator accelerates upward with a net force of 2725 N.

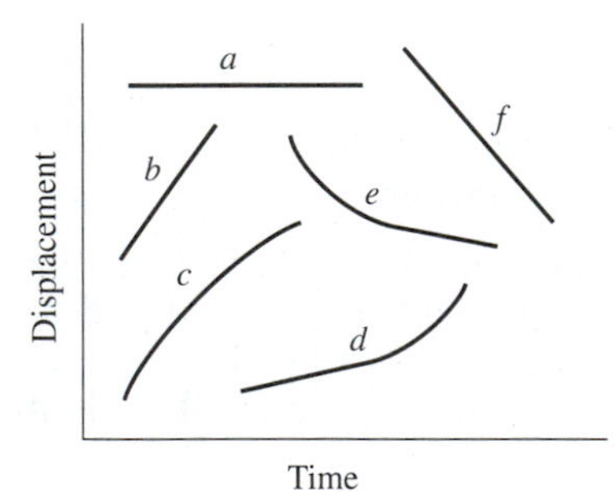

Figure 6–15: Plot of the motion of six automobiles.

9. An adult human has a head weighing $1\bar{0}$ lb. Someone walking at 3.0 mph (4.4 ft/s) strikes her head on an overhanging obstruction, bringing her head to a stop in 5.0 ms. What retarding force acts on her head?

10. A jet engine develops a thrust of 5.0×10^4 N, accelerating the plane by 9.0 m/s^2. If the air drag is 1.2×10^4 N, what is the mass of the vehicle?

11. Figure 6–15 is a plot of displacement versus time for six cars.
    A. Which curves show acceleration?
    B. Which curves show deceleration?

12. Table 6–1 shows the position of a motor scooter at different times. Does an unbalanced force act on the scooter at any time?

*13. An oil-drilling rig lowers a section of piping with the drill attached with a combined weight of 1230 lb at a speed of 8.0 ft/s. The brake is applied, bringing the assemblage to a stop in 3.6 ft. What is the force on the brake drum?

Table 6–1: Position (displacement) of a motor scooter at different times.

DISPLACEMENT (m)	TIME (s)
0.0	0.0
8.2	1.0
16.4	2.0
24.6	3.0
34.6	4.0
46.4	5.0
60.0	6.0

***14.** An automobile strikes a tree, coming to rest in 0.10 s.

A. If the car was traveling 18.3 m/s and the driver's mass is 50 kg, estimate the force exerted on the driver by the seat harness.

B. Assume the driver neglected to fasten his seat belt. Instead of coming to rest with the car he is catapulted into the windshield, coming to a stop in 5 ms. What is the force on the driver?

***15.** A 15-g bullet is accelerated down a 6$\bar{0}$-cm rifle barrel with a force of 1.0×10^2 N. What is the muzzle velocity of the bullet?

Section 6.3:

16. What is the mass of a 60$\bar{0}$0-lb truck in slugs? In kilograms?

17. What are the weights of the following masses?

A. A 9.1-kg computer monitor.
B. A 0.093-slug textbook.
C. A 54-kg ingot of aluminum.
D. A 625-slug earthmover with a load.
E. A 37-g computer card.

18. Calculate the mass of each of the following weights.

A. A 3$\bar{0}$0-lb I beam.
B. A 142-N drive shaft.
C. A 0.44-N drill bit.
D. A 13,200-lb turbine.

***19.** On the surface of Mars the gravitational pull is 0.38 that of earth's surface gravity.

A. An astronaut weighs 190 lb on earth. What is his weight on Mars?

B. If the astronaut falls on the surface of Mars, what will be his free-fall acceleration?

**20. The gravitational acceleration on the moon is one-sixth that on the earth. A rock is measured to be 60 N here on earth. It is again measured on the moon, first with a spring balance and then with a balance scale. What would each scale read?

Section 6.5:

21. A tire is balanced by placing a $3\bar{0}0$-g weight on the rim 35.5 cm from the center of the wheel and on the opposite side of the off-center mass. If the tire has a mass of 18.2 kg, how far off center was the unbalanced mass of the wheel?

22. An 8.0-lb weight is 2.0 ft from an unknown weight. The balance point is 7.9 in from the 8.0-lb weight. What is the weight of the unknown?

23. The range of a triple-beam balance can be extended by hanging counterweights at the end of the beam arm. The distance from the balance pan to the knife edge support is $1\bar{0}$ cm. The beam end is 25 cm from the knife edge. What mass must the counterweight be to balance off 1000 g resting in the pan?

*24. The center of the moon is 60.25 earth radii from the center of earth. The mass of the moon is 0.123 earth masses. Find how far from the center of the earth is the center of mass of the earth-moon system. Express your answer in earth radii.

**25. A $15\bar{0}$-lb camper is at the end of a $10\bar{0}$-lb canoe 15 ft from the pier (see Figure 6–16). The center of mass of the 16 ft long canoe is at its center.

A. Where is the center of mass of the camper-canoe system relative to the camper?

B. Where is the center of mass of the camper-canoe system relative to the pier?

The camper moves from the end of the canoe nearest the pier to the far end.

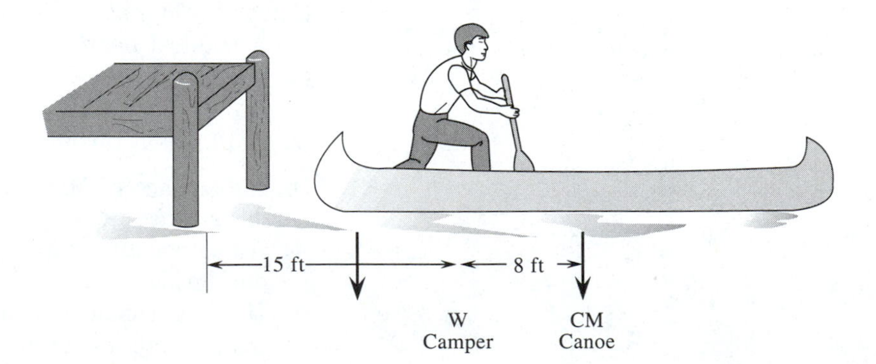

Figure 6–16: *A camper moves from one end of a canoe to the other. W, weight; CM, center of mass.*

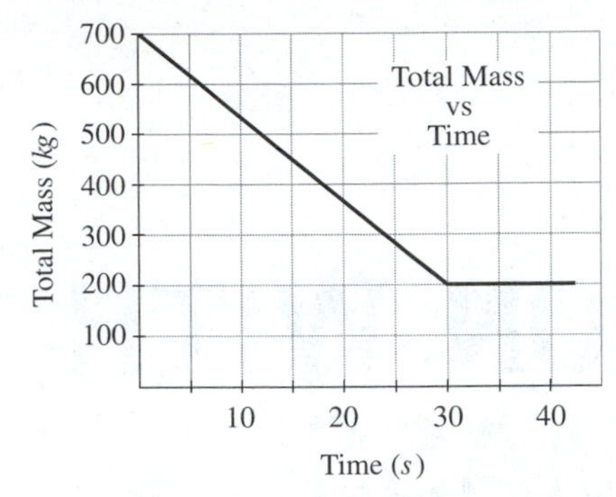

Figure 6–17: A plot of the total mass of a rocket versus time as it burns fuel.

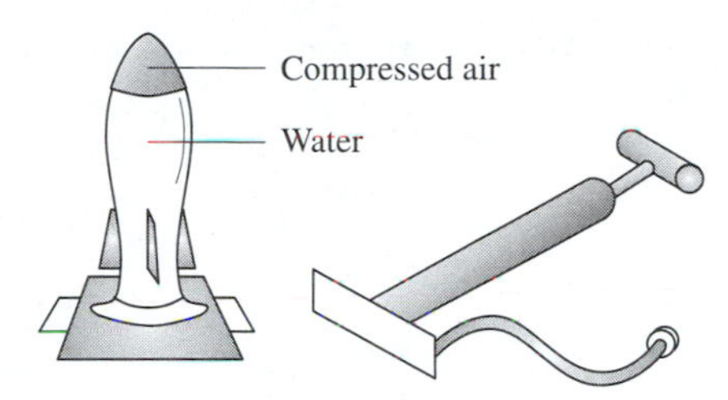

Figure 6–18: A toy water rocket.

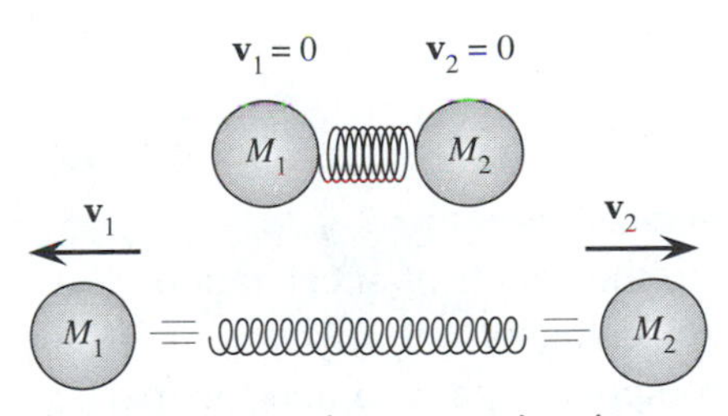

Figure 6–19: A coil spring is placed between two balls and then is released. **v**, *velocity;* M, *mass.*

C. If the canoe is initially at rest describe the motion of the center of mass of the system. Neglect friction.

D. How far away from the pier is the near end of the canoe now?

Section 6.6:

26. Calculate the momentum of the following objects.

A. A 20,000-lb semitrailer truck traveling at 88 ft/s.

B. A 52-kg person jogging at 13 ft/s.

C. An 11.0-g bullet with a speed of 375 m/s.

27. An industrial pump delivers 300 liters (l) of water per minute at an outlet speed of 15 m/s. What net force must the pump exert on the water? (The density of water is 1.00 kg/l.)

28. A machine gun fires 1200 rounds of 50.0-g bullets a minute with a muzzle velocity of 550 m/s. What is the average recoil force?

29. Figure 6–17 shows a plot of the total mass of a rocket versus time. What is the thrust of the rocket during the first 30.0 s of flight if the speed of the exhaust is 420 m/s?

30. A toy water rocket operates with compressed air (see Figure 6–18). The rocket is filled with water. A tire pump is used to build up pressure over the water. If 150 g of water is forced out of the rocket in 0.80 s at an average velocity of 10 m/s, what is the thrust of the rocket?

***31.** A compressed spring is placed between two balls at rest as shown in Figure 6–19. The spring is released. One of the balls has a mass of 50 g and is accelerated to a final velocity of 120 m/s. The final velocity of the second ball is 100 m/s.

A. What happens to the center of mass of each of the two balls?

B. What is the mass of the second ball?

C. Calculate the final momentum of each of the two balls.

D. What is the vector sum of the final momentums of the two-ball system?

***32.** A venting fan must deliver 8.0×10^3 ft^3/min (cfm) of air against a static resistance of 10.5 lb. If the linear velocity of air at the fan outlet is 26 ft/s, calculate the total thrust the fan must develop. (The weight density of air is 0.080 lb/ft^3.)

Chapter 7

ACCELERATION WITH CONSTANT MASS

This engraving shows Elisha Graves Otis demonstrating his new elevator—the first to use a braking system for safety. If you were caught in a free-falling elevator, could you do anything to save yourself? Read about acceleration in this chapter to find out what your options are. (Photograph courtesy of the Otis Elevator Company.)

OBJECTIVES

In this chapter you will learn:

- how to solve force problems involving acceleration
- that the effects of gravity are equivalent to the effects of acceleration
- to apply strategies to solve inclined plane problems
- how to break a composite system into component parts to find internal forces
- applications of accelerometers

Time Line for Chapter 7

370 BC	Aristotle states free fall is accelerated motion.
1590	Galileo refutes Aristotelean physics.
1686	Isaac Newton publishes the *Principia, De Motu,* the motion of bodies.
1743	Jean le Rond d'Alembert describes equilibrium of internal forces inside a system and develops equilibrium mechanics.
1784	George Atwood finds an accurate value of g.
1854	Elisha G. Otis demonstrates his elevator brake at the World's Fair.

In this and the following three chapters, we will look at different kinds of problems that can be solved using the principles of Newton's laws of dynamics set forth in Chapter 6.

7.1 VERTICAL ACCELERATION

Here are a couple of problems. Although they look different, they have a lot in common. A student enters an elevator carrying a 1.0-kg mass hung on a spring scale. The elevator accelerates upward at 2.0 m/s^2. The spring changes from a weight of 9.8 N to a larger value. What is the new weight?

At a construction site a derrick lifts a large steel bucket of concrete. The combined weight of the bucket and concrete is 1800 lb. If the bucket is lifted with an acceleration of 6.0 ft/s^2, what is the tension on the derrick's cable?

Both problems involve an upward acceleration. This means that there must be a net upward force. Let us draw the forces on the objects, showing only the external forces acting on them. These kinds of diagrams are called **free-body diagrams.**

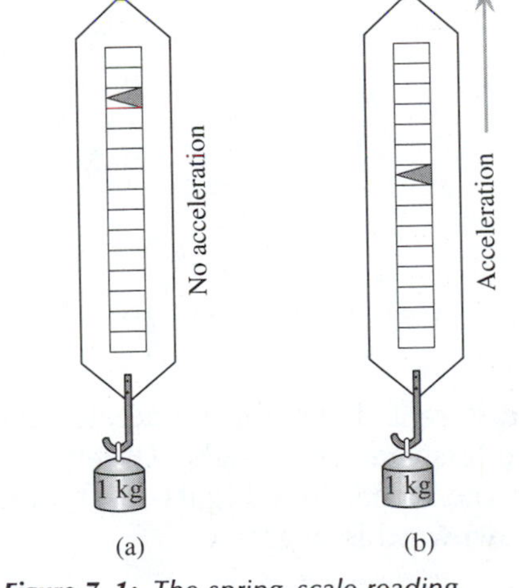

Figure 7–1: *The spring-scale reading increases with an upward acceleration.*

Figure 7–3a shows the forces on the 1.0-kg mass. The weight (**W**) of the mass acts downward. The tension (**T**) in the spring acts upward on the mass. Figure 7–3b shows the forces on the bucket. Tension (T) in the cable acts up on the bucket. The weight of the bucket and its contents act downward. The same vector diagram can be used to represent both problems. Let us use upward as the positive direction. We can write a general equation to represent the forces in both systems using Figure 7–4a as a guide. Here are the component equations:

$$\mathrm{T} - \mathrm{W} = m\,\mathrm{a}$$

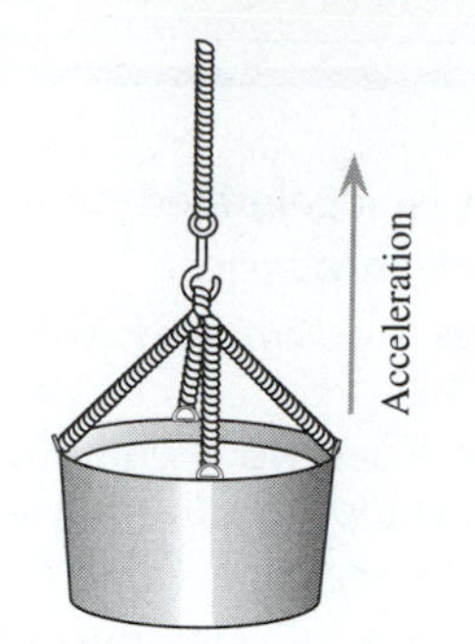

Figure 7–2: A bucket of concrete is accelerated upward.

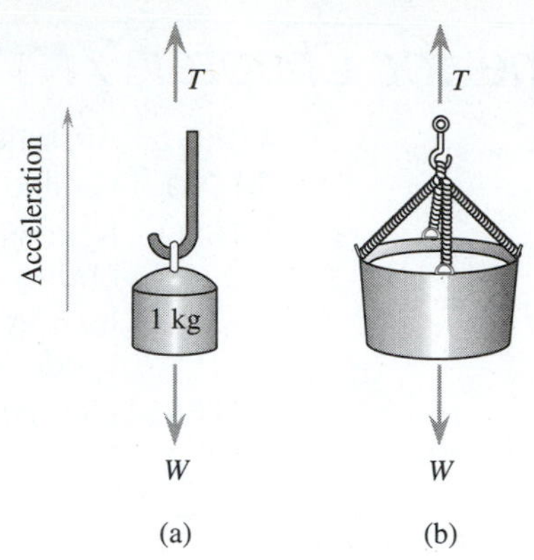

Figure 7–3: Free-body diagrams of the two systems. (a) The 1.0-kg mass on a spring. (b) The bucket of concrete. T, *tension in the spring;* ***W,*** *weight.*

The weight can be expressed in terms of mass:

$$T - (m\,g) = m\,a$$

or

$$T = m\,(g + a) \qquad \text{upward acceleration}$$

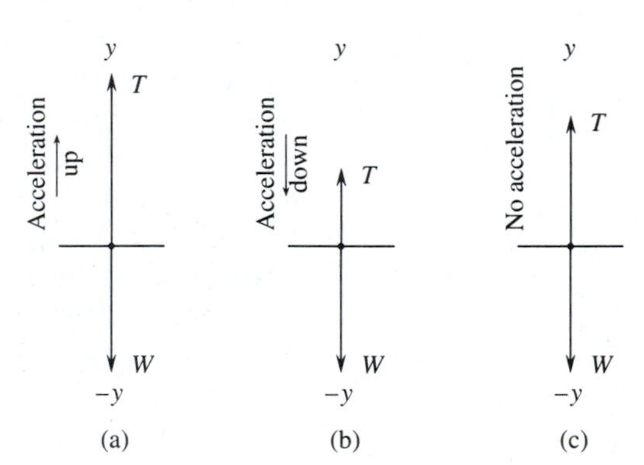

Figure 7–4: The tension on a loaded spring varies with the vertical acceleration.

The diagram would be slightly different if the objects accelerated downward. The tension would be less than the weight. Otherwise, the relationship is pretty much the same. See Figure 7–4b; and remember upward is positive, downward is negative:

$$T - (m\,g) = m\,(-a)$$

or

$$T = m\,(g - a) \qquad \text{downward acceleration}$$

Of course, the tension must equal the weight if there is no acceleration, as depicted in Figure 7–4c.

$$T = m\,g \qquad \text{no acceleration}$$

All three situations may be summarized in one *vector* equation where boldface letters are vector quantities.

$$\mathbf{T} + (m\,\mathbf{g}) = m\,\mathbf{a}$$

or

$$\mathbf{T} = m\,(\mathbf{a} - \mathbf{g}) \qquad \textbf{(Eq. 7–1)}$$

In the first problem the reading of the spring scale depends on the tension in the spring. As the elevator accelerates upward, the scale will register a weight different from that registered when there was no acceleration. The **apparent weight** includes the effects of the motion of the elevator. The apparent weight ($\mathbf{W}_a$) is in the opposite direction of the tension.

$$\mathbf{W}_a = -\mathbf{T}$$

or

$$\mathbf{W}_a = m\,(\mathbf{g} - \mathbf{a}) \qquad \textbf{(Eq. 7–2)}$$

where $\mathbf{W}_a$ is the apparent weight of the vertically accelerated object. In the case of the derrick, $\mathbf{W}_a$ is the apparent weight of the bucket and its contents experienced by the cable.

One usually knows when an elevator is moving. The vibrations and noise of the cables and the relative motion of the car to the elevator shaft seen through the crack between the doors suggest the motion of the elevator. These observations require gathering information from outside the elevator car. If the passenger's ears were plugged, if the doors fit tightly, and if the elevator moved smoothly without vibration, there would be no proof that the elevator was accelerating. If the 1.0-kg mass were not marked, and if we had not glanced at the spring scale until acceleration was under way, there would be no way of knowing whether or not the weight read by the scale was real or apparent. The effects of acceleration are equivalent to gravitational effects. This idea, called **equivalence,** was first suggested by Albert Einstein.

Elevator Evolution

Early elevators for tall buildings, circa 1850, used hydraulic motors and pulley systems with a free-hanging cage held up by a strong rope. Ropes wore and broke; nothing but terra firma stopped the elevator's fall if an accident occurred. Not too many people were anxious to use elevators.

Along came Elisha G. Otis. He designed a system with ratchets, pieces of steel with an upward-facing sawtooth pattern, placed on the walls of the shaft. Spring-loaded pawls, steel rods with a tooth on the end, were installed on the cage. The tension of the rope held back the pawls. When the rope tension went down to practically nothing, the pawls snapped outward and caught on the ratchets.

Otis demonstrated his elevator brake at the 1854 New York Fair. He stood in an elevator installed with his invention while someone cut the rope. The elevator fell a few feet and then stopped. Otis, the elevator man, had established his fortune.

(Photograph courtesy of the Otis Elevator Company.)

Today we use more reliable steel cables and electric winches.

In the unlikely situation that you find yourself in a free-falling elevator, will jumping up just before the elevator hits the bottom of the shaft save your life?

There's a problem here. If you jump in the air, you have to come down again. If you toss a ball upward, it comes back to your hand with the same speed downward that it left your hand going upward. In the elevator, you'd slow down your fall for maybe a second. In the meantime the elevator has come to a stop. When you hit the floor of the elevator, it will be at the same speed the cage hit the bottom of the shaft. Tough luck!

Someone in a closed system cannot distinguish between the effects produced by a gravitational field and the effects caused by the acceleration of the system.

The apparent weight observed by the student is an example of equivalence. The apparent weight is equivalent weight.

One way NASA prepares astronauts for the weightlessness of space takes advantage of equivalence. Trainees are placed in the padded cabin of a large plane. The plane goes into a controlled dive maintaining a downward acceleration of **g**. For approximately 30 s, future astronauts can experience the weightlessness they will encounter in space.

Now let us solve the two problems.

☐ **EXAMPLE PROBLEM 7–1: THE ELEVATOR**

The mass (m) = 1.00 kg. The acceleration (**a**) = 2.0 m/s^2, upward. Find the weight of the mass.

■ ***SOLUTION***

Up is taken as the positive direction.

$$\mathbf{W} = m\,(\mathbf{g} - \mathbf{a})$$

$$\mathbf{W} = 1.00\text{ kg}\,(-9.8 - 2.0)\text{m/s}^2$$

$$\mathbf{W} = -11.8\text{ N}$$

The negative sign indicates that the weight acts downward.

Motion and Inertia

Aristotle developed ideas of motion with no apparent thought to friction. This led him to the idea that an object moving in a horizontal direction needed the aid of some sort of push to keep it going. Roll a pencil across a tabletop; it slows to a stop. He also noticed heavy things, in general, do fall faster than lighter ones. Feathers fall more slowly than stones. His laws were based on observation of everyday events.

Medieval scientists were able to be more abstract. They could visualize a system with no friction. They dreamed up the idea of impetus. A moving body would continue to move in the absence of an applied force because it had a property called *impetus*.

In 1586, Simon Stevinius performed an experiment that showed falling bodies of the same weight had the same motion, contrary to Aristotle's rules of motion.

Very soon Galileo performed several experiments with motion. He slowed down the "falling" motion by rolling objects down an inclined plane. (Incidentally, the records kept by Galileo are located at Pisa. They show no evidence that he actually performed the famous cannonball experiment on the leaning tower.)

Galileo's experiments convinced him the following things are true.

1. Any object moving along a horizontal plane would continue at a constant speed without the help of an external force. Objects had impetus.
2. Objects with different weights and different materials will all fall the same in a vacuum.
3. Bodies rolling down an inclined plane undergo uniform acceleration. The distance traveled is proportional to time squared (t^2).

Isaac Newton expanded Galileo's work on motion. He replaced the idea of impetus with the more general idea of inertia. Impetus kept moving objects in motion. Inertia was a property of both moving objects and objects at rest. Inertia resisted the change of motion no matter what the object's state of motion might be. Newton then drew up mathematical rules to show how force changed motion and how forces interacted.

EXAMPLE PROBLEM 7–2: THE DERRICK

The unaccelerated weight of the bucket of concrete is 1800 lb. The acceleration is upward at 6.0 ft/s². Find the tension in the cable.

SOLUTION

First find the mass of the bucket and its contents.

$$m = \frac{\mathbf{W}}{\mathbf{g}}$$

$$m = \frac{1800 \text{ lb}}{32 \text{ ft/s}^2}$$

$$m = 56.3 \text{ slug}$$

Use Equation 7–1 to find the tension in the cable.

$$\mathbf{T} = m(\mathbf{a} - \mathbf{g})$$

$$\mathbf{T} = 56.3 \text{ slug } [6.0 - (-32.0)] \text{ ft/s}^2$$

$$\mathbf{T} = 2100 \text{ lb}$$

7.2 INCLINED PLANES

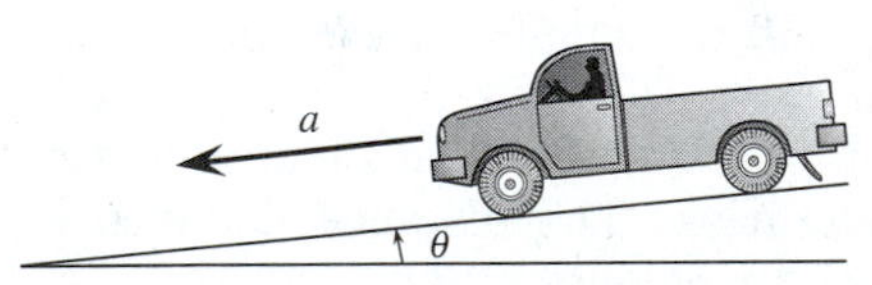

Figure 7–5: A truck accelerates down an incline. (The angle is exaggerated.) a, acceleration.

In Chapter 6 we looked at some problems with horizontal motion. In the first section of this chapter we tackled vertical motion. Now we need to look at something in between—motion along an incline. We will analyze a specific problem to find a strategy for solving this kind of problem.

A pickup truck is parked on a hill with a grade of 8.8% (the ratio of rise: run is 8.8:100, or tan θ is 0.088, where θ is the angle of the incline). The truck rolls down the hill. Neglecting friction, what is the acceleration of the truck?

First, draw the forces on the truck. Gravity will act toward the center of the earth, and the pavement will push upward **normal,** or perpendicular, to the surface of the hill. The normal force will balance some of the weight. Figure 7–6 shows the forces acting on the truck. Figure 7–7 is a simplified diagram showing only the two vectors. The vector sum of these two forces must be the net force pushing the truck down the incline. This net force must be parallel to the incline because we know the net force is in the same direction as the resultant acceleration.

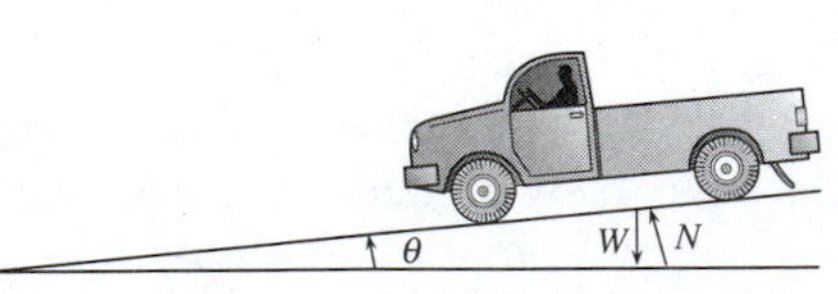

Figure 7–6: Two forces act on the pickup truck.

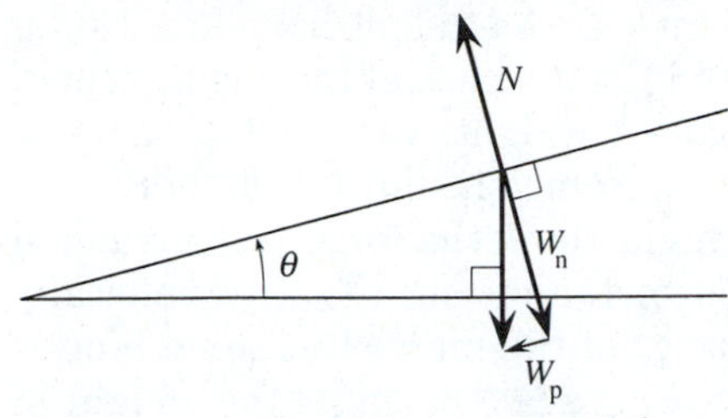

Figure 7–7: *A simplified version of the forces on the truck.* ***N,*** *normal force;* ***W,*** *weight.*

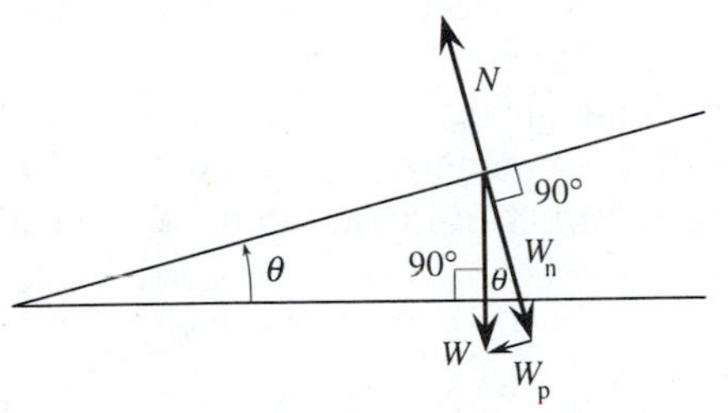

Figure 7–8: *Resolution of the weight into parallel and perpendicular components to the incline.*

To solve inclined plane problems, convert the forces into components parallel and perpendicular to the incline. The perpendicular forces will usually add to zero. Only the forces parallel to the incline will cause acceleration.

The angle between the weight vector and the normal force is the same angle as the incline. According to Euclidean geometry, if both rays of two angles cross at right angles, the two angles must be the same (see Figure 7–8). The normal component of weight is

$$W_n = |\mathbf{W}|\cos\theta$$

The component of weight parallel to the incline is

$$W_p = |\mathbf{W}|\sin\theta$$

where θ is the angle of the incline measured from the horizontal. Now we can determine the acceleration of the truck.

□ **EXAMPLE PROBLEM 7–3: THE ROLLING TRUCK**

A truck rolls down an 8.8% grade. What is the acceleration?

■ ***SOLUTION***

Find the angle of the incline.

$$\tan\theta = 0.088$$
$$\theta = 5.0°$$

The parallel component along the incline (W_p) will cause the acceleration.

$$W_p = m\,a$$

In terms of the weight of the truck, the parallel component is also

$$W_p = m\,g\sin\theta$$

Combine the two equations.

$$m\,a = m\,g\sin\theta$$
$$a = 9.8\ \text{m/s}^2 \times \sin 5.0°$$
$$a = 0.85\ \text{m/s}^2$$

Let us extend Example Problem 7–3. A student holding a 1.0-kg mass hung on a spring scale is riding in the back of the pickup truck. In the accelerated elevator apparent weight was not the same as the weight of the mass at rest. The same should be true here.

Figure 7–9 indicates how we might think the forces are arranged on the 1.0-kg mass. The tension in the spring (**T**) acts vertically upward to balance the weight (m **g**) of the mass. But something is wrong here. The tension in the spring scale balances the weight of the mass, giving a vector sum of zero for the net force. If the net force on the 1.0-kg mass were zero, the mass would not be accelerated down the hill with the spring, the student, and the truck. The diagram in Figure 7–9 must be wrong. One of the forces must act at an angle to give a component of force parallel to the incline. The pull of gravity cannot change; it acts vertically. The spring scale must pull at an angle to supply the needed parallel component (see Figure 7–10). The vector sum of the spring tension acting at an angle and m **g** gives the correct acceleration.

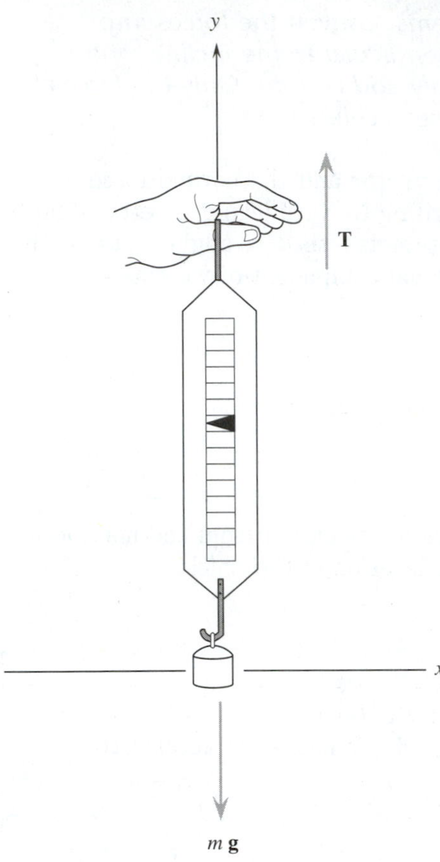

Figure 7–9: An improper view of the forces on the accelerated mass. ***T****, tension;* m ***g****, weight.*

$$\mathbf{T} + (m\,\mathbf{g}) = m\,\mathbf{a}$$

The equivalent weight ($\mathbf{W}_a$) read by the spring scale is in the opposite direction to **T**.

$$-\mathbf{W}_a + (m\,\mathbf{g}) = m\,\mathbf{a}$$

or

$$\mathbf{W}_a = m\,(\mathbf{g} - \mathbf{a})$$

We have already seen this vector equation in Section 7.1 for vertical motion. Equivalence applies to any accelerated system. To find the equivalent weight of an object, subtract the net force from the gravitational force using vectors. Figure 7–11 shows vector diagrams used to calculate the equivalent weight in three situations.

If a mug of coffee is accelerated, the liquid arranges itself so that the surface is normal to the equivalent weight vector (see Figure 7–12). The change in height (Δh) at the edge of the container can be magnified by pouring the fluid into a shallow container.

A liquid will seek a surface normal to the equivalent weight vector that causes the fluid to slosh back and forth in a partially filled container. For example, two rectangular gas cans are in the back of a pickup truck. One is completely filled; the other is two-thirds full. At rest, the full can will tip over more easily because it has a higher center of mass. If the truck accelerates unevenly, however, the partially filled can will probably tip over first because of the liquid's motion.

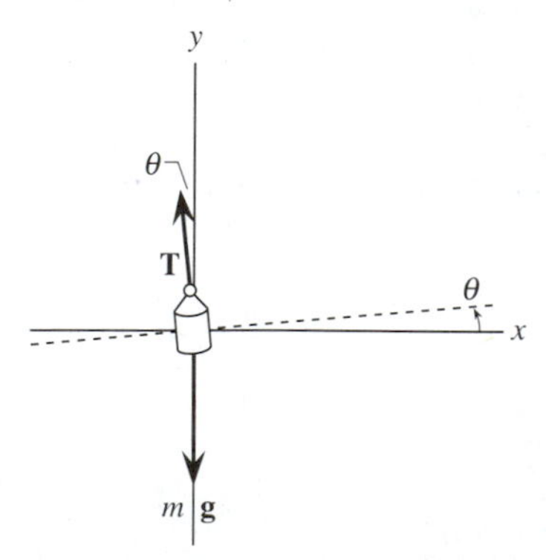

Figure 7–10: The correct view of the forces on the accelerated mass.

Here are two more examples of acceleration along an incline.

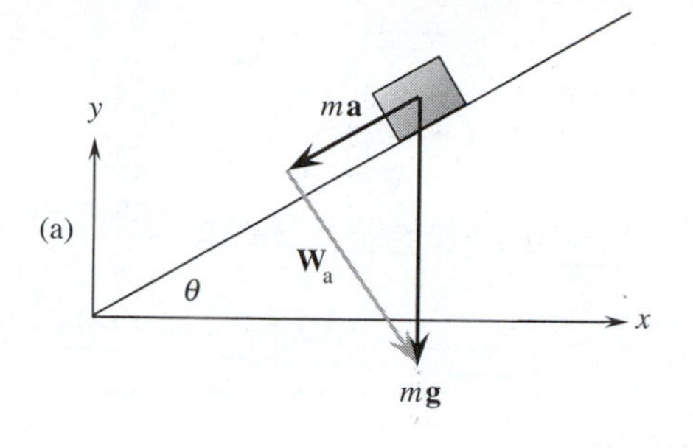

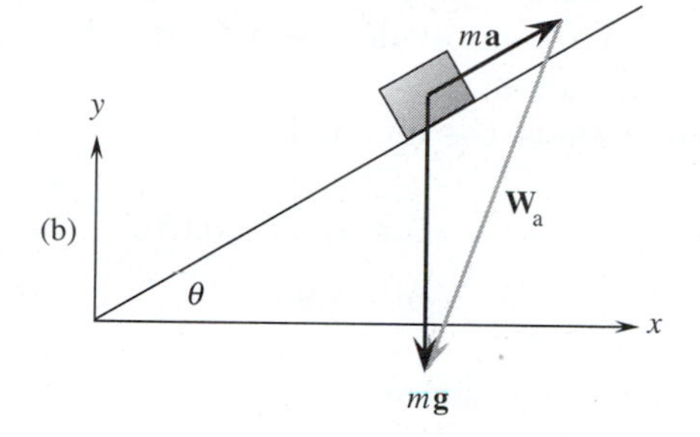

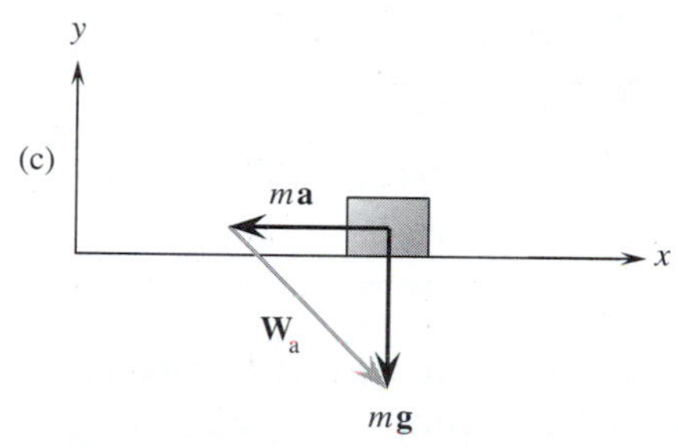

Figure 7–11: *Vector diagrams for calculating apparent weight* (**W_a**)*. (a) Acceleration down an incline. (b) Acceleration up an incline. (c) Horizontal acceleration.*

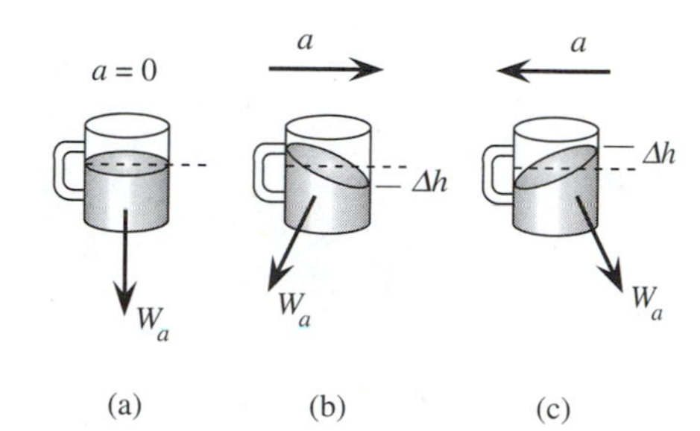

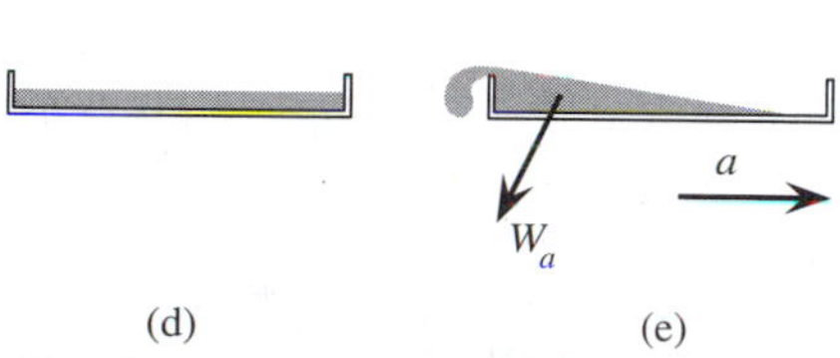

Figure 7–12: *(a – c) The coffee in a mug aligns itself with the surface perpendicular to the equivalent weight vector. Δh, change in height of the level of the liquid. (d – e) In a shallow pan, a small acceleration will cause the liquid to spill over the edge.*

☐ **EXAMPLE PROBLEM 7–4: WORK SHIFT IN THE MINES**

At the end of the workday, miners pile into two mining cars pulled along the rails by a small diesel engine. The engine accelerates at 5.0 ft/s^2, pulling the 260-slug train up a 7.0% grade as it begins its journey to the elevators. With what force (**F**) must the track push against the drive wheels of the engine if the frictional drag (**f**) is 300 lb?

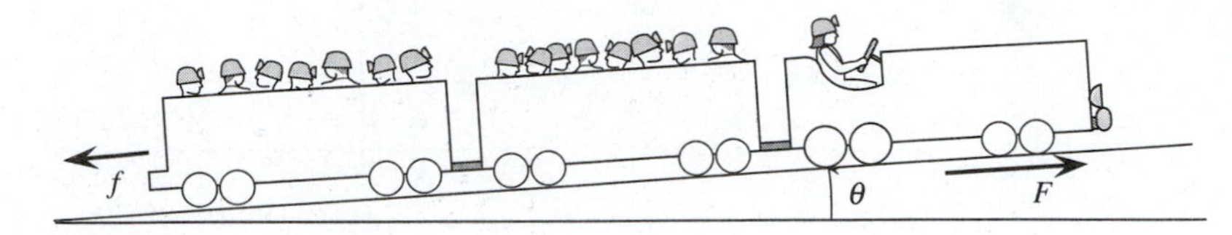

Figure 7–13: A load of miners accelerate up an incline.

■ *SOLUTION*

The strategy, of course, is to find the net force parallel to the incline. We then divide the net force by the system's mass to find the acceleration.

The angle of the incline is

$$\theta = \tan^{-1}(0.070)$$
$$\theta = 4.0^\circ$$

Take up as positive.

$$\mathbf{F} - \mathbf{f} - (m\,\mathbf{g}\sin\theta) = m\,\mathbf{a}$$
$$\mathbf{F} = \mathbf{f} + m(\mathbf{g}\sin\theta + \mathbf{a})$$
$$\mathbf{F} = 300 \text{ lb} + 260 \text{ slug}\,(32\sin 4 + 5)\text{ ft/s}^2$$
$$\mathbf{F} = 2.2 \times 10^3 \text{ lb}$$

Here is a more complicated problem with more than one force not parallel to the incline.

□ **EXAMPLE PROBLEM 7–5: THE WATER-SKIER**

A 120-lb water-skier accelerates up a 20° ramp at 3.0 ft/s^2. If the frictional drag of the skis along the incline is 20 lb, what is the tension in the horizontal tow rope?

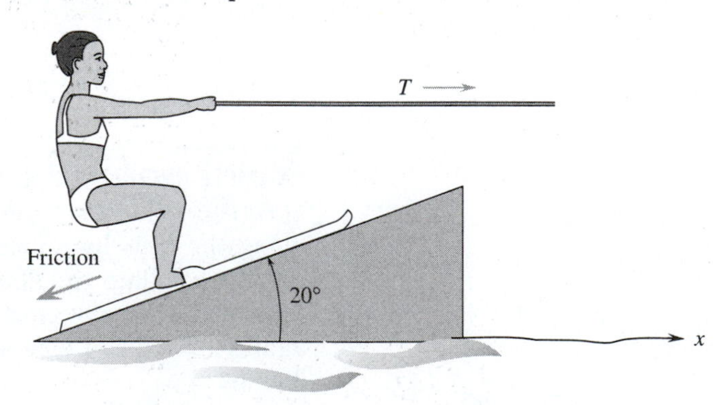

Figure 7–14: A water-skier accelerates up a ramp.

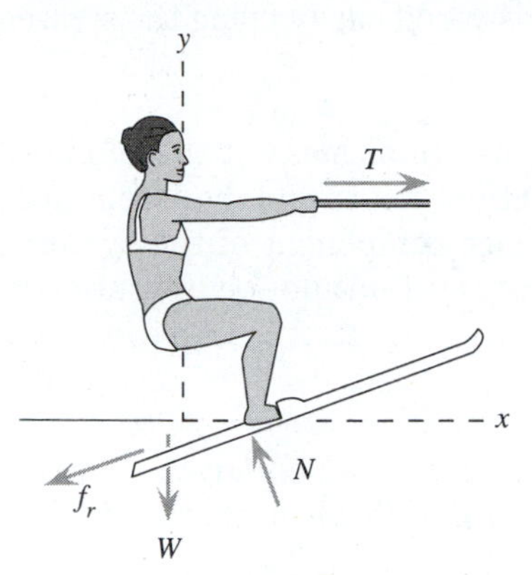

Figure 7–15: *The forces on the water-skier.*

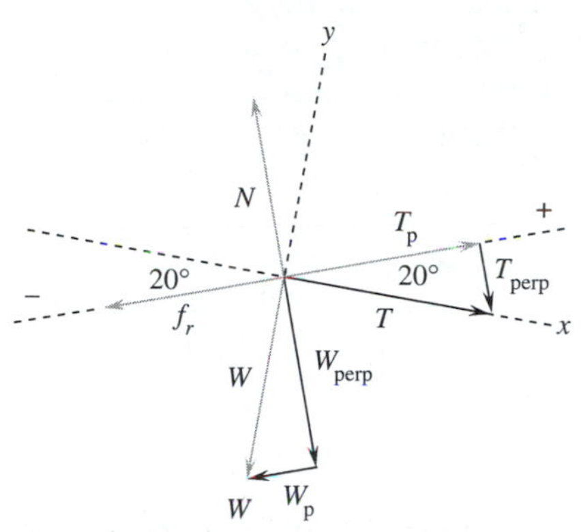

Figure 7–16: *Resolution of forces into components parallel and perpendicular to an inclined plane.*

■ ***SOLUTION***

Look at the forces on the skier (see Figure 7–15). The frictional drag ($\mathbf{f}_r$) is down along the incline. The weight (**W**) is vertical, the tension in the tow rope (**T**) pulls forward horizontally, and the normal force (**N**) prevents the skier from falling through the ramp.

We will use the standard strategy for inclined planes: break the forces into components parallel and perpendicular to the incline. Take up as the positive direction. Figure 7–16 shows that there are three forces along the incline.

$$\mathbf{T}_{\text{para}} - \mathbf{f}_r - \mathbf{W}_{\text{para}} = m\,\mathbf{a}$$

or

$$(\mathbf{T}\cos\theta) - \mathbf{f}_r - (\mathbf{W}\sin\theta) = m\,\mathbf{a}$$

Solve the equation for **T**.

$$\mathbf{T} = \frac{(m\,\mathbf{a}) + \mathbf{f}_r + (\mathbf{W}\sin 20°)}{\cos 20°}$$

The mass of the skier is $m = \mathbf{W}/\mathbf{g}$.

$$m = 3.75 \text{ slug}$$

Substitute the values into the equation.

$$\mathbf{T} = \frac{[(3.75 \text{ slug})(3.0 \text{ ft/s}^2)] + 20 \text{ lb} + [(120 \text{ lb})(0.342)]}{0.949}$$

$$\mathbf{T} = 77 \text{ lb}$$

7.3 INTERNAL FORCES ON COMPOSITE SYSTEMS

A cable elevator has a counterweight to balance the weight of the elevator car. As the elevator moves up, the counterweight moves down. An engineer wants to know what tensions the cables must withstand. A railroad engine pulls 27 cars behind it. The couplings must be designed to withstand the forces involved. In the power train of a large truck, the universal joint couples the drive shaft with the output shaft of the transmission. The joint must be constructed to stand up to the load.

Usually the external forces acting on a composite system are known. Questions concerning the internal forces acting on one of the components of the system can be answered easily using the strategy below.

1. If the acceleration is unknown, use total mass of the composite system and the external forces acting on it to find the acceleration.
2. Draw a free-body diagram for each component of the system.
3. Treat each component as a separate system and solve for unknown forces.

Hint: *In some systems different parts of the system may move in different directions. An elevator car and its counterweight are an example. Use either clockwise or counterclockwise motion to define the positive direction.*

☐ **EXAMPLE PROBLEM 7–6: THE CART**

A string is looped over a frictionless plastic pulley. One end of the string is connected to a 1.2-kg cart resting on a horizontal tabletop. On the other end of the string, a 300-g mass is hung. The frictional drag on the cart is 2.0 N. What is the tension on the string?

■ ***SOLUTION***

Find the acceleration of the entire system first. Choose clockwise as the positive direction. The mass of the assembly is $m_1 + m_2$.

$$F_{net} = (m_1 g) - f_r = (m_1 + m_2) a$$

The tension in the string is an internal force holding the two parts of the system together. Tension will not appear in the equation of the composite system. Solve the equation for a.

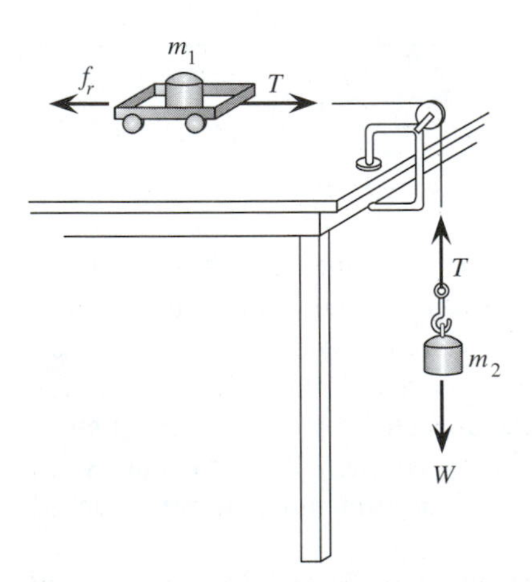

Figure 7–17: A hung mass pulls a cart along the tabletop.

$$a = \frac{(m_1 g) - f_r}{m_1 + m_2}$$

Substitute the values into the equation.

$$a = \frac{[(0.300 \text{ kg})(9.8 \text{ m/s}^2)] - 2.0 \text{ N}}{0.300 \text{ kg} + 1.2 \text{ kg}}$$

$$a = 0.63 \text{ m/s}^2$$

Now we can find the tension by looking at the forces on one of the parts of the system. It does not matter which part of the system we choose.

For the cart:

$$T - f_r = m_2 a$$

$$T = (m_2 a) + f_r$$

$$T = [(1.2 \text{ kg})(0.63 \text{ m/s}^2)] + 2.0 \text{ N}$$

$$T = 2.8 \text{ N}$$

As a check on our calculations find the tension in the string attached to the hung weight.

For the hung weight:

$$(m_1 g) - T = m_1 a$$

$$T = m_1 (g - a)$$

$$T = 0.300 \text{ kg } (9.8 - 0.63) \text{ m/s}^2$$

$$T = 2.8 \text{ N}$$

In Example Problem 7–6 the pulley was not considered part of the system. It was used as a device to convert a vertical force into a horizontal force without changing its magnitude. Bearings have frictional resistance. Wheels have inertia. A force is needed to make a wheel spin faster. The effect of a lightweight pulley with low friction will probably be small compared with the two-figure accuracy of the calculation. In this case it was all right to ignore the pulley. If a heavy pulley with worn bearings had been used, however, the experimental results would have been substantially different from our prediction. We would have had to resort to a more complicated analysis including the forces on the pulley. Remember, calculations are merely estimates of what is likely to happen under a certain set of conditions. Ignoring small factors helps to simplify calculations. Such shortcuts are fine as long as the predictions are satisfactory.

In some two-body systems—e.g., a counterweighted elevator—we would like to have friction in the pulley. We want to control the motion by a motor rather than let gravity control it. In the next example problem the pulley is replaced by a winch.

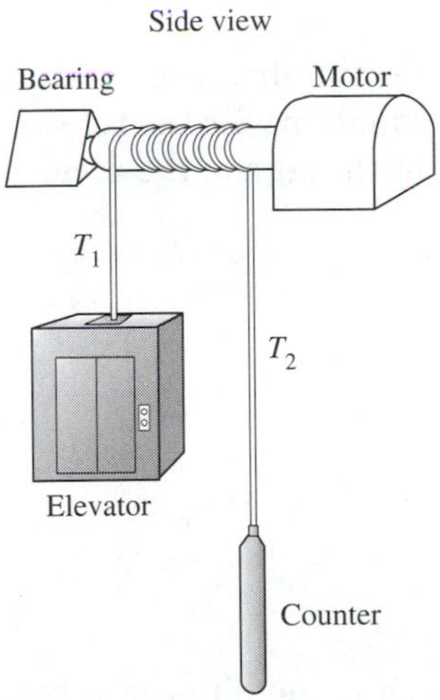

Figure 7–18: *(Side view) An elevator with a countermass.*

EXAMPLE PROBLEM 7–7: AN ELEVATOR WITH A COUNTERMASS

A loaded cable elevator has a mass of 2920 kg. At the top of the shaft, the steel cable winds a few times around the drum of a winch. From the free end a 2150-kg countermass is hung (see Figure 7–18). The frictional forces are high enough to prevent the cable from slipping on the drum. Tension $\mathbf{T}_1$ is the tension in the cable connected to the elevator car; $\mathbf{T}_2$ is the tension in the cable connected to the countermass. Because there is a great deal of friction, the two tensions will be different.

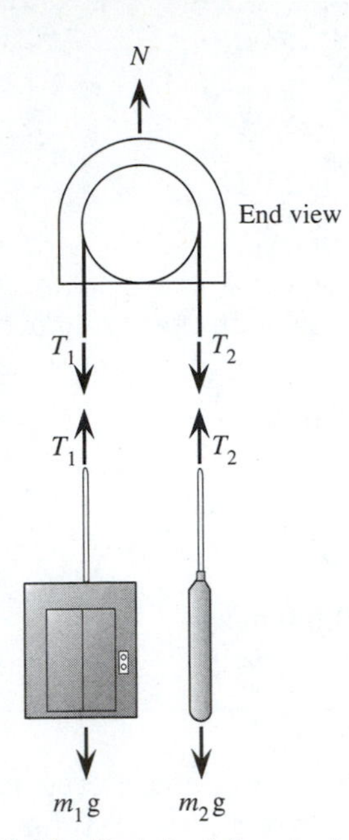

Figure 7–19: *(End view) Free-body diagrams of the elevator system.*

A. What are $\mathbf{T}_1$ and $\mathbf{T}_2$ when the elevator accelerates upward at 1.5 m/s^2?

B. What are the two tensions when the elevator descends at 1.5 m/s^2?

C. What net driving force must the winch exert on the cables when the elevator accelerates upward? When the elevator accelerates downward?

■ ***SOLUTION***

Figure 7–19 shows the free-body diagrams for the three parts of the system: elevator, countermass, and winch. Choose forces that cause a counterclockwise rotation of the drum to be positive.

A. Write Newton's second law equations for the elevator and for the countermass. Notice the acceleration causes a clockwise motion of the drum.

$$(m_1\,\mathrm{g}) - \mathrm{T}_1 = m_1\,(-\mathrm{a}) \qquad \text{(elevator)}$$

$$\mathrm{T}_2 - (m_2\,\mathrm{g}) = m_2\,(-\mathrm{a}) \qquad \text{(countermass)}$$

Solve the equations for the tensions.

$$\mathrm{T}_1 = m_1\,(\mathrm{g} + \mathrm{a}) \qquad \text{(elevator)}$$

$$\mathrm{T}_2 = m_2\,(\mathrm{g} - \mathrm{a}) \qquad \text{(countermass)}$$

$$\mathrm{T}_1 = 2920\ \mathrm{kg}\ (9.8 + 1.5)\ \mathrm{m/s^2}$$

$$\mathrm{T}_1 = 3.3 \times 10^4\ \mathrm{N}$$

$$\mathrm{T}_2 = 2150\ \mathrm{kg}\ (9.8 - 1.5)\ \mathrm{m/s^2}$$

$$\mathrm{T}_2 = 1.8 \times 10^4\ \mathrm{N}$$

B. In Part B, the acceleration causes the drum to rotate counterclockwise (+ direction). The equations for the tensions are the same as those in Part A except the sign of the acceleration is changed.

$$\mathrm{T}_1 = m_1\,(\mathrm{g} - \mathrm{a})$$

and

$$\mathrm{T}_2 = m_2\,(\mathrm{g} + \mathrm{a})$$

$$\mathrm{T}_1 = 2.4 \times 10^4\ \mathrm{N}$$

$$\mathrm{T}_2 = 2.4 \times 10^4\ \mathrm{N}$$

C. The driving force ($\mathbf{F}_d$) of the winch is the difference between the two tensions ($\mathrm{T}_2 - \mathrm{T}_1 = [m_1 + m_2]\,\mathrm{a}$). It must be negative (clockwise) to lift the elevator.

For upward acceleration

$$-F_d = T_1 - T_2$$

$$F_d = -(3.3 \times 10^4 \text{ N}) - (1.8 \times 10^4 \text{ N})$$

$$F_d = -1.5 \times 10^4 \text{ N}$$

For downward acceleration

$$F_d = (2.4 \times 10^4 \text{ N}) - (2.4 \times 10^4 \text{ N})$$

$$F_d = 0.0$$

The extra weight of the elevator is just enough to cause the 1.5 m/s^2 acceleration.

☐ EXAMPLE PROBLEM 7–8: THE HOIST

A cable weighing 0.30 lb/ft is used by a hoist to lift a 280-lb load. For an upward acceleration of 6.0 ft/s^2, what is the tension in the cable just above the load? At 120 ft above the load?

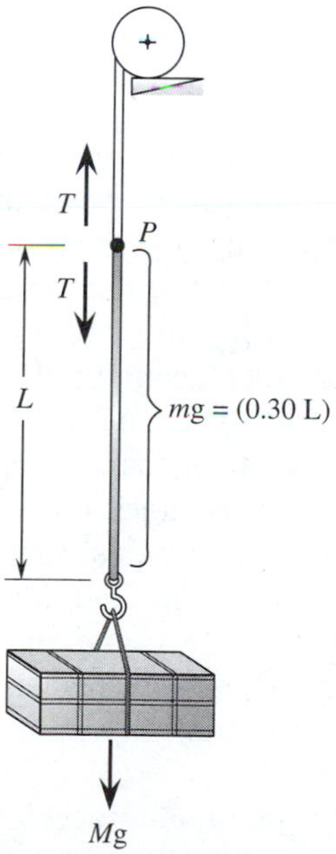

■ *SOLUTION*

Quite often the mass of a supporting cable is assumed to be small and is ignored in finding the tension in a system. For a heavy cable, the strategy of finding internal forces can be used to find the tension anywhere along the cable if we know how much the cable weighs per unit length.

At any point on the cable, the weight of the load plus the weight of cable below the point must be supported by the tension (see Figure 7–20). For a length L, the weight of the cable is $W_c = 0.30\,L$ lb/ft. Take upward as positive.

The mass of the load (M) is:

$$M = \frac{W}{g} = \frac{280 \text{ lb}}{32 \text{ ft/s}^2} = 8.75 \text{ slug}$$

The mass (m) of the cable of length L is:

$$m = \frac{(0.30\,L)\text{lb/ft}}{32 \text{ ft/s}^2} = (0.0094\,L) \text{ slug/ft}$$

The tension at point P must be large enough to accelerate both the load and connecting cable. Newton's dynamics equation is

Figure 7–20: *The tension at point P must support the load (M* ***g****) and the weight (m* ***g****) of a length of cable L long.*

tension − load − weight of cable = mass of system × acceleration

$$T - (M\,g) - (m\,g) = (M + m)\,a$$

$$T = [(M + m)\,g] + [(M + m)\,a] = (M + m)(g + a)$$

or

$$T = [(8.75 \text{ slug}) + (0.0094\,L) \text{ slug/ft})]\,(32 + 6.0)\text{ft/s}^2$$

The mass of the cable below the point of interest is simply added to the system. At the point just above the load $L = 0$.

$$T = (8.75 \text{ slug})(38 \text{ ft/s}^2)$$

$$T = 330 \text{ lb}$$

At a point 120 ft above the load, $m = 0.0094$ slug/ft × 120 ft, or 1.13 slug.

$$T = (8.75 + 1.13)\text{slug} \times 38 \text{ ft/s}^2$$

$$T = 380 \text{ lb}$$

7.4 APPLICATIONS OF ACCELEROMETERS

Aviation

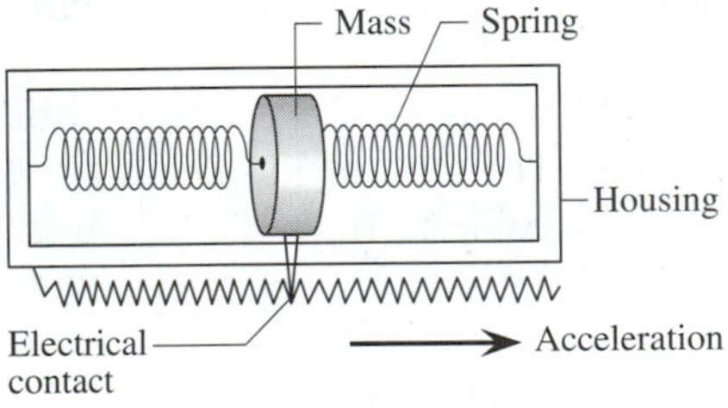

Figure 7–21: *A spring-loaded accelerometer.*

Devices to measure acceleration in airplanes come in a variety of designs. One type, a spring-loaded accelerometer, is shown in Figure 7–21. A relatively heavy mass is supported by low-friction guides. Carefully calibrated springs hold the mass in the center position. When an acceleration occurs, the mass will slide relative to the housing. An electrical contact indicates the displacement of the mass. The displacement in turn is proportional to the acceleration. As an airplane is accelerated forward, the front spring is stretched; the rear one is compressed, indicating a forward acceleration. When the plane is throttled back, the mass is thrown into a position forward of the equilibrium point. At a constant speed, the inertial force of the mass acting on the springs is, of course, zero, and the mass is held in the center position by the springs. This is all consistent with Newton's second law: the net force on a system is proportional to the acceleration.

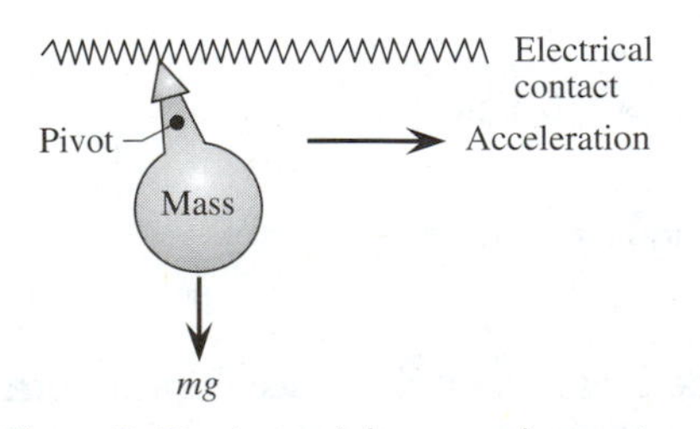

Figure 7–22: *A pendulous accelerometer.*

Another kind of accelerometer is the pendulous accelerometer (Figure 7–22). A mass is mounted off-center from a pivot point. This operates a lot like the spring balance described in Section 7.2. The difference is that the pendulum is mounted horizontally so that the weight of the mass will not affect its motion. Only inertial effects are detected.

Automobiles and Buildings

Accelerometers have other uses than in aviation. Some automobiles now have accelerometers. Information is fed to minicomputers, which adjust the suspension system. Rapidly reacting gas-filled shock absorbers give the car a smoother ride than can be obtained by conventional shocks.

Accelerometers are beginning to be used in tall buildings. An old building such as the Empire State Building is constructed with heavy, cross-braced steel and a great deal of concrete. In a stiff wind, it sways back and forth with an amplitude of about 3 in. More modern structures use half as much steel in the frame to enclose the same amount of volume. Replacing concrete with aluminum and Plexiglas skins saves construction costs and leads to more flexible buildings. The top floors of the World Trade Center have a horizontal motion with an amplitude of about 3 ft on windy days. Office workers have suffered motion sickness because of this. Several techniques have been developed to deal with flexible buildings.

The Citicorp Center located in New York City uses a "tuned mass damper" located on the 59th story of the 960 ft tall building. Sensitive accelerometers detect building motion. If the acceleration exceeds 3 m**g** (three thousandths of the acceleration of gravity), a hydraulic system moves a 400-ton block of concrete at right angles to the motion of the building with a period matching the natural frequency of the building. This reduces the building sway by 45%.

A different technique is being designed at the University of Buffalo. Cables, called tendons, are stretched diagonally across the structure. When motion is sensed by detectors, motors tighten selected tendons and relax others. In this way, the building is braced against motion in the same way that your arm stiffens when you tighten your tendons.

Inertial Guidance Systems

If an accelerometer is tipped, its weight will move the mass, and it will give an erroneous reading of acceleration. This problem occurs in airplanes and rockets. The solution is to place the accelerometer on a table whose surface is always at right angles to the pull of gravity. Such a surface is called an **artificial horizon.**

Place two accelerometers on the artificial horizon at right angles. One measures acceleration north and south; the other, east and west. We have created one component of an inertial guidance system. We also need a clock and a computer. The clock tells the computer how long the vehicle has been accelerated in a particular direction. The computer assumes a constant velocity in the time interval between accelerations. With the measurements of time, velocity, and acceleration, the computer can calculate the horizontal displacement of the vehicle.

However, as the plane moves, the artificial horizon tilts relative to its orientation at the beginning of the trip. If the tilting motion of the artificial horizon can also be recorded by the computer, displacements can be calculated for any place on or near the earth. This problem can be solved by using a gyroscope, the last component needed to make an inertial guidance system. Once a gyroscope starts spinning its spin axis keeps the same orientation in space. The spin axis can be used as a fixed reference direction. In an inertial guidance system, the orientation of the gyroscope is set at the beginning of the trip. Changes of the tilt of the artificial horizon relative to this direction can be continuously fed to the computer. We now have a complete inertial guidance system.

SUMMARY

The effects of acceleration are equivalent to the effects of gravity. In a closed system, these effects cannot be distinguished. The equivalent weight of a mass in an accelerated system is

$$\mathbf{W}_a = m(\mathbf{g} - \mathbf{a})$$

To solve problems involving inclined planes convert forces into components parallel and perpendicular to the plane. Only the parallel components can cause acceleration.

To solve problems involving internal forces, first find the acceleration of the composite system if it is not known. Then break the composite into individual systems to find the internal forces.

KEY TERMS

If you can explain these terms to a friend or classmate, you understand their meaning. If you cannot explain the terms, you should reread the sections in which they are discussed.

apparent weight
artificial horizon
equivalence
free-body diagrams
normal

EXERCISES

Section 7.1:

1. A physics student does not believe in equivalence. The student carries a 1.0-kg mass, a spring scale, a tape measure, a stop watch, and a triple-beam balance into an elevator. The following experiments are performed.
   A. The 1.0-kg mass is hung from the spring scale. The weight is 11.8 N.

B. The mass is measured on the triple-beam balance.
C. The mass is dropped to measure its acceleration as it falls to the floor of the elevator.

All the experiments are performed with the same acceleration of the elevator. What are the results of experiments B and C?

2. A thin cable has just enough strength to suspend a heavy load from it without breaking, if the load is lifted very slowly. However, if the load, initially at rest, is pulled up with a rapid motion the cable will break. Why?

3. A light line will support a maximum load of 140 lb without breaking. Is it possible for a person weighing 160 lb to slide down the rope without breaking it? Explain why or why not.

4. A heavy spring scale is placed between a load and the cable of a hoist. Compare the scale readings of the following situations with the reading when the load is at rest.
A. The load is accelerated upward.
B. The load has a rapid upward speed.
C. The load comes to a stop in its upward motion.
D. The load accelerates downward.

Section 7.2:

5. The gas gauge in an automobile usually uses a float in the gas tank. When the fuel level in the tank is low, the gauge reading changes when the car speeds up or slows down. Why?

6. Sometimes a plane and its pilot undergo accelerations a few times that of gravity. These accelerations are commonly expressed in multiples of **g**. A **g** is the earth's gravitational acceleration. A 2-**g** acceleration is 64 ft/s^2 or 19.6 m/s^2.

As an F-14 fighter accelerates, the pilot experiences a 7-**g** acceleration. Estimate forces caused by acceleration on the following.

A. The force with which the 80-kg pilot is pushed against the seat.
B. The equivalent weight of a 250-kg wing.
C. The mass of the pilot's head is 4.5 kg. With what force will his head be pushed against his neck and shoulders?

Figure 7–23: *A lawnmower is pushed up a hill.*

7. A road grader rests on an incline with an 8.8% grade ($\tan^{-1}[0.088] = 5.0°$). Which of the following is the part of the grader's weight (**W**) supported by the normal force: **(a)** zero, **(b) W**, **(c) W** sin 5°, or **(d) W** cos 5°?

8. A lawnmower is pushed up a hill (see Figure 7–23). Must the normal force on the mower be less than the weight? Explain why or why not.

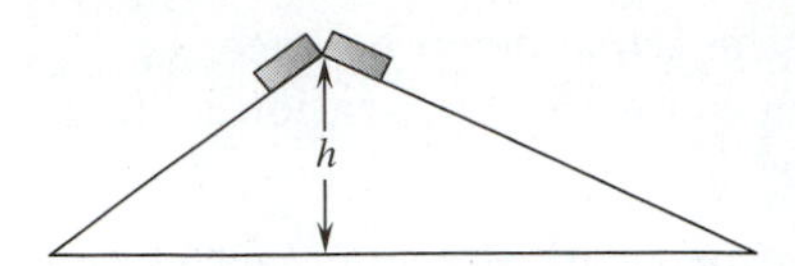

Figure 7–24: Two inclined planes with the same height (h), but different slopes.

9. Two frictionless inclined planes have the same height, but different angles (see Figure 7–24). Which of the following statements is (are) true? Explain why or why not.
   A. A block sliding down the steeper incline will reach the bottom first.
   B. A block sliding down the steeper incline will have a larger acceleration.
   C. A block sliding down the steeper incline will have a larger speed at the bottom.
   D. A more massive block will move down one of the inclines more quickly than a less massive block.

Section 7.3:

10. In Figure 7–17 a hung weight accelerates a cart along the tabletop.
    A. Will the tension in the string be zero if the frictional force is zero?
    B. What maximum tension can the string have? What is the acceleration at maximum tension?

11. In Figure 7–25 two trunks are connected by a rope. Forces $\mathbf{F}_1$ and $\mathbf{F}_2$ pull the trunks in opposite directions. $\mathbf{T}$ is the tension in the rope. The positive direction is to the right, and $\mathbf{F}_1 > \mathbf{F}_2$. Which of the following is true?

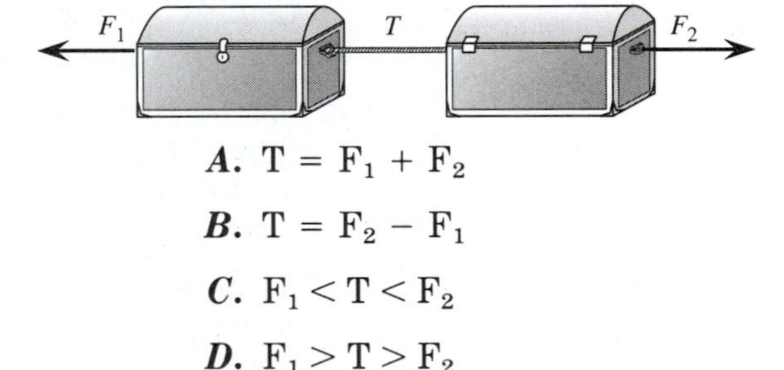

Figure 7–25: Two forces pull in opposite directions on the trunks tied together with a rope.

A. $T = F_1 + F_2$

B. $T = F_2 - F_1$

C. $F_1 < T < F_2$

D. $F_1 > T > F_2$

PROBLEMS

Section 7.2:

1. A 1.0-slug mass undergoes vertical motion. Calculate the apparent weight of the mass in each case.
   A. Upward at 4.0 ft/s.
   B. Upward at 4.0 ft/s^2.
   C. Downward at 4.0 ft/s^2.

2. A 1.0-kg mass is hung from a spring balance. Find the direction and magnitude of the vertical accelerations corresponding to the following readings of the scale.
   A. 9.8 N
   B. 10.8 N
   C. 7.6 N

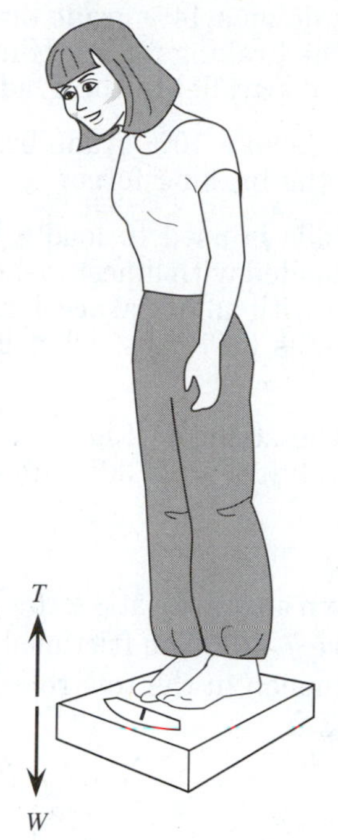

Figure 7–26: Measuring weight with a bathroom scale.

3. A woman who normally has a weight of 490 N weighs herself in an elevator on a bathroom scale. What is the reading of the scale under the following conditions?
   A. The elevator accelerates upward at 2.0 m/s^2.
   B. The acceleration is 2.0 m/s^2 downward.
   C. The elevator moves down with a speed of 2.0 m/s.

4. A rope is made from towels and bed linen to escape from the third story of a burning building. The rope will support a maximum of 670 N without breaking. With what minimum acceleration must an 800-N man slide down the rope so it will not break?

5. What thrust must a 27.0-ton rocket have to give it an acceleration of 5.0 ft/s^2?

6. A 2700-lb slab of marble is lifted vertically out of a rock quarry by a winch. The operator applies tension to the cable rapidly, causing the rock to reach a velocity of 2.6 ft/s in the first 0.80 ft of height. What is the tension in the cable just above the rock?

Section 7.2:

7. A 75-kg skier swoops down a 45° slope. The frictional drag is 400 N. What is the magnitude of the acceleration of the skier?

8. Find the normal and parallel components of a 10-lb weight:
   A. on a 10° incline
   B. on a 45° incline
   C. on a 60° incline
   D. on a 90° incline

9. Find the magnitude of acceleration of a block sliding down a frictionless incline:
   A. with an angle of 30°.
   B. with an angle of 127°.
   C. with an angle of 53°.

10. What would be the angle (θ) of the string in Figure 7–28 if the system is accelerating horizontally at 5.0 m/s^2?

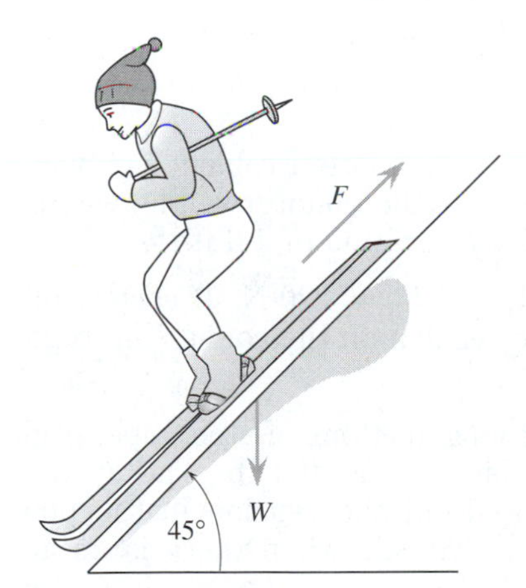

Figure 7–27: Accelerating down a ski slope.

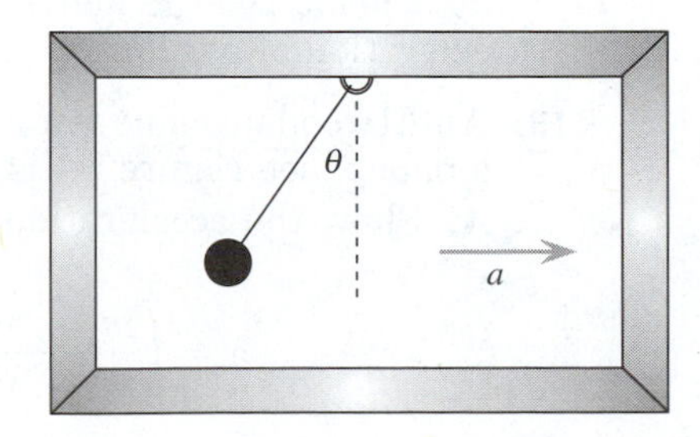

Figure 7–28: A bob on a string is used to find the acceleration.

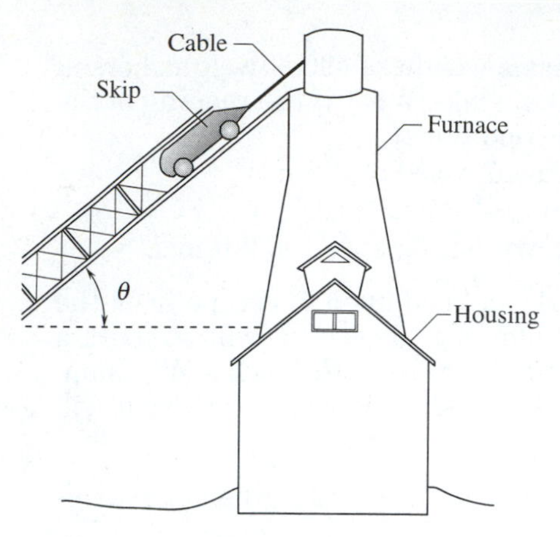

Figure 7–29: A skip is pulled up an inclined track to load a blast furnace.

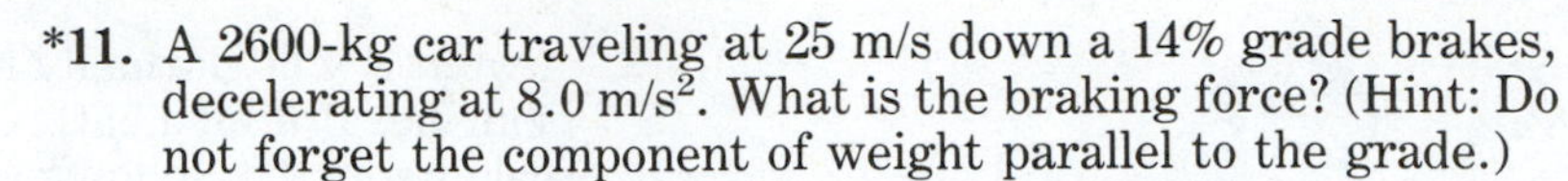

***11.** A 2600-kg car traveling at 25 m/s down a 14% grade brakes, decelerating at 8.0 m/s². What is the braking force? (Hint: Do not forget the component of weight parallel to the grade.)

***12.** A 5600-lb car traveling at 80 ft/s down a 10% grade brakes, coming to rest in 350 ft. What is the braking force?

***13.** A cable car (a skip) guided by rails is used to load a blast furnace (see Figure 7–29). A skip loaded with limestone, coke, and iron ore starts up a 60° incline with an initial acceleration of 2.5 ft/s². If the loaded car weighs 0.87 tons, what is the tension on the cable?

****14.** What would be the angle (θ) of the string in Figure 7–28 if the system were accelerating at 5.0 ft/s² up a 30° hill?

Section 7.3:

15. A 1600-lb airplane accelerates down a runway at 3.2 ft/s² with an 800-lb glider in tow (see Figure 7–30). The frictional drag on the glider is 40 lb. Find the tension in the tow rope.

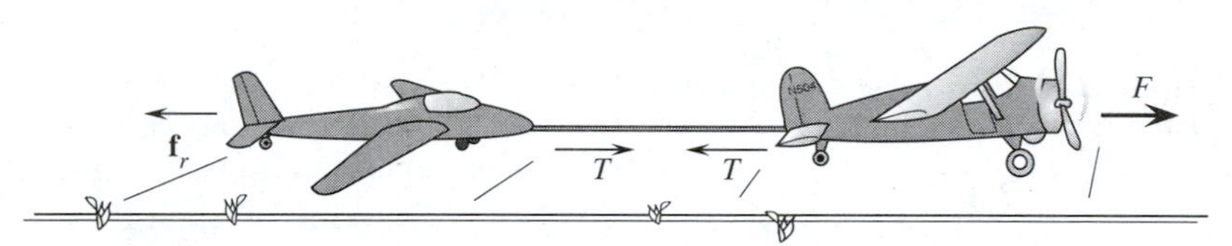

Figure 7–30: An airplane tows a glider along a runway.

16. Find the tensions $\mathbf{T}_1$ and $\mathbf{T}_2$ for Example Problem 7–7 when the elevator weighs 6400 lb and the counterweight weighs 5600 lb. Assume an upward acceleration of 4.6 ft/s².

17. In Figure 7–25, $\mathbf{F}_1 = 350$ N, and $\mathbf{F}_2 = 220$ N. If each trunk has a mass of 18 kg, find the tension on the connecting rope. (Ignore friction.)

***18.** An oil-drilling rig uses 40 ft long sections of steel pipe, each weighing 300 lb. The drill bit weighs 400 lb. The system is hauled up to change bits. Find the tension in the pipe at a point 2000 ft above the drill bit when there is an acceleration of 5.0 ft/s².

***19.** An Atwood machine is a weight and counterweight hung over a pulley (see Figure 7–31). Assume the pulley has no friction.
A. Show the acceleration of the system is:

$$a = \left(\frac{m_1 - m_2}{m_1 + m_2}\right) g$$

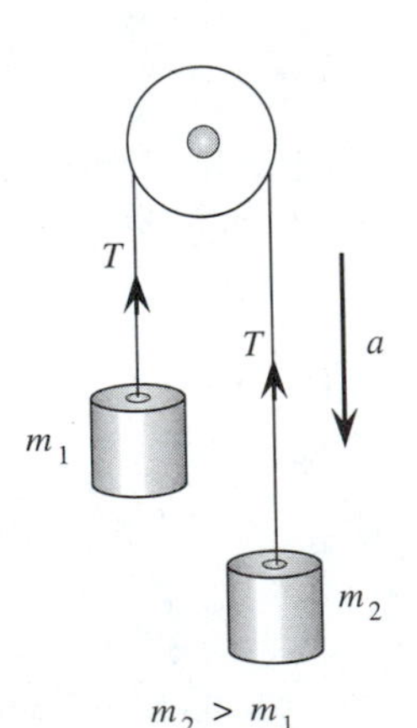

Figure 7–31: An Atwood machine.

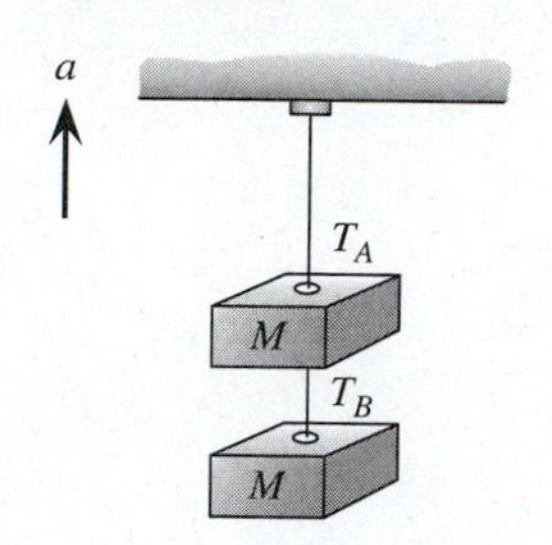

Figure 7–32: *Two blocks are hung from the ceiling of an accelerating elevator.*

B. Find the acceleration and the tension in the string when $m_1 = 10\bar{0}0$ g and $m_2 = 11\bar{0}0$ g.

***20.** Two 5.0-kg blocks are hung in tandem from the ceiling of an elevator (see Figure 7–32). What are the tensions $\mathbf{T_A}$ and $\mathbf{T_B}$ on the strings if there is an upward acceleration of 2.0 m/s^2?

Chapter 8

KINDS OF FORCES

Imagine that you could bore a hole through the center of the earth and jump through it. How would the earth's gravitational forces affect your voyage? (Photograph courtesy of NASA.)

OBJECTIVES

In this chapter you will learn:

- how static and sliding frictional forces behave
- to solve problems involving frictional forces
- to solve problems involving viscous friction
- to apply Hooke's law to forces directly proportional to distance
- to solve problems involving universal gravitation
- to recognize the kinds of forces that occur in everyday situations

Time Line for Chapter 8

1686 Isaac Newton publishes the *Principia, De Motu,* the motion of bodies.

1638 Galileo describes the effects of friction on motion.

1676 Robert Hooke finds the force acting on a stretched string is proportional to displacement.

1680 Isaac Newton calculates an inverse square law of gravity will give the planets elliptical orbits.

1781 Coulomb's "Theory of Simple Machines" studies the effects of friction.

Prediction may not be the only goal of science, but it is certainly a major one. Engineers designing an automobile need to know how much force an engine must deliver to give a car a desired performance on the road. The designers want to be able to predict the net output of their engine before they build it. The engineer in charge of the valve system needs to understand the forces on the rocker arms and valve stems.

Many forces obey patterns. If we recognize these patterns, we can anticipate the sizes and directions of forces in a variety of situations. This makes it possible to predict the behavior of a system or of a piece of machinery before it is fabricated. This chapter explores the predictable patterns of some kinds of forces.

8.1 FORCES PROPORTIONAL TO THE NORMAL FORCE

Lay a book on the table in front of you. Push gently on the book with your fingertips, increasing the force until the book moves. Lay a second book on top of the first. Repeat the experiment. Try three or four books. Notice that as the number of books in the stack increases, the force required to start the books in motion increases. Here are a couple of variations on the experiment.

See if it takes more force to start the books in motion than to keep them sliding across the table. If the bottom book is a hard cover, replace it with a paperback. See if this changes the force needed to move the books.

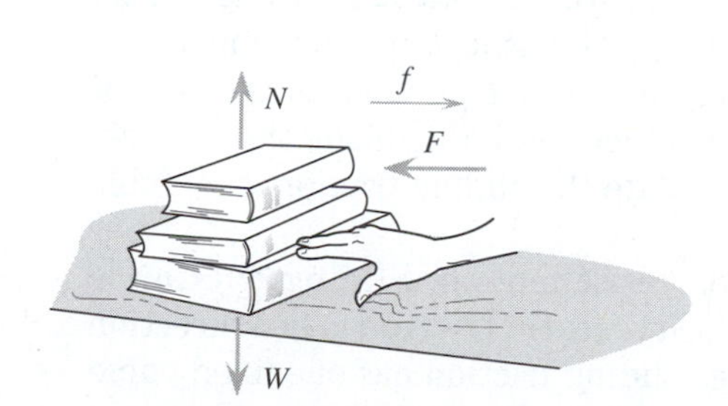

Figure 8–1: *A force is needed to start a stack of books in motion.*

We should be able to draw some conclusions about the forces acting on surfaces in contact. When we push a pile of books along a surface we notice several things. These observations are generally true for any two surfaces in contact.

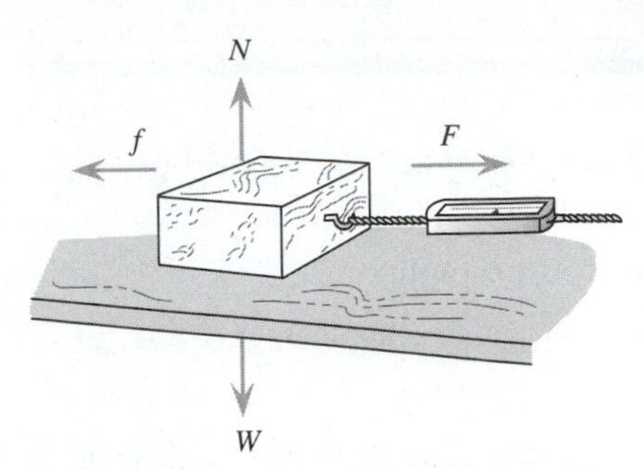

Figure 8–2: Spring scales can be used to measure the resisting frictional force.

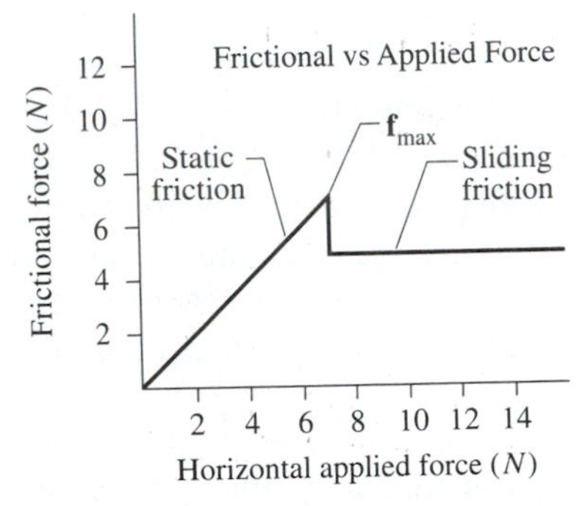

Figure 8–3: Friction versus applied force. A smaller force ($\mathbf{f}_{sliding}$) is needed to maintain an object at a constant speed once the maximum static friction ($\mathbf{f}_{max}$) has been overcome.

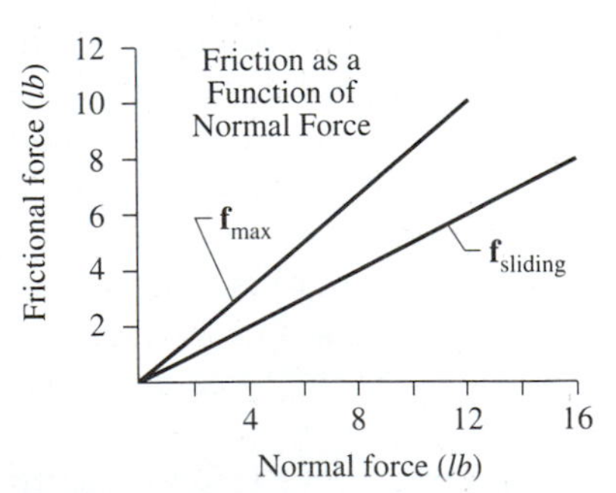

Figure 8–4: Friction as a function of normal force. As the normal force increases, the frictional force increases proportionally.

A. When an object is at rest, the applied force is balanced by some resisting force up to some maximum value. If the maximum resisting force is exceeded, motion will occur.

B. The maximum resisting force needed to move a stationary object increases as the force pushing the two surfaces together (normal force) increases.

C. The maximum resisting force depends on the nature of the surfaces. The roughness of the surfaces, the kind of materials used, temperature and lubricants may change the resisting force.

D. Once the object starts to slide along the surface, the resisting force decreases.

E. The area of contact of the two surfaces has little effect on the resisting forces.

In the laboratory, spring scales can be used to quantify the relationship between the normal force on a stationary object and the resisting force necessary to move it (see Figure 8–2). A block of wood is pulled along a horizontal surface by a spring scale. The resisting force (friction) balances the pull of the string up to some maximum force ($\mathbf{f}_{max}$). When this force is exceeded, the block begins to slide. A smaller force ($\mathbf{f}_{sliding}$) is needed to maintain the block at a constant speed (Figure 8–3). When the normal force pushing the two surfaces together increases, both $\mathbf{f}_{max}$ and $\mathbf{f}_{sliding}$ will increase. Figure 8–4 shows the relationship between the normal force and the frictional force.

Both the maximum static friction and the sliding friction are directly proportional to the normal force. The slope of a friction **(f)** versus normal force plot is called the **coefficient of friction** (μ).

$$\mathbf{f} = \mu \mathbf{N} \qquad \textbf{(Eq. 8–1)}$$

The coefficient of friction takes into consideration the properties of the surface, such as roughness and type of material.

Two coefficients are needed for each pair of surfaces. The static coefficient is used to calculate the maximum static frictional force. If the applied force is less than $\mathbf{f}_{max}$ the object will not move and the friction will be equal to, but opposite in direction to, the sum of the other applied forces. Once the object is sliding along the surface, a second coefficient is used to calculate the sliding friction (see Table 8–1).

Both static and sliding friction resist motion. **Sliding friction** is in a direction opposite to the velocity; **static friction** is in a direction opposite to the net applied force. Sliding friction has one fixed value for a particular combination of materials. Static friction will have a value between zero and $\mathbf{f}_{max}$.

Table 8–1: Coefficients of friction (μ).

MATERIAL	STATIC FRICTION*	SLIDING FRICTION
Rubber on		
dry concrete	1.02	0.80
wet concrete	0.80	0.5–0.7
ice, at 0°C	—	0.10
Steel on steel	0.74	0.57
Soft wood on wood	0.40	0.20
Waxed ski on dry snow at:		
0°C	—	0.04
−10°C	—	0.18
−40°C	—	0.40

* —Data not available.

Equation 8–1 is true for most common situations. However, it is an experimental equation; under extreme conditions it may not be valid. The coefficients assume the surfaces do not penetrate. If rough concrete blocks are dragged along a soft pine board, the coefficients may increase for high normal forces because the blocks begin to gouge the board. In some cases, the area *may* affect the frictional force. An example is ice skates on ice. The frictional force is less for sharper skates. Another example is a knife drawn across a cardboard surface.

□ **EXAMPLE PROBLEM 8–1: MOVING A CRATE**

The static coefficient of friction between a 120-lb crate and a concrete floor is 0.60. The sliding coefficient is 0.4.

A. A lateral force of $6\bar{0}$ lb is applied to the stationary crate. Find its acceleration.

B. What is the acceleration for an $8\bar{0}$-lb horizontal force?

■ ***SOLUTION***

Before calculating the acceleration, check to see if there is enough force to start the crate moving. Since the crate is on a level surface, the normal force is the weight. The maximum static force of friction that can be developed is:

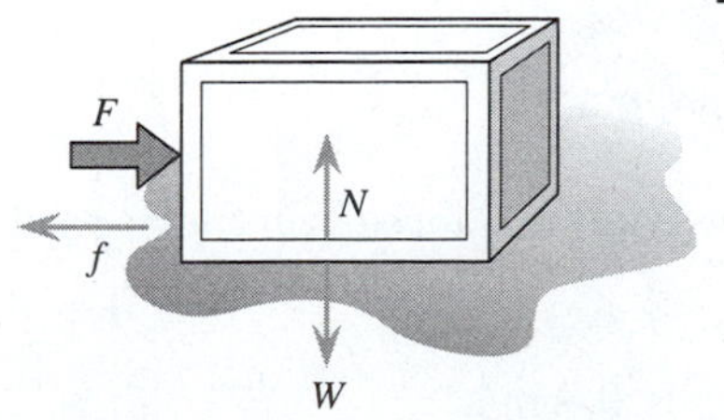

Figure 8–5: *The friction between the crate and the floor resists motion.*

$$\mathbf{f}_{max} = \mu \mathbf{N}$$

$$\mathbf{f}_{max} = 0.60 \times 120 \text{ lb}$$

$$\mathbf{f}_{max} = 72 \text{ lb}$$

A. The $6\bar{0}$-lb force is not enough to overcome static friction. Since the crate was initially at rest, its acceleration is zero.

B. The $8\bar{0}$-lb force is larger than the maximum static friction. The net force on the crate is the difference between the applied force and the *sliding* frictional resistance.

The mass of the crate is:

$$m = \frac{W}{g}$$

$$m = \frac{120 \text{ lb}}{32 \text{ ft/s}^2}$$

$$m = 3.75 \text{ slug}$$

$$F_{net} = F - f$$

$$F_{net} = F - (\mu N) = m\,a$$

$$a = \frac{F - (\mu N)}{m}$$

$$a = \frac{8\bar{0} \text{ lb} - 0.4\bar{0}\,(120 \text{ lb})}{3.75 \text{ slug}}$$

$$a = 8.5 \text{ ft/s}^2$$

☐ **EXAMPLE PROBLEM 8–2: THE SKIDDING CAR**

A $16\bar{0}0$-kg car traveling at a speed of 25 m/s brakes on a dry concrete highway and skids to a halt. If the coefficient of friction between the tires and the road is 0.80, what distance does the braking car travel?

■ ***SOLUTION***

Use Newton's second law to find the acceleration. The only force acting on the car is the frictional sliding force.

$$F_{net} = -\mu N = -\mu\,(m\,g)$$

The negative sign indicates that the frictional force is in the opposite direction to the car's motion.

$$F_{net} = -0.80 \times 1600 \text{ kg} \times 9.8 \text{ m/s}^2$$

$$F_{net} = -1.25 \times 10^4 \text{ N}$$

$$a = \frac{F_{net}}{m}$$

$$a = \frac{-1.25 \times 10^4 \text{ N}}{1600 \text{ kg}}$$

$$a = -7.8 \text{ m/s}^2$$

Now use one of the equations describing linear motion to find the distance, where v_f = final speed and v_0 = initial speed.

$$2 a s = v_f^2 - v_0^2$$

$$s = \frac{v_f^2 - v_0^2}{2 a}$$

$$s = \frac{0^2 - (25 \text{ m/s})^2}{2(-7.8 \text{ m/s}^2)}$$

$$s = 4\bar{0} \text{ m}$$

This distance is nearly half the length of a football field. Notice the braking distance is proportional to the square of the initial velocity. (v_f will always be zero because the car comes to a stop.)

$$\Delta s \propto v_0^2$$

If the car had been traveling at twice the speed, the braking distance would have been 160 m.

☐ **EXAMPLE PROBLEM 8–3: FRICTION ON AN INCLINE**

What minimum force must a bulldozer exert on a 1.2-ton boulder to push it up a $2\bar{0}°$ hill if the coefficient of friction between the boulder and the ground is 0.80?

■ ***SOLUTION***

The bulldozer must overcome the component of weight parallel to the incline and the maximum static frictional force to start the boulder moving. The acceleration is zero.

The parallel component of weight is:

$$W_{parallel} = m \text{ g} \sin \theta$$

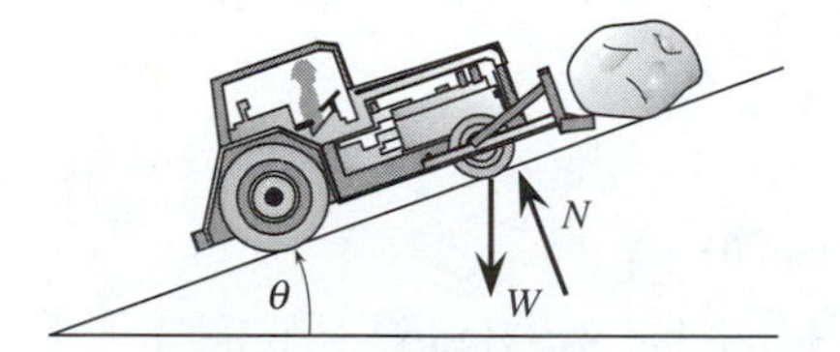

Figure 8–6: *A bulldozer pushes a boulder up an incline.*

The normal force is:

$$N = m \text{ g} \cos \theta$$

$$F - f_{max} - W_{parallel} = 0$$

$$F = f_{max} + W_{parallel}$$

$$F = (\mu m \, g \cos \theta) + (m \, g \sin \theta)$$

$$F = m \, g \, (\mu \cos \theta + \sin \theta)$$

$$F = 1.2 \text{ ton} \times \{2000 \text{ lb/ton} \, [(0.80 \times 0.940) + 0.342]\}$$

$$F = 2.6 \times 10^3 \text{ lb}$$

☐ **EXAMPLE PROBLEM 8–4: THE WHEELBARROW**

A workman pushes a 55-kg wheelbarrow load of concrete mix up a 15° incline by applying a horizontal force of $30\bar{0}$ N. The coefficient of friction is 0.15. To find the acceleration break the problem into different parts.

A. What is the normal force?
B. What is the frictional force?
C. Find the component of the applied force parallel to the ramp.
D. Find the component of weight parallel to the ramp.
E. What is the acceleration of the wheelbarrow?

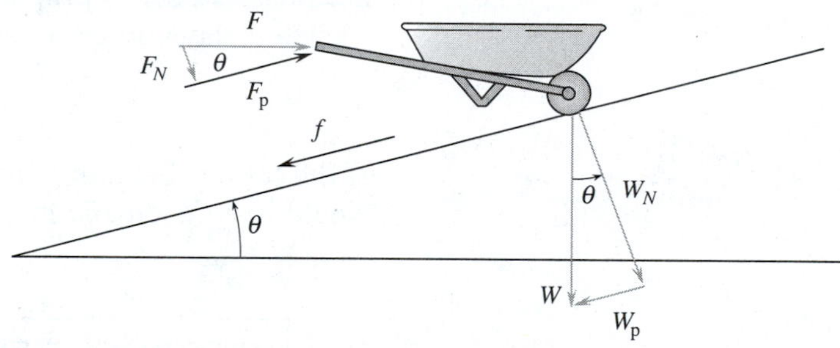

Figure 8–7: A horizontal force pushes a loaded wheelbarrow up an incline.

■ *SOLUTION*

A. The normal force:

The normal force has two parts. Part of the weight of the wheelbarrow pushes down on the incline and part of the applied horizontal force also pushes down. The ramp must push upward with a normal force equal to the sum of these two components (see Figure 8–7).

$$N = F_N + W_N$$

$$N = (F \sin \theta) + (m \, g \cos \theta)$$

$$N = (30\bar{0} \text{ N} \times 0.259) + (55 \text{ kg} \times 9.8 \text{ m/s}^2 \times 0.966)$$

$$N = 598 \text{ N}$$

B. The frictional force:

Since we have the normal force, calculation of the frictional force is easy.

$$f = \mu N$$
$$f = 0.15 \times 598 \text{ N}$$
$$f = 9\bar{0} \text{ N}$$

C. The component of applied force parallel to the ramp ($\mathbf{F}_p$):

$$F_p = F \cos \theta$$
$$F_p = 30\bar{0} \text{ N} \times 0.966$$
$$F_p = 29\bar{0} \text{ N}$$

D. The parallel components of the weight ($\mathbf{W}_p$):

$$W_p = m \, g \sin \theta$$
$$W_p = 55 \text{ kg} \times 9.8 \text{ m/s}^2 \times 0.259$$
$$W_p = 140 \text{ N}$$

E. The acceleration:

Look only at the forces along the incline to calculate the acceleration. Take up as the positive direction. The acceleration and parallel component of the applied force are positive. The parallel component of the weight and the resisting friction are negative.

$$F_p - f - W_p = m \, a$$

or

$$a = \frac{F_p - f - W_p}{m}$$
$$a = \frac{29\bar{0} \text{ N} - 90 \text{ N} - 140 \text{ N}}{55 \text{ kg}}$$
$$a = 1.1 \text{ m/s}^2$$

Example Problem 8–4 looks long and cumbersome, but it illustrates the general approach to solving frictional force problems, summarized below.

Falling Through the Earth

Suppose we could blow a hole through the center of the earth with some sort of annihilator cannon borrowed from the set of a sci-fi movie. You peer down the hole that begins in Portland, Maine, hoping to see friends in Canberra, Australia. Being the stumble-bum you are, you fall in.

Since this is going to be a flight of fancy, we'll make a few simplifying assumptions. We'll assume the hole is lined with a thermally insulating material that will keep the molten rock that's down there from seeping in and frying you to a crisp. We'll also ignore air friction.

Down you go, accelerating to faster and faster speeds. At the end of one minute, you're falling at a speed close to 1,900 ft/s, or 1,300 mph, and you've traveled nearly 11 miles. At the end of two minutes, your speed is 3,840 ft/s. You're still only 44 miles on your way to visit your friends in Canberra and you wonder if you should have brought a lunch with you.

As time passes, you begin to notice something. Your speed is continuing to increase, but at a slower rate. The mass of the earth below you continues to exert a gravitational force downward, but you now have a sizable amount of the earth's mass above you. That part of the earth has a gravitational pull upward. The net gravitational force is less than on the planet's surface.

As you pass through the center of the earth, half of the mass is below you toward Canberra and half of the earth's mass is above you toward Portland. Your acceleration at the center of the earth is zero. The mass of the two halves of the earth exert equal, but oppositely directed, gravitational pulls.

As you proceed past the center of the earth toward Canberra, you discover there's more mass behind you than in front of you. Your velocity begins to decrease. As you reach the planet's surface at Canberra, your velocity has fallen to zero. If someone doesn't grab you at this point, you'll fall back into the hole and retrace your trip in the opposite direction.

Photograph courtesy of NASA

A. Find all the forces normal to the surfaces of contact and sum them algebraically. Forces that tend to push the surfaces will increase the normal force (+). Forces that tend to push the surfaces apart decrease the normal force (−).

B. Once the normal force is known, calculate the frictional force. The frictional force will always be parallel to the contact surfaces.

C. Find all the remaining components of forces parallel to the surface of contact.

D. Apply Newton's second equation of dynamics to find the unknowns.

8.2 VELOCITY-DEPENDENT FORCES

As an object moves through a fluid, a viscous drag opposes the motion. This viscous force depends upon:

A. the shape of the object,
B. the size of the object,
C. the density of the fluid,
D. the **viscosity** of the fluid (a measure of how easily the fluid flows), and
E. the speed of the object *relative to the fluid*. (If both the fluid and the object move, the combined motion has to be considered.)

The equations to describe motion developed in Chapter 4 ignored viscous friction. They applied only to dense objects traveling a short distance through the atmosphere. When a baseball player hits a fly ball to centerfield, the ball's trajectory is different from the symmetric path predicted by friction-free equations. Viscous friction causes the ball to fall short (see Figure 8–8).

Experiments have shown that the viscous force acting on an object moving through air is proportional to the square of its velocity, and it is also proportional to the cross-sectional area of the object and to air density.

$$\mathrm{f} = \left(\frac{1}{2}\right) C\,\rho\,A\,v^2 \qquad \textbf{(Eq. 8–2)}$$

where C is a unitless proportionality constant that involves the effects of the shape of the object (called the **drag coefficient**); ρ is air density (1.3 kg/m^3 at sea level); A is the cross-sectional area, and v is the velocity of the object relative to the air. C has a value between 0 and 1.0. For a rough sphere the drag coefficient is 0.40. For more irregularly shaped objects the value is closer to one.

Frictional equations are empirical. That is, they are derived from experiment rather than from basic physical principles. Usually they are good for only a modest range of conditions. At extremely large velocities, such as the reentry of a satellite or the occurrence of a

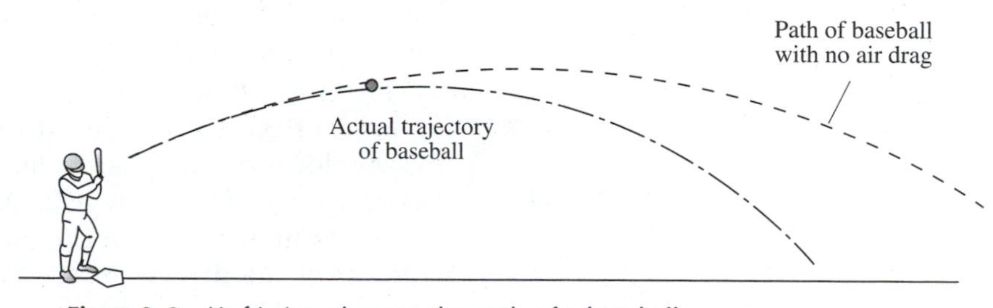

Figure 8–8: *Air friction changes the path of a baseball.*

meteor in the earth's atmosphere, the friction may be proportional to a higher power of velocity ($\mathbf{f} \propto \mathbf{v}^3$). Objects falling in a liquid rather than a gas usually obey a frictional force law that is directly proportional to the velocity ($\mathbf{f} \propto \mathbf{v}$). When the flow of the fluid around an object is restricted—e.g., when an object falls through a fluid-filled pipe—the relationship is even more complicated. Despite these cautions, Equation 8–2 can be applied to several situations.

EXAMPLE PROBLEM 8–5: AIR RESISTANCE ON A CAR

A small car has a cross-sectional area of 36 ft². Estimate the air drag on the car at 45 mph and at 60 mph. Assume a drag coefficient of 0.50. (In British units the density of air is 2.5×10^{-3} slug/ft³.)

SOLUTION

Express the speed in feet per second.

$$45 \text{ mph} \times 1.47 \frac{\text{ft/s}}{\text{mph}} = 66 \text{ ft/s}$$

$$60 \text{ mph} \times 1.47 \frac{\text{ft/s}}{\text{mph}} = 88 \text{ ft/s}$$

$$\mathrm{f} = \frac{1}{2} C \rho A \mathbf{v}^2$$

At 45 mph

$$\mathrm{f} = \frac{1}{2} \times 0.50 \times 2.5 \times 10^{-3} \text{ slug/ft}^3 \times 36 \text{ ft}^2 \times (66 \text{ ft/s})^2$$

$$\mathrm{f} = 98 \text{ lb}$$

At 60 mph

$$\mathrm{f} = \frac{1}{2} \times 0.50 \times 2.5 \times 10^{-3} \text{ slug/ft}^3 \times 36 \text{ ft}^2 \times (88 \text{ ft/s})^2$$

$$\mathrm{f} = 170 \text{ lb}$$

Air friction is an important factor in controlling an automobile's speed. As the velocity of the vehicle increases, the frictional drag also increases. Eventually the frictional forces will balance the driving force of the wheels, resulting in no acceleration. If the driver presses down on the gas pedal, the driving force will exceed the frictional drag until a new balance is found at a higher speed. If the driver lets up partway on the gas pedal, the frictional drag will be larger than the driving force. The balance of forces is found at a lower speed. Without velocity-dependent air friction it would be difficult to control a moving vehicle.

Things falling through the atmosphere often reach a **terminal velocity,** the point at which a falling object's weight is balanced by an upward frictional air drag. Raindrops fall several thousand feet, but usually strike the earth at less than 10 mph. The forces on the raindrop are shown in Figure 8–9. If the two forces balance, the raindrop falls at a constant speed (terminal speed, $\mathbf{v}_t$).

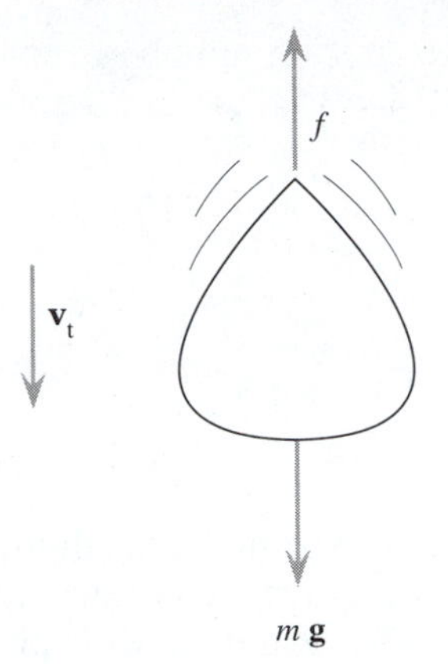

Figure 8–9: The velocity-dependent air friction (***f***) *balances the weight* (m **g**) *of the falling raindrop at the terminal velocity* (***v***$_t$)*.*

$$F_{net} = (m\,g) - \frac{1}{2}(C\,\rho\,A\,v_t^2) = 0$$

The equation can be solved for the terminal velocity.

$$v_t = \left(\frac{2\,m\,g}{C\,\rho\,A}\right)^{1/2} \qquad \textbf{(Eq. 8–3)}$$

The terminal velocity of an object falling near the surface of the earth is proportional to the square root of the ratio of mass to cross-sectional area ($v_t \propto [m/A]^{1/2}$). The other values in the expression are constants.

□ **EXAMPLE PROBLEM 8–6: THE PARACHUTE DROP**

A team of geologists is taking core samples in a mountainous region for a mining company. An airplane drops two identical 70-kg packages of supplies by parachute. One parachute fails to open. The effective area of the packages is 0.25 m^2 with a drag coefficient of 0.50. The parachute has a cross section of 15 m^2 and a drag coefficient of 1.0. Both packages reach terminal speed before striking the ground. With what speed does each package land?

■ ***SOLUTION***

A. No parachute:

$$v_t = \left(\frac{2\,m\,g}{C\,\rho\,A}\right)^{1/2}$$

$$v_t = \left(\frac{2 \times 70\text{ kg} \times 9.8\text{ m/s}^2}{0.50 \times 1.3\text{ kg/m}^3 \times 0.25\text{ m}^2}\right)^{1/2}$$

$$v_t = 92\text{ m/s}$$

B. Package with a parachute:

Use the same equation with the different values for C and A.

$$v_t = \left(\frac{2 \times 70\ \text{kg} \times 9.8\ \text{m/s}^2}{1.0 \times 1.3\ \text{kg/m}^3 \times 15\ \text{m}^2}\right)^{1/2}$$

$$v_t = 8.4\ \text{m/s}$$

8.3 FORCE PROPORTIONAL TO DISPLACEMENT

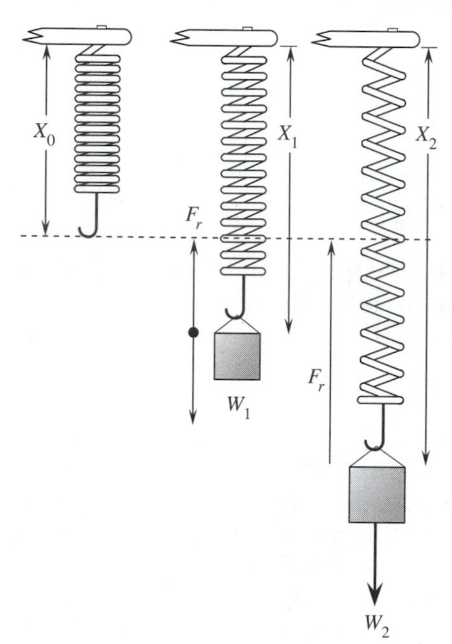

Figure 8–10: *The restoring force (**F**$_r$) balances the weight hung on the spring. Weight **W**$_2$ is larger than weight **W**$_1$.*

A force is needed to stretch or compress a spring. A small force will produce a small change in length; a larger force will cause a larger change. If you stretch a spring or rubber band, it will pull back. The reacting force tries to restore the object to its original shape or **equilibrium position.**

A spring is hung from a horizontal bar (see Figure 8–10). The spring hangs down with the free end at its equilibrium position (X_0). A weight (**W**) is hung on the spring. With the added weight, the spring stretches until the **restoring force** (**F**$_r$) balances the weight at a new length (X_1). The displacement of the spring (change of length) is ($X_1 - X_0$). If more weight is added, a larger displacement will be produced ($X_2 - X_0$). The displacement is always measured from the original equilibrium position.

If several different weights are hung on the spring, the displacement may be plotted against the restoring force. The restoring force will have the same size as the hung weight, but in the opposite direction. If the spring weighs much less than the weights, the graph will look like Figure 8–11. The straight line of the graph indicates that the force is directly proportional to the displacement. The slope of the line is the proportionality constant (k). We can write the equation for the restoring force of the spring.

$$F_r = -k\,(X - X_0) \qquad \textbf{(Eq. 8–4)}$$

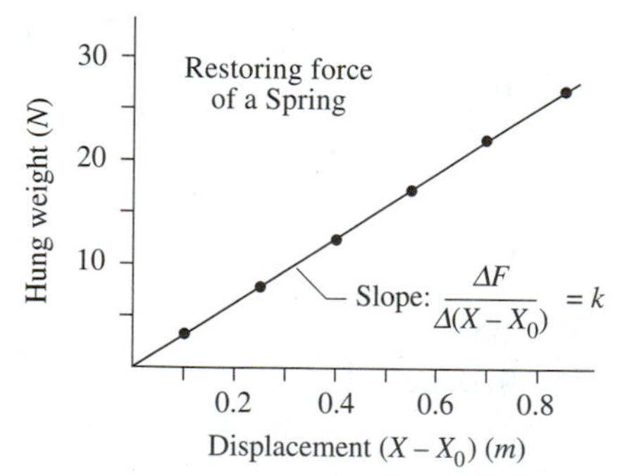

Figure 8–11: *The restoring force of a spring is directly proportional to its displacement. The slope of the curve is called the stiffness of the spring.*

This relationship is called **Hooke's law.** The negative sign indicates that the restoring force is opposite in direction to the displacement. If the spring is stretched downward, it will react with an upward restoring force. If the spring is compressed, it will push outward to resist the compression. In general, the restoring force tries to prevent an increase of the displacement.

The proportionality constant (k) found from the slope of the plot is called the **stiffness** of the spring. Sometimes it is called the **spring constant** or the **force constant.** For example, a screen-door spring has a stiffness of 50 N/m, while the coil suspension on an automobile would have a k value close to 4×10^4 N/m.

☐ **EXAMPLE PROBLEM 8–7: THE SPRING SCALE**

Since the displacement of a spring is proportional to the weight hung on it, a spring can be used to construct a weight scale. A 1.0-lb weight stretches a spring by 0.75 in.

A. What is the force constant of the spring in pounds per foot?
B. How much is the spring stretched by a 7.3-lb weight?
C. An unknown weight stretches the spring 7.0 in. How large is the weight?

■ ***SOLUTION***

A. The restoring force is in the opposite direction to the weight.

$$\mathrm{F_r} = -\mathrm{W} = -k\,\Delta X$$

or

$$k = \frac{\mathrm{W}}{\Delta X}$$

$$k = \frac{1.0\ \mathrm{lb}}{0.75\ \mathrm{in} \times 1.0\ \mathrm{ft}/12\ \mathrm{in}}$$

$$k = 16\ \mathrm{lb/ft}$$

B. A 7.3-lb weight downward will result in a restoring force of −7.3 lb (upward). Remember, Equation 8–4 is for the restoring force, not for the weight.

$$X = \frac{-\mathrm{F_r}}{k}$$

$$X = \frac{-(-7.3\ \mathrm{lb})}{16\ \mathrm{lb/ft}}$$

$$X = 0.46\ \mathrm{ft}$$

C. If we use the weight rather than the restoring force, the minus sign disappears from the equation because the weight and restoring force are an action-reaction pair of forces.

$$\mathrm{W} = k\,\Delta X$$

$$\mathrm{W} = 16\ \mathrm{lb/ft} \times 7.0\ \mathrm{in} \times \frac{1.0\ \mathrm{ft}}{12\ \mathrm{in}}$$

$$\mathrm{W} = 9.3\ \mathrm{lb}$$

□ **EXAMPLE PROBLEM 8–8: THE TRUCK SUSPENSION**

A small truck has a coil suspension with a combined spring constant of 3.0×10^5 N/m. By how much do the coils compress when a load of 700 kg is placed on the truck?

■ ***SOLUTION***

$$\Delta X = \frac{W}{k}$$

$$\Delta X = \frac{700 \text{ kg} \times 9.8 \text{ m/s}^2}{3.0 \times 10^5 \text{ N/m}}$$

$$\Delta X = 2.3 \times 10^{-2} \text{ m or about 1 in}$$

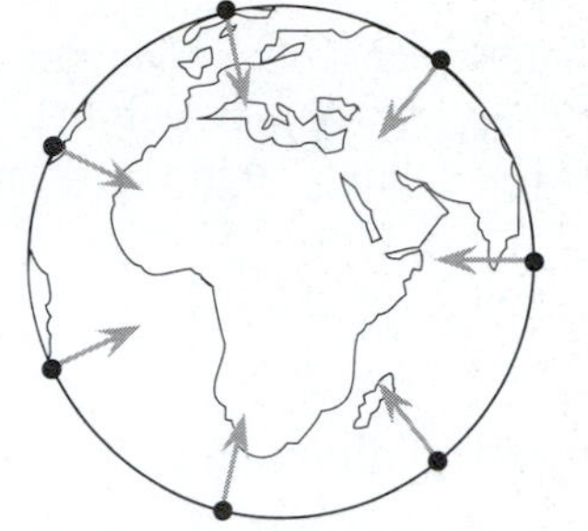

Figure 8–12: *The weights of objects on the surface of the earth act radially inward to the earth's center.*

8.4 GRAVITY: A FIELD FORCE

If 1-kg test masses were placed in different locations around the surface of the earth, their weights would act radially inward (see Figure 8–12). If these 1-kg masses were located farther away from the planet, the gravitational pull on them would be less, but still radially inward (see Figure 8–13).

The gravitational force acting on a mass, its weight, depends on two things: the mass's distance from the center of the earth and its size. The gravitational pull is twice as large on a 2-kg mass as on a 1-kg mass.

A vector quantity that depended only on the position of an object and not on its size would make it possible to represent the effect of gravity at some location using only a single set of vectors independent of mass. Weight will not do. For each different-sized object we would have a different-sized weight vector. But by dividing weight by mass, we get what we want—a quantity independent of mass. This quantity is called the **gravitational field.**

$$\mathbf{g} = \frac{\mathbf{F}}{m} \qquad \textbf{(Eq. 8–5)}$$

The gravitational field is nothing more than the acceleration caused by gravity. To find the gravitational force acting on a mass at any point simply multiply the gravitational field by the mass. The product is force.

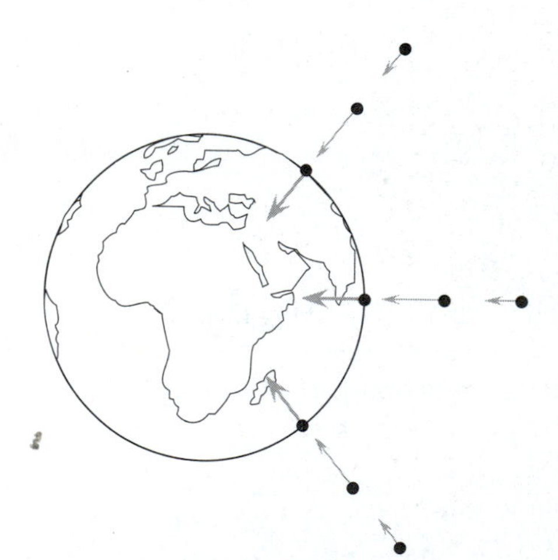

Figure 8–13: *The gravitational force on a mass decreases with increasing distance from the earth.*

EXAMPLE PROBLEM 8–9: GRAVITATIONAL FIELD

On the surface of the earth the gravitational field is 9.8 m/s². What is the gravitational force (weight) on a 3.0-kg object on the earth's surface?

SOLUTION

$$\mathbf{W}(\text{eight}) = m(\text{ass}) \times \mathbf{g}(\text{ravitational field})$$

$$\mathbf{W} = 3.0\ \text{kg} \times 9.8\ \text{m/s}^2$$

$$\mathbf{W} = 3\bar{0}\ \text{N}$$

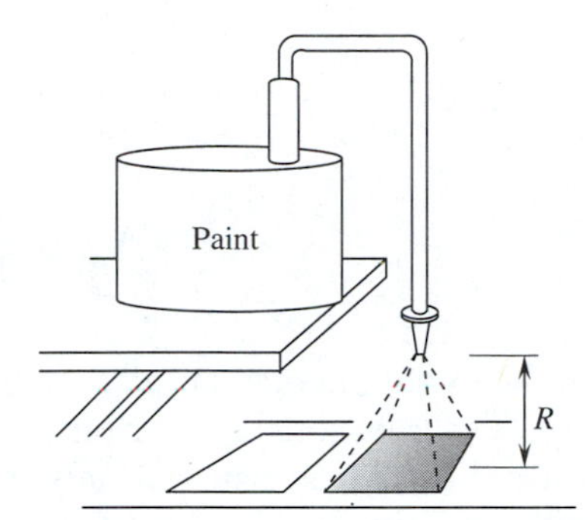

Figure 8–14: *A machine to spray enamel on tiles.*

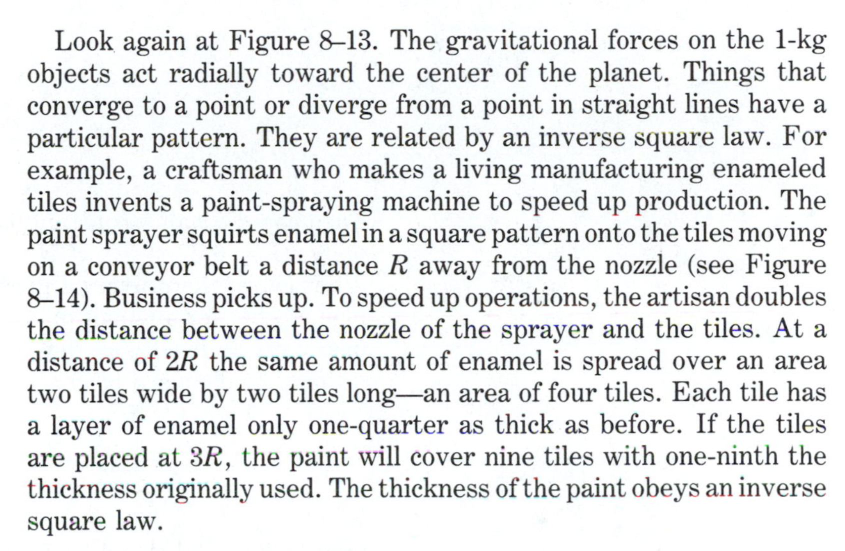

Look again at Figure 8–13. The gravitational forces on the 1-kg objects act radially toward the center of the planet. Things that converge to a point or diverge from a point in straight lines have a particular pattern. They are related by an inverse square law. For example, a craftsman who makes a living manufacturing enameled tiles invents a paint-spraying machine to speed up production. The paint sprayer squirts enamel in a square pattern onto the tiles moving on a conveyor belt a distance R away from the nozzle (see Figure 8–14). Business picks up. To speed up operations, the artisan doubles the distance between the nozzle of the sprayer and the tiles. At a distance of $2R$ the same amount of enamel is spread over an area two tiles wide by two tiles long—an area of four tiles. Each tile has a layer of enamel only one-quarter as thick as before. If the tiles are placed at $3R$, the paint will cover nine tiles with one-ninth the thickness originally used. The thickness of the paint obeys an inverse square law.

$$\textit{enamel thickness} \propto \frac{1}{R^2}$$

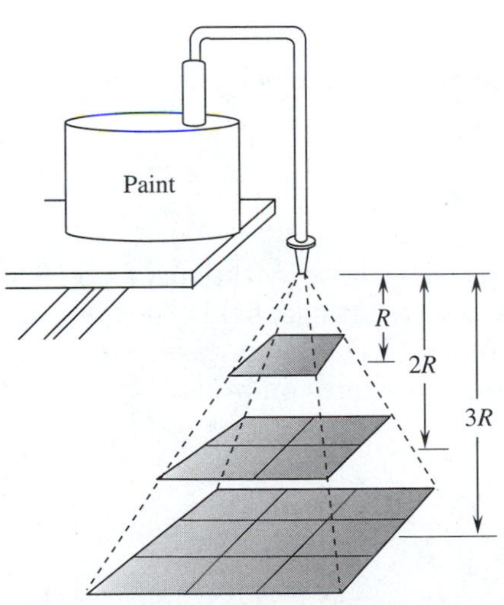

Figure 8–15: *As the distance (R) between the spray nozzle and the tile increases, the same amount of paint is spread over a larger area.*

The gravitational field around a planet behaves somewhat like the paint. The lines of action are in straight lines from a single point. This suggests that the strength of the gravitational field, like the paint, obeys an inverse square law.

$$|\mathbf{g}| \propto \frac{1}{R^2}$$

The strength of the gravitational field is proportional to the mass of the planet (m_1). Around a larger planet, such as Jupiter, the gravitational field is stronger at the equivalent distance from its center. A smaller mass such as the moon has a smaller gravitational field. We can include planet mass in the proportion.

$$|\mathbf{g}| \propto \frac{m_1}{R^2}$$

We can make this an equation by introducing a proportionality constant (G).

$$|\mathbf{g}| = \frac{G\, m_1}{R^2}$$

The size of the gravitational force on a mass (m_2) near a planet is:

$$|\mathbf{F}| = m_2 \times \mathrm{g}$$

or

$$|\mathbf{F}| = G\frac{m_1\, m_2}{R^2} \qquad \textbf{(Eq. 8–6)}$$

The proportionality constant (G) can be found experimentally. In SI units, $G = 6.67 \times 10^{-11}\ \mathrm{N \cdot m^2/kg^2}$. The relationship in Equation 8–6 is called the **universal gravitational law.** It is true not only for planets, but for any pair of objects in the universe. Notice the symmetry of the masses in the equation. We could consider the gravitational field caused by mass m_1.

$$|\mathbf{g}_1| = \frac{\mathrm{F}}{m_2} = \frac{G\, m_1}{R^2}$$

or we could find the strength of the gravitational field of m_2.

$$|\mathbf{g}_2| = \frac{\mathrm{F}}{m_1} = \frac{G\, m_2}{R^2}$$

You may wonder which mass is causing the gravitational force. The answer is both of them. Each has a gravitational field. They interact with an action-reaction pair of forces. The force m_1 exerts on m_2 is $\mathbf{F}_{1\cdot 2} = m_2\, \mathbf{g}_1$ toward m_1. The force m_2 exerts on m_1 is $\mathbf{F}_{2\cdot 1} = m_1\, \mathbf{g}_2$ toward m_2.

$$|\mathbf{F}_{1\cdot 2}| = |-\mathbf{F}_{2\cdot 1}| = \frac{G\, m_1\, m_2}{R^2}$$

where R is the distance between the centers of mass of the two objects.

When a raindrop falls, there is an action-reaction pair of forces.

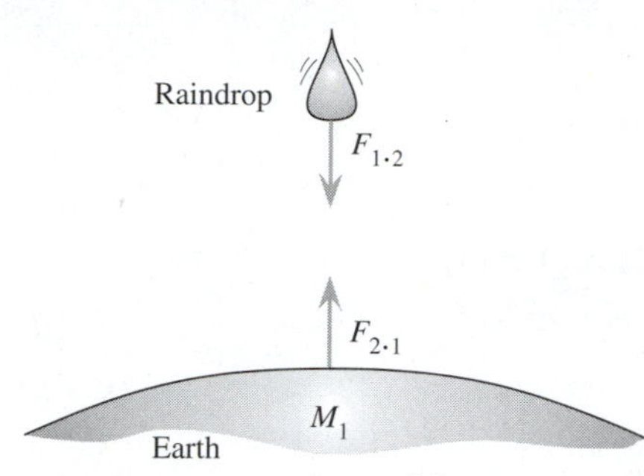

Figure 8–16: *Gravitational forces occur in interacting pairs.*

The earth exerts a force down on the raindrop, trying to pull it to the center of the earth. At the same time, the raindrop exerts an equal, but upward-directed, gravitational force on the earth. The motion of the earth is not noticed because the earth's inertia is much larger.

There is an attractive, action-reaction pair of forces between any two objects. The strength of the force is given by the universal gravitational law.

$$F = G\frac{m_1 m_2}{R^2}$$

Now is a good time to point out that there are two kinds of mass. Newton's second law deals with **inertial mass,** which is involved when an unbalanced force changes the motion of an object. Inertial mass resists the change. The other kind of mass is **gravitational mass.** It is not involved with motion. When a rock is placed on a triple-beam balance, no motion occurs. You are measuring the mass related to gravitational forces. Several experiments to find a difference between these two kinds of mass have shown, to an accuracy of one part in 10^{11}, that inertial mass and gravitational mass are the same. That extraordinary coincidence is the basis of Einstein's equivalence described in Chapter 7.

□ **EXAMPLE PROBLEM 8–10: MOON MASS**

The strength of the moon's gravitational field is 1.62 m/s². Its diameter (D) is 3.48×10^3 km. Estimate the mass of the moon.

■ ***SOLUTION***

The distance from the surface of the moon to its center of gravity is its radius.

$$R = \frac{D}{2}$$

$$R = \frac{3.48 \times 10^3 \text{ km} \times 10^3 \text{ m/km}}{2}$$

$$R = 1.74 \times 10^6 \text{ m}$$

The gravitational field is:

$$|\mathbf{g}| = \frac{G\,m}{R^2}$$

or

$$m = \frac{|g|\, R^2}{G}$$

$$m = \frac{(1.62 \text{ m/s}^2)\,(1.74 \times 10^6 \text{ m})^2}{6.67 \times 10^{-11} \text{ N}\cdot\text{m}^2/\text{kg}^2}$$

$$m = 7.35 \times 10^{22} \text{ kg}$$

☐ **EXAMPLE PROBLEM 8–11: GRAVITATIONAL PULL OF SUN AND MOON**

The mass of the sun is 2.7×10^7 times the mass of the moon. The sun's average distance from earth is 1.50×10^8 km. The moon's average distance from earth is 3.84×10^5 km. Find the ratio of the gravitational pull of the sun on the earth to the pull of the moon.

■ ***SOLUTION***

Take the ratio of the two forces. Cancel like terms.

$$\frac{F_s}{F_m} = \frac{(G\, m_s\, m_e)/R_s^{\,2}}{(G\, m_m\, m_e)/R_m^{\,2}}$$

Since we are taking a ratio, units are not important as long as they are consistent.

$$\frac{F_s}{F_m} = \left(\frac{2.7 \times 10^7 \text{ moon masses}}{1.0 \text{ moon mass}}\right) \times \left(\frac{3.84 \times 10^5 \text{ km}^2}{1.50 \times 10^8 \text{ km}^2}\right)^2$$

$$\frac{F_s}{F_m} = 180$$

☐ **EXAMPLE PROBLEM 8–12: FORCE ON A SATELLITE**

A satellite used to assess mineral resources on the earth's surface is placed into orbit one earth radius above the planet. If it weighed 1200 lb before launch, what gravitational force does the earth exert on it in orbit?

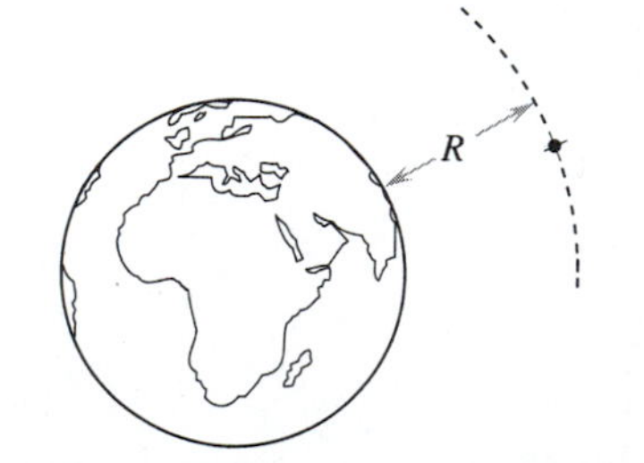

Figure 8–17: *A satellite in orbit one earth radius above the surface of the earth.*

■ ***SOLUTION***

Let $\mathbf{F}_1$ be the weight of the satellite on earth. $\mathbf{F}_2$ is the gravitational force on the satellite in orbit. M is the mass of earth, and m the mass of the satellite. R is the earth's radius.

$$\frac{F_2}{F_1} = \frac{G\, m\, M/(2R)^2}{G\, m\, M/R^2}$$

$$\frac{F_2}{F_1} = \frac{R^2}{(2R)^2} = \frac{R^2}{4\,R^2}$$

$$F_2 = 0.25 \times 1200 \text{ lb}$$

$$F_2 = 300 \text{ lb}$$

8.5 APPLICATIONS

Clutches

A clutch is a device used to transmit power from a motor to a drive shaft. Its essential feature is the ability to disengage the power without shutting off the motor. Many clutches use friction as the driving force. Figure 8–18 shows the essential features of a disk clutch used in a standard-transmission automobile.

A driven clutchplate, situated between the flywheel and a pressure plate, is lined with a material that has a high coefficient of friction. When the clutch is engaged a pressure plate pushes the clutchplate against a third plate fastened to the back of the flywheel. All three plates rotate, tied together by frictional forces. Because the clutchplate is connected to the driven shaft by a spline, the clutchplate can move back and forth freely, and rotary motion is transmitted to the driven shaft. A throw-out collar is used to disengage the clutch. When the collar is pushed forward it levers back the pressure plate.

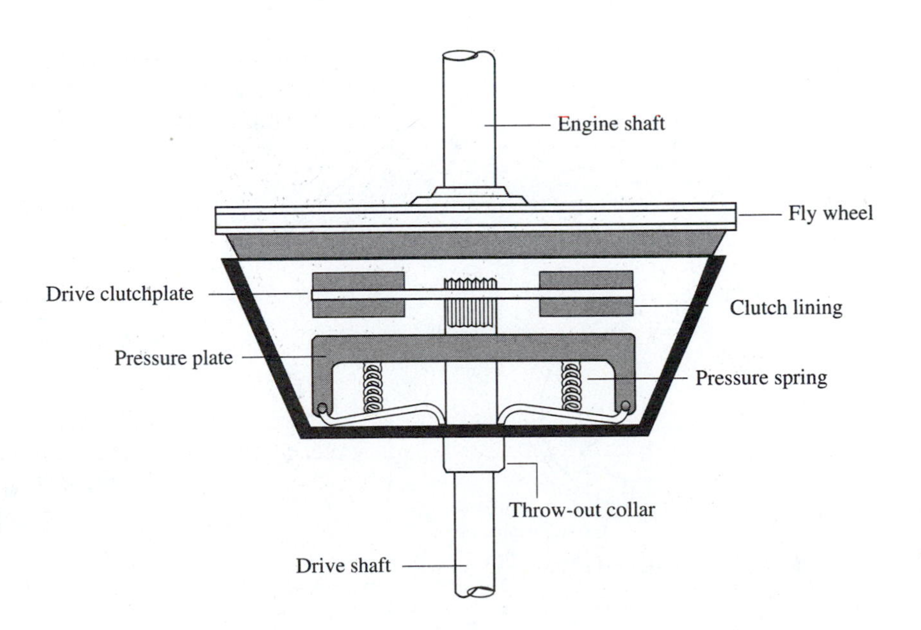

Figure 8–18: *A disengaged clutch on an automobile. When the clutch is engaged, the pressure plate presses the driven clutchplate against the back of the flywheel.*

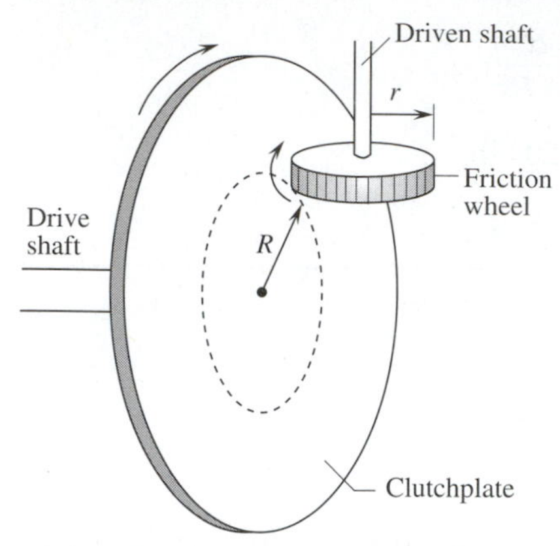

Figure 8–19: A variable-speed friction clutch. The driven shaft rotates faster as it moves to the outside of the clutchplate.

Thus the normal force between the clutchplate and the flywheel is reduced drastically and power transmission is cut off.

Figure 8–19 shows a variable-speed clutch sometimes used with small electric motors. A large rubberized disk is attached to the motor's drive shaft. The driven shaft has a smaller disk attached to it. The rim of the smaller disk makes contact with the larger disk, causing it to rotate. As the large disk makes one rotation, the driven disk travels over a distance (S) of $2\pi R$. Expressed in the radius of the smaller disk, this distance is $\mathbf{N}(2\pi r)$, where **N** is the number of turns. If the large disk makes one turn, the smaller disk must make $\mathbf{N} = R/r$ turns. Usually the driven shaft can be adjusted to make contact on the driving disk at any R. By increasing R, the rotational speed of the driven shaft is increased. When the driven shaft moves toward the center of the larger disk, its speed approaches zero. If the small disk is placed below the center of the driving wheel, its rotation is reversed.

Brakes

Many cars use drum brakes. A simplified diagram of a drum brake is shown in Figure 8–20. Brake shoes are attached to a fixed mounting inside the rotating drum. The two shoes are connected with an adjustable pivot at one end. At the other end is a hydraulic cylinder. When pressure inside the cylinder is increased, it expands, pushing the linings of the shoes against the inside rim of the rotating drum. Frictional force then slows down the drum.

Air Deflectors

For a fast-moving vehicle on the level, much of the force generated by the motor goes into overcoming air resistance. Figure 8–21 shows a device used by semitrailer trucks to reduce air drag. The blunt end of the trailer causes air drag coefficients in the range of 0.7–0.8 for this kind of truck. Air deflectors mounted on the cab divert the

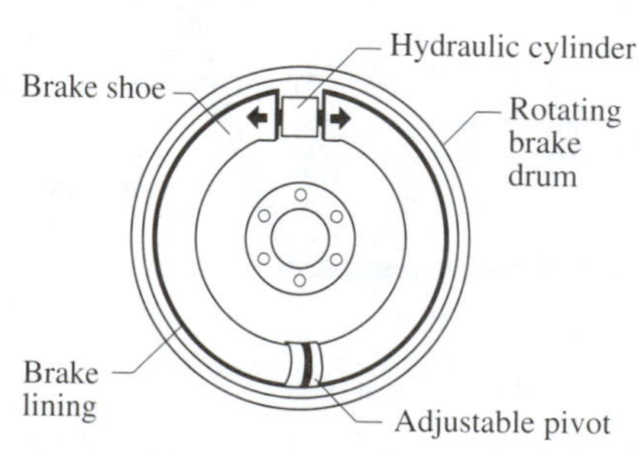

Figure 8–20: Simplified diagram of a drum brake.

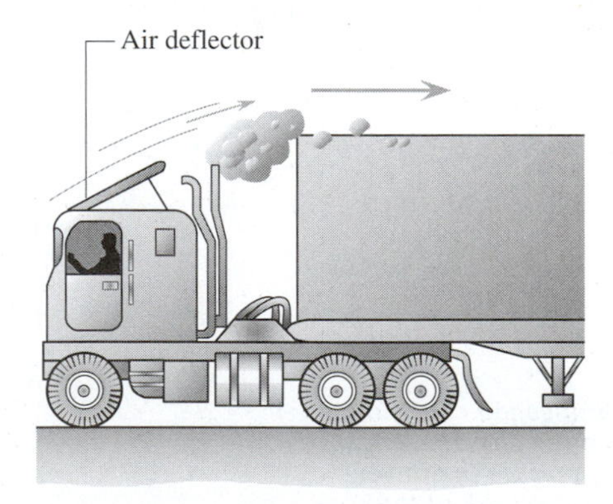

Figure 8–21: Air deflector on the cab of a trailer truck reduces air drag.

airstream over the trailer rather than directly into the flat plane of its front end. Air deflectors reduce the drag coefficient from 11% to 21% with an average of 17%. The amount of reduction depends on the truck cab design and the speed of the truck. Many truckers have found they can reduce fuel consumption up to 10% with this simple device.

Intermolecular Forces

The forces that bind molecules depend on position. When a compressional force acts on a material, strong repulsive forces resist the attempt to push the molecules closer together. Another set of positional forces attracts molecules, pulling the molecules together. The sum of the repulsive and attractive forces balances at separation distances in the order of 10^{-10} m for most materials. These equilibrium distances are called the **interatomic distances.**

The **cohesive forces** are the net attractive forces that hold molecules of a material together. In general, cohesive forces are less in a liquid than in a solid. The strength of cohesive forces in a gas is much less than in a liquid or solid.

The tendency for one material to cling to another material is called **adhesion.** Adhesive forces increase as the two materials are pushed closer together. If you push down harder on a graphite pencil, the line drawn on the paper is clearer. Pliable materials, such as chewing gum or a liquid, will tend to stick to a solid surface more readily than would another solid.

SUMMARY

Static friction acts parallel to surfaces in contact. It opposes the net force parallel to the surface. The maximum static friction that can be developed is $f = \mu N$, where μ is the static coefficient of friction and N is the force normal to the contact surfaces.

Sliding friction is in a direction opposite to its motion. Its magnitude is $f = \mu N$, where μ is the sliding coefficient of friction. Sliding friction has one fixed value. Static friction may react with a value of zero to $\mathbf{f}_{max}$ depending on the applied forces parallel to the contact surface.

The viscous drag of fluids on a moving object depends on the object's velocity relative to the fluid. In liquids, the viscous friction is proportional to velocity. For objects moving through a gas at a moderate speed, the viscous force has a magnitude of $f = C \rho A v^2/2$, where C is the drag coefficient, ρ is the density of the gas, v is the velocity of the object relative to the gas, and A is the cross-sectional area.

A terminal speed is reached by a falling object when the frictional force balances its weight. In a gas, the terminal velocity is $v_t = (2\,m\,g/C\,\rho\,A)^{1/2}$.

Hooke's law tells us that the restoring force of a spring or material under tension or compression is directly proportional to the displacement. The proportionality constant is called stiffness, or the force constant, $\mathbf{F} = -k\,\Delta X$, where the minus sign indicates that the restoring force is opposite in direction to the displacement.

Any two masses exert equal, but oppositely directed, attractive forces on each other (universal gravitation). The magnitude of the forces is $F = (G\,m_1\,m_2)/R^2$, where R is the distance between the two centers of gravity and m_1 and m_2 are the masses. $G = 6.67 \times 10^{-11}\ \text{N} \cdot \text{m}^2/\text{kg}^2$. The gravitational field surrounding a mass is $\mathbf{g} = \mathbf{F}/m$.

KEY TERMS

If you can explain the following terms to a friend or classmate, you understand their meaning. If you cannot explain the terms, you should reread the sections in which they are discussed.

adhesion
coefficiency of friction
cohesive forces
drag coefficient
equilibrium position
force constant
gravitational field
gravitational mass
Hooke's law
inertial mass
interatomic distances
inverse law of gravity
restoring force
sliding friction
spring constant
static friction
stiffness
terminal velocities
universal gravitational law
viscosity

EXERCISES

Section 8.1:

1. A 2$\bar{0}$-lb wooden chair is slid across a wooden floor. First it is *pushed* by a force ($\mathbf{F}_1$) at an angle of 3$\bar{0}$° above the floor. Later it is *pulled* by a force ($\mathbf{F}_2$) also at an angle of 3$\bar{0}$°. In both cases, the force just balances friction (see Figure 8–22). Are the magnitudes of the forces the same or different? Explain.

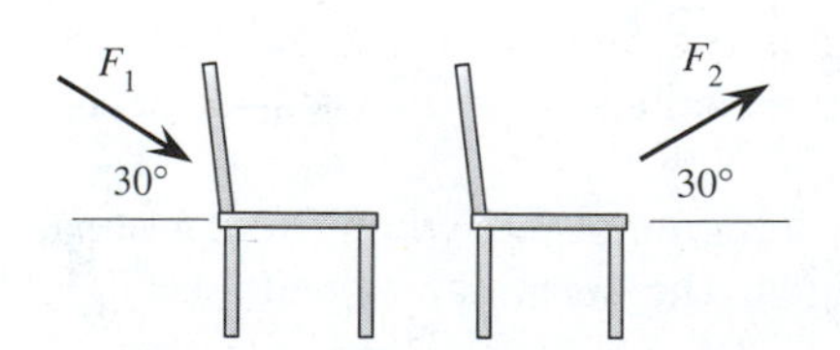

Figure 8–22: *Two ways to move a chair along a floor by a nonparallel force.*

2. Ice is slippery. In a Minnesota winter are icy roads more of a hazard during some months than others? Explain.
3. A rotating tire contacts the road statically. The portion of tire in contact with the road rises while another section descends to make contact without sliding. When a car goes into a skid, why must the car's speed be reduced to gain control?
4. Ball bearings or roller bearings are often put between a rotating shaft and its housing. Why does this reduce friction?

Section 8.2:

5. Two identical cars travel at the same speed on the same road surface. One travels into the wind; the other, with the wind. Is the air friction on the two cars the same? Why or why not?
6. Why does a downhill skier assume a "tuck" position during a race?
7. What happens to a sky diver's terminal velocity as he falls from 15,000 ft to 2000 ft? As the sky diver assumes a spread-eagle position?
8. Why does a snowflake have a smaller terminal velocity than a raindrop with the same mass?
9. If Galileo's leaning tower experiment were done with a ping-pong ball and a golf ball would the results be different? Why or why not?

Section 8.3:

10. Figure 8–23 shows the displacement of some different objects under tension or compression. Which plots represent objects obeying Hooke's law? Explain why.

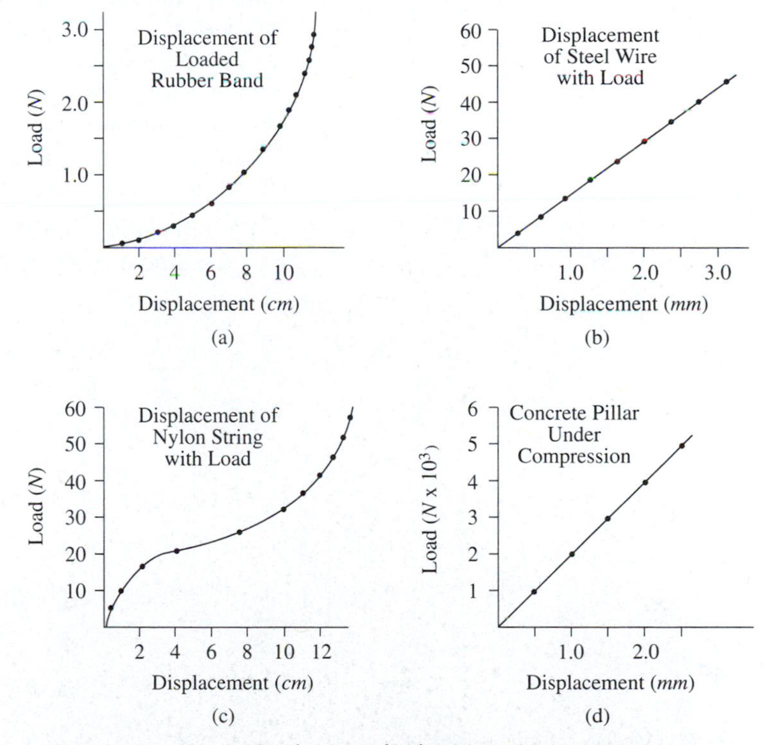

Figure 8–23: *Plots of load versus displacement for four situations.*

11. If two springs were connected end to end, would the stiffness of this compound spring be larger or smaller than the stiffness of a single spring?

12. If two springs were connected side by side, would the stiffness of the combination be larger or smaller than that of a single spring?

Section 8.4:

13. The earth's gravitational field on its surface is 32 ft/s^2. What is the strength of the earth's gravitational field two earth radii from its center? Three radii?

14. If the earth had half its present diameter and twice its present mass, how many times larger or smaller would your weight be?

15. If Equation 8–6 were valid *inside* a mass, what would be the gravitational force at the center of gravity?

16. Why does the inverse law of gravity lead to the formation of spherical planets rather than square or cylindrical ones?

PROBLEMS

Section 8.1:

1. Find the normal force and the maximum static friction for a 2.40-kg mass in the following situations. Assume a static coefficient of friction of 0.30.
   A. On a horizontal surface.
   B. On an incline making an angle of 30° with the horizontal.
   C. On an incline making an angle of 70° with the horizontal.
   D. On a vertical surface.

2. A 12$\bar{0}$0-lb slab of marble is taken out of a quarry (see Figure 8–24). Assume a coefficient of sliding friction of 0.25.

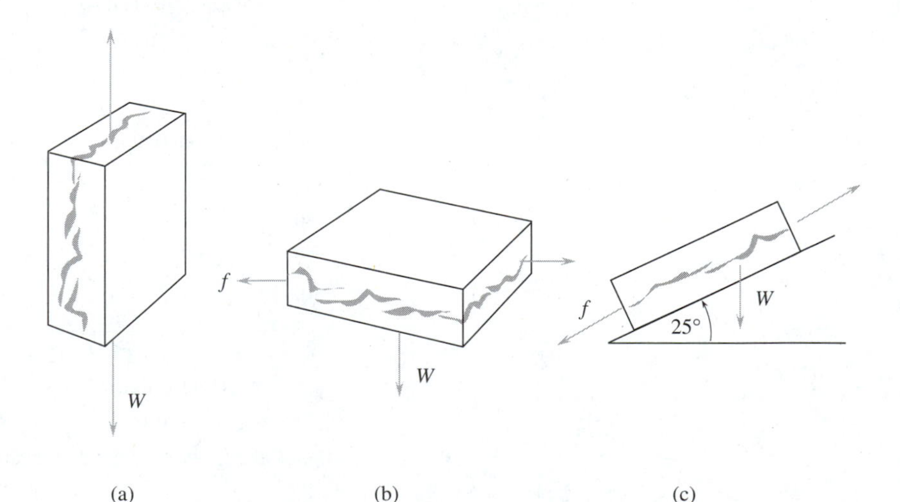

Figure 8–24: *A heavy marble slab is: (a) lifted vertically, (b) dragged horizontally, and (c) pulled up a 25° incline.*

A. The slab is first lifted out of the excavation vertically. What is the tension on the cable?
B. The slab is dragged along the ground. What is the tension on the cable?
C. The slab is loaded on a truck by pulling it up a 20° incline. What is the tension on the cable?

3. A 70.0-lb wooden crate is moved along a wooden floor at constant speed as shown in Figure 8–25. Find the sliding frictional force for each method. (See Table 8–1.)

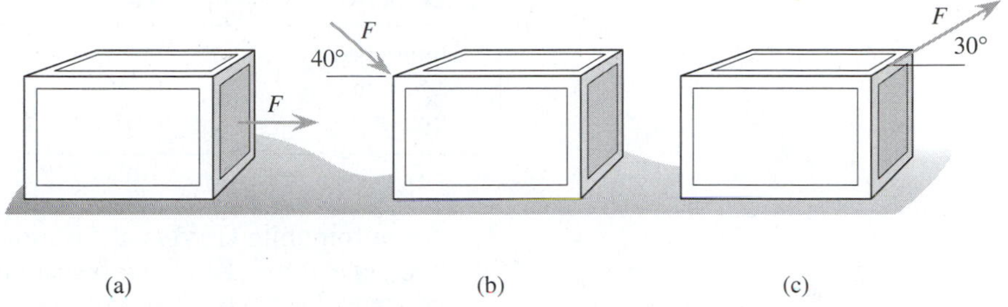

Figure 8–25: Three methods of moving a crate along a wooden floor.

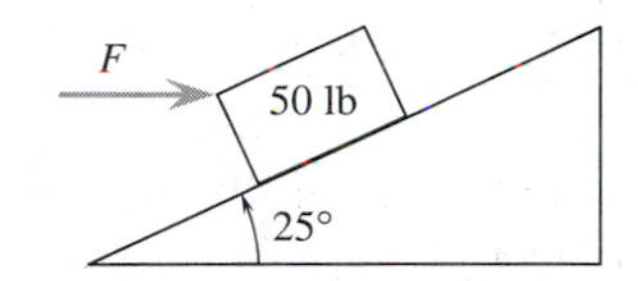

Figure 8–26: Diagram for Problem 8. A horizontal force pushes a crate up an incline.

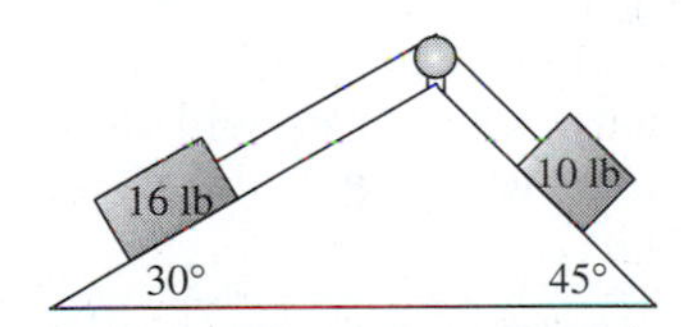

Figure 8–27: Diagram for Problem 9. Two blocks are connected by a string looped through a pulley.

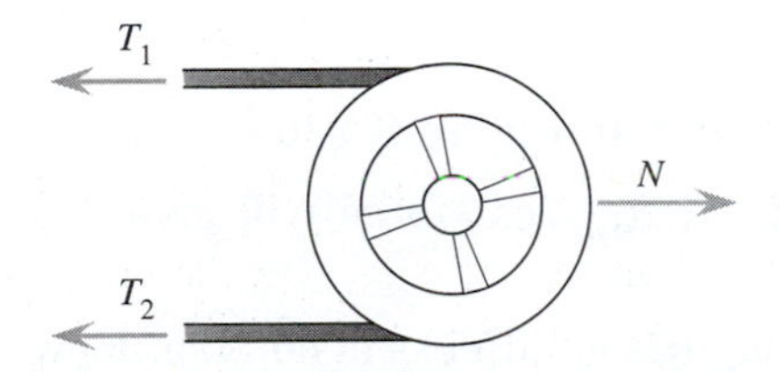

Figure 8–28: Diagram for Problem 10. The forces on a pulley driven by a V-belt.

4. Determine whether or not a 25-kg bundle of shingles will slide off a roof with a slope of 40° if the static coefficient of friction is 0.80.

5. Find the sliding friction of a $7\bar{0}$-kg skier hurtling down a 45° slope on waxed skis when the temperature is $-1\bar{0}$°C.

6. Find the braking distance of a $46\bar{0}0$-lb car skidding to a stop on ice at 0.0°C. Assume an initial speed of $4\bar{0}$ mph.

7. A block is placed on an incline. The incline is tilted upward to an angle θ_s at which the block begins to slide. Show the static coefficient of friction if $\mu = \tan \theta_s$.

8. A horizontal force pushes a 50-lb wooden crate up a 25° wooden incline (see Figure 8–26). Calculate:
   A. the force required to start the crate moving up the incline.
   B. the force required to keep the crate moving up the incline at a constant speed.

9. If the blocks in Figure 8–27 are initially at rest, will they start to move? The static coefficient of friction between the blocks and the inclines is 0.30.

10. For the V-belt drive shown in Figure 8–28, T_1 is $14\bar{0}$ N and T_2 is 45 N.
    A. What is the normal force the pulley exerts on the belt?
    B. What is the net driving force the belt delivers to the pulley?
    C. If the static coefficient of friction between the V-belt and the pulley is 0.70, will the belt slip or not?

Section 8.2:

11. Different-sized raindrops fall at different rates. Find the terminal velocities of the raindrops in Table 8–2. The density of air is 1.3 kg/m^3 at sea level. Use a drag coefficient of 0.40.

Table 8–2: Raindrops.

PRECIPITATION	RADIUS (cm)	MASS (kg)	CROSS-SECTIONAL AREA (m^2)
Thunderstorm	0.40	2.7×10^{-4}	5.0×10^{-5}
Light rain	0.10	4.2×10^{-6}	3.1×10^{-6}
Drizzle	0.01	4.2×10^{-9}	3.1×10^{-8}

12. An automobile travels at 55 mph. The car has a cross-sectional area of 2.9 m^2. Find the frictional drag on the car. The density of air is 2.5×10^{-3} slug/ft^3.

13. The car in Problem 12 travels into a 15-mph head wind. What is the new frictional drag? (The density of air is 2.5×10^{-3} slug/ft^3.)

14. A baseball has a radius of 3.66 cm and a mass of 0.145 kg.
    A. Find the viscous force on a baseball hurled horizontally at a speed of 90 mph. Air density is 1.3 kg/m^3. Assume a drag coefficient of 0.50.
    B. What is the horizontal deceleration of the baseball?

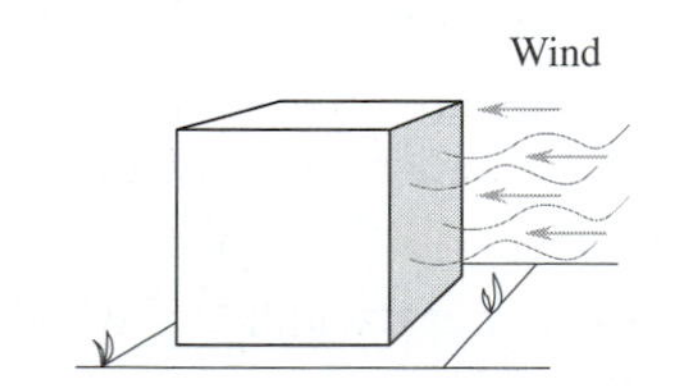

Figure 8–29: Diagram for Problem 16. Wind pushes on a square box.

15. A carpenter's apprentice carries a 4.0 ft $\times$ 8.0 ft sheet of plywood in a 25-mph wind. If the apprentice holds the sheet of plywood so the flat face is into the wind, with what force will the wind push against the plywood? Use a drag coefficient of 0.90 and an air density of 2.5×10^{-3} slug/ft^3.

16. A 1.80-kg square box 0.6 m on a side sits on a concrete sidewalk (see Figure 8–29). The static coefficient of friction between the box and the sidewalk is 0.60. Determine by calculation whether or not the box will slide along the sidewalk in an 8.0 m/s wind. Use a drag coefficient of 0.90. (The density of air is 1.3 kg/m^3.)

Section 8.3:

17. Find the stiffness of the steel wire in Figure 8–23b.

18. A spring has a stiffness of $2\bar{0}$ lb/ft. What force will give it a displacement of 1.0 in?

19. An unloaded spring 0.40 m long has a stiffness of $5\bar{0}$ N/m. How long will it be when a mass of 1.5 kg is suspended from it?

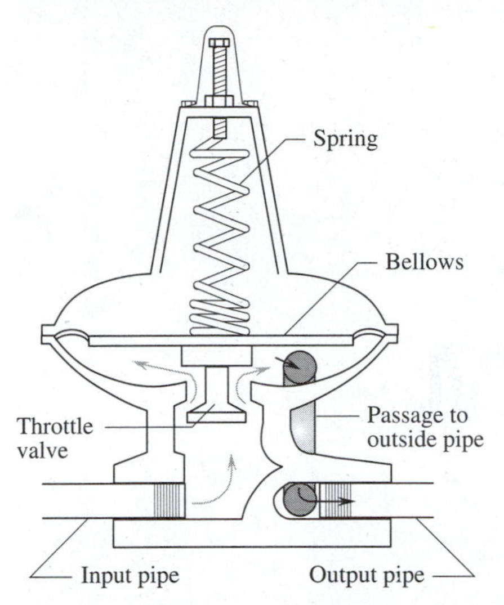

Figure 8–30: Diagram for Problem 20. A flow regulator. Fluid pushing against the bellows compresses the spring. When the spring is compressed the throttle valve closes.

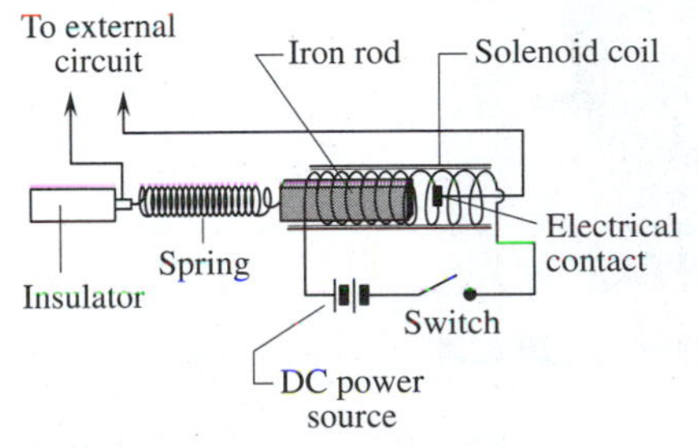

Figure 8–31: When the switch is closed a magnetic force inside the solenoid coil pulls the iron rod in against the electrical contact closing an external circuit. When the switch is opened the spring pulls the rod out of the coil, breaking contact.

20. The compressional spring in the flow regulator shown in Figure 8–30 has a force constant of 2500 N/m. With what force must the fluid push against the bellows to cause the throttle valve to close by 1.0 mm?

21. Figure 8–31 shows a simplified relay. When current passes through the solenoid coil a magnetic force is produced to pull the iron rod into the solenoid. The rod makes electrical contact, closing a second electrical circuit. When the current stops flowing through the coil, the spring pulls the rod away from the contact. If a magnetic force of 0.25 N moves the rod 3.0 mm, what is the stiffness of the spring?

22. A spring has a length of 1.2 ft with a load of 2.0 lb. When its load is 4.0 lb, it is 1.4 ft long.

A. What is the spring's force constant?

B. What is the length of the spring with no load?

Section 8.4:

23. What is the gravitational attraction between two 1.0-kg masses separated by 1.0 m?

24. Calculate the gravitational attraction between a 60-kg filing cabinet and a 75-kg physics professor.

25. Mars has a mass of 6.4×10^{23} kg. Its diameter is 6790 km. With what acceleration would a carpenter's hammer fall on the surface of Mars?

26. The mass of the sun is 2.0×10^{30} kg. The average distance between the sun and Mars is 2.28×10^{8} km. What is the gravitational force between Mars and the sun?

27. At what distance from the earth will a 16-lb wood maul have a weight of 2.0 lb? Express your answer in earth radii.

28. A spaceship travels between the earth and the moon in a straight line. At what distance from the earth will the gravitational pull of the moon equally balance the pull of the earth on the spaceship? The mass of the earth is 6.0×10^{24} kg; the moon, 7.35×10^{22} kg. The earth-moon distance is 3.84×10^{5} kg.

Chapter 9

EQUILIBRIUM

This unusual kite was flown in Bermuda, where kite-flying is a tradition on Good Friday. The concepts in this chapter will help you learn how to get a kite in the air and keep it flying. (Photograph courtesy of Capital Newspapers/the Times Union.*)*

OBJECTIVES

In this chapter you will learn:

- the difference between concurrent and nonconcurrent forces
- about sets of forces that will *not* cause the linear acceleration of an object
- how to calculate torques
- about sets of torques that will *not* cause rotational acceleration of an object
- how to find the center of gravity of an object
- to calculate the forces and torques acting on a system in total equilibrium
- to distinguish among stable, unstable, and neutral states of equilibrium
- how to apply the idea of equilibrium to a variety of situations

Time Line for Chapter 9

391 BC	Chinese artisan Lu Pan invents the first kite.
1686	Isaac Newton publishes the *Principia, De Motu,* the motion of bodies.
1743	Jean le Rond d'Alembert describes equilibrium of internal forces inside a system and develops equilibrium mechanics.
1761	John Rennie builds the Waterloo and London bridges.
1881	J. Willard Gibbs develops a system of three-dimensional vectors.

When all of the forces acting on something add up to zero, they are in **equilibrium,** which was discussed briefly in Chapters 7 and 8. If an object is at rest or if it has a constant velocity, we know the forces acting on it add up to zero. A drive shaft rotating at a constant speed; the matrix of rafters, sills, and braces that constitute a gable roof; and the stress exerted on a power pole by its transmission lines are all examples of systems in equilibrium. If the forces are in equilibrium, a new question arises. What are the size and direction of an unknown force in the system? The study of forces in unaccelerated systems is called **statics.**

9.1 THE FIRST CONDITION OF MECHANICAL EQUILIBRIUM

Systems of forces can be classified into two categories: concurrent and nonconcurrent forces. Figure 9–1a shows one kind of force sys-

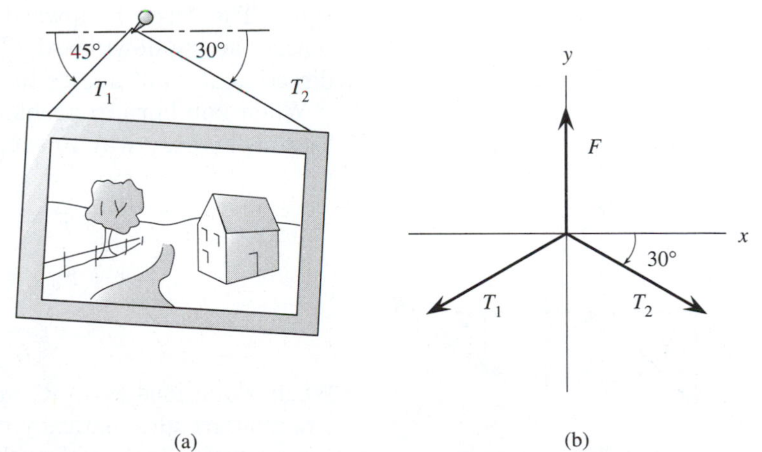

Figure 9–1: *Concurrent forces.* (a) *A picture hung on a nail.* (b) *Free-body diagram. Forces act through a common point.*

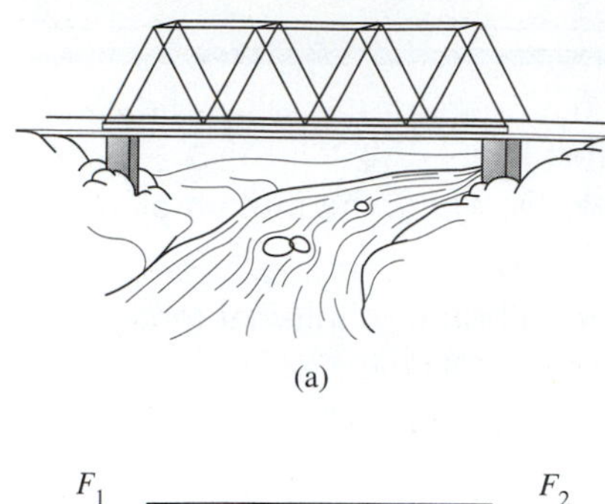
(a)

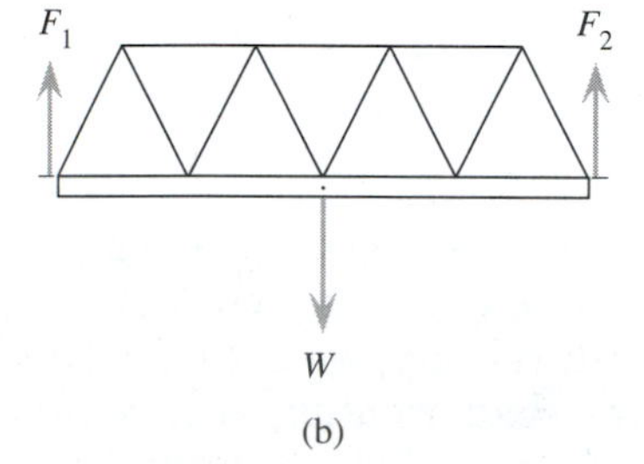

(b)

Figure 9–2: *Nonconcurrent forces. (a) A bridge. (b) Free-body diagram. Forces do not act through a common point.*

tem. A picture is hung from a nail. The wire has three forces acting on it. The nail pushes upward on the wire, and the picture pulls downward causing the two sections of wire to be under tension, $\mathbf{T}_1$ and $\mathbf{T}_2$. Figure 9–1b shows the vector diagram for the hung picture. The three forces all pass through one common point. Forces that pass through a single point are called **concurrent forces.**

Figure 9–2a shows a bridge held up by two abutments. The free-body diagram is shown in Figure 9–2b. The weight of the bridge acts downward, while the abutments push upward on the ends of the span. These forces do **not** pass through a common point. Forces that do not pass through a common point are called **nonconcurrent forces.**

For the moment we will consider only concurrent forces. If an object has no linear acceleration, it is in equilibrium with respect to forces. In other words, the sum of the forces acting on the object is zero.

$$\sum \mathbf{F} = 0 \qquad \textbf{(Eq. 9–1)}$$

This vector equation describes the **first condition of mechanical equilibrium.** The component forces must add up to zero if the forces are balanced.

$$\sum F_x = 0$$
$$\sum F_y = 0 \qquad \textbf{(Eq. 9–2)}$$
$$\sum F_z = 0$$

Figure 9–3 shows the components of the forces T_1 and T_2 in Figure 9–1b. Force T_{1x}, to the left, must balance force T_{2x}, acting to the right. The force F upward must balance the two components of tension acting downward, T_{1y} and T_{2y}. There are no forces in the z direction at right angles to the xy plane.

When equilibrium problems are expressed in component form, there is an easy way to set up the force equations.

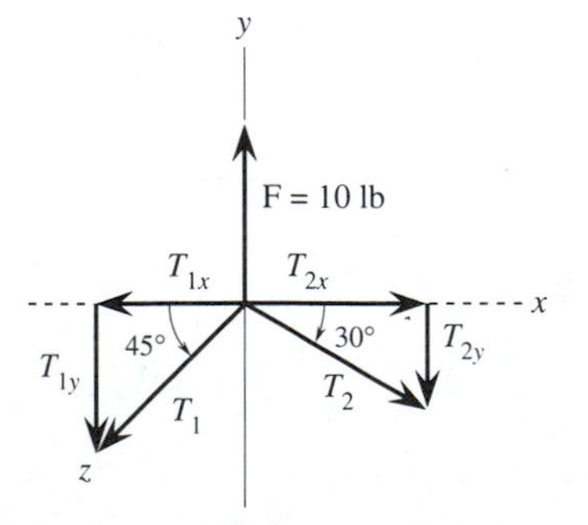

Figure 9–3: *Vectors can be resolved into horizontal and vertical components.*

$$F_{up} = F_{down} \qquad \textbf{(Eq. 9–3A)}$$

$$F_{left} = F_{right} \qquad \textbf{(Eq. 9–3B)}$$

$$F_{in} = F_{out} \qquad \textbf{(Eq. 9–3C)}$$

When Equations 9–3A–C are used, the negative signs indicating direction are automatically controlled. Simply place the magnitude of each component on the proper side of the equations. No negative signs are needed.

EXAMPLE PROBLEM 9–1: THE HANGING PICTURE

A $1\bar{0}$-lb picture is hung from a nail by a string. The angles the string makes with the horizontal are shown in Figure 9–3. Find the tensions in the string.

SOLUTION

We know that the force the nail exerts upward on the string must balance the weight of the picture. The vertical components of the two tensions act downward.

$$F_{up} = T_{1y} + T_{2y}$$

$$1\bar{0}\text{ lb} = (T_1 \sin 45°) + (T_2 \sin 3\bar{0}°)$$

or

$$(0.707\ T_1) + (.500\ T_2) = 1\bar{0}\text{ lb}$$

The horizontal components of tension are equal since there are no other horizontal forces.

$$F_{left} = F_{right}$$

$$T_{1x} = T_{2x}$$

$$T_1 \cos 45° = T_2 \cos 3\bar{0}°$$

$$0.707\ T_1 = 0.816\ T_1$$

Solve T_2 in terms of T_1.

$$T_2 = 1.414\ T_1$$

T_2 can be eliminated from the equation of vertical components by substitution.

$$(0.707\ T_1) + [0.866\ (1.414\ T_1)] = 1\bar{0}\text{ lb}$$

$$(0.707\ T_1) + [0.500\ (0.816\ T_1)] = 1\bar{0}\text{ lb}$$

or

$$T_1 = 9.\bar{0}\text{ lb}$$

T_2 is $0.816\ T_1$.

$$T_2 = 7.3\text{ lb}$$

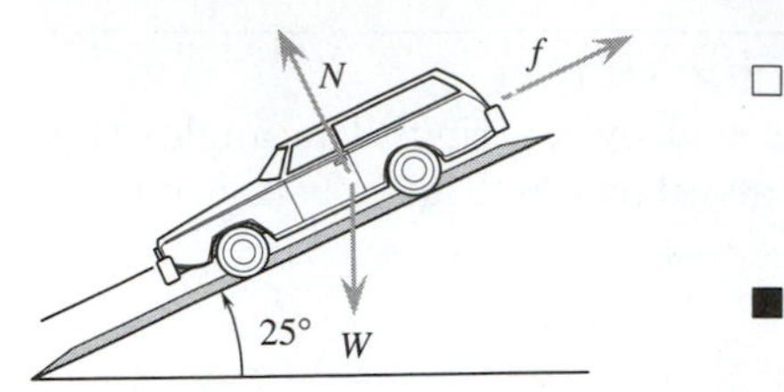

Figure 9–4: Forces on a car parked on an incline.

EXAMPLE PROBLEM 9–2: THE PARKED CAR

An 1800-kg car is parked on a 25° hill. What static frictional force prevents the car from sliding down the hill?

SOLUTION

Figure 9–4 shows the forces on the car. **N** is the normal force. The frictional force (**f**) is parallel to the hill, and the weight (**W**) acts downward.

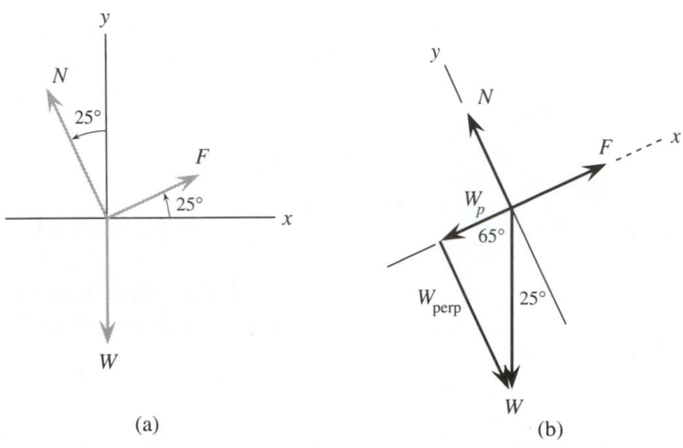

Figure 9–5: Two ways to draw the coordinate axes. (a) A horizontal coordinate system. (b) The coordinate axis is parallel to **f.**

Figure 9–5 shows two ways to draw the vector diagram. In Figure 9–5a the x axis is horizontal. In order to solve the problem we must calculate the normal force as well as the frictional force. Figure 9–5b shows a simpler approach. The x and y axes are rotated 25°. The frictional force is now parallel to the x axis. It is balanced by the parallel component of weight (W_P). We no longer need to know the value of **N.**

$$f = W_P$$

$$f = (m\,g) \sin 25°$$

$$f = (1800\text{ kg} \times 9.8\text{ m/s}^2) \times 0.423$$

$$f = 7500\text{ N}$$

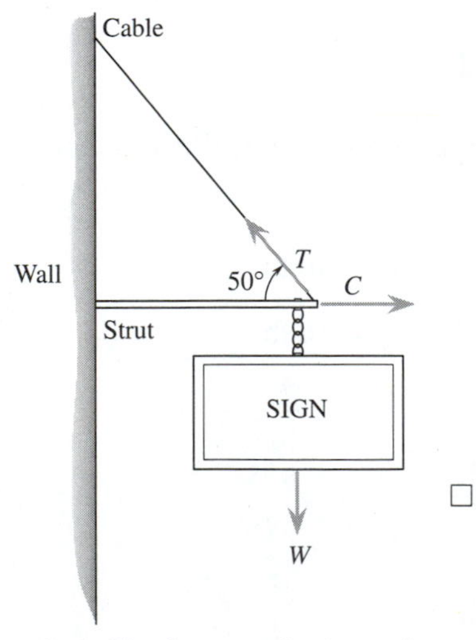

Figure 9–6: The forces acting on a sign.

EXAMPLE PROBLEM 9–3: THE SIGN

A 40-lb sign is hung from a cable supported by a light, horizontal strut (see Figure 9–6). Find the tension in the cable and the compressional force (**C**) on the strut. The strut pushes outward horizontally. The term **compressional force** refers to the fact that the cable pushes back with an inward-reacting force that will compress the strut.

■ *SOLUTION*

Write the component force equations.

$$F_{up} = \mathbf{F}_{down}$$

$$T_y = \mathbf{W}$$

$$T \sin 5\bar{0}^\circ = 4\bar{0} \text{ lb}$$

$$T = \frac{4\bar{0} \text{ lb}}{0.766}$$

$$T = 52 \text{ lb}$$

$$F_{right} = \mathbf{F}_{left}$$

$$C = \mathbf{T} \cos 5\bar{0}^\circ$$

$$C = 52 \text{ lb} \times 0.643$$

$$C = 33 \text{ lb}$$

9.2 TORQUES

Place a book on a flat surface. With your fingertips on opposite sides of the book at the center, push inward (see Figure 9–7). Not much happens. You are exerting equal but oppositely directed concurrent forces. The system is in equilibrium. Now push inward on diagonal corners (see Figure 9–8). You are now creating a pair of noncon-current forces. The book rotates even though the forces are of equal size. Look at Figure 9–8b. If the upper force were pulling at point b, rather than pushing at point a, there would be the same vertical separation between forces. If we attached a string at point c and

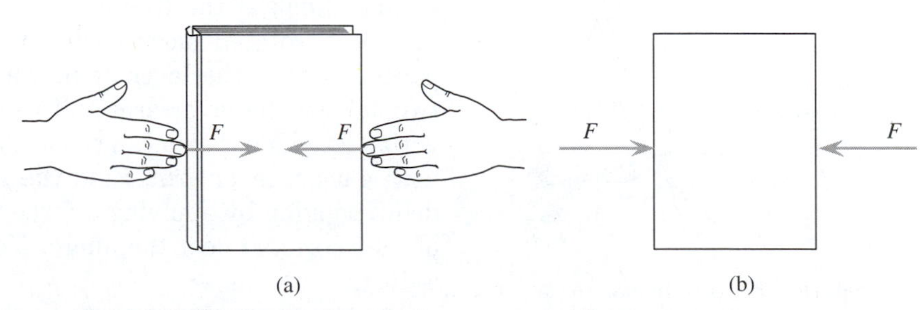

Figure 9–7: *A pair of action-reaction concurrent forces.*

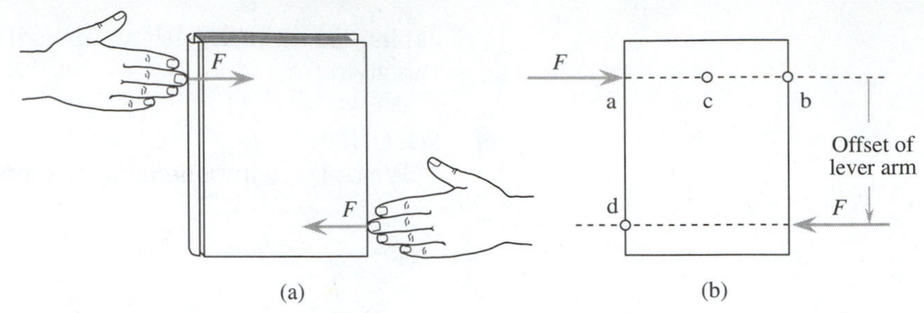

Figure 9–8: A pair of action-reaction nonconcurrent forces.

pulled on it, we would get the same rotational effect as before. Sliding the bottom force to point d would result in the same tendency to rotate.

If we draw a line along the direction of an applied force, the force can be considered as being applied anywhere along that line. Mathematically we get the same tendency of the object to rotate. This is a **line of action.** The dashed lines in Figures 9–8 and 9–9 are lines of action. Push again on the book. This time push inward with the fingers held near the center of the book, but offset by only an inch or two (see Figure 9–9). The book will again rotate, but not as easily as before.

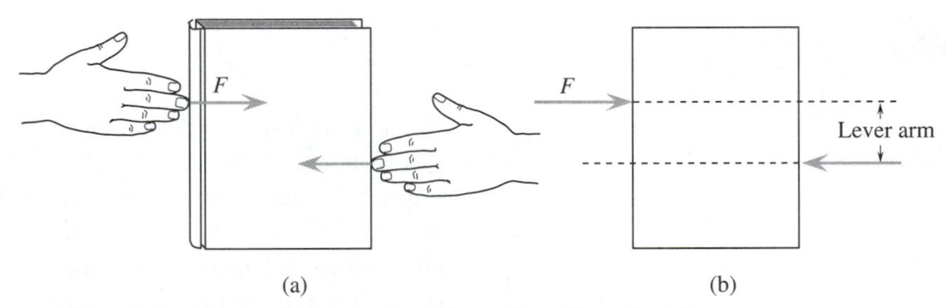

Figure 9–9: The book does not rotate easily with a short lever arm.

Apparently nonconcurrent forces can cause something to spin even though the forces add up to zero. Forces that cause rotation exert **torque.** Look at the free-body diagrams (Figures 9–8b and 9–9b). The book rotated more easily when the **offset** between the two parallel forces, the **lever arm,** was larger. The torque depends on two things: the lever arm and the size of the force. A small force with a large lever arm can cause the same rotation as a larger force with a short lever arm. Find the minimum force needed to start a door swinging by applying a force near the doorknob, at the center of the door, and near the hinge. Your result would look like Figure 9–10.

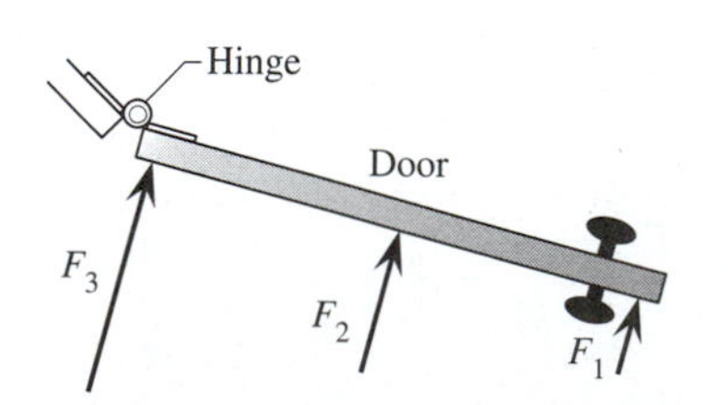

Figure 9–10: The force needed to swing a door is largest near the hinge.

Try one more experiment. Push on the edge of the door with a

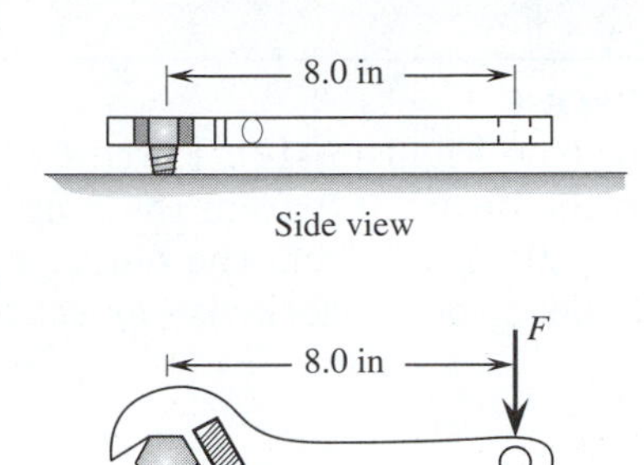

Figure 9–11: *The torque on a wrench.*

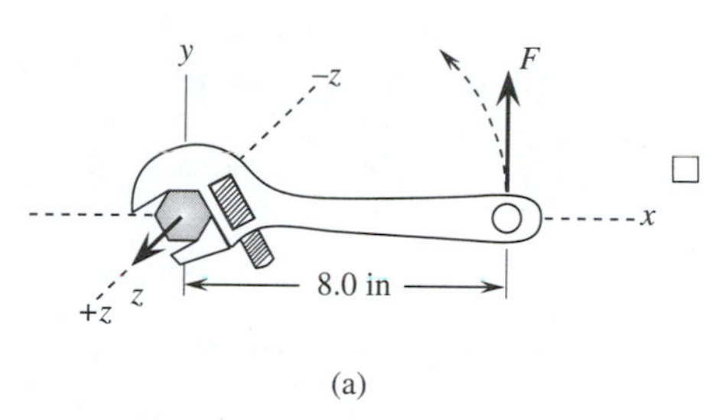

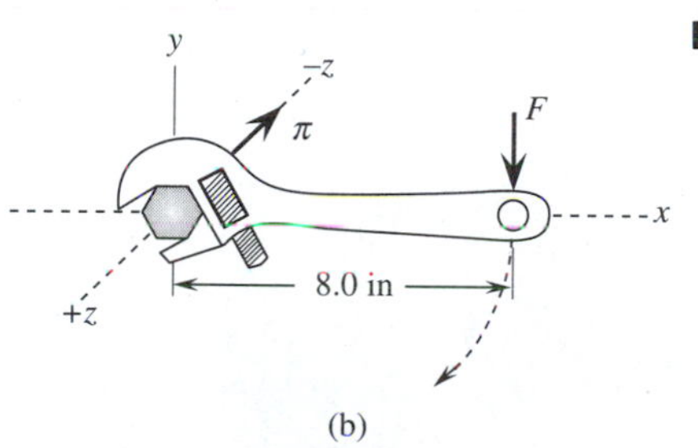

Figure 9–12: *Positive and negative torques. (a) A counterclockwise torque moves the bolt in the +z direction. (b) A clockwise rotation advances the bolt in the −z direction.*

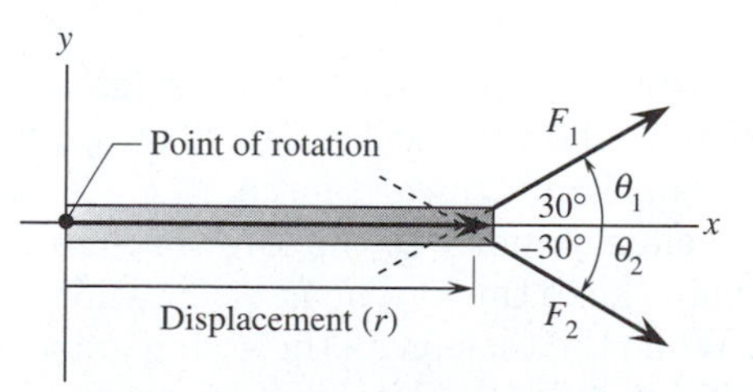

Figure 9–13: *When measured from the lever arm to the force, the angle is positive for a counterclockwise torque and negative for a clockwise torque.*

force directed at the hinge. Only forces at right angles to the lever arm, in this case the door, are involved with rotation. If the force is applied at a slant, only the component at right angles to the lever arm is involved with the torque. We can summarize our results in an equation. τ is the symbol for torque.

$$\tau = F\,r \sin\theta \qquad \textbf{(Eq. 9–4)}$$

where **F** is the force, **r** is the displacement between the rotational axis and the point at which the force is applied (in other words, the lever arm), and θ is the angle between the displacement and the force. You may notice that this is the vector cross product.

$$\boldsymbol{\tau} = \mathbf{r} \times \mathbf{F} \qquad \textbf{(Eq. 9–4A)}$$

EXAMPLE PROBLEM 9–4: THE LAG BOLT

A lag bolt is a large wood screw that has a head shaped so that a wrench rather than a screwdriver can be used. To turn a lag bolt, a force of 15 lb is exerted at the end of an 8.0-in wrench at right angles (see Figure 9–11). How large is the torque?

SOLUTION

$$\textit{torque} = \textit{force} \times \textit{lever arm} \times \sin\theta$$

$$\tau = 15\text{ lb} \times 8.0\text{ in} \times \frac{1.0\text{ ft}}{12.0\text{ in}} \times \sin 90^\circ$$

$$\tau = 1\bar{0}\text{ lb}\cdot\text{ft}$$

Torque has a direction. If the bolt in Example Problem 9–4 is turned clockwise, it advances into the wood. If the torque is counterclockwise, the bolt backs out of the wood. We need to keep track of the direction.

Figure 9–12 shows a three-dimensional right-handed coordinate system. Normally the $+z$ direction is drawn coming out of the page. In Figure 9–12a a counterclockwise torque is applied. The bolt moves out of the page in the $+z$ direction. In Figure 9–12b the torque is clockwise. The bolt moves inward in the $-z$ direction. Use the following convention to indicate the direction of a torque. Counterclockwise torques are positive. Clockwise torques are negative.

Here is another way to figure out if a torque is positive or negative. The angle θ is measured from the lever arm (r) to the force (**F**). Look at Figure 9–13. Force $\mathbf{F}_1$ exerts a counterclockwise torque. Angle θ_1 is positive. Force $\mathbf{F}_2$ exerts a clockwise torque. Angle θ_2 is negative.

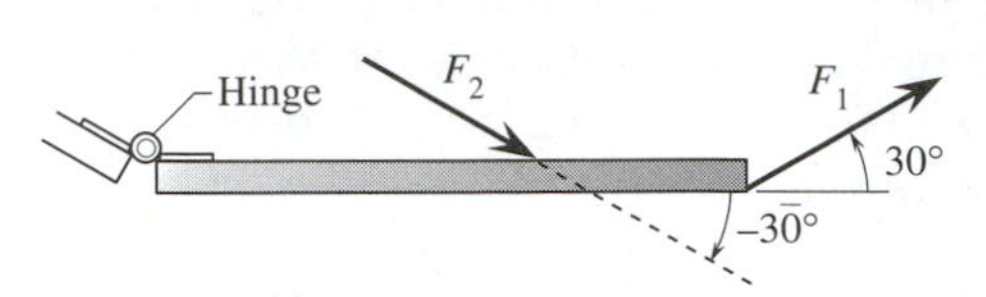

Figure 9–14: *(Top view) Nonconcurrent forces on a door.*

EXAMPLE PROBLEM 9–5: A DOOR TORQUE

Two forces are exerted on a door (see Figure 9–14). Force $\mathbf{F}_1$ is 15 lb at an angle of $3\bar{0}°$ pushing on the door 2.0 ft from the hinge. Force $\mathbf{F}_2$ is applied at an angle of $-3\bar{0}°$, 1.3 ft from the hinge. $\mathbf{F}_2$ has a magnitude of $3\bar{0}$ lb. Will the door rotate clockwise or counterclockwise?

SOLUTION

We can add the torques keeping track of − and + signs to find the net torque.

$$\tau_{net} = (\mathbf{F}_1 \times r_1 \times \sin \theta_1) + (\mathbf{F}_2 \times r_2 \times \sin \theta_2)$$

$$\tau_{net} = (15 \text{ lb x } 2.0 \text{ ft} \times \sin 3\bar{0}°) + [3.0 \text{ lb} \times 1.3 \text{ ft} \times \sin(-3\bar{0}°)]$$

$$\tau_{net} = +15 \text{ ft} \cdot \text{lb} - 19.5 \text{ ft} \cdot \text{lb}$$

$$\tau_{net} = 4.5 \text{ ft} \cdot \text{lb}$$

The net torque is +4.5 ft · lb in the counterclockwise direction.

How Kites Fly

Kites have been around for a long time. A Chinese artisan named Lu Pan invented the first kite in 391 B.C.

A kite in flight is a study of balanced forces.

The air passing over the curved top of the kite creates an upward force called lift. The amount of lift depends on the speed of the air over the kite. The stronger the wind, the stronger the lift will be. Running to get the kite in the air is one way to increase the lift. Pulling on the string when the kite is in the air is another way to get a larger lift. This pulls the kite into the wind. The speed of the air over the kite goes up. The tail acts as a drag holding the bottom of the kite down so the wind blows across the correct part of the kite. The lift is balanced partly by the weight of the kite acting downward. Wind striking the kite pushes it forward.

The tension on the string balances the forward push of the wind, holding the kite motionless. The string has other functions. It has a downward component that, along with the weight of the kite, balances the lift. If the tension is slackened as more string is played out, the kite will climb. The lift becomes larger than the downward force.

(Photograph courtesy of Capital Newspapers/*The Times Union.)*

The string also adds stability to the kite through the harness. It helps keep the kite in an orientation for the best lift. When the tension on the string gets too weak, the wind will flip the kite so it offers the least air resistance. In this position, the lift is greatly reduced and the kite falls.

9.3 CENTER OF GRAVITY

In Section 6.6 we found how to find the center of mass of an object. In the presence of a uniform gravitational field this is also the **center of gravity.** When torques acting on an object are calculated, the weight of the object behaves as though it were located at the center of gravity. The weight of an object acting at the center of gravity can cause a torque.

☐ **EXAMPLE PROBLEM 9–6: TORQUE ON A METER STICK**

A $5\bar{0}$-g meter stick is supported on a finger at the $6\bar{0}$-cm mark. What is the size of the torque causing the meter stick to rotate around the finger (see Figure 9–15)?

■ ***SOLUTION***

We assume that the meter stick has a uniform density. The center of gravity is located at the geometric center of the stick at the $5\bar{0}$-cm mark. The meter stick will rotate around the $6\bar{0}$-cm mark, the only point of support. The displacement (r) is ($6\bar{0}$ cm $-$ $5\bar{0}$ cm), or $1\bar{0}$ cm. Convert the units into SI units and use the torque equation. The angle between the lever arm and the force is $9\bar{0}°$; $\sin 9\bar{0}° = 1.00$. The force in Newtons is the weight of the stick, 0.050 kg × 9.8 m/s^2.

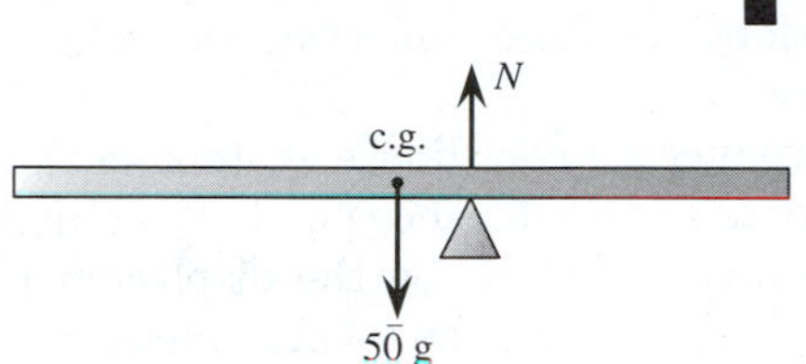

Figure 9–15: *The center of gravity (c.g.) of a meter stick can produce a torque. **N,** normal force.*

$$\tau = 0.050 \text{ kg} \times 9.8 \text{ m/s}^2 \times 0.100 \text{ m}$$

$$\tau = 0.049 \text{ N} \cdot \text{m}$$

There is a simple method of finding the center of gravity of something that does not have a uniform density or that has an odd shape. Hold a book between the thumb and index finger as in Figure 9–16a. The center of gravity causes a torque. This makes the book

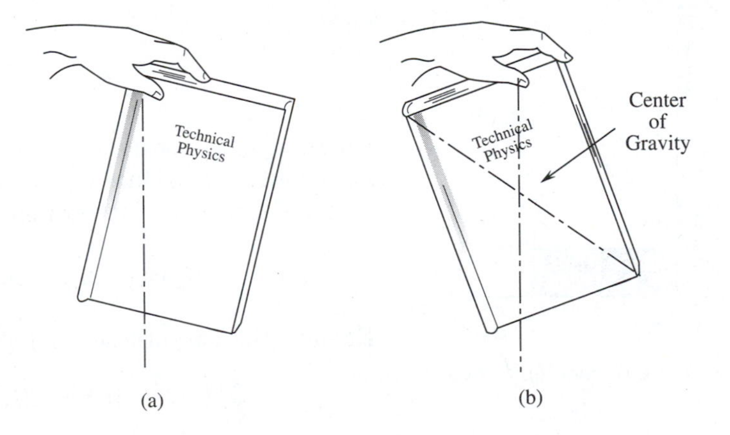

Figure 9–16: *(a) The center of gravity falls below the point of support. (b) The center of gravity is located where the two lines cross.*

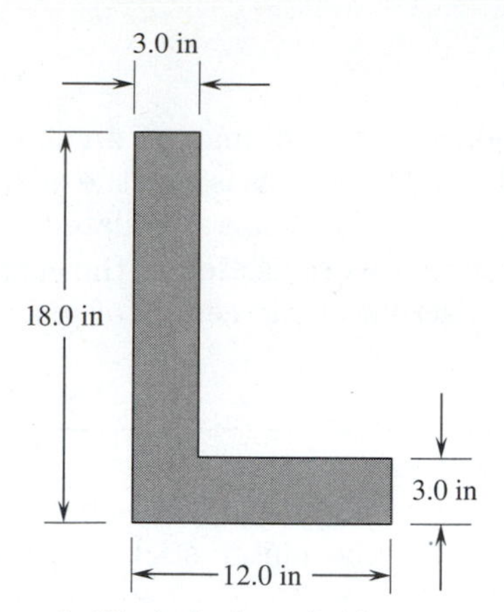

Figure 9–17: *An L-shaped object.*

rotate. Before long the center of gravity lies below the point of support. In this position, the torque on the book is zero because the angle between the force and the displacement is zero. The displacement and the line of action of the weight are parallel ($\sin \bar{0}° = 0.00$). Draw a vertical line from the point of support downward. We know the center of gravity must lie somewhere along this line.

Next hold the book by another corner and repeat the process (see Figure 9–16b). The center of gravity is located where the two lines cross. You should be able to balance the book at that point. This method works for objects of any shape.

We can calculate the center of gravity of an object that has a uniform density, but lacks symmetry. Figure 9–17 shows the L shape of the cross section of a piece of angle iron. We can think of the L as being made of two rectangles. The centers of gravity of the rectangles are at their geometric centers. The center of gravity of the whole L must lie somewhere along the line connecting the centers of gravity of the two rectangles.

In Figure 9–18 we have introduced a coordinate system so that we can express distance relative to some reference point. The object will balance at the center of gravity. Let $\mathbf{R}_1$ be the displacement from the center of gravity to M_1. Let $\mathbf{R}_2$ be the displacement from the center of gravity to M_2. These displacements can be broken into components. Let us look at the x components, x_1 and x_2.

We can use the balance equation (Equation 9–4). The L would balance at the center of gravity. The torques caused by the two parts of the L must balance.

$$M_1 x_1' = M_2 x_2'$$

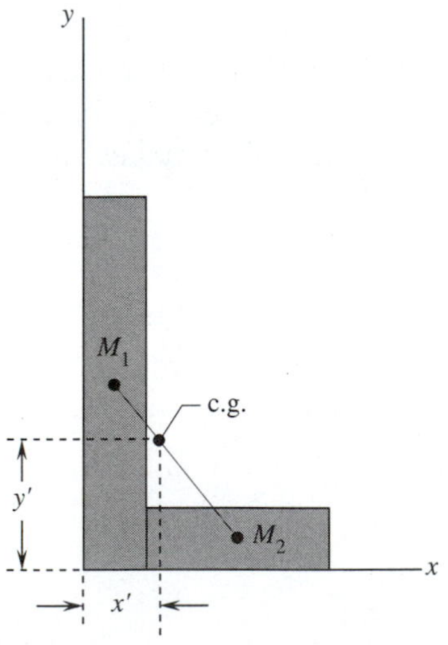

In terms of the coordinate reference point:

$$x_1' = x_c - x_1$$

and

$$x_2' = x_c + x_2$$

where x_c is the position of the center of gravity. Substitute these values into the balance equation. This gives us the balance point in terms of the coordinate system.

$$M_1 (x_c - x_1) = M_2 (x_2 + x_c)$$

Remove the parentheses and group the terms with x_c on the left.

$$x_c (M_1 + M_2) = (M_1 x_1) + (M_2 x_2)$$

Figure 9–18: *The L is made up of two rectangles.*

or

$$x_c = \frac{(M_1\,x_1) + (M_2\,x_2)}{M_1 + M_2}$$

The pattern seems to be: *add up the products of mass and position and then divide by the total mass*. This will give the position of the center of gravity. Although this equation has been developed for only two masses, the pattern works for three or more masses. We can put this pattern in a general form. For i masses the center of mass is:

$$x_c = \frac{\Sigma\, m_i\, x_i}{\Sigma\, m_i} \qquad \textbf{(Eq. 9–5A)}$$

The same equation can be developed for other components.

$$y_c = \frac{\Sigma\, m_i\, y_i}{\Sigma\, m_i} \qquad \textbf{(Eq. 9–5B)}$$

$$z_c = \frac{\Sigma\, m_i\, z_i}{\Sigma\, m_i} \qquad \textbf{(Eq. 9–5C)}$$

Notice that Equations 9–5A–C find the center of mass rather than the center of gravity. To find the center of gravity, weight should be used rather than mass. In most cases, the gravitational field (**g**) is a constant and cancels out of the equation. In most everyday situations the center of mass and the center of gravity coincide.

Notice also that these equations are nothing more than a weighted average. If tests have a weight of 10 times that of a quiz, and you have a test grade of 86 and a quiz grade of 100, you can use Equation 9–5A to find the average, if x's are grades, and m's are weights.

$$x_c = \frac{(10 \times 86) + (1 \times 100)}{10 + 1}$$

$$x_c = 87.3$$

☐ **EXAMPLE PROBLEM 9–7: CENTER OF GRAVITY OF L**

Find the center of gravity of the L-shaped bracket shown in Figures 9–17 and 9–18. The object is 1.00 in thick and has a constant density.

■ ***SOLUTION***

Since the bracket is uniformly thick and has a constant density we can use a shortcut in calculating the center of mass.

$$mass = density\ (\rho) \times volume = density \times area \times thickness$$

or

$$m = \rho\, A\, t$$

Equation 9–5A can be rewritten.

$$x_c = \frac{\Sigma\, \rho\, A_i\, t\, x_i}{\Sigma\, \rho\, A_i\, t}$$

Since ρ and t are constant they can be brought outside the summation.

$$x_c = \frac{t\, \rho\, \Sigma\, A_i\, x_i}{t\, \rho\, \Sigma\, A_i} = \frac{\Sigma\, A_i\, x_i}{\Sigma\, A_i}$$

Density and thickness cancel out of the equation. In this special case, the center of mass depends only on the distribution of areas.

Find the coordinates of M_1 and M_2 in Figure 9–18. The center of masses of the two rectangles will be at the geometric centers.

$$x_1 = \frac{(0.0 + 3.0)\text{ in}}{2} = 1.5\text{ in}$$

$$y_1 = \frac{(0.0 + 18.0)\text{ in}}{2} = 9.0\text{ in}$$

$$x_2 = \frac{(3.0 + 12.0)\text{ in}}{2} = 7.5\text{ in}$$

$$y_2 = \frac{(0.0 + 3.0)\text{ in}}{2} = 1.5\text{ in}$$

The areas are:

$$A_1 = 3.0\text{ in} \times 18.0\text{ in} = 54\text{ in}^2$$

$$A_2 = 3.0\text{ in} \times 12.0\text{ in} = 36\text{ in}^2$$

The coordinates of the center of gravity are:

$$x_c = \frac{(A_1\, x_1) + (A_2\, x_2)}{A_1 + A_2}$$

$$x_c = \frac{(54\text{ in}^2 \times 1.5\text{ in}) + (36\text{ in}^2 \times 7.5\text{ in})}{54\text{ in}^2 + 36\text{ in}^2}$$

$$x_c = 3.9\text{ in}$$

$$y_c = \frac{(54 \text{ in}^2 \times 9.0 \text{ in}) + (36 \text{ in}^2 \times 1.5 \text{ in})}{54 \text{ in}^2 + 36 \text{ in}^2}$$

$$y_c = 6.0 \text{ in}$$

Figure 9–19 shows the location of the center of gravity. Notice it is outside the L.

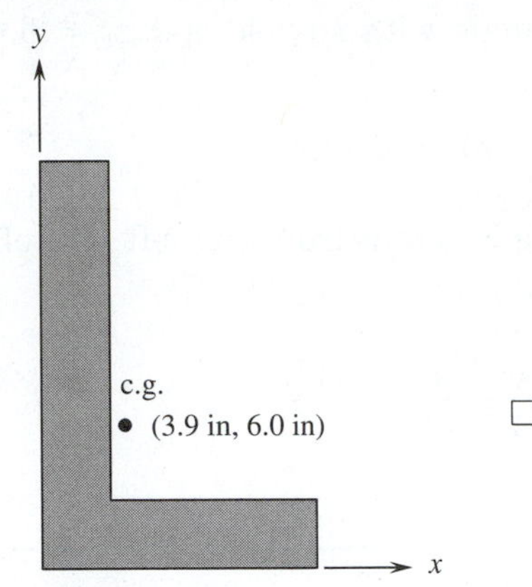

Figure 9–19: The center of gravity (c.g.) is outside the object.

EXAMPLE PROBLEM 9–8: CENTER OF GRAVITY OF A TIE-ROD

Find the center of gravity of the tie-rod shown in Figure 9–20. The sphere on the end is part of a ball-and-socket joint. The sphere has a mass of 200 g; the rod, 500 g. Where along the tie-rod would it balance? Assume uniform density.

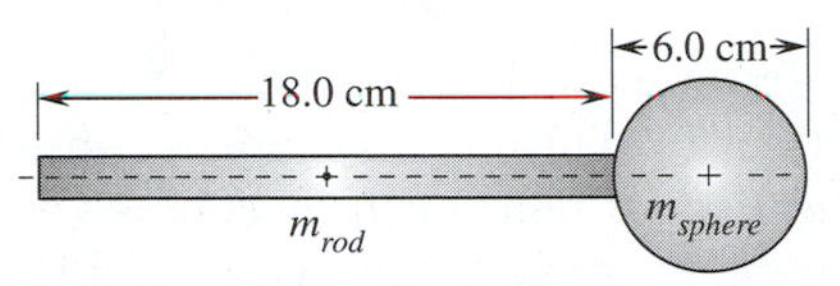

Figure 9–20: A tie-rod, made of a rod and sphere.

SOLUTION

The center of gravity of the sphere is at its center. The center of gravity of the rod lies halfway along its axis.

$$x_{\text{rod}} = \frac{(0.0 + 18.0) \text{ cm}}{2} = 9.0 \text{ cm}$$

$$x_{\text{sphere}} = (18.0 + 3.0) \text{ cm} = 21.0 \text{ cm}$$

$$x_c = \frac{(500 \text{ g} \times 9.0 \text{ cm}) + (200 \text{ g} \times 21.0 \text{ cm})}{500 \text{ g} + 200 \text{ g}}$$

$$x_c = 12.4 \text{ cm}$$

EXAMPLE PROBLEM 9–9: THE HOLE

Figure 9–21 shows a different kind of center of mass problem. A plate is altered by punching a 3.0 in diameter hole out of a sheet of stock with a uniform thickness.

Instead of adding the center of mass of the two objects, we must subtract the hole from the rectangle. Because the circular hole is symmetrically placed along the y axis, we know the vertical component of the center of mass is $y_c = 3.0$ in from the bottom of the plate. We need only find the x position. The mass is proportional to area as it was in Example Problem 9–7.

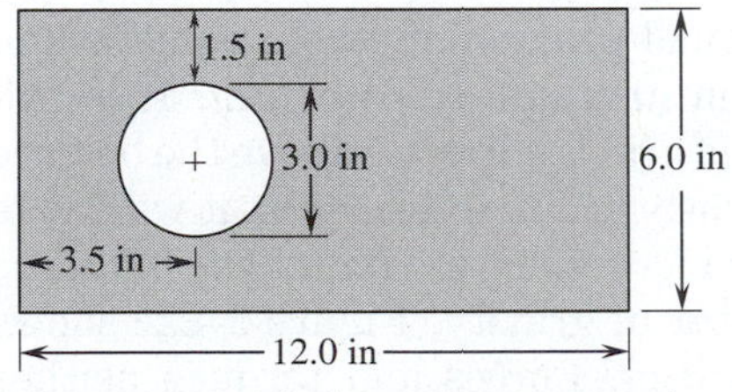

Figure 9–21: A hole cut in a rectangular plate shifts the center of gravity.

$$x_c = \frac{(x_{\text{rect}} \times A_{\text{rect}}) - (x_{\text{circle}} \times A_{\text{circle}})}{A_{\text{rect}} + A_{\text{circle}}}$$

The area of the rectangle is:

$$A_{\text{rect}} = \textit{length} \times \textit{width} = 12.0 \text{ in} \times 6.0 \text{ in} = 72 \text{ in}^2$$

The center of mass of the rectangle with no hole is $x_{rect} = 6.0$ in. The area of the hole is:

$$A_{circle} = \pi R^2 = \pi (1.5 \text{ in})^2 = 7.1 \text{ in}^2$$

The center of mass of the circle is 3.5 in from the left end of the rectangle.

$$x_c = \frac{(6.0 \text{ in} \times 72 \text{ in}^2) - (3.5 \text{ in} \times 7.1 \text{ in}^2)}{72 \text{ in}^2 - 7.1 \text{ in}^2}$$

$$x_c = 6.3 \text{ in}$$

9.4 TOTAL MECHANICAL EQUILIBRIUM

In Section 9.1 we found that there was no linear acceleration on something if the sum of the forces is zero. This means that the center of mass of the object is not accelerated. If nonconcurrent forces act on the object, there may be an angular acceleration (α) around the center of mass. In order to have rotational equilibrium, the net torque must add up to zero. This is called **the second condition of mechanical equilibrium.**

$$\Sigma \tau = 0 \qquad \textbf{(Eq. 9–6)}$$

For a system to be in **total mechanical equilibrium** it must satisfy both Equation 9–1, no linear acceleration, and Equation 9–6, no rotational acceleration.

$$\Sigma \mathbf{F} = 0 \quad \text{and} \quad \Sigma \tau = 0 \qquad \textbf{Total Mechanical Equilibrium}$$

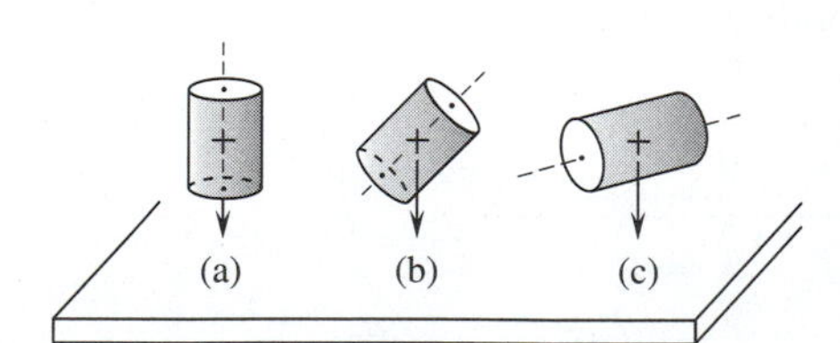

Figure 9–22: A cylinder in three states of equilibrium. (a) Stable. (b) Unstable. (c) Neutral.

Equilibrium systems generally fall into three categories: stable, unstable, and neutral. Figure 9–22a shows a cylinder in **stable equilibrium.** If the cylinder is tilted slightly away from equilibrium, the torque exerted by the center of gravity resists the change. The cylinder will fall back to its original position of stable equilibrium. Figure 9–22b shows a cylinder in **unstable equilibrium.** It is balanced on its edge with the center of gravity directly above the balance point. If the cylinder is tilted slightly in either direction, it will move away from this equilibrium at an increasing rate under the influence of the torque caused by the center of gravity. Figure 9–22c shows the cylinder in **neutral equilibrium.** Forces and torques neither resist nor assist change. A small rotation will not result in a return to the original position. A small rotation will not result in the cylinder rolling away at an *increasing* rate.

☐ **EXAMPLE PROBLEM 9–10: A BRIDGE**

An 8.0-ton truck is $2\bar{0}$ ft from the left end of a $3\bar{0}$-ton bridge with a $7\bar{0}$-ft span (see Figure 9–23). What forces are exerted by the two bridge abutments?

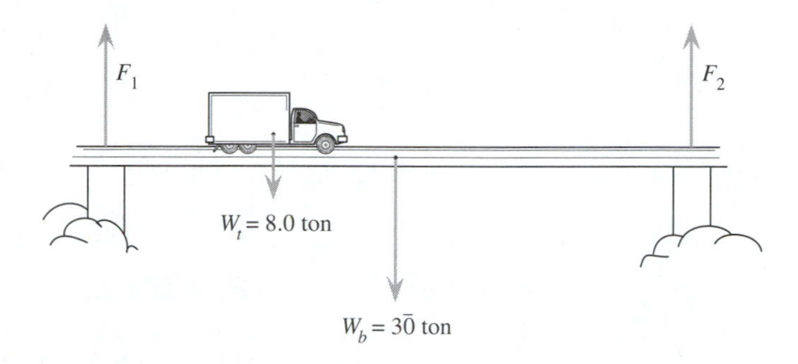

Figure 9–23: *Forces on a bridge supporting a truck.*

■ ***SOLUTION***

The bridge is symmetric, so the center of gravity is halfway across it, 35 ft from either end. Because the bridge does not move, it is in total equilibrium.

$$\textit{forces up} = \textit{forces down}$$
$$F_1 + F_2 = W_t + W_b$$
$$F_1 + F_2 = 8.0 \text{ ton} + 3\bar{0} \text{ ton}$$
$$F_1 + F_2 = 38 \text{ ton}$$

Because there are two unknowns, $\mathbf{F_1}$ and $\mathbf{F_2}$, we need two independent equations. For the second equation we will use a torque equation. The bridge is in rotational equilibrium; therefore *the sum of the torques must add up to zero around* ***any*** *point on the bridge.* We are free to choose any point on the bridge as the "hinge" for our torque equation. A good strategy is to choose the position of one of the unknowns. The lever arm for the unknown will be zero. The force's torque will not appear in the equation. We will choose the left-hand side of the bridge as the rotational axis.

In order to visualize whether a force will cause a clockwise or a counterclockwise torque look at each force separately. Choose the point of rotation. If it were the only force, in which direction would the bridge rotate around the point of rotation?

$\mathbf{W_t}$ would cause the bridge to rotate clockwise around the left-hand abutment if it were the only force.

$\mathbf{W_b}$ would rotate clockwise around the left abutment.

$\mathbf{F_2}$ would cause a counterclockwise rotation around the left end of the bridge.

$\mathbf{F_1}$ has no torque because its lever arm is zero.

counterclockwise torques = *clockwise torques*

$$7\bar{0}\ \text{ft} \times \text{F}_2 = (2\bar{0}\ \text{ft} \times 8.0\ \text{ton}) + (35\ \text{ft} \times 2\bar{0}\ \text{ton})$$

$$\text{F}_2 = \frac{860\ \text{ft} \cdot \text{ton}}{7\bar{0}\ \text{ft}}$$

$$\text{F}_2 = 12\ \text{ton}$$

From the equilibrium of forces we have

$$\mathbf{F_1} = 38\ \text{ton} - \mathbf{F_2} = 38\ \text{ton} - 12\ \text{ton}$$

$$\mathbf{F_1} = 26\ \text{ton}$$

☐ **EXAMPLE PROBLEM 9–11: A LADDER PROBLEM**

A $4\bar{0}$-lb wooden ladder is $2\bar{0}$ ft long and has a center of gravity 9.0 ft from the bottom. The ladder leans against a smooth wall, making an angle of $4\bar{0}°$ with the ground. The coefficient of friction between the bottom of the ladder and the ground is 0.80. How far up the ladder can a 180-lb painter climb without the ladder slipping (X)?

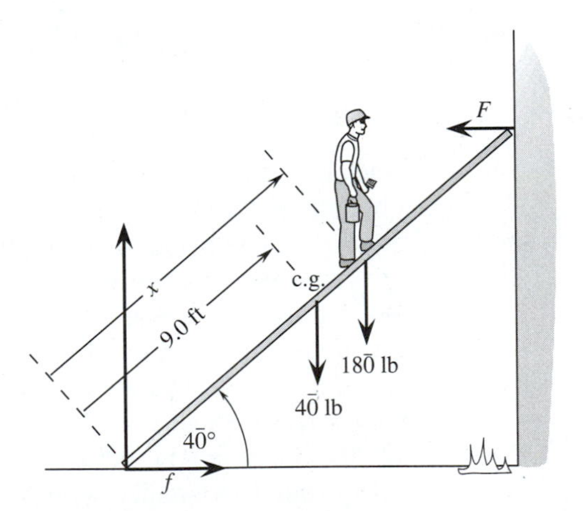

Figure 9–24: *Forces on a ladder. The horizontal force of the wall (**F**) acting on the ladder must be balanced by a frictional force (**f**) at the base.*

■ ***SOLUTION***

Because the wall is smooth, it cannot have a component of friction parallel to the wall. F is a horizontal force to the left. The only other horizontal force is the frictional force on the bottom of the ladder.

Assume the frictional force (f) is maximum. N is the normal force on the ladder.

$$F_{left} = F_{right}$$

$$F = f = \mu N = 0.80\,(18\bar{0}\text{ lb} + 4\bar{0}\text{ lb})$$

$$F = 180\text{ lb}$$

We know F is the maximum force the ladder can have without slipping. Take the torques around the bottom of the ladder.

counterclockwise torques = *clockwise torques*

$$176\text{ lb} \times 2\bar{0}\text{ ft} \times \sin 4\bar{0}^\circ = (9\text{ ft} \times 4\bar{0}\text{ lb} \times \sin 5\bar{0}^\circ) + (18\bar{0}\text{ lb} \times X \times \sin 50^\circ)$$

$$X = 14\text{ ft}$$

When the painter climbs past the last level, the ladder will slip. The ladder would be much safer with a steeper angle.

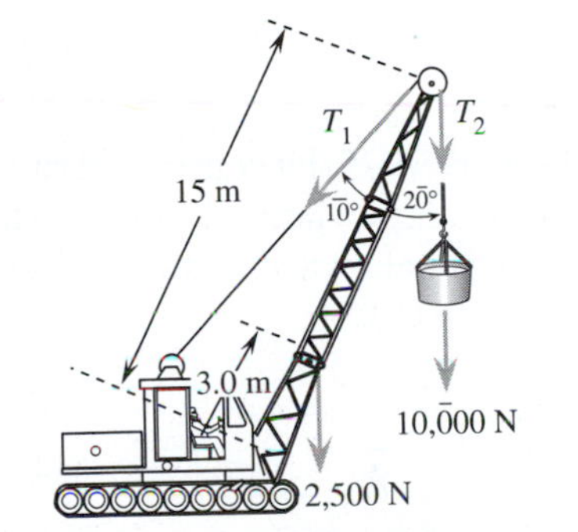

Figure 9–25: A crane lifts a heavy load.

☐ **EXAMPLE PROBLEM 9–12: THE CRANE**

Figure 9–25 shows a crane lifting a 10,$\bar{0}$00-N load (W_l). The boom has a weight of 25$\bar{0}$0 N (W_b). Find the tension in the cable and the compressional force on the boom.

■ ***SOLUTION***

First draw a free-body diagram (see Figure 9–26).

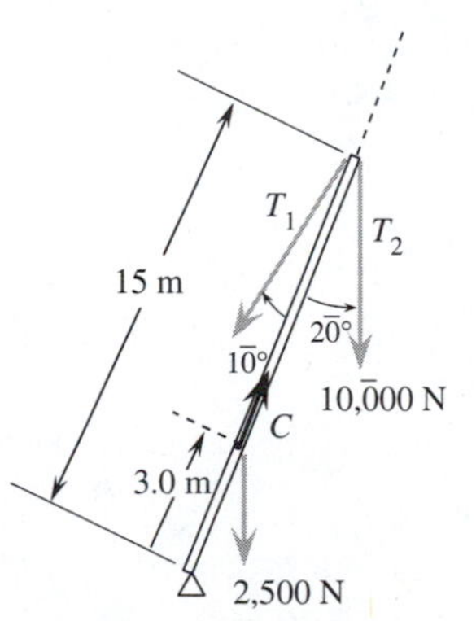

Figure 9–26: Free-body diagram for the crane.

The second condition of mechanical equilibrium can be used to find the tension on the cable.

counterclockwise torques = *clockwise torques*

$$15\text{ m} \times T_1 \times \sin 1\bar{0}^\circ = [15\text{ m} \times (1.00 \times 10^4\text{ N}) \times \sin 15^\circ] + [3.0\text{ m} \times (2.50 \times 10^3\text{ N}) \times \sin 15^\circ]$$

$$T_1 = 1.6 \times 10^4\text{ N}$$

We can get the compressional force **(C)** from the equilibrium of vertical forces.

forces up = *forces down*

$$C_y = T_y + W_b + W_l$$

$$C \sin 75^\circ = [1.6 \times 10^4 (\sin 65^\circ)] + (2.50 \times 10^3) + (1.00 \times 10^4)$$

$$C = 2.7 \times 10^4\text{ N}$$

9.5 EQUILIBRIUM SYSTEMS

In this section we will look at the application of equilibrium in some special kinds of systems.

The Stability of a Sailboat

A sailboat is a system in stable equilibrium. Figure 9–27a shows a sailboat when there is no wind. The buoyant force ($\mathbf{F}_b$) is equal, but opposite in direction, to the weight of the ship. Figure 9–27b shows the sailboat in a fresh breeze. X marks the rotational axis of the boat. The torque caused by the horizontal wind is balanced by a countertorque from the center of gravity of the boat. An increase in wind will cause the boat to tilt further. The angle between the wind ($\mathbf{F}_w$) and the sail decreases from 90°, while the lever arm between the weight of the boat acting at the center of gravity and the buoyant force increases. A new equilibrium is reached.

It is essential that the center of gravity of the boat be below the rotational axis. In the era of large wooden sailing ships it was a common practice to lay rocks (ballast) at the bottom of the ship over the keel to ensure stability.

Body Mechanics

The erect human body is an unstable system. If the center of gravity moves away from a position over the feet, balance is lost. When someone bends over, the hips are thrust backward to maintain the center of gravity over the feet (Figure 9–28).

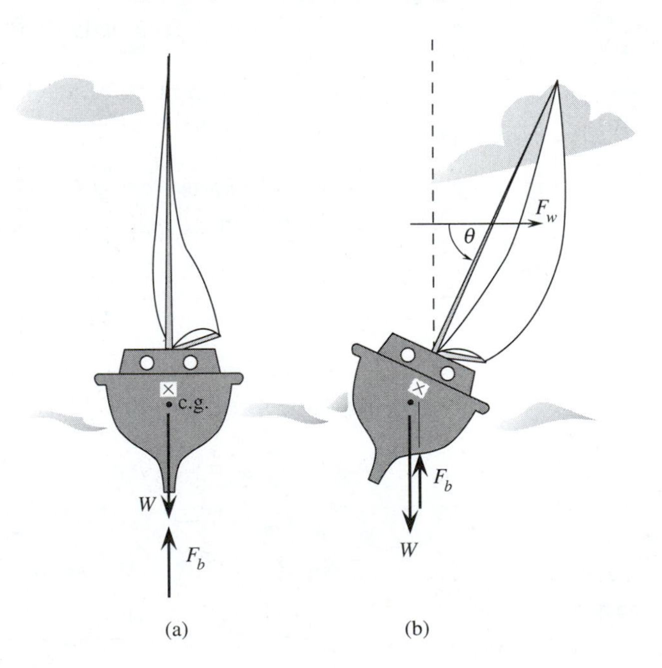

Figure 9–27: *Equilibrium of a sailboat. (a) No wind. (b) With a fresh breeze.*

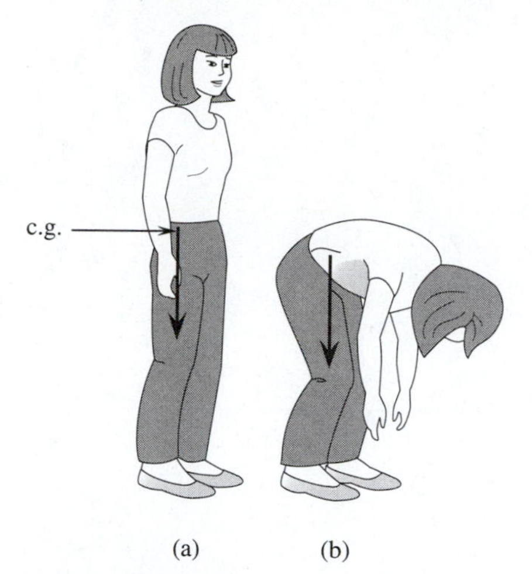

Figure 9–28: When someone bends, the center of gravity (c.g.) is maintained over the feet.

Figure 9–29 shows the torques around the backbone. The center of gravity of the body is located in front of the backbone. The weight is balanced by the muscles in the back. The vertebrae must support both forces. The tension in the back muscles is larger than the weight because of the short lever arm. Figure 9–29b shows the forces on someone who is overweight. The lever arm for the center of gravity increases more rapidly than the lever arm for the back muscles as the person gains weight.

Here are some example values. For the situation in Figure 9–29a, if the weight is 120 lb, the lever arm for the center of gravity is 3.0 in, and the lever arm for the back muscles is 2.0 in, then the tension on the muscles is 180 lb. For Figure 9–29b, if the weight is 200 lb, the lever arm for the center of gravity is 6.5 in, and the lever arm for the back muscles is 2.5 in, then the tension on the back muscles is 500 lbs. It is no wonder then that overweight people commonly complain of lower back pains.

Figure 9–30 shows the forces on the foot when the weight is on the toes. The muscles in the back of the leg are attached to the heel bone by a tendon called the Achilles tendon. The tendon pushes upward while the bone pushes downward with a compressional force to balance the two upward forces. Typically the force on the tendon is two to three times the weight; in action sports the tension can be much larger. A torn Achilles tendon is a fairly common sports injury—particularly among people who are not in proper physical condition.

Transmission Lines

The angle of attachment of a transmission line to a pole is rather small. As a result, the tension on the line is much larger than the

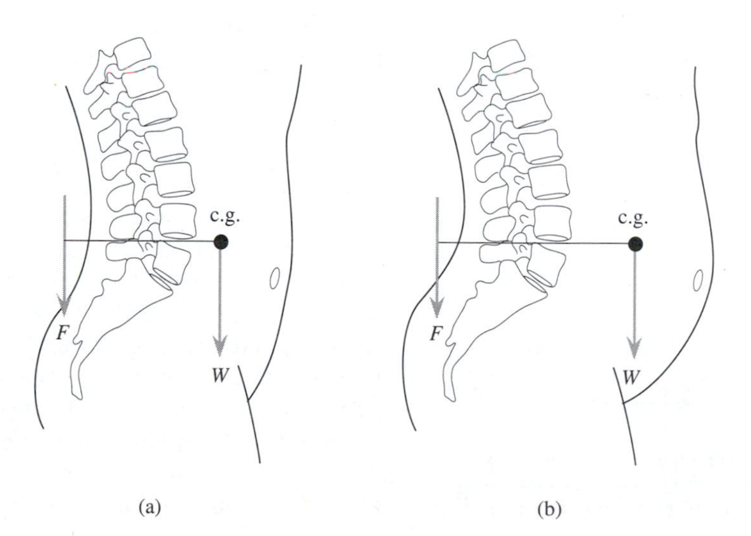

Figure 9–29: Torques around the spinal column. (a) A person of normal weight. (b) An overweight person.

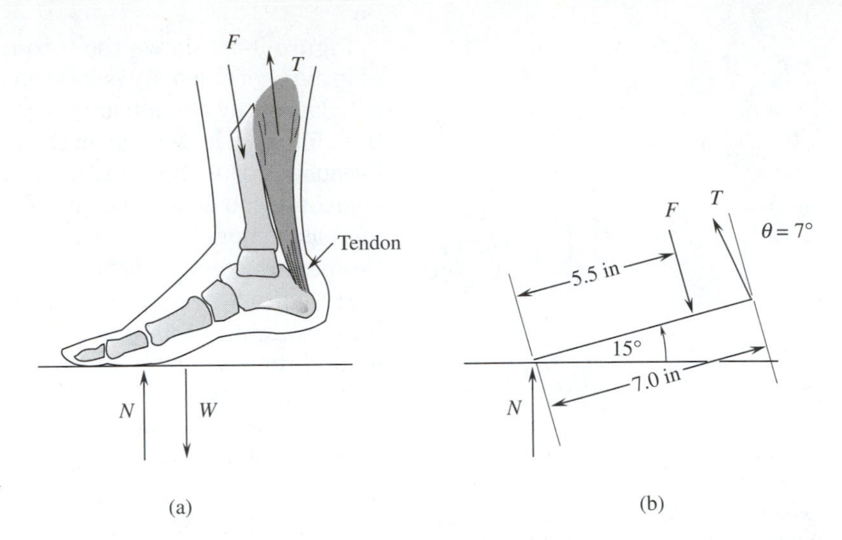

Figure 9–30: *(a) The forces on the foot when standing on the toes. (b) Free-body diagram.*

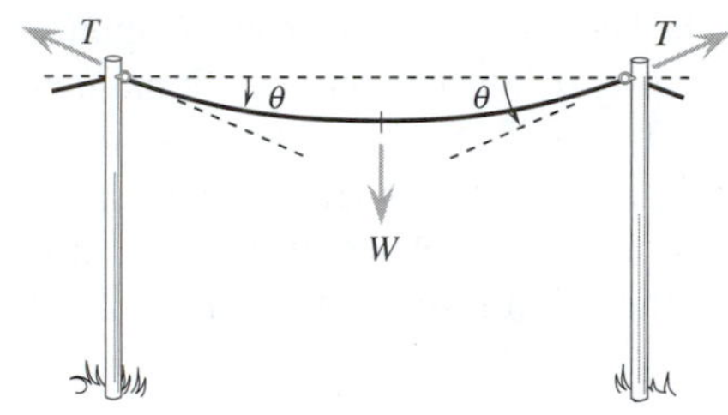

Figure 9–31: *Tension on a power line.*

weight of the cable (see Figure 9–31). If the weight of the cable is 400 N, each pole will support half the weight with the vertical component of the tension ($\mathbf{T}_y$). The tension is inversely proportional to the sine of the angle ($\mathbf{T} = \mathbf{W}/[2 \sin \theta]$). For an angle of 4.0° this gives a tension of 2900 N. In cold weather, the line will contract, decreasing the angle and increasing the tension. The added weight of a coating of ice on the power line can increase the tension beyond its yield point. If the two sections of transmission line attached to the pole are not in a straight line there will be a resultant force (**R**) (Figure 9–32). This force will cause a large torque around the base of the pole. If no force is added to the pole to balance **R,** the pole will look like the one shown in Figure 9–33. Either a guy wire or a strut is commonly used to balance **R.** A guy wire, which is cheaper and easier to install, is used most often.

Raising the Roof

Figure 9–34a shows a common gable roof construction for a small building. A board is used as a spine to which the roof rafters are nailed. One rafter acts as a strut to hold the upper end of the opposing rafter. The horizontal component of the force exerted by one rafter on the other must be balanced by an inward force on the bottom end. The reacting force of the rafter will eventually cause the wall to be pushed outward. Figure 9–34b shows a simple solution to this problem. A collar board connects the two rafters. The tension in the collar, rather than the wall, supplies the needed horizontal force.

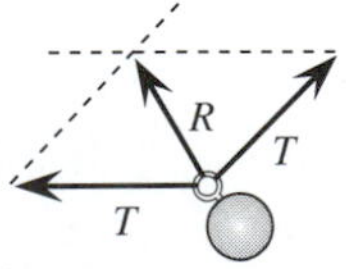

Figure 9–32: *(Top view) When the direction of the power line changes, there is a resultant force* (***R***) *on the pole.*

Figure 9–35 shows a more ambitious way to construct the roof. The sills for the attic replace the collar boards. Struts are used to prevent sag in the roof. Stretchers attach the sills to the roof near

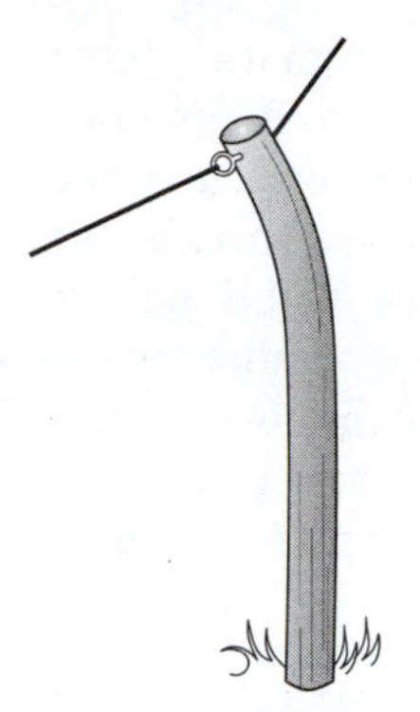

Figure 9–33: The results of a net torque on a power pole.

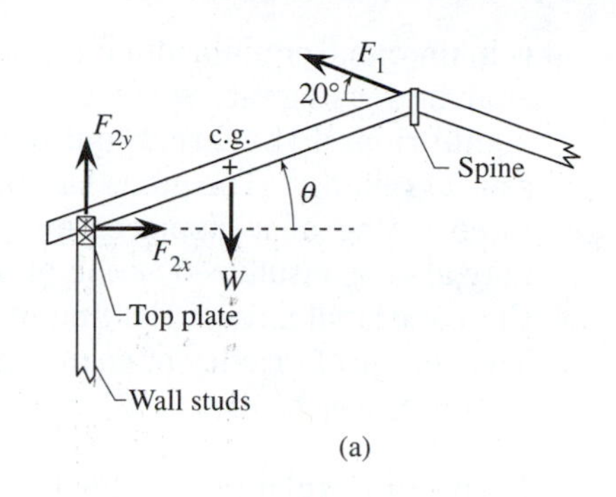

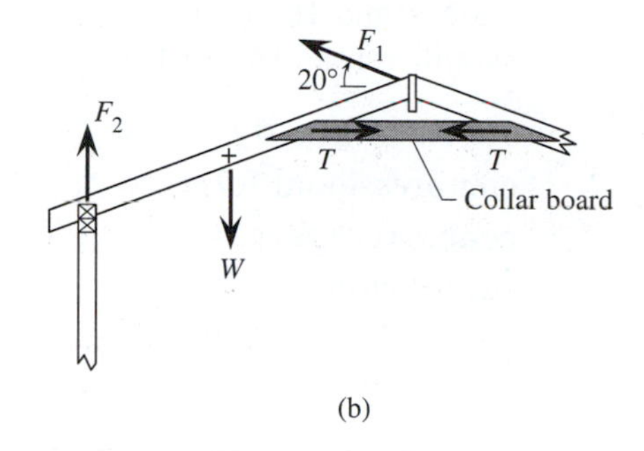

Figure 9–34: (a) The forces on a roof rafter. The wall must supply a horizontal force to balance the horizontal component of $\mathbf{F_1}$*. (b) The tension in the collar board balances the horizontal component of* $\mathbf{F_1}$*.*

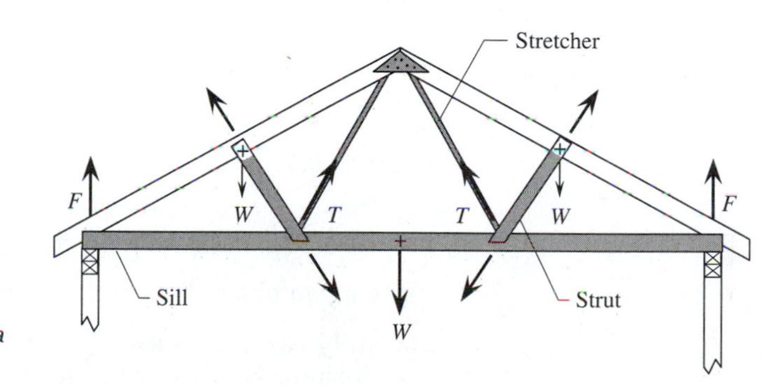

Figure 9–35: A standard way of framing a rigid roof.

the peak. This helps support the floor sills. Usually these units are assembled in one piece before they are placed on the walls.

SUMMARY

The first condition of mechanical equilibrium states that if the sum of the forces acting on a system is zero, there will be no linear acceleration of the system. The second condition of mechanical equilibrium states that if the sum of the torques acting on a system is zero, there will be no angular acceleration of the system. For a system to be in total mechanical equilibrium both the first and second conditions must be satisfied. A system is in stable equilibrium if it returns to its original condition when the system is slightly altered. A system is in unstable equilibrium if it moves away from its original condition at an increasing rate when it is slightly altered. A system is in neutral equilibrium if the forces and torques acting on it neither resist nor assist a change. Torque is the product of the vertical component of force acting at a displacement r from the rotational axis. Torque may also be visualized as the product of perpendicular distance from the rotational axis to the line of action of the force times the force. The center of gravity of an assembly of objects is usually equivalent to the center of mass. It can be found using the balance equation.

KEY TERMS

If you can explain the following terms to a friend or classmate, you understand their meaning. If you cannot explain the terms, you should reread the sections in which they are discussed.

center of gravity
compressional force
concurrent forces
equilibrium
conditions of mechanical equilibrium
lever arm
line of action
neutral equilibrium
nonconcurrent forces
offset
stable equilibrium
statics
torque
total mechanical equilibrium
unstable equilibrium

EXERCISES

Section 9.1:

1. Can a moving automobile satisfy the first condition of equilibrium? Explain why or why not.
2. Can an object be in equilibrium if only one force is acting on it?
3. Only two forces act on a book that is in total mechanical equilibrium. Explain what must be true about the forces.
4. A rope is pulled taut with equal but opposite forces on the two ends. Why is the tension not equal to zero?
5. An airplane travels at a constant speed in level flight with a heading of due north. Is it in equilibrium? Draw a free-body diagram of the forces acting on it.

Section 9.2:

6. Why do doors not have handles on the hinge side?
7. Can concurrent forces cause a torque?
8. Draw a free-body diagram showing the forces involved in opening a jar of peanut butter.

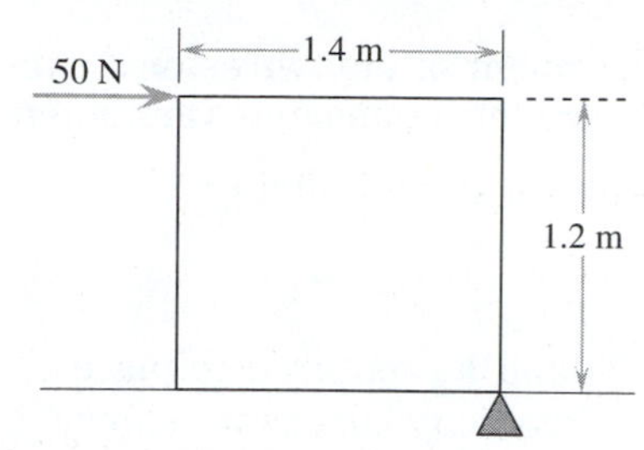

Figure 9–36: Diagram for Exercise 10. Find the torque produced by the $5\bar{0}$-N force.

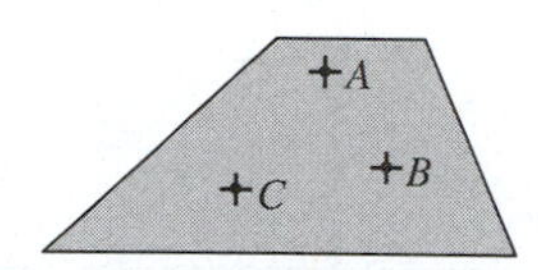

Figure 9–37: Diagram for Exercise 14. Is A, B, or C the most likely location for the center of gravity?

9. Why are the rear wheels on a farm tractor much larger than the front wheels?

10. A force of $5\bar{0}$ N is used to push the box in Figure 9–36. Calculate the torque the force exerts around the front bottom edge of the box.

11. When loading a trailer it is good practice to put the load over the wheels of the trailer rather than at the back end. Why?

Section 9.3:

12. Where is the center of gravity of: a baseball? a doughnut? a length of pipe with a circular cross section? a pipe wrench?

13. In a particular physics class, hour exams make up $6\bar{0}$% of the final grade; the lab average, 25%; and quizzes, 15%. A student has an exam average of 87, a lab average of 96, and a quiz average of 73. What is the final grade?

14. For the trapezoid shown in Figure 9–37 is the center of gravity most likely to be at point A, B, or C?

Section 9.4:

15. Can a spinning object be in mechanical equilibrium?

16. Is the moon in equilibrium? Is the earth-moon system in equilibrium? Draw a free-body diagram for each system to support your answer.

17. "Weebles wobble, but don't fall down." This is the claim of a toy manufacturer. A weeble is a wooden figure with a steel ball inserted in the hemispherical base. Discuss the stability of a weeble.

18. What kind of stability do the following items have resting on a level tabletop? A coffee can resting on its side. A carpenter's hammer balanced on its head. An orange. An upright saltshaker. A soda straw lying on its side. A soda straw standing on end.

19. How are equilibriums of torque involved with: a pipe wrench? a nutcracker? a bumper jack? a screwdriver?

Section 9.5:

20. Some sailboats have a keel weighted with steel. Why?

21. No matter how taut a clothesline may be it will sag when a wet sheet is hung on it. Why?

22. Stand with your heels against a wall. Bend over and touch your toes. Explain what happens.

23. Examine power lines for the placement of guy wires or struts. How are corner posts on barbed-wire fences and gates arranged?

24. What is the function of buttresses on stone buildings?

PROBLEMS

Section 9.1:

1. Find the magnitude of forces $\mathbf{F}_1$ and $\mathbf{F}_2$ required to place the systems in equilibrium in the free-body diagrams shown in Figure 9–38.

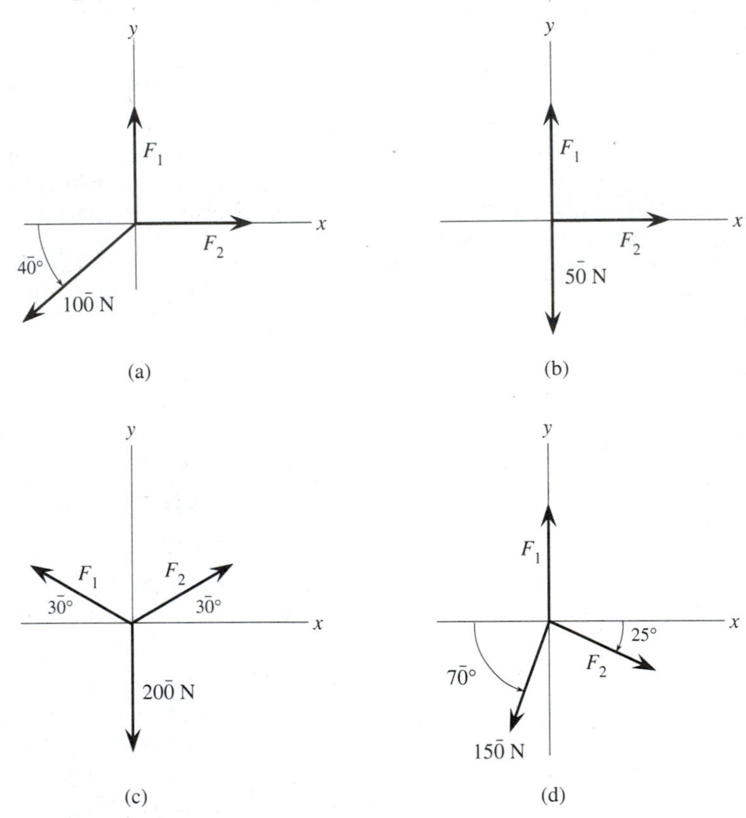

Figure 9–38: *Diagrams for Problem 1.*

2. A 3$\bar{0}$0-N traffic light is suspended halfway over a street by a cable. The cable makes an angle of 8.0° with respect to the horizontal where it is attached to the two supporting poles at each end of the cable. What is the tension in the cable?

3. A 1.5-kg wooden box on a level wooden table is pulled at a constant velocity by a mass hung over a pulley at the edge of the table (see Figure 9–39). How large is the hung mass? See Table 8–1.

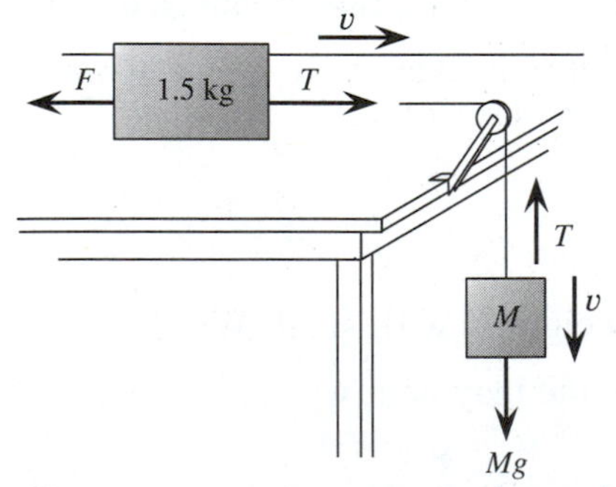

Figure 9–39: *Diagram for Problem 3. Mass (M) moves a box at a constant rate.*

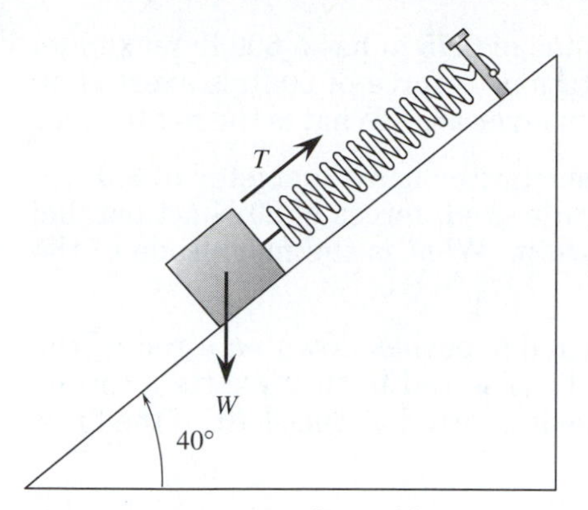

Figure 9–40: Diagram for Problem 4. Weighted spring on a 40° incline.

***4.** The spring in Figure 9–40 has a stiffness of $5\bar{0}$ N/m and an equilibrium length of $6\bar{0}$ cm. What is its length when the attached weight is $4\bar{0}$ N? (Ignore friction.)

***5.** A parent pulls two children in tandem on sleds (see Figure 9–41). The coefficient of friction between the sleds and the snow is 0.10. The child in the front sled weighs $6\bar{0}$ lb; the other, 55 lb. With what force must the parent pull along the rope?

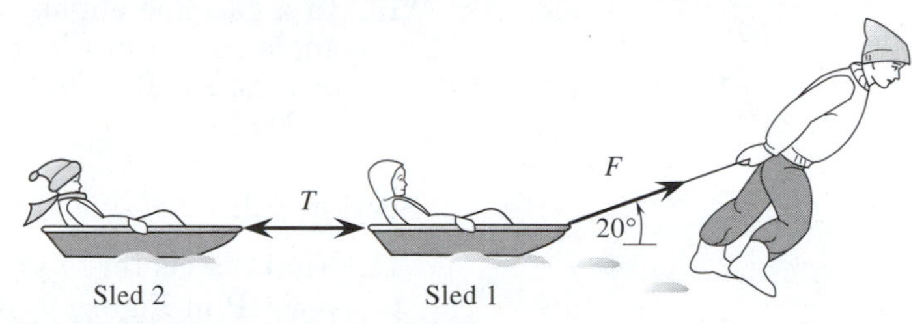

Figure 9–41: Diagram for Problem 5.

***6.** A homeowner mows the lawn pushing an 85-lb lawnmower at an angle of $4\bar{0}°$ above the horizontal. The coefficient between the mower and the lawn is 0.70. What is the magnitude of the force along the handle?

Section 9.2:

7. Find the net torque for each of the free-body diagrams shown in Figure 9–42.

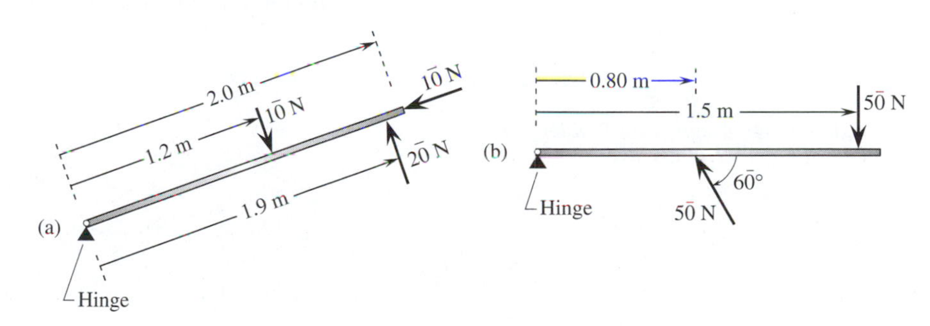

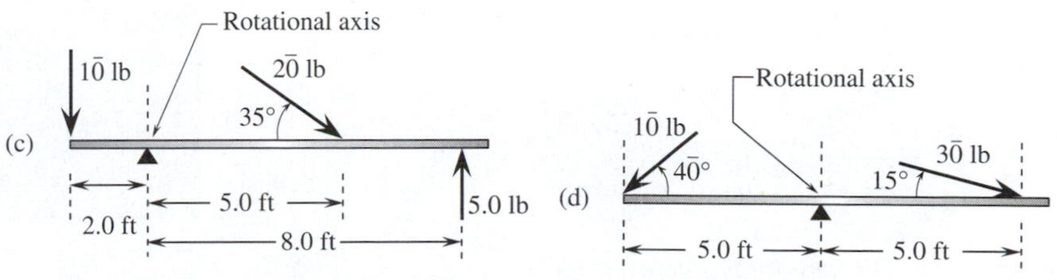

Figure 9–42: Diagrams for Problem 7. Find the net torques.

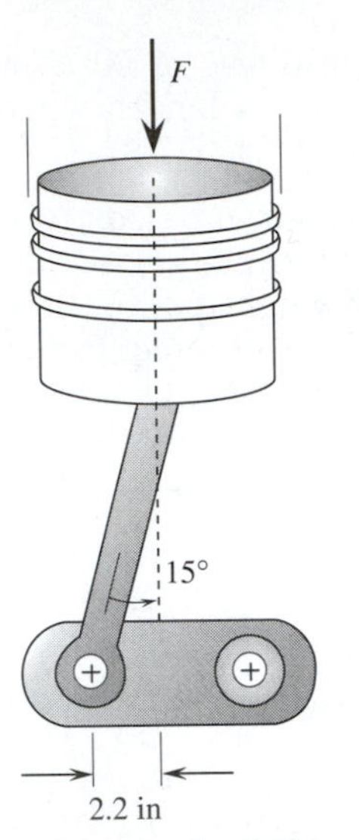

Figure 9–43: *Diagram for Problem 10.*

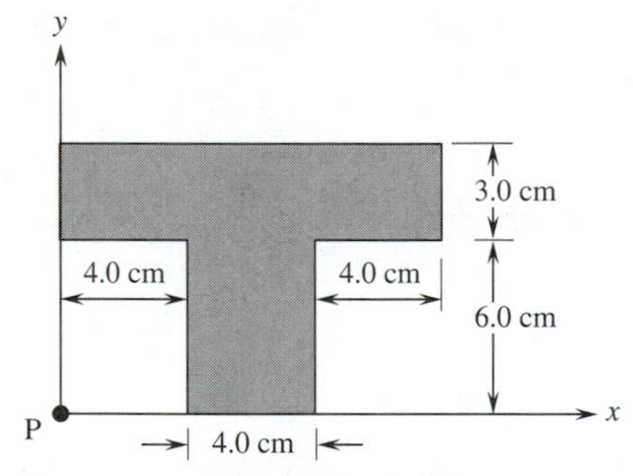

Figure 9–44: *Diagram for Problem 11. Find the center of gravity.*

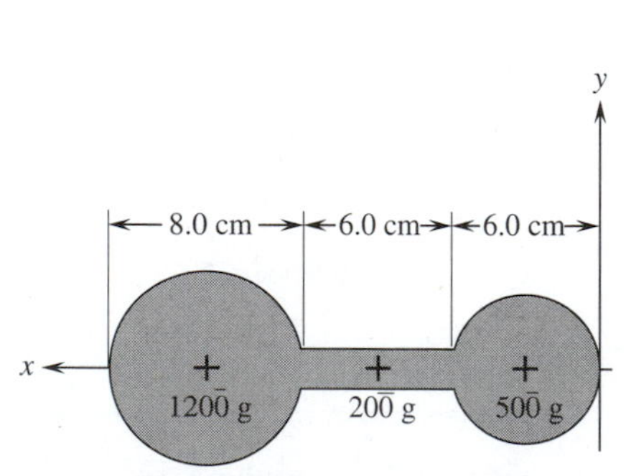

Figure 9–45: *Diagram for Problem 13. Find the center of gravity.*

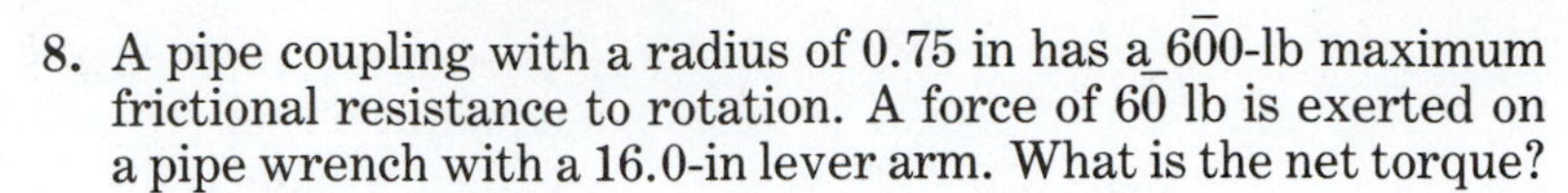

8. A pipe coupling with a radius of 0.75 in has a $6\bar{0}0$-lb maximum frictional resistance to rotation. A force of $6\bar{0}$ lb is exerted on a pipe wrench with a 16.0-in lever arm. What is the net torque?

9. The cover of a jar of peanut butter has a diameter of 8.0 cm. Two equal, but oppositely directed, forces of $2\bar{0}$ N act parallel to the rim of the lid to turn it. What is the magnitude of the applied torque?

***10.** In a gasoline engine, a cylinder pushes down on a rod at the angle shown in Figure 9–43. The rod in turn exerts a torque on a camshaft. What torque is exerted on the shaft, if the force is 300 lb?

Section 9.3:

11. Find the center of gravity of the T-shaped bracket relative to point P in Figure 9–44.

12. A piece of wood 2.0 in wide, 2.0 in thick, and 10.0 in long stands on end. If it is tilted 10°, will it fall over? (Hint: Is the line of action of the weight at the center of gravity, inside or outside the base?)

13. A $5\bar{0}0$-g sphere with a 3.0-cm radius, and a $12\bar{0}0$-g sphere with a 4.0-cm radius are connected by a $20\bar{0}$-g uniform rod (see Figure 9–45). Where is the center of gravity of the system relative to the outside edge of the 500-g sphere?

14. A 3.0-cm diameter circle is cut out of a 10.0-cm square piece of boiler plate (see Figure 9–46). Find the center of gravity.

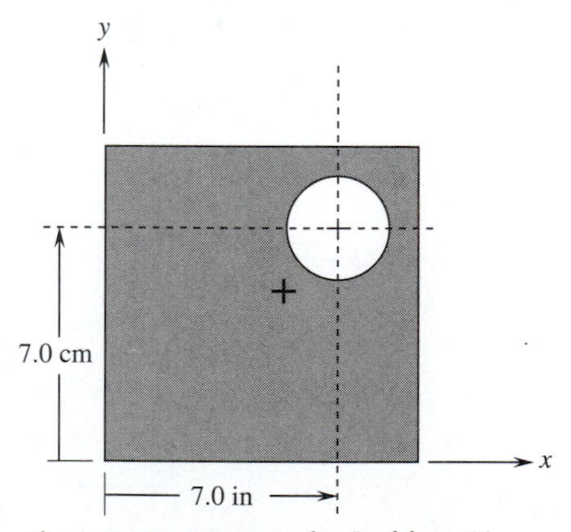

Figure 9–46: *Diagram for Problem 14.*

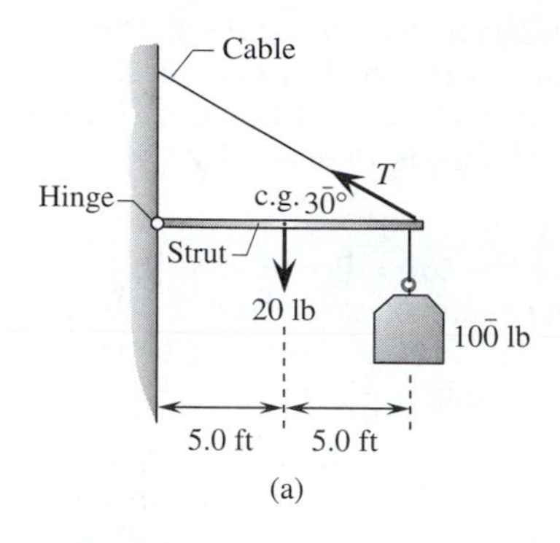

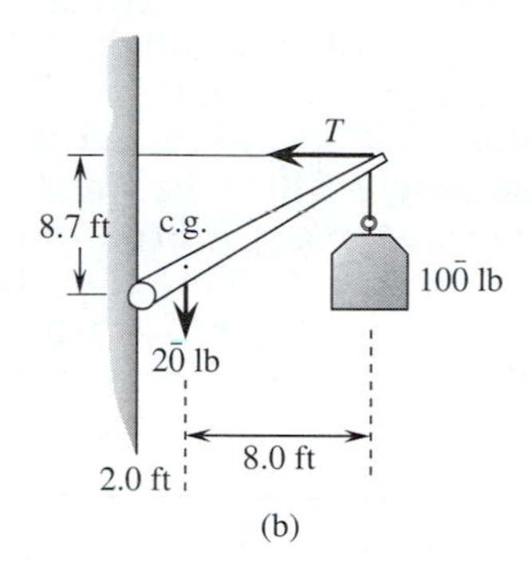

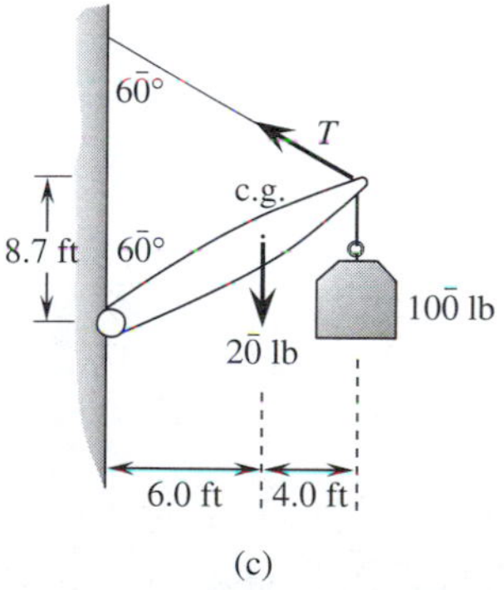

Figure 9–47: Diagrams for Problem 17.

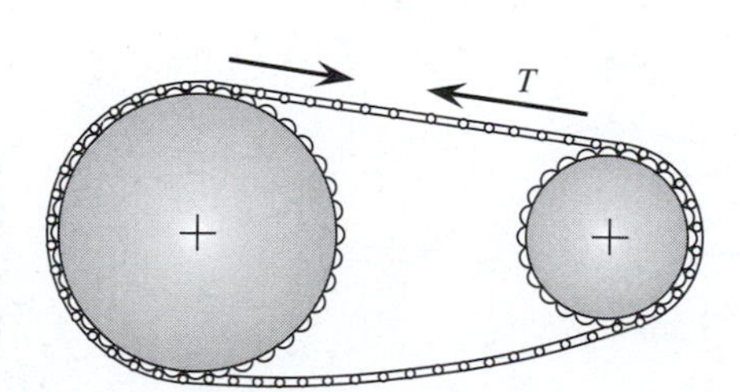

Figure 9–48: Diagram for Problem 18. Chain-and-sprocket wheel drive.

Section 9.4:

15. Two carpenters carry a uniform 6̄0-lb plank that is 24 ft long. Sam supports his end 5.0 ft from the end. Wilma holds the plank 1.0 ft from the other end. How much of the plank's weight does each carry?
16. A uniform 8.0-m plank weighing 140 N lies on a platform with 3.5 m jutting off the platform. How far out on the plank from the edge of the platform can a 35-N cat walk without tipping the plank?
17. Find the tension on the cable and the horizontal and vertical forces on the hinge on each of the hinged poles in Figure 9–47.
18. Figure 9–48 shows a chain-and-sprocket wheel drive. If the radius of the large wheel is 14.0 cm and the radius of the small wheel is 3.0 cm, find the ratio of the torques on the two wheels. Assume the tension in the bottom length of chain is zero.
19. A pulldozer is a device used in some body shops to straighten automobile bodies. The car is fastened onto a heavy steel frame of I beams. Hydraulic devices are used to pull parts of the body into the correct shape. Figure 9–49a shows a car mounted to straighten a fender. Figure 9–49b shows a side view of the hinged vertical I beam and hydraulic piston. If angle θ is 35° and $\mathbf{F}_1$ exerted by the hydraulic cylinder is 30̄0 lb, find forces $\mathbf{F}_2$ and $\mathbf{F}_3$ and the torque around point P.

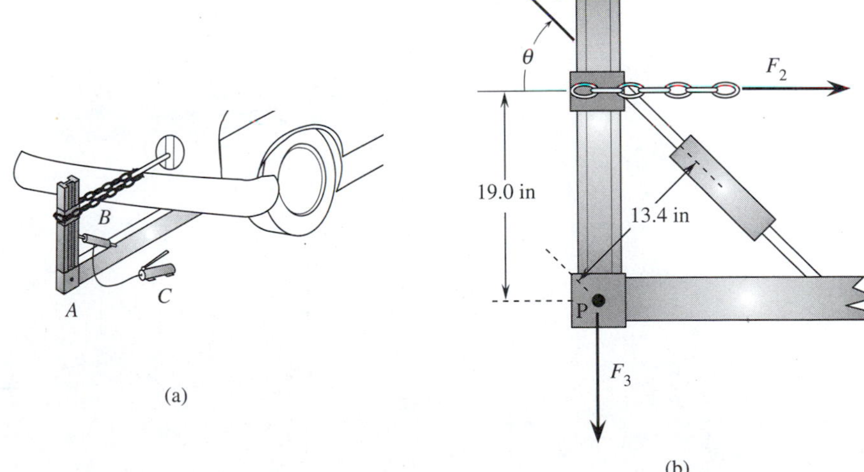

Figure 9–49: A pulldozer. (a) An automobile mounted to straighten a fender. (b) (Side view) The forces on the vertical I beam.

***20.** A 16-ft, 25-lb ladder leans against a smooth wall with the base 4.0 ft from the wall. The center of gravity of the ladder is 7.0 ft from its base. A 17$\bar{0}$-lb painter stands 13 ft from the base. Find the vertical and horizontal forces exerted by the ground on the ladder.

***21.** A horizontal force is applied to the top edge of a 9$\bar{0}$-lb cabinet, 15 in wide and 5.0 ft tall. The center of gravity is 2.5 ft above the base. If the static coefficient of friction between the cabinet and floor is 0.30, will the filing cabinet tip over?

***22.** Figure 9–50 shows a 16.0-ft plank weighing 3$\bar{0}$ lb, supported 3.0 ft from one end by a 6.0 ft tall fence. A 5.5 ft tall stack of crates supports the plank 2.5 ft from the other end. A 1-ft^2 box weighing 8.0 lbs rests on one end. Spitford, who weighs 4.0 lb, stands on the center of the box. Killer, 6 in inside the fence, weighs 20 lb. Tiger applies a force of 6$\bar{0}$ lbs down and to the right 1.0 ft from the fence. Moose weighs 14$\bar{0}$ lb; he stands 4.0 ft to the right of the fence. Sam exerts a force of 75 lb downward to the left at an angle of 70° at a position 3.0 ft to the left of the crates. Archie weighs 10$\bar{0}$ lb, with his center of gravity 8.0 in from the end of the right end of the plank. The coefficient of friction is large enough to prevent the plank from sliding.

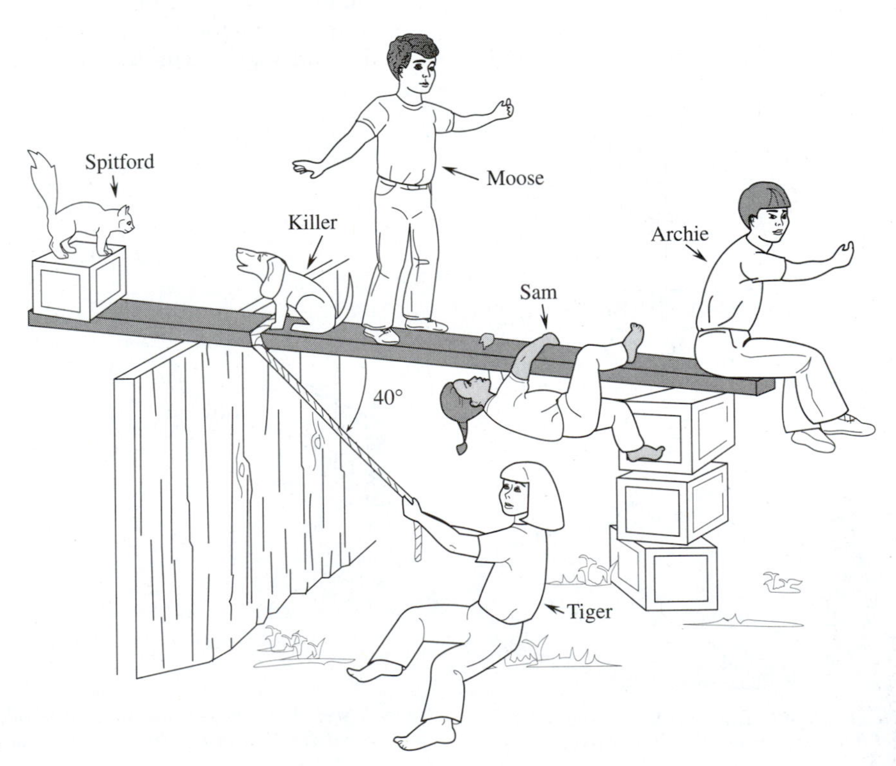

Figure 9–50: *Diagram for Problem 22.*

A. What is the sum of the forces on the plank?
B. What is the sum of the torque on the plank around the fence?

Section 9.5:

23. In Figure 9–27, the center of gravity of a 2.0-ton sailboat is 2.0 ft below the point of rotation. The wind acts as if it were pushing on the sail 12 ft above the rotational axis. With what force does the wind push on the sails if the boat tilts $1\bar{0}°$?

24. The cable in Figure 9–31 has a weight of 400 N. The colder winter weather causes the cable to contract, reducing the angle to 2.5°. An ice storm deposits 295 N of ice on the line. Find the new tension on the cable.

25. For the situation shown in Figure 9–34a calculate the force (F_1) on a 30-lb uniform rafter 24 ft long with a roof angle of 35°.

***26.** The shoulder muscles connect with the upper arm bone at an angle of 13° at a position 5.0 in from the shoulder joint. The center of gravity of the arm is located at the elbow. The average arm weight is 7.0 lb. In Figure 9–51, the arm is fully extended horizontally, holding a 3.0-lb weight in the hand. Find the tension on the shoulder muscle.

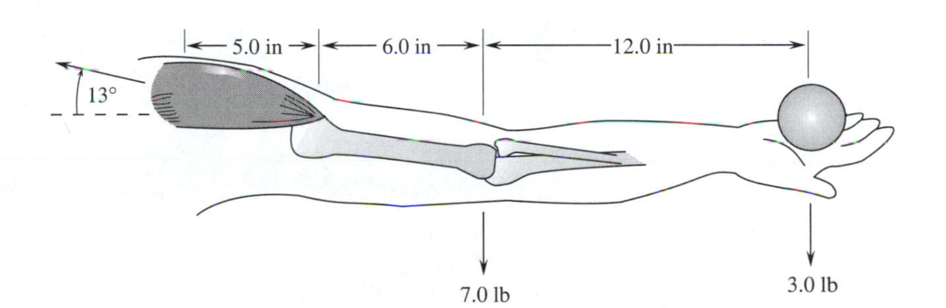

Figure 9–51: Diagram for Problem 26. Forces on an extended arm.

***27.** in Figure 9–30 find the tension on the leg muscle (**T**) and the compressional force on the bone (**C**) for a 200-lb person.

Chapter 10

CENTRIPETAL FORCE

This photograph of Hurricane Ellen was taken during a 1973 Skylab mission. In this chapter, you will read about the Coriolis effect, which helps keep hurricanes spinning. (Photograph courtesy of NASA.)

OBJECTIVES

In this chapter you will learn:

- how centripetal acceleration changes the direction of an object's velocity without changing its speed
- to calculate the centripetal acceleration for an object with angular motion
- to distinguish between centripetal force and centrifugal acceleration
- to calculate the centripetal force acting on objects moving in horizontal and vertical circles
- to apply centripetal force to a variety of situations

Time Line for Chapter 10

1666 Isaac Newton discovers that the centrifugal force a body in circular motion exerts is proportional to the inverse square of the radius of that body's path.

1673 Christian Huygens develops laws for centripetal force.

1674 Robert Hooke suggests there is a balance between the earth's centripetal force and the sun's gravitational force on earth.

1686 Isaac Newton publishes the *Principia,* enunciating his laws of motion.

1835 Gustave-Gaspart Coriolis describes the Coriolis effect.

1851 Leon Foucault performs an experiment in a church to show that the earth rotates.

In Chapter 5 we found that a spinning object had a rotational acceleration (α) if its angular speed changed. If the spin rate is constant then α is zero. Does this mean that there is no acceleration on something that is rotating at a constant speed?

Newton's first law indicates there should be a force and corresponding acceleration on *any* rotating body. If we make a chalk mark on a tire and then set the tire spinning at a constant rate, the mark will move in a circular path. Since the piece of the tire marked by the chalk is not moving in a straight line, an unbalanced force must be acting on it. If there is an unbalanced force, Newton's second law tells us there is an acceleration. In this chapter we will look at force and acceleration on objects rotating at a constant speed.

10.1 CENTRIPETAL ACCELERATION

A weather satellite orbits the earth at an altitude high enough to make the air drag of the outer atmosphere negligible. The force of gravity pulls downward on it; its velocity is at right angles to the force of gravity (see Figure 10–1). The gravitational force will, of course, cause the weather satellite to fall. Over a short period of time it falls a distance s. At the same time, its horizontal velocity moves at a right angle to the gravitational force. It is no closer to the earth than it was before. Gravity is still pulling at right angles to its velocity. Repeated combinations of gravitational pull and tangential velocity cause the satellite to move in a circular orbit. We can say it "falls" around the earth. Let us write an equation to describe this motion.

A satellite orbits the earth at a constant speed in a circle of radius

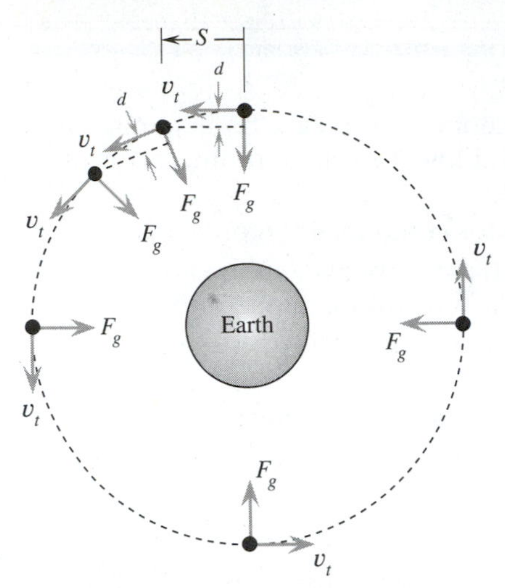

Figure 10–1: *A weather satellite "falls" around the earth. As the satellite falls a distance (d) inward, its tangential velocity moves it at right angles to d.*

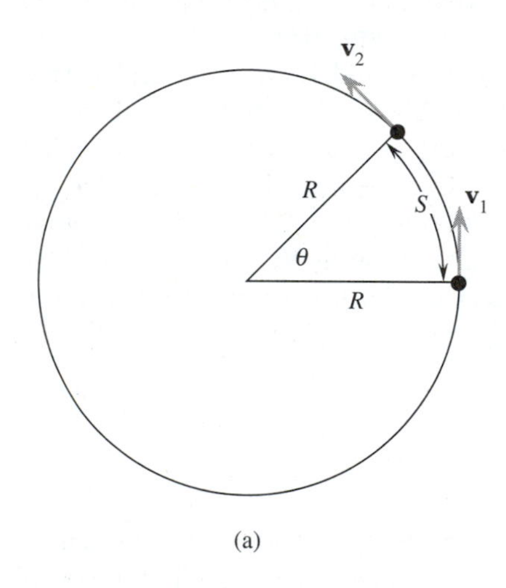

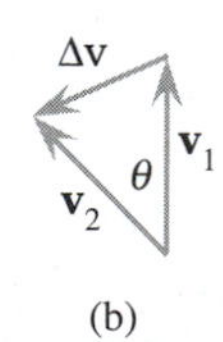

Figure 10–2: *(a) An object moves in a circular path at a constant speed. The object travels an arc length s as it moves through an angle Θ. (b)* **Δv** *is added to* $\mathbf{v}_1$ *to produce* $\mathbf{v}_2$*, a velocity with the same speed, but a different direction.*

R (see Figure 10–2). The satellite has a linear speed of **v**. Although the speed is constant, its direction is changing. We have a change of the vector quantity, velocity. $\mathbf{v}_1$ is the velocity of the satellite at one point in the circular path. $\mathbf{v}_2$ is its velocity a little bit later when it has traveled an arc length of s. The vector difference between these two velocities divided by the time interval will give us the acceleration. (Acceleration is the time rate of change of velocity.) Figure 10–2b shows the subtraction of $\mathbf{v}_2$ from $\mathbf{v}_1$. We have two similar triangles. The angle θ in Figure 10–2a is

$$\theta = \frac{s}{R}$$

In Figure 10–2b angle θ is

$$\theta = \frac{\Delta \text{v}}{\text{v}}$$

where v is the magnitude of either $\mathbf{v}_1$ or $\mathbf{v}_2$.
We can set these two expressions for the angle equal to each other and solve for Δv.

$$\Delta\text{v} = \frac{\text{v}\, s}{R}$$

Divide the equation by Δt, the time interval to travel the distance s, to get the magnitude of the acceleration.

$$\text{a}_c = \frac{\text{v}\, s}{R\Delta\text{t}}$$

Since the speed v is s/t, the equation can be rewritten as:

$$\text{a}_c = \frac{\text{v}^2}{R} \qquad \textbf{(Eq. 10–1)}$$

The acceleration is in the direction of the change of velocity. From Figure 10–2b we see that the acceleration is directed in toward the center of the circular path. An unbalanced force must be acting in this direction. In the case of the satellite the gravitational pull of the earth acts on it. The acceleration that changes the direction of a velocity vector without changing its speed is called a **centripetal acceleration.**

EXAMPLE PROBLEM 10–1: A BOB ON A STRING

A string is attached to a bob. It is then whirled in a horizontal circle at a constant speed with a radius of $6\bar{0}$ cm. The bob makes $8\bar{0}$ complete revolutions (N) in 1 min.

A. Find the magnitude of the tangential velocity of the bob.
B. Find the centripetal acceleration on the bob.

SOLUTION

A. Tangential velocity:

The distance traveled in one revolution is $2\pi R$. The total distance traveled in 1 min is N $(2\pi R)$, where N is the number of revolutions made in one minute.

$$s = 8\bar{0} \times 2\pi \times 0.60 \text{ m}$$

$$s = 302 \text{ m}$$

The tangential velocity is s/t.

$$v = \frac{302 \text{ m}}{60 \text{ s}}$$

$$v = 5.0 \text{ m/s}$$

B. Centripetal acceleration:

$$a_c = \frac{v^2}{R}$$

$$a_c = \frac{(5.0 \text{ m/s})^2}{0.60 \text{ m}}$$

$$a_c = 42 \text{ m/s}^2$$

Equation 10–1 can be written in terms of angular velocity rather than linear acceleration. Recall $v = \omega R$.

$$a_c = \frac{(\omega R)^2}{R} \qquad \textbf{(Eq. 10–1A)}$$

$$a_c = \omega^2 R$$

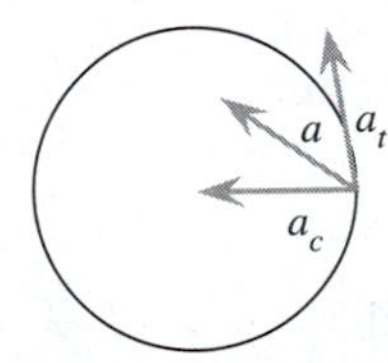

Figure 10–3: *Tangential acceleration* (a_t) *is perpendicular to the centripetal acceleration* (a_c). *The Pythagorean theorem can be used to find the net acceleration* (a).

When a rotating object speeds up or slows down there are two accelerations acting on it. A tangential acceleration changes the speed, and a centripetal acceleration changes the direction of the velocity (see Figure 10–3). The two accelerations are at right angles.

The Pythagorean theorem can be used to find the combined acceleration.

$$a^2 = a_c^2 + a_t^2$$

EXAMPLE PROBLEM 10–2: FLYWHEEL

A 20.0 in diameter flywheel accelerates from rest to 80.0 rad/s in 3.0 s. Find the combined acceleration at the moment it has an angular velocity of 10.0 rad/s.

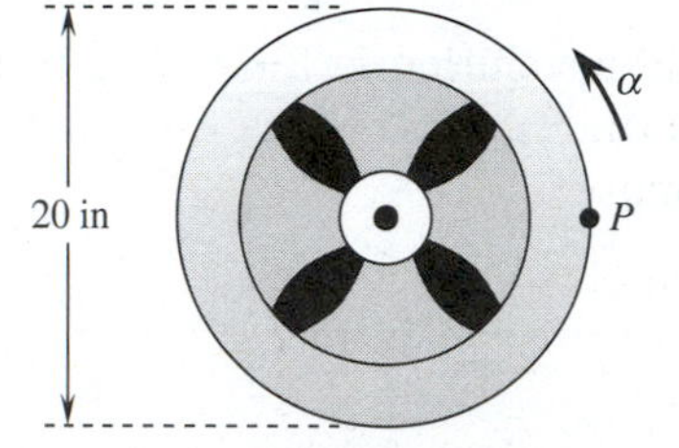

Figure 10–4: An accelerating flywheel.

SOLUTION

First find the accelerations. Calculate the angular acceleration.

$$\alpha = \frac{\omega_f - \omega_0}{t}$$

$$\alpha = \frac{(80.0 - 0.0) \text{ rad/s}}{3.0 \text{ s}}$$

$$\alpha = 26.7 \text{ rad/s}^2$$

Find the tangential acceleration from the angular acceleration.

$$a_t = \alpha R$$
$$a_t = 26.7 \text{ rad/s}^2 \times 0.833 \text{ ft}$$
$$a_t = 22.2 \text{ ft/s}^2$$

Equation 10–1A can be used to find the centripetal acceleration.

$$a_c = \omega^2 R$$

$$a_c = (10.0 \text{ rad/s})^2 \times 0.833 \text{ ft}$$

$$a_c = 83.3 \text{ ft/s}^2$$

The combined acceleration can now be found.

$$a = (a_t^2 + a_c^2)^{1/2}$$

$$a = [(22.2 \text{ ft/s}^2)^2 + (83.3 \text{ ft/s}^2)^2]^{1/2}$$

$$a = 86.2 \text{ ft/s}^2$$

We can find the angle θ between the radial direction and the acceleration.

$$\theta = \tan^{-1}\left(\frac{a_t}{a_c}\right)$$

$$\theta = \tan^{-1}\frac{(22.2\ \text{ft/s}^2)}{86.2\ \text{ft/s}^2}$$

$$\theta = 14.4°$$

10.2 CENTRIPETAL FORCES

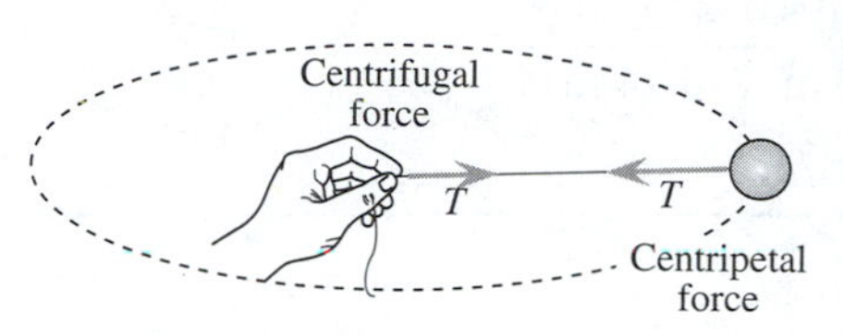

***Figure 10–5:** A centripetal force acts on a bob pulling it inward. A centrifugal force is the reacting force pulling outward on the hand.*

Tie a bob on a piece of string. Swing it over your head in a horizontal orbit. The tension you feel on the string is *not* the centripetal force changing the direction of the bob's motion. It is the force the bob exerts on the string. As always, forces come in pairs. The force exerted on the bob is the **centripetal force.** The reacting force of the bob is called the **centrifugal force** (see Figure 10–5).

There is some confusion concerning these forces. In everyday use the word *centrifugal* is sometimes used when *centripetal* is meant. For example, a centrifuge is often used in the chemistry laboratory to separate precipitates from solution. It is the centripetal force, not the centrifugal force, that acts on the particles.

If a centripetal force acts on something, it will create a centripetal acceleration. We can rewrite Newton's second law for this situation. Let $\hat{\mathrm{r}}$ be a vector with a magnitude of one, pointing to the center of the circle. For constant mass:

$$\mathbf{F}_{\text{net}} = m\,\mathbf{a}_c$$

$$\mathbf{F}_{\text{net}} = \frac{m\,\mathrm{v}^2\hat{\mathrm{r}}}{R}$$

$$\mathbf{F}_{\text{net}} = m\,\omega^2 R\hat{\mathrm{r}} \qquad \textbf{(Eq. 10–2)}$$

The centripetal force may be a single force or the combination of several forces. The important thing is that they act at right angles to the velocity of the object under consideration.

☐ **EXAMPLE PROBLEM 10–3: WEIGHT ON A SPRING**

A spring's unloaded length is 14 cm. A 400-g mass is placed on the spring. The spring and weight are swung in a circle with an angular speed of 12 rad/s. If the radius of the circle is 0.40 m, what is the stiffness of the spring?

■ *SOLUTION*

The tension on the spring acts as the centripetal force. It pulls the mass toward the center of the circle. We will take this direction as positive.

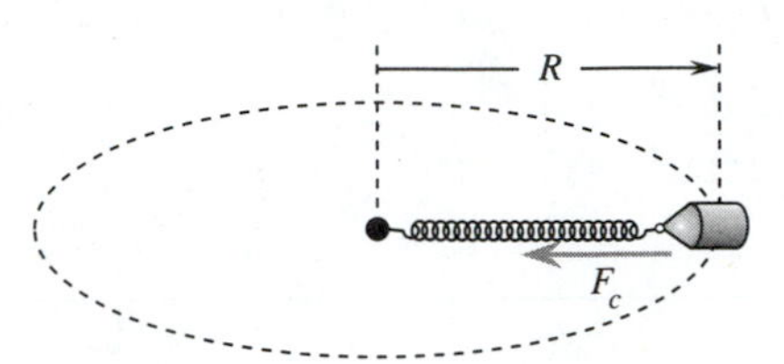

Figure 10–6: *A spring supplies the centripetal force to keep a mass orbiting in a circle.*

$$F_{net} = k\,\Delta x = m\,\omega^2 R$$

or

$$k = m\,\frac{\omega^2 R}{\Delta x}$$

Entering the values we get:

$$k = 0.400\ \text{kg}\,\frac{(12\ \text{rad/s})^2 \times 0.40\ \text{m}}{(0.40\ \text{m} - 0.14\ \text{m})}$$

$$k = 89\ \text{kg/s}^2 \text{ or } 89\ \text{N/m}$$

10.3 FORCES IN A VERTICAL CIRCLE

A pail of water can be swung in a vertical circle without the water falling out. Some roller coasters have vertical loops built into the tracks. The car can travel through the loop without the passengers falling out. Centripetal force makes these possible. Let's take a look at the forces on the water undergoing circular motion in the vertical plane.

Figure 10–7 shows the forces acting on water in a pail. The bottom of the pail pushes on the water toward the center of the circle. Gravity acts downward. We'll take toward the center of the circle as positive.

Case A: The Top of the Circle

At the top of the circle, the weight of the water and the bottom of the pail combine to create the centripetal force toward the center of the circle.

$$F + (m\,g) = \frac{m\,v^2}{R}$$

or

$$F = m\left(\frac{v^2}{R} - g\right)$$

Let us look at the possible values for **F**.

1. *F is positive*. The centripetal acceleration is larger than **g**. The water will stay in the pail. The bottom of the pail pushes against the water.
2. *F is zero*. The centripetal force is **g**. The bottom of the pail does not push against the water. At this special condition, $v = (g\,R)^{1/2}$. This is the minimum velocity the pail may have without the water spilling out.
3. *F is negative*. The centripetal equation is not valid for this case. There is no way the bottom of the pail can exert an upward force on the water. The water spills out.

Case B: Halfway Down the Circle

The gravitational force is tangent to the circle. No component of **g** acts as a centripetal force. Only the force of the bottom of the pail acts inward.

$$F = \frac{m\,v^2}{R}$$

Case C: At the Bottom of the Circle

The weight acts outward in the negative direction.

$$F - (m\,g) = \frac{m\,v^2}{R}$$

or

$$F = m\left(\frac{v^2}{R} + g\right)$$

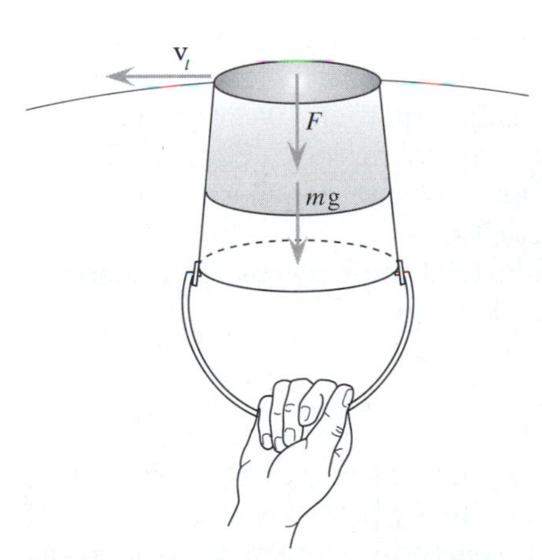

*Figure 10–7: A bucket is swung in a vertical circle. At the top of the circle the centripetal force has two parts acting toward the center of the circle: the weight of the water (m **g**) and the force exerted by the bottom of the bucket on the water.*

The pail pushes upward with a force large enough to hold the weight of the water and supply the centripetal force, too.

Generally, the analysis of vertical loop motion with gravity for other situations follows this same outline. Aircraft often undergo maneuvers involving circular motion. They must be designed to withstand the centripetal forces that occur. The F-14 fighter is designed to routinely operate under 8-g stresses. Unfortunately pilots have not been redesigned as well. The heart is built to operate near normal conditions. Centripetal forces act on all parts of the body including the circulatory system. When a pilot pulls out of a dive or makes a horizontal turn, the blood is forced toward the feet. If there are large **g** forces, the heart cannot create enough pressure to push the blood back into the upper body. Blackouts occur. This has been blamed for some of the crashes of F-14 fighters.

10.4 APPLICATIONS

The Centrifuge

In many chemical reactions precipitates result. The small particles will usually settle under the influence of gravity over a long period of time. The process can be speeded up by using a centrifuge (see Figure 10–8). A tilted test tube whirls at a rapid rate. The gravitational force acts down. The buoyant force of the liquid on the particles acts upward. These two forces partly cancel. The small net unbalanced force causes the particles to fall to the bottom of the test tube. The motion is resisted by a velocity-dependent friction. The centripetal force acts horizontally.

Assume the precipitate has a density larger than the solution. A volume V of precipitate will have a mass of

$$m = V\rho$$

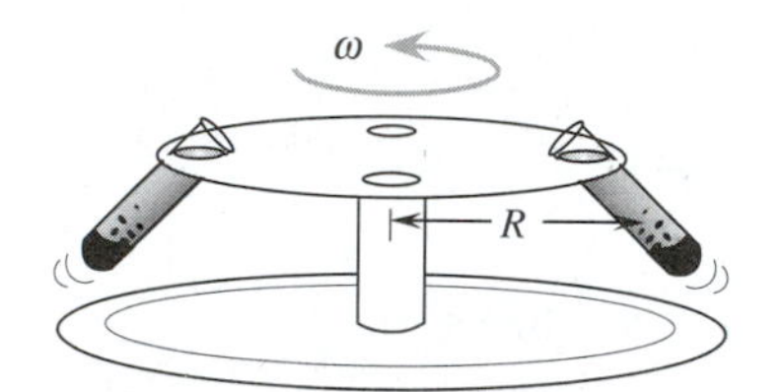

Figure 10–8: The centrifuge. A differential centripetal force separates substances of different densities.

Let ρ_1 be the density of the precipitate and ρ_2 be the density of the solution. The difference between the centripetal forces required to keep the particles and solution in the same orbit is:

$$F_c = \frac{V\,v^2(\rho_1 - \rho_2)}{R} \qquad \textbf{(Eq. 10–3)}$$

The centripetal force is proportional to the density. The denser particles cannot be held in orbit as easily as the solution. They pile up at the end of the test tube. The same idea can be used to separate two liquids, such as milk and cream. Equation 10–3 does not take into account the details of the centripetal force. In Section 17.6 we will look at the buoyant forces and velocity-dependent friction to calculate the settling rate of particles.

Centrifugation has several industrial applications. It is used in cyclonic separators and centrifugal casting.

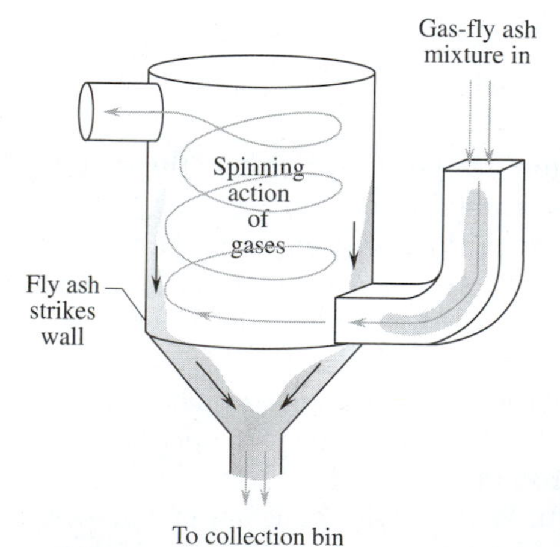

Figure 10–9: The cyclonic separator. Fly ash is thrown against the wall of the separator and falls into a collection bin.

The Cyclonic Separator

Some industries generate a large amount of particles, or fly ash, in the exhaust fumes of the plants. Heavy particles on the order of 100 μm diameter will settle to the earth in a few minutes in areas around the plant. Smaller particles on the order of 1 μm diameter may float in the atmosphere, borne by air currents for a few months. Usually they are washed out of the atmosphere by rain droplets before they settle. The fly ash usually contains SO_2, which reacts

Hurricanes and the Coriolis Effect

A baseball pitcher is standing on top of a large sphere that is spinning counterclockwise. A catcher at the "equator" of this large ball is held in place by some mysterious force. The pitcher throws the ball directly at the catcher. While the ball is in the air, the sphere continues to rotate. The catcher rotates with the sphere. The ball seems to be diverted to the right as it travels toward the catcher. Standing outside of the system, we can see that the ball is actually traveling in a straight line. If the baseball players on the sphere do not know that their sphere is spinning, it will seem to them that some force is diverting the ball to the right.

Photograph courtesy of NASA.

We live on a large sphere called earth. Moving bodies in the Northern Hemisphere have their velocity diverted to the right because of the earth's spin. This effect, known as the Coriolis effect, was first described in 1835 by Gustave-Gaspart Coriolis. In 1851 Leon Foucault used a pendulum to prove the earth spins. The plane of the pendulum rotated according to the predictions of Coriolis.

Large air masses moving on the face of the earth are influenced by the Coriolis effect. North of the equator, air rushes toward a large low-pressure area in the atmosphere. As the winds approach the center, the Coriolis effect causes the wind to be diverted to the right of its motion. This causes a spiral of counterclockwise-moving air. At the center of the spiral, air is pushed upward. If the air is over warm tropical oceans, moisture is picked up from the water. As the water condenses at high altitudes, additional heat is added to the air, driving it up higher. A chimney effect results. Additional air is drawn to the center, and a tropical storm is produced. If the water is warm enough and the Coriolis effect is strong enough, a hurricane develops.

An analysis of the Coriolis effect shows that the effect is strongest at the poles of spinning spheres. The effect approaches zero toward the equator.

Hurricanes depend strongly on two things: warm oceans and the Coriolis effect. Rarely does a hurricane develop below a latitude of 5°. The Coriolis effect is too weak. If the climatic models predicting a warming of the earth over the next century are correct, we can expect stronger hurricanes at higher latitudes.

---

with water to form sulfuric acid causing acid rains downwind from the source.

One method of removing the fly ash from exhaust fumes is to use a cyclonic separator (see Figure 10–9). Particle-laden air is forced through a cylindrical column in a spiral motion. Since the particles have a larger density than air, they are forced to the outside wall of the cylinder. They strike the wall and fall into a collection bin. This method removes some of the heavier fly ash—particles with diameters of 5 μm or more. Other techniques must be used to remove the smaller particles.

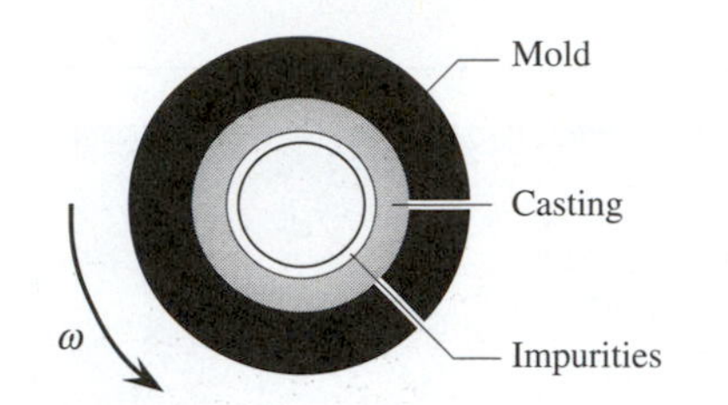

Figure 10–10: Centrifugal casting. Denser metal materials are driven against the mold. Less-dense impurities collect inside the casting.

Centrifugal Casting

Most casting techniques use gravity to push the molten metal against a mold. A higher quality casting can be obtained by using centripetal forces. Molten metal is poured into a cylindrical mold rotating with speeds high enough to produce gs around 70 times the acceleration of gravity. The dense metals are forced against the inside of the cylinder, where the mold is located. Less-dense impurities are pushed toward the center of the cylinder (see Figure 10–10).

The Centrifugal Clutch

A centrifugal clutch is used in a variety of machines run by one-cylinder engines. Examples are snow blowers, snowmobiles, and different off-road vehicles. The clutch will engage only when the motor is running at a high rpm rate (see Figure 10–11). Two slotted D-shaped friction shoes ride out on studs when the centripetal force becomes large enough to overcome the inward force of the garter spring. The friction shoes engage the drum, which transmits power to the rest of the drive train.

Centripetal Forces on the Highway

Whenever an automobile changes its direction, a centripetal force acts on it. Figure 10–12a shows a car traveling over the top of a hill with a vertical radius of R. Two forces act on the vehicle. N acts upward and weight acts downward. We will use the usual convention of labeling into the circle as the positive direction. The centripetal force equation is:

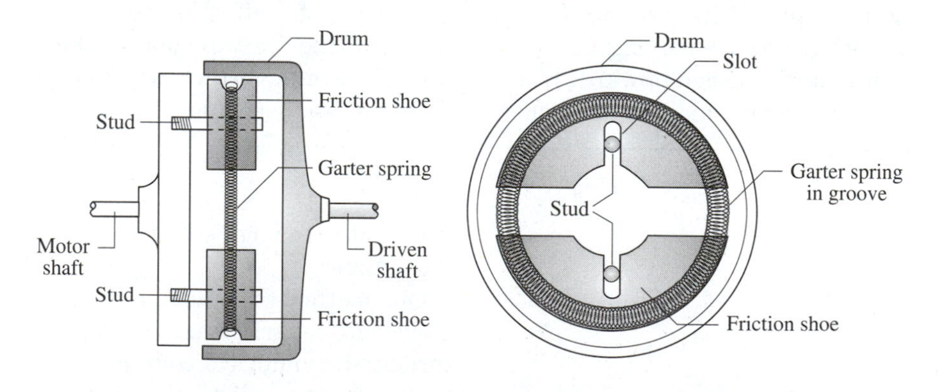

Figure 10–11: Centrifugal clutch. As the motor speeds up, the garter spring cannot supply enough centripetal force to hold the friction shoes in the center. The shoes ride out on studs to make contact with the drum.

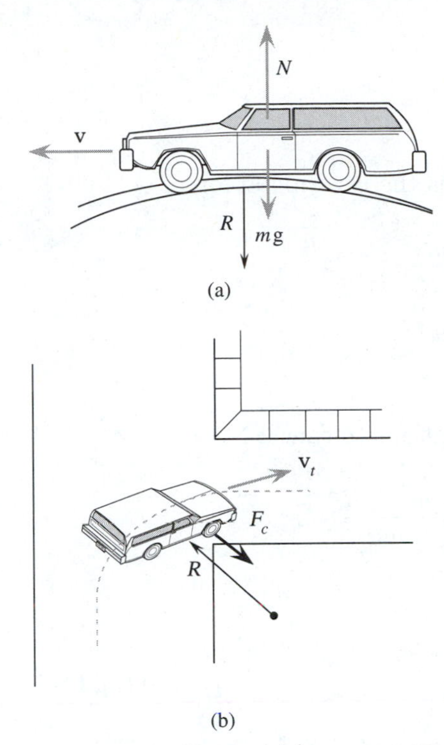

Figure 10–12: *Centripetal forces on a car. (a) Forces on a car topping the crest of a hill. (b) A car rounding a corner on a level street.*

$$-\mathrm{N} + (m\,\mathrm{g}) = \frac{m\,\mathrm{v}^2}{R}$$

$$\mathrm{N} = m\left(\mathrm{g} - \frac{\mathrm{v}^2}{R}\right)$$

As long as the acceleration of gravity is larger than the centripetal force, the car will stay on the highway. The limiting condition occurs when the gravitational acceleration is just equal to the centripetal acceleration.

$$0 = m\left(\frac{\mathrm{v}^2}{R} - \mathrm{g}\right)$$

or

$$\mathrm{v} = (\mathrm{g}\,R)^{1/2}$$

If the velocity is larger than this quantity, the car will leave the ground as it goes over the crest of the hill.

Figure 10–12b shows a car turning a corner on a level street. The turning radius is R. Static friction is the centripetal force. The maximum friction force is μmg. We can use the centripetal force equation to determine the maximum frictional force acting on the car to hold it in the turn.

$$\mu m\,\mathrm{g} = \frac{m\,\mathrm{v}^2}{R}$$

The maximum speed the car can have to round the corner without skidding is:

$$\mathrm{v} = (\mu\mathrm{g}\,\mathrm{R})^{1/2}$$

Highways are much safer when it is not necessary to depend on friction for turning. A banked turn uses part of the normal force on the car to supply the needed centripetal force to round the curve. Figure 10–13 shows the forces on a car rounding a curve of radius R. In cross section, the road is built as an inclined plane with the bottom of the incline pointed to the center of the turn. The vertical component of the normal force (**N**) must balance the weight of the car.

$$\mathrm{N}_y = m\,\mathrm{g}$$

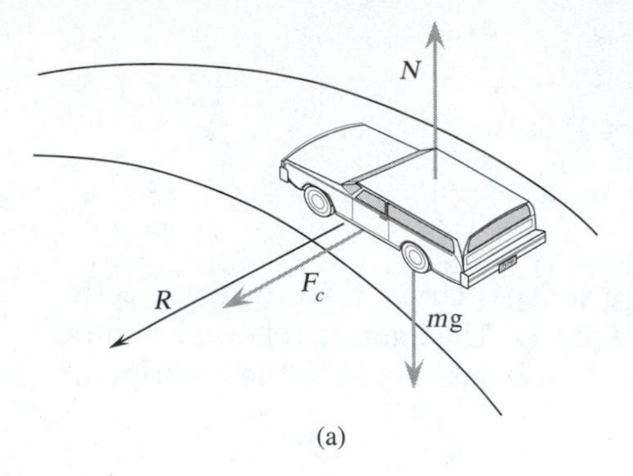

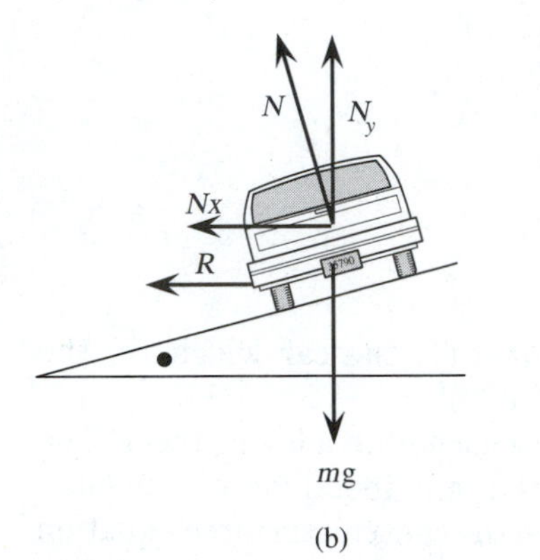

Figure 10–13: A banked highway. (a) Perspective view of the forces on the car. (b) The same forces seen in a cross-sectional view.

or

$$\text{N} \cos \theta = m\,\text{g}$$

The horizontal component of **N** is the centripetal force.

$$\text{N}_x = \frac{m\,\text{v}^2}{R}$$

$$\text{N} \sin \theta = \frac{m\,\text{v}^2}{R}$$

Take the ratio of these two equations to cancel **N.**

$$\frac{\text{N}_x}{\text{N}_y} = \frac{\text{N} \sin \theta}{\text{N} \cos \theta} = \frac{(m\,\text{v}^2)/R}{m\,\text{g}}$$

or

$$\tan \theta = \frac{\text{v}^2}{R\,\text{g}} \qquad \textbf{(Eq. 10–4)}$$

This equation gives the proper banking angle when the design speed of the highway and the turning radius of the curve are known.

□ **EXAMPLE PROBLEM 10–4: THE BANKED HIGHWAY**

A highway has a design speed of 55 mph. What is the banking angle of a curve with a turning radius of 800 ft?

■ ***SOLUTION***

$$\tan \theta = \frac{\text{v}^2}{R\,\text{g}}$$

$$\tan \theta = \frac{(55 \text{ mi/h} \times 1.47 \text{ ft/s/mi/h})^2}{800 \text{ ft} \times 32 \text{ ft/s}^2}$$

$$\theta = \tan^{-1}(0.255)$$

$$\theta = 14°$$

Satellite and Planetary Motion

Gravity is the only force acting on a satellite orbiting around the earth or on a planet around the sun. We can write the equation for a satellite in circular orbit. M is the large mass, and m is the mass of the satellite, and G is the proportionality constant for universal gravitation introduced in Equation 8–6. $G = 6.67 \times 10^{-11}\ \mathrm{N \cdot m^2/kg^2}$.

$$\mathbf{F}_c = \frac{m\,v^2}{R}$$

or

$$\frac{G\,m\,M}{R^2} = \frac{m\,v^2}{R}$$

Cancel like quantities.

$$v^2 = \frac{GM}{R} \qquad \textbf{(Eq. 10–5)}$$

The speed of the satellite depends on two things: the mass of the object it orbits and the radius of its orbit. Let us find how its period, the time for one orbit, is related to these quantities.

The distance traveled in one complete circle is $2\pi R$ in a time interval of T. The speed is:

$$v = \frac{2\pi R}{T}$$

Substitute this expression into Equation 10–5.

$$\left(\frac{2\pi R}{T}\right)^2 = \frac{G\,M}{R}$$

$$T^2 = \frac{(4\pi^2)\,R^3}{G\,M} \qquad \textbf{(Eq. 10–6)}$$

The period squared is directly proportional to the cube of the radius of the orbit. This principle is Kepler's third law of planetary motion. Johannes Kepler spent a good deal of his life discovering the laws of motion of planets. He had neither algebra nor Newton's laws of motion to help him. Geometry and a large volume of data were his only aids. He formulated three laws from his observations, known as **Kepler's laws of planetary motion.**

1. **The planets travel in an elliptical orbit with the sun at one focus.** An ellipse can be drawn easily; place a piece of paper on

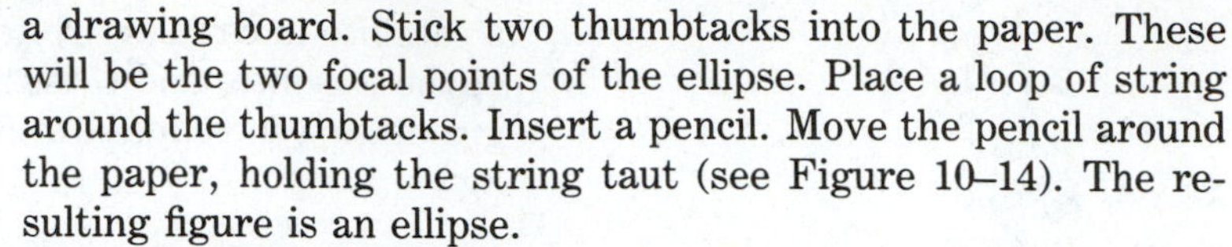

a drawing board. Stick two thumbtacks into the paper. These will be the two focal points of the ellipse. Place a loop of string around the thumbtacks. Insert a pencil. Move the pencil around the paper, holding the string taut (see Figure 10–14). The resulting figure is an ellipse.

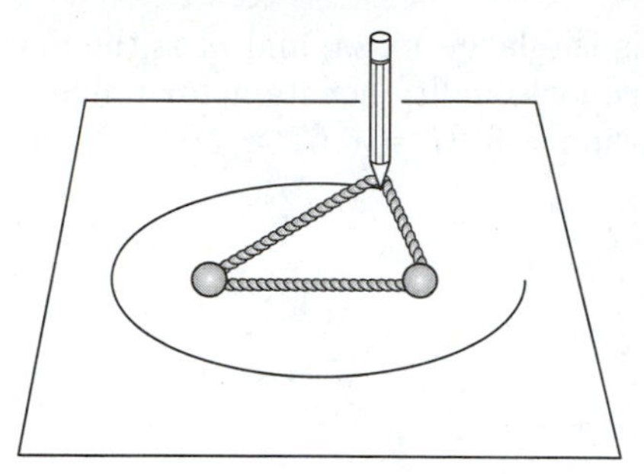

Figure 10–14: A method for drawing an ellipse with a string and two thumbtacks.

2. **The straight line that connects the sun and a planet sweeps out equal areas during equal time intervals.** We can see why this might be true by looking at the gravitational forces acting on a planet (see Figure 10–15). As the planet approaches the sun, part of the gravitational force is tangent to the orbit speeding it up. As it moves away, the tangential component slows down the planet. When it is far away, it moves slowly. When the distance is small, it moves rapidly.

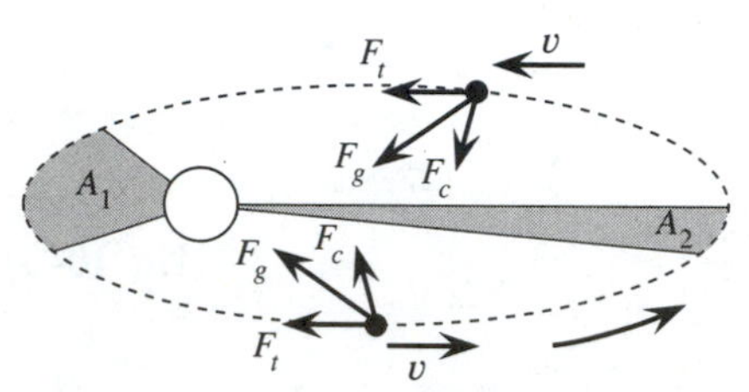

Figure 10–15: In an elliptical orbit, gravitational forces speed up a planet as it approaches the sun and slow it down as it moves away from the sun. Areas swept out during equal time periods have the same area ($A_1 = A_2$).

3. **The square of the period (*T*) is proportional to the cube of the radius of orbit (*R*).** Equation 10–5 summarizes the last relationship found by Kepler. In an elliptical orbit, R is half the major axis. The major axis is the longest line of the ellipse passing through the focal points (see Figure 10–16).

Care should be taken in using Equation 10–5. It assumes the satellite mass is much smaller than the mass it orbits. If m and M are similar in size, the two objects will orbit around their center of mass. The masses of both objects enter into the equation.

☐ **EXAMPLE PROBLEM 10–5: SPUTNIK**

On October 4, 1957, the Russians placed earth's first manmade satellite into an elliptical orbit. Its closest approach to the earth, or perigee, was 228 km; its farthest point, or apogee, was 947 km. Calculate its period of orbit in minutes.

■ *SOLUTION*

Express the distances from the earth relative to its center. The radius of the earth is $628\bar{0}$ km.

$$\text{perigee} = 628\bar{0}\text{ km} + 228\text{ km} = 6508\text{ km}$$

$$\text{apogee} = 628\bar{0}\text{ km} + 947\text{ km} = 7227\text{ km}$$

Half the major axis is then

$$R = \frac{(6508\text{ km} + 7227\text{ km})}{2}$$

$$R = 6867\text{ km}$$

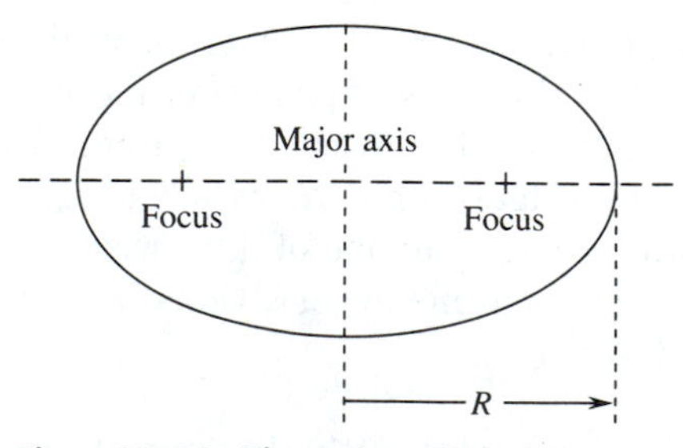

Figure 10–16: The major axis is the longest line of an ellipse.

Use Equation 10–5 to find the period. The mass of the earth is 5.99×10^{24} kg.

$$T^2 = \frac{(4\pi^2)\, R^3}{G\, M}$$

$$T^2 = \frac{4\,(\pi)^2\,(6.867 \times 10^6 \text{ m})^3}{(5.99 \times 10^{24} \text{ kg})\,(6.67 \times 10^{-11} \text{ N} \cdot \text{m}^2/\text{kg}^2)}$$

$$T^2 = 3.272 \times 10^7 \text{ s}^2$$

$$T = 5660 \text{ s} \times \frac{1}{60} \text{ min/s} = 94 \text{ min}$$

SUMMARY

A centripetal acceleration changes the direction of a velocity vector without changing its magnitude. The centripetal acceleration is directed inward along the line of curvature. Its magnitude is $\omega^2 R$.

Whenever an object changes direction, a centripetal force is acting upon it. The magnitude of the centripetal force is $m\, \omega^2 R$. The net centripetal force is the sum of the forces acting at right angles to the direction of an object's motion.

Centripetal accelerations act at right angles to the tangential acceleration, the component of acceleration that changes the speed of an object. The combined acceleration is found using the Pythagorean theorem.

KEY TERMS

If you can explain the following terms to a friend or classmate, you understand their meaning. If you cannot explain the terms, you should reread the sections in which they are discussed.

centripetal acceleration
centripetal force
centrifugal force
Kepler's laws of planetary motion

EXERCISES

Section 10.1:

1. For each of the following situations determine which accelerations are involved: a centripetal acceleration only; a tangential acceleration only; both a centripetal acceleration and a tangential acceleration; neither a centripetal nor a tangential acceleration.
   - ***A.*** An automobile makes a right-hand turn at constant speed.
   - ***B.*** A flywheel initially at rest accelerates to 180 rpm.
   - ***C.*** A pickup truck slows as it reaches the top of a hill and then speeds up.
   - ***D.*** An electron orbits the nucleus of an atom in an elliptical orbit.
   - ***E.*** A weather satellite orbits the earth in a circular path.

F. A child on a sled accelerates down a hill that has a constant slope.

2. in Chapter 4 we analyzed projectile motion using linear equations. Discuss whether or not a centripetal acceleration is involved in projectile motion.
3. A motorcycle rounds a curve with a centripetal acceleration of 12 ft/s^2. What is the acceleration if:
   A. the radius of the curve is doubled.
   B. the speed of the motorcycle is doubled.
   C. the radius of the curve is tripled and the speed is doubled.

Section 10.2:

4. Why does a lariat shape itself into a circular loop when spun in a circle?
5. Why does mud fly off a rotating tire? Describe the motion of the mud.
6. Figure 10–17 shows a ride often found at county fairs. As a central shaft rotates at an increasing speed, the chairs swing out, increasing θ. Why must the angle increase with the increasing speed of the central shaft?

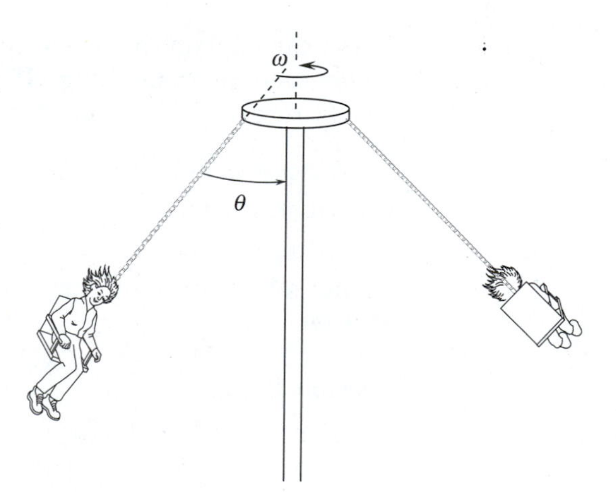

Figure 10–17: *Chain chair ride. As the angular speed (ω) increases the angle (θ) also increases.*

7. How does the spin-dry cycle of a washing machine work?

Section 10.3:

8. A stunt pilot performs a vertical loop at a constant speed. Where in the loop are the forces on the plane and pilot the largest?
9. Tie a bob to a piece of string. Rotate it in a vertical circle. Explain why the speed is not constant.

Section 10.4:

10. Under icy conditions is a car more likely to skid rounding a curve on the open highway or turning a corner on a city street? Explain your answer.
11. On a speedway, the curves are banked steeper on the outside than on the inside. Why?
12. A centrifuge is used to separate cream and milk. Is the cream skimmed from the inside or from the outside of the rotating cylinder?
13. In 1970, the Apollo 13 experienced an explosion on its way to the moon. The command module was damaged. Much of the fuel was lost. Rather than turn around and immediately return, the Apollo mission orbited the moon before returning. Why?
14. Comets orbit the sun in highly elliptical orbits. Why are they seen for only a small fraction of their period of orbit?
15. Io, the inner large moon of Jupiter, orbits the planet in approximately 42.5 h. It is 4.22×10^5 km from the planet. Europa is another moon of Jupiter 6.71×10^5 km from the planet. Estimate Europa's period of orbit.

PROBLEMS

Section 10.1:

1. Find the centripetal acceleration of a motorcycle rounding a curve with a radius of 450 ft at 45 mph.
2. **A.** What is the centripetal acceleration at the earth's equator?
   B. What is the centripetal acceleration at the North Pole?
   C. What percentage change occurs in the weight of an object moved from the North Pole to the equator?
3. What is the centripetal acceleration of the rim of a 12 in diameter 33-$\frac{1}{3}$ rpm record? What is the centripetal acceleration 3.0 in from the center?
4. A 16.0 in diameter flywheel accelerates from $10\bar{0}0$ rpm to $34\bar{0}0$ rpm in 8.2 s. At $30\bar{0}0$ rpm:
   A. what is the tangential acceleration?
   B. what is the centripetal acceleration?
   C. what are the magnitude and direction of the combined acceleration?
5. A bob on the end of a $2\bar{0}$-cm string makes $10\bar{0}$ complete rotations in 1 min in a horizontal circle. What is the centripetal acceleration of the bob?

Section 10.2:

6. A $20\bar{0}$-g mass is twirled by a string in a horizontal circle with a radius of 25 cm. What is the tension on the string if the angular speed is 2.3 rad/s?

7. A 4200-lb pickup truck turns a corner on a level road at $2\bar{0}$ mph. If the radius of curvature of the turn is $5\bar{0}$ ft, what is the frictional force between the pavement and the truck's tires?

** 8. The moon orbits the earth in 27 d 07 h 43 min. This is an angular speed of 2.66×10^{-6} rad/s. Use the centripetal force equation to estimate the earth-moon distance. (Hint: The centripetal force acting on the moon is the universal gravitational force.)

Section 10.3:

9. A $20\bar{0}$-g mass is swung on a string in a vertical circle with a radius of $4\bar{0}$ cm.
   - ***A.*** At the top of the circle, the tangential velocity is 3.0 m/s. What is the tension on the string?
   - ***B.*** At the bottom of the circle, the tangential velocity is 3.4 m/s. What is the tension on the string in this case?

10. A jet fighter with a speed of $50\bar{0}$ mph (730 ft/s) pulls out of a dive with a turning radius of $60\bar{0}0$ ft.
    - ***A.*** What is the centripetal acceleration?
    - ***B.*** What is the equivalent weight of the pilot in gs?

11. The crest of a hill on a rural road has a radius of 100 ft (see Figure 10–18). What is the maximum speed in miles per hour at which a pickup truck can go over the hill without leaving the road?

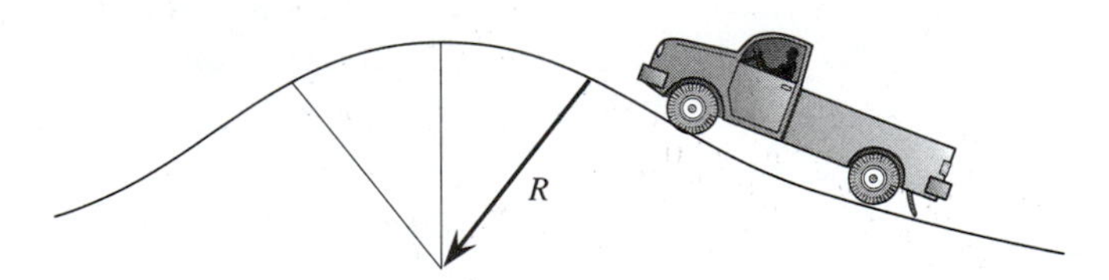

Figure 10–18: *A pickup truck goes over the crest of a hill.*

12. The maximum force a particular string can withstand without breaking is $8\bar{0}$ N. A 0.40-kg bob on a 0.60-m length of string is swung in a vertical circle with a slow, steady, increasing tangential speed.
    - ***A.*** Where will the bob most likely be in the circle when the string breaks?
    - ***B.*** What is the tangential speed of the bob when the string breaks?

Section 10.4:

13. A centrifugal casting is made using a mold with a 12-in inside diameter. What rotational speed in revolutions per minute must be used to cause a 70-g force?

14. A car rounds a corner on an unbanked road at 25 mph with a turning radius of $6\bar{0}$ ft. What must be the minimum coefficient of friction between the tires and the road to keep the car from skidding?

15. A car rounds a curve with a turning radius of $3\bar{0}$ m on an unbanked road. If the coefficient of static friction between the tires and the road is 0.85, what is the maximum speed the car can have without skidding?

16. A curve has a radius of curvature of $12\bar{0}0$ ft. If the highway is designed for 60-mph traffic, what is the banking angle?

17. A highway is designed for $8\bar{0}$ km/h traffic. A curve is banked at 8.0°. What is the radius of curvature?

18. Plot period versus distance to the 3/2 power using the data in Table 10–1. From the slope of the line, calculate the mass of the sun.

Table 10–1: Planetary orbits.

PLANET	AVERAGE DISTANCE FROM THE SUN ($\times 10^{10}$ m)	ORBITAL PERIOD ($\times 10^{7}$ s)
Mercury	5.79	0.761
Venus	10.8	1.94
Earth	15.0	3.16
Mars	22.8	5.94
Jupiter	77.8	37.4
Saturn	143	93.0
Uranus	287	265
Neptune	450	520

*19. The average distance between the earth and the sun (1.496×10^{8} km) is called an astronomical unit (AU). The earth is 1.00 AU from the sun. Show that, if the sun-planet distance (R) for other planets is expressed in AUs, the period of orbit (T) in earth years is simply $T = (R)^{3/2}$.

*20. A precipitate in water has a density of 1.12 g/cm^3 and an average particle diameter of 1.0×10^{-2} cm. A sample of this mixture is placed in a centrifuge operating at 3400 rpm. What is the difference between centripetal force on a particle and on the water at a radius of 6.0 cm? At 9.0 cm?

***21.** Io, one of the moons of Jupiter, has an average orbital distance of 4.22×10^8 m and a period of 2.55×10^3 min.

A. From these data determine the mass of Jupiter.

B. How many times more massive is Jupiter than earth? (Earth has a mass of 5.97×10^{24} kg.)

****22.** How far above the surface of the earth must a satellite orbit in order to maintain a fixed position in the sky over the United States?

Chapter 11

IMPULSE AND MOMENTUM

How does a cat land on its feet? As you read about momentum in this chapter, you will see how the cat uses some of the principles of physics to right itself. (Photograph courtesy of J. E. Frederickson, Department of Physics, California State University at Long Beach.)

OBJECTIVES

In this chapter you will learn:

- about the relationship between impulse and momentum
- to calculate the linear momentum of a moving mass
- to solve collision problems using the conservation of linear momentum
- to calculate the angular momentum of a spinning object
- about the relationship between torque and angular momentum
- to find the moment of inertia of a rotating object
- to use conservation of angular momentum to solve some types of rotation problems
- to identify some applications of linear momentum and angular momentum

Time Line for Chapter 11

1350	Jean Buridan advances medieval concepts of impetus.
1590	Galileo Galilei refutes Aristoteean physics.
1668	John Wallis formulates conservation of momentum.
1686	Isaac Newton publishes the *Principia, De Motu,* the motion of bodies.
1744	Pierre de Maupertuis advances the idea of least action: nature moves in a direction so as to keep the product of force, distance, and time at a minimum.

In a baseball game a heavily muscled first baseman comes to bat. His powerful muscles ripple as he swats the ball against the right-field fence for a double. Next to bat is the shortstop. He's puny compared with the first baseman. Shortstops are designed for agility, not strength. Yet the shortstop is able to hit a home run over the left-field fence using agility rather than strength. A spinning flywheel on a one-cylinder engine keeps the piston moving between power strokes. A sailboat uses the wind to tack against a fresh breeze. The boat moves into the wind, progressing along a course opposite the flow of air. Water rushes down a penstock. The direction of water flow is diverted by the blades of a turbine, causing the turbine to spin.

These are a few of many situations that are best analyzed by impulse and momentum. In this chapter we will examine impulse and momentum by rearranging Newton's second law. In this new form, it becomes a powerful tool for solving problems involving the interaction of two or more masses.

11.1 IMPULSE AND MOMENTUM

We introduced the idea of momentum in Section 6.6 when we were looking at the general form of Newton's second law. Momentum is the product of mass times velocity.

$$\mathbf{p} = m\,\mathbf{v}$$

Notice momentum is a vector quantity. The direction is the same as the velocity of the mass.

The general form of Newton's second law is usually written using momentum. *Force is equal to the time rate of change of momentum.* In Chapter 6, we used the special case in most cases, where mass is constant ($\mathbf{F} = m\,\mathbf{a}$). The general form is:

$$\mathbf{F} = \frac{\Delta \mathbf{p}}{\Delta t}$$

where $\Delta \mathbf{p} = (m_f \mathbf{v}_f) - (m_i \mathbf{v}_i)$. The equation can be altered by multiplying the force acting on the object by the interval of time in which it acts.

$$\mathbf{F}\, \Delta t = \Delta \mathbf{p} \qquad \textbf{(Eq. 11–1)}$$

The left-hand side of the equation is called **impulse.** Impulse is the product of the force acting on something times the length of time the force acts. Impulse is a vector quantity that has the same direction as the force.

$$impulse = \mathbf{F}\, \Delta t \qquad \textbf{(Eq. 11–2)}$$

Equation 11–1 tells us the change of momentum experienced by a mass is dependent upon two things: the size of the applied force and how long the force acts. An experienced baseball player knows the ball will go farther if the ball is pulled. The batter does not break the wrists until the bat is already in contact with the ball. The bat is in contact with the ball for a longer period of time than for a normal swing. The ball receives a larger impulse, which means a larger change of momentum. Equation 11–1 may be expressed this way:

impulse = the change of momentum

☐ **EXAMPLE PROBLEM 11–1: THE FOUR-TON TRUCK**

An $8\bar{0}00$-lb truck travels at a speed of 40.0 ft/s.

A. What is the magnitude of the momentum of the truck?

B. With what speed must a $32\bar{0}0$-lb car travel to have the same momentum?

■ ***SOLUTION***

A. The mass of the truck is $m = \mathbf{W/g}$.

$$\mathrm{p} = m\,\mathrm{v}$$

$$\mathrm{p} = \frac{\mathrm{W\,v}}{\mathrm{g}}$$

$$\mathrm{p} = \frac{8\bar{0}00 \text{ lb} \times 40.0 \text{ ft/s}}{32.0 \text{ ft/s}^2}$$

$$\mathrm{p} = 1.00 \times 10^4 \text{ slug} \cdot \text{ft/s}$$

B. The mass of the car is $10\bar{0}$ slug.

$$v = \frac{p}{m}$$

$$v = \frac{1.00 \times 10^4 \text{ slug} \cdot \text{ft/s}}{10\bar{0} \text{ slug}}$$

$$v = 10\bar{0} \text{ ft/s}$$

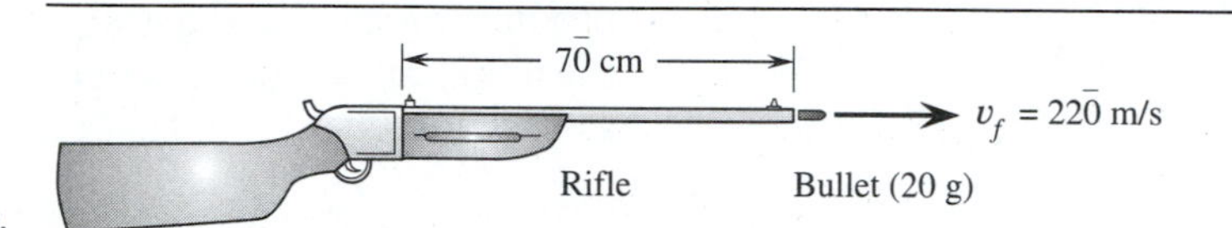

Figure 11–1: *A rifle fires a $2\bar{0}$-g bullet.*

□ **EXAMPLE PROBLEM 11–2: THE RIFLE**

A rifle with a 70-cm barrel fires a $2\bar{0}$-g bullet with a muzzle velocity of $22\bar{0}$ m/s. Find:

A. the initial momentum of the bullet.
B. the final momentum of the bullet.
C. the impulse acting on the bullet.
D. the magnitude of the average force acting on the bullet.

■ ***SOLUTION***

A. The bullet seated in the chamber has a velocity of zero. Take the forward direction as positive.

$$p_1 = m\,v_1 = 0 \text{ kg} \cdot \text{m/s}$$

B.

$$p_f = m\,v_f = 0.020 \text{ kg} \times 220 \text{ m/s}$$

$$p_f = 4.4 \text{ kg} \cdot \text{m/s}$$

C.

$$impulse = \Delta p = (4.4 - 0) \text{ kg} \cdot \text{m/s}$$

$$impulse = 4.4 \text{ N} \cdot \text{s} \ (1.00 \text{ N} = 1.00 \text{ kg} \cdot \text{m/s}^2)$$

D. We need to find the time of contact to calculate the average force. The bullet is accelerated down the barrel by the force of hot expanding gases. Find the time for the bullet to travel the barrel length. We'll make the not-quite-correct assumption that the acceleration is constant. The acceleration is:

$$a = \frac{v_f^2 - v_i^2}{2\,s}$$

$$a = \frac{(220\ m/s)^2 - (0\ m/s)^2}{2 \times 0.70\ m}$$

$$a = 3.46 \times 10^4\ m/s^2$$

The time of contact is then

$$t = \frac{v_f - v_i}{a}$$

$$t = \frac{220\ m/s - 0\ m/s}{3.46 \times 10^4}$$

$$t = 6.36 \times 10^{-3}\ s$$

From the definition of impulse we have:

$$F = \frac{\Delta p}{\Delta t} = \frac{4.4\ N \cdot s}{6.36 \times 10^{-3}\ s}$$

$$F = 690\ N$$

In Part D of Example Problem 11–2 we made an assumption about the force acting on the bullet. We assumed a constant force to give us a constant acceleration. What we would really expect to see is a rapidly increasing force when the gunpowder ignites, followed by a decrease as the bullet moves to leave space for the gas to expand. Friction acts between the barrel and the accelerating bullet. If the barrel is very long, frictional forces will exceed the force of the expanding gas. A rifle has an optimum length that will give the highest muzzle velocity for a particular bullet design.

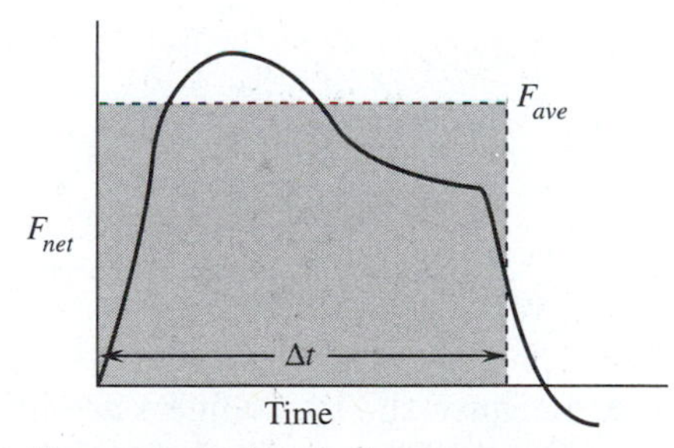

Figure 11–2: The net force acting on the bullet varies with time. The dashed line represents the average constant force. The area under the dashed line is the same as the area under the more complicated curve.

Figure 11–2 shows a plot of net force on the bullet with time. The impulse is the area under the curve. The horizontal dashed line represents the average force we calculated. The area under this horizontal line has the same area (impulse) as the more complicated curve.

In most kinds of collisions we can find the initial and final momentums. We often have an estimate of the period of contact. The impulse is calculated from the momentum data. When we use Equation 11–1 we find an *average* force for the reaction. At any instant, the actual forces may be larger or smaller than the calculated average.

11.2 CONSERVATION OF MOMENTUM

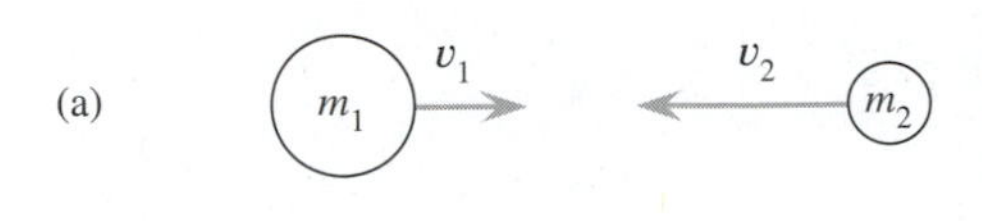

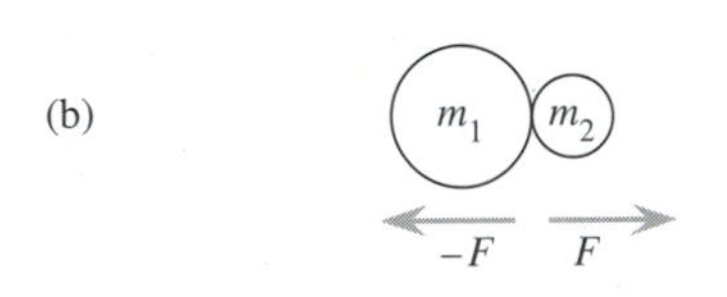

Figure 11–3: *A linear collision. (a) Mass m_1 and mass m_2 approach each other with velocities $\mathbf{v}_1$ and $\mathbf{v}_2$. (b) The masses collide with equal and opposite forces acting on them. (c) The masses rebound with new velocities, $\mathbf{u}_1$ and $\mathbf{u}_2$.*

Two automobiles run into each other head on. A rifle fires a bullet; the gun kicks backward. The path of a free electron is diverted by the electrical forces between the electron and nucleus of an atom. A stream of water is deflected by the blades of a turbine, causing the turbine to spin. A bomb bursts, sending fragments in several directions.

These are all examples of collisions. A **collision** is a process in which two or more objects interact, causing a change of momentum of the individual masses. First let's look at linear collisions involving only one dimension (see Figure 11–3).

A mass, m_1, with a velocity $\mathbf{v}_1$ to the right collides with another mass, m_2, moving in the opposite direction with a velocity of $\mathbf{v}_2$. In this example, m_1 is larger than m_2. After the collision both masses move with velocities $\mathbf{u}_1$ and $\mathbf{u}_2$. The initial momentums before the collision are:

$$\mathbf{p}_{1i} = m_1 \mathbf{v}_1$$

$$\mathbf{p}_{2i} = m_2 \mathbf{v}_2$$

The final momentums after the collision are:

$$\mathbf{p}_{1f} = m_1 \mathbf{u}_1$$

$$\mathbf{p}_{2f} = m_2 \mathbf{u}_2$$

During the collision each mass receives an impulse changing the momentum of the mass. For mass m_1:

$$\mathbf{F}_1 \Delta t = \Delta \mathbf{p}_1 = \mathbf{p}_{1f} - \mathbf{p}_{1i}$$

$$\mathbf{F}_1 \Delta t = (m_1 \mathbf{u}_1) - (m_1 \mathbf{v}_1)$$

For mass m_2:

$$\mathbf{F}_2 \Delta t = \Delta \mathbf{p}_2 = \mathbf{p}_{2f} - \mathbf{p}_{2i}$$

$$\mathbf{F}_2 \Delta t = (m_2 \mathbf{u}_2) - (m_2 \mathbf{v}_2)$$

There is no subscript on the time Δt because the two objects are in contact for the same period of time. $\mathbf{F}_1$ and $\mathbf{F}_2$ are a pair of action-reaction forces. Each mass pushes on the other with an equal but oppositely directed force ($\mathbf{F}_2 = -\mathbf{F}_1$). We may write this as equal but oppositely directed impulses.

$$\mathbf{F}_2 \Delta t = -\mathbf{F}_1 \Delta t$$

Now substitute the corresponding changes of momentum for the impulses.

$$(m_2\,\mathbf{u}_2) - (m_2\,\mathbf{v}_2) = -[(m_1\,\mathbf{u}_1) - (m_1\,\mathbf{v}_1)]$$

We see that each mass undergoes an equal, but oppositely directed, change of momentum. Let us combine the terms to eliminate all of the minus signs.

$$(m_1\,\mathbf{u}_1) + (m_2\,\mathbf{u}_2) = (m_1\,\mathbf{v}_1) + (m_2\,\mathbf{v}_2) \qquad \textbf{(Eq. 11–3)}$$

Notice everything on the left-hand side of the equation is related to momentum *after* the collision. The right-hand side contains only initial momentums. Here's another way of expressing this result.

The total momentum of a system before a collision is equal to the total momentum after the collision.

The above rule is called the **conservation of momentum.** Each thing involved in the collision may give up some of its individual momentum to the other object. It might receive some momentum in the collision from the other object. What one gives up, the other gains. It is simply an *exchange* of momentum. If the momentum of each particle is added up before the collision, it will equal the total of the momentum of the individual masses after collision. (When calculating totals remember that momentum is a vector.) Let's try some example problems using this idea.

☐ **EXAMPLE PROBLEM 11–3: ANOTHER RIFLE**

A 3.20-kg rifle fires a $2\underline{0}$-g bullet with a muzzle velocity of 310 m/s. What is the recoil velocity of the rifle? (See Figure 11–4.)

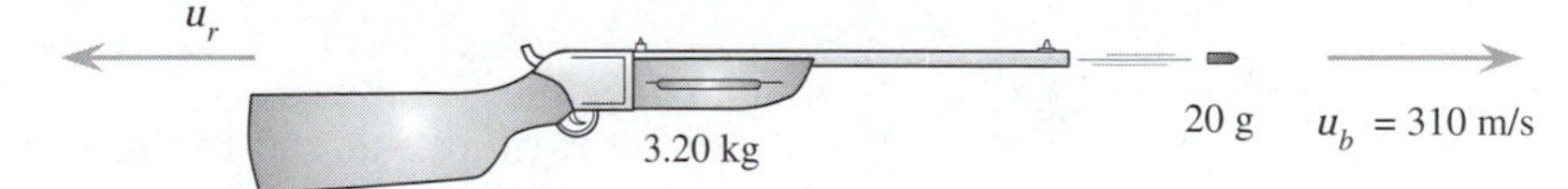

Figure 11–4: *A rifle recoils when it fires a bullet.*

■ *SOLUTION*

Initially the rifle and the bullet are at rest.

$$\mathbf{p}_r = \mathbf{p}_b = 0 \text{ kg} \cdot \text{m/s}$$

The law of conservation of momentum states that:

$$\textit{initial momentum} = \textit{final momentum}$$

$$0 = (m_r \mathbf{u}_r) + (m_b \mathbf{u}_b)$$

$$\mathrm{u}_r = -\left(\frac{m_b \mathrm{u}_b}{m_r}\right)$$

$$\mathrm{u}_r = -\left[\frac{(0.020 \text{ kg} \times 310 \text{ m/s})}{3.2 \text{ kg}}\right]$$

$$\mathrm{u}_r = -1.9 \text{ m/s}$$

The minus sign indicates that the gun moves in a direction opposite to the bullet. This is a typical recoil problem. If the initial momentum is zero, the two objects must move in opposite directions to maintain conservation of momentum.

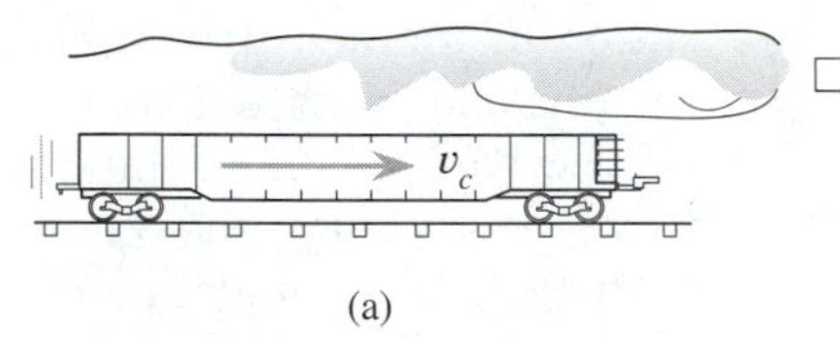

(a)

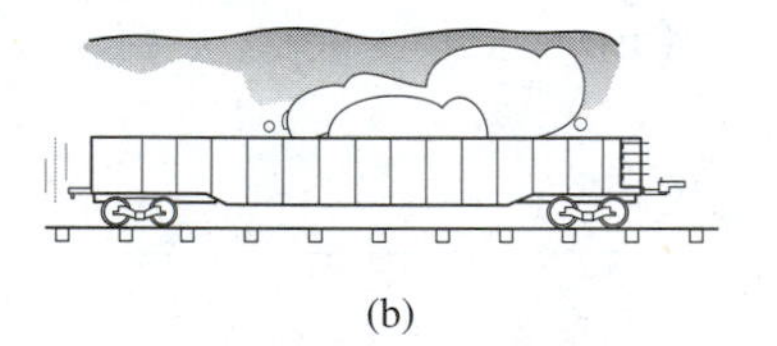

(b)

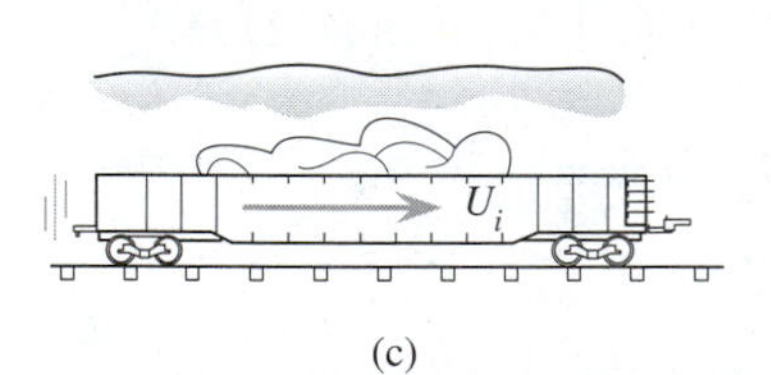

(c)

Figure 11–5: *(a) A hopper car moves along a level track. (b) A mass of snow falls into the car. (c) The car and snow move at a new velocity.*

□ **EXAMPLE PROBLEM 11–4: THE HOPPER CAR**

A $2\bar{0}$-ton hopper car moves along a level siding at a constant speed of 12 ft/s. Suddenly an overhanging bank of snow collapses, dumping 8.0 tons of snow in the hopper car. What is the new velocity of the car?

■ ***SOLUTION***

The snow has no initial momentum along the horizontal direction before the collision. After the collision, the snow moves at the same speed as the car.

$$\textit{initial momentum} = \textit{final momentum}$$

$$(m_c \mathbf{v}_c) + 0 = (m_c + m_s)\,\mathbf{U}$$

$$\mathbf{U} = \frac{m_c \mathbf{v}_c}{m_c + m_s}$$

The ratio of masses is the same as the ratio of weights, so we can state that

$$\mathbf{U} = \frac{2\bar{0} \text{ ton} \times 12 \text{ ft/s}}{(20 + 8)\text{ton}}$$

$$\mathbf{U} = 8.6 \text{ ft/s}$$

□ **EXAMPLE PROBLEM 11–5: THE BILLIARD BALLS**

A billiard cue ball traveling at 5.0 m/s strikes a stationary, identical ball. After the collision, the cue ball has a velocity of 3.0 m/s and is traveling at an angle of $6\bar{0}°$ to the right of its original path. The other

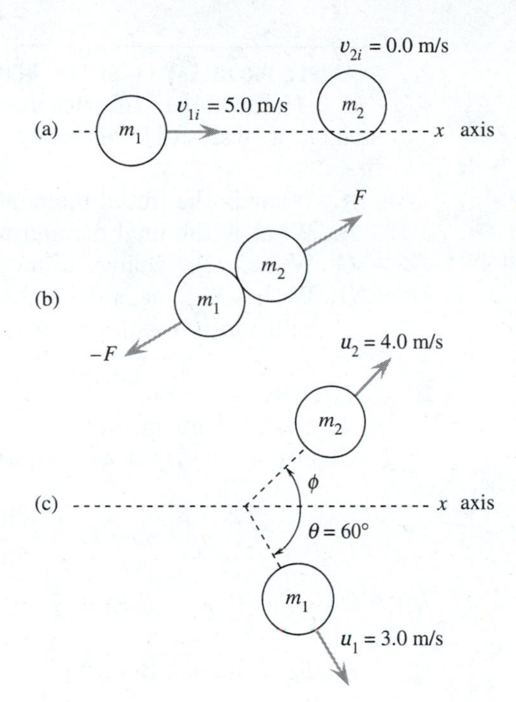

Figure 11–6: *A two-dimensional collision. (a) Mass m_1 approaches mass m_2, initially at rest. (b) Equal but opposite forces act on the masses for the same length of time. (c) Both masses move away from the collision site with components of momentum along the y axis.*

ball has a speed of 4.0 m/s. At what angle does it travel? (See Figure 11–6.)

■ *SOLUTION*

Remember, momentum is a vector quantity. Look at the x components of momentum. Before the collision, there was no horizontal component of momentum. Measure the angles from the x axis. Choose to the right as positive.

$$\text{initial } x \text{ momentum } (\mathbf{u}_1) = \text{final } x \text{ momentum } (\mathbf{u}_2)$$

$$mv_1 = (\text{m u}_1 \cos\theta) + (\text{m u}_2 \cos\phi)$$

Mass cancels out of the equation.

$$\cos\phi = -\frac{v_1 - u_1 \cos 6}{\text{u}_2}$$

$$\cos\phi = -\frac{5.0 \text{ m/s} - 3.0 \text{ m/s} \cos 60°}{4.0 \text{ m/s}}$$

$$\cos\phi = 0.875$$

$$\phi = 29°$$

□ **EXAMPLE PROBLEM 11–6: THE BOUNCING BALL**

A 0.15-lb rubber ball with an initial velocity of $4\bar{0}$ ft/s to the right strikes a wall and rebounds to the left with a speed of $3\bar{0}$ ft/s.

A. What is the initial momentum of the ball?
B. What is the final momentum of the ball?
C. What is the change of momentum of the ball?
D. With what mass does the ball exchange momentum in this collision to maintain conservation of momentum?

■ ***SOLUTION***

A. Initial momentum:
Choose to the right as the positive direction.

$$\mathbf{p}_1 = m\,\mathbf{v} = \frac{0.15 \text{ lb} \times 4\bar{0} \text{ ft/s}}{32.0 \text{ ft/s}^2}$$

$$\mathbf{p}_i = 0.19 \text{ slug} \cdot \text{ft/s}$$

B. Final momentum:

$$\mathbf{p}_f = m\,\mathbf{u} = \frac{0.15 \text{ lb} \times (-30 \text{ ft/s})}{32.0 \text{ ft/s}^2}$$

$$\mathbf{p}_f = -0.14 \text{ slug} \cdot \text{ft/s}$$

C. Change of momentum:

$$\mathbf{p} = \mathbf{p}_f - \mathbf{p}_i = -(0.141 \text{ slug} \cdot \text{ft/s}) - (0.188 \text{ slug} \cdot \text{ft/s})$$

$$\mathbf{p} = -0.33 \text{ slug} \cdot \text{ft/s}$$

D. The two action-reaction forces are between the ball and the wall with which it collides. The change of momentum of the wall is equal to the change of momentum of the ball, but in the opposite direction. Because the masses of the wall and the earth to which it is attached are very large, the change of velocity of its mass is not measurable.

11.3 ANGULAR MOMENTUM

In the last section we found that the linear momentum of a system was unchanged unless an *external* force acted upon it. Although internal forces allowed an exchange of momentum among the various parts of a system, the total momentum of the system was constant.

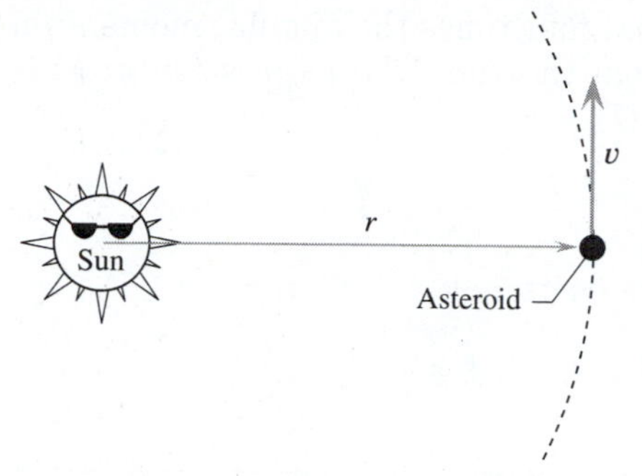

Figure 11–7: *An asteroid has a circular orbit around the sun.*

Look at Figure 11–7. An asteroid rotates around the sun in a circular path. The linear momentum of the asteroid is continually changing. Although the speed of the object is constant, its direction is changing. Newton's second law in the general form shows that the external force causing this change of linear momentum is centripetal force.

$$F_c = \frac{\Delta p}{\Delta t}$$

Since this is a case of rotational motion, let us rewrite the momentum in terms of angular speed.

$$v = r\omega$$

The force equation now looks like this.

$$F_c = \frac{\Delta(m\,r\omega)}{\Delta t}$$

Rotational motion is affected by torques. Recall $\tau = \mathbf{F}\,r$ where r is the perpendicular lever arm.

$$\tau = F_c r = \frac{\Delta\,(m\,r^2\,\omega)}{\Delta t}$$

We will define $\tau\Delta t$ as the **angular impulse.** Consistent with the development of the linear form of Newton's second law, the expression inside the parentheses is called **angular momentum (L).** We now have a relationship for rotation that is similar in form to the impulse-momentum equation for linear motion.

angular impulse = the change of angular momentum

$$\tau\Delta t = \Delta \mathbf{L} \qquad \textbf{(Eq. 11–4)}$$

where $|\mathbf{L}| = m\,r^2\omega$.

In order for the angular momentum of something to change, an unbalanced *external* force must act on it. In the case of our asteroid, there is no unbalanced torque. Because it is traveling at a constant speed in a circular path, the only force acting on it is the centripetal force supplied by gravity. The force is radial. The angle between the force and the lever arm is zero. There is no change of angular momentum in the circular path. The orbital radius, mass, and angular speed are all constant.

Notice that the angular momentum depends not only on the size of the mass, but also on how far the mass is from the spin axis. An asteroid 4.0 AU from the sun has four times the angular momentum of an asteroid only 2.0 AU from the sun. The expression $m\,r^2$ is called the **moment of inertia** (I).

$$I = m\,r^2$$

The angular momentum can be written:

$$\mathrm{L} = I\omega$$

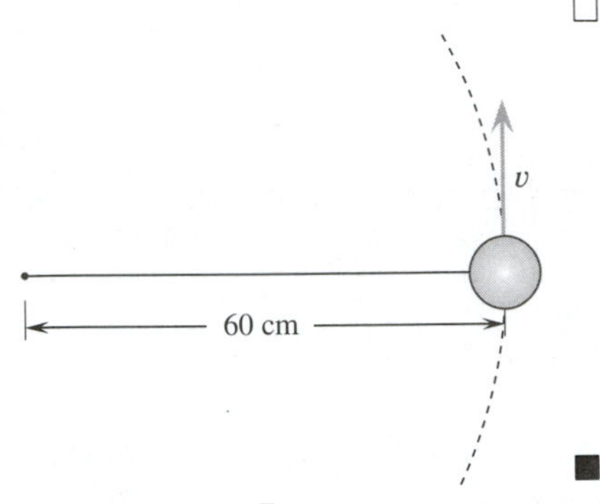

Figure 11–8: A 20$\bar{0}$-g bob is whirled in a 6$\bar{0}$-cm circle.

□ **EXAMPLE PROBLEM 11–7: A BOB ON A STRING**

A 20$\bar{0}$-g bob on a string is whirling on a 6$\bar{0}$-cm circle at an angular speed of 10 rad/s. A torque is applied to increase the angular speed to 12 rad/s in 2.0 s (see Figure 11–8).

A. What is the moment of inertia of the bob?
B. What is the initial angular momentum?
C. What is the final angular momentum?
D. What is the change of momentum?
E. What average torque was applied to the bob?

■ ***SOLUTION***

A. Moment of inertia:

$$I = m\,r^2$$

$$I = 0.200\text{ kg} \times (0.60\text{ m})^2$$

$$I = 0.072\text{ kg}\cdot\text{m}^2$$

B. Initial angular momentum:

$$\mathrm{L_i} = I\,\omega_\mathrm{i}$$

$$\mathrm{L_i} = 0.072\text{ kg}\cdot\text{m}^2 \times 10\text{ rad/s}$$

$$\mathrm{L_i} = 0.72\text{ kg}\cdot\text{m}^2\text{/s}$$

C. Final angular momentum:

$$\mathrm{L_f} = I\,\omega_\mathrm{f}$$

$$\mathrm{L_f} = 0.072\text{ kg}\cdot\text{m}^2 \times 12\text{ rad/s}$$

$$\mathrm{L_f} = 0.86\text{ kg}\cdot\text{m}^2\text{/s}$$

D. The change of the angular momentum:

$$\mathrm{L} = \mathrm{L_f} - \mathrm{L_i}$$

$$\mathrm{L} = (0.86 - 0.72)\text{ kg}\cdot\text{m}^2\text{/s}$$

$$\mathrm{L} = 0.14\text{ kg}\cdot\text{m}^2\text{/s}$$

E. The average torque:

$$\tau = \frac{(\Delta L)}{\Delta t}$$

$$\tau = \frac{0.14 \text{ kg} \cdot \text{m}^2/\text{s}}{2.0 \text{ s}}$$

$$\tau = 0.070 \text{ N} \cdot \text{m}$$

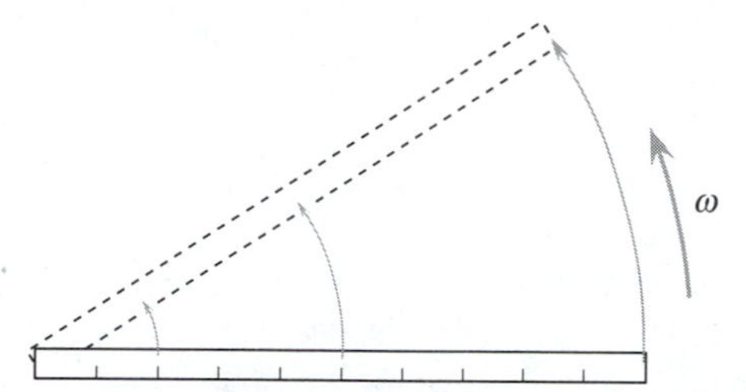

Figure 11–9: *A meter stick is rotated around one end.*

So far we have looked at systems in which the mass can be viewed as a point mass orbiting in a circle. Let's look at a situation in which an object has an extended mass.

A meter stick is swung from one end in a circle (see Figure 11–9). All parts of the meter stick have the same angular speed. The outside edge of the meter stick travels through a larger circle than the middle part. The 10-cm mark moves in yet a smaller circle. Since angular momentum depends on radius, different parts of the stick have different angular momentums.

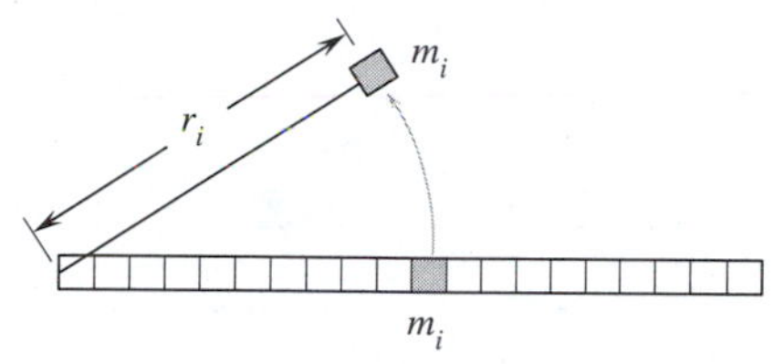

Figure 11–10: *The meter stick can be divided into small masses. Each mass,* m_i*, orbits at a radius of* r_i*.*

We can find the total angular momentum of the stick by dividing the stick into smaller parts. The momentums of the different parts can be added to find the total momentum of the stick (see Figure 11–10).

Assume the stick has a uniform density. If we divide the stick into equal-sized lengths, each part will have the same mass. The total angular momentum is then:

$$\mathbf{L} = (m\, r_1{}^2\omega) + (m\, r_2{}^2\omega) + (\ldots m\, r_n{}^2\omega)$$

$$\mathbf{L} = [(m\, r_1{}^2) + (m\, r_2{}^2) + (\ldots m\, r_n{}^2)]\omega$$

$$\mathbf{L} = \left(\sum_{i=1}^{n} m\, r_i{}^2\right)\omega = I\omega$$

where I is the moment of inertia of the stick. $\left(I = \sum_{i=1}^{n} m\, r_i{}^2.\right)$

A mathematical device called calculus can be used to find the moment of inertia of a slender stick. $I = \frac{1}{3}(m\, R^2)$ where R is the length of the stick. If the stick is twirled around its center, the moment of inertia will be different from the moment of inertia if the stick is swung from one end. The end will not move through as large a circle. In this case, the moment of inertia is $I = \frac{1}{12}(m\, R^2)$.

The moment of inertia of an object depends on the distribution of mass around the spin axis. It is proportional to $m\, R^2$.

$$I = k\, m\, R^2 \qquad \textbf{(Eq. 11–5)}$$

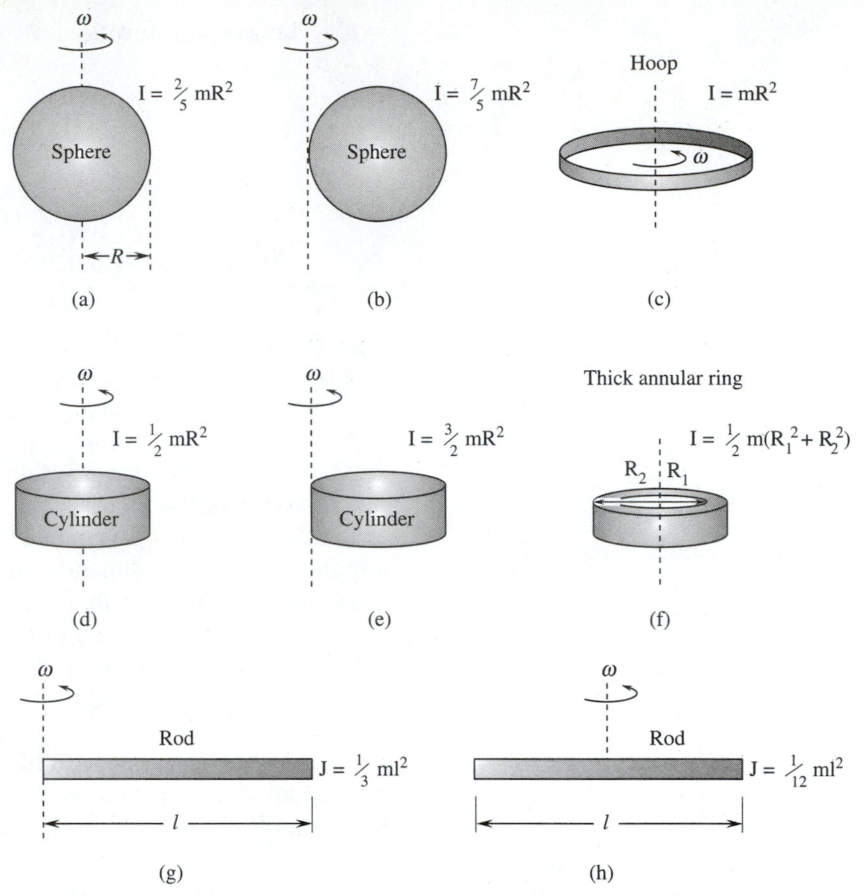

Figure 11–11: *The moments of inertia of some simple shapes. Notice the moment of inertia is larger if the shape is not spun on an axis through its center of mass.*

where k depends on the shape of the object and the location of the spin axis. The following factors affect the moment of inertia.

A. The shape of the object affects k.
B. The size of the object affects R.
C. The mass of the object affects m.
D. The location of the spin axis affects k.
E. Variation in density will affect k.

Figure 11–11 shows the moments of inertia for some simple shapes with constant density. In Chapter 12 we will find an experimental way to find the moment of inertia for a more complicated object or one with nonuniform density.

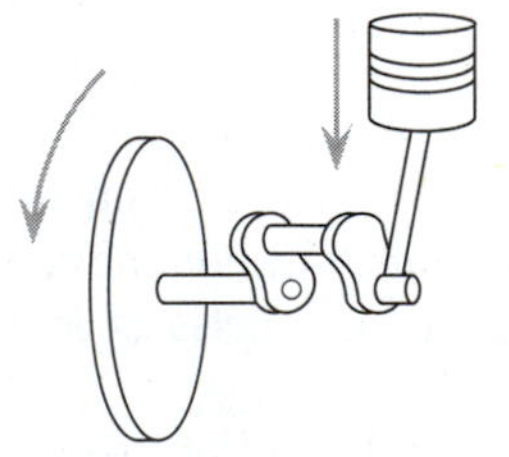

Figure 11–12: *The momentum of a flywheel keeps the piston moving between power strokes.*

Angular momentum and angular motion are major factors in many kinds of machinery. A piston engine takes advantage of angular momentum (see Figure 11–12). In a four-stroke engine only one stroke in four supplies power to the cam case. The angular momentum is given to a flywheel during the power stroke. A small momentum decrease is then converted to torque to drive the cam case. The rotary motion is converted by the cams into linear motion to move the piston and valves.

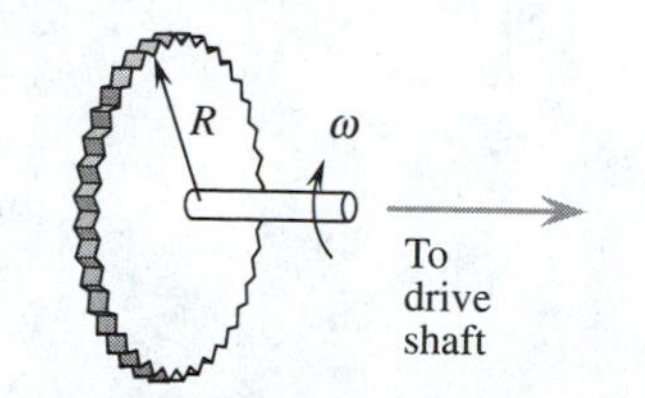

Figure 11–13: A disk-shaped flywheel.

EXAMPLE PROBLEM 11–8: THE FLYWHEEL

A small motor has a 2.50-kg disk-shaped flywheel with a radius of 20.0 cm. How much angular momentum does the flywheel have when the motor operates at $30\bar{0}0$ rpm (see Figure 11–13)?

SOLUTION

In radians per second the rotational speed of the flywheel is:

$$\omega = 3\bar{0}\bar{0}0 \frac{\text{rev}}{\text{min}} \times 2\pi \frac{\text{rad}}{\text{rev}} \times \frac{1 \text{ min}}{60 \text{ s}} = 314 \frac{\text{rad}}{\text{s}}$$

From Figure 11–11 the moment of inertia of the disk-shaped flywheel is:

$$I = \tfrac{1}{2}(mR^2)$$

$$I = \tfrac{1}{2} \times 2.50 \text{ kg} \times (0.200 \text{ m})^2$$

$$I = 0.0500 \text{ kg} \cdot \text{m}^2$$

The magnitude of the angular momentum is:

$$L = I\omega$$

$$L = 314 \text{ rad/s} \times 0.0500 \text{ kg} \cdot \text{m}^2\text{/s}$$

$$L = 15.7 \text{ kg} \cdot \text{m}^2\text{/s}$$

11.4: CONSERVATION OF ANGULAR MOMENTUM

In Section 11.2 we found there is no change of total linear momentum on a system unless there is an unbalanced external force. Angular momentum obeys a similar rule.

The angular momentum of a system is constant unless an unbalanced external torque acts on it.

Why a Cat Lands on Its Feet

A cat can be held upside down by the legs and released with no initial angular momentum. The cat twirls around and lands on its feet. This has puzzled lots of people for a long time because this would seem to violate the conservation of angular momentum.

Fortunately cats are smarter than people. The cat rotates its tail in the opposite direction from its body. This maintains a total rotational momentum of zero.

There are breeds of cats, like the manx, that have no tail. They too are able to land on their feet. As you can see in the series of photographs on page 263, tailless cats rotate one of their hind feet in the opposite direction from the body. This is a more complicated solution, but it does the same job as the missing tail.

(Photograph courtesy of J. E. Frederickson, Department of Physics, California State University at Long Beach.)

Let us see why this might be true. Look at Figure 11–14. Two freely spinning disks on the ends of shafts are rotating with different angular speeds of ω_1 and ω_2. Their moments of inertia are I_1 and I_2. The two disks are brought together. Frictional torques bring them to a common rotational speed of ω_f. The torques are equal in magnitude, but opposite in direction. One disk is caused to spin faster. The other is slowed down.

$$\tau_1 \Delta t = (I_1 \omega_f) - (I_1 \omega_1)$$

$$\tau_2 \Delta t = (I_2 \omega_f) - (I_2 \omega_2)$$

Since the angular impulses are equal in size, but opposite in direction:

$$\tau_1 \Delta t = -\tau_2 \Delta t$$

or

$$(I_1 \omega_f) - (I_1 \omega_1) = -(I_2 \omega_f) - (I_2 \omega_2)$$

Rearrange the terms of the equation to eliminate the minus signs.

initial angular momentum = *final angular momentum*

$$(I_1 \omega_1) + (I_2 \omega_2) = (I_1 \omega_f) + (I_2 \omega_f)$$

In this example the spin axes are along the same line of action. Notice that the moments of inertia of the two objects are additive.

$$(I_1 \omega_1) + (I_2 \omega_2) = (I_1 + I_2)\,\omega_f \qquad \textbf{(Eq. 11–6)}$$

EXAMPLE PROBLEM 11–9: TWO SPINNING DISKS

The two disks in Figure 11–13 have initial angular speeds of $\omega_1 = 80$ rad/s and $\omega_2 = 120$ rad/s. The moments of inertia are $I_1 = 0.024$ kg · m^2 and $I_2 = 0.016$ kg · m^2. Find the final angular speed if:

A. the two disks initially spin counterclockwise.

B. disk 1 initially spins counterclockwise and disk 2 spins clockwise.

SOLUTION

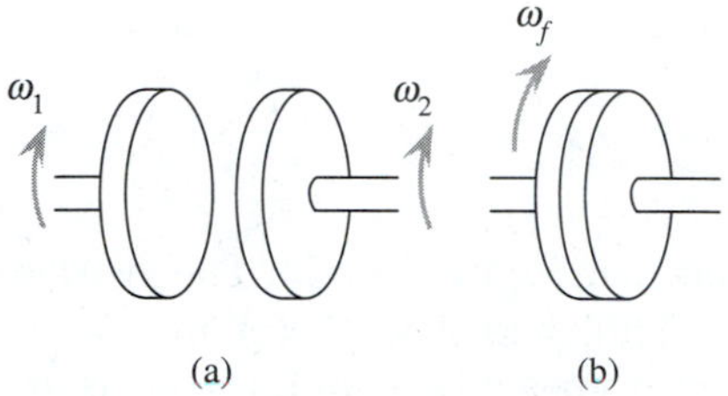

Like torque, the vector for angular momentum is along the spin axis. It obeys the same right-hand screw rule (see Figure 11–14). A clockwise rotation will advance a screw into the page; a counterclockwise rotation will advance a screw outward. A counterclockwise momentum is generally taken as positive.

A. Both initial spins are positive.

$$\textit{initial angular momentum} = \textit{final angular momentum}$$

$$(I_1\,\omega_1) + (I_2\,\omega_2) = (I_1 + I_2)\,\omega_f$$

$$\omega_f = \frac{(I_1\,\omega_1) + (I_2\,\omega_2)}{(I_1 + I_2)}$$

$$\omega_f = \frac{(0.024\ \text{kg}\cdot\text{m}^2 \times 80\ \text{rad/s}) + (0.016\ \text{kg}\cdot\text{m}^2 \times 120\ \text{rad/s})}{(0.024 + 0.016)\ \text{kg}\cdot\text{m}^2}$$

$$\omega_f = 96\ \text{rad/s}$$

B. Disk 1 has a positive spin. Disk 2 has a negative spin.

$$\omega_f = \frac{(0.024\ \text{kg}\cdot\text{m}^2 \times 80\ \text{rad/s}) + [(0.016\ \text{kg}\cdot\text{m}^2)\,(-120\ \text{rad/s})]}{(0.024 + 0.016)\ \text{kg}\cdot\text{m}^2]}$$

$$\omega_f = 0.00\ \text{rad/s}$$

Figure 11–14: *(a) Two disks rotate freely at different angular speeds. (b) The two disks are brought together. Friction forces them to rotate at the same final angular speed.*

The angular momentum of an object depends on three things: the direction of the spin axis, the angular speed, and the distribution of mass around the spin axis (moment of inertia). The analysis can be very complicated for systems in which the spin axes are not aligned along the same line of action. We will not cover this type of system in this text. Let us look at a problem involving only the last two factors affecting angular momentum.

If the moment of inertia of a spinning object changes and there is no external torque acting on it, then the angular velocity must change. The product of the moment of inertia (I) and angular speed (ω) is a constant.

$$I_i \omega_i = I_f \omega_f \quad \text{(Eq. 11–7)}$$

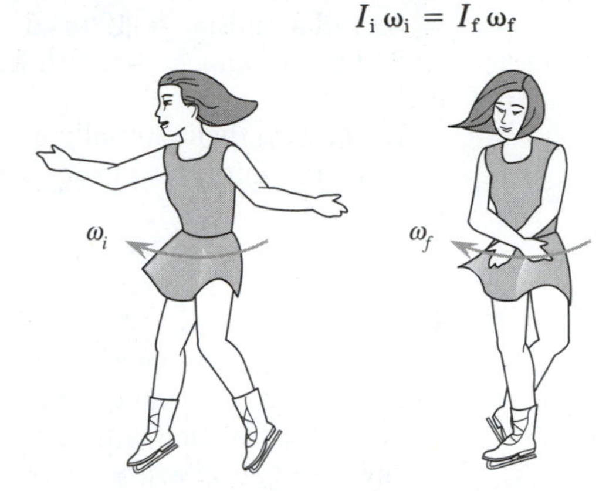

Figure 11–15: A spinning skater. (a) As the skater goes into the spin, the arms are outstretched. (b) The arms are brought across the body to increase the spin rate.

A figure skater uses this principle (see Figure 11–15). The average weight of an adult human arm is 7 lb, with the center of gravity located near the elbow. As the skater goes into a spin, the arms are initially outward as momentum is built up. The arms are then brought across the chest to reduce the moment of inertia. The skater spins faster, maintaining angular momentum. To stop the spin, the arms are again spread outward to slow down the rotation. A skate is then used as a brake.

□ **EXAMPLE PROBLEM 11–10: THE SKATER**

A figure skater goes into a spin of 6.0 rad/s with the arms extended with an initial moment of inertia of 1.20 slug · ft^2. She brings her arms inward, reducing the moment of inertia to 0.40 slug · ft^2. Estimate the new spin rate.

■ ***SOLUTION***

Solve Equation 11–7 for ω_f.

$$\omega_f = \frac{\omega_i I_i}{I_f}$$

$$\omega_f = \frac{6.0 \text{ rad/s} \times 1.20 \text{ slug} \cdot \text{ft}^2}{0.40 \text{ slug} \cdot \text{ft}^2}$$

$$\omega_f = 18 \text{ rad/s}$$

11.5 APPLICATIONS

The Hydraulic Ram

The hydraulic ram in Figure 11–16 works if the device can be located below the water source. The momentum of the water entering the

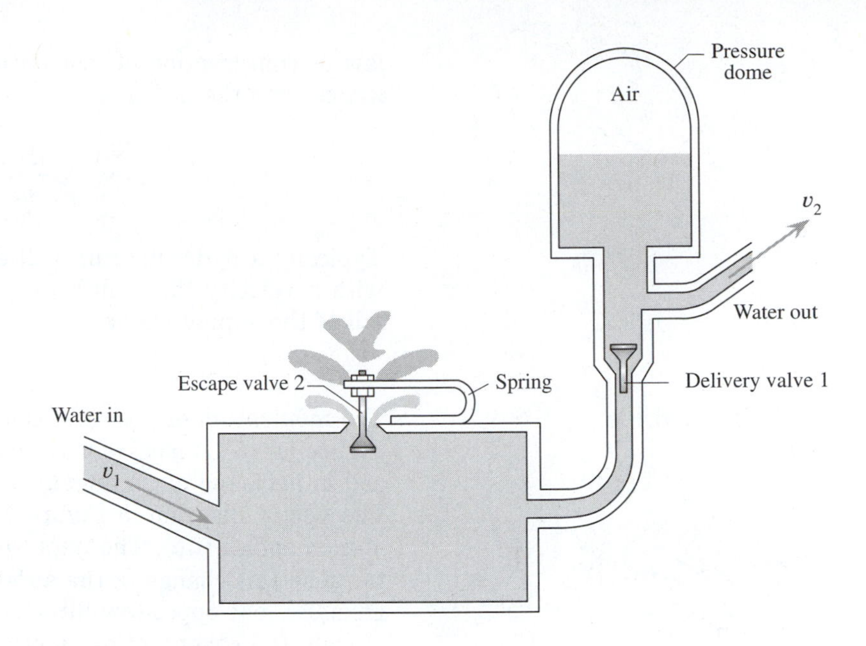

Figure 11–16: *A hydraulic ram. The momentum of the incoming water is used to force out a smaller mass of water at a larger speed.*

ram at one side is used to drive a smaller mass of water out the other side at a higher velocity. When water enters the ram on the supply side, valve 1 is closed and valve 2 is open. The momentum of the water closes valve 2 as the chamber fills. The water then pushes valve 1 open. Water surges into the dome, compressing the air. As the back-pressure from the compressed air increases, valve 1 is forced shut. The compressed air then forces the water in the dome out the outlet pipe. While this is happening there is a back-surge of water in the chamber and the supply pipe. Valve 2 opens, releasing waste water. The cycle starts over again. This cycle is repeated from 25 to 100 times a minute. The frequency depends on the height of the water source and the design of the ram.

The mechanics of the collision inside the ram seems a bit complicated. Figure 11–17 shows a simpler view. We will ignore the details of the collision and focus on the water entering and leaving the hydraulic ram. A mass of water (M_1) enters the ram with an initial speed of v_1. In a collision process, a smaller mass of water (m_2) is driven out the other side with a final speed (v_2). Using the

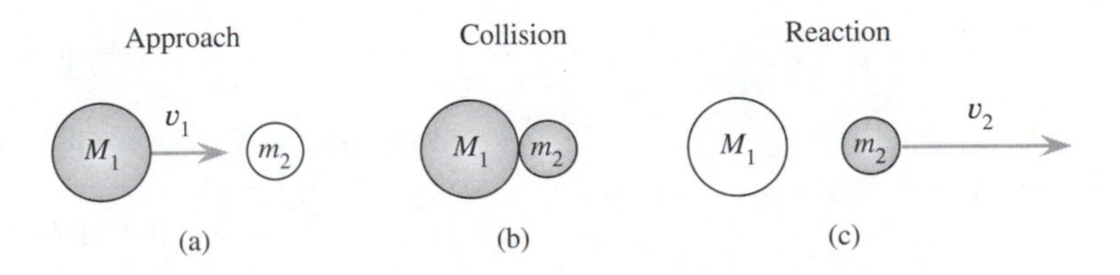

Figure 11–17: *The hydraulic ram can be visualized as a collision between a large mass and a small mass initially at rest.*

law of conservation of momentum, we see that the final speed is larger than the initial one.

$$v_2 = \frac{M_1 v_1}{m_2}$$

Typically a hydraulic ram will deliver 10% of the incoming water with a velocity that will raise the water a distance five times the fall of the supply water.

The Wind Turbine

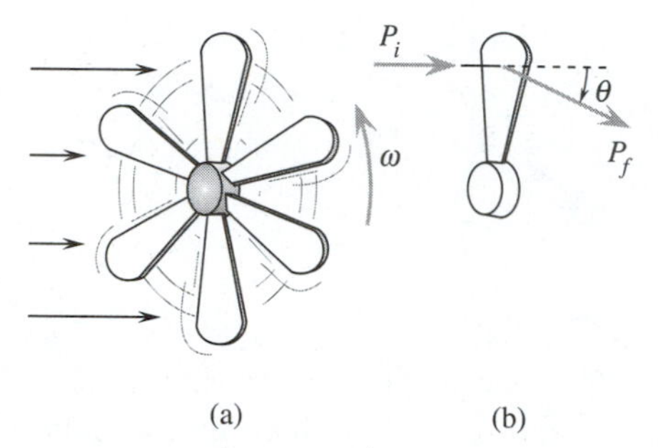

Figure 11–18: *A wind turbine. (a) Moving air causes the blades of the turbine to turn. (b) A blade of the turbine deflects the wind.*

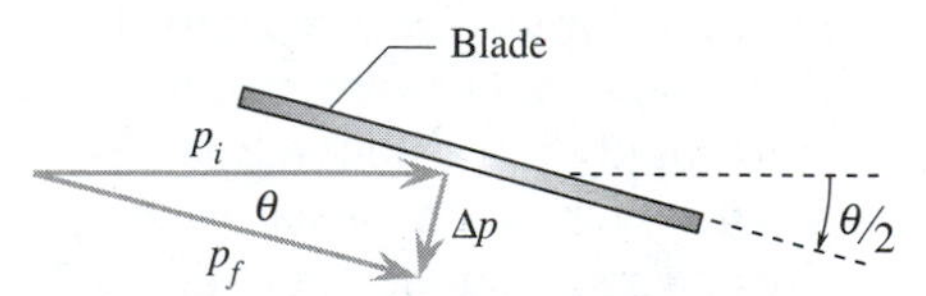

Figure 11–19: *Δ **p** is the change of momentum of the air caused by the blade of the turbine.*

The momentum of the wind causes a wind turbine to operate (see Figure 11–18). A mass of moving air (m) strikes the blade of a turbine and is deflected at an angle (θ) from its original direction of flow. The vector diagram in Figure 11–19 shows that there is a change of momentum (**Δp**). The impulse delivered by the blade on the air to cause this change of the air's momentum is **F**Δt. The air exerts an equal, but oppositely directed, impulse on the blade.

Only the change of momentum in the y direction will cause the blade to turn on its axis. This is the same as the y component of the final momentum.

$$\Delta p_y = \Delta p_f \sin \theta$$

The y component of the impulse acting on the blade is:

$$F_y \Delta t = - p_f \sin \theta$$

Let us assume that the wind is deflected without changing its speed. In this case, the final momentum will have the same magnitude as the initial momentum. Express the mass of air in terms of its density.

$$m = \rho V$$

where V is the volume of air striking the blade during a time interval Δt. If the wind has a speed of **v**, a column of air **v** Δt long will strike the blade of area A during the time Δt (see Figure 11–20).

$$m = \rho A v \Delta t$$

The magnitude of the wind's momentum is:

$$p = \rho A v^2 \Delta t$$

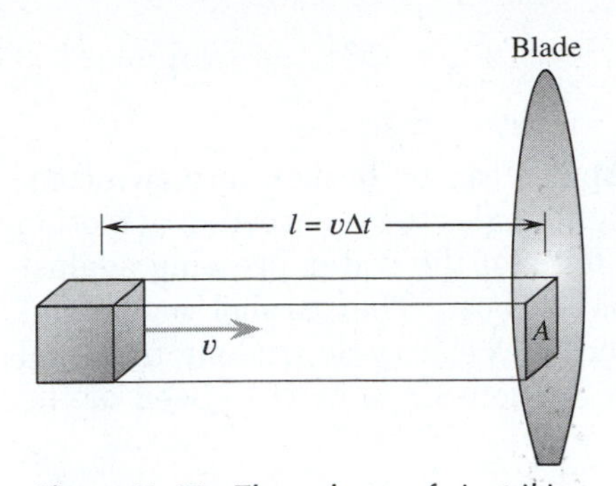

*Figure 11–20: The volume of air striking an area A of the turbine in time Δ t is a rectangular block. The block has a cross-sectional area A and length **v** Δ t.*

and the y component of the impulse acting on the turbine blade is:

$$F_y \, \Delta t = \rho \, A \, v^2 \, \Delta t \sin \theta$$

The Δts cancel out, giving us the component of force causing the blade to turn.

$$F_y = \rho \, A \, v^2 \sin \theta$$

The force is proportional to the square of the wind velocity. The pitch of the blade will determine the angle (θ). Notice in Figure 11–19 that the pitch is half the angle of θ.

□ **EXAMPLE PROBLEM 11–11: THE WIND TURBINE**

A wind turbine blade has an area of 0.60 m² and a pitch of $2\bar{0}°$. What turning force is exerted on the blade by a $3\bar{0}$ km/h wind? (The density of air is 1.20 kg/m³.)

■ ***SOLUTION***

The angle θ is twice the angle of pitch: $\theta = 4\bar{0}°$.
Convert the wind speed to SI units:

$$\mathbf{v} = 3\bar{0} \text{ km/h} \times 0.278 \frac{\text{m/s}}{\text{km/h}} = 8.34 \text{ m/s}$$

$$F_y = \rho \, A \, v^2 \sin \theta$$

$$F_y = 1.2 \frac{\text{kg}}{\text{m}^3} \times 0.6 \text{ m}^2 \, (8.34 \text{ m/s})^2 \sin 4\bar{0}°$$

$$F_y = 32 \text{ N}$$

Tacking against the Wind

In nautical terms a tack is a course set obliquely against the wind. By using a zigzag course to port (left) and starboard (right) a sailboat can advance against the wind (see Figure 11–21).

The vector nature of momentum makes this process possible. The wind is deflected by the sail. This causes an impulse to act on the

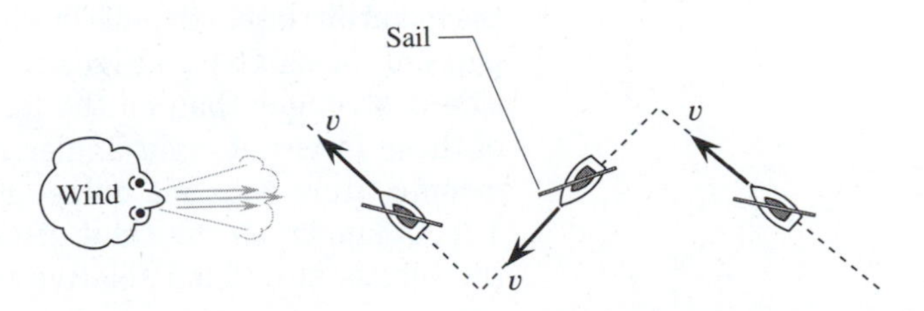

Figure 11–21: Tacking against the wind. Notice the angles of the sail.

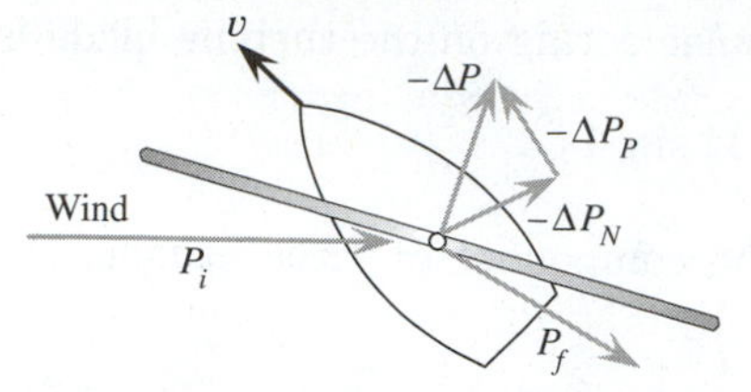

Figure 11–22: Wind on the sail has a component parallel to the boat's forward direction, $-\Delta P_p$. The perpendicular component is balanced by the water pushing against the side of the hull, $-P_N$.

sail (see Figure 11–22). The impulse can be broken into two components. The y component, at right angles to the direction of motion of the boat, is balanced by the force of the water pressing against the side of the hull and keel of the boat. The parallel component gives the ship its forward velocity. Velocity-dependent frictional forces of the type described in Section 8.2 limit the speed of the boat.

The Earth-Moon System

Gravitational forces act between the moon and the earth. These are internal forces in the earth-moon system. The total angular momentum of the earth-moon system should be conserved.

The moon has very little spin motion. It always has the same side facing the earth, so that it rotates once during one complete orbit of the earth, approximately 27.3 days. Its angular momentum is mostly in its orbital motion.

The earth's angular momentum is mostly found in its spin. The earth and moon orbit around their common center of gravity. The center of gravity of the system is much closer to the earth because the earth is much more massive than the moon. The earth travels through an orbit with a radius a little less than one earth radius in 27.3 days. The angular momentum associated with this is very small compared with its spin angular momentum.

There is little error in assuming the earth's angular momentum is caused only by its spin and the moon's angular momentum is found only in its orbital motion. The total angular momentum of the system is then the spin angular momentum of the earth plus the orbital angular momentum of the moon.

$$\mathbf{L} = (I_e\, \omega_e) + (m_m R^2\, \omega_m)$$

where I_e is the moment of inertia of the earth, ω_e is the spin rate of the earth, m_m is the mass of the moon, R is the distance between the earth and the moon, and ω_m is the angular velocity of the moon's orbit.

Because the moon is close to the earth, the gravitational force of the moon on the earth pulls in slightly different directions on different parts of the earth (see Figure 11–23). The pull of gravity on the near side is stronger than on the far side of the earth. The components of these forces at right angles to the line connecting the earth and moon centers causes a bulge on the near and far side of the earth. This is known as the **tidal effect** of the moon. As the earth spins, the bulges stay fixed relative to the line connecting the earth and

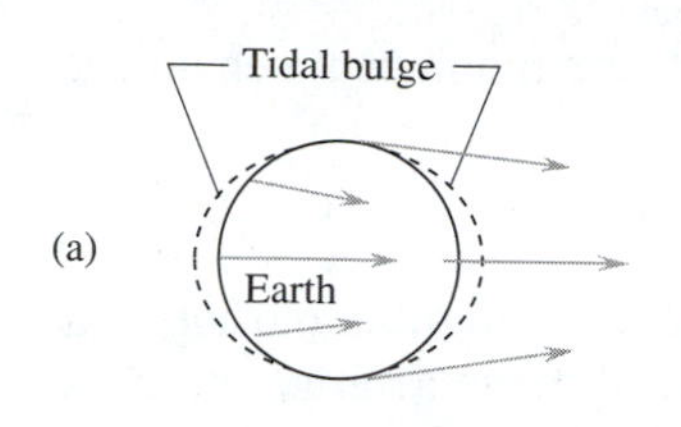

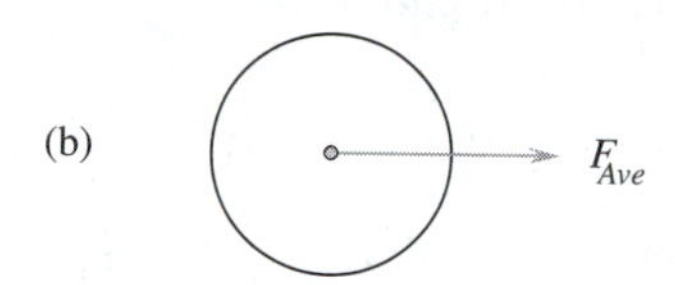

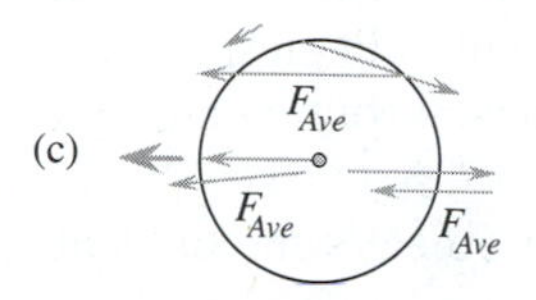

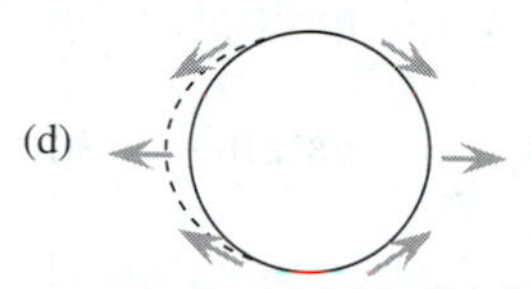

Figure 11–23: *(a) Gravity attracts different parts of the earth with different strengths and directions. (b) The average gravitational force of attraction acts at the center of gravity. (c) The tidal force can be found by vector subtraction of the average gravitational force from the gravitational force at each location on the earth. (d) The net tidal force.*

the moon. To an earth dweller these tides sweep across the planet. Actually the planet is spinning underneath the nearly stationary bulges. It does not matter which way you look at it; the effect is the same. The tides move relative to the surface of the earth sweeping eastward with the daily rotation of the earth.

Frictional forces between the tidal bulges and the earth act as a brake, slowing down the earth. The length of the day is increasing by 2×10^{-3} s each century. This does not sound like much, but the earth has been around for over four billion years. The study of growth bands of fossilized coral reefs indicates that the periods of daylight were shorter millions of years ago. According to these data, the day was around 21 h long 500 million years ago, in agreement with the tidal friction theory.

If the earth is slowing down, its angular momentum is decreasing. The moon's angular momentum must increase to maintain a constant total angular momentum for the earth-moon system. The angular momentum of the moon is:

$$m_m R^2 \omega = m_m R \mathbf{v}$$

We can use Equation 10–5 to express the tangential velocity of the moon in terms of the mass of the earth and the earth-moon radius.

$$\mathbf{v} = \left(\frac{G M}{R}\right)^{1/2}$$

Substitute this into the angular momentum equation.

$$\textit{angular momentum of the moon} = m_m R \left(\frac{G M}{R}\right)^{1/2}$$

or

$$\mathbf{L}_{moon} = m_m (G M)^{1/2} \times R^{1/2}$$

The only variable on the right-hand side of the equation is the earth-moon distance. The moon's orbital angular momentum is proportional to $R^{1/2}$. As the earth slows down, the moon spirals outward moving farther away from the earth.

SUMMARY

Impulse is the product of the force acting on a system multiplied by the interval of time through which it acts.

$$\text{impulse} = \mathbf{F} \, \Delta \, t$$

The momentum (**p**) of an object is the product of its mass and velocity.

$$\text{momentum} = \mathbf{p} = \text{m } \mathbf{v}$$

When an external impulse acts on a system, its momentum changes.

$$\mathbf{F} \, \Delta \, t = \Delta \, \mathbf{p} = m_f \, \mathbf{v}_f - m_i \, \mathbf{v}_i$$

The moment of inertia (I) of something with angular motion is a function of mass (m) and the distribution of mass around the axis of rotation. For a point mass a distance (R) from the axis

$$I = m \, R^2.$$

For an extended body the moment of inertia is calculated by finding the sum of the moments of inertia of all the small masses (m_i) at distances R_i from the rotational axis. I depends not only on the mass of the object, but also on the distribution of mass around the spin axis.

Angular momentum (**L**) is the product of an object's moment of inertia and its angular velocity.

Angular impulse is the torque acting on something multiplied by the period of action.

When an angular impulse acts on a system, the system's total angular momentum will change.

The law of conservation of momentum states that if no external force acts on a system, the total momentum of the system is unchanged.

Momentum is conserved in all collisions. If no external torque acts on a system, the total angular momentum of the system remains unchanged.

$$I_{1i} \, \omega_{1i} + I_{2i} \, \omega_{2i} = I_{1f} \, \omega_{1f} + I_{2f} \, \omega_{2f}$$

KEY TERMS

If you can explain the following terms to a friend or classmate, you understand their meaning. If you cannot explain the terms, you should reread the sections in which they are discussed.

angular impulse
angular momentum
collisions
conservation of momentum
impulse
moment of inertia
tidal effect

EXERCISES

Section 11.1:

1. Will a large force always give a larger impulse than a small force? Explain why or why not.

2. A large mass and a small mass collide. Assume no external forces. Which of the following statements is (are) correct?
   A. The larger mass exerts a larger force on the smaller mass.
   B. The smaller mass undergoes a larger change of velocity.
   C. The larger mass exerts a larger impulse on the smaller mass.
   D. The smaller mass undergoes a larger change of momentum.
3. In which of the following systems is linear momentum most likely to be unchanged? What are the external impulses?
   A. A wrench falls a short distance in air.
   B. A motor scooter rounds a corner at a constant speed.
   C. A piston is driven by a gasoline explosion.
   D. A hockey puck slides along the ice.
4. A rubber ball and a ball of mud with the same mass are thrown at a wall. They have the same velocity. The mud sticks; the ball rebounds.
   A. Do they undergo the same change of momentum?
   B. Do they exert the same impulse on the wall?
5. Explain why a counterweight on a cable elevator reduces the impulse needed by the winch to increase the speed of the elevator car. Use an impulse-momentum argument.

Section 11.2:

6. When mud is thrown at a wall and it sticks, what happens to the linear momentum?
7. Large rockets are usually designed in stages. In flight, booster rockets are ejected from the final stage. Give two reasons why this is done.
8. Two identical cars traveling at the same speed have a head-on collision. Are the impulses on each car larger than for a single car striking a massive concrete abutment? Explain why or why not. Use the law of conservation of momentum to explain your answer.
9. Dynamite is used to break a boulder. The rock breaks into three pieces. One piece is hurled due east; another piece is thrown in a due west direction. Which of the following situations is (are) impossible according to the law of conservation of momentum?
   A. The third piece has a velocity of zero.
   B. The third piece has a velocity due west.
   C. The third piece has a velocity due south.
10. What happens to the center of mass of the boulder in Exercise 9?
11. Two identical cars travel at the same speed. One is traveling east; the other, north. They collide at an intersection, and remain joined together after the collision. In what direction do they travel after the collision?

Section 11.3:

12. Why does the front end of a car tend to tip downward during an emergency braking situation?
13. Two flywheels have the same mass and radius. One has the shape of a uniform disk; the other has most of its mass on the rim. If they are rotating at the same angular speed, do they have the same angular momentum? Explain why or why not.
14. Forces of the same size act on identical disks for the same amount of time (same linear impulse). Are the angular impulses necessarily the same for the two disks? Explain why or why not.
15. Why are the brake shoes on disk brakes near the rim of the brake rather than near the center?
16. In what direction is the rotational axis of:
    - ***A.*** a car rounding an unbanked curve?
    - ***B.*** a car going over the top of a hill?

Section 11.4:

17. Why do helicopters have a small rotor on the tail?
18. What has conservation of angular momentum to do with seasonal changes here on earth?
19. Many machines convert linear momentum to angular momentum or vice versa. Figure 11–12 shows a flywheel and crankshaft connected to a piston by a piston rod. Are angular and linear momentum conserved in this system? If not, where are the external forces and torques?
20. Why does a diver gather his body up into a tight ball while somersaulting overboard?

Section 11.5:

21. Why must the mass of water entering a hydraulic ram be smaller than the water delivered by the ram?
22. Why can a hydraulic ram not be used in a well?
23. Neglecting other factors, what pitch will give the largest force for the wind turbine shown in Figure 11–18?
24. The sun exerts a tidal force on the earth that is about half as strong as the lunar tidal force. How might this affect the earth's orbit around the sun?

PROBLEMS

Section 11.1:

1. Figure 11–24 shows the impulse imparted to a $40\bar{0}$-g mass.
    - ***A.*** From the figure determine the magnitude of the impulse.

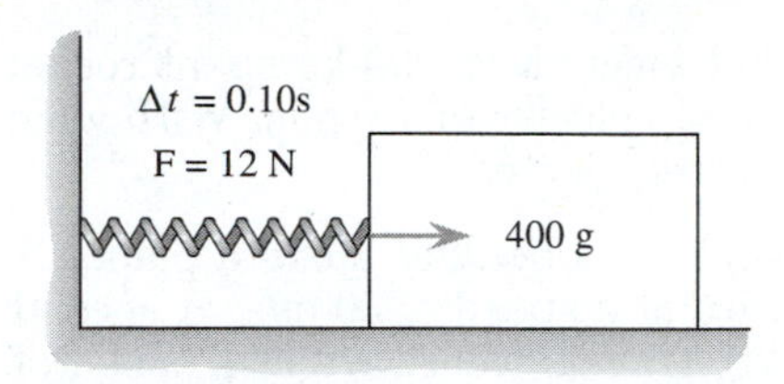

Figure 11–24: *A compressed spring exerts an average net force of 12 N for 0.10 s on a $4\bar{0}0$-g mass initially at rest.*

B. What is the change of momentum of the mass?
C. What is the change of velocity of the $4\bar{0}0$-g mass?

2. A fan in an air conditioning system takes in $2\bar{0}$ ft^3 of air per second (1.5 lb of air per second) and gives it a velocity of 30 ft/s. Assume the air is initially at rest.
   A. What is the change of momentum of the air per second?
   B. Neglecting friction, what impulse must the motor deliver to push 60 ft^3 of air into the conduits?

3. A $3\bar{0}$-N · s impulse will give a 250-g mass what velocity?

4. A $5\bar{0}$-g golf ball is driven off a tee. Its speed as it leaves the tee is 45 m/s.
   A. What is the magnitude of the change of momentum of the ball?
   B. If the period of impact is 5.0×10^{-4}s, what is the magnitude of the average force acting on the ball?

5. A $48\bar{0}0$-lb car traveling at 45 mph comes to a stop in 6.5 s.
   A. What is the change of momentum of the car?
   B. What is the average braking force?

6. A $22\bar{0}0$-lb cable elevator car initially at rest is counterweighted by $16\bar{0}0$ lb. What is the change of momentum of the counterweighted elevator when it is accelerated to a speed of 5.0 ft/s?

*7. Water travels through a pipe at a linear speed of 5.5 ft/s, delivering 2.5 lb of water per second at the end of the pipe. The water must flow through a right-angle elbow. What is the magnitude of the average force the water exerts on the elbow?

*8. A rifle fires a $4\bar{0}$-g bullet with a muzzle velocity of 1200 ft/s.
   A. What impulse acts on the bullet?
   B. If the barrel is 28 in long, what average force acts on the bullet?

Section 11.2:

9. A 9.0-lb rifle fires a 0.044-lb bullet with a muzzle velocity of $90\bar{0}$ ft/s. What is the recoil velocity of the rifle?

10. A $2\bar{0}$-ton hopper car is moving along a siding with a speed of 14 mph. Fifteen tons of crushed lime is dumped into the car from a hopper directly overhead. What is the final speed of the loaded hopper car?

11. A $12,\bar{0}00$-lb truck traveling at $5\bar{0}$ mph eastward has a head-on collision with a $40\bar{0}0$-lb car traveling westward at 65 mph. If the vehicles stick together after the collision, what are the direction and speed of the vehicles immediately after the collision?

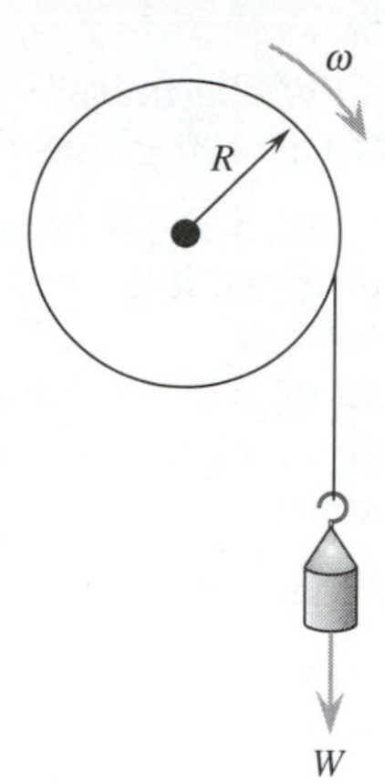

Figure 11–25: *A hung weight causes a pulley to rotate.*

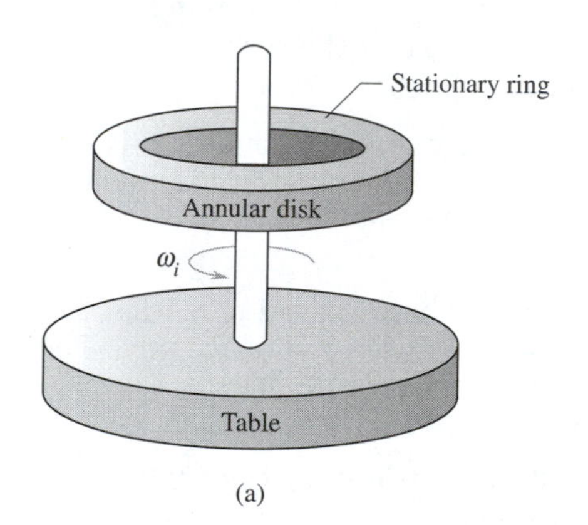

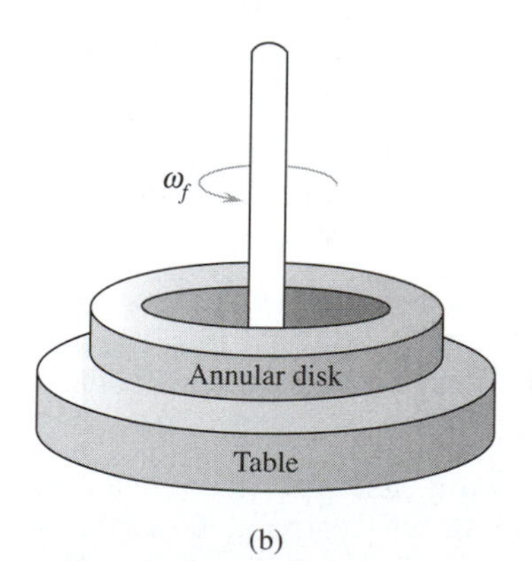

Figure 11–26: (a) *An annular ring is dropped onto a rotating table.* (b) *The ring and table rotate together.*

***12.** A 20,0$\bar{0}$0-kg final-stage rocket ejects a 10,0$\bar{0}$0-kg spent rocket booster, giving it an increase of velocity of 7.0 m/s. With what backward speed was the booster ejected?

***13.** An explosion sends three pieces of a boulder along a plane. A 12-kg piece is thrown eastward at a speed of 5$\bar{0}$ m/s. A second piece, having a mass of 18 kg, travels westward at a speed of 4$\bar{0}$ ft/s. What are the direction and speed of the third piece if it has a mass of 8.0 kg?

***14.** A billiard ball with an initial velocity of 4.0 m/s strikes a second billiard ball at rest. The first ball is deflected at an angle of 3$\bar{0}$° from its original direction. It has a speed of 2.0 m/s. What are the components of the velocity of the second ball?

Section 11.3:

15. A 4.0-kg flywheel has a radius of 25 cm and a disk shape. Find its moment of inertia.

16. An angular impulse of 32 N · m · s acts on a 3$\bar{0}$ cm diameter sphere. The sphere's mass is 2.5 kg. If the sphere was initially at rest, what is its final angular speed around its center?

17. Which will produce a larger angular impulse: a 1$\bar{0}$-lb tangential force acting at 7.0 in from a spin axis for 2.0 s, or a 5.0-lb tangential force acting at 10.0 in from a spin axis for 3.0 s?

18. Find the moment of inertia of the earth, assuming it is a uniform sphere. ($M_e = 5.98 \times 10^{24}$ kg; $R_e = 6.38 \times 10^6$ m.)

19. A string is wound around a pulley with a moment of inertia of 0.45 slug · ft^2 and a 6.0-in radius (see Figure 11–25). A 0.50-lb weight is hung from the free end of the string. The weight is released. At the end of 3.0 s what will be the angular speed of the pulley if it was initially at rest?

Section 11.4:

20. A skater goes into a spin. The initial moment of inertia is 2.2 slug · ft^2, and the initial angular speed is 6.0 rad/s. The skater lowers his arms to reduce the moment of inertia to 0.60 slug · ft^2. What is the new spin rate?

21. Two disks at the end of axles are initially spinning freely in opposite directions. They are brought in contact so that friction causes them to spin together. What is the final angular speed? ($I_1 = 0.80$ slug · ft^2; $\omega_1 = 35$ rad/s; $I_2 = 1.20$ slug · ft^2; $\omega_2 = -40$ rad/s.)

22. A stationary annular disk of mass 2.3 kg has an inside radius of 8.0 cm and an outside radius of 12.0 cm. It is dropped onto a rotating table. The center of the table and the center of the annular disk coincide (see Figure 11–26). If the table has an initial angular speed of 55 rad/s and a moment of inertia of 0.065 $kg \cdot m^2$, what is the final angular speed of the combination?

23. A frictionless turntable with a moment of inertia of 1.13×10^{-2} $kg \cdot m^2$ is spinning at 5.0 rad/s. A 400-g mass is dropped onto the turntable 8.0 cm from the center. What is the new angular speed of the turntable? (Hint: The 400-g mass acts as a point mass with an orbit of 8.0 cm.)

*24. When a star the size of the sun converts all of its hydrogen into helium it collapses into a white dwarf, a small, very dense sphere about the size of the earth. Assume a star with an effective radius of 1.00×10^9 m collapses into a white dwarf with a radius of only 1.00×10^6 m. A star with planets takes approximately one month to make one rotation, a spin rate of 2.4×10^{-6} rad/s. Use these data to estimate the spin rate of a white dwarf.

Chapter 12

WORK, POWER, AND ENERGY

Before you take your next roller coaster ride, read this chapter to find out how much potential energy your car needs to get through the loop! (Photograph courtesy of the Great Escape Fun Park, Lake George, NY.)

OBJECTIVES

In this chapter you will learn how to:

- calculate the work performed by a constant force that moves an object through a displacement
- graphically determine the work done by a varying force
- find the work done by torques
- determine power, the time rate of doing work
- find the kinetic energy of a moving object
- calculate potential energies for a variety of systems
- find the heat generated by work done against friction
- associate work with changes of energy
- relate work and energy to gas consumption of an automobile

Time Line for Chapter 12

1748 Mikhail Vasilievich Lomonosov, a Russian scientist, advances the ideas of conservation of energy and conservation of mass. Lomonosov's ideas are slow to reach Europe.

1807 Thomas Young independently introduces the concept of energy and uses the term *energy* for the first time.

1824 Nicholas Carnot shows that work can be done when heat moves from a higher temperature to a lower temperature.

1842 Julius Mayer is the first to state the principle of the conservation of energy and heat and recognizes heat is a form of energy.

1847 Hermann von Helmholtz develops ideas on the conservation of heat and energy.

1847 James Prescott Joule measures the transformation of mechanical energy into heat.

1853 Gustave-Gaspart Coriolis coins the term *kinetic energy* in a paper explaining mechanical action.

1853 William Rankine introduces the idea of energy of position, or potential energy.

To excavate for a building foundation, heavy equipment must exert a force to lift the dirt from the hole. A force is applied through a distance. In order to shape sheet steel into an automobile fender, a die exerts a force on the metal, pushing it into the correct shape. A balloon is expanded when air is blown into it. Increased pressure pushes the sides of the balloon outward. Again, force acts through a distance. When you make a telephone call, electrons move along a wire. Their motion is caused by an electrical force.

In order to communicate or to change the size or shape of something or to transport something from one place to another we must apply a force through some distance. The application of a force through a distance is called **work.** We must do work to influence our world in any way. In this chapter we will define work, and learn to determine how fast work can be done and how it can be stored.

12.1 WORK WITH A CONSTANT FORCE

Try pushing a chair across the floor. First push parallel to the floor. This is easy. Next push the chair the same distance using an angle close to 90° to the floor. This method requires more force to move the chair. Finally push straight down on the chair as hard as you

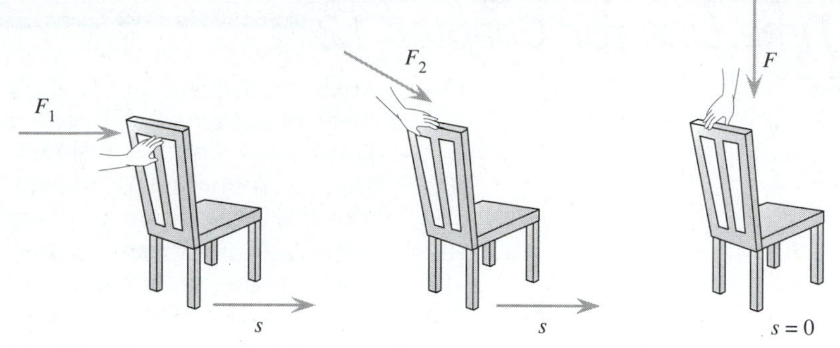

Figure 12–1: *More force is needed to push a chair as the angle between the force (**F**) and the distance moved (s) approaches 90°.*

can. No matter how hard you push, the chair will not move (see Figure 12–1).

In the first two cases you did some work. You applied a force through a distance. In the last case, you did no work. Even though your face may have gotten red from the exertion, the chair did not move. Two things have to happen to perform work: a force must be applied, and something has to move.

There is another factor involved with work. When you pushed down at an angle, the vertical part of the force tried to push the chair into the floor. The vertical component of force did not help move the chair. Only the horizontal component of the force pushed the chair along the floor. When we are calculating how much work is done, we should take into account only the part of the force that causes the change of position. In the case of the chair it is the horizontal component of the force you exerted on the chair. Let us put these ideas in the form of an equation.

work = *force parallel to the distance moved* × *the distance*

$$W = \mathbf{F}_{\text{para}}\, s$$

Work actually involves displacement rather than distance. The change of position is in a particular direction. If we know the angle (θ) between the applied force and the direction of the displacement, we can write the work equation this way.

$$W = |\mathbf{F}|\,|\mathbf{s}|\cos\theta$$

or

$$W = \mathbf{F} \cdot \mathbf{s} \qquad \textbf{(Eq. 12–1)}$$

Work is a scalar quantity even though it is the dot product of two vectors: force and displacement.

□ **EXAMPLE PROBLEM 12–1: THE DRILLING RIG**

The winch of a drilling rig exerts a force of $12\bar{0}0$ lb to lift the drill bit and piping $2\bar{0}$ ft. How much work is done?

■ ***SOLUTION***

The force and displacement are in the same direction, upward. The angle between the applied force and the displacement is zero. Cos 0.00 = 1.00.

$$W = \mathbf{F}\,\mathbf{s}\cos 0.00^0$$

$$W = 12\bar{0}0\text{ lb} \times 2\bar{0}\text{ ft} \times 1.00$$

$$W = 2.4 \times 10^4\text{ ft} \cdot \text{lb}$$

Units

In British units there is no special unit for work. It is expressed in foot · pounds (ft · lb). In SI units, there is a special unit for work. It is called a **joule** (pronounced jewel). One joule (J) of work is performed when 1 N of force moves through 1 m (1 J = 1 N · m).

□ **EXAMPLE PROBLEM 12–2: THE PLANER**

A carpenter pushes wooden stock through a planer with shaper blades to form molding (see Figure 12–2). A force of 50.0 N is exerted on the carpenter at an angle of 35.0° above the horizontal to push a 5.00-m length of stock through the planer. How much work is done by the carpenter?

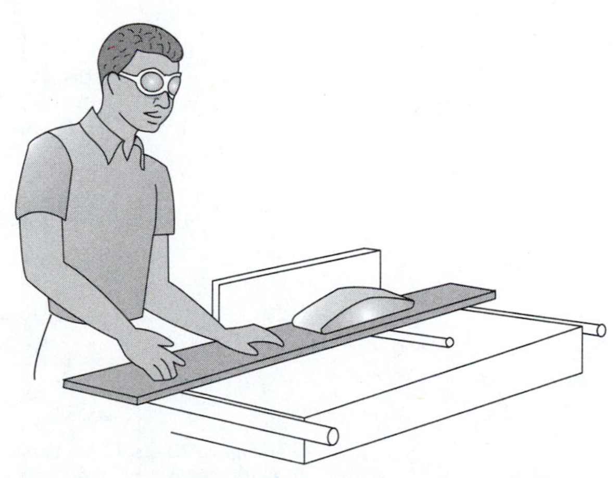

Figure 12–2: *A carpenter pushes a board through a planer.*

■ *SOLUTION*

The carpenter must push a distance of 5.00 m to get the stock through the planer.

$$W = \mathbf{F}\ \mathbf{s} \cos 35.0°$$

$$W = 50.0 \text{ N} \times 5.00 \text{ m} \times 0.819$$

$$W = 205 \text{ N} \cdot \text{m}$$

$$W = 205 \text{ J}$$

12.2 WORK WITH VARIABLE FORCES

In many situations the force doing work is not constant. When a spring is stretched, the force increases proportionally to the change of length of the spring ($\mathbf{F} = k\ \Delta\ x$) (see Figure 12–3). More and more work is required to increase the spring's length by each additional centimeter. The average force needed to stretch the spring 1.0 cm is 2.0 N. An average force of 6.0 N is needed to increase the length of the spring from 1.0 cm to 2.0 cm. To stretch the spring from 2.0 cm to 3.0 cm, an average force of 10.0 N is needed. We can find the total work done in stretching the spring from 0 to 3.0 cm by adding up the work needed for each centimeter, indicated by the shaded areas in Figure 12–3.

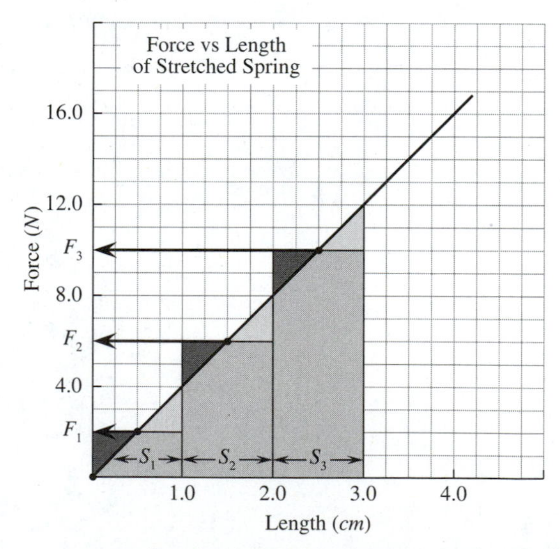

Figure 12–3: *Force versus length of a stretched spring. As the displacement increases, the average force to lengthen the spring by one additional centimeter increases.*

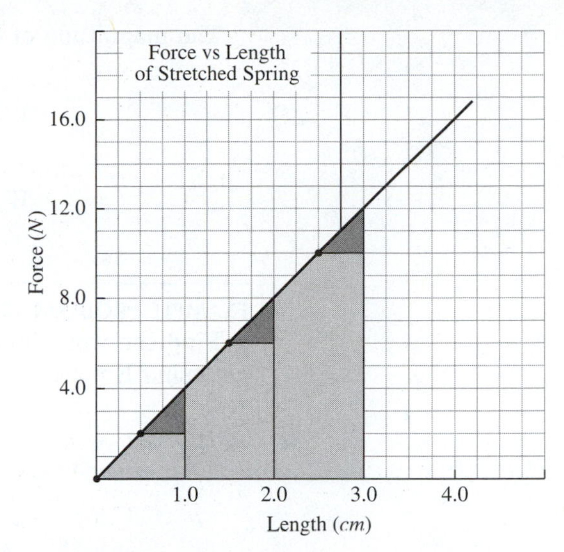

Figure 12–4: *The area under the plot of force versus length is work. The work is represented by the total shaded area: a triangle with a base of 3.0 cm and a height of 12.0 N.*

$$W(\text{total}) = W_1 + W_2 + W_3$$

$$W = (\mathbf{F}_1\ \mathbf{s}_1) + (\mathbf{F}_2 \cdot \mathbf{s}_2) + (\mathbf{F}_3 \cdot \mathbf{s}_3)$$

$$W = (2.0\ \text{N} \times 1.0\ \text{cm}) + (6.0\ \text{N} \times 1.0\ \text{cm}) + (10.0\ \text{N} \times 1.0\ \text{cm})$$

$$W = 18\ \text{N} \cdot \text{cm} = 0.18\ \text{N} \cdot \text{m}$$

$$W = 0.18\ \text{J}$$

There is an easier way to calculate the work. If you look carefully at the shaded areas in Figure 12–3, you will notice that the heavily shaded triangles above the force versus length plot are the same size and shape as the unshaded triangles below the plot. We can move the dark triangles to fill in the unshaded triangles to get Figure 12–4. The area under the plot of force versus length in Figure 12–4 represents the work done. The area is a triangle with a base of Δx and a height of f.

$$\textit{work done} = \textit{area under the}\ \text{f}\ \textit{vs.}\ \Delta\ \text{x}\ \textit{plot}$$

$$W = \frac{1}{2}(b\ h)$$

$$W = \frac{1}{2}(\Delta\ x\ \text{f})$$

The magnitude of the force to stretch a spring is $f = k \Delta x$.

$$W = \frac{1}{2}[\Delta x (k \Delta x)]$$

$$W = \frac{1}{2}(k \Delta x^2) \qquad \textbf{(Eq. 12–2)}$$

☐ **EXAMPLE PROBLEM 12–3: THE SPRING**

Find the work done to stretch the spring in Figure 12–4 by 3.0 cm using Equation 12–2.

■ ***SOLUTION***

First we need the spring constant k. This is the slope of the plot.

$$k = \frac{f_2 - f_1}{x_2 - x_1}$$

$$k = \frac{(12.0 - 0)\text{ N}}{(3.0 - 0)\text{ cm}}$$

$$k = 4.0\text{ N/cm}$$

We can now use Equation 12–2.

$$W = \frac{1}{2}(k \Delta x^2)$$

$$W = \frac{1}{2}[4.0\text{ N/cm }(3.0\text{ cm})^2]$$

$$W = 18\text{ N} \cdot \text{cm}$$

$$W = 0.18\text{ J}$$

The variable force of stretching a spring plotted out to be a triangle on a force versus distance plot. Some other forces may have a different shape when plotted against distance. The area under any force versus distance plot has units of work. We can find the work performed by finding this area.

☐ **EXAMPLE PROBLEM 12–4: THE CAR**

Figure 12–5 represents the frictional forces exerted on an automobile during a short trip. Estimate the total work done.

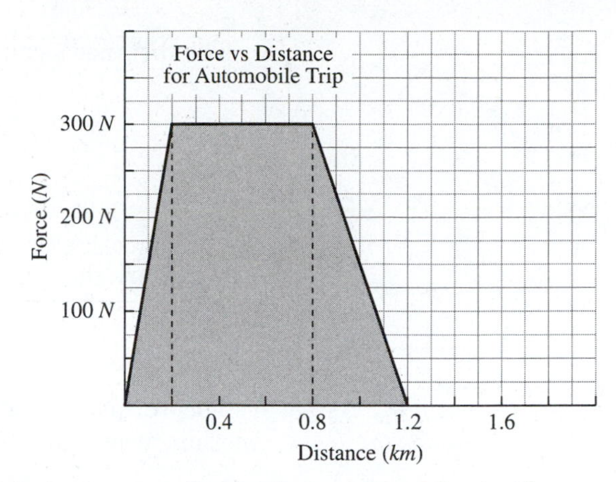

Figure 12–5: *Force versus distance for an automobile trip. The area under the curve is the work done* **(F · s).**

■ ***SOLUTION***

The area under the force versus distance curve can be broken into three parts. The area from 0.00 km to 0.20 is a triangle. The area from 0.20 to 0.80 km is a rectangle, and the area from 0.80 km to 1.2 km is another triangle. The total area will give us the work.

$$W = \frac{1}{2}[30\bar{0}\text{ N }(0.20 - 0)\text{ km}] + [30\bar{0}\text{ N }(0.80 - 0.20)\text{ km}]$$

$$+ \left[\frac{1}{2}(30\bar{0}\text{ N})(1.20 - 0.80)\text{ km}\right]$$

$$W = 270\text{ N}\cdot\text{km} = 270\text{ kJ}$$

$$W = 2.7 \times 10^5\text{ J}$$

12.3 WORK IN ROTATING SYSTEMS

Torques can perform work, too. Look at Figure 12–6. A force perpendicular to the door is used to turn the door through an angle θ. The force moves with the door. It travels a distance **s**. The work done is:

$$W = \mathbf{F} \cdot \mathbf{s}$$

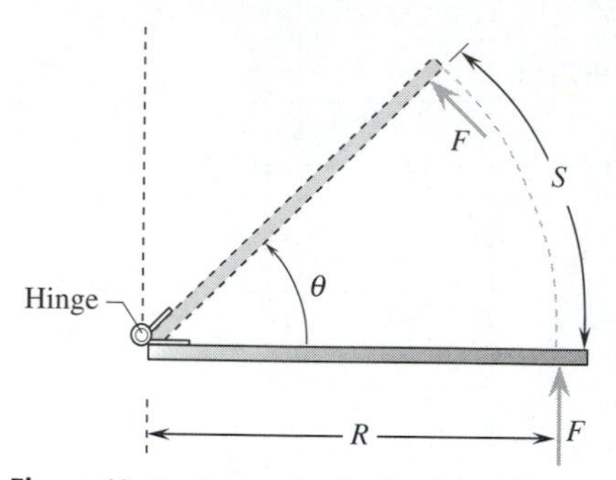

*Figure 12–6: A perpendicular force (**F**) acts through a distance (**s**). This is the same as the torque (**R × F**) moving the door through an angle of 90°.*

The distance s can be expressed in terms of the angle of rotation.

$$\theta = \frac{s}{R}$$

$$s = R\,\theta$$

The work done is:

$$W = \mathbf{F} \cdot R\,\theta$$

The torque acting on the door is $\mathbf{F} \times R$. In terms of angular motion, work can be written as:

$$W = \tau \cdot \theta \qquad \textbf{(Eq. 12–3)}$$

☐ **EXAMPLE PROBLEM 12–5: THE DOOR**

A force of 10.0 lb acting at an angle of $\bar{7}0°$ on a door 2.50 ft from the hinge swings the door through $\bar{9}0°$ (see Figure 12–7).

A. Calculate the work done using Equation 12–1.

B. Calculate the work done using Equation 12–3.

■ ***SOLUTION***

A. The component of the force moving parallel to the path **s** is:

$$F_p = \mathbf{F} \sin \bar{7}0°$$

$$F_p = 10.0 \text{ lb } (0.940)$$

$$F_p = 9.4 \text{ lb}$$

The path length is:

$$s = R\,\theta$$

$$s = 2.50 \text{ ft} \times \frac{\pi}{2}$$

$$s = 3.93 \text{ ft}$$

The work done is:

$$W = F_p\, s$$

$$W = 9.4 \text{ lb} \times 3.93 \text{ ft}$$

$$W = 37 \text{ ft} \cdot \text{lb}$$

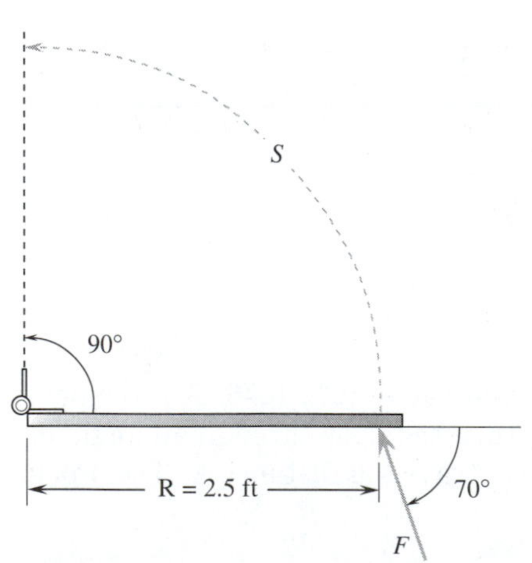

Figure 12–7: A force acting at an angle of 70° swings the door through an angle of 90°.

B. The magnitude of the torque is:

$$\tau = F\,R \sin 70^\circ$$
$$\tau = 10.0 \text{ lb} \times 2.50 \text{ ft} \times 0.940$$
$$\tau = 23.5 \text{ ft} \cdot \text{lb}$$

The door moves through an angle of $\pi/2$ radians. The work is then:

$$W = \tau \cdot \boldsymbol{\theta}$$
$$W = 23.5 \text{ lb} \cdot \text{ft} \times 1.57 \text{ rad}$$
$$W = 37 \text{ ft} \cdot \text{lb}$$

□ **EXAMPLE PROBLEM 12–6: THE WINCH**

A winch lifts a load of 322 kg. If the drum has a diameter of 80.0 cm, how much work must the motor do to rotate the drum through 12.0 revolutions? See Figure 12–8.

■ ***SOLUTION***

The magnitude of the torque the motor must exert is:

$$\tau = m\,g\,R \sin 90^\circ$$
$$\tau = 322 \text{ kg} \times 9.80 \text{ m/s}^2 \times 0.400 \text{ m} \times 1.00$$
$$\tau = 1.26 \times 10^3 \text{ N} \cdot \text{m}$$

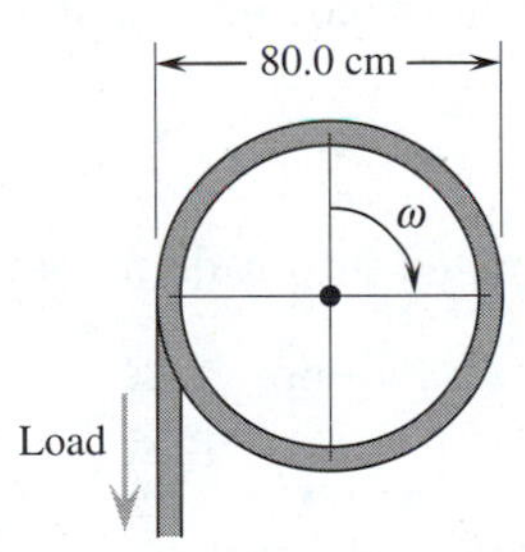

Figure 12–8: *An 80.0 cm diameter winch barrel rotates 12 revolutions in lifting a 322-kg load.*

Since there are 2π radians in one revolution, the total angle through which the torque must rotate is:

$$\theta = 2\,\pi\,\frac{\text{rad}}{\text{rev}} \times 12.0 \text{ rev}$$
$$\theta = 75.4 \text{ rad}$$

The work done by the motor is:

$$W = \tau\,\theta$$
$$W = 1.26 \times 10^3 \text{ N} \cdot \text{m} \times 75.4 \text{ rad}$$
$$W = 9.50 \times 10^4 \text{ J}$$

12.4 POWER

If properly geared, a winch powered by a lightweight motor can lift the same load as a heavy-duty system, but it will take longer to perform the work. It makes more sense to rate machines by how fast they can perform work rather than by how much work they do.

Power is the time rate of doing work.

$$P = \frac{W}{t} = \frac{\mathbf{F} \cdot \mathbf{s}}{t} = \frac{\boldsymbol{\tau} \cdot \boldsymbol{\theta}}{t} \qquad \textbf{(Eq. 12–4)}$$

We can express power in a slightly different way. Velocity is the time rate of change of distance ($\mathbf{v} = s/t$). Angular speed is the time rate of change of angular speed ($\omega = \theta/t$). We can rewrite power in terms of speed.

$$P = \mathbf{F} \cdot \mathbf{v}$$

or

$$P = \boldsymbol{\tau} \cdot \boldsymbol{\omega} \qquad \textbf{(Eq. 12–4A)}$$

Units

In British units power is often expressed in **horsepower** (hp). 1.00 hp = 550 ft · lb/s. The SI unit of power is the **watt** (W). 1.00 W = 1.00 J/s. Thus, 1 hp = 746 W.

☐ **EXAMPLE PROBLEM 12–7: THE AUTOMOBILE ENGINE**

An automobile engine develops a torque of 225 ft · lb when the engine is operating at 2500 rpm. What horsepower is delivered by the engine?

■ ***SOLUTION***

First we must convert the angular speed into radians per second.

$$2500 \frac{\text{rev}}{\text{min}} \times \frac{1.00 \text{ min}}{60.0 \text{ s}} \times 2\pi \frac{\text{rad}}{\text{rev}} = 262 \frac{\text{rad}}{\text{s}}$$

The power according to Equation 12–4A is:

$$P = \tau\,\omega$$

$$P = 225 \text{ ft} \cdot \text{lb} \times 262 \text{ rad/s}$$

$$P = (5.90 \times 10^4 \text{ ft} \cdot \text{lb/s}) \times \frac{1 \text{ hp}}{550 \text{ ft} \cdot \text{lb/s}}$$

$$P = 107 \text{ hp}$$

☐ **EXAMPLE PROBLEM 12–8: A CRANE**

A crane delivers 4.20 kW of power while lifting a 525-kg girder 40.0 m at a building site. How much time will it take the crane to lift the girder?

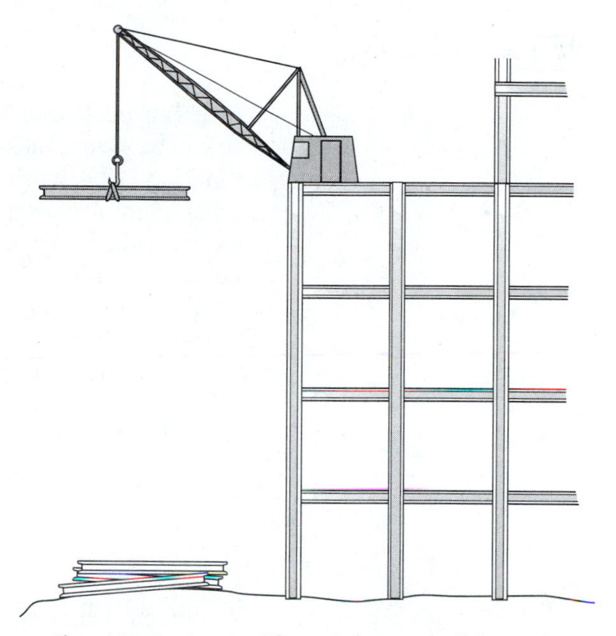

Figure 12–9: *A crane lifts a girder at a building site.*

■ ***SOLUTION***

The tension on the cable is equal to the weight of the girder.

$$\mathbf{F} = m\,\mathbf{g} = 525 \text{ kg} \times 9.80 \text{ m/s}^2$$
$$|\mathbf{F}| = 5.14 \times 10^3 \text{ N}$$

The work done is then:

$$W = \mathbf{F}\,\mathbf{s}$$
$$W = (5.14 \times 10^3 \text{ N}) \times 40.0 \text{ m}$$
$$W = 2.06 \times 10^5 \text{ J}$$

The crane can perform 4.20×10^3 J of work in 1 s. The required time is the total work done divided by the power.

$$t = \frac{W}{p}$$

$$t = \frac{2.06 \times 10^5\ \text{J}}{4.20 \times 10^3\ \text{J/s}}$$

$$t = 49.0\ \text{s}$$

12.5 KINETIC ENERGY

Work can be stored in something for later use. One way to store work is to give a mass motion. Work is done on a flywheel to set it in motion. The work stored in the wheel's rotary motion pushes valves open or closed and moves the pistons up and down between power strokes. Work is done on a bowling ball when it is thrown. The work stored in the ball's motion acts on the bowling pins at the end of the alley. Work stored in motion is called **kinetic energy.** For the moment we will look at linear kinetic energy. We will examine rotational kinetic energy in Chapter 13.

If a constant net force acts on a mass over a distance, the mass will be accelerated. Let us use Newton's second law for constant mass.

$$\mathbf{F}_{net} = m\,\mathbf{a}$$

$$W_{net} = \mathbf{F} \cdot \mathbf{s} = m\,\mathbf{a} \cdot \mathbf{s}$$

Recall from Chapter 4 for constant acceleration $2\,\mathbf{a} \cdot \mathbf{s} = \text{v}_f^2 - \text{v}_i^2$.

$$W_{net} = m\frac{(\text{v}_f^2 - \text{v}_f^2)}{2} \qquad \textbf{(Eq. 12–5)}$$

$$W_{net} = \left(\frac{\text{m v}_f^2}{2}\right) - \frac{m\,v_i^2}{2}$$

We will identify the terms $(m\text{v}^2/2)$ as kinetic energy. Equation 12–5 shows us that the work put into the motion of a system is equal to the difference between the initial and final kinetic energies. Work done on a system can change its kinetic energy.

☐ **EXAMPLE PROBLEM 12–9: THE SNOWMOBILE**

A $3\bar{2}0$-kg snowmobile is traveling at a speed of 8.0 m/s. An accelerating force of $1\bar{2}0$ N acts on the sled while it travels 100 m.

Figure 12–10: Work stored in the bowling ball's motion knocks down the pins.

A. How much work is done on the snowmobile?
B. What is the snowmobile's initial kinetic energy?
C. What is the snowmobile's final kinetic energy?
D. What is the snowmobile's final speed?

■ ***SOLUTION***

A. Work:

$$W = \mathbf{F} \cdot \mathbf{s}$$
$$W = 12\bar{0}\text{ N} \times 10\bar{0}\text{ m}$$
$$W = 1.20 \times 10^4\text{ J}$$

B. Initial kinetic energy (KE):

$$KE_1 = \frac{m\,v^2}{2}$$
$$KE_1 = \frac{32\bar{0}\text{ kg} \times (8.0\text{ m/s})^2}{2}$$
$$KE_1 = 1.0 \times 10^4\text{ J}$$

C. Final kinetic energy:
The final kinetic energy is easily found by adding the work done on the snowmobile to the initial kinetic energy.

$$KE_f = KE_1 + W$$
$$KE_f = (1.0 \times 10^4\text{ J}) + (1.20 \times 10^4\text{ J})$$
$$KE_f = 2.2 \times 10^4\text{ J}$$

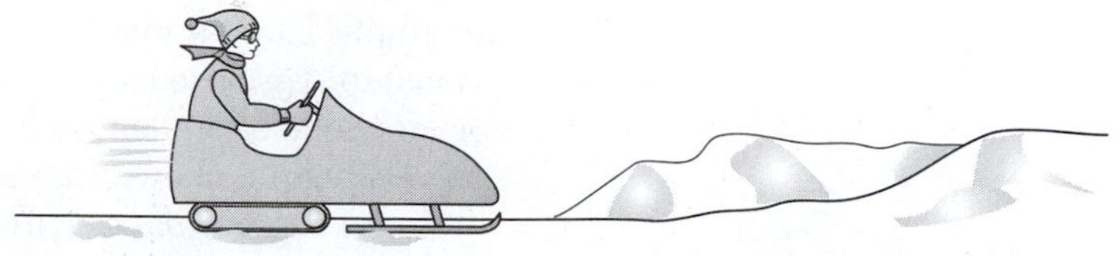

Figure 12–11: A snowmobile accelerates on level ground. Work is done to change its speed.

D. Final speed:

$$KE = \frac{m\,v_f^2}{2}$$

or

$$v_f = \left(\frac{2\,KE}{m}\right)^{1/2}$$

$$v_f = \left(\frac{2 \times 2.2 \times 10^4\ \text{J}}{320\ \text{kg}}\right)^{1/2}$$

$$v_f = 12\ \text{m/s}$$

12.6 POTENTIAL ENERGY

Another way to store work is to change the position of something. In Section 12.2 we saw that work is done when a spring is stretched or compressed from the equilibrium length. The position of one end of the spring changed relative to the other. The work stored in the spring may be used later on. For example, work stored in a screen-door hinge will close the door for you after you have opened it. The spring in a pressure-relief valve will close the valve when the pressure is reduced. Work that is stored in the position of something is called **potential energy.**

When work is done on an object to compress it or stretch it, **elastic potential energy** is stored in it. If the object obeys Hooke's law, as a spring does, the work stored in the object is given by equation 12–2.

$$PE = \frac{1}{2}(k\,\Delta\,x^2) \qquad \textbf{(Eq. 12–6)}$$

Many things that can be stretched elastically do not obey Hooke's law. A hunting bow, a rubber band, and leaf springs on a truck, for example, have a more complicated force versus displacement relationship. In these cases the work stored in the objects can be found from the area under their force versus displacement curves using the technique shown in Section 12.2.

Some objects have **gravitational potential energy.** The weight of something near the surface of a planet is fairly constant. To lift

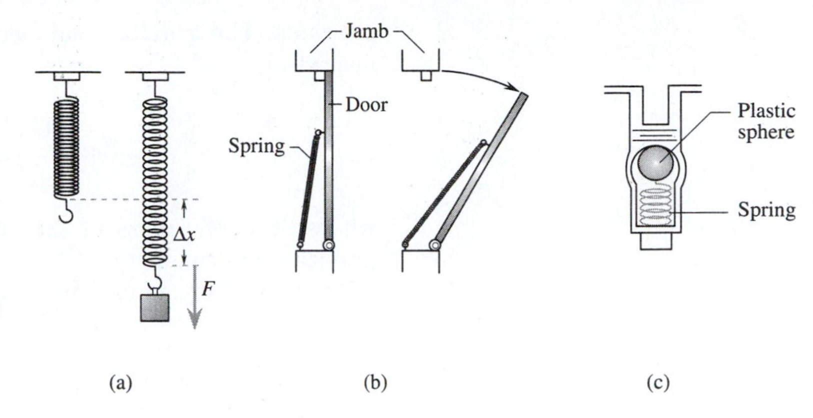

Figure 12–12: Springs store work in different situations. (a) Work is done to stretch a spring. (b) Work stored in a stretched screen-door spring closes the door after it has been opened. (c) In a pressure-relief valve, pressure compresses a spring. The work stored in the spring closes the valve when the pressure decreases.

a book, a force must be exerted to balance out its weight. The work done in lifting the book through a displacement (**s**) is:

$$W = \mathbf{F} \cdot \mathbf{s} = m\,g\,h$$

The gravitational potential energy (stored work) of an object lifted a short distance is:

$$PE = m\,g\,h \qquad \textbf{(Eq. 12–7)}$$

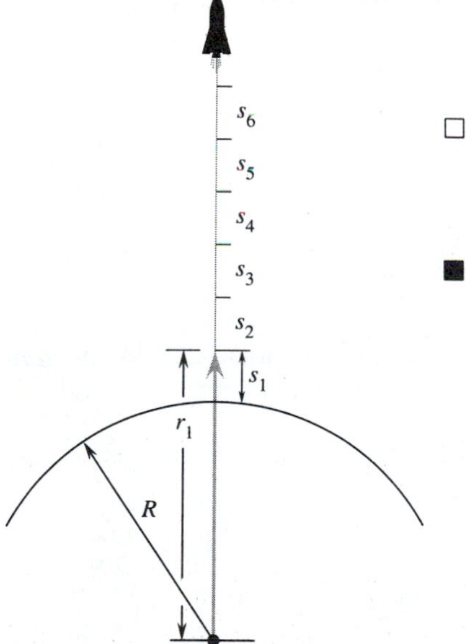

Figure 12–13: The gravitational force acting on an ascending rocket decreases with the increasing distance from the center of the earth. The intervals traveled $(s_i) = r_i - r_{i-1}$.

□ **EXAMPLE PROBLEM 12–10: THE TRUCK**

An 8340-lb truck climbs a $55\bar{0}$ ft high hill. How much gravitational potential energy is stored in the truck in climbing the hill?

■ ***SOLUTION***

$$PE = m\,g\,h$$
$$PE = 8340 \text{ lb} \times 55\bar{0} \text{ ft}$$
$$PE = 4.59 \times 10^6 \text{ ft} \cdot \text{lb}$$

When a shuttle is launched or when a satellite is lifted from earth into orbit by a rocket, the vertical distance traveled is no longer trivial compared with the radius of the earth. We have to take into consideration the inverse square law of universal gravitational forces. As the rocket rises away from the earth, the gravitational force decreases. Less and less work is required to raise the rocket a specified distance (see Figure 12–13).

In order to calculate the work done against gravity in lifting the rocket of mass m into orbit, break the distance up into smaller intervals. The gravitational force on the surface of the planet has a magnitude:

$$\mathbf{F}_0 = \frac{G\,M\,m}{R^2}$$

where M is the mass of the earth, R is its radius, and G is the gravitational constant.

At the end of the first interval the magnitude of the force is

$$\mathbf{F}_1 = \frac{G\,M\,m}{r^2}$$

We will use an average square distance for the interval of $R - r_1$ to estimate the average force for the interval. This approximation is good as long as the interval $(r_1 - R)$ is very small compared with the distance R or r_1. $r^2 = Rr_1$

$$\mathrm{F} = \frac{G\,M\,m}{R\,r_1}$$

The work done against gravity for this interval is:

$$W_1 = \mathbf{F} \cdot \mathbf{s}$$

$$W_1 = \frac{G\,M\,m}{R\,r_1}(r_1 - R)$$

$$W_1 = G\,M\,m\left(\frac{1}{R} - \frac{1}{r_1}\right)$$

The work for the remaining intervals can be calculated in the same way.

$$W_2 = G\,M\,m\left(\frac{1}{r_2} - \frac{1}{r_1}\right)$$

$$W_3 = G\,M\,m\left(\frac{1}{r_3} - \frac{1}{r_2}\right)$$

$$\vdots$$

$$W_n = G\,M\,m\left(\frac{1}{r_{n-1}} - \frac{1}{r}\right)$$

We can find the total work by adding up the work for the smaller intervals.

$$W = W_1 + W_2 + W_3 \ldots W_n$$

$$W = G\,M\,m\left(\frac{1}{R} - \frac{1}{r_1} + \frac{1}{r_1} - \frac{1}{r_2} + \frac{1}{r_2} - \frac{1}{r_3} \cdots \frac{1}{r_{n-1}} - \frac{1}{r}\right)$$

All the terms in the parentheses cancel out except for the first and the last one. This leaves us with:

$$W = G\,M\,m\left(\frac{1}{R} - \frac{1}{r}\right)$$

This gives us the gravitational potential energy stored in the mass m when it is moved a distance r away from the planet's surface.

$$PE = G\,M\,m\left(\frac{1}{R} - \frac{1}{r}\right) \qquad \textbf{(Eq. 12–8)}$$

□ **EXAMPLE PROBLEM 12–11: THE SATELLITE**

A 455-kg satellite is placed in orbit $82\bar{0}$ km above the earth's surface. Calculate its gravitational potential energy relative to the earth's surface.

■ ***SOLUTION***

The earth's radius (R) is 6380 km. Its mass is 5.98×10^{24} kg. $G = 6.67 \times 10^{-11}\ \text{N} \cdot \text{m}^2/\text{kg}$. The distance between the center of the earth and the satellite is:

$$r = R + \mathbf{s}$$

$$r = (6.38 \times 10^6\ \text{m}) + (8.20 \times 10^5\ \text{m})$$

$$r = 7.20 \times 10^6\ \text{m}$$

We can use Equation 12–8.

$$PE = G\,M\,m\left(\frac{1}{R} - \frac{1}{r}\right)$$

$$PE = (6.67 \times 10^{-11}\ \text{N} \cdot \text{m/kg}^2)\,(5.98 \times 10^{24}\ \text{kg})\,(455\ \text{kg})$$

$$\left(\frac{1}{6.38 \times 10^6\ \text{m}} - \frac{1}{7.20 \times 10^6\ \text{m}}\right)$$

$$PE = 3.24 \times 10^9\ \text{J}$$

Let us check this answer with the one we would have gotten from Equation 12–7. In that equation we assume the vertical distance is small compared with the earth's radius, so that the force is constant.

$$PE = m\,\mathbf{g}\,h$$

$$PE = 455\ \text{kg} \times 9.8\ \text{m/s}^2 \times (8.20 \times 10^5\ \text{m})$$

$$PE = 3.66 \times 10^9\ \text{J}$$

Equation 12–7 overestimates the gravitational potential energy by 13%.

12.7 WORK AND HEAT

When a drill bores a hole in steel, much of the work done by the drill is against friction. This work against friction does not change the kinetic or potential energy of the piece of steel. Instead, the work goes into increasing the random vibrational motion of individual atoms and molecules. This is kinetic energy on the atomic, or *microscopic*, level. We cannot see the motion of the individual atoms the way we can see the chunk of steel move when it has kinetic energy. What we do notice is an increase of temperature of the drill bit and metal around the hole.

Work done against friction produces heat. **Heat** is kinetic and potential energy on a microscopic scale. In other words, heat is another form of energy. It is a bit different from the potential and kinetic energies discussed so far. When energy is given to something on a large, or *macroscopic*, level, we can force that object to give up all the potential and kinetic energy we stored in it. We call this **mechanical energy** to distinguish it from heat. When work is stored in heat, we can never get all of the energy back. In some cases, heat engines can be used to regain some, but not all, of the work. Some of the energy dissipates in the random motion of atoms.

If we know the nature of the frictional forces, we can calculate how much heat (Q) will be produced by determining the amount of work done against friction.

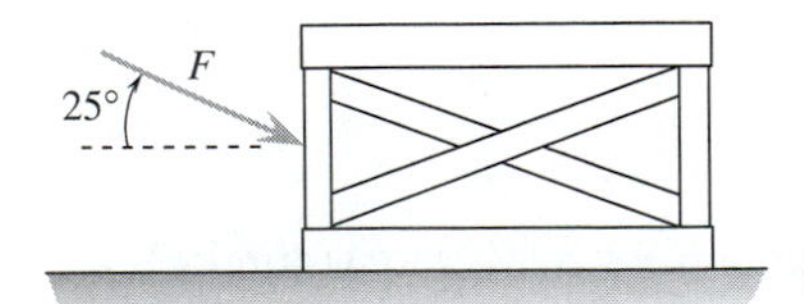

Figure 12–14: A crate is pushed across the floor with a force acting 25° above the horizontal.

EXAMPLE PROBLEM 12–12: THE CRATE

An 85-kg crate is pushed 9.50 m across a floor by a 290-N force acting 25° above the horizontal (see Figure 12–14). Find the heat produced.

■ *SOLUTION*

The horizontal component of the applied force acts against friction. All the work is done against friction. We will use the symbol Q to represent heat.

$$Q = W = \mathbf{F}\, s \cos 25°$$
$$Q = 290 \text{ N} \times 9.50 \text{ m} \times 0.9063$$
$$Q = 2.5 \times 10^3 \text{ J}$$

□ **EXAMPLE PROBLEM 12–13: THE DRILL**

A drill operating at 0.25 hp takes 23 s to bore a hole in a piece of thick steel plate. If 95% of the work goes into overcoming friction, how much heat is generated?

■ *SOLUTION*

First, find the work done.

$$W = P\,t$$
$$W = \frac{1}{4} \text{ hp} \times \frac{550 \text{ ft} \cdot \text{lb/s}}{\text{hp}} \times 23 \text{ s}$$
$$W = 3.16 \times 10^3 \text{ ft} \cdot \text{lb}$$

The heat generated is 95% of this quantity.

$$Q = 0.95\, W$$
$$Q = 0.95 \times 3.16 \times 10^3 \text{ ft} \cdot \text{lb}$$
$$Q = 3.0 \times 10^3 \text{ ft} \cdot \text{lb}$$

12.8 WORK AND ENERGY

Much of what we have learned in this chapter can be summarized in one simple idea. When work is done on a system, the kinetic energy of the system can be changed; the potential energy of the system can be changed; or heat can be generated. One or any combination of these things can happen.

$$W = \Delta KE + \Delta PE + Q \qquad \textbf{(Eq. 12–9)}$$

Notice that the work done depends on the *change* of kinetic energy. This change can be either positive (an increase) or negative (a de-

How Roller Coasters Loop the Loop

A roller coaster is a marvelous example of conservation of energy. A cable pulls the coaster car to the highest point in the track. The work done by the cable gives gravitational potential energy to the car and its occupants. The car is released and rolls down and up a track and down again, exchanging potential energy for kinetic energy and back again.

Newer roller coasters have loops built into the track. How high above the top of the loop must the car start to make it through the loop without falling off the track?

In Chapter 10 we looked at vertical loops. At the top of a vertical loop the minimum speed is related to the radius (R) of the loop by $\mathbf{v}^2 = R\ \mathbf{g}$. We can substitute $R\ \mathbf{g}$ for $\mathbf{v}^2$ in the expression for kinetic energy in the conservation of energy equation. This tells us that if the car starts just a half radius above the top of the loop, it will make it through the loop. However, if a design engineer used only this factor, a major disaster would occur. We need to consider some other factors.

First, most loops are really a spiral, or helix. The component of the car's velocity that moves it along the axis of the spiral does not participate in the circular motion. It is not part of the centripetal acceleration. The car will have to be started farther up on the track to make up this difference.

Another factor is the rotation of the wheels. Part of the kinetic energy of the wheels goes into spinning the wheels, not moving the car along the track. This added energy must be considered in the total energy of the coaster car.

(Photograph courtesy of the Great Escape Fun Park, Lake George, NY.)

Finally, some of the initial work put into the system is lost to rolling friction and air friction. Mechanical energy is converted into heat. The rolling friction of the bearings will slow the car down. Air friction is greatest when the car is traveling the fastest. Roller coaster aficionados who like to hold their hands in the air as if to say, "Look Ma, no hands," increase the drag coefficient of the car.

When we consider all these factors we see that the practical starting height of a coaster car is quite a bit higher than what we might predict using the conservation of energy rule for a simple vertical loop.

---

crease). During the power stroke of a one-cylinder engine, work is done on the flywheel, increasing its kinetic energy. No energy is stored as potential energy in this case. A small amount of work is lost to friction.

$$W_{\text{in}} = +\Delta KE + Q_1$$

During the rest of the cycle, work stored in the rotating flywheel is used to move the pistons and valves. The flywheel slows down as kinetic energy is drawn from it. Again there are some losses to friction. The flywheel must supply enough energy to overcome friction as well as perform useful mechanical work.

$$W_{\text{out}} + Q_2 = -\Delta KE$$

When energy is added to a system, positive (+) work is done. When energy is removed from a system negative (−) work is done.

Keeping track of the work and energy of a system is a lot like banking. The work is the deposits and withdrawals. The potential and kinetic energies are the different accounts that can be used, and heat is a surcharge on your banking activities.

☐ **EXAMPLE PROBLEM 12–14: THE MOTOR SCOOTER**

A $30\bar{0}$-kg motor scooter is initially at rest halfway up a hill 60.0 m high. The scooter climbs the hill, reaching a speed of 50.0 km/h at the crest. If the scooter did $20\bar{0}$ kJ of work, how much work was done against friction?

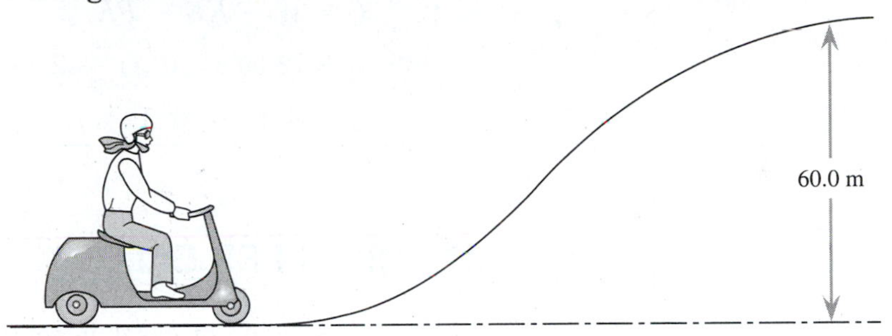

Figure 12–15: *A motor scooter accelerates up a 60.0-m hill.*

■ ***SOLUTION***

We can use Equation 12–9, but first we must find the changes of kinetic energy and potential energy.

Kinetic energy:

The initial kinetic energy is zero.

$$KE_f = \frac{1}{2}(m\,\mathbf{v}^2)$$

$$KE_f = \frac{1}{2}\left[30\bar{0}\text{ kg}\left(50.0\,\frac{\text{km}}{\text{h}} \times 1000\,\frac{\text{m}}{\text{km}} \times \frac{1\text{ h}}{3600\text{ s}}\right)^2\right]$$

$$KE_f = 2.89 \times 10^4\text{ J}$$

The change of kinetic energy is:

$$KE = KE_f - KE_i$$

$$KE = (2.89 \times 10^4) - 0\text{ J}$$

$$KE = 2.89 \times 10^4\text{ J}$$

At normal driving speeds, a radial tire will improve gas mileage. A lighter car reduces the normal force. Frictional losses are further reduced.

Potential energy:

$$PE = PE_f - PE_i$$

$$PE = (m\,g\,h_f) - (m\,g\,h_i) = m\,g\,(h_f - h_i)$$

$$PE = 30\bar{0}\text{ kg} \times 9.80\text{ m/s}^2\,(60.0 - 30.0)\text{ m}$$

$$PE = 8.82 \times 10^4\text{ J}$$

Solve Equation 12–9 for heat.

$$Q = W - KE - PE$$

$$Q = (2.00 \times 10^5\text{ J}) - (2.89 \times 10^4\text{ J}) - (8.82 \times 10^4\text{ J})$$

$$Q = 8.29 \times 10^4\text{ J}$$

12.9 APPLICATION: MORE MILES PER GALLON

A US gallon of gasoline contains 1.3×10^8 J (9.5×10^7 ft · lb) of chemical potential energy. Exhaust heat, cooling-system heat losses, and internal friction of the engine dissipate about 70% of the available energy. Certain accessories are necessary to maintain the engine. These include a cooling fan, hydraulic and water pumps, and an alternator. These accessories use up another 5% of the available energy. (If the car has an air conditioner, 10%.) The drive train, made up of the clutch transmission, drive shaft, and differential, loses another 10%. Roughly 15% of the available potential energy of the fuel reaches the drive wheels of the car to propel it forward. This is 2.0×10^7 J or 1.4×10^7 ft · lb.

The work done by the drive wheels falls into four categories:

1. acceleration of the car,
2. change of potential energy,
3. work done against rolling friction, and
4. work done against aerodynamic drag.

Let us look at these four factors and the way they affect gas consumption.

The first two factors involve the kinetic and potential energy of the car as it accelerates and climbs hills. The combined change of mechanical energy is:

$$W = \frac{1}{2}\{[m(v_f^2 - v_i^2)] + m\,g\,(h_f - h_i)\}$$

or

$$\frac{W}{m} = \frac{1}{2}\{[(v_f^2 - v_i^2) + g\,(h_f - h_i)]\}$$

Work done to change the mechanical energy of the car is directly proportional to the mass of the car. A lighter car will use less energy to accelerate and climb hills. Gas mileage is improved by making a car smaller and replacing steel parts by aluminum.

The third factor involves the flex of the tires. At normal driving speeds the rolling friction can be treated in the same way as sliding friction (see Table 12–1). The frictional force is directly proportional to the normal force. At high speeds, this is no longer true. The frictional losses become velocity-dependent. For moderate speeds the rolling friction is:

$$f = \mu m\,g$$

The last factor, aerodynamic drag, does not depend on mass. The shape and size of the car are important. Table 12–2 shows some drag coefficients (C) for different shapes. An egg shape has the lowest drag coefficient. A convertible with the top down has the highest. The aerodynamic drag is:

$$f = \frac{1}{2}(C\,A\,\rho\,v^2)$$

The drag coefficient can be reduced by designing the car with simple, rounded surfaces. From Table 12–2 you can see why the hatchback has become a more popular design than the blocky shape with a long trunk. Again, smaller cars win out in gas economy. Although weight is not a factor here, the cross-sectional area is. A third factor in reducing air drag depends on the driver. Faster speeds increase the

Table 12–1: Rolling resistance (lb/1000 lb load) of tires.

SPEED (mph)	RADIAL TIRE	BIAS-BELTED TIRE	BIAS TIRE
20	13	16	18
40	13	16	18
60	14	17	18
80	17	18	22
90	23	23	26

Table 12–1: Rolling resistance of tires. Resistance is listed per (lb/1000 lb load). Example: At 20 mph a 2000-lb car will have 2 × 16 lb or 32 lb of rolling resistance for bias-belted tires.

Table 12–2: Drag coefficients for different shapes of passenger cars. Egg shapes have the lowest drag; convertibles the highest.

Table 12–2. Drag coefficients for different shapes of passenger cars.

DRAG COEFFICIENT	SHAPE
0.1	
0.2	
0.3	
0.4	
0.6	
0.9	

air friction. Table 12–3 shows typical percentages of energy use by a medium-weight car traveling on the level at a constant speed for three speeds. When a car travels a distance (s) on a level road, the total work done by the drive wheels against friction is:

$$W = (\mu\, m\, g\, s) + \frac{1}{2}(C\, A\, \rho\, v^2\, s) \qquad \textbf{(Eq. 12–10)}$$

Table 12–3: Percents of energy used as a function of speed by a medium-weight car. The percentages refer to energy or power delivered by the engine. The efficiency of the motor is not considered.

Table 12–3: Percents of energy (or power) used as a function of speed by a medium-weight car.

	PERCENT AVAILABLE ENERGY USED BY:			
SPEED (mph)	ACCESSORIES	TRANSMISSION	ROLLING FRICTION	AERODYNAMIC DRAG
20	21	14	55	10
40	12	12	44	32
60	9	11	30	50

NOTE: Values indicate the percent usage of power delivered by the engine. They do not include the engine efficiency.

□ EXAMPLE PROBLEM 12–15: RADIAL TIRES

A $30\bar{0}0$-lb car with bias tires and a $20\bar{0}0$-lb compact with radial tires both travel $40{,}0\bar{0}0$ mi at an average speed of $4\bar{0}$ mph. What is the difference in gasoline consumption attributable to rolling friction for the two cars? (1 US gallon of gasoline contains 9.5×10^7 ft · lb of chemical energy.)

■ *SOLUTION*

The frictional forces can be obtained from Table 12–1. The rolling friction of the $30\bar{0}0$-lb car is:

$$f_a = \frac{18 \text{ lb}}{1000 \text{ lb}} \times 30\bar{0}0 \text{ lb load} = 54 \text{ lb}$$

The rolling friction of the $20\bar{0}0$-lb car is:

$$f_b \frac{13 \text{ lb}}{1000 \text{ lb}} \times 20\bar{0}0 \text{ lb load} = 26 \text{ lb}$$

The difference in the work against rolling friction done by the two cars is:

$$W = W_a - W_b = (f_a - f_b)\,\mathbf{s}$$

$$W = [(54 \text{ lb} - 26 \text{ lb})\, 40{,}0\bar{0}0 \text{ mi}] \times 5280 \frac{\text{ft}}{\text{mi}}$$

$$W = \frac{5.9 \times 10^9 \text{ ft} \cdot \text{lb}}{9.5 \times 10^7 \text{ ft} \cdot \text{lb/gal}} = 62 \text{ gal}$$

The difference in gas consumption is 62 gallons.

SUMMARY

Work must be performed to alter a system. Work is the product of force times the distance through which the force acts. In the British system work is measured in foot · pounds. In SI, it is measured in joules.

Power is the time rate of doing work. It is often expressed in horsepower in the British system. The SI unit of power is the watt.

$$P = W/t = \frac{Fs}{t} = Fv = \tau\omega \qquad \textbf{Power}$$

$$1.00 \text{ watt} = 1.00 \text{ joule/s} = 0.738 \text{ ft} \cdot \text{lb/s}$$

$$1 \text{ hp} = 550 \text{ ft} \cdot \text{lb/s}$$

Energy is stored work. There are three forms of energy. Kinetic energy is work stored in motion. Potential energy is work stored in position. Heat is work stored in random atomic motion and position. When work is done on a system, the kinetic energy of the system may change, the potential energy may change, or heat can be produced.

$$W = \Delta KE + \Delta PE + Q$$ **Work and Energy**

The linear kinetic energy of a system is $\frac{1}{2}(m\,v^2)$. Potential energy may take several forms depending on the nature of the force interacting with the system. It can take the form of elastic potential energy when work is done to stretch or compress an object. In other situations it can take the form of gravitational potential energy, involving short distances or interplanetary distances.

$$PE = G\,mM\left(\frac{1}{R} - \frac{1}{r}\right)$$ **Gravitational Potential Energy (Planetary)**

$$PE = \frac{1}{2}k\,\Delta x^2$$ **Elastic Potential Energy**

The work done against sliding or rolling friction is:

$$W = \mu N \cdot s$$

where N is the normal force.

The work done against aerodynamic drag is:

$$W = \frac{1}{2}C\,A\,\rho\,v^2\,s$$

KEY TERMS

If you can explain the following terms to a friend or classmate, you understand their meaning. If you cannot explain the terms, you should reread the sections in which they are discussed.

elastic potential energy
gravitational potential energy
heat
horsepower
joule
kinetic energy
mechanical energy
potential energy
power
watt
work

EXERCISES

Section 12.1:

1. Is it possible to exert a force on something without doing work? Explain why or why not.
2. As a truck rounds a curve at a constant speed, a centripetal force acts on it. Does the centripetal force do work on the vehicle? Explain.

3. A laboratory assistant moves a cart that is initially at rest. If the maximum static frictional force is 16 lb, how much work does the assistant do against static friction?

Section 12.2:

4. If $1\bar{0}$ J. of work are needed to stretch a spring $1\bar{0}$ cm from the equilibrium length, how much work is needed to stretch the same spring $2\bar{0}$ cm from its equilibrium length?

5. If a *constant* force moves something, the work done is directly proportional to the distance moved. If the force is *not constant*, as in the case of a spring, is the work necessarily directly proportional to the distance moved? Explain.

Section 12.3:

6. Twice as much force must be exerted to open a door from the middle as that needed to open it from the handle side. Is twice as much work done?

7. Is it possible to exert a torque on a system without doing work?

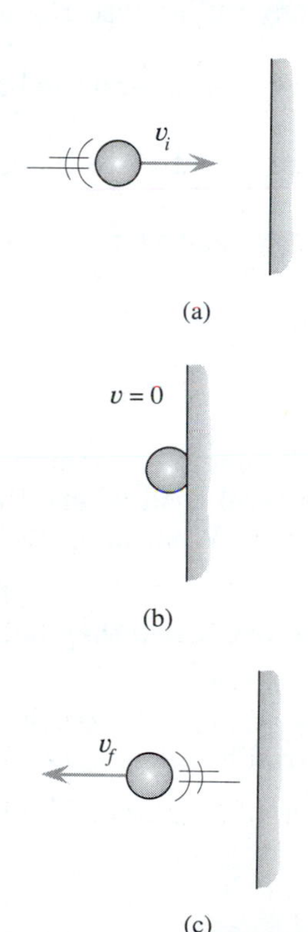

Figure 12–16: A ball collides against a wall. (a) The ball has a velocity ($+\mathbf{v}$) to the right. (b) The ball comes to rest at the wall. (c) The ball rebounds with a velocity ($-\mathbf{v}$) to the left. The total change of momentum is $\Delta \mathbf{p} = [m(-\mathbf{v})] - [m(+\mathbf{v})] = -2\,(m\,\mathbf{v})$.

Section 12.4:

8. It takes only $3\bar{0}$ hp to maintain a medium-sized automobile at 60 mph on a level road. This is the power to overcome rolling friction, air drag, and internal friction of the power train and to operate accessories. Why are engines used that can deliver over 125 hp?

9. The unit used for electric power is usually watts or kilowatts. If a light bulb is rated at $1\bar{0}0$ W, how is the power used?

10. An electric motor is rated at 0.25 hp. Express this in kilowatts.

11. Electricity is bought in kilowatt hours. Is this a unit of power or of energy?

Section 12.5:

12. Does a baseball pitcher expend twice as much energy to deliver a baseball at $9\bar{0}$ mph as at 45 mph? Explain your answer.

13. Can kinetic energy be negative? Explain.

14. An $8\bar{0}$-g ball with an initial horizontal velocity of $+18.0$ m/s strikes a wall and rebounds with a final velocity of -18.0 m/s (see Figure 12–16). The change of linear momentum is $-2\,(mv)$ or 2.88 kg · m/s. What is the change of kinetic energy?

15. Car A has a mass of $15\bar{0}0$ kg and a speed of $3\bar{0}$ m/s. Car B has a speed of only 28 m/s.

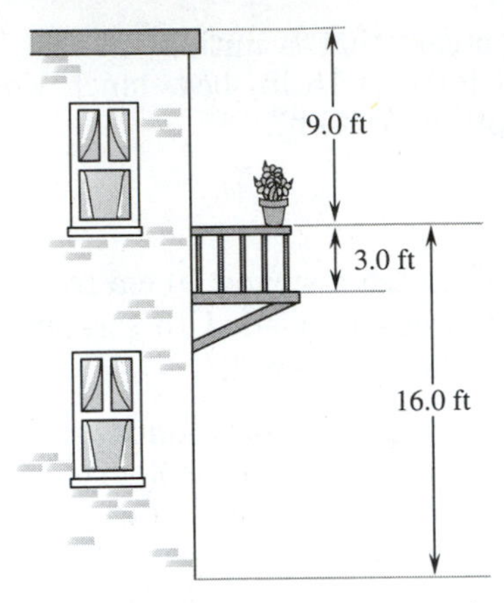

Figure 12–17: The potential energy of a potted plant depends on the frame of reference. The potential energy is different relative to the ground, to the roof, and to the floor of the balcony.

A. What is the kinetic energy of car A relative to the road?
B. What is the kinetic energy of car A relative to car B as they approach each other head on?
C. What is the kinetic energy of car A relative to car B as it approaches car B from behind?
D. Why would there be less damage done in a rear-end collision between cars A and B than in a head-on collision?

Section 12.6:

16. A spring stores 3.0 J of potential energy when it is displaced 4.0 cm from its equilibrium length. How much energy is stored when it is displaced by 8.0 cm? By 12.0 cm?

17. How much potential energy relative to the earth would a 1.0-kg mass have if it were very far away from the planet? ($r = \infty$; $M = 5.98 \times 10^{24}$ kg; $G = 6.67 \times 10^{-11}$ N $\cdot$ m^2/kg^2; $R = 6.38 \times 10^6$ m.)

18. A 3.7-lb potted plant rests on a balcony railing (see Figure 12–17).
    A. What is the plant's potential energy relative to the balcony floor 3.0 ft below the railing?
    B. What is the plant's potential energy relative to the ground 16.0 ft below the railing?
    C. What is the plant's potential energy relative to the roof 9.0 ft *above* the plant?

19. Can potential energy ever be negative?

Section 12.7:

20. Work must be done by an automobile engine to maintain the automobile at a constant speed on a level road where there is no change of kinetic or potential energy. What happens to the work done by the engine?

21. Why do basketball players get floor burns when they fall while running?

22. Less power is delivered to the drive wheels of a truck at the end of the power train than is delivered by the engine. Why?

23. Why is it a bad idea to ride the clutch of an automobile with a standard transmission?

Section 12.8:

24. The efficiency of a machine is the ratio of the mechanical work done by the machine divided by the work put into the machine. Use Equation 12–9 to explain why a 100% efficient machine is not possible.

25. A rock is hurled upward in the air. After the rock leaves the hand, no external work is done on the rock. As it moves upward, its kinetic energy decreases. What happens to the kinetic energy?

26. Why does a small meteoroid falling through the earth's atmosphere vaporize before it hits the ground?

PROBLEMS

Section 12.1:

1. Calculate the work done for the following forces and displacements. The angle is the angle between the applied force and the displacement.

FORCE	ANGLE	DISPLACEMENT
12.0 lb	0.00°	23.0 ft
12.0 lb	30.0°	23.0 ft
12.0 lb	90.0°	23.0 ft
8.50 N	53.0°	40.0 m
40.0 N	53.0°	8.5 m
36.6 N	20.0°	38.7 m

2. How much work must a winch do to lift a 425-lb load 11.2 ft?

3. A $12\bar{0}$-N chest is pushed 7.8 m by a parallel force of 45 N. How much work is done?

4. A cyclist exerts an average force of $2\bar{0}$ lb against the pedals of a bicycle as he travels 1.2 mi. How many foot · pounds of work does the cyclist perform?

5. Convert the following work to the indicated units.
   A. $10\bar{0}$ J = ? ft · lb
   B. 272 ft · lb = ? J
   C. 230 J = ? ft · lb

6. Which is larger, $23\bar{0}$ J or $20\bar{0}$ ft · lb?

7. Neglecting friction, how much work in foot · pounds can be done by 1.00 kW · h of electricity? How many joules?

8. An $8\bar{0}$-N box is pushed across a floor with a coefficient of friction of 0.30 between the box and floor. If 288 J of work is done, how far is the box moved?

Section 12.2:

9. A spring has a stiffness of $23\bar{0}$ N/m. How much work must be done to stretch the spring 8.5 cm?

10. A spring has a stiffness of 26.0 lb/ft. How much work must be done to stretch the spring from 4.0 in from its equilibrium length to 8.0 in?

11. Estimate the work done for the plots of force versus distance shown in Figure 12–18.

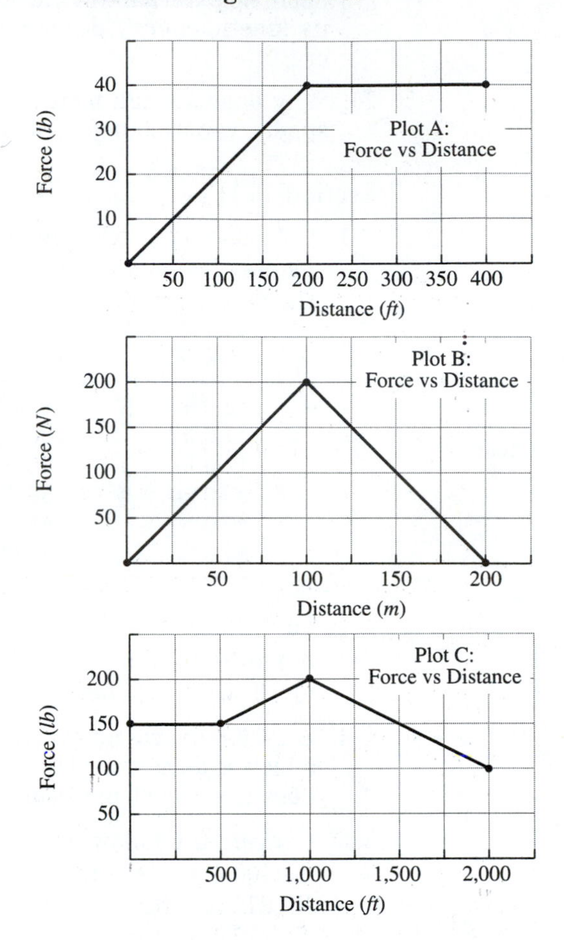

Figure 12–18: *Graphs for Problem 11.*

12. Figure 12–19 shows the plot of force versus displacement for a spring. Plot work versus distance for the spring.

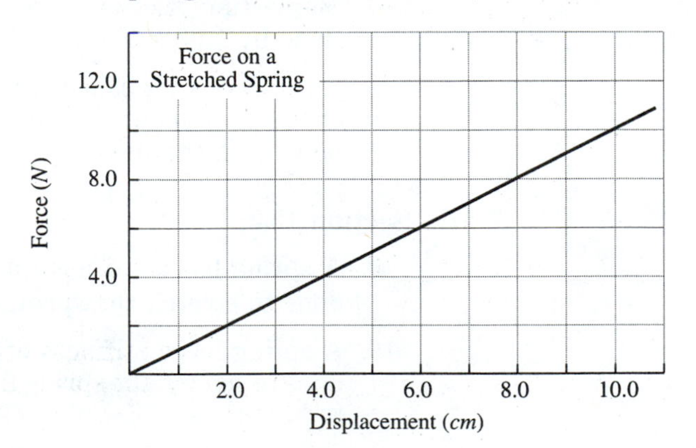

Figure 12–19: *Graph for Problem 12. Force on a stretched spring.*

Section 12.3:

13. Find the work done by the following pairs of torques and angular displacements.

TORQUE	ANGULAR DISPLACEMENT
23.0 ft · lb	12.0 rad
19.5 ft · lb	465°
12.7 ft · lb	2.20 rev
76.4 N · m	45.0°
35.0 N · m	23.2 rad

14. A force of 2.0 lb exerted perpendicular to the outer edge of a 32-in door is needed to swing the door through an angle of 90°. How much work is done in foot · pounds?

15. A table saw uses a blade with a 21.0 cm diameter to cut through a plank that resists with a force of $3\bar{0}$ N. If the blade spins at 600 rpm and it takes 2.0 s to make the cut, how many joules of work is done? (Hint: Assume the force is on the outer edge of the blade.)

16. An electric motor exerts a torque of $11\bar{0}$ ft · lb on its power take-off pulley. How much work is done for one rotation of the pulley?

Section 12.4:

17. Make the indicated unit conversions.
    A. 6.05×10^3 ft · lb/s = ? hp
    B. 1.7 hp = ? ft · lb/s
    C. 1.23 kW = ? W
    D. $23\bar{0}$ W = ? ft · lb/s
    E. 55 hp = ? kW

18. How much power must a winch deliver in horsepower to lift an 825-lb load 50.0 ft in 7.5 s?

19. How many joules of energy is used by a 30-W amplifier in 1 min? How many foot · pounds?

20. How many kilowatts are needed to lift a 230-kg load 15.0 m in 19.5 s?

21. A light car traveling at 45 mph on a level road must overcome 98 lb of frictional force. What horsepower must be delivered to the drive wheels?

22. At 65 mph, a compact car must overcome $16\bar{0}$ lb of frictional force. What horsepower must be delivered to the drive wheels?

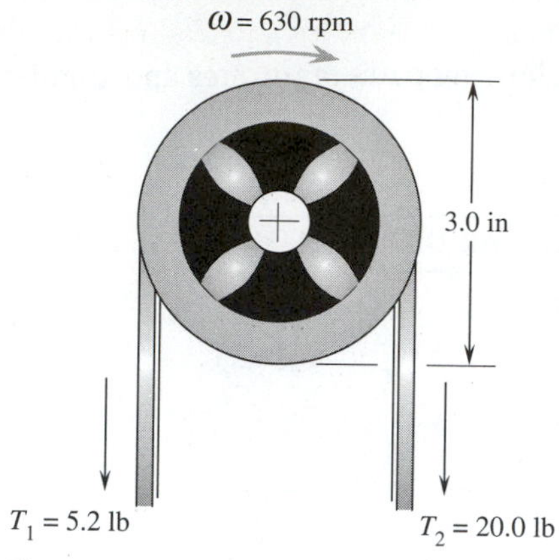

Figure 12–20: *Diagram for Problem 24.*

23. An electric motor exerts a torque of 0.16 N · m at $12\bar{0}0$ rpm. What power is delivered in watts?

***24.** The 3.0 in diameter pulley shown in Figure 12–20 is driven at $63\bar{0}$ rpm. What power is transmitted to the pulley in foot · pounds per second? In horsepower?

Section 12.5:

25. Calculate the kinetic energy of the following objects:

A. A nitrogen molecule with a mass of 4.65×10^{-26} kg and a speed of 3.20×10^{7} m/s.
B. A 12.3-g bullet with a velocity of 347 m/s.
C. A 0.37-lb ball hurled at 124 ft/s.
D. A 16.0-lb ball traveling at 33.0 ft/s.
E. A 360-kg bull charging at 24 ft/s.
F. A 432-ton train moving at 8.5 ft/s.

26. Ignoring friction, how much work must be done to accelerate a 4.52×10^{3} kg truck from 13.4 m/s to 22.8 m/s?

27. A 175-lb water-skier has a kinetic energy of 2470 ft · lb. How fast is the skier moving?

Section 12.6

28. A 145-lb mountain climber moves from an elevation of $56\bar{0}0$ ft to $63\bar{0}0$ ft.

A. What is the mountain climber's potential energy at the lower level with respect to sea level?
B. What is the climber's potential energy at the higher level with respect to sea level?
C. How much work is done in climbing to the higher level?
D. If the climber moves back down from the $63\bar{0}0$-ft level to the $56\bar{0}0$-ft level, what is the change of the climber's potential energy?

29. How much potential energy is stored in a spring stretched 12.0 cm from its equilibrium length if it has a stiffness of $2\bar{0}$ N/m?

30. What is the potential energy relative to the earth's surface of a 325-kg satellite 7.20×10^{5} m above the earth's surface? (See Exercise 17 for earth constants.)

31. Use the short distance approximation ($PE = m\,\mathbf{g}\,h$) to recalculate the potential energy of the satellite in Problem 30. What is the percentage error introduced by this approximation?

Section 12.7:

32. How many joules are generated by the brake drums of a $32\bar{0}0$-lb car as it brakes to a stop if its initial speed was 32.0 m/s?

33. A crate is pushed 32.0 ft across a floor by a parallel force of 12.0 lb. How many foot · pounds of work is done?

34. A motorist coasts down a 5000 ft high mountain using the brakes to control the car's speed. If the car weighs 2800 lb, how much heat must be dissipated by the brakes? Why would using low gear reduce the chances of burning out the brakes?

Section 12.8:

35. A pump lifts 2.30 ft^3 of water per minute out of a 178 ft deep well. The pump operates at 0.85 hp to perform this task. The density of water is 62.5 lb/ft^3.
    A. How much work is done by the pump in 1 min?
    B. What is the increase of potential energy of the water?
    C. How much of the work done by the pump is lost to fluid friction?

36. An 80% efficient electric motor does 8.98×10^3 J of work in 1 min.
    A. How much work must be supplied to the motor by the electric power source?
    B. How many watts of power are drawn by the motor?
    C. What is the power rating of the motor in horsepower?

37. In a processing plant, crushed zinc ore is carried up a 12 ft high incline by a conveyor belt delivering 34.5 lb/s to the processing plant. The friction resistance of the conveyor belt is 68 lb at a linear speed of 3.0 ft/s. How much work per second must be supplied to the belt?

Section 12.9:

38. A 3200-lb car accelerates from rest to 88 ft/s while climbing a 200 ft high hill.
    A. How much work is done to increase its kinetic energy?
    B. How much work is done to increase its potential energy?
    C. If only 14% of the energy from the fuel is transmitted to the drive wheels of the car, what fraction of a gallon of gasoline is consumed to perform this mechanical work? (Neglect friction.) (1.00 US gallon of gasoline = 9.5×10^7 ft · lb.)

39. How much horsepower must be used to overcome rolling friction for a 3800-lb automobile traveling at 60 mph (88 ft/s) with bias-belted tires? (See Table 12–1.)

40. A convertible has a cross-sectional area of 23.6 ft^2. The density of air is 8.11×10^{-2} lb/ft^3. How many foot · pounds of work are done against air drag in a 55-mi trip at 60 mph? (See Table 12–2.)

41. A medium-sized car has an engine that is 18.0% efficient in fuel usage. Rolling friction is 45 lb at $6\bar{0}$ mph.

A. After energy losses to accessories and transmission, what is the efficiency of the car at $6\bar{0}$ mph?

B. Use a ratio between rolling friction and aerodynamic drag to find the aerodynamic drag in pounds.

C. How much work is done against aerodynamic drag and rolling friction combined in 1 mi?

D. Estimate the distance the car could travel on the level on 1 gallon of gas at $6\bar{0}$ mph.

Chapter 13

CONSERVATION OF ENERGY

This power plant is a good example of the efficient use of electricity. The station uses surplus electric power from other sources to pump water uphill at night. The water is stored in a reservoir. During peak daytime power demands, the water flows back downhill and turns turbine-generators to generate electricity. (Photograph courtesy of New York Power Authority.)

OBJECTIVES

In this chapter you will learn:

- how to calculate rotational kinetic energy
- how to determine the moment of inertia for axes parallel to the spin axis through the center of gravity of an object
- that energy is conserved in linear motion and rotational motion when no work is done
- how to find the motion of rolling objects using the law of conservation of energy for both linear and rotational motion
- how to apply the laws of conservation of momentum and conservation of energy to solve collision problems

Time Line for Chapter 13

1748 Mikhail Vasilievich Lomonosov advances the ideas of conservation of energy and conservation of mass.

1807 Thomas Young introduces the concept of energy and uses the term for the first time.

1824 Nicholas Carnot shows work can be done by heat moving from a higher temperature to a lower temperature.

1842 Julius Mayer is the first to state the principles of conservation of energy and heat.

1847 Hermann von Helmholtz develops ideas on the conservation of heat and energy

1847 James Prescott Joule measures transformation of mechanical energy into heat.

1853 Coriolis coins the term *kinetic energy* in a paper explaining mechanical action.

1853 William Rankine introduces the idea of energy of position or potential energy

---

Make a pendulum from a piece of string and something heavy for a bob. Hold the end of the string. Pull the bob up at an angle of about 60°, keeping the string taut (see Figure 13–1). You do work to raise the bob. Release the stationary bob. It falls, gaining speed until it passes through the bottom of the swing. As it swings up on the other

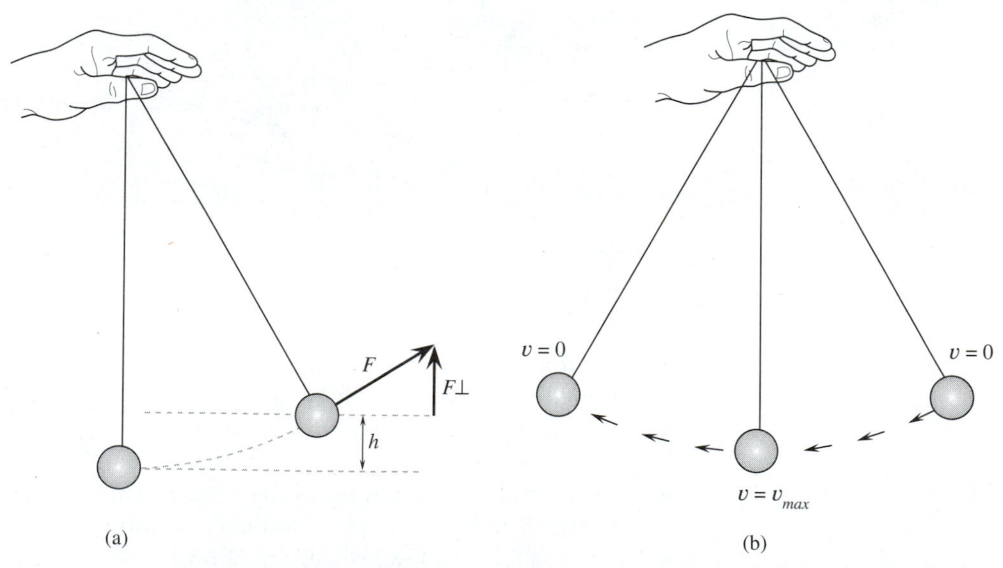

Figure 13–1: *(a) Work is done by a force* **(F)** *to raise the bob to a height* (h). *(b) The initial potential energy stored in the bob oscillates between kinetic and potential energy.*

side of the cycle, it slows down, coming to a stop before falling back. The bob will swing back and forth until friction slowly brings it to a stop.

Initially, gravitational potential energy was stored in the pendulum when it was raised to its starting position. Once the bob was released, no additional work was done. The motion of the pendulum can be described in terms of the initial energy oscillating between potential and kinetic energy with a slow but steady transfer of mechanical energy into heat.

Like the pendulum, many kinds of systems can be analyzed by keeping track of the energy stored in them. Roller coasters, electrical circuits, nuclear power plants, colliding billiard balls, gasoline engines, and falling objects all can be described in terms of energy transfer.

Before we do energy analysis we need to look at another form of kinetic energy. Work is required to make a flywheel turn. The flywheel spins without its center of gravity moving from one place to another. It has no linear kinetic energy of the type described in Chapter 12. The work is stored in rotational kinetic energy.

13.1 ROTATIONAL KINETIC ENERGY

See Figure 13–2. Two masses tied to a string are whirled in a circular orbit. We can calculate the combined kinetic energy of the two masses in terms of their angular speed. The kinetic energy of the two masses is:

$$KE = \left[\frac{1}{2}(m_1\,v_1^2)\right] + \left[\frac{1}{2}(m_2\,v_2^2)\right]$$

Speeds v_1 and v_2 are:

$$v_1 = R_1\,\omega$$

$$v_2 = R_2\,\omega$$

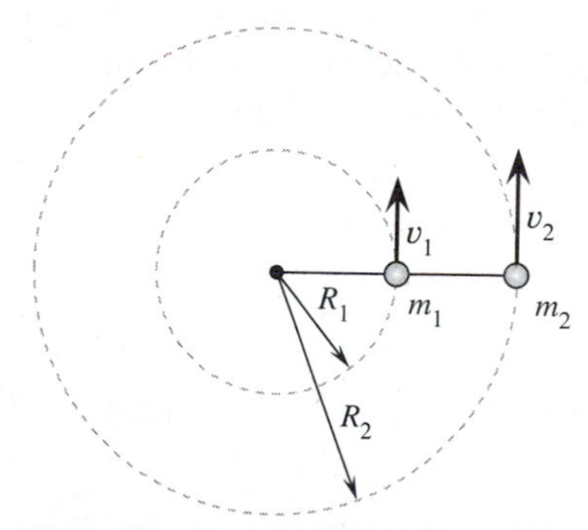

Figure 13–2: *Two masses are tied to a string at different radii. They are whirled in a circle.*

Substitute these velocities into the energy equation.

$$KE = \frac{1}{2}[(m_1\,R_1^2) + (m_2\,R_2^2)]\,\omega^2$$

The expression in brackets is the moment of inertia first encountered in Chapter 11 for angular momentum. The rotational kinetic energy of the orbiting masses can be written compactly.

Energy Storage

Electric power companies have a problem; the demand for electricity changes quite a lot during the day. At night there is a minimal need for electricity. After dawn human activity increases and so does the demand for electric power, peaking in the early afternoon. As the workday ends, demand decreases and slowly fades to the base demand in the wee hours of the morning. Some electrical generating systems cannot respond rapidly to the quickly changing power demand. Systems such as coal-burning steam generators are slow to adjust.

One solution is to store some of the unused nighttime energy as potential or kinetic energy. Pumping stations, like the one at Niagara Falls, store potential energy. The unused nighttime energy from the generating plant is used to pump water uphill to a water reservoir. During peak demand, the water passes back into the river through a hydroelectric generating system. This system works very well, but not all power plants have a lot of water nearby to pump. Some power plants store excess energy in the kinetic energy of flywheels.

The amount of energy stored in a rotating wheel is proportional to the square of the angular speed. If the spin rate is doubled, four times as much energy is stored. The outer part of the wheel has a greater tangential speed and, therefore, more kinetic energy than the part of the wheel near the axle. Doubling the diameter of the wheel will quadruple the amount of energy. So wheels should be large. Mass also affects energy. Doubling the mass yields twice as much energy.

Another design factor is centripetal force. Like rotational kinetic energy, centripetal force depends on spin rate, mass, and radius. The pieces of the wheel near the rim have the greatest centripetal force acting on them. A layer of atoms on the outside surface must have a centripetal force exerted on it from neighboring atoms a bit closer to the axle. The strength of the material becomes important. If the stress between the layers, expressed in units of pressure, is less than the pressure needed to supply the needed centripetal force, the flywheel will break.

You might think that steel would be the ideal material for a flywheel. It is not. The strength of steel is typically 6×10^4 psi. Glass has a strength of 3×10^{10} psi—a million times stronger than steel. We can make larger flywheels and spin them faster by making super flywheels out of laminated fiberglass.

Photo Courtesy of New York Power Authority

$$KE_{\text{rot}} = \frac{1}{2}(I\,\omega^2) \qquad \textbf{(Eq. 13–1)}$$

You will find moments of inertia for some simple shapes in Figure 11–11. You may want to review Section 11.3 to help you recall how to calculate moments of inertia.

EXAMPLE PROBLEM 13–1: A FLYWHEEL

A disk-shaped flywheel has a mass of 2.45 kg and a radius of 23.9 cm. How much rotational kinetic energy does the wheel have when it spins at 1270 rpm? See Figure 13–3.

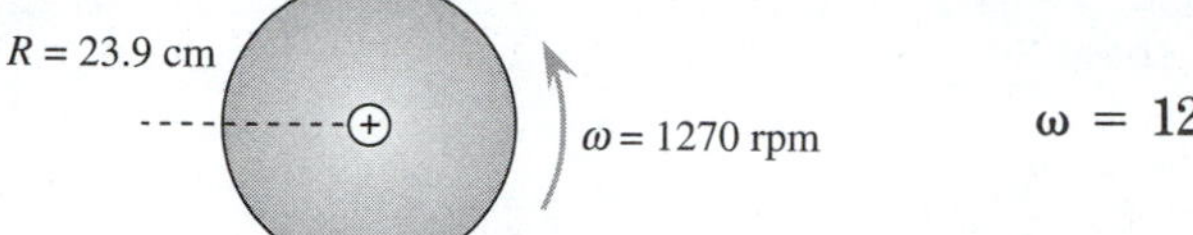

Figure 13–3: A disk-shaped flywheel stores energy in its rotational motion.

SOLUTION

First convert the angular speed to radians per second.

$$\omega = 1270\,\frac{\text{rev}}{\text{min}} \times \frac{1\text{ min}}{60\text{ s}} \times 2\,\pi\,\frac{\text{rad}}{\text{rev}} = 133\,\frac{\text{rad}}{\text{s}}$$

The moment of inertia for a disk spinning around its center is:

$$I = \frac{1}{2}(m\,R^2) = 0.5 \times 2.45\text{ kg} \times (0.239\text{ m})^2$$

$$I = 0.0697\text{ kg}\cdot\text{m}^2$$

The rotational kinetic energy is then:

$$KE_{(\text{rot})} = \frac{1}{2}(I\,\omega^2)$$

$$KE_{(\text{rot})} = 0.5 \times 0.0697\text{ kg}\cdot\text{m}^2 \times \left(133\,\frac{\text{rad}}{\text{s}}\right)^2$$

$$KE_{(\text{rot})} = 6.16 \times 10^2\text{ J}$$

13.2 PARALLEL SPIN AXES

The spin axis in Example Problem 13–1 is at the center of mass of the disk. Sometimes machine parts rotate around a spin axis that is not at the center of mass. In such cases Equation 13–2, the **parallel axis theorem,** can be used. **D** is the distance between the center of mass and a parallel spin axis. I_0 is the moment of inertia of the object around its center of mass.

$$I = I_0 + (m\,D^2) \qquad \textbf{(Eq. 13–2)}$$

Notice, if the distance D is much larger than the average radius of the object, I_0 will be small compared with the second term. A small bob is tied to the end of a long piece of string of length D. When the bob is twirled in a circle, the shape of the bob is not important. Its moment of inertia is simply $m\,D^2$.

☐ **EXAMPLE PROBLEM 13–2: PARALLEL AXIS OF A SPHERE**

A sphere has a moment of inertia of 2/5 ($m R^2$) around its center of mass (see Figure 11–11). Find the moment of inertia for parallel spin axes at distances R, $2R$, and $10R$ from the center of mass of the sphere.

■ ***SOLUTION***

A. If $D = R$, then

$$I = \frac{2}{5} m R^2 + m R^2$$

$$I = \frac{7}{5} (m R^2)$$

This result agrees with the value shown in Figure 11–11.

B. If D = $2R$, then

$$I = \frac{2}{5} m R^2 + m(2R)^2$$

$$I = 4.4 (m R^2)$$

C. If $D = 10R$, then

$$I = \frac{2}{5} m R^2 + m(10R)^2$$

$$I = 100.4 (m R^2) \text{ or } I = 100 (m R^2)$$

Notice that for $D = 10R$, the approximation of $I = m D^2$ is good to three significant figures.

13.3 TRANSFORMATIONS FOR ROTATIONAL DYNAMICS

In Chapter 5 we found that the equations describing rotational motion were similar to the equations for linear motion. Transform equations could be used to convert linear equations into analogous rotational equations. We can do the same thing with dynamics. Newton's second law for rotation, angular momentum, and rotational kinetic energy all have a form that is similar to the linear equations (see Table 13–1).

Table 13–1: Transformations between linear equations and rotational equations.

MOTION	
Transform	**Transform equation**
$s \rightarrow \theta$	$s = R\,\theta$
$v \rightarrow \omega$	$v = R\,\omega$
$a \rightarrow \alpha$	$a = R\,\alpha$
Linear equation	**Rotational analog**
$\bar{v} = \dfrac{s_f - s_i}{t}$	$\bar{\omega} = \dfrac{\theta_f - \theta_i}{t}$
$\bar{v} = \dfrac{v_f + v_i}{2}$	$\bar{\omega} = \dfrac{\omega_f + \omega_i}{2}$
$v_f = v_i + (a\,t)$	$\omega_f = \omega_i + (\alpha\,t)$
$s = (v_i\,t) + \left(\dfrac{a\,t^2}{2}\right)$	$\theta = (\omega_i\,t) + \left(\dfrac{\alpha\,t^2}{2}\right)$
$2\,\mathbf{a}\,s = v_f^2 - v_i^2$	$2\,\alpha\,\theta = \omega_f^2 - \omega_i^2$
DYNAMICS	
Transform	**Transform equation**
$\mathbf{F} \rightarrow \boldsymbol{\tau}$	$\boldsymbol{\tau} = \mathbf{r} \times \mathbf{F}$
$\mathbf{p} \rightarrow \mathbf{L}$	$\mathbf{L} = \mathbf{r} \times \mathbf{p}$
$m \rightarrow I$	$I = \sum_i (m_i\,r_i^2)$
Linear equation	**Rotational analog**
$\mathbf{F} = \dfrac{\Delta\,(m\,\mathbf{v})}{t}$	$\boldsymbol{\tau} = \dfrac{\Delta\,(I\,\boldsymbol{\omega})}{t}$
$\mathbf{p} = m\mathbf{v}$	$\mathbf{L} = I\,\boldsymbol{\omega}$
$KE = \dfrac{m\,v^2}{2}$	$KE_{rot} = \dfrac{I\,\omega^2}{2}$
$W = \mathbf{F} \cdot \mathbf{s}$	$W = \boldsymbol{\tau} \cdot \boldsymbol{\theta}$

13.4 CONSERVATION OF ENERGY WITH LINEAR MOTION

Work must be done to change the total energy of a system. When work is done on a system the potential energy may change, the kinetic energy (linear and rotational) may change, or heat may be generated.

$$W = \Delta\,PE + \Delta\left[\frac{1}{2}(m\,v^2)\right] + \Delta\left[\frac{1}{2}(I\,\omega^2)\right] + Q$$

If no work is done on a system, the total work cannot change. The form of the energy may change. Some potential energy may change to kinetic energy. Frictional forces may change kinetic energy to heat. Mechanical devices may convert rotational kinetic energy into linear kinetic energy.

One gallon of gasoline contains 1.3×10^8 J of chemical potential energy. When this one gallon is burned to run a car, much of the energy shows up as heat in the engine and exhaust; another fraction of the energy is used to run accessories to maintain the engine; and another portion shows up as heat from frictional losses in the power train and wind resistance. Finally there is the mechanical change of kinetic and potential energy of the car as it changes speed and climbs hills. When we add all of these energies in terms of potential energy changes, kinetic energy changes, and heat, we find the same total energy we started with—1.3×10^8 J. We say that energy is conserved. The total energy of an isolated system is constant.

Look at the work-energy equation above. If no work is done on the system, the left-hand side is zero.

$$0 = \Delta KE + \Delta PE + Q$$

$$0 = (KE_f - KE_i) + (PE_f - PE_i) + Q$$

Rearranging the equation with the initial energies on the left gives us the equation for **conservation of energy.**

$$KE_i + PE_i = KE_f + PE_f + Q \qquad \textbf{(Eq. 13–3)}$$

The sum of the kinetic and potential energies of a system is called the **mechanical energy** of the system. When no external work is done, the initial mechanical energy of a system is equal to the final mechanical energy of the system plus whatever heat is generated. The total energy of the system is unchanged.

initial total energy = final total energy = constant

This gives us a powerful relationship to solve a variety of problems. If we can find the total energy of a system at some time, we can determine what the total energy is at some later time.

☐ **EXAMPLE PROBLEM 13–3: THE PARKED CAR**

A $32\bar{0}0$-lb car is parked $10\bar{0}$ ft above the base of a hill. The brakes give way. The car rolls down the hill, reaching a speed of 70.0 ft/s at the bottom. How much heat was generated by frictional forces? See Figure 13–4.

■ ***SOLUTION***

Initially, the car's kinetic energy was zero. The total energy (E_t) was tied up in potential energy.

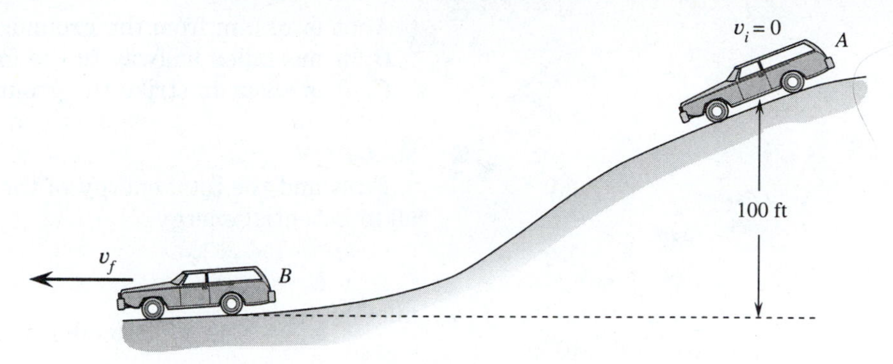

Figure 13–4: A car initially at rest rolls down a hill. Its potential energy is converted to kinetic energy and heat.

$$E_t = PE = m\,g\,h$$
$$E_t = 32\bar{0}0 \text{ lb} \times 10\bar{0} \text{ ft}$$
$$E_t = 3.20 \times 10^5 \text{ ft} \cdot \text{lb}$$

At the bottom of the hill no gravitational potential energy is left. The final energy is in the form of kinetic energy plus heat.

$$E_t = KE + Q$$

Solve the equation for Q.

$$Q = E_t - KE$$
$$Q = (3.20 \times 10^5 \text{ ft} \cdot \text{lb}) - \left\{\frac{1}{2}\left[\left(\frac{3200 \text{ lb}}{32 \text{ ft/s}^2}\right) \times (70.0 \text{ ft/s})^2\right]\right\}$$
$$Q = [(3.20 \times 10^5) - (2.45 \times 10^5)] \text{ ft} \cdot \text{lb}$$
$$Q = 7.5 \times 10^4 \text{ ft} \cdot \text{lb}$$

In some situations, the heat generated in a system is small compared with the mechanical energy. In such a case, a good approximation can be made by ignoring the heat term.

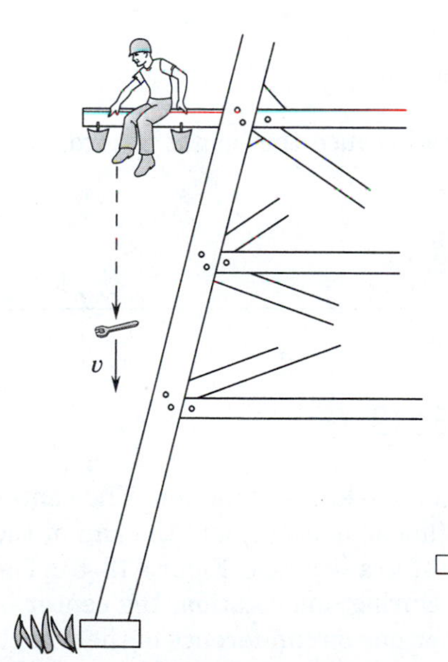

Figure 13–5: A rigger drops a wrench from a great height.

☐ **EXAMPLE PROBLEM 13–4: THE FALLING WRENCH**

A rigger is assembling a high-voltage transmission tower. The rigger drops a 248-g wrench from a height of 32.6 m (see Figure 13–5). Ignoring air friction, find the potential, kinetic, and total energy of the wrench when:

A. it is 23.0 m from the ground.
B. it has fallen halfway to the ground.
C. it is about to strike the ground.

■ ***SOLUTION***

First find the total energy of the system. Initially, the energy is all in potential energy.

$$E_t = m\,g\,h$$

$$E_t = 0.248\text{ kg} \times 9.80\text{ m/s}^2 \times 32.6\text{ m}$$

$$E_t = 79.2\text{ J} = \text{a constant for the problem}$$

A.

$$PE = m\,g\,h$$

$$PE = 0.248\text{ kg} \times 9.80\text{ m/s}^2 \times 23.0\text{ m}$$

$$PE = 55.7\text{ J}$$

$$KE = E_t - PE$$

$$KE = 78.9\text{ J} - 55.7\text{ J}$$

$$KE = 23.2\text{ J}$$

B. The wrench has lost half its potential energy. Half of the energy is converted to kinetic energy.

$$PE = KE = \frac{79.2}{2}\text{J} = 39.6\text{ J}$$

C. The potential energy is zero because the height is zero.

$$PE = 0$$

$$KE = E_t = 79.2\text{ J}$$

13.5 CONSERVATION OF ENERGY IN ROLLING SYSTEMS

A wheel rolling along a surface has two kinds of motion. The center of mass moves in a straight line (linear motion), and the rim of the wheel rotates around the center of gravity (see Figure 13–6). The two kinds of motion are related. During one rotation, the center of mass travels a distance $s = 2\pi R$, or one circumference of the wheel. At the same time, the wheel rotates through one revolution of $\theta = 2\,\pi$. Combining these two relationships we get:

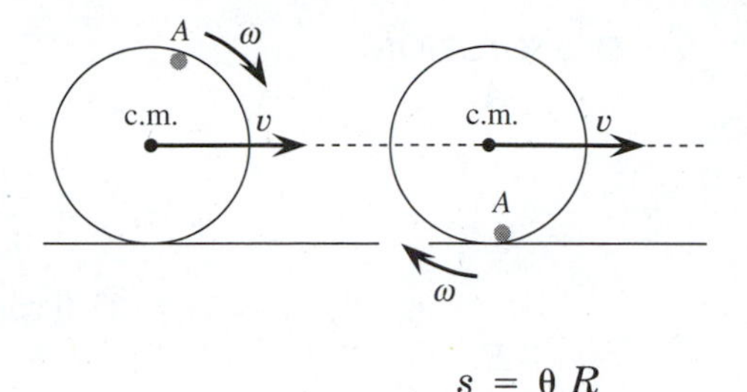

Figure 13–6: *A rolling wheel has two kinds of motion. The center of mass (c.m.) moves in a straight line with a velocity* ***v****. Points on the wheel, such as point A, rotate around the moving center of mass with an angular speed ω.*

$$s = \theta R$$

Dividing by the time required to make the rotation gives us an equation that relates the linear speed with the rotational speed.

$$\frac{s}{t} = \frac{\theta R}{t}$$

or

$$v = \omega R \qquad \textbf{(Eq. 13–4)}$$

When something is rolling without slipping, Equation 13–4 can be used to convert between rotational kinetic energy and linear kinetic energy.

A rolling object has both linear and rotational kinetic energy. How much of the energy is invested in rotational motion depends on the moment of inertia, or shape of the object.

☐ **EXAMPLE PROBLEM 13–5: THE ROLLING DRUM**

A steel drum of chemicals with a radius of 38.4 cm rolls off a moving truck. The drum has a moment of inertia of 4.76 kg · m^2 and a mass of 56.8 kg. If the drum rolls with a linear speed of 14.2 m/s, what is its:

A. rotational speed in radians per second?
B. linear kinetic energy?
C. rotational kinetic energy?
D. total kinetic energy?

Figure 13–7: *A steel drum falls off a moving truck. It moves foward with a linear velocity* ***v****.*

■ *SOLUTION*

A.

$$\omega = \frac{v}{R} = \frac{14.2 \text{ m/s}}{0.384 \text{ m}}$$

$$\omega = 37.0 \text{ rad/s}$$

B.

$$KE = \frac{1}{2}(m\,V^2) = 0.5 \times 56.8 \text{ kg} \times (14.2 \text{ m/s})^2$$

$$KE = 5.73 \times 10^3 \text{ J}$$

C.

$$KE_{(rot)} = \frac{1}{2}(I\,\omega^2) = 0.5 \times 4.76 \text{ kg} \cdot \text{m}^2 \times (37.0 \text{ rad/s})^2$$

$$KE_{(rot)} = 0.880 \times 10^3 \text{ J}$$

D.

$$E_t = KE + KE_{(rot)} = (5.73 + 0.880) \times 10^3 \text{ J}$$

$$E_t = 6.61 \times 10^3 \text{ J}$$

The *fraction* of kinetic energy involved with rotational motion does *not* depend on mass or radius. Only shape affects the fraction of kinetic energy in rotation. In Example Problem 13–6 it seems as though there is not enough information, but the final linear velocities are independent of mass and radius.

□ **EXAMPLE PROBLEM 13–6: THE RACE**

A cylinder, a sphere, and a hoop roll down a 3.46 ft high incline starting from rest. What is the speed of each object as it reaches the bottom of the incline? See Figure 13–8.

■ *SOLUTION*

The moments of inertia for the three shapes have the form:

$$I = k\,m\,R^2$$

The constant k is the shape factor of the objects. We can find the constants (k) from Figure 11–11.

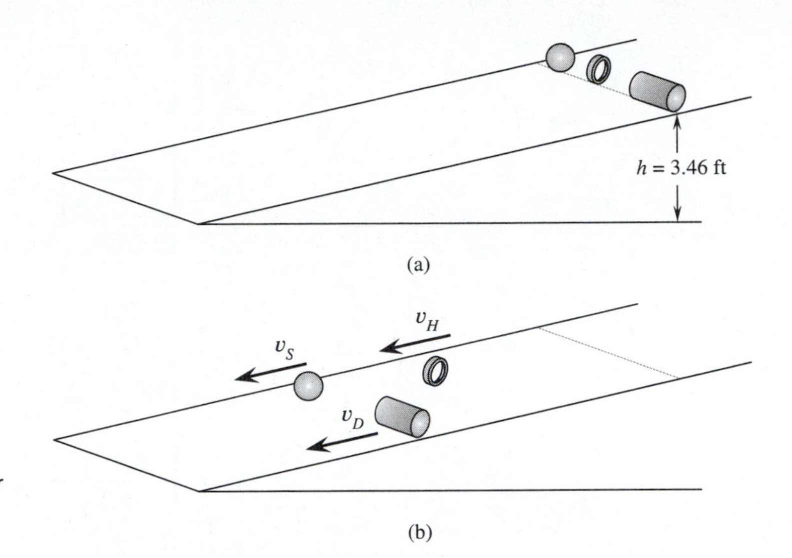

Figure 13–8: *A disk, a sphere, and a cylinder roll down an incline. The linear speed depends on their moments of inertia.*

$$\text{Sphere:} \quad k_s = \frac{2}{5}$$

$$\text{Cylinder:} \quad k_d = \frac{1}{2}$$

$$\text{Hoop:} \quad k_h = 1$$

The total energy in each case is the initial potential energy.

$$E_t = m\ g\ h$$

At the bottom of the incline all the total energy is invested in kinetic energy.

$$E_t = \left[\frac{1}{2}(m\ v^2)\right] + \left[\frac{1}{2}(I\ \omega^2)\right]$$

$$m\ \mathbf{g}\ h = [0.5\ (m\ v^2)] + \left[0.5\ (k\ m\ R^2)\left(\frac{v}{R}\right)^2\right]$$

Mass (m) and radius (R) cancel. Solve the equation for v.

$$v = \left[\frac{2\ g\ h}{1 + k}\right]^{1/2}$$

$$v = \left[\frac{2\ (32\ \text{ft/s}^2 \times 3.46\ \text{ft})}{1 + k}\right]^{1/2}$$

$$v = \left[\frac{221}{1 + k}\right]^{1/2} \text{ ft/s}$$

$$v_d = \left[\frac{221}{1 + 0.5}\right]^{1/2} \text{ ft/s}$$

$$v_d = 12.1 \text{ ft/s}$$

$$v_s = \left[\frac{221}{1 + 0.4}\right]^{1/2} \text{ ft/s}$$

$$v_s = 12.6 \text{ ft/s}$$

$$v_h = \left[\frac{221}{1 + 1}\right]^{1/2} \text{ ft/s}$$

$$v_h = 10.5 \text{ ft/s}$$

Notice that the hoop is the slowest of the three objects because half of its kinetic energy is in the form of rotational energy. The sphere, with the smallest moment of inertia, has the fastest linear speed.

13.6 APPLICATION: COLLISIONS

In Chapter 11 we found that *momentum* is conserved in *all* kinds of collisions. *Mechanical energy* is not necessarily conserved. Some or all of the mechanical energy may be converted to heat in a collision. Collisions are usually sorted into three categories.

Perfectly Elastic Collisions

In a perfectly elastic collision, mechanical energy is conserved. In nature, this kind of collision rarely happens. On the atomic scale, some kinds of atomic and molecular collisions conserve mechanical energy. On the large scale some collisions generate a very small amount of heat compared with the mechanical energy of the system. We can solve such problems by assuming conservation of energy without introducing large errors in the results. The collision of billiard balls is an example.

Perfectly Inelastic Collisions

In perfectly inelastic collisions the colliding objects stay together. They move with the same final velocity after the collision. Mechanical energy is not conserved. A bullet fired into a wooden block is an example of this kind of collision.

Imperfectly Elastic Collisions

Most collisions are imperfectly elastic collisions. Mechanical energy is not conserved, and the objects do not stay together after the collision.

☐ **EXAMPLE PROBLEM 13–7: PERFECTLY ELASTIC COLLISION**

A bowling ball traveling at a velocity v_1 on the return rack strikes another identical stationary bowling ball. What are the velocities of the two balls after the collision?

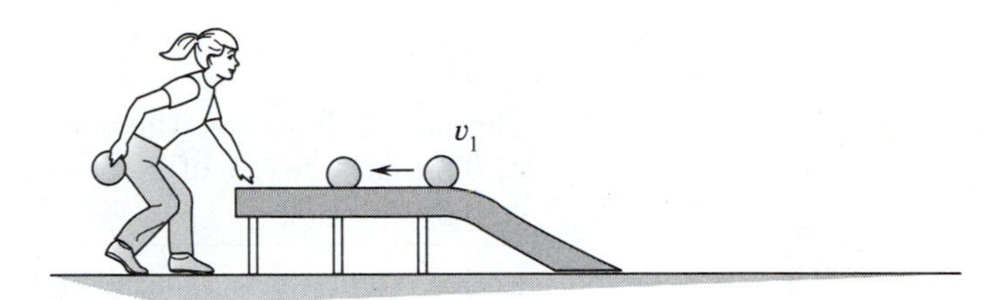

Figure 13–9: *A moving bowling ball collides with an identical ball at rest on the return rack.*

■ ***SOLUTION***

Momentum must be conserved. (In the equations below **u** = velocity after the collision.)

$$(m\ \mathrm{v}_1) + 0 = (m\ \mathrm{u}_1) + (m\ \mathrm{u}_2)$$

Since the balls are identical, mass cancels.

$$\mathrm{v}_1 = \mathrm{u}_1 + \mathrm{u}_2 \qquad \textbf{(Eq. 13–5)}$$

We will assume mechanical energy is conserved.

$$\left[\frac{1}{2}(m\ \mathrm{v}_1^2)\right] + 0 = \left[\frac{1}{2}(m\ \mathrm{u}_1^2)\right] + \left[\frac{1}{2}(m\ \mathrm{u}_2^2)\right]$$

or

$$\mathrm{v}_1^2 = \mathrm{u}_1^2 + \mathrm{u}_2^2 \qquad \textbf{(Eq. 13–6)}$$

We can square Equation 13–5.

$$\mathrm{v}_1^2 = \mathrm{u}_1^2 + [2\,(\mathrm{u}_1\ \mathrm{u}_2)] + \mathrm{u}_2^2 \qquad \textbf{(Eq. 13–5A)}$$

Compare Equation 13–5A with Equation 13–6. The two equations describe the same collision. The middle term on the right-hand side of Equation 13–5A must be zero in order to conserve both momentum and mechanical energy.

$$2\,(u_1\,u_2) = 0$$

Either u_1 or u_2 must be zero. If $\mathbf{u_2}$ is zero, no collision has occurred. Neither ball shows a change of motion.

$$u_1 = 0$$

From the momentum equation we see:

$$u_2 = v_1.$$

The first ball strikes the second, transferring all its momentum to it. This is a result of conserving both momentum and mechanical energy.

☐ **EXAMPLE PROBLEM 13–8: PERFECTLY INELASTIC COLLISION**

A 2.3-g bullet is fired horizontally into a 1.248-kg block of wood hung from a string (see Figure 13–10). The block of wood swings upward reaching a height of 6.3 cm.

A. Estimate the velocity of the bullet before the collision.

B. Estimate the fraction of mechanical energy converted to heat.

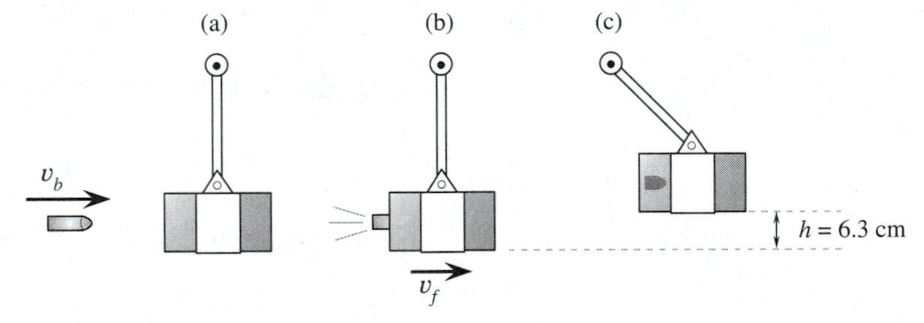

Figure 13–10: *A bullet strikes a suspended block of wood. The embedded bullet and block are driven back and upward to a maximum height of h.*

■ ***SOLUTION***

A. Speed of the bullet:

Although mechanical energy is not conserved in the collision, momentum is conserved. The bullet and block have the same final velocity of $\mathbf{V_f}$ after the collision.

$$(m\,v_b) + 0 = (m + M)V_f$$

or

$$v_b = \frac{(m + M)}{m}\,V_f$$

After the collision mechanical energy is conserved. The kinetic energy that the block and the bullet had right after the collision is transformed into potential energy at the top of the swing.

$$\frac{1}{2}[(m + M)\,v_f^2] = (m + M)\,g\,h$$

$$V_f = [2\,(g\,h)]^{1/2}$$

Combine the energy and momentum equations.

$$v_b = \frac{(m + M)}{m}[2(g\,h)]^{1/2}$$

$$v_b = \frac{(0.0023 + 1.248)\text{ kg}}{0.0023\text{ kg}} \times [2\,(9.80\text{ m/s}^2 \times 0.053\text{ m})]^{1/2}$$

$$v_b = 554\text{ m/s}$$

B. Mechanical energy converted to heat:

Before the collision, the mechanical energy was all in the bullet.

$$KE_i = \frac{1}{2}(m\,v_b^2)$$

$$KE_i = 0.5 \times 0.0023\text{ kg} \times [(554\text{ m/s})^2]$$

$$KE_i = 353\text{ J}$$

After the collision there was enough mechanical energy to swing the block and bullet to a height of 6.3 cm.

$$PE_f = (m + M)\,g\,h$$

$$PE_f = (0.0023 + 1.248)\text{ kg} \times 9.80\text{ m/s}^2 \times 0.063\text{ m}$$

$$PE_f = 0.77\text{ J}$$

$$Q = KE_i - PE_f = (353 - 0.77)\text{ J}$$

$$Q = 352\text{ J}$$

The percentage of mechanical energy converted to heat is:

$$Q\% = \frac{352\text{ J}}{353\text{ J}} \times 100$$

$$Q\% = 99.7\%$$

EXAMPLE PROBLEM 13–9: IMPERFECTLY ELASTIC COLLISION

A rubber ball is dropped from a height of 6.00 ft onto a concrete floor. The ball bounces to a maximum height of 5.34 ft. What is the ratio (e) of the ball's velocity just after the collision divided by its velocity just before the collision?

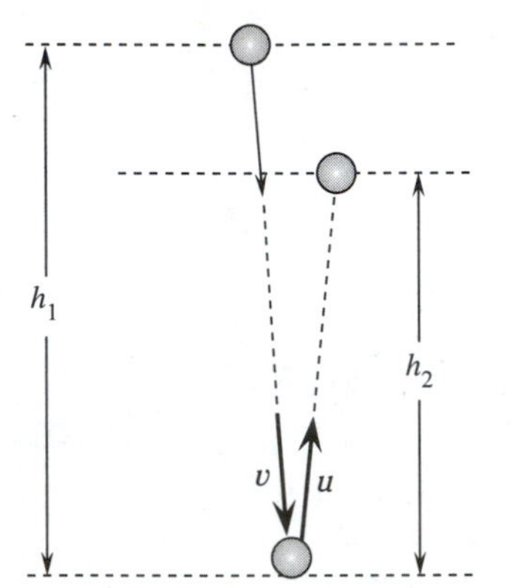

Figure 13–11: *A rubber ball is dropped from a height h_1. It rebounds to a lower height h_2.*

SOLUTION

Before the collision:

$$\frac{1}{2}(m\,v^2) = m\,g\,h_1$$

After the collision:

$$\frac{1}{2}(m\,u^2) = m\,g\,h_2$$

Divide one equation by the other.

$$\frac{u^2}{v^2} = \frac{h_2}{h_1}$$

$$e = \frac{u}{v} = \left(\frac{h_2}{h_1}\right)^{1/2}$$

$$e = 0.943$$

The ratio e, called the **coefficient of restitution,** is the ratio of the relative speed of approach divided by the speed of separation for two objects colliding head on. In this problem one object, the floor, did not move. In general cases:

$$e = \frac{u_2 - u_1}{v_1 - v_2}$$

SUMMARY

The rotational kinetic energy of a spinning object is one-half the product of the moment of inertia times the angular momentum squared. When a round object rolls without slipping, it has both linear and rotational kinetic energy. For a rolling object of radius R the rotational motion is the linear speed divided by the object's radius.

$$\omega = \frac{v}{R}$$

The moment of inertia of an object rotated around an axis parallel to, and a distance D from, the center of mass can be found by the parallel axis theorem. It is the moment of inertia for rotation around the center of mass plus the mass of the object times D^2.

$$I = I_0 + m D^2 \qquad \textbf{Parallel Axis Theorem}$$

If no external work is done on a system, then the total energy of the system is unchanged. The energy may vary in type—kinetic energy, potential energy, and heat—but the total of all types of energy in the system has a fixed value. This principle is called the law of conservation of energy.

$$\textbf{Initial Energy = Final Energy = Constant}$$
$$KE_i + PE_i = KE_f + PE_f + Q$$

Mechanical energy (the sum of potential and kinetic energy in a system) is not necessarily conserved in a collision. Some energy may be lost as heat. But the total final energy (mechanical plus heat) will equal the total initial energy.

Collisions may be sorted into three types. In a perfectly elastic collision, mechanical energy is conserved. In a perfectly inelastic collision the objects stay together after the collision. Mechanical energy is *not* conserved. Mechanical energy is also *not* conserved in an imperfectly elastic collision. The objects do not stay together after the collision.

KEY TERMS

If you can explain the following terms to a friend or classmate, you understand their meaning. If you cannot explain the terms, you should reread the sections in which they are discussed.

coefficient of restitution
conservation of energy
mechanical energy
parallel axis theorem

EXERCISES

Section 13.1:

1. Flywheel A has twice the radius and half the mass of flywheel B. If flywheel B spins twice as fast as A, does it have more or less rotational kinetic energy? Both wheels have the same shape.
2. A sphere, a hoop, and a disk have the same radius, mass, and rotational speed. Which has the largest rotational kinetic energy? The smallest?
3. What effect does the speed of a car have on the fraction of kinetic energy stored in the tires?
4. A racing bike has thin lightweight wheels. Why would it be easier to accelerate than another bike with the same weight, but lighter frame and heavier wheels?

Section 13.2:

5. If an object rotates around an axis a distance D away from the center of mass, we found in Example Problem 13–2 that if $D >> R$, the moment of inertia can be approximated by $m\ D^2$ with very little error. The moon orbits around the earth at a distance (R_m) of 60.3 earth radii. The mass of the moon (M_m) is 1/81.3 earth masses. What percentage of the earth-moon system's moment of inertia is given by the approximation $I = M_m\ R_m{}^2$?

6. A hoop spun around its center has a moment of inertia of $m\ R^2$. Show that its moment of inertia is $2\ (m\ R^2)$ if it is rotated around a point on the rim.

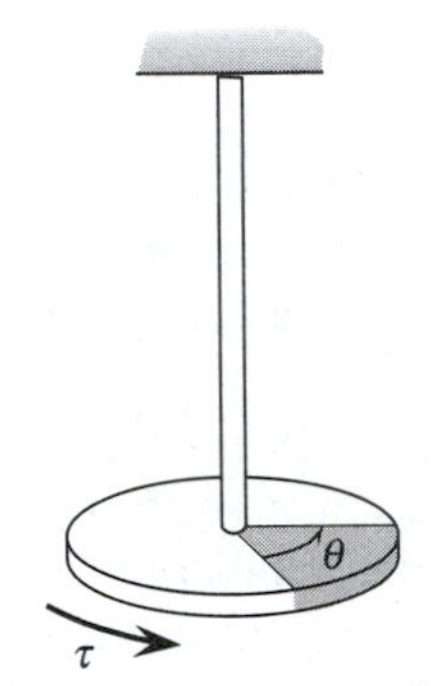

Figure 13–12: A torque, τ, on the attached disk twists the rod through an angle, θ.

Section 13.3:

7. The centripetal force of a bob on a long string is: $F_c = (m\ v^2)/R$. Use the transform equations between linear and rotational motion to show that the centripetal force on the bob is $\mathbf{F}_c = I\ \omega^2/R$.

8. A weight is fastened to the end of a rod (see Figure 13–12). The weight is twisted through an angle (θ) by a torque (τ). The torque and angular displacement are related by: $\tau = -C\theta$, where C is a proportionality constant. Use the transform equations to show that this is the rotational analog to Hooke's law. (Hint: R is constant.)

*9. From Hooke's law, we found that the potential energy stored in a stretched spring is $PE = 1/2\ (k\ \Delta\ x^2)$. Use the analog equations to find the expression for the potential energy stored in the torsion bar in Exercise 8.

Section 13.4:

10. A child on a swing is pulled back and released.
    A. The kinetic energy of the swing increases as the swing approaches the lowest point in the arc. Where does the kinetic energy come from?
    B. After the swing passes through the lowest point in the arc, it loses kinetic energy. Where does the kinetic energy go?
    C. After a period of time, the swing stops moving and hangs vertically. What happened to the mechanical energy?

11. Give some examples of situations in which gravitational potential energy is mostly converted into:
    A. elastic potential energy.
    B. kinetic energy.
    C. heat.

Figure 13–13: *(a) The swing is drawn back to a height* h. *(b) The swing has its maximum speed at the lowest point. (c) After a period of time the swing hangs motionless.*

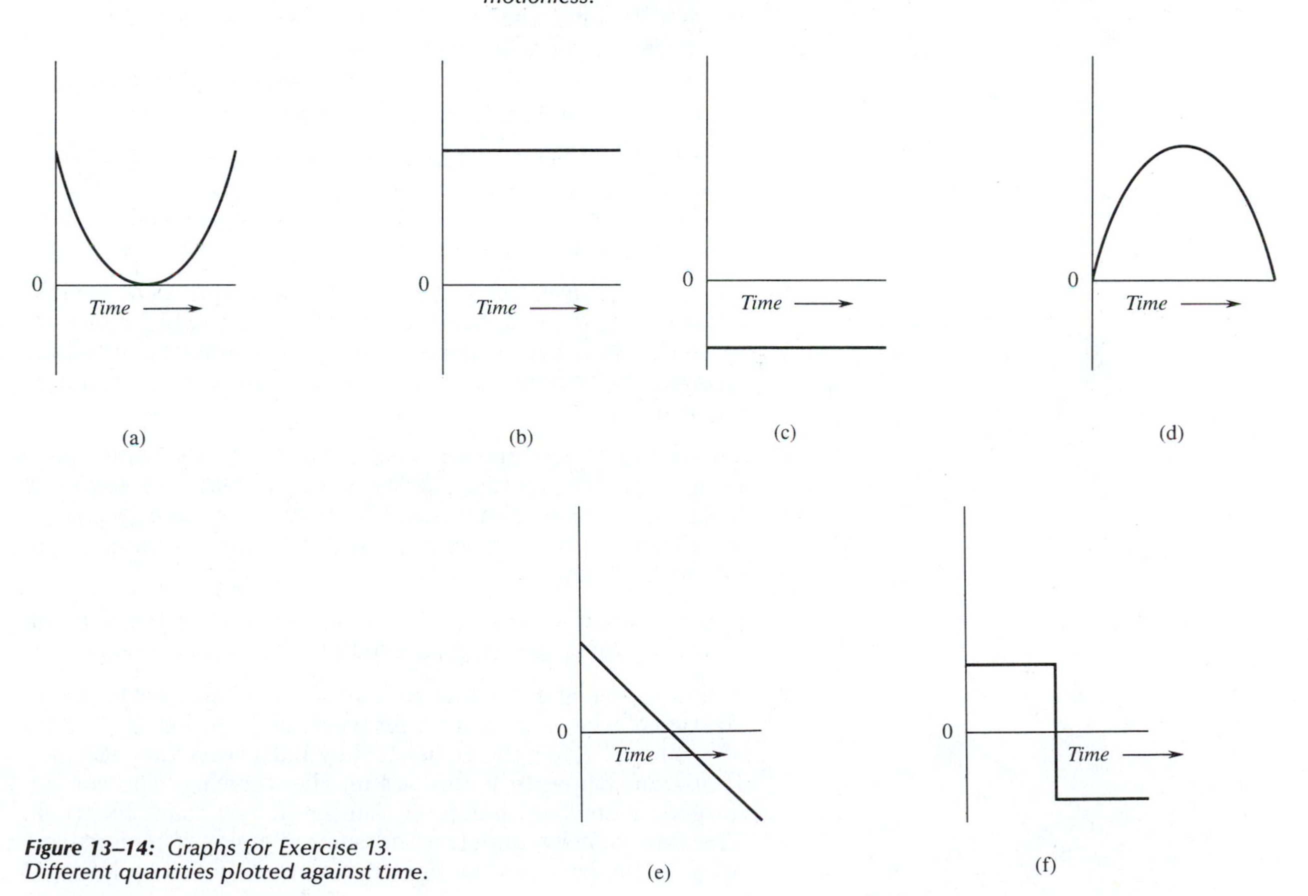

Figure 13–14: *Graphs for Exercise 13. Different quantities plotted against time.*

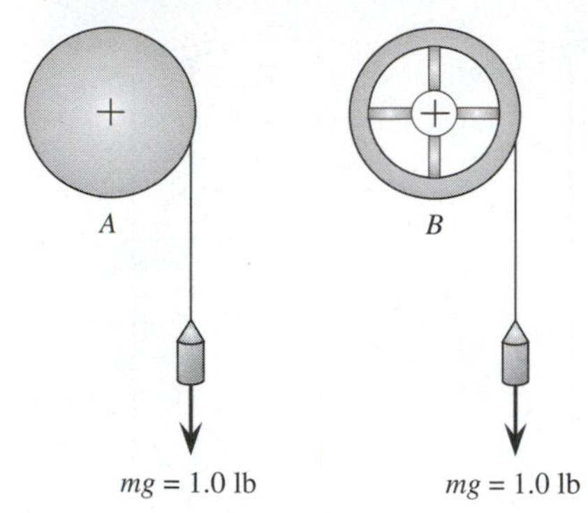

Figure 13–15: *Two pulleys have the same radius and mass, but different shapes. Identical weights are hung from the pulleys.*

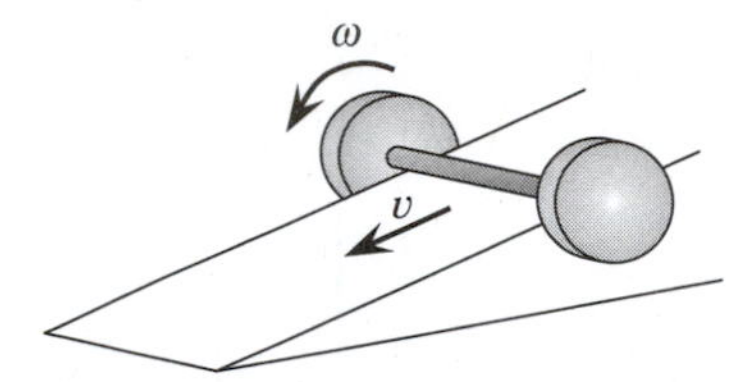

Figure 13–16: *A dowel with a wheel on each end rolls down an incline.*

12. Use a conservation of energy argument to explain why a comet has its greatest velocity at its closest approach to the sun.

13. Figure 13–14 represents some different quantities plotted against time for a stone thrown vertically into the air. Which graph best represents the:
    A. kinetic energy of the stone?
    B. potential energy of the stone?
    C. total energy of the stone?
    D. acceleration of the stone?

Section 13.5:

14. A 1.0-lb weight is hung from a string wrapped around pulley A. A second 1.0-lb weight is hung from a string wrapped around pulley B (see Figure 13–15). The two pulleys have the same radius and mass. Pulley A is a solid disk; pulley B has most of its mass on the rim. The two weights are released. They fall through the same distance of 5.0 ft.
    A. Which weight will have the larger linear speed?
    B. Which pulley will be spinning faster?
    C. Which pulley will have the larger rotational kinetic energy?
    D. Compare the total energy of the two systems.

15. A dowel has a larger diameter wheel on each end. The dowel rests on a central rail and rolls at a constant speed (see Figure 13–16). At the end of the rail, the wheels make contact with the tabletop. The linear speed of the system increases. Why?

16. An elevator is counterbalanced by a weight equal to the weight of the elevator car. Use conservation of energy arguments to show that less work is done in lifting the counterbalanced car than would be done to lift an identical car with no counterbalance.

17. Two identical cars approach each other with the same speed relative to the highway. They come to rest in a perfectly inelastic collision. Show that the mechanical energy change for either car is no more than if it had run into a concrete bridge abutment.

18. Is *mechanical* energy conserved only in perfectly elastic collisions? Is *total* energy conserved in other kinds of collisions?

19. A light passenger car traveling at $4\bar{0}$ m/s has a perfectly inelastic collision with a heavier truck with an initial velocity of $-4\bar{0}$ m/s. After the collision they both travel at $-2\bar{0}$ m/s. Equal and opposite forces act on the vehicles. The car undergoes a smaller change of kinetic energy than the truck. The two vehicles undergo the same change of momentum. Why is the driver in the light car more likely to be injured?

PROBLEMS

Figure 13–17: *A pulley that is a composite of six rods rotated around their ends and a hoop rotated around its center.*

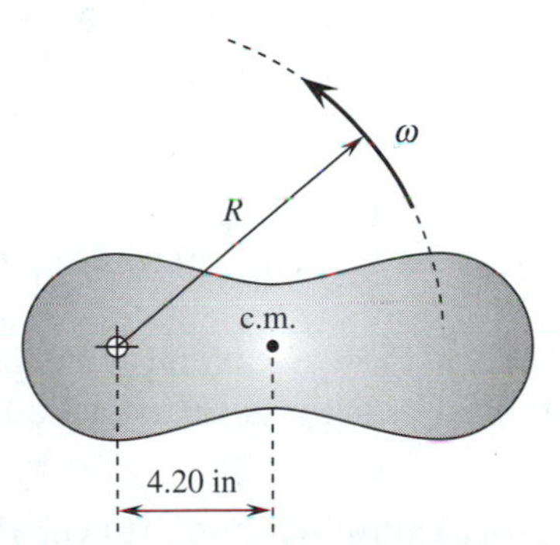

Figure 13–18.: *Diagram for Problem 5. A machine part is rotated 4.20 in from its center of mass.*

Section 13.1:

1. A mounted tire has a moment of inertia of 0.831 slug · ft^2. Calculate its rotational kinetic energy when it is spinning at:
   A. 230 rpm.
   B. 5.3 rev/s.
   C. 37.8 rad/s.

2. A flywheel accelerates from 54 rad/s to 63 rad/s. What is the change of rotational kinetic energy if its moment of inertia is 0.344 kg · m^2?

*3. A roller in a sheet-metal processing plant has a diameter of 1.34 m and a mass of 6.5 metric tons (1 metric ton = 1000 kg). Assume the roller is a solid cylinder.
   A. What is the moment of inertia of the roller?
   B. What is its kinetic energy when it spins at 3.54 rad/s?

**4. A pulley has a circular steel rim with a diameter of 8.22 in and a weight of 2.70 lb. It is supported by six uniform spokes. Each spoke has a mass of 0.320 lb (see Figure 13–17). How much kinetic energy is stored in the pulley when it rotates at 612 rpm?

Section 13.2:

5. The machine part shown in Figure 13–18 is used to convert rotational motion to linear motion in a large lift pump. Its moment of inertia around the center of gravity is 0.136 slug · ft^2. Its weight is 9.70 lb, and its axle is mounted 4.20 in from the center of mass. What is its moment of inertia around the axle?

6. What is the moment of inertia around a point halfway between the center of a solid disk and its rim? The disk has a mass of 1.42 kg and a diameter of 17.8 cm.

Section 13.4:

7. A 3.2-lb weight is fastened to a spring with a stiffness of 23.2 lb/ft. The spring is stretched horizontally 4.4 in from the equilibrium position and then released. What is the velocity of the weight as it passes through the equilibrium point? Assume the surface it slides along is horizontal and has negligible friction.

8. A 236-kg satellite has an altitude of 89 km and an orbital speed of 7.88 km/s. What is the total mechanical energy of the satellite relative to the earth's surface? (Earth radius = 6.37×10^3 km; earth mass = 5.99×10^{24} km.)

9. A 3820-lb car initially at rest coasts down a hill $20\bar{0}$ ft high.
   A. What is its kinetic energy at the bottom of the hill?
   B. What is its velocity?

10. The roller coaster shown in Figure 13–19 has a speed of 16.2 m/s as it passes point A. By the time it reaches point B, 10,000 ft · lb of energy have been converted to heat by frictional forces. What is the car's speed at point B if the coaster car weighs 1730 lb?

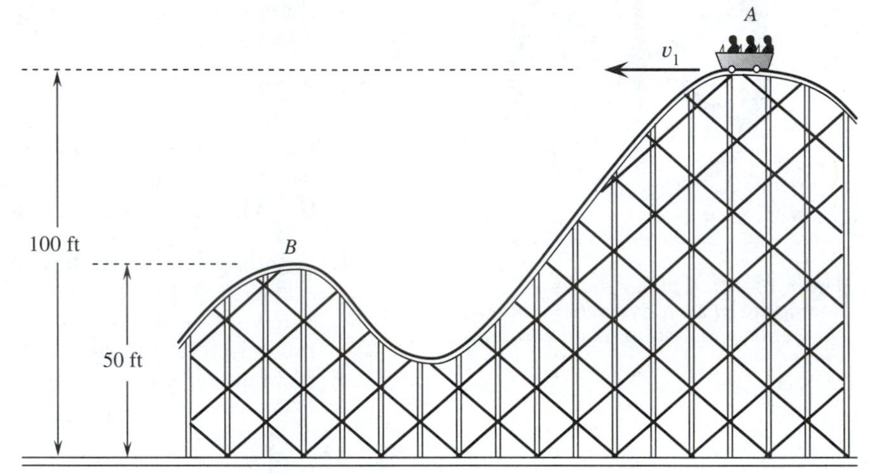

Figure 13–19: *A roller coaster car coasts from point A to point B. The car's potential and kinetic energy change. Friction produces heat.*

11. A derrick lifts a 230-kg wrecking ball 4.30 m above a roadbed.
    A. What is the potential energy of the ball relative to the road?
    B. When the ball has fallen to a height of 1.95 m, what percentage of the initial potential energy is transformed to kinetic energy?

12. A 24-lb cement block slides down a ramp 16 ft high. Its velocity at the bottom is 18.7 ft/s. How much energy was converted to heat?

*13. A simple pendulum is made from a steel bob and a 1.00 m long piece of string. The bob is pulled back through an angle of 35° and then released. What is its speed at the lowest point of the swing? See Figure 13–20.

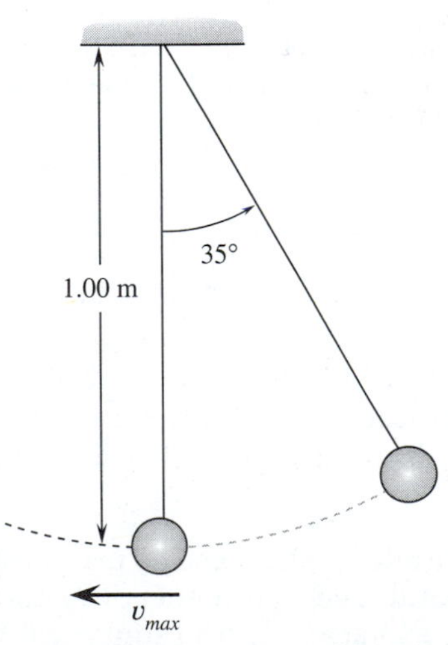

Figure 13–20: *Diagram for Problem 13.*

Section 13.5:

14. A mounted truck tire with a moment of inertia of 1.26 kg · m^2 and a mass of 28.1 kg rolls from rest down an incline 4.66 m high. The tire's radius is 30.0 cm. Assume no frictional losses. What is the speed of the tire at the bottom of the hill?

15. A hula hoop with a weight of 0.87 lb and a diameter of 30.4 in rolls with a linear speed of 15 ft/s. What is its
   A. linear kinetic energy?
   B. rotational kinetic energy?
   C. total kinetic energy?

16. A string is wrapped around a pulley with a moment of inertia of $0.023\ \text{kg}\cdot\text{m}^2$. On the free end of the string, a 50-g mass is attached. When the mass has fallen 1.2 m, what is the rotational speed of the pulley?

17. A 12-lb, 10 in diameter bowling ball has a velocity of 17 ft/s. What is its kinetic energy?

18. A flywheel with a moment of inertia of $1.48\ \text{kg}\cdot\text{m}^2$ is spinning freely at 2460 rpm. It is suddenly connected to a drive shaft by a clutching system. The moment of inertia of this second assembly is $0.94\ \text{kg}\cdot\text{m}^2$. The final angular speed of the combined system is 1500 rpm. How much kinetic energy was converted to heat by frictional forces on the clutch?

*19. Typically, each of the four tires on an automobile has a radius of 14 in and a weight of about 40 lb. What percentage of the kinetic energy of a 3060-lb car traveling at 80.0 ft/s is in the rotational energy of the wheels?

*20. A uniform vertical rod 2.24 m long is hinged at the bottom. When it is released, it falls with a rotational motion. Its center of mass falls 1.12 m (see Figure 13–21). Just before the rod hits the floor:
   A. What is the rotational kinetic energy of the rod?
   B. What is the tangential speed of the center of the rod?
   C. What is the tangential speed of the free end?
   D. If the rod were unhinged and dropped from a horizontal

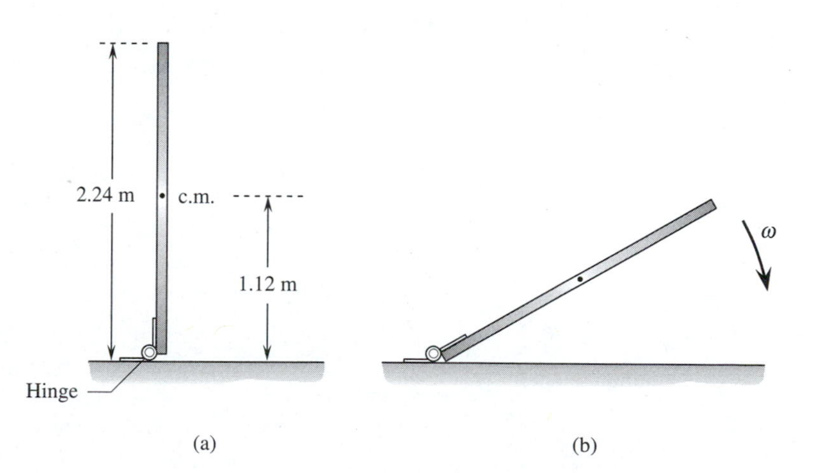

Figure 13–21: Diagram for Problem 20. A hinged rod falls. cm., center of mass.

position from a height of 1.12 m, would the final velocity of the rod be more or less than the tangential velocity of the hinged rod?

Section 13.6:

21. A tennis ball with a coefficient of restitution of 0.78 strikes a wall with a speed of 93 ft/s. With what speed will it rebound?

22. A rubber ball with a coefficient of restitution of 0.83 is dropped from a height of 8.24 ft. How high will it bounce?

23. A 3240-lb car initially traveling at 68 ft/s has a head-on collision with a 6780-lb pickup truck traveling at -76 ft/s. After the collision, the two vehicles stick together.

A. What is the final velocity of the vehicles right after the collision?

B. How much kinetic energy is converted to heat?

24. A $23\bar{0}$-g glider on an air track traveling at 34 cm/s strikes a 115-g glider at rest. If the collision is perfectly elastic, find the final velocities.

***25.** A 56-g mass traveling at 8.9 m/s has a perfectly elastic head-on collision with a 78-g mass at rest. What are the final velocities of the two masses?

***26.** A 27.4-g arrow traveling horizontally at 34.0 m/s strikes a stationary 1.08-kg block resting on a level floor. The sliding coefficient between the floor and the block is 0.30.

A. What is the velocity of the arrow and block after the collision?

B. How much kinetic energy do they have immediately after the collision?

C. How far will the block slide along the floor before coming to rest?

Chapter 14

MACHINES

This backhoe is not 100% efficient—no machine is. But it can accomplish this task much more efficiently than people with shovels could. It uses simple machines to gain a mechanical advantage. (Photograph courtesy of Deere and Company, Moline, IL.)

OBJECTIVES

In this chapter you will learn:

- that machines obey the law of conservation of energy
- how to calculate the efficiency of a machine
- how to find the ideal mechanical advantage and the actual mechanical advantage of a machine
- how to analyze machines containing an inclined plane
- how to analyze machines containing levers
- how to analyze machines containing pulleys
- how to calculate the ideal mechanical advantage and efficiency of composite systems
- how to find the gear ratios, speed ratios, and torque ratios for power drives

Time Line for Chapter 14

2,400,000 B.C.	Early humans develop stone tools.
25,000 B.C.	The bow and arrow are invented.
5000 B.C.	Sailing ships are used in Mesopotamia.
3500 B.C.	Wheeled carts are used in Sumer.
420 B.C.	Archytas of Tarentum develops theory of the pulley.
250 B.C.	Archimedes develops the principle of the lever and other simple machines.
250 B.C.	Philon of Byzantium designs a chain drive.
200 B.C.	The use of gears leads to waterwheels for irrigation.
100 B.C.	The Chinese invent the crank handle.
7 B.C.	Ko Yu invents the wheelbarrow.
1648	John Wilkins writes a book on the basic principles of machines.
1656	Christian Huygens develops a pendulum that keeps accurate time.
1702	A steam engine is invented that pumps water out of coal mines.

An athlete can generate about 0.20 hp for a period of a few minutes. With much exertion a weight lifter can press 250 lb. Most of us are not capable of accomplishing these feats unaided. By using machines we can multiply the force that humans can exert. With a block and tackle, a single person can lift a 450-lb motor from a car. Machines can also make things move faster. The gearing system in a bicycle makes it possible to pedal faster than we can run. Machines can use different kinds of energy. A gasoline engine uses chemical potential energy to do work. An electric motor may use hydroelectric energy. Machines can be used to move heat from one place to another. For example, in a refrigerator heat is moved from the inside of an insulated box to the outside.

In this chapter we will examine how machines perform mechanical work. We will look at some different kinds of simple machines and find out how they work. Machines that transfer heat will be discussed in a later chapter.

14.1 EFFICIENCY

Machines obey the law of conservation of energy. No more work can be obtained from a machine than is put into it. Much of the chemical potential energy of an automobile engine's fuel is converted to heat; some of it is converted to useful mechanical work. No more work can be done by the engine than is supplied by the fuel. A block and tackle can lift a heavy load by exerting a larger force on the load than the operator puts into the system, but the operator exerts

In Search of Time

We can easily measure acceleration and velocity in the physics laboratory. We can use stop clocks with a precision of $\frac{1}{100}$ second. We may elect to use strobe lights to freeze time and distance on photographs. Another choice we have is to use heat tapes and 60 Hz sparks from a high-voltage transformer.

If it's so easy to measure motion, why did it take thousands of years for scientists to unravel the puzzle of motion? Aristotle knew falling objects accelerated, but had only a hazy idea of the nature of acceleration. Galileo was able to find that the distance traveled by a ball rolling down an inclined plane was proportional to time squared. Isaac Newton came up with his laws of motion and general gravitational force without being able to measure gravitational acceleration. George Atwood was the first to find the acceleration of gravity in 1784.

Early scientists were hampered in their studies of motion because they had no good way to measure time. Many advances in science and technology have come with parallel advances in timing devices.

The earliest timing devices were sundials. Then came hourglasses and water clocks. In a water clock water flowed from one container into a lower container. In some a float in the lower vessel moved an indicator to show how much time had elapsed. One of the problems with these devices was that the flow rate of water was not constant.

During the Middle Ages the first mechanical clocks appeared. A falling weight turned a toothed gear. The rate of rotation was slowed down by an escapement. This was a special pawl that oscillated between two gears, allowing the gears to advance by only one tooth as it oscillated through one back-and-forth motion. This gives the tick to a clock. The movement of the escapement was not uniform. Clocks had only a single hand to indicate the hour.

Galileo added a pendulum to the clock to make the escapement move at a more regular pace, but the simple pendulum he and his son used was not completely regular. In 1656 Christian Huygens developed a more complicated pendulum that kept accurate time. A minute hand was added to the clock.

A pendulum must swing on a vertical plane. It cannot be moved around easily. Eventually coil springs replaced the falling weights to power clocks. The pendulum was replaced by a balance wheel, which is a torsion pendulum. The balance wheel didn't use gravity to determine its period. This meant the clock could lie in any position. Easily portable clocks, or watches, were created. Eventually a second hand was possible.

This is why two thousand years passed between the time Aristotle recognized that falling bodies accelerate and George Atwood was able to measure the acceleration of gravity using a pulley, counterweights, and a state-of-the-art clock.

---

a force through a larger distance. The input force must overcome friction as well as lift the load. Less mechanical work is done by the block and tackle than is put into the machine.

$$\textit{input work (or energy)} = \textit{useful mechanical work} + \textit{heat}$$

$$W_{in} = W_{out} + Q \qquad \textbf{(Eq. 14–1)}$$

Since frictional forces are nearly always present, the work done by a machine is less than the work put into the machine. We can estimate how well the machine converts input work or energy into useful mechanical work by dividing the work out (W_{out}) by the work put into the system (W_{in}). This ratio is called the **efficiency** of the machine.

$$eff = \frac{W_{out}}{W_{in}} \qquad \textbf{(Eq. 14–2)}$$

Often the efficiency is expressed in percentage.

$$\%\ eff = \frac{W_{out}}{W_{in}} \times 100 \qquad \textbf{(Eq. 14–2A)}$$

☐ **EXAMPLE PROBLEM 14–1: THE GASOLINE ENGINE**

A gasoline engine converts chemical potential energy into useful work. A particular engine performs 1.2×10^7 ft · lb of useful mechanical work for each gallon of gasoline. What is its efficiency?

■ ***SOLUTION***

One gallon of gasoline contains 9.5×10^7 ft · lb of potential energy.

$$eff = \frac{W_{out}}{W_{in}}$$

$$eff = \frac{1.2 \times 10^7 \text{ ft} \cdot \text{lb}}{9.5 \times 10^7 \text{ ft} \cdot \text{lb}}$$

$$eff = 0.13$$

Notice that efficiency has no units. It is a ratio of work divided by work.

☐ **EXAMPLE PROBLEM 14–2: THE BLOCK AND TACKLE**

A mechanic pulls a 236-kg motor using a block and tackle. The mechanic pulls 6.70 m of rope through the block and tackle to lift the motor 83 cm. If the input force exerted by the mechanic is $42\overline{0}$ N, what is the percentage efficiency of the block and tackle?

Figure 14–1: *An automobile mechanic uses a block and tackle to pull a motor.*

■ ***SOLUTION***

Letting F = force and s distance, the work done by the mechanic is:

$$W_{in} = F_{in}\, s_{in}$$

$$W_{in} = 42\bar{0}\ \text{N} \times 6.70\ \text{m}$$

$$W_{in} = 2810\ \text{J}$$

The machine lifts a weight a distance h. This is the useful mechanical output work.

$$W_{out} = F_{out}\, s_{out} = m\, g\, h$$

$$W_{out} = 236\ \text{kg} \times 9.8\ \text{m/s}^2 \times 0.83\ \text{m}$$

$$W_{out} = 1920\ \text{J}$$

$$\%\, eff = \frac{W_{out}}{W_{in}} \times 100$$

$$\%\, eff = \frac{1920\ \text{J}}{2810\ \text{J}} \times 100$$

$$\%\, eff = 68\%$$

14.2 MECHANICAL ADVANTAGE

One of the advantages of using a machine is that it can multiply the input force. As in Example Problem 14–2, the output force can be larger than the effort put into the device. Remember part of the input force is doing work against friction; the output force is the force that does useful mechanical work.

$$W_{in} = W_{out} + Q$$

$$\mathbf{F}_{in}\, s_{in} = \mathbf{F}_{out}\, s_{out} + Q$$

A ratio of the output force **(resistance)** to the input force **(effort)** measures the actual multiplication of force. This ratio is called the **actual mechanical advantage (*AMA*).**

$$AMA = \frac{\mathbf{F}_{out}}{\mathbf{F}_{in}} \qquad \textbf{(Eq. 14–3)}$$

If we lubricate the machine, we can reduce the frictional forces. Less effort will be needed to overcome friction. The denominator in the *AMA* ratio ($\mathbf{F}_{in}$) will get smaller, and the actual mechanical advantage will increase. If we could eliminate all of the frictional forces, we would have a 100% efficient machine. All of the work put into the device would go into mechanical work.

$$W_{in} = W_{out} \qquad \text{(no friction)}$$

$$\mathbf{F}_{in}\, s_{in} = \mathbf{F}_{out}\, s_{out}$$

In this ideal situation, the ratio of force-out to force-in can be expressed in terms of distance. We can see this by rearranging the work equation.

$$\frac{\mathbf{F}_{out}}{\mathbf{F}_{in}} = \frac{s_{in}}{s_{out}}$$

This is the highest ratio of forces we could theoretically obtain if there were no friction in the machine. The ratio of the distance the input force must travel divided by the distance the output force travels is called the **ideal mechanical advantage (*IMA*).**

$$IMA = \frac{s_{in}}{s_{out}} \qquad \textbf{(Eq. 14–4)}$$

The *IMA* depends only on the geometry of the machine, not on the actual forces. Later, we will see that the *IMA* often can be determined by simply looking at the design of a machine.

Why Use Machines?

No machine is 100% efficient. We get less work out of a machine than we put into it. Why do we use machines if we lose energy?

A simple machine such as a lever gives us two possible advantages. One advantage is a multiplication of force. If we make the input distance greater than the output distance, we can exert a small force to move a heavy object. Archimedes bragged, "Show me a place to stand and I'll move the world."

Another advantage of a simple machine is speed. If we shorten the input distance on our lever, we can move something faster. A shovel is an example of this kind of lever. When we throw a shovelful of gravel, the end of the handle becomes the pivot point. The hand slides up the handle to shorten the input distance to give the blade an increased speed.

Machines let you use something other than your own muscle power to do work for you. Some early machines, such as plows, mowers, and grain mills, traded human muscle power for beast muscle power. But oxen and horses needed to be fed, so more work had to be done to raise crops and tend the animals.

A better idea is to use solar power to do work. Most of the energy that we use to run machines comes from the sun. Wind and running water are created by solar energy. Windmills and waterwheels are simple devices that tap solar energy. Kinetic energy does the work.

(Photograph courtesy of Deere and Company, Moline, IL.)

Steam engines were the first machines to use solar energy in the form of heat. The solar energy stored in fossil fuels and biomass is used to create heat. The heat creates steam. The steam does work on a piston. The first practical use of a heat machine was the "Miner's Friend," a steam engine invented in 1702 to pump water out of coal mines. Most modern engines use electricity or fossil fuels. These engines also convert solar energy into usable work.

So why do we use machines? To multiply effort or speed, and to harness energy from the sun.

The function of a machine is related to the IMA. Usually the following is true:

If $IMA > 1$, then either the force or torque is increased.

If $IMA = 1$, then power is transferred.

If $IMA < 1$, then either speed or distance is increased.

The efficiency of a device can be found from the IMA and AMA.

$$\frac{AMA}{IMA} = \frac{F_{out}/F_{in}}{s_{in}/s_{out}} = \frac{F_{out}\, s_{out}}{F_{in}\, s_{in}} = \frac{W_{out}}{W_{in}}$$

$$eff = \frac{AMA}{IMA} \qquad \textbf{(Eq. 14–5)}$$

☐ **EXAMPLE PROBLEM 14–3: THE BLOCK AND TACKLE REVISITED**

Calculate the *AMA* and *IMA* of the block and tackle system given in Example Problem 14–2. Check to see whether the efficiency is equal to the ratio of the *AMA* and *IMA*.

■ ***SOLUTION***

The force-in is $42\bar{0}$ N. The force-out is m g.

$$AMA = \frac{F_{out}}{F_{in}} \qquad IMA = \frac{s_{in}}{s_{out}}$$

$$AMA = \frac{236 \text{ kg} \times 9.80 \text{ m/s}^2}{42\bar{0} \text{ N}} \qquad IMA = \frac{6.70 \text{ m}}{0.83 \text{ m}}$$

$$AMA = 5.51 \qquad IMA = 8.1$$

$$eff = \frac{AMA}{IMA}$$

$$eff = \frac{5.51}{8.1}$$

$$eff = 0.68$$

14.3 THE INCLINED PLANE

The Basic Incline

The inclined plane is perhaps the simplest of all machines (see Figure 14–2). A force exerted parallel to the incline lifts a weight a vertical distance h.

The load travels the length of the incline (s_{in}). The load is lifted a distance h (s_{out}) against gravity, giving it potential energy. The *IMA* is:

$$IMA = \frac{\text{length of incline}}{\text{height of incline}}$$

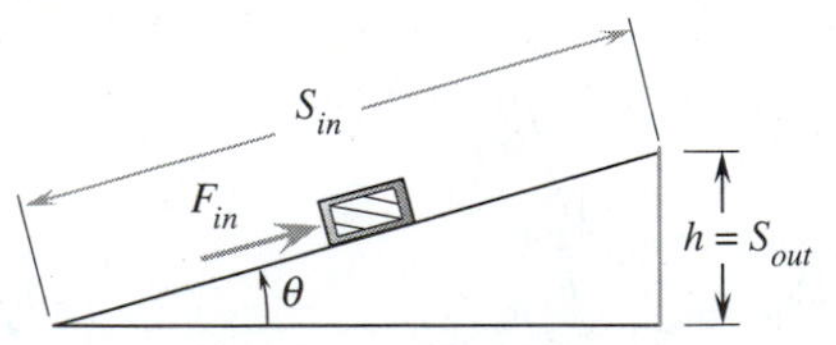

Figure 14–2: An inclined plane is used to lift a weight a vertical distance (h) by applying a force parallel to the incline.

Notice, if the angle of the incline is θ, then the *IMA* is the reciprocal of the sine function.

$$IMA = \frac{1}{\sin \theta}$$

EXAMPLE PROBLEM 14–4: AN INCLINE

A parallel force of 53.0 lb is needed to push a 162-lb crate up a 12.0° incline. What is the efficiency of the incline?

SOLUTION

$$IMA = \frac{1}{\sin 12.0°} \qquad AMA = \frac{F_{out}}{F_{in}}$$

$$IMA = \frac{1}{0.2079} = 4.81 \qquad AMA = \frac{162 \text{ lb}}{53.0 \text{ lb}} = 3.06$$

$$eff = \frac{AMA}{IMA} = \frac{3.06}{4.81}$$

$$eff = 0.64$$

The Wedge

One variation of the inclined plane is the wedge (see Figure 14–3). A wedge is used to separate a material into two portions. A vertical force on the base of the wedge drives the wedge into the material. The horizontal component of force exerted by the wedge on the material pushes the material apart.

The ideal mechanical advantage is the ratio of the length of the wedge (L) divided by the base thickness of the wedge.

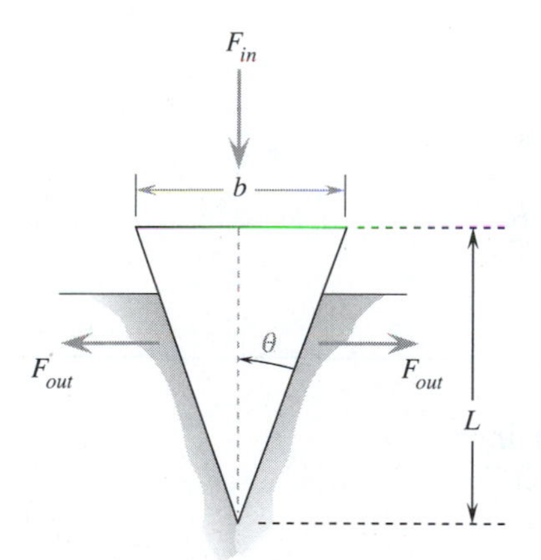

Figure 14–3: A wedge is driven into a material, separating it into two parts. The IMA is base width (b) times the length of the wedge (L).

$$IMA = \frac{s_{in}}{s_{out}} = \frac{L}{b}$$

The wedge is usually formed from two inclined planes forming a symmetric triangle. The IMA can be expressed in terms of the inclined plane angle. The "height" of one incline is $b/2$:

$$\tan \theta = \frac{b/2}{L}$$

$$IMA = \frac{L}{b} = \frac{1}{2 \tan \theta}$$

The AMA for a wedge is usually much smaller than the IMA. The material squeezes the wedge. Large frictional forces keep it from popping out of the material. Figure 14–4 shows some of the many tools that have the form of a wedge.

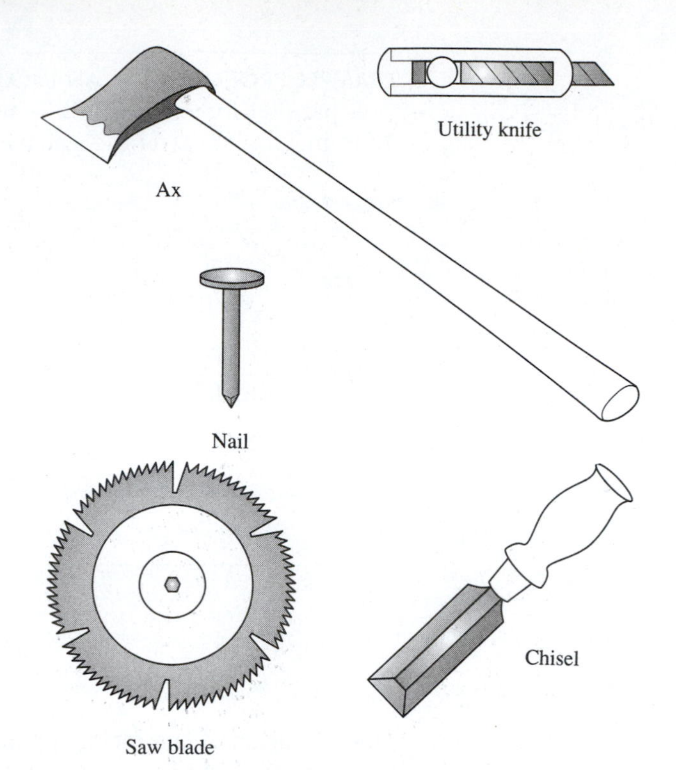

Figure 14–4: *Different kinds of wedges. A nail has a wedge point. Most cutting tools have wedge-shaped edges.*

The Screw

A screw is an inclined plane wrapped around a cylinder. The **pitch** of the screw is the distance between adjacent thread loops of the incline (see Figure 14–5). During one rotation, the screw advances forward a distance equal to the pitch (s_{out}).

s_{in} is the circumference of the cylinder.

$$IMA = \frac{2\,\pi\,R}{p}$$

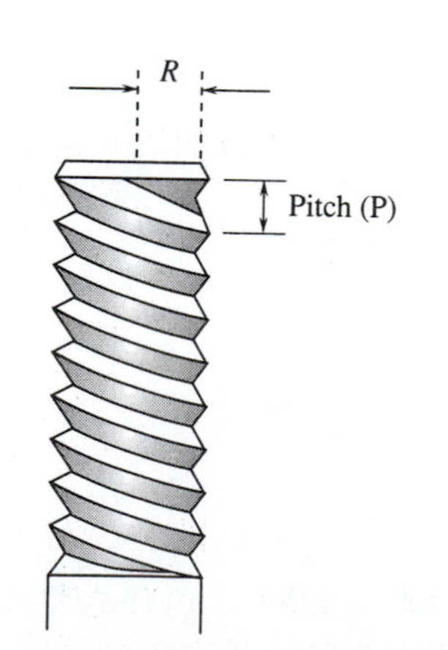

Figure 14–5: *A screw is an inclined plane wrapped around a cylinder. The pitch is the separation of adjacent threads.*

□ **EXAMPLE PROBLEM 14–5: A BOLT**

A $\frac{1}{4}$-in bolt has 12 threads per inch.

A. What is the pitch of the bolt?

B. What is its *IMA?*

■ ***SOLUTION***

A.

$$p = \frac{1.00\text{ in}}{12\text{ threads}}$$

$$p = 0.0833\text{ in}$$

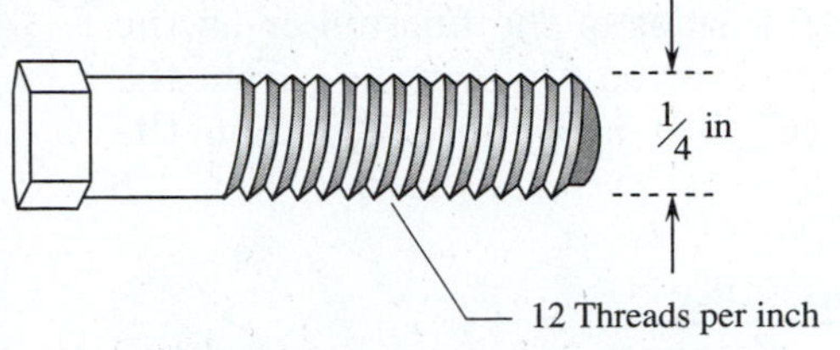

Figure 14–6: *A ¼-in machine bolt. The bolt size refers to the diameter of the cylinder, not the radius.*

B.

$$IMA = \frac{2\pi R}{p}$$

$$IMA = \frac{2\pi \times 0.250 \text{ in}}{0.0833 \text{ in}}$$

$$IMA = 18.8$$

14.4 THE LEVER

The Basic Lever

A pair of pliers, a wheelbarrow, scissors, and a wrecking bar have something in common. They all are levers. A lever uses two forces around a point of rotation, called a **fulcrum,** to generate a mechanical advantage (see Figure 14–7).

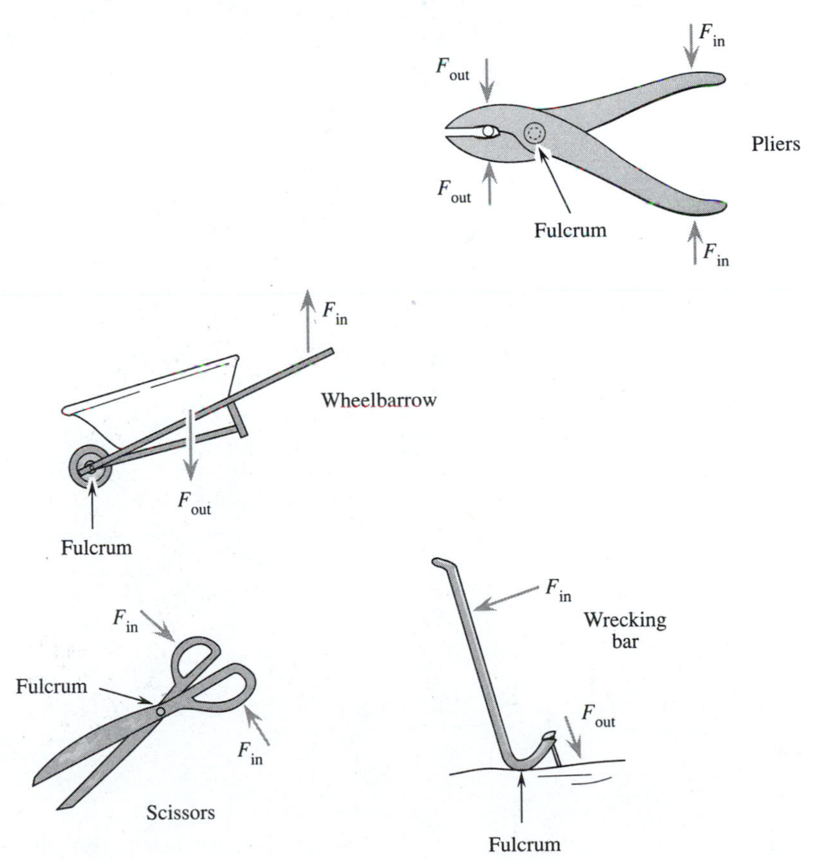

Figure 14–7: *An assortment of levers. Most lever-shaped tools multiply the input force.*

The two forces may be on opposite sides of the fulcrum or on the same side. An effort force ($\mathbf{F}_{in}$) is exerted a distance r_{in} from the fulcrum. The resistance force ($\mathbf{F}_{out}$) is a distance r_{out} from the fulcrum.

$$IMA = \frac{\textit{effort distance}}{\textit{resistance distance}}$$

$$IMA = \frac{r_{in}}{r_{out}}$$

Because levers involve rotation around a fixed point rather than sliding surfaces, friction is low, and the efficiencies are high. The *AMA* is nearly the same as the *IMA*.

$$F_{in} \times r_{in} = F_{out} \times r_{out}$$

You may recognize this as the equation for the equilibrium of torques (see Chapter 9).

$$\tau_{in} = \tau_{out}$$

EXAMPLE PROBLEM 14–6: THE BRUSH HOOK

An apple tree is trimmed with a brush hook (see Figure 14–8). A branch is between the blades 2.1 from the hinge. A force of 63 lb is exerted on the handle 19 in away from the fulcrum. Estimate the size of the force that the blades exert on the branch.

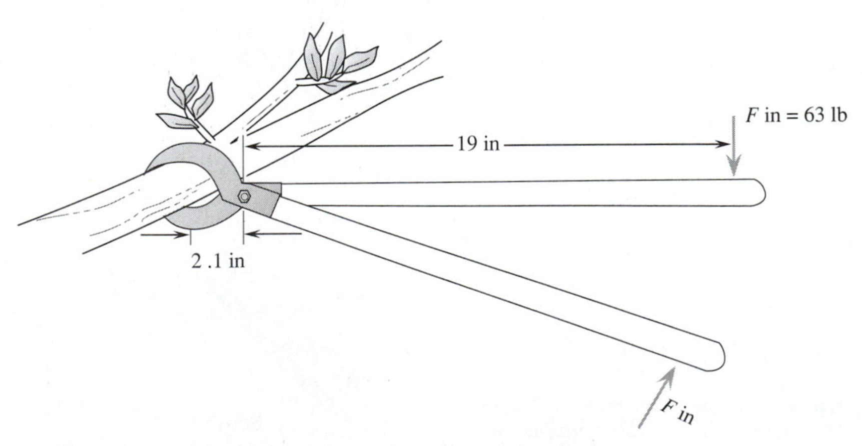

Figure 14–8: A brush hook is used to prune a tree limb.

■ *SOLUTION*

$$F_{out} \times r_{out} = F_{in} \times r_{in}$$

$$F_{out} = \frac{63 \text{ lb} \times 19 \text{ in}}{2.1 \text{ in}}$$

$$F_{out} = 570 \text{ lb}$$

The Wheel and Axle

Figure 14–9 shows a wheel and axle. During one rotation, the effort force travels through a distance $2\ \pi\ r_{in}$. The resisting force travels through a distance of $2\ \pi\ r_{out}$.

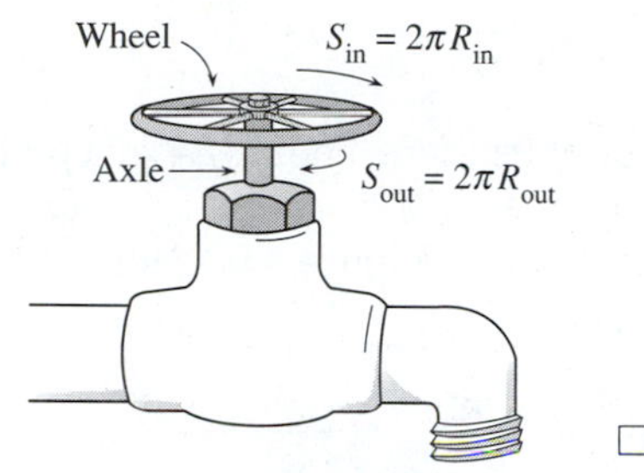

Figure 14–9: *A wheel and axle on a water tap. A force exerted on the rim of the wheel travels through a larger circle than the resisting force on the axle.*

$$IMA = \frac{2\ \pi\ r_{in}}{2\ \pi\ r_{out}}$$

$$IMA = \frac{r_{in}}{r_{out}}$$

□ **EXAMPLE PROBLEM 14–7: THE TORQUE WRENCH**

A bolt with a diameter of 0.95 cm is to be tightened by a torque wrench to a setting of 19.0 N · m. What is the resisting force of the bolt at this setting?

■ *SOLUTION*

$$\tau_{out} = \tau_{in}$$

$$F_{out}\, s_{out} = \tau_{in}$$

$$F_{out} = \frac{\tau_{in}}{s_{out}}$$

$$F_{out} = \frac{19.0 \text{ N} \cdot \text{m}}{9.5 \times 10^{-3} \text{ m}}$$

$$F_{out} = 2.0 \times 10^{3} \text{ N}$$

14.5 PULLEY SYSTEMS

Figure 14–10 shows a simple pulley system. A load is attached to the pulley. The pulley is supported by two strands of rope. Each strand supports half the weight. If there is no friction, someone can

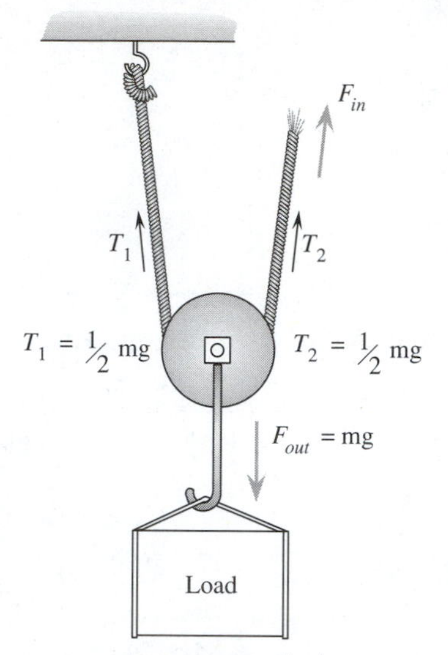

Figure 14–10: *Each strand of rope attached to the load pulley supports half the weight of the load.*

lift the load by pulling on the free end of the rope with a force equal to only half the weight of the load. However, as the rope is pulled through the pulley, both strands of rope must be shortened to lift the load. Each strand will be shortened by half the input distance. A person pulls with a force equal to only half, but the load is lifted only half the input distance. Work in, equals work out.

If there are three strands connected to the load pulley, each supports one-third of the load and each must be shortened by one-third the input distance to lift the load (see Figure 14–11).

In general, the *IMA* of a pulley system is equal to the number of strands (n) connected to the **load pulley** (the pulley to which the load is attached).

$$IMA = n$$

Usually, pulley systems have a lot of friction. The *AMA* of a pulley system will be substantially less than the *IMA*.

Care should be taken in counting strands. Figure 14–12 shows two

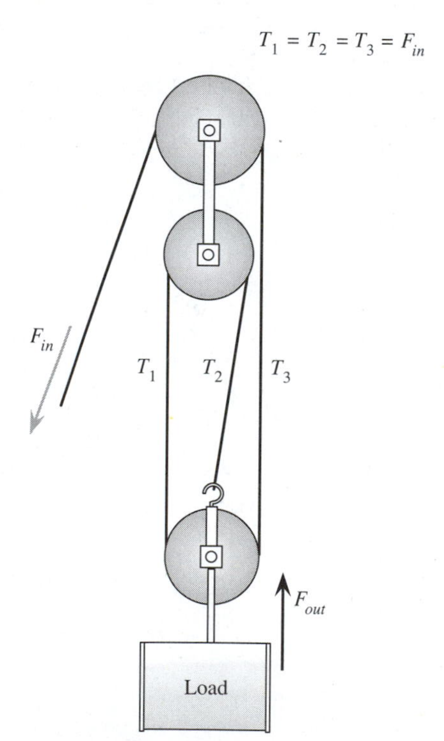

Figure 14–11: *The load is supported by three strands of rope. If there were no friction, F_{in} would equal one-third of the load.*

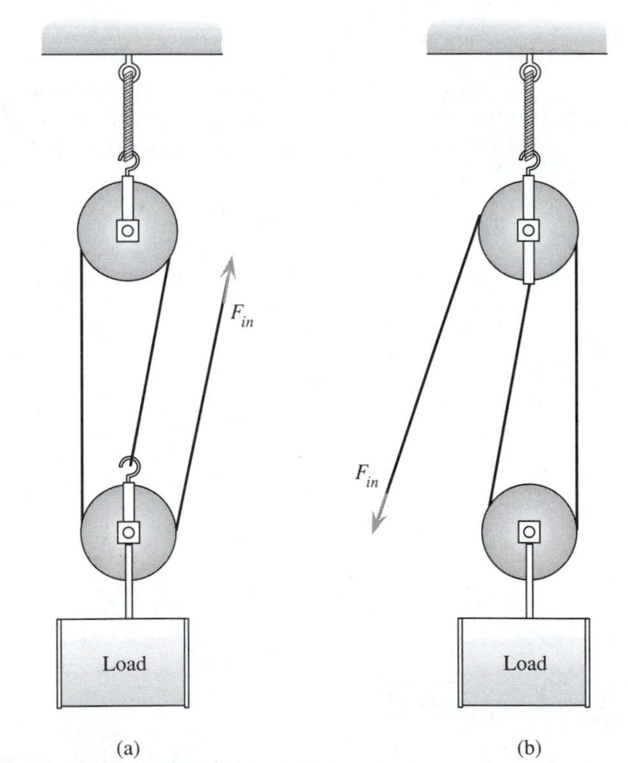

Figure 14–12: *Two pulleys with different* IMAs. *(a) The load is supported by three strands;* IMA = 3. *(b) The load is supported by only two strands;* IMA = 2.

pulley systems that look very much alike. System A has an *IMA* of 3. In system B, only two of the strands are connected to the load pulley. Its *IMA* is 2.

14.6 COMPOSITE SYSTEMS

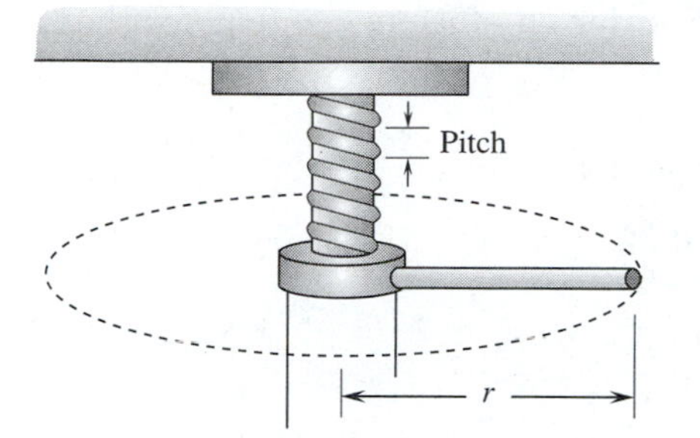

Figure 14–13: *A jack post is a composite machine: a wheel and axle plus a screw.*

Many systems are composites of the kinds of simpler machines we have already discussed. They are often called **compound machines.** For instance, a jack post is a hybrid of a wheel and axle mated with a screw. A steel lever arm is used to rotate a threaded column (see Figure 14–13). During one rotation, the handle goes through a distance (s_{in}) of $2\ \pi\ r_l$, where r_l is the lever arm of the handle. The threaded column advances by the amount of the thread pitch (p). The overall *IMA* is then

$$IMA = \frac{2\ \pi\ r_l}{p}$$

Here is another way to look at the same system. A force exerted on the handle of the lever rotates the threaded column. For one rotation, the force on the lever travels through a distance $2\ \pi\ r_l$ causing the column to rotate through a distance $2\ \pi\ r_c$, and the column advances by the separation of the threads. The output distance of the first system becomes the input distance for the second system.

$$IMA = \frac{2\ \pi\ r_l}{2\ \pi\ r_c} \times \frac{2\ \pi\ r_c}{p}$$

$$IMA = IMA_1 \times IMA_2$$

In composite systems the total *IMA* can be found by multiplying the *IMA*s of the different parts of the systems. A similar analysis will show that the *AMA* and efficiencies can be found in the same way.

$$AMA = AMA_1 \times AMA_2 \ldots AMA_n$$

$$eff = eff_1 \times eff_2 \ldots eff_n$$

☐ **EXAMPLE PROBLEM 14–8: A NINE-STRAND SYSTEM**

At a quarry, a 1700-lb slab of marble is dragged up a 23° slope by a block and tackle with nine rope strands connected to the load pulley. The efficiency of the pulley system is 0.63. Forty-eight percent of the work done on the incline is against friction.

A. What is the overall efficiency of the system?
B. What is the overall *IMA* of the pulley system and incline?
C. What minimum force must be exerted on the block and tackle to move the slab?

■ *SOLUTION*

A. Efficiency:

Since 48% of the work on the incline is done against friction, 52% is left to do useful mechanical work, for an efficiency of 0.52 for the incline.

$$eff = eff_1 \times eff_2$$

$$eff = 0.63 \times 0.52$$

$$eff = 0.33$$

B. *IMA:*

$$IMA = IMA_1 \times IMA_2$$

$$IMA = n \times \frac{1}{\sin 23^\circ}$$

$$IMA = 9 \times \frac{1}{0.391}$$

$$IMA = 23$$

C. Input force:

$$AMA = \frac{F_{out}}{F_{in}} = eff \times IMA$$

$$F_{in} = \frac{F_{out}}{eff \times IMA}$$

$$F_{in} = \frac{1700 \text{ lb}}{0.33 \times 23}$$

$$F_{in} = 220 \text{ lb}$$

14.7 POWER DRIVES

Basic Ideas

Since power is the time rate of doing work, power transmission can be analyzed in the same fashion as work. Let us look at efficiency and mechanical advantage for power transmission.

$$eff = \frac{W_{out}}{W_{in}} = \frac{W_{out}/time}{W_{in}/time}$$

$$eff = \frac{power\ out}{power\ in}$$

$$IMA = \frac{s_{in}}{s_{out}} = \frac{s_{in}/time}{s_{out}/time}$$

$$IMA = \frac{v_{in}}{v_{out}}$$

where v = speed.

Most power trains operate on rotational motion using belt drives, chain-and-sprocket drives, or a gearing system. Therefore, we can analyze power trains in terms of rotational motion. Tangential velocity of a gear can be transformed into rotational motion by the equation $v = r\,\omega$, where r is the radius of the gear or pulley. The efficiency is then:

$$eff = \frac{power\ out}{power\ in} = \frac{F_{out}\, v_{out}}{F_{in}\, v_{in}}$$

$$eff = \frac{F_{out}\, r_{out}\, \omega_{out}}{F_{in}\, r_{in}\, \omega_{in}}$$

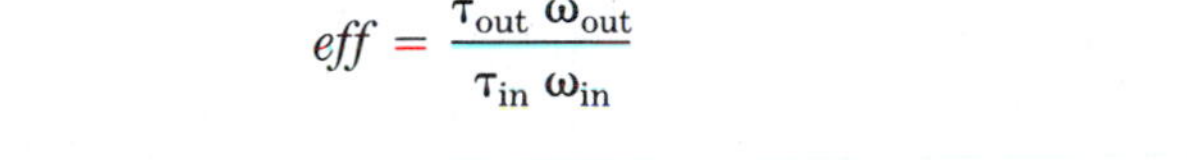

$$eff = \frac{\tau_{out}\, \omega_{out}}{\tau_{in}\, \omega_{in}}$$

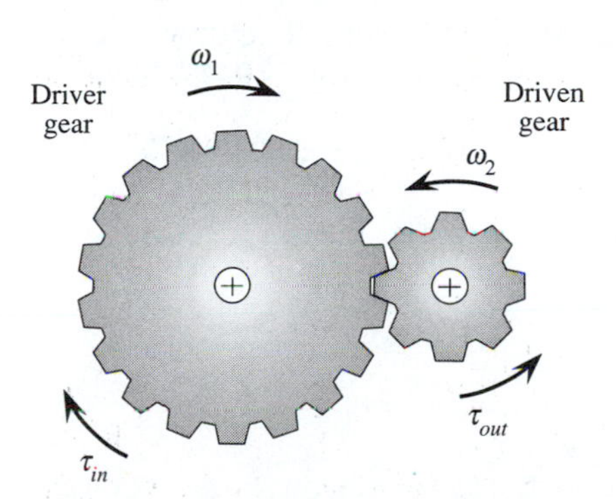

Figure 14–14: *The driver gear does work on the driven gear. The relative rotational speeds of the gears depend on the gear ratio.*

The mechanical advantages can be expressed in terms of rotation.

$$IMA = \frac{\omega_{in}}{\omega_{out}}$$

$$AMA = \frac{\tau_{out}}{\tau_{in}}$$

Gear Drives

Gears are used to transmit rotational power. The input gear that causes the motion is called the **driver gear.** The gear to which the power is transferred is called the **driven gear.** The teeth on the driver gear and those on the driven gear must be the same size so that they will mesh properly. If the distance between two adjacent teeth is d, then the circumference of a gear can be written in terms of a multiple of d. The number of teeth is proportional to the radius of the gear for a fixed value of d (see Figure 14–14). The circumference of a gear with n teeth is:

$$C = 2\pi R = n\,d$$

The angular speed in terms of revolutions per second is:

$$\omega = \frac{2\pi R}{t} = \frac{n\,d}{t}$$

When two rotating gears are meshed, they must have the same tangential speed.

$$v_1 = v_2$$

$$\omega_1 r_1 = \omega_2 r_2$$

This can be rearranged to give us the *IMA* for the gearing system.

$$IMA = \frac{\omega_1}{\omega_2} = \frac{r_2}{r_1} = \frac{n_2}{n_1} = \textit{gear ratio}$$

EXAMPLE PROBLEM 14–9: A GEAR DRIVE

For the gear drive shown in Figure 14–15 a driver gear with eight teeth has an input power of 1.26 kW at 1420 rpm. The driven gear has 18 teeth and a radius of 10.40 cm. The gear system is 94% efficient.

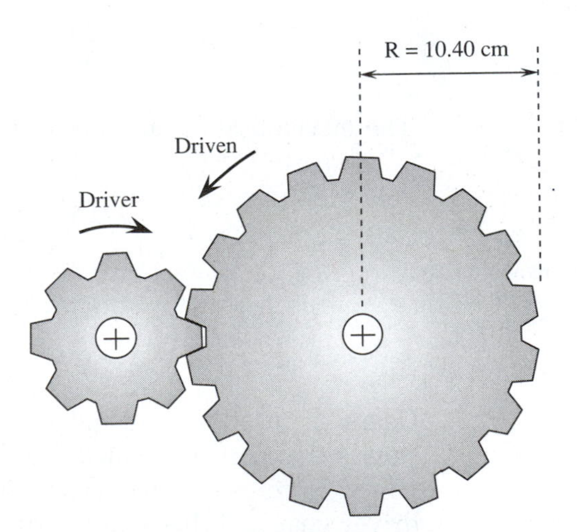

Figure 14–15: *A small driver gear (eight teeth) turns a larger gear (18 teeth). Because the two gears need to have the same size teeth to mesh properly, the gear ratio can be found by counting teeth.*

A. What power does the driven gear transmit to its axle?
B. What is the radius of the driver gear?
C. What is the rotational speed of the driven gear?

D. What is the input torque?
E. What is the output torque?
F. What is the *IMA?*
G. What is the *AMA?*

■ *SOLUTION*

A. Power transmission:

$$\textit{power out} = \textit{power in} \times \textit{efficiency}$$

$$P_{out} = 1.26\ \text{kW} \times 0.94$$

$$P_{out} = 1.2\ \text{kW}$$

B. R_{in}:

$$R_{in} = R_{out} \times \frac{n_{in}}{n_{out}}$$

$$R_{in} = 10.40\ \text{cm} \times \frac{8}{18}$$

$$R_{in} = 4.62\ \text{cm}$$

C. Angular speed of driven gear:

$$\omega_{out} = \omega_{in} \times \frac{n_{in}}{n_{out}}$$

$$\omega_{out} = 1420\ \text{rpm} \times \frac{8}{18}$$

$$\omega_{out} = 631\ \text{rpm} = 66.0\ \text{rad/s}$$

D. Torque in:

$$P_{in} = \tau_{in} \times \omega_{in}$$

$$\tau_{in} = \frac{P_{in}}{\omega_{in}} = \frac{1.26\ \text{kW} \times 1000\ \text{W/kW}}{1420\ \text{rpm} \times 2\pi\ \ \text{rad/rev} \times 1/60\ \text{min/s}}$$

$$\tau_{in} = 8.47\ \text{N} \cdot \text{m}$$

E. Torque out:

$$\tau_{out} = \frac{\text{P}_{out}}{\omega_{out}}$$

$$\tau_{out} = \frac{1.26 \times 10^3\ \text{W}}{66.0\ \text{rad/s}}$$

$$\tau_{out} = 19.1\ \text{N} \cdot \text{m}$$

F. Ideal mechanical advantage:

$$IMA = \frac{n_{\text{in}}}{n_{\text{out}}} = \frac{8}{18}$$

$$IMA = 0.444$$

G. Actual mechanical advantage:

$$AMA = eff \times IMA$$

$$AMA = 0.94 \times 0.444$$

$$AMA = 0.42$$

Belt Drives

Belt drives or chain-and-sprocket drives are analyzed in the same fashion as gear drives. In a belt drive, there is a larger chance for slippage. For this reason their efficiencies are lower than gear drives.

The belt drive shown in Figure 14–16 transmits torque between the two pulleys. Since the diameter of the pulleys is given more often than the radius, the ideal mechanical advantage can be given in terms of the ratio of the diameters. This ratio is sometimes called the **speed ratio.**

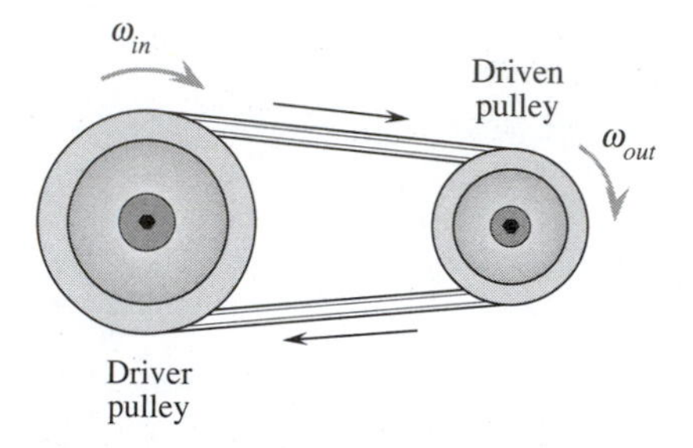

Figure 14–16: *A driver pulley (input) turns a driven pulley (output). The speed ratio is inversely proportional to the diameters of the pulleys.*

$$IMA = \frac{D_{\text{out}}}{D_{\text{in}}} = \frac{\omega_{\text{in}}}{\omega_{\text{out}}} = \textit{speed ratio}$$

□ **EXAMPLE PROBLEM 14–10: A BELT DRIVE**

A variable-speed belt drive with step pulleys is used on a drill press (see Figure 14–17). The belt is set on the 2.2 in diameter level of the driver pulley and on the 10.4 in diameter step of the driven pulley. If the driver pulley spins at 2370 rpm, what is the rotational speed of the driven pulley?

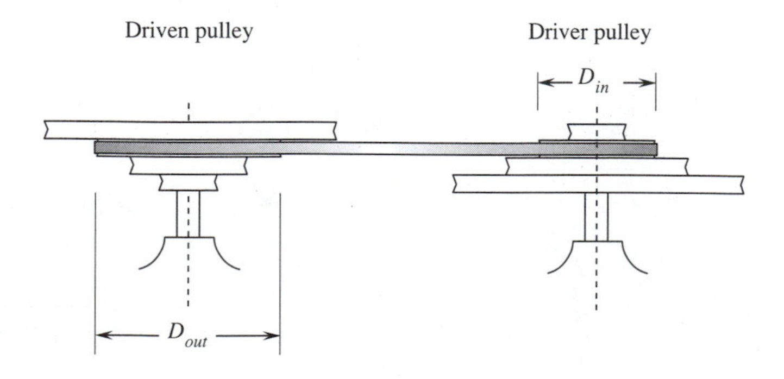

Figure 14–17: *A variable-speed pulley drive. Sets of pulleys have common axes. The speed ratio is changed by slipping the belt from one set of pulleys to another set.*

■ *SOLUTION*

The speed ratio (sr) is:

$$sr = \frac{D_{out}}{D_{in}} = \frac{2.2 \text{ in}}{10.4 \text{ in}}$$

$$sr = 0.21$$

$$\omega_{out} = sr \times \omega_{in}$$

$$\omega_{out} = 0.21 \times 2370 \text{ rpm}$$

$$\omega_{out} = 500 \text{ rpm}$$

SUMMARY

All machines obey the law of conservation of energy: *work in* = *work out* + *heat*. The efficiency of a machine is the fraction of input work that is converted by the machine into useful mechanical work. Efficiency may be expressed as work or power out divided by work or power in.

The input force (effort) does work against friction and produces useful mechanical work. The output force (resistance) is only that part of the force that produces useful mechanical work. The input distance is the distance the input force travels. The output distance is the distance the output force travels.

The actual mechanical advantage (AMA) of a machine is the ratio of force-out to force-in. This is the actual multiplication factor the machine gives to the force. For rotational systems AMA may be expressed in torques.

$$AMA = \frac{F_{out}}{F_{in}} = \frac{\tau_{out}}{\tau_{in}}$$

The ideal mechanical advantage (IMA) is the mechanical advantage the machine would have if there were no friction (*work in* = *work out*). The IMA can be expressed in terms of the geometry of the machine.

$$IMA = \frac{s_{in}}{s_{out}}$$

The efficiency of a machine is equivalent to the ratio of the actual mechanical advantage divided by the ideal mechanical advantage.

$$eff = \frac{AMA}{IMA}$$

For compound machines the efficiencies or mechanical advantages may be multiplied together to get the composite efficiency or mechanical advantages.

$$eff = eff_1 \times eff_2 \times \ldots eff_n$$

$$AMA = AMA_1 \times AMA_2 \times \ldots AMA_n$$

$$IMA = IMA_1 \times IMA_2 \times \ldots IMA_n$$

The *IMA* may have some special expressions for specific kinds of machines. For an inclined plane it is $1/\sin \theta$; for a screw, $\pi D/p$; for a wedge, L/b; a wheel and axle, R_{in}/R_{out}; a pulley system, n (the number of strands attached to the load pulley). For power drives:

$$IMA = \frac{v_{in}}{v_{out}} = \frac{\omega_{in}}{\omega_{out}}$$

Gear ratio for gear drives:

$$IMA = \frac{n_{out}}{n_{in}}$$

Belt drives:

$$IMA = \frac{r_{out}}{r_{in}}$$

KEY TERMS

If you can explain the following terms to a friend or classmate, you understand their meaning. If you cannot explain the terms, you should reread the sections in which they are discussed.

actual mechanical advantage (*AMA*)
compound machine
driven gear
driver gear
efficiency
effort
fulcrum
ideal mechanical advantage (*IMA*)
load pulley
pitch
resistance
speed ratio

EXERCISES

Section 14.1:

1. Can a machine have an efficiency larger than one? Explain.
2. Some devices, such as disk brakes, have efficiencies of zero. Can you think of two other devices that convert mechanical work entirely to heat?

Section 14.2:

3. Some machines and drives have mechanical advantages of less than one. What is gained by such a machine?
4. If the *AMA* is equal to the *IMA* of a device, what can you say about its efficiency? About the heat produced by the machine?
5. Is it possible for the *AMA* to be larger than the *IMA?* Explain why or why not.

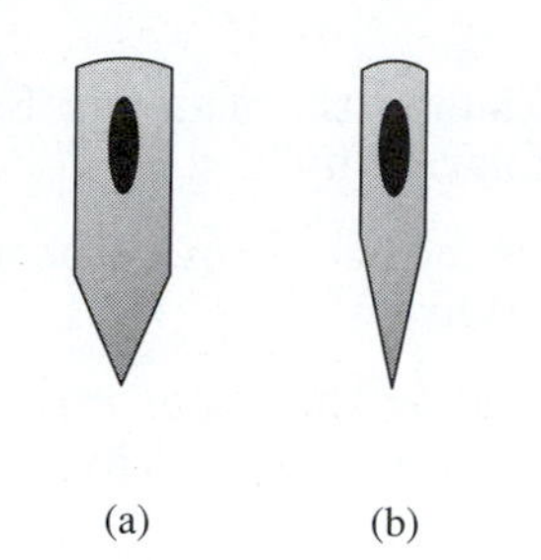

Figure 14–18: *(a) A wood maul is a blunt, heavy wedge. (b) An ax is a narrow, sharp wedge.*

Section 14.3:

6. A wood maul, a blunt wedge mounted on a handle, is often used instead of an ax to split wood. The ax is a thinner, sharp wedge with a higher mechanical advantage (see Figure 14–18). Why is the wood maul preferred for splitting wood?
7. Machine screws have blunt ends. Wood screws have pointed ends. Why?
8. It is easier to slide a heavy crate up a loading ramp onto a truck than to lift the crate. Is less work done by using the ramp?
9. As the angle of an incline decreases, the *IMA* increases. What happens to the efficiency of the incline? (Hint: Look at the change of the normal force.)

Section 14.4:

10. Figure 14–19 shows some different kinds of levers.

Figure 14–19: *Different kinds of lever systems.*

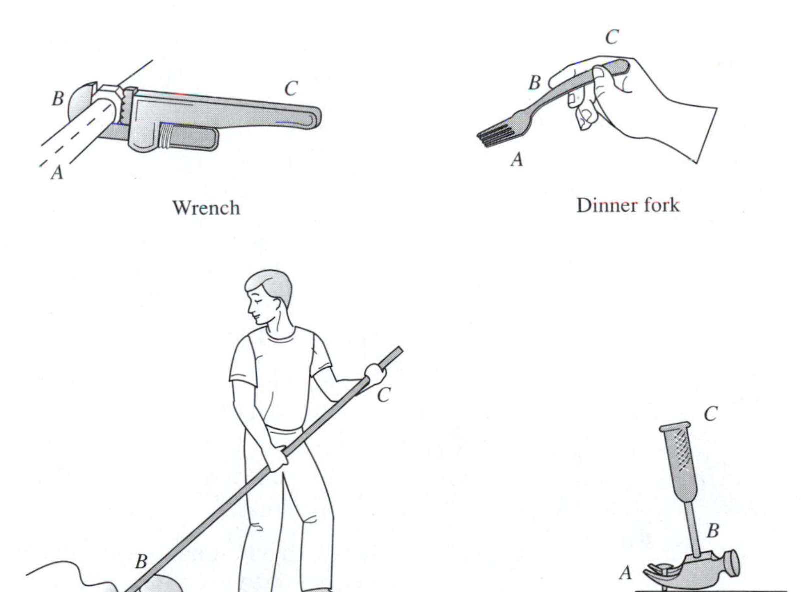

A. Identify the location of the fulcrum in each case.

B. For each diagram, identify the effort force and the resistance force.

C. Which diagrams show a lever system with an *IMA* of less than one?

11. Rented trailers have a notice cautioning the user not to place the load at the back end of the trailer. Why?

12. A plumber slips a length of pipe over the handle of a pipe wrench in order to loosen a rusted union. Why?

13. It is safer to lift a heavy load by bending the knees rather than bending over and lifting with the back muscles. Why?

Section 14.5:

14. Determine the *IMA* for the pulley systems shown in Figure 14–20.

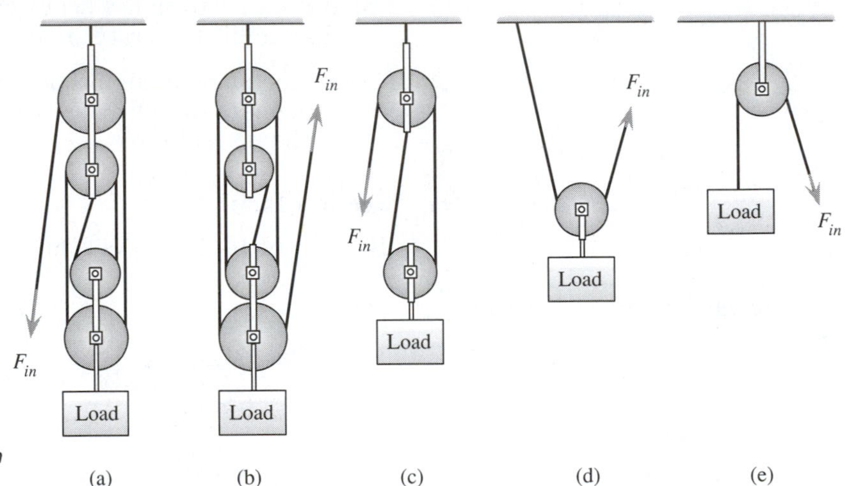

Figure 14–20: *Diagram for Exercise 14. An assortment of pulley systems.*

Section 14.6:

15. An electric motor operates at an efficiency of 0.91 to run a belt drive with an efficiency of 0.73. The belt drive turns a gearing system that rotates a small cement mixer. The gearing system has an efficiency of 0.94. What is the overall efficiency of the machine?

16. A driver gear with nine teeth transmits a torque to an intermediate gear (idler gear) with 17 teeth. The idler in turn drives the driven gear. The driven gear has 24 teeth (see Figure 14–21). What is the overall *IMA* of the gearing system?

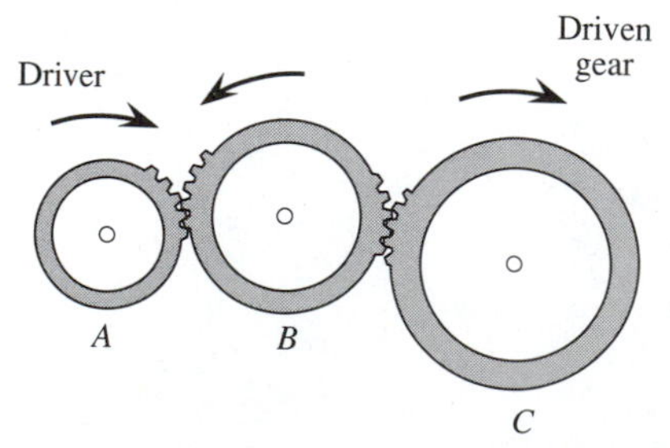

Figure 14–21: *Diagram for Exercise 16. A three-gear system.*

Section 14.7:

17. In a standard-transmission car, when you shift from first gear to second gear, are you increasing the gear ratio or decreasing it?

18. A drill press has a variable belt drive. To drill copper, the bit should rotate at a low speed. Should the drive pulley have a smaller diameter or a larger diameter than the driven pulley?

19. Sometimes people think they are getting more *power* when they use a high gear ratio. If the motor operates at the same revolutions per minute for two different gear ratios, the power output of the motor is the same. What does increase with the gear ratio?

20. Does an increase of the gear ratio cause the driven gear to go faster or slower?

PROBLEMS

Section 14.1:

1. An electric motor draws 0.520 kW/h of energy while performing 0.438 kW/h of work. What is the efficiency of the motor?

2. An automobile engine delivers 64.2 hp to the power train. If the drive wheels of the car receive 51.5 hp, what is the efficiency of the power train of the car?

3. A pulley system is used to raise a 97.2-kg load 13.2 m. If the system is 57% efficient, how much work must be put into the system?

4. A pickup truck consumes 1.3 gallons of gasoline to perform 1.46×10^7 ft · lb of work. What is the efficiency of the truck?

Section 14.2:

5. A force of 143 N is required to push a 440-N weight up an incline. The input distance is 14.6 m; the output distance is 2.7 m.
   A. What is the *IMA?*
   B. What is the *AMA?*
   C. What is the efficiency?

6. A force of 63 lb must be exerted on a pulley system to raise a 345-lb load. If the *IMA* of the system is 9.0, what is the efficiency of the pulley system?

7. An engine delivers a torque of 153 ft · lb to a drive train. If the drive train transmits a torque of 62.0 ft · lb:
   A. What is the *AMA* of the power train?
   B. If the power train is 89% efficient, what is the *IMA?*

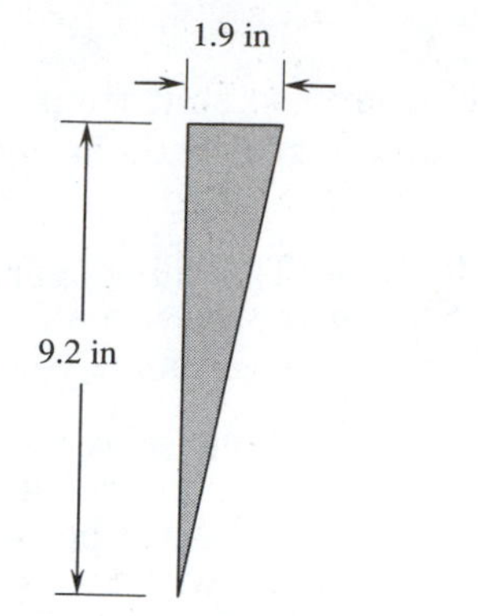

Figure 14–22: Diagram for Problem 10. A simple wedge.

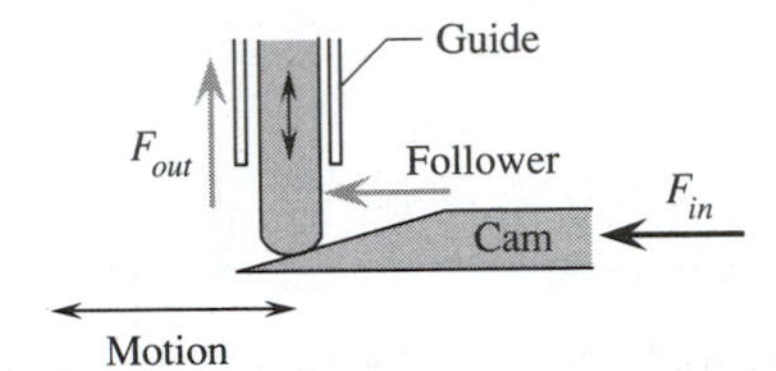

Figure 14–23: A wedge-shaped cam moves a follower at right angles to the motion of the wedge.

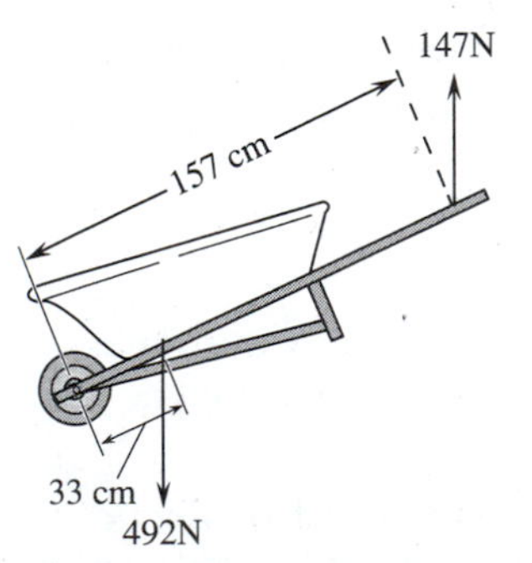

Figure 14–24: Diagram for Problem 15. A wheelbarrow.

Section 14.3:

8. A force of 62.5 N is needed to push a 184-N box up a 15° incline. What is the *AMA?* The *IMA?*
9. What is the IMA of a 3̄0° ramp?
10. A wedge has a length of 9.2 in and a base width of 1.9 in (see Figure 14–22). What is its *IMA?*
11. A wedge-shaped cam is used in a manufacturing process. The cam moves back and forth through a distance of 1.20 in, moving a follower (see Figure 14–23). The force on the cam is 530 N with a wedge angle of 18°. What force is transmitted to the follower? (Assume negligible friction.)
12. A ¼-in bolt has 12 threads per inch.
    A. What is the pitch of the thread?
    B. What is the *IMA* of the bolt?
13. A screw has a pitch of 1.3 mm and a radius of 2.4 mm. What is its *IMA?*

Section 14.4:

14. A bolt is held 0.80 in from the fulcrum of a pair of pliers. A force of 23.0 lb is exerted on the handles of the pliers 3.4 in from the fulcrum. What force is exerted on the bolt? What is the mechanical advantage of the pliers? (Assume negligible friction.)
15. For the wheelbarrow in Figure 14–24, estimate the *AMA* and *IMA*.
16. A hand winch has a handle length of 63.0 cm and a barrel radius of 10.2 cm. What is its *IMA?*
17. A force of 43.0 lb is exerted with a 10.0-in wrench to turn a ⅜-in machine bolt.
    A. What torque acts on the bolt?
    B. What is the ideal mechanical advantage of the wrench?

Section 14.5:

18. What are the *IMA*s of the pulley systems shown in Figure 14–25?
19. A chain hoist has seven chains attached to the load. A force of 72 lb is exerted to lift a 358-lb load 4.2 ft.
    A. What length of chain is pulled through the hoist?
    B. What is the *AMA* of the hoist?
    C. What is the percentage efficiency of the hoist?

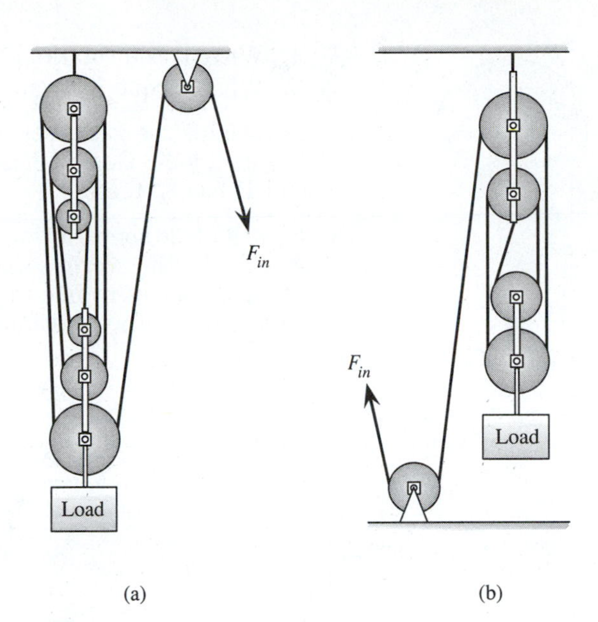

Figure 14–25: Diagram for Problem 18.

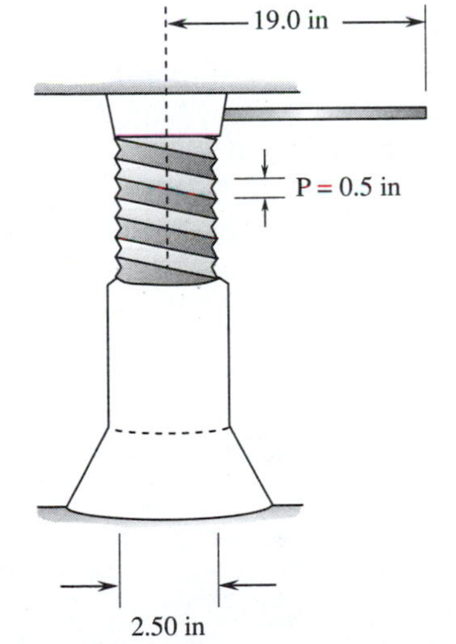

Figure 14–26: Diagram for Problem 20. A house jack.

Section 14.6:

20. The central column of a house jack is a 2.50 in diameter threaded cylinder with a pitch of 0.50 in (see Figure 14–26). If a handle with a length of 19.0 inches is used on the jack, what is the *IMA?*

21. A hand winch with a barrel diameter of 23.0 cm and a handle length of 37.0 cm is used to drag a stone slab up a 15° slope.
   A. What is the *IMA* of the system?
   B. If the overall *AMA* is 5.9, what is the efficiency?

***22.** Water from a storage reservoir is used to run a 37% efficient hydroelectric generator that distributes electrical power through an 89% efficient transmission line. Electricity is used to operate an 82% efficient electric motor using a 77% efficient V-belt drive to operate a planer. What fraction of the original gravitational potential energy of the water in the reservoir is lost to heat?

Section 14.7:

23. A driver gear with 14 teeth turns a gear with 32 teeth.
   A. What is the gear ratio?
   B. What is the speed ratio?

24. A belt drive has an efficiency of 0.86 at 1240 rpm. The driver pulley has a diameter of 4.0 in. The driven pulley has a diameter of 1.0 in. The input torque is 5.86 N · m.

A. What is the rotational speed of the output pulley?

B. What torque is transmitted to the driven pulley?

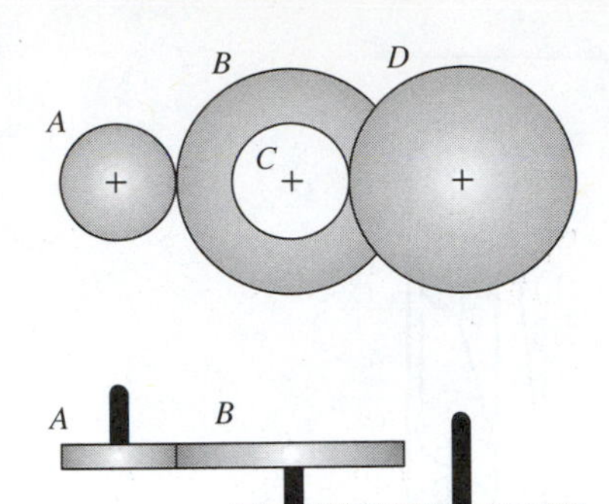

Figure 14–27: *Diagram for Problems 25 and 27. A four-gear system. Gears B and C are on the same axle.*

25. Determine the gear ratio (*IMA*) for the gear system shown in Figure 14–27. Gear A has 84 teeth, B has 32 teeth, C has 84, and D has 32 teeth.

26. Figure 14–28 shows a worm gear (a threaded shaft) and a worm wheel. As the worm gear rotates one revolution, the worm wheel advances by one tooth. If the worm gear rotates at 2800 rpm, what is the rotational speed of the worm wheel if it has 35 teeth?

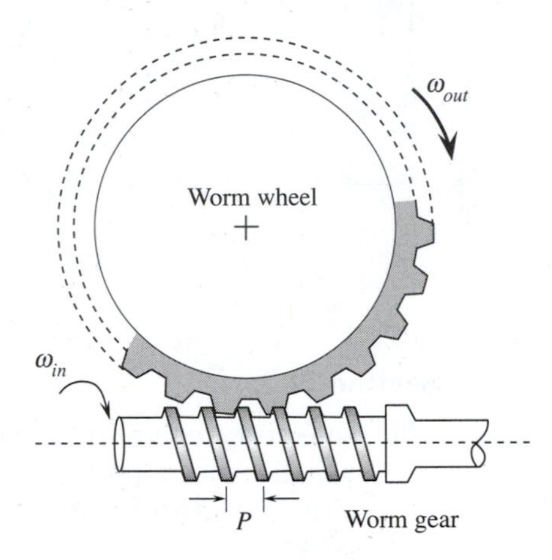

Figure 14–28: *The wheel turns by one tooth for each complete rotation of the worm gear. A worm gear and wheel can be used for a large speed reduction.*

27. The driver gear, A, in Figure 14–27 has an angular speed of 1430 rpm. What is the angular speed of gear D?

Chapter 15

MECHANICAL PROPERTIES OF SOLIDS

Could King Kong really have been this big? A careful study of the physics of Kong suggests that his proportions are all wrong. (Photograph courtesy of Turner Entertainment Company.)

OBJECTIVES

In this chapter you will learn to:

- distinguish among solids, liquids, and gases
- identify a material constant
- find the density and specific gravity of materials
- calculate Young's modulus from stress and strain data
- find the change in cross section of a stressed bar or wire
- calculate the shear modulus from shear stress and shear strain
- understand the properties of ductility, toughness, and hardness

Time Line for Chapter 15

4000 B.C.	Egyptians use copper.
3200 B.C.	Egyptians use bronze.
1400 B.C.	Iron is used extensively by Hittites.
190	Chinese porcelain is developed.
210	Concrete is used in Italy.
400	Chinese make steel.
1380	Cast iron is available in Europe.

Materials usually exist in one of three states: solid, liquid, or gas. The atoms or molecules making up a **solid** material are closely packed (see Figure 15–1a). They cannot move around easily in the material. The lump of material has a definite shape, which cannot be changed easily.

The atoms or molecules making up a **liquid** material are close together, but they can move freely relative to each other (see Figure 15–1b). A sample of material changes shape easily. When it is placed in a container, it will take the shape of the container.

A **gas** is made up of widely spaced atoms or molecules moving freely throughout the material (see Figure 15–1c). Because the molecules are widely spaced, there is little interaction among them. Like a liquid, gases have no fixed shape. They will conform to the shape of the container holding them. Gases are easily compressed, while liquids compress with great difficulty.

In this and the following chapter we will examine the behavior and some of the properties of solids and liquids. This chapter deals mostly with solids.

A civil engineer designs a suspension bridge. Concrete is chosen for the abutments, but steel cables are chosen for the suspension. The selection depends on the behavior of the materials with the kind of forces existing in the different parts of the bridge.

In an effort to meet Environmental Protection Agency (EPA) standards for gasoline mileage an automotive engineer replaces some of the parts of a car that were originally steel with lighter-weight aluminum and plastic.

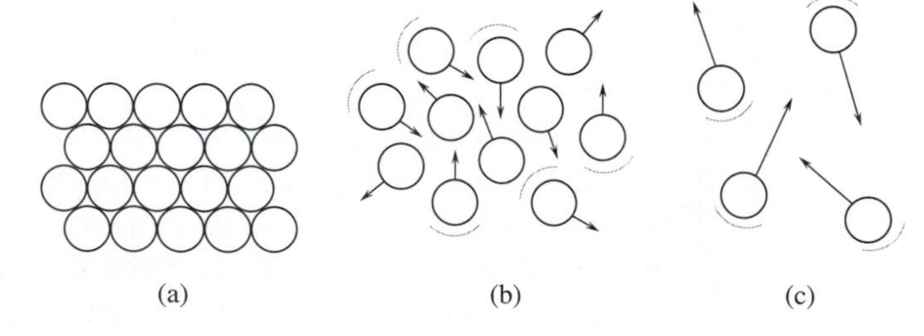

Figure 15–1: *Three states of matter. (a) Solid. Closely spaced atoms have fixed positions. (b) Liquid. Closely packed atoms move relative to each other. (c) Gas. Widely spaced atoms move with little interaction.*

The casing for a $\frac{1}{2}$-in electric drill must be lightweight and an electrical insulator. In addition it must be sturdy. The casing should not break when it is dropped. The designer selects an impact plastic.

Engineers choose building materials for specific jobs according to their properties, ease of fabrication, and cost. Let us look at some of the mechanical properties of materials.

15.1 MATERIAL CONSTANTS

We want to classify materials according to their behavior under specific kinds of external influences such as force and temperature. If we can perform an experiment that generates a number, we can sort the materials according to the size of the number.

We must be careful. We want a number that does *not* depend on the size or shape of the sample of material we are testing. The number must depend only on the kind of material we choose. A number that fits this condition is called a **material constant.**

You have already encountered a material constant—the coefficient of friction. When two surfaces are in contact, the ratio of the frictional force to the normal force is reasonably constant. The ratio depends only on the kinds of materials that are in contact, not on the shape or size of the surfaces. The ratio is a material constant. We can tabulate the ratio of frictional force to the normal force from the results of tests on a variety of pairs of surfaces. The tables can be used to calculate frictional forces in specific situations. Notice that the material constant is the ratio between an effect and its cause: the frictional force and a normal force.

Geometry has no effect at all on the coefficients of friction of surfaces. This is unusual. More often than not, geometric factors *do* influence the result of an experiment. When geometric factors affect the results of testing, we can still generate a material constant by dividing by the geometric factors that influence the result. We can find the effect per unit volume, per unit area, or per unit length.

A material constant is usually an effect: cause ratio per unit geometric factor that depends only on the kind of material used. The coefficient of friction is an example where the effect is a frictional force caused by a normal force. This formula seems a bit broad. Let us look at some examples.

15.2 DENSITY AND SPECIFIC GRAVITY

Density was considered in Chapter 3 in the discussion of proportion. The weight of a material was proportional to its volume. The proportionality constant between weight and volume is called **weight density** (D_w).

$$D_w = \frac{mg}{V} \qquad \textbf{(Eq. 15–1)}$$

Weight density is the ratio between a material's weight and its volume. Here is another way to look at the material constant called density.

Figure 15–2 shows four samples of material. Each sample is the same size, 1 ft^3. Since the volumes are all the same, the difference of weight depends only on the material. Any sample of copper with a volume of 1.00 ft^3 will have a weight of 555 lb. Any 1.00-ft^3 lump of aluminum will have a weight of 169 lb.

Since the volumes in Figure 15–2 are exactly 1 ft^3, the densities are numerically the same as the weights.

Weight density *is the weight of a unit volume of a material.*

To find the weight of a sample of material that is larger or smaller than 1 ft^3, multiply the unit size weight (density) by the number of volume units of the substance in the sample.

In SI, kilograms are used rather than pounds. Kilograms are units of mass, not weight. **Mass density** (d) is nearly always used for metric units.

$$d = \frac{m}{V} \qquad \textbf{(Eq. 15–2)}$$

Table 15–1 shows both mass and weight densities for some common substances.

Everyone has a good sense of the density of water. You have lifted different-sized containers of water in different forms—a glass of water, a liter of soda, or a full coffeepot. The numbers in Table 15–1 would make more intuitive sense if we compared the density of materials with the density of water. The ratio of the density of a substance to the density of water is called **specific gravity.**

$$\textit{specific gravity} = \frac{\textit{density of material}}{\textit{density of water}}$$

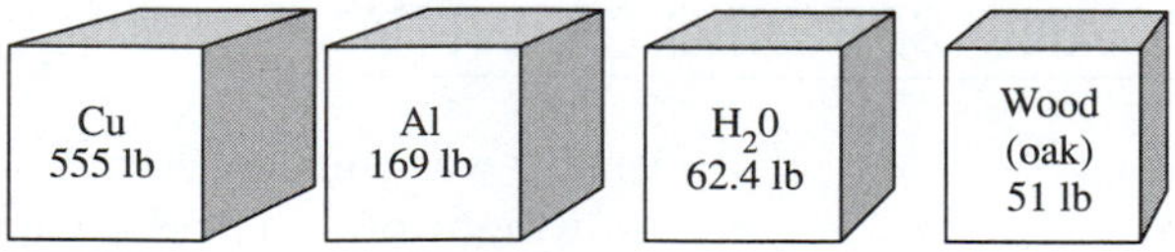

Cu
555 lb

Al
169 lb

H_2O
62.4 lb

Wood
(oak)
51 lb

Figure 15–2: *Density. Materials with a unit volume of 1 ft^3 have characteristic weights.*

Table 15–1: Density for selected materials.

MATERIAL	WEIGHT DENSITY (lb/ft^3)	MASS DENSITY (kg/m^3)
Solids		
Aluminum	169	2700
Brass	540	$86\bar{0}0$
Concrete	145	$23\bar{0}0$
Copper	555	8890
Ice	57	920
Steel	$49\bar{0}$	7850
Wood (oak)	51	$81\bar{0}$
Wood (pine)	26	$42\bar{0}$
Liquids		
Ethyl alcohol	49	$79\bar{0}$
Gasoline	42	$68\bar{0}$
Seawater	64.0	1025
Turpentine	54	$87\bar{0}$
Fresh water	62.4	1000
Gases		
Air	0.0807	1.29
Carbon dioxide	0.1234	1.96
Helium	0.0111	0.18
Hydrogen	0.0056	0.090
Propane	0.1254	2.02

Note: Values are for 20°C (68°F) and standard atmospheric pressure.

The ratio has no units and simply indicates how many times more dense or less dense the substance is than water. Table 15–2 lists some common specific gravities. It is much easier to read than the array of densities shown in Table 15–1.

□ **EXAMPLE PROBLEM 15–1: THE I BEAM**

What is the weight of a 54.5 ft long steel I beam with the cross section shown in Figure 15–3?

■ ***SOLUTION***

The volume is the cross-sectional area (A) times the length (l). The cross section is made up of three rectangles. Let w = width.

$$A = w\,[2(L_1 + L_2)]$$

$$A = 1.20 \text{ in } [(2 \times 14.0 \text{ in}) + 11.8 \text{ in}]$$

Table 15–2: Specific gravity for selected materials.

MATERIAL	SPECIFIC GRAVITY	MATERIAL	SPECIFIC GRAVITY	MATERIAL	SPECIFIC GRAVITY
Solids		**Liquids**		**Gases**	
Aluminum	2.7	Ethyl alcohol	0.79	Air	0.00125
Brass	8.6	Gasoline	0.68	Carbon dioxide	0.00196
Concrete	2.3	Seawater	1.02	Helium	0.00018
Copper	8.89	Turpentine	0.87	Hydrogen	0.000090
Ice	0.92			Propane	0.00202
Steel	7.85				
Wood (oak)	0.81				
Wood (pine)	0.42				

Note: Values are for 20°C (68°F) and standard atmospheric pressure.

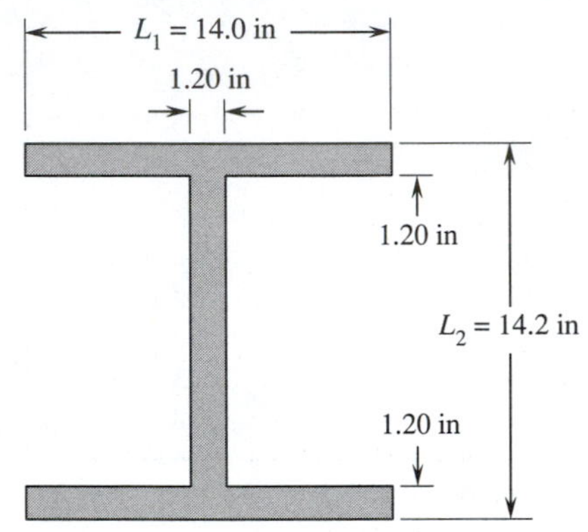

Figure 15–3: *A steel I beam.*

$$A = 47.76 \text{ in}^2 \times \frac{1.00 \text{ ft}^2}{144 \text{ in}^2}$$

$$A = 0.332 \text{ ft}^2$$

$$V = A \times l$$

$$V = 0.332 \text{ ft}^2 \times 54.5 \text{ ft}$$

$$V = 18.1 \text{ ft}^3$$

$$weight = D_w \times V$$

$$W = 49\bar{0} \text{ lb/ft}^3 \times 18.1 \text{ ft}^3$$

$$W = 8870 \text{ lb}$$

☐ **EXAMPLE PROBLEM 15–2: A FORENSIC PROBLEM WITH GLASS**

A small glass shard is found near the scene of a crime. There are three sources of broken glass at the crime with specific gravities of soda glass (2.32), plate glass (2.50), and optical glass (2.65). The piece is found to have a volume of 5.12 cm³ and a mass of 12.8 g. Is it possible that the piece of glass is from one of the sources at the scene?

■ ***SOLUTION***

Find the density or specific gravity of the sample.

$$d = \frac{m}{V}$$

$$d = \frac{12.8 \text{ g}}{5.12 \text{ cm}^3}$$

$$d = 2.50 \text{ g/cm}^3$$

Since the density of water is 1.00 g/cm^3, the density in centimeter, gram, second units is numerically equal to specific gravity.

$$\text{specific gravity} = 2.50$$

The shard is plate glass. It *may* have come from one of the sources at the scene of the crime. Further testing is needed to make sure it is the same plate glass found at the scene and is not from some external source.

King Kong, Mice, and Elephants

In the movies, King Kong had a hand large enough to hold a full-grown woman. For a human-sized gorilla we might expect a hand length of 7–8 in. Assuming a hand length of about 6 ft for King Kong yields a size factor of about 10. So King Kong was about 60 ft tall.

If the linear length of an object changes by a factor of 10, then the volume changes by a factor of 10 × 10 × 10, or 1000. The weight of the oversized gorilla would be not 10, but 1000 times as great as the weight of a man in a monkey suit.

The skeleton of a mammal is designed to support the weight of its muscle, blood, and viscera. If King Kong's bones were proportionately 10 times as large as a human-sized gorilla, then he had a problem. The cross-sectional area of one of King Kong's bones would be proportionately larger than a human's by a factor of a hundred. This means that the stress on the bone would be 10 times as great.

Bone, of course, has an elastic limit. A bone could easily fail during normal activity. So heavier animals have thicker bones. In this case, the diameter of King Kong's bones would be 30 times as thick as the bones of an average gorilla. He would look more like a hippopotamus than a primate. The fraction of a mammal's total weight made up of bone increases as the mammal's size increases. For example, only 3.6% of the weight of a mouse is invested in bone structures; of rabbits, 7.4%; of humans, 13%, and of elephants, 29%. King Kong would have had to invest half his weight in bone structure. Imagine an elephant-shaped monkey climbing the Empire State Building.

(Photograph courtesy of Turner Entertainment Company.)

15.3 HOOKE'S LAW REVISITED

We can perform an experiment to test Hooke's law for strands of metal wire. The restoring force should be proportional to the change of length of the wire (see Section 8.3). We cut several different lengths of copper wire from the same spool and stretch the pieces by hanging weights on the end. Each wire has the same cross-sectional area and is composed of the same material (see Figure 15–4). When the weight is removed the wire goes back to its original length—the same behavior as a spring. Figure 15–5 shows a plot of four samples of wire. Each wire obeys Hooke's law; force plotted against change of length is a straight line.

Notice that the longest wire has the largest change of length and the shortest wire the least. The amount of stretch may be proportional to the length of the wire. Rather than plotting the total change of length, let us plot the fractional change ($\Delta L/L$). The fractional change of length is called **strain.**

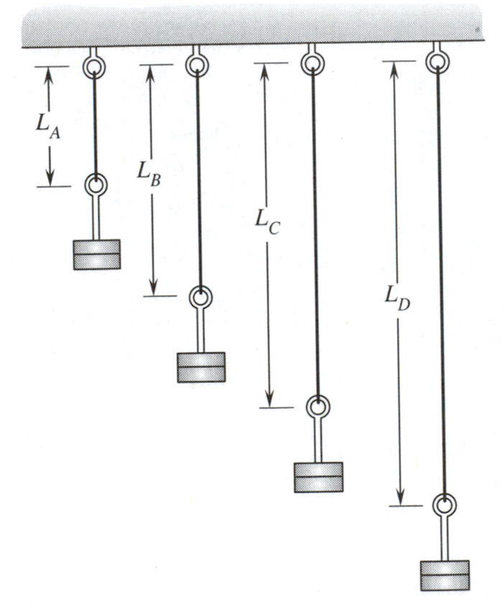

Figure 15–4: Weights are hung on copper wires of different lengths. All wires have the same cross-sectional area.

$$e = \frac{\Delta L}{L} \qquad \textbf{(Eq. 15–3)}$$

Figure 15–6 shows load plotted against strain. The data for all four wires fall on the same straight line. While the total change of the length of the wire depends on its length, strain does not.

Let us extend our experiment by doing additional testing with copper wires of different diameters. When we plot the load against strain we get the results shown in Figure 15–7. Hooke's law is obeyed, but we have different slopes for different cross-sectional areas of copper.

Notice that the strain is the largest for the smallest cross section, the smallest for the largest cross section. We will try plotting not the load, but the load per cross-sectional area (F/A). Force per unit area is called **stress.**

$$S = \frac{F}{A} \qquad \textbf{(Eq. 15–4)}$$

We can try stretching copper wires and rods of different lengths and diameters. We also try compressing rods of copper by pushing in on the two ends. A plot of stress versus strain gives us the result shown in Figure 15–8. The slope of the graph is always the same. The only way we can change the slope of the stress versus strain curve is to change to a new material such as iron or brass. Different

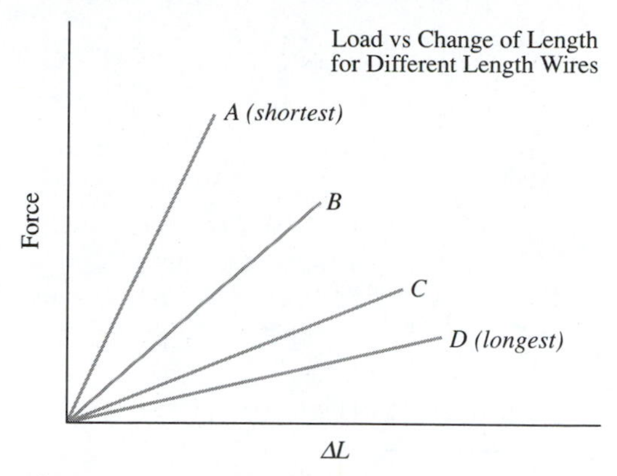

Figure 15–5: A plot of load versus stretch for copper wires that have different lengths, but the same cross section. The shortest wire has the largest slope; the longest, the smallest.

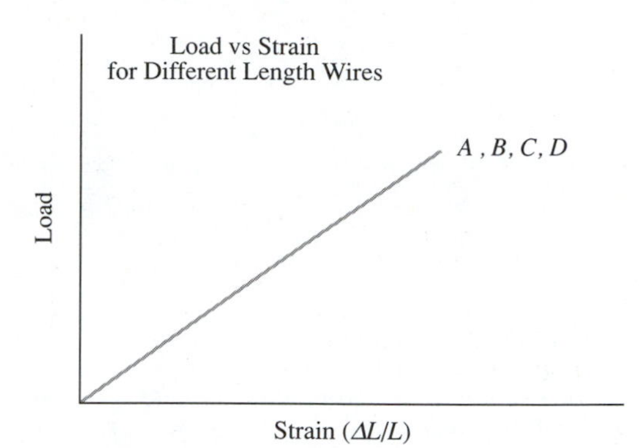

Figure 15–6: The data for Figure 15–5 using fractional change of length (strain). The data for all four wires fall on the same straight line.

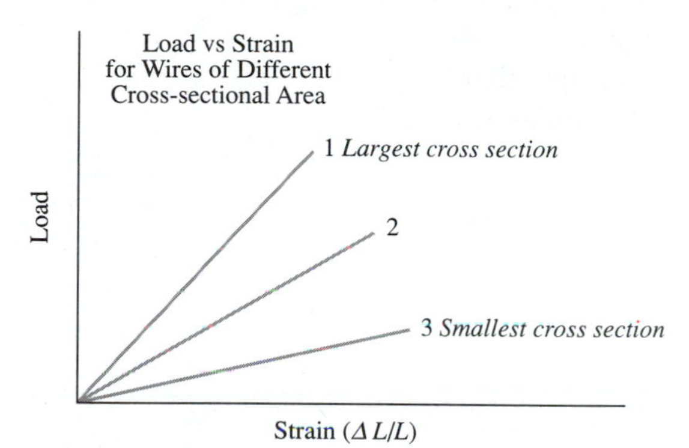

Figure 15–7: The force on wires with different cross-sectional areas plotted against strain. The wire with the largest cross section has the steepest slope.

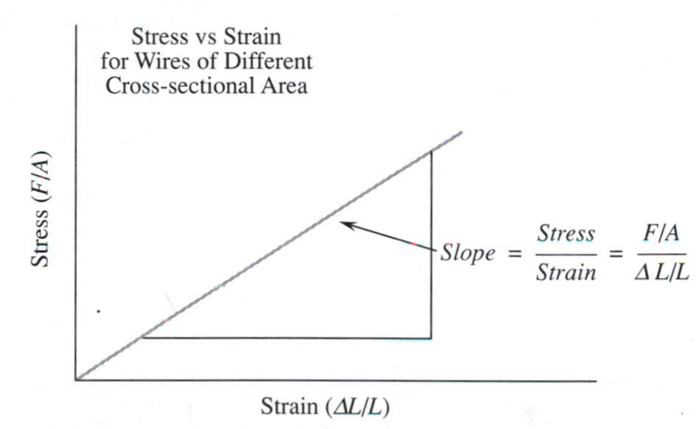

Figure 15–8: The data for Figure 15–7 plotted using force per unit area (stress). All the data fall on the same straight line. The slope of the stress-strain curve is a material constant.

materials have different slopes (see Figure 15–9).

The slope of a stress versus strain curve is a material constant. The slope does not depend on the geometry of the sample. It depends only on the material.

The slope of a stress-strain curve that obeys Hooke's law is called the **elastic modulus,** or **Young's modulus** (Y).

$$Y = \frac{stress}{strain} = \frac{F/A}{\Delta L/L} \qquad \textbf{(Eq. 15–5)}$$

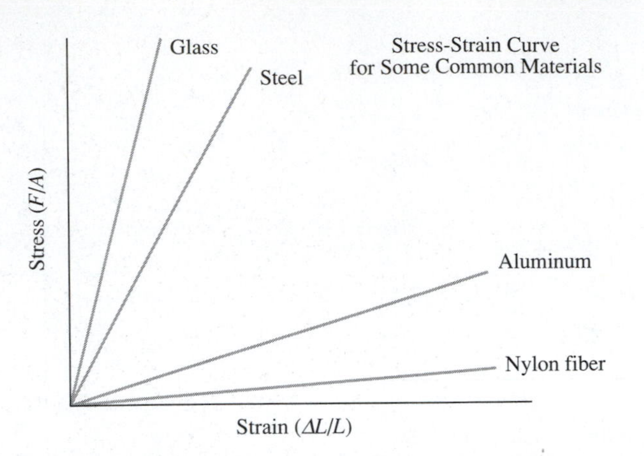

Figure 15–9: *Stress-strain curves for some common materials. The steeper the slope (Young's modulus) the more difficult it is to stretch the material.*

□ **EXAMPLE PROBLEM 15–3: THE CONCRETE PILLAR**

A solid concrete pillar is used to support part of a building. The length of the unloaded post is 4.220 m and its diameter is 23.8 cm. When the building is completed the pillar is compressed to 4.211 m. Young's modulus for concrete is $1.40 \times 10^9\ \text{N/m}^2$.

A. What strain does the pillar undergo?

B. What is the stress on the pillar?

C. What weight does the pillar support?

■ ***SOLUTION***

A. Strain:

$$e = \frac{\Delta L}{L}$$

$$e = \frac{4.211\ \text{m} - 4.220\ \text{m}}{4.220}$$

$$e = -2.133 \times 10^{-3}$$

B. Stress:

$$S = Y e$$

$$S = (1.40 \times 10^9\ \text{N/m}^2)(-2.133 \times 10^{-3})$$

$$S = -2.98 \times 10^6\ \text{N/m}^2$$

C. Load:

$$\text{F} = S A = S \times (\pi r^2)$$

$$\text{F} = -2.98 \times 10^6\ \text{N/m}^2 \times 3.1416 \times (0.119\ \text{m})^2$$

$$\text{F} = -1.33 \times 10^5\ \text{N}$$

The minus sign indicates a compressional force.

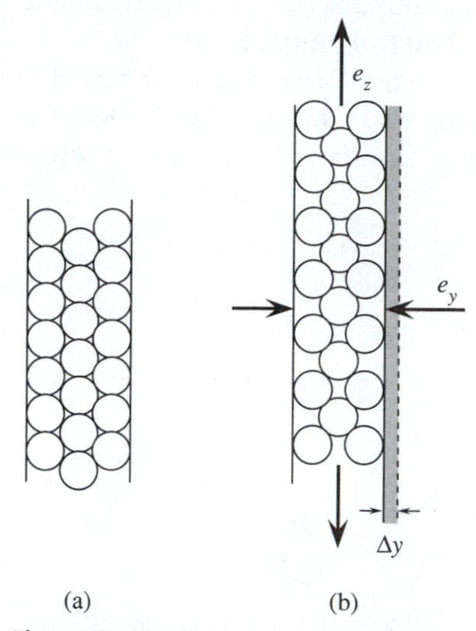

Figure 15–10: **A.** *Atoms in an unstressed material. As the atoms are stretched farther apart by a tensile strain (e_z) along the axis of the wire, the atoms in the cross section fill in the gaps. This causes a lateral strain (e_y).*

We know that substances are made up of atoms. We can visualize the material as being made up of small spheres with diameters on the order of 10^{-10} m. Each sphere is attracted to its neighbors. When we stretch a wire with a **tensile stress** (a stress along the wire's axis), the atoms are pulled farther apart (see Figure 15–10). This makes it possible for the atoms to crowd closer together on the plane at right angles to the tensile stress. We would expect the cross section of the wire to decrease. This is what is observed experimentally. How much lateral strain occurs depends on the properties of the material.

If the tensile stress (e_z) is along the z axis, then the lateral stress can be measured along the y or the x axis. For a homogeneous material, the ratio of lateral stress divided by the tensile stress is a constant. The proportion is called **Poisson's ratio.** The Greek letter nu (ν) is used to represent it.

$$\nu = \frac{-e_y}{e_z} \qquad \textbf{(Eq. 15–6)}$$

The negative sign indicates that one strain decreases while the other increases. In compression, the length of the material will decrease while the cross section will increase. Poisson's ratio for most materials has a value between 0.25 and 0.50. Table 15–3 shows this ratio for some common substances.

Table 15–3: Elastic constants for selected solids.

MATERIAL	YOUNG'S MODULUS psi (× 10^6)	YOUNG'S MODULUS N/m² (x 10^{10})	POISSON'S RATIO	YIELD STRENGTH* psi (× 10^3)	YIELD STRENGTH* N/m² (× 10^8)	ULTIMATE STRENGTH* psi (× 10^3)	ULTIMATE STRENGTH* N/m² (× 10^8)
Aluminum	11	7.0	0.33	19	1.3	21	1.4
Brass	13	9.0	0.34	55	3.8	67	4.7
Copper	16	11	0.34	6	0.4	31	2.2
Plate glass	10	7.0	0.34	na	na	na	na
Rubber (vulcanized)	0.000014	0.000020	0.5	na	na	na	na
Steel	29	20	0.28				

Note: Approximate values for room temperature. The history of the material may affect its strength.
* na = not available; blank

EXAMPLE PROBLEM 15–4: THE STRETCHED RUBBER CORD

A rubber cord with a diameter of 1.8 mm is used to attach a rubber ball to a paddle. When the ball is hit with the paddle, the length of the cord increases by 60%. What is the diameter of the stretched cord?

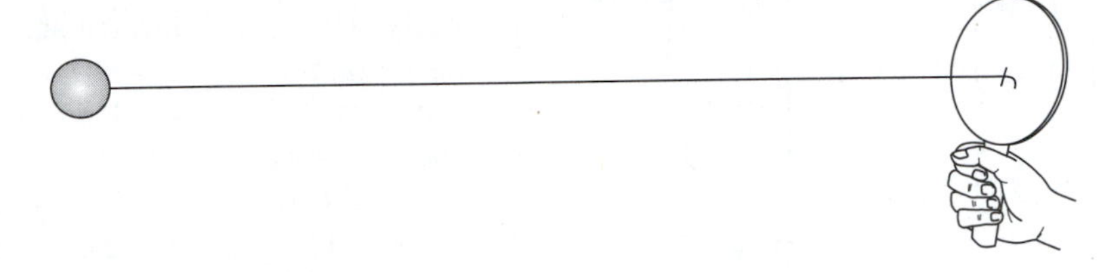

Figure 15–11: A rubber cord is stretched.

SOLUTION

$$\nu = \frac{-e_y}{e_z} = \frac{-\Delta D/D}{e_z}$$

or

$$\Delta D = -\nu \times D \times e_z$$

From Table 15–3 we see that Poisson's ratio for rubber is 0.5.

$$\Delta D = -0.50 \times 1.80 \text{ mm} \times 0.60$$

$$\Delta D = -0.54 \text{ mm}$$

$$D_{\text{stretched}} = D + \Delta D$$

$$D_{\text{stretched}} = 1.8 \text{ mm} + (-0.54 \text{ mm})$$

$$D_{\text{stretched}} = 1.3 \text{ mm}$$

15.4 SHEAR

A tensile stress is a *unilateral* stress. The stress exerted on the material and the resulting strain are along a single axis. We can exert nonconcurrent forces on a material to cause a torque. This is called a *bilateral* stress.

Look at Figure 15–12. A force F_s is exerted along the top surface of the material. The bottom surface is fixed. The force per unit area is stress, but in this case the force is exerted parallel to the surface

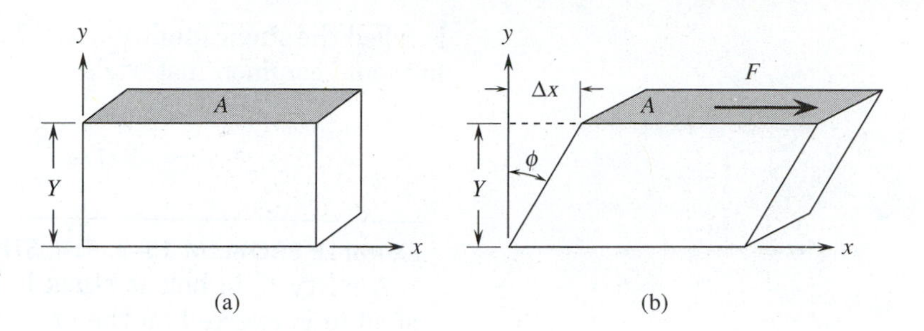

Figure 15–12: Shear strain. A force acts along the top surface. The surface is displaced a distance Δx. The distance between the top surface and the fixed bottom is y. The angle ϕ in radians is the shear strain. ($\sigma = \phi = \Delta x/y$.)

rather than perpendicular to it. This kind of stress is called **shear stress** (S).

$$S = \frac{F_s}{A_s} \quad \textbf{(Eq. 15–7)}$$

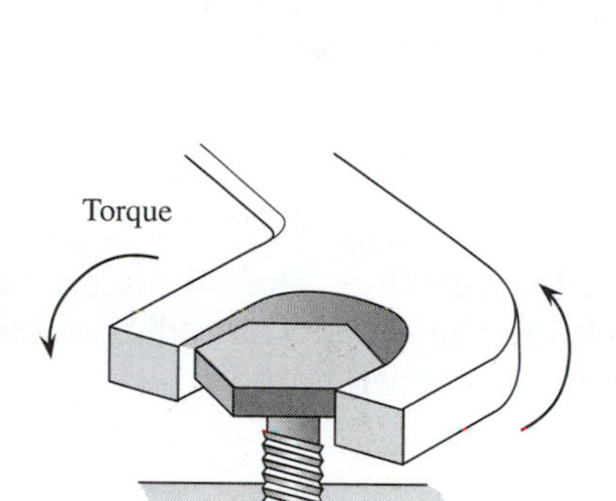

Figure 15–13: A torque is exerted by a wrench. A shear strain results in the shank of the bolt.

The top layer of material can move a distance x. Layers between the stressed surface and the fixed surface move proportionately. The ratio $\Delta x/y$ is called the **shear strain** (σ).

The shear strain is the tangent of angle ϕ. For small angles, we can use the approximation: $\tan \phi = \phi$. ϕ must be expressed in radians.

$$\sigma = \phi = \frac{\Delta x}{y} \quad \textbf{(Eq. 15–8)}$$

Expressing shear strain as an angle makes sense in the situation shown in Figure 15–13. Shear strains frequently occur when torques are exerted.

The ratio of shear stress to shear strain is a material constant. It

Table 15–4: Shear modulus for selected solids.

MATERIAL	SHEAR MODULUS psi ($\times 10^6$)	N/m^2 ($\times 10^{10}$)
Aluminum	3.4	2.4
Brass	5.1	3.5
Copper	6.1	4.2
Lead	0.78	0.54
Plate glass	6.0	4.1
Rubber (vulcanized)	0.00004	0.00003
Steel	12	8.0

Note: Approximate values for room temperature.

is called the **shear modulus** (G). Table 15–4 shows the shear modulus for some common materials.

$$G = \frac{S}{\sigma} \qquad \textbf{(Eq. 15–9)}$$

☐ **EXAMPLE PROBLEM 15–5: THE STICKY BOLT**

A rusty ¼-in bolt is stuck in a threaded mounting. A force (F_w) of $4\bar{0}$ lb is exerted on the end of a $1\bar{0}$-in wrench.

A. What is the shear stress on the shank of the bolt?

B. What shear strain develops on the bolt shank?

■ ***SOLUTION***

A. Shear stress:

First find the shearing force on the bolt (F_b). The torque on the wrench and the torque on the bolt are the same. The radius of the bolt (R_b) is ⅛ in. The lever arm of the wrench is $1\bar{0}$ in.

$$F_b R_b = F_w R_w$$

$$F_b = \frac{F_w R_w}{R_b}$$

or

$$F_b = \frac{4\bar{0} \text{ lb} \times 1\bar{0} \text{ in}}{0.125 \text{ in}}$$

$$F_b = 3200 \text{ lb}$$

$$S = \frac{F_b}{A} = \frac{F_b}{\pi R_b^2}$$

$$S = \frac{3200 \text{ lb}}{3.1416 \times (0.125)^2}$$

$$S = 6.52 \times 10^4 \text{ lb/in}^2$$

$$S = 6.52 \times 10^4 \text{ psi}$$

B. Shear strain:

$$\sigma = \frac{S}{G}$$

$$\sigma = \frac{6.52 \times 10^4 \text{ psi}}{12 \times 10^6 \text{ psi}}$$

$$\sigma = 5.43 \times 10^{-3} \text{ radians}$$

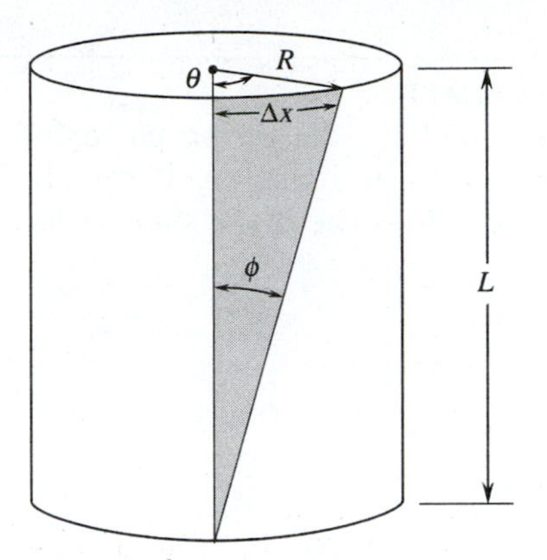

Figure 15–14: The shear angle (ϕ) is related to the angle of twist (θ) of a cylinder. When the angles are expressed in radians they have a common arc length (Δx).

Sometimes the angle of twist (θ) is needed. A torque twists the top of a cylinder, while the bottom of the cylinder remains fixed (see Figure 15–14). The top moves through an angle $\theta = \Delta x/R$. The shear strain is $\phi = \Delta x/L$, where L is the length of the cylinder. The two angles are then related by:

$$\Delta x = \theta R = \phi L$$

or

$$\theta = \frac{\phi L}{R} \qquad \textbf{(Eq. 15–10)}$$

This can be written in terms of the area and length of the cylinder. The shear stress is F_s/A_s. The area is $A = \pi R^2$. If we assume the force acts over the entire cross section, the lever arm of the torque varies from zero to R with an average lever arm of $R/2$. The average torque is $\tau = F(R/2)$. The force in terms of torque is $F = 2(\tau/R)$. From this we can write the shear stress in terms of torque.

$$\textit{shear stress} = \frac{F_s}{A} = \frac{2\,\tau/R}{\pi R^2}$$

or

$$\textit{shear stress} = \frac{2\tau}{\pi R^3} = G\sigma = \frac{G\,\theta R}{L}$$

Solve for θ.

$$\theta = \frac{2\tau L}{\pi G R^4} \qquad \textbf{(Eq. 15–11)}$$

☐ **EXAMPLE PROBLEM 15–6: THE STICKY BOLT AGAIN**

If the bolt in Example Problem 15–5 has 2.3 in free to twist, through what angle does the head of the bolt twist?

■ ***SOLUTION***

Since we have the shear strain, we can use Eq. 15–10.

$$\theta = \frac{\phi L}{R}$$

$$\theta = \frac{5.67 \times 10^{-3}\ \text{rad} \times 2.3\ \text{in}}{0.125\ \text{in}}$$

$$\theta = 0.104\ \text{rad, or about } 6^\circ$$

☐ **EXAMPLE PROBLEM 15–7: THE DRIVE SHAFT**

An automobile engine operates at 114 hp, delivering power to a steel drive shaft 7.75 ft long and 1.88 in in diameter. If the shaft turns at 920 rpm, through what angle does the drive shaft twist?

■ ***SOLUTION***

$$L = 7.75 \text{ ft}$$

$$R = 0.94 \text{ in} \times \frac{1 \text{ ft}}{12 \text{ in}} = 0.0783 \text{ ft}$$

$$G = 12 \times 10^6 \text{ lb/in}^2 \times \frac{144 \text{ in}^2}{1 \text{ ft}^2} = 1.73 \times 10^9 \text{ lb/ft}^2$$

We can calculate the torque from the power.

$$\textit{power} = \tau\omega$$

$$\tau = \frac{\textit{power}}{\omega}$$

$$\tau = \frac{114 \text{ hp} \times 550 \text{ ft} \cdot \text{lb/s} \cdot \text{hp}}{920 \frac{\text{rev}}{\text{min}} \times \frac{2\pi \text{ rad}}{\text{rev}} \times \frac{1 \text{ min}}{60 \text{ s}}}$$

$$\tau = 651 \text{ ft} \cdot \text{lb}$$

Use Equation 15–11.

$$\theta = \frac{2\tau L}{\pi G R^4}$$

$$\theta = \frac{2 \times 651 \text{ ft} \cdot \text{lb/s} \times 7.75 \text{ ft}}{3.14 \times 1.73 \times 10^9 \times (0.0783 \text{ ft})^4}$$

$$\theta = 0.0494 \text{ rad or } 2.8°$$

15.5 OTHER MECHANICAL PROPERTIES

Ductility

Ductility is the permanent strain at fracture. When a sample of material is stretched until it fails, the permanent strain that has occurred is the measure of ductility.

Ductility is usually expressed in percent elongation. Curve D in Figure 15–15 has a large percent elongation. Curve A has a ductility

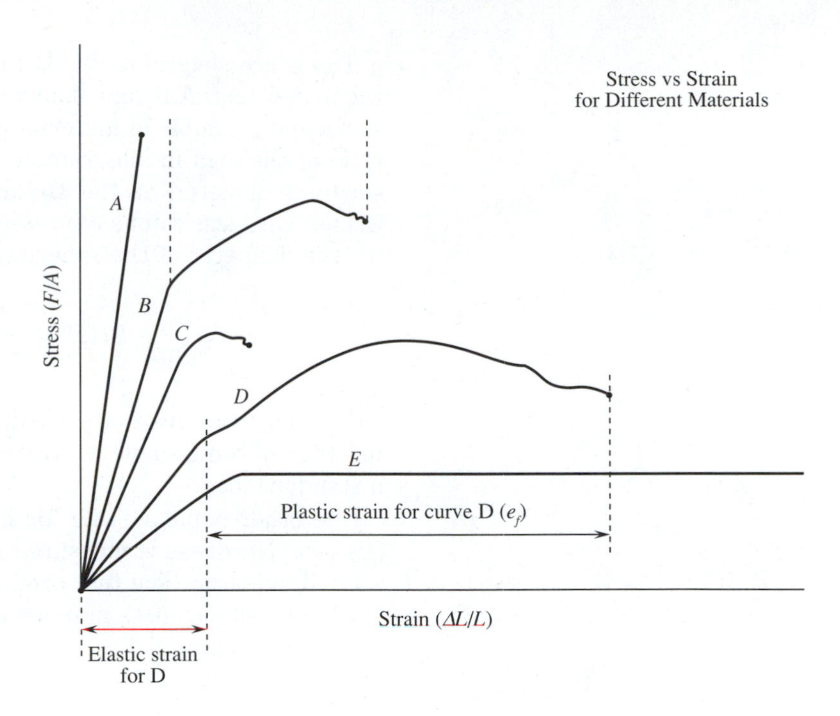

Figure 15–15: Stress versus strain for materials with different ductilities.

of zero. This is a *brittle* material. It will fracture without entering the plastic phase. Curve E is a perfectly elastic material. It can be drawn out into a long filament.

$$\%\ elongation = \frac{\Delta L_f}{L} \times 100 \qquad \textbf{(Eq. 15–12)}$$

Toughness

The area under the stress-strain curve has units of energy per unit volume ($\mathbf{F}/A \times \Delta L/L = work/V$). **Toughness** is the area under the stress-strain curve up to the point of failure. It is the energy per volume needed to cause failure. Look at Figure 15–15. The material in curve B has a larger ultimate strength than the material in curve C, but material C is tougher than B. The ultimate strength is the maximum stress of the material. This is usually in the plastic region.

Hardness

Hardness is a measure of resistance to plastic indentation. For ductile materials (this includes everything in Figure 15–15 except A) there is a relationship between the strength of the material and its hardness. The greater the hardness, the greater the strength will be.

There are several methods for testing hardness. One method is the Brinell test. A 10 mm diameter sphere is pressed into the polished surface of a sample of material with a load of 3000 kg for 30 s. The ratio of the load in kilograms to the area of contact of the dent in square millimeters is the **Brinell hardness number** (BHN). The Brinell hardness can be expressed in terms of the depth of the dent (t), the diameter of the sphere (D), and the load (**F**).

$$BHN = \frac{\mathrm{F}}{\pi D\, t} \qquad \textbf{(Eq. 15–13)}$$

Other hardness tests are used, but the procedure is similar. An indenter of some shape is pressed into the surface of a sample with a standard load.

One other popular scale for hardness is the *Rockwell hardness* (R) test. Hardness is measured in terms of depth of penetration of a small indenter. The test produces a convenient dial reading, but the Rockwell hardness number has a complicated relationship with strength of material.

Table 15–5: Mechanical properties of solids.

PROPERTY	SYMBOL	DEFINITION
Ductility (% elongation)	e_f	Plastic strain at failure $(L_f - L)/L$
Hardness	BHN or R	Resistance to plastic indentation.
Toughness		Energy per volume for failure by fracture. This is the area under the stress-strain curve.
Ultimate strength	s_u	Maximum stress the material can withstand
Yield strength	s_y	Stress at which plastic deformation begins
Young's modulus	Y	Stress/elastic strain (s/e)

☐ **EXAMPLE PROBLEM 15–8: DUCTILITY**

An aluminum alloy undergoes tensile testing. A 50-mm gauge length is marked on the sample. After stressing the sample to the breaking point, the gauge markings are found to be 124 mm apart. What is the ductility?

■ *SOLUTION*

$$\%\ elongation = \frac{\Delta L}{L} \times 100$$

$$\%\ elongation = \frac{(124 - 5\bar{0})\ \text{mm} \times 100}{5\bar{0}\ \text{mm}}$$

$$\%\ elongation = 150$$

SUMMARY

Substances usually occur in one of three states: solid, liquid, or gas. A material constant measures a property of a material. It is a cause-and-effect relationship that is independent of the dimensions or shape of the material. Some material constants of substances are density and specific gravity.

Density is the ratio of mass (or weight) to volume of a substance. It is the mass (or weight) of a unit volume of the material. Specific gravity is the ratio of the density of a material to the density of water. Strain is the fractional change of length of a stressed material. Stress is the force per unit area acting on an object.

The ratio of stress to strain is constant for a given material in the plastic region. Young's modulus (elastic modulus) is the ratio of tensile or compressional stress to strain. Poisson's ratio is the ratio of the lateral strain to the tensile strain of a material. It is used to calculate the change of cross-sectional area of a stressed object.

Shear is caused by a bilateral or shear stress (S) when nonconcurrent forces act on parallel surfaces of an object. The shear strain (σ) can be measured as an angle in radians. The ratio of shear stress to shear strain is called the shear modulus (G).

Some mechanical properties of materials are ductility, toughness, and hardness. Table 15–5 summarizes some of the mechanical properties of solids.

Strains are either **elastic,** meaning the material regains its original dimensions when the stress is removed, or they are **plastic.** A plastic strain causes a permanent deformation of the material. The material will not regain its original dimensions when the stress is removed. The **yield strength** is the stress on the stress-strain curve at which plastic behavior begins.

KEY TERMS

If you can explain the following terms to a friend or classmate, you understand their meaning. If you cannot explain the terms, you should reread the sections in which they are discussed.

Brinell hardness number (BHN)
density
ductility
elastic strains
elastic modulus

gas
hardness
liquid
mass density
material constant
plastic strains
Poisson's ratio
shear modulus
shear strain
shear stress
solid
specific gravity
strain
stress
tensile stress
toughness
ultimate strength
weight density
yield strength
Young's modulus

EXERCISES

Section 15.1:

1. Which of the following are *material* constants? For each constant explain why or why not.
   A. Coefficient of friction
   B. Spring stiffness
   C. Drag coefficient
   D. Density

2. Which of the following are *material* constants? For each constant explain why or why not.
   A. Stress
   B. Shear modulus
   C. Hardness
   D. Ductility

Section 15.2:

3. The weight density of water is 62.5 lb/ft^3. Is the weight density of water in the following units numerically larger or smaller than 62.5? Answer without doing a numerical conversion. (1 US gallon = 0.134 ft^3.)
   A. lb/yd^3
   B. lb/in^3
   C. lb/gallon

4. The specific gravity of aluminum is 2.70. What would 1 ft^3 of aluminum weigh in pounds?

5. The mass density of water is $100\bar{0}$ kg/m^3. The specific gravity of ice is 0.92. What would be the mass of 1 m^3 of ice in kilograms?

6. A student measures sand into a graduated cylinder that is on a triple-beam balance. The student plots the mass reading of the scale against the volume of sand (see Figure 15–16).
   A. What does the y intercept represent?
   B. What is the average density of the sand?

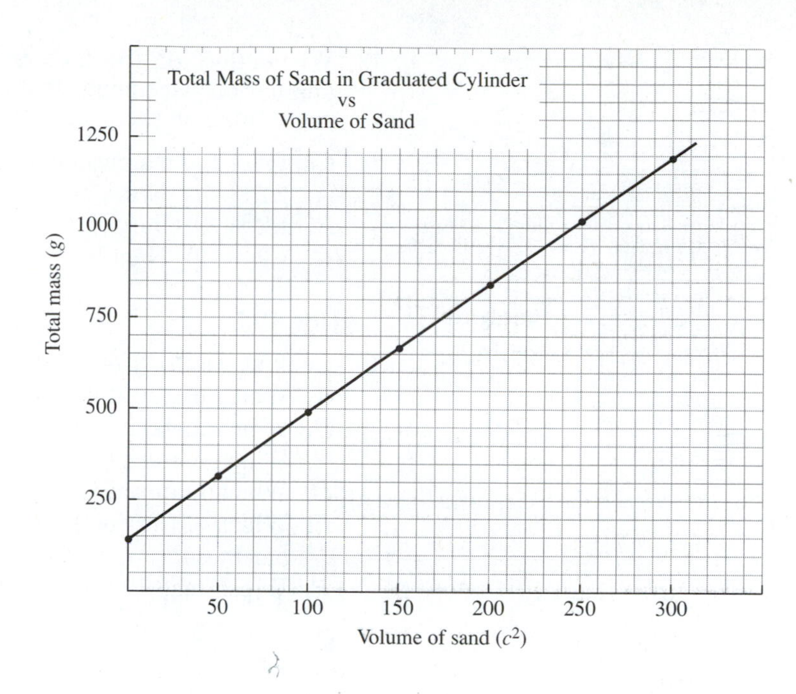

Figure 15–16: Diagram for Exercise 6. Total mass of sand in graduated cylinder versus volume of sand.

Section 15.3:

7. Look up the word *tensile* in a dictionary. What does it mean?
8. If a material has a high value for Young's modulus, is it easy to stretch? Explain.
9. Two materials have the same value of Young's modulus, but material A has a larger Poisson's ratio than material B. What difference in behavior will there be between the two materials under tensile stress? Under compression?
10. Two pillars made of different materials stand beside each other with a girder lying level across them so that they undergo the same strain (see Figure 15–17). The unstressed dimensions of the materials are the same. If pillar A has a Young's modulus that is two-thirds that of pillar B, will they both be supporting the same amount of load? Explain why or why not.

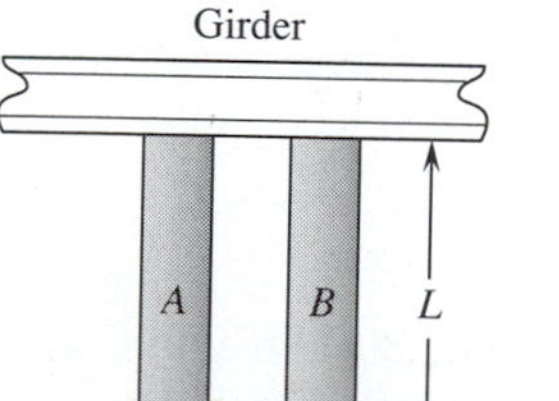

Figure 15–17: Diagram for Exercise 10. Load pillars have the same dimensions, but are made of different materials.

Section 15.4:

11. Identical rods of copper and aluminum undergo the same shear stress. Which will be twisted through the larger angle? Why?
12. What effect do the following have on the amount of shear strain (σ) of a threaded rod? (See Figure 15–18.)
    A. The length of the rod.
    B. The diameter of the rod.
    C. The shear modulus.

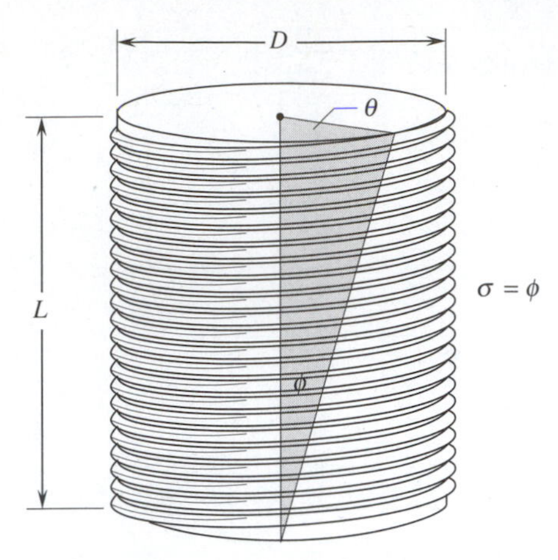

Figure 15–18: Diagram for Exercises 12 and 13.

13. What effect do the factors listed in Exercise 12 have on the angular displacement (θ) of the free end of a rod? This is not the same thing as the shear strain. (See Figure 15–18.)

14. Compare the magnitude of a shear stress of 1.2×10^4 psi with:
    A. 1.2×10^4 N/m^2.
    B. 1.2×10^4 lb/ft^2.
    C. 1.2×10^4 N/cm^2.

Section 15.5:

15. Look at Figure 15–19.
    A. Which material is a brittle material?
    B. Which material has the largest ultimate strength?
    C. Which material can best be drawn into a thin filament?
    D. Which material is the toughest?
    E. Which of the ductile materials is likely to have the greatest hardness?
    F. Which material has the greatest percent elongation?

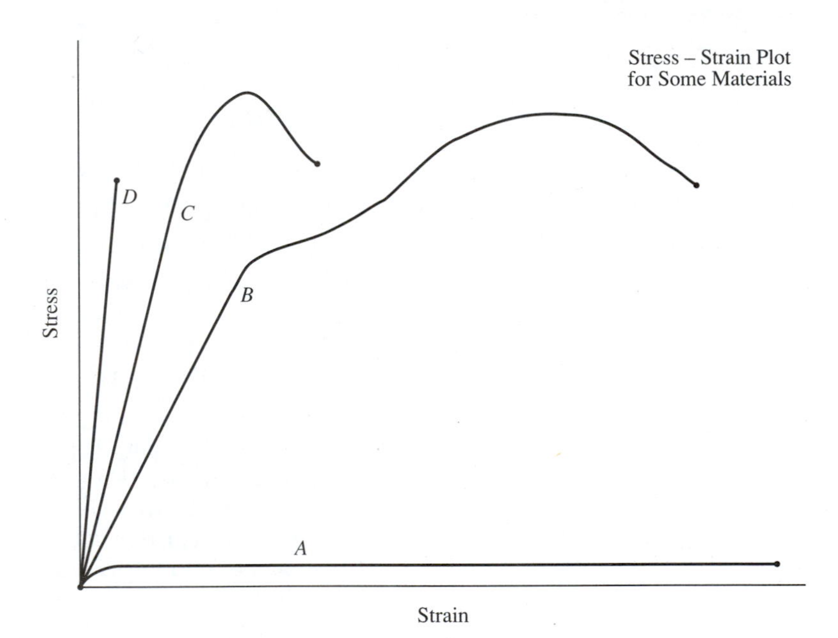

Figure 15–19: Diagram for Exercise 15. Stress-strain plot for some materials.

PROBLEMS

Section 15.2:

1. A slab of limestone is 12.7 in thick, 26.7 in wide, and 53.6 in long. The slab weighs 657 lb. What is the density of the limestone slab?

2. How heavy is the water in a completely filled 20-gallon aquarium? (1 US gallon = 0.134 ft^3.)

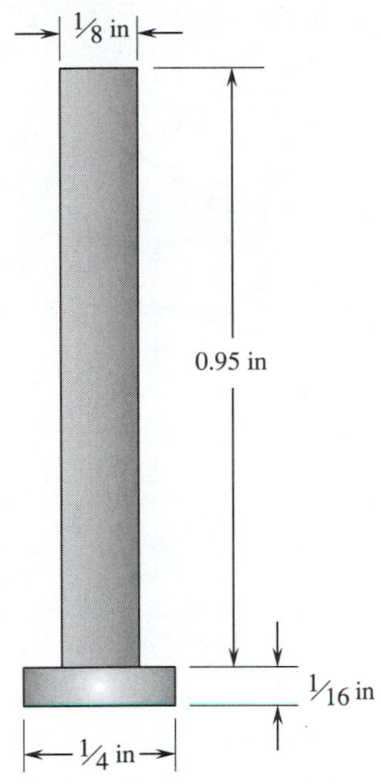

Figure 15–20: Diagram for Problem 5. Specifications for an aluminum rivet.

3. What is the specific gravity of a material with a volume of 0.124 m^3 and a mass of 524 g?

*4. A copper bar has a diameter of 5.12 cm and a mass of 14.6 kg. How long is the bar in centimeters?

**5. The cost for shipping a package to a specified destination by a particular carrier is $0.56/lb. Sixty thousand aluminum rivets with the specifications shown in Figure 15–20 are to be shipped in a 9.8-lb crate. Estimate the shipping charges.

Section 15.3:

6. A load of 3.00×10^5 N compresses a 2.42-m metal post by 1.00 cm. If the cross section of the post is 7.88×10^{-4} m^2, find:
   A. the strain.
   B. the stress.
   C. Young's modulus for the metal.

7. A 120-lb load is hung by a 12.4 ft long steel wire with a diameter of 0.187 in (see Figure 15–21).
   A. What is the tensile stress on the wire?
   B. What is the strain?
   C. What is the increase in length?
   D. What is the decrease in diameter?

8. A no. 18 copper wire has a diameter of 0.04 in. Originally it is 9.7 ft long. By how much does its length change under a load of 37 lb?

9. A 3.72 m long wire with a diameter of 1.42 mm increases in length by 1.22 mm when a tensile force of 128 N is applied to it. What is Young's modulus for the material?

10. A 0.300 in diameter wire decreases in diameter by 0.005 in when it undergoes a tensile strain of 0.42%. What is Poisson's ratio for the material from which the wire is made?

11. Find the Young's modulus for the material shown in Figure 15–22.

Section 15.4:

12. A 54-lb force is exerted on the end of a 12-in lug wrench in an attempt to loosen the nuts on a wheel (see Figure 15–23). If the wheel has ½-in steel studs:
   A. what is the shear stress on the studs?
   B. what is the shear strain?

13. A torque is exerted on a 3.27 m long copper wire with a diameter of 1.2 mm. One end of the wire is fixed. If the free end of the wire twists through an angle of 38°, what is the shear strain?

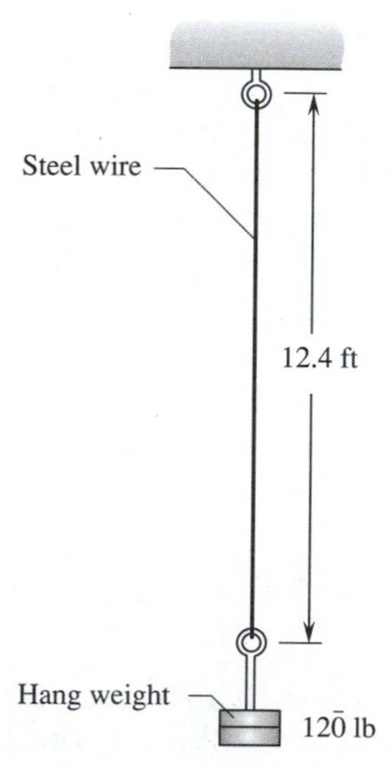

Figure 15–21: Diagram for Problem 7.

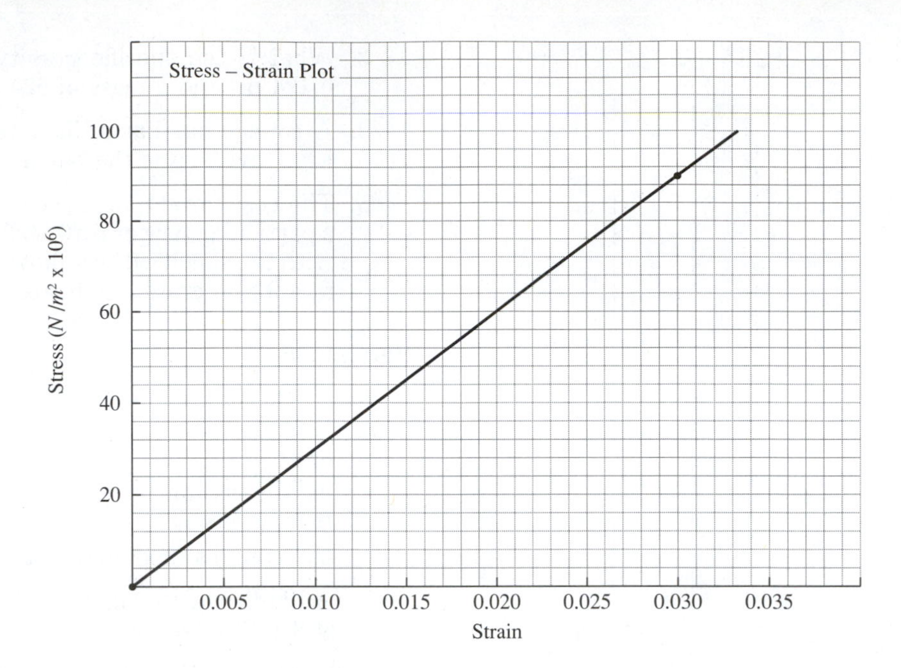

Figure 15–22: *Diagram for Problem 11. Stress-strain plot.*

14. A net torque of 328 ft · lb acts on the end of a steel rod 8.4 ft long and 1.00 in in diameter.

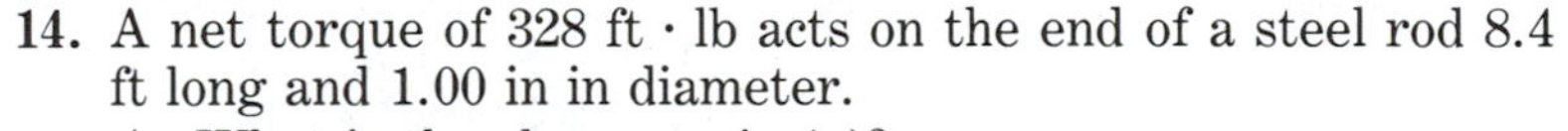

A. What is the shear strain (σ)?

B. What is the shear stress?

C. Through what angle (θ) does the rod twist?

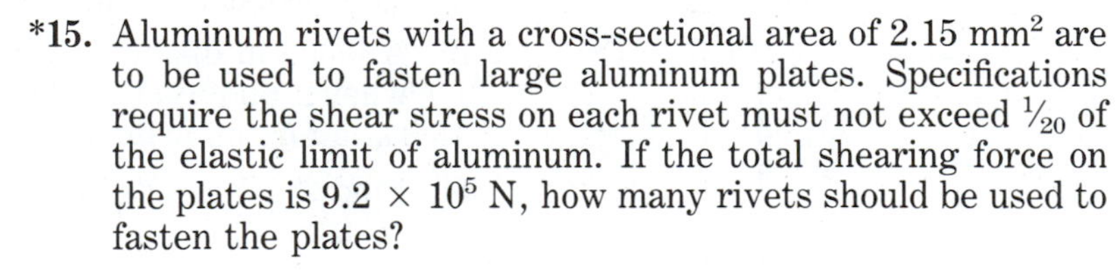

***15.** Aluminum rivets with a cross-sectional area of 2.15 mm^2 are to be used to fasten large aluminum plates. Specifications require the shear stress on each rivet must not exceed $\frac{1}{20}$ of the elastic limit of aluminum. If the total shearing force on the plates is 9.2×10^5 N, how many rivets should be used to fasten the plates?

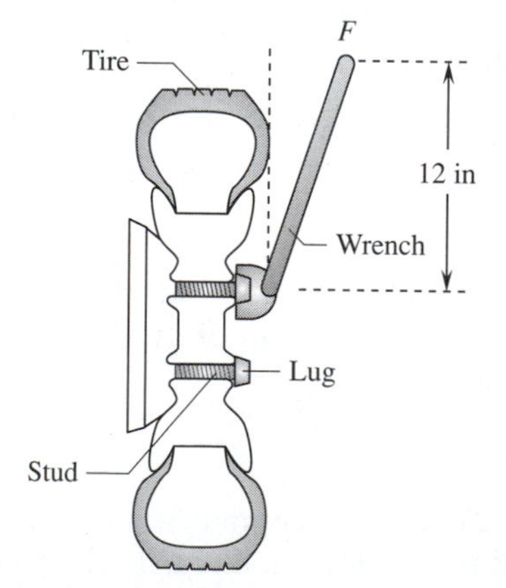

Figure 15–23: *Diagram for Problem 12. A lug wrench is used on a wheel.*

Section 15.5:

16. A wire initially 1.20 m long is stretched to 1.67 m at fracture. What is the percent elongation of the material?

17. 30–70 brass (30% zinc and 70% copper) has an elongation of 60%. To what length can a piece of brass wire be stretched if its initial length is 12.3 ft?

18. A sample of steel is given a standard Brinell test. The dent is 9.2 mm in diameter and 6.9 mm deep. What is the Brinell hardness number (BHN)?

19. For the two materials shown in Figure 15–24:

A. which is more ductile?

B. which has the larger ultimate strength?

C. estimate the toughness of the two materials.

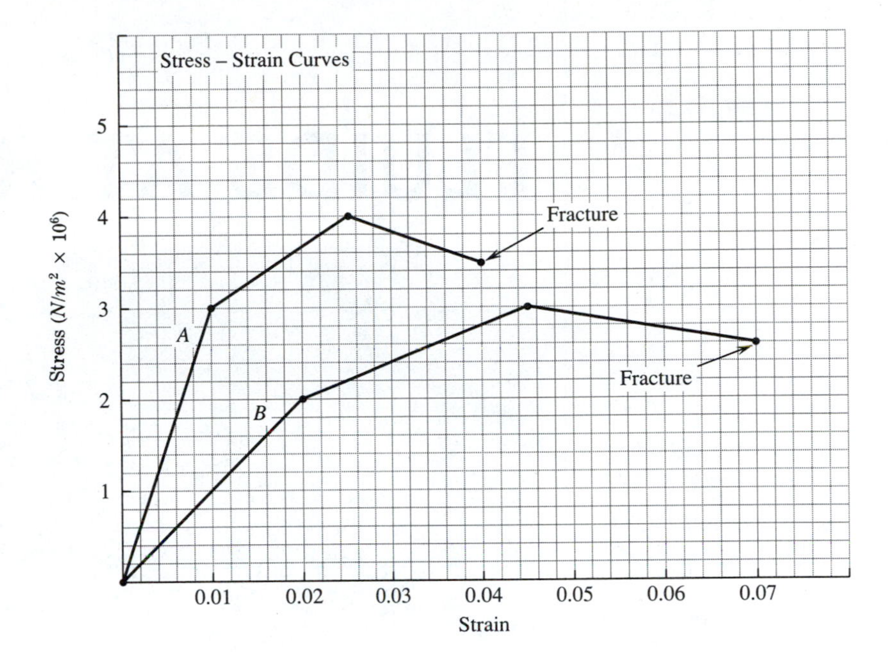

Figure 15–24: *Diagram for Problem 19. Stress-strain curves for two materials.*

Chapter 16

FLUIDS AT REST

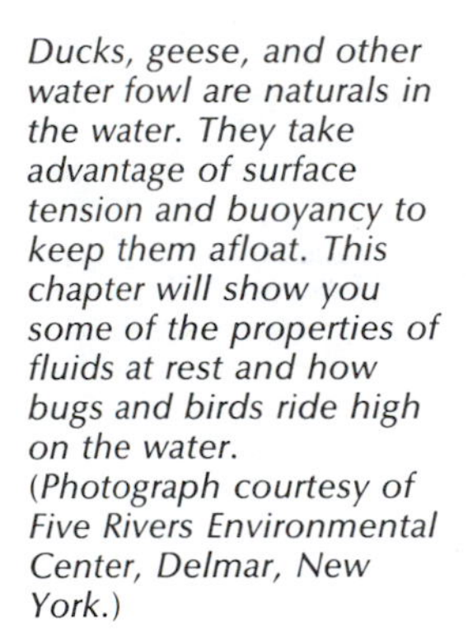

Ducks, geese, and other water fowl are naturals in the water. They take advantage of surface tension and buoyancy to keep them afloat. This chapter will show you some of the properties of fluids at rest and how bugs and birds ride high on the water. (Photograph courtesy of Five Rivers Environmental Center, Delmar, New York.)

OBJECTIVES

In this chapter you will learn how:

- to identify fluids
- to find the pressure a fluid exerts on its container
- pressure changes with depth in a fluid
- to calculate buoyant forces (Archimedes' principle)
- a change of pressure is transmitted through a confined fluid (Pascal's principle)
- to calculate the mechanical advantage and efficiency of a hydraulic system
- to calculate the bulk modulus and compressibility of a fluid
- to identify isotropic substances

Time Line for Chapter 16

1490 Capillary action is studied by Leonardo da Vinci.
1586 Simon Stevinus of Bruges finds that the pressure exerted by a column of liquid depends upon the height of the column.
1612 Galileo Galilei develops ideas on hydrostatics.
1643 Evangelista Torricelli invents the mercury barometer.
1663 Blaise Pascal publishes a paper presenting Pascal's principle.

A heavy rock is lifted out of a river by a light rope. As the rock clears the surface of the water, the tension on the rope increases. The rope breaks. A small force is exerted on the small input piston of a hydraulic lift. The output piston lifts the front end of a heavy truck. A scuba diver experiences increasing forces on her body as she swims down to deeper water. Robot submarines must be used for very great depths. These are effects of static fluids. The study of fluids at rest is called **hydrostatics.**

A **fluid** is a substance that flows. It offers little resistance to the forces that change its shape. Both gases and liquids are fluids. Because liquids do not have much space between atoms, they do not compress easily. When forces are exerted upon them, their density remains fairly constant. Gases, on the other hand, have a lot of empty space between atoms. They easily compress into smaller volumes.

Some of the equations we will develop in this chapter will assume that the fluid has a constant density. This means that the equations will give good numerical results only for liquids and will only qualitatively describe the behavior of gases. We will say more about gases in Chapter 20.

16.1 PRESSURE

When a pencil is placed in a glass, the force of its weight is supported at two well-defined points: where the pencil lead touches the bottom of the glass and where the pencil leans against the rim (see Figure 16–1). When a fluid is placed in a glass, it flows until it pushes against the sides of the glass. A reacting force stops the flow. The molecules of water pile up on top of each other, taking the shape of the container. The bottom of the glass with straight, vertical sides holds up the weight of water. The molecules push against the glass wherever they are in contact with it. Forces are not localized as they are in the case of the pencil in the glass.

The kinds of forces exerted by a stationary fluid are called **hy-**

drostatic forces. It is easiest to describe hydrostatic forces in terms of force per unit area. This quantity is called **pressure (P).**

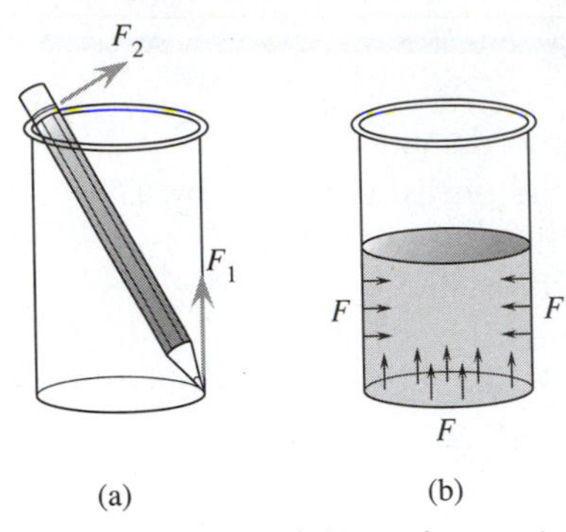

Figure 16–1: *(a) A solid in a glass retains its shape. Forces act only at the two points of contact. (b) A fluid takes the shape of the container. Forces act between the fluid and glass at all surfaces of contact.*

$$pressure = \frac{total\ hydrostatic\ force}{area}$$

$$P = \frac{F}{A} \qquad \textbf{(Eq. 16–1)}$$

Several kinds of units are used to describe pressure. Here are a few of them.

Metric units	1 pascal (Pa) = 1 N/m²
British units	1 lb/in² (psi) = 6895 Pa
Other units	1 torr = 1 mm of Hg (mercury) at 0° C. 1 torr = 133.3 Pa is used in vacuum systems. It originates from the era when mercury barometers and pressure gauges were used.
	1 millibar = 100 N/m² = 100 Pa. Millibars are sometimes used in meteorology.
	1 standard atmosphere (atm) = 14.7 psi = 1.013×10^5 Pa. This is the pressure of the earth's atmosphere at sea level at room temperature.

☐ **EXAMPLE PROBLEM 16–1: THE GLASS OF WATER**

A glass with straight sides contains 0.58 lb of water. The inside diameter of the glass is 2.2 in. What pressure is exerted by the weight of the water on the bottom of the glass?

■ ***SOLUTION***

The area of the bottom of the glass is:

$$A = \pi r^2$$
$$A = 3.14 \times (1.1\ \text{in})^2$$
$$A = 3.8\ \text{in}^2$$

The bottom of the glass must support the weight of the water. The pressure is:

$$P = \frac{F}{A}$$
$$P = \frac{0.58\ \text{lb}}{3.8\ \text{in}^2}$$
$$P = 0.15\ \text{lb/in}^2$$

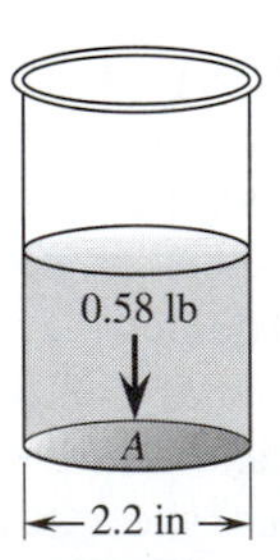

Figure 16–2: *The bottom of a container with vertical sides supports the weight of the liquid.*

16.2 PROPERTIES OF FLUID PRESSURE

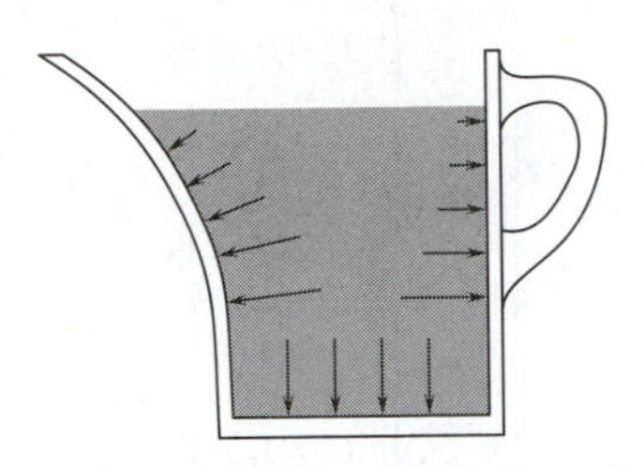

Figure 16–3: Because a fluid at rest does not support a shear stress, pressure is always perpendicular to the surface of the container that holds it.

Fluid pressure has the same units as stress, but it behaves differently. A solid resists shearing stresses while changing shape. A liquid at rest will easily change shape under a shear stress. Even a viscous liquid such as honey cannot maintain a resistance to shear for very long as a solid can. In equilibrium, a static liquid or gas has no shearing forces. This means that where it is in contact with a surface the forces must be at right angles to the surface (see Figure 16–3).

A fluid exerts forces at right angles to any surface in contact with it.

The pressure of water contained in a cup pushes outward on the walls and bottom of the cup everywhere that the water is in contact with the cup. **Hydrostatic pressure,** the pressure caused by a fluid at rest, cannot have a component of force parallel to its container wall.

Fluid pressure has another interesting property. If a cork is completely immersed in water, the water will exert forces on the cork in all directions. The fluid will press down on the top of the cork, up on the bottom, and inward on the sides. This happens because a fluid exerts a force normal to any surface with which it comes in contact. The orientation of the surface is not important (see Figure 16–4).

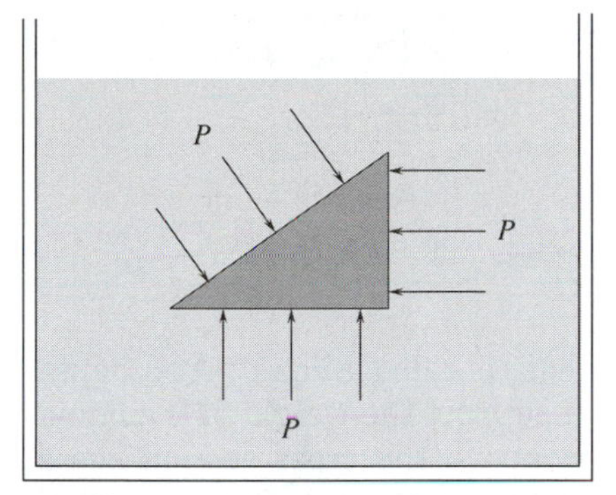

Figure 16–4: Pressure at points in a fluid acts equally in all directions.

Forces are exerted equally in all directions at points in a fluid.

Figure 16–5 shows an experiment with agribusiness tomatoes, developed for uniformity in size and shape to make machine handling possible. If we place one tomato at the bottom of a pipe, the bottom must support the weight of the tomato. If a tomato weighs 0.25 lb, then the force at the bottom is 0.25 lb. If we add a tomato, the bottom must support two tomato weights, or 0.50 lb. The bottom tomato needs to hold up only the added tomato; the force between the first and second tomato is only 0.25 lb. If we add several tomatoes, the force between each tomato is just enough to hold up the weight of the tomatoes above it. The weight of all the tomatoes must be borne by the bottom of the pipe.

Figure 16–6 is a plot of the forces between tomatoes as a function of depth. The relationship is a direct proportion.

Molecules piled in a container behave in the same way. Each molecule must support the weight of the molecules above it. Liquids, like agribusiness tomatoes, do not easily compress. We can assume a constant density of the liquid.

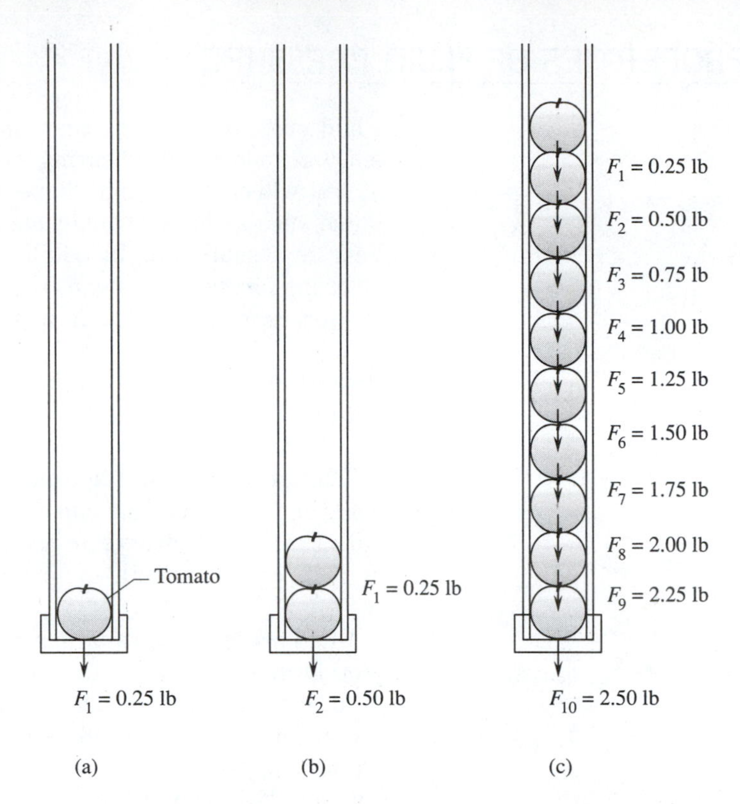

Figure 16–5: *Uniform tomatoes are placed in a pipe. (a) The bottom of the pipe supports the weight of one tomato. (b) The bottom supports the weight of two tomatoes. The bottom tomato supports the weight of the second tomato. (c) The bottom of the pipe supports the weight of all 10 tomatoes. Each tomato must support the ones above it.*

Figure 16–7 depicts a cylinder that contains a liquid. At a depth (h) a cross section (A) of liquid must support the weight of the liquid above it. The weight of the liquid above the cross section has a volume $A \times h$. If its density is ρ, then the weight (**w**) of the supported volume is:

$$\mathrm{w} = \mathrm{g}\,\rho\,V$$

$$\mathrm{w} = \mathrm{g}\,\rho\,A\,h$$

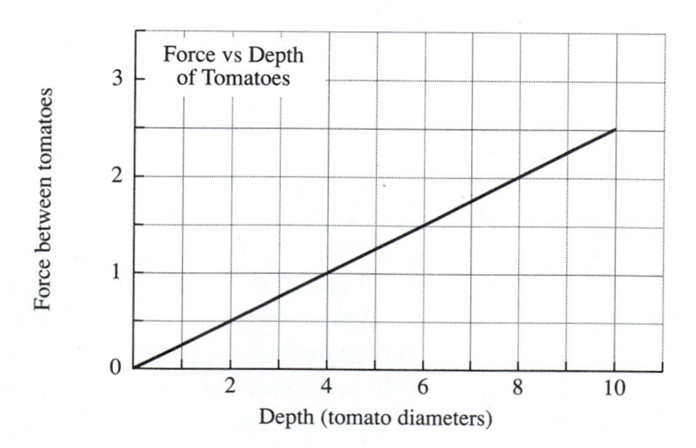

Figure 16–6: *A plot of force versus depth for the stack of tomatoes shown in Figure 16–5c.*

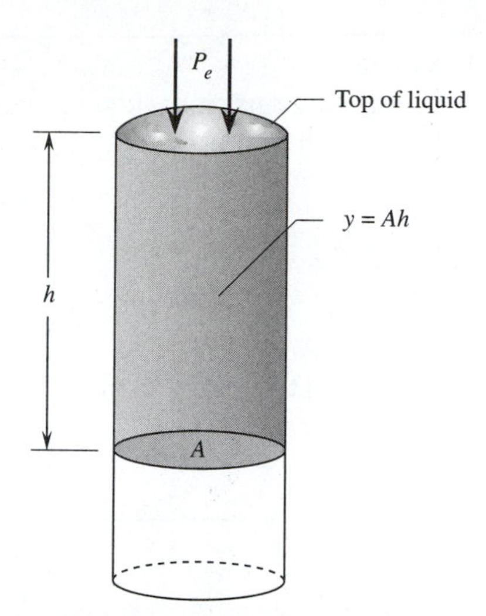

Figure 16–7: *The pressure (**F**/A) at a depth h depends on the weight of the liquid above it (ρ g A h) and the external pressure (**P_e**) pushing down on the top of the fluid.*

The pressure on the cross section (A) of fluid is weight divided by the cross section.

$$P = \frac{w}{A} = \frac{g \rho A h}{A}$$

$$P = g \rho h$$

$$P = D h$$

The increase in pressure with depth is directly proportional to the weight density of the fluid.

If we want to find the total pressure at a depth h in a fluid, we need to add on any pressure existing at the surface. In Figure 16–7, the liquid is in a cylinder with a tight-fitting, lightweight membrane on top. A weight sits on the membrane. A cross section of water at a depth h must support the weight of the water above it plus the added weight on the membrane. The external pressure exerted by the membrane on the top surface of the water is P_e. We can find P_e by dividing the added weight by the cross section.

$$P = (g \rho h) + P_e$$

or **(Eq. 16–2)**

$$P = (D h) + P_e$$

Equation 16–2 works well for most liquids. It does not work well with gases. Since gases are easily compressed, as we move deeper into a column of gas the density of the gas increases. A unit volume of gas near the bottom of the column weighs more than a unit volume near the top. Pressure *does* increase with depth, but not linearly. Figure 16–8 shows the variation of pressure of the earth's atmosphere for the first few miles above the ground.

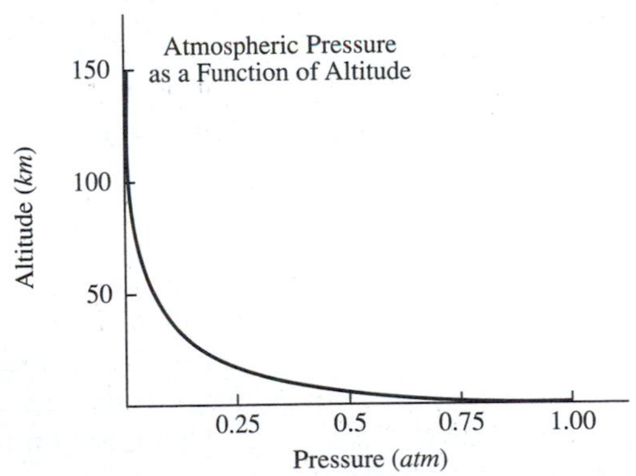

Figure 16–8: *A plot of atmospheric pressure as a function of height above the earth's surface. In this case, pressure is not described by Equation 16–2.*

EXAMPLE PROBLEM 16–2: THE PUMP VALVE

A lift pump lifts water a height of 12.4 m against a back pressure of 1.30×10^5 Pa. What is the pressure on the pump valve?

SOLUTION

The density of water is 1.00×10^3 kg/m³.

$$P = (g\,\rho\,h) + P_e$$

$$P = (1.00 \times 10^3\ \text{kg/m}^3 \times 9.8\ \text{m/s}^2 \times 12.4\ \text{m}) + (1.30 \times 10^5\ \text{Pa})$$

$$P = (1.21 \times 10^5\ \text{Pa}) + (1.30 \times 10^5\ \text{Pa})$$

$$P = 2.51 \times 10^5\ \text{Pa}$$

16.3 BUOYANT FORCES

Let us look at the forces acting on something immersed in a fluid (see Figure 16–9). A block with a cross-sectional area A and height $= (h_2 - h_1)$ is submerged in a liquid with a density of ρ_f. Pressure pushes on all sides of the block. Remember force is a vector quantity. We will take to the right and upward as the positive directions.

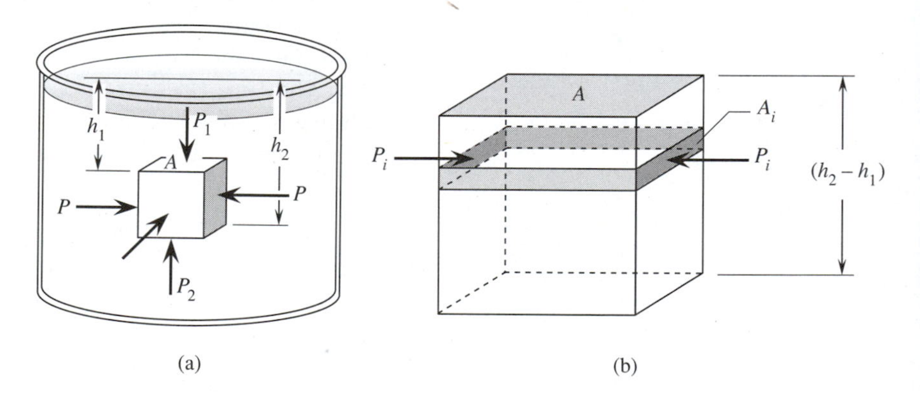

Figure 16–9: *Pressure acting on a block of material immersed in a fluid. (a) Pressure* P_2 *is larger than pressure* P_1 *because the bottom of the block is at a deeper depth. (b) Strips on the side of the block at the same depth have equal but opposite forces* ($F_i = P_iA_i$) *acting on them.*

The forces on the sides of the block cancel out. For each strip of the block (A_i) at a depth of h_i there are equal-sized forces on opposite sides of the block. The top and bottom forces do not have the same magnitude because they are at different depths.

The force on the bottom of the block pushing up is:

$$F_2 = A\,P_2 = A\,[(\rho_f\,g\,h_2) + P_e]$$

Water Bugs and Sinking Ducks

(Photograph courtesy of Five Rivers Environmental Center, Delmar, New York.)

Water bugs are able to walk on the surface of a pond because water molecules have a very strong attraction to each other. In the depth of the water, each molecule is equally attracted by molecules in all directions. At the surface, the top layers of molecules have attractive forces pulling down, but no forces pulling up. This makes a few layers of compressed molecules on the surface of the water. This surface layer, which acts like a stretched rubber sheet, is called the surface tension layer.

The surface tension layer resists attempts to break through the surface layer of molecules. The amount of weight held up by the surface is proportional to the length of contact of the object on the surface (*force* [**F**] = surface tension [**T**] × length [*L*].)

The surface tension for water is 72×10^{-2} N/m. Suppose we place a 4.0 cm long needle on the water. The surface tension will hold up to $\mathbf{F} = 0.04 \text{ m} \times 72 \times 10^{-2} \text{ N/m} = 0.029 \text{ N}$. If the needle weighs less than this, it will float. Razor blades float well, too, because the numerous holes provide a lot of contact length.

Water bugs usually have furry feet, which increases the amount of contact surface. Each whisker adds length.

Liquids other than water have molecules that have a strong mutual attraction. Oil molecules are strongly attracted to each other. We say there is a strong adhesive force. Water and oil do not mix because the oil molecules have a stronger attraction to other oil molecules than to water molecules.

Ducks coat their feathers with oil. Water does not penetrate the fluffy layer of feathers surrounding the duck. The air trapped in the feathers makes the duck very buoyant. Ducks usually ride high on the water. Detergents placed in the water destroy the surface tension; the water can penetrate the feathers. The duck will lose much of its extra buoyancy from the trapped air and will sink deeper into the water.

The force on the top of the block pushing down is:

$$F_1 = -A\,P_1 = -A[(\rho_f\, g\, h_1) + P_e]$$

The sum of these two forces is:

$$F_2 - F_1 = \rho_f\, g\, A\,(h_2 - h_1)$$

$A\,(h_2 - h_1)$ is the volume of the rectangular block. The net force the fluid exerts on the block is:

$$F_B = \rho_f\, g\, V \qquad \textbf{(Eq. 16–3)}$$

Equation 16–3 is sometimes called **Archimedes' principle.** When something is immersed in a fluid, the fluid exerts a force upward on the object. The force depends on the weight density of the fluid and on the volume of fluid that is displaced. It does *not* depend on the density of the immersed object.

Archimedes' principle has many applications. One application is an industrial process called **flotation.** It is used to enrich low-grade ores such as zinc blende (zinc sulfide), copper pyrite (copper sulfide), and galena (lead sulfide). The ore is finely crushed and mixed with water and a small percentage of oil in a tank. The oil adheres to the sulphide metal compounds, but not to the stony material. Air bubbles are blown into the bottom of the tank. Only the oily particles—the sulfides—adhere to the bubbles. The metal ore is brought to the surface where it can be skimmed off.

□ **EXAMPLE PROBLEM 16–3: THE ROCK**

A rock has a density of 296 lb/ft³ and a weight of 124 lb. It lies at the bottom of a pond. What minimum force is required to lift the rock to the surface?

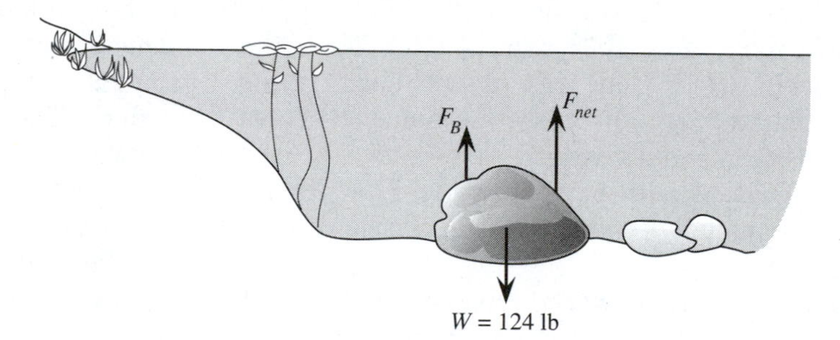

Figure 16–10: *The forces acting on a rock at the bottom of a pond. A buoyant force (F_B) helps lift the weight of the rock. Its apparent weight in water is F_{net}.*

■ ***SOLUTION***

The weight density of fresh water is 62.5 lb/ft³. The volume of the rock is:

$$V = \frac{W}{D} = \frac{124 \text{ lb}}{296 \text{ lb/ft}^3}$$

$$V = 0.419 \text{ ft}^3$$

The minimum force required to lift the rock is the difference between the weight of the rock in air and the buoyant force.

$$F_{net} = W - F_B = W - (D\,V)$$

$$F_{net} = 124 \text{ lb} - (62.5 \text{ lb/ft}^3 \times 0.419 \text{ ft}^3)$$

$$F_{net} = 97.8 \text{ lb}$$

EXAMPLE PROBLEM 16–4: THE SHIP

An oceangoing cargo ship has a cross-sectional area (A) of 218 m^2. It loads on 223 metric tons (223×10^3 kg) of cargo. By how much does the water line rise on the ship?

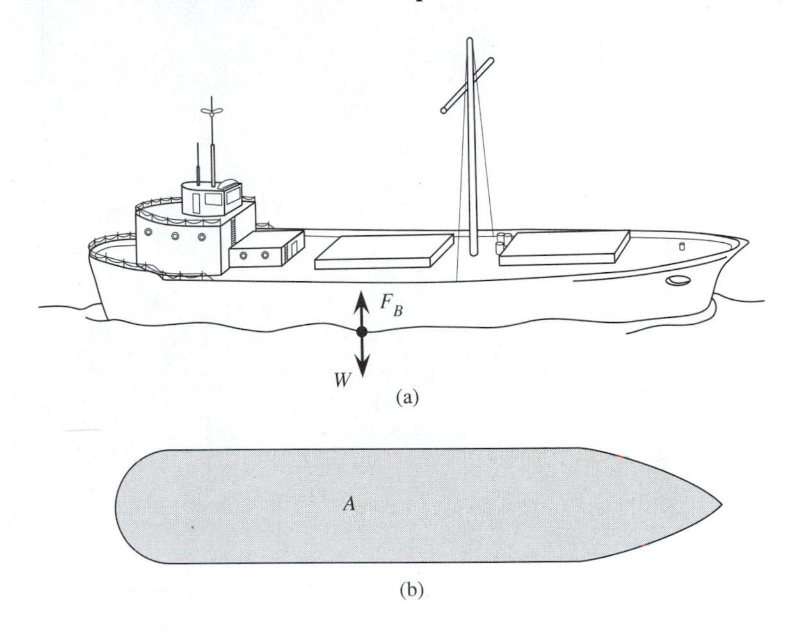

Figure 16–11: *As a cargo ship loads on cargo it sinks farther into the water. If h is the distance the ship sinks, the volume of water displaced is V = A h. A is the cross-sectional area of the ship.*

SOLUTION

The ship will sink farther into the water until the increase in buoyant force matches the added weight (W). The density of seawater is 1.03×10^3 kg/m^3. The change of volume is $V = A\,h$.

$$W = F_B = \rho_f\, g\, V = \rho_f\, g\, A\, h$$

or

$$h = \frac{W}{\rho_f\, g\, A}$$

$$h = \frac{223 \times 10^3 \text{ kg} \times 9.80 \text{ m/s}^2}{1.03 \times 10^3 \text{ kg} \times 9.80 \text{ m/s}^2 \times 218 \text{ m}^2}$$

$$h = 0.993 \text{ m}$$

16.4 PASCAL'S PRINCIPLE

Equation 16–2 describes the pressure with depth for a column of liquid. The shape of the column is not indicated in the equation. The pressure depends only on the depth of the liquid, the density of the

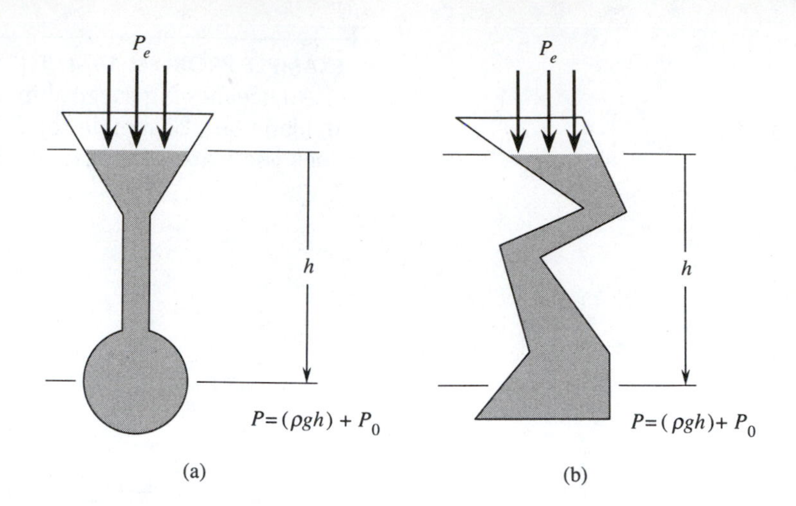

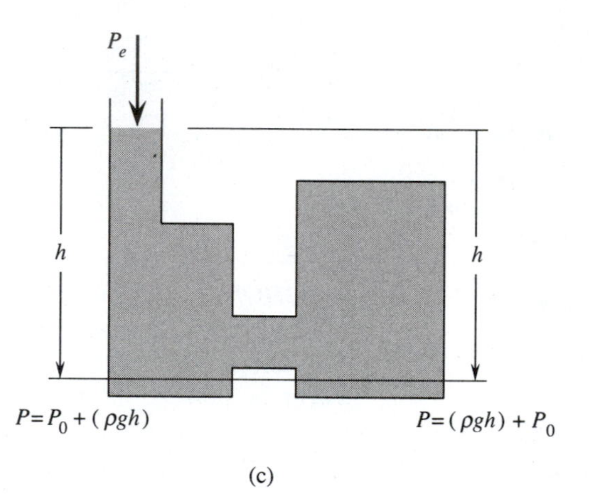

Figure 16–12: *(a,b) The pressure at a depth* h *in a liquid does not depend on the shape of the container. It depends only on the external pressure* (P_e) *at the top of the liquid and* h. *(c) The pressure in the side compartment is the same at depth* h *as in the main compartment.*

liquid, and the external pressure $\mathbf{P_e}$. Figure 16–12 shows some different-shaped containers. They all have the same pressure at depth h since the top ends are open to the atmosphere. The external pressure in each case is atmospheric pressure, P_0. The pressure inside the liquid is:

$$P_1 = (\rho\ g\ h) + P_0$$

Let us replace the open end of a column with a lightweight piston. The atmosphere still transmits its pressure through the piston (see Figure 16–13a). There is no change of the pressure anywhere in the column. It is still $(\rho\ g\ h) + P_0$. In Figure 16–13b we will increase the external pressure by pressing down on the piston. The increase

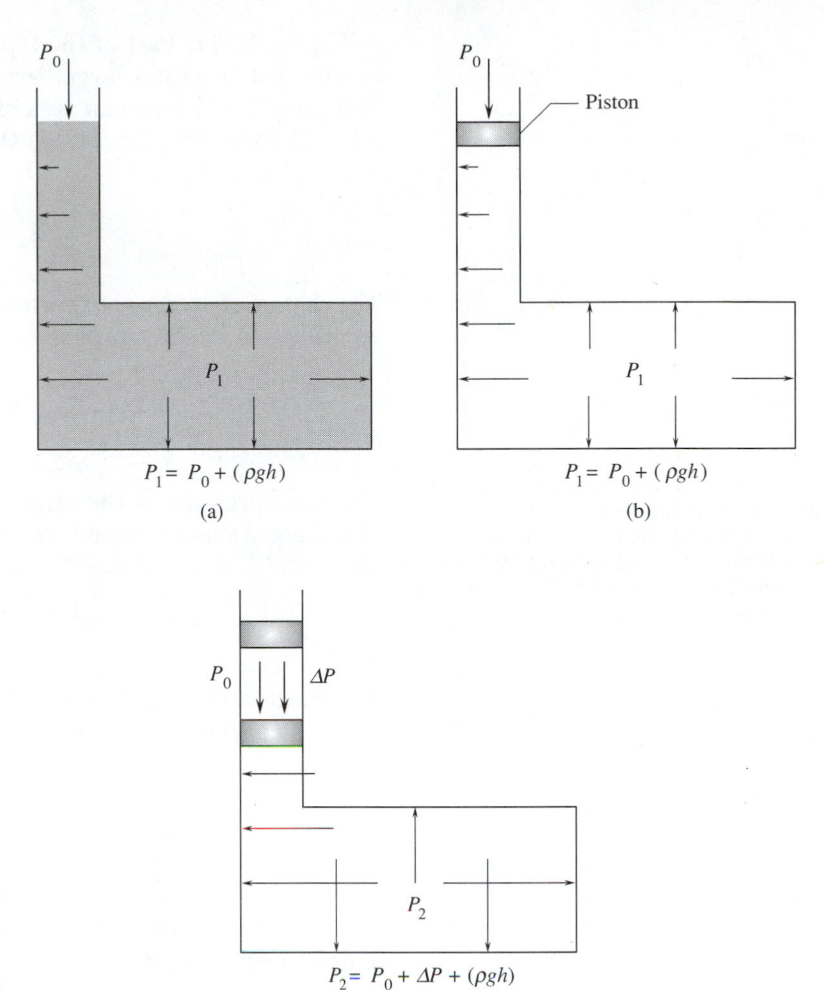

Figure 16–13: *(a) Atmospheric pressure pushes down on the top of the liquid. (b) A lightweight piston is placed on top of the fluid. Since its weight is negligible there is no change of pressure in the confined liquid. (c) An added external pressure is exerted on the piston. The added pressure is transmitted throughout the vessel.*

of pressure is Δ P. The total external pressure is now $P_e = P_0 + \Delta P$. The pressure at any depth is:

$$P_2 = (\rho\ g\ h) + P_0 + \Delta P$$

The change in pressure at any point is:

$$P_2 - P_1 = \Delta P$$

The *change* of pressure exerted on a confined liquid is transmitted to every point in the liquid. This rule is called **Pascal's principle.**

Pascal's principle applies to all fluids including gases. We can use it to make a machine that will give us a mechanical advantage. Look

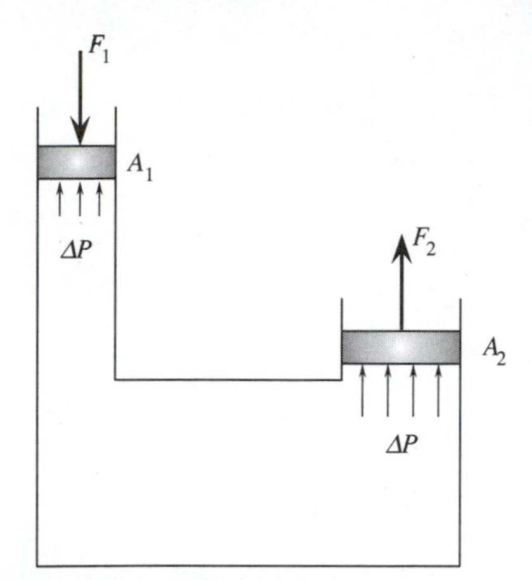

Figure 16–14: *The change of pressure (ΔP) is transmitted uniformly throughout the confined liquid. Force F_2 is larger than force F_1 because the second piston has a larger area. $F_1 = \Delta P \times A_1$. $F_2 = \Delta P \times A_2$.*

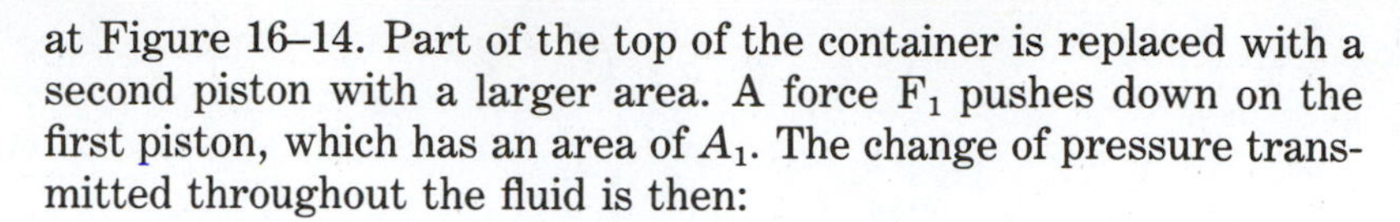
at Figure 16–14. Part of the top of the container is replaced with a second piston with a larger area. A force F_1 pushes down on the first piston, which has an area of A_1. The change of pressure transmitted throughout the fluid is then:

$$\Delta P = \frac{F_1}{A_1}$$

The change of pressure experienced by the second piston is the same according to Pascal's principle.

$$\Delta P = \frac{F_2}{A_2}$$

Since the pressure is the same on both pistons, the left-hand sides of the equations are equal, so we can say

$$\frac{F_1}{A_1} = \frac{F_2}{A_2}$$

The ratio of output force divided by the input force gives us the actual mechanical advantage.

$$\frac{F_2}{F_1} = \frac{A_2}{A_1} \qquad \textbf{(Eq. 16–4)}$$

We have assumed an ideal situation. No work was lost to friction. In a real case, some of the work done by $\mathbf{F_1}$ goes into heat. For a hydraulic system we can write:

$$AMA = \frac{F_2}{F_1}$$

$$IMA = \frac{A_2}{A_1}$$

$$eff = \frac{F_2/F_1}{A_2/A_1} = \frac{P_2}{P_1}$$

The efficiency is the ratio of the output pressure to the input pressure.

Let us look at the conservation of work and energy for this system. Work is done by F_1 as the first piston moves a distance S_1. The work done by the second piston is $F_2 \times S_2$ (see Figure 16–15).

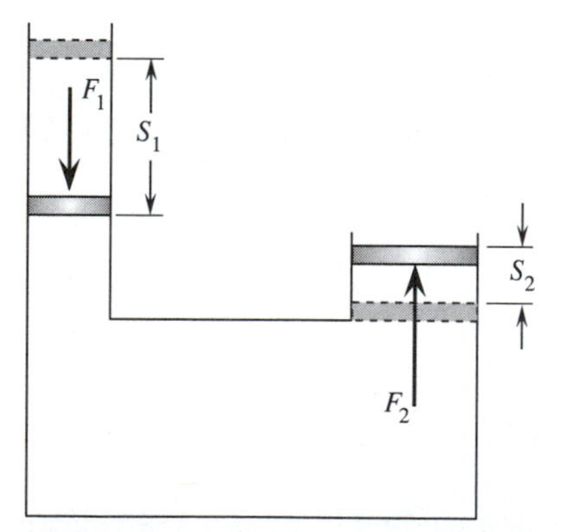

Figure 16–15: *The two pistons do the same amount of work. The smaller force (F_1) moves through a larger distance than the larger force (F_2).*

$$work\ in = work\ out + heat$$
$$F_1 S_1 = (F_2 S_2) + Q$$

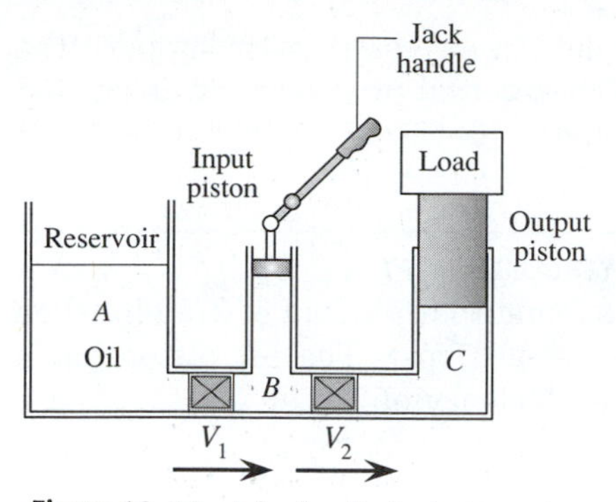

Figure 16–16: A hydraulic jack. V_1 and V_2 are one-way valves. They open only when the pressure to the left of the valve is larger.

The relationship between force and pressure is: $F = P A$.

$$P_1 A_1 S_1 = (P_2 A_2 S_2) + Q$$

If there is no frictional force $Q = 0$.

For a fluid that is incompressible, such as a liquid, the volumes of fluid moved by the pistons must be the same.

$$V = A_1 S_1 = A_2 S_2$$

$$S_1 = \frac{A_2 S_2}{A_1}$$

The distance each piston moves is inversely proportional to its cross-sectional area. When the system has a large mechanical advantage, the smaller piston must move a much larger distance than the output piston. In a hydraulic jack this is accomplished by using valves. Figure 16–16 shows a simplified diagram. Valves V_1 and V_2 are one-way valves. If the pressure in the reservoir (P_A) is less than the pressure in region B (P_B), the valve V_1 will close. If P_A is greater than P_B, the valve will open. Valve V_2 opens only when the pressure in region B is larger than in chamber C ($P_B > P_C$).

On the downstroke of the input piston, V_2 opens and V_1 closes. Fluid is forced into chamber C, increasing the pressure on the output piston. On the upstroke of the input piston, V_2 closes and V_1 opens. Fluid is drawn from the reservoir, refilling region B. The system is now ready for another downstroke. The total distance S_1 is the sum of all the downstrokes.

Many hydraulic systems use a motor-operated piston instead of a handle. Fork lifts, blade control of bulldozers, power steering in automobiles, cherry pickers, and dump trucks all use the same principle.

Some systems use both liquids and gases. The lift found in a

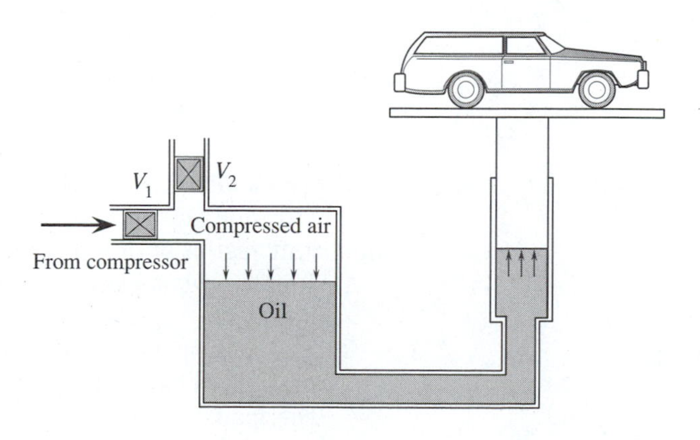

Figure 16–17: An air-oil hydraulic lift. A compressor exerts the external pressure on the liquid. Valve V_1 lets air in from the compressor. Valve V_2 is used to bleed air from the system to lower the lift.

commercial garage is an example. An air compressor supplies the necessary external pressure. The external pressure is transmitted directly to the reservoir (see Figure 16–17).

☐ EXAMPLE PROBLEM 16–5: THE HYDRAULIC LIFT

A hydraulic lift in a garage is found to lift a load of 3750 lb when the compressor exerts a pressure of 37 psi. The lift piston has a diameter of 12.0 in. What is the efficiency of the lift?

■ *SOLUTION*

The area of the output piston is:

$$A = \pi R^2$$

$$A = 3.14 \times (6.0 \text{ in})^2$$

$$A = 110 \text{ in}^2$$

The output pressure is the cross-sectional area of the piston divided into the load.

$$P_2 = \frac{F}{A}$$

$$P_2 = \frac{3750 \text{ lb}}{110 \text{ in}^2}$$

$$P_2 = 33 \text{ psi}$$

The efficiency is then:

$$eff = \frac{P_2}{P_1}$$

$$eff = \frac{33 \text{ psi}}{37 \text{ psi}}$$

$$eff = 0.89$$

☐ EXAMPLE PROBLEM 16–6: THE JACK

A hydraulic jack has an input piston with a 0.80 cm diameter and an output piston with a diameter of 4.60 cm.

A. Ideally, what force is exerted on the input piston to lift a 500-kg load?

B. The jack operates with a handle with a lever arm with an *IMA* of 12.2. What force is exerted at the end of the handle to lift the load?

C. What is the overall *AMA* of the machine if the force on the handle is 14.2 N?

■ *SOLUTION*

A. Input force on the piston:

$$F_1 = F_2 \times \frac{A_1}{A_2} = (m\ g) \times \frac{\pi\ (D_1/2)^2}{\pi\ (D_2/2)^2}$$

$$F_1 = 500\ \text{kg} \times 9.80\ \text{m/s}^2 \times \frac{(0.80\ \text{cm})^2}{(4.60\ \text{cm})^2}$$

$$F_1 = 150\ \text{N}$$

B. Ideal input force on handle:

$$F_{in} = \frac{F_{out}}{IMA}$$

$$F_{in} = \frac{150\ \text{N}}{12.2}$$

$$F_{in} = 12\ \text{N}$$

C. Total *AMA:*

$$AMA = \frac{F_{out}}{F_{in}} = \frac{500\ \text{kg} \times 9.80\ \text{m/s}^2}{14.2\ \text{N}}$$

$$AMA = 345$$

16.5 COMPRESSIBILITY

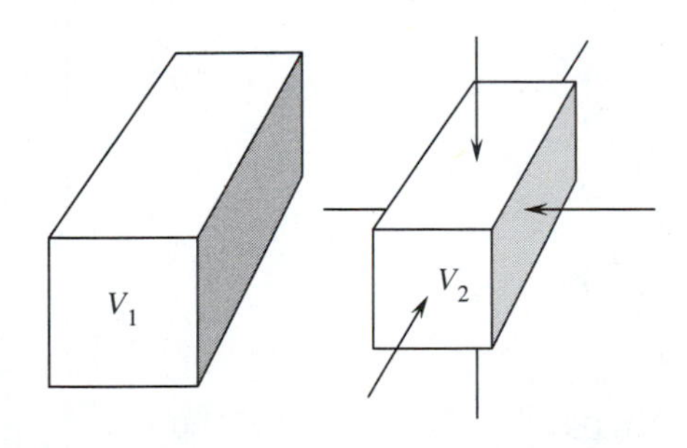

$\Delta V = V_1 - V_2$

Figure 16–18: *Hydrostatic pressure acts on all surfaces of a material, decreasing its volume.*

Any material under hydrostatic pressure will be compressed. The pressure pushes inward normal to the surface of a solid immersed in a fluid (see Figure 16–18). As the pressure increases, the volume of the object will decrease. The pressure causes a volume strain, $\Delta V/V$. Liquids will also compress under hydrostatic pressure.

We can find a material constant in the same way we developed Young's modulus and shear modulus. We will call this the **bulk modulus.** Let A be the total surface of the object and F be the total force pushing inward.

$$\textit{bulk modulus}\ (B) = \frac{\textit{stress on volume}}{\textit{volume strain}}$$

$$B = \left(\frac{F/A}{\Delta V/V}\right)$$

F/A is nothing more than the fluid pressure (**P**) acting on the volume.

$$B = -\left(\frac{P}{\Delta V/V}\right) \qquad \textbf{(Eq. 16–5)}$$

The minus sign indicates that the volume decreases while the pressure increases.

The reciprocal of the bulk modulus measures the fractional change of volume with pressure. This quantity is called the **compressibility** (k) of the material.

$$k = \frac{1}{B} = -\left(\frac{\Delta V/V}{P}\right) \qquad \textbf{(Eq. 16–6)}$$

Table 16–1 gives values of bulk modulus for some common materials.

An **isotropic** material is a material whose properties are independent of direction. A single crystal of brass is not isotropic. A piece of brass made up of millions of small crystals randomly oriented is isotropic. Most polycrystalline materials with random crystal orientation are isotropic.

We now have four material constants for solids: Young's modulus, shear modulus, Poisson's ratio, and bulk modulus. These constants are not all independent for isotropic materials. We can write some of the constants in terms of the other. For example, the bulk modulus

Table 16–1: Bulk modulus for selected materials.

	BULK MODULUS	
MATERIAL	**N/m² (x 10^{10})**	**psi (x 10^6)**
Solids		
Aluminum	6.9	10
Brass	5.9	8.5
Copper	12	17
Lead	0.77	1.1
Steel	16	23
Glass	3.7	5.2
Liquids		
Ethyl alcohol	0.091	0.13
Glycerine	0.48	0.67
Water	0.20	0.29

Note: Values are for near room temperature; varying temperature and extreme pressure will alter them.

(B) and shear modulus (G) can be calculated from Young's modulus (Y) and Poisson's ratio (ν). This is useful if you cannot find data on all four constants. (Be careful, though. These relationships are good only for isotropic materials.)

$$B = \frac{Y}{3(1 - 2\nu)}$$

$$G = \frac{Y}{2(1 + \nu)}$$

☐ **EXAMPLE PROBLEM 16–7: HYDRAULIC OIL**

A hydraulic system uses 0.60 m^3 of hydraulic oil with a bulk modulus of 1.88×10^9 N/m^2.

A. What is the compressibility of the oil?

B. What will be the change of volume under a pressure of 2.00 $\times 10^6$ N/m^2? (That is roughly a pressure of 20 atm or 290 psi.)

■ ***SOLUTION***

A. Compressibility:

$$k = \frac{1}{B} = \frac{1}{1.88 \times 10^9 \text{ N/m}^2}$$

$$k = 5.32 \times 10^{-10} \text{ m}^2\text{/N}$$

B. Change of volume:

$$k = -\left(\frac{\Delta V/V}{P}\right)$$

or

$$\Delta V = -k P V$$

$$\Delta V = -5.32 \times 10^{-10} \text{ m}^2\text{/N} \times 2.00 \times 10^6 \text{ N/m}^2 \times 0.60 \text{ m}^3$$

$$\Delta V = -0.00064 \text{ m}^3$$

The decrease in volume would hardly be noticed. That is why we often treat liquids as incompressible fluids and assume a constant density under pressure.

SUMMARY

The study of fluids at rest is called *hydrostatics*. Hydrostatic forces can be most easily described in terms of pressure. The pressure exerted on a surface is equal to the total force divided by the total area. The most common units of pressure are the pascal (1Pa =

1 N/m^2); psi (1 lb/in^2 = 6895 Pa); and the standard atmosphere (atm) = 14.7 psi = 1.013×10^5 Pa).

A static fluid exerts a force at right angles to surfaces that it contacts. At any point inside a fluid, forces are exerted equally in all directions. The pressure at a depth h in a fluid is the sum of the pressure caused by the weight of the fluid above it plus any external pressure existing at the surface.

The buoyant force acting on an object immersed in fluid depends on the volume of fluid displaced by the object and the density of the fluid. This effect is called Archimedes' principle.

When the external pressure acting on a fluid changes, the change is transmitted uniformly throughout the fluid. This is called Pascal's principle.

The bulk modulus (B) of a material is the ratio of hydrostatic pressure (P) acting on the material to the volume strain ($\Delta V/V$). Compressibility (k) is the reciprocal of the bulk modulus.

Young's modulus (Y), the shear modulus (G), Poisson's ratio (ν), and the bulk modulus (B) are dependent material constants for isotropic materials (materials whose properties are independent of direction). If two of the material constants are known the other two can be calculated.

KEY TERMS

If you can explain the following terms to a friend or classmate, you understand their meaning. If you cannot explain the terms, you should reread the sections in which they are discussed.

Archimedes' principle
bulk modulus
compressibility
flotation
fluid
hydrostatic forces
hydrostatic pressure
hydrostatics
isotropic
Pascal's principle
pressure

EXERCISES

Section 16.1:

1. One atmosphere is equivalent to how many millibars?
2. The term *fluid* includes what states of matter?
3. Two glasses with straight sides are shown in Figure 16–19. They contain the same volume of water, but glass B has a larger diameter.
   A. Does the weight of the water exert the same force on the bottom of both glasses? Explain.
   B. Does the water exert the same pressure on the bottom of both glasses? Explain.

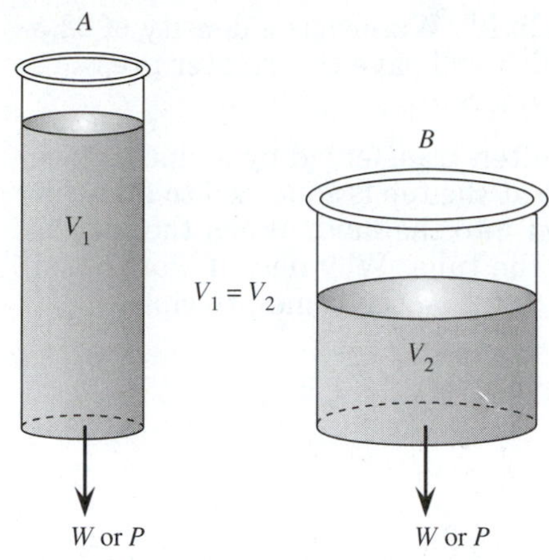

Figure 16–19: Containers with vertical sides hold the same volume of liquid. Container B has a larger cross section than container A.

4. When a car falls into a pond it is almost impossible to open a car door until the inside fills with water. Why?

Section 16.2:

5. Figure 16–20 shows three containers that hold the same height of fluid. Is the pressure at the bottom of each container the same or different? Explain.

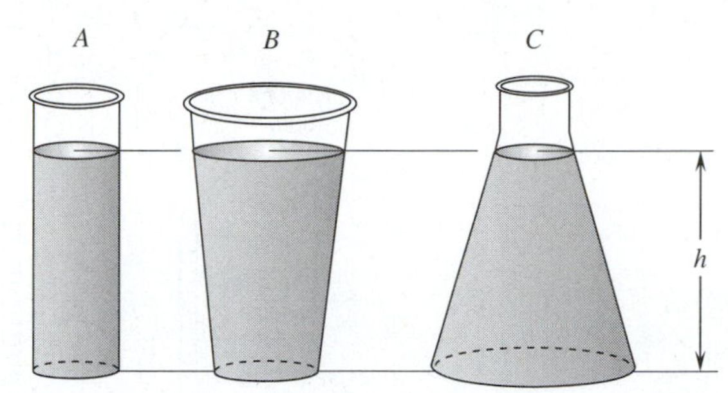

Figure 16–20: Three containers hold the same height of fluid.

6. A full glass of water is inverted on a smooth tabletop (see Figure 16–21). A vacuum is created at the top of the water column in the inverted glass. Why does the water not flow out?

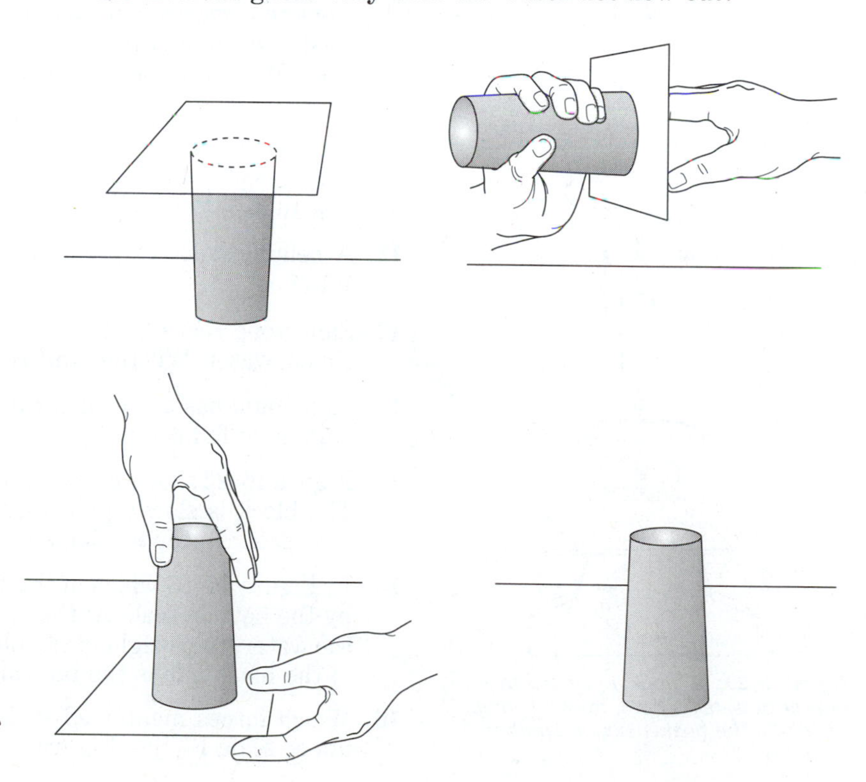

Figure 16–21: A piece of cardboard is used to invert a glass of water without spilling. Water does not flow out of the inverted glass.

7. Alcohol has a density of 49.4 lb/ft^3. Water has a density of 62.5 lb/ft^3. At a depth of 1.2 ft which will have the greater pressure if they are both in open containers?

8. In the laboratory, fluids are often transferred by a pipette (see Figure 16–22). A rubber bulb at the top is squeezed to force air out. The open end is inserted into the fluid. When the bulb is released, fluid is drawn into the tube. Why does it work? List at least two other tools that work on the same principle.

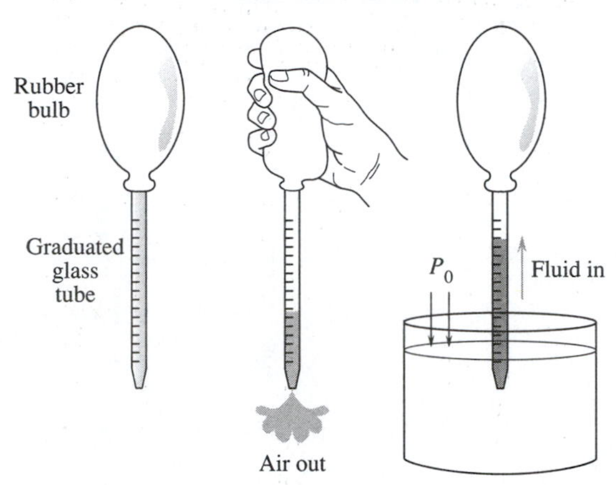

Figure 16–22: *A pipette is used to transfer liquids.*

9. Remote cameras on board an expensive robot submarine were used to examine the wreckage of the Titanic lying on the ocean floor. Why were scuba divers not used for the job?

10. There are two main reasons why dams have a broader base than top. What are they?

Section 16.3:

11. A helium balloon is limited in the height to which it will rise. Why?

12. Sandstone rocks have an average specific gravity of 2.7. They sink in water. Will the same rocks float or sink in liquid mercury?

13. Aluminum has a specific gravity of 2.7. Explain how a sheet of aluminum foil can float.

14. Figure 16–23 shows a metal block immersed in a beaker of water. The block is slowly pulled out of the water. What changes, if any, occur in the readings of scales A and B?

15. In Figure 16–23 which of the following forces is (are) supported by the bottom scale A: the weight of the beaker, the weight of the water, the weight of the block, the buoyant force, the weight of the block minus the buoyant force.

16. Which forces mentioned in Exercise 15 are supported by the upper scale B? See Figure 16–23.

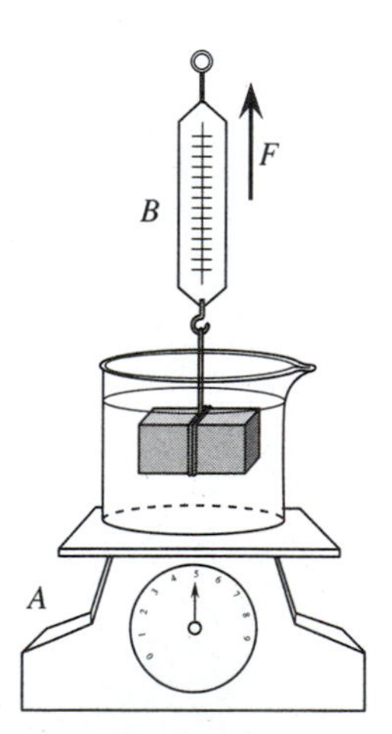

Figure 16–23: *A block immersed in a beaker of water is hung from a spring scale (B). The beaker sits on another scale (A).*

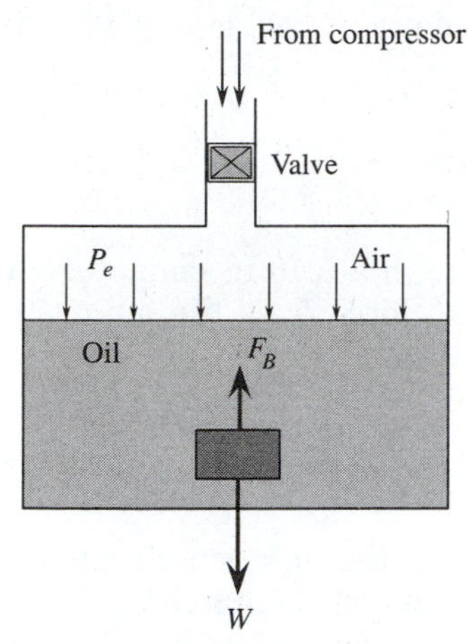

Figure 16–24: A block of material is immersed in the oil of a hydraulic system.

17. As the block in Figure 16–23 is pulled out of the beaker of water, does the pressure on the bottom of the beaker change? Why or why not?

18. Aluminum has a density of 2700 kg/m³. Brass has a density of 8600 kg/m³. A 4-kg block of aluminum and a 4-kg block of brass are immersed in gasoline, which has a density of 670 kg/m³. Will the same size buoyant forces act on the two blocks? Why or why not?

19. Identical volumes of aluminum and brass are immersed in water. Will the aluminum have a larger buoyant force acting on it? Why or why not?

Section 16.4:

20. According to Pascal's principle is the *total* pressure the same at all points in a hydraulic system no matter what the depth?

21. Assume oil is incompressible. A block of solid material is immersed in oil in a closed tank (see Figure 16–24). Compressed air is pumped into the area above the oil level. Will the buoyant force on the block be affected?

22. The diameter of an output piston in a hydraulic lift is doubled. By what factor will the ideal mechanical advantage increase?

23. A hydraulic lift has a mechanical advantage of 24. The input piston moves through an effective distance of 12 in. Through what distance does the output piston move?

Section 16.5:

24. If a material is easily compressed will it have a large or small bulk modulus?

25. What is meant by the term *hydrostatic pressure?* Give an example of hydrostatic pressure. Give an example of pressure that is not hydrostatic.

PROBLEMS

Section 16.1:

1. Find the pressure in the indicated units for the following data:

A.	***B.***	***C.***
F = 24.3 lb	F = 24.5 lb	F = 2.08 $\times 10^5$ N
A = 23.7 in²	A = 0.67 ft²	A = 1.32 m²
P = ? psi	P = ? psi	P = ? Pa

D. $F = 1.23 \times 10^3$ N

$A = 23.4\ cm^2$

$P = ?$ Pa

E. $F = 89.6$ lb

$A = 2.33\ ft^2$

$P = ?$ atm

F. $F = 8.45 \times 10^5$

$A = 0.930\ m^2$ N

$P = ?$ atm

2. A 2-lb coffee can is 6.00 in high with a 5.10 in diameter. At a pressure of 1 atm with what total force does the air push inward on the lateral surface of the can?

3. A force of 214 lb is exerted on a piston by expanding gases in its cylinder, which has a diameter of 2.08 in. What pressure acts on the piston in pounds per square inch? In atmospheres?

4. A pressure of 3.09×10^5 Pa is exerted on a submarine hatch 82.0 cm in diameter. What force acts on the hatch?

Section 16.2:

5. A diving shelter is lowered to a depth of 34 m in the Pacific Ocean (see Figure 16–25). Aquanauts can live in the shelter and work in the surrounding waters without decompressing. In order for the divers to freely enter and leave the shelter the atmospheric pressure in the shelter must match the pressure of the surrounding water to keep the seawater out. What is this absolute pressure?

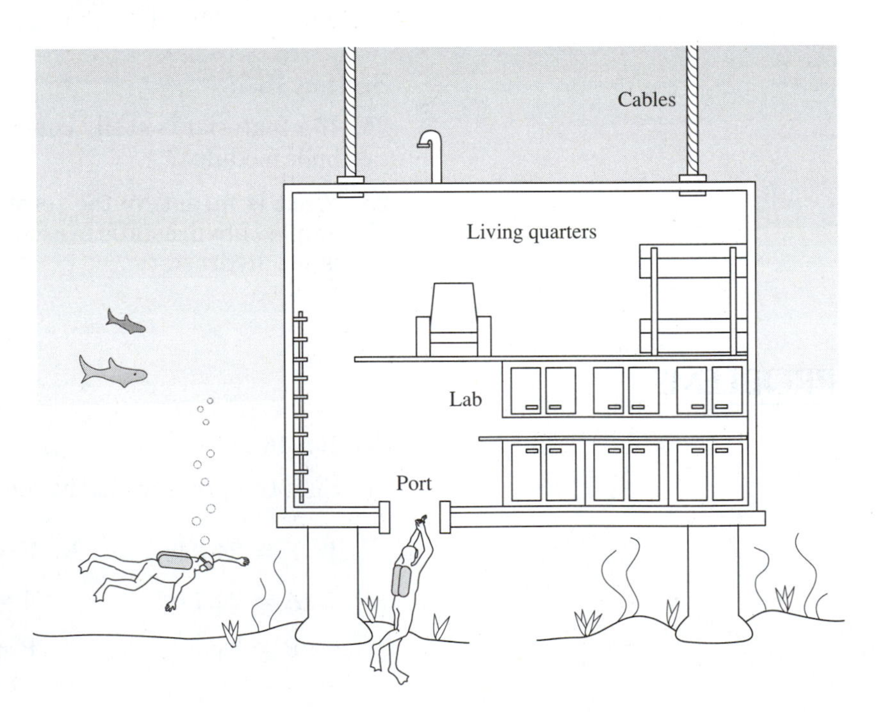

Figure 16–25: *Air pressure in the undersea shelter must be high enough to keep water from entering.*

6. What pressure in pounds per square inch does water exert on a dam at a depth of 27.3 ft?
7. ***A.*** What is the pressure at the bottom of a column of mercury 74 cm high? (The density of mercury is 13.6×10^3 kg/m^3.)
   B. What column height of water in meters is required to produce the same pressure? The density of water is 1.00×10^3 kg/m^3.
8. Ethyl alcohol has a specific gravity of 0.81. What is the pressure at the bottom of a 34.0 cm high, closed cylinder full of ethyl alcohol? Assume no air pressure at the top of the cylinder.

Section 16.3:

9. Ice has a density of $91\bar{0}$ kg/m^3. Kerosene has a density of $82\bar{0}$ kg/m^3. What is the apparent weight of a cube of ice $1\bar{0}$ cm on a side immersed in kerosene? Will ice float in kerosene?
10. What is the buoyant force on a 6.0 in long cylinder of aluminum with a 6.0-in diameter when it is immersed in water?
11. A spherical helium balloon has a 10.0-in diameter at room temperature and 1 at of pressure. What buoyant force acts on it in air? (The density of air is 0.0807 lb/ft^3; the density of helium is 0.0111 lb/ft^3.)
12. An ice floe (density = 57.2 lb/ft^3) has a cross-sectional area of 34.2 ft^2 (see Figure 16–26). It floats in seawater (density = 64.0 lb/ft^3). How much additional weight can it support without sinking below the surface of the water?

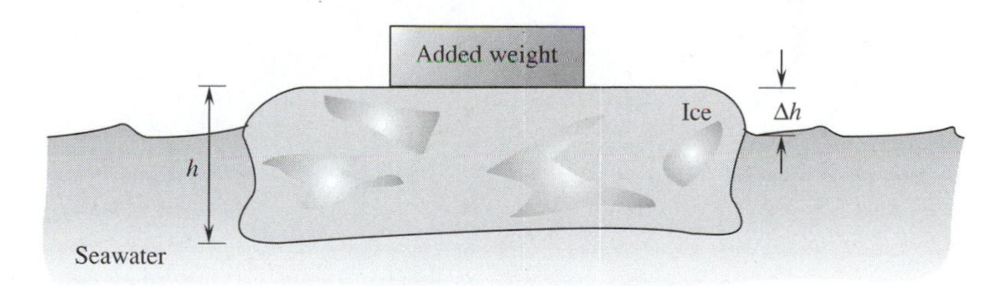

Figure 16–26: *An ice floe has a thickness h. Find the minimum added weight required to reduce Δh to zero.*

Section 16.4:

13. A press is used to compact trash. The diameters of the pistons are 1.00 in and 16.5 in. If a force of 6500 lb is exerted on the trash by the larger piston, how much force is exerted on the smaller piston?
14. The lift piston in a service garage has a diameter of 23.0 cm. What air pressure is needed to lift a load of $94\bar{0}$ kg?

15. A hydraulic system used on a dump truck delivers a force of 5300 lb. The system is 89% efficient. The input cylinder is 0.90 in in diameter; the output piston is 4.60 in in diameter.
    A. What is the *IMA?*
    B. What is the *AMA?*
    C. What force must be exerted on the input piston?

16. A hydraulic elevator is lifted 6.30 m. The input piston has a diameter of 6.80 cm; the output piston, a diameter of 30.0 cm. The input piston has a stroke length of 21.0 cm. How many strokes are taken by the smaller piston to lift the elevator?

Section 16.5: Compressibility

17. Steel has a Young's modulus of 2.0×10^{11} N/m^2 and a Poisson's ratio of 0.28. Estimate the bulk modulus of steel.

18. Copper has a Young's modulus of 16×10^6 psi and a Poisson's ratio of 0.34. Estimate its shear modulus.

19. Steel has a bulk modulus of 23×10^6 psi. What would be the change of volume of a 18.0-in long cylinder of steel with a diameter of 4.32-in when it is under a hydrostatic pressure of 500 atmospheres?

20. If lead has a bulk modulus of 0.77×10^{10} N/m^2, what is its compressibility?

21. Water has a compressibility of 34.5×10^{-6} (psi)$^{-1}$. What is its bulk modulus.

*22. Seawater has a compressibility of 49×10^{-11} m^2/N and a density of 1.03×10^3 kg/m^3. Estimate the density of seawater at a depth of 2.30 km?

Chapter 17

FLUIDS IN MOTION

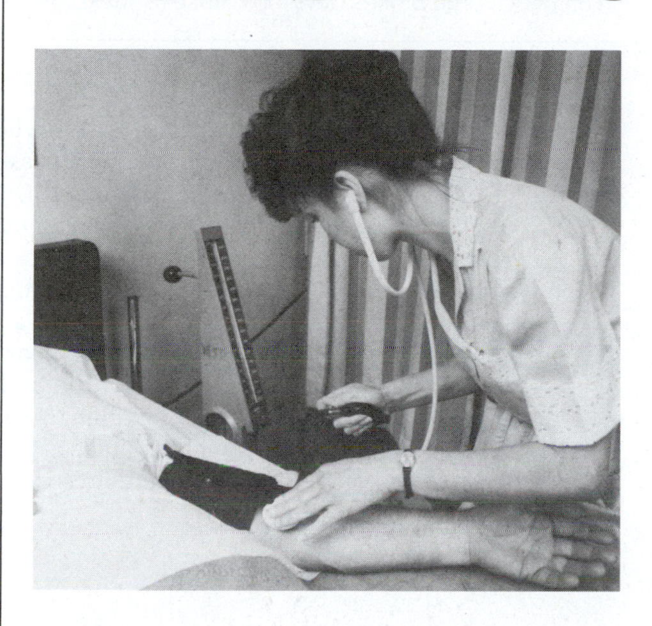

This nurse is taking a patient's blood pressure. What she records will depend in part on how the Bernoulli effect is working in the patient's blood vessels. As you read about fluids in motion, you will learn about the Bernoulli effect and how it works inside you. (Photograph reprinted with permission from Nursing Assistant, *5th edition, by Caldwell and Hegner. Copyright © 1989 by Delmar Publishers Inc.)*

In this chapter you will learn:

- how to calculate volumetric flow in a pipe for an incompressible fluid
- how work and energy per unit volume behave in a pipe (Bernoulli's principle)
- to distinguish between absolute pressure and gauge pressure
- to apply Bernoulli's principle to a variety of situations
- how to distinguish between turbulent flow and laminar flow
- to identify the factors that influence the flow of a viscous fluid through a pipe (Poiseuille's law)
- to apply the ideas of viscous flow to a variety of systems

Time Line for Chapter 17

100 B.C.	Illyrians (in Yugoslavia) use water power to grind corn.
90 B.C.	Parachutes are mentioned in a Chinese historical record.
600	Windmills are used in Persia.
1640	Evangelista Torricelli applies Galileo Galilei's laws of physics to the motion of fluids.
1738	Daniel Bernoulli relates pressure and velocity in moving fluids.
1797	Jean Poiseuille studies the flow of viscous fluids through narrow tubes to describe blood flow through the body.
1851	George Stokes develops Stokes's law.

Water flows from a fire hydrant through a hose. As the water passes through a nozzle it speeds up, making it possible to project the stream of water three stories onto a flaming roof. A tornado passes over a wooden frame house. The house explodes *outward* and the roof rises upward. The stand pipe for a village water supply is placed on a hill at the edge of town. Water is pumped from the treatment plant to this highest point in the community. In a tall building, the water flowing from a water faucet on the top floor has a slower speed than the water flowing from a faucet on the bottom floor. A plastic pipe collapses when the flow rate of fluid in the tubing becomes too large.

So far, we have looked only at fluids at rest. Let us look at moving fluids. The study of fluids in motion is called **hydrodynamics.**

17.1 VOLUME FLOW

Look at Figure 17–1. An incompressible fluid passes through a tube that has a changing cross section. The same volume of fluid comes out of one end of the tube that flowed into the other end. If V_1 is the volume of fluid that entered one end of the pipe, and V_2 is the volume of fluid that exited the other end, then $V_1 = V_2$. We would expect the volume flow rate in such a case to be constant.

During some interval of time, the volume of fluid passing some point is V/t. This is the **volumetric flow rate.** During time t the fluid

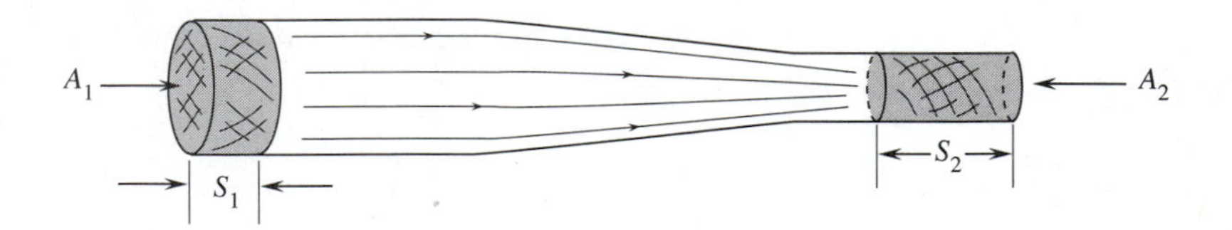

Figure 17–1: *The volume of fluid entering one end of the pipe per unit time equals the volume coming out the other end during the same time interval.*

with a cross-sectional area A_1 moves a distance s_1 at one end of the pipe, and fluid with a cross section of A_2 moves a distance s_2 at the other end. The volume of the two cylinders formed must be the same for a constant volumetric flow rate.

$$\frac{V_1}{t} = \frac{V_2}{t}$$

$$\frac{A_1\, s_1}{t} = \frac{A_2\, s_2}{t}$$

Distance divided by time is speed ($v = s/t$).

$$A_1\, v_1 = A_2\, v_2 \qquad \textbf{(Eq. 17–1)}$$

The linear speed of the fluid is inversely proportional to the cross-sectional area of the tube. Because the opening of the nozzle on a fire hose is small compared with the hose, the water speeds up as it exits. Thus it is possible to project the water a fairly long distance.

□ **EXAMPLE PROBLEM 17–1: THE PIPE REDUCER**

Water flows through a 1-in pipe with a linear speed of 2.3 ft/s. It passes through a reducer into a ½-in pipe.

A. What is the volumetric flow rate (R) in gallons per minute?

B. What is the linear speed of the water in the ½-in pipe?

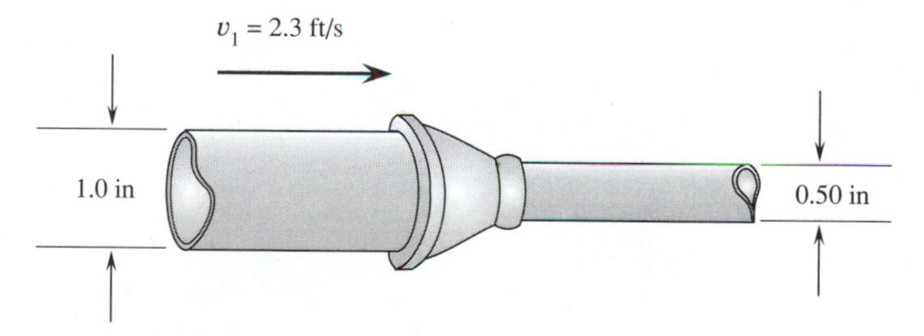

Figure 17–2: *A reducer adapts a ½-in pipe to a 1-in pipe.*

■ *SOLUTION*

A. Volumetric flow rate:

$$R = V/t = A\, v$$

$$R = [\pi(D/2)^2] \times v$$

$$R = 3.14\, (0.50 \text{ in})^2 \times 2.3 \text{ ft/s} \times 12 \text{ in/ft}$$

$$R = 21.7 \text{ in}^3\text{/s} \times 60 \text{ s/min} \times \frac{1 \text{ gal}}{231 \text{ in}^3}$$

$$R = 5.6 \text{ gal/min}$$

B. Linear flow rate in the ½-in pipe:

$$v_2 = v_1 \frac{A_1}{A_2} = v_1 \frac{\pi (D_1/2)^2}{\pi (D_2/2)^2}$$

$$v_2 = v_1 \left(\frac{D_1}{D_2}\right)^2$$

$$v_2 = 2.3 \text{ ft/s} \times \left(\frac{1.00}{0.50}\right)^2$$

$$v_2 = 9.2 \text{ ft/s}$$

17.2 BERNOULLI'S PRINCIPLE

The volume of fluid traveling through a pipe is conserved as long as the fluid does not compress under pressure. Pressure works on the fluid to move it through the pipe. Let us use the work-energy equation to describe the fluid. We will assume there are no frictional forces between the fluid and the pipe or between different parts of the fluid (see Figure 17–3).

Force F_1 pushes the fluid to the right. Force F_2 is a resisting force set up by back-pressure in the pipe. The fluid at point 1 travels a distance S_1 during a unit of time. At point 2, the fluid travels a distance S_2 during the same time interval. The net work done is:

$$\textit{net work} = (F_1 S_1) - (F_2 S_2)$$

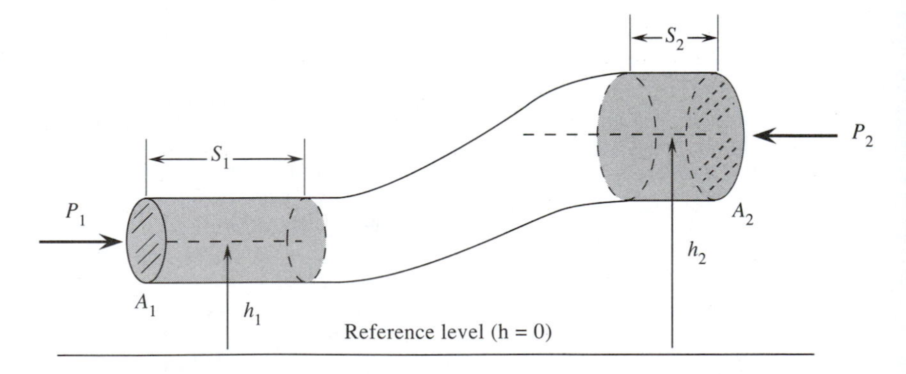

Figure 17–3: *Pressure P_1 does work on the fluid as it pushes the fluid to the right. Back-pressure P_2 does negative work by resisting the motion. If there is no friction, the net work done by the two pressures will change the potential and kinetic energy of the fluid.*

The net work will cause a change in the kinetic and potential energy.

$$\textit{net work} = \Delta KE + \Delta PE$$

$$(F_1 S_1) - (F_2 S_2) = \left[\frac{1}{2} m\, v_2^2 - \frac{1}{2} m\, v_1^2\right] + [(m\, g\, h_2) - (m\, g\, h_1)]$$

Bernoulli and You

You might not have considered how Bernoulli's principle can work inside you. A fat-rich, fast-food diet of hot dogs, burgers, and fries can lead to atherosclerosis. Fat deposits called plaque build up on the walls of large arteries. The plaque hardens the artery so it becomes less flexible. The plaque also partially blocks the passage of blood through the vessel. In order to maintain an adequate blood flow, the heart must exert a greater pressure. The result is a patient with high blood pressure.

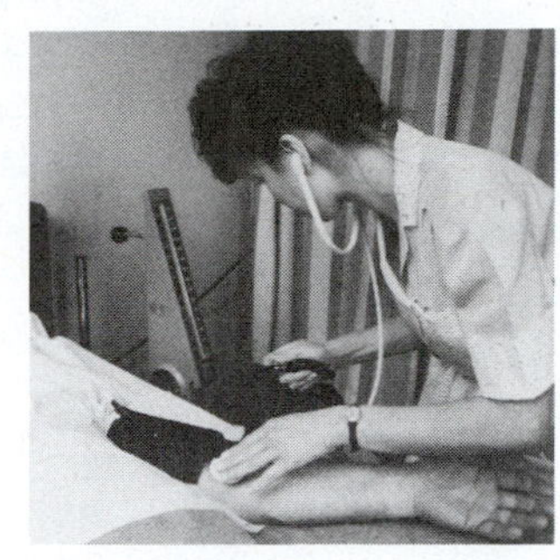

(Photograph reprinted with permission from Nursing Assistant, *5th edition, by Caldwell and Hegner. Copyright © 1989 by Delmar Publishers Inc.)*

Plaque and Bernoulli's principle are involved with strokes. The blood flows faster past an obstruction formed by the plaque. The pressure on the surface of the plaque away from the vessel wall is lower than the normal pressure in an unobstructed vessel. The blood forms slow-moving eddies just in front of and behind the plaque deposit where it attaches to the artery wall. Here the pressure is higher than normal for the blood vessel. The high pressure at the base of the plaque combines with the low pressure of the flow over the plaque to pry the deposit loose. The deposit flows downstream, where it may block a vital artery and cause a stroke.

The forces can be expressed in terms of pressure.

$$F_1 = P_1 A_1 \text{ and } F_2 = P_2 A_2$$

The left-hand part of the equation becomes:

$$(F_1 S_1) - (F_2 S_2) = (P_1 A_1 S_1) - (P_2 A_2 S_2)$$

The volume of fluid (V) flowing through the pipe per unit time is constant.

$$A_1 S_1 = A_2 S_2 = V$$

$$(P_1 - P_2)\,V = \left[\frac{1}{2} m\, v_2^{\,2} - \frac{1}{2} m\, v_1^{\,2}\right] + (m\, g\, h_2) - (m\, g\, h_1)$$

Divide the equation by volume. Volume will cancel out on the left side. Each term on the right will have a factor (ρ), mass divided by volume, or density ($\rho = m/V$). Rearrange the equation to eliminate the negative signs.

$$P_1 + \left[\frac{1}{2} \rho\, v_1^{\,2}\right] + (\rho\, g\, h_1) = P_2 + \left[\frac{1}{2} \rho\, v_2^{\,2}\right] + (\rho\, g\, h_2) \qquad \textbf{(Eq. 17–2)}$$

The sum of the pressure, the kinetic energy per unit volume, and the potential energy per unit volume is a constant anywhere in the pipe. It is a work-energy conservation relationship per unit volume.

This relationship is called **Bernoulli's equation.** It is a powerful tool for analyzing fluid flow. Let us look at an example problem to see how it works.

□ **EXAMPLE PROBLEM 17–2: THE PRESSURE PUMP**

A household plumbing system pumps water from a well into a pressure tank in the basement, creating a pressure of 11.0 psi above atmospheric pressure in the compressed air trapped in the tank (see Figure 17–4). A faucet is opened 15.3 ft above the tank. With what linear speed does water come out of the faucet?

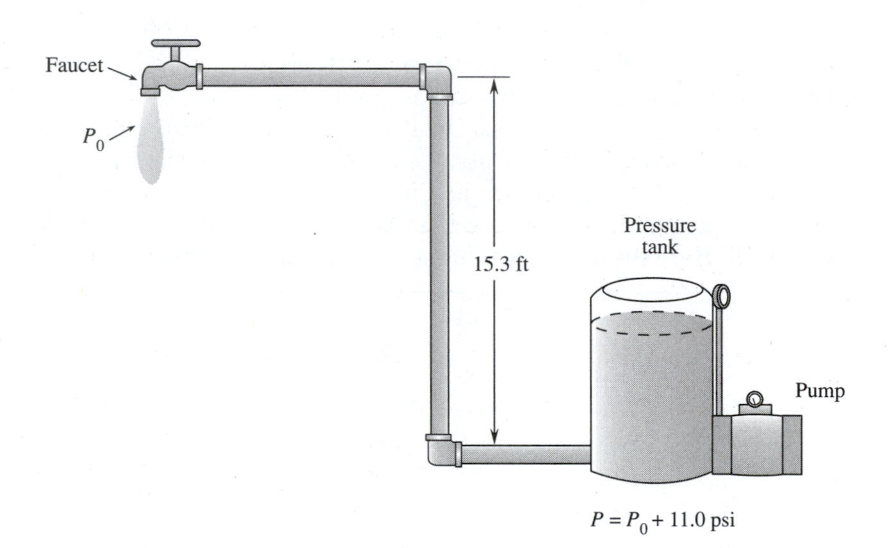

***Figure 17–4:** Pressure at the tank forces the water upward and produces a flow rate at the faucet.*

■ ***SOLUTION***

Take point 1 as the water level in the pressure tank. Point 2 is the open faucet.

Point 1:

The pressure is atmospheric pressure + 11.0 psi ($P_0 + P_g$), where P_g is the gauge pressure. The velocity of the water is negligibly small ($v = 0$). The height of the water relative to the pressure tank is 0.

Point 2:

The atmosphere pushes inward at the open faucet ($P_2 = P_0$). The linear speed of the water is v. The height (h) of the faucet is 15.3 ft.

Substitute these data into Bernoulli's equation.

$$P_o + P_g + 0 + 0 = P_o + \frac{1}{2}(\rho v^2) + (\rho g h)$$

Solve the equation for v.

$$v = \left\{2\left[\left(\frac{P_g}{\rho}\right) - (g\,h)\right]\right\}^{1/2}$$

The mass density of water is 62.5 lb/ft^3/32 ft/s^2 = 1.95 slug/ft^3.

$$v = \left\{2\left[\left(\frac{11.0 \text{ lb/in}^2 \times 144 \text{ in}^2/\text{ft}^2}{1.95 \text{ slug/ft}^3}\right) - \left(32\,\frac{\text{ft}}{\text{s}^2} \times 15.3 \text{ ft}\right)\right]\right\}^{1/2}$$

$$v = 25.4 \text{ ft/s}$$

17.3 ABSOLUTE PRESSURE AND GAUGE PRESSURE

Fluids are usually transported through tubes open at one end when in use to the atmosphere. A hose on a fire department pumper, a portable compressed-air sprayer, and a municipal water system are examples. In such systems, the velocity of water coming out of the system depends in part on the amount of pressure above atmospheric pressure. Most pressure gauges measure the difference between the pressure of the system and the surrounding atmospheric pressure. This kind of pressure is called **gauge pressure** ($\mathbf{P_g}$). The total pressure, atmospheric pressure plus gauge pressure, is called **absolute pressure.**

absolute pressure = *atmospheric pressure* + *gauge pressure* **(Eq. 17–3)**

$$P_a = P_o + P_g$$

Often the distinction between gauge pressure and absolute pressure is indicated in the units when US engineering units are used. Gauge pressure is often given in units of psig, and absolute pressure is written as psia.

A container holds a fluid. Absolute pressure is caused by the molecules of the fluid bombarding the sides of the container. If the fluid is a gas, removing some of the molecules with a vacuum pump will reduce the absolute pressure inside the container, because there will be fewer molecules striking the sides. If we remove *all* the molecules, the absolute pressure will be zero. There will be no more molecules to bump up against the container. This condition is a perfect vacuum. We cannot reduce the pressure further. Absolute pressure has a minimum value of zero; it cannot be negative. Gauge pressure can have a negative value. Assume we have a perfect vacuum (P_a = 0.00). According to Equation 17–3:

$$0 = P_o + P_g \qquad \text{or}$$

$$P_g = -P_o$$

The gauge pressure equals minus the atmospheric pressure for a perfect vacuum. If $P_a = (1/3)P_o$, then $P_g = -(2/3)P_o$.

In Equation 17–3 the atmospheric pressure (P_o) is *not necessarily* a standard atmosphere. It is whatever the pressure happens to be for the atmosphere surrounding the system.

☐ **EXAMPLE PROBLEM 17–3: ABSOLUTE AND GAUGE PRESSURE**

What is the absolute pressure in the water tank in Example Problem 17–2?

■ ***SOLUTION***

The gauge pressure is given in the problem ($P_g = 11.0$ psi). Atmospheric pressure is 14.7 psi. The absolute pressure is:

$$P_a = P_o + P_g$$

$$P_a = 14.7 \text{ psi} + 11.0 \text{ psi}$$

$$P_a = 25.7 \text{ psi}$$

☐ **EXAMPLE PROBLEM 7–4: THE PUMP**

A water pump for a portable water purification system develops a pressure of 1.3×10^5 Pa. Find the gauge pressure of the pump for the following situations.

A. The pump is used at sea level ($P_o = 1.0 \times 10^5$ Pa).

B. The pump is used on a mountaintop ($P_o = 0.9 \times 10^5$ Pa).

C. The pump is used in a scuba diver's decompression chamber ($P_o = 1.5 \times 10^5$ Pa).

■ ***SOLUTION***

$$P_g = P_a - P_o$$

A.

$$P_g = (1.3 \times 10^5 \text{ Pa}) - (1.0 \times 10^5 \text{ Pa})$$

$$P_g = 0.3 \times 10^5 \text{ Pa}$$

B.

$$P_g = (1.3 \times 10^5 \text{ Pa}) - (0.9 \times 10^5 \text{ Pa})$$

$$P_g = (0.4 \times 10^5 \text{ Pa})$$

C.

$$P_g = (1.3 \times 10^5 \text{ Pa}) - (1.5 \times 10^5 \text{ Pa})$$

$$P_g = -0.2 \times 10^5 \text{ Pa}$$

In case C the pump does not develop enough absolute pressure to overcome atmospheric pressure. It will not work.

17.4 APPLICATIONS OF BERNOULLI'S PRINCIPLE

The height H of fluid stored above an outlet is called the **head.** The head is related to the potential energy of the top of the column of liquid. If there is a large head, there will be a rapid flow rate. Potential energy will be converted to kinetic energy. Municipal water systems pump water into stand pipes to create a head. Dams are used to create a head for hydroelectric generation. The speed of the water hitting the power turbines depends on the level of the water above the dam (see Figure 17–5).

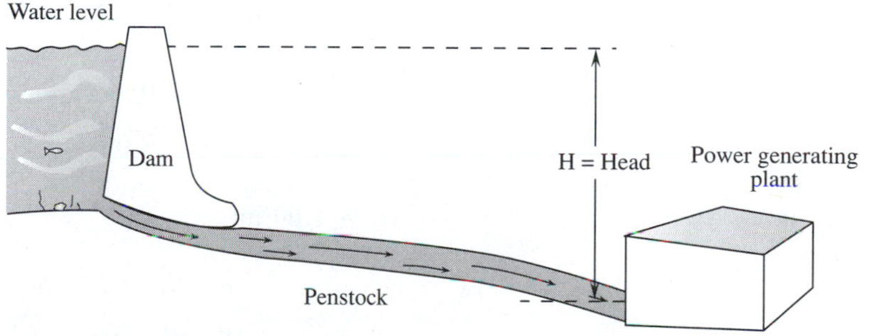

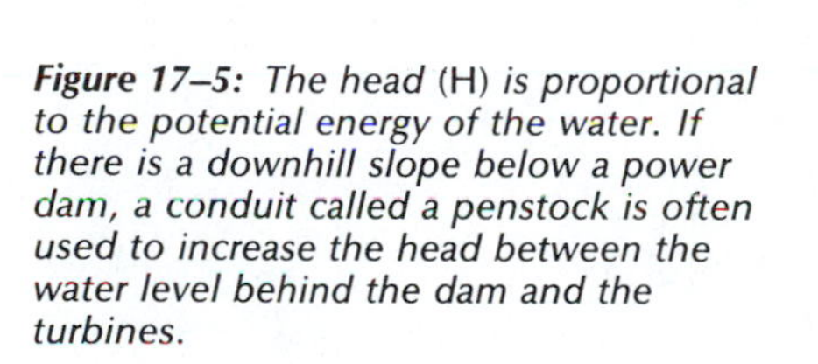

Figure 17–5: *The head (H) is proportional to the potential energy of the water. If there is a downhill slope below a power dam, a conduit called a penstock is often used to increase the head between the water level behind the dam and the turbines.*

☐ **EXAMPLE PROBLEM 17–5: THE MECHANICAL COW**

Cafeterias often use a milk dispenser that consists of a plastic bag of milk inside a stainless steel box (see Figure 17–6). At the bottom of the milk bag is a flexible plastic tube that passes through a spring-loaded dispenser handle. The handle squeezes the tube shut when it is not in use.

A. Find an expression of the linear speed of milk flowing out of the dispenser as a function of the height of the milk above the spigot.

B. Find the speed of the milk for heights of 1.00 ft and 1.00 in.

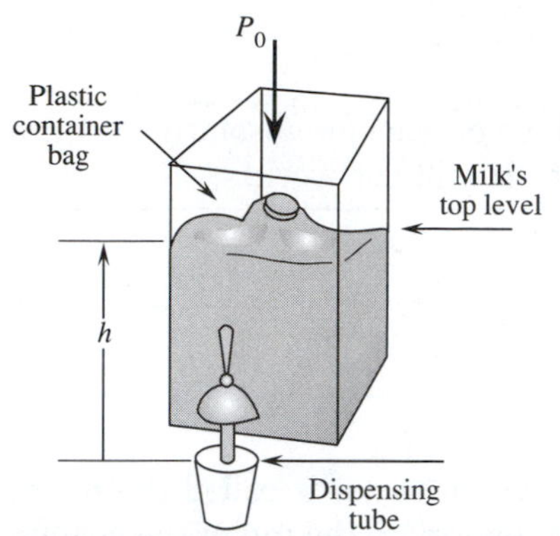

Figure 17–6: A mechanical cow. How fast the cup fills up depends on the height of the column of milk.

■ *SOLUTION*

A. The equation:

Take point 1 as the top of the milk column and point 2 as the end of the spigot.

Point 1:

The atmosphere presses down with a pressure P_o. The column of milk goes down slowly ($v_1 = 0$). The milk is a height H above the spigot.

Point 2:

The atmosphere presses inward on the milk flowing out of the spigot with a pressure of P_o. The milk coming out of the spigot has a linear speed v. The height relative to the end of the spigot is zero.

Substitute these data into Bernoulli's equation.

$$P_o + 0 + (\rho\, g\, H) = P_o + \left[\frac{1}{2}(\rho\, v^2)\right] + 0$$

Solve the equation for v.

$$v = (2\, g\, H)^{1/2}$$

B. The linear flow rate:

$H = 1.00$ ft:

$$v = (2 \times 32\ \text{ft/s}^2 \times 1.0\ \text{ft})^{1/2}$$
$$v = 8.0\ \text{ft/s}$$

$H = 1.00$ in:

$$v = (2 \times 32\ \text{ft/s}^2 \times 0.0833\ \text{ft})^{1/2}$$
$$v = 2.3\ \text{ft/s}$$

Pressure differences in a system can result from differences in the speed or kinetic energy of a fluid. A paint sprayer is based on this principle. Compressed air passes over the top of a tube inserted into the paint container (see Figure 17–7). The pressure inside the holding tank and at the bottom of the tube is close to atmospheric pressure. The pressure is depressed at the top by the airstream ($P_o - 1/2\ \rho\ v^2$). The difference of pressure between the bottom of the tube and the top is $\Delta P = 1/2\ (\rho\ v^2)$. This pressure difference forces the paint upward into the airstream.

Airplane wings are designed to form an airfoil (see Figure 17–8). As the wing planes through the air, air must travel over the top of

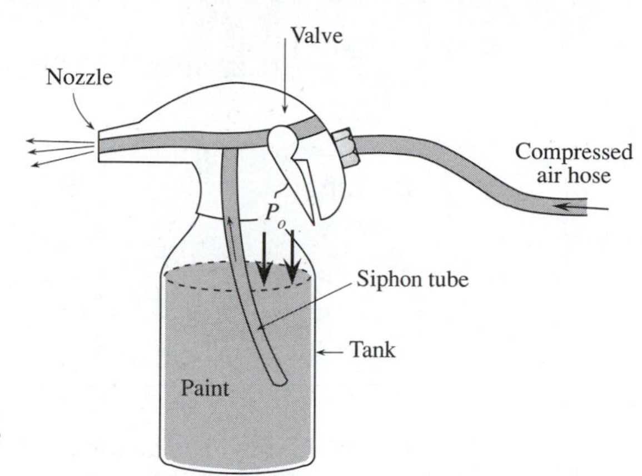

Figure 17–7: A paint sprayer. Air passing over the top of the siphon tube reduces the pressure at the top of the tube. Atmospheric pressure forces paint up the tube into the airstream.

the wing and under the wing during the same time. Because the distance over the top of the wing is longer, the air that passes over the top is faster. A pressure difference is created, giving lift to the airplane.

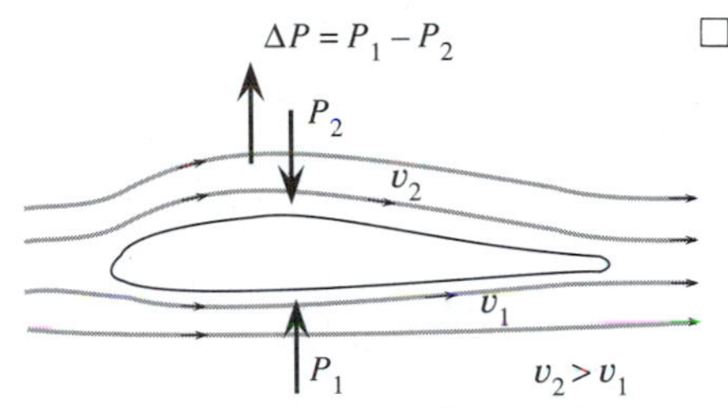

Figure 17–8: A section of an airplane wing. Air must travel a greater distance over the top of the wing than below the wing as the wing moves through the air. The pressure differential created by the moving air pushes up on the wing.

EXAMPLE PROBLEM 17–6: THE AIRPLANE WING

An airplane wing has an area of 10.3 m². When the airplane is traveling at 192 km/h, air passes over the top of the wing with an average speed of 59.2 m/s and under the wing at 55.7 m/s. The density of air is 1.29 kg/m³.

A. What is the pressure difference between the top and bottom of the wing?

B. How much lift will be exerted upward on the wing?

SOLUTION

A. Pressure difference:

Top of wing:

The pressure is P_1. The speed is v_1 or 59.2 m/s. The height is essentially 0. Wings are thin, and air density is small.

Bottom of wing:

The pressure is P_2. The speed of air is v_2 or 55.7 m/s. The height is 0.

Substitute the data into Bernoulli's equation.

$$P_1 + \left[\frac{1}{2}(\rho\, v_1^2)\right] + 0 = P_2 + \left[\frac{1}{2}(\rho\, v_2^2)\right] + 0$$

$$P = P_2 - P_1 = \frac{1}{2}[\rho\,(v_1^2 - v_2^2)]$$

$$P = \frac{1}{2}\{1.29 \text{ kg/m}^3 [(59.2 \text{ m/s})^2 - (55.7 \text{ m/s})^2]\}$$

$$P = 259 \text{ N/m}^2$$

B. Lift:

$$F = P A$$

$$F = 259 \text{ N/m}^2 \times 10.3 \text{ m}^2$$

$$F = 2.67 \times 10^3 \text{ N}$$

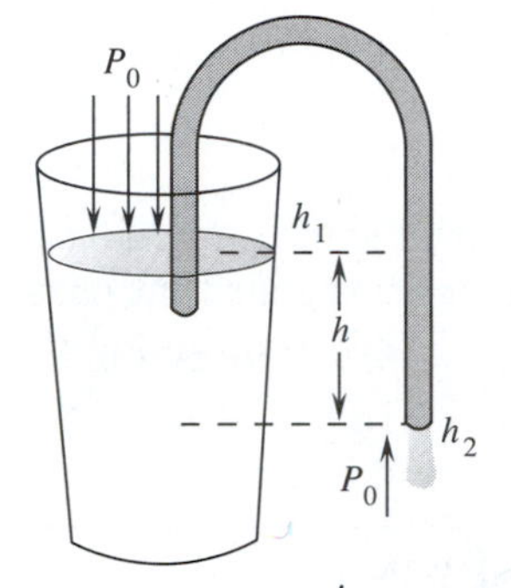

Figure 17–9: A siphon. Fluid flowing out of the siphon at P_1 creates a partial vacuum in the tube.

Figure 17–9 shows a siphon. A partial vacuum must be formed in the tube to start the flow. The same principle is used to make fluids flow in a soda straw. Let us look at points 1 and 2 in the diagram.

Point 1: Air pushes down on the top of the fluid (P_0). Velocity, $v_1 = 0$. Height above the bottom of the tube $= h$.

Point 2: Atmospheric pressure acts on the end of the tube (P_0). The velocity of the fluid is v_2. The height is 0.

Place these data into Bernoulli's equation.

$$P_0 + 0 + (\rho g h) = P_0 + \frac{1}{2}(\rho v^2) + 0$$

$$v = (2 g h)^{1/2}$$

As long as the height of the liquid is above the bottom end of the siphon the fluid will flow.

17.5 VISCOUS FLUIDS

When we derived Bernoulli's equation from the work-energy relationship, we ignored friction. The equation works well for fluids that flow easily over short distances. When applied to gases and water transported short distances through large-diameter pipes, the equation gives reasonably accurate results. The relationship between pressure and potential and kinetic energy per unit volume is still qualitatively true in systems that have fluid friction.

There are two modes of fluid flow. When eddy currents are present, the fluid is said to have **turbulent flow** (see Figure 17–10). Turbulent flow is difficult to analyze. When layers of a fluid slip over each other without eddy currents, the fluid is said to have **laminar flow** (see Figure 17–11). Laminar flow involves shear in a viscous fluid.

Fluid moves along a fixed surface such as the inside wall of a pipe.

Figure 17–10: Turbulent flow. Eddy currents increase frictional losses in a pipe.

Figure 17–11: Laminar flow. In laminar flow, there are no eddy currents. Each successive layer of fluid slides over the layer adjacent to it.

Motion is negligible against the pipe wall. The next layer of fluid moves over the bottom layer with a greater speed. The next layer moves even faster. The speed (v) increases linearly with distance (l) from the fixed bottom layer. This ratio is the time rate of change of shear strain (R).

$$R = \Delta v / \Delta l$$

A shearing stress (F/A) acts on each layer of fluid to force it over the layer below it. This is similar to the shear experienced by a solid, except here we have a flowing material. We use speed instead of displacement for the strain.

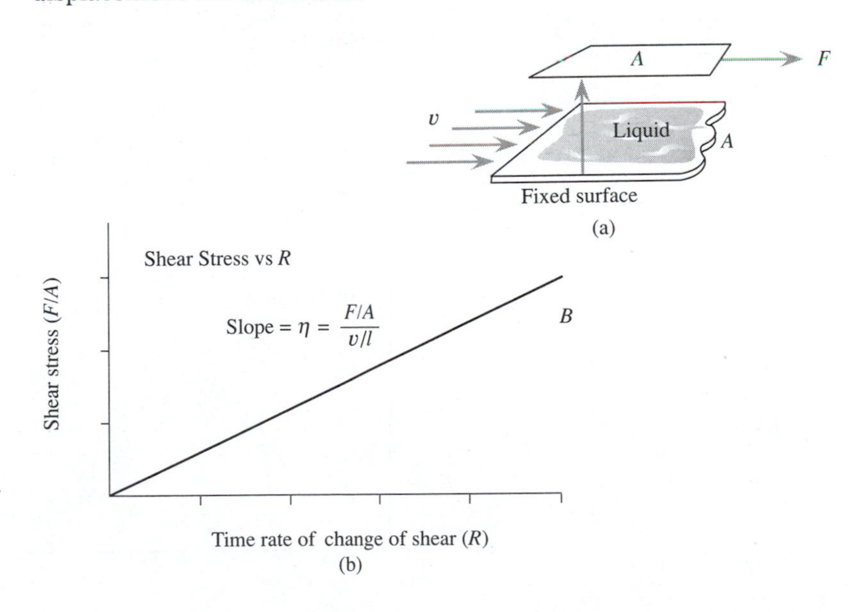

Figure 17–12: Vicosity. (a) The shear stress (F/A) acting on a layer of fluid causes a time rate of change of shear strain ($R = \Delta v/\Delta l$). (b) The proportionality constant between shear stress and time rate of shear strain is called viscosity (η).

If we plot stress and change of shear for a liquid undergoing laminar flow, we get a plot similar to that shown in Figure 17–12. The slope of the curve (η) is called the **viscosity** of the fluid. Viscous forces show up only when the fluid is moving. Viscosity is highly temperature-dependent in most fluids.

$$\eta = \frac{F/A}{v/l} = \frac{\textit{shear stress}}{\textit{time rate of change of shear strain}} \quad \textbf{(Eq. 17–4)}$$

The basic unit of viscosity in SI is the **poiseuille** ($1\ N \cdot s/m^2$). Most engineering handbooks use a smaller unit called a **poise.** A poise is still an inconveniently large unit for some applications. Often viscosity is expressed in centipoise.

- 1 poiseuille (Pl) = $1\ N \cdot s/m^2$
- 1 poise (p) = 10^{-1} Pl
- 1 centipoise (cp) = 10^{-2} p = $10^{-3}\ N \cdot s/m^2$

Table 17–1 gives typical values of viscosity for some common liquids. The viscosity of liquids usually decreases with increasing tem-

Table 17–1: Viscosity of some common liquids.

LIQUID	TEMPERATURE (°C)	VISCOSITY ($N \cdot s/m^2$)
Ethyl alcohol	20	1.2×10^{-3}
	30	1.0×10^{-3}
Blood plasma	37	1.3×10^{-3}
Glucose	22	9.1×10^{12}
	30	6.6×10^{10}
	40	2.8×10^{8}
Glycerin	20	1.5
	30	0.63
Olive oil	20	8.4×10^{-2}
	30	6.0×10^{-2}
Sulfuric acid	20	2.5×10^{-2}
	30	1.6×10^{-2}
Water	0	1.8×10^{-3}
	20	1.0×10^{-3}
	40	6.5×10^{-4}
	60	4.7×10^{-4}
	80	3.5×10^{-4}
	100	2.8×10^{-4}

Table 17–2: Viscosity of some gases, at 1 atm of pressure.

GAS	VISCOSITY ($N \cdot s/m^2 \times 10^{-5}$) at: 0°C	40°C	80°C	100°C
Air	1.70	1.90	2.12	—
Carbon dioxide	1.39	1.57	1.76	1.86
Hydrogen	0.83	0.90	0.98	1.01
Methane	1.02	1.14	1.27	1.33
Sulfur dioxide	—	1.43	1.52	1.61

perature. Gases behave differently. Hot gases have higher viscosities than cool gases (see Table 17–2).

17.6 APPLICATIONS: VISCOUS FLOW

Liquid Flow in Pipes

Figure 17–13 shows laminar flow of a viscous fluid through a section of pipe. The speed of the fluid is highest at the center of the pipe. It is nearly zero at the edges. A calculation can be made of the volume flow of a liquid in a pipe with a circular cross section. The resulting equation is called **Poiseuille's law.**

$$\frac{V}{t} = \frac{\pi R^4 (P_1 - P_2)}{8 \eta L} \qquad \textbf{(Eq. 17–5)}$$

The volume rate of flow is:
A. inversely proportional to the viscosity (η),
B. directly proportional to the fourth power of the radius (R),
C. inversely proportional to the length of the pipe (L), and
D. directly proportional to the pressure difference ($P_1 - P_2$).

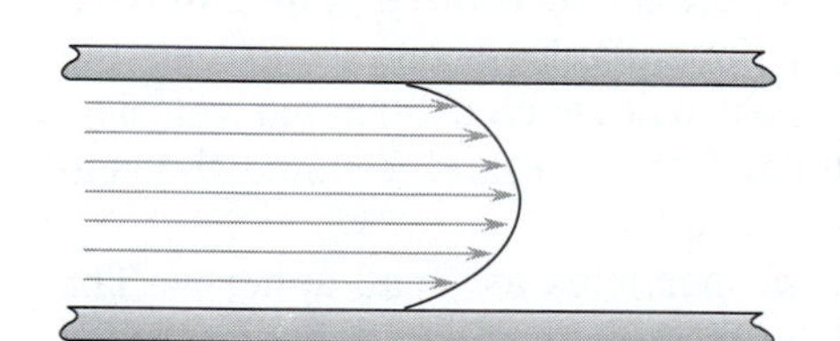

Figure 17–13: *Laminar flow in a circular pipe. The flow rate is greatest at the center of the pipe. The speed of the fluid approaches zero as it comes into contact with the pipe.*

The ratio of the pressure difference to the length of the pipe is called the pressure gradient. As the pressure gradient increases, the flow rate will also increase.

$$\text{Pressure gradient} = \frac{(P_2 - P_1)}{L}$$

EXAMPLE PROBLEM 17–7: THE FOOD PLANT

At a food-manufacturing plant a 2.50 cm diameter pipe is used to transfer soya oil from a holding tank to a mixing vat. The pipe is to deliver 1.2 liters of oil per second through a 12.3 m long pipe. What must be the gauge pressure at the holding tank end of the pipe? At room temperature, soya oil has a viscosity of 69 cp.

SOLUTION

Solve Poiseuille's law for the pressure difference and input the data using SI units.

$$\eta = 69 \text{ cp} = 0.069 \text{ N} \cdot \text{s/m}^2$$

$$R = 1.25 \text{ cm} = 0.0125 \text{ m}$$

$$\frac{V}{t} = 1.2 \text{ l/s} = 1.2 \times 10^{-3} \text{ m}^3\text{/s}$$

$$L = 12.3 \text{ m}$$

$$P_g = P_1 - P_2 = \frac{8\,\eta\, L\,(V/t)}{\pi R^4}$$

$$P_g = \frac{8 \times 0.069 \text{ N} \cdot \text{s/m}^2 \times 12.3 \text{ m} \times 1.2 \times 10^{-3} \text{ m}^3\text{/s}}{3.14 \times (0.0125 \text{ m})^4}$$

$$P_g = 1.1 \times 10^5 \text{ N/m}^2$$

$$P_g = 7.6 \text{ psi}$$

Fluids often have small lumps of solids suspended in them. Precipitates from a chemical reaction, clay particles in a river, and waste particles in waste water are examples.

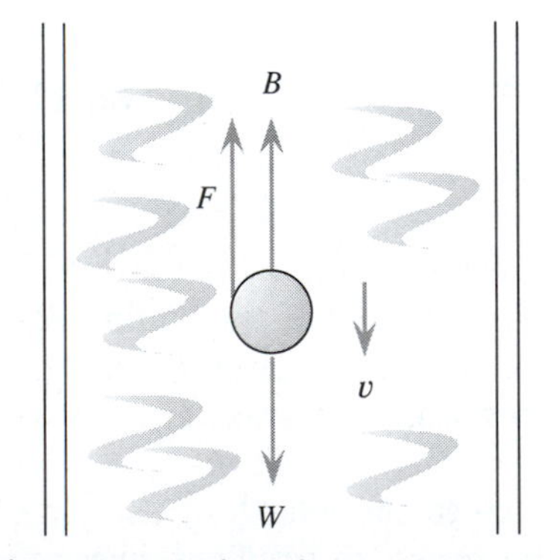

*Figure 17–14: Three forces act on a mass falling through a fluid. Its weight **(W)** acts downward. A buoyant force (B) acts upward. The third force is a viscous force (F) acting in a direction opposite the particle's motion.*

Three forces act on particles in a fluid (see Figure 17–14). A buoyant force acts upward. The weight of the particle acts downward. If the density of the solid is greater than the density of the fluid, the particle will sink under the influence of an unbalanced force. For a moving particle, there will be a velocity-dependent viscous drag resisting the motion. Eventually the particle will reach a terminal velocity. The terminal velocity for small particles in a fluid is called the **sedimentation rate.**

For simplicity we can treat the particles as small spheres. The terminal velocities of spheres, large and small, have been studied carefully. In a liquid, the viscous drag is proportional to the velocity. Other factors affecting the force are the viscosity of the liquid and the radius of the sphere. These can be combined into an equation called **Stokes's law.**

$$F = 6 \pi \eta r v_t \qquad \textbf{(Eq. 17–6)}$$

This is the form of the third force acting on the particle falling through a liquid. At the sedimentation rate the three forces will be in equilibrium.

$$\textit{buoyant force} + \textit{viscous drag} = \textit{weight} \qquad \textbf{(Eq. 17–7)}$$

Let us express the weight and buoyant force in terms of the density of the solid (ρ_s) and the density of the fluid (ρ_f).

$$w = m\,g = V \rho_s\, g = \frac{4}{3} \pi r^3 \rho_s\, g$$

$$F_B = V \rho_f\, g = \frac{4}{3} \pi r^3 \rho_f\, g$$

The equilibrium of forces is:

$$\frac{4}{3} \pi r^3 \rho_f\, g + 6 \pi \eta r v_t = \frac{4}{3} (\pi r^3 g \rho_s)$$

We can solve this equation for the terminal velocity.

$$v_t = \frac{2\,[r^2 g\,(\rho_s - \rho_f)]}{9\,\eta} \qquad \textbf{(Eq. 17–8)}$$

This is the sedimentation rate. In addition to liquids, this equation can be used for very small particles in a gas where the sedimentation rate is small. The equation is used to predict settling rates of fly ash and small liquid particles in still air. For larger particles in a gas, the equations found in Chapter 10 should be used.

Of course, not all particles are spheres. A more general form can be developed for the sedimentation rate. Small particles moving slowly through a viscous fluid have a force acting on them that is proportional to the velocity. The magnitude of the force is:

$$F = k\,v$$

where k is a constant depending on the shape of the particle and the viscosity of the fluid. For a sphere, $k = 6 \pi \eta r$. We can rewrite Equation 17–7.

$$(V \rho_f\, g) + (k\, v_t) = V \rho_s\, g$$

or

$$v_t = \frac{(\rho_s - \rho_f)\,V\,g}{k} \qquad \textbf{(Eq. 17–9)}$$

k can be found for a particular sized particle by experiment. The terminal velocity of other particles of the same shape in the same fluid can then be predicted.

☐ **EXAMPLE PROBLEM 17–8: AIR POLLUTION**

The particulate matter in the air falls roughly into two sizes. Coarse particles are between 2 and 10 μm (1 μm = 10^{-6} m). This is mostly dust from roads and particles from agricultural activities, ocean spray, and volcanic activity. Finer particles in the range of 0.1–1 μm come primarily from combustion. This includes soot, tar, fly ash, and droplets such as sulfuric acid. Find the sedimentation rate of the coarse particles (6.0 μm) and the fine particles (0.50 μm). Assume a spherical shape and a density twice that of water (2.0×10^3 kg/m³). The viscosity of air at 20°C is 1.81×10^{-4} poise.

■ ***SOLUTION***

A. Coarse particles:

$$v_t = \frac{2\,(\rho\,g\,r^2)}{9\,\eta}$$

$$v_t = \frac{2 \times (2 \times 10^3\ \text{kg/m}^2) \times 9.8\ \text{m/s}^2 \times (6 \times 10^{-6}\ \text{m})^2}{9 \times (1.81 \times 10^{-5}\ \text{N} \cdot \text{s/m}^2)}$$

$$v_t = 6.0 \times 10^{-5}$$

B. Fine particles:

The sedimentation rate is proportional to the square of the radius.

$$v_t' = v_t \left(\frac{R'}{R}\right)^2$$

$$v_t' = 8.7 \times 10^{-3}\ \text{m/s} \times \left(\frac{0.50\ \mu\text{m}}{6.0\ \mu\text{m}}\right)^2$$

$$v_t' = 7.3 \times 10^{-5}\ \text{m/s}$$

The coarser particles would settle a distance of 1 m in a couple of minutes. The finer and more toxic particles would need about 4–5 h to settle the same distance in still air. Wind currents would further delay the sedimentation. Sulfuric acid droplets would have a density close to that of water. Their sedimentation rate would be even slower. The fine particles stay mixed with the air until they are washed out with rain droplets, creating acid rain.

The Centrifuge

Sedimentation rates can be increased by taking advantage of Einstein's equivalence. If a sample is whirled rapidly in a centrifuge, the centripetal acceleration behaves just like gravity (see Chapter 7). If the centripetal acceleration is quite a bit larger than the acceleration of gravity, we can replace g in Equations 17–8 and 17–9 with the centripetal acceleration ($\omega^2 R$).

$$v_t = 2\,\frac{(\omega^2 R)\,[r^2\,(\rho_s - \rho_f)]}{9\,\eta}$$

and

$$v_t = \frac{(\rho_s - \rho_f)\,[V\,(\omega^2 R)]}{k}$$

EXAMPLE PROBLEM 17–9: RED BLOOD CELLS

Under the normal influence of gravity, red blood cells settle in blood plasma at a rate of 5 cm/h. The sample is placed in a high-speed centrifuge that rotates at 20,$\bar{0}$00 rpm. What is the sedimentation rate in the centrifuge, assuming the average distance of the test tube is 12 cm from the center of rotation?

SOLUTION

First find the centripetal acceleration.

$$a_c = \omega^2 R$$

$$a_c = \frac{(20{,}\bar{0}00\text{ rpm} \times 2\pi\text{ rad/rev})^2 \times 0.12\text{ m}}{(60\text{ s/min}^2)}$$

$$a_c = 5.26 \times 10^5\text{ m/s}^2$$

$$v_t' = v_t\left(\frac{a_c}{g}\right)$$

$$v_t' = 5\text{ cm/h} \times \frac{1\text{ h}}{3600\text{ s}} \times \frac{5.26 \times 10^5\text{ m/s}^2}{9.8\text{ m/s}^2}$$

$$v_t' = 7 \times 10^1\text{ cm/s}$$

By using the centrifuge a process that normally would take hours can be done in a second.

SUMMARY

The study of fluids in motion is called hydrodynamics. The volumetric flow rate of an incompressible fluid in a pipe is constant. Work-energy of a fluid in a pipe is conserved if the fluid does not compress under pressure. For an incompressible fluid, the conservation of work and energy per unit volume is expressed by Bernoulli's equation.

Absolute pressure is the gauge pressure plus atmospheric pressure. The absolute pressure is the total pressure on the fluid. It is always positive. For a pipe subject to atmospheric pressure, the flow rate is determined by the gauge pressure.

When layers of a fluid flow smoothly over each other, the fluid is said to have laminar flow. If eddy currents exist in a fluid, it undergoes turbulent flow. Viscosity (η) is a measure of the friction between layers of a fluid in laminar flow. It is defined as the ratio of shear stress divided by the time rate of change of shear strain. Viscosity is usually measured in poise or centipoise. Poiseuille's equation describes the volume flow of a viscous fluid with laminar flow through a horizontal pipe.

An object falling (or rising) through a viscous fluid has a viscous force resisting the motion. The object will reach a terminal speed (v_t) under the influence of three forces: the object's weight, a buoyant force, and viscous drag. The terminal velocity is often called the sedimentation rate.

KEY TERMS

If you can explain the following terms to a friend or classmate, you understand their meaning. If you cannot explain the terms, you should reread the sections in which they are discussed.

absolute pressure	**Poiseuille's law**
Bernoulli's equation	**pressure gradient**
gauge pressure	**sedimentation rate**
head	**Stokes's law**
hydrodynamics	**turbulent flow**
laminar flow	**viscosity**
poise	**volumetric flow rate**
poiseuille	

EXERCISES

Section 17.1:

1. Explain the following terms in your own words.

   A. Volumetric flow rate
   B. Bernoulli's principle
   C. Absolute pressure
   D. Gauge pressure
   E. Turbulent flow
   F. Laminar flow
   G. Viscosity
   H. Sedimentation rate

Figure 17–15: *Diagram for Exercise 3. An engineer's idea to improve the efficiency of a wind turbine.*

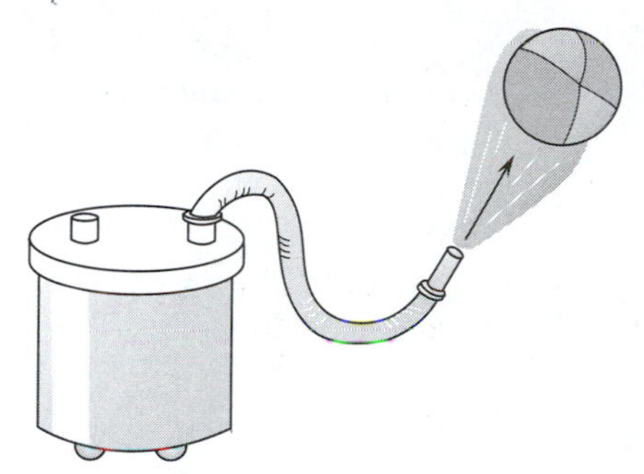

Figure 17–16: *Diagram for Exercise 5. Air flowing from the blower on a shop vacuum keeps the beach ball in the air.*

2. Water flowing through a main pipe passes through a Y connection, distributing the flow of water into two feeder pipes.
   A. If the feeder pipes have the same diameter as the main pipe, will the linear speed through the feeder pipes be faster or slower than through the main pipe?
   B. If the diameter of the feeder pipes is half that of the main, will the linear speed through the feeder pipes be faster or slower than through the main pipe?
   C. Under what conditions would the linear speed through the feeder pipes be the same as through the main pipe?
3. An engineer plans to construct a wind turbine with a funnel to direct the wind onto the turbine (see Figure 17–15). What advantages would there be in adding the funnel to the turbine? What problems might exist?

Section 17.2:

4. A gale wind causes a window to break. The glass shards fly outward rather than into the room. Why?
5. In a department store display, a beach ball is held in an airstream (see Figure 17–16). Draw a diagram to show the forces acting on the ball.
6. Wind blowing over the top of a chimney will cause smoke to be drawn up the chimney more rapidly. Why?
7. The normal force the roadway exerts on an automobile is usually less when the automobile is traveling at a high speed than when it is at rest. Why?

Section 17.3:

8. What is the absolute pressure inside a flat tire? What is the gauge pressure?
9. Under what conditions can there exist a negative gauge pressure? A negative absolute pressure?
10. Does a barometer read absolute or gauge pressure?

Sections 17.5, 17.6:

11. At a refinery, two pipelines have the same diameter, but pipe A has twice the length of pipe B. Petroleum is pumped through the two pipes with the same pressure difference at the ends of the pipes. The ratio of volume flow rate of pipe A to pipe B (V_a/V_b) is:

A. 2	***C.*** 0.5
B. 1	***D.*** 0.25

12. The **pressure gradient** across a pipe is the difference between the pressure at the ends of the pipe divided by the length of the pipe (see Figure 17–17). Which of the following pipes has the largest pressure gradient?
   A. A pipe 23 m long with a pressure difference of 460 Pa between the ends of the pipe.
   B. A 42 m long pipe with a pressure difference of 900 Pa between the ends of the pipe.
   C. A pipe 32 m long with a pressure of 1120 Pa at one end and a pressure of 220 Pa at the other.

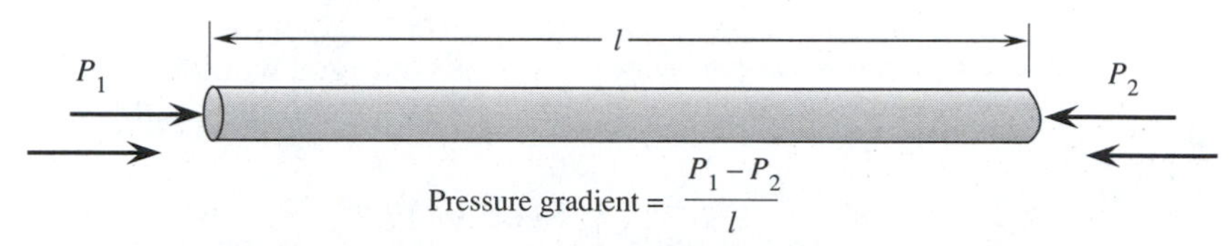

Figure 17–17: *Diagram for Exercise 12. The pressure gradient in a pipe is the change of pressure per unit length.*

13. Two pipes have the same pressure gradient. Pipe A has a 0.5-in diameter; pipe B, a 1.0-in diameter. When lubricating oil is forced through the pipes, the ratio of volumetric flow rates (V_B/V_A) is:
   A. 0.5
   B. 2
   C. 4
   D. 8
   E. 16

14. Why will Poiseuille's law not work for a fluid with a viscosity of zero?

15. If the rotational speed of a centrifuge is doubled, by what factor will the sedimentation rate increase?

*16. Galileo Galilei is reported to have dropped two cannonballs of different weights from the Leaning Tower of Pisa to show that heavier masses do not fall faster than light masses. It is said that the two cannonballs hit the ground at the same time. (A search of his extensive papers at Pisa show no written record of the event.) The experiment is to be recreated, but styrofoam balls are to be used instead of cannonballs. Ball A has twice the diameter of ball B.
   A. Which ball will have the greater viscous force acting on it?
   B. Which ball will have the larger terminal velocity?
   C. What will be the ratios of the terminal velocities?
   D. Which ball will hit the ground first?

PROBLEMS

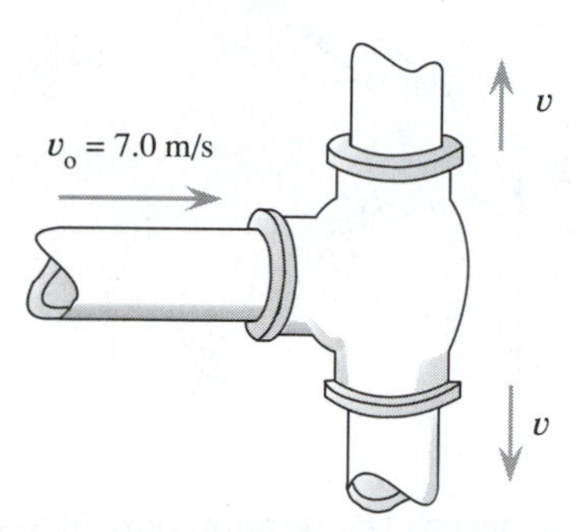

Figure 17–18: Diagram for Problem 4. Water flows into a plumbing T. Fluid flows out the two branches at the same speed (v).

Section 17.1:

1. The area of a ventilation duct reduces from 1.24 ft^2 to 0.78 ft^2. If the linear speed of air in the large-diameter duct is 24 ft/s, what is the speed of air in the smaller duct?
2. Water passes through a reducing nibble from a 1.0-in pipe to a $\frac{3}{8}$-in pipe. What is the linear speed of the water in the $\frac{3}{8}$-in pipe compared with its speed in the 1.0-in pipe?
3. The speed of water in a 1.3 cm diameter pipe is 8.2 m/s. How many liters of water pass through the pipe in 1 min?
4. A 2.5 cm diameter pipe is connected by a T to two other pipes of the same diameter. If the linear speed of water approaching the T is 7.0 m/s, find the flow rate in the other two pipes leaving the T. Assume the flow rate is the same in the two exit pipes. (See Figure 17–18.)

*5. A 2.0 in diameter hose delivers water at a rate of 2.1 ft^3/min (cfm). What is the linear speed of the water in the hose? Give your answer in feet per second.

Section 17.2:

6. At point A in a level pipe the pressure is 24 psia and the speed of the fluid is 120 in/s. At point B the pressure is 14 psia. If the fluid has a density of 43 lb/ft^3, what is its speed at point B?
7. The water level in a large-diameter, open storage tank is 22.0 m above a small, open pipe. With what speed does the water exit the pipe?
8. During a commercial break of a television broadcast of *The Wizard of Oz*, the public works department of a large city notes that the water level in its municipal storage tanks drops 17.3 ft. This corresponds to how much of a decrease of pressure in the water mains?
9. An irrigation pump produces 15 psi above atmospheric pressure to lift water 23.0 ft from a river to water a cornfield. With what speed does water flow through the irrigation pipes in the field?
10. A penstock has a head of 21 m. With what velocity will water strike the turbines at the lower end of the penstock?
11. A jet pump produces a pressure 23 psi above atmospheric pressure. How high a column of water can it support in an open pipe?

*12. Wind moves across a windowpane at $\overline{2}0$ m/s (see Figure 17–19).

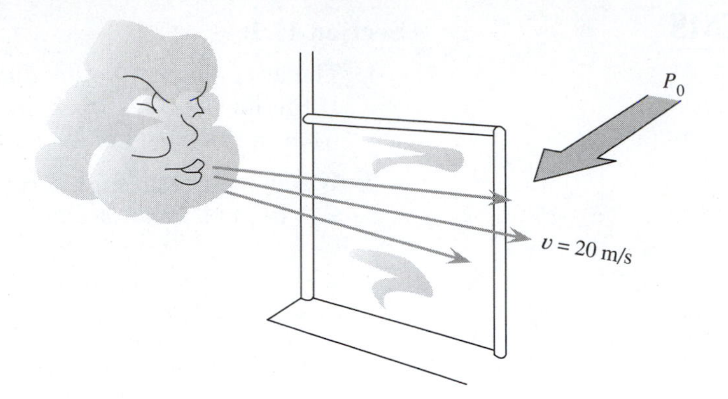

Figure 17–19: Diagram for Problem 12. Old Man Winter blows air past a windowpane, creating a pressure difference between the moving air and the static air inside the building.

A. What is the drop in pressure across the windowpane caused by the air's motion? (Hint: Compare the pressure of moving air with atmospheric pressure for still air.)

B. If the windowpane has an area of 1.20 m^2, what net force pushes out on the window?

Section 17.3:

13. A rotary vacuum pump produces an absolute pressure of $900\bar{0}$ Pa. What is the gauge pressure if atmospheric pressure is 1.013×10^5 Pa?

14. The gauge on an air compressor reads $4\bar{0}$ psig. What is the absolute pressure? (Assume standard atmosphere.)

15. A gauge pressure of -14.2 psig corresponds to what absolute pressure when atmospheric pressure is 14.2 psi?

17. A pressure scale reads an absolute pressure of 2.02×10^5 Pa. What is the corresponding gauge pressure for a standard atmosphere?

Section 17.4:

17. A water storage tank stands on top of a factory building with a water level 12.3 m above a faucet in the building. What is the linear speed of water flowing out of the faucet?

18. A fuel oil storage tank is filled to a height of 22.5 ft above an outlet. If the outlet valve is opened, what will be the linear speed of the oil flowing out?

19. An airplane wing is designed to have 23 lb lift for each square foot of wing. If air moves under the wing at $22\bar{0}$ ft/s, with what speed must it move over the top of the wing to give the required lift? (The density of air is 0.081 lb/ft^3.)

20. A siphon is used to skim cream off the top of a vat of whole milk. The bottom of the siphon tube is 0.70 m below the liquid

level. With what linear speed will the cream come out of the siphon?

*21. A $^3\!/_8$ in diameter rubber hose is used to siphon gasoline from an automobile gas tank. The level of the gas in the tank is 1.3 ft above the bottom end of the tube. How much gasoline will be siphoned off in 1 min?

Section 17.5:

22. An aluminum plate $3\bar{0}$ cm × 15 cm is pulled over a fixed, smooth surface by a hanging mass of $6\bar{0}$ g (see Figure 17–20). If there is 0.50 mm of oil with a viscosity of 2.3 poise between the surfaces, what is the terminal speed of the plate?

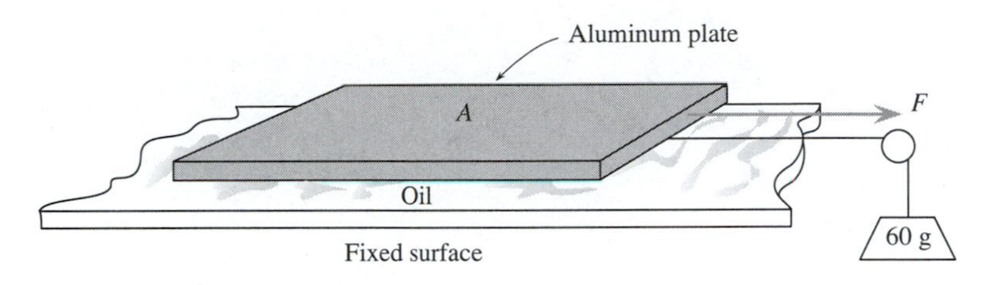

Figure 17–20: Diagram for Problem 22. An aluminum plate is drawn across a thin layer of viscous oil.

23. The wind speed near the ground is 7.0 mph. At an altitude of 200 ft, the wind speed is 13.5 mph. Estimate the time rate of change of shear strain (R) for the wind.

24. For the situation in Problem 22 what force is required to pull the plate at a terminal speed of 11.0 cm/s?

Section 17.6:

25. At $5\bar{0}$°F a gauge pressure of 10 psi forces 2.1 gal/min of formic acid (viscosity = 2.26 cp) through a pipe. What will be the flow rate at 86°F where formic acid has a viscosity of 1.46 cp with the same pressure?

26. A pipeline used to transfer chemicals at a petroleum-processing plant is replaced with one that has twice the diameter of the original and twice the length, and the pressure difference between the ends of the new pipe is twice that of the original. By what factor has the volume flow of the liquid changed?

27. A 2.00-mm steel ball with a density of 7.8×10^3 kg/m^3 is dropped into a container of glycerin (density = 1.26×10^3 kg/m^3 and viscosity = 0.83 Pl). What will be its terminal velocity?

28. An oil has a viscosity of 0.20 Pl. What is its volume flow rate when forced through a 40.0 m long pipe with a 2.50-cm inside radius under a gauge pressure difference of 6.0×10^4 Pa between the ends of the pipe?

29. What is the sedimentation rate in air of fly ash with a density of 2.5×10^3 kg/m^3 and diameter of 30 μm? (The viscosity of air is 1.8×10^{-4} p.)

30. A precipitate from a chemical reaction settles at a rate of 1.2 cm/h. What will be the sedimentation rate in a centrifuge that rotates at 22,000 rpm with the average radius of the sample 10 cm from the axis of rotation?

Chapter 18

TEMPERATURE AND MATTER

This aerial photograph shows a portion of the Outer Banks of North Carolina. Some scientists believe that the greenhouse effect will contribute to rising sea levels in coming years. Imagine what could happen to this fragile coastal area if predictions about global warming are accurate. (Photograph courtesy of NASA.)

OBJECTIVES

In this chapter you will learn:

- the distinction among temperature, thermal energy, and heat
- how to measure temperature using Celsius and Fahrenheit thermometers
- how to convert between Fahrenheit and Celsius temperature scales
- that thermal expansion is uniform in nature
- to calculate linear expansion and area expansion for solids
- to calculate volume expansion for liquids and solids
- that there is an absolute zero temperature at which molecular motion is at a minimum
- that absolute temperature, absolute pressure, and volume are simply related for an ideal gas

Time Line for Chapter 18

100	Hero of Alexandria studies the thermal expansion of air.
1592	Galileo Galilei builds a thermoscope, the first crude thermometer. It uses the volume expansion of air to measure temperature.
1662	Robert Boyle discovers that pressure and volume are inversely proportional for a gas at constant temperature.
1714	Gabriel Daniel Fahrenheit makes a mercury thermometer.
1742	Anders Celsius invents the Celsius temperature scale.
1787	Jacques Alexandre Charles finds that the thermal expansion of gases is proportional to temperature at constant pressure.
1802	Joseph-Louis Guy-Lussac finds that the pressure of a gas confined to a fixed volume will increase proportionally to temperature.
1852	Henri-Victor Regnault determines absolute zero.

As the temperature in a living room decreases, a bimetal strip in the thermostat bends, making an electrical contact; the furnace turns on. Iron heated to a high temperature starts to glow. At first it is dull red. As it gets hotter it changes to orange, then yellow. It is white just before it melts. Water fills a beaker to the brim. As the beaker is heated, some of the water overflows and dribbles down the side of the container. A Boy Scout tosses a sealed can of tomatoes onto a fire during his first camp-out. The pressure builds up inside the can until it bursts.

Many physical properties change with temperature. In this chapter, we will learn how some of these phenomena are used to measure temperature. We will also look at different temperature scales and compare them. Finally, we will examine how the sizes of things change when their temperatures are altered.

18.1 TEMPERATURE, THERMAL ENERGY, AND HEAT

We need to tell the difference between three different terms: thermal energy, heat, and temperature. When we were discussing work and energy, we found that work can cause three things to happen: kinetic energy can change, potential energy can change, or heat can be generated. Under the influence of frictional forces work is converted into kinetic and potential energy on the atomic scale. Let us perform a thought experiment to expand this idea.

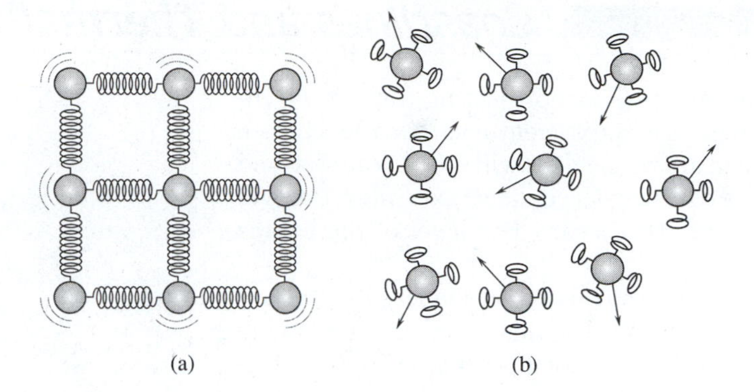

Figure 18–1: Potential energy is used to break interatomic bonds. (a) Atoms in a solid can vibrate, but they are held in fixed positions. (b) In a liquid atoms move about freely.

Imagine placing a saucepan full of ice and water on an electric stove. We know the heating element is hot because it glows and we can feel the heat rising from the burner. When we put the pan on the burner we expect heat to enter the pan and its contents. The reading on a thermometer placed in the ice-water mixture does not change. It stays at the melting temperature of ice as the ice melts. Something must be happening to the heat energy going into the ice and water.

Figure 18–1 will help you to visualize what is happening on the molecular level. In a solid such as ice, the molecules are attracted strongly to each other by intermolecular forces. These forces act as bonds holding each molecule in a fixed position relative to its neighbors. The molecules can oscillate and bounce around as if they were connected by springs, but they cannot change position.

At the melting temperature, enough work is done on the bond to stretch the bond between molecules to the breaking point. The molecules are still attracted to each other, but they are now free to move around. Since this is a process involving the positions of the molecules relative to each other, the heat is increasing the potential energy of the molecules.

After all the ice melts, there are no more molecular bonds to be broken. As heat enters the water, the average speed of the molecules increases, and the scale reading on the thermometer increases. Apparently temperature has something to do with kinetic energy on the atomic scale and nothing to do with molecular potential energy.

Temperature *is a measure of the amount of kinetic energy of individual atoms or molecules in a substance.*

All of the potential and kinetic energy contained in all the atoms of a substance is called thermal energy. A bathtub full of lukewarm water would probably have more thermal energy than a steaming

Coastlines and Thermal Expansion

Some meteorologists are projecting a warming climate through the next century. As the climate warms, ice in alpine glaciers will melt. Water stored in ice sheets covering Greenland and Antarctica will be dumped into the ocean. The level of the oceans will rise.

The most important factor in the rise of sea level is thermal expansion. Liquids, such as the liquid in the reservoir of a thermometer, expand as temperature increases. The extra volume shows up as a rising cylinder of fluid in the stem of the thermometer. The salty ocean waters are expanding in the same way.

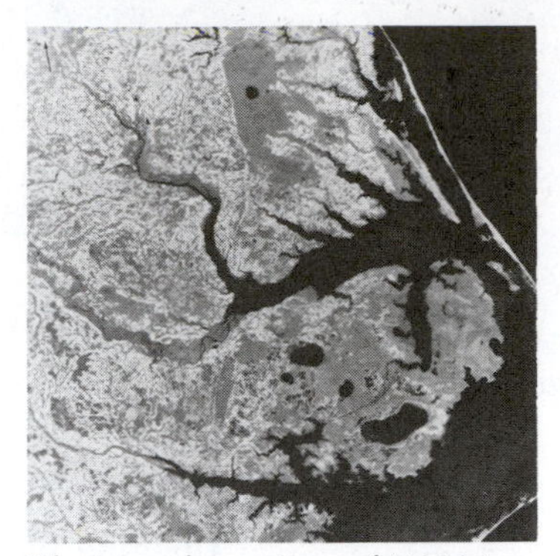

(Photograph courtesy of NASA.)

The calculation of expansion of water is complicated. Water expands very little near the freezing point. At higher temperatures, such as 25°C, the expansion rate is greater. A one-degree temperature increase in tropical ocean water will cause a much larger increase of volume than a one-degree temperature increase in cold arctic water. With computers, a reasonable estimate can be made.

The world's weather records show that the average temperature of the earth has increased 0.5°C, or 1.0°F, over the past 100 years. During that same time the records of seacoast harbors show an average increase in water level of 10 cm, or 4 in. Average estimates of the earth's temperature increase over the next century forecast 5°C of warming. This would correspond to a sea-level rise of about 1 m. Thermal expansion will be the major factor in this effect.

---

hot cup of soup. The soup has a higher temperature—that is, each atom has a higher kinetic energy than the atoms in the bathwater—but there are many more molecules in the bathwater. When the individual energies of the atoms are added up, the bathwater will have a larger total energy. For this reason, even a chilly iceberg could have more thermal energy than the cup of soup.

It is very difficult to determine exactly how much thermal energy something contains. It depends upon the way the atoms are bonded together, the ways the atoms are free to move, the kinds of imperfections in the structure, and what impurities are present. We can measure only the amount of thermal energy being added to or subtracted from the object. This is called heat.

Thermal *energy is total kinetic and potential energy on the atomic scale.*

Heat *is thermal energy in transit from one place to another.*

18.2 MEASURING TEMPERATURE

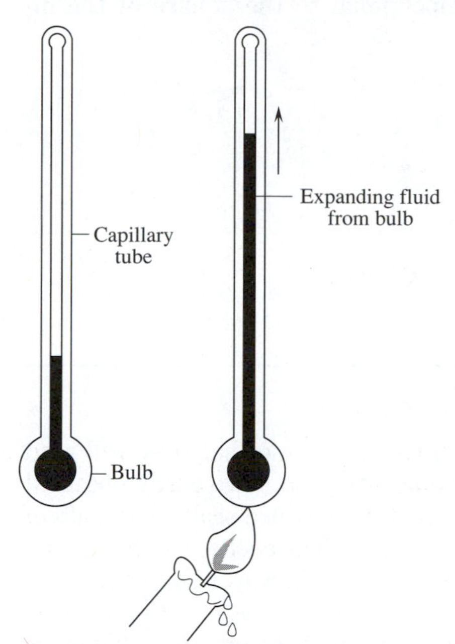

Figure 18–2: *A thermometer. The bulb has a reservoir of fluid. When the fluid is heated it expands up a narrow capillary tube.*

The physical properties of materials often change with temperature. These changes can be used to measure temperature. The dimensions of most materials are altered by temperature changes. This is the most popular thermal property for measuring temperature.

Many liquids increase in volume when their temperature is raised. A liquid thermometer consists of a reservoir (bulb) at the bottom of a capillary tube. Small changes in the volume of fluid are magnified by the narrowness of the tube. Since the tube has a very small cross section, fluid is forced along the length of the tube as the volume of fluid in the bulb changes (see Figure 18–2).

For common temperature ranges, the change of volume of the fluid in the reservoir is directly proportional to the temperature change.

$$\Delta V = k \, \Delta T$$

where k is a constant depending on the dimensions of the reservoir and the kind of fluid in the thermometer.

The change of length of fluid in the tube depends on the tube's cross section. $\Delta V = A \, \Delta L$. The temperature change is proportional to the change in length of the column of fluid.

$$\Delta L = \frac{k}{A} \Delta T$$

The amount of change of length of the column will be different for different thermometers, depending on the dimensions of the bulb and the tube and on the fluid. Mercury expands differently from alcohol.

☐ **EXAMPLE PROBLEM 18–1: TWO THERMOMETERS**

Two thermometers are identical except thermometer B has a capillary tube with half the diameter of thermometer A. The column of fluid in thermometer A moves upward 8.0 cm for a temperature change. How far up the capillary tube will fluid move in thermometer B for the same temperature change?

■ *SOLUTION*

The ratio of diameters is proportional to the square of the diameters ($A = \pi D^2/4$).

$$\frac{A_A}{A_B} = \frac{(2D)^2/4}{D^2/4} = 4$$

$$\frac{L_B}{L_A} = \frac{(k/A_B)\,\Delta T}{(k/A_A)\,\Delta T} = 4$$

$$\Delta L_B = 4\,\Delta L_A = 4 \times 8.0 \text{ cm}$$

$$\Delta L_B = 32 \text{ cm}$$

We need to standardize the thermometers so that they will read the same temperatures for the same situation. Two easily obtained temperatures are used to make the temperature scale. The bulb of the thermometer is immersed in ice water. The position of the column is marked. The thermometer is then placed in water boiling at standard atmosphere. (The boiling temperature of water varies with air pressure.) A second mark is made on the thermometer. This gives us two known temperature readings (see Figure 18–3). The two marks are inconveniently far apart. We can break this distance into smaller equally spaced divisions. One common way to mark the scales is to divide the distance into 100 equal parts. The ice temperature is labeled 0°. The boiling temperature is labeled 100°. This is a Celsius thermometer.

Figure 18–3: A thermometer scale is calibrated by finding the height of fluid for the ice point and for the boiling point of water.

On a Celsius *scale, one degree Celsius is 1/100 the difference between the freezing temperature and the boiling temperature of water.*

Another scale divides the distance between the two marks by 180. The ice temperature is 32° and the boiling point is 212°. This is a Fahrenheit thermometer.

On a Fahrenheit *scale, one degree Fahrenheit is 1/180 the difference between the freezing temperature and the boiling temperature of water.*

The spacing of the marks is arbitrary. We could invent our own scale by using different divisions and placing different labels at the two fixed points.

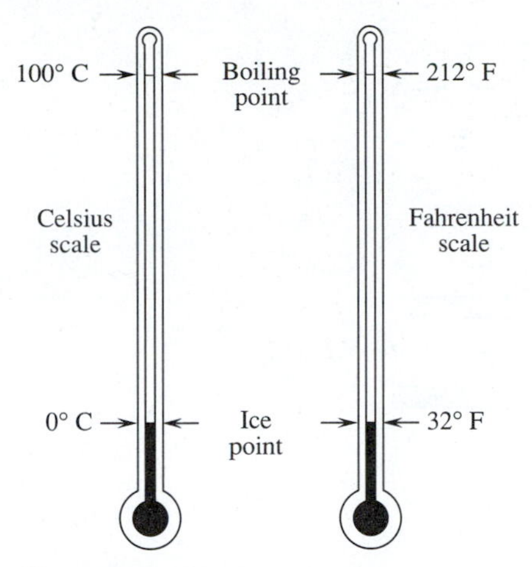

Figure 18–4: Comparison of the Celsius scale and the Fahrenheit scale.

We can easily compare readings from two temperature scales. Figure 18–4 shows a Fahrenheit scale next to a Celsius scale. There are 180 Fahrenheit degrees between the two fixed points for water. There are only 100 Celsius degrees to represent the same temperature change. The ratio of degrees is:

$$\frac{180°\text{F}}{100°\text{C}} = \frac{9°\text{F}}{5°\text{C}}$$

A 9° Fahrenheit change of temperature is the same as a 5° change on the Celsius scale. We need to correct for the ice temperature. To convert from a Celsius reading to a Fahrenheit reading we need to multiply by the ratio and then add 32°.

$$\text{F} = \frac{9}{5}\text{C} + 32 \qquad \textbf{(Eq. 18–1)}$$

□ **EXAMPLE PROBLEM 18–2: TEMPERATURE CONVERSION**

Most people have a normal body temperature of 98.6°F. What is this temperature on the Celsius scale?

■ ***SOLUTION***

Solve Equation 18–1 for C.

$$\text{C} = \frac{5}{9}(\text{F} - 32.0)$$

$$\text{C} = \frac{5}{9}(98.6 - 32.0)$$

$$\text{C} = 37.0°\text{C}$$

18.3 LINEAR EXPANSION

Most materials increase in size when they are heated. Figure 18–5 shows a rod at four different temperatures. Its change of length is proportional to the change of temperature ($\Delta L \propto \Delta T$).

If different lengths of the same material are used, the change of length varies. Longer pieces expand more than shorter pieces (see

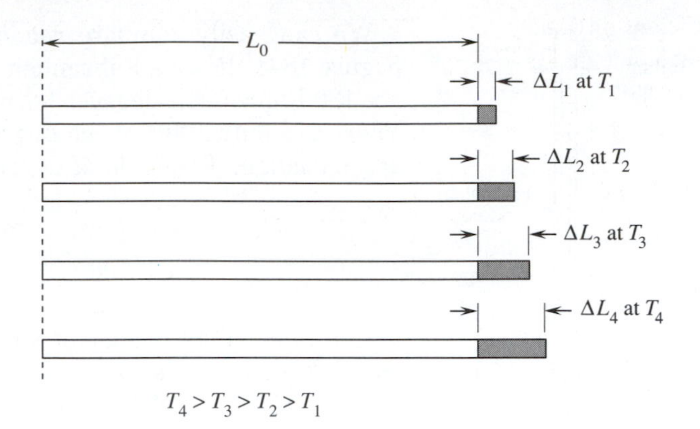

Figure 18–5: *The change of length of a rod (Δ L) is proportional to the change of temperature (Δ T).*

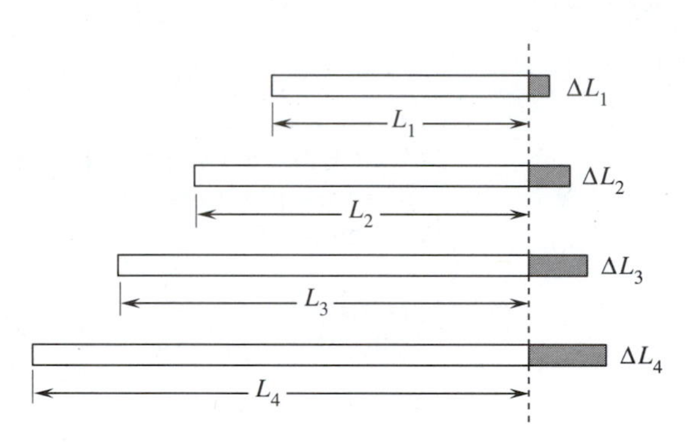

Figure 18–6: *Four rods of the same material experience the same change of temperature.* $\Delta L \propto L$.

Figure 18–6). The change of length is proportional to the original length ($\Delta L \propto L$). Another way of writing this is: $\Delta L/L$ = constant.

We can combine this information to get a single equation. ∝ is the combined proportionality constant. It is called the **coefficient of linear expansion.**

$$\Delta L/L = \propto \Delta T \qquad \textbf{(Eq. 18–2)}$$

The fractional change of length ($\Delta L/L$) is directly proportional to the temperature change. $\Delta L/L$ is called the **thermal strain.** The thermal strain plotted against temperature change plots into a straight line. The slope of the graph is the coefficient of linear expansion (∝). Different materials have different ∝s. ∝ is a material constant. Table 18–1 lists the coefficient of linear expansion for some common solids.

Table 18–1: Coefficient of linear expansion for some common solids.

SOLID	$\times 10^{-5}/°C$	$\times 10^{-5}/°F$
Metals		
Aluminum	2.4	1.3
Brass	1.8	1.0
Copper	1.7	0.94
Iron	1.2	0.66
Lead	3.0	1.7
Steel	1.2	0.66
Zinc	2.6	1.4
Ceramics		
Brick		
fireclay	0.45	0.25
building	0.90	0.50
Concrete	1.3	0.72
Glass		
plate	0.90	0.50
pyrex	0.27	0.15
silica	0.05	0.03
Polymers		
Nylon	10	5.6
Polystyrene	6.3	3.5
Polyvinylidene chloride	19	10.5
Rubber (vulcanized)	8.1	4.5

☐ **EXAMPLE PROBLEM 18–3: THE RAILROAD TRACK**

Steel rails 30.0 ft long are laid on a railroad track with a gap of 0.20 in between each pair of rails. If the temperature is 52°F when the rails are laid, what will be the width of the gap when the temperature is 91°F?

■ ***SOLUTION***

The increase of length of the rails is:

$$\Delta L = 0.67 \times 10^{-5}/°\text{F} \times 30.0 \text{ ft} \times 12 \text{ in/ft} \times (91°\text{F} - 52°\text{F})$$
$$\Delta L = 0.094 \text{ in}$$

The width of the gap at 91°F is:

$$W = 0.20 \text{ in} - 0.094 \text{ in}$$
$$W = 0.11 \text{ in}$$

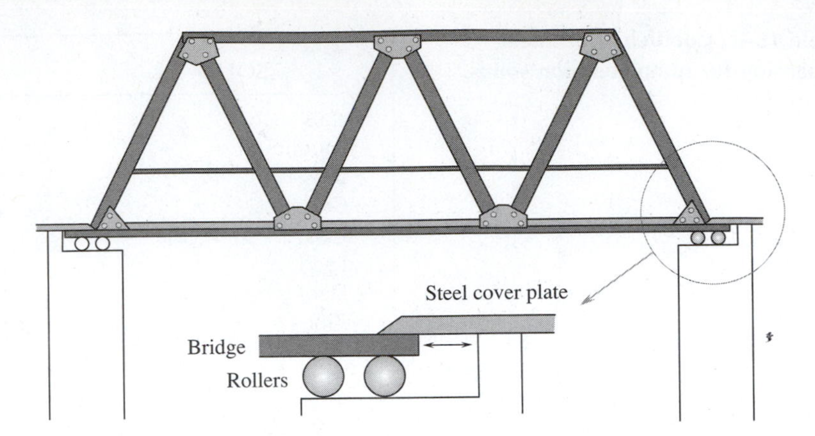

Figure 18–7: One kind of expansion joint for a steel bridge.

Thermal expansion must be taken into account when designing structures that will undergo temperature cycling. Bridges undergo thermal expansion. Figure 18–7 shows one solution to the problem. One end of the bridge rests on rollers. A cover plate lies over the rollers to let traffic pass. Power lines are hung with enough slack to prevent the thermal shrinkage caused by cold weather from increasing the tension on the lines to the breaking point. Sidewalks are poured in blocks with expansion joints between the sections.

□ **EXAMPLE PROBLEM 18–4: THERMAL STRESS**

A 19.2 m long steel beam used as a support member for a bridge abuts two granite banks (see Figure 18–8). When the temperature increases by 18°C, what force will be exerted on the beam? The beam has a cross section of 1.23×10^{-3} m².

■ ***SOLUTION***

If the beam were allowed to expand, its thermal strain would be:

$$\frac{\Delta L}{L} = \propto \Delta T$$

The granite exerts a compressional stress on the beam to prevent the expansion.

$$\frac{F}{A} = Y \frac{\Delta L}{L}$$

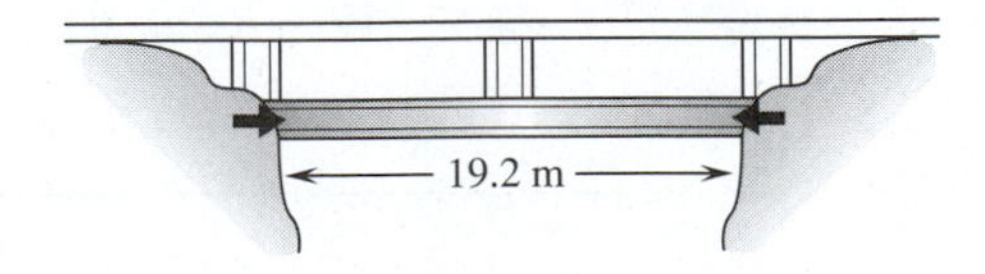

Figure 18–8: Rock walls prevent a steel beam's expansion. Compressional thermal stresses occur.

where Y is Young's modulus. The strain is the same in both equations if the beam is prevented from expanding.

$$\frac{F}{A} = Y \propto \Delta T$$

or

$$F = Y A \propto \Delta T$$

$$F = (20 \times 10^{10}\ \text{N/m}^2)\ (1.23 \times 10^{-3}\text{m}^2)\ (1.2 \times 10^{-5}/^\circ\text{C}) \times 18^\circ\text{C}$$

$$F = 5.3 \times 10^4\ \text{N}$$

18.4 AREA EXPANSION

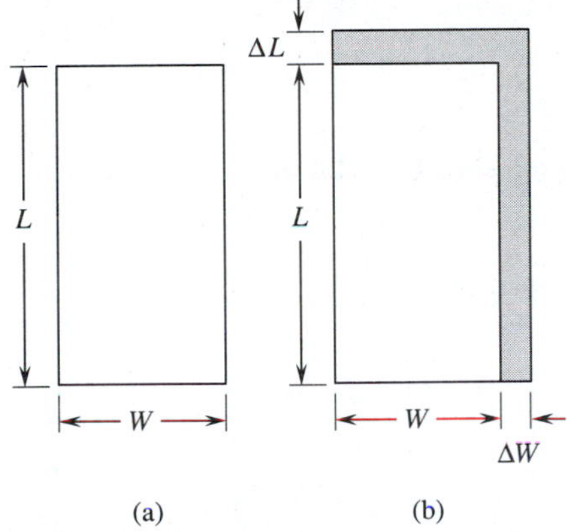

Figure 18–9: *Thermal expansion of a plate with an initial area of A = L W.*

Occasionally we run into problems involving change of the dimensions of a plate or cross section. This can be handled easily using the coefficient of linear expansion. Look at Figure 18–9. A plate has an initial length L and width W. When it is heated, the new length is $(L + \Delta L)$ and the width is $(W + \Delta W)$. The new area is:

$$A + \Delta A = (L + \Delta L)\,(W + \Delta W)$$

$$A + \Delta A = (L\,W + \Delta L\,W) + (L\,\Delta W + \Delta L\,\Delta W)$$

The last term on the right is much smaller than the other terms, so we can ignore it. Subtract the area from each side of the equation $(A = L\,W)$.

$$\Delta A = (\Delta L\,W) + (L\,\Delta W)$$

Divide by area to get the area strain.

$$\frac{\Delta A}{A} = \frac{\Delta L\,W}{L\,W} + \frac{L\,\Delta W}{L\,W}$$

or

$$\frac{\Delta A}{A} = \frac{\Delta L}{L} + \frac{\Delta W}{W}$$

For linear expansion we have:

$$\frac{\Delta L}{L} = \propto \Delta T \quad \text{and} \quad \frac{\Delta W}{W} = \propto \Delta T$$

Substitute these expressions into the equation for area strain.

$$\frac{\Delta A}{A} = 2 \propto \Delta T \qquad \textbf{(Eq. 18–3)}$$

The area thermal strain is directly proportional to the temperature change. The only difference is we must use twice the coefficient of linear expansion.

☐ **EXAMPLE PROBLEM 18–5: THE CYLINDER**

An aluminum cylinder has a diameter of 8.240 cm at 23°C.

A. What is its diameter at 45°C?
B. What is its increase in area?

■ ***SOLUTION***

A.

$$\Delta D = \propto D_0 \Delta T$$

$$\Delta D = (2.4 \times 10^{-5}/°C)(8.240 \text{ cm})(45°C - 23°C)$$

$$\Delta D = 0.004 \text{ cm}$$

$$D = D_0 + \Delta D = 8.240 \text{ cm} + 0.004 \text{ cm}$$

$$D = 8.244 \text{ cm}$$

B.

$$A = \left(\frac{\pi}{4}\right) D_0^2$$

$$\Delta A = 2 \propto \Delta T A = 2 \propto \Delta T \left(\frac{\pi}{4}\right) D_0^2$$

$$\Delta A = 2(2.4 \times 10^{-5}/°C)(22°C)\left(\frac{\pi}{4}\right)(8.240 \text{ cm})^2$$

$$\Delta A = 5.6 \times 10^{-2} \text{ cm}^2$$

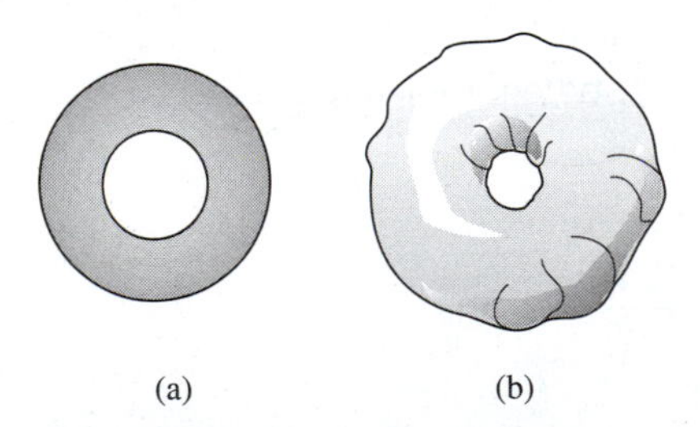

Figure 18–10: *Nonuniform expansion. (a) A doughnut is stamped out of dough. (b) The cooked doughnut has swelled. Different parts of the doughnut expand at different rates.*

We have to be careful in the way we think of thermal expansion. Some people mistakenly think of doughnuts. When the ring shapes cut from dough are dropped into hot cooking oil they swell up; part of the hole in the center is lost. This is an example of **nonuniform expansion** (see Figure 18–10). Thermal expansion does *not* behave this way.

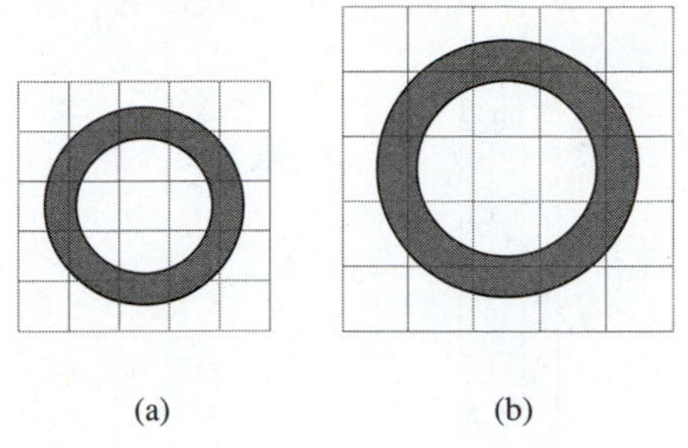

Figure 18–11: Uniform expansion. All parts of the object expand proportionally like a photographic enlargement.

Thermal expansion is uniform.

In thermal expansion all dimensions increase equally. A good analogy is a photograph enlargement (see Figure 18–11). All parts of the object are increased proportionally. If there is a hole in a plate, the hole will increase as the temperature goes up. Think of photographic enlargements, not doughnuts.

EXAMPLE PROBLEM 18–6: THE COMPRESSION FIT

At 26°C, a steel ring has an inside diameter of 9.197 cm. It is to be fitted around a cylinder with a diameter of 9.200 cm. Compressional forces will hold it in position. To what minimum temperature must the ring be heated to slip over the cylinder?

SOLUTION

As the ring is heated, it will expand uniformly. The size of the hole will increase. The inside diameter of the ring must at least match the diameter of the cylinder.

$$\Delta T = \frac{\Delta D}{D \propto}$$

$$\Delta T = \frac{(9.200 - 9.197)\text{ cm}}{9.200\text{ cm} \times 1.2 \times 10^{-5}/^\circ\text{C}}$$

$$\Delta T = 27^\circ\text{C}$$

$$T_f = T_i + \Delta T = 26^\circ\text{C} + 27^\circ\text{C}$$

$$T_f = 53^\circ\text{C}$$

18.5 VOLUME EXPANSION OF SOLIDS AND LIQUIDS

The volume expansion of solids is handled a lot like area expansion (see Figure 18–12). A rectangular solid has a volume of $V = L\,W\,H$. It expands to a new volume $V + \Delta V = (L + \Delta L)(W + \Delta W)(H + \Delta H)$. It is left as an exercise to derive the equation below, showing that the change of volume is:

$$\frac{\Delta V}{V} = 3 \propto \Delta T \qquad \textbf{(Eq. 18–4)}$$

Because liquids have no definite shape, they have only bulk properties; there are no linear expansions. Instead, a coefficient of volume

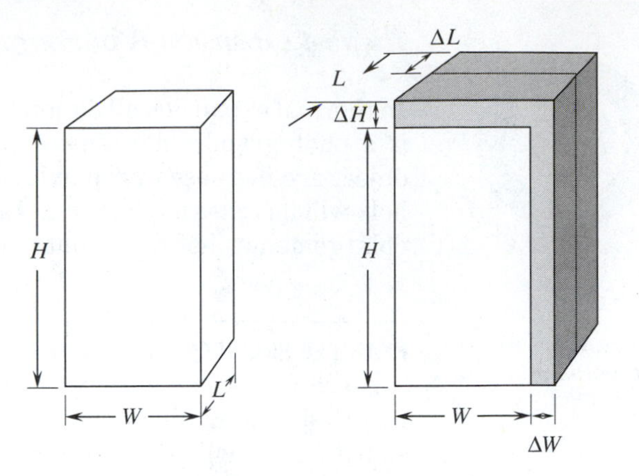

Figure 18–12: *A rectangular volume expands.*

Table 18–2: Coefficient of volume expansion for some common liquids.

LIQUID	$\times 10^{-4}/°C$	$\times 10^{-4}/°F$
Ethyl alcohol	11	6.1
Benzene	12.4	6.9
Gasoline	10.8	6.0
Glycerin	5.3	2.9
Sulfuric acid	5.8	3.2
Turpentine	9.4	5.3
Water (at 20°)	2.1	1.2

expansion (β) is given. Table 18–2 lists the coefficient of volume expansion for some common liquids. As with solids, the volume strain is directly proportional to the temperature change.

$$\frac{\Delta V}{V} = \beta \, \Delta T \qquad \textbf{(Eq. 18–5)}$$

□ **EXAMPLE PROBLEM 18–7: DIFFERENTIAL EXPANSION**

An aluminum cup with a 200-ml volume is filled to the brim with glycerin at 24°C. If the cup is heated to 46°C, how much fluid will overflow the cup?

■ ***SOLUTION***

This is a differential expansion problem. Both the container and the fluid will expand. The difference in the expanded volumes will equal the overflow. Let ΔV_f be the change of volume of the fluid and ΔV_c equal the change of volume of the container. V_0 is the initial volume of the liquid and container.

$$overflow = \Delta V$$
$$= \Delta V_f - \Delta V_c$$
$$\Delta V = (\beta_f \Delta T V_0) - (3 \propto_c \Delta T V_0)$$
$$\Delta V = (\beta_f - 3 \propto) \Delta T V_0$$
$$\Delta V = [(53 \times 10^{-5}) - (3 \times 2.4 \times 10^{-5})°C] \times (46°C - 24°C) \times 220 \text{ ml}$$
$$\Delta V = 2.2 \text{ ml}$$

18.6 IDEAL GASES

The thermal behavior of gases needs to be handled differently from that of solids or liquids. At pressures of less than 10 atm most gases have a simple relationship between temperature, pressure, and volume. These are called **ideal gases.**

Look at Figure 18–13. A metal sphere contains a sample of gas. Attached to the sphere is an absolute pressure gauge. The sphere is dipped into solutions at different temperatures. The absolute pressure of the gas is plotted against temperature to give us line A.

More gas is pumped into the sphere to give us a higher initial pressure. It is again dipped into different temperature solutions. The pressure as a function of temperature plots into line B.

The experiment is repeated one more time after evacuating some of the gas to lower the initial pressure to a point below that of the other two trials. The result is line C.

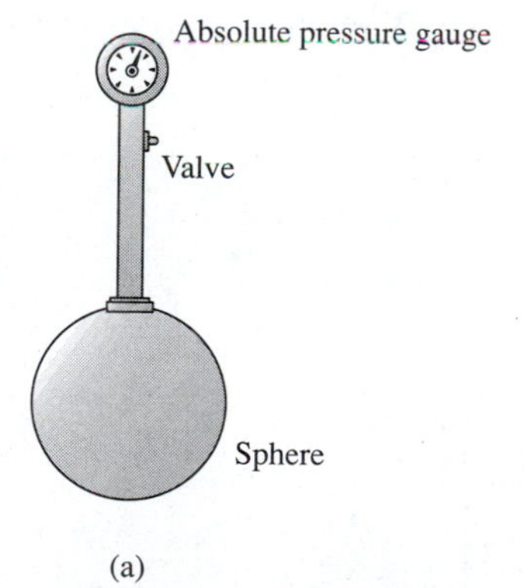

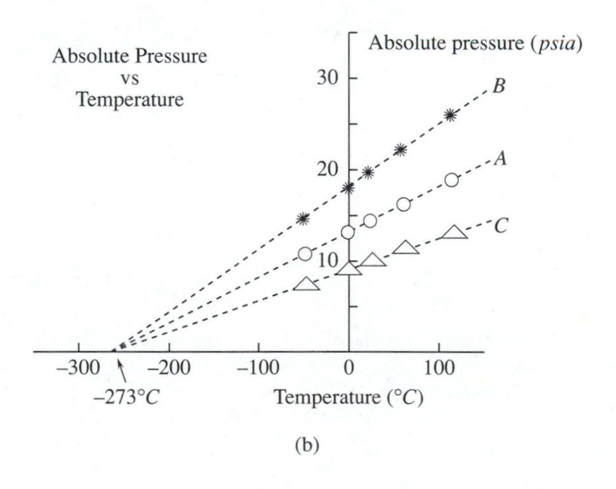

Figure 18–13: *(a) A sphere full of gas is connected to an absolute pressure gauge. (b) A graph of absolute pressure versus Celsius temperature for three different volumes of gas in the bulb.*

Notice that we get a straight line in each case. Each line can be extrapolated to the same spot on the temperature axis. It does not matter whether we use air, neon, hydrogen, or any other ideal gas. The result is the same. We get a straight line that is dependent only on the initial pressure and the temperature. All pressure-temperature lines meet at −273°C.

If we define a new temperature scale with −273°C as the zero of the scale, the absolute pressure will be directly proportional to the new temperature scale. Since there are no negative readings on this new scale it is called an **absolute temperature scale.**

For Celsius degrees of temperature, the absolute temperature is called the **kelvin scale (K).** A Celsius reading can be converted to a kelvin reading by:

$$K = C + 273 \qquad \textbf{(Eq. 18–6)}$$

The kelvin temperature of 0°K is called **absolute zero.** If the gases did not liquefy, this would be the temperature at which the pressure would be zero. We will have more to say about absolute zero later.

If a Fahrenheit scale had been used in the experiment, the pressure-temperature lines would have all met at −460°F. This is the Fahrenheit equivalent of −273°C. We can make an absolute temperature scale for Fahrenheit degrees by adding $46\bar{0}$ to the Fahrenheit reading. This absolute temperature scale is called the **Rankine scale.**

$$R = F + 46\bar{0} \qquad \textbf{(Eq. 18–7)}$$

Figure 18–14 shows some equivalent temperatures on the four temperature scales we now know: the Celsius scale (C), the kelvin scale (K), the Rankine scale (R), and the Fahrenheit scale (F).

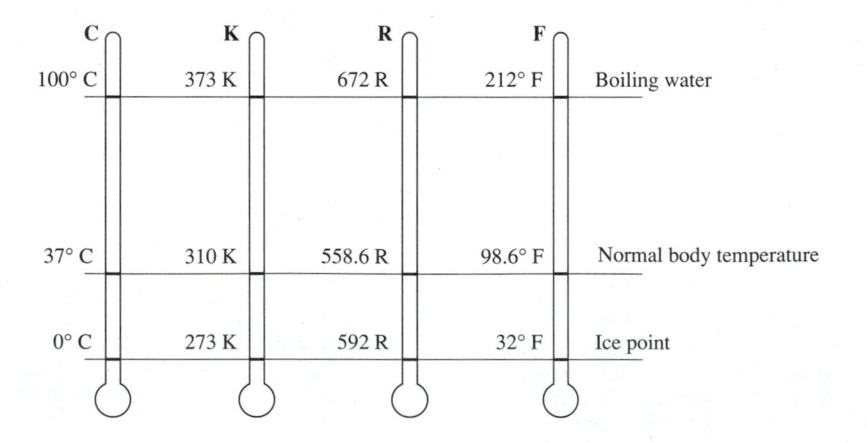

Figure 18–14: *Comparison of temperatures using four different temperature scales. C, Celsius; K, kelvin; R, Rankine; F, Fahrenheit.*

EXAMPLE PROBLEM 18–8: ABSOLUTE TEMPERATURES

Lead melts at 327°C. What is the melting temperature of lead on the kelvin scale? On the Fahrenheit scale? On the Rankine scale?

SOLUTION

A.

$$K = 327°C + 273$$

$$K = 60\bar{0}\ K$$

B.

$$F = \frac{9}{5}C + 32$$

$$F = \frac{9}{5}(327) + 32$$

$$F = 621°F$$

C.

$$R = 621°F + 46\bar{0}$$

$$R = 1081\ R$$

The data for Figure 18–13 can be replotted with the absolute pressure graphed against absolute temperature. This produces a graph that looks like Figure 18–15. The absolute pressure is directly proportional to the absolute temperature.

$$P_{abs} = k\,T_{abs}$$

where k is a proportionality constant.

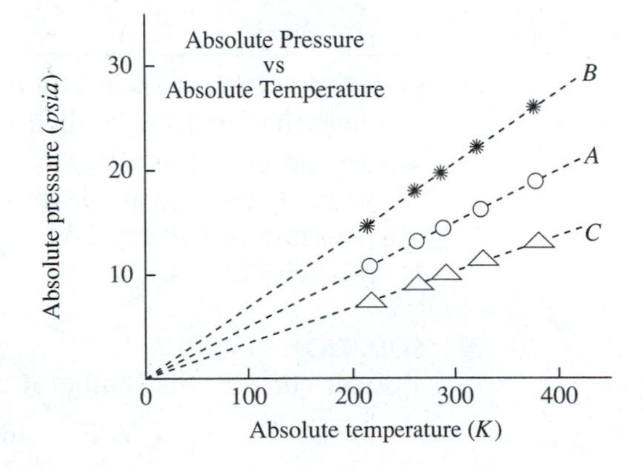

Figure 18–15: *Absolute pressure versus absolute temperature. The data from Figure 18–13b are replotted, using absolute temperature ($P \propto T$).*

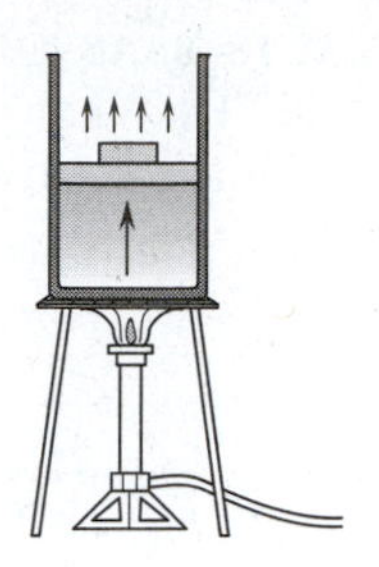

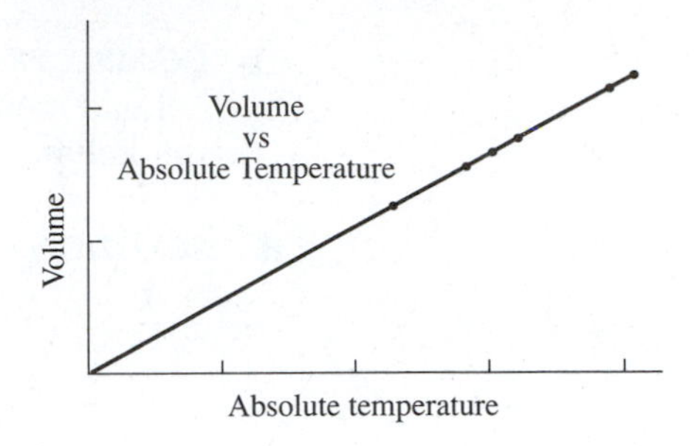

Figure 18–16: The volume of gas in a cylinder versus absolute temperature (V ∝ T).

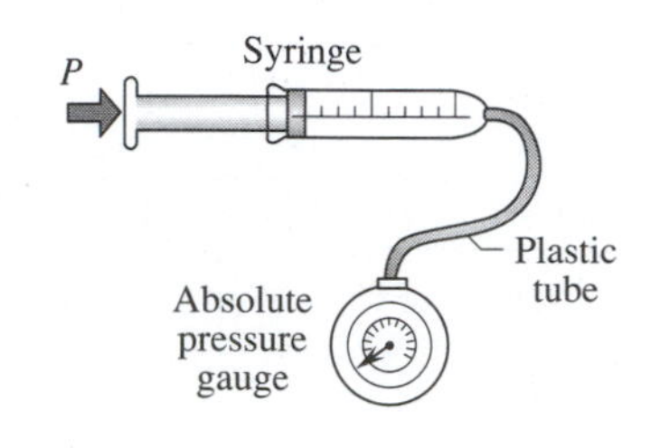

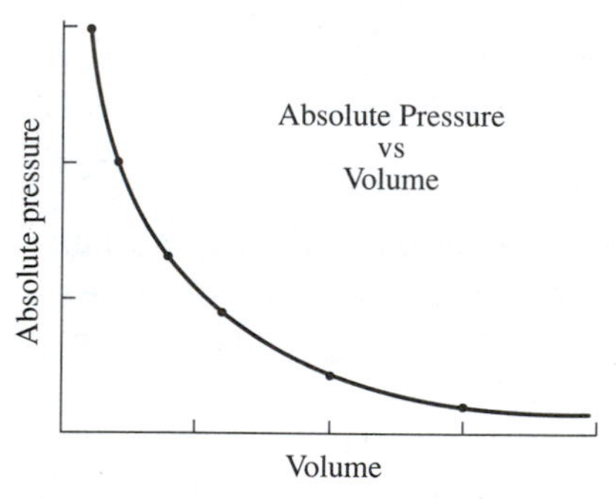

Figure 18–17: The volume of gas in a syringe versus absolute pressure (P ∝ 1/V).

Let us perform a different experiment. Look at Figure 18–16. We will place gas in a friction-free piston. As we heat the gas, it will expand against a constant pressure. The variables are volume and temperature. When we plot volume against *absolute* temperature, we again get a direct proportion. The volume extrapolates to zero at absolute zero.

$$V = k\,T_{\text{abs}}$$

Now let us compare absolute pressure against volume. We will connect an absolute pressure gauge to one end of a plastic tube. At the other end of the tube we will attach a syringe (see Figure 18–17). As we press on the syringe plunger, the pressure increases as the volume of gas decreases. A graph of the data has a hyperbolic shape, as we would expect for an inverse relationship: $V = k/\mathrm{P}$ or $V\,\mathrm{P} = k$.

The information from the three experiments can be combined into one equation called the **ideal gas law.**

$$\mathrm{P}\,V = k\,T \qquad \textbf{(Eq. 18–8)}$$

P and T must be expressed in absolute scales. k is a constant as long as no gas is added or subtracted from the sample.

□ **EXAMPLE PROBLEM 18–9: THE WEATHER BALLOON**

A weather balloon is filled with 985 ft^3 of helium on the ground where the pressure is 1.00 atm and the temperature is 56°F. It is released. It rises to an elevation where the temperature is 5°F and the pressure is 0.40 atm. What is the volume of the gas in the balloon at this height?

■ ***SOLUTION***

First convert the temperature to the Rankine scale.

$$T_1 = 56°\text{F} + 46\bar{0} = 516\ \text{R}$$

$$T_2 = 5°\ \text{F} + 46\bar{0} = 465\ \text{R}$$

Since no gas is added to or subtracted from the sample, k is constant in the ideal gas law equation. It has the same value for both locations of the balloon.

$$k = \frac{P_1 V_1}{T_1} = \frac{P_2 V_2}{T_2}$$

or

$$V_2 = \frac{P_1 T_2 V_1}{P_2 T_1}$$

$$V_2 = \frac{(1.00 \text{ atm})(465 \text{ R})(985 \text{ ft}^3)}{(0.40 \text{ atm})(516 \text{ R})}$$

$$V_2 = 2.2 \times 10^3 \text{ ft}^3$$

18.7 OTHER WAYS TO MEASURE TEMPERATURE

Volume expansion is not the only way to measure temperature. Here is a list of some other techniques.

The Thermistor

Semiconducting diodes undergo changes in electrical resistance as their temperature changes. Unlike metals, the resistance of semiconductors decreases as the temperature goes up. Over a small range of temperatures a diode will undergo a very rapid decrease. Currents passing through a diode can be measured directly, and the information can be fed directly into a computer, an electronically managed control device, or a recorder.

Diodes constructed specifically to measure temperature are called **thermistors.** Different diodes are designed to measure different temperature ranges. Figure 18–18 shows a plot of resistance of a thermistor designed to operate near room temperature.

The Thermocouple

A **thermocouple** is a junction between two pieces of wire made of different metals (see Figure 18–19). If the temperature of the junction (T_H) is higher (or lower) than the temperature of a reference junction (T_0), a small voltage difference occurs. The reference junction usually is dipped into a water-ice mixture, producing a reference temperature of 0°C. The potential difference is small; it is in the range of millivolts. The potential difference is not linear. A chart of temperature versus voltage is used to find the temperature. Another technique is to feed the information into a computer circuit to get a digital readout.

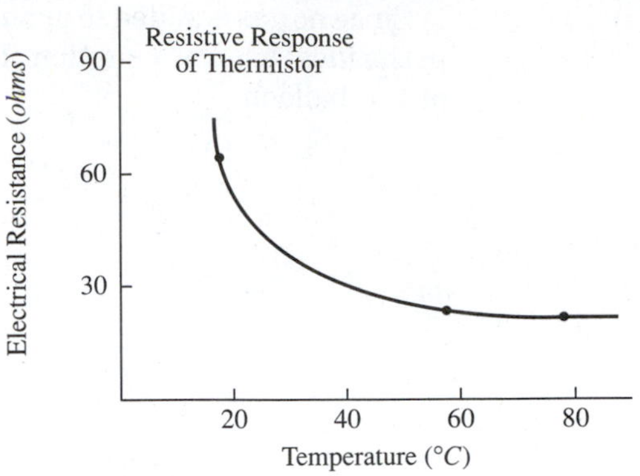

Figure 18–18: *The electrical resistance of a thermistor decreases as temperature increases.*

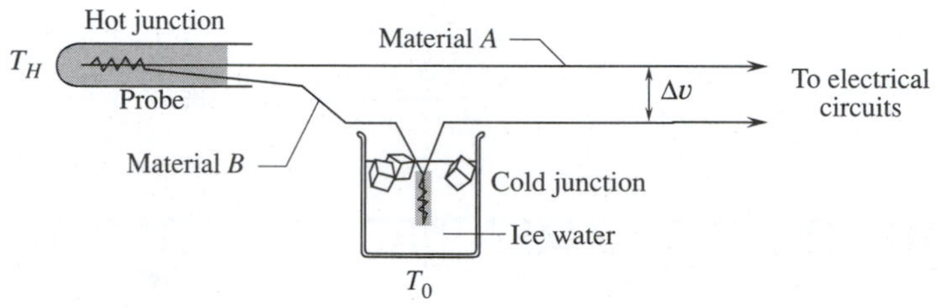

Figure 18–19: *A thermocouple is made by welding two different metals to form a junction. A small electrical potential occurs between the hot junction and the cold reference junction.*

Thermocouples can measure temperatures over a wide range. The same thermocouple can measure the temperature of liquid nitrogen and the melting point of aluminum. The reference junction and the hot junction can be several feet apart. For example, the hot probe can be in a furnace and the cold junction and electrical circuitry can be some distance away.

Pyrometers

At any temperature above absolute zero objects radiate heat energy. Around room temperature, most of the radiation is in long wavelengths called **infrared (IR) radiation** (see Figure 18–20). As the temperature rises, more and more of the radiation occurs in shorter wavelengths. At very high temperatures, a large proportion is in the visible light range. The object begins to glow. At first, most of the visible light is in the longer wavelengths, which we see as red. As the temperature continues to rise, more of the energy is emitted at shorter wavelengths. The object turns orange. At yet higher temperatures, it turns white. For solids and liquids, the color depends only on the temperature. It does not depend on the kind of material.

The change in wavelength of the emitted radiation can be used to measure temperature. If the object is hot enough to glow, an **optical pyrometer** can be used. For lower temperatures, an infrared scanner is used to sample and analyze the invisible infrared (IR) radiation. This is an **infrared pyrometer.**

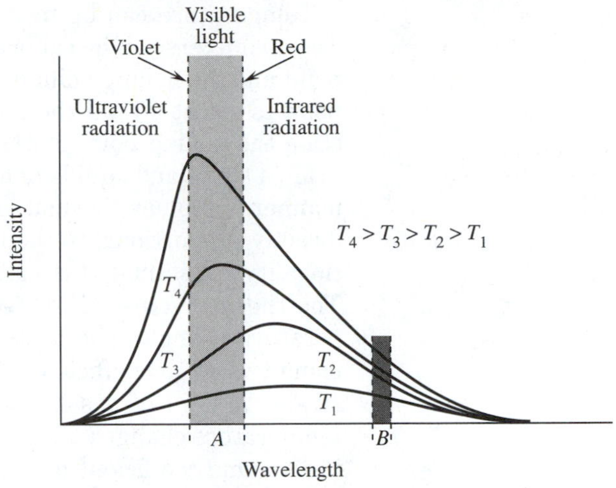

Figure 18–20: A plot of the intensity of radiant energy for an object at four different temperatures. The peak intensity shifts to shorter wavelengths as temperature increases. (A) The visible wavelengths. (B) A small region of infrared radiation that might be used by an infrared pyrometer.

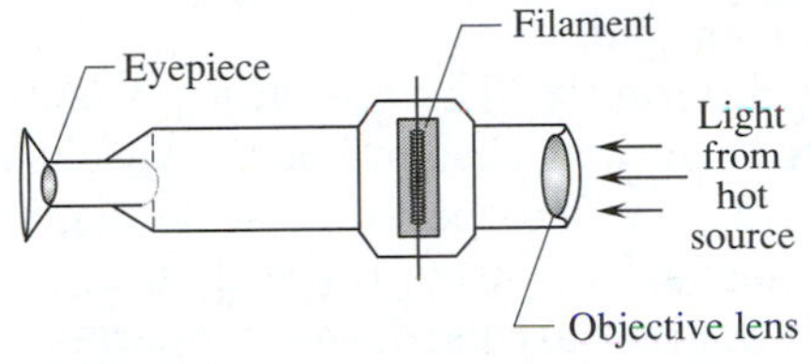

Figure 18–21: An optical pyrometer. A glowing filament is seen against a background of light from the material being measured.

1. Optical pyrometer. In an optical pyrometer, a standardized wire filament is heated by electrical current until it glows. The filament is enclosed in a glass tube. In turn the tube is placed in the field of a viewing scope (see Figure 18–21). The pyrometer is held so that the filament is seen against a background consisting of the object to be measured, such as the inside of a furnace. The electrical current is adjusted until the color of the filament matches the background. A properly calibrated scale between temperature and electrical current through the filament makes it possible to read the temperature.
2. Infrared pyrometer. Infrared (IR) pyrometers are now used more often than optical pyrometers. IR pyrometers view a narrow range of wavelengths in the infrared region (see Figure 18–20). The field of view of the pyrometer must be completely filled by radiation from the sample (see Figure 18–22). The radiation intensity is converted into electrical impulses, which can be passed on to a microprocessor for analysis. The intensity of the radiation from the source in the selected bandwidth is related to the temperature.

   IR pyrometers are designed to operate over a range of a few hundred degrees. Commercial infrared pyrometers are available with ranges reaching temperatures as low as 0°F and as high as 3600°F.

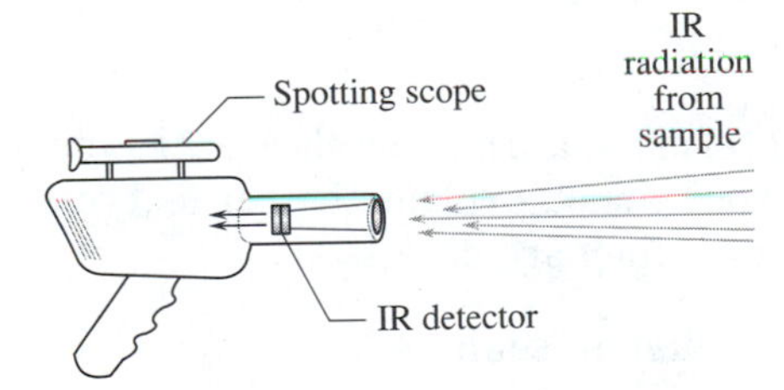

Figure 18–22: An infrared (IR) pyrometer. An IR detector measures the intensity of infrared radiation.

SUMMARY

Three important terms are used in discussing temperature and matter. Temperature is a measure of the average kinetic energy of individual atoms in a material. Thermal energy is the sum of all the individual potential and kinetic energies of the atoms in a material. Heat is thermal energy moving from one place to another.

Temperature can be measured by devices called thermometers. Thermometers can be calibrated using two fixed points: the melting point and the boiling point of water. On the Celsius scale the melting point is set at 0° and the boiling point at 100°. On the Fahrenheit scale the boiling point is 212° and the melting point is 32°.

Most solids and liquids react to changes in temperature in a similar manner. In linear thermal expansion, the thermal strain ($\Delta L/L$) is directly proportional to the temperature change (ΔT). The proportionality constant (α) is called the coefficient of linear expansion. The thermal stress (F/A) is also directly proportional to the temperature change. The area expansion of a solid can be calculated using twice the coefficient of linear expansion. The thermal volume strain ($\Delta V/V$) of a solid or liquid is directly proportional to the temperature change (ΔT). The proportionality constant (β) is called the volume coefficient of thermal expansion. For a solid, the value of β is three times the coefficient of linear expansion. Thermal expansion is uniform. All dimensions of the object change proportionately, as when a photograph is enlarged.

Gases are treated somewhat differently. The absolute pressure and volume of a gas extrapolate to zero at absolute zero. Absolute zero is −273°C or −460°F. An absolute temperature scale begins at absolute zero. If Fahrenheit degrees are used, the scale is called the Rankine scale (R). If Celsius degrees are used, the temperature scale is called the kelvin scale (K).

For pressures less than 10 atm gases obey the ideal gas law ($P V = k T$). Pressure must be expressed in absolute pressure, and temperature must be given on an absolute temperature scale (kelvin or Rankine). k is a constant depending on the number of gas molecules in the sample. For a confined gas the ideal gas law can be written without the proportionality constant.

KEY TERMS

If you can explain the following terms to a friend or classmate, you understand their meaning. If you cannot explain the terms, you should reread the sections in which they are discussed.

absolute temperature scale
absolute zero
Celsius
coefficient of linear expansion
Fahrenheit
heat
ideal gases
ideal gas law
infrared pyrometer
infrared (IR) radiation
kelvin scale (K)
nonuniform expansion
optical pyrometer
Rankine scale
temperature
thermal energy
thermal strain
thermistors
thermocouple

EXERCISES

Section 18.1:

1. Define the terms thermal energy, heat, and temperature. How are they different?
2. Ice is placed in a glass of room-temperature lemonade. The ice melts.
   A. What gains thermal energy?
   B. What loses thermal energy?
   C. What is the heat in this situation?
3. Compare a steaming-hot bowl of soup with a 500-lb cake of ice at the freezing temperature of water.
   A. Do the atoms in the soup have a larger or smaller average kinetic energy than the atoms in the ice? Why?
   B. Which has the greater thermal energy? Why?
4. The temperature of a freshly formed steel ingot decreases as it approaches room temperature. What can you say about the thermal energy of the ingot? About the average kinetic energy of atoms in the ingot?

Section 18.2:

5. Which is larger: a temperature change of 6°C or a temperature change of 11°F?
6. Why will immersing only the high-temperature end of a mercury thermometer in a fluid give you an inaccurate reading of the fluid's temperature?
7. Why is it necessary to dip only the bulb of a thermometer into a fluid to get a reading of its temperature?

Section 18.3:

8. Steel rod A is 3.5 ft long. Steel rod B is 6.0 ft long. They undergo the same temperature increase.
   A. Do they have the same thermal strain? Explain.
   B. Do they have the same change of length? Explain.
9. A bimetal strip made of copper and steel is formed by fusing the two metals together (see Figure 18–23). When the strip is heated, it bends. The strip can be used to operate a dial thermometer or to activate contacts in a thermostat. The strip is initially straight. It is heated 10°F. Which part of Figure 18–23 best shows the shape of the heated bimetal strip: A, B, or C? Why?
10. A rod of iron compound and a rod of copper compound are wedged between thick granite walls so they cannot expand. The rods are the same length (see Figure 18–24). If they undergo the same temperature increase, will they undergo the same thermal stress? Why or why not?

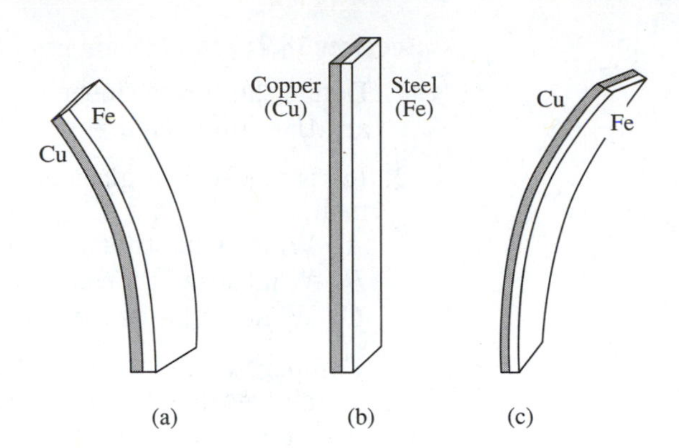

Figure 18–23: *Diagram for Exercise 9. A bimetal strip.*

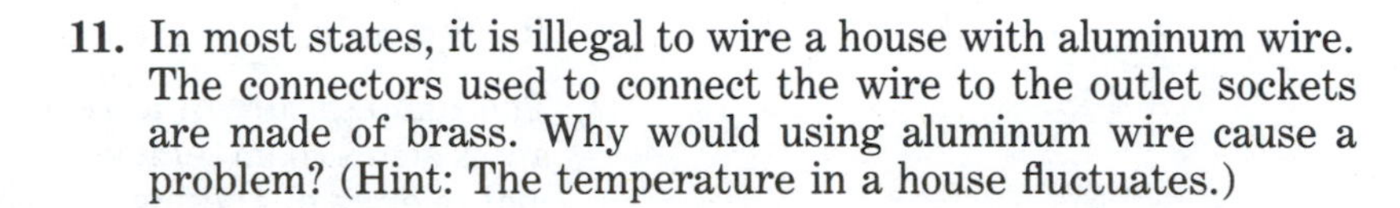

11. In most states, it is illegal to wire a house with aluminum wire. The connectors used to connect the wire to the outlet sockets are made of brass. Why would using aluminum wire cause a problem? (Hint: The temperature in a house fluctuates.)

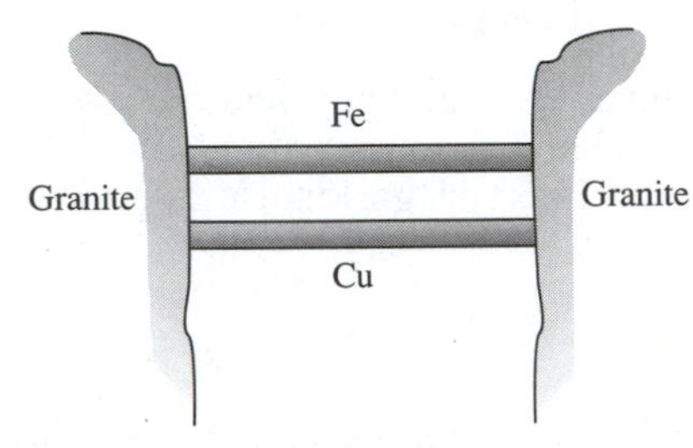

Figure 18–24: *Diagram for Exercise 10. Bars of iron compound and copper compound are wedged between granite walls.*

Section 18.4:

12. A $^3\!/_8$-in hole is drilled into a steel plate at room temperature. If the plate is heated to $4\bar{0}$°C, will the hole be larger or smaller? Why?

13. Roofing nails are usually coated to increase the friction between the nail and the wood. In the northern states, uncoated roofing nails have a tendency to pop loose on cold winter nights. Why?

Section 18.5:

14. Two thermometers are identical except thermometer A has a bulb that contains twice as much liquid as the bulb in thermometer B. Will there be any difference in the behavior of the two thermometers? Explain.

15. Compare the thermal expansion coefficients in Tables 18–1 and 18–2. Remember, for solids $\beta = 3\alpha$. In general, which has the larger volume expansion coefficient: inorganic solids or liquids?

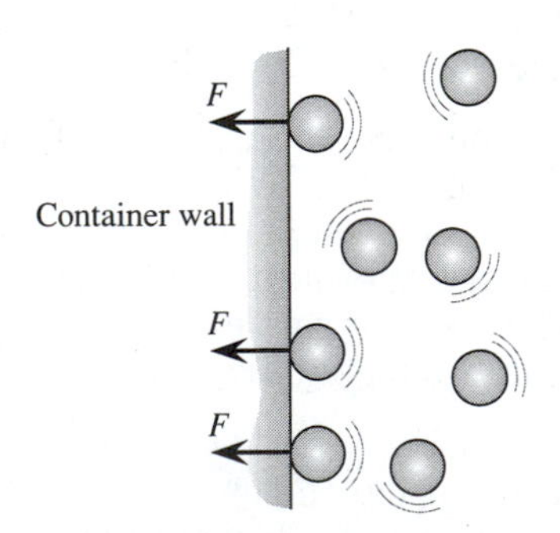

Figure 18–25: *Diagram for Exercise 17. Molecules collide with a wall to create pressure.*

Section 18.6:

16. Can an ideal gas be cooled to $-4\bar{0}$°F? To $-4\bar{0}$°K? Explain.

17. Gas pressure is produced as the gas molecules strike the sides of the container in which the gas is enclosed (see Figure 18–25). If the absolute pressure of a gas is zero at a temperature of absolute zero, what can you guess concerning the kinetic energy of molecules at absolute zero?

18. Which is the larger temperature change: a $1\bar{0}$° change on the Rankine scale or a $1\bar{0}$° change on the kelvin scale?

19. A balloon has the same volume at two different temperatures. How is this possible?

20. A student solves a pressure-temperature problem using the ideal gas law. The student gets an answer of $\mathbf{P}_2 = -4.6$ psia. Why is this an impossible answer?

PROBLEMS

Section 18.2:

1. Convert the following Fahrenheit temperature measurements to Celsius degrees.
   A. 674°F
   B. −27°F
   C. 0°F
   D. $10\bar{0}$°F

2. Convert the following Celsius temperature measurements to Fahrenheit degrees.
   A. 32°C
   B. $-4\bar{0}$°C
   C. 724°C
   D. −193°C

3. An air conditioner reduces the temperature of a computer laboratory by 17°F. How much of a temperature change is this in Celsius degrees?

4. The heater in an automobile warms the air in the car by 18°C. What is this temperature change on the Fahrenheit scale?

Section 18.3:

5. What will be the change of length of a 1.22 m long brass rod when it undergoes a temperature change of 14°C?

6. A concrete pillar is 23.0 ft high at 78°F. How much shorter is it at a temperature of −12°F?

7. A concrete highway is made of slabs of concrete 18.0 ft long. How wide should the expansion joints be to allow for a temperature change of 100°F?

8. A 50.0-ft measuring tape gives an accurate reading at 20°C. If the tape is made from a vinyl polymer with a coefficient of linear expansion of 1.9×10^{-4}/°C, what will be the error in its length at 0°C? Give your answer in inches.

9. A high-voltage transmission line is strung with aluminum wire in the summer when the temperature is 85°F. The length of wire between towers is 280.0 ft. What will be the change of length of the wire between towers in the winter when the temperature drops to 8°F?

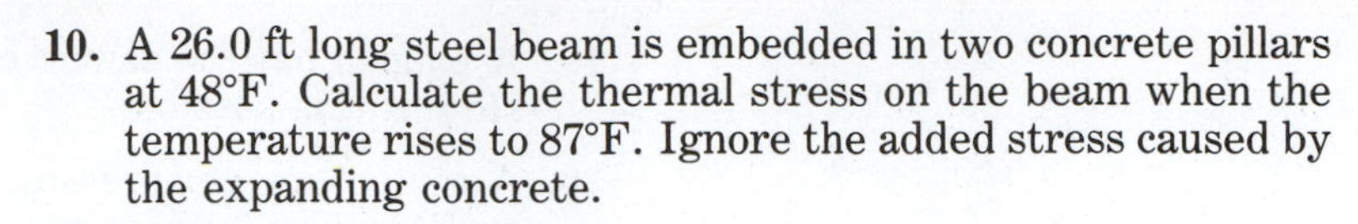

10. A 26.0 ft long steel beam is embedded in two concrete pillars at 48°F. Calculate the thermal stress on the beam when the temperature rises to 87°F. Ignore the added stress caused by the expanding concrete.

Section 18.4:

11. A sheet of steel is 4.00 ft × 8.00 ft at 68°F. What is its increase of area when it is heated to 110°F?

12. A hole with an area of 1.200 cm is drilled in a brass plate at 25°C. What will be the area of the hole at a temperature of −20°C?

13. A brass sphere has a diameter of 2.400 cm. At 20°C, it just fits through a hole drilled in a brass plate. If they are both heated to 100°C, how much must the hole be increased for the sphere to again pass through the hole?

14. The gap in the split ring shown in Figure 18–26 is 1.00 mm. If the radius of the ring is 12.0 cm, at what temperature must the ring be in order to close the hole?

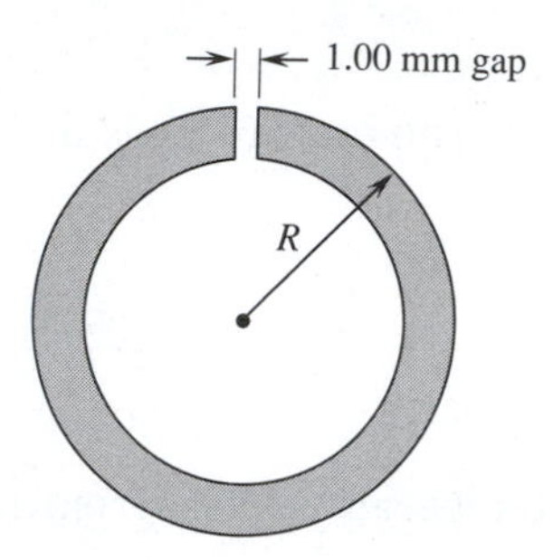

Figure 18–26: *Diagram for Problem 14. A split ring.*

Section 18.5:

15. A polyvinylidene chloride float in a toilet tank displaces 122 in^3 of water at 50.0°F. How much water will it displace at 87.0°F?

16. A polystyrene cup holds 532 ml of water at a temperature of 22°C. How many milliliters will it hold at a temperature of 89°C?

17. A tanker delivers gasoline to a service station in the winter when the temperature of the air and the tanker is −12°F. The deliveryman pumps $15\bar{0}0$ gallons of gas into an underground storage tank where the ground temperature is 48°F. The gas expands as it is warmed in the storage tank. How many extra gallons of gasoline does the service station have to sell?

18. A $23\bar{0}$-ml pyrex cup is filled to the brim with ethyl alcohol at 0°C. How much fluid overflows the cup when it is heated to 32°C?

*19. Derive Equation 18–4.

**20. Show that the hydrostatic pressure required to prevent a fluid from undergoing thermal expansion is $\mathbf{P} = -B\,\beta\,\Delta T$, where B is the bulk modulus of the fluid.

Section 18.6:

21. Convert the following temperatures to the Rankine scale.
    A. 230°F
    B. −32°F
    C. $1\bar{0}$°C

22. Convert the following temperatures to the Fahrenheit scale.
A. $10\bar{0}$ R
B. $50\bar{0}$ R
C. $46\bar{0}$ K

23. Convert the following temperatures to the kelvin scale.
A. 527°C
B. $-18\bar{0}$°C
C. 68°F

24. Convert the following temperatures to the Celsius scale.
A. 12 K
B. $62\bar{0}$ K
C. $52\bar{0}$ R

25. An air compressor produces a gauge pressure of $3\bar{0}$ psia. What is the volume of $32\bar{0}$ ft^3 of air initially at an absolute pressure of 1 atm after it has passed through the compressor into a tank? (see Figure 18–27.) Assume no temperature change.

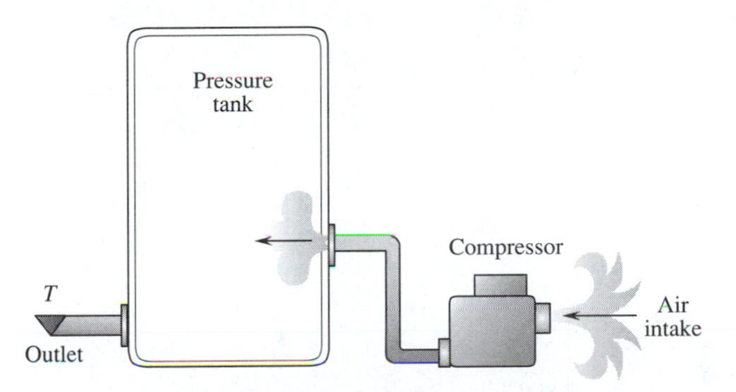

Figure 18–27: Diagram for Problem 25. An air compressor fills a tank.

26. Air is sealed in a container at 25°C at a pressure of 1.4×10^5 Pa. The container is dropped into a vat of boiling water. What is the new pressure in the container?

27. Helium is put into a weather balloon at 1.0 atm and a temperature of 24°C. The balloon is released and rises to an elevation where the temperature is $-3\bar{0}$°C and the volume of gas has expanded to 1.8 times its original volume. What is the air pressure at this height?

***28.** A bubble 2.40 in in diameter is released by a scuba diver swimming at a depth of 22.0 ft in a freshwater lake. What is the diameter of the bubble as it reaches the surface of the water? (Assume constant temperature.)

Chapter 19

ATOMIC MODEL OF GASES

Smoke stacks and the burning of tropical forests are major sources of atmospheric particles. Some particles are so small they are suspended in the atmosphere until they are washed out by raindrops. (Photograph by Brent Miller and Sonya Stang.)

OBJECTIVES

In this chapter you will learn how:

- atoms are constructed
- to calculate the mass of a molecule
- to find the number of molecules in a mole (Avogadro's number)
- to find the number of moles of a compound in a given mass
- to use the general equation for an ideal gas
- to calculate the root mean square speed and the kinetic energy of molecules using the theory of kinetic gases
- the speeds of molecules are distributed at any absolute temperature (Maxwell distribution)

Time Line for Chapter 19

450 B.C.	Leucippus of Miletus first presents the idea of indivisible particles.
1649	Pierre Gassendi's study of Epicurus's works leads him to believe matter is made of atoms.
1803	John Dalton presents his atomic theory of matter.
1811	Amedeo Avogadro finds that equal volumes of gas under the same pressure contain equal numbers of molecules.
1827	Robert Brown discovers that small particles in a liquid have a random motion.
1844	Ludwig Boltzmann develops the kinetic theory of gases using statistical mechanics.
1851	William Thomson, Lord Kelvin, proposes the idea of absolute zero.
1905	Albert Einstein explains Brownian movement. This is thought to be the first proof that atoms exist.

In Chapter 18 we learned that the kinetic energy of molecules in a gas is directly proportional to the absolute temperature. In this chapter we will discuss that idea further. We will first look at the structure of atoms and molecules and then develop a model to describe the motion of the molecules. We will then apply the model to diffusion.

19.1 ATOMIC AND MOLECULAR MASSES

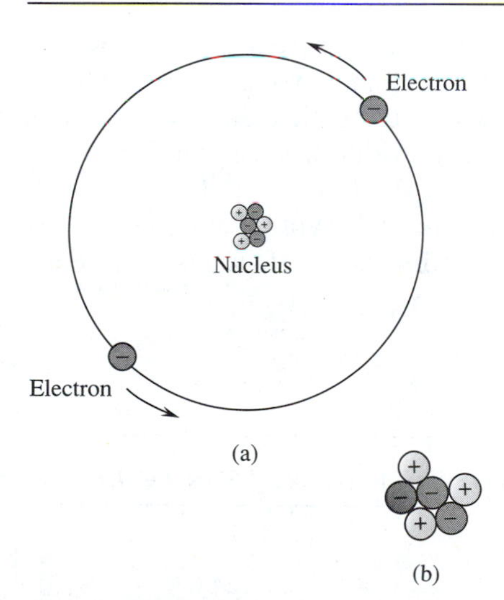

Figure 19–1: *The structure of an atom. (a) Negatively charged electrons orbit a massive nucleus. (b) The nucleus is composed of neutrons (shaded spheres) and positively charged protons (+).*

A simple model of an atom looks a lot like a solar system. In our solar system, planets are held in orbits around a much more massive body, the sun, by a gravitational force. Much of the solar system is empty space. Most of its mass is concentrated in the sun.

In an atom, negatively charged particles called **electrons** orbit around a much more massive nucleus. Electrostatic force holds the electrons in orbit. Most of the atom is empty space. Most of the mass is concentrated in the atom's nucleus (see Figure 19–1).

The nucleus is made up of two kinds of massive particles called **nucleons. Protons** have a positive charge equal to the charge of the electrons. In a complete atom, the number of electrons equals the number of protons. The chemical behavior of the atom is determined by the array of orbiting electrons. Indirectly, the protons specify the chemical nature of the atom. The number of protons in the nucleus is called the **atomic number (*Z*).**

$$\textit{atomic number}\ (Z) = \textit{number of protons}$$

The other particle found in the nucleus is called a **neutron.** The neutron has a mass very close to that of the proton, but it does not have an electric charge. The mass of an atom is determined by the total number of nucleons in the atom: the number of neutrons and protons. The total number of nucleons is called the **mass number (A).**

$$\text{mass number } (A) = \text{number of protons} + \text{number of neutrons}$$

Two atoms may be chemically identical (same number of protons), but may have different mass numbers (different number of neutrons). Such atoms are called **isotopes.**

We can write a notation to distinguish isotopes. Place the mass number as a superscript and the atomic number as a subscript before the chemical symbol. For example, three isotopes of oxygen are $^{18}_{8}O$, $^{17}_{8}O$, and $^{16}_{8}O$. Each isotope has eight protons. The number of neutrons in each isotope is found by subtracting the atomic number from the mass number. The numbers of neutrons are 10, 9, and 8, respectively.

$^{12}_{6}C$ is the isotope of carbon used as a standard to measure atomic masses. One-twelfth of the mass of $^{12}_{6}C$ is called an **atomic mass unit (u).** In terms of atomic mass units, we can write the masses of particles that make up an atom.

- mass of proton (m_p) = 1.0073 u
- mass of neutron (m_n) = 1.0087 u
- mass of electron (m_e) = 0.000549 u

In nature, the chemical elements have isotopes. For example, chlorine has two isotopes. The ratio of the two isotopes is always the same (see Table 19–1). The average atomic mass for chlorine is 35.453 u. We need to use this average value when solving problems involving chemical compounds.

Some gases—e.g., helium, neon, argon, krypton, and xenon—are composed of single atoms **(monatomic gases).** Most gases are **com-**

Table 19–1: Isotopes of chlorine (Cl).

	ATOMIC MASS (u)	RELATIVE ABUNDANCE (%)
$^{35}_{17}Cl$	34.97	75.4
$^{36}_{17}Cl$	35.97	24.6
Average atomic mass (u)	35.45	

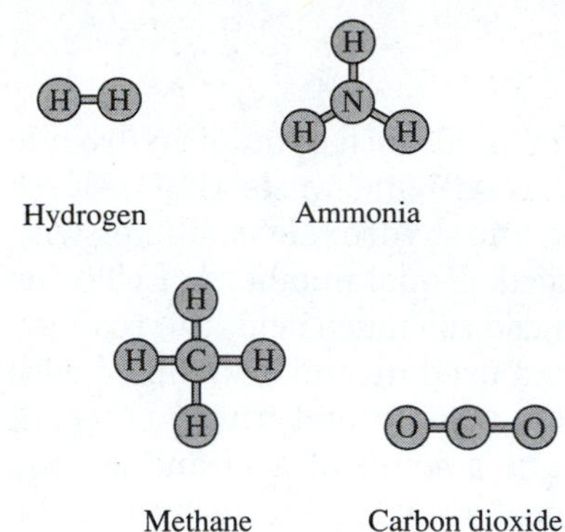

Figure 19–2: *The composition of the molecules of four gases.*

Table 19–2: Average atomic mass of selected elements.

ELEMENT	SYMBOL	AVERAGE ATOMIC MASS (u)
Carbon	C	12.011
Chlorine	Cl	35.45
Fluorine	F	19.00
Helium	He	4.0028
Hydrogen	H	1.0080
Neon	Ne	20.18
Nitrogen	N	14.01
Oxygen	O	16.00
Sulfur	S	32.06

pounds composed of two or more atoms chemically bonded by interatomic forces to form a molecule. Examples are hydrogen (H_2), ammonia (NH_3), carbon dioxide (CO_2), and methane (CH_4) (see Figure 19–2).

We can find the average atomic mass of a gas molecule by adding up the average masses of the atoms of which it is composed. Table 19–2 lists the average atomic masses of some of the elements that occur in common gases.

☐ **EXAMPLE PROBLEM 19–1: MOLECULAR MASS OF CARBON DIOXIDE**

Find the molecular mass of carbon dioxide.

■ ***SOLUTION***

Carbon dioxide (CO_2) has one atom of carbon and two atoms of oxygen. From Table 19–2:

$$1\,\mathrm{C} = 1 \times 12.01\ \mathrm{u} = 12.01\ \mathrm{u}$$

plus

$$2\,\mathrm{O} = 2 \times 16.00\ \mathrm{u} = 32.00\ \mathrm{u}$$

$$\text{Total molecular mass} = 44.01\ \mathrm{u}$$

19.2 COUNTING ATOMS

Hydrochloric acid (HCl) can be combined with sodium hydroxide (NaOH) to form common table salt (NaCl) and water (H_2O). Each atom of the sodium (Na) from the sodium hydroxide combines with one atom of chlorine (Cl) from the acid. Equal numbers of chlorine and sodium atoms are needed. If we use too much acid, the reaction will stop when all the sodium has been used up and acid mixed with water and salt will be left over. If we could find a way to count atoms, we could use exactly the right amount of acid and sodium hydroxide. There would be no extra acid, no extra sodium hydroxide.

Let us relate numbers of molecules with mass. One molecule of HCl has a mass of 36.46 u. Two molecules of HCl have twice the mass of one molecule: 72.92 u. One hundred molecules of HCl have a mass of 3646 u.

One molecule of sodium hydroxide has a mass of 40.00 u. Two molecules have a mass of 80.00 u, and 100 molecules have a mass of 4000 u.

The total mass of a chemical (M) is equal to the molecular mass (m) of the substance times the number of molecules (N). If we have the total mass we can find the number of molecules by dividing by the molecular mass.

$$\textit{number of molecules } (N) = \frac{\textit{total mass of the substance } (M)}{\textit{molecular mass } (m)}$$

Chemists have found a convenient measure for the number of molecules: the **mole.** The mole is the mass in grams equivalent to the molecular mass of a substance. For example, the gram-equivalent mass of NaOH is 40.00 g. The gram-equivalent mass of HCl is 36.46 g. Each represents the same number of molecules. In both cases, the ratio of total mass in grams divided by molecular mass is the same.

Careful measurements have been made using a variety of tech-

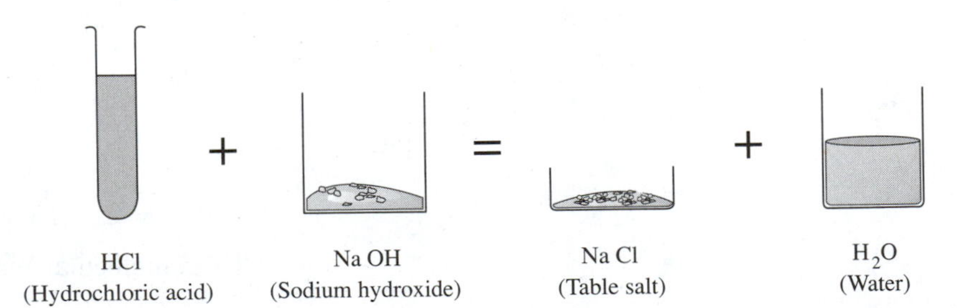

Figure 19–3: *Hydrochloric acid (HCl) combines with sodium hydroxide (NaOH) to form table salt (NaCl) and water (H_2O).*

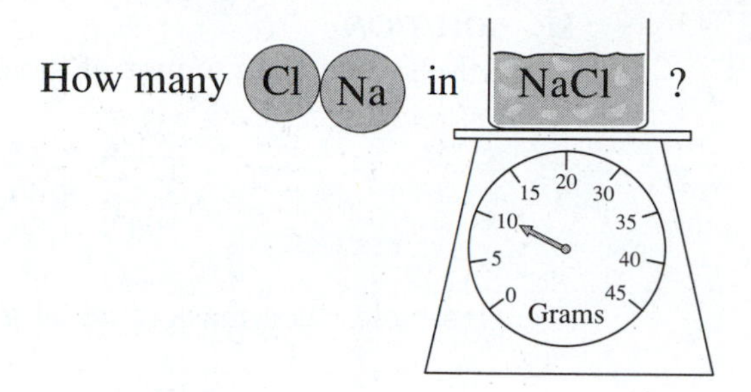

Figure 19–4: Diagram for Example Problem 19–3.

niques to find the number of molecules in one mole. This number of molecules is called Avogadro's number (N_a).

$$1 \text{ mole} = N_a = 6.023 \times 10^{23} \text{ molecules}$$

If we measured mass in kilograms, we would have 1000 times as many molecules.

$$1 \text{ kilomole} = 6.023 \times 10^{26} \text{ molecules}$$

□ **EXAMPLE PROBLEM 19–2: CONVERSION OF ATOMIC MASS UNITS TO METRIC UNITS**

Find the conversion factor between atomic mass units and kilograms.

■ ***SOLUTION***

Assume we have 1 kilomole of particles; each particle has a mass of 1.00 u. The one kilomole is 1.00 kg.

$$M = N_a \times m$$

$$m = \frac{M}{N_a}$$

$$1.00 \text{ u} = \frac{1.00 \text{ kg}}{6.023 \times 10^{26} \text{ particles}}$$

$$1.00 \text{ u} = 1.66 \times 10^{-27} \text{ kg}$$

Avogadro's number is used to convert between metric units and atomic mass units.

□ **EXAMPLE PROBLEM 19–3: TABLE SALT**

How many molecules are there in 10.0 g of table salt (NaCl)? (Na = 22.99 u; Cl = 35.45 u.)

■ *SOLUTION*

First find the number of moles of NaCl. The molecular mass is:

$$\begin{aligned} Na &= 22.99 \text{ u} \\ Cl &= 35.45 \text{ u} \\ NaCl &= 58.44 \text{ u} \end{aligned}$$

One mole has a mass of 58.44 g. The number of moles (n) is:

$$n = \frac{10.0 \text{ g}}{58.44 \text{ g}} = 0.171 \text{ mole}$$

We can get the number of molecules by multiplying by Avogadro's number.

$$\textit{number of molecules} = N_a \times n$$

$$\textit{number of molecules} = 6.023 \times 10^{23} \text{ molecules/mole} \times 0.171 \text{ mole}$$

$$\textit{number of molecules} = 1.03 \times 10^{23} \text{ molecules}$$

Because gases have very small densities, it is difficult to measure their masses. Fortunately, chemists have found that under standard conditions of temperature and pressure (0°C and 1 atm pressure), 1 mole of any gas occupies the same volume (see Figure 19–5). This volume is called the **molar volume.**

$$1 \text{ molar volume} = 22.4 \text{ liters} = 1 \text{ mole of gas}$$

One liter is 10^3 cm^3. A molar volume is 2.24×10^4 cm^3. The cube root of this number is the size of a cubic volume containing 1 mole of gas under standard conditions: roughly 28 cm, or 11 in. A cubic box 11 in on a side will hold 1 mole of gas.

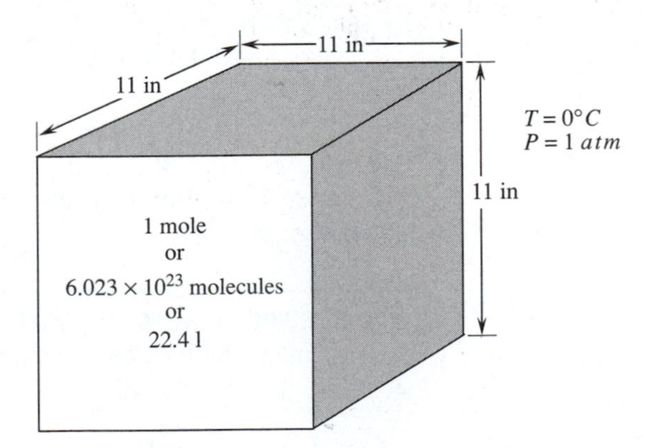

Figure 19–5: *One mole of gas fills a cube 11 in on a side at standard conditions of pressure and temperature.*

□ **EXAMPLE PROBLEM 19–4: A ROOM FULL OF AIR**

Air is chiefly a mixture of nitrogen and oxygen with an average molecular mass of 28.8 u.

A. How many moles of air are in a room with a volume of 126 m^3?

B. What is the mass of the air?

($1\ m^3 = 10^3$ l. Assume standard conditions.)

■ *SOLUTION*

A.

$$n = \frac{126 \times 10^3\ \text{l}}{22.4\ \text{l/mole}}$$

$$n = 5.63 \times 10^3\ \text{moles}$$

B.

$$M = n \times m$$

$$M = 5.63 \times 10^3\ \text{moles} \times 28.8\ \text{g/mole}$$

$$M = 1.62 \times 10^5\ \text{g, or 162 kg}$$

19.3 THE IDEAL GAS LAW REVISITED

In Section 18–6 we found that the equation relating pressure, volume, and temperature for an ideal gas could be written as:

$$P\,V = \textit{constant} \times T$$

where the constant depended on the amount of gas in our sample. If no gas was added or subtracted from our sample the constant did not change. Let us see what happens when we express the amount of gas in moles.

Let R equal the value of the constant when we have 1 mole of gas at standard conditions ($T = 0°\text{C}$ and $P = 1$ atm). The volume for 1 mole will be 22.4 l under these conditions.

$$R = \frac{P\,V}{T}$$

$$R = \frac{(1.00\ \text{atm})(22.4\ \text{l/mole})}{273\ \text{K}}$$

$$R = 0.0821\ \text{atm} \cdot \text{l/mole} \cdot \text{K}$$

The constant for 2.0 moles under standard conditions is twice the molar volume. The pressure and temperature will be the same.

$$constant = \frac{PV}{T} = \frac{(1.00 \text{ atm})(2 \times 22.4 \text{ l})}{273 \text{ K}} = 2R$$

The constant for any number of moles (n) is

$$constant = \frac{PV}{T} = n\,R$$

because the volume can always be expressed in multiples of the molar volume.

The constant R is known as the **universal gas constant.** It has the same value for any ideal gas. If we use SI units (pressure in pascals, volume in cubic meters, and kelvin temperature scale), the value of R is 8.314 J/mole · K.

$$R = 8.314 \text{ J/mole} \cdot \text{K} = 0.0821 \text{ atm} \cdot \text{l/mole} \cdot \text{K} = \textit{universal gas constant}$$

If pressure is given in pounds per square inch, the molar volume in cubic inches, and temperature on the Rankine scale, we can calculate the gas constant in British units.

$$R = 40.84 \text{ lb} \cdot \text{in/mole} \cdot \text{R or } 3.403 \text{ ft} \cdot \text{lb/mole} \cdot \text{R}$$

We can now rewrite the ideal gas law in terms of the number of moles of gas and the universal gas constant. We no longer need to find the proportionality constant experimentally.

$$P\,V = n\,R\,T \qquad \textbf{(Eq. 19–1)}$$

□ **EXAMPLE PROBLEM 19–5: THE ENGINE CYLINDER**

An engine cylinder has a maximum volume of 422 ml during the intake stroke. Before compression, the pressure is equal to 1 atm. Assume the engine has a temperature of 60°C. How many moles of oxygen enter the cylinder? (Air is roughly 21% oxygen.)

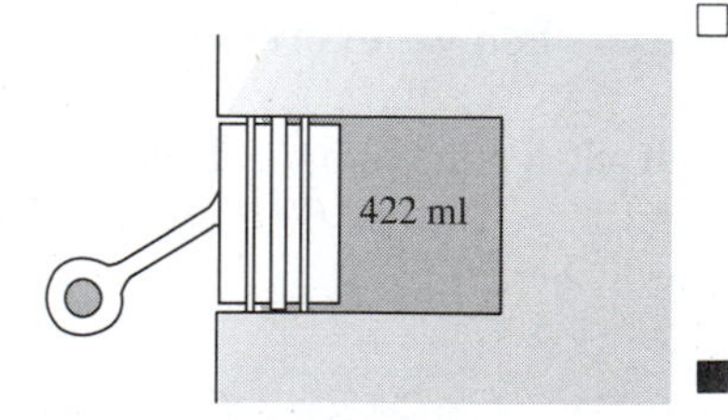

Figure 19–6: *The maximum volume of a cylinder is 422 ml.*

■ ***SOLUTION***

$$T = 60°\text{C} + 273 = 333 \text{ K}$$

$$P = 1.00 \text{ atm}$$

$$V = 0.422 \text{ l}$$

Solve the ideal gas equation for n.

$$n_{air} = \frac{P\,V}{R\,T} = \frac{(1.00 \text{ atm})(0.422 \text{ l})}{(0.0821 \text{ atm} \cdot \text{l/mol} \cdot \text{K})\ (333 \text{ K})}$$
$$n_{air} = 0.0154 \text{ moles}$$

This gives us the number of moles of air. Only 21% is oxygen.

$$n_{O_2} = 0.21 \times n_{air} = 0.21 \times 0.0154 \text{ mole}$$
$$n_{O_2} = 3.2 \times 10^{-3} \text{ mole}$$

19.4 KINETIC THEORY OF GASES

We can visualize a gas as being made up of widely spaced molecules undergoing continuous motion. From time to time they bump into the sides of the container, creating pressure. We will make some simplifying assumptions for this model of the gas.

1. **The molecules do not interact with each other.** We will assume that the molecules are very small, and there is a vast space between them. The probability of one molecule bumping into another is very small. Further, we will assume that there are no attractive or repulsive forces between molecules.
2. **Molecules undergo perfectly elastic collisions.** When molecules collide with the walls of the container, no energy is lost in the collision.
3. **There is no preferred direction of motion.** The molecules are likely to move in any direction. For a large number of molecules, this means that the average velocity in any direction should be the same as that in any other direction.

We will first look at what happens to one molecule of mass m (see Figure 19–7). A molecule with a component of velocity v_x in the $-x$

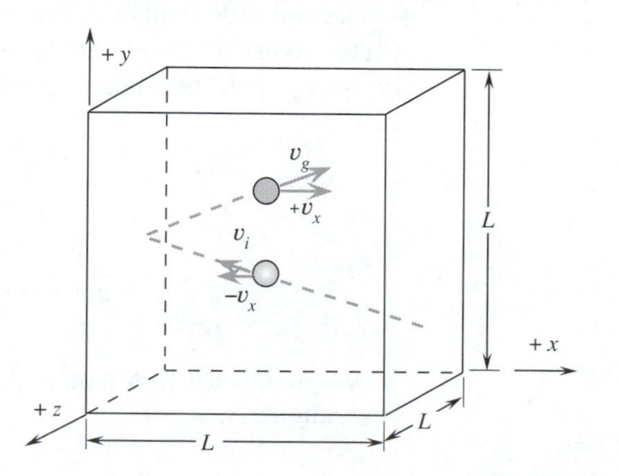

Figure 19–7: *A gas molecule bounces back and forth inside a cubic container.*

direction strikes the wall of a cubic container L long on a side. The molecule rebounds in a perfectly elastic collision. It undergoes a change of momentum. Initially, it is traveling in the negative direction. After the collision with the wall, it rebounds in the plus direction, with no loss of kinetic energy.

$$\Delta P_x = (m\, v_x) - (-m\, v_x) = 2\, m\, v_x$$

The molecule must travel to the other wall and back again, a distance of $2L$, before it hits the same wall again. The time required for this trip is:

$$t = \frac{2L}{v_x}$$

The force the molecule exerts on the wall is then:

$$F = \frac{\Delta P}{\Delta t} = \frac{2\, m\, v_x}{2L/v_x} = \frac{m\, v_x^{\,2}}{L}$$

The total force N molecules will exert on the wall if they are traveling an *average* speed of $\mathbf{v}_x$ is N times as large.

$$F_{\text{total}} = N\frac{(m\, v_x^{\,2})}{L}$$

The area of the wall is $A = L^2$. The pressure exerted on the wall is:

$$P = \frac{F_{\text{total}}}{A} = \frac{N\,(m\, v_x^{\,2})}{L^3} = N\frac{(m\, v_x^{\,2})}{V} \qquad \textbf{(Eq. 19–2)}$$

where V is the volume of the cube.

The molecules will have the same average velocity in any direction. On the average, we expect the components of motion to be the same ($\mathbf{v}_x = \mathbf{v}_y = \mathbf{v}_z$) because of purely random motion.

$$v_{\text{avg}}^{\;2} = v_x^{\,2} + v_y^{\,2} + v_z^{\,2} = 3v_x^{\,2}$$

or

$$v_x^{\,2} = \frac{v_{\text{avg}}^{\;2}}{3}$$

If we substitute this expression into Equation 19–2 and do some rearranging, we get:

$$P V = \frac{N\,(m\,v^2)}{3} = \frac{2\,N}{3}\left(\frac{m\,v^2}{2}\right)$$

$m\,v^2/2$ is the translational kinetic energy of a gas molecule. Compare this with Equation 19–1. The left-hand sides of the equations are the same. We can equate the right-hand sides to find a relationship between the kinetic energy of a molecule and absolute temperature.

$$\frac{2\,N}{3}\left(\frac{m\,v^2}{2}\right) = n\,R\,T$$

This can be simplified by considering 1 mole of gas. For $n = 1$, the number of molecules in the container will be Avogadro's number (N_a). Solving for the average kinetic energy of molecules:

$$KE = \frac{3\,R\,T}{2\,N_a}$$

Aerosols and Brownian Movement

In Example Problem 17–8 we discussed the terminal speed of particles in the atmosphere. The smaller the particle became, the longer it took to settle out of the atmosphere. If a particle is very small, it will not settle at all because of an effect called Brownian movement.

In 1827, a Scottish botanist, Robert Brown, was peering through his microscope. He discovered that small particles suspended in a liquid had random, jerky motion. People puzzled over this for years. In 1905, Albert Einstein calculated that the motion of the particles was caused by the random motion of molecules. Brownian movement, as it is now called, is caused by molecules bumping into the small particles. The particles recoil from the collision. This is what Brown saw through his microscope. Einstein's analysis was hailed as the first proof of the existence of atoms.

(Photograph by Brent Miller and Sonya Stang.)

When tiny smoke particles reach a size smaller than the average distance between molecules in the air, they experience Brownian movement. Random recoil holds the aerosols suspended in still air. About 20% of the aerosols in the atmosphere are small enough for this to occur.

The two constants, R and N_a, can be combined to form a single constant (k) called **Boltzmann's constant.**

$$k = \frac{R}{N_a} = \frac{8.314 \text{ J/(mole} \cdot \text{K)}}{6.023 \times 10^{23} \text{ molecules/mole}}$$
$$k = 1.38 \times 10^{-23} \text{ J/K}$$

The kinetic energy of a molecule is then:

$$KE = \frac{3}{2}(k\, T) \qquad \textbf{(Eq. 19–3)}$$

We can also find the speed of gas molecules.

$$KE = \frac{1}{2}(m \text{ v}^2) = \frac{3}{2}(k\, T)$$
$$\text{v}_{rms} = \left(\frac{3kT}{m}\right)^{1/2} \qquad \textbf{(Eq. 19–4)}$$

This is called the **root mean square (rms) speed.** Since absolute temperature is proportional to kinetic energy, we are taking the square root of the kinetic energy $\propto \mathbf{v}^2$. In effect, we are adding up the squares of the velocities to calculate the root mean square speed rather than adding up the velocities themselves.

☐ **EXAMPLE PROBLEM 19–6: AIR SPEED**

Calculate the kinetic energies and root mean square speeds of nitrogen (N_2) and oxygen (O_2) at a temperature of 27°C.

■ ***SOLUTION***

$$T = 27°\text{C} + 273 = 300 \text{ K}$$
$$2\, m_N = 2\,(14.00 \text{ u}) \times 1.66 \times 10^{-27} \text{ kg/u} = 4.65 \times 10^{-26} \text{ kg}$$
$$2\, m_O = 2\,(16.00 \text{ u}) \times 1.66 \times 10^{-27} \text{ kg/u} = 5.31 \times 10^{-26} \text{ kg}$$

A. Nitrogen:

$$KE = \frac{3}{2}(k\, T)$$
$$KE = 1.5\,(1.38 \times 10^{-23} \text{ J/K})\,(300 \text{ K})$$
$$KE = 6.21 \times 10^{-21} \text{ J}$$

$$v_{rms} = \left(\frac{3\,k\,T}{m}\right)^{1/2}$$

$$v_{rms} = \left(\frac{3 \times 1.38 \times 10^{-23}\,\text{J/K} \times 300\,\text{K}}{4.65 \times 10^{-26}\,\text{kg}}\right)^{1/2}$$

$$v_{rms} = 517\,\text{m/s}$$

B. Oxygen:

$$KE = 6.21 \times 10^{-21}\,\text{J}$$

$$v_{rms} = \left(\frac{3 \times 1.38 \times 10^{-23}\,\text{J/K} \times 300\,\text{K}}{5.31 \times 10^{-26}\,\text{kg}}\right)^{1/2}$$

$$v_{rms} = 484\,\text{m/s}$$

19.5 DISTRIBUTION OF MOLECULAR VELOCITIES

The kinetic theory developed in Section 19.4 gives us some information concerning the *average* behavior of molecules in a gas. It is reasonable to expect some of the molecules to move faster and others slower than the rms speed. An experiment can be devised to find the distribution of the molecular speeds.

In Figure 19–8 a furnace (F) supplies a stream of hot molecules. They pass through a pair of pinholes (P) to produce a narrow beam. The beam then passes through a slit (S_1) on a rotating disk. The beam travels parallel to an axle through a distance L. It encounters a second disk (S_2). If the time to travel the length L is just right, the second slit will be lined up so that the molecules can pass through to a detector (D). Only those molecules that have the correct speed

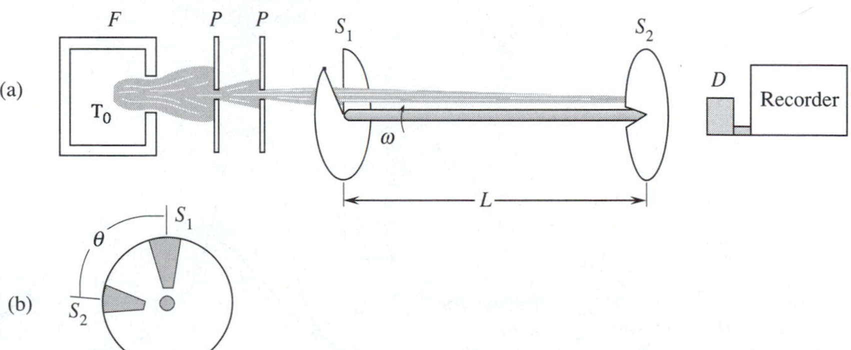

Figure 19–8: *(a) Rotating disks measure the speed distribution of gas molecules. Slits S_1 and S_2 act as shutters. (b) θ is the offset angle of the rotating slits.*

to travel the distance L in time to reach the second slit can reach the detector.

If **v** is the velocity of a molecule, the time to travel between slits is:

$$t = \frac{L}{v}$$

θ is the angular offset of the second slit, and ω is the rotational speed of the disks. The time (T) needed for the second slit to line up with the beam coming through the first slit is:

$$T = \frac{\theta}{\omega}$$

Only those molecules that have a speed for which $t = T$ will pass through the second slit to the detector. These molecules have a speed of:

$$v = \frac{L\omega}{\theta}$$

By changing the angular speed of the disks, molecules with different velocities can be measured. Figure 19–9 shows a plot of relative beam intensity versus the speed of molecules at two fixed temperatures of the furnace. Three different average speeds are shown on the plot. The **most probable speed** is the peak of the distribution. More molecules have this speed than any other. The **mean average speed** is calculated by adding up the individual speeds and then dividing by the total number of molecules. The root mean

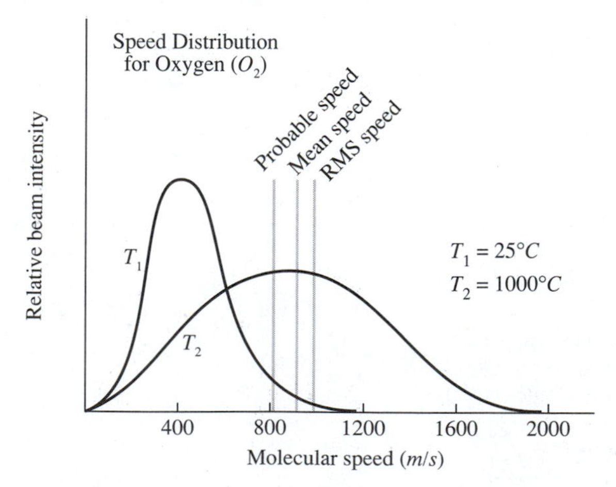

Figure 19–9: *The velocity distribution for molecules of oxygen at two different temperatures. At 1000°C the probable speed is 813 m/s, the mean speed is 918 m/s, and the rms speed is 996 m/s.*

square (rms) speed is identical to that calculated in Equation 19–4.

In a mole of gas, a few molecules have nearly no speed at all. Most of the molecules have a speed near that of the average speed. A small number of molecules in the exponential tail of the distribution curve have velocities much higher than the average. This type of distribution is called a **Maxwell distribution.**

SUMMARY

An atom is made of negatively charged electrons orbiting a much more massive nucleus. The nucleus is composed of two kinds of nucleons: positively charged protons, which indirectly determine the chemical behavior of the atom, and neutrons, which carry no charge. Isotopes are atoms that have the same number of protons, but different numbers of neutrons.

The mass of atoms is often expressed in atomic mass units (u). One atomic mass unit is one-twelfth the mass of carbon 12 ($^{12}_{6}C$).

A mole is the mass of a compound in grams numerically equal to the molecule's atomic mass in atomic mass units. The number of molecules in one mole is Avogadro's number (N_a), 6.023×10^{23} molecules. One mole of a gas at 0°C and 1 atm of pressure occupies a volume of 22.4 l.

The ideal gas law can be written in the form $PV = n R T$, where n is the number of moles of gas and R is the universal gas constant.

According to the kinetic theory of ideal gases the average translational kinetic energy of a gas molecule is $3(kT)/2$, where k is Boltzmann's constant. k is the ratio between the universal gas constant and Avogadro's number.

The root mean square (rms) speed of a molecule is $(3\ kT/m)^{1/2}$. The distribution of speeds of a gas molecule follows a Maxwell distribution (see Figure 19–9).

KEY TERMS

If you can explain the following terms to a friend or classmate, you understand their meaning. If you cannot explain the terms, you should reread the sections in which they are discussed.

atomic mass unit (u)
atomic number (*z*)
Boltzmann's constant
compounds
electrons
isotopes
mass number (*A*)
Maxwell distribution
mean average speed
molar volume
mole
monatomic gases
most probable speed
neutron
nucleons
protons
root mean square (rms) speed
universal gas constant

EXERCISES

Section 19.1:

1. Explain the following terms in your own words.
   A. Nucleon *B.* Isotope *C.* Atomic number
   D. Compound *E.* Mass number *F.* Atomic mass unit

2. How many neutrons do the following atoms have?
   A. ${}^{13}_{6}C$
   B. ${}^{21}_{9}F$
   C. ${}^{34}_{16}S$
   D. ${}^{238}_{92}U$

3. An atom has an atomic number of 10 and contains 11 neutrons.
   A. What is the mass number of this atom?
   B. How many protons does it have?
   C. How many nucleons does it have?

Section 19.2:

4. Explain the following terms in your own words.
   A. Mole
   B. Avogadro's number
   C. Molar volume
   D. Gram-equivalent mass

5. The mass of one molecule of carbon dioxide (CO_2) is 44.01 u. At 0°C and 1.00 atm of pressure a sample of carbon dioxide occupies a volume of 48.8 l.
   A. How many moles of carbon dioxide is this?
   B. How many grams of gas are in the sample?
   C. How many molecules of CO_2 are in the 48.8 l?

Section 19.3:

6. What is the advantage of breaking the proportionality constant in the ideal gas law into two parts: n, the number of moles of gas; and R, the universal gas constant?

7. Gasoline consists mostly of octane, C_8H_{18}. In a gasoline engine, 2 molecules of octane in a gaseous state combine with 25 molecules of oxygen to form 16 molecules of carbon dioxide (CO_2) and 18 molecules of water (H_2O) when the gas mixture is ignited. Heat liberated in this reaction raises the temperature of the gas (See Figure 19–10).
   A. How does the heat liberated in this reaction affect the pressure in the cylinder?
   B. After the reaction, are there fewer or more gas molecules in the cylinder? What effect does this have on the pressure in the cylinder?

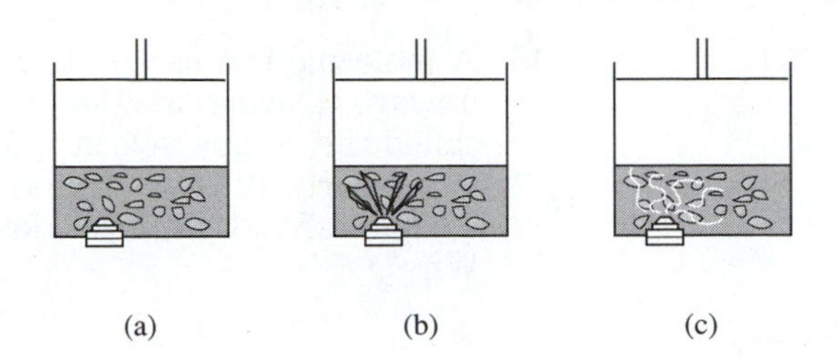

Figure 19–10: (a) A mixture of air and gasoline (octane) is compressed. (b) The mixture is ignited with a spark. (c) Oxygen and octane react to form water and carbon dioxide.

8. A mole (6.023×10^{23} molecules) of gas occupies 22.4 l of volume at 0°C and 1 atm of pressure. Would 22.4 l contain fewer, the same, or more molecules under the following conditions?
   A. A pressure of 1.00 atm and a temperature of 273 K.
   B. A pressure of 14.7 lb/in^2 and a temperature of 350 R.
   C. A pressure of 1.00 atm and a temperature of 80°C.
   D. A pressure of 2.3 atm and a temperature of 0°C.

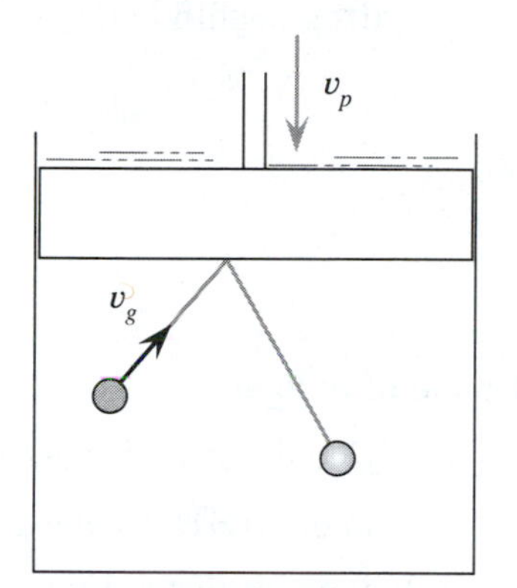

Figure 19–11: Diagram for Exercise 10. A gas is compressed. A molecule has a velocity v_g before striking the piston and moves away at velocity v_p.

Section 19.4:

9. Two containers have the same temperature and pressure. One container is filled with carbon dioxide (CO_2). The other is filled with carbon monoxide (CO).
   A. Compare the kinetic energy of the two kinds of molecules.
   B. Compare the velocities of the two kinds of molecules.

10. Figure 19–11 shows a cylinder with a piston compressing the gas. In this situation,
    A. will molecules colliding with the piston have a greater, smaller, or unchanged speed after the collision?
    B. how is the kinetic energy of the molecules affected by the collision?
    C. what effect might this have on the temperature of the gas?

11. Figure 19–12 shows a cylinder with a piston expanding a gas.
    A. Will molecules colliding with the piston have a greater, smaller, or unchanged speed after the collision?
    B. How is the kinetic energy of the molecules affected?
    C. What effect might this have on the temperature of the gas?

12. What variables affect the kinetic energy of gas molecules?

13. What variables affect the speed of gas molecules?

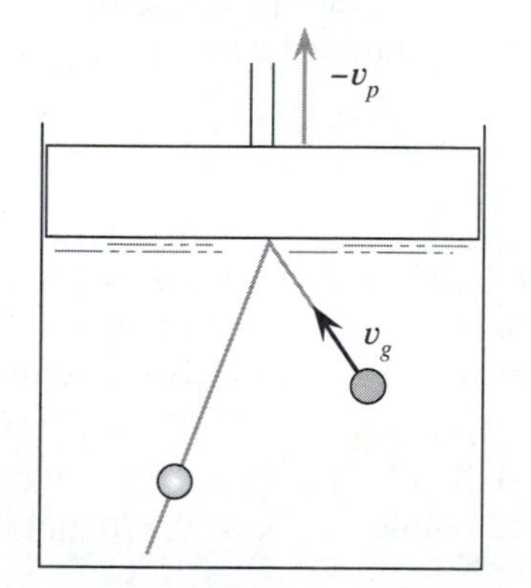

Figure 19–12: Diagram for Exercise 11. A gas is expanded. A molecule has a velocity v_g before colliding with the piston and moves away at velocity $-v_p$.

Section 19.5:

14. The v_{rms} energy of a molecule is calculated for 300 K for oxygen molecules (refer to Figure 19–9).
    A. Is the rms speed larger or smaller than the mean average speed?
    B. Is the v_{rms} the most probable speed a molecule would have?

15. A molecule can escape from a planet if it has enough velocity upward to overcome the gravitational pull of the planet. This is called the escape velocity. Mars has an escape velocity of 5100 m/s, nearly 10 times the average speed of molecules in its atmosphere, and yet it has lost much of its original atmosphere. Why?

PROBLEMS

Section 19.1:

1. Use Table 19–2 to find the average atomic mass of the following molecules.
   A. Carbon monoxide (CO) ***B.*** Ammonia (NH_3)
   C. Methane (CH_4) ***D.*** Freon (CF_2Cl_2)
   E. Sulfur dioxide (SO_2) ***F.*** Hydrogen sulfide (H_2S)

2. $^{35}_{17}Cl$ has
   A. how many protons?
   B. what total number of nucleons?
   C. how many neutrons?

Section 19.2:

3. How many *atoms* are in 2.34 mole of oxygen?

4. What is the mass of 34.5 l of SO_2 at 0°C and 1 atm of pressure.

5. What is the mass in kilograms of 12 ammonia (NH_3) molecules?

6. What volume of space will 2.10×10^{24} molecules of hydrogen gas occupy at standard conditions of pressure and temperature?

* 7. Neon is a monatomic gas. It does not chemically bond with itself as oxygen and nitrogen do. What is the density of neon in grams per cubic centimeter at standard conditions?

Section 19.3:

8. Find the number of moles of gas for the following data.

A.	***B.***	***C.***
$P = 1.3$ atm	$P = 3\bar{0}$ lb/in^2	$P = 2.77 \times$ Pa
$V = 82.2$ l	$V = 23\bar{0}$ ft^3	$V = 3.2$ m^3
$T = 46$°C	$T = 51\bar{0}$ R	$T = 24\bar{0}$ K

9. Find the volume of the gas for the following sets of data.

A.	***B.***	***C.***
$P = 0.45$ atm	$P = 9\bar{0}$ lb/in^2	$P = 3.21 \times 10^5$ Pa
$n = 12.4$ mole	$n = 3.20$ mole	$n = 0.870$ mole
$T = 42\bar{0}$ K	$T = 9\bar{0}$°F	$T = 45$°C

10. Argon from a 0.38-ml chamber at 1 atm is vented into an evacuated light bulb with a volume of 198 ml to produce a low gas pressure. The pressure from the inert gas will inhibit

evaporation of the tungsten filament. If the pressure remains constant, what will be the pressure in the light bulb?

11. A tank of liquid propane (C_3H_8) develops a leak. The chemical evaporates into the air as a gas.
 A. What is the molecular mass of propane?
 B. If 2.53 kg of propane escapes, how many moles escape?
 C. How many molecules of propane are released?

12. A monatomic gas such as argon does not transmit heat as well as diatomic gases such as nitrogen and oxygen. A glass patio door has a sealed double glazing 38 in × 84 in. There is ½ in between the plates. Air between the glass plates is to be evacuated and replaced by argon at 1 atm and a temperature of 25°C. How many moles of argon are needed?

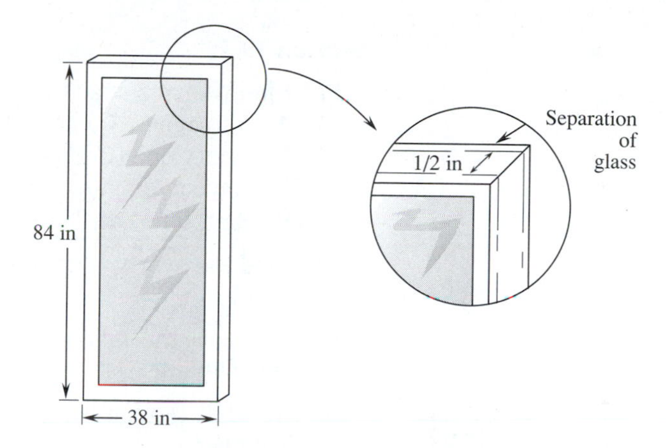

Figure 19–13: *Diagram for Problem 12. Argon is used to replace air in the space between the double glazing of a patio glass door.*

13. At 32°C and an absolute pressure of 0.80 atm, 12.3 moles of helium are introduced into a plastic balloon. What is the volume of the balloon?

*14. How many grams of steam (H_2O) are there in a steam pipe with a length of 32.0 ft and a radius of 4.00 in at 215°F and 15.0 lb/in^2 pressure?

*15. A welder uses a tank that contains 19.2 lb of acetylene (C_2H_2) (see Figure 19–14). How many cubic feet of gas is this at a pressure of 14.7 lb/in^2 and a temperature of 76°F? (22.4 l = 0.791 ft^3, and 1.00 kg = 2.21 lb.)

**16. A liquid-oxygen plant produces 2300 l of liquid oxygen from air. The density of liquid oxygen is 1.14 g/cm^3. Air contains only 21% oxygen. How many liters of air at a temperature of 20°C and a pressure of 0.98 atm must be processed to produce the liquid oxygen?

Figure 19–14: Diagram for Problem 15. A welder uses compressed acetylene. If the gas is released into the atmosphere it will occupy several cubic feet of space.

Section 19.4:

17. Find the rms speed of the following gases.
    A. Nitrogen (N_2) at 320 K
    B. Acetylene (C_2H_2) at −23°C
    C. Carbon dioxide (CO_2) at 120°C
    D. Argon (A) at 260 K
18. Find the average kinetic energy of the gases given in Problem 17.
19. Calculate the value of Boltzmann's constant (k) in foot · pounds per Rankine.
20. At what temperature will oxygen (O_2) have an rms speed of 530 m/s?
21. Oxygen (O_2) has an rms speed of 530 m/s at 87°C. At what temperature will hydrogen (H_2) have the same rms speed?

Chapter 20

WORK AND HEAT

Speed skater Eric Heiden won five gold medals during the 1980 winter Olympics in Lake Placid. He also became a successful cyclist to stay in shape during the summer. If Eric works equally hard at each sport, when will he burn more calories, in winter or in summer? (Photographs courtesy of the Journal/Sentinel, *Milwaukee, WI.)*

OBJECTIVES

In this chapter you will learn:

- to relate temperature change with molecular kinetic energy
- how to calculate heat transfer from temperature changes
- how to convert between units of heat and units of work
- how to predict how much heat is needed to increase an object's temperature by a specified amount
- to apply the law of conservation of energy to thermally insulated systems
- to calculate the heat released when something burns
- to calculate the heat released when a substance changes state
- how vapor pressure changes with temperature
- why evaporation and boiling occur

Time Line for Chapter 20

1761 Joseph Black discovers latent heat in phase changes of water.
1781 Johan Carl Wilke develops the idea of specific heat.
1842 Julius Mayer is the first to state the principle of conservation of energy and heat.
1847 Hermann von Helmholtz develops the laws of conservation of heat and energy.
1847 James Prescott Joule measures the transformation of mechanical energy into heat.

We know work done on a system can cause three things to happen: there may be a change in mechanical potential energy; the mechanical kinetic energy of the system may change; and heat may be generated. If there is no change of mechanical energy, work creates heat. Heat is thermal, potential, and mechanical energy on the atomic level. Thermal energy appears in many ways.

In a gasoline engine, fuel is oxidized. Molecular potential energy is changed into molecular kinetic energy that is capable of driving a piston. Thermal energy is converted to mechanical work (see Figure 20–1). In a refrigerator, freon expands. The work needed to cause the expansion is taken from the kinetic energy of the gas molecules. The temperature of the gas decreases. Again thermal energy performs work (see Figure 20–2). An ice cube melts in a glass of soda, causing the drink to cool. The ice gains potential energy at the expense of the kinetic energy of the molecules in the soda. One kind of thermal energy is converted into another form of thermal energy (see Figure 20–3). In an evaporator air conditioner hot, dry air is

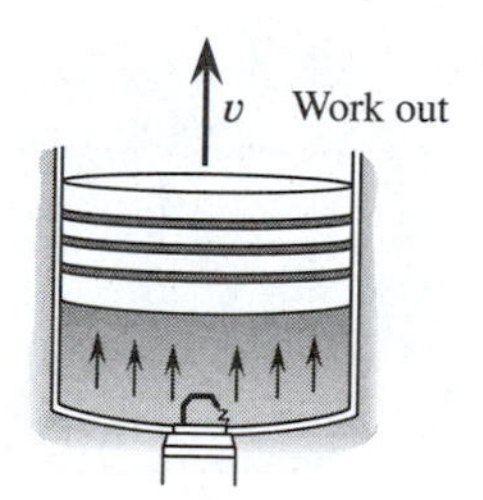

Figure 20–1: *The heat from burning fuel does work on a piston.*

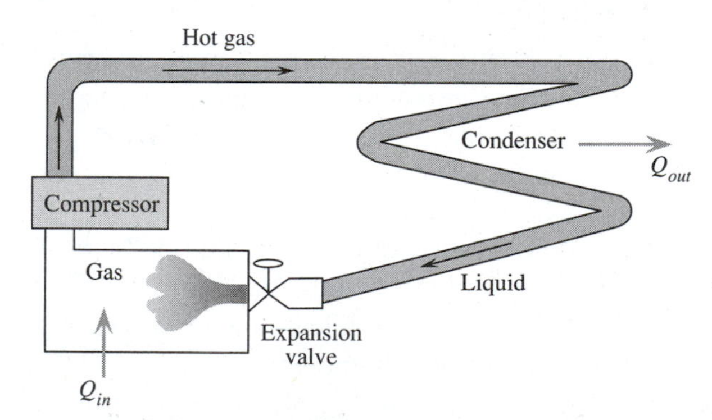

Figure 20–2: *A compressor does work to maintain a partial vacuum over liquid freon. The evaporation of freon draws heat from the surrounding region.*

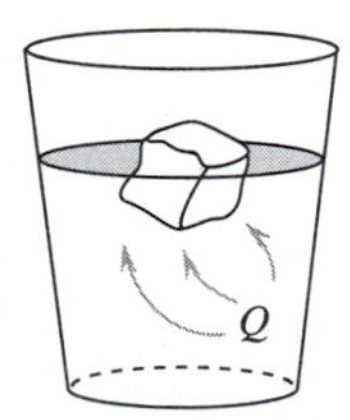

Figure 20–3: *The heat needed to melt an ice cube comes from the surrounding liquid.*

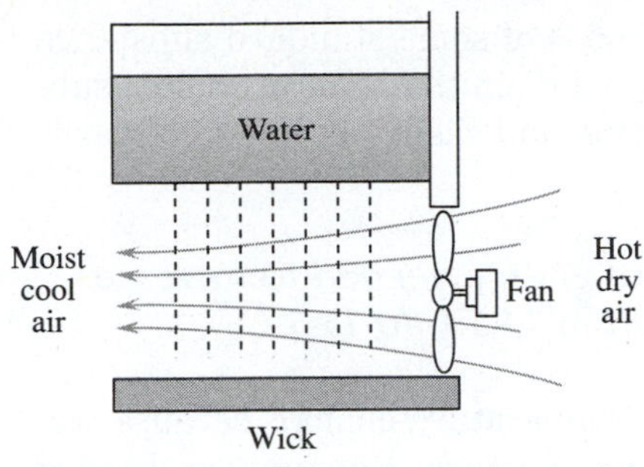

Figure 20–4: *Air blown over a damp wick causes evaporation of liquid, resulting in cooler air.*

passed over a wick saturated with water. The water vaporizes, cooling the air. The water gains potential energy as it changes from a liquid to a gas. The gain in potential energy by the water is obtained from the kinetic energy of the surrounding air molecules. Again, one kind of thermal energy is converted into another (see Figure 20–4).

The work-energy equation applies to atoms and molecules just the same as it applies to larger mechanical systems. In this chapter we will explore the relationship between work and thermal energy. We will look at different forms of thermal energy and develop a conservation of energy equation for thermal systems.

20.1 WORK AND HEAT

We will use Q to represent heat. The thermal energy a liquid, solid, or gas gains is proportional to its increase in temperature.

$$Q \propto \Delta T$$

We can visualize this increase of thermal energy as an increase in kinetic energy of the molecules making up the substance. As the temperature increases, the molecules move faster. In a gas, the molecules gain translational speed. In a solid, the molecules vibrate more rapidly in a fixed position.

The thermal energy of a substance is the sum of all the individual energies of the molecules and atoms making up the material. The more molecules there are to vibrate or move about, the larger the change of thermal energy for the same temperature change will be. Much less energy is needed to increase the temperature of a cup of cocoa by 50°F than to increase the temperature of water in a 50-gal hot-water heater by the same amount.

We can count the number of molecules in something by measuring its mass. One hundred grams of water have twice as many molecules as 50 g of water. The gain of thermal energy is proportional to the mass of the material as well as to its gain of temperature.

$$Q \propto m\,\Delta T$$

We can convert this proportionality into an equation by introducing a constant (c).

$$Q = c\,m\,\Delta T \qquad \textbf{(Eq. 20–1)}$$

This equation provides the basis for the unit of energy called a calorie. Heat (energy) can be related to temperature change by defining the energy required to heat one unit mass of some standard substance by one degree. This is an operational definition. The standard substance should be familiar to everyone and easily obtained. Water is the logical standard.

One calorie (cal) is the amount of energy needed to increase the temperature of 1 g of water from 14.5°C to 15.5°C.

We have specified a particular temperature change because water's density varies slightly with temperature. Remember, thermal energy is proportional to the total number of molecules. We have used mass as a method to measure the number of molecules in the material. If density changes, so will the number of molecules.

Using this definition, we can find the units for the proportionality constant (c). For water:

$$c = \frac{Q}{m\,\Delta T} = \frac{1\text{ cal}}{1\text{ g}\cdot 1\text{°C}} = 1.00\text{ cal/g}\cdot\text{°C}$$

☐ **EXAMPLE PROBLEM 20–1: HEATING WATER**

How many calories are needed to heat 57.0 g of water from 24°C to 88°C?

■ ***SOLUTION***

$$Q = c\,m(T_f - T_0)$$

$$Q = 1.00\text{ cal/g}\cdot\text{°C} \times 57.0\text{ g}\,(88\text{°C} - 24\text{°C})$$

$$Q = 3.6 \times 10^3\text{ cal}$$

A calorie is a rather small unit of energy. Often kilocalories are used.

One kilocalorie (kcal) is the amount of heat required to heat 1 kg of water from 14.5°C to 15.5°C.

The British thermal unit (Btu) is the unit of heat used in the British system of measurement. It too can be defined using Equation 20–1.

One British thermal unit (Btu) is the amount of heat required to increase the temperature of 1 lb of water from 63°F to 64°F.

For water, $c = 1.00$ Btu/lb · °F.

One pound is approximately 454 g. A change of one Celsius degree

Calories

Early theories of matter, following the ideas of Aristotle, included heat and cold as two of the elements that made up matter. Two scientists that lived at the time of Galileo, Francis Bacon and Rene Descartes, suggested that heat had something to do with local motion. At that time, the atomic theory of matter was in its infancy; it did not have universal support.

In the early 1700s, a Dutch physician named Hermann Boerhaave strongly supported the idea that heat was a weightless, invisible fluid that flowed from one object to another. This fluid was eventually called *caloric*.

Throughout the eighteenth century there was a controversy between the caloric theory and the local motion theory. Meanwhile, the atomic theory of matter, which tended to support the local motion theory, began to gain approval.

In 1798, Count Rumford (Benjamin Thompson) described how the motion performed to bore cannons created heat. This was another piece of evidence supporting the motion theory of heat.

In the mid-1800s Julius Mayer and James Prescott Joule independently developed the idea that heat and mechanical energy were simply two different forms of the same thing. Heat is energy. Joule was able to measure the conversion between units of heat and units of mechanical work. This finally put to rest any remaining support for the old caloric theory. All that remains of that theory is the name *calorie* for a unit of heat.

---

is the same as a change of 1.8 Fahrenheit degrees. We can use these values to find conversion equations among the units of thermal energy.

$$1 \text{ Btu} = 1 \text{ lb(water)} \times 1°\text{F} \times \frac{454 \text{ g/lb}}{1.8°\text{F}/°\text{C}}$$

$$1 \text{ Btu} = 252 \text{ cal} = 0.252 \text{ kcal}$$

$$1 \text{ kcal} = 3.97 \text{ Btu}$$

☐ **EXAMPLE PROBLEM 20–2: HEATING WATER THE BRITISH WAY**

One pound of diesel fuel supplies 19,500 Btu of heat. How many gallons of water can be heated 15.0°F by 1 lb of diesel fuel? (1.00 gal water = 8.38 lb.)

■ ***SOLUTION***

$$m = \frac{Q}{c\,\Delta T} = \frac{19{,}500 \text{ Btu}}{1 \text{ Btu/lb} \cdot °\text{F} \times 15.0°\text{F}}$$

$$m = 13\bar{0}0 \text{ lb}$$

The volume of water in gallons is then:

$$V = \frac{13\bar{0}0 \text{ lb}}{8.38 \text{ lb/gal}}$$

$$V = 155 \text{ gal}$$

British thermal units and calories measure energy in terms of thermal effects. This is handy when we want to see how much thermal energy is transferred from one place to another as heat. In some situations we need a conversion between units of heat and units of work.

Assume we know the annual heating load of a building in British thermal units. The building is heated electrically. In order to calculate the heating costs, the heating load needs to be expressed in kilowatt hours.

We know how much thermal energy is released when gasoline is burned. An engineer designing an automobile engine needs to find out how much work is done on the pistons to power the vehicle. Again a conversion between work units and thermal units would be useful.

Heat is defined in terms of the average kinetic energy of molecules. Work is defined in terms of force and distance on the macroscale. These are two entirely different processes. There is no simple conversion. The conversion constant needs to be found experimentally.

We can use the work-energy equation. A measured amount of work in joules can be done to heat water. The increase of thermal energy of the water can be found with a thermometer so we can express the work in terms of calories. The two kinds of energy units will be equal and related by a proportionality constant (J) called the **mechanical equivalent of heat.**

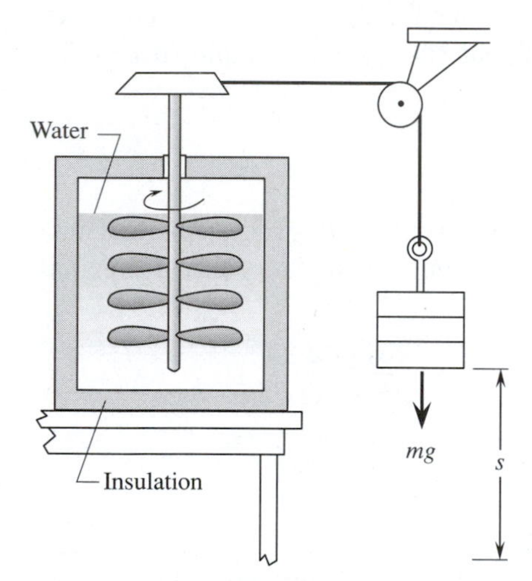

Figure 20–5: A weight (mg) falls a vertical distance (s). The falling weight does work against liquid friction. The water is heated by agitation.

$$\textit{work}\ (\text{joules}) = J \times \textit{thermal energy}\ (\text{calories})$$

Figure 20–5 shows an apparatus for performing the experiment. Weights ($m\,\mathbf{g}$) are dropped N times through a distance (s). The work performed is $W = m\,\mathbf{g}\,(N\,s)$. The work turns paddles in the water. The water is heated by the agitation. Its change of thermal energy is $Q = c\,m\,\Delta T$. The ratio of work divided by equivalent heat generated is the mechanical equivalent of heat (J).

$$J = \frac{W}{Q} = 4.186\ \text{J/cal}$$

In British units the mechanical equivalent of heat is:

$$J = 778\ \text{ft} \cdot \text{lb/Btu}$$

☐ **EXAMPLE PROBLEM 20–3: THE ELECTRIC HAIR DRYER**

Through resistive heating, an electric hair dryer converts 950 J of electric energy into heat for each second of operation. How much heat is this in calories?

■ *SOLUTION*

$$Q = \frac{W}{J}$$

$$Q = \frac{950\ \text{J/s}}{4.186\ \text{J/cal}}$$

$$Q = 227\ \text{cal/s}$$

□ **EXAMPLE PROBLEM 20–4: THE LAWN MOWER**

A pound of gasoline releases 19,$\bar{0}$00 Btu of thermal energy when it is burned. A lawn mower converts 12% of the thermal energy of the gasoline into useful work. How much useful work is done by the mower when 1 quart of gasoline is burned? (The density of gasoline is 42.0 lb/ft^3. One quart has a volume of 0.0334 ft^3.)

■ *SOLUTION*

The heat liberated from the gasoline is:

$$Q = \textit{density} \times \textit{volume} \times \text{heat/pound}$$

$$Q = 42.0\ \text{lb/ft}^3 \times 0.0334\ \text{ft}^3 \times 19{,}\bar{0}00\ \text{Btu/lb}$$

$$Q = 2.66 \times 10^4\ \text{Btu}$$

Only 12% of the thermal energy is converted into work by the mower. The useful work is:

$$W = J\,Q \times 0.12$$

$$W = 778\ \text{ft} \cdot \text{lb/Btu} \times 2.66 \times 10^4\ \text{Btu} \times 0.12$$

$$W = 2.5 \times 10^6\ \text{ft} \cdot \text{lb}$$

20.2 SPECIFIC HEAT

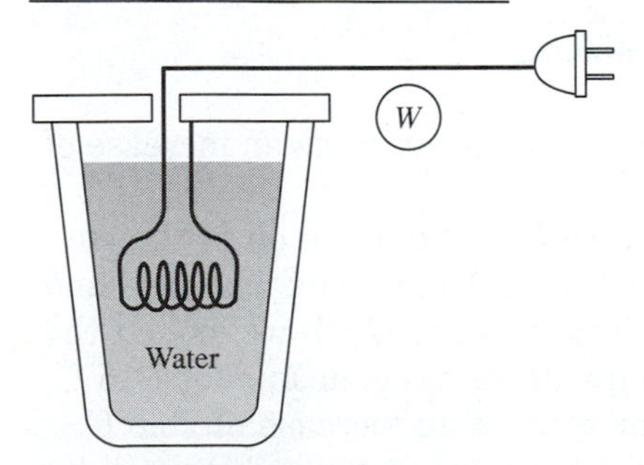

Figure 20–6: *Water in an insulated cup is heated by an electric coil.*

If we place 50.0 g of water in a polystyrene cup and heat it with a 40.0-W electrical coil for 1 min, we can calculate the amount of heat going into the water (see Figure 20–6). We will assume the amount of heat absorbed by the cup is trivial. A watt is a joule per second.

$$Q = \frac{W}{J} = \frac{40.0\ \text{J/s} \times 60.0\ \text{s}}{4.186\ \text{J/cal}} = 573\ \text{cal}$$

The Mechanical Equivalent of Heat, Climate and Diet

In 1843, James Prescott Joule presented to the public the results of his experiments showing that mechanical work could be converted into heat. He built an apparatus that used the potential energy of falling weights to increase the temperature of water. In his experiments, he found a conversion factor between heat and work. The conversion factor is called the mechanical equivalent of heat.

Joule was not the first to suggest that heat is a form of energy. Julius Mayer, a German medical doctor working in Java, noticed that the blood in the veins of people in the tropics was a brighter red than blood in the veins of people living in cooler Europe. Mayer argued that less chemical energy was needed in the warm tropics to maintain body heat. The heat released from the body came from oxidation. Chemical energy is converted to heat, he argued. Therefore, heat is energy. He made this suggestion in 1842, just a year before Joule's experimental results were released.

(Photograph courtesy of the Journal/Sentinal, *Milwaukee, WI.)*

This also explains why many people have less of an appetite during humid, hot summer days. The body needs less food to maintain an appropriate body temperature.

This all suggests that, if you want to lose weight, winter is the best time to do it. Lower the thermostat and find something fun to do in the cool outdoors.

We can use Equation 20–1, the heat equation, to predict the change of temperature of the water.

$$\Delta T = \frac{Q}{c\,m} = \frac{573\text{ cal}}{(1.00\text{ cal/g}\cdot{}^\circ\text{C})(50.0\text{ g})} = 11.5^\circ\text{C}$$

We are not surprised when the thermometer records an increase of over 11°C.

Let us repeat the experiment with 50.0 g of olive oil in the cup. We heat the oil with the same heater for 1 min. Five hundred and seventy-three calories of heat go into the cup. When we look at the thermometer, we discover the temperature has gone up, not 11.5°C, but much more. The thermometer records an increase of 28.8°C.

Less heat is required to increase the temperature of olive oil by one degree than is needed to increase the temperature of water by the same amount. Let c(oil) equal the amount of heat required to increase the temperature of 1 g of oil by 1°C. Let us look at the heat equation. We used the same amount of heat in each experiment.

$$Q(\text{water}) = Q(\text{olive oil})$$

The right-hand sides of the equations must be equal, too.

$$1 \text{ cal/g} \cdot {}^\circ\text{C} \times 50.0 \text{ g} \times 11.5^\circ\text{C} = \text{c(oil)} \times 50.0 \text{ g} \times 28.8^\circ\text{C}$$
$$\text{c(oil)} = 0.40 \text{ cal/g} \cdot {}^\circ\text{C}$$

Only 40% as much heat is required to heat a gram of olive oil by one degree compared with water.

In general, different substances require different amounts of heat energy to increase their temperature by one degree. These values can be found experimentally. In each case, the proportionality constant in the heat equation will have a different value. The proportionality constant (c) is called the specific heat.

The specific heat (c) *of a substance is the amount of heat required to increase the temperature of one unit mass of the substance by one degree.*

The specific heat of water is 1.00 cal/(g · °C) = 1.00 Btu/(lb · °F). The specific heat of olive oil is 0.40 cal/(g · °C) = 0.40 Btu/(lb · °F). Notice the numerical value of the specific heat is independent of the kinds of heat and temperature units used. Table 20–1 lists the specific heats for some common substances.

Table 20–1: Specific heat for selected materials at 25°C.

MATERIAL	SPECIFIC HEAT	MATERIAL	SPECIFIC HEAT
Solids		Solids	
Aluminum	0.22	Concrete	0.19
Copper	0.092	Glass	0.2
Gold	0.031	Granite	0.11
Iron (steel)	0.11	Ice	0.50
Lead	0.038	Paraffin	0.50
Tin	0.052	Rubber	0.42
Zinc	0.093	Wood	0.22
Liquids		Gases	
Alcohol (ethanol)	0.59	Air (dry)	0.24
Benzene	0.41	Helium	1.24
Ether	0.51	Hydrogen	3.41
Mercury	0.033	Nitrogen	0.25
Olive oil	0.40	Oxygen	0.22
Water	1.00	Steam	0.48

Note: Units are cal/(g · °C), kcal/(kg · °C), or Btu/(lb · °F). The values of the specific heats are the same in all three sets of units.

□ **EXAMPLE PROBLEM 20–5: HEATING A BEAKER OF OIL**

One hundred and sixty-eight grams of olive oil are heated in a 123-g beaker from 24°C to 67°C. How much heat is gained by the beaker and oil?

■ ***SOLUTION***

$$Q = Q(\text{oil}) + Q(\text{glass})$$

$$Q = [c(\text{oil})\, m_o\, \Delta T] + [c(\text{glass})\, m_g\, \Delta T]$$

$$Q = [(0.40\ \text{cal/g}\cdot{}^\circ\text{C} \times 168\ \text{g}) + (0.20\ \text{cal/g}\cdot{}^\circ\text{C} \times 123\ \text{g})]\,(67^\circ\text{C} - 24^\circ\text{C})$$

$$Q = 3.9 \times 10^3\ \text{cal}$$

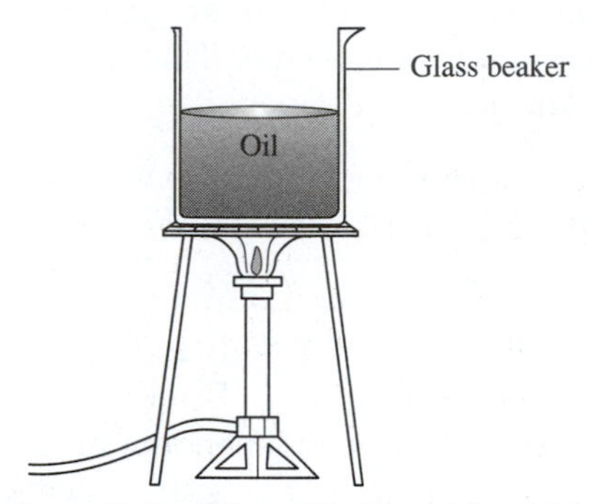

Figure 20–7: *Olive oil is heated in a glass beaker.*

We use many objects that are composed of several different kinds of material with different specific heats. Rather than consider the masses and specific heats of all the things making up the object, it is often more useful to use heat capacity. **Heat capacity** is simply the heat required to increase the temperature of an object by one degree. Mass is part of the heat capacity. Objects that have more mass will tend to have more heat capacity. A tub of bathwater has a much higher heat capacity than a cup of coffee. We will use C to indicate heat capacity. When an object with heat capacity C increases by a temperature change of ΔT, the heat gain (Q) is:

$$Q = C\, \Delta T$$

□ **EXAMPLE PROBLEM 20–6: THE SOLAR PORCH**

A homeowner decides to insulate and enclose a south-facing porch for passive solar heating. The solar gain is calculated to be 1.83×10^5 Btu/day. What heat capacity should the porch have so that the temperature of the porch does not increase by more than $3\bar{0}$°F?

■ ***SOLUTION***

$$Q = C\, \Delta T$$

$$C = \frac{Q}{\Delta T} = \frac{1.83 \times 10^5\ \text{Btu/day}}{3\bar{0}^\circ\text{F/day}}$$

$$C = 6100\ \text{Btu/}^\circ\text{F}$$

If the heat capacity of the porch is much less than 6100 Btu/°F, the homeowner can increase the capacity of the porch by adding water tanks or cement blocks.

Figure 20–8: A homeowner makes a solar collector by enclosing a south-facing porch.

20.3 CALORIMETRY

Heat is thermal energy in transit. If we have a system in which heat can neither enter from the outside nor leave the system, heat will be conserved. Anything inside the system that loses heat must pass the heat on to something else inside the system.

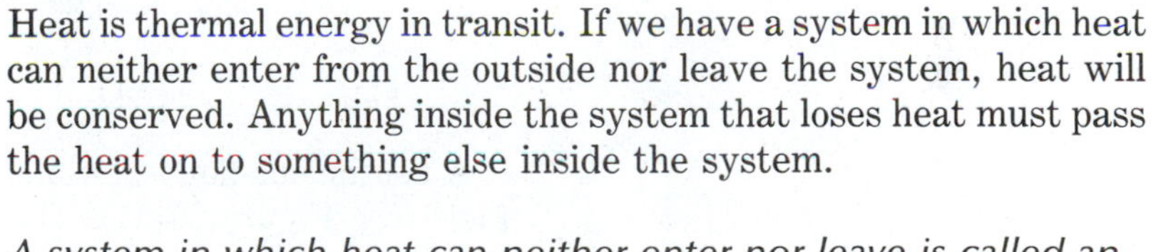

A system in which heat can neither enter nor leave is called an adiabatic system.

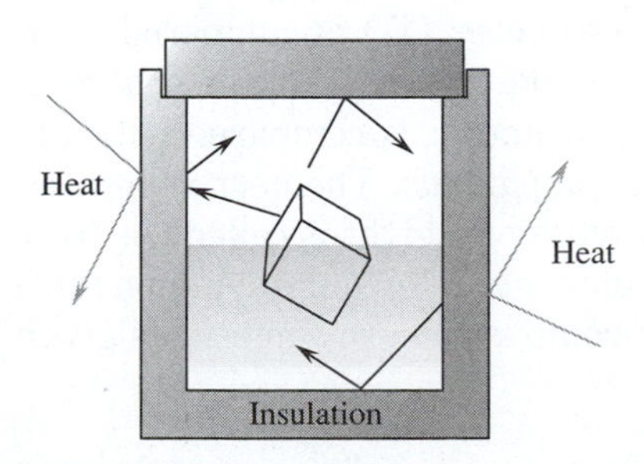

Figure 20–9: In an adiabatic system heat can neither enter nor leave.

It is difficult to devise a perfectly adiabatic container. Some things come close. A vacuum thermos bottle and a capped, thick-walled styrofoam cup leak heat, but they can be used as nearly adiabatic containers for experiments of short duration.

An adiabatic container used to measure the transfer of heat from one thing to another is called a **calorimeter.** Objects of different temperatures are placed in it. In a short period of time, hot objects will transfer heat to cooler objects. A heat equation can be set up using the principle of energy conservation. All the heat gained by things in the system must be heat lost by other things in the insulated system.

$$\textit{sum of heat gains} = \textit{sum of heat losses}$$

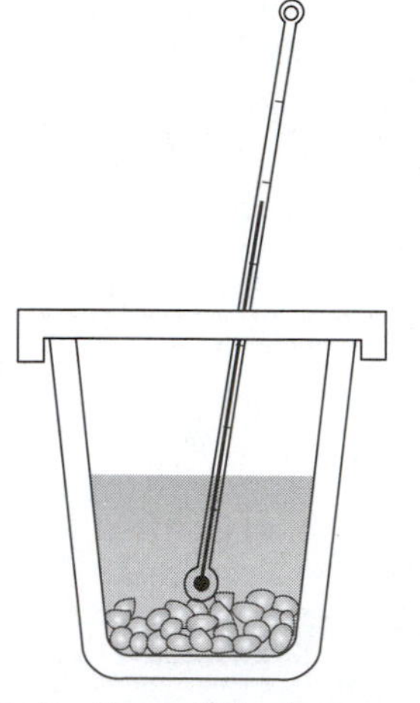

Figure 20–10: *Hot pebbles heat water in an insulated cup.*

☐ **EXAMPLE PROBLEM 20–7: MEASURING THE SPECIFIC HEAT OF DOLOMITE**

A calorimeter consists of a thick-walled styrofoam cup with a styrofoam cap. A thermometer sticks through the cap (see Figure 20–10). In the cup is 32.4 g of water at 24°C. The thermometer has a heat capacity of 2.7 cal/°C. The experimeter drops 146 g of dolomite pebbles heated to 97°C into the calorimeter. The final temperature of the mixture is 59°C. What is the specific heat of dolomite?

■ *SOLUTION*

The water and thermometer increase in temperature. They gain heat. The dolomite decreases in temperature. It loses heat.

$$\textit{heat gain} = \textit{heat loss}$$

$$Q(\text{water}) + Q(\text{thermometer}) = Q(\text{dolomite})$$

$$[(c_w\, m_w) + (C_{th})](T_f - T_1) = c_d\, m_d\, (T_2 - T_f)$$

$$[(1.0\text{ cal/g}\cdot°\text{C} \times 32.4\text{ g}) + (2.7\text{ cal/°C})]\,(59°\text{C} - 24°\text{C}) = c_d(146\text{ g})(97°\text{C} - 59°\text{C})$$

$$c_d = \frac{1228\text{ cal}}{5548\text{ g}\cdot°\text{C}}$$

$$c_d = 0.221\text{ cal/g}\cdot°\text{C}$$

A continuous-flow calorimeter is used to find the heat content of fuels (see Figure 20–11). The fuel is burned in the calorimeter. A steady flow of water passes through a heat exchanger. The temperature of water entering the calorimeter (T_o) is compared with the temperature of water leaving the calorimeter (T_f). The heat gain in the water flowing through the calorimeter is compared with the mass or volume of fuel burned in the apparatus. The heat generated by oxidation of a chemical per unit mass or weight is called the **heat of combustion.** Values of heat of combustion for some common fuels are shown in Table 20–2. Heats of combustion are sometimes given in **therms.**

$$1\text{ therm} = 10^5\text{ Btu}$$

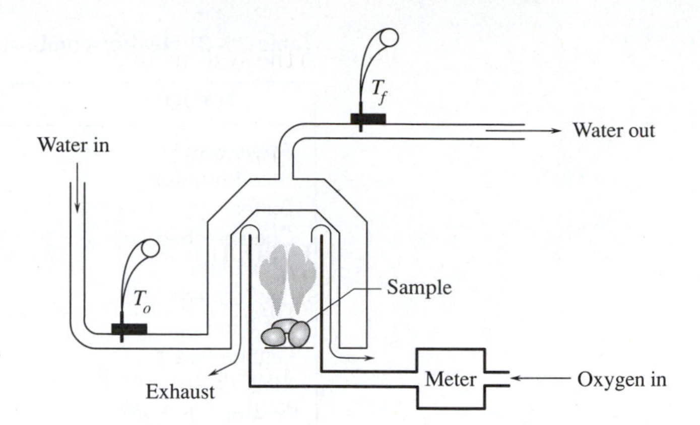

Figure 20–11: *The heat of combustion of a sample of material is measured by a continuous-flow calorimeter. The temperature of water flowing through the heat exchanger is monitored by thermocouples.*

Food Calories are heats of combustion. The values are found by burning food in a calorimeter in the presence of pure oxygen. Some heats of combustion for food are shown in Table 20–3. Notice the food Calorie (Cal) is spelled with a capital *C*. When a nutritionist talks about a Calorie, he or she is really talking about a thermal *kilo*calorie.

1 food Calorie (Cal) = 1 thermal kilocalorie (kcal)

Table 20–2: Heat of combustion for selected fuels.

MATERIAL	KCAL/KG	BTU/LB	KCAL/M³	BTU/FT³
Solids				
Charcoal	8,100	14,600		
Coal	7,800	14,000		
Wood	4,400	8,000		
Liquids				
Diesel oil	10,700	19,300		
Ethyl alcohol	7,800	14,000		
Fuel oil	10,800	19,400		
Gasoline	11,300	20,300		
Kerosene	11,000	19,800		
Gases at 0° and 1 atm				
Acetylene	12,000	22,000	12,900	1,450
Hydrogen	29,000	52,000	2,400	270
Natural gas	13,000	23,000	13,000	1,460
Propane	12,000	22,000	20,000	2,250

Table 20–3: Heat of combustion for selected foods.

FOOD	CAL/G (KCAL/G)
Apple (raw)	0.64
Bread (white)	2.66
Butter	7.95
Cheese (cheddar)	3.93
Chocolate	5.70
Egg (boiled)	1.62
Fat (lard)	9.30
Lettuce (leaf)	0.20
Meat (lean)	0.27
Potatoes (boiled)	0.97
Rice (cooked)	1.12
Sugar (granulated)	3.94

20.4 CHANGE OF STATE

Latent Heat of Vaporization

When a liquid is heated, its temperature increases until it reaches the boiling point. As it changes state from a liquid to a gas, the temperature tends to remain constant.

The thermal energy going into the liquid does not change the kinetic energy of the molecules. We know this because the temperature does not change. Instead, thermal energy does two things: first, the attractive bonds between molecules are broken, and second, work is done on the molecules to space them farther apart (see Figure 20–12).

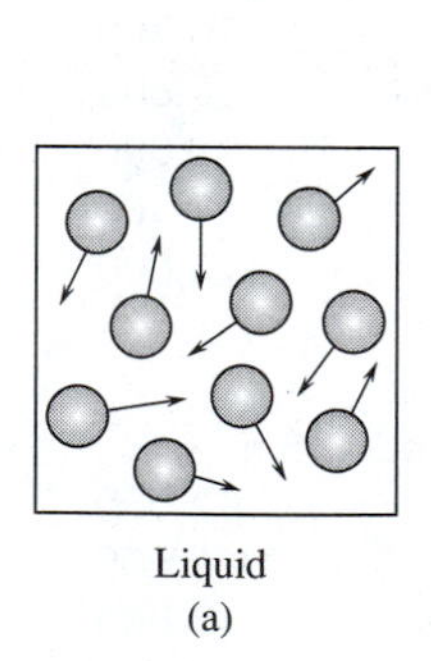

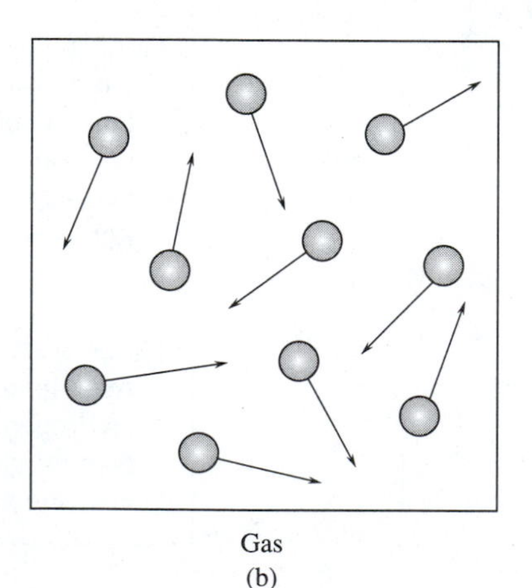

Figure 20–12: Work is done in moving the molecules farther apart as a material changes from a liquid to a gas.

Table 20–4: Latent heat of some common materials at 1 atm.

MATERIAL	MELTING POINT °C	(°F)	BOILING POINT °C	(°F)	L_f KCAL/KG	L_f BTU/LB	L_v KCAL/KG	L_v BTU/LB
Ammonia	−75	−102	−33	−28	108	195	327	465
Ethyl alcohol	−144	−173	78	172	25	45	204	367
Lead	330	620	1620	2950	5.9	10.6	208	315
Oxygen	−218	−362	−182	−297	3.3	5.9	51	92
Water	0	32	100	212	80	144	540	970

The amount of energy required to completely vaporize a liquid depends on the number of molecules in the liquid. We can estimate the number of molecules by measuring the mass of the liquid.

The amount of heat required to transform one unit of mass of a liquid into a gas is called the **latent heat of vaporization.** The heat needed to change a liquid to a vapor is the latent heat of vaporization (L_v) times the liquid's mass (m). Table 20–4 lists the latent heat of vaporization for some common substances.

$$Q = m L_v \quad \textbf{(Eq. 20–2)}$$

□ **EXAMPLE PROBLEM 20–8: MAPLE SYRUP**

Maple syrup is made by boiling much of the water content from maple sap in an evaporator pan. Typically, 24 gal of water must be removed from the sap to produce 1 gal of syrup. Fuel oil releases 19,$\bar{0}$00 Btu/lb. One gallon of water weighs 8.35 lb. Estimate the number of pounds of fuel oil needed to produce 1 gal of maple syrup.

■ ***SOLUTION***

Find the mass of water.

$$m = volume \times density$$

$$m = 24 \text{ gal} \times 8.35 \text{ lb/gal} = 2\bar{0}0 \text{ lb}$$

Figure 20–13: *Heat from burning fuel oil evaporates much of the water in maple sap to produce syrup.*

The heat needed to vaporize the water is:

$$Q = m\,L_v$$

$$Q = 2\bar{0}0 \text{ lb} \times 970 \text{ Btu/lb}$$

$$Q = 1.9 \times 10^5 \text{ Btu}$$

The weight of fuel oil (**W**) needed is then:

$$\text{W} = \frac{\textit{total heat}}{\textit{heat per pound of fuel}}$$

$$\text{W} = \frac{1.9 \times 10^5 \text{ Btu}}{1.90 \times 10^4 \text{ Btu/lb}}$$

$$\text{W} = 1\bar{0} \text{ lb}$$

Latent Heat of Fusion

As a solid melts, the temperature of the solid-liquid mixture stays constant during the process. Thermal energy goes into changing the potential energy of molecules. Because little work is needed to separate the molecules as they melt, less energy is required for a change of state between liquids and solids than is needed between liquids and gases. The amount of heat required to transform one unit mass or weight of a solid into a liquid is the **latent heat of fusion** (L_f). Table 20–4 lists the latent heat of fusion for some common materials. The total heat needed to transform a mass (m) of substance is:

$$Q = m\,L_f$$

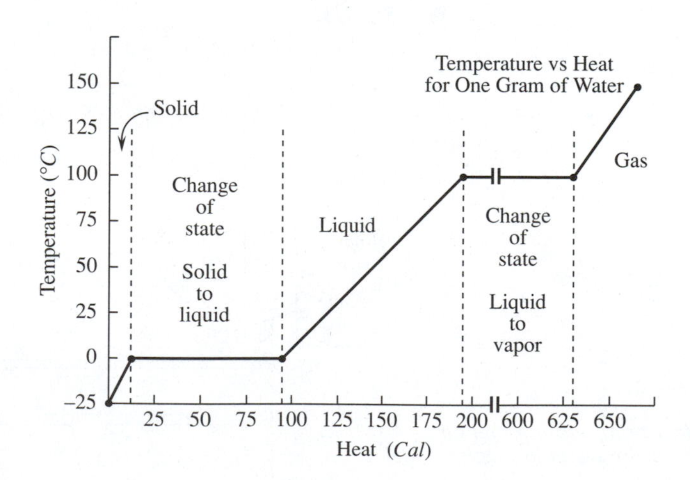

Figure 20–14: Different states of matter of a substance usually have different specific heats. The specific heats are the slopes in the temperature versus heat curve for 1 g of water. Notice the truncated heat scale between 200 cal and 600 cal.

A substance usually has different specific heats in its various states—solid, liquid, and gas. For example, the specific heat of ice is 0.50 cal/g · °C; of water, 1.00 cal/g · °C; of steam, 0.48 cal/g · °C. Figure 20–14 shows a plot of temperature versus heat gained for water. Notice the different slopes for the different states of matter.

☐ **EXAMPLE PROBLEM 20–9: ICE TO STEAM**

How much heat is needed to change 45.0 g of ice at −16°C to steam at 120°C and 1 atm of pressure?

■ ***SOLUTION***

Q_1 = heat to raise temperature of ice to 0°C

Q_2 = heat to melt ice

Q_3 = heat to raise the temperature of water to the boiling point

Q_4 = heat to vaporize water

Q_5 = heat to raise temperature of steam to 120°C

The total heat equals the sum of the five processes.

$$Q = Q_1 + Q_2 + Q_3 + Q_4 + Q_5$$

$$Q = m\,[(c_i\,T_1) + L_f + (c_w\,T_2) + L_v + (c_s\,T_3)]$$

$$Q = 45.0\text{ g }(\{0.50\text{ cal/g}\cdot{}^\circ\text{C }[0^\circ\text{C} - (-16^\circ\text{C})]\} + 80\text{ cal/g} + [1.00\text{ cal/g}\cdot{}^\circ\text{C }(100^\circ\text{C} - 0^\circ\text{C})] + 540\text{ cal/g} + [0.48\text{ cal/g}\cdot{}^\circ\text{C }(120^\circ\text{C} - 100^\circ\text{C})])$$

$$Q = 45.0\text{ g }(8.0\text{ cal} + 80.0\text{ cal} + 100.0\text{ cal} + 540.0\text{ cal} + 9.6\text{ cal})$$

$$Q = 3.32 \times 10^4\text{ cal}$$

20.5 VAPOR PRESSURE, EVAPORATION, BOILING

Vapor Pressure

The vibrational kinetic energy of liquid molecules has a distribution similar to that of gas molecules (see Figure 20–15). A small number of molecules at the surface of the fluid have enough energy to overcome the attractive forces of the nearby molecules and escape.

Let us place water in a closed container and create a vacuum in the space above the liquid (see Figure 20–16). Water molecules will escape from the fluid. In a short time, there will be water vapor in the space above the water.

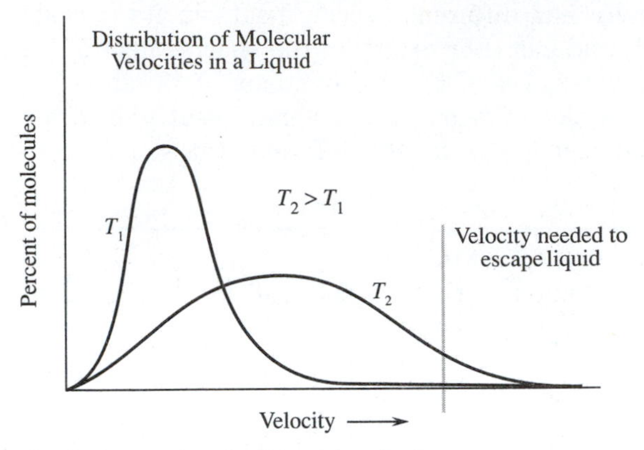

Figure 20–15: In a liquid, molecules have a wide distribution of velocities. At any given temperature, a few molecules will have enough kinetic energy to escape the liquid.

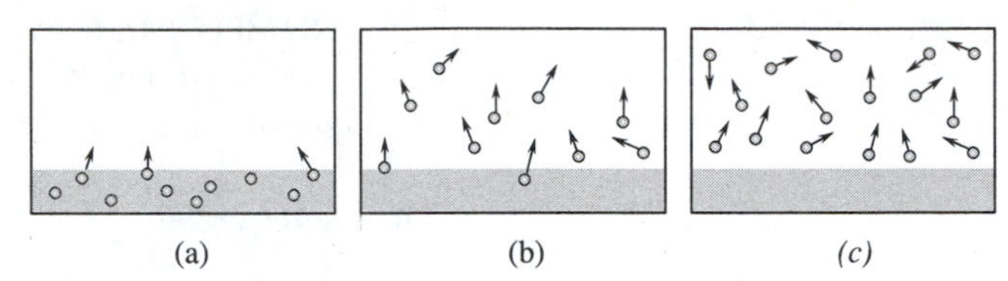

Figure 20–16: (a) A vacuum is formed over a liquid. (b) Molecules escape into the vacuum, creating a partial pressure. (c) An equilibrium is reached when the number of gas molecules captured by the water matches the number of molecules evaporating. This equilibrium pressure is the vapor pressure.

As the pressure of the vapor increases, more and more of the molecules in the vapor will collide with the liquid surface and become trapped. Eventually the number of molecules returning to the liquid will equal the number of molecules escaping from it. The pressure over the water will remain constant. If we increase the temperature of the water, more molecules will have enough energy to escape the water. The pressure of the vapor will increase until a new equilibrium is reached between molecules leaving and entering the liquid. If the temperature is decreased, fewer molecules will have enough energy to escape the liquid phase. More molecules will enter the liquid than will leave it. The density of molecules in the vapor will be reduced. A lower equilibrium pressure will be reached. The pressure of a vapor in equilibrium with its liquid state at a given temperature is called the **vapor pressure.**

Evaporation

If we use an open container rather than a closed container in a dry atmosphere, some of the molecules will escape into the open air. This will reduce the density of molecules in the gas over the water. More molecules will escape the liquid than will enter it. Eventually all of the liquid will be evaporated. If the air already has water molecules in it, some of them will enter the container and collide with the water. This will slow the evaporation process.

An equilibrium will exist if the density of molecules in the air matches the density of water molecules at the vapor pressure of the water. There will be no net gain of molecules in the air. The liquid water will not evaporate. When the water pressure of water vapor in the air reaches the vapor pressure, the air is **saturated.** Table

Table 20–5: Saturated water vapor pressure.

TEMPERA-TURE °0	°F	ABSOLUTE PRESSURE Psi	Pa	MASS OF VAPOR IN AIR G/M³ of Air
−15	5	0.028	1.91×10^2	1.5
0	32	0.089	6.10×10^2	4.85
10	50	0.178	1.23×10^3	9.41
20	68	0.338	2.33×10^3	17.3
30	86	0.615	4.24×10^3	30.4
40	104	1.07	7.37×10^3	51.2
50	122	1.78	1.23×10^4	83.0
60	140	2.86	1.99×10^4	130
70	158	4.52	3.11×10^4	197
80	176	6.86	4.73×10^4	292
90	194	10.2	7.00×10^4	422
100	212	14.7	1.01×10^5	—
105	221	17.5	1.21×10^5	—
110	230	20.8	1.43×10^5	—
115	239	24.5	1.69×10^5	—
120	248	28.8	1.98×10^5	—
125	257	33.7	2.32×10^5	—

20–5 shows the mass of water in saturated air for a range of temperatures.

Often the actual pressure of water in the air, called the **partial pressure,** is less than the saturated values given in Table 20–5. The ratio of the partial pressure of water in the air to the saturated pressure is called the **relative humidity (RH).** It is usually expressed in percentage.

$$RH = 100 \times \frac{partial\ pressure}{saturated\ vapor\ pressure}$$

The partial pressure is proportional to the mass of water in a volume of air. We can view relative humidity as the amount of water in the air compared with the amount of water the air could hold.

Notice that in Table 20–5 the mass of water per unit volume decreases as the temperature decreases. If there are 17.3 g/m^3 of water in the air at 30°C, the relative humidity is 17.3 g/m^3/30.4 g/m^3. This gives a relative humidity of 57%. If the air is cooled to 20°C, the relative humidity will increase to 100% for the same amount of water per cubic meter. If the air cools below this point, water will condense out of the air. Water droplets will form. We call the temperature at which the partial vapor equals the saturated vapor pressure the **dew point.**

Condensing vapors give off heat.

Moist air can cause a problem for air conditioners. If the air is saturated, water will condense as the air cools. Latent heat will be given off by the condensing water. This extra heat must be removed by the air conditioner along with the heat needed to cool the air.

Evaporating liquids absorb heat.

Figure 20–17 shows a water chiller that operates on evaporation. A vacuum system removes water vapor and maintains a partial vacuum over the water in the tank. Warm water is sprayed into the tank. As the droplets fall, they are cooled by evaporation. Chilled air is taken out of the bottom of the tank. A float-operated valve allows make-up water into the tank to replace the water lost by evaporation.

The latent heats of vaporization listed in Table 20–4 are listed only for the boiling point of the liquids at 1 atm. Evaporation can occur at temperatures below the normal boiling point. As a pan of

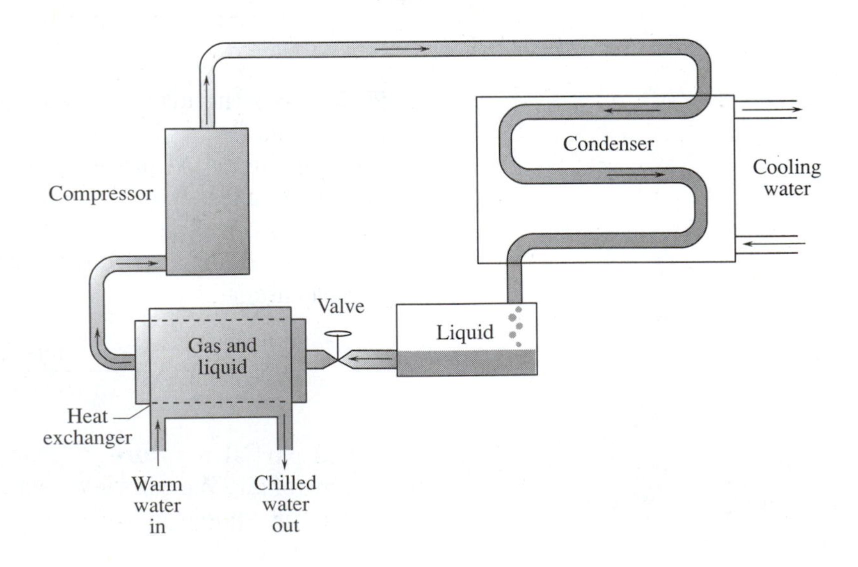

Figure 20–17: *A compressor maintains a partial pressure over a liquid refrigerant. As the liquid evaporates, heat is drawn from the heat exchanger, chilling the water.*

water is heated, steam can be seen rising above the water at temperatures well below the boiling point. For evaporation at one standard atmosphere, we can estimate the latent heat for evaporation.

If the liquid evaporates at temperature T_e, and then the vapor temperature is raised to the normal boiling point of the liquid (T_b), the molecules should have the same energy as molecules that changed state by boiling. The final energy is independent of the processes.

Let L_e be the latent heat of the evaporation of a fluid at temperature T_e. c_v is the specific heat of the vapor, and c_l is the specific heat of the liquid.

Q(heat to evaporate fluid + heat vapor to boiling temperature) = Q(heat to raise fluid to boiling temperature and vaporize)

$$(m\,L_e) + [c_v\,m\,(T_b - T_e)] = [c_l\,m\,(T_b - T_e)] + (m\,L_v)$$

$$L_e = L_v + [(c_l - c_v)\,(T_b - T_e)]$$

☐ **EXAMPLE PROBLEM 20–10: EVAPORATION OF WATER**

At 1 atm of pressure, water evaporates at 80°C. What is the latent heat of evaporation at this temperature?

■ ***SOLUTION***

$$L_e = 54\bar{0}\text{ cal/g} + [(1.00 - 0.48)\text{ cal/g}\cdot{}^\circ\text{C}\,(100^\circ\text{C} - 80^\circ\text{C})]$$

$$L_e = 550\text{ cal/g}$$

Boiling

Bubbles can form in a liquid if the vapor pressure is equal to the surrounding water. If the water pressure were more than the vapor pressure, the water would press in on the wall of the bubble and crush it. Usually the heating surface is beneath the liquid. Bubbles form on cracks and irregularities of the surface. The bubbles grow until they have enough buoyant force to break away from the surface and move upward.

At 1 atm of pressure, boiling will usually occur near the surface of water at 212°F (100°C). If boiling is extremely rapid, the temperature can be a bit higher. Increased boiling temperatures have been measured for situations in which the volume of a liquid is filled with 75% bubbles or more. A pressurized system, such as a pressurized boiler or a pressure cooker, makes it possible for water to boil at higher temperatures than it can at 1 atm. At high altitudes, where the atmospheric pressure is low, water will boil at temperatures below normal. Water can be boiled at room temperature if it is placed under

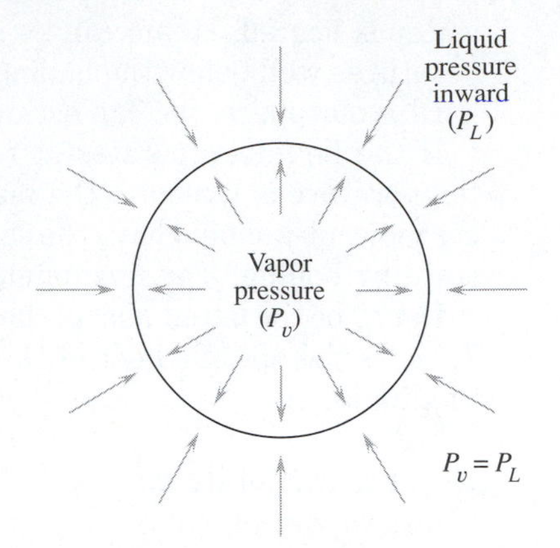

Figure 20–18: The vapor pressure P_v of the gas in a bubble keeps it from being crushed. As the bubble rises, it expands to maintain a pressure outward equal to the reduced pressure inward caused by the water (P_L). $P_v = P_L$.

a bell jar and a partial vacuum is created over the water with a vacuum pump.

□ **EXAMPLE PROBLEM 20–11: THE PRESSURE COOKER**

A cannery needs to process some canned goods at a temperature of at least 239°F. What minimum gauge pressure (P_g) must the cannery maintain in the pressure cookers?

■ ***SOLUTION***

From Table 20–5 the absolute vapor pressure is 24.5 psi at 239°F.

$$P_g = P_a - P_o$$

$$P_g = 24.5 \text{ psia} - 14.7 \text{ psi}$$

$$P_g = 9.8 \text{ psig}$$

The cannery will probably use a gauge pressure of 10 psig.

20.6 CHANGE OF STATE AND PRESSURE

Figure 20–19 is a typical plot of pressure versus temperature for a substance. The lines separate the different states of matter. At low pressures, the substance is a gas; at low temperatures, the substance is a solid. For a range of intermediate pressures and temperatures, the substance is a liquid.

Line AB separates the solid and gas phases. This is called the **sublimation line.** At low temperatures and pressures, a solid may turn into a gas without passing through the liquid state. This process is called **sublimation.** Solid carbon dioxide (dry ice) sublimates into vapor at normal atmospheric pressures.

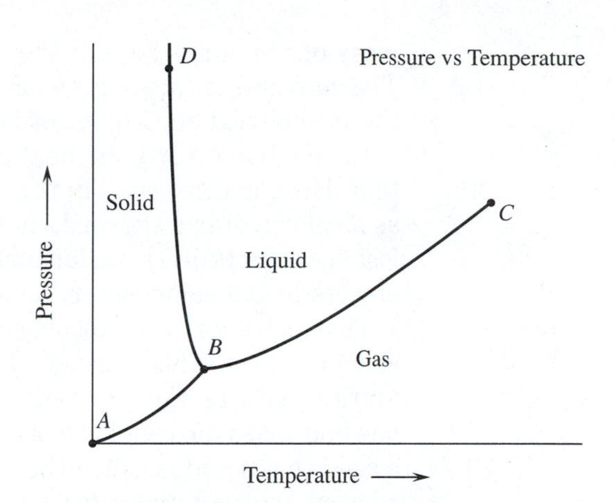

Figure 20–19: A pressure versus temperature curve. The lines indicate the combinations of pressure and temperature where changes of state occur. ***Line AB.*** *Transformation from solid to gas.* ***Line BD.*** *Conversion between solid and liquid.* ***Line BC.*** *Conversion between liquid and vapor.* ***Point B.*** *The triple point, where all three states of matter—solid, liquid, and gas—can exist in equilibrium.*

Line BD separates the solid and liquid phases. For the substance shown in the diagram, the melting point will be lower if the solid is put under pressure.

Line BC is between the liquid and gas phases. As pressure increases, the boiling temperature of the substance also increases. The liquid-gas transition is much more pressure-sensitive than the solid-liquid transition.

Point B is called the **triple point.** It is the only combination of pressure and temperature where all three phases of matter can exist in equilibrium. That is, they can coexist over a long period of time without being converted into one of the other phases. The triple point temperature (T_3) is used to calibrate thermometers. The triple point temperature is 273.16 K or 0°C for water.

Point C is called the **critical point.** As the pressure on a gas increases, the molecules are packed closer and closer together. Eventually, the molecules are so closely packed that we can no longer distinguish between the liquid phase and the tightly compressed vapor phase. For pressures and temperatures above the critical point, there is no latent heat of vaporization because no transition occurs.

SUMMARY

Heat (energy) is measured in terms that define the energy required to heat a standard unit of water by one degree. One calorie is the amount of heat required to raise the temperature of 1 g of water by one Celsius degree. One British thermal unit (Btu) is the amount of heat required to raise the temperature of 1 lb of water by one Fahrenheit degree.

The specific heat (c) of a substance is the amount of heat required to change the temperature of a unit mass of the substance by one degree. The heat capacity (C) of an object is the amount of heat required to change the temperature of its entire mass by one degree. The heat (Q) gained or lost by a substance is proportional to the mass of the material and the temperature change it experiences. The conversion factor between work units and heat units is called the mechanical equivalent of heat (J).

In an adiabatic system heat can neither escape nor enter the system. Heat is conserved in the system. In an adiabatic system such as a calorimeter a thermal energy equation may be used. The heat lost by objects in the calorimeter equals the heat gained by other objects in the calorimeter.

Thermal energy is absorbed or given off by a substance as it changes state. This is called the latent heat. No change of temperature occurs as the substance changes state. The amount of heat one unit mass (or weight) of a substance must absorb to change from a solid to a liquid is called the latent heat of fusion (L_f). The amount of heat one unit mass (or weight) of a substance must absorb to change from a liquid to a gas is called the latent heat of vaporization (L_v). The amount of heat released when one unit mass (or weight) of a substance is burned is called the heat of combustion (L_c).

The pressure of a gas in equilibrium with its liquid state is called the vapor pressure. The vapor pressure is highly dependent on temperature. When the pressure of water in the air is equal to the vapor pressure, the air is saturated; evaporation cannot occur. Boiling occurs in a liquid when the vapor pressure equals the pressure in the liquid.

KEY TERMS

If you can explain the following terms to a friend or classmate, you understand their meaning. If you cannot explain the terms, you should reread the sections in which they are discussed.

adiabatic system
British thermal unit (Btu)
calorie
calorimeter
critical point
dew point
heat capacity
heat of combustion
kilocalorie
latent heat of fusion
latent heat of vaporization
mechanical equivalent of heat
partial pressure
relative humidity (*RH*)
saturated
specific heat (*c*)
sublimation
sublimation line
therm
triple point
vapor pressure

EXERCISES

Section 20.1:

1. Explain the following terms in your own words:

   A. British thermal unit
   B. Mechanical equivalent of heat
   C. Kilocalorie

2. Does it take more heat to increase the temperature of 1.00 kg of water by 1.00°C than to increase the temperature of 1.00 lb of water by 3.50°?

3. Water strikes rocks after going over a 154-ft waterfall. Is the temperature of the water at the bottom of the falls likely to be slightly warmer than at the top? Why?

Section 20.2:

4. Explain the difference between heat capacity and specific heat.

5. Table 20–6 shows the densities and corresponding specific heats of some metals. Brass has a density 8.5 g/cm^3. Its specific heat would probably be between what two metals from Table 20–6?

Table 20–6: Density and specific heat of selected metals.

METAL	DENSITY (g/cm^3)	SPECIFIC HEAT
Aluminum	2.7	0.22
Copper	8.9	0.093
Iron	7.5	0.11
Lead	11.3	0.030
Titanium	4.5	0.125

Section 20.3:

6. Explain what is meant by an adiabatic system.

7. Ice at 32°F, water at 56°F, and lead shot at 400°F are placed in an aluminum calorimeter cup at 74°F. The final temperature of the mixture is 63°F. What items gain heat? What items lose heat?

8. One kilogram of ice at 0°C is placed in a calorimeter of negligible heat capacity; 5 g of steam at 100°C are added. Estimate the final temperature of the mixture.

Section 20.4:

9. A burn caused by steam at 100°C can cause more damage than a scald with hot water at the same temperature. Why?

10. Explain why freezing is a warming process and melting is a cooling process.

11. Salt interacts with the surface of ice to cause it to melt. A brine solution can have a freezing point as low as 0°F. Use this in-

formation to explain how salt can be added to ice at 32°F in an ice cream freezer to create a brine-ice mixture at a temperature of 10°F, 22°F below the starting temperature.

12. Before they had central heating, on cold winter days people used to put tubs of water in the cellar where food was kept, to keep canned goods and root vegetables from freezing. How would this help?

Section 20.5:

13. When you leave a swimming pool you feel cool. Why?

14. How would you explain to a homemaker why it is difficult to remove the lid from the pressure cooker when it has cooled?

15. Why do wet clothes hung outdoors dry better on a windy day than on a calm day?

16. Why does it take longer to boil potatoes at a high altitude than at sea level?

17. A food processor wants to reduce the water content in fruit juices to form a frozen concentrate. Boiling the juice in a cooker was tried, but produced an unpleasant "cooked" taste. Suggest a process that will solve the problem.

PROBLEMS

Section 20.1:

1. How much heat is required to raise the temperature of 1.56 lb of water from 67.0°F to 86.0°F?

2. Two hundred and sixty-eight calories will raise 23.0 g of water initially at 23.0°C to what final temperature?

3. Perform the following unit conversions:

   A. 456 Btu to kcal
   B. 456 cal to Btu
   C. 4890 cal to kcal
   D. 4.3×10^6 kcal to therm

4. A 4800-lb car coasts down a 5430 ft high mountain. The driver uses the brakes to maintain a constant speed rather than using the engine compression by operating in low gear. Estimate how much heat is generated by the brakes in British thermal units.

5. A 214-ton train traveling at 68 ft/s brakes to a stop. How much heat is generated between the rails and the wheels of the train? Give your answer in therms.

6. A wind turbine is connected to a paddle wheel used to agitate a tank of water (see Figure 20–20). The tank holds 1.8 m^3 of water. The turbine generates an average of 8.2 kW of power. How many minutes will it take to increase the temperature of the water by 15°C?

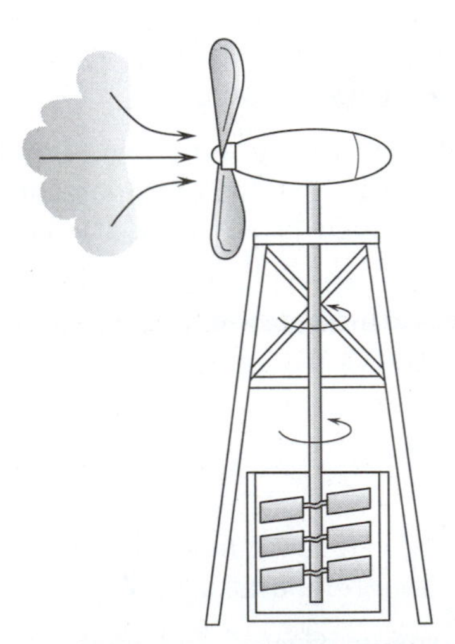

Figure 20–20: *Diagram for Problem 6. A wind turbine uses paddle wheels to convert the kinetic energy of the wind into heat.*

7. A gasoline engine is used to generate electricity. If the system is 15% efficient, how many joules of electrical energy are generated by 1 gal of gasoline? (The density of gasoline is 69.0 kg/m^3. 1 gal = 3.79 l.)

Section 20.2:

8. The temperature of 64.8 g of rubber O rings is increased 11.5°C by 328 cal. What is the specific heat of the rubber?
9. What is the heat capacity of a 432-lb steel engine block?
10. How many calories are needed to raise the temperature of 1.20 l of olive oil from 23°C to 155°C? (The density of olive oil is 0.918 g/cm^3.)
11. An empty electrical hot-water heater has a heat capacity of 15.5 Btu/°F.
    - **A.** What is the heat capacity of the heater when it contains 35 gal of water? (1 gal = 0.134 ft^3)
    - **B.** Assuming the heater is perfectly insulated, how many minutes will it take to heat 35 gal of water in the heater from 50°F to 110°F with a 1.2 kW heating element? (Hint: 1 W = 0.737 ft · lb/s)

Section 20.3:

12. A styrofoam cup like the one in Example Problem 20–7 contains 120 g of copper at 98°C and 35.0 g of water at 27°C. What is the final temperature of the mixture?
13. A calorimeter has a heat capacity of 0.035 Btu/°F and contains 0.110 lb of water at 71°F. Sandstone pebbles with a weight of 0.23 lb and temperature of 190°F are dropped into the calorimeter. The final temperature of the mixture is 88°F. What is the specific heat of sandstone?
14. A fuel is burned in a continuous-flow calorimeter. The inlet temperature of the water is 69°F and the outlet temperature is 106°F. If 168 lb of water pass through the calorimeter while 0.450 lb of fuel is burned, what is the heat of combustion of the fuel?
15. A slice of bread contains about 100 Calories. The human body has an average specific heat of 0.85 kcal/(kg · °C). If all of the energy in a slice of bread were to go into heat, by how many Celsius degrees would the temperature of a 54-kg person increase?
16. How many pounds of fuel oil are needed to increase the temperature of a home by 12°F, if the building has an overall heat capacity of 3.7×10^3 Btu?

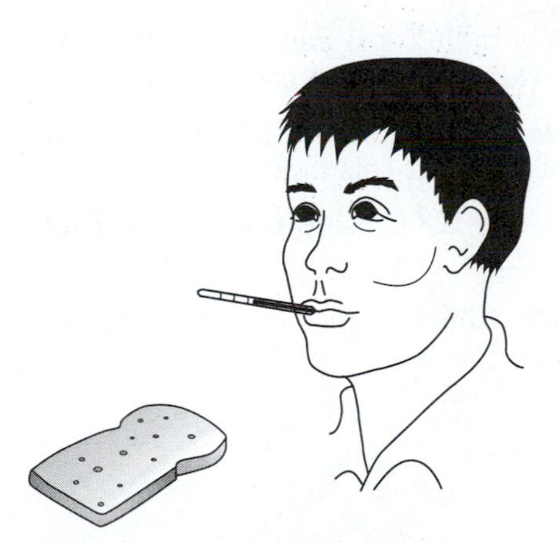

Figure 20–21: *Diagram for Problem 15. The heat of combustion of a slice of bread is converted into body heat.*

Section 20.4:

17. A styrofoam calorimeter contains 55.0 g of water at 27°C; 12.3 g of ice at 0.0°C are added to the water. What is the final temperature of the resulting mixture?

18. How much heat must be removed from 21.0 lb of water at 54°F to produce ice at 5.0°F?

19. How much heat is needed to raise the temperature of 70 g of ice at −20°C to steam at 125°C.

20. How many gallons of water initially at 75°F can be converted to steam at 212°F by a pound of fuel oil? (1 gal of water has a weight of 8.56 lb.)

Section 20.5:

21. Estimate the latent heat of evaporation of water at 35°C.

22. Calculate the relative humidity when the partial pressure of air is 2.33×10^3 Pa:

    A. at 30°C.
    B. at 20°C.

23. On a dry, hot summer day a baseball pitcher sweats off 9.5 lbs during a game. Assume a latent heat of evaporation of 1043 Btu/lb. How much body heat did the pitcher lose by evaporation?

24. Evaporator air conditioners are used in the southwestern United States, where the humidity is very low. A fan passes dry air past a wick soaked with water. Air has a density of 1.29 kg/m³ and a specific heat of 0.24 kcal/(kg · °C). How many kilograms of water will be evaporated to cool 200 m³ of air by 10°C? Assume a latent heat of evaporation of 582 kcal/kg.

25. A pressure cooker operates with 15.0 psi gauge pressure. At what temperature does the water boil in the cooker?

26. On a mountaintop the atmospheric pressure is 0.89 atm. What is the boiling temperature of water at this location?

27. The caldron that heats a geyser is 128 ft below the surface level of the geyser (see Figure 20–22).

    A. What is the vapor pressure of bubbles formed at this depth?
    B. What is the boiling temperature?

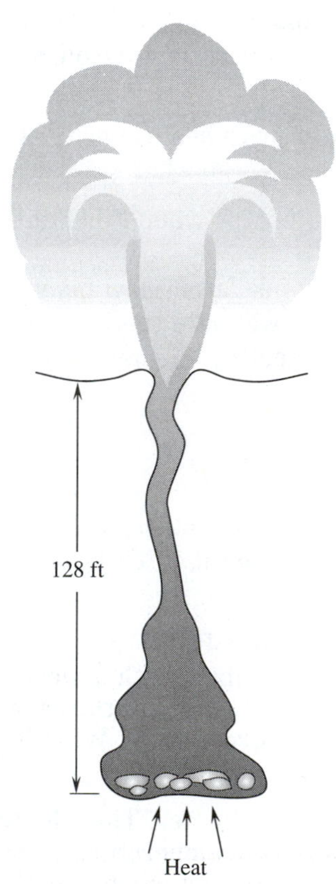

Figure 20–22: Diagram for Problem 27. In a geyser, water is heated deep in the ground to cause a percolating effect.

Chapter 21

THERMODYNAMICS

Why does soda fizz? Does hot water freeze sooner than cold water? Thermodynamics deals with such questions through study of the relationship between heat and energy. (Photograph courtesy of the Coca-Cola Company.)

OBJECTIVES

In this chapter you will learn:

- how work, heat, and change of internal energy are related (first law of thermodynamics)
- to calculate engine efficiency by keeping track of heat entering and leaving the engine
- to calculate thermodynamic work using pressure and volume change
- to analyze P-V diagrams
- that no engine operating in a cycle can convert heat entirely into work (second law of thermodynamics)
- that energy flows in a direction to increase disorder (entropy)
- to calculate entropy for simple systems
- to calculate the maximum efficiency an engine can have operating between two fixed temperatures (Carnot cycle)
- to find the coefficient of performance (*COP*) for refrigerators

Time Line for Chapter 21

1824 Nicolas Carnot shows that work can be done when heat moves from a higher temperature to a lower temperature.

1851 William Thomson, Lord Kelvin, develops the second law of thermodynamics from Carnot's work.

1858 Rudolph Clausius states the second law of thermodynamics. Entropy, or disorder, always increases in a closed system.

A fuel is burned in an internal combustion engine to create high-temperature heat. The heat does some work. Waste heat is ejected through the exhaust and cooling system. Not all of the heat is converted to work. The engine's efficiency is less than 100%.

A refrigerator uses electrical energy to run a compressor. Work is done by the compressor to evaporate a coolant. The evaporation process removes heat from inside the refrigerator and ejects heat outside the refrigerator. Frictional forces in the compressor create additional heat. The only part of the work done by the compressor moves heat from inside the refrigerator to outside.

Thermodynamics is the study of the relationship between heat and work. Thermodynamics makes it possible to determine the upper limits on the efficiencies of engines, refrigerators, and heat-related processes.

In this chapter we will outline some of the basic ideas involved with thermodynamics.

21.1 THE FIRST LAW OF THERMODYNAMICS

Thermal energy is sometimes called **internal energy** (U). When ice melts, its internal energy is increased. The molecules making up the ice gain potential energy to change state. If the water is then heated to a higher temperature, the molecules gain kinetic energy; their internal energy is again increased.

The **first law of thermodynamics** is simply a restatement of the energy-work equation. A quantity of heat (Q) can be used to do two things: perform some work (W) and change the internal energy of a system (ΔU).

$$Q = W + \Delta U \qquad \textbf{(Eq. 21–1)}$$

☐ **EXAMPLE PROBLEM 21–1: A DIESEL ENGINE**

A diesel engine performs 1.83×10^6 ft · lb of work while burning 1.00 lb of fuel. What is the internal change of energy of the exhaust and coolant in British thermal units?

■ *SOLUTION*

From Table 20–2 we see that 1.00 lb of diesel fuel creates 19,300 Btu when burned. This is the heat going into the system.

$$\Delta U = Q - W$$

$$\Delta U = 19{,}300 \text{ Btu} - \frac{1.83 \times 10^6 \text{ ft} \cdot \text{lb}}{788 \text{ ft} \cdot \text{lb/Btu}}$$

$$\Delta U = 19{,}300 \text{ Btu} - 2{,}320 \text{ Btu}$$

$$\Delta U = 17{,}000 \text{ Btu}$$

21.2 ENGINE EFFICIENCY

In most situations we are interested in the amount of work we can obtain from the energy we put into a system. If a station wagon burns a gallon of gasoline (Q_{in}), we would like to know how much useful work (W) can be obtained from it. We can view the change of internal energy as waste heat (Q_{out}). We can rearrange Equation 21–1 to express the useful work in terms of heat. The work done is:

$$W = Q_{in} - Q_{out}$$

We can express the efficiency of the engine in terms of heat. The efficiency is 100 times the ratio of work performed divided by the energy put into the machine.

$$eff = 100 \frac{W}{Q_{in}}$$

$$eff = 100 \frac{(Q_{in} - Q_{out})}{Q_{in}}$$

$$eff = 100 \left[1 - \frac{Q_{out}}{Q_{in}}\right] \qquad \textbf{(Eq. 21–2)}$$

□ **EXAMPLE PROBLEM 21–2: EFFICIENCY OF DIESEL ENGINE**

Find the efficiency of the engine in Example Problem 21–1.

■ *SOLUTION*

$$eff = 100 \left[1 - \frac{Q_{out}}{Q_{in}}\right]$$

$$eff = 100 \left[1 - \frac{17{,}000 \text{ Btu}}{19{,}300 \text{ Btu}}\right]$$

$$eff = 12\%$$

21.3 THERMODYNAMIC WORK

Look at Figure 21–1. A piston fits neatly into a cylinder filled with air. The air is heated; it expands, pushing the piston upward to maintain a constant pressure on the piston. The piston has a cross-sectional area of A. The force acting on the piston is $\mathbf{F} = \mathbf{P}\,A$ as it moves upward through a displacement of Δs. The work done is:

$$W = \mathrm{F}\,\Delta s = \mathrm{P}\,A\,\Delta s$$

The change of volume (ΔV) of the air is $A\,\Delta s$. The work done by the heat is then:

$$W = \mathrm{P}\,\Delta V \qquad \textbf{(Eq. 21–3)}$$

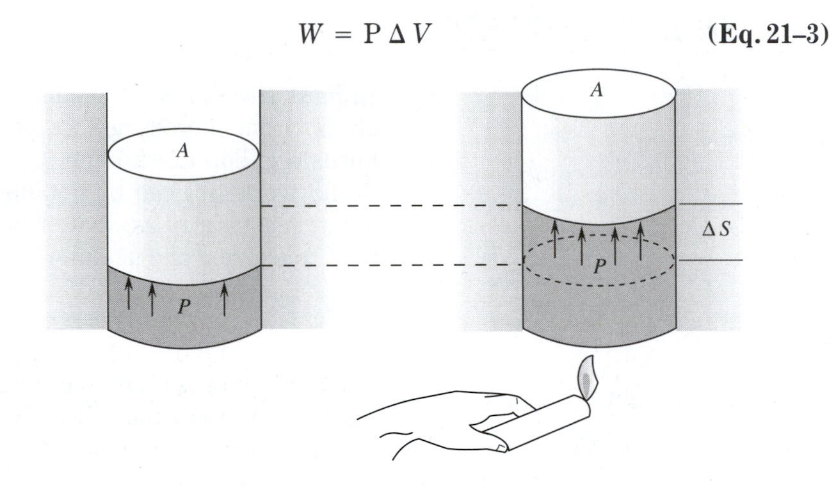

Figure 21–1: *The pressure of heated gas pushes a piston through a distance Δ s. Work is performed on the piston.*

☐ **EXAMPLE PROBLEM 21–3: EXPANDING GAS**

Gas is heated in a cylinder like the one shown in Figure 21–1. If the piston has a cross-sectional area of 82 cm^2 and moves upward 5.60 cm at 1.00 atm of pressure, how much work is done?

■ ***SOLUTION***

$$\mathrm{P} = 1.00\ \mathrm{atm} = 1.013 \times 10^5\ \mathrm{Pa}$$

$$\Delta V = 82\ \mathrm{cm}^2 \times 5.60\ \mathrm{cm} = 4.59 \times 10^2\ \mathrm{cm}^3 = 4.59 \times 10^{-4}\ \mathrm{m}^3$$

$$W = \mathrm{P}\,\Delta V = 1.013 \times 10^5\ \mathrm{N/m}^2 \times 4.59 \times 10^{-4}\ \mathrm{m}^3$$

$$W = 46\ \mathrm{J}$$

21.4 P-V DIAGRAMS

Figure 21–2 shows a specially designed cylinder. The stroke of the piston is limited by changes in the diameter of the cylinder. It is

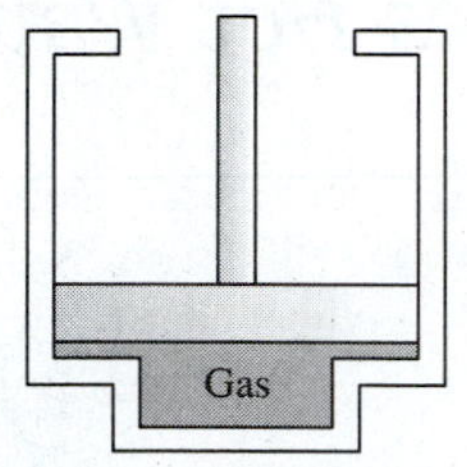

Figure 21–2: *A piston rests on the bottom lip of a cylinder. Gas is trapped below it.*

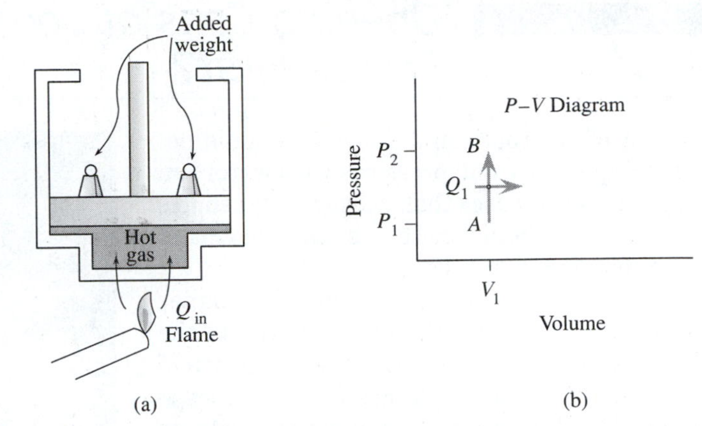

Figure 21–3: (a) *Heat added to the gas increases the pressure of the gas without creating enough force to raise the piston.* (b) *P-V diagram. Pressure versus volume in the cylinder.*

narrower at the bottom and top. We will follow it through one cycle, keeping track of the work done in each part and the heat added to and extracted from the system.

At the beginning of the cycle the gas is heated in the cylinder. Weights are added to the top of the piston (Figure 21–3a). The pressure is not high enough to lift the piston. As we add heat, the pressure of the gas will increase with no change of volume. Figure 21–3b shows a plot of pressure versus volume in the cylinder. This part of the cycle plots into a vertical line (A-B).

Eventually the pressure of the gas is enough to move the cylinder. We will continue to add heat slowly so that the gas will expand at constant pressure. This part of the cycle is a horizontal line (Figure 21–4b, line B-C).

Now we will remove heat from the cylinder. The pressure of the gas will decrease. We will take away some of the weight on the

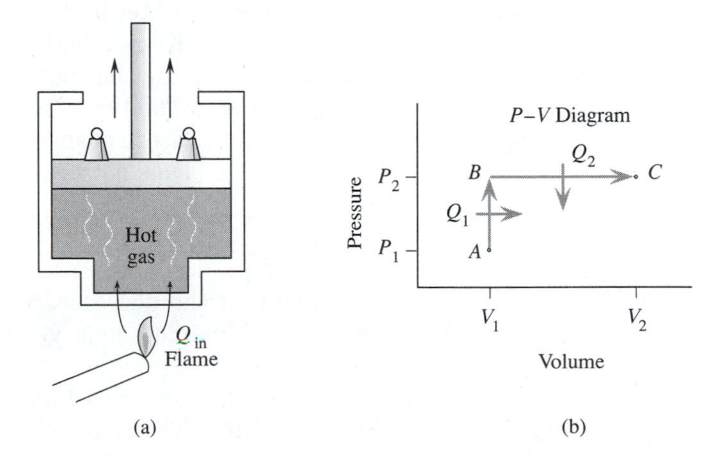

Figure 21–4: (a) *Heat is added slowly.* (b) *P-V diagram. The pressure of the gas remains constant as the volume increases (line B-C).*

Opening Cans of Soda, or, Does Hot Water Freeze First?

The amount of gas that can dissolve in a liquid depends on temperature and pressure. Cool water can hold more dissolved gases than warm water can. As the pressure on a liquid increases, the amount of gas it can dissolve increases.

A can of soda is a liquid-gas solution. The higher pressure in the can keeps the gas dissolved in the liquid until you open the can. At the lower atmospheric pressure, some of the gas needs to escape from the soda. The escaping gas gives the soda its effervescence.

Before you open it, tap the side of the can. Small bubbles sticking to the side of the can will break loose and float to the airspace at the top of the can. When you open the can, the extra pressurized air will escape with a swoosh.

Small bubbles on the side of the can or trapped in the fluid increase in size when the pressure over the soda falls when you open the can. They boil up and make a froth. Shaking the can before opening it catches more small gas bubbles in the liquid. The pressure inside a warm can of carbonated drink will be higher than the pressure in a cold can. The higher pressure accentuates the foaming process.

Dissolved gas in a liquid affects some of its properties. In general, the boiling temperature is increased and the freezing temperature is decreased in proportion to the mass of dissolved gas. This fact has led to the question, "Does hot water freeze faster than cold water?"

Here is an experiment you can do if you live in a cold climate or have a deep freezer handy. Pour leftover coffee into two mugs. Heat one of the mugs in the microwave and then place the two mugs at a sub-zero temperature. You will not be surprised to find that the cold coffee starts to freeze first.

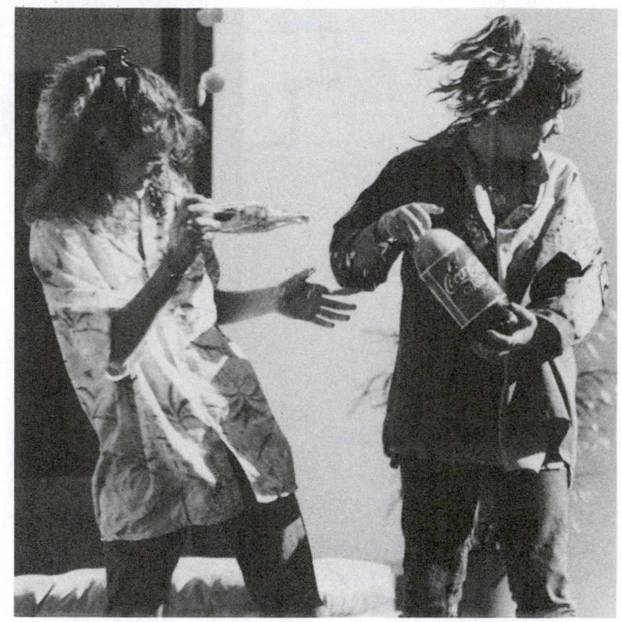

(Photograph courtesy of the Coca-Cola Company.)

On the other hand, let us consider cold- and hot-water pipes in a poorly closed cellar on a cold winter evening. The water in the hot-water pipe has been heated. It has less dissolved gas in it than the cold-water pipe. As the cellar slowly cools at night, the water in the hot-water pipe will freeze first because it has a higher freezing temperature. This does not say that hot water freezes faster than cold water. The two pipes are going through the same temperature changes. The chances are that the water that was not heated has more dissolved gas in it.

piston so that the decreased pressure of the gas can still hold the piston up. This plots as a vertical line moving downward (Figure 21–5, line C-D).

The gas continues to cool, but there are no weights to be removed from the piston. The piston slowly sinks back to its original position

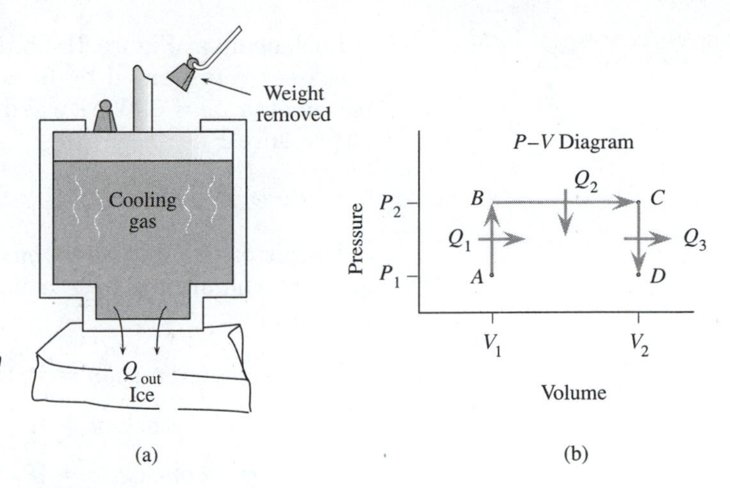

Figure 21–5: (a) *Heat is removed from the gas. Weight is carefully removed from the piston so that the piston will not fall.* (b) *P-V diagram. The pressure of the gas decreases, while the volume remains the same (line C-D).*

(Figure 21–6, line D-A). We are back to the starting position of the cycle.

The heat added to the cylinder in one cycle is $Q_{in} = Q_1 + Q_2$. The heat extracted in one cycle is $Q_{out} = Q_3 + Q_4$. The quantity of heat used by the cylinder is the difference between the heat added and the heat extracted from the cylinder.

$$\Delta Q = Q_{in} - Q_{out}$$

At the end of one complete cycle, the gas has the same temperature, pressure, and volume as it had at the beginning of the cycle. That is, the gas has not changed its internal energy. The difference in heat must have gone into work. This is in agreement with Section 21.2.

$$W = \Delta Q = Q_{in} - Q_{out}$$

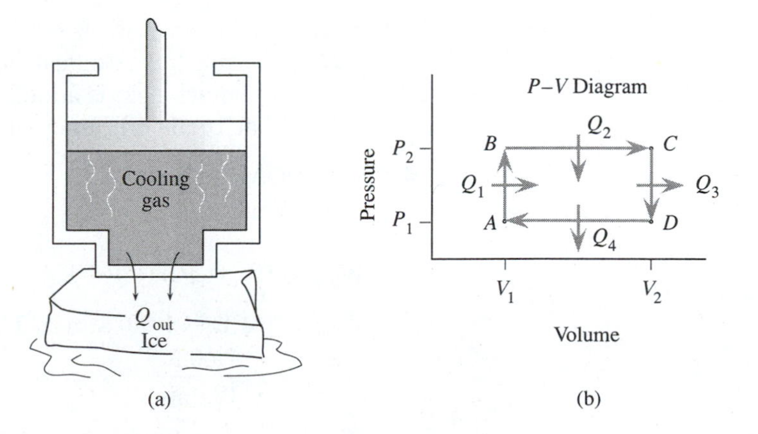

Figure 21–6: (a) *The unweighted piston moves downward as the gas continues to cool, returning to its original position.* (b) *P-V diagram.*

Look again at Figure 21–6b. No mechanical work was done in the processes represented by lines A-B and C-D. There were no displacements; $\Delta s = 0$. Work was done *on the piston* as the gas expanded at pressure P_2.

$$\textit{work done on piston by gas} = P_2 (V_2 - V_1) = \textit{work out}$$

The piston did work *on the gas* as the piston compressed the cooling gas. The volume was decreasing. This will give us a negative value for the work.

$$\textit{work done by piston on gas} = P_1 (V_1 - V_2)$$

$$\textit{work in} = -P_1 (V_2 - V_1)$$

$$\textit{net work in one cycle} = [P_2 (V_2 - V_1)] - [P_1 (V_2 - V_1)]$$

$$W_{net} = (P_2 - P_1)(V_2 - V_1)$$

The net work is the enclosed area on the P-V diagram. We used a special cycle, so pressure and volume plotted out into horizontal and vertical lines. This, of course, is not true for a real engine. However, the enclosed area of the P-V diagram for any cyclic process is the net work done by the process.

The enclosed area on a P-V diagram is the net work done in one cycle.

☐ **EXAMPLE PROBLEM 21–4: P-V DIAGRAM**

The quantities in Figure 21–6b have the following values:

$$P_1 = 1.00 \text{ atm} \qquad V_1 = 8.90 \text{ in}^3$$

$$P_2 = 31.0 \text{ atm} \qquad V_2 = 46.1 \text{ in}^3$$

$$Q_3 = 7.2 \text{ Btu} \qquad Q_4 = 5.3 \text{ Btu}$$

A. How much work is done in one cycle?
B. How much waste heat is extracted during each cycle (Q_{out})?
C. How much heat is added to the system during one cycle (Q_{in})?
D. What is the efficiency of the system?

■ ***SOLUTION***

A. Work:

$$W_{net} = (P_2 - P_1)(V_2 - V_1)$$

$$W_{net} = (31.0 - 1.00) \text{ atm} \times 14.7 \text{ lb/in}^2 \times (46.1 - 8.90) \text{ in}^3 \times \frac{1.00 \text{ ft}}{12.0 \text{ in}}$$

$$W_{net} = 1.37 \times 10^3 \text{ ft} \cdot \text{lb, or } 1.73 \text{ Btu}$$

B. Exhaust heat:

$$Q_{out} = Q_3 + Q_4$$
$$Q_{out} = 7.2\text{ Btu} + 5.3\text{ Btu}$$
$$Q_{out} = 12.5\text{ Btu}$$

C. Heat in:

$$Q_{in} = W + Q_{out}$$
$$Q_{in} = 1.73\text{ Btu} + 12.5\text{ Btu}$$
$$Q_{in} = 14.2\text{ Btu}$$

D. Efficiency:
Here are two ways to calculate the efficiency.

$$eff = \left[1 - \frac{Q_{out}}{Q_{in}}\right] \times 100 \qquad eff = \left(\frac{W}{Q_{in}}\right) 100$$

$$eff = \left[1 - \frac{12.5\text{ Btu}}{14.2\text{ Btu}}\right] \qquad eff = \left(\frac{1.73\text{ Btu}}{14.2\text{ Btu}}\right) 100$$

$$eff = 12\% \qquad eff = 12\%$$

On your calculator, the efficiency will differ in the third significant figure for the two methods of calculation. This difference is caused by round-off error.

21.5 THE SECOND LAW OF THERMODYNAMICS

Rub your hands together briskly. Work is changed into heat. You can feel the heat generated by the work you perform. In this process, the work done turns completely into heat, or internal energy. No mechanical energy is gained or lost by the action.

$$work = heat \qquad \textbf{(Eq. 21–4)}$$

It is easy to demonstrate a process in which work is transformed completely into heat, but it is difficult to demonstrate a process in which energy or heat is transformed *completely* into work. In fact, no one has ever found such a process. The conversion of work into heat is an *irreversible* process. Equation 21–4 cannot be read backwards. Yes, heat can do some work, but a quantity of heat or energy cannot be completely converted to work. There is always some waste heat left over.

$$heat = work + waste\ heat \text{ (or, } change\ of\ internal\ energy\text{)}$$

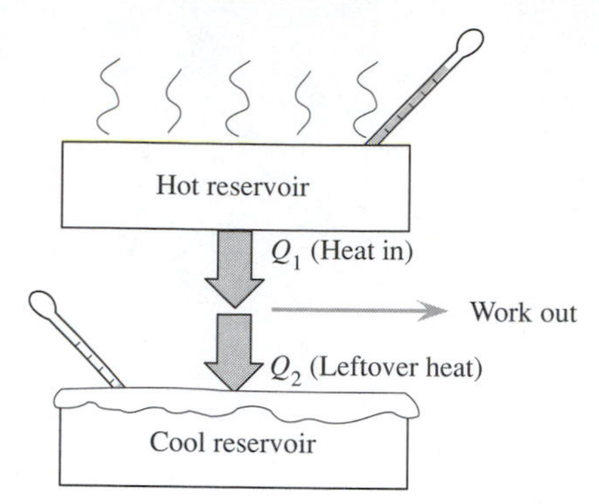

Figure 21–7: *Work is done by some of the heat flowing from a hot place to a cool place.*

This principle is called the **second law of thermodynamics.** It puts some limits on what can be done with the first law of thermodynamics.

Heat flows from something with a high temperature toward something with a low temperature. In an internal combustion machine, fuel is burned at a high temperature. Waste heat is deposited into the atmosphere at a cooler temperature.

Work can be done only when heat flows. This means that we need a temperature difference between two things. We will call the high-temperature source the "hot reservoir," and the place where the exhaust heat goes the "cool reservoir." In a heat engine, work is done by heat moving from a hot reservoir to a cool reservoir (see Figure 21–7). According to the second law of thermodynamics, not all of the heat moving between the reservoirs can be converted into work.

It is impossible to construct an engine that will continuously take a quantity of heat and convert it completely into work.

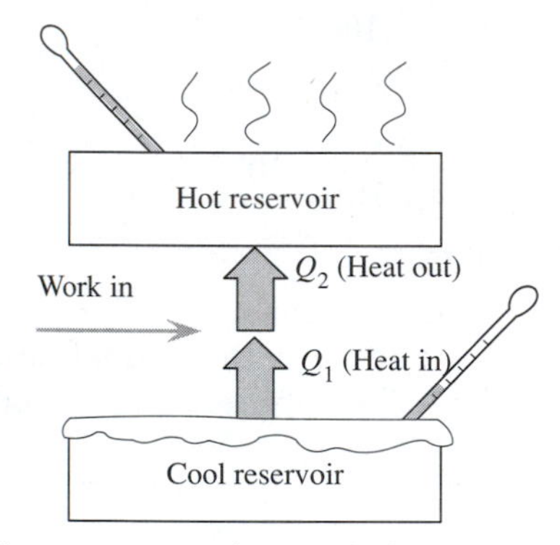

Figure 21–8: *Work is needed to move heat from a cool place to an area with a higher temperature.*

A refrigerator forces heat flow from a cool reservoir to a hot reservoir by performing work on the substance carrying the heat. The most common type of refrigerator uses a compressor that creates a partial vacuum in the evaporator and removes vapor. The compressor does the necessary work. Heat is removed from a cool region surrounding the evaporator and is extracted at a higher temperature at the condenser. If the compressor is turned off, evaporation will stop when the vapor in the evaporator chamber becomes saturated (see Figure 21–8). For refrigerators, the second law of thermodynamics can be stated this way:

It is impossible to construct a machine that will continuously move heat from a cool reservoir to a hot reservoir without doing work.

21.6 ENTROPY: DISORDERLY CONDUCT

Nature—the universe—tends toward disorder. Octane, the principal component of gasoline, is a molecule with 8 carbon atoms and 18 hydrogen atoms. When it is burned, it combines with oxygen to make several smaller, less organized molecules of water and carbon dioxide.

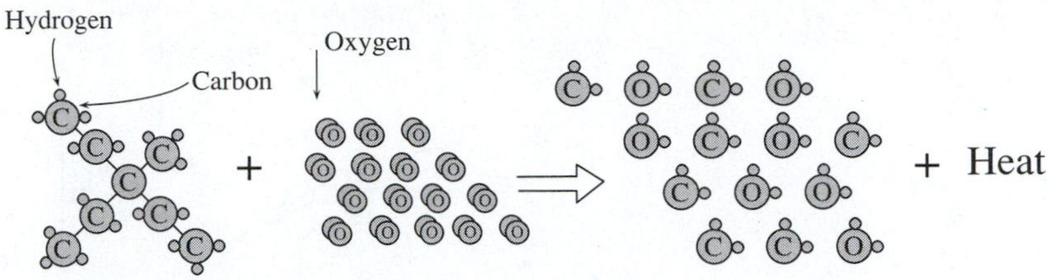

Figure 21–9: A highly organized hydrocarbon is burned, producing heat and smaller, less organized molecules.

A tank is partitioned, with oxygen on one side of the partition and nitrogen on the other. When the partition is removed, the gases diffuse into a uniform mixture of gas molecules.

A fresh deck of playing cards falls, spewing all over the floor. When they are picked up, they are no longer in order of suit or number.

The valve on a tank of acetylene is opened in a vacuum chamber. The gas expands into the chamber. The acetylene becomes less organized in the sense that it is less localized.

Entropy is a measure of disorder. The less entropy something has the more order it has. In nature, energy flows in a direction to increase entropy. Processes flow in a direction of increased random disorder.

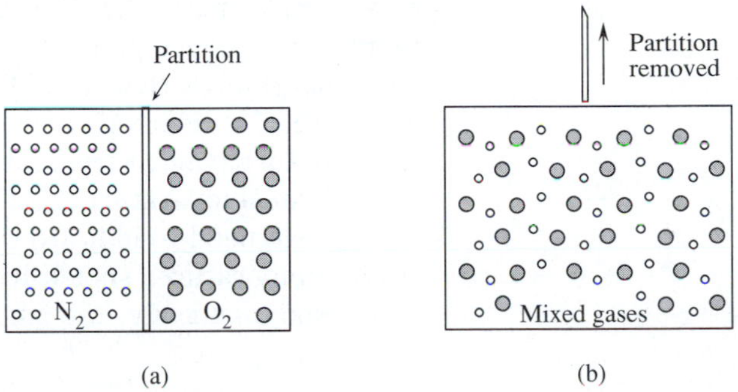

Figure 21–10: (a) A partition separates compartments containing pure gases. (b) The partition is removed and the gases diffuse, mixing thoroughly.

Figure 21–11: A new deck of cards is in order. If the cards fall on the floor, they will have a random order when they are picked up.

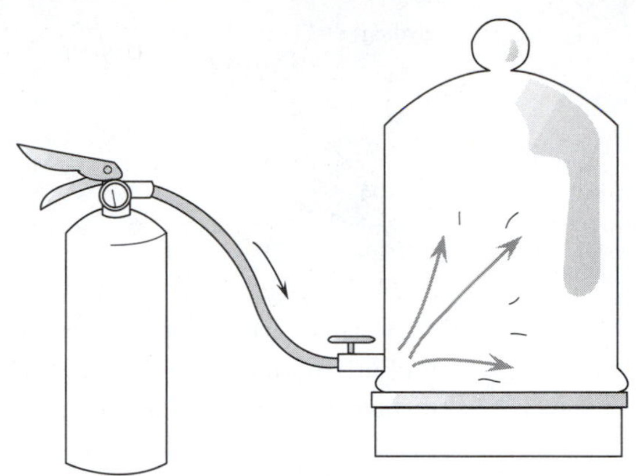

Figure 21–12: *Gas from a compressed gas tank expands into a vacuum. The molecules become less ordered because they are not in as small a space as before.*

A high-grade energy source has low entropy. When a high-grade energy source is burned in an automobile, its entropy increases as it changes into a high-temperature gas under pressure (see Figure 21–13). The gas expands and cools as it pushes a piston. Its entropy again increases. The exhaust gases are discharged into the atmosphere, cooling some more and mixing with the air in yet a more disordered state. It then has too much entropy to be very useful as an energy source.

For processes involving heat the *change* of entropy can be expressed mathematically. The more heat something gains or loses, the greater the change in entropy will be. For example, a liquid is more disorganized than a solid. As we add heat to a piece of ice, we melt more of it. Entropy increases as the added heat increases.

Heat naturally flows from a high-temperature source to something with a lower temperature. If energy flows in a direction of increasing entropy, we need to indicate the effect of temperature on entropy. Lower temperatures should have higher entropy.

Thermodynamically, the change of entropy (S) is defined as:

$$\Delta S = \frac{\Delta Q}{T} \qquad \textbf{(Eq. 21–5)}$$

where T is an absolute temperature.

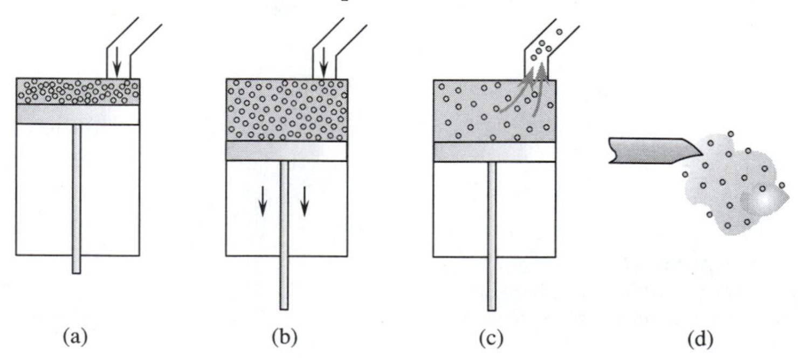

Figure 21–13: *Increased entropy occurs in each step of operating an internal combustion engine.* (a) *Gasoline burns. Large molecules are converted to smaller, less organized molecules.* (b) *The gas becomes less localized as it expands and its temperature decreases.* (c), (d) *The gas expands, cools, and mixes with the atmosphere.*

Notice that Equation 21–4 gives only the *change* of entropy. It is difficult to know what the *total* entropy of something is unless its total history is known in great detail. Two apparently identical pieces of copper may have different amounts of total entropy. The grain sizes of the metal making up the two samples may be slightly different; they may fit together better in one piece. One may have more dissolved impurities or missing atoms in the lattice. All these things will affect the total entropy of the two pieces of copper. It is difficult to know all these things. What we do know is that they will undergo the same change of entropy if they are given the same amount of heat at the same absolute temperature.

☐ **EXAMPLE PROBLEM 21–5: MELTING ICE**

An ice cube with a mass of 23 g absorbs heat from the tabletop on which it rests. Calculate its change of entropy as it melts at 0.0°C.

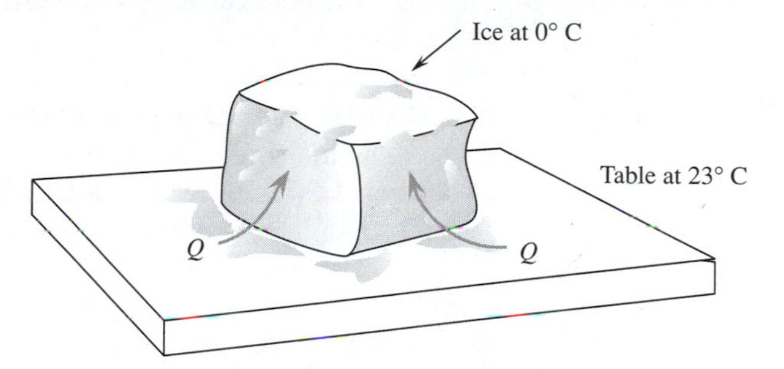

Figure 21–14: *Ice melts on a tabletop. Heat is drawn from the warmer table to the ice.*

■ ***SOLUTION***

The absolute temperature is $T = 0°\text{C} + 273 = 273\ \text{K}$. The gain in heat is $Q = m\,L_f = 23\ \text{g} \times 80\ \text{cal/g} = 1840\ \text{cal}$.

$$\Delta S = \frac{\Delta Q}{T}$$

$$\Delta S = \frac{1840\ \text{cal}}{273\ \text{K}}$$

$$\Delta S = 6.74\ \text{cal/K}$$

While you were looking over Example Problem 21–5, it may have occurred to you that the heat needed to melt the ice must have come from the table. The entropy of the table must have changed. Assume the table has a large heat capacity, so it remains at a constant temperature of 23°C (296 K). Its change of entropy is:

$$\Delta S = \frac{-1840\ \text{cal}}{296\ \text{K}} = -6.22\ \text{cal/K}$$

The negative sign indicates that the table *lost* 1840 cal of heat instead of gaining it. The decrease in entropy does not violate the idea of entropy flow. What matters is the total change of entropy: the entropy change of a system plus the change of entropy of its surroundings. When a thermodynamic process occurs, there must be a net increase in the universe. If we look at individual parts of the universe, we will find that some things gain and some lose entropy. If we add up all the entropy changes, there will be a net gain. We can write this as an equation.

$$\Delta S(\text{system}) + \Delta S(\text{surroundings}) \geqslant 0 \qquad \textbf{(Eq. 21–6)}$$

For the case of the melting ice cube ΔS(system) is the entropy change of the ice and ΔS(surroundings) is the entropy change of the table.

$$+6.74\ \text{cal/K} - 6.22\ \text{cal/K} = +0.52\ \text{cal/K} \geqslant 0$$

□ **EXAMPLE PROBLEM 21–6: CHANGE OF STATE FOR LEAD**

The melting point of lead is 621°F. What change of entropy occurs when 1.20 lb of lead solidifies by giving off heat to the surrounding air that has a temperature of 86°F? The heat of fusion for lead is 10.6 Btu/lb.

■ ***SOLUTION***

Lead:

$$Q_1 = -(m\,L_f) = -(1.20\ \text{lb} \times 10.6\ \text{Btu/lb}) = -12.7\ \text{Btu}$$

$$T_1 = 621°\text{F} + 460 = 1081\ \text{R}$$

Air:

$$Q_2 = +12.7\ \text{Btu}$$

$$T_2 = 86°\text{F} + 460 = 546\ \text{R}$$

$$S(\text{total}) = S(\text{lead}) + S(\text{air})$$

$$S(\text{total}) = \frac{Q_1}{T_1} + \frac{Q_2}{T_2}$$

$$S(\text{total}) = -\frac{12.7\ \text{Btu}}{1081\ \text{R}} + \frac{12.7\ \text{Btu}}{546\ \text{R}}$$

$$S(\text{total}) = +0.0115\ \text{Btu/R or } +9.07\ \text{ft} \cdot \text{lb/R}$$

21.7 ENTROPY, THE SECOND LAW OF THERMODYNAMICS, AND ABSOLUTE ZERO

In Section 21.5 we reduced the operation of a heat engine to essentials. Heat is gained from a hot source such as steam, burning fuel, or an electrical heater. Some work is done. A piston may be moved or a turbine blade rotated. Exhaust heat is then ejected at a cooler temperature (see Figure 21–15).

Let's look at the entropy of a heat engine. Energy is extracted from the fuel (Q_{in}) at some hot temperature (T_{in}). Leftover heat (Q_{out}) is discharged into the surroundings at a lower exhaust temperature (T_{out}). The net change of entropy cannot be negative,

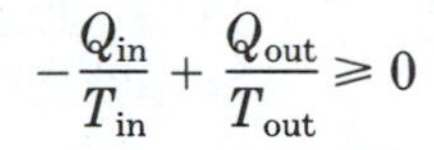

$$-\frac{Q_{in}}{T_{in}} + \frac{Q_{out}}{T_{out}} \geqslant 0$$

or

$$\frac{Q_{out}}{T_{out}} \geqslant \frac{Q_{in}}{T_{in}}$$

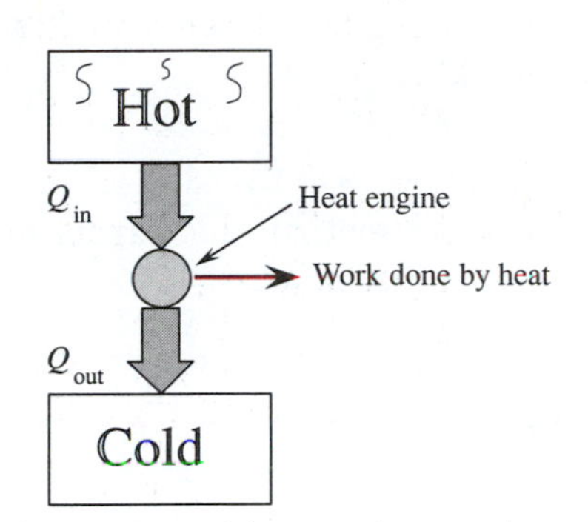

Figure 21–15: *A heat engine uses heat moving from a hot region to a cool region to perform work.*

This is what we expect. The exhausted fuel has a higher entropy than the initial fuel. We can rearrange this inequality.

$$\frac{Q_{out}}{Q_{in}} \geqslant \frac{T_{out}}{T_{in}}$$

We can use this relationship to find the best possible efficiency a continuously operating heat engine could have. If we multiply by a minus sign the inequality reverses. (If $3 > 2$, then $-3 < -2$.)

$$-\frac{Q_{out}}{Q_{in}} \leqslant -\frac{T_{out}}{T_{in}}$$

If we add one to both sides of the relationship we have the efficiency of the engine.

$$eff = 1 - \frac{Q_{out}}{Q_{in}} < 1 - \frac{T_{out}}{T_{in}}$$

The expression on the extreme right is called the **Carnot efficiency.** It is the best efficiency an engine can obtain when operating in a cycle. This theoretical efficiency depends only upon the absolute temperature of input heat and the exhaust.

$$eff(\text{Carnot}) = 1 - \frac{T_{out}}{T_{in}} \qquad \text{(Eq. 21–7)}$$

Equation 21–7 is the second law of thermodynamics in a different form. At first it looks as though we could have a 100% efficient machine if we could devise a system that would operate at an exhaust temperature of absolute zero. Unfortunately, because it takes an infinite amount of work to cool something to this temperature, a 100% efficient engine operating in a cycle is theoretically impossible.

□ **EXAMPLE PROBLEM 21–7: THE FABULOUS MACHINE**

An inventor enters the U.S. patent office with plans for a new highly efficient engine. It is a modified Wankel engine with superchargers, special fuel injectors, and a radically new ignition system. It burns fuel at 1450°F and produces an exhaust temperature of 180°F. The wide-eyed inventor claims an engine efficiency of 72%. Should the inventor be granted a patent?

■ *SOLUTION*

The engine cannot have an efficiency better than the Carnot efficiency. For the data given, the Carnot efficiency is:

$$eff(\text{Carnot}) = 1 - \frac{(180°\text{F} + 460)}{(1450°\text{F} + 460)}$$

$$eff(\text{Carnot}) = 66\%$$

Show the inventor the door. The fabulous machine is not thermodynamically possible.

The inability to reach absolute zero is the **third law of thermodynamics.**

It is impossible to cool something to absolute zero.

The three laws of thermodynamics can be crudely stated as: **(first law)** there it is; **(second law)** you can get it if you can get there; **(third law)** you can't get there. "It," of course, means an equivalent amount of work for the heat put into an engine.

21.8 THE REFRIGERATION CYCLE

A refrigerator uses an evaporation-condensation cycle. A liquid evaporates in the cooled region, absorbing heat from the surroundings. Outside the cooled region, vapor ejects heat in a condenser as it returns to a liquid state. In effect, heat is moved from a cool region

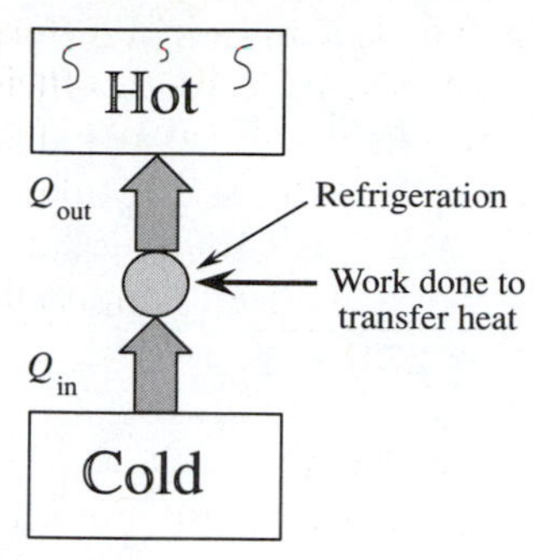

Figure 21–16: Work must be done to transfer heat from a cool region to a warmer region.

to a warmer region by doing work to maintain the evaporation cycle (see Figure 21–16).

Instead of getting work from heat, work is used to move heat from one place to another. The performance of a refrigerator can best be defined in terms of the amount of heat transferred for the work put into the system. This ratio is called the **coefficient of performance (*COP*).**

$$COP = \frac{\textit{heat transferred}}{\textit{work done}} = \frac{Q_{in}}{W} \qquad \textbf{(Eq. 21–8)}$$

For a cyclic process the work done is the difference between heat in and heat out.

$$W = Q_{out} - Q_{in}$$

Substitute this into Equation 21–8.

$$COP = \frac{Q_{in}}{Q_{out} - Q_{in}} = \frac{1}{Q_{out}/Q_{in} - 1}$$

From the equations for entropy we know that $Q_{out}/Q_{in} \geqslant T_{out}/T_{in}$. The best possible coefficient of performance can be expressed in terms of the input and output temperatures of the refrigerator. This is called a **Carnot refrigerator cycle.**

$$COP(\text{Carnot}) = \frac{1}{T_{out}/T_{in} - 1}$$

□ **EXAMPLE PROBLEM 21–8: A REFRIGERATOR**

A refrigerator does 10 J of work to extract 14 cal of heat from a freezer at 0.0°C. If the condenser coil has an effective temperature of 34°C:

A. What is the actual *COP* of the unit?

B. What is the theoretical (Carnot) *COP* of the unit?

■ ***SOLUTION***

A.

$$COP(\text{actual}) = \frac{Q_{in}}{W}$$

$$COP(\text{actual}) = \frac{14\text{ cal} \times 4.186\text{ J/cal}}{10\text{ J}}$$

$$COP(\text{actual}) = 5.9$$

B.

$$COP(\text{Carnot}) = \frac{1}{T_{out}/T_{in} - 1}$$

$$COP(\text{Carnot}) = \frac{1}{(34°\text{C} + 273)/(0°\text{C} + 273) - 1}$$

$$COP(\text{Carnot}) = 8.0$$

21.9 APPLICATIONS

Refrigeration Equipment

The Compressor Refrigerator

The most commonly used refrigeration cycle uses a compressor (see Figure 21–17).

Liquid with a high vapor pressure at low temperatures, such as freon, passes through a throttle valve (A) from a region of high pressure to a region of low pressure. Low pressure causes rapid evaporation in the evaporator (B). The heat needed for the change of state is drawn from the region surrounding the evaporator. A compressor (C) draws off the freon vapor, maintaining a partial vacuum in the evaporator. The hot vapor passes through a condenser (D). In a household refrigerator, heat is passed out into the air outside the unit. Larger units will use cooling water in the condenser. After passing through the condenser, the freon is again in the liquid state. It is collected in a tank (E) ready to pass through the throttle valve to start the cycle again.

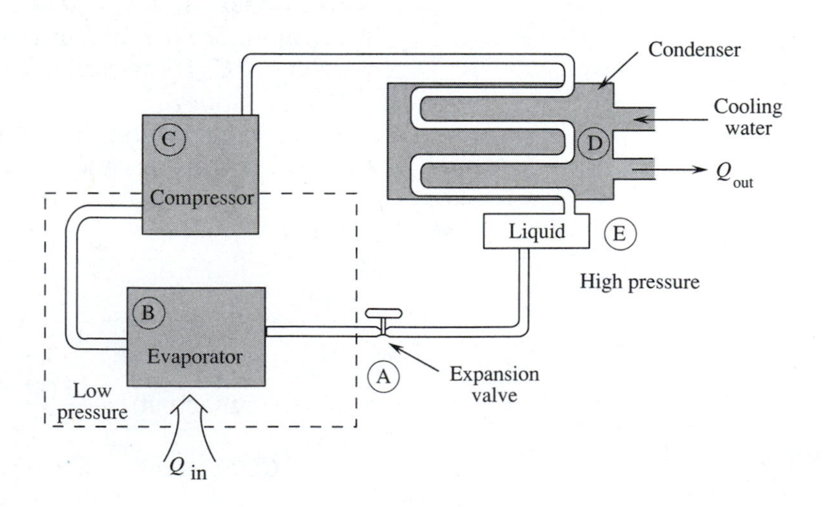

Figure 21–17: The operation of a mechanical refrigerator.

Absorption Refrigerators

A refrigerator that operates on propane or natural gas does not use a compressor. It uses the absorption cycle shown in Figure 21–18.

In the evaporator (A), ammonia evaporates, drawing heat from the surrounding region. The ammonia vapor mixes with hydrogen gas that is already in the evaporator. The density of the combined gases is greater than the density of saturated ammonia alone. The gas mixture sinks to a collector (C) and then to an absorber (D) that contains water. The ammonia vapor is easily dissolved in the water, but the hydrogen gas is not. As the gases pass through the absorber they are separated. The ammonia is absorbed by the water, and cooled hydrogen gas is recycled to the evaporator. The water and dissolved ammonia continue on to the generator (E). Here heat is applied to the water to cause a percolating effect. The water is lifted to the rectifier (F).

The rectifier separates the ammonia from the water. The vapor pressure of the ammonia in the heated water is much higher than the vapor pressure of the water. The vapor over the rectifier is mostly ammonia. The vapor rises to the condenser (G). Heat is removed to return the ammonia to the liquid state. The ammonia then drips down into the evaporator to start the cycle again. In the meantime water flows out of the rectifier into the absorber, where it is cooled so it can absorb additional ammonia.

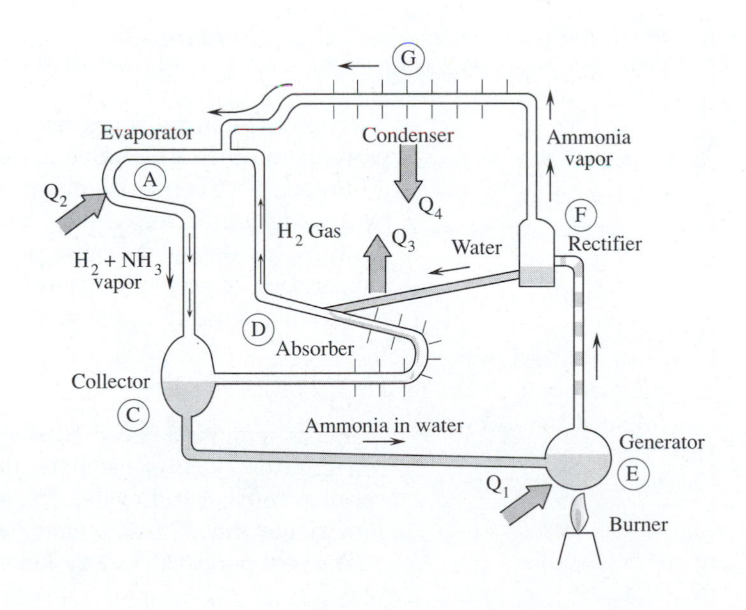

Figure 21–18: *The absorption refrigerator. This system does not use a compressor. The heat used to separate ammonia and water is the input energy.*

The absorption refrigerator has three cycles going at once.

1. Hydrogen gas is cycled through the evaporator and absorber, gaining and losing heat. Hydrogen gains heat in the evaporator and loses heat in the absorber. It increases the density of vapor in the evaporator so the hydrogen-ammonia vapor mixture will sink into the absorber.
2. Ammonia is cycled in a complex way. It gains heat as it changes state in the evaporator, loses some heat in the absorber, gains heat again in the generator, and finally loses heat in the condenser as it returns to the liquid state. The ammonia acts as the refrigerant for the device.
3. Water gains heat in the generator and cools in the absorber. It acts as an absorbent for the ammonia.

Figure 21–18 shows a schematic of the heat flow. Heat enters at the generator (Q_1) at a high temperature (T_1) and at the evaporator (Q_2) at a low temperature (T_2). Heat leaves the system at room temperature (T_3) at the absorber (Q_3) and the condenser (Q_4).

Conservation of energy requires that the heat gained equals the heat lost.

$$Q_1 + Q_2 - Q_3 - Q_4 = 0$$

The heat put in at the generator (Q_1) is the energy needed to drive the system. This represents the work input. The heat extracted from the freezing compartment is the heat entering the evaporator (Q_2). The coefficient of performance is:

$$COP = \frac{\textit{heat extracted}}{\textit{energy in}} = \frac{Q_2}{Q_1}$$

An absorption refrigerator has no moving parts. Neither a compressor nor a throttle valve is needed because there is no pressure difference between parts of the system. The system can be driven by a variety of external heat sources. All that is needed is a temperature hot enough to cause percolation in the generator and rapid evaporation of ammonia from the water. Designs have been made to use solar energy to drive absorption air-cooling systems.

Heat Pumps

A heat pump is a reverse refrigerator. The exhaust heat from the refrigerator cycle is used for heating. Figure 21–19 shows a reversible refrigerator cycle. The system can be converted into an air conditioner during the summer and a heat pump in the winter.

The heat pump can be an efficient way to use electricity. An electric

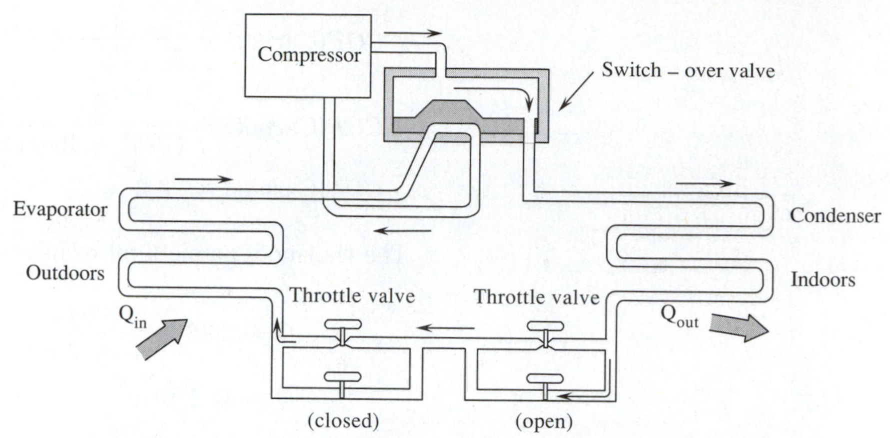

Figure 21–19: A reversible refrigerator cycle is used as a heat pump. By reversing the cycle, the same machine can be used as an air conditioner in the summer.

heater using resistive heating directly converts electrical energy into heat ($Q = W/J$). The heat delivered is the heat equivalent of the electrical work done.

A heat pump extracts heat from a source and delivers it to the heated area. The heat transferred can be larger than the heat equivalent of the work done. As long as the coefficient of performance is larger than one, there is a net gain.

$$\textit{heat delivered} = COP \times \frac{\textit{electrical work}}{\text{J}}$$

The coefficient of performance depends on the temperature difference between the source and the heated area. Heat pumps cost quite a bit more than resistive heaters. Initial costs must be considered in the overall costs of heating.

□ **EXAMPLE PROBLEM 21–9: USING A HEAT PUMP**

A heat pump operating on outside air is used to maintain the inside temperature of a building at 70°F. Estimate the number of British thermal units delivered by the pump per kilojoule of work when:

A. the outside temperature is 45°F.
B. the outside temperature is −20°F.

■ ***SOLUTION***

A. At 45°F:

We can use a Carnot refrigerator estimate to get the theoretical high ratio of heat to work. The actual heat pump will not perform as well.

$$COP(\text{Carnot}) = \frac{1}{T_{out}/T_{in} - 1}$$

$$COP(\text{Carnot}) = \frac{1}{(70°F + 460)/(45°F + 460) - 1}$$

$$COP(\text{Carnot}) = 20.2$$

The thermodynamic limit of heat delivered by the pump is:

$$Q = 20.2 \times \frac{1000 \text{ J}}{4.186 \text{ J/cal}} \times \frac{1 \text{ Btu}}{252 \text{ cal}}$$

$$Q = 19.1 \text{ Btu}$$

B. At −20°F:

$$COP = \frac{1}{(70°F + 460)/(-20°F + 460) - 1} = 4.9$$

$$Q = 4.9 \times \frac{1000 \text{ J}}{4.186 \text{ J/cal}} \times \frac{1 \text{ Btu}}{252 \text{ cal}}$$

$$Q = 4.6 \text{ Btu}$$

Resistive heating would deliver 0.94 Btu for the same amount of electrical energy.

In some of the northern states, heat pumps operating on outside air are marginally cost-effective. One way to make them more effective is to cool water circulating through plastic pipes buried in the lawn below the frost line, where the ground temperature is around 45°F to 50°F. Other schemes use heat pumps to recover heat from "gray water," the waste water from washers, sinks, tubs, and so on.

Engines

The Four-cycle Engine

A four-cycle engine is commonly used in automobiles. Figure 21–20 shows the four parts of the cycle along with a corresponding P-V diagram. This operation is called the **Otto cycle.** In the following description of its operation, the letters in parentheses represent segments of the P-V diagram.

Stroke 1: Intake (A to B). The intake valve is open. As the piston moves backward the fuel-air mixture from the carburetor is drawn into the piston.

Stroke 2: Compression (B to D). The intake valve closes. The

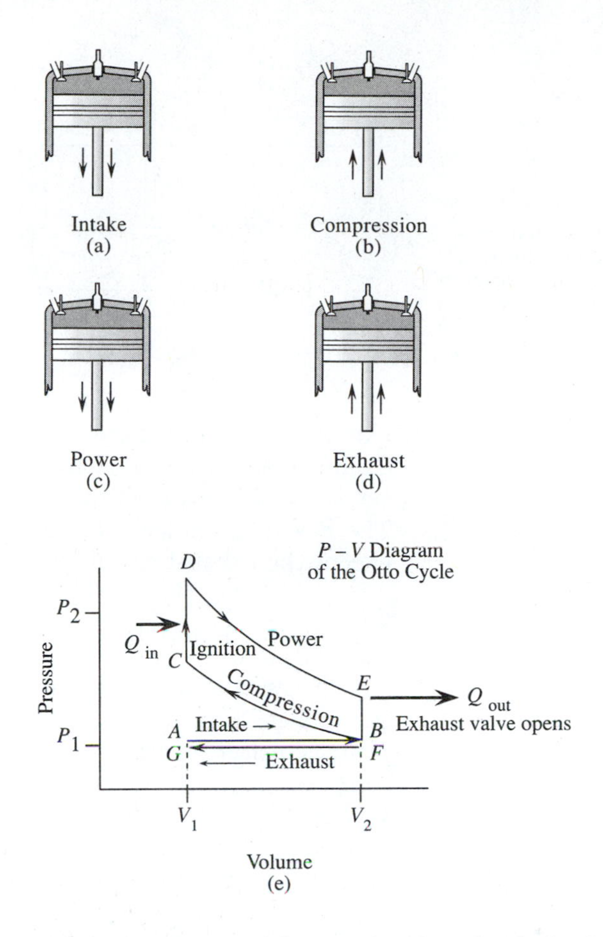

***Figure 21–20:** (a–d) The cycles of a four-stroke engine. (e) P-V diagram of the Otto cycle.*

piston moves forward, compressing the fuel mixture (B to C). The pressure and temperature in the cylinder increase. At the end of the compression stroke, the fast-burning fuel is ignited. Pressure and temperature in the cylinder abruptly increase (C to D).

Stroke 3: Power Stroke (D to E). The increased pressure on the piston drives it backward, delivering work to the drive shaft (D to E). As the gas in the cylinder expands, its temperature decreases.

Stroke 4: Exhaust. The exhaust valve opens. The pressure in the cylinder drops quickly (E to F) as the exhaust components escape through the exhaust manifold. The piston drives forward to push out the remaining exhaust (F to G). The cycle is ready to start again.

The Two-cycle Engine

A four-cycle engine needs a camshaft to operate the valve system. This is cumbersome in a small motor, and only one stroke in four is a power stroke. A two-cycle engine eliminates these two problems.

Instead of using valves in the combustion chamber, ports are drilled in the cylinder wall. As the piston reaches the end of the

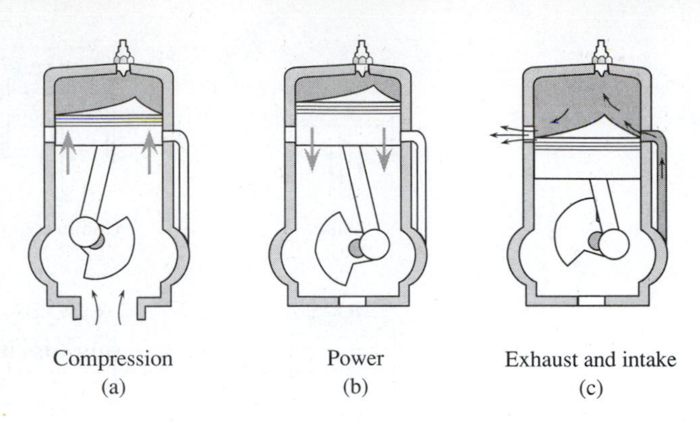

Figure 21–21: *The operation of a two-stroke engine. (a) During compression the air-fuel mixture enters the crankcase. (b) The power stroke. (c) At the end of the power stroke, the exhaust is vented on one side of the cylinder; fuel and air are drawn into the other side.*

power stroke, the ports are exposed (see Figure 21–21). The exhaust exits one set of ports. A fresh fuel mixture is blown into another set of ports. The fuel-air mixture is compressed on the return stroke. The P-V diagram looks much like the one for a four-stroke engine except the exhaust and intake strokes are missing.

The Diesel Engine

If the gas-air mixture in a gasoline engine is compressed too much, the fuel mixture will get hot enough to cause ignition before the piston finishes its compression stroke. This **preignition** limits the compression ratio of gasoline engines. The **compression ratio** is the volume of the cylinder at the beginning of the compression stroke divided by the volume at the end of the stroke (V_2/V_1 on the P-V diagram). Typically, for an automobile engine the compression ratio

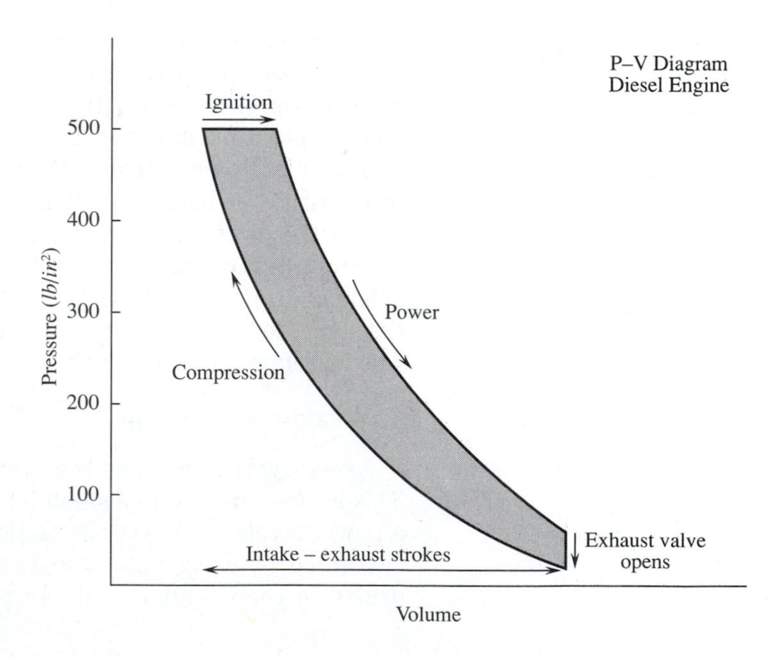

Figure 21–22: *P-V diagram for a high-compression diesel engine.*

is 4.6. The efficiency of the motor could be increased by increasing the compression ratio.

The diesel engine makes larger compression ratios possible by not adding the fuel to the cylinder until the compression stroke is completed. With large compression ratios, an ignition system is not necessary. The heat of compression ignites the fuel. Typically, diesel engines use compression ratios of 15 to 20.

Figure 21–22 shows the P-V diagram for a diesel engine. The piston begins to move backward after the compression stroke is completed. A slow-burning fuel is sprayed into the combustion chamber at a rate that will keep a constant pressure in the cylinder (the horizontal line at the top of the P-V diagram). The power stroke follows.

SUMMARY

Thermodynamics is the study of the relationship between heat and work with an emphasis on application to machines. The first law of thermodynamics states that a quantity of heat (Q) can be used to do work (W) and change the internal energy (ΔU) of a system. Thermodynamic work at constant pressure is defined as $W = P \Delta V$.

The second law of thermodynamics states that the first law is *irreversible*. Work can be totally converted to heat, but heat cannot be totally converted to work. In terms of machines it is stated as follows:

1. It is impossible to construct a continuously operating engine that will take a quantity of heat and convert it entirely to work.
2. It is impossible to construct a continuously operating refrigerator that will move heat from a low-temperature reservoir to a high-temperature reservoir without work being performed.

The second law can be expressed in terms of entropy. Entropy is a measure of disorder. Mathematically, the change of entropy (ΔS) of a thermodynamic system is defined as $\Delta S = \Delta Q/T$. Natural processes proceed in a direction to increase the total entropy of the universe.

The third law of thermodynamics states that it is impossible to cool a system to absolute zero.

The efficiency (*eff*) of an engine is the percentage work performed by the amount of heat (or energy) put into the engine. The most efficient engine possible is a Carnot engine. Its efficiency is expressed in terms of the absolute temperatures of heat-in and heat-out. Real engines will have a lower efficiency.

The performance of a refrigerator is expressed in terms of the coefficient of performance (COP). It is the ratio of heat transferred from the low-temperature area of the refrigerator (evaporator) divided by the amount of work done (or energy input) by the machine. A Carnot refrigerator has the largest possible coefficient of performance. It is expressed in terms of the absolute temperatures of heat taken in and exhaust heat.

KEY TERMS

If you can explain the following terms to a friend or classmate, you understand their meaning. If you cannot explain the terms, you should reread the sections in which they are discussed.

Carnot efficiency
Carnot refrigerator cycle
coefficient of performance (COP)
compression ratio
entropy
internal energy (U)
laws of thermodynamics
Otto cycle
preignition
thermodynamics

EXERCISES

Section 21.1:

1. What is the change of internal energy when
   A. 10 g of ice melt?
   B. 1.2 lb of water change to steam?
   C. 130 g of water are heated by 10°C?
2. A diesel engine burns fuel to perform work.
   A. Identify the quantity of heat going into the engine.
   B. Identify the change of internal energy.

Section 21.2:

3. What is the efficiency of an engine that does no work?
4. An engine burns an amount of fuel that produces 1.74×10^3 Btu. The exhaust heat is 1.43×10^3 Btu. How much work is done?
5. An engine is 20% efficient. What percentage of the energy put into the system shows up as exhaust heat?

Section 21.3:

6. Gas in a cylinder expands at constant pressure. Does the gas do work, or is work being done on the gas? Explain.
7. A volume of gas is compressed at constant pressure. Is work done on the gas, or is the gas doing work? Explain.
8. A gas expands adiabatically. No energy is added to the system to do the work to cause the expansion. Where does the energy producing the expansion come from?
9. Solid carbon dioxide cakes are made by allowing CO_2 to escape from high-pressure tanks. A cloth bag tied over the nozzle of the tank collects the solid CO_2 (see Figure 21–23). Why does this system work?

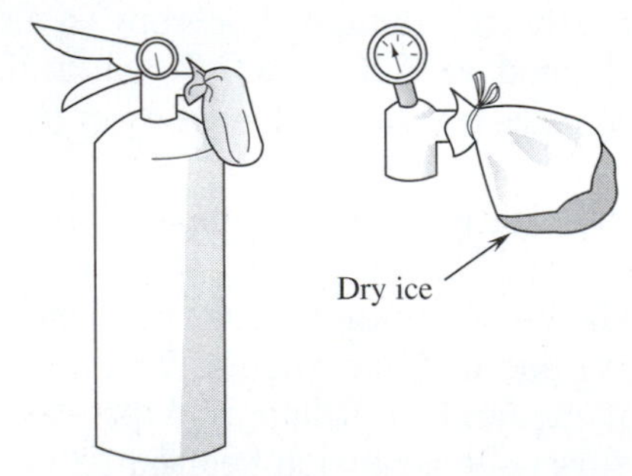

Figure 21–23: Diagram for Exercise 9. A bag is tied over the nozzle of a tank of compressed carbon dioxide. When the valve is opened dry ice forms inside the bag.

Section 21.4:

10. Figure 21–24 shows a P-V diagram for a cylinder in an engine. What area represents:
    A. the work done by expanding hot gases?
    B. the work done on the gas by the piston?
    C. the net work done in one cycle?

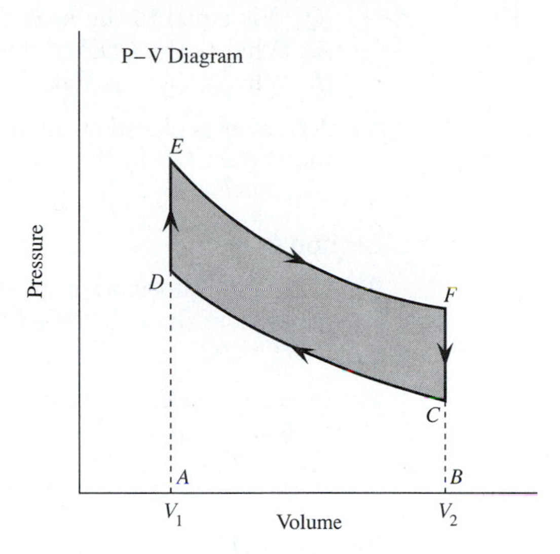

Figure 21–24: *P-V diagram for Exercise 10.*

Section 21.5:

11. In your own words, explain what is meant by an irreversible process.
12. Why is it not thermodynamically possible to build a perpetual motion machine?
13. Why is it not possible to construct a 100% efficient machine?

Section 21.6:

14. Explain the term *entropy* in your own words.
15. Ice cream freezes in an ice-cream maker. The ice cream undergoes a *decrease* in entropy. Why can we say entropy increases in all processes?
16. Explain why the idea that entropy increases in thermodynamic processes leads to the idea that heat flows from a higher temperature to a lower temperature.

Section 21.7:

17. In order for a Carnot engine to be 100% efficient what temperature must the exhaust be?

18. If the exhaust temperature is greater than the input temperature:
    A. what is true about the efficiency of the machine?
    B. what kind of machine do you have?

Section 21.8:

19. An inventor designs a refrigerator in which the exhaust heat (Q_{out}) is equal to the heat absorbed by the evaporator (Q_{in}).
    A. What is the *COP* of this machine?
    B. Why is this thermodynamically impossible?

20. A freezer is placed on an uninsulated back porch. Would it have the same *COP* in the summer as it had in the winter? Why or why not?

Section 21.9:

21. From the information given in Section 21.9 on absorption refrigerators show that the *COP* for an absorption refrigerator is: $COP = (Q_3 + Q_4/Q_1) - 1$.

22. In most situations, the heat delivered by a heat pump is greater than that supplied by resistive heating for the same amount of electrical work. Can you think of some reasons why resistive electric heaters are more commonly used?

Section 21.10:

23. What are the advantages of using a two-cycle engine in cases where only one cylinder is to be used?

24. Why are slow-burning fuels used in diesel engines rather than fast-burning fuels such as gasoline?

PROBLEMS

Section 21.1:

1. A quantity of heat melts 1.30 lb of ice at 32°F. What is the change of internal energy (ΔU) of the ice?

2. An engine burns fuel, creating 1.20×10^3 Btu of heat; 1.05×10^3 Btu of heat is exhausted through the manifold and cooling system. How much work does the engine perform in foot · pounds?

3. An engine performs 1.76×10^4 J of work. The exhaust heat is 2.10×10^4 cal. What amount of heat from fuel went into the engine?

Section 21.2:

4. What is the efficiency of the engine in Problem 2?

5. A natural gas furnace burns 1.00 ft^3 of natural gas while supplying a house with 1100 Btu of heat. What fraction of the input heat of the fuel is delivered to the house? (See Table 20–2.)

6. An engine performs 2.30 kJ of work. The input heat is 3.20 kcal. What is the engine's efficiency?

7. A $\frac{1}{4}$-hp electric motor uses $22\bar{0}$ W of power to operate at the rated level. What is the efficiency of the motor?

Section 21.3:

8. How many foot · pounds of work are done on an air-fuel mixture when it is compressed in a cylinder from 35.3 in^3 to 7.70 in^3 at an effective pressure of 62 lb/in^2?

9. How much work is done by a gas as it expands from 1.00 m^3 to 4.00 m^3 at a constant pressure of 2.30×10^5 Pa?

10. A piston has a stroke of 10.0 cm and a diameter of 10.2 cm. If the effective pressure in the cylinder is 4.6×10^5 Pa, how much work is done on the piston during a power stroke of the piston? (See Figure 21–25).

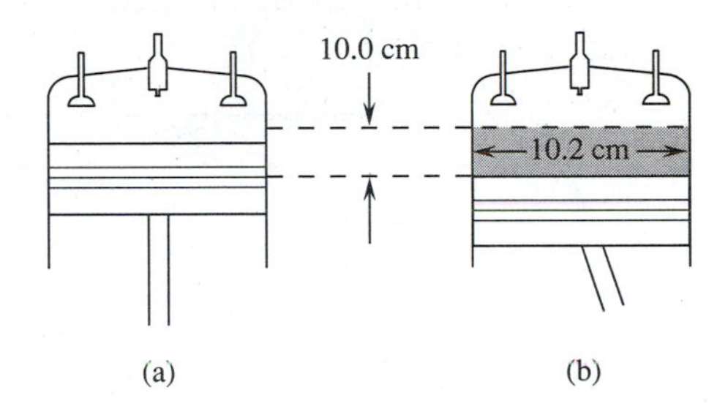

Figure 21–25: Diagram for Problem 10. Determine the work done on the piston during a power stroke.

Section 21.4:

11. Estimate the net work done in foot · pounds for the cyle shown on the P-V diagram in Figure 21–26.

12. According to Figure 21–26:
    A. how much work is done on the air-fuel mixture to compress it?
    B. how much work is done on the piston while the hot burned fuel expands?

13. If the engine shown in Figure 21–26 has an efficiency of 18%, how many British thermal units are supplied by the fuel during one cycle?

14. Estimate the net work done by the engine shown in Figure 21–27.

Section 21.6:

15. Natural gas burns at a temperature of about $12\bar{0}0$°F. What is the increase of entropy when 1 ft^3 of natural gas is burned? (See Table 20–2 for heat of combustion.)

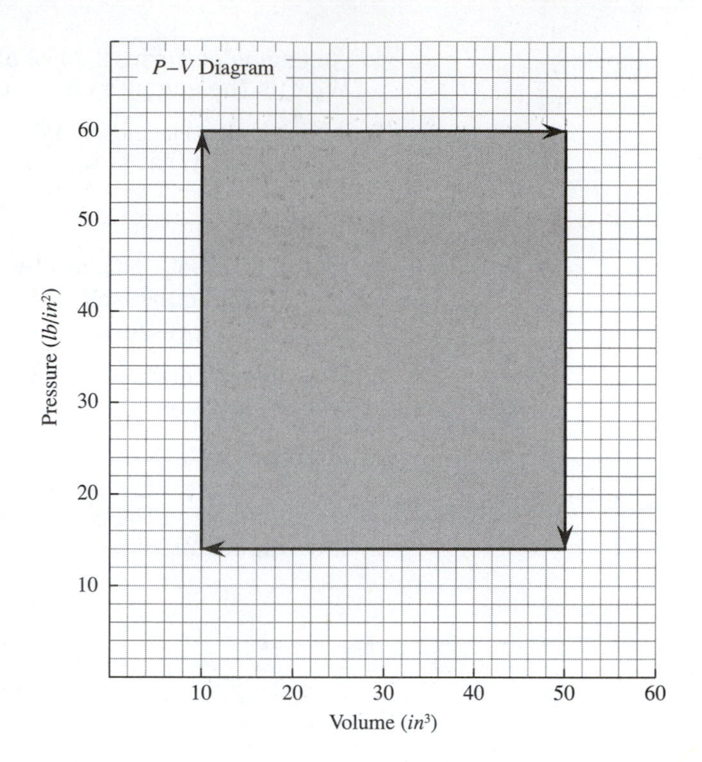

Figure 21–26: *P-V diagram for Problems 11, 12, and 13.*

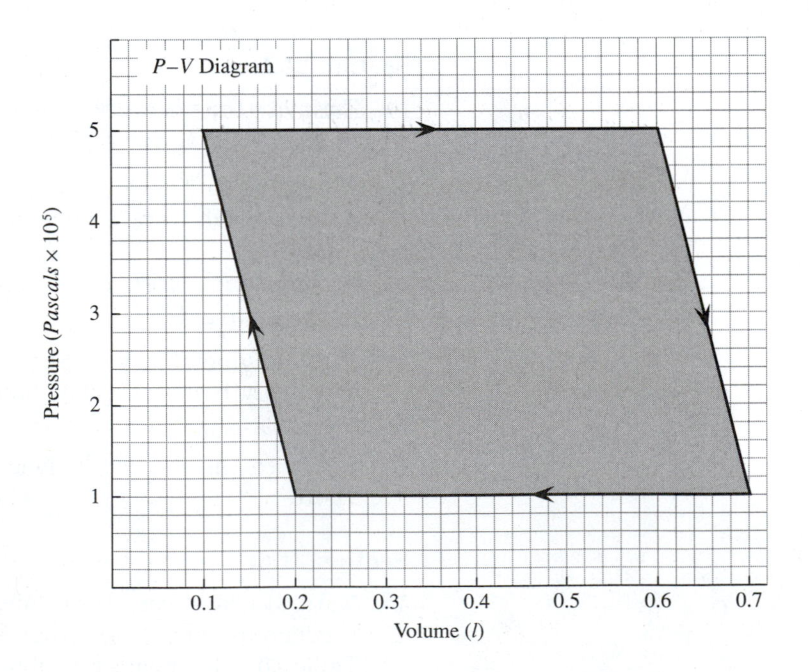

Figure 21–27: *P-V diagram for Problem 14. Determine the work done.*

16. Fifty-six grams of ice melt at 0.0°C. What is the increase of entropy?

17. At 10$\bar{0}$°C, 82.3 g of water is converted to steam. What is the change of entropy of the water?

18. In the freezing compartment of a refrigerator, 0.24 lb of water is changed into ice cubes at 32°F. The lost heat is absorbed by the evaporator coils of the refrigerator at −5°F.
    A. What is the decrease of entropy of the water?
    B. What is the gain of entropy of the freon in the coils?
    C. What is the total change of entropy?

Section 21.7:

19. What is the maximum efficiency for a motor that burns gasoline at 70$\bar{0}$°F and has an exhaust temperature of 15$\bar{0}$°F?

20. Calculate the maximum efficiency of an engine that burns kerosene at 26$\bar{0}$°C and has an exhaust temperature of 5$\bar{0}$°C.

Section 21.8:

21. The compressor in a refrigerator does 3.24 kJ of work while absorbing 2.02 kcal in the evaporator. What is the *COP* for the refrigeration cycle?

22. A freezer maintains a temperature of −15°F in the evaporator and exhausts heat at 7$\bar{0}$°F. What is the maximum *COP* the freezer can have?

Section 21.9:

23. A gas-absorption refrigerator takes in 49.2 Btu from the flame at the generator. It ejects 59.3 Btu through the absorber and 90.9 Btu through the condenser.
    A. How much heat is taken in at the evaporator?
    B. What is the *COP* of this refrigerator?

24. Estimate the maximum rate that heat can be transferred by a heat pump from outside air at −3°C into a room at 23°C using a 300-W compressor.

25. A piston engine has a piston with a stroke of L and a cross-sectional area of A. N is the number of strokes per minute of the piston operating at an effective pressure of P. Show that the horsepower generated by the piston is numerically equal to: $\text{hp} = (\text{P}\ L\ A\ N)/33{,}000$.

26. Use the results of Problem 25 to calculate the horsepower generated by a six-cylinder engine with a 3.825-in bore, a 4.00-in stroke, and an effective pressure of 67.2 lb/in^2 when it is operating at 1700 rpm.

Chapter 22

HEAT TRANSFER

This man is walking on a bed of wood embers with a temperature of 1200°F. Many people try "fire-walking" and report that they feel no pain, suffer no burns. Is fire-walking the triumph of mind over matter? As you read about heat transfer in this chapter, you will discover how Professor Taylor was able to make this trek across the coals smiling. (Photograph courtesy of Professor John Taylor, University of Colorado at Boulder, Department of Physics.)

OBJECTIVES

In this chapter you will learn:

- how heat moves through a solid (conduction)
- about steady-state heat flow
- to calculate the amount of heat conducted through a surface under steady-state conditions
- how to use *R* values and *U* values to calculate heat flow
- how heat is transferred by the movement of masses of a fluid (convection)
- to calculate convective heat flow for some simple cases
- how heat is transferred by radiant energy
- about the effectiveness of surfaces in absorbing or emitting radiation (emissivity)
- how the peak wavelength of a radiator changes with temperature (Wein's law)
- why solar collectors work

Time Line for Chapter 22

1792	Pierre Prevost shows that radiation is exchanged between objects.
1804	John Leslie shows that radiant heat has the same properties as light.
1879	Josef Stefan finds that the total radiation from a body is proportional to the fourth power of the absolute temperature.
1893	Wilhelm Wien finds that the peak wavelength of radiation from a black body is inversely proportional to the absolute temperature.
1962	Henry Hess identifies sea-floor spreading with convection currents in the earth's mantle.

Trying to capture and store useful heat is like trying to carry water in a leaky bucket. Both will eventually be lost from the container. A hot mug of coffee placed on a table will eventually cool down to room temperature. Ice cubes stored in a thermos jug will slowly melt. The composition and structure of the jug determine how fast heat will leak in. An oil furnace heats a house in the winter. Heat leaks out through the windows, the walls, and the roof. Cold air infiltrating through cracks must be heated to maintain a comfortable temperature. Insulation, weather stripping, thermal curtains, and caulking can prevent some of the leaks, but heat will still be lost, although at a slower rate.

In this chapter we will look at the ways in which heat moves from a warmer to a cooler location. We will also discuss the greenhouse effect, a way to trap solar energy.

22.1 CONDUCTION

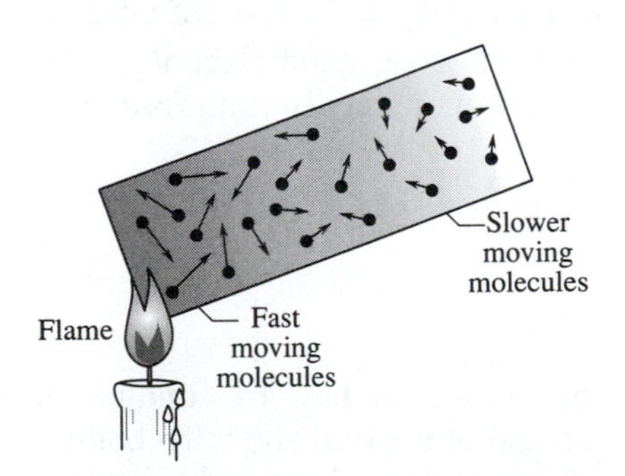

Figure 22–1: *The kinetic energy of heated molecules is transferred to the cool end of the rod by molecular collision.*

When one end of a metal rod is placed in a flame, heat moves slowly along the rod from the hot end toward the cool end. If we place rings of paraffin along the rod we can see the progress of the heat flow. The top of the rings melt and the pieces of wax fall, beginning with the one nearest the flame and moving upward. We can use a molecular model to describe what is happening (see Figure 22–1).

Remember, the higher the temperature of an object, the more energy individual atoms contain. The molecules on the end of the rod in the flame have much more kinetic energy than the molecules on the other end. They have higher kinetic energies than their neighbors. They vibrate rapidly, colliding with the atoms next to them. Kinetic energy is transferred to the slower moving neighbors. These in turn start to vibrate faster, hitting slower moving atoms on the other side. This process proceeds on down the rod. Eventually, some

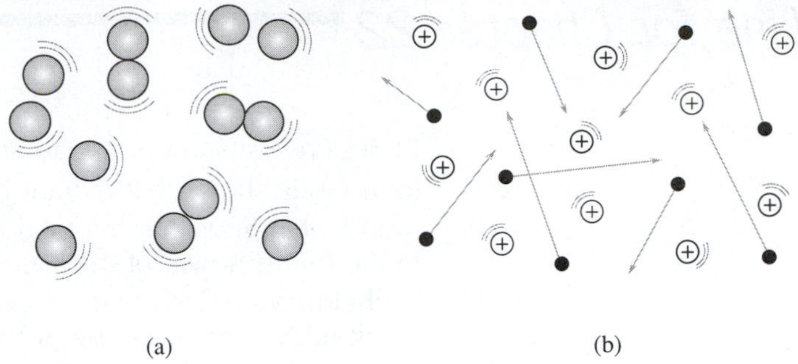

(a)

(b)

Figure 22–2: (a) *Nonmetals transfer kinetic energy when their particles collide with each other. The particles can collide with only their nearest neighbors.* (b) *In a metal, free electrons can strike metal ions farther away than nearest neighbors.*

of the kinetic energy is transferred the full length of the rod, heating up atoms on the cool end. This process is called thermal conduction.

Thermal conduction *is a process in which thermal energy is transferred in a substance from particle to particle by collision.*

All materials conduct heat. Some are better thermal conductors than others. Porous polymer material such as styrofoam, vermiculite, and polystyrene are poor conductors. Ceramics, such as glass, stone, and concrete, are moderately good thermal conductors. Metals, on the other hand, are excellent conductors of heat.

The particles making up a solid are frozen into fixed positions, making a lattice structure (see Figure 22–2a). They can only vibrate around their assigned site, bumping into their nearest neighbor. In most materials, the electrons are held by chemical bonds and are attached to specific atoms making up the lattice. The electrons cannot move around.

Metals are different (see Figure 22–2b). For each atom, one or more negatively charged electrons are free to move around a lattice of positive metal ions. These free electrons can strike atoms farther away than a nearest neighbor. This mobility speeds up the transfer of heat. These same free electrons participate in the transfer of electricity through a metal. As a general rule, good electrical conductors are good thermal conductors because they contain free electrons.

22.2 STEADY HEAT FLOW

Let us design an experiment to measure heat flow by conduction (see Figure 22–3). We will heat one end of a metal rod with boiling water. The other end will be placed in a jacket with running water. If we measure the temperature difference of water flowing into the jacket and the water flowing out we can find the amount of heat the

FIRE-WALKING

Fire-walking is a skill that shows the difference between temperature and heat. Fire walkers cross a bed of hot coals on their bare feet without getting burned. The coals can be as hot as 1500°F, but little heat is transferred to the bare skin. Heat, not temperature, causes burns.

The amount of heat transferred to the feet of a fire walker depends on a number of factors. Usually wood coals are used in fire-walking. Heat conduction of wood coals is rather low compared with that of metals. Consequently, only heat near the surface of the ember is transferred to the feet. In addition, the coals have a small heat capacity. Only a small amount of heat is contained near the surface of the embers even though the temperature is high. These same factors make a cold rug seem comfortably warm compared with a metal surface at the same temperature.

Fire walkers do not stand still. Their bare feet are in contact with the embers for a fairly short period of time. The total heat conducted to the walker's feet is proportional to the period of contact.

(Photograph courtesy of Professor John Taylor, University of Colorado at Boulder, Department of Physics.)

Fire-walking is explained by physics. It is not a demonstration of mind over matter.

water will pick up from the rod. We will cover the rod with an insulating material so that the heat lost to the air will be negligible. Thermometers will measure the temperature of the rod at different places along its length.

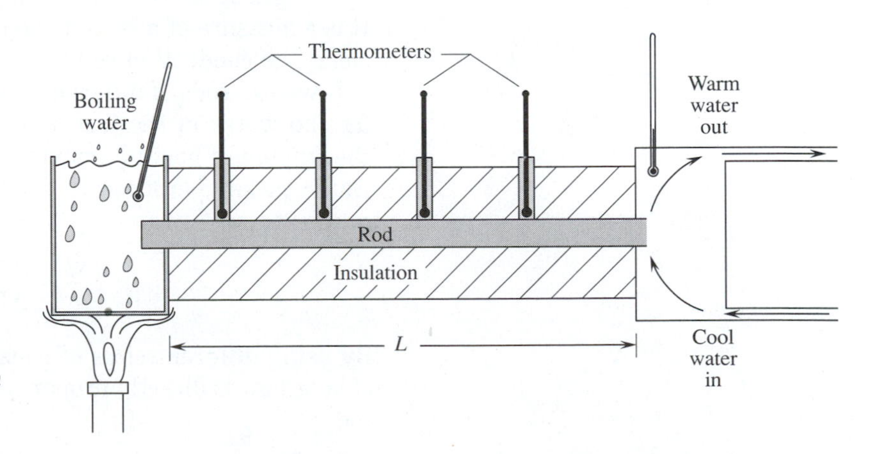

Figure 22–3: *An apparatus for measuring the flow of heat by conduction through a rod. Boiling water heats one end of the rod. Water cools the other end.*

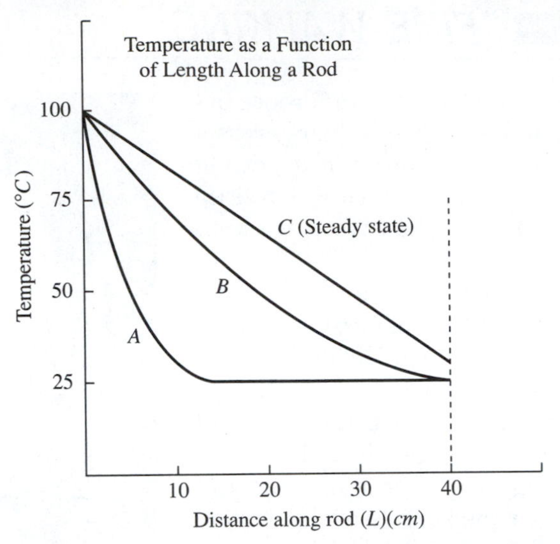

Figure 22–4: *Temperature versus length along the rod in Figure 22–3. At first, heat transferred by conduction does not immediately reach the cool end (curve A). A little later, heat begins to flow out the cool end (curve B). Much later, heat flows through the rod at a steady rate (curve C).*

When we first start the apparatus, the end of the rod in the boiling water will heat up quickly, but it will take time for thermal conduction to carry the heat very far along the rod (see Figure 22–4, curve A). A little later, some thermal energy has nearly reached the other end of the rod, but the water in the jacket has not picked up any heat (Figure 22–4, curve B). If we let the apparatus run for a long time, we will get the temperature versus length profile shown by curve C. As long as the temperatures at the two ends of the rod do not change, the slope of the straight line will remain constant. There will be a steady flow of heat from the hot end of the rod to the cool end. The water in the jacket will collect the same amount of heat for each minute of operation. This situation is called **steady-state heat flow.**

The magnitude of the slope of curve C is $\Delta T/\Delta L$, where ΔT is the temperature difference of the ends of the rod ($T_h - T_c$) and ΔL is the length of the rod. This slope is called the **temperature gradient.** It is a measure of how fast the temperature changes per unit length along the conduction rod.

If we use rods of different lengths and try different temperatures for the water in the jacket on the cool end, we will discover that the amount of heat (Q) moving from the boiling water to the cooling water jacket per time (t) is directly proportional to the thermal gradient.

$$\frac{Q}{t} \propto \frac{\Delta T}{\Delta L}$$

By using different sizes of rods we will also discover that the rate of heat flow is directly proportional to the cross-sectional area of the rod.

$$\frac{Q}{t} \propto A$$

These proportionality relationships can be combined into one equation where k is a proportionality constant. We will use H to represent the time rate of heat flow ($\Delta Q/\Delta t$).

$$H = \frac{\Delta Q}{\Delta t} = k\, A\, \frac{\Delta T}{\Delta L} \qquad \textbf{(Eq. 22–1)}$$

Rods with the same dimensions and the same thermal gradient, but made of different materials, conduct heat at different rates. The proportionality constant (k) called **thermal conductivity,** is a material constant. We can find the values of k for different materials by experiment.

Table 22–1 shows the thermal conductivities for some selected materials. Notice that in British units the cross-sectional area is in square feet, while the distance in the temperature gradient is given in inches.

Table 22–1: Thermal conductivity (k) for selected materials near room temperature.

	kcal/(m • s • °C)	(Btu • in)/(ft² • h • °F)
Metals		
Aluminum	5.0×10^{-2}	1480
Brass	2.6×10^{-2}	750
Copper	9.2×10^{-2}	2640
Steel (iron)	1.1×10^{-2}	320
Gases		
Air	5.5×10^{-6}	0.16
Helium	3.3×10^{-5}	0.95
Nitrogen	5.6×10^{-6}	0.16
Other Materials		
Asbestos	1.4×10^{-4}	4.2
Brick	1.7×10^{-4}	5.0
Concrete	4.1×10^{-4}	12
Fiberglass (loose batt)	9.0×10^{-5}	2.4
Glass	2.5×10^{-4}	7.3
Gypsum board	8.4×10^{-5}	2.4
Ice	5.2×10^{-4}	15
Styrofoam	2.4×10^{-6}	0.07
Water	1.4×10^{-4}	4.2
Wood (oak)	4.0×10^{-5}	1.1
Wood (pine)	2.9×10^{-5}	0.84

☐ EXAMPLE PROBLEM 22–1: THE CONCRETE WALL

A concrete wall is 6.0 in thick, $4\bar{0}$ ft long, and 8.0 ft tall. The temperature is 65°F on one side of the wall and 29°F on the other side. Find:

A. The temperature gradient.

B. The rate of heat flow (H).

C. The number of British thermal units passing through the wall in one day.

■ SOLUTION

A. The temperature gradient is:

$$\frac{\Delta T}{\Delta L} = \frac{(65°\text{F} - 29°\text{F})}{6.0 \text{ in}}$$

$$\frac{\Delta T}{\Delta L} = 6.0°\text{F/in}$$

B. The heat flow rate is found using Equation 22–1.

$$H = k\,A\,\frac{\Delta T}{\Delta L}$$

$$H = (12 \text{ Btu} \cdot \text{in})/(\text{ft}^2 \cdot \text{h} \cdot °\text{F}) \times (4\bar{0} \text{ ft} \times 8.0 \text{ ft}) \times 6.0°\text{F/in}$$

$$H = 2.3 \times 10^4 \text{ Btu/h}$$

C. Btu per day:

$$H = \frac{\Delta Q}{\Delta t} \qquad \text{or} \qquad Q = H \times \Delta t$$

$$Q = (2.3 \times 10^4 \text{ Btu/h}) \times 24 \text{ h}$$

$$Q = 5.5 \times 10^5 \text{ Btu}$$

☐ EXAMPLE PROBLEM 22–2: A BRASS BAR

The temperature gradient through a brass bar is measured to be 15°C/m. If the bar has a 10.0-cm diameter, how much heat will flow through the rod in 1 h? Assume no losses along the length of the rod.

■ SOLUTION

$$area = A = \frac{\pi D^2}{4} = \frac{\pi\,(0.10 \text{ m})^2}{4} = 7.85 \times 10^{-3} \text{ m}^2$$

$$time = t = 1\text{ h} \times 60\,\frac{\text{min}}{\text{h}} \times 60\,\frac{\text{s}}{\text{min}} = 3.6 \times 10^3\text{ s}$$

$$Q = (2.6 \times 10^{-2}\text{ kcal/m} \cdot \text{s} \cdot {}^\circ\text{C}) \times (7.85 \times 10^{-3}\text{ m}^2) \times (15^\circ\text{C/m}) \times (3.6 \times 10^3\text{ s})$$

$$Q = 11\text{ kcal}$$

Heat is thermal energy. Because of the conservation of energy law we would expect the amount of heat going in one end of a thermal conductor to equal the energy coming out the other end as long as no heat is lost in between. This was the assumption made when we developed Equation 22–1. We can also use this assumption to calculate the heat flow through a wall made up of more than one material. For a steady-state thermal system

$$H(\text{in}) = H(\text{out})$$

□ **EXAMPLE PROBLEM 22–3: THE INSULATED WALL**

The wall in Example Problem 22–1 has been insulated with 1.0 in of styrofoam placed on the outside of the wall. Assume all the other conditions are the same as before.

A. Find the temperature of the wall at the surface of contact between the two materials.

B. Find the temperature gradient for each material.

C. Find the heat passing through the insulated wall in one day and compare this with the heat lost through the uninsulated wall.

■ ***SOLUTION***

A. The same amount of heat flows through the two materials.

$$H(\text{concrete}) = H(\text{styrofoam})$$

Let T_a be the temperature at the interface of the materials.

$$k_1\,\frac{(T_1 - T_a)}{L_1} = k_2\,\frac{(T_a - T_2)}{L_2}$$

$$(12\text{ Btu} \cdot \text{in})/(\text{ft}^2 \cdot \text{h} \cdot {}^\circ\text{F}) \times \frac{(65^\circ\text{F} - T_a)}{6.0\text{ in}}$$

$$= 0.07\,(\text{Btu} \cdot \text{in})/(\text{ft}^2 \cdot \text{h} \cdot {}^\circ\text{F}) \times \frac{(T_a - 29^\circ\text{F})}{1.0\text{ in}} = 2\,(65^\circ\text{F}) - 2T_a$$

$$= 0.07\,T_a - 0.07\,(29^\circ\text{F})$$

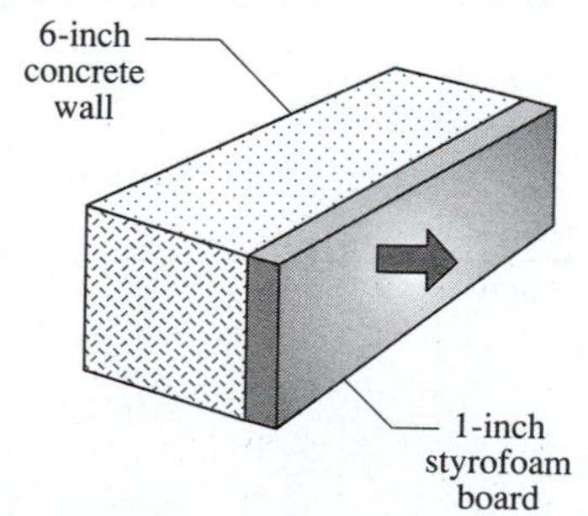

Figure 22–5: Heat flows through an insulated cement wall.

$$2.07\ T_a = 132$$

$$T_a = 63.8°F$$

B. Now we can find the thermal gradients for the two materials. Concrete:

$$\frac{\Delta T_1}{\Delta L_1} = \frac{(65°F - 63.8°F)}{6.0 \text{ in}}$$

$$\frac{\Delta T_1}{\Delta L_1} = 0.20°F/\text{in}$$

Styrofoam:

$$\frac{\Delta T_2}{\Delta L_2} = \frac{(63.8°F - 29°F)}{1.0 \text{ in}}$$

$$\frac{\Delta T_2}{\Delta L_2} = 34.8°F/\text{in}$$

C. Now that we have a temperature gradient we can calculate the heat flow. Find the flow through the styrofoam.

$$H = (0.07\ (\text{Btu} \cdot \text{in})/(\text{ft}^2 \cdot \text{h} \cdot °F) \times (320\ \text{ft}^2) \times (34.8°F/\text{in})$$

$$H = 780\ \text{Btu/h}$$

In one day, the heat flowing through the wall is:

$$Q = H \times t = 780\ \text{Btu/h} \times 24\ \text{h}$$

$$Q = 1.87 \times 10^4\ \text{Btu (insulated wall)}$$

For the uninsulated wall, 5.5×10^5 Btu's flow through the wall in a day.

$$\frac{Q(\text{insulated})}{Q(\text{uninsulated})} = \frac{1}{30}$$

22.3 *U* VALUES AND *R* VALUES

Example Problem 22–3 can be solved in a simpler manner. We really do not need to know the temperature at the interface of the styrofoam and the concrete. The equations for heat flow through the two materials can be added to eliminate the intermediate temperature. (Warning: We are going to make things more complicated before we make them easier!)

For the concrete:

$$H = k_1 \frac{A\,(T_1 - T_a)}{\Delta L_1} \quad \text{or} \quad \frac{H\,\Delta L_1}{k_1 A} = T_1 - T_a$$

For the styrofoam:

$$H = k_2 \frac{A\,(T_a - T_2)}{\Delta L_2} \quad \text{or} \quad \frac{H\,\Delta L_2}{k_2 A} = T_a - T_2$$

Add the equations:

$$\frac{H}{A}\left(\frac{\Delta L_1}{k_1} + \frac{\Delta L_2}{k_2}\right) = T_1 - T_2$$

Now, we will introduce a **conduction coefficient (*U*).** The conduction coefficient (sometimes called the ***U* value**) is defined as:

$$\frac{1}{U_i} = \frac{\Delta L_i}{k_i} \quad \text{or} \quad U_i = \frac{k_i}{\Delta L_i}$$

where $i \neq 1, 2, 3 \ldots$ The conduction coefficients for the materials in Example Problem 22–3 are for concrete:

$$U_1 = \frac{12 \text{ Btu} \cdot \text{in}/(\text{ft}^2 \cdot \text{h} \cdot {}^\circ\text{F})}{6.0 \text{ in}}$$

$$U_1 = 2.0 \text{ Btu}/(\text{ft}^2 \cdot \text{h} \cdot {}^\circ\text{F})$$

and for styrofoam:

$$U_2 = \frac{0.07 \text{ Btu} \cdot \text{in} \cdot (\text{ft}^2 \cdot \text{h} \cdot {}^\circ\text{F})}{1.0 \text{ in}}$$

$$U_2 = 0.07 \text{ Btu}/(\text{ft}^2 \cdot \text{h} \cdot {}^\circ\text{F})$$

Look at the equation we got by adding the heat flow through the two parts of the wall to eliminate T_a. The factor in the parentheses is related to an overall **transmission coefficient.**

$$\frac{1}{U} = \sum \frac{1}{U_1} \qquad \textbf{(Eq. 22–2)}$$

The overall transmission coefficient for the wall is:

$$\frac{1}{U} = \frac{1}{U_1} + \frac{1}{U_2}$$

$$\frac{1}{U} = \frac{1}{2.0\ \text{Btu/(ft}^2 \cdot \text{h} \cdot {}^\circ\text{F)}} + \frac{1}{0.07\ \text{Btu/(ft}^2 \cdot \text{h} \cdot {}^\circ\text{F)}}$$

$$\frac{1}{U} = 14.8\ (\text{ft}^2 \cdot \text{h} \cdot {}^\circ\text{F)/Btu}$$

$$U = 0.0676\ \text{Btu/(ft}^2 \cdot \text{h} \cdot {}^\circ\text{F)}$$

Again referring to the equation for the complex wall:

$$\frac{H}{A}\left(\frac{1}{U}\right) = (T_1 - T_2)$$

In general if the overall coefficient of transmission is known, the heat flow equation can be written as:

$$H = U\,A\,(T_1 - T_2) \qquad \textbf{(Eq. 22–3)}$$

The rate of heat flow through the insulated styrofoam wall is:

$$H = 0.0676\ \text{Btu/(ft}^2 \cdot \text{h} \cdot {}^\circ\text{F)} \times 320\ \text{ft}^2\ (65^\circ\text{F} - 29^\circ\text{F})$$

$$H = 780\ \text{Btu/h}$$

This answer agrees with Example Problem 22–3, but it seems to be a rather complicated way to solve the problem. Now let us make things simple.

In the building trades, materials come in standard sizes and thicknesses. Usually, R values are given for the materials. The R stands for **thermal resistance.** An ***R* value** is the reciprocal of the transmission coefficient.

$$R = \frac{1}{U} \qquad \textbf{(Eq. 22–4)}$$

In order to find the overall transmission coefficient simply add up the R values and take the reciprocal.

$$R(\text{overall}) = R_1 + R_2 \ldots$$

$$U(\text{overall}) = \frac{1}{R(\text{overall})}$$

Table 22–2 lists the R values for some common building materials.

Table 22–2: Thermal resistance (*R* value) for common building materials.

MATERIAL	FOR LISTED THICKNESS ($ft^2 \cdot h \cdot °F/Btu$)	PER INCH THICKNESS ($ft^2 \cdot h \cdot °F/Btu \cdot in$)
Brick	—	0.20
Cement block		
8 in	1.1	—
12 in	1.3	—
Felt (vapor-permeable)	0.06	—
Fiberglass (3-½-in batt)	10.5	3.0
Glass		
insulated double	1.6	—
single-pane	0.4	—
Gypsum board	0.31	0.90
Plywood (⅜-in)	0.47	1.25
Stucco	—	0.20
Styrofoam	—	14.3
Wood clapboards	0.9	—

□ **EXAMPLE PROBLEM 22–4: *R* VALUE FOR A WALL**

Find the overall *R* value and transmission coefficient for a house wall consisting of:

A. gypsum board (⅜-in),
B. fiberglass insulation (3½-in batt),
C. plywood (⅜-in),
D. vapor-permeable felt, and
E. brick (3-in).

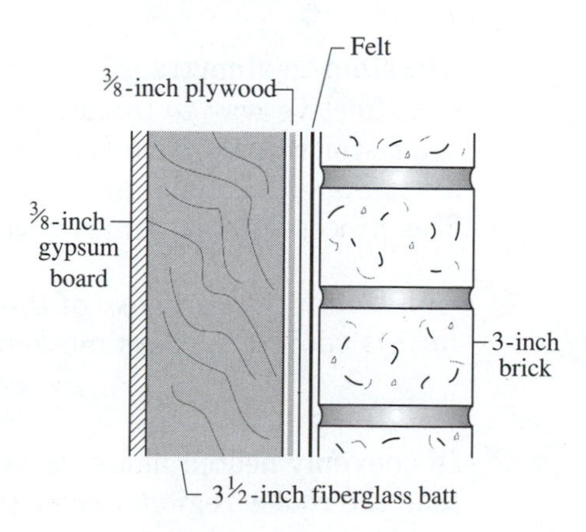

Figure 22–6: The composite R value can be found for a wall made of several building materials.

■ *SOLUTION*

The R value for the brick is:

$$R(\text{brick}) = 0.20\ (\text{ft}^2 \cdot \text{h} \cdot {}^\circ\text{F})/(\text{Btu} \cdot \text{in}) \times 3.0\ \text{in} = 0.60\ (\text{ft}^2 \cdot \text{h} \cdot {}^\circ\text{F})/\text{Btu}$$

The rest of the R values can be read directly from Table 22–2.

$$R = R(\text{gypsum}) + R(\text{fiberglass}) + R(\text{plywood}) + R(\text{felt}) + R(\text{brick})$$

$$R = 0.31 + 10.5 + 0.47 + 0.06 + 0.60\ (\text{ft}^2 \cdot \text{h} \cdot {}^\circ\text{F})/\text{Btu}$$

$$R = 11.9\ (\text{ft}^2 \cdot \text{h} \cdot {}^\circ\text{F})/\text{Btu}$$

$$U = \frac{1}{R} = \frac{1\ \text{Btu}}{11.9 \cdot (\text{ft}^2 \cdot \text{h} \cdot {}^\circ\text{F})}$$

$$U = 0.084\ \text{Btu}/(\text{ft}^2 \cdot \text{h} \cdot {}^\circ\text{F})$$

□ **EXAMPLE PROBLEM 22–5: HEAT FLOW THROUGH A HOUSE WALL**

How much heat is lost in a day through a 230-ft² section of wall with an overall R value of 14 if the effective temperature difference between the inside and outside faces of the wall is 47°F?

■ *SOLUTION*

$$U = \frac{1}{\text{R}} = \frac{1}{14} = 0.0714\ \text{Btu}/(\text{ft}^2 \cdot \text{h} \cdot {}^\circ\text{F})$$

$$Q = H \times t = (U\,A\,\Delta\,T) \times t$$

$$Q = 0.0714\ \text{Btu}/(\text{ft}^2 \cdot \text{h} \cdot {}^\circ\text{F}) \times 230\ \text{ft}^2 \times 47^\circ\text{F} \times 24\ \text{h}$$

$$Q = 1.85 \times 10^4\ \text{Btu}$$

22.4 CONVECTION

The atom-to-atom transfer of thermal energy by conduction is a slow but effective way to transfer heat through a solid. In fluids, there is a faster way to move heat from one place to another: simply take a volume of hot fluid and move it along with the heat it contains. This process of heat transfer is known as convection.

Convection *is a process of thermal energy transport in which heat is carried by mass movement within a fluid.*

Natural Convection

In unevenly heated fluids, hotter portions of the fluid expand more than the cooler regions. The expansion causes a decreased density.

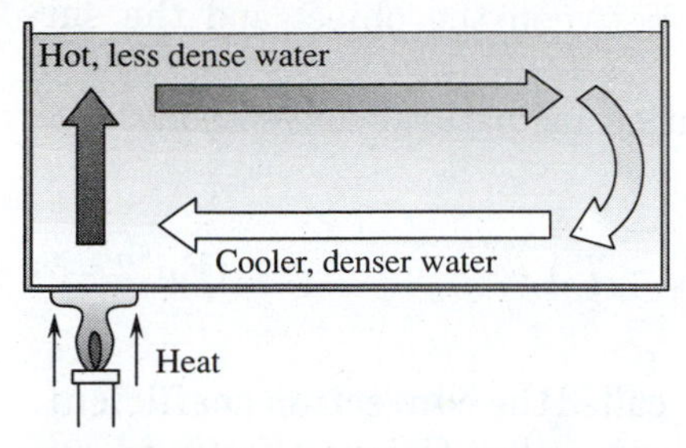

Figure 22–7: *Buoyant forces lift hot, less dense fluid. The rising water pushes fluid aside at the top of the column. Cooler, more dense water moves in to occupy the space left by the rising water. A convection current is formed.*

Buoyant forces push the hotter, less dense fluid upward. A **convection cell** is built up (see Figure 22–7).

Sea-land breezes are caused by convection. On a sunny day, the warm air rising above a hot beach is replaced by heavier cool air from over the water. At night the land cools faster and becomes cooler than the water. The breeze reverses. Air rises above the warmer water and is replaced by cooler air from over the land.

Thermosiphoning solar hot-water heaters use natural convection for their operation. Figure 22–8 shows a simple design for this type of heater. Sunlight warms the top of the collector. Natural convection moves the heated water upward to the storage area. Cooler water is drawn in below the baffle to complete the cycle.

Forced Convection

Natural convection is too slow to adequately heat and ventilate a large building or to supply fresh air to an underground mining operation. Pumps and fans are used to force fluids through pipes and ventilation ducts more rapidly. When pumps or fans are used to aid the movement of a fluid we have forced convection.

Convection Equation

The heat flow by forced convection can be complicated. Fortunately there is a simple relationship that can be used for natural convection for many technical applications.

A solid material of temperature T_s and surface area A is placed in air at temperature T_a (see Figure 22–9). An air film moves along the surface of the object, picking up heat. The film moves back to the body of air, mixes with it, and deposits the heat it picked up.

Experiments indicate that the thickness of the film has little bearing on the rate at which heat flows from the solid surface into the body of air. The rate of heat flow depends only on the area of contact

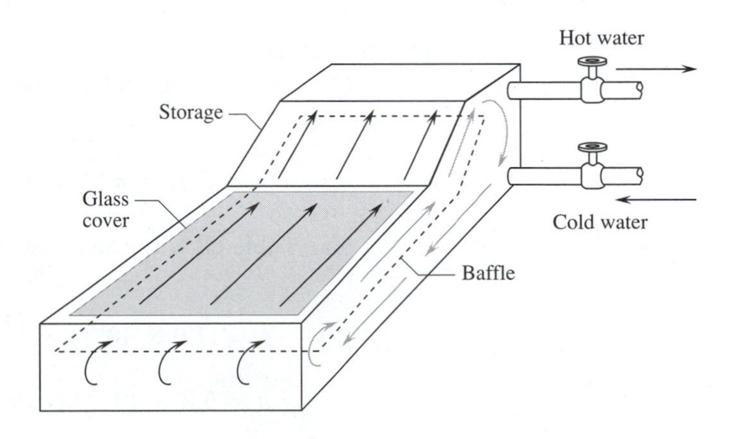

Figure 22–8: *A thermosiphoning solar collector is used to heat water. The system operates on natural convection currents.*

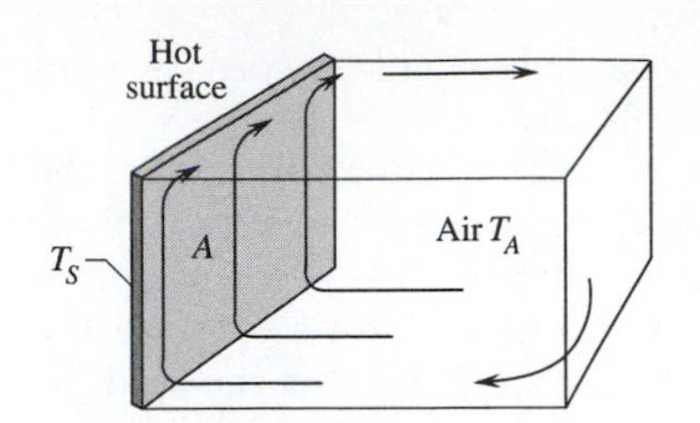

Figure 22–9: *Air at temperature T_a picks up heat from an air film moving across a surface with a temperature T_s.*

and the temperature difference between the object and the surrounding air.

We can write a heat flow equation for natural convection.

$$H = \frac{Q}{t} = h\,A\,(T_s - T_a) \qquad \textbf{(Eq. 22–5)}$$

The proportionality constant (h) is called the **convection coefficient.** It is slightly temperature-dependent. Other factors affecting h are the roughness of the surface, the viscosity of the fluid, and the orientation of the surface. Rough surfaces will impede the flow of the air film and will reduce the rate of heat transfer. Table 22–3 lists convection coefficients for some common geometries.

Table 22–3: Convection coefficient (*h*) for still air and smooth surfaces.

SURFACE	kcal/(m² · s · °C)
Vertical plate	$4.24 \times 10^{-4}\,(\Delta T)^{1/4}$
Horizontal plate	
facing up	$5.95 \times 10^{-4}\,(\Delta T)^{1/4}$
facing down	$3.14 \times 10^{-4}\,(\Delta T)^{1/4}$
Vertical or horizontal	
Pipe of diameter D	$1.0 \times 10^{-3}\,(\Delta T/D)^{1/4}$

□ **EXAMPLE PROBLEM 22–6: CONVECTION FLOW AROUND A STEAM PIPE**

A vertical steam pipe has a diameter of $1\bar{0}$ cm and a length of 4.0 m. If the surface temperature of the pipe is 95°C and the room temperature is 23°C, how much heat per second is lost by the pipe by convection?

■ ***SOLUTION***

First find the area of the pipe. The area is the circumference times length.

$$A = \pi D L = \pi\,(0.10\text{ m})\,(4.0\text{ m}) = 1.26\text{ m}^2$$

Use Table 22–3 to calculate the convection coefficient.

$$h = (1.0 \times 10^{-3}) \times \left(\frac{95°\text{C} - 23°\text{C}}{0.10\text{ m}}\right)^{1/4}$$

$$h = 5.2 \times 10^{-3}\text{ kcal/m}^2 \cdot \text{s} \cdot °\text{C}$$

The rate of heat flow is:

$$H = h\,A\,\Delta\,T$$

$$H = (5.2 \times 10^{-3}) \times (1.26\ \text{m}^2) \times 72^\circ\text{C}$$

$$H = 0.47\ \text{kcal/s}$$

22.5 RADIATION

Radio waves, microwaves, infrared waves, visible light, ultraviolet waves, X rays, and gamma rays are all **electromagnetic radiation.** They interact differently with materials and can come from different sources, but they are pretty much the same thing. They all travel at the speed of light. The major difference among these kinds of electromagnetic waves is size of the wavelengths. Radio waves have the longest wavelength. AM radio waves can have wavelengths that are miles long. Microwaves are around a centimeter in length. Visible light comes in waves a little shorter than the thickness of a soap-bubble film. X rays are about the size of atoms. Gamma rays have wavelengths that are even smaller.

The term *radiation* refers to energy transfer by electromagnetic waves. When heat is transferred by conduction or convection some medium must be present. Either atoms must collide with other atoms or a fluid must be available to circulate. Radiation is different; electromagnetic radiation can pass through a vacuum. No atoms are needed. No fluid is needed. Most of the energy the earth receives from the sun is in the form of electromagnetic radiation.

Thermal radiation *is a process in which heat is transferred from one place to another by electromagnetic waves.*

The rate at which radiant heat is gained or lost by something depends on several factors. These factors are discussed below.

Surface Area

Like other heat transfer processes the flow of heat is directly proportional to surface area (A). The greater surface area an object has, the more rapidly it will transfer heat by radiation.

Absolute Temperature

Temperatures must be expressed in Kelvin or Rankine degrees. The gain or loss of radiant heat depends on the *absolute* temperature. The transfer of energy is proportional to the fourth power of the temperature (T^4).

The Properties of the Surface: Emissivity

Different surfaces absorb or radiate thermal energy with varying efficiencies. In general, a dull black surface is a better emitter of radiation than a shiny silver one. It will also be a better absorber of radiant energy. There is an interesting relationship in the way a surface absorbs and emits radiation.

For any particular surface the efficiency of absorption of incident radiation is the same as the efficiency of emission. The best possible absorber (or emitter) is a surface that absorbs all of the incident radiation striking it. Such a surface is called a **black body.** A black body is a perfect absorber or emitter of radiant energy.

The effectiveness of the absorption of a surface is indicated by comparing it with a black body under the same conditions. The ratio of the radiation absorbed by the surface divided by the radiation absorbed by a black body is called **emissivity** (ϵ).

$$\epsilon = \frac{\textit{radiation absorbed by material}}{\textit{radiation absorbed by black body}}$$

A perfect absorber would have an emissivity of 1.00. A perfect reflector (no radiation absorbed) would have an emissivity of 0.00. Table 22–4 lists the emissivities of some common materials.

Radiation comes in a variety of wavelengths. A good absorber must absorb a wide range of wavelengths, not just visible light. Sometimes a surface that reflects visible light may be a better overall absorber than a darker surface. Compare the emissivity of fresh snow and graphite on Table 22–4. Appearances can be deceiving.

Table 22–4: Emissivity (ϵ) of some common materials.

SURFACE	EMISSIVITY
Aluminum foil	0.05
Black tar paper	0.93
Concrete	0.88
Dry sand	0.90
Flat black paint	0.88
Fresh snow	0.82
Galvanized steel	0.13
Granite	0.44
Graphite	0.41
Red brick	0.92
Water	0.96
White paint	0.91
White plaster	0.91

Temperature of the Surroundings

At the same time that an object is emitting radiation it is absorbing radiation from its surroundings. Emission depends on the temperature of the object's surface (T); absorption depends on the temperature of the surroundings (T_s). The net gain or loss of heat is the difference between these two processes.

Radiant Heat Equation

All the factors affecting heat transfer can be put into a proportionality relationship. The heat radiated by a surface (A) is:

$$H(\text{loss}) = \frac{Q}{t} \propto (-\epsilon\, A\, T^4)$$

We can convert this into an equation by introducing a proportionality constant (σ). This constant is known as the **Stefan-Boltzmann constant.** It is a true physical constant. It is independent of material and geometry of the radiating object.

$$\sigma = 5.67 \times 10^{-8}\,\text{J/m}^2 \cdot \text{s} \cdot \text{K}^4$$

$$H(\text{loss}) = -(\sigma\, \epsilon\, A\, T^4)$$

The heat gained by the surface from radiation from the surroundings striking it is:

$$H(\text{gain}) = +(\sigma\, \epsilon\, A\, T_s^{\,4})$$

The net heat gained or lost is the difference between the heat absorbed and the heat emitted by the surface.

$$H(\text{net}) = \sigma\, \epsilon\, A\, (T_s^{\,4} - T^4) \qquad \textbf{(Eq. 22–6)}$$

□ **EXAMPLE PROBLEM 22–7: RADIATION HEAT LOSS OF STEAM PIPE**

Compare the radiation heat loss with the natural convection heat loss of the steam pipe in Example Problem 22–6. Assume an emissivity of 0.80.

■ ***SOLUTION***

First convert temperatures to absolute values.

$$T_s = 23°\text{C} + 273 = 296\ \text{K}$$

$$T = 95°\text{C} + 273 = 368\ \text{K}$$

From Example Problem 22–6:

$$A = 1.26 \text{ m}^2$$

$$H = \sigma \epsilon A (T_s^4 - T^4)$$

$$H = 0.80 \times 5.67 \times 10^{-8} \text{ J/m}^2 \cdot \text{s} \cdot \text{K}^4 \times 1.26 \text{ m}^2 [(296 \text{ K})^4 - (368 \text{ K})^4]$$

$$H = -609 \text{ J/s} \times \frac{1 \text{ kcal}}{4.186 \times 10^3 \text{ J}}$$

$$H = 0.15 \text{ kcal/s}$$

The radiation losses of the steam pipe are about ⅓ the heat convection losses.

22.6 THE GREENHOUSE EFFECT

Any surface with a temperature above absolute zero will radiate energy. That is why we need to use absolute temperatures in Equation 22–6. The distribution of wavelengths depends on the temperature. Figure 22–10 shows the distribution of wavelengths at different temperatures for a black body, or perfect emitter. Notice that the peak of the distribution of wavelengths shifts with temperature. A hot object will emit more shortwave radiation than a cooler object.

There is a simple relationship between the peak intensity (λ_{max}) and the absolute temperature of the radiating surface. The peak intensity is inversely proportional to the temperature. This relationship is known as **Wein's law.**

$$\lambda_{max} T = constant \qquad \textbf{(Eq. 22–7)}$$

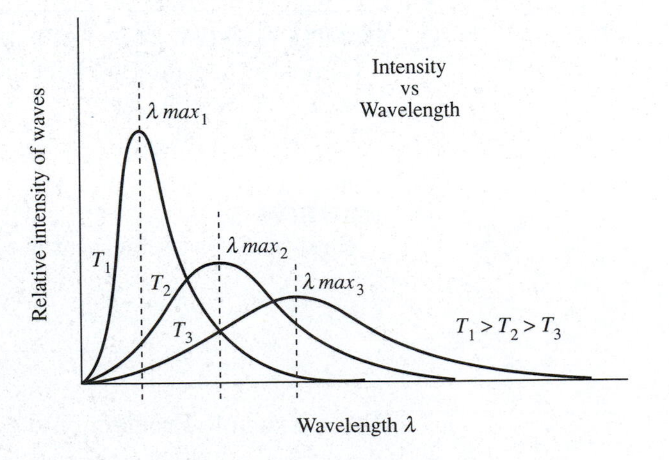

Figure 22–10: *Intensity versus wavelength. The wavelength peaks of radiation intensity occur at longer wavelengths for cooler surfaces.*

EXAMPLE PROBLEM 22–8: WEIN'S LAW

At a temperature of $60\bar{0}0$ K the peak intensity of emission of radiation is at a wavelength of about 5×10^{-7} m. What is the peak wavelength for a surface with a temperature of 35°C? (Note: $60\bar{0}0$ K is the approximate surface temperature of the sun and 5.0×10^{-7} corresponds to the wavelength of visible yellow light.)

SOLUTION

$$T_1 = 60\bar{0}0 \text{ K}$$

$$T_2 = 35°\text{C} + 273 = 308 \text{ K}$$

$$\lambda_1 = 5.0 \times 10^{-7} \text{ m}$$

$$\lambda_2 = ?$$

$$\lambda_1 T_1 = \lambda_2 T_2 = \textit{constant}$$

$$\lambda_2 = \frac{T_1 \lambda_1}{T_2}$$

$$\lambda_2 = \frac{60\bar{0}0 \text{ K}}{308 \text{ K}} 5.0 \times 10^{-7} \text{m}$$

$$\lambda_2 = 9.7 \times 10^{-6} \text{ m}$$

λ_2 is in the infrared range. It is invisible to the eye, but it will give a sensation of warmth.

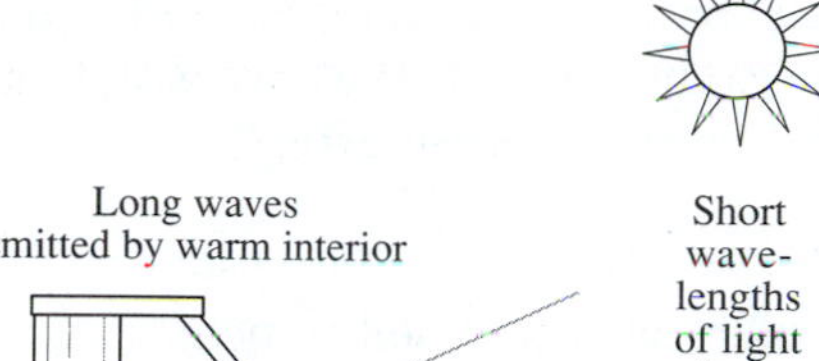

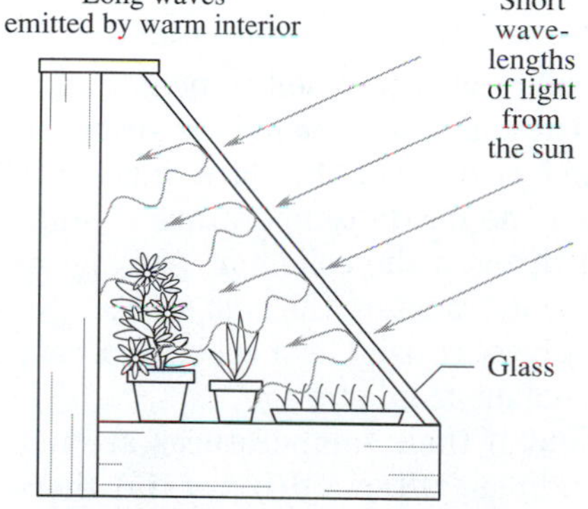

Figure 22–11: Short wavelengths of light from the hot sun easily penetrate the glass wall of a greenhouse and are absorbed. The warmed materials in the enclosure emit longer infrared radiation. The longer wavelengths cannot get through the glass. Radiant heat is trapped in the greenhouse.

Various wavelengths are absorbed, reflected, or transmitted in a different manner by different materials. Radio waves can easily penetrate a wall of wood and gypsum board. A portable radio works quite well inside a wooden frame house. On the other hand, metals are opaque to radio waves. Metals have emissivities that are very low. Most of the radio waves are reflected; the rest are absorbed in the metal. In order to operate a radio in a car the antenna should be on the outside for good reception.

A single material can behave differently for different wavelengths. Glass is transparent to visible light, but it is opaque to infrared radiation. Visible light can pass through the glass. Infrared radiation sees the same pane of glass as a blank wall; it can't get through. This behavior can be used to make a radiant heat trap.

The Greenhouse Effect

Figure 22–11 shows a greenhouse. The surface of the sun is about 6000 K. Much of its radiation is in the visible light range. Much of

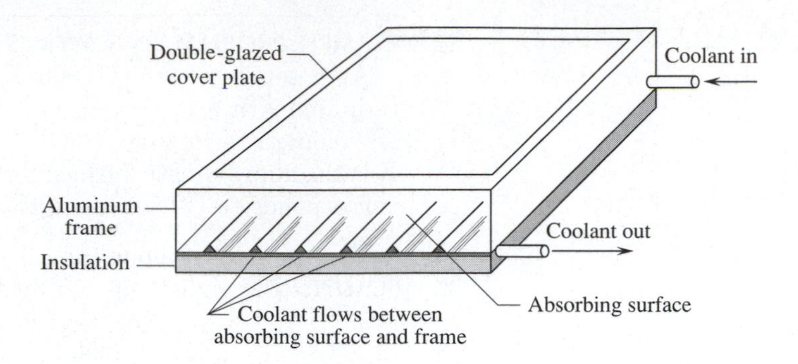

Figure 22–12: A typical flat-plate collector used to heat water using the greenhouse effect.

the radiant energy from the sun can pass through the glass walls of the greenhouse.

The light entering the greenhouse is absorbed by the surfaces inside the enclosure. The objects inside the greenhouse will also emit radiation, but at longer wavelengths according to Wein's law. These longer wavelengths cannot get through the windows. They are trapped inside the greenhouse.

Short wavelengths of radiant energy continue to enter the greenhouse, warming it as the light is absorbed. The infrared radiation from materials inside the enclosure cannot get out. We have a radiant heat trap. This situation is called the **greenhouse effect.**

Solar Collectors

Figure 22–12 shows a flat-plate solar collector used to heat water. A cover glass allows sunlight into the collector. The light is absorbed by a dark surface. The reradiated heat from the dark surface is trapped inside the collector. When the inside temperature is sufficiently high, a coolant is pumped through the collector, picking up heat by conduction. Forced convection transfers the heat to storage. Since an antifreeze such as ethyl glycol is used as a coolant, a heat exchanger is used to isolate the coolant from storage.

Solar collectors are more efficient if their temperatures are not much different from their surroundings. Often a **differential thermostat** is used to decide when to pass coolant through the collector. It monitors the temperature of the storage area and compares it with the temperature of the collector. If the collector is at least 15°F warmer than the storage area, the pumps are turned on. When the temperature difference drops to 5°F, the pumps turn off.

Flat-plate collectors can also use air as a coolant. Fans circulate air through the space between the collector plate and absorbing surface. The warmed air is carried through ducts to a bin filled with rocks approximately 1 in in diameter. As the air passes through the cavities between the stones, heat is absorbed by the rock. When heat is needed to warm the building, a fan passes cool air through the rock to pick up heat and distributes it throughout the house.

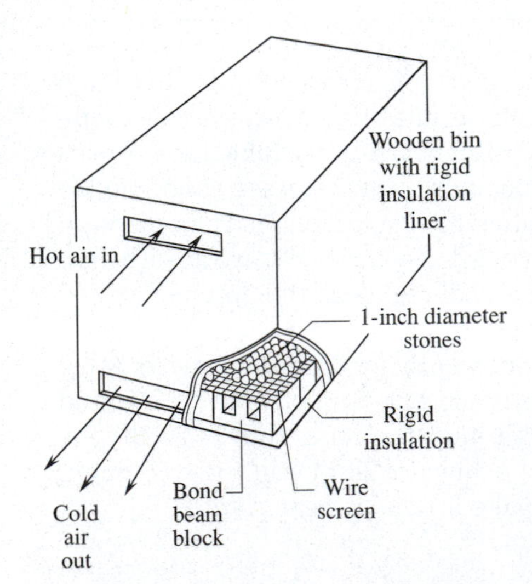

Figure 22–13: Details of the construction of a rock storage box used to store heat from a hot-air collector. Fans reverse the direction of airflow to draw heat from the box.

SUMMARY

Heat is transferred by three different processes: conduction, convection, and radiation. Thermal conduction is a process in which thermal energy is transferred by collision of atoms in a substance.

The heat flow rate by thermal conduction is proportional to the cross-sectional area (A) and the temperature gradient ($\Delta T/\Delta L$). The proportionality constant (k) is called thermal conductivity.

If the sides of a conductor of heat are insulated so that thermal energy is not lost along the length of the conductor, the rate of heat flow into the high-temperature side of the conductor equals the rate of flow of heat coming out the other side. Energy is conserved. $H(\text{in}) = H(\text{out})$.

For a conductor made of layers of different materials, an overall thermal conduction coefficient (U) can be found using R values. R values for the different materials are added up to give the total effective R value. The conduction coefficient is the reciprocal of the effective R value. An R value of a material is its thickness divided by its thermal conductivity.

While conduction is the major form of thermal transport in a solid, convection is the major form of heat transport in a volume of fluid. Thermal convection is a process in which masses of fluid move from one place to another carrying heat. For natural convection near a surface of area (A) and temperature T_a with surrounding fluid of temperature T_s the heat flow equation is: $H = Q/t = h\,A\,(T_s - T_a)$, where h is the convection coefficient.

Thermal radiation is a process in which heat is transported by electromagnetic waves. Radiant energy can travel through a vacuum. Unlike conduction and convection, no medium is necessary. A perfect absorber is called a black body. *Emissivity* (ϵ) is the ratio of the absorption of a surface compared with the absorption of a black body. A perfect absorber has an emissivity of 1.00; a perfect reflector, an emissivity of 0.00. A surface has the same emissivity for emission as it has for absorption of radiant energy.

The rate of heat transfer for an object with area (A) and emitting heat at the absolute temperature T while receiving radiant heat from a source of absolute temperature T_s is:

$$H = \frac{Q}{t} = \sigma\,\epsilon\,A\,(T_s^4 - T^4)$$

where σ is the Stefan-Boltzmann constant.

Something with a temperature above absolute zero emits a wide range of electromagnetic wavelengths. The hotter the object, the shorter will be the wavelength. The peak of the distribution of wavelengths (λ_{max}) can be found using Wein's law ($\lambda_{max}\,T = \textit{constant}$).

A glassed-in area will allow visible light waves to pass through the glass and be absorbed by the material inside. The material inside the glassed-in area will emit longer wavelengths according to Wein's law. These longer wavelengths cannot escape through the glass. They are trapped inside. This is called the greenhouse effect.

KEY TERMS

If you can explain the following terms to a friend or classmate, you understand their meaning. If you cannot explain the terms, you should reread the sections in which they are discussed.

black body
conduction
conduction coefficient (*U*)
convection
convection coefficient
differential thermostat
electromagnetic radiation
emissivity
greenhouse effect
radiation
***R* value**
steady-state heat flow
Stefan-Boltzmann constant
thermal conductivity
temperature gradient
thermal resistance
transmission coefficient
***U* value**
Wein's law

EXERCISES

Sections 22.1, 22.2:

1. A hardwood floor seems colder to a bare foot than a rug at the same temperature. Explain why.
2. On a frigid winter day, a child touches the wooden slats of a sled with her tongue. Nothing happens. The child then touches the steel runner of the sled with her tongue and it becomes frozen fast to the runner. Why does this happen?
3. The thickness of wall A is 4.0 in. The inside temperature is 65°F and the outside temperature is 45°F. Wall B is 2.0 in thick, and the temperature inside is 70°F, while the outside temperature is 80°F.
   A. Compare the size of the temperature gradients for the two walls.
   B. In which direction will heat travel for each wall?
   C. Assume the two walls are both solid concrete and have the

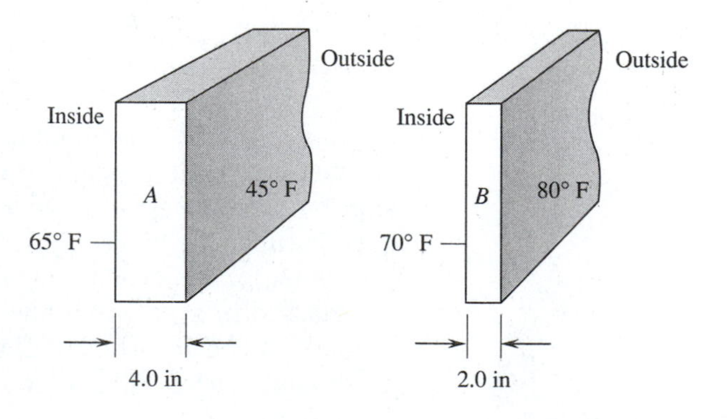

Figure 22–14: *Diagram for Exercise 3. Two walls are made of the same material. They differ in thickness and temperature difference across the walls.*

same cross-sectional area. Will more heat pass through wall B than wall A? Explain.

4. Heat passes through a wall with uniform properties. How will the rate of heat flow change if:
   A. the thickness of the wall is doubled?
   B. the temperature difference across the wall is doubled?
   C. the area of the wall is doubled?
   D. the area is doubled and the thickness is tripled?
   E. the temperature difference is doubled, the thickness is halved, and the area is reduced to one-quarter the original area?

Section 22.3:

5. A uniform, 3.0 in thick wall has an R value of 5. The amount of heat flowing through the wall is 16×10^4 Btu/h. What will be the heat flow rate if the R value of the wall is increased to 10?

6. Calculate the heat flow rate for the wall in Exercise 4 for R values of 5, 10, 20, 40, and 80. Plot heat flow rate versus R value.
   A. How much is the heat loss reduced by changing the R value from 5 to 10?
   B. How much is the heat loss reduced by changing the R value from 40 to 80?

7. How thick must the wall in Exercise 4 be to reduce the heat loss to zero?

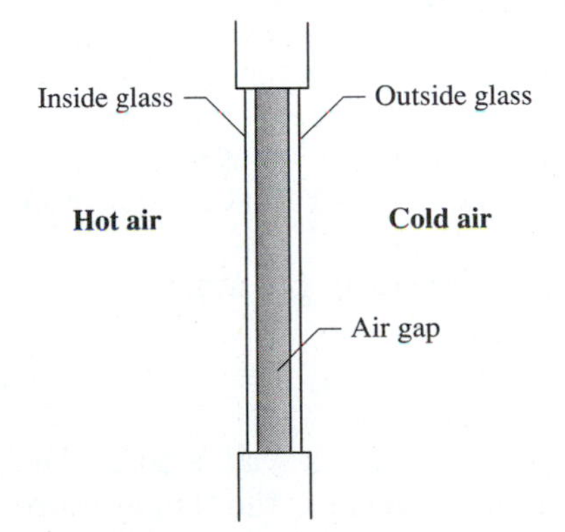

Figure 22–15: Diagram for Exercise 9. A cross section of a double-glazed windowpane.

Section 22.4:

8. The temperature of a living room is 65°F on a midwinter evening, yet frost forms on the *inside* of the windows. Explain why this is possible.

9. Figure 22–15 shows the cross section of a double glazed window. Copy the diagram and sketch in the convection currents on the inside of the window, the outside, and the space between the panes.

10. A highly effective thermos bottle has a vacuum between the inside and outer walls. A cheap plastic thermos has an air gap. Why is the vacuum more effective?

11. Glass is not a very good thermal insulator, but spun fiberglass batts are commonly used for wall insulation. Why?

Section 22.5:

12. For wavelengths in the visible range or longer, a silvered mirror reflects about 95% of the incoming radiation. Estimate the emissivity of the mirror.

13. An object such as this book at room temperature is well above absolute zero; it must continuously emit radiant heat. Why does it not cool below freezing?

14. Why do baseboard radiators have fins? Why are there fins on the cylinder housings of small two-cycle engines?

15. Describe the heat gains and heat losses of a fireplace. Is a fireplace an efficient way to heat a room?

Section 22.6:

16. In the visible light spectrum, red has the longest wavelength and blue the shortest wavelength.
    A. Explain how astronomers can use Wein's law to estimate the surface temperature of stars.
    B. Order the following stars according to increasing temperature: red supergiant, blue supergiant, yellow main sequence.

17. Why are solar collectors more efficient on a hot day than on a cold day?

18. Flat-plate collectors that are used for space heating of a building are usually tilted more toward the horizon than flat-plate collectors used for heating domestic hot water. Can you think of a reason why?

19. Ozone absorbs short ultraviolet radiation. The behavior of radiation with carbon dioxide is similar to its behavior with glass. Explain what changes in climate are likely for the following situations.
    A. The ozone layer high in the atmosphere is depleted.
    B. The concentration of carbon dioxide in the atmosphere increases.
    C. There is an increase of dust particles in the atmosphere.

PROBLEMS

Sections 22.1, 22.2:

1. The temperature on one side of a 3.5 in thick wall is $8\bar{0}$°F. The temperature on the other side is 45°F. What is the temperature gradient across the wall?

2. The thermal gradient across a concrete wall is 5.20°C/cm.
    A. What is the thermal gradient in Celsius degrees per meter?
    B. What is the rate of heat flow (*H*) per square meter through the wall?

3. The bottom of an aluminum tea kettle has an area of 59 in^2 and a thickness of $\frac{1}{16}$ in. It sits on a burner with a temperature of $65\bar{0}$°F. Find the rate of heat flow (*H*) when the temperature of the water in the kettle is 82°F.

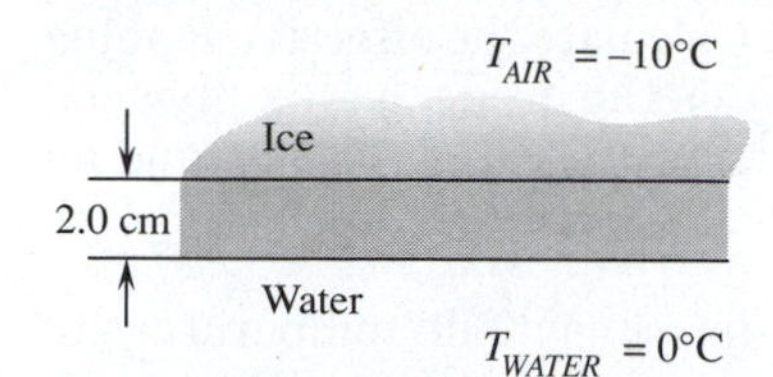

Figure 22–16: *Diagram for Problem 6. Water loses heat through the ice, causing new ice to form on the bottom side.*

4. How much heat is conducted in 2 h through a single-glazed 32 in × 80 in patio door that is ⅛ in thick? (Assume an inside temperature of 55°F and an outside temperature of 23°F.)

5. What is the temperature in the interface of a 5.0-in concrete wall with 1.0 in of styrofoam insulation on the outside? (Inside temperature is 70°F; outside 42°F.)

6. A layer of ice on a pond is 2.0 cm thick. The water temperature under the ice is 0°C, and the air temperature above the ice is −10°C. How many grams of ice will be formed for each square meter of ice in 12 h?

Section 22.3:

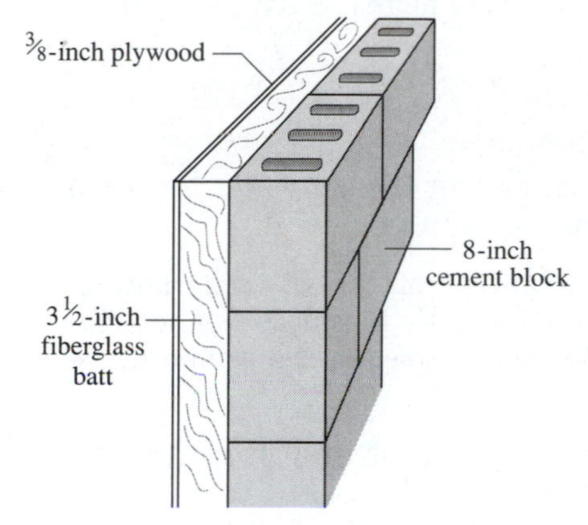

Figure 22–17: *Diagram for Problem 7. Find the R value and conduction coefficient (U) for a composite wall.*

7. A cellar wall in a raised ranch home consists of 8-in cement blocks, 3-½-in fiberglass batts, and ⅜-in plywood.
   A. What is the overall R value for the wall?
   B. What is the wall's conduction coefficient?

8. A wall has an R value of 13.5 and a temperature difference of 18.3°F. If the area of the wall is 450 ft², how many British thermal units will flow through the wall in 12 h?

9. The effective temperature gradient across an insulated double-glazed window is 32°F. If the area of the window is 30 in × 72 in, how much heat is lost through the window in one day? (Use R values.)

Section 22.4:

10. A 2.5 cm diameter hot-water pipe has an outside temperature of 50.0°C and a length of 12 m. How much heat is convected in 1.2 h to the air surrounding the pipe at a temperature of 9.0°C?

11. The inside temperature of a 0.92-m² windowpane is 7°C in a room with an average temperature of 22°C. How much heat is convected from the room in 2.0 h?

12. The convective air film moving over a surface creates an effective R value. For convection, the R value is defined as $R = 1/h$, where h is the convection coefficient. Table 22–5 lists the R value

Table 22–5: *R* values for convective air film for a vertical wall with average roughness and temperature.

CONDITION OF AIR	*R* VALUE (ft² · h · °F/Btu)
Still air	0.65
7-½-mph wind	0.25
16-mph wind	0.17

for smooth, vertical, flat surfaces. Calculate the effective R value of an uninsulated wall with still air on the inside, a $^3/_8$-in plywood inside surface, a still air gap, a clapboard outer surface, and an air film, in a 7-$^1/_2$ mph wind.

13. An uninsulated hot-water heater has an outside temperature of 31°C in a basement with a mean temperature of 13°C. The heater has a length of 1.4m and a diameter of 0.62 m.
    A. What is the value of the convection coefficient? (Treat it as a pipe.)
    B. How much heat is lost by convection in a 30-day month?
    C. What is the work equivalent in kilojoules?
    D. How much electrical energy in kilowatt hours is needed to replace the lost heat? (A kilowatt hour is 1 kW $\times$ 3600 s.)

Section 22.5:

14. Calculate the net radiation heat loss for the hot-water heater in Problem 13. Assume an emissivity of 0.91.

15. The sun is a sphere with a diameter of 1.39×10^9m and an effective surface temperature of 6000 K. Estimate the heat radiated by the sun in 1 s. Interstellar space has a temperature of 4 K.

16. Calculate the net radiation loss of heat for the hot-water pipe in Problem 10. Use an emissivity of 0.82.

Section 22.6:

17. Use the data in Example Problem 22–8 to estimate the peak wavelength of radiation emitted by ice at 0°C.

18. A fuel burns with a peak wavelength of 3.7×10^{-6} m. What is the temperature of the fuel? Use the data from Example Problem 22–8 in your calculation.

19. Hot plate collectors in a solar energy system work best when they are not allowed to heat to high temperatures. Compare the radiation heat loss per square meter of the cover glass of a hot plate collector held at an effective temperature of 67°C by a differential thermostat with that of a collector with a plate that heats to 89°C. Assume an air temperature of 18°C and an emissivity of 1.00 for both situations.

20. Solar energy is often expressed in langleys (1 langley = 3.69 Btu/ft^2). How many gallons of water can be heated in one day from 60°F to 110°F by a solar collector with an area of 32.0 ft^2 and an efficiency of 75%? Assume the radiation energy for one day is 130 langleys. (One gallon of water has a weight of 8.4 lb.)

Chapter 23

VIBRATIONAL MOTION

With the right set of physical conditions, you can hear wires singing in the wind. In the next few chapters, you will begin to learn about the physics of sound production. (Photograph courtesy of New York Power Authority.)

OBJECTIVES

In this chapter you will learn:

- the behavior of a simple harmonic motion
- that simple harmonic oscillators obey Hooke's law
- to calculate the restoring force acting on a simple harmonic oscillator
- that the acceleration of a simple harmonic oscillator is proportional to displacement
- that the motion of a simple harmonic oscillator behaves like the vertical components of the motion of a particle moving in a circle
- to calculate the frequency and period for simple harmonic motion
- to calculate the energy stored in a simple harmonic oscillator

Time Line for Chapter 23

1676	Robert Hooke discovers that the tension on a string is directly proportional to displacement.
1679	De La Hautefeuille invents the seismograph.
1747	Jean d'Alembert develops a wave theory of vibrating strings.
1809	Ernst Chladni finds that sand figures can be formed on the surface of a vibrating metal sheet.
1863	Augustus Love discovers surface earthquake waves (L waves) moving through the earth's crust.
1935	Charles Richter develops a scale to measure earthquakes.
1940	Tacoma Narrows bridge fails from wind-induced vibrations.

We are surrounded by vibrational motion. A plucked guitar string oscillates. A tree limb sways rhythmically in the wind. A sewing machine needle repeatedly plunges in and out of a fabric. The diaphragms of a woofer and a tweeter oscillate to the sound of rock music.

Different kinds of waves are caused by vibrational motion. The guitar string and the speaker form sound waves in the air. A vibrating electric charge can cause electromagnetic waves to occur.

Often vibrational motion is coupled with rotational motion. The up and down motion of the piston in a gasoline engine is tied to the rotational motion of the crankshaft. The vibrational motion of a sewing machine needle comes from a gearing system connected to the rotating shaft of an electric motor.

In this chapter we will look at vibrational motion and its connection with rotational motion. In the following chapters we will look at the waves caused by vibration and how they behave.

23.1 HARMONIC MOTION

Figure 23–1 shows some naturally oscillating systems. A ball rolls from side to side in a bowl. A pendulum swings to and fro. A bob bounces up and down on a spring. A plucked string vibrates. A metal rod pulled to one side and released oscillates back and forth. A mass on a twisted wire rotates first clockwise then counterclockwise over and over again.

These systems have some things in common. Their motion is called **simple harmonic motion.** Let us look at some of the common factors.

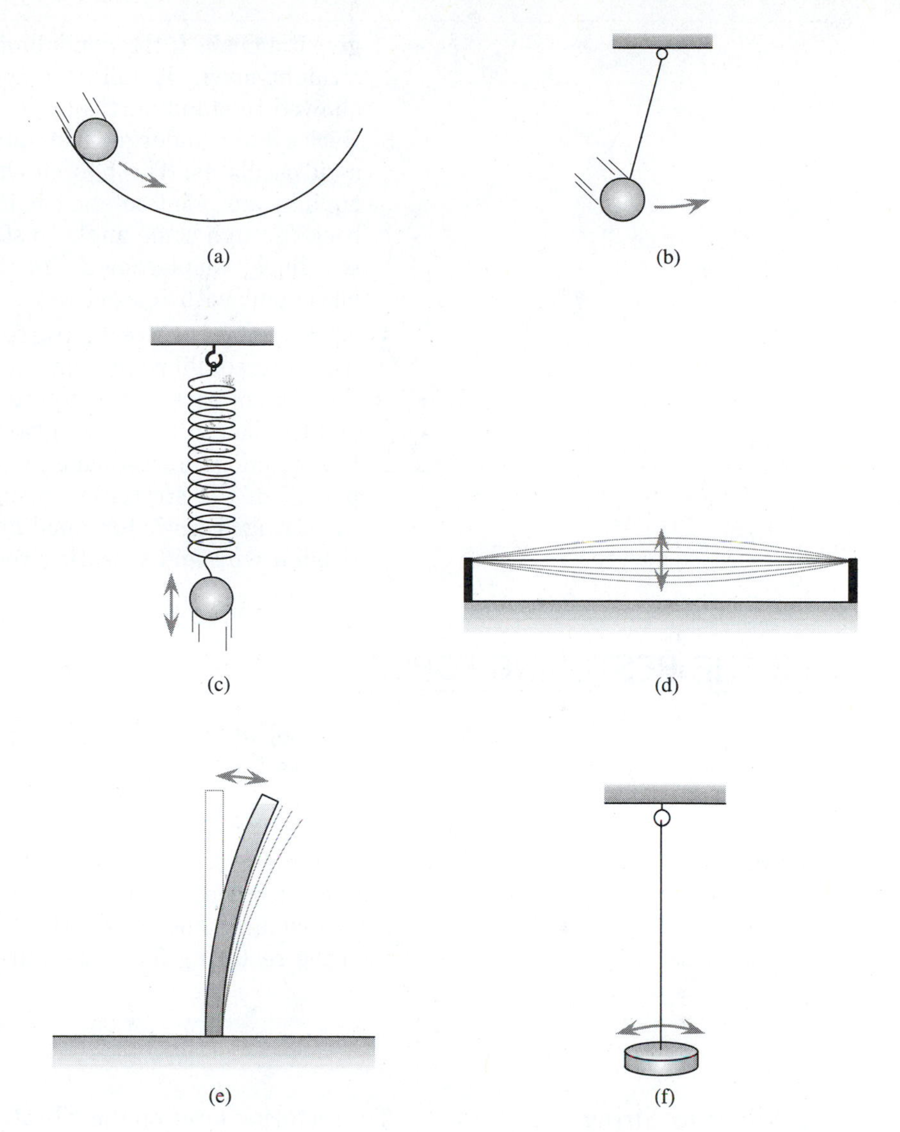

Figure 23–1: Different kinds of systems that vibrate with simple harmonic motion. (a) *A ball rolls back and forth in a bowl.* (b) *A pendulum swings.* (c) *A mass bobs up and down on a spring.* (d) *A stressed string vibrates.* (e) *The free end of a rod vibrates.* (f) *A mass on a twisted wire oscillates.*

1. In each of the systems an object goes through a repeating **cycle.** The ball rolls from one side of the dish, returns to its starting point, and starts its motion over again. The bob bounces upward, compressing the spring, falls back, stretching the spring until it stops its downward plunge, and then starts the motion over again. The mass on the twisted wire turns clockwise, reverses its direction, and rotates counterclockwise until it comes to a stop. It reverses its direction and starts the clockwise motion again.
2. Each system has an **equilibrium point.** If the ball is allowed to rest at the bottom of the bowl, it will remain at rest unless we

give it a push. If the pendulum is allowed to hang motionless and straight down, it will no longer move. If the top of the rod is allowed to stand vertically, it will not start to vibrate.

3. Each system undergoes a **displacement** from its equilibrium point as it oscillates. The bob on the spring is pulled down from the equilibrium point to start it in motion. The pendulum is pulled back through some angle to start it in motion. In one case there is a linear displacement, in the other there is an angular displacement with respect to the equilibrium point.
4. Each system has a **restoring force.** When each system is displaced from the equilibrium point, a restoring force attempts to move the system back to the equilibrium condition. For the pendulum and the ball in the bowl, gravity is the restoring force. Hooke's law supplies the restoring force for the spring. The vertical component of the stretched spring is the restoring force acting on the string. Elastic torsional forces act as restoring forces on the twisted wire and vibrating rod.

23.2 THE RESTORING FORCE

Let us look at the nature of the restoring force for simple harmonic oscillators.

The Spring

A spring obeys Hooke's law. The restoring force is proportional to the displacement from equilibrium and is in the opposite direction to the displacement (see Section 8.3). If x_0 is the equilibrium position, then the restoring force of a spring is:

$$F = -k\,(x - x_0) = -k\,\Delta x$$

The Vibrating String

The restoring force on the vibrating string is the vertical component of tension. Figure 23–2a is a simplified diagram of the stretched string. The center of the string under tension is displaced upward

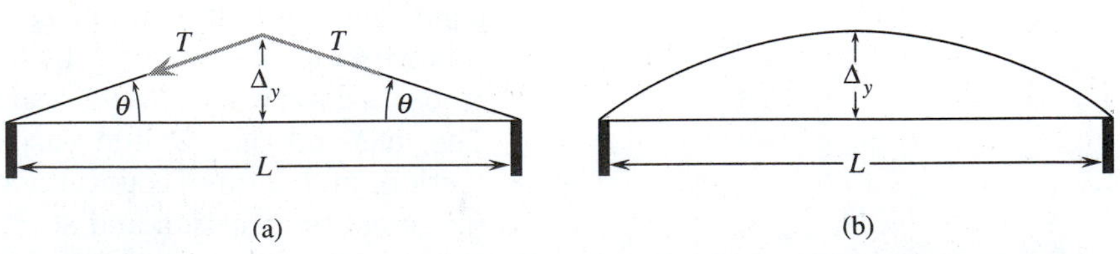

Figure 23–2: The vertical component of the tension of a stretched string is the restoring force.

by Δy. Using the small angle approximation, $\sin\theta = \tan\theta = \theta$, the component of tension (F_y) acting downward on a string L long is:

$$F_y = -2\,T\sin\theta = \frac{-2\,T\,\Delta y}{L/2}$$

Figure 23–2b shows the real shape of a vibrating string. The vertical tension along the string is not constant. An averaging technique can be used to find the expression of the vertical force in terms of T, y, and L.

$$F_y = -\left(\frac{\pi^2 T\,\Delta y}{L}\right)$$

If we lump the constants together, we get an equation that looks like Hooke's law.

$$F_y = -K\Delta y$$

where

$$K = \frac{\pi^2 T}{L}.$$

The Pendulum

Figure 23–3 shows the forces acting on a simple pendulum. The pendulum, with a length L, is displaced horizontally by Δx from the equilibrium position. The horizontal component of the string tension acts in the opposite direction to restore the bob.

First find the tension on the string caused by the weight of the bob. We can use components of force.

$$T_y = T\cos\theta = m\,g$$

$$T = \frac{m\,g}{\cos\theta}$$

$$F_x = -T_x = -T\sin\theta$$

We have taken the displacement to be positive, so the force is negative since it acts in the opposite direction. Combining these two equations we get:

$$F_x = \frac{-m\,g\sin\theta}{\cos\theta} = m\,g\tan\theta$$

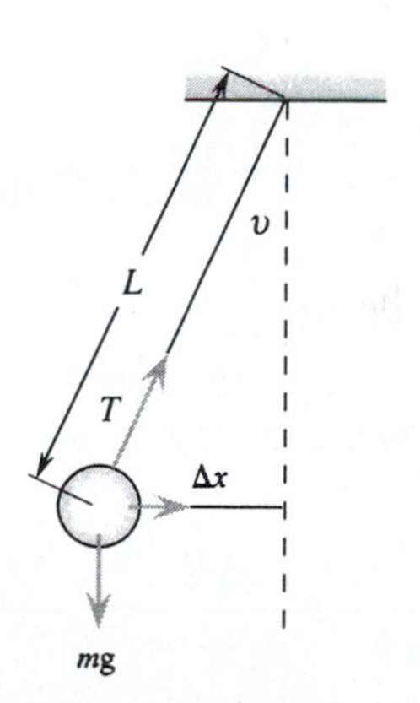

Figure 23–3: *The horizontal component of the tension on the string acts as the restoring force for a simple pendulum.*

We can use the small angle approximation $\tan \theta = \sin \theta = \theta$. Then

$$\tan \theta = \frac{\Delta x}{L}$$

$$F_x = \frac{-m\, g\, \Delta x}{L}$$

Again we get an equation that looks like Hooke's law if we represent the constants by one general symbol. Let $K = (m\, g)/L$.

$$F_x = -K\, \Delta x$$

The General Force Equation for Simple Harmonic Motion

The other cases shown in Figure 23–1 are much the same. The analysis for the ball in the bowl is similar to that for the pendulum if the bowl has a circular cross section. The restoring force of the twisted wire is proportional to the twist angle. The proportionality constant depends on the length of the wire and the shear modulus (review Section 15.4). The vibrating rod has a similar relationship.

The important thing to recognize is that harmonic oscillators obey Hooke's law. The proportionality constant may depend on a variety of physical factors concerning a specific system, but it is just a constant. The restoring force for any harmonic oscillator can be reduced to:

$$F = -K\, \Delta x \qquad \textbf{(Eq. 23–1)}$$

Let us apply Newton's second law of motion. Assume we have a constant mass.

$$F = m\, a = -K\, \Delta x$$

then

$$\frac{a}{(-\Delta x)} = \frac{K}{m} = \textit{constant} \qquad \textbf{(Eq. 23–2)}$$

For simple harmonic motion, the acceleration is proportional to the displacement from the equilibrium. The more the center of gravity is displaced from the equilibrium point the larger will be the acceleration of the object.

□ **EXAMPLE PROBLEM 23–1: RESTORING FORCE ON A STRETCHED STRING**

What is the restoring force of an 80.0 cm-long guitar string stretched 1.00 cm from its equilibrium position when it is under a tension of 65.0 N?

■ *SOLUTION*

$$K = \frac{\pi^2 T}{L}$$

$$F = -K\,\Delta x = \frac{-\pi^2 T \Delta x}{L}$$

$$F = -\left[\frac{(3.14)^2 \times 65.0\ \text{N} \times 1.00\ \text{cm}}{80.0\ \text{cm}}\right]$$

$$F = -8.02\ \text{N}$$

23.3 SIMPLE HARMONIC MOTION AND ROTATION

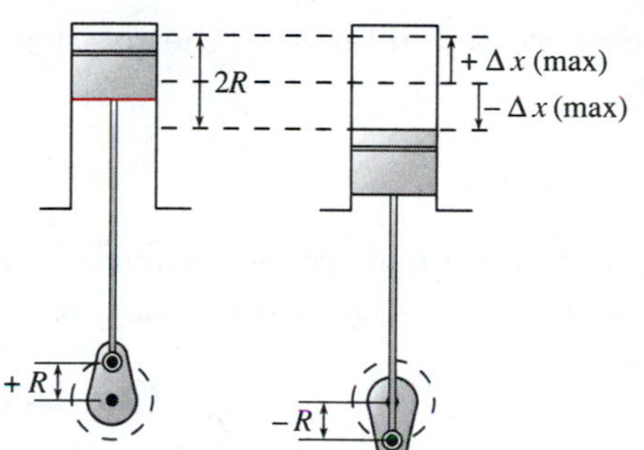

Figure 23–4: The diameter of the circle traveled by the end of the rod equals the total displacement of the piston.

Many mechanical systems couple rotational motion with vibrational motion. Look at Figure 23–4. A piston undergoes vibrational motion up and down while the crankshaft rotates. The piston and the crankshaft have the same frequency, since they are tied together by a rod. The forces causing the motion are complex, but the motion of the piston is harmonic.

The rod is connected a distance R from the center of rotation of the crankshaft. The stroke of the piston will be $2R$. When the rod is at the bottom of the circle, the piston is at the bottom of the stroke. When the rod is at the top of the circular motion, the piston is at the top of the stroke.

We will label the midpoint (x axis) as zero (see Figure 23–5). The piston will have a maximum displacement of $+R$ at the top of the stroke and a displacement of $-R$ at the bottom of the stroke (see Figure 23–4). The angle θ measures the position of the rod as it moves around the crankshaft. $\theta = 0°$ as it starts at the equilibrium position. $\theta = 90°$ as the piston reaches the top of the stroke. $\theta =$

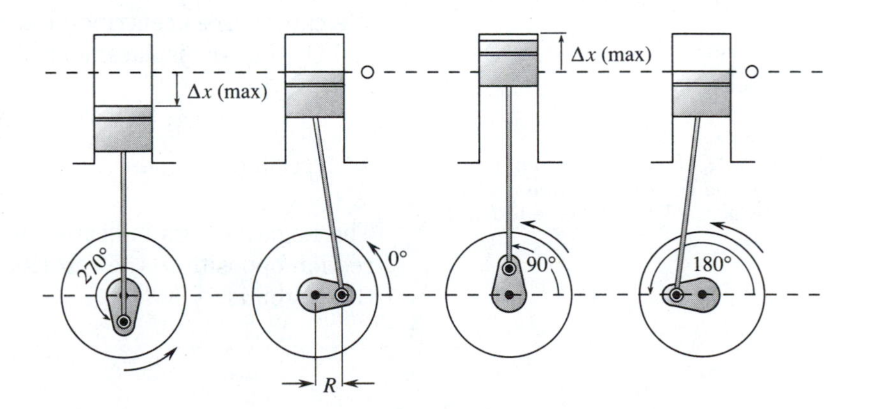

Figure 23–5: The rotational motion of the end of the rod connected to the crankshaft is converted to linear harmonic motion of the piston.

180° as the piston is halfway down the stroke, and $\theta = 270°$ as the piston reaches the bottom of the stroke.

The displacement of the piston is the vertical component of R.

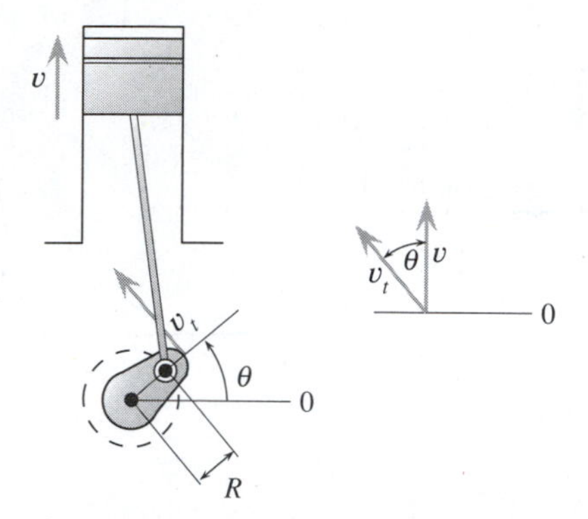

Figure 23–6: *The velocity of the piston is the vertical component of the tangential velocity of the rod end.*

$$\Delta x = R \sin \theta$$

Since $R = \Delta x(\max)$, for displacement for harmonic motion, we can write:

$$x = x(\max) \sin \theta \qquad \textbf{Eq. 23–3}$$

The velocity of the piston is the vertical component of the tangential velocity ($\mathbf{v}_t$) (see Figure 23–6). At the top and bottom of the piston's stroke, the linear velocity is zero as the piston reverses direction.

$$v = v_t \cos \theta$$

The tangential velocity can be expressed in terms of the rotational speed of the rotating rod end.

$$v_t = R\omega = \Delta x(\max)\,\omega$$

The relationship between angular speed and frequency (f) is $\omega = 2\pi f$. In terms of frequency, the linear velocity of the piston is:

$$v = 2\pi f \Delta x(\max) \cos \theta \qquad \textbf{(Eq. 23–4)}$$

The acceleration of the piston is the vertical component of the centripetal acceleration of the rotating rod end (see Figure 23–7). The acceleration will be the greatest at the top and bottom of the piston's stroke. Notice that the acceleration is in the opposite direction of the linear displacement.

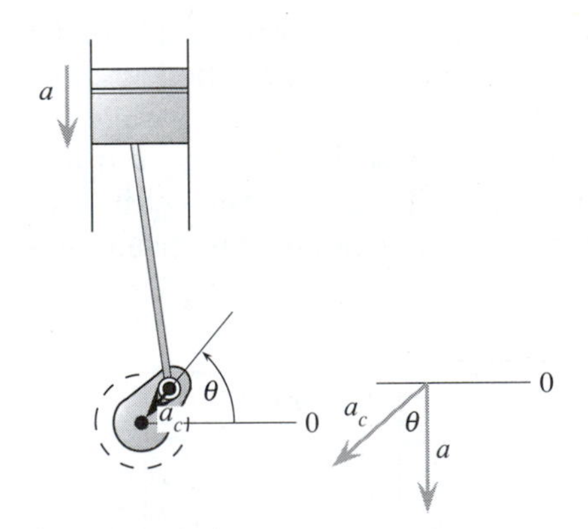

Figure 23–7: *The acceleration of the piston is the vertical component of the centripetal acceleration of the rod end.*

$$a = a_c \sin \theta$$

We can express centripetal acceleration in terms of the frequency and the linear displacement of the piston.

$$a_c = R\omega^2 = -\Delta x\,(2\pi f)^2$$

$$a_c = -4\,\pi^2 f^2 \Delta x$$

The negative sign indicates that the linear displacement is in a direction opposite to the centripetal acceleration. The acceleration of the piston is:

$$a = -4\pi^2 f^2\,\Delta x \sin \theta \qquad \textbf{(Eq. 23–5)}$$

☐ **EXAMPLE PROBLEM 23–2: THE PISTON**

A piston with a 4.00-in stroke operates at $12\bar{0}0$ rpm. Find:

A. the maximum linear speed of the piston in feet per second.

B. the maximum linear acceleration in feet per second squared.

■ ***SOLUTION***

The stroke is $2R$. Therefore, Δx = stroke/2 or $\Delta x = 2.00$ in

$$x = 2.00 \text{ in} \times \frac{1 \text{ ft}}{12 \text{ in}} = 0.167 \text{ ft}$$

$$f = \frac{12\bar{0}0 \text{ cycles}}{\text{min}} \times \frac{1 \text{ min}}{60 \text{ s}} = \frac{20.0 \text{ cycles}}{\text{s}}$$

A. v(max) occurs when $\cos \theta = 1.00$.

$$\text{v(max)} = 2\pi f \Delta x \times \cos \theta$$

$$\text{v(max)} = 2\pi \times \frac{20.0 \text{ cycle}}{\text{s}} \times 0.167 \text{ ft} \times 1.00$$

$$\text{v(max)} = 21.0 \text{ ft/s}$$

B. a(max) occurs when $\sin \theta = 1.00$. We will ignore the negative sign since we are interested only in the size of the acceleration.

$$\text{a(max)} = 4\pi^2 f^2 \Delta x \sin \theta$$

$$\text{a(max)} = 4\pi^2 \left(\frac{20.0 \text{ cycle}}{\text{s}}\right)^2 \times 0.167 \text{ ft} \times 1.00$$

$$\text{a(max)} = 2.64 \times 10^3 \text{ ft/s}^2$$

23.4 FREQUENCY OF A SIMPLE HARMONIC OSCILLATOR

We can see that the motion of a piston is closely related to the rotary motion of the crankshaft. Complex forces act on the piston to drive it. It is a **driven oscillator.** Things like pendulums and springs respond to a simple restoring force obeying Hooke's law. These are **natural oscillators.**

The up and down motion of a mass on a spring is the same as that of a piston. The spring is not connected to a rotating shaft, but we can analyze its motion in the same way. It can be described as the vertical projection of a **reference circle** with a radius equal to the maximum displacement of the spring and rotating at the same fre-

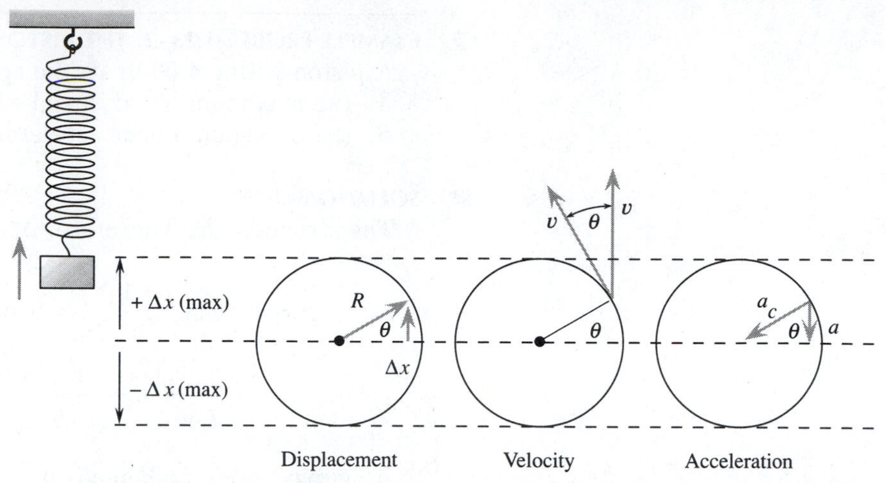

Figure 23–8: *The motion of a loaded spring can be analyzed using a reference circle.*

quency (see Figure 23–8). Mathematically the motion of the spring is the same as the motion of a piston.

For a natural oscillator obeying Hooke's law we can find some additional information. The centripetal acceleration seen in the reference circle is related to the angular velocity by:

$$a_c = R\omega^2$$

or

$$\omega = \left(\frac{a_c}{R}\right)^{1/2}$$

In terms of the linear motion of the oscillator:

$$\omega = \left(\frac{-a \sin\theta}{\Delta x \sin\theta}\right)^{1/2}$$

or

$$\omega = \left(\frac{-a}{\Delta x}\right)^{1/2}$$

Since $f = \omega/2\pi$, the frequency is:

$$f = \frac{1}{2\pi}\left(\frac{-a}{\Delta x}\right)^{1/2}$$

From section 23.2,

$$\frac{-a}{\Delta x} = \frac{K}{m}$$

For an oscillator obeying Hooke's law:

$$f = \frac{1}{2\pi}\left(\frac{K}{m}\right)^{1/2} \qquad \textbf{(Eq. 23–6)}$$

If we know the mass of a system and the nature of the proportionality constant K, we can predict the frequency of an oscillating system. K for a spring is the stiffness k. When a mass m is hung on a spring it will tend to vibrate with a frequency of:

$$f(\text{spring}) = \frac{1}{2\pi}\left(\frac{k}{m}\right)^{1/2}$$

K for a simple pendulum with a length of L is (mg/L). The frequency is:

$$f(\text{pendulum}) = \frac{1}{2\pi}\left(\frac{\cancel{m}g/L}{\cancel{m}}\right)^{1/2}$$

or

$$f(\text{pendulum}) = \frac{1}{2\pi}\left(\frac{g}{L}\right)^{1/2}$$

EXAMPLE PROBLEM 23–3: THE FREQUENCY OF A GUITAR STRING

A 30.0-cm long segment of guitar string is held between the bridge and a fret. The length of string has a mass of 3.80 g. If it is under a tension of 125.0 N, at what frequency will it vibrate when it is plucked?

SOLUTION

K for a vibrating string is:

$$K = \frac{\pi^2 T}{L}$$

The frequency is:

$$f = \frac{1}{2\pi}\left(\frac{K}{m}\right)^{1/2}$$

$$f(\text{string}) = \frac{1}{2\pi}\left(\frac{\pi^2 T}{m L}\right)^{1/2}$$

$$f(\text{string}) = \frac{1}{2\pi}\left(\frac{\pi^2 \times 125\text{N}}{3.80 \times 10^{-3}\,\text{kg} \times 0.300\ \text{m}}\right)^{1/2}$$

$$f(\text{string}) = 166\ \text{Hz}$$

(Hz = hertz, a unit of frequency equal to one cycle per second.)

The restoring force for a twisted wire has the form:

$$\mathbf{F} = -\frac{G\,C\,\theta}{L}$$

Singing Wires

Solids passing along the surface of a fluid can form whirlpools or eddy currents. A canoe paddle passed along the surface of water forms an eddy current on each side of the paddle. One eddy current spins clockwise; the other counterclockwise. It is not surprising that the eddies spin in opposite directions, because of the principle of conservation of angular momentum. The eddy currents are evenly spaced.

Similar whirlpools make wind whistle when it moves past an obstruction. Air moving past something, such as a wire, will leave a trail of eddy currents much like the whirlpools formed by a canoe paddle. Behind the wire is a series of alternating clockwise and counterclockwise whirlpools that gives the air behind the wire a transverse vibration.

A stretched wire has a natural frequency. If the wind has a frequency matching one of the overtones of the wire, a resonance between the wire and the wind occurs. A particularly loud vibration develops, and the wire sings.

The frequency of the vibration of air behind the wire is directly proportional to the speed of the air past the wire. Higher wind speeds will cause a higher tone. This frequency also depends on the thickness of the wire. The frequency is inversely proportional to the thickness of the wire; a thicker wire will cause a lower tone.

(Photograph courtesy of New York Power Authority.)

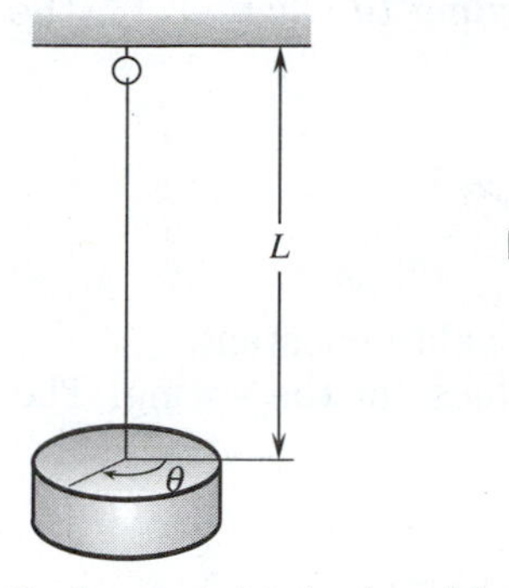

Figure 23–9: *A mass twists back and forth on a wire.*

where θ is the angle of twist, G is the shear modulus, L is the length of wire, and C is a constant.

EXAMPLE PROBLEM 23–4: THE TWISTED WIRE

A mass is hung on a wire (see Figure 23–9). The wire is given an initial twist. The frequency of rotation of the mass is found to be 0.62 cycles/s.

A. In the force equation, identify the symbol representing displacement.

B. What symbols represent the general proportionality constant K for Hooke's law?

C. What would be the frequency if the wire's length were cut in half?

SOLUTION

A. The displacement is the angle of twist (θ).

B.

$$K = \frac{G\,C}{L}$$

C.

$$\frac{f_2}{f_1} = \frac{1/(2\pi) \quad (K_2/m)^{1/2}}{1/(2\pi) \quad (K_1/m)^{1/2}}$$

or

$$\frac{f_2}{f_1} = \left(\frac{K_2}{K_1}\right)^{1/2}$$

$$f_2 = f_1 \times \left(\frac{G\,C/L_2}{G\,C/L_1}\right)^{1/2} = f_1 \left(\frac{L_1}{L_2}\right)^{1/2}$$

$$f_2 = 0.62\text{ Hz}\left(\frac{L}{L/2}\right)^{1/2} = 0.62\text{ Hz} \times (2.0)^{1/2}$$

$$f_2 = 0.88\text{ Hz}$$

23.5 ENERGY TRANSFER IN A SIMPLE HARMONIC OSCILLATOR

Work is done on a harmonic oscillator to start it in motion. A pendulum bob is pulled back and released. A spring is stretched and released. A wire is twisted. In all three cases, the work goes into storing initial potential energy in the system. Since harmonic oscillators obey Hooke's law, according to Chapter 13, the potential energy is in the form of:

$$PE = \frac{K}{2}(\Delta x^2)$$

where K is the general proportionality constant.

For a spring, $K = k$, the stiffness of the spring. The potential energy is:

$$PE(\text{spring}) = \frac{k}{2}(\Delta x^2)$$

For a stretched string, $K = \pi^2\text{T}/L$ at a small angle.

$$PE(\text{string}) = \frac{\pi^2\text{T}}{2L}\Delta y^2$$

For a simple pendulum, $K = (m\text{ g})/L$ at a small angle.

$$PE(\text{pendulum}) = \frac{m\text{ g}}{2L}(\Delta x^2)$$

where Δx is the horizontal displacement. It is left as an exercise to show that this is equivalent to $PE = m\text{ g}\,\Delta y$, where Δy is the vertical displacement.

The work stored in an oscillator is dependent on the square of maximum displacement of the system. The size of the maximum

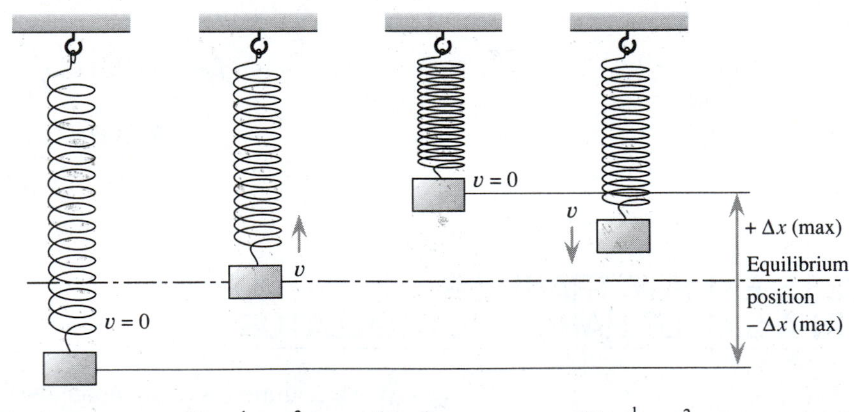

Figure 23–10: *As a mass bobs up and down on a spring, its energy oscillates between kinetic and potential energy.*

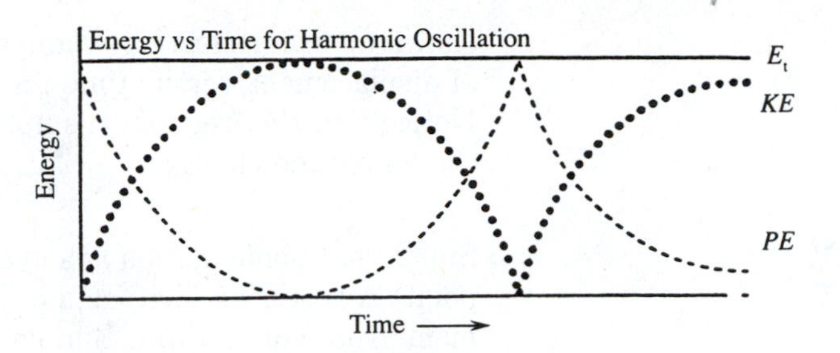

Figure 23–11: *Energy versus time in a harmonic oscillator. The sum of the kinetic energy (dotted line KE) and potential energy (dashed line PE) is a constant total energy (solid line E_t).*

displacement is also called the **amplitude.** The total amount of energy is proportional to the amplitude squared of the oscillator.

$$\textit{total energy} \propto \textit{amplitude}^2$$

Figure 23–10 shows a mass on a stretched spring. When it is released, the energy is initially all potential energy. As the mass approaches the equilibrium point, its potential energy decreases as it speeds up. The lost potential energy is converted to kinetic energy. As it passes through the equilibrium point, its energy is all kinetic energy. After the mass passes the equilibrium point, kinetic energy is used to compress the spring, storing up potential energy. At the top of the motion, all the kinetic energy is used up. It has been converted to potential energy. A restoring force accelerates the mass back toward the equilibrium position, starting the potential-to-kinetic energy transfer over again.

We can view harmonic motion as a system in which energy oscillates between potential and kinetic energy. Figure 23–11 shows a plot of kinetic and potential energy as a function of time in a harmonic oscillator. The sum of the potential energy and the kinetic energy is the total energy of the system. It is constant as long as there are no frictional losses.

When there is friction in the system, some of the kinetic and potential energy is converted to heat. The loss of mechanical energy

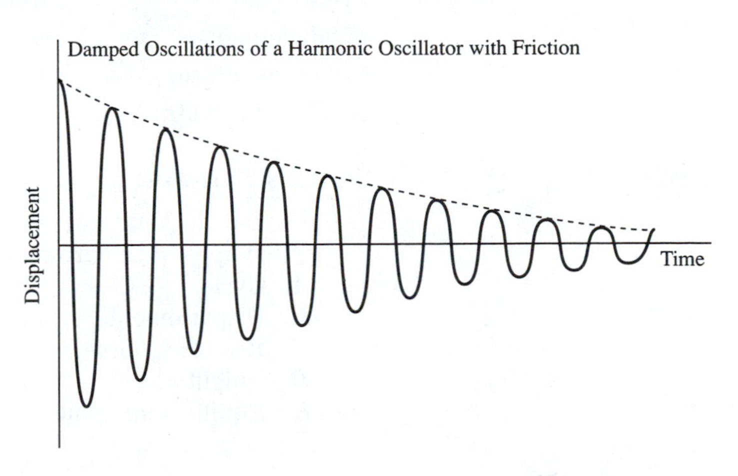

Figure 23–12: *Damped oscillations of a harmonic oscillator with friction. Mechanical energy lost to friction reduces the amplitude of the motion but does not change the frequency.*

shows up as a reduction in amplitude. Figure 23–12 shows the plot of displacement versus time for a harmonic oscillator with friction. Notice that the frequency is not affected by the energy losses; only the amplitude changes.

SUMMARY

Simple harmonic motion is a cyclic vibration around an equilibrium point. A restoring force acts in a direction opposite to the displacement from equilibrium. Simple harmonic oscillators obey Hooke's law: $F = -K\Delta x$, where Δx is the displacement and K is a proportionality constant depending on the geometry and physical properties of the oscillator. $K = k$ (stiffness) for a coil spring; $K = (m\,\mathbf{g})/L$ for a simple pendulum at small angles; $K = \pi^2\,\mathbf{T}/L$ for a vibrating wire under tension; $K = G\,C/L$ for a twisted wire or torsion bar.

The ratio of acceleration to displacement of a simple harmonic oscillator is constant. The cyclic motion of a harmonic oscillator behaves mathematically like the vertical components of motion of a particle moving in a circle at constant speed. The frequency of a harmonic oscillator is related to the proportionality constant (K) in Hooke's law by:

$$f = \frac{1}{2\pi}\left(\frac{K}{m}\right)^{1/2}$$

The energy stored in a harmonic oscillator is proportional to the square of the amplitude. Frictional forces reduce the amplitude of a harmonic oscillator without altering its frequency.

KEY TERMS

If you can explain the following terms to a friend or classmate, you understand their meaning. If you cannot explain the terms, you should reread the sections in which they are discussed.

amplitude
cycle
displacement
driven oscillator
equilibrium point
natural oscillators
reference circle
restoring force
simple harmonic motion

EXERCISES

Sections 23.1, 23.2:

1. Explain in your own words what the following terms mean when applied to harmonic motion.
   A. Cycle.
   B. Displacement.
   C. Restoring force.
   D. Amplitude.
   E. Equilibrium point.

2. Will the restoring force of a simple pendulum increase, decrease, or remain unchanged if the following things are increased? Assume an amplitude of 5°.
   A. The mass of the bob.
   B. The length of the string.
   C. The thickness of the string. (Assume negligible mass.)
   D. The lateral displacement of the bob.
3. How will the restoring force of a violin string be affected by an increase of the following things?
   A. The tension in the string.
   B. The mass of the string.
   C. The length of the string.

Section 23.3:

4. A harmonic oscillator is at its largest displacement. What can you say about its speed and acceleration?
5. A pendulum is passing through its equilibrium point. What can you say about its displacement, velocity, and acceleration?

Section 23.4:

6. Why are the strings on a guitar used to produce lower tones thicker than the ones used to produce high-pitched tones?
7. The restoring force of a simple pendulum is directly proportional to the mass of the bob. How does the mass of the bob affect the frequency of a simple pendulum?
8. The strings on guitar A are longer than the ones on guitar B. In order to tune an E string on guitar A, will the tension on the string need to be larger or smaller than on the E string on guitar B?
9. A pendulum clock is too fast. Should the pendulum be lengthened or shortened to correct the clock's speed?
10. A piano is in a room in which the temperature drops by 20°F. Will the pitch (frequency) of the piano wires increase or decrease?
11. The natural vibrational frequency of a quartz crystal can be used to regulate the frequency of electrical circuits. The proportionality constant (K) for Hooke's law for a quartz crystal is: $K = C\,A\,Y/L$, where Y is Young's modulus, L is the dimension of the crystal in the direction of the displacement, A is the cross-sectional area, and C is a numerical constant (see Figure 23–13).
    A. Write an expression for the frequency of a quartz crystal.
    B. What effect will decreasing the thickness (L) of the crystal have on the natural frequency of the crystal?

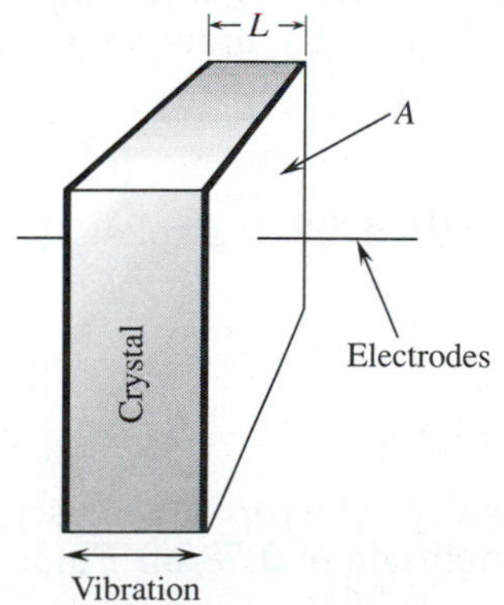

Figure 23–13: *Diagram for Exercise 11. The vibration of a quartz crystal interacts with an electric circuit.*

Section 23.5:

12. The amplitude of a simple pendulum is doubled. How will the following be affected?
    A. The total energy of the pendulum.
    B. The velocity of the bob as it passes through the equilibrium point.
    C. The frequency of the pendulum.
    D. The restoring force of the pendulum at maximum amplitude.
    E. The acceleration of the bob as it passes through the equilibrium point.

13. A mass hung on a coil spring bobs up and down, mechanical energy is lost to frictional forces. How does the lost energy affect the following?
    A. Amplitude.
    B. Maximum speed of the mass.
    C. Frequency.
    D. Maximum acceleration of the mass.
    E. The period of the harmonic motion.

PROBLEMS

Sections 23.1, 23.2:

1. What force is needed to stretch a spring with a stiffness of 2.0 lb/in by 4.3 in from the equilibrium condition?

2. A 1.30-kg bob hangs from a string 1.40 m long. What lateral displacement will result when a horizontal force of 0.50 N is exerted on the bob?

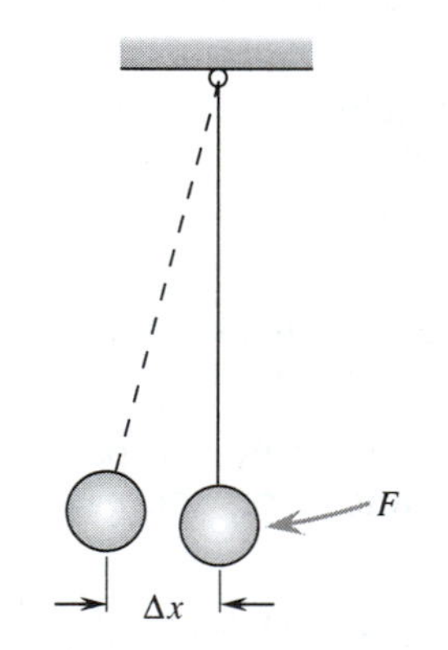

Figure 23–14: *Diagram for Problem 2.*

3. What force is required to give a 14-in ukulele string under 32 lb pressure a vertical displacement of 0.50 in?

4. A force of 128 lb is needed to twist a 2.30 ft long torsion bar through an angle of 6.0°. If the bar were 3.30 ft long, what force would be needed to twist it through an angle of 5.0°?

Section 23.3:

5. A gasoline engine has a piston with a 4.2-in stroke. If it operates at 1500 rpm, find:
    A. the amplitude of the piston.
    B. the piston's maximum speed.
    C. the piston's maximum acceleration.

6. On a windy day, tall buildings sway. The top of a skyscraper has a period of 170 s and an amplitude of 0.78 m. Find:
    A. the frequency.
    B. the maximum velocity.
    C. the maximum acceleration.

7. A large, double-bladed ceiling fan in a restaurant rotates with a period 1.60 s. The shadow of the blades cast on the wall by the late afternoon sun describes two harmonic oscillators. The maximum length of the shadow of one blade is 3.25 ft. Do a harmonic motion analysis of the motion of the shadow of the blade tip.
   A. What is the frequency?
   B. What is the maximum velocity?
   C. What is the maximum acceleration?

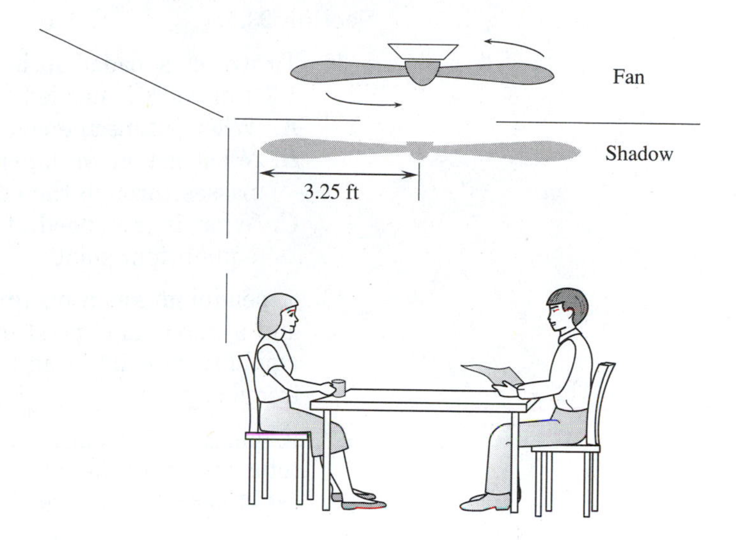

Figure 23–15: *Diagram for Problem 7. The shadow of a fan blade tip undergoes linear harmonic motion.*

Section 23.4:

8. A mass of $5\bar{0}0$ g is hung from a spring with a stiffness of 0.10 cm/kg. What is the period of the spring?

9. A $46\bar{0}0$-lb automobile with worn shocks oscillates as it hits a pothole. If the stiffness of the suspension springs is 5.1×10^3 lb/ft, what is the frequency of the car's up and down motion?

10. A child is on a swing with an effective length of 9.5 ft. How long will it take the child to make one complete cycle back and forth (see Figure 23–16)?

11. A $23\bar{0}$-g mass is hung by a length of twisted fishline. The mass spins with a period of 3.5 s. What would be its period if the length of the line were doubled?

12. The string of a hunting bow is under 90 lb tension and is 44 in long. The weight of the string is 0.122 lb. When an arrow is fired, what will be the vibrational frequency of the string after the arrow is released?

13. An 80.0-cm wire has a tension of 64.2 N and a mass of 4.3 g. What will be its vibrational frequency?

*14. A derrick lifts a heavy load at an angle slightly off from the vertical. The load starts to swing back and forth as a pendulum.
   A. What is the period of the load when the cable has an effective length of 18.2 m?
   B. What is the period when the effective length is 7.0 m?
   C. If the amplitude of the harmonic motion stays the same, what happens to the maximum acceleration and maximum speed of the load as it is lifted?

Section 23.5:

15. To stretch a loaded spring 6.5 cm from the equilibrium point, 1.7 J of work is needed. The mass on the spring is 245 g.
   A. What potential energy is stored in the spring?
   B. What maximum kinetic energy does the mass have as it passes through the equilibrium point?
   C. What is the speed of the mass as it travels through the equilibrium point?

16. A pendulum has a maximum lateral displacement of 14.0 in and a maximum speed of 134 in/s as it passes through the equilibrium point. Use the results of Problem 14 to find its frequency.

17. A guitar string vibrates with an amplitude of 0.50 cm. The total energy of the vibrating string is 4.0 J. If the amplitude is increased to 1.00 cm, what will be its new total energy?

Figure 23–16: *Diagram for Problem 10. Find the period of a child on a swing.*

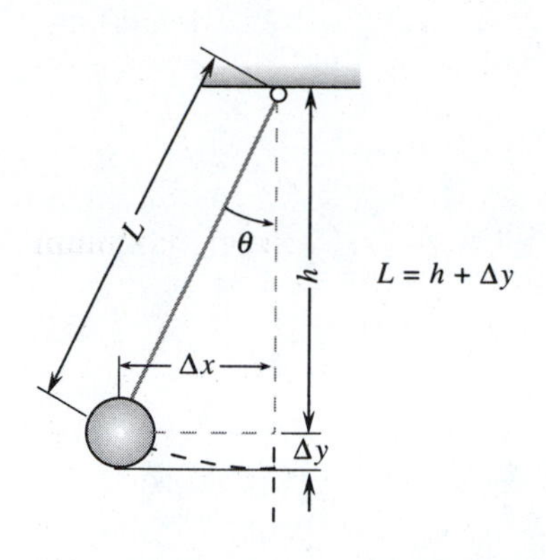

Figure 23–17: *Diagram for Problem 19.*

***18.** For a harmonic oscillator show that:

A. $$\frac{K}{m} = \frac{v(\max)^2}{x(\max)^2}$$

B. $$f = \frac{1}{2\pi}\frac{v(\max)}{x(\max)}$$

****19.** See Figure 23–17. Δy is the vertical displacement of a pendulum; Δx, the horizontal displacement. Use the small angle approximations $\sin\theta = \theta$ and $\cos\theta = (1 - \theta^2/2)$ to show that:

$$PE = m\,g\,\Delta y = \frac{m\,g}{2L}(\Delta x^2)$$

Chapter 24

WAVES

This photo shows some of the destruction borne by the Cypress Structure during the 1989 earthquake in San Francisco. This upper level of the viaduct collapsed onto the lower level. In this chapter, you will learn how waves inside the earth can create such disaster above the crust. (Photograph by Michael Macor; courtesy of the Oakland Tribune.)

OBJECTIVES

In this chapter you will learn:

- that energy can be transported by waves
- how to identify the amplitude, wavelength, and frequency of a wave
- to calculate the phase velocity of a wave
- how rays are related to wave fronts
- to determine how energy is related to frequency and amplitude of a mechanical wave
- how two or more waves can be superposed
- that mechanical waves undergo a 180° phase shift when reflected from a fixed boundary
- how resonant waves are produced and to calculate their wavelengths and frequencies
- to distinguish between longitudinal (compressional) waves and transverse waves
- how the velocity of a mechanical wave is related to the mechanical properties of the material through which it moves

Time Line for Chapter 24

1618 Francesco Grimaldi discovers diffraction patterns with light, leading him to believe that light is a wave.

1690 Christiaan Huygens publishes a work describing a wave theory of light.

1746 Leonhard Euler develops a mathematical wave theory of refraction.

1801 Johann Ritter discovers ultraviolet radiation.

1866 August Kundt designs a method to measure the speed of sound in gases.

In April, 1860, the Russell, Majors and Waddell stagecoach company established the famed Pony Express route. Relays of horses and riders moved $\frac{1}{2}$-oz mail parcels along the 1838-mi route from St. Joseph, Missouri, to Sacramento, California. Occasionally, the goal of 10-day service was achieved. In October, 1861, the first transcontinental telegraph line went into operation, making it possible to send messages across the United States in less than one day. In October, 1861, the Pony Express went out of business.

Today we have choices for sending information. We can transport a letter or parcel by the U.S. Postal Service or by one of several messenger services. Some achieve the goal of 24-h service. Information gets from one place to another by transporting a particle of some sort.

We can also pick up a telephone and send information across the United States in less than a minute. Information is transmitted by electronic waves. No particle is moved over the length of the information path. The information is superimposed on the waves.

Energy, like information, can be transmitted by either a particle or a wave. Kinetic energy can be moved from one place to another by a particle. A bullet transmits kinetic energy from the muzzle of a rifle to a target. Potential energy can also be transported from one place to another by particles. Petroleum can be transmitted along a pipeline. Dry cell batteries can be sent by parcel post.

Energy can be transmitted by waves. We receive solar energy from the sun by electromagnetic waves. Earthquakes send out vibrational waves that can move mountains and topple buildings miles from the site of the earthquake's epicenter.

In this chapter, we will study the basic properties of waves. In the next few chapters we will look at specific kinds of waves and how they can be used.

24.1 PHASE VELOCITY

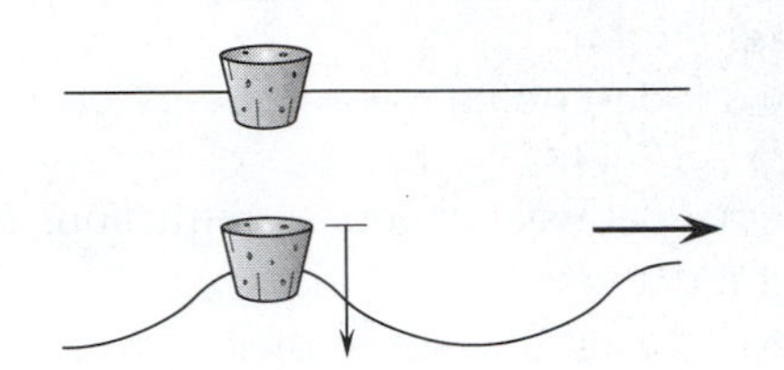

Figure 24–1: Ripples cause a floating cork to vibrate up and down.

Figure 24–1 shows a cork floating in a pool of water. If the water is not disturbed, the cork will not move. If there are ripples on the surface of the water, the cork will bob up and down as it rides on the wave. When the cork falls into a trough, it falls below the level it has in calm water (equilibrium point). Crests lift it above the level it had in calm water. Its maximum displacement above or below the equilibrium point is the **amplitude** of the wave (see Figure 24–2).

The frequency of oscillation of the cork depends on the number of crests that pass through its location during some time period. If N is the number of crests passing the cork during the time t, the frequency of the wave is:

$$f = \frac{N}{t} \qquad \textbf{(Eq. 24–1)}$$

The frequency of a wave is the number of waves passing a fixed point in a unit of time.

As we move along a wave we notice a repeating pattern. The shape from crest to crest is repeated over and over again. The length of this repeating pattern is called the **wavelength** (λ) (see Figure 24–2).

The distance between adjacent peaks of a wave is called the wavelength (λ).

Water ripples move on the surface of the water. The water does not move along with the wave. Instead, the surface of the water bobs up and down like the cork. This is characteristic of a wave. The wave moves *through* a medium. The medium carries the wave, but in most cases it does not move with the wave.

We can measure a wave's speed through a medium by keeping track of the motion of a particular wave crest. This motion is called the phase velocity. We are not concerned with the motion of the water surface up or down, but only with the speed of the crest along the water's surface (see Figure 24–3).

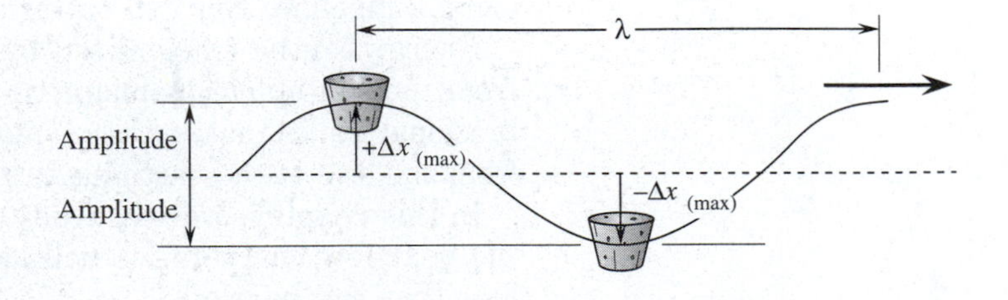

***Figure 24–2:** The amplitude of a wave is the maximum displacement from equilibrium. The wavelength (λ) is the distance between peaks.*

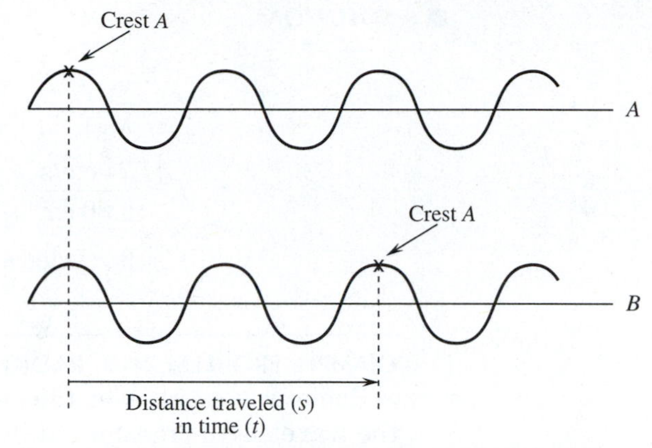

Figure 24–3: *The phase velocity is the speed with which a wave peak moves.*

The phase velocity of a wave is the velocity a peak moves in the direction of the motion of the wave.

The speed of the crest is the distance (s) it moves along the water divided by the time (t) it takes to travel that distance.

$$v = \frac{s}{t}$$

The distance can be expressed in units of wavelengths (λ). If the distance traveled is N wavelengths:

$$s = N\lambda$$

or

$$v = \frac{N\lambda}{t}$$

But N/t is the frequency of the wave, so the phase velocity can be expressed as:

$$v = f\lambda \qquad \textbf{(Eq. 24–2)}$$

☐ **EXAMPLE PROBLEM 24–1: WATER RIPPLES**

Water ripples with a wavelength of 0.80 cm have a phase speed of 74 cm/s. What is the frequency of the ripples?

■ *SOLUTION*

$$f = \frac{v}{\lambda}$$

$$f = \frac{74 \text{ cm/s}}{0.80 \text{ cm}}$$

$$f = 93 \text{ cycles/s} = 93 \text{ Hz}$$

☐ **EXAMPLE PROBLEM 24–2: RADIO WAVES**

Radio waves travel at the speed of light (186,000 mi/s). What is the wavelength transmitted by an AM radio station that operates at 860 kHz?

■ *SOLUTION*

$$\lambda = \frac{v}{f}$$

$$\lambda = \frac{186{,}000 \text{ mi/s}}{8.6 \times 10^5 \text{ cycles/s}}$$

$\lambda = 0.22$ mi, roughly a quarter of a mile.

24.2 WAVE FRONTS AND RAYS

Look at Figure 24–4. A rock is dropped in a pond. Ripples radiate outward from the disturbance. The wave peaks move out in circles of increasing radii. Figure 24–4a shows a cross section of the wave disturbance. Figure 24–4b is a sketch of the wave viewed from over-

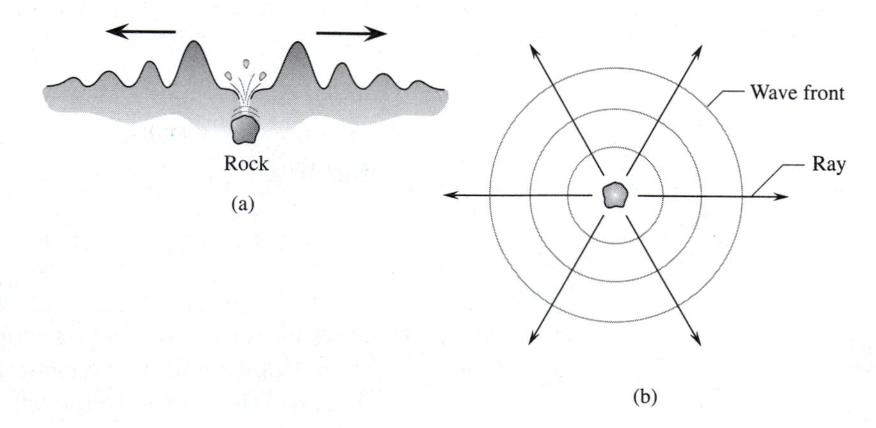

Figure 24–4: (a) (*Side view*) *Ripples moving away from a disturbance on the water's surface.* (b) (*Top view*) *Ripples moving away from a point source. The concentric circles are the wave peaks (wave fronts), and the rays point in the direction of motion of the wave fronts.*

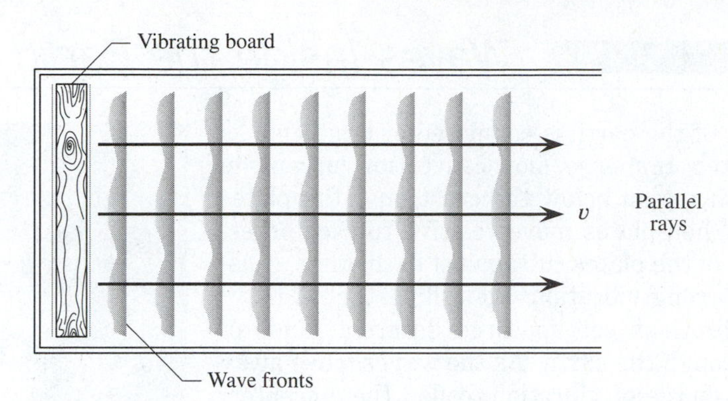

Figure 24–5: A vibrating board in a narrow trough creates parallel waves.

head. The circles represent the wave peaks. The circles are called **wave fronts.** They show the location of the waves. The rays in the diagram show the direction of motion of the wave fronts. Notice that they are perpendicular to the wave fronts.

Figure 24–5 shows a set of waves sent out by a vibrating board in a long narrow pan. In this case, the wave fronts are parallel. Waves that travel with parallel wave fronts are called **plane waves.** Again, the direction of motion is perpendicular to the wave fronts. In general we can say:

Rays *are always perpendicular to wave fronts.*

24.3 WAVES AND ENERGY

Energy, Amplitude, and Frequency

In Figure 24–1 the cork bobbing on water is a harmonic oscillator. The energy it receives is from the water wave. The maximum velocity of the oscillator depends on the amplitude and frequency (refer to Section 23.3).

$$\text{v(max)} = 2\pi f \, \Delta x(\text{max})$$

where Δx is the amplitude.

Let us use A to represent the maximum amplitude, Δx(max). The maximum kinetic energy of the cork with mass m is:

$$KE(\text{max}) = \frac{m}{2}(\text{v}^2) = \frac{m}{2}(2\pi f A)^2$$

$$KE(\text{max}) = 2\,m\pi^2 f^2 A^2$$

Since we are finding the *maximum* kinetic energy of an oscillator, we are determining the total energy, E. The energy imparted to

Waves Inside the Earth

(*Photograph by Michael Macor; courtesy of the* Oakland Tribune.)

The crust of the earth is composed of large chunks called plates. Immense, slow convection currents in the molten region below the crust push the plates around. When plates move relative to each other, the edges of the plates rub against each other, causing very strong vibrations we call earthquakes.

The vibrations set up three different kinds of waves through the earth. As the waves move away from the source of vibration, called the epicenter, their intensity decreases.

One kind of wave is a surface wave called an L wave, or Love wave. These were first identified by Augustus Love in 1863. L waves are much like the rolling surface waves on the ocean. In the ocean, water under the wave rolls in a circular motion. L waves do the same thing. They cause a rotating motion of the ground that causes tall buildings to sway. By observing the speed of these waves, scientists have been able to deduce the thickness of the earth's crust.

Another kind of earthquake wave is a P wave. These are longitudinal waves that push and pull material in the direction they travel. In effect they are sound waves.

The third kind of wave is an S wave. These are transverse waves. They cause material to vibrate at right angles relative to their direction of travel.

As the waves travel through the interior of the earth, differences in density affect the direction and speed with which they travel. They are refracted by the earth's materials. P waves move easily through liquids, but S waves can move only through a solid.

A network of seismic stations has been set up around the world to monitor the waves as they return to the earth's surface. Measurements of their reflection and refraction have made it possible to develop a complex model of the layers of solid and liquid in the earth's interior.

---

the cork by the wave is proportional to the square of the frequency and the square of the amplitude. The amplitude determines the distance up and down the cork must travel in one cycle, and the frequency specifies how quickly a cycle must be finished. Together they determine the average vibrational speed of the cork.

The energy of the wave itself is related to frequency and amplitude in the same way. This is true whether we discuss waves on a stretched string, or sound waves, or any other kind of *mechanical* wave. The frequency dependence of energy for electromagnetic radiation is slightly different.

$$E \propto f^2 A^2 \qquad \textbf{(Eq. 24–3)}$$

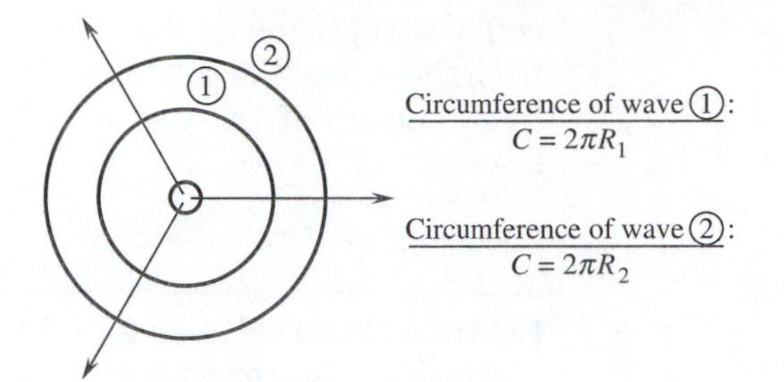

Figure 24–6: If there is no friction, when the wave front is in position 2 it will have the same total energy that it had at position 1.

Energy and Wave Fronts

As circular ripples spread out on water from a point disturbance, the amplitude decreases as the radius of the wave fronts increases. This occurs because energy is conserved along a wave front. Look at Figure 24–6. The total energy along the wave front is the energy per unit length of front times the length of the front.

The energy per unit length of a two-dimensional wave is called the **linear intensity (I)** of the wave.

$$\textit{linear intensity} = \frac{\textit{energy}}{\textit{length of two-dimensional wave}}$$

At wave front 1:

$$E = kf^2 A_1^2(2\pi R_1) = I_1(2\pi R_1)$$

where k is a proportionality constant.

A little later the wave front has spread out to circle 2. If there have been no losses caused by frictional forces, the wave front should have the same energy. The energy is spread over a larger circle. The total energy of the front is:

$$E = kf^2 A_2^2 (2\pi R_2) = I_2 (2\pi R_2)$$

The two expressions represent the same amount of energy. We can equate the two equations.

$$I_2 R_2 = I_1 R_1 = constant \qquad \textbf{(Eq. 24–4)}$$

or

$$A_2^2 R_2 = A_1^2 R_1$$

For two-dimensional waves the amplitude decreases with the square root of the distance traveled.

The reverse effect can be seen in a cup of coffee. Set a cup of coffee abruptly on a table. The vibration of the cup will send waves from the side of the cup toward the center. The amplitude often gets large enough at the center of the cup to project a drop of coffee upward.

☐ **EXAMPLE PROBLEM 24–3: RIPPLES AGAIN**

Ripples on water emanating from the site of a dropped stone have an amplitude of 1.2 cm at a distance of 80 cm from the center of disturbance. What is the amplitude 2.0 m from the center?

■ ***SOLUTION***

$$A_2 = A_1 \left(\frac{R_1}{R_2}\right)^{1/2}$$

$$A_2 = 1.2 \text{ cm} \left(\frac{0.80 \text{ m}}{2.0 \text{ m}}\right)^{1/2}$$

$$A_2 = 0.76 \text{ cm}$$

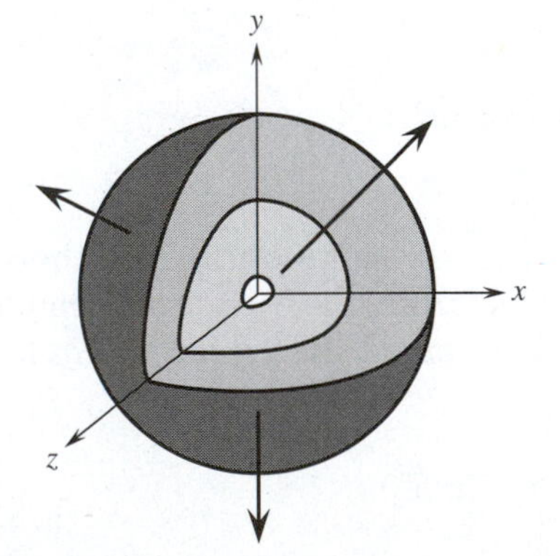

Figure 24–7: *As a wave spreads out in three dimensions from a point source, the wave fronts form spherical shells. In the absence of friction, each shell contains the same total energy.*

Sound waves and electromagnetic waves usually travel in three-dimensional waves. Wave fronts spread out from a point source in spherical shells (see Figure 24–7). In this case, it makes more sense to express energy per unit area of wave front.

$$\textit{intensity} = \frac{\textit{energy}}{\textit{area of three-dimensional wave}}$$

The total energy of a spherical wave front is:

$$\textit{energy} = \textit{intensity} \times \textit{area}$$

$$E = I\,(4\pi R^2)$$

We can use the same analysis used for the two-dimensional wave.

$$I_2 R_2^2 = I_1 R_1^2 \qquad \textbf{(Eq. 24–5)}$$

One-dimensional waves are waves such as the vibrations on a stretched string or plane waves such as those shown in Figure 24–5. The dimensions of the wave are fixed. If there are no frictional forces, intensity (or amplitude) will not change as the wave moves from its source.

□ **EXAMPLE PROBLEM 24–4: THE RADIO STATION**

A local radio station transmits 15,000 W. What is the intensity of the signal 15 km from the station?

■ ***SOLUTION***

$$I = \frac{E}{4\pi R^2}$$

$$I = \frac{15{,}000 \text{ W}}{4\pi (1.5 \times 10^4 \text{ m})^2}$$

$$I = 5.3 \times 10^{-6} \text{W/m}^2$$

24.4 SUPERPOSITION PRINCIPLE

Look at Figure 24–8. Two people send wave pulses along a string. One pulse moves to the right; the other, to the left. The waves interact when they meet at the center. The amplitude of the combined wave is the sum of the single waves. After the waves pass each other, they continue on at their original size.

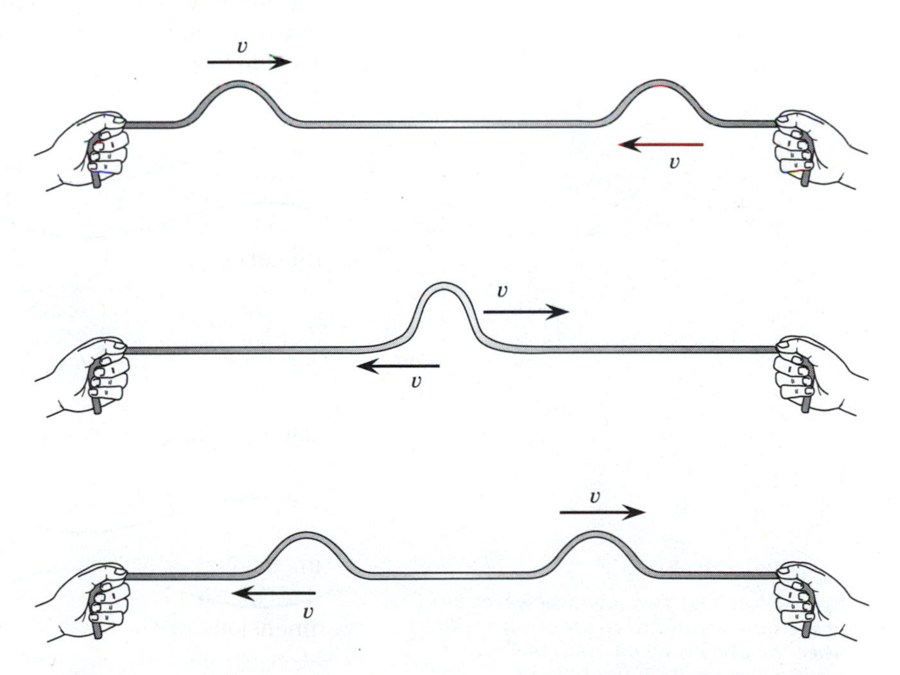

Figure 24–8: *When two "up" wave pulses meet, their amplitude is doubled.*

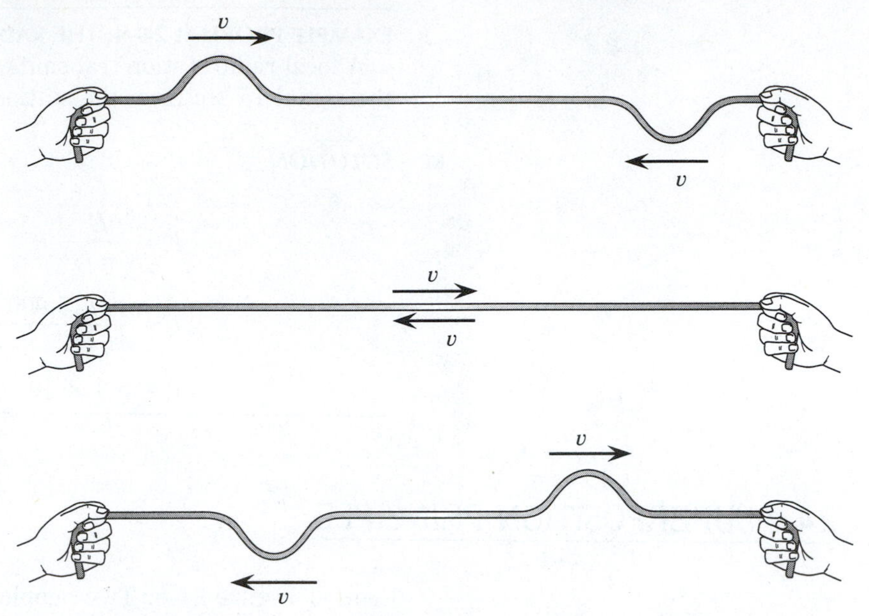

Figure 24–9: When an "up" pulse meets a "down" pulse, their amplitudes cancel.

Figure 24–9 shows a similar experiment. The difference is that one wave pulse is up and the other is down. When they meet, the net amplitude is no net displacement. The waves cancel out. After the waves pass each other, they continue on at the same amplitude as before they met.

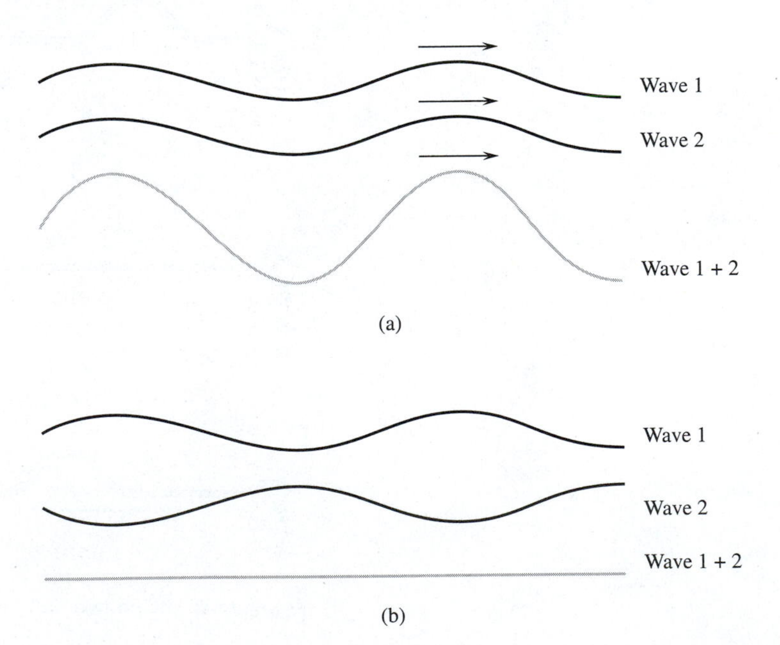

Figure 24–10: (a) *Two identical waves in phase superimpose to create a wave with twice the original amplitude.* (b) *Two identical waves 180° out of phase superimpose to cancel.*

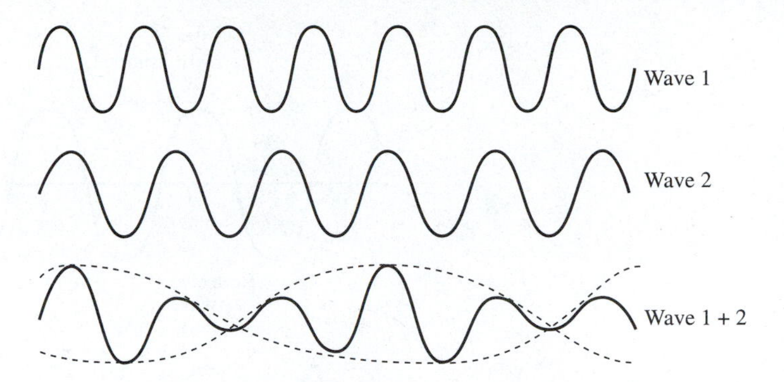

Figure 24–11: *Two waves of nearly the same wavelength combine to make wave groups.*

These two experiments give us an idea of how waves interact. Waves combine by adding amplitudes using vector algebra. The net amplitude at any point of two or more waves is the vector sum of the amplitudes. This is the **superposition principle.**

The net amplitude of the combination of two or more waves is the vector sum of the displacements.

The superposition principle applies to waves no matter what direction of motion they have. Figure 24–10 shows pairs of waves traveling in the same direction. The same rule applies.

If two waves are traveling in the same direction with nearly the same frequency, the superposition forms **wave groups** (see Figure 24–11). A longer wavelength is superimposed on the average wavelength of the individual waves. The bow waves created by a boat behave this way. The longer wave pattern is called the **group wave.**

24.5 REFLECTION OF WAVES

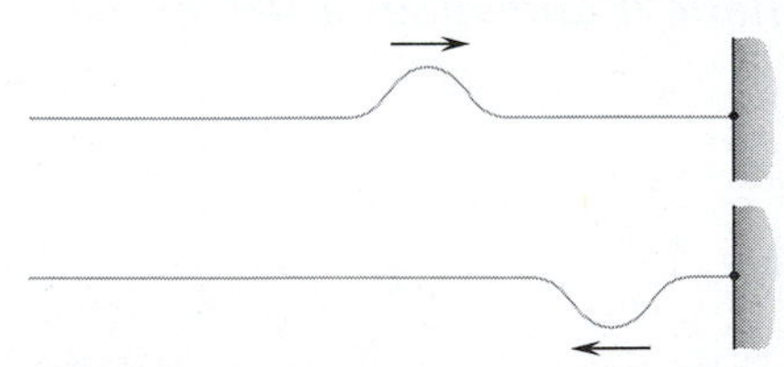

Figure 24–12: *A wave pulse is reflected by the fixed end of a string. Notice that the reflected wave is upside down.*

Look at Figure 24–12. A wave pulse can be made to move along a string fixed at one end by giving the free end a rapid up-and-down motion. The wave front moves along the string until it collides with the wall at the fixed end. The wave can no longer move forward. The energy of the wave is conserved by rebounding. The wave moves back along the string. The rebounding wave is called a **reflected wave.**

The fixed end of the string cannot move. A fixed point for wave motion is called a **node.** This puts a condition on the reflecting wave. The sum of the amplitudes of the incoming wave and the reflected wave must be zero. The reflected wave must have the same amplitude as the incoming wave, but a displacement in the opposite direction.

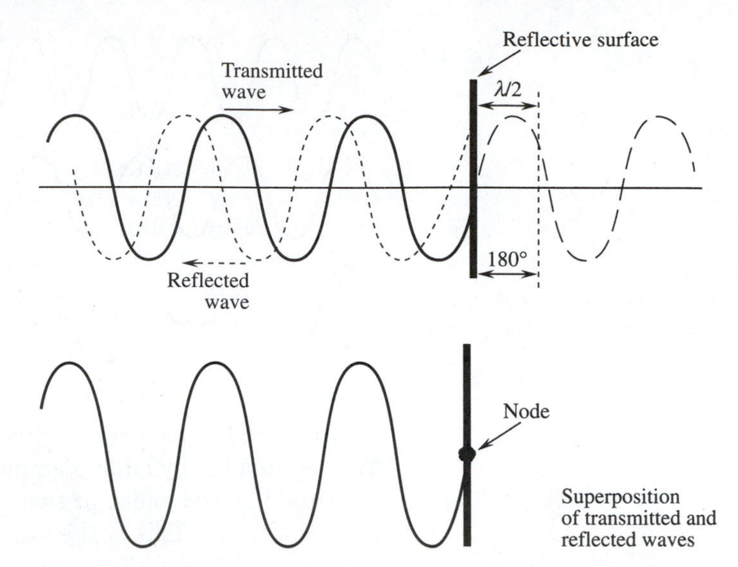

Figure 24–13: Superposition of reflected and transmitted waves. The reflected wave undergoes a 180° phase shift during reflection. The transmitted wave and reflected wave combine to make a wave that always has a node at the reflection point.

In Figure 24–13, the dashed line represents the reflected wave; the dotted line is the shape the incoming wave would have if it were possible for it to continue past the fixed end of the string, which is a node. If we measure exactly half a wavelength (180°) past the point of reflection, we find the displacement needed for the reflected wave. This will work no matter what displacement the incoming wave has.

In the diagram there is a discontinuity between the reflected wave and the incoming wave. There is no need to worry. What we actually observe is the superposition of the two waves. This is a continuous wave with a zero amplitude at the fixed end of the string.

This property of reflection is true for any wave. Water waves, sound waves, electromagnetic waves, and any other kind of wave undergo reflection with a 180° phase shift if there is a node at the point of reflection.

When a wave is reflected at a node it undergoes a 180° phase shift.

24.6 RESONANCE

Figure 24–14 shows a wave pulse moving along a string with both ends fixed. It is reflected first at one end (node) and then, later, at the other end. At first, the pulse is up; after the first reflection, it is down. When it is reflected the second time, at the other end, the pulse is turned upright into its original orientation.

Figure 24–15 shows an impossible situation. We are trying to reflect a continuous train of waves off both end nodes of a string. A train of waves moves to the right and undergoes the first reflection. When the waves return to the starting point, there is no way that the amplitude of the two waves can cancel at what should be a node. If a wave of this sort is sent down a string, the reflected wave and the transmitted wave will shortly interfere with each other. The string will wobble with no particular pattern, and the wave energy will be turned into internal energy.

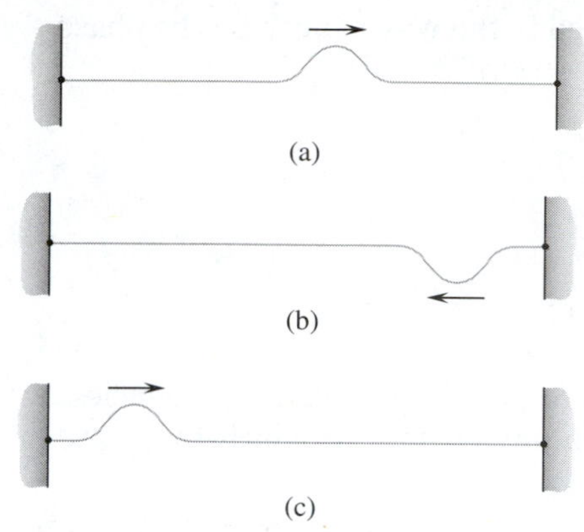

Figure 24–14: A pulse is reflected at both ends of a string. After the second reflection (c), it is again in the upward position.

Figure 24–16 shows cases of **resonance.** Reflection can occur at both nodes. At each end of the string, the wave is flipped over by a 180° phase shift, to make an exact match with the wave going the other way.

Notice that the length of the string is related to the wavelength. The longest wave that will reflect at both ends of the string has a wavelength of twice the length of the string. This is the **fundamental wavelength.** Wavelengths of one-half, one-third, and one-fourth wavelength are also reflected at both ends of the string. This corresponds with an integer number of wavelengths in a distance of 2 L. These are called the **overtones** of the fundamental wavelength. In general, we can say that the wavelengths corresponding to resonance are:

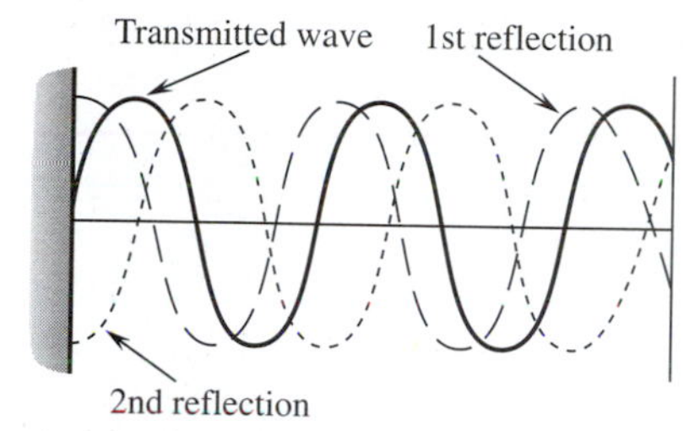

Figure 24–15: For most wavelengths, a multiple reflected wave will interfere with itself in a destructive way.

$$N\lambda = 2L$$

where $N = 1, 2, 3 \ldots$, or

$$\lambda = \frac{2L}{N} \qquad \textbf{(Eq. 24–6)}$$

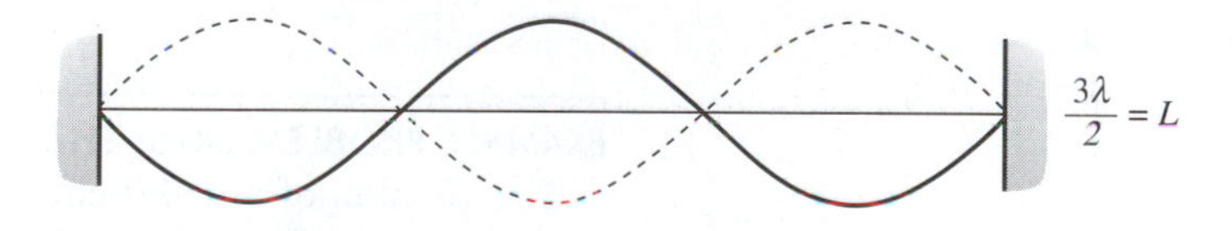

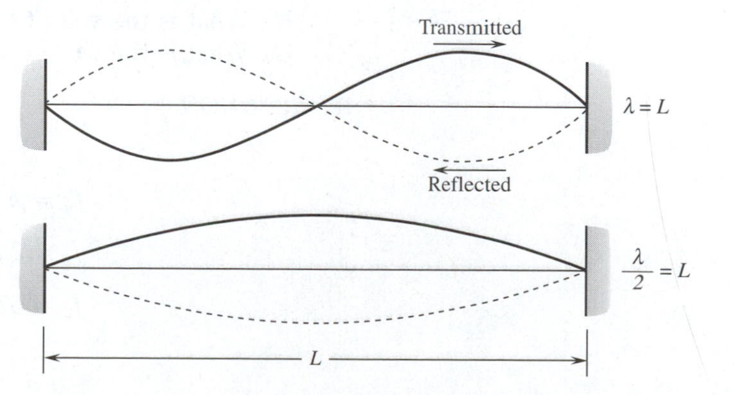

Figure 24–16: Cases where the distance (L) between reflection points is a multiple of half wavelengths. After the second reflection, the wave is in phase with the original transmitted wave. This is the resonance condition.

The frequency of any wave is related to the wavelength by the phase velocity.

$$f\lambda = \mathrm{v}$$

or

$$f = \frac{\mathrm{v}}{\lambda}$$

We can substitute Equation 24–6 to get the resonant frequencies. This equation is good for waves in any system in which there are nodes at both ends.

$$f = \frac{N\,\mathrm{v}}{2\,L} \qquad \textbf{(Eq. 24–7)}$$

The overtone frequencies can be expressed in terms of the fundamental frequency (f_1):

$$f_1 = \frac{\mathrm{v}}{2\,L}$$

and the overtones are:

$$f_n = n\,f_1$$

where $n = 2, 3, 4. . . .$

□ **EXAMPLE PROBLEM 24–5: THE GUITAR**

The tension of an 80.0 cm long guitar string is adjusted to have a fundamental frequency of 288 Hz.

A. What are the first two overtones?
B. What is the velocity of waves on the string?
C. What is the wavelength of the first overtone?

■ ***SOLUTION***

A.

$$f_n = n\,f_1$$

$$f_2 = 2\,f_1 = 2\;(288\ \mathrm{Hz})$$

$$f_2 = 576\ \mathrm{Hz}\ \text{(first overtone)}$$

$$f_3 = 3\,(288\text{ Hz})$$

$$f_3 = 864\text{ Hz (second overtone)}$$

B.

$$v = 2\,L\,f/N,\text{ for } N = 1:$$

$$v = 2\,(0.800\text{ m})\,(288\text{ cycle/s})$$

$$v = 461\text{ m/s}$$

C.

$$\lambda_2 = \frac{2\,L}{2} = \frac{2\,(80.0\text{ cm})}{2}$$

$$\lambda_2 = 80.0\text{ cm}$$

24.7 KINDS OF WAVES

Transverse Waves

Water waves, waves along a stressed string, light, and other electromagnetic waves are transverse waves. The amplitude oscillates in a direction at right angles to the direction of motion (see Figure 24–17). This kind of wave is called a **transverse wave.**

Tie one end of a string to a door handle. Or, if you have a stringed instrument, tighten a string and pluck it. As it vibrates, look at it from different angles. You will notice it is not only vibrating up and

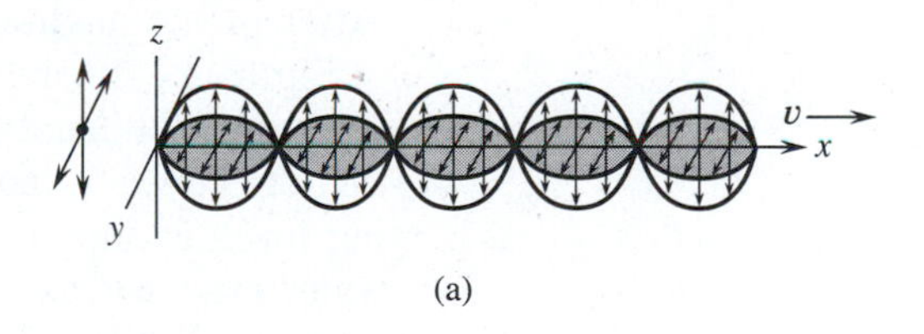

(a)

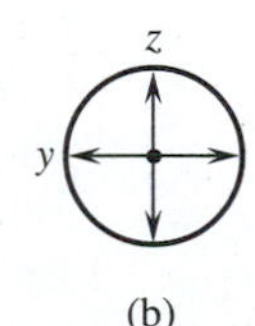

(b)

Figure 24–17: A transverse wave vibrates at right angles to the phase velocity. (a) A perspective view. (b) A view looking along the x axis.

down, but sideways, too. This is the way most transverse waves behave.

Sometimes waves vibrate only along one transverse axis. If you have something heavier, such as a piece of clothesline, again tie one end to something. You can send waves along the line with an up-and-down motion with your hand. The amplitude of the waves you create is along the vertical axis only. You have created what is called a **polarized wave.** Transverse waves can be polarized either vertically or horizontally. Figure 24–18 shows some polarized and partially polarized transverse waves.

There are several situations in which transverse waves are polarized. In Chapter 26 we will look at some cases for light. Light reflected from a surface is usually partially polarized. Some materials will transmit light vibration along one direction; other materials split a light wave into two separate polarized waves traveling at different directions through the material.

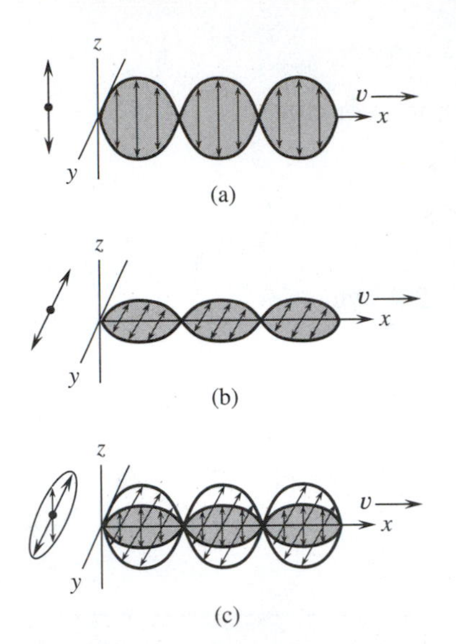

Figure 24–18: Polarized waves moving in the x direction. (a) Vibration along the z axis. (b) Vibration along the y axis. (c) Partially polarized wave. The amplitude of the wave is reduced along the z axis.

Longitudinal Waves

Electromagnetic waves can travel through all sorts of things: solids such as glass or plastic, clear liquids, and gases. They can even pass through a vacuum. Electromagnetic waves supply their own medium on which waves move.

Mechanical waves are different. They need a medium. Ripples on water depend on the properties of water. No water means no ripple. Sound cannot move through a vacuum. It needs a medium to carry the wave. The phase speed and wavelength depend upon the properties of the medium through which the mechanical waves move.

Figure 24–19 shows a transverse wave moving along a string. As it moves, the front end of the wave moves upward, while the back end of the wave moves downward. This is a shearing process. The ing force is supplied by the tension on the string. All mechanical transverse waves depend on shearing forces to move through a medium. While solids have a shear modulus, liquids and gases do not. A different kind of wave moves through these materials.

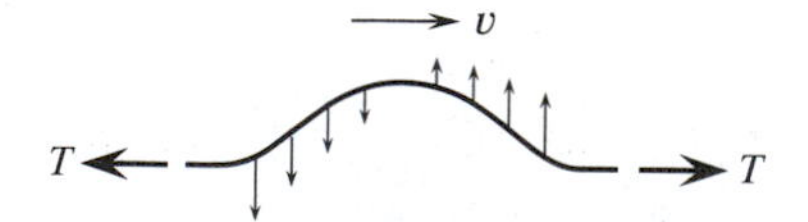

Figure 24–19: As a transverse wave moves along a string, the front end rises, while the back end subsides. The string's tension causes the needed shear stress.

If we place a Slinky toy on a table and move the end of the Slinky back and forth we create a **compressional wave** (see Figure 24–20). When we move the end of the spring forward, the coils of the spring are compressed. The momentum moves the compression along the spring. When we move the end of the coil back, the coils are pulled farther apart. As we repeat this process several times, alternate compression and rarifications occur. These compressed and sepa-

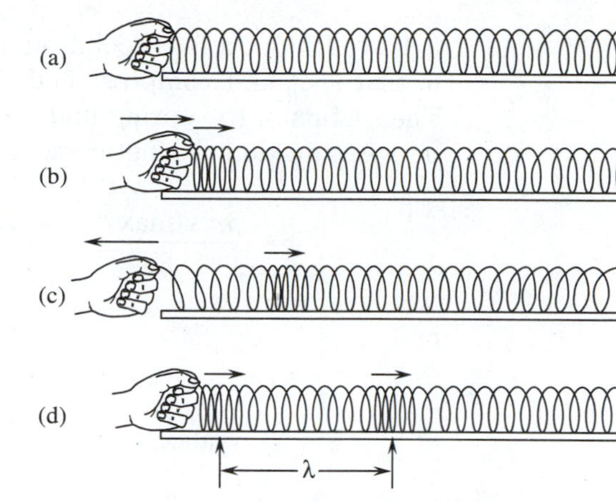

Figure 24–20: (a) *One end of a Slinky is held in the hand.* (b) *The end of the Slinky is pushed forward to create a compression.* (c) *The compression continues forward along the Slinky as the end is pulled back.* (d) *The next compressional wave is formed. The wavelength* (λ) *is the distance between compressions.*

rated sections of the coils move down the length of the spring. This compressional wave is also called a **longitudinal wave.** The vibrations are in the same direction as the wave velocity.

Compressional waves are the only kind of mechanical waves that can move through all three states of matter: solid, liquid, and gas. Sound waves are compressional waves. In air, a vibrating source causes the air around it to be first compressed and then "rarified," like the coils in the slinky. These variations in air density spread out from the source, creating sound (see Figure 24–21).

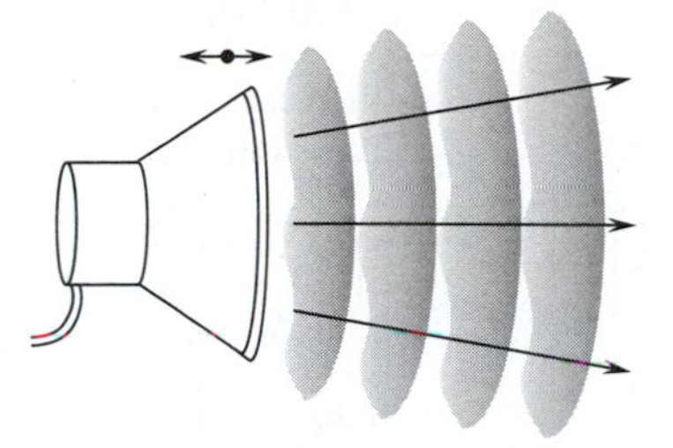

Figure 24–21: Sound waves made by a speaker are compressional waves that cause a variation in the density of the air.

Phase Speeds of Waves in Different Mediums

Figure 24–22 shows two long springs. We can send a compressional wave pulse along a spring by oscillating one end back and forth. If the spring is stiff (elastic property), the pulse will move rapidly. If the spring is soft (low k), the pulse will move slowly. A large elastic property enhances the motion of mechanical waves. If the mass (inertial property) of the spring is large, the acceleration of the waves will be small compared with that for a lighter spring with the same stiffness. A large inertial property hinders the motion of mechanical waves. A look at the relationship between kinetic energy and potential energy may give us a better understanding of the relationship between phase speed and the inertial and elastic properties.

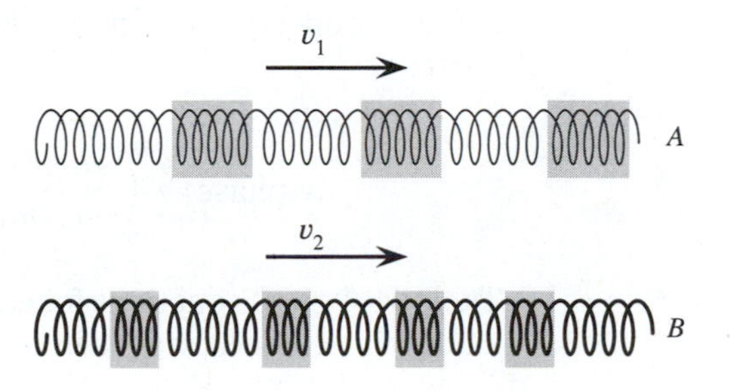

Figure 24–22: If springs A and B have the same spring constant, waves will move more slowly in the more massive spring B.

If we look at a fixed point on a spring, we will see that the coils in that spot first compress and then expand as the wave goes by. The section of the spring undergoes harmonic motion. It oscillates between maximum kinetic energy and maximum potential energy.

$$\frac{m\,\mathrm{v(max)}^2}{2} = \frac{k\,\Delta x(\mathrm{max})^2}{2}$$

or

$$\mathrm{v(max)} = \left(\frac{k}{m}\right)^{1/2} \Delta x(\mathrm{max})$$

Table 24–1: Phase speeds formulas for mechanical waves in different mediums.

KIND OF WAVE AND MEDIUM	PHASE SPEED (v)	VARIABLES
Compressional wave in gases and liquids	$\left(\frac{B}{\rho}\right)^{1/2}$	B = bulk modulus ρ = mass density
Longitudinal wave in solids	$\left[\frac{(B + 4S)/3}{\rho}\right]^{1/2}$	B = bulk modulus S = shear modulus ρ = mass density
Transverse wave in solids	$\left(\frac{S}{\rho}\right)^{1/2}$	S = shear modulus ρ = mass density
Longitudinal wave in a thin rod	$\left(\frac{Y}{\rho}\right)^{1/2}$	Y = Young's modulus ρ = mass density
Transverse wave on a string	$\left(\frac{\mathbf{T}}{\mu}\right)^{1/2}$	$\mathbf{T}$ = tension μ = linear mass density

For the same amplitude, the maximum velocity of the spring's vibration will be directly proportional to the square root of the elastic property (k) and inversely proportional to the square root of the inertial property (m). In general, for any mechanical wave we can write:

$$\mathbf{v}(\mathrm{phase}) \propto \left(\frac{\textit{elastic property}}{\textit{inertial property}}\right)^{1/2}$$

Table 24–1 gives equations for calculating the phase speed of different kinds of mechanical waves in terms of elastic and inertial properties.

□ **EXAMPLE PROBLEM 24–6: SOUND WAVES IN A STEEL RAIL**

At what speed does sound travel along a steel rail?

■ *SOLUTION*

From Table 24–1, the speed of a compressional wave along a rod is:

$$v = \left(\frac{Y}{\rho}\right)^{1/2}$$

The density of steel is: $\rho = 7.8 \times 10^3$ kg/m^3. Young's modulus is: $Y = 2\bar{0} \times 10^{10}$ N/m^2.

$$v = \left(\frac{2\bar{0} \times 10^{10}\ \text{N/m}^2}{7.8 \times 10^3\ \text{kg/m}^3}\right)^{1/2}$$

$$v = 5.1 \times 10^3\ \text{m/s}$$

SUMMARY

Energy and information can be transported by waves. The frequency of a wave is the number of wave crests passing a point during an interval of time. The amplitude of a wave is the maximum displacement of the wave, or peak height. The wavelength of a wave is the distance between adjacent peaks of a wave. The phase velocity is the speed with which a wave peak moves. It is related to wavelength and frequency by: $v = f\lambda$.

Rays indicate the direction of motion of a wave front. They are always perpendicular to wave fronts. Plane waves have parallel wave fronts.

The energy of a mechanical wave is dependent upon the frequency and amplitude of the wave. The linear intensity of a two-dimensional wave is energy per unit length of wave front. For a three-dimensional wave, the intensity is energy per unit area of wave front. The intensity of a wave changes to maintain conservation of energy along a wave front.

The net amplitude of the combination of two or more waves is the vector sum of the waves' displacements. This is called the superposition principle.

When mechanical waves are reflected from a fixed boundary they undergo a 180° phase shift. A node is a point along the path of a combination of transmitted and reflected waves where there is no motion. Resonance can occur in a system with a node at both ends.

The phase speed of a mechanical wave depends upon the elastic and inertial properties of the medium through which it moves.

There are two kinds of waves. Transverse waves vibrate at right angles to the direction of motion of a wave. If the transverse wave vibrates along only one vertical axis it is a polarized wave. Longitudinal waves are compressional waves. They vibrate in the same direction as the wave's motion. Longitudinal waves are the only kind of *mechanical* waves that can move through a liquid or gas.

Electromagnetic waves are transverse waves that are *not* mechanical waves. They can pass through selected solids, liquids, and gases as well as through a vacuum.

KEY TERMS

If you can explain the following terms to a friend or classmate, you understand their meaning. If you cannot explain the terms, you should reread the sections in which they are discussed.

amplitude
compressional wave
frequency
fundamental wavelength
group wave
linear intensity
longitudinal wave
node
overtones
phase velocity
plane waves
polarized wave
rays
reflected wave
resonance
superposition principle
transverse wave
wave fronts
wave groups
wavelengths

EXERCISES

Section 24.1:

1. Give some examples in which energy is transported from one place to another by waves.
2. Describe an experiment using everyday objects to find the speed of sound.
3. For the wave shown in Figure 24–23 find:
   A. the amplitude.
   B. the wavelength.
   C. the frequency if the phase speed is 19.0 m/s.

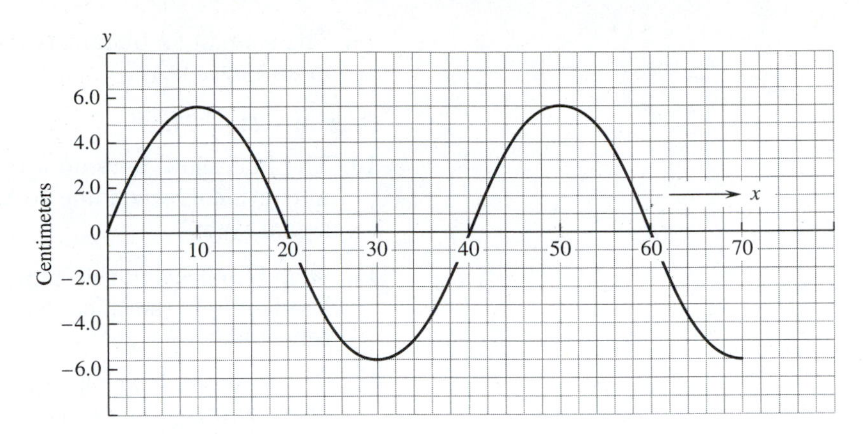

Figure 24–23: Diagram for Exercise 3.

Section 24.2:

4. Copy the diagrams of wave fronts in Figure 24–24 and draw rays on the diagrams.

5. In Figure 24–24 what diagrams correspond to plane waves?

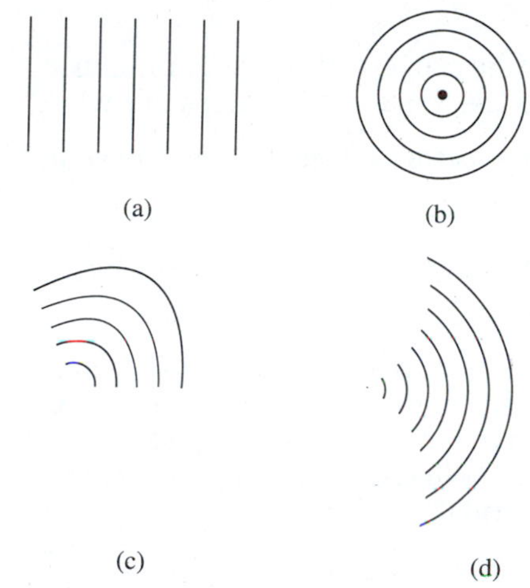

Figure 24–24: Diagrams for Exercises 4 and 5.

Section 24.3:

6. Ripples on the surface of a pond cause a cork floating on its surface to bob up and down. By what factor will the maximum kinetic energy of the cork change if:
   A. the frequency of the ripples is doubled?
   B. The amplitude of the ripples is doubled?
   C. The wavelength of the ripples is doubled?

7. A stone is dropped in a pond, creating both ripples and sound waves. For which kind of wave does the intensity fall off more rapidly with distance? Explain.

Section 24.4:

8. Plot superposition waves resulting from the waves shown in Figure 24–25.

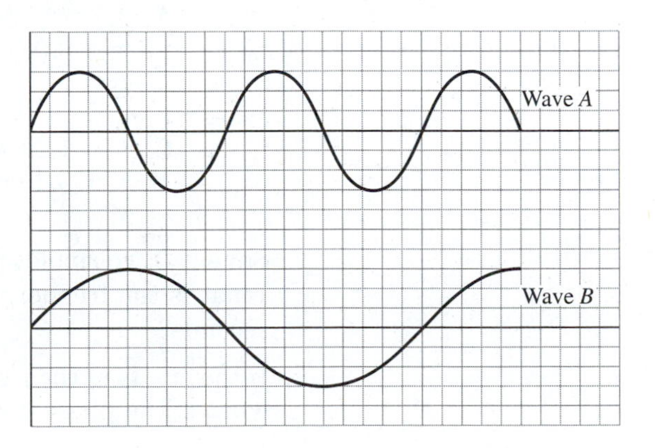

Figure 24–25: Diagram for Exercise 8. Find the superposition of the two waves.

9. The sound of a distant twin-engine plane with propellers fades in and out. Why?

Sections 24.5, 24.6:

10. The horizontal bars on a television antenna are designed with a length corresponding to half the wavelength of the receiving signal. Why?

11. Soap bubbles and thin films of oil tend to reflect specific colors. Explain how the thickness of the film determines what color is reflected.

12. On an organ, long pipes produce bass notes, while short pipes produce treble notes. Explain why.

Section 24.7:

13. When an earthquake occurs, the tremor is felt before it is heard. Explain why.

14. The distance to a lightning stroke can be estimated by measuring the time between the flash and the thunder. Why does this work?

15. How do you think the following factors affect the speed of sound in air?
   A. Temperature
   B. Humidity (Water molecules are less massive than molecules of nitrogen and oxygen.)

16. Seismologists have found that compressional waves (P waves) caused by earthquakes pass through the center of the earth, while transverse waves (S waves) from the same quake do not pass through the earth's center. What can be deduced about the earth's core from this observation?

PROBLEMS

Section 24.1:

1. Find the missing quantities for the following sets of data:

   A. $\lambda = 23$ cm
   $v = 3.0 \times 10^8$ m/s
   $f = ?$

   B. $f = 288$ Hz
   $v = 1090$ ft/s
   $\lambda = ?$

   C. $f = 1.23 \times 10^4$ Hz
   $\lambda = 3.20$ m
   $v = ?$

   D. $f = 5.6 \times 10^6$ Hz
   $\lambda = 0.034$ ft
   $v = ?$

2. Seventy-two ripples pass a frog sleeping on a lily pad in 3.4 min. What is the frequency in hertz and the period of the ripples in seconds?

3. If the crests of the ripples in Problem 2 are spaced 8.1 cm apart, what is the phase speed of the ripples?

4. A 440-Hz note is sounded by a clarinet. If the speed of sound in air is 340 m/s, what is the wavelength of the note in air?

Section 24.3:

5. A rock is dropped into a calm pool of water. The ripples have an amplitude of 0.80 in at a distance of 3.0 ft from the disturbance. What will be their amplitude 10.0 ft from the source?
6. A point source of light emits 53 W of light. What will be the light's intensity in watts per square foot at a distance of 21.2 ft from the source?
7. A public address speaker delivers 210 W of sound energy to the surrounding air. What is the sound intensity in watts per square meter at a distance of 12.4 m from the speaker?
8. The intensity of light striking the outer atmosphere of the earth is 1.35 k/m^2. What will be the intensity of radiation from the sun on Saturn, 9.2 AU from the sun? (The radius of the earth's orbit is 1 astronomical unit [AU]. The earth is 1.00 AU from the sun.)

Section 24.6:

9. A television antenna has a horizontal resonance rod 28 in long. Electromagnetic waves in air travel at 186,000 mi/s. Estimate the resonating fundamental frequency of the television signal for which the antenna is designed.
10. The phase speed of a transverse wave along a violin A string is 288 m/s. The fundamental frequency is 440 Hz.
    - ***A.*** What is the fundamental wavelength of the string?
    - ***B.*** Estimate the length of a violin string.
    - ***C.*** The phase speed of sound in air is 434 m/s. Find the wavelength of the tone in air.
11. A hollow tube is closed at both ends and has a length of 19.5 in. Find the resonant wavelengths and frequencies for the fundamental and the first harmonic wavelengths oscillating in the air enclosed in the tube. Take the speed of sound to be 1110 ft/s.
12. A 1.26 ft long string has a node at each end.
    - ***A.*** Find the first three overtones if the fundamental wavelength is 80.0 Hz.
    - ***B.*** Calculate the phase speed of a transverse wave on the spring.
13. The second harmonic of light with a frequency of 5.6×10^{14} Hz undergoes a resonant reflection between the surfaces of the soap film forming a soap bubble. It makes the bubble look green (see Figure 24–26). Estimate the thickness of the soap film. The phase speed of light in the soap film is 2.3×10^{8} m/s.

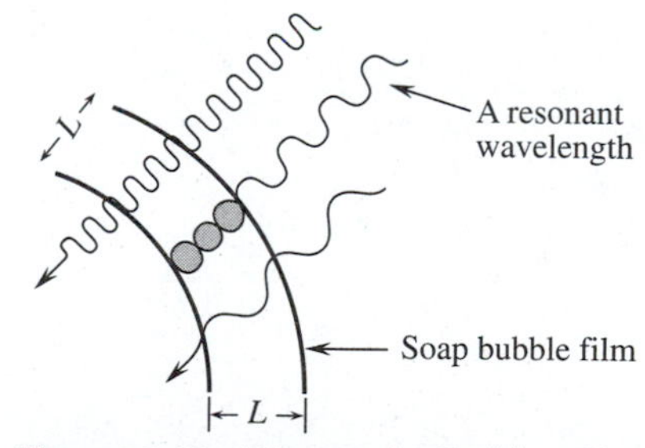

Figure 24–26: *Diagram for Problem 13. Resonant waves undergo multiple reflections in a soap film. Wavelengths not meeting the resonant conditions pass through the film without being reflected back and forth in the film.*

Section 24.7:

14. A string has a linear mass density of 1.72 g/cm and is under a tension of 116 N. What is the phase speed of a transverse wave on the string?

15. Copper has a Young's modulus of 16×10^6 psi, a shear modulus of 6.1×10^6 psi, and a bulk modulus of 17×10^6 psi. Its mass density is 17.3 slug/ft^3.

- **A.** Find the speed of sound in a long, thin rod of copper.
- ***B.*** With what speed will a compressional wave travel through a large block of copper?
- ***C.*** With what speed will a transverse wave travel through a large block of copper?

16. Water has a weight density of 62.5 lb/ft^3 and a bulk modulus of 2.9×10^5 psi. What is the speed of sound through water?

17. At room temperature, the speed of sound in air is roughly 340 m/s and has a density of 1.29 kg/m^3. Estimate the bulk modulus of air.

Chapter 25

SOUND

Is it true that more men sing in the shower than women? Read about the physics of sound and you will begin to see how some acoustical environments can flatter the human voice.

OBJECTIVES

In this chapter you will learn:

- how the speed of sound is related to the mechanical properties of the medium through which it moves
- how the speed of sound in a gas is related to temperature
- that sound resonates with reflections at both free and fixed boundaries
- how the voice is produced
- to measure intensity and pressure of sound
- to calculate the frequency shifts caused by the relative motion between a sound source and an observer
- how beats are created
- how ultrasonics may be applied in a variety of useful ways

Time Line for Chapter 25

1705 Francis Hauksbee shows that sound cannot travel through a vacuum.
1842 Christian Doppler discovers the Doppler effect.
1915 Paul Langevin invents sonar.

A trumpet sings. A dog hears a high-frequency whistle that humans cannot sense. A fishing trawler locates a catch by sonar. A technician finds defects in a steel casting using ultrasound equipment. A medical team uses echocardiographic equipment to locate a congenital heart disorder in a patient.

In the broadest sense of the word, sound is not limited to the frequency range of hearing of the human ear. It is any compressional wave. The human range of hearing is from about 16 Hz to 20,000 Hz. For many people the frequency range is narrower. Frequencies above the human hearing range are called **ultrasound.** There is a broad range of technical applications for these higher-frequency compressional waves.

In this chapter we will look at the behavior of sound, human perception of sound, and some of the technical applications of sound waves.

25.1 THE SPEED OF SOUND

In Chapter 24 we learned that the speed of sound (longitudinal, or compressional, waves) is the one kind of mechanical wave that can pass through any medium. In a long slender rod the speed of the wave depends on Young's modulus (Y) and density (ρ) (see Table 24–1).

$$\text{v(sound)} = \left(\frac{Y}{\rho}\right)^{1/2} \qquad \textbf{(Eq. 25–1)}$$

In an extended solid, compressional waves can spread out rather than be constrained to move in a straight line. In this case, the speed of sound depends on both the bulk modulus (B) and the shear modulus (S) as well as density (ρ).

$$\text{v(sound)} = \left[\frac{(B + 4S)/3}{\rho}\right]^{1/2} \qquad \textbf{(Eq. 25–2)}$$

A gas or liquid cannot support a shear force. For these mediums the speed depends only on density and the bulk modulus.

$$\text{v(sound)} = \left(\frac{B}{\rho}\right)^{1/2} \qquad \textbf{(Eq. 25–3)}$$

In most cases we hear sound transmitted through the air. Let us take a closer look at Equation 25–3. In most situations, sound travels through a gas adiabatically; that is, when the sound passes through a medium, the medium neither gains nor loses heat. For this situation the bulk modulus of an ideal gas is proportional to the pressure by a factor of 1.4. We can rewrite Equation 25–3 in terms of pressure.

$$\frac{1.4\,\text{P}}{\rho}$$

It would seem that the speed of sound would increase if the pressure of a gas increased. However, if the pressure increases, the density will increase, also. Let us see how volume affects the speed. Since density is m/V, we can write:

$$\text{v} = \left(\frac{(1.4\,\text{P})\,V}{m}\right)^{1/2}$$

We can eliminate both the pressure and volume by using the ideal gas law.

$$\text{P}\,V = N\,R\,T$$

where N is the number of kilomoles, $R = 8314$ J/(kmole · K), and T is the absolute temperature. Substitute this into the speed of sound relationship.

$$\text{v} = \left(\frac{1.4\,NRT}{m}\right)^{1/2}$$

For one kilomole of gas molecules, $N = 1$ and m is the mass (M) of one kilomole of gas in kilograms. So the speed of sound in an ideal gas is:

$$\text{v} = \left(\frac{1.4\,R\,T}{M}\right)^{1/2} \qquad \textbf{(Eq. 25–4)}$$

☐ **EXAMPLE PROBLEM 25–1: THE SPEED OF SOUND IN AIR**

The mass of a kilomole of air is 29.0 kg. Find the speed of sound at 25.0°C in air in feet per second.

■ ***SOLUTION***

$$T = C + 273 = 25°\text{C} + 273$$
$$T = 298 \text{ K}$$

$$v = \left(\frac{1.4\, R\, T}{M}\right)^{1/2}$$

$$v = \left(\frac{1.4 \times 8314 \text{ J/(kmole} \cdot \text{K)} \times 298 \text{ K}}{29 \text{ kg/kmole}}\right)^{1/2}$$

$$v = 345.8 \text{ m/s} \times 3.28 \text{ ft/m}$$

$$v = 1130 \text{ ft/s}$$

☐ **EXAMPLE PROBLEM 25–2: HELIUM**

The kilogram equivalent mass of helium is 4.00 kg. How many times as fast will sound travel in helium as in air? Assume identical temperatures.

■ ***SOLUTION***

$$\frac{v(\text{He})}{v(\text{air})} = \left[\frac{(1.4\, R\, T)/M(\text{He})}{(1.4\, R\, T)/M(\text{air})}\right]^{1/2} = \left[\frac{M(\text{air})}{M(\text{He})}\right]^{1/2}$$

$$\frac{v(\text{He})}{v(\text{air})} = \left(\frac{29 \text{ kg/kmole}}{4 \text{ kg/kmole}}\right)^{1/2}$$

$$\frac{v(\text{He})}{v(\text{air})} = 2.7 \text{ times as fast}$$

25.2 SOUND RESONANCE

In Chapter 24 we found that resonance can occur on a string when both ends are fixed. A string has transverse waves, but the same condition applies to compressional waves. A column of fluid (liquid or gas) with a length L is closed on both ends.

For this situation, the resonant wavelengths are:

$$\lambda = \frac{2L}{N}$$

where $N = 1, 2, 3, 4. \ldots$

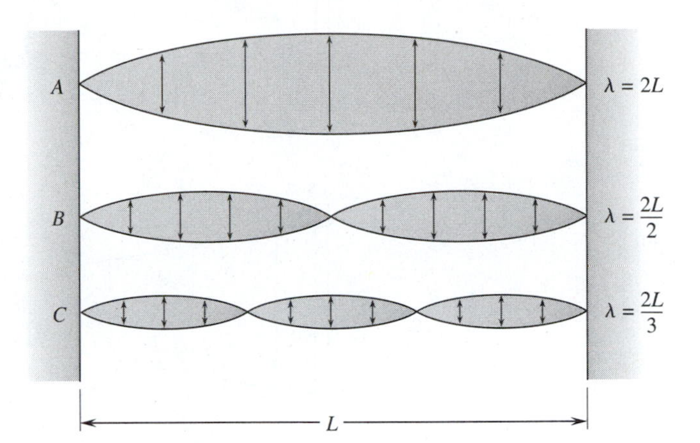

Figure 25–1: *Resonance for a string fixed at both ends. The ends must be nodes.* ***A.*** *Fundamental wavelength.* ***B.*** *First overtone.* ***C.*** *Second overtone.*

Many musical instruments, such as a clarinet and a trombone, resonate, but they are not closed columns. The resonant sound comes out of the open end of the instrument. There must be another resonant condition.

Figure 25–2 shows a string with one fixed end (A). The other end is looped loosely around a vertical bar at B. Point B can freely vibrate up and down. There is no need for a node to exist here during reflection. In fact, we expect the string to vibrate vertically as waves are reflected. Waves can be reflected at point B without destructive interference if it is a point of maximum amplitude. The reflected wave can undergo a 180° phase shift and still be in step with the transmitted wave. A point that has no motion is called a **node**; a point of maximum amplitude is an **antinode.**

Reflection can support resonance in two ways. Either the reflection point is a node or it is an antinode. If the reflection point is a fixed position, such as the closed end of a pipe or the end of a string tied to a surface, it is a node. If at the reflection point the medium can move freely, as at the open end of a pipe or the free end of a string, the reflection point is an antinode. Figure 25–3 shows some resonant conditions for a variety of end configurations for pipes.

☐ **EXAMPLE PROBLEM 25–3: KUNDT'S TUBE**

A 92.0 cm long aluminum rod is clamped at the center. One end extends into a glass tube with cork dust along the tube. The aluminum

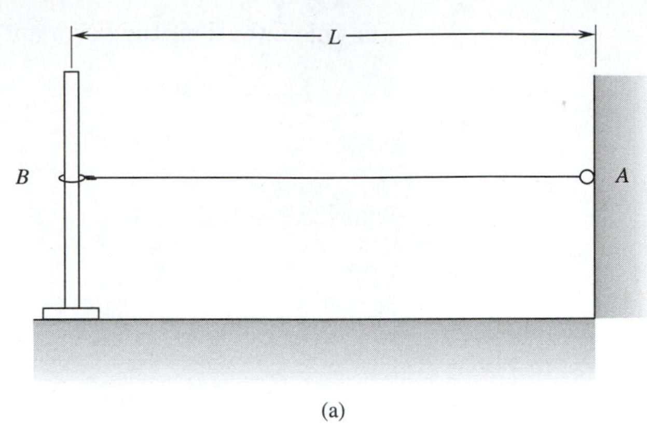

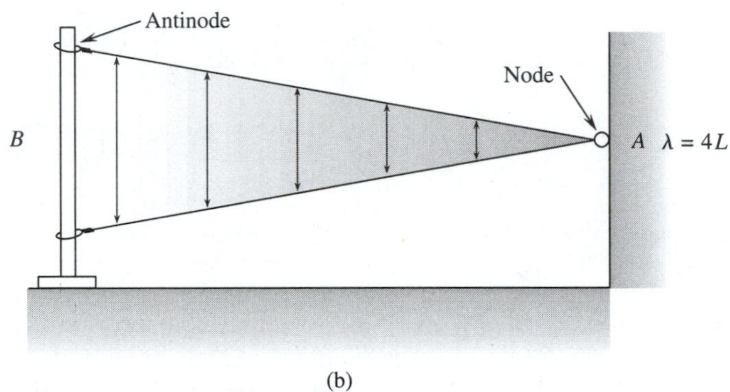

Figure 25–2: A string has a fixed end (A) and an end free to vibrate up and down (B). For resonance, end A is a node and end B is an antinode.

rod is stroked with a rosin-impregnated cloth to make it vibrate. The position of the glass tube is adjusted until resonance occurs. Cork dust piles up 6.2 cm apart at the nodes of the sound wave in the glass tubing. If the speed of sound in air is 343 m/s, find:

A. the resonant wavelength in air.
B. the frequency of sound in air and in the aluminum.
C. the speed of sound in aluminum.

■ *SOLUTION*

A. The distance (s) between nodes is a half wavelength.

$$\lambda(\text{air}) = 2\,s = 2 \times 0.062 \text{ m}$$

$$\lambda(\text{air}) = 0.124 \text{ m}$$

B. The air column and the rod will resonate at the same frequency (f).

$$f = \frac{\text{v(air)}}{\lambda(\text{air})} = \frac{343 \text{ m/s}}{0.124 \text{ m}}$$

$$f = 2770 \text{ Hz}$$

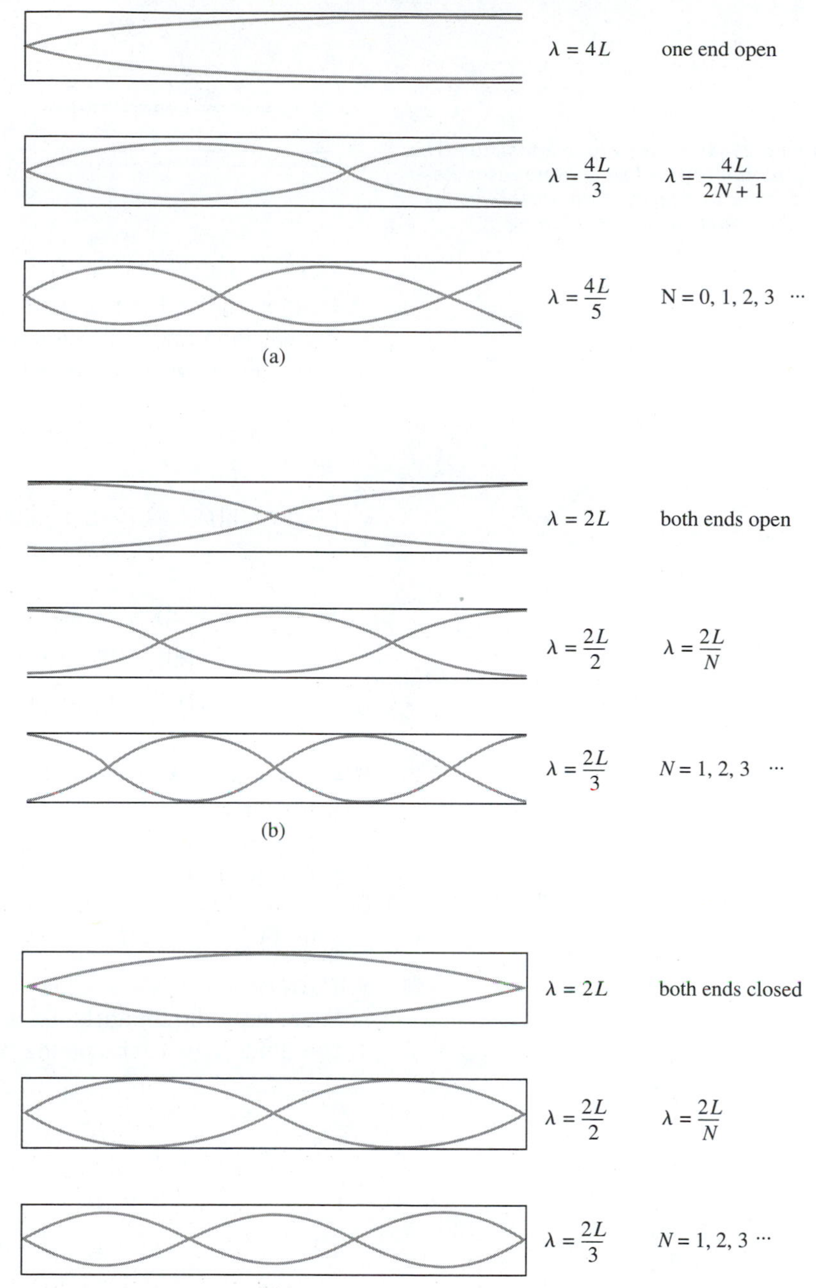

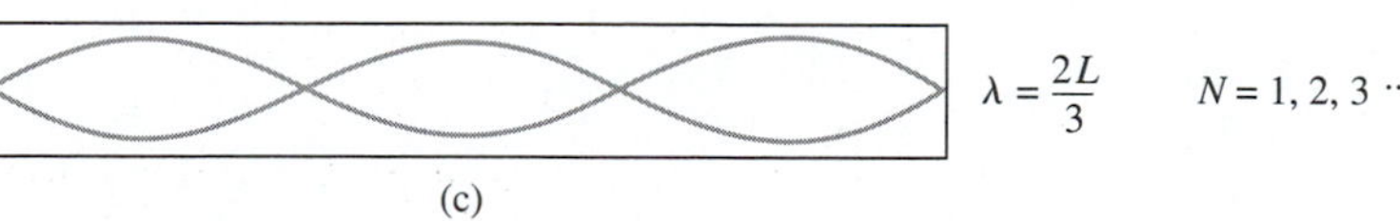

Figure 25–3: *Resonance for different end conditions for a hollow pipe.* (a) *One end closed and one end open.* (b) *Both ends open.* (c) *Both ends closed.*

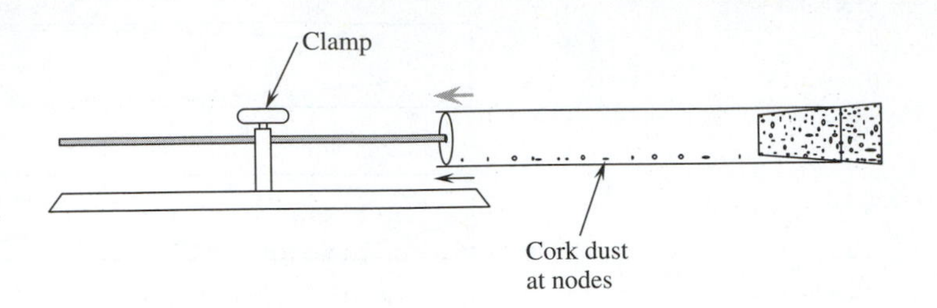

Figure 25–4: Kundt's tube apparatus. The aluminum rod and air column resonate at the same frequency. Cork dust piles up at the nodes in the air column.

C. Since the rod is clamped in the center, the center of the rod must be a node. The ends are free to vibrate (see Figure 25–4). The longest wavelength satisfying these conditions is:

$$\frac{\lambda(\text{al})}{2} = L$$

$$\lambda(\text{al}) = 2L = 2 \times 0.92\text{ m} = 1.84\text{ m}$$

$$v(\text{al}) = f\lambda(\text{al})$$

$$v(\text{al}) = 2770\text{ Hz} \times 1.84\text{ m}$$

$$v(\text{al}) = 51\bar{0}0\text{ m/s}$$

□ **EXAMPLE PROBLEM 25–4: THE SODA BOTTLE**

A teenager blows across the top of a partly filled bottle of soda pop, creating a resonant tone (see Figure 25–5). If the distance from the mouth of the bottle to the remaining fluid is 5.3 in, find the fundamental resonant wavelength and frequency. Take the speed of sound to be 1124 ft/s.

■ ***SOLUTION***

There must be an antinode at the mouth of the bottle and a node at the fluid. This corresponds to a quarter wavelength.

Figure 25–5: The resonance wavelength of a soda pop bottle is four times the distance from the rim of the bottle to the liquid level.

$$\frac{\lambda}{4} = L$$

$$\lambda = 4L$$

$$\lambda = 4 \times 5.3\text{ in}$$

$$\lambda = 21.2\text{ in or } 1.77\text{ ft}$$

$$f = \frac{v}{\lambda} = \frac{1124\text{ ft/s}}{1.77\text{ ft}}$$

$$f = 640\text{ Hz}$$

25.3 THE VOICE

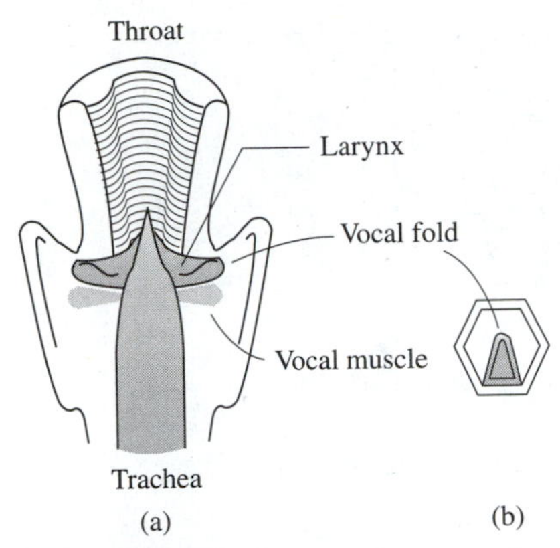

***Figure 25–6:** The larynx is between the throat and the trachea (air tube to the lungs). Tension on the vocal folds and the size of the opening are controlled by the vocal muscles.* (a) *Longitudinal section of the throat and larnyx.* (b) *Top view.*

Many stringed instruments, violins and banjos for example, consist of a vibrating string and a resonance box. The oscillations of the string are reinforced by the resonance cavity. The human voice is based on the same principle.

Figure 25–6 shows a diagram of the larynx. The larynx is a hollow tube located above the windpipe (trachea) that contains two muscular folds called vocal cords. When wind is forced through the larynx, the vocal cords vibrate much like the string on a guitar. The pitch of the vocal cords depends on the thickness of the cords and the tension. Men usually have lower-pitched voices than women because they have thicker, more massive vocal cords. The pitch can be raised by increasing the tension. When the vocal cords are completely relaxed no sound is produced.

The vibrating vocal cord can produce little sound alone. The larynx along with other cavities—the mouth, the throat, and the nasal and sinus cavities—all resonate to amplify the sound (see Figure 25–7). The size and shape of these cavities determine the characteristics of a particular voice.

25.4 SOUND INTENSITY

The lowest intensity of sound that is detected by the human ear is about 10^{-12} W/m^2. Sound begins to be painful at intensities of 100 W/m^2. Hearing is most sensitive at 3500 Hz. For lower and higher frequencies, sound must have greater intensity to be heard. The threshold of hearing for a 60-Hz sound is about 10^{-7} W/m^2. For 10,000 Hz the threshold is 10^{-10} W/m^2.

Like many human senses, the sense of loudness is logarithmic. We can sense a very large range of intensity because our senses perceive not the intensity, but its logarithm. If the intensity of a sound increases by 100-fold we perceive an increase of loudness of only fourfold.

We can use the threshold of hearing (I_0), an intensity of 10^{-12} W/m^2, as a standard to compare other intensities. The comparison will be made by taking a ratio. For example, if I is 10^{-3} W/m^2, its intensity compared with I_0 is:

$$\frac{I}{I_0} = \frac{10^{-3}\ \text{W/m}^2}{10^{-12}\ \text{W/m}^2} = 10^9$$

***Figure 25–7:** The throat, sinus cavities, mouth, and nasal passages act as resonating cavities.*

The intensity I is a billion times greater than the threshold intensity of hearing. Units cancel. The measurement of sound levels has no

SINGING IN THE BATHTUB

More men sing in the shower than women. There is a reasonable physical explanation for this: resonance. A typical enclosed tub-shower combination is about 5 ft long, producing a fundamental frequency of 10 ft. Sound travels at speeds near 1100 ft/s, giving a frequency near 110 Hz, well below middle C. A frequency of 110 Hz and a few of the lower overtones bounce back and forth in standing waves between the walls, making these tones louder.

A deep voice has a rich bass resonance in the shower. Higher-frequency resonances are not amplified as strongly. If a soprano would like to hear the same pleasing amplification, she should invest in a much shorter shower stall.

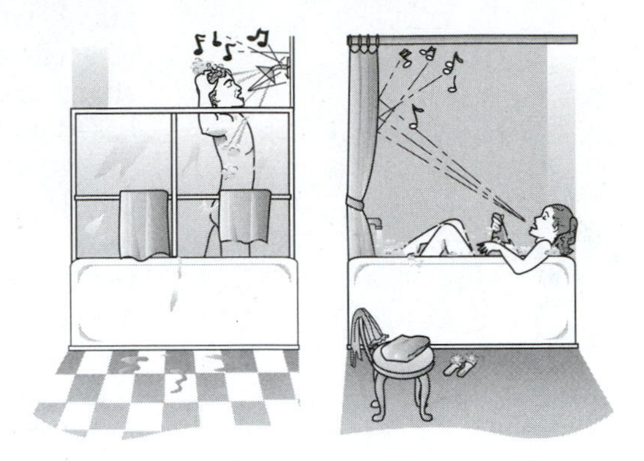

units. The exponent, or power of 10, is called the **sound intensity level.** Recall that $\log(10^x) = x$. The sound level defined in this way is called a **bel (b).**

$$\textit{sound intensity level in bels} = \log\left(\frac{I}{I_0}\right)$$

A bel is a fairly large change of sound level. Most often one-tenth of a bel is used to express sound levels. A tenth of a bel is a **decibel (db).**

$$1 \text{ bel} = 10 \text{ decibels (db)}$$

$$1 \text{ db} = 10 \log\left(\frac{I}{I_0}\right) \qquad \textbf{(Eq. 25–5)}$$

Table 25–1 shows some sound intensity levels in decibels.

When sound waves strike something a pressure is exerted. Microphones react to the variation in pressure produced by sound. Older style microphones used carbon detectors. The variation in pressure caused a variation in the electrical resistance of the carbon. Other microphones allow pressure to vary the electrical voltage across a crystal.

The pressure caused by a sound wave is proportional to the amplitude, not to the intensity.

Table 25–1: Sound intensity levels.

SOUND INTENSITY LEVEL (db)	INTENSITY (W/m^2)	EXAMPLE
0	10^{-12}	Threshold of hearing
10	10^{-11}	Rustling of leaves
20	10^{-10}	Whisper
40	10^{-8}	Quiet home
60	10^{-6}	Normal conversation
80	10^{-4}	Inside of car in traffic
100	10^{-2}	Machine shop
120	10^{0}	Threshold of pain, jackhammer
140	10^{2}	Jet airplane 30 m away

$$I \propto A^2 \propto \mathrm{P}^2$$

or

$$I = k\,\mathrm{P}^2$$

where k is a proportionality constant. We can compare pressures relative to a standard pressure P_0 related to the threshold intensity.

$$I_0 = k\,\mathrm{P}_0^{\,2}$$

$$\mathrm{P}_0 = 2 \times 10^{-5}\ \mathrm{N/m^2}$$

Sound intensity levels can be expressed in terms of the pressure created by sound waves.

$$\mathrm{db} = 10 \log \left(\frac{I}{I_0}\right)$$

$$\mathrm{db} = 10 \log \left(\frac{k\,\mathrm{P}^2}{k\,\mathrm{P}_0^{\,2}}\right)$$

$$\mathrm{db} = 20 \log \left(\frac{\mathrm{P}}{\mathrm{P}_0}\right) \qquad \textbf{(Eq. 25–6)}$$

□ **EXAMPLE PROBLEM 25–5: SOUND LEVEL**

Find the sound level for a sound intensity of $2.8 \times 10^{-3}\ \mathrm{W/m^2}$.

■ ***SOLUTION***

$$\mathrm{db} = 10 \log \left(\frac{I}{I_0}\right)$$

$$db = 10 \log \left(\frac{2.8 \times 10^{-3} \text{ W/m}^2}{10^{-12} \text{ W/m}^2} \right)$$

$$db = 10 \log 2.8 \times 10^9$$

$$db = 94.5$$

☐ **EXAMPLE PROBLEM 25–6: A SECURITY SYSTEM**

The security system of an industrial file room is activated by sound levels larger than 20 db (Figure 25–8). What sound pressure corresponds to this level?

Figure 25–8: *A security system is activated by sound.*

■ ***SOLUTION***

$$db = 20 \log \left(\frac{P}{P_0} \right)$$

$$\log \left(\frac{P}{P_0} \right) = \frac{db}{20}$$

$$P = P_0 \times 10^{db/20}$$

$$P = (2 \times 10^{-5} \text{ N/m}^2)(10^{20/20})$$

$$P = 2 \times 10^{-5} \text{ N/m}$$

25.5 THE DOPPLER EFFECT AND BEATS

An automobile races along a highway at a constant speed. It overtakes a hitchhiker walking along the shoulder of the road. The pitch

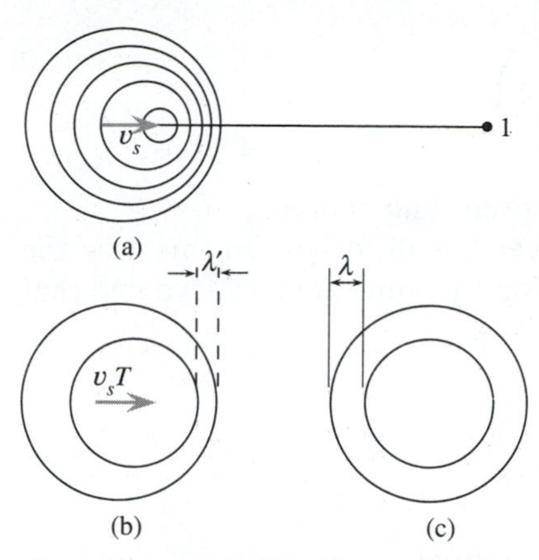

Figure 25–9: *Doppler effect caused by a moving sound source. (a) A source moves toward a stationary observer. Wave fronts pile up in the direction of motion. (b) Measurement of λ′, the altered wavelength in the direction of motion. (c) The unshifted wave length (λ) of a stationary source.*

of the car's motor and the whine of its tires change to a lower frequency as the car passes the pedestrian. This is an example of the **Doppler effect,** or **Doppler shift.**

The Doppler shift, or the frequency shift caused by motion, is a general property of waves. The behavior of the Doppler shift is a bit different for light waves than for mechanical waves such as sound. In this chapter, we will discuss the Doppler shift for sound waves. In Chapter 28, we will look at the Doppler effect for electromagnetic waves.

If a sound source moves toward or away from an observer, the frequency of the sound changes. Look at Figure 25–9. A sound source moves toward observer 1. It is moving as it sends out wave peaks. T is the period of the sound, the time elapsed between wave peaks. If the source moves toward 1 at a speed $\mathbf{v}_s$, it will travel a distance $S = \mathbf{v}_s T$ for each wavelength it emits. The waves traveling toward 1 are closer together by this amount. The wavelength (λ') observed by 1 is:

$$\lambda' = \lambda - (v_s T)$$

Remember, frequency and period are reciprocals:

$$f = \frac{1}{T}$$

So $f\lambda = v$ can be written as $T = \lambda/v$.

Observer 1 knows the speed of sound is v and sees the waves arrive with a wavelength λ'. Observer 1 divides the observed wavelength by the speed of sound to find the observed period (T').

$$\frac{\lambda'}{v} = \frac{\lambda - (v_s T)}{v}$$

or

$$T' = T - \frac{v_s T}{v}$$

Algebra gives us:

$$\frac{T}{T'} = \frac{v}{v - v_s}$$

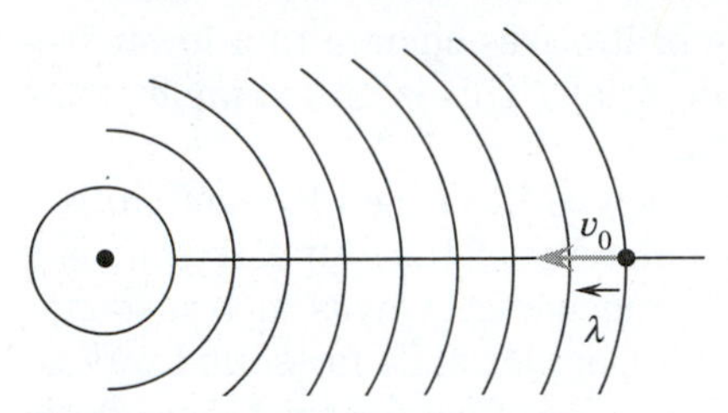

Figure 25–10: *An observer moving toward a stationary source encounters more wavelengths per unit time than if she were not moving.*

since $f = 1/T$

$$f' = f\left(\frac{v}{v - v_s}\right)$$

for the frequency shift for the approaching source.

The frequency heard by observer 2 is different. In this case the wave peaks are farther apart. Using the same analysis, we find that frequency observed at B is:

$$f' = f\left(\frac{v}{v + v_s}\right)$$

the frequency shift for a receding source.

Next look at Figure 25–10. An observer with speed **v** approaches a stationary sound source with a speed of v_0. The speed of the sound waves relative to the observer is:

$$v' = v + v_0$$

The observer hears a frequency $f' = \mathbf{v}'/\lambda$.

$$f' = \frac{(v + v_0)}{\lambda}$$

The wavelength in terms of the frequency of the stationary source is $\lambda = v/f$. Substituting this into the equation, we get

$$f' = f\frac{(v + v_0)}{v}$$

for the frequency shift for an approaching observer.

If the observer is moving away from the source, the relative speed is $v - v_0$. The frequency heard is:

$$f' = f\frac{(v - v_0)}{v}$$

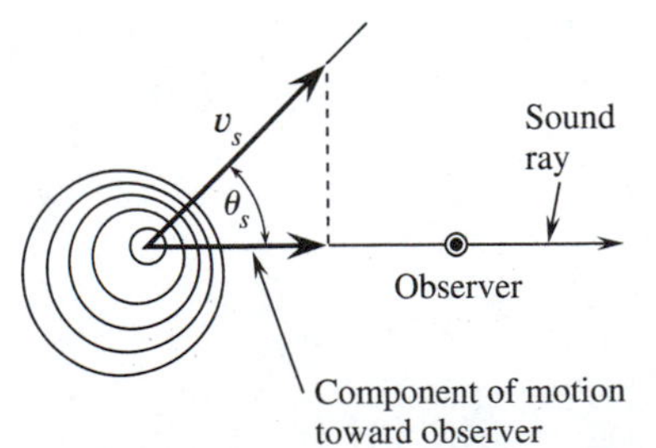

Figure 25–11: *A source is moving at an angle θ_s relative to the sound ray traveling from the source to the observer. Only the component of motion along the ray affects the Doppler frequency.*

We have assumed that the observer and the source are moving directly toward each other. In each equation the important factor is motion relative to the sound waves. In Figure 25–11 the waves move from the source outward. Rays show the motion of the wave fronts. A source moves at an angle θ_s relative to the ray connecting the source and observer. The component of the velocity acting along the ray is $\mathbf{v}_s \cos \theta_s$. Only this component of the velocity will alter the frequency of sound heard by the observer. The Doppler shift for a moving source and stationary observer is:

$$f' = f\left(\frac{v}{v - (v_s \cos \theta_s)}\right) \quad \textbf{(Eq. 25–7)}$$

If the source is moving toward the observer, $\cos \theta_s = 1$. If the source is moving directly away, $\cos \theta_s = -1$. One equation covers both the case of recession and of approach. We can find the Doppler shift for any direction of motion.

Figure 25–12 shows the case of a moving observer. Again we measure the angle from the ray connecting the source and the observer. Only the component along the ray is involved with the observer's motion relative to the sound. The observer's motion relative to the ray is $v_0 \cos \theta_0$. The Doppler shift for a moving observer and a stationary source is:

$$f' = f\left(\frac{v - (v_0 \cos \theta_0)}{v}\right) \quad \textbf{(Eq. 25–8)}$$

We can combine Equations 25–7 and 25–8 to get a general equation for the Doppler effect.

$$f' = f\left(\frac{v - (v_0 \cos \theta_0)}{v - (v_s \cos \theta_s)}\right) \quad \textbf{(Eq. 25–9)}$$

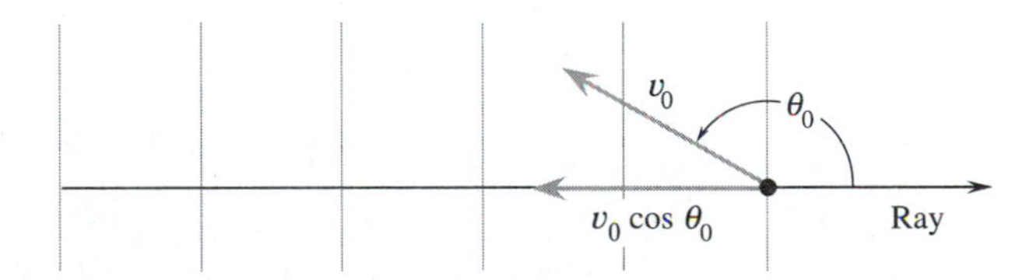

Figure 25–12: *Only the component of the observer's motion parallel to the sound ray from the source is involved with the Doppler shift.*

☐ **EXAMPLE PROBLEM 25–7: THE PARALLEL HORSEPERSONS**

A person riding a horse toots a trumpet while heading northeast. A second rider is riding along on a parallel course east of the first rider (see Figure 25–13). Show that the second rider does not hear a Doppler shift.

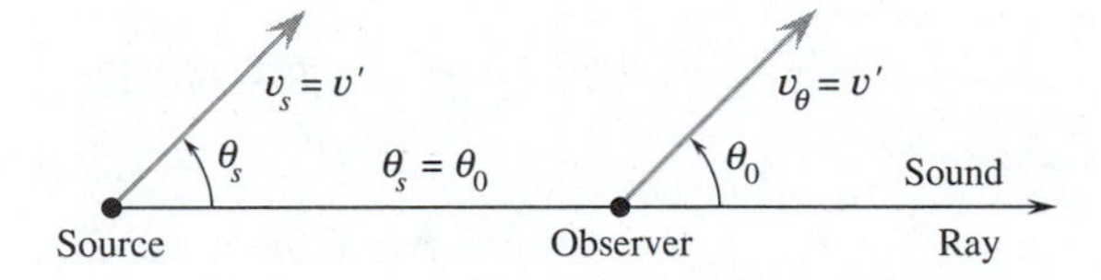

Figure 25–13: *The motion of a source and the motion of an observer are parallel and equal in magnitude.*

■ *SOLUTION*

$$\theta_s = \theta_0 \qquad \text{and } v_s = v_0 = v'$$

Substitute these values into Equation 25–9.

$$f' = f\left(\frac{v - (v' \cos \theta)}{v - (v' \cos \theta)}\right)$$

or

$$f' = f \qquad \text{no Doppler shift}$$

□ **EXAMPLE PROBLEM 25–8: A DOPPLER FLOWMETER**

Echo-ultrasonic flowmeters can be used in fluids that contain things, such as undissolved solids, particles, or bubbles, that will reflect the sonic wave. High-frequency signals are alternately transmitted and received. Assume that a Doppler flowmeter transmits a frequency of 1.80 MHz (megahertz) at an angle of 25° to a fluid. The speed of sound in the fluid is 3230 ft/s. What frequency shift would be found in the echo for a 12.0 ft/s flow rate?

■ *SOLUTION*

The transmitter-receiver is the stationary observer; the fluid is the source.

$$f = 1.80 \times 10^6$$

$$v = 3230 \text{ ft/s}$$

$$v_s = 12.0 \text{ ft/s}$$

$$\theta_s = 25°$$

$$v_0 = 0.0$$

$$f' = f\left[\frac{v - (v_0 \cos \theta_0)}{v - (v_s \cos \theta_s)}\right]$$

$$f' = 1.80 \times 10^6 \text{ Hz} \frac{(3230 \text{ ft/s} - 0)}{3230 \text{ ft/s} - (12.0 \text{ ft/s} \times \cos 25°)}$$

$$f' = 1{,}806{,}081 \text{ Hz}$$

$$f = f' - f = 1{,}806{,}081 - 1{,}800{,}000$$

$$f = 6.08 \times 10^3 \text{ Hz}$$

In the Doppler effect, frequency is shifted by relative motion between a sound source and an observer. There is another way that sound can be shifted slightly in frequency. In Section 24.4, we discussed the superposition of waves. When two waves have nearly the same frequency or wavelength, they superimpose to make a wave group. The wave group has the average frequency of the two waves.

$$f_{avg} = \frac{f_1 + f_2}{2}$$

The average frequency is modulated to form the group pattern shown in Figure 24–11. If the wave is a sound wave, we will hear the average tone fade in and out with a beat frequency (f_b) equal to the difference of the two tones.

$$f_b = f_1 - f_2$$

Musicians use beat frequencies to tune instruments. One way to tune a guitar is to strum the same note on adjacent strings. If the strings are out of tune, beats will be heard. The guitarist adjusts the tension of the untuned string until the beats disappear.

☐ **EXAMPLE PROBLEM 25–9: BEATS**

One clarinet produces a fundamental frequency of 440 Hz, and another clarinet produces a frequency of 446 Hz as its player attempts to produce the same note as the first clarinetist.

A. Find the average frequency of the two instruments played together.

B. What is the beat frequency?

■ ***SOLUTION***

A.

$$f_{avg} = \frac{f_1 + f_2}{2}$$

$$f_{avg} = \frac{446\text{ Hz} + 440\text{ Hz}}{2} = 443\text{ Hz}$$

B.

$$f_b = f_2 - f_1 = 6\text{ Hz}$$

25.6 ULTRASONICS APPLICATIONS

Ultrasound is a compressional wave with a frequency higher than the upper range of human hearing. Most industrial applications use frequencies from 1 MHz to 20 MHz. Below are a few of the many applications of ultrasound.

Sonar

Except at very low frequencies, electromagnetic waves do not penetrate water very well. In this medium, sound works better for communications and location. The word *sonar* is an acronym for the expression "sound navigation and ranging." In the broadest sense, **sonar** is a branch of acoustics concerned with the use of the ocean as a medium for sound waves.

The most popular use of sonar is echo ranging. A transmitter emits sound pulses and then listens for echos (see Figure 25–14). If T is the time between transmission and reception of an echo (transit time), v is the speed of sound in water, and L is the distance between the sonar gear and the echo source, then the total path traveled by the sound is $2L$.

$$L = \frac{\mathrm{v}\,T}{2} \qquad \textbf{(Eq. 25–10)}$$

Direction-sensitive receivers are usually used to pinpoint the echo source. Ultrasound waves can penetrate silt and mud. Geologists use sonar to search out geological structures on the continental shelf that are probable sources of petroleum.

Echocardiography

Echocardiography is one of several medical applications of ultrasonics. Sound waves are used to form images. Sound waves are

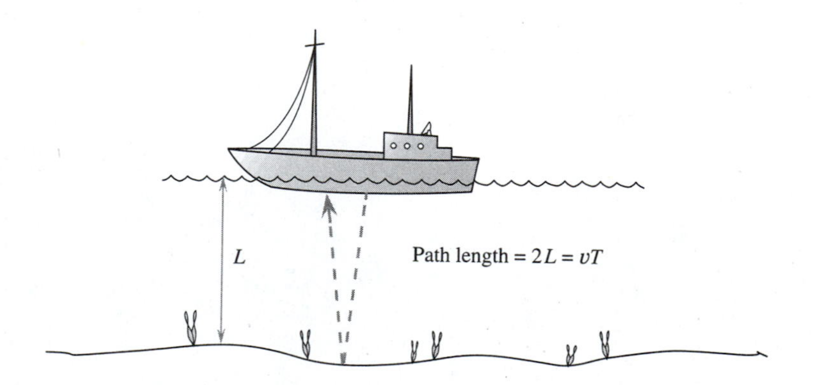

Figure 25–14: *In echo ranging, the distance traveled by sound in a round-trip is* 2L *at a constant speed* ***v****. The time of transit is* T.

narrowly focused on the region of the heart, forming a wedge of sound. The echos are plotted against time on a cathode ray tube to get a cross-sectional image. A three-dimensional image can be built up by a computer from a series of two-dimensional slices. Congenital heart defects can be evaluated with little discomfort or danger to the patient.

Thickness Gauging

Ultrasound frequencies between 1 MHz and 15 MHz are used for some industrial processes such as monitoring the thickness coating on surfaces. Ultrasound thickness gauges primarily use one of two techniques.

1. The **pulse-echo method** is similar to sonar. Short, single-frequency pulses are transmitted. The sound waves bounce off from the surfaces of the material being monitored. Equation 25–10 is used to find the thickness of the material.
2. **Ultrasonic resonance thickness gauges** transmit a continuous signal with a varying frequency. When the thickness matches half a wavelength or overtone, a standing wave is set up in the material. If the thickness is L, the speed of the sound is v, and the resonant fundamental frequency is F, then

$$L = \frac{v}{2F} \qquad \textbf{(Eq. 25–11)}$$

□ **EXAMPLE PROBLEM 25–10: COATED PAPER**

The coating of high-gloss paper is monitored by an ultrasonic resonance thickness gauge. The speed of sound in the coating is 752 m/s. A fundamental resonance occurs at 12.5 MHz. What is the thickness of the coating?

■ ***SOLUTION***

$$L = \frac{v}{2F}$$

$$L = \frac{752\ \text{m/s}}{2 \times 1.25 \times 10^{7}\ \text{Hz}}$$

$$L = 3.01 \times 10^{-5}\ \text{m}$$

$$L = 0.030\ \text{mm}$$

SUMMARY

Sound is compressional (longitudinal) waves. In the absence of friction, sound travels adiabatically. Heat is neither gained nor lost by the medium through which the wave travels.

The speed of sound in an ideal gas is:

$$v = \left(\frac{1.4\,R\,T}{M}\right)^{1/2}$$

where T is the absolute temperature, R is the universal gas constant (8314 J/[kmole · K]), and M is the kilogram equivalent mass of the gas.

Sound waves may resonate in a gas column or in a solid (see Figure 25–15). The conditions for resonance are:

1. If the reflection point is fixed, such as a closed end of a column or clamped end of a rod, the resonant wave has a node at the reflection point.
2. If the reflection point is free to move, such as an open-ended gas column or unrestricted end of a rod, the reflection point is an antinode.
3. A node exists anywhere a rod is clamped.

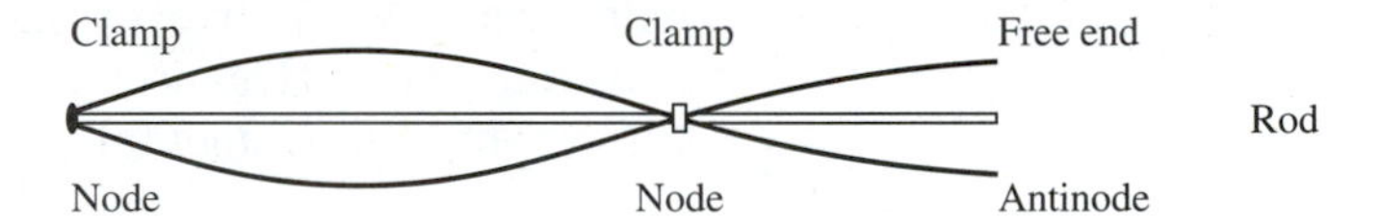

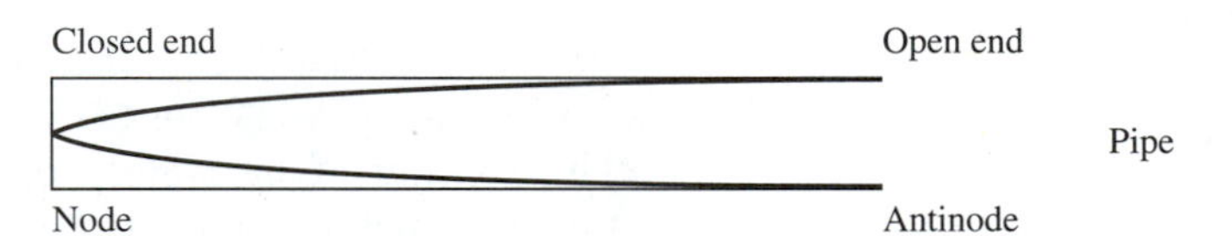

Figure 25–15: *Resonance conditions:* ***1.*** *Clamped rod ends or closed pipe ends are nodes.* ***2.*** *Open pipe ends or free rod ends are anitnodes.* ***3.*** *Clamped points are nodes.*

Sound intensity level is often expressed by a logarithmic scale known as a decibel scale (db). The threshold of hearing is usually taken as a threshold intensity (I_0) of 10^{-12} W/m^2. The sound intensity level in decibels is:

$$\text{db} = 10 \log \left(\frac{I}{I_0}\right)$$

The pressure caused by sound waves is proportional to the amplitude of a compressional wave ($I \propto A^2$). The sound level in terms of pressure is:

$$\text{db} = 20 \log \left(\frac{P}{P_0}\right)$$

where $P_0 = 2 \times 10^{-5}$ N/m^2.

Frequency shifts occur when there is relative motion between an observer and a sound source. This is known as the Doppler effect, or Doppler shift. The Doppler-shifted frequency (f') is related to the frequency of the unshifted source (f) by:

$$f' = f\frac{(v - v_0 \cos \theta_0)}{(v - v_s \cos \theta_s)}$$

where v is the speed of sound, v_0 is the observer's speed, and v_s is the source's speed. Angles θ_0 and θ_s are measured from the ray traveling from the source to the observer (see Figure 25–11).

Ultrasound is compressional waves with frequencies greater than the upper limit of human hearing. Ultrasound has several technical applications, such as sonar (echo ranging), imagery, and resonance thickness measurements.

KEY TERMS

If you can explain the following terms to a friend or classmate, you understand their meaning. If you cannot explain the terms, you should reread the sections in which they are discussed.

antinode
bel (b)
decibel (db)
Doppler effect
Doppler shift
echocardiography
node
pulse-echo method
sonar
sound intensity level
ultrasonic resonance thickness gauges
ultrasound

EXERCISES

Section 25.1:

1. How is the speed of sound in air affected by:
   A. an increase of temperature?
   B. an increase of humidity (increased H_2O)?
   C. a decrease of pressure at constant temperature?
2. Compared with the speed of sound in air would the speed of sound be faster or slower in:
   A. carbon dioxide (CO_2)?
   B. hydrogen (H_2)?
   C. steam (H_2O)?
3. Figure 25–16 shows stress-strain curves for three different steels. A is a brittle steel, B is a moderately tough steel, and C is a soft steel. They all have approximately the same density. Rods are made from the three steels. In which material will the speed of sound be the slowest? In which will it be the fastest?

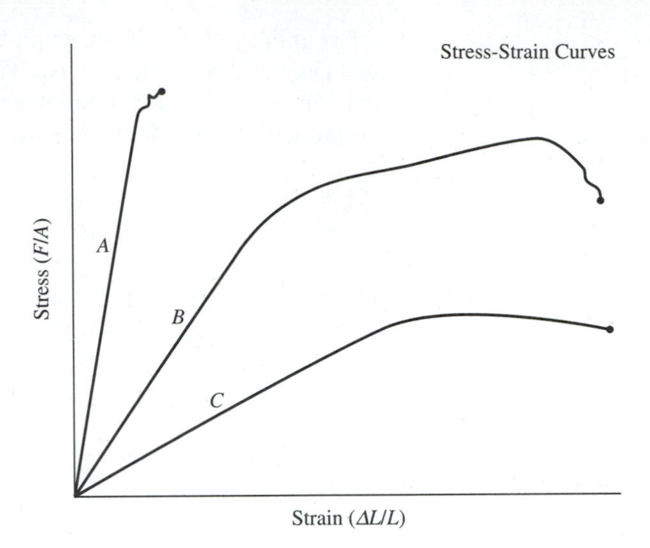

Figure 25–16: Diagram for Exercise 3. Stress-strain curves showing the elastic properties of three steels.

Section 25.2:

4. Draw diagrams to show that the fundamental resonant wavelength for a rod clamped at both ends is the same as for a rod clamped in the middle with the ends free to vibrate.

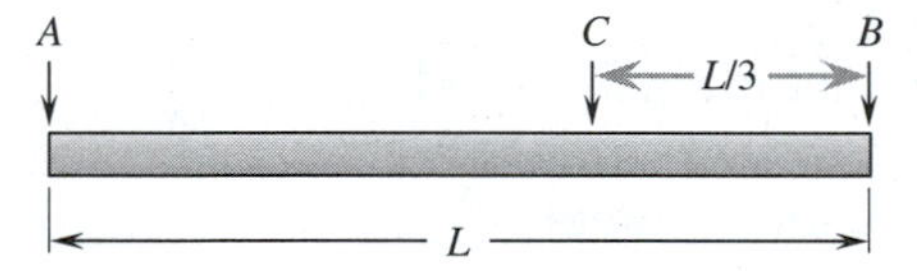

Figure 25–17: Diagram for Exercise 5.

5. Figure 25–17 shows a rod of length L with three possible clamping positions: end A, end B, and a distance $L/3$ from end B at C. Draw diagrams to verify the following situations.
   A. The rod is clamped at A and C. B is free to vibrate. The fundamental resonant wavelength is $4L/3$.
   B. The rod is clamped at all three positions. The fundamental resonant wavelength is $2L/3$.
   C. The rod is clamped at B and C. A is free to vibrate. The fundamental resonant wavelength is $8L/3$.

6. Draw a diagram to show that the fundamental wavelength and first two overtones for a pipe of length L that is open at one end and closed at the other end are $\lambda_1 = 4\text{L}$, $\lambda_2 = 4L/3$, and $\lambda_3 = 4L/5$.

7. Resonant tones can be produced by blowing across the lip of a soda bottle. If the distance from the liquid to the lip is 1.00 in, what is the wavelength of the fundamental resonance?

8. Some of the soda is drunk from the bottle in Exercise 7. The bottle is now only half full. Is the resonant frequency higher or lower than it was before?

9. Most wind instruments can be tuned by moving the mouthpiece or a section of tubing in or out to adjust the length of the instrument. A marching band rehearses indoors tuning to concert B-flat. The band then goes outside where the temperature is lower (a cool fall day). To retune to B-flat should the instruments

be lengthened or shortened? (Hint: Consider the change of the speed of sound.)

Section 25.4:

10. In what units are decibels expressed? Why?
11. If the intensity of a sound source increases 100-fold, by how many bels does it increase? By how many decibels?
12. Why is there a factor of 20 rather than a factor of 10 in Equation 25–6?
13. The intensity of a sound wave is increased by a factor of four. By what factor is the pressure of the sound wave increased?

Section 25.5:

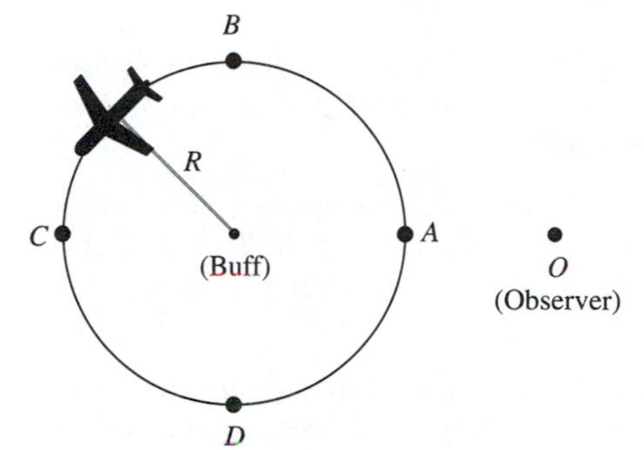

Figure 25–18: *A model airplane orbits around a model airplane buff. Sounds heard by an observer (O) outside the orbit are different from those heard by the model airplane buff.*

14. See Figure 25–18. A model airplane buff (X) operates a radio-control model airplane. The plane circles B at a constant radius. Observer (O) stands some distance away.
    - ***A.*** What does the observer notice concerning the tone of the model plane engine between A and B, B and C, C and D, D and A?
    - ***B.*** What does B hear?
15. Show that Equations 25–7 and 25–8 are special cases of Equation 25–9 where $\mathbf{v}_0$ or $\mathbf{v}_s$ are, respectively, zero.
16. From Equation 25–7, show that the equations of approach and recession are obtained when $\theta_s = 0°$ and $\theta_s = 180°$.
17. Will wind cause a Doppler shift of the frequency of a sound source? Why or why not?
18. A person drives rapidly toward a cliff. The driver toots the car horn. A moment later an echo returns. Will the echo have a higher pitch, lower pitch, or the same pitch as the original toot? Explain.

PROBLEMS

Section 25.1:

1. At what temperature does the speed of sound in air equal 350 m/s?
2. What is the speed of sound in air at 30°C?
3. At what temperature does the speed of sound in helium (He) equal 350 m/s?
4. How many times as fast is the speed of sound in hydrogen (H_2) than in air? Assume the same temperature for the gases.
5. Aluminum has a density of 5.27 slug/ft^3 and a Young's modulus of 1.1×10^7 psi. What is the speed of sound traveling the length of an aluminum rod?
6. Aluminum has a bulk modulus of 1.0×10^7 psi and a shear modulus of 3.4×10^6 psi. What is the speed of sound traveling through a large block of aluminum?

7. Ethyl alcohol has a density of 790 kg/m^3 and a bulk modulus of 9.1×10^8 N/m^2. Fresh water has a density of 1000 kg/m^3 and a bulk modulus of 2.0×10^9 N/m^2. Find the ratio of the speed of sound in ethyl alcohol compared with its speed in fresh water.

8. Show that Equations 25–1 and 25–4 are dimensionally correct.

Section 25.2:

9. A 1.00-m long hollow tube is open at both ends. What is its fundamental wavelength at 20°C?

10. The tube in Problem 9 is filled with air at 25.0°. What is the fundamental frequency of the resonance?

11. The tube in Problem 9 is filled with a liquid in which the speed of sound is 1400 m/s. What is the fundamental resonant frequency of the tube for this case?

12. A 32.0-in long metal rod is clamped in the center. The ends are free to oscillate. If the speed of sound is 4320 ft/s in the rod, what is the fundamental resonant frequency of compressional waves in the rod?

13. A 39.0-in hollow tube is open at one end and closed on the other. Air resonates in the tube at a fundamental frequency of 87.0 Hz.
    A. What is the fundamental resonant wavelength?
    B. What is the speed of sound in this situation?

*14. An 84-cm long rod is clamped at 21.0 cm from one end. Find the fundamental wavelength if:
    A. the ends are free to vibrate.
    B. the ends are fixed.

Section 25.4:

15. What is the sound level in decibels for an intensity of 2.5×10^{-4} W/m^2?

16. What sound intensity corresponds to 100 db?

17. A rock band produces sound at the 130-db level. What pressure is exerted on the listeners?

18. When a sound intensity increases by a factor of 1000, what is the increase in decibels?

Section 25.5:

19. A bat chirps with an 80,000-Hz frequency as it flies toward a tree at 50.0 ft/s (see Figures 25–19). What is the frequency of the echo? Take the speed of sound as 1125 ft/s. The tree acts as a mirror reflecting the sound of an approaching source.

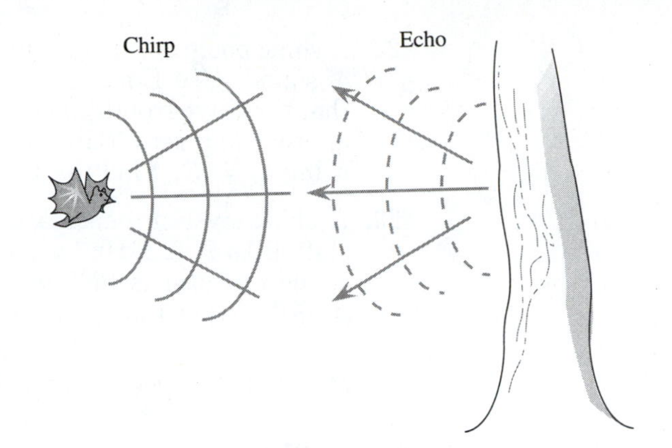

Figure 25–19: *Diagram for Problem 19. A bat uses echo ranging to fly at night. The Doppler shift of the echo depends on the bat's speed.*

20. A dolphin emits a 67,0̄00-Hz frequency as it swims under water at 50 ft/s toward an ultrasonic microphone. What frequency is picked up by the microphone? (The speed of sound in water is 4600 ft/s.)

21. A car horn blares at 324 Hz in still air. The speed of sound in air is 340̄ m/s. See Figure 25–20. Find the frequency heard by an observer when the car is moving at a speed of 27 m/s when the car is

A. traveling directly toward the observer.
B. approaching at an angle of 3̄0°.
C. receding at an angle of 135°.
D. moving directly away from the observer.

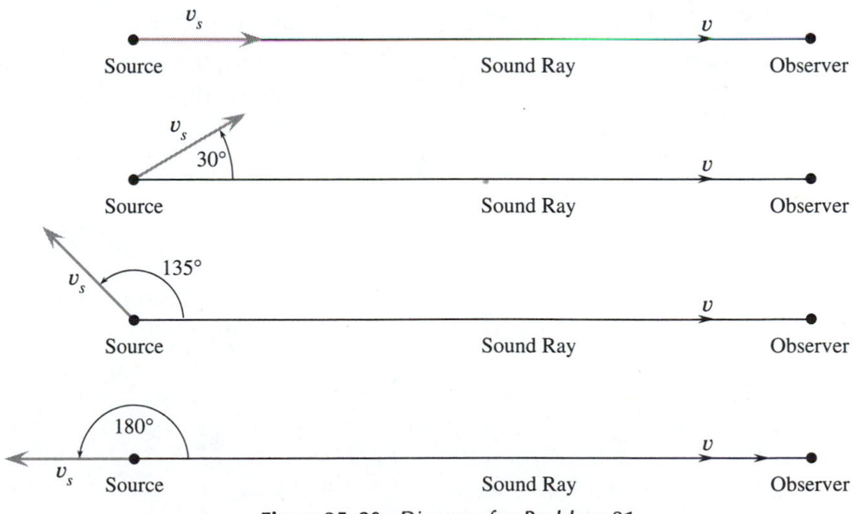

Figure 25–20: *Diagram for Problem 21.*

22. A sonic boom occurs as a jet airplane passes through the speed of sound. Use Equation 25–9 with $\mathbf{v}_s = \mathbf{v}$ and $\mathbf{v}_0 = 0$ to show that the wave fronts pile up in front of the jet with a wavelength approaching zero. (Hint: Substitute wavelength for frequency using $\lambda f = C$. Find the ratio of λ'/λ.)

*23. A sonar system emits a frequency of 9.20 MHz. An echo is shifted to 9.22 MHz by an approaching object. The speed of sound in water is 1400 m/s.

A. Show that the speed of the approaching object is $\mathbf{v}_s = \mathbf{v}(1 - f/f')$.

B. Find the speed of approach of the object.

Section 25.6:

24. The signal from a sonar echo-ranging system has a transit time of 1.24 s as it bounces off the ocean bottom. Assume a speed of sound of 4600 ft/s. How deep is the ocean at this point?

25. An ultrasonic resonance thickness gauge is used to monitor the thickness of sheet aluminum. Specifications require the material to be rolled to a thickness of 3/64 in with a tolerance of 5%. The speed of sound in aluminum is 1.67×10^4 ft/s. What are the upper and lower range of resonant frequencies that correspond to an acceptable thickness?

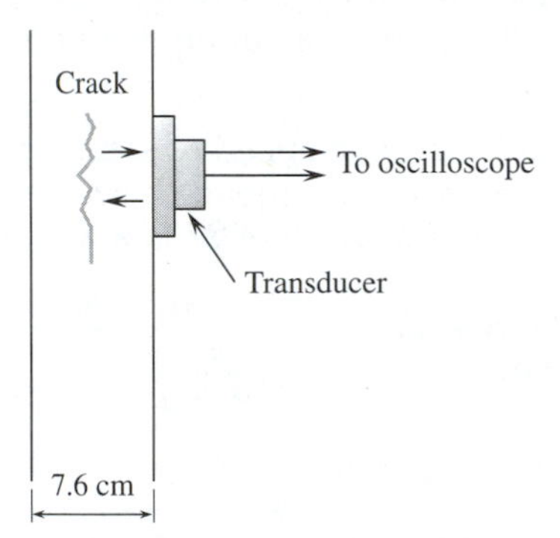

Figure 25–21: *Diagram for Problem 26. Ultrasound is used to locate defects in a steel girder.*

*26. Ultrasound echo ranging is used to find defects in concrete and steel structures. Ultrasonic waves will reflect from any surface, including interior cracks and cavities (see Figure 25–21). The speed of sound in steel is 5800 m/s. The steel bridge girder is 0.076 m thick.

A. In one area of the girder the transit time is $1.05\ 10^{-5}$ s for an ultrasound signal. How far is the crack from the surface of the girder?

B. In another area the transit time is 2.63×10^{-5} s. Interpret this result.

Chapter 26

LIGHT WAVES

How far away are the stars in this cluster? Scientists use their understanding of the properties of light waves to measure the distances to the stars. (*Photograph courtesy of NASA.*)

OBJECTIVES

In this chapter you will learn:

- about the full spectrum of electromagnetic radiation
- how the speed of light varies as it passes through different materials
- how light changes direction as it passes through a surface between two different materials (refraction)
- that different wavelengths are refracted at different angles as they pass through a surface between two different materials (dispersion)
- that the energy associated with light is directly proportional to its frequency
- to distinguish among illumination, luminous flux, and luminous intensity
- to calculate the effects of distance and surface orientation on illumination

Time Line for Chapter 26

1675 Ole Romer estimates the speed of light from Jupiter's eclipses.
1760 Johann Lambert investigates light reflected from the planets.
1801 Thomas Young rediscovers light interference.
1849 Hippolyte Fizeau measures the speed of light using rotating mirrors.
1926 Albert Michelson measures the speed of light to seven significant figures.

Light is a tiny frequency range of the spectrum of electromagnetic waves. Other electromagnetic waves have longer or shorter wavelengths. They may penetrate materials light cannot pass through, or they might be absorbed by other materials that are transparent to light. They differ mostly in wavelength. The general properties of light discussed in this chapter are very similar for most of the rest of the electromagnetic spectrum.

26.1 THE ELECTROMAGNETIC SPECTRUM

Figure 26-1 shows the range of the electromagnetic spectrum. Much of the spectrum has wavelengths longer than light. **AM radio waves** have very long wavelengths—from a few meters to several miles long. **Television** and **FM radio waves** are carried on shorter waves. **Microwaves,** used for telephone communications and microwave ovens, have wavelengths in the millimeter-centimeter range. **Infrared radiation,** which we sense as radiant heat, falls in between visible light and microwaves.

The **visible light** spectrum falls between wavelengths of 4×10^{-7} m and 7×10^{-7} m. Violet light has the shortest wavelength, and red light has the longest. The lower end of the visible spectrum merges into **ultraviolet light.** These are the wavelengths that cause suntans and sunburn.

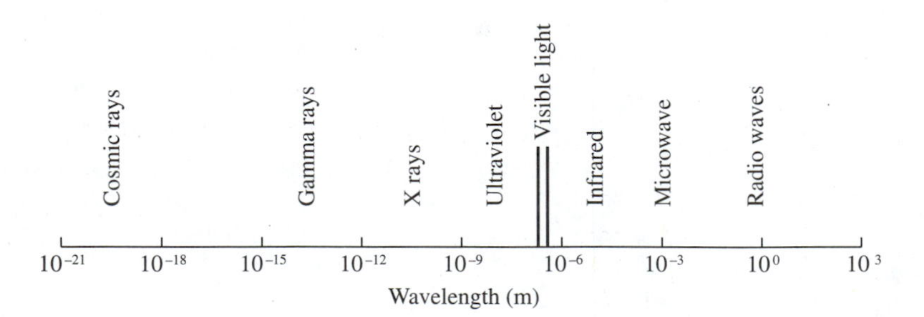

Figure 26–1: Visible light is a very small region of the full electromagnetic spectrum.

At shorter wavelengths, around 10^{-10} m, **X rays** occur. These wavelengths are close to the diameter of atoms. X rays are often used to study the atomic structure of materials. Shorter waves called **gamma rays** are emitted from the nuclei of atoms. They emanate from one of the smallest structures we know. The shortest wavelengths of the electromagnetic spectrum are called **cosmic rays.** They seem to come primarily from large structures far from our solar system.

26.2 THE SPEED OF LIGHT

Speed of Light in a Vacuum

All electromagnetic waves including light have the same speed through a vacuum. The speed of light, symbolized by the letter c, is one of the few true physical constants. The speed of light in a vacuum is

$$c = 3.00 \times 10^8 \text{ m/s}$$

$$c = 1.86 \times 10^5 \text{ mi/s}$$

Speed of Light in a Substance

When light or other electromagnetic waves pass through some material such as glass, plastic, or water, their speed is slower than in a vacuum. The ratio of the speed of light in a vacuum divided by the speed of light in a material is called the **index of refraction (n).**

$$n = \frac{c}{\text{v}} \qquad \textbf{(Eq. 26–1)}$$

Table 26–1 lists the average index of refraction for visible light for some materials. The index will be different for wavelengths larger or smaller than visible light.

□ **EXAMPLE PROBLEM 26–1: SPEED OF LIGHT IN WATER**

The index of refraction for fresh water is $n = 1.33$. What is the speed of light in water?

■ ***SOLUTION***

$$n = \frac{c}{\text{v}}, \text{ or v} = \frac{c}{n}$$

$$\text{v} = \frac{3.00 \times 10^8 \text{ m/s}}{1.33}$$

$$\text{v} = 2.26 \times 10^8 \text{ m/s}$$

Measuring the Distance to Stars

On a clear, dark night the sky is a canopy of twinkling stars. Ancients thought that the stars were painted on the inside of a crystal sphere rotated by some supreme power. As you look up, you can see why they might have thought that. To the naked eye, there is little to indicate that one star is any closer than another.

Astronomers have several schemes to measure stellar distance. One technique is based on the varying brightness of certain kinds of stars. Many stars, called variable stars, do not give steady light as the sun does. A Cepheid variable is a variable star that has the curious habit of pulsating at a fixed frequency of a few days. Its average brightness is related to the rate of pulsation.

(Photograph courtesy of NASA.)

Astronomers have made charts of period and brightness of Cepheids, indicating how bright the stars would be at a standard distance of 3.26 light years from the earth.

When a Cepheid star is found in a distant cluster of stars, the inverse square law of illumination is used to determine the distance from the earth to the Cepheid. The brightness of the distant star is compared with the brightness it would have at the standard distance. This ratio tells us how far away the star is and, in turn, the distance to the cluster of stars in which it is found.

Table 26–1: Index of refraction for selected materials.

MATERIAL	INDEX OF REFRACTION*
Acetone	1.36
Air (standard conditions)	1.00029
Chloroform	1.44
Diamond	2.42
Ethyl alcohol	1.36
Fluorite	1.44
Glass	
crown	1.50
dense flint	1.65
light flint	1.58
Halite (rock salt)	1.54
Ice	1.31
Octane	1.40
Paraffin	1.43
Quartz (fused)	1.46
Sodium carbonate	1.54
Water	1.33

* Index of refraction for yellow light (wavelength = 5.9×10^{-7} m).

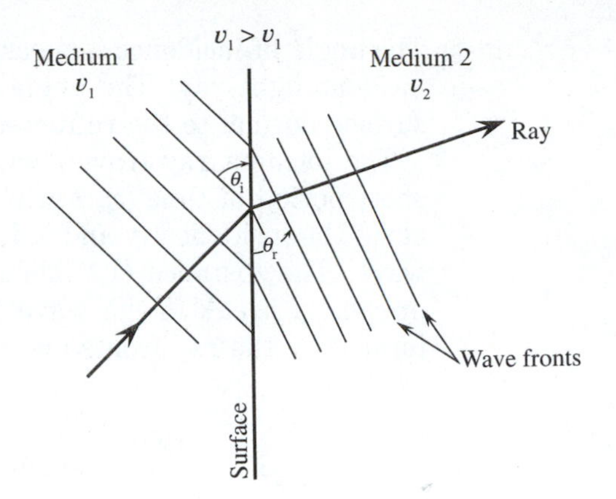

Figure 26–2: Wave fronts bend as they pass through a surface between two materials with different indices of refraction. This makes it possible for the fronts to be continuous as their speed changes.

Refraction

When a marching band turns a corner, the musicians on the inside of the turn move slowly and those on the outside move faster to keep the line of marchers straight. Light passing through a surface between two mediums behaves in a similar way.

Figure 26–2 shows wave fronts passing through a surface between two mediums such as water and glass. The **angle of incidence** (θ_i) is the angle at which the incoming waves strike the surface. The part of the wave front that has entered the material moves at a slower speed. In order to prevent a break in the wave front, waves move at a smaller angle to the surface in the second material. This angle is called the **angle of refraction** (θ_r).

Refraction *is the change of direction of a wave front as its speed changes.*

Figure 26–3 shows the light rays as well as the wave fronts at the surfaces. Remember, a light wave is at right angles to wave fronts.

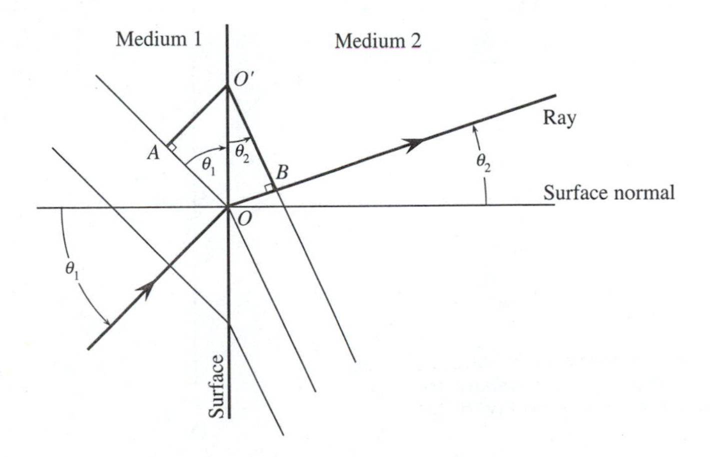

Figure 26–3: Triangles AOO′ and BOO′ are used to derive Snell's law.

The angle of incidence is measured from the surface normal to the incident light ray. The angle of refraction is measured from the surface normal to the refracted ray.

The incident ray travels with a speed of v_1 at an angle θ_1. In a short period of time (t), a wave front moves a distance $AO' = v_1 t$ along the incident ray and a distance $OB = v_2 t$ along the refracted wave. The distance OO′ along the surface must be the same to prevent a break in the wave front. From the two right triangles formed by the ray fronts and rays we have:

$$OO' = \frac{v_1 t}{\sin \theta_1} = \frac{v_2 t}{\sin \theta_2}$$

Time cancels. Rearrange the equation.

$$\frac{\sin \theta_1}{v_1} = \frac{\sin \theta_2}{v_2}$$

We can multiply both sides of the equation by c.

$$\frac{c}{v_1} \sin \theta_1 = \frac{c}{v_2} \sin \theta_2$$

Substitute n (see Equation 26–1) for c/v on each side of the equation.

$$n_1 \sin \theta_1 = n_2 \sin \theta_2 \qquad \textbf{(Eq. 26–2)}$$

Equation 26–2 is known as Snell's law. It tells us that the amount a light beam is bent as it passes through a surface depends on the index of refraction of the two materials on either side of the surface.

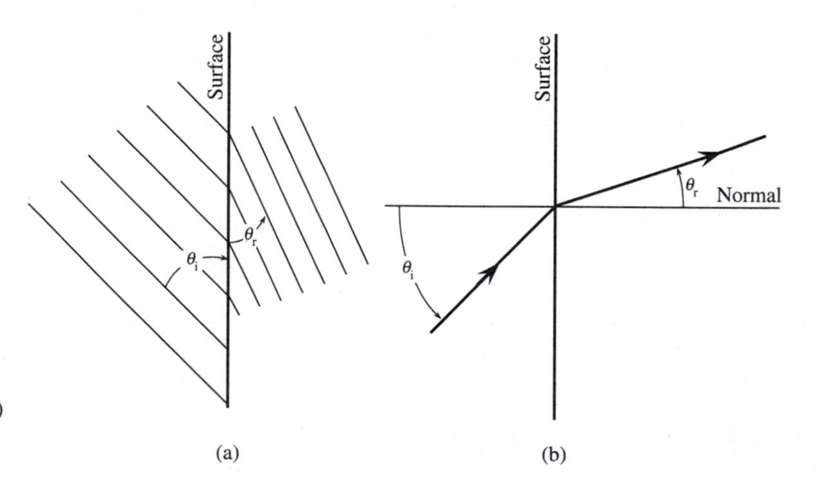

Figure 26–4: *(a) The behavior of wave fronts passing between two materials. (b) The same behavior diagrammed with light rays.*

It is easier to draw ray diagrams than to draw wave fronts. In a ray diagram, angles are measured from the surface normal. Figure 26–4 shows refraction diagrammed in two equivalent ways. Figure 26–4a shows the behavior of the wave fronts. Figure 26–4b shows the same thing diagrammed with rays. In optics, ray diagrams are usually used rather than wave front diagrams. We will follow this practice.

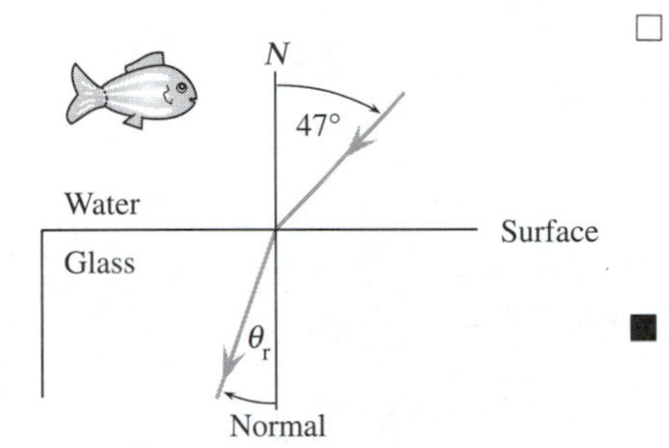

Figure 26–5: *Diagram for Example Problem 26–2.*

EXAMPLE PROBLEM 26–2: REFRACTION

A light ray traveling through water strikes a piece of glass at an incident angle of 47.0° (see Figure 26–5). If the index of refraction for water is 1.33 and for the glass it is 1.56, what is the angle of refraction?

SOLUTION

$$n_1 \sin \theta_1 = n_2 \sin \theta_2$$

$$\sin \theta_2 = \frac{n_1}{n_2} \sin \theta_1$$

$$\sin \theta_2 = \frac{1.33 \times 0.7314}{1.56}$$

$$\sin \theta_2 = 0.6236$$

$$\theta_2 = \sin^{-1}(0.6236)$$

$$\theta_2 = 38.6°$$

Dispersion

The speed of electromagnetic radiation is the same for all wavelengths in a vacuum, but not for radiation traveling through a medium. In Figure 26–6, the index of refraction for three different kinds of optical glass is plotted against wavelength around the visible light region. We notice three things.

1. The index of refraction increases as the wavelength decreases.
2. The *rate* of increase is greater for shorter wavelengths.
3. When different materials are compared in the same wavelength range, the ones with higher indices of refraction tend toward a steeper slope.

Light from a black body such as the sun or a hot incandescent light bulb is a mixture of many different wavelengths. A full-spec-

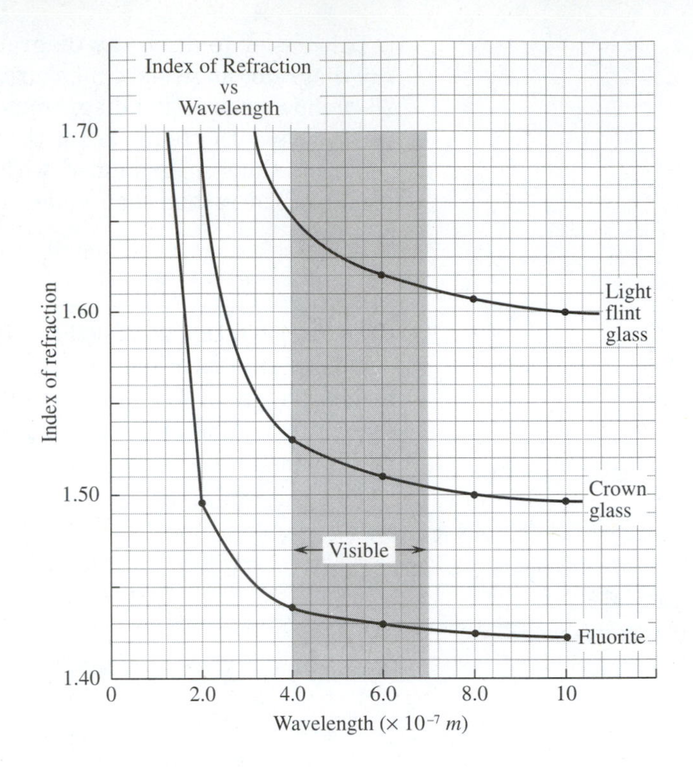

Figure 26–6: *Index of refraction versus wavelength.*

trum light beam from a black-body source is called **white light.** When white light is refracted, different wavelengths are refracted in slightly different directions. Shorter wavelengths undergo a greater change of direction.

Figure 26–7 shows white light passing through a prism. Different wavelengths travel in different directions after two refractions. We see the black-body spectrum—a rainbow. This effect is called **dispersion.**

Dispersion *is the angular divergence of electromagnetic waves according to wavelength.*

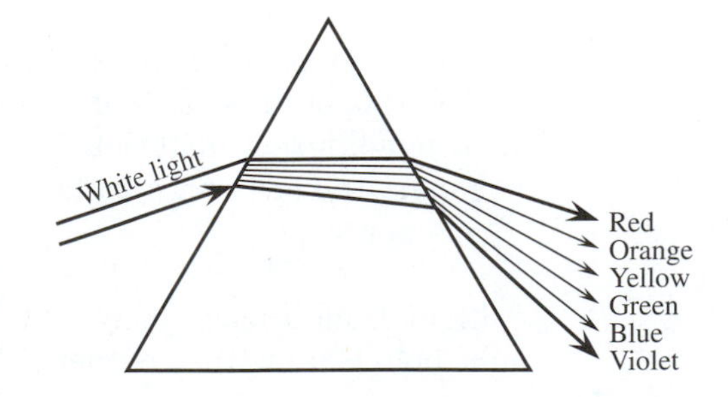

Figure 26–7: *Dispersion. The index of refraction is slightly different for different wavelengths of light. Shorter wavelengths are bent more by the prism than longer wavelengths.*

26.3 PHOTONS

Electromagnetic waves are emitted in **wave trains,** or wave groups, similar to the one shown in Figure 26–8. The length of the wave train depends on the frequency of the wave. Visible light has wave trains about 1 m long. The waves in the train have a very small variation in frequency. This causes a single "beat" similar to the beats described for sound in Chapter 25.

In situations where group waves behave more like particles than wave groups, they are called **photons.** A hot light source such as a light bulb continually emits photons with a wide range of frequencies.

The energy of a photon or electromagnetic wave is directly proportional to its frequency (f). Low-frequency (long-wavelength) photons carry less energy than high-frequency (short-wavelength) photons. The proportionality constant is called **Planck's constant (h).**

$$E = hf \qquad \textbf{(Eq. 26–3A)}$$

where E = energy and $h = 6.626 \times 10^{-34}\ \mathrm{J \cdot s}$.

Planck's constant is a very small number. This means that a beam of light detectable by the eye must contain a very large number of wave trains or photons. If n is the number of photons transmitted by a light beam, the energy of the beam is:

$$E = nhf \qquad \textbf{(Eq. 26–3B)}$$

Figure 26–8: *A wave group, or wave train. Visible light is emitted in group waves (photons), about 1 m long.*

☐ **EXAMPLE PROBLEM 26–3: PHOTON ENERGY**

How much energy is in a single photon of violet light with a wavelength of 4.2×10^{-7} m in a vacuum?

■ ***SOLUTION***

First find the frequency.

$$f = \frac{c}{\lambda} = \frac{3.0 \times 10^{8}\ \mathrm{m/s}}{4.2 \times 10^{-7}\ \mathrm{m}}$$

$$f = 7.14 \times 10^{14}\ \mathrm{Hz}$$

$$E = hf$$

$$E = 6.628 \times 10^{-34}\ \mathrm{J \cdot s} \times 7.14 \times 10^{14}\ \mathrm{Hz}$$

$$E = 4.7 \times 10^{-19}\ \mathrm{J}$$

□ **EXAMPLE PROBLEM 26–4: THE NUMBER OF PHOTONS IN A LIGHT BEAM**

How many violet photons with a frequency of 7.14×10^{14} Hz are needed per second to supply 2.1 W of radiant energy?

■ *SOLUTION*

$$P = \frac{E}{t} = \frac{n\,hf}{t}$$

$$\frac{n}{t} = \frac{P}{hf}$$

$$\frac{n}{t} = \frac{2.1\ \text{W}}{6.628 \times 10^{-34}\ \text{J} \cdot \text{s} \times 7.14 \times 10^{14}\ \text{Hz}}$$

$$\frac{n}{t} = 4.5 \times 10^{18}\ \text{photons/s}$$

26.4 ILLUMINATION

Three quantities are used to measure the quantity of light emitted from a source or falling on a surface. These quantities are luminous flux (F), illumination (E), and luminous intensity (I).

Illumination and Distance

In the absence of exceptionally strong gravitational fields electromagnetic waves travel in a straight line. Figure 26–9 shows a point source of electromagnetic radiation. All of the photons emitted from the source pass through a sphere of radius R_1 surrounding the source. The same number of photons pass through another sphere with a larger radius, R_2. The flow of photons is called the **flux (F).** The same total flux of photons passes through the two concentric spheres.

Since the same number of photons passes through the two spheres per unit time, we can write:

$$F_1 = F_2 \qquad \textbf{(Eq. 26–4)}$$

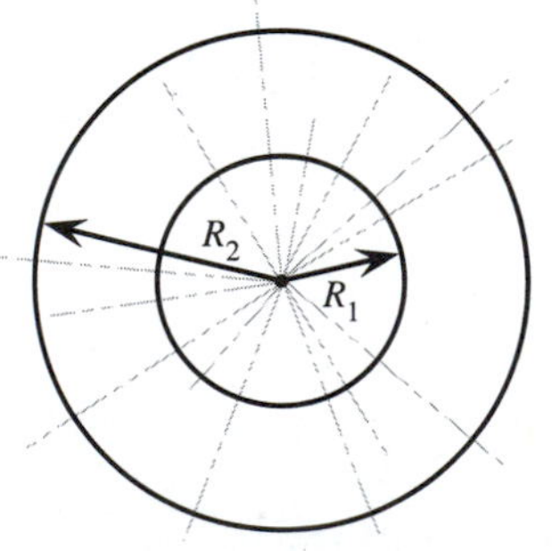

Figure 26–9: *The same number of photons emitted by a point source passes through two concentric spheres.*

Illumination (E) is the amount of flux striking a surface per unit area.

$$E = \frac{F}{A} \qquad \textbf{(Eq. 26–5)}$$

The area of a sphere is $A = 4\pi R^2$. Then

$$F = 4\pi R^2 E$$

We can substitute this into Equation 26–4.

$$4\pi R_1^{\,2} E_1 = 4\pi R_2^{\,2} E_2$$

or

$$E_2 = \frac{R_1^{\,2}}{R_2^{\,2}} E_1$$

The illumination falls off as an inverse square of the distance. For a single surface we can write

$$E = k\frac{F}{R^2}$$

where k is a proportionality constant.

Illumination and Orientation

The illumination of a surface depends on the orientation of the surface. Figure 26–10 shows two surfaces far away from a light source. The light hits surface A at right angles. The same flux is spread over a wider surface B tilted at an angle (θ) to the light rays. From the diagram we see that

$$\frac{A_{\mathrm{A}}}{A_{\mathrm{B}}} = \cos\theta$$

or

$$A_{\mathrm{B}} = \frac{A_{\mathrm{A}}}{\cos\theta}$$

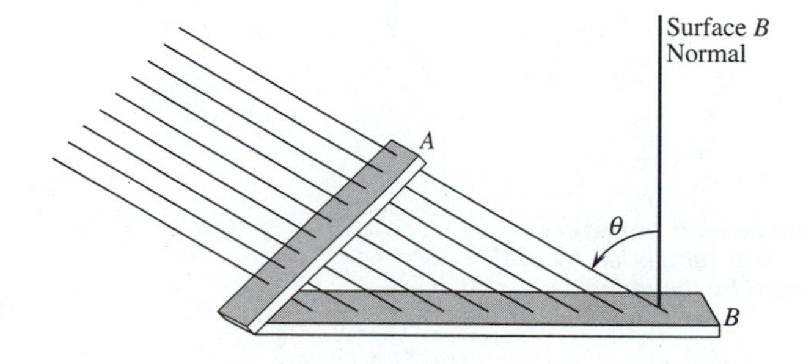

Figure 26–10: *The same number of photons strikes areas A and B. Area B has fewer photons per unit area striking it. It has a lower illumination.*

We can use this to compare the illumination of a tilted surface with one that is not tilted.

$$E(\text{untilted}) = \frac{F}{A_A}$$

$$E(\text{tilted}) = \frac{F}{A_B} = \frac{F}{A_A/\cos\theta}$$

$$E(\text{tilted}) = E(\text{untilted}) \times \cos\theta$$

This gives us the effect of orientation of a surface. In general, we can write:

$$E = k\frac{F\cos\theta}{R^2} \qquad \textbf{(Eq. 26–6)}$$

Intensity

In two dimensions we defined an angle called a radian. It is the arc length of a circle (S) divided by the radius (R). There are 2π radians in a circle (see Figure 26–11).

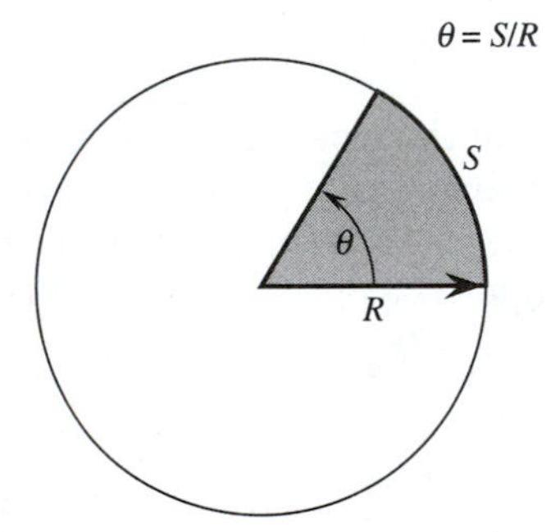

Figure 26–11: *A radian is the arc length subtended by angle θ to the radius of the circle.*

$$\text{radian} = \frac{S}{R}$$

For a complete circle $S = 2\pi R$. The number of radians in a circle is:

$$\frac{S}{R} = \frac{2\pi R}{R} = 2\pi$$

We can define a **solid angle** for three dimensions in a similar way. Look at Figure 26–12. A solid angle forms a cone intersecting the surface of a sphere. Instead of a distance along the circumference

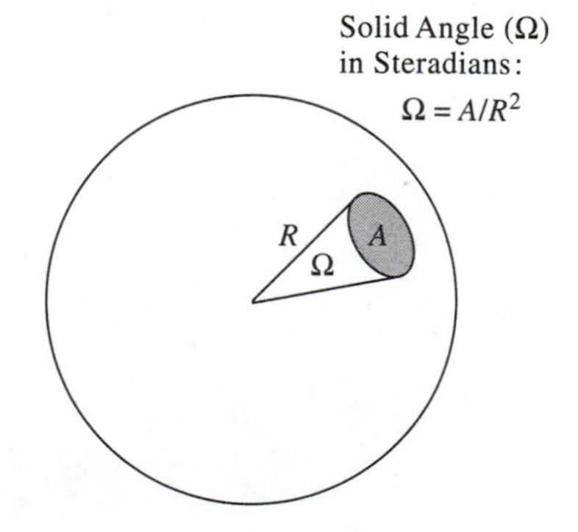

Figure 26–12: *A steradian is the ratio of the surface of a sphere subtended by the solid angle Ω divided by the radius squared.*

of a circle, the solid angle encloses an area on the surface of a sphere. The ratio of the area enclosed by the solid angle divided by R^2 is a unitless quantity analogous to the radian. It is called a **steradian (sr).**

$$\textit{solid angle in steradians}\ (\text{sr}) = \frac{A}{R^2}$$

We can calculate the number of steradians in a complete sphere. The area of a sphere is $4\pi R^2$.

$$\textit{steradians in a complete sphere} = \frac{4\pi R^2}{R^2}$$

$$\textit{steradians in a sphere} = 4\pi\ \text{sr}$$

We can now define luminous intensity (I) in terms of steradians.

Luminous intensity *is the luminous flux passing through a solid angle of 1 sr.*

Since there are 4π sr in a complete sphere, the intensity (I) is proportional to the total luminous flux (F) from a point source of light by:

$$F = 4\pi I \qquad \textbf{(Eq. 26–7)}$$

If we use luminous intensity (I) rather than luminous flux, the constant $k = 1$ in Equation 26–6.

$$E = \frac{I \cos \theta}{R^2} \qquad \textbf{(Eq. 26–8)}$$

Units

An engineer designing a solar collecting system is interested in the total radiant energy of sunlight. The engineer will measure the total electromagnetic energy flux in terms of langleys per unit time. The designer is interested in how much radiant heat she can collect in an hour or day. She would probably express solar illumination in langleys per day.

$$1\ \text{langley} = 1\ \text{cal/cm}^2 = 3.687\ \text{Btu/ft}^2$$

Only a small part of the radiant energy from the sun or from an incandescent lamp is in the visible light range. An architect is interested only in this part of the radiation. The design concern is

Table 26–2: Typical luminous intensity for gas-filled tungsten-filament light bulbs.

POWER RATING (W)	LUMINOUS INTENSITY (cd)	LUMINOUS FLUX (lm)
25	20.7	260
50	55.0	695
100	125	1580
200	290	3640
500	800	10,000

proper lighting, not total heat energy. Let us define the luminous intensity that is concerned only with visible light.

Luminous intensity (I) is expressed in **candela (cd).** One candela is $\frac{1}{60}$ of the black-body radiation per steradian from a 1-cm^2 area of a standard light source. The standard source is radiation from platinum at its melting point (2043 K). The fraction $\frac{1}{60}$ is the portion of black-body radiation in the visible range at this temperature.

Table 26–2 shows typical luminous intensities for ordinary tungsten-filament light bulbs. Roughly 1 cd is produced per watt of electrical power. Fluorescent lighting is more efficient. About 4 cd/W are produced.

The total luminous flux (F) can be found using Equation 26–7. The unit of luminous flux is the **lumen** (lm).

$$1 \text{ lm} = \frac{1 \text{ cd}}{4\pi \text{ sr}}$$

Illumination (E) is expressed in lumens per unit area.

$$1 \text{ lux (lx)} = 1 \text{ lm/m}^2$$

$$1 \text{ foot} \cdot \text{candle (ft} \cdot \text{c)} = 1 \text{ lm/ft}^2$$

□ **EXAMPLE PROBLEM 26–5: THE SOLAR COLLECTOR**

A homeowner wants to design a passive solar collector. He finds that for January the solar illumination on a horizontal flat surface is 64 langleys per day with an effective altitude of 28° for the sun (see Figure 26–13). How many British thermal units will strike an 8.0 ft × 4.0 ft vertical glass surface?

■ ***SOLUTION***

First find the illumination on a surface oriented with a zero-degree normal to the sun's rays (E_0) (see Figure 26–13a). The normal of the horizontal surface makes an angle of 62°.

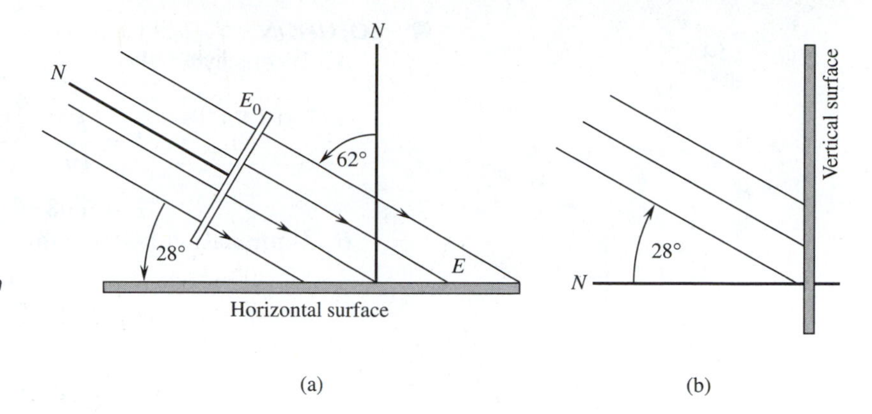

Figure 26–13: (a) *Horizontal surface. The sun makes an angle of 62° with the surface normal. E_0 is the maximum illumination for a surface orientation with the surface normal parallel to the sun's rays.* (b) *Vertical surface. The surface normal makes an angle of 28° with respect to the light rays.*

$$E(\text{horizontal surface}) = E_0 \cos 62°$$

$$E_0 = \frac{E(\text{horizontal surface})}{\cos 62°}$$

$$E_0 = \frac{64 \text{ langley/d}}{0.4695} = 136 \text{ langley/d}$$

Look at Figure 26–13b. The angle between the vertical normal and the light rays is 28°.

$$E(\text{vertical}) = E_0 \cos 28°$$

$$E(\text{vertical}) = 136 \text{ langley/d} \times 0.8829$$

$$E(\text{vertical}) = 120 \text{ langley/d}$$

We now have the solar illumination on the vertical plate. We want the total flux striking the sheet of glass.

$$F = E \times A$$

$$F = 120 \text{ langley/d} \times 8.0 \text{ ft} \times 4.0 \text{ ft} \times 3.687 \text{ Btu/langley/ft}^2$$

$$F = 1.4 \times 10^4 \text{ Btu}$$

EXAMPLE PROBLEM 26–6: A LIGHT BULB OVER A DESK

A light bulb is rated at 695 lm. See Figure 26–14.

A. What is the light intensity (I) of the bulb?

B. What is the illumination on a desktop 6.5 ft below the bulb?

C. If the desk is moved 8.0 ft horizontally, what will be the illumination on the desktop?

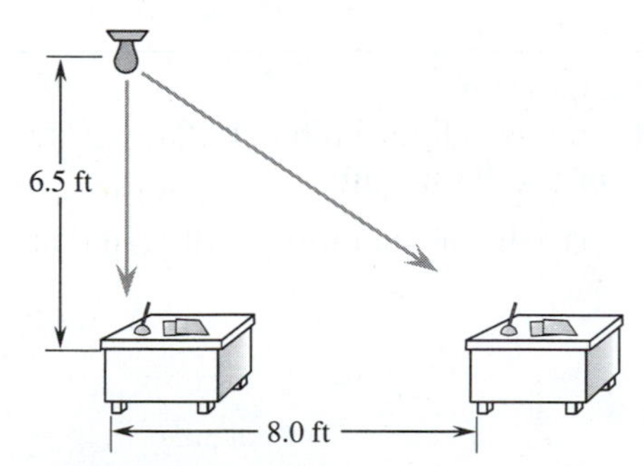

Figure 26–14: Diagram for Example Problem 26–6.

■ *SOLUTION*

A. Intensity:

$$I = \frac{F}{4\pi} = \frac{695 \text{ lm}}{4\pi}$$

$$I = 55.3 \text{ cd}$$

B. Illumination below bulb:

$$E = \frac{I \cos \theta}{R^2}$$

$$E = \frac{55.3 \text{ cd} \times 1.00}{(6.5 \text{ ft})^2}$$

$$E = 1.3 \text{ ft} \cdot \text{c}$$

C. Illumination at an angle:

Moving the desk changes two things. The desk is farther away, and the light hits the desktop at an angle. We can find the new distance using the Pythagorean theorem.

$$R^2 = (6.5 \text{ ft})^2 + (8.0 \text{ ft})^2$$

$$R^2 = 106 \text{ ft}^2$$

The angle is the inverse tangent.

$$\theta = \tan^{-1}\left(\frac{8.0 \text{ ft}}{6.5 \text{ ft}}\right) = 51°$$

$$E = \frac{I \cos \theta}{R^2}$$

$$E = \frac{55.3 \text{ cd} \times \cos 51°}{106 \text{ ft}^2}$$

$$E = 0.33 \text{ ft} \cdot \text{c}$$

□ **EXAMPLE PROBLEM 25–7: LUMINOUS FLUX**

The illumination 3.2 m directly below a light bulb is 16.2 lux. Find the intensity and luminous flux of the light bulb.

■ *SOLUTION*

$$I = \frac{R^2 E}{\cos \theta}$$

$$I = \frac{(3.2 \text{ m})^2 \times 16.2 \text{ lx}}{1.00}$$

$$I = 166 \text{ cd}$$

$$F = 4\pi I$$

$$F = 2.09 \times 10^3 \text{ lm}$$

SUMMARY

Visible light is a narrow band of the electromagnetic spectrum. Light and other electromagnetic waves travel in a vacuum at a speed (c) of 3.00×10^8 m/s. In a material, light travels at a slower speed. The ratio of the speed of light in a vacuum divided by its speed in a material is called the index of refraction (n):

$$n = \frac{c}{v}$$

When light passes through the surface between two materials its direction changes. If it moves from a material with a low index of refraction into a material with a higher index, the ray will be bent toward the surface normal. If θ_i is the angle incident to the surface and θ_r is the angle of refraction, then, according to Snell's law:

$$n_i \sin \theta_i = n_r \sin \theta_r$$

The index of refraction varies with frequency. Different frequencies or colors of light will be refracted at different angles. This is called dispersion.

Visible light travels in wave groups, or wave trains, about a meter long. An individual wave train is called a photon. The radiant energy of an individual photon is proportional to its frequency.

$$E = hf$$

where $h = 6.626 \times 10^{-34}$ J · s, Planck's constant.

Three things—luminous intensity, luminous flux, and illumination—are used to measure a quantity of light. The luminous intensity (I) is measured in candelas (cd). One candela is $\frac{1}{60}$ the intensity of radiation from a 1-cm^2 area of a black body at the melting temperature of platinum. It is a measure of the number of photons passing through a solid angle of 1 sr/s. The luminous flux (F) is the total number of photons from a source per second. It is measured in lumens (lm).

$$F = 4\pi I$$

The illumination (E) is the number of photons striking a surface per unit area per second. Since light usually travels in a straight line, illumination obeys an inverse square law.

$$E = \frac{I \cos \theta}{R^2}$$

where θ is the angle between light rays from the source and the surface normal, and R is the distance between the source and the illuminated surface. Illumination is measured in either foot · candles or lux.

KEY TERMS

If you can explain the following terms to a friend or classmate, you understand their meaning. If you cannot explain the terms, you should reread the sections in which they are discussed.

AM radio waves
angle of incidence
angle of refraction
candela (cd)
cosmic rays
dispersion
flux (F)
FM radio waves
gamma rays
illumination (I)
index of refraction
infrared radiation
lumen
luminous intensity
microwaves
photons
Planck's constant (h)
refraction
Snell's law
solid angle
steradian (sr)
television waves
ultraviolet light
visible light
wave trains
white light
X rays

EXERCISES

Section 26.1:

1. Which of the following waves have wavelengths longer than visible light?

   A. AM radio waves **B.** X rays **C.** Microwaves
   D. Ultraviolet waves **E.** Gamma rays **F.** Infrared waves

2. Explain the following terms in your own words.

   A. Electromagnetic spectrum **B.** Refraction
   C. Dispersion **D.** Photons
   E. Luminous flux **F.** Luminous intensity
   G. Illumination **H.** Steradian

Section 26.2:

3. A wet yellow polka-dot bikini has the same color in water as it has in air. The index of refraction of water is 1.33. The index for air is close to 1.00. What can you say about the speed of light in water compared with its speed in air? About the frequency

of light in water compared with air? About the wavelength of yellow light in water compared with air?

4. As light passes from air ($n_1 = 1.00$) into glass ($n_2 = 1.50$) is violet light bent more toward the surface normal than red light? Violet light has the shorter wavelength.

5. Light passes from glass into air. Do light waves bend away from or toward the surface normal? Explain.

Section 26.3:

6. Arrange the following waves according to the amount of energy of a single photon. The most energetic first; the least energetic last.

   A. FM radio waves **B.** Ultraviolet radiation **C.** Red light
   D. Blue light **E.** Microwaves **F.** Infrared light

7. If the energy of a photon is directly proportional to its frequency, how is its energy related to its wavelength?

Section 26.4:

8. One surface has an illumination of 1 lux. Another surface has an illumination of 1 ft · c. Which surface is more brightly lit?

9. How many steradians are there in a hemisphere?

10. A foot · candle is a British unit of illumination. A lux is a metric unit. Is a lumen a British unit or a metric unit? Explain.

11. At the grocery store, light bulbs are sometimes rated in terms of lux at 1 m. In other cases, the rating is given in lumens. If the same light bulb were rated in both lux and lumens, which quantity would be numerically larger?

PROBLEMS

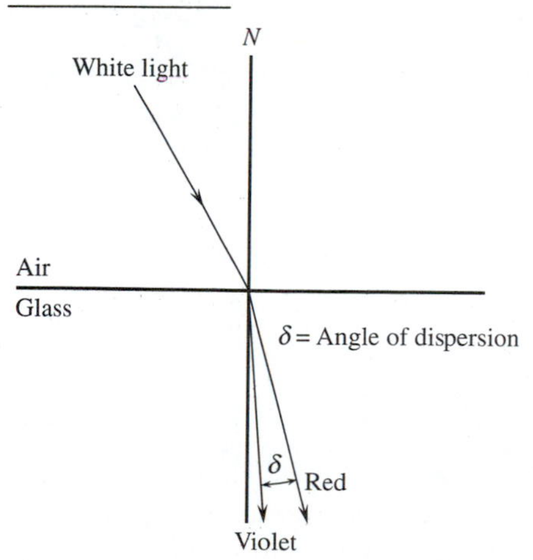

Figure 26–15: Diagram for Problem 5.

Section 26.2:

1. Light in air strikes at an angle of 58° on the surface of glass with an index of refraction of 1.63. What is the refraction angle?

2. What is the speed of light in diamond ($n = 2.42$)?

3. Light in air is incident at an angle of 45.0° on a glass plate. The light is refracted at an angle of 29.2°. What is the index of refraction of the material?

4. The average distance between the earth and the sun is 9.3×10^7 mi. How long does it take light to travel from the sun to earth?

5. Crown glass has an index of refraction of 1.52 for red light and an index of 1.54 for violet light. White light strikes crown glass at an angle of 37.0°. What is the *difference* between the angles of refraction for these frequencies of light? This is the angle of dispersion (see Figure 26–15).

6. Light traveling through water ($n_i = 1.33$) strikes a glass surface ($n_r = 1.63$) at an angle of 32.0°. Find the angle of refraction.

7. Show that the wavelength of light in a medium (λ') is related to its wavelength in a vacuum (λ) by: $\lambda' = \lambda/n$, where n is the medium's index of refraction.

Section 26.3:

8. Find the energy of the following photons:
   A. An X ray with a frequency of 2.50×10^{18} Hz
   B. A radio wave with a wavelength of 2.34 km
   C. Visible light at a wavelength of 6.22×10^{-7} m
   D. A cosmic ray with a frequency of 1.30×10^{24} Hz

9. A laser delivers 2.3×10^{-2} W of radiant power at a wavelength of 6.2×10^{-7} m. How many photons per second does the laser emit?

Section 26.4:

10. Find the conversion factor between lux and foot · candles.

11. The illumination 5.3 m below a street lamp is 620 lux.
    A. Calculate the luminous intensity of the street lamp.
    B. Find the luminous flux of the lamp.

12. Find the illumination on a horizontal surface 7.0 ft below a 1650-lm lamp and displaced 12.0 ft horizontally (see Figure 26–16).

13. What would be the illumination for the surface in Problem 12 if the surface were vertical rather than horizontal?

14. Find the maximum illumination 2.3 m from a 700-lm lamp.

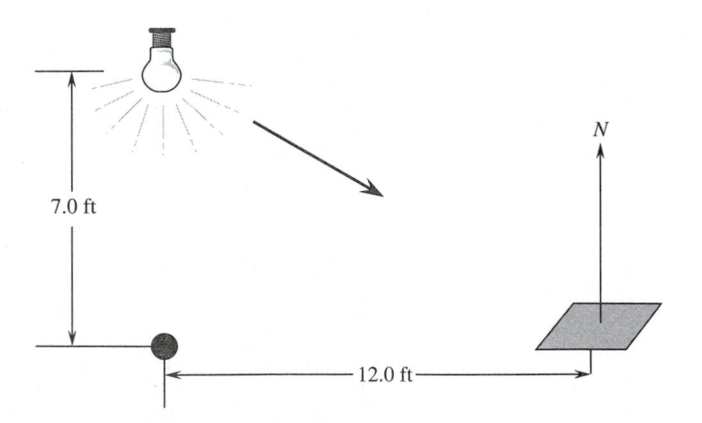

Figure 26–16: *Diagram for Problem 12.*

Chapter 27

GEOMETRIC OPTICS

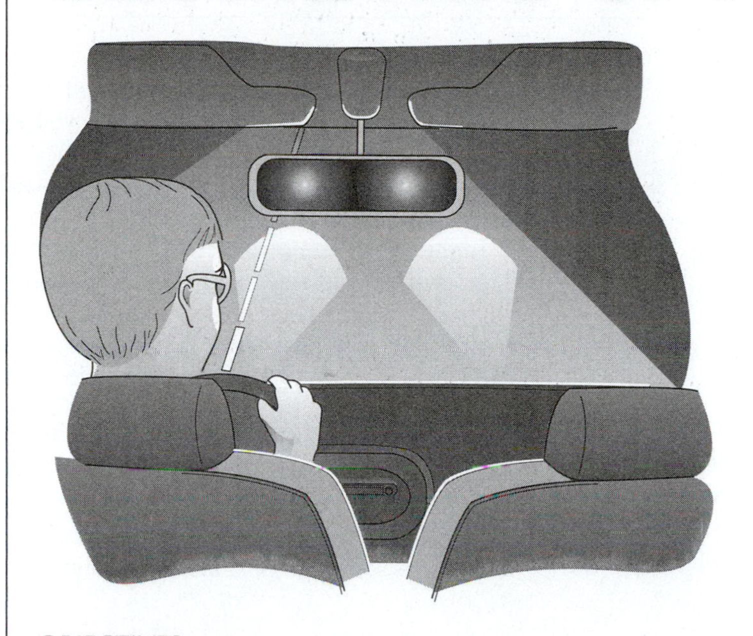

A rearview mirror that adjusts for day or night driving is an everyday application of geometric optics. This chapter explains how the behavior of light affects the images it creates.

OBJECTIVES

In this chapter you will learn:

- about the behavior of light reflected from flat surfaces
- how images are formed in a plane mirror
- to distinguish between real and virtual images
- how some light rays cannot escape from a material with a high index of refraction
- how images are formed by spherical mirrors
- to calculate image distances, object distances, and magnifications of spherical mirrors
- to find the apparent depth of objects immersed in materials with optical indices greater than air
- to calculate the image distances, object distances, and magnifications of thin lenses
- how optical fibers transmit light signals
- how to find angular magnification for a variety of optical instruments

Time Line for Chapter 27

1106 Robert Grosseteste experiments with lenses and mirrors.
1270 Witelo writes a paper dealing with reflection, refraction, and geometric optics.
1608 Hans Lippershey invents the telescope.
1663 James Gregory describes a reflecting telescope.
1668 Isaac Newton invents a reflecting telescope.
1704 Isaac Newton publishes *Optics*.
1827 William Rowan Hamilton publishes *Theory of Systems of Rays*.

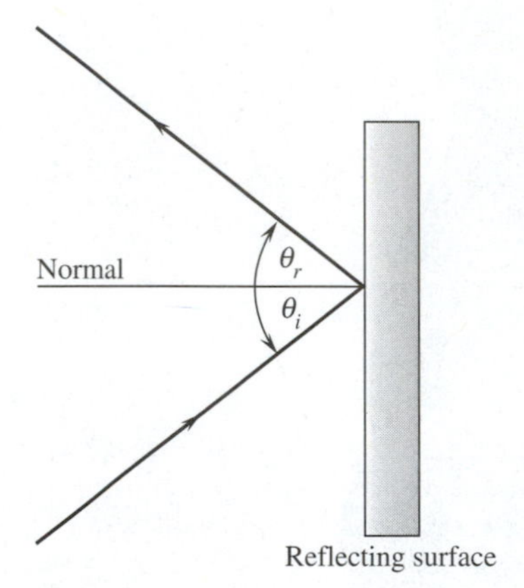

Figure 27–1: Angles of incidence and angles of reflection are measured from the surface normal.

You prepare for the day by looking at your image in the bathroom mirror. As you wait for your toast, you see a somewhat different image of yourself on the polished surface of the toaster. In the car, a rearview mirror lets you check the traffic behind you without turning around.

Light rays can be reflected by a smooth surface to form images. The location and size of the image depend on the curvature of the surface. If light rays pass through a surface, the rays will be refracted. This is another way images can be formed. Your view of the world is through a lens. Lenses in your eyes focus an image on the lining of your eye. If the natural focus is not good, the focus is adjusted by adding glasses or contact lenses.

The study of the path of light rays without regard for their wave nature is called **geometric optics.** In this chapter, we will trace the path of light rays through different devices such as mirrors, prisms, and lenses. We will also look at a few optical instruments.

27.1 REFLECTION

The Law of Reflection

See Figure 27–1. In geometric optics angles are measured from the surface normal. The **angle of incidence** (θ_i) is the angle the incoming ray makes with the normal. The **angle of reflection** (θ_r) is the angle the reflected ray makes with the normal.

The **law of reflection** is simple. When a light ray bounces off a reflecting surface the angle of incidence equals the angle of reflection.

$$\theta_i = \theta_r \qquad \textbf{(Eq. 27–1)}$$

Figure 27–2 shows a tilted arrow in front of a plane (flat surface) mirror. Light bounces off from the head of the arrow in all directions.

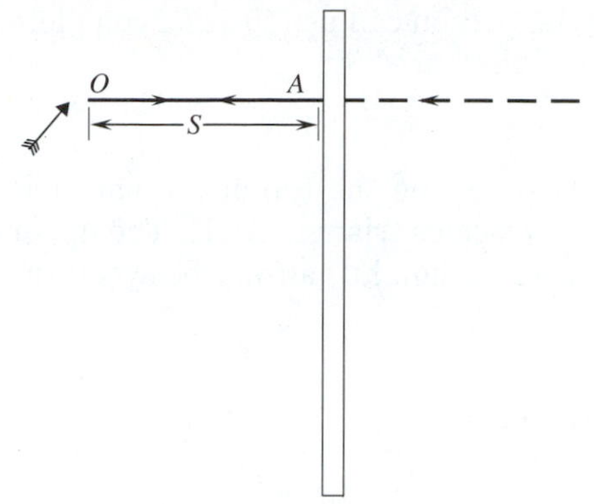

Figure 27–2: *Light rays with an incident angle of zero degrees (0°) are reflected straight back. The reflected ray appears to be coming from behind the mirror.*

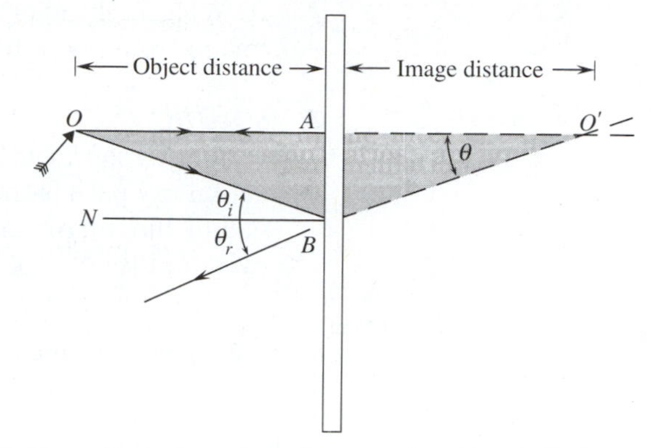

Figure 27–3: *Two reflected rays from the arrowhead locate its image.*

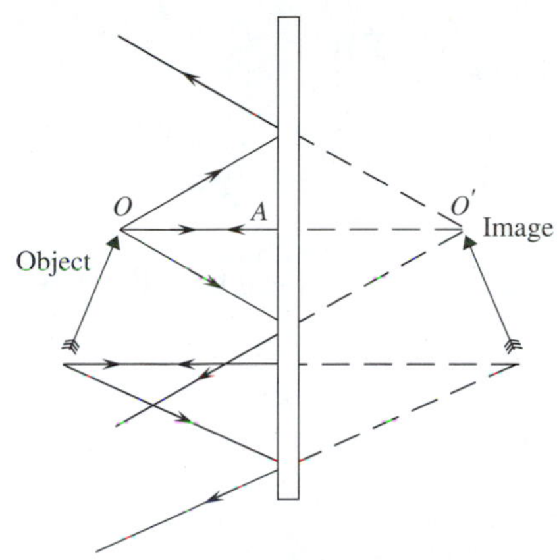

Figure 27–4: *The complete image of the arrow can be located by tracing rays from different parts of the object.*

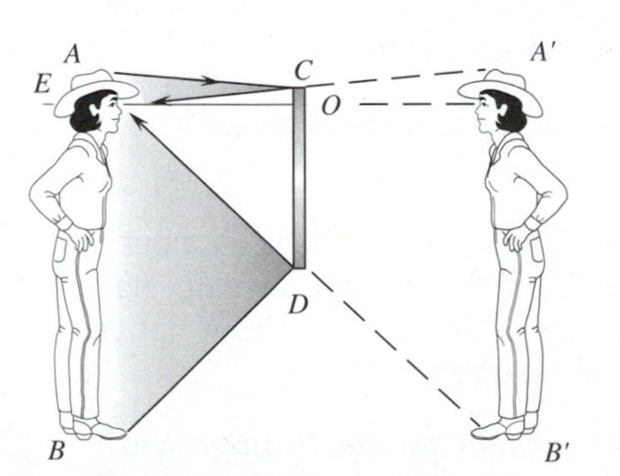

Figure 27–5: *Distance C-D is the minimum mirror length that will give a full-length image.*

The arrow, the object, acts as a light source. Its distance S away from the mirror is the **object distance.**

An image can be located by tracing the path of two or more rays. Rays parallel to the surface normal will be reflected straight back (ray O-A). From an observer's point of view, the reflected ray appears to be coming from some point behind the mirror along the line indicated by dashes.

Ray O-B has an incident angle θ_i (see Figure 27–3). After the reflection, the ray appears to be coming from some point along the dashed line extended behind the mirror. The two dashed lines intercept at point O′. This is where the image of the arrowhead will appear. The distance from the mirror to O′ is called the **image distance.**

Triangles OAB and O′AB are similar right triangles with a common side (AB). The two triangles are the same shape and size. Corresponding sides must be equal. For a plane mirror the image distance and the object distance are the same. The image is formed behind the mirror a distance equal to the object's distance in front of the mirror.

$$p \text{ (object distance)} = q \text{ (image distance) for a plane mirror}$$

Figure 27–4 shows some additional rays drawn from the arrow. For each part of the arrow, the image distance equals the object distance.

☐ **EXAMPLE PROBLEM 27–1: THE CASE OF THE MINIMAL MIRROR**

A person wearing a Stetson hat stands in front of a mirror (see Figure 27–5). The distance from the eyes to the top of the hat is 1.10 ft. The distance from the eyes to the sole of the boots is 5.60

ft. What is the shortest mirror the person can use to see a complete image from boots to the crown of the hat?

■ *SOLUTION*

The person looks along line E-A′ to see the top of the hat. The actual ray path is the sides of the isosceles triangle ACE. The upper end of the mirror must have a vertical height halfway between the eyes and the top of the hat.

$$\text{CE(top edge of mirror relative to eyes)} = \frac{\text{EA}}{2}$$

$$\text{CE} = \frac{1.10\text{ ft}}{2} = 0.55\text{ ft}$$

Another isosceles triangle (EDB) is formed in viewing the boots.

$$\text{ED (lower edge of mirror relative to eyes)} = \frac{\text{EB}}{2}$$

$$\text{ED} = \frac{5.60}{2}\text{ ft} = 2.80\text{ ft}$$

The total length of the mirror CD is the sum of these two distances.

$$\text{CD} = \text{CE} + \text{ED}$$

$$\text{CD} = 0.55\text{ ft} + 2.80\text{ ft}$$

$$\text{CD} = 3.35\text{ ft}$$

Kinds of Images

Light rays do not actually pass through the point on a plane mirror where the image is formed. They just appear to be coming from the image. An image will not be projected on a screen placed behind the mirror. This kind of image is a virtual image.

A virtual image *is an image formed at a point through which light rays do* not *actually pass.*

A movie projector casts a bright image on a screen. A camera focuses a light image on a piece of film. Light rays actually pass through the point at which the image is formed by these optical devices. This kind of image is a real image.

A real image *is formed when light actually passes through the point at which the image is formed.*

Real images can be projected onto a screen; virtual images cannot.

Coefficient of Reflection

When light hits a surface three things can happen (see Figure 27–6):

1. the light can be reflected;
2. the light can be absorbed; and
3. the light can be transmitted.

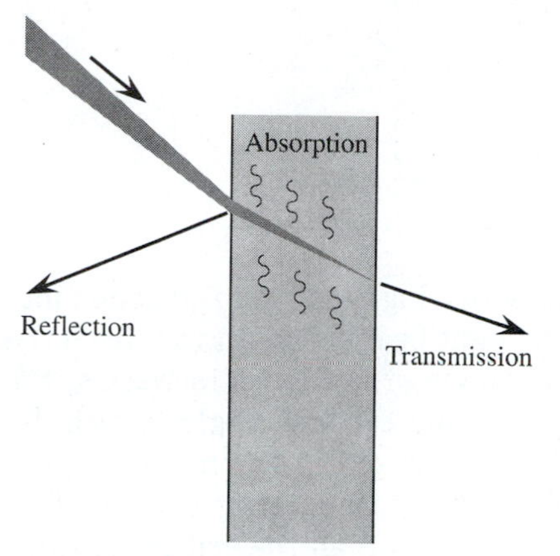

Figure 27–6: When a light beam strikes a surface three things can happen: reflection, absorption, and transmission.

The fraction of the light that is absorbed or transmitted depends on the wavelength. A yellow enameled surface absorbs most of the wavelengths except yellow wavelengths, which are reflected. We perceive the surface as yellow (see Figure 27–7a). A piece of yellow glass absorbs wavelengths other than yellow. Only the yellow light is transmitted through the glass (see Figure 27–7b). Many materials such as polished aluminum or silver reflect a wide range of wavelengths. Other materials such as black construction paper absorb most wavelengths.

The ratio of the intensity of light reflected by a surface divided by the intensity of the incoming light is called the **coefficient of reflection.** The coefficient of a perfect reflector is 1.00; of a perfect absorber or transmitter, 0.00.

The coefficient of reflection involves different wavelengths than emissivity. Emissivity involves absorption or emittance of all black-body wavelengths. The coefficient of reflection involves only the visible light range. Table 27–1 shows some coefficients of reflection for incandescent light on selected materials.

Internal Reflection

The materials in Table 27–1 are not perfect reflectors. Aluminum is used for mirrors more often than silver even though its coefficient of reflection is poorer. Silver has a higher coefficient but it easily forms sulfides when exposed to the air.

Under special conditions transparent materials can be made to be perfect reflectors, reflecting surfaces with a coefficient of reflection of 1.000. Let us take another look at Snell's law.

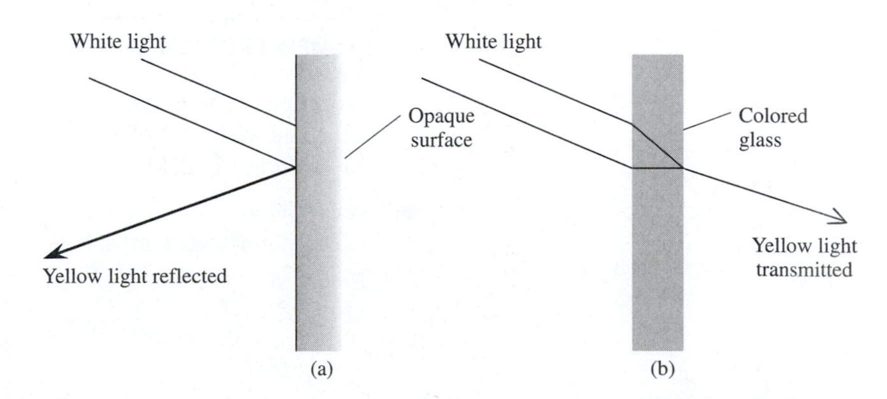

Figure 27–7: White light contains the full spectrum. (a) When it strikes a yellow opaque surface, all the wavelengths except yellow are absorbed. (b) When white light passes through a yellow filter, only the yellow light is transmitted.

Table 27–1: **Coefficient of reflection for incandescent light on various materials.**

MATERIAL	COEFFICIENT OF REFLECTION
Aluminum	
polished	0.69
on glass	0.84
Black paper	0.05
Copper	0.83
Magnesium oxide	0.98
Silver	0.92
Snow	0.93
Steel	0.55

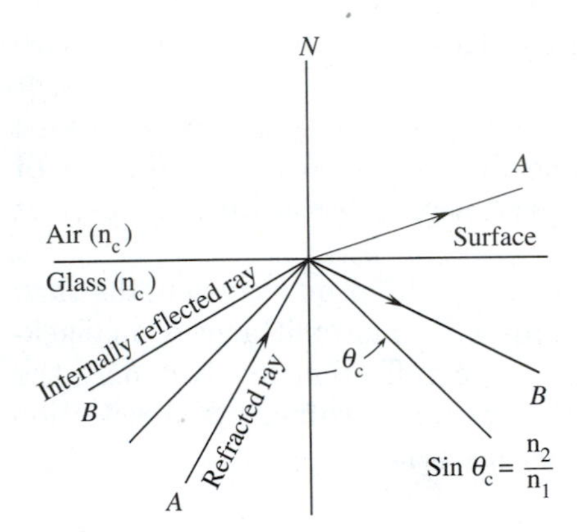

Figure 27–8: *For angles greater than the critical angle (θ_c) light rays will undergo internal reflection (ray B) rather than refraction (ray A).*

If light travels from a material with a high index of refraction into a material with a lower index, the light beam is bent away from the normal (see Figure 27–8). As the angle of incidence increases, the refraction angle also increases. At some **critical angle,** which depends on the relative indices of refraction of the two materials, the refraction angle is 90°. The refracted ray runs along the interface of the two materials at angles a little less than the critical angle.

For angles greater than the critical angle, according to Snell's law, the ray cannot get out of the material. The angle of refraction is greater than 90°, but, instead of being refracted at the surface, the light ray is *reflected* at the surface. This kind of reflection is total; 100% of the light hitting the surface is reflected back into the material. This is called **internal reflection.**

If n_1 is the index of refraction of the first material, and n_2 is the index of refraction for the second material, the critical angle (θ_c) at which internal reflection begins can be found by setting $\theta_2 = 90°$.

$$n_1 \sin \theta_c = n_2 \sin 90°$$

$$\sin \theta_c = \frac{n_2}{n_1} \qquad \textbf{(Eq. 27–2)}$$

□ **EXAMPLE PROBLEM 27–2: THE CRITICAL ANGLE FOR DIAMOND IN WATER**

Compare the critical angle of light rays striking a diamond-water interface with that of a diamond-air interface. The index of refraction for diamond is 2.43; for water, 1.33; and for air, 1.00.

■ ***SOLUTION***

Diamond-water interface:

$$\sin \theta_c = \frac{1.33}{2.43}$$

$$\theta_c = 33.2°$$

Diamond-air interface:

$$\sin \theta_c = \frac{1.00}{2.43}$$

$$\theta_c = 24.3°$$

The critical angle is smaller for the diamond-air interface.

27.2 SPHERICAL MIRRORS

Convex Spherical Mirrors

A shiny Christmas tree ornament is an example of a spherical mirror. A convex spherical mirror is the outer surface or section of the outer surface of a sphere. A light ray traveling from an object toward the center of the sphere will be reflected straight back toward the object. The incident angle is zero (see Figure 27–9).

The **optical axis** connects the surface of the object with the center

Rearview Mirrors and "Red-eye"

An automobile's rearview mirror has a wedge-shaped cross section. The front surface is clear glass; the back surface is silvered.

When the mirror is adjusted for daytime driving, light passing through the rear window passes through the glass wedge. It is then reflected from the silvered back surface into the eyes of the driver. Ninety to ninety-five percent of the incident light is reflected.

At night, the glare of following headlights can be very distracting. The mirror is switched to the nighttime position. The silvered surface is now in a position to reflect light coming from the ceiling of the car, which is usually dark. Now the clear glass front of the wedge is at an angle to reflect light from the rear window into the driver's eyes. The intensity of the light is greatly reduced. The same principle applies when you use the reflection from the plate glass of a store display case to comb your hair.

"Red-eye" is the bane of color photographers. The effect frequently occurs with cameras that have built-in flashes. The flash is close to the objective lens of the camera. Red-eye usually occurs when the subject is in subdued light, so the pupils are dilated. The flash is reflected by the back lining of the eye. The dense population of blood vessels located there gives the reflection its red color.

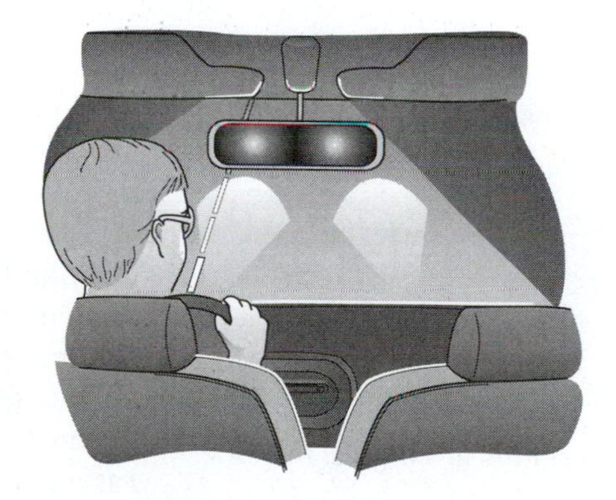

Red-eye can be eliminated by using natural lighting—remove the batteries from the camera—or by using an external flash held farther away from the camera lens.

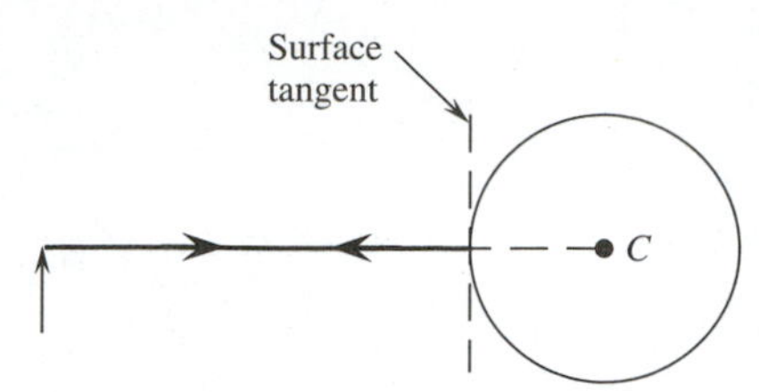

Figure 27–9: Light rays traveling toward the center of curvature of a sphere have an incident angle of 0°.

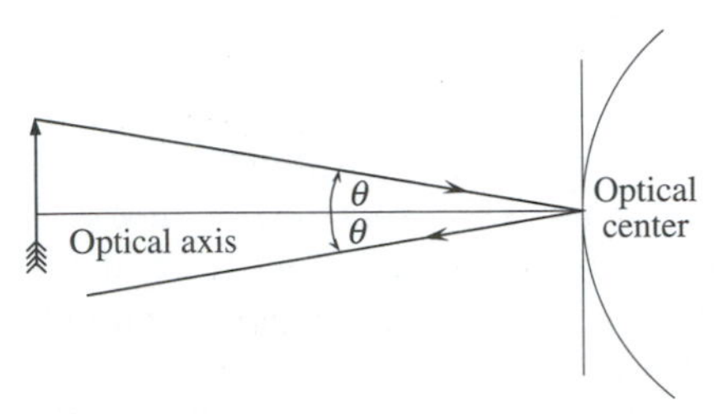

Figure 27–10: The optical axis connects the object and the center of curvature.

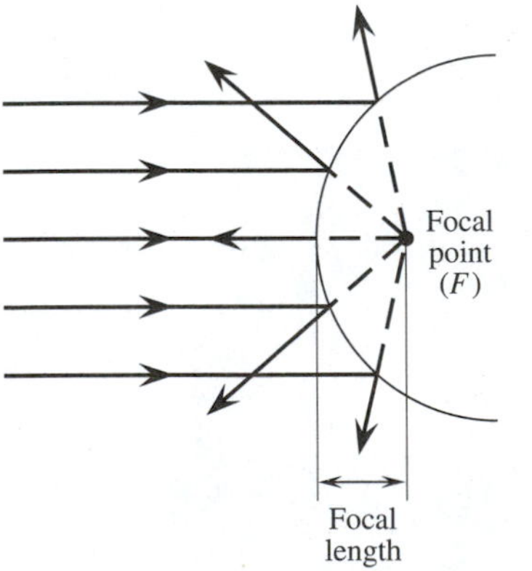

Figure 27–11: Parallel light rays from a distant object form an image at the focal point.

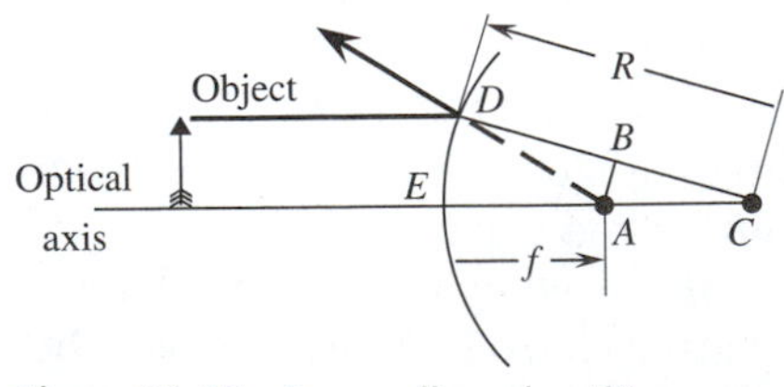

Figure 27–12: For small angles DB = BC = f = R/2. The focal length is half the radius of the sphere.

of the sphere with a line normal to the sphere. The point where the optical axis touches the sphere is known as the **optical center** (see Figure 27–10). A ray moving from the object to the optical center will have equal angles above and below the optical center.

The rays from a distant object, such as the sun or a cloud, travel parallel to each other. If we point the optical axis in the direction of the distant source, the rays will be parallel to the optical axis. The distance from the sphere to the source of the image of a distant object, ideally at infinity, is the **focal point** (see Figure 27–11). The **focal length** is the distance from the surface of the mirror to the focal point.

An object that is not at infinity will still have a ray that is parallel to the optical axis. After it is reflected, the ray will appear to have come from the focal point (see Figure 27–12). If the parallel ray is close to the optical axis, the angles of incidence and reflection will be small.

The distance from the center of the sphere to the point where the ray strikes the mirror's surface is R. Line AB bisects the isosceles triangle ADC. The projection of the leg of the triangle ABD onto the optical axis is $R/2 \times \cos \theta_r$. For small angles, $\cos \theta_r = 1.00$. The length EC is $R = f + R/2$. The focal length is:

$$f = -\left(\frac{R}{2}\right) \qquad \textbf{(Eq. 27–3)}$$

The minus sign indicates that the focal point is behind the reflecting surface. Remember this is true only for rays close to the optical axis, because we used a small angle approximation.

If we trace two or more of the rays coming from the object, we can find the location of the image. The image will be where the rays appear to cross after they are reflected. Figure 27–13A shows how the image can be found by tracing two rays. P is the height of the object; Q, the height of the image. The two shaded triangles are proportional; corresponding sides are proportional. We can find the ratio of the height of the image to the height of the object. This ratio is called the **magnification.** We will measure distances above the optical axis as positive. For a mirror, distances measured behind the mirror are negative. The magnification is proportional to the image distance ($-q$) and the object distance (p).

$$\textit{magnification}\ (M) = \frac{Q}{P} = \frac{-q}{p} \qquad \textbf{(Eq. 27–4)}$$

Figure 27–13b is the same diagram as 27–13a, but two other similar triangles have been shaded. Again we can take the ratio of sides:

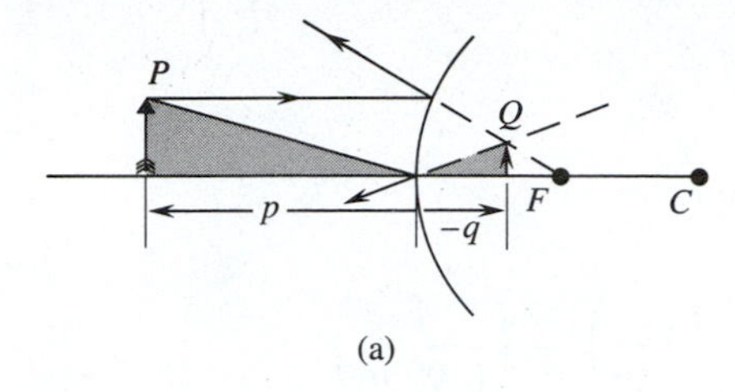

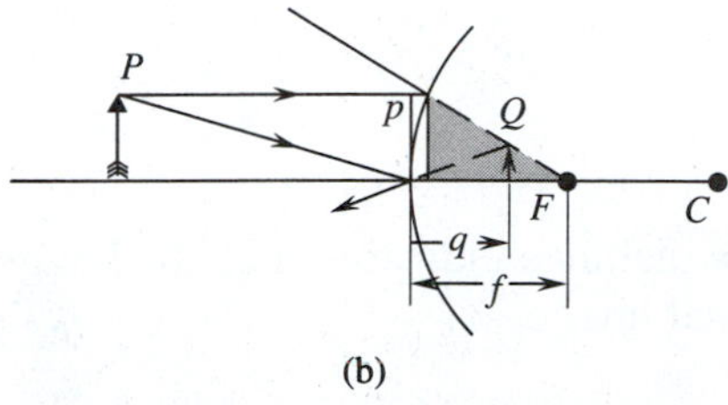

Figure 27–13: (a) *The shaded triangles are proportional. The object height* (P) *and object length* (p) *are proportional to the image height* (Q) *and the image length* (q). (b) *The shaded similar triangles make it possible to express the magnification in terms of focal length and image length.*

$$\frac{Q}{P} = \frac{-(f-q)}{-f}$$

We can set this ratio equal to Equation 27–4:

$$\frac{Q}{P} = \frac{f-q}{f} = \frac{-q}{p}$$

or

$$\frac{q}{p} = \frac{q-f}{f} = \left(\frac{q}{f}\right) - 1$$

If we divide the equation by q we get the spherical mirror equation:

$$\frac{1}{p} = \left(\frac{1}{f}\right) - \left(\frac{1}{q}\right)$$

or

$$\left(\frac{1}{p}\right) + \left(\frac{1}{q}\right) = \frac{1}{f} \qquad \textbf{(Eq. 27–5)}$$

If we know the focal length of the mirror and the object distance we can find where the image will be formed by Equation 27–5. The magnification can then be found using Equation 27–4.

□ **EXAMPLE PROBLEM 27–3: THE CHRISTMAS TREE ORNAMENT**

A shiny spherical Christmas tree ornament has a diameter of 3.0 in.

A. What is the focal length of the sphere?
B. What is the image distance of a cat's paw 5.0 in from the ball?
C. Is the image real or virtual?
D. If the real paw is 0.60 in wide, how wide will the paw in the image be?

■ ***SOLUTION***

A. Focal length:

$$f = \frac{-R}{2} = \frac{-D}{4} = \frac{-3.00 \text{ in}}{4}$$

$$f = -0.75 \text{ in}$$

B. Image distance:

$$\left(\frac{1}{p}\right) + \left(\frac{1}{q}\right) = \frac{1}{f}$$

$$\frac{1}{q} = \left(\frac{1}{f}\right) - \left(\frac{1}{p}\right) = \left(\frac{1}{-0.75\ \text{in}}\right) - \left(\frac{1}{5.00\ \text{in}}\right)$$

$$\frac{1}{q} = (-1.333 - 0.200)\ \text{in}^{-1} = -1.533\ \text{in}^{-1}$$

$$q = -0.65\ \text{in}$$

C. Image type:

The negative sign on the image distance indicates that the image is behind the mirror. It is a virtual image.

D. Image size:

$$Q = P \times \left(\frac{-q}{p}\right)$$

$$Q = 0.60\ \text{in} \times \left[\frac{-(-0.65\ \text{in})}{5.00\ \text{in}}\right]$$

$$Q = 0.08\ \text{in}$$

Concave Spherical Mirrors

Figure 27–14 shows a concave spherical mirror. We have merely turned a section of a sphere around so that the light reflects on the inside surface of the sphere.

Since we have chosen to call directions in front of the mirror positive, the focal length of the concave mirror is $+R/2$. The image formed in Figure 27–14 is real (light actually passes through the image) and inverted (the image is upside down).

We can find similar triangles in the ray diagram and find the same ratios as we did for the convex mirror. We get the same proportions.

$$\textit{magnification}\ (M) = \frac{Q}{P} = \frac{-q}{p}$$

$$\left(\frac{1}{p}\right) + \left(\frac{1}{q}\right) = \frac{1}{f}$$

What mathematically distinguishes a convex mirror from a concave mirror is the focal length. Otherwise the equations are the same.

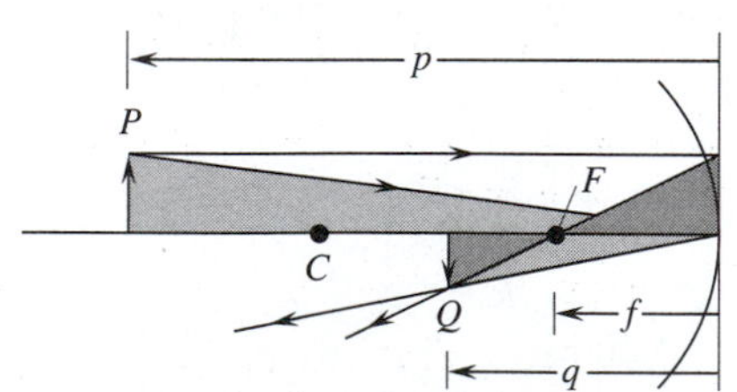

Figure 27–14: *The focal point of a concave mirror is in front of the mirror. Real images can be formed.*

$$f(\text{concave}) = \frac{+R}{2}$$

$$f(\text{convex}) = \frac{-R}{2}$$

□ **EXAMPLE PROBLEM 27–4: A CONCAVE MIRROR**

A concave mirror has a radius of 25.0 cm (see Figure 27–15). For an object 70.0 cm from the mirror find:

A. the image distance.
B. the magnification.
C. the nature of the image.

■ ***SOLUTION***

A. Image distance:

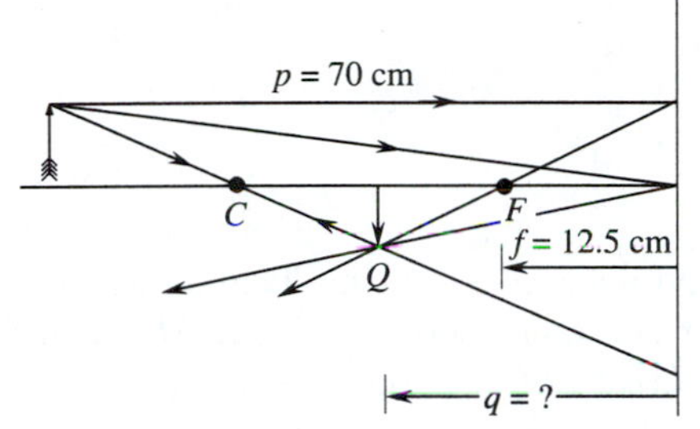

Figure 27–15: *Diagram for Example Problem 27–4. A concave mirror.*

$$\frac{1}{q} = \left(\frac{1}{f}\right) - \left(\frac{1}{p}\right)$$

$$\frac{1}{q} = \frac{1}{12.5 \text{ cm}} - \frac{1}{70.0 \text{ cm}} = 0.0657 \text{ cm}^{-1}$$

$$q = 15.2207 \text{ cm}$$

B. Magnification:

$$M = \frac{-q}{p} = \frac{-15.2207 \text{ cm}}{70.0 \text{ cm}}$$

$$M = -0.2174$$

C. Nature of the image:

The image is inverted (negative magnification), smaller than the object ($M < 1$), and real (q is positive, or in front of the mirror).

□ **EXAMPLE PROBLEM 27–5: THE MAKEUP MIRROR**

A concave spherical mirror with a focal length of 60.0 cm is used as makeup mirror. A facial pimple is 28.0 cm from the mirror. Find:

A. the image distance.
B. the magnification.
C. the nature of the image.

■ ***SOLUTION***

A. Image distance:

$$\frac{1}{q} = \left(\frac{1}{f}\right) - \left(\frac{1}{p}\right)$$

$$\frac{1}{q} = \frac{1}{60.0\text{ cm}} - \frac{1}{28.0\text{ cm}}$$

$$q = -52.5\text{ cm}$$

B. Magnification:

$$M = \frac{-q}{p} = \frac{-(-52.5\text{ cm})}{28.0\text{ cm}}$$

$$M = +1.88$$

C. Nature of the image:

The image is virtual (negative image distance; it is behind the mirror), larger than the object ($M > 1$), and erect (M is +).

27.3 MORE ON REFRACTION

Apparent Depth

Refraction can fool you when you estimate depth. Place a coin in a glass of water and look down at the coin (see Figure 27–16). Both the depth of the coin and the bottom of the glass seem to be less than you expected.

Letting L be the true depth and L' the apparent depth, from the diagram we see:

$$\frac{D}{L} = \tan\theta_1$$

and

$$\frac{D}{L'} = \tan\theta_2$$

Combine the equations to eliminate D.

$$L\tan\theta_1 = L'\tan\theta_2$$

For small angles, $\tan\theta_1 = \sin\theta_1$, and $\tan\theta_2 = \sin\theta_2$.

$$L' = L\frac{\sin\theta_1}{\sin\theta_2}$$

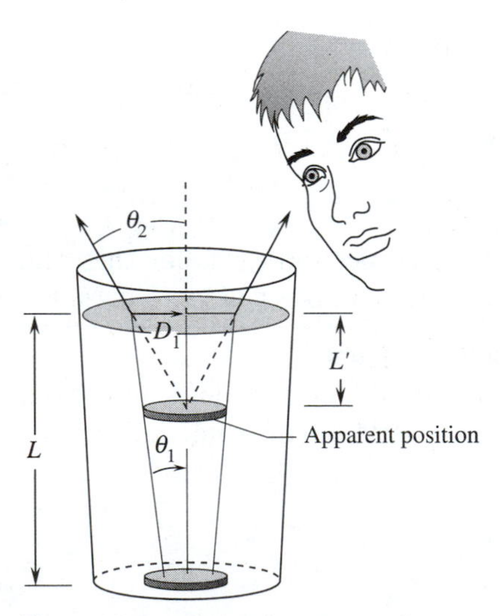

Figure 27–16: Refraction causes the apparent depth of a coin in a glass to be less than expected.

We can use Snell's law to replace the ratio of sines with the ratio of indices of refraction to determine the **apparent depth.**

$$L' = L\frac{n_2}{n_1} \qquad \textbf{(Eq. 27–6)}$$

☐ **EXAMPLE PROBLEM 27–6: APPARENT DEPTH**

A person inside a canoe looks over the side and estimates the apparent depth to be 3.0 ft. If this estimate is reasonably correct, what is the actual depth? (The index of refraction for water is 1.33.)

■ ***SOLUTION***

$$L = L'\frac{n_1}{n_2}$$

$$L = (3.0\text{ ft}) \times \frac{1.33}{1.00}$$

$$L = 4.0\text{ ft}$$

Prism

Figure 27–17a shows a light ray passing through a prism. The ray deviates from its original direction by refraction. The angle between the original direction and the final direction is called the **angle of deviation** (D). Notice that the ray deviates toward the base of the prism.

Figure 27–17b shows two ways to combine two prisms: base to base or apex to apex. When the prisms are base to base the light rays converge, or come together. When the prisms are apex to apex the rays diverge; they are bent away from each other.

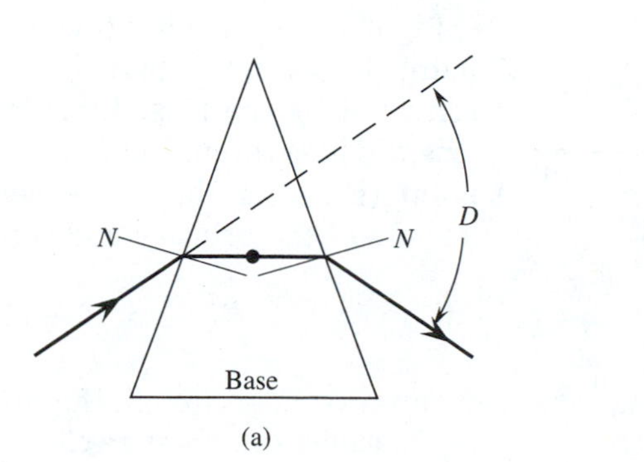

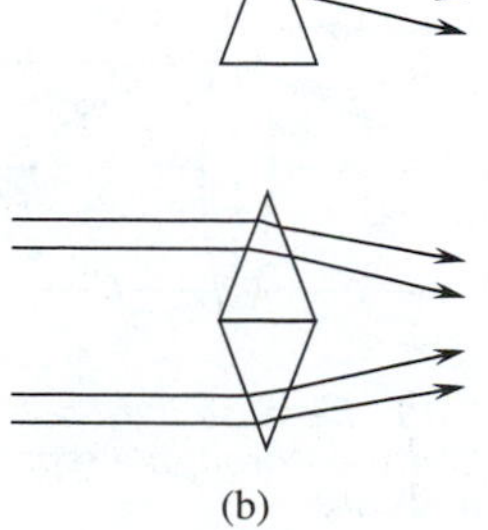

Figure 27–17: Prisms. (a) A light ray is bent toward the base of the prism. (b) Arrangements of two prisms for divergence and convergence of light rays.

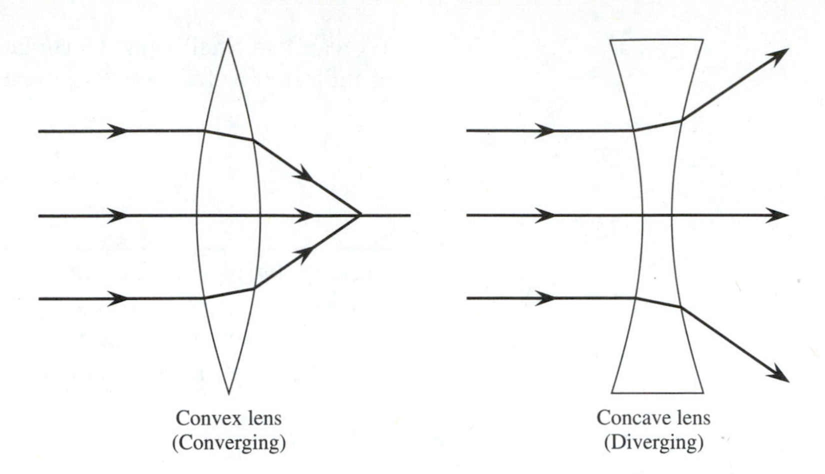

Figure 27–18: *Convex lenses cause convergence of light rays. Concave lenses cause divergence.*

If we replace the flat-faced prisms with ones that have curved surfaces, we have created lenses (see Figure 27–18). Lenses that are fat in the middle are called convex lenses. They will cause light to converge. Lenses that are thin in the center are called concave lenses. Light rays refracted through them will diverge.

27.4 THIN SPHERICAL LENSES

A **thin lens** is a lens with a focal length much longer than the actual thickness of the lens. We do not need to trace the path of a light ray inside such a lens to get a good estimate of the location of an image. An image is formed when light passes through a lens, rather than when it is reflected back as for a mirror. For lenses, we will call image distances positive if they are on the side of the lens opposite the object.

Because light can pass through a lens in either direction, there are two focal points—one on each side of the lens (see Figure 27–19). We will say that a focal length is positive if it is a converging (convex) lens; negative, if it is a diverging (concave) lens.

Light rays traveling toward or coming from the **primary focal point** (F) will travel parallel to the optical axis after they have been refracted by the lens. Light rays traveling parallel to the optical axis will pass through or appear to pass through the **secondary focal point** (F′) after they have been refracted by the lens. The focal lengths of the primary and secondary focal points are the same.

$$f = f'$$

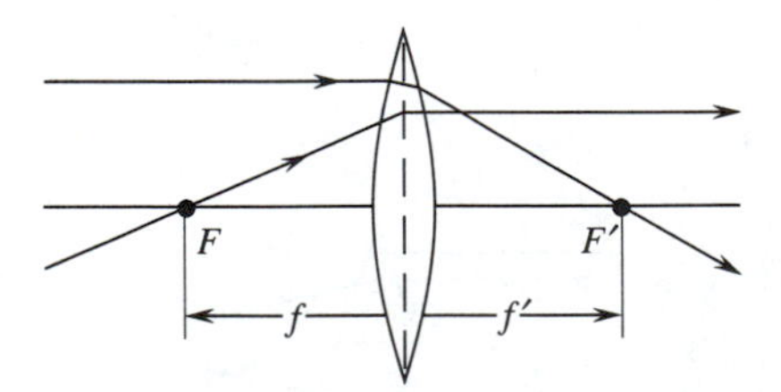

Figure 27–19: *Because light rays can travel either way through a lens, there are two focal points.*

A ray traveling through the center of the lens will not deviate. This ray is called the **chief ray.**

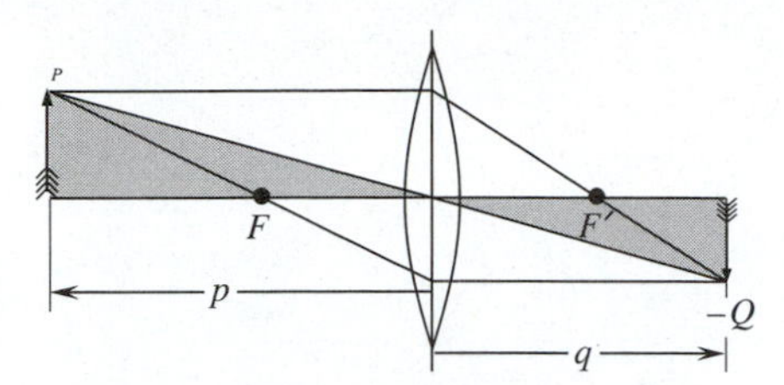

Figure 27–20: The shaded triangles are similar. Magnification is proportional to the ratio of image length to object length.

Figure 27–20 shows the three rays—the chief ray and the rays through the two focal points—for a converging lens. Two similar triangles are shaded. Notice that the height of the image ($-Q$) is negative. We can find the magnification in terms of the image length and object length.

$$\frac{-Q}{P} = \frac{q}{p}$$

or

$$M = \frac{Q}{P} = \frac{-q}{p} \qquad \textbf{(Eq. 27–7)}$$

Notice that magnification is defined in the same way as for a spherical mirror.

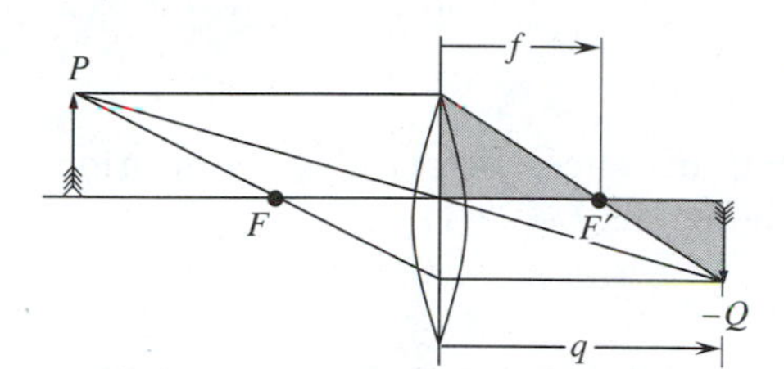

Figure 27–21: The shaded areas are similar. They are used to derive the thin lens equation.

Figure 27–21 is the same ray diagram with two other shaded similar triangles. Using proportional sides we get:

$$\frac{-Q}{P} = \frac{q-f}{f}$$

or

$$\frac{Q}{P} = \frac{-q}{p} = \frac{f-q}{f}$$

Rearranging the expression we get the thin lens equation:

$$\frac{1}{p} + \frac{1}{q} = \frac{1}{f} \qquad \textbf{(Eq. 27–8)}$$

The thin lens equation is the same as the spherical mirror equation, but be careful. The conventions of plus and minus signs are reversed for reflection and refraction for the image distance. A positive image is behind the lens and a negative image is in front of the mirror.

☐ **EXAMPLE PROBLEM 27–7: CONVEX LENS**

A lizard is 18.0 cm from a convex lens with a 12.7-cm focal length. Find:

A. the image distance.
B. the magnification.
C. the nature of the image.

■ *SOLUTION*

A. Image length:

$$\frac{1}{q} = \frac{1}{f} - \frac{1}{p}$$

$$\frac{1}{q} = \frac{1}{12.7 \text{ cm}} - \frac{1}{18.0 \text{ cm}}$$

$$q = 43.1 \text{ cm}$$

B. Magnification:

$$M = \frac{-q}{p} = \frac{-43.1 \text{ cm}}{18.0 \text{ cm}}$$

$$M = -2.4$$

C. Nature of the image:

The image is real (positive image distance), inverted (magnification is negative), and larger than the object ($M > 1$).

□ **EXAMPLE PROBLEM 27–8: CONCAVE LENS**

An 8.0 in diameter ball is 23.0 in from a concave lens. The erect image of the ball has a diameter of 2.9 in. Find the image distance.

■ *SOLUTION*

$$M = \frac{Q}{P} = \frac{-q}{p}$$

The image is erect, so Q is positive.

$$q = \frac{-(Q \times p)}{P}$$

$$q = \frac{-(2.9 \text{ in} \times 23.0 \text{ in})}{8.0}$$

$$q = -8.3 \text{ cm}$$

(The image is virtual, since q is negative.)

27.5 APPLICATIONS: OPTICAL DEVICES

Optical Fibers

Optical fibers take advantage of internal reflection (see Figure 27–22). A fiber is built with a central core of glass with a high index of refraction surrounded by cladding with a lower index. As the light ray travels along the fiber, it is repeatedly reflected from the interface of the core and the cladding. The cladding ensures few pits or imperfections on the surface of the central core.

Two fiber designs are popular. The step index fiber design is shown in Figure 27–22a. The core and cladding have fixed indices of refraction. The light moves down the fiber in a zigzag fashion. Another design, called graded index (GRIN) optical fibers, has a varying index of refraction in the core, with the highest index at the center. This design helps to focus the light beam. It travels down the fiber in a sine wave pattern rather than a zigzag (see Figure 27–22b).

Numerical Aperture

Figure 27–23 shows the **capture angle** ($\boldsymbol{\theta_c}$) for an optical fiber. This is the maximum angle at which the light ray can strike the surface and be reflected internally. The angle is measured from the surface rather than from the surface normal. The capture angle is related to the critical angle of internal refraction by:

$$\sin \theta = \cos \theta_c$$

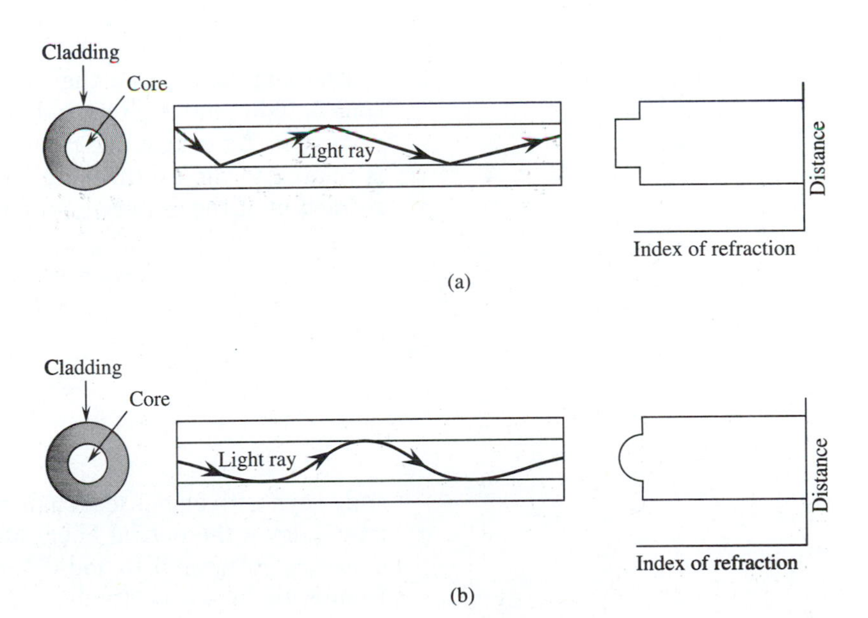

Figure 27–22: *Two kinds of optical fibers. (a) Step index optical fiber. (b) Graded index (GRIN) optical fiber.*

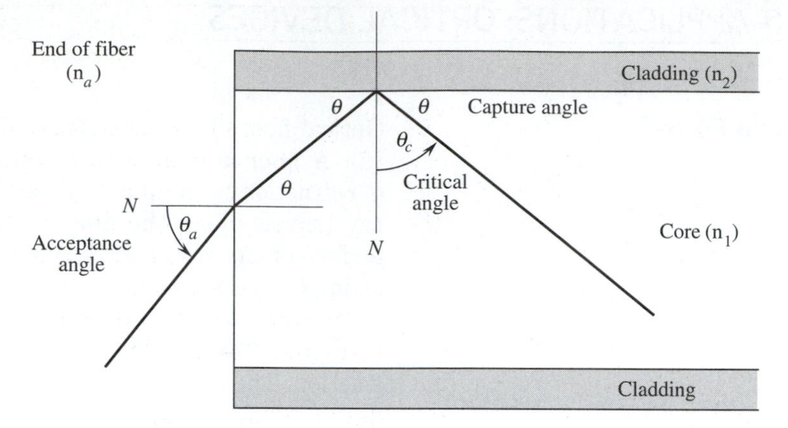

Figure 27–23: *The capture angle (θ) is the complement of the critical angle for internal reflection (θ_c). θ_a is the acceptance angle. Light rays must have angles smaller than θ_a to undergo internal reflection in the fiber.*

We can use the trigonometric identity $\cos^2\theta = 1 - \sin^2\theta$.

$$\sin\theta = (1 - \sin^2\theta_c)^{1/2}$$

If n_1 is the index of refraction of the core, and n_2 is the index for the cladding, then:

$$\sin\theta_c = \frac{n_2}{n_1}$$

By substitution we get:

$$n_1 \sin\theta = (n_1{}^2 - n_2{}^2)^{1/2} \qquad \textbf{(Eq. 27–9)}$$

This quantity, called the **numerical aperture,** is related to the cone of light entering one end of the fiber that will undergo internal reflection. Look again at Figure 27–23. The **acceptance angle,** θ_a, is related to the capture angle by Snell's law. If n_a is the index of refraction of the material outside the end of the fiber, then:

$$n_1 \sin\theta = n_a \sin\theta_a$$

or

$$\sin\theta_a = \frac{n_1}{n_a}\sin\theta$$

Only rays with angles less than or equal to the acceptance angle will travel down the optical fiber. Most optical fibers have a numerical aperture between 0.15 and 0.4, corresponding to capture angles of from 9° to 24°, respectively.

EXAMPLE PROBLEM 27–9: NUMERICAL APERTURE

The core of an optical fiber in air has an index of refraction of 1.62 and cladding with an index of 1.57. Calculate the numerical (NA) aperture, capture angle (θ), and acceptance angle (θ_a) for the fiber.

SOLUTION

$$NA = (n_1^2 - n_2^2)^{1/2}$$

$$NA = (1.62^2 - 1.57^2)^{1/2}$$

$$NA = 0.399 = n_1 \sin \theta$$

$$\theta = \sin^{-1}\left(\frac{0.399}{1.62}\right)$$

$$\theta = 14.3°$$

$$NA = 0.399 = n_a \sin \theta_a$$

$$\theta_a = \sin^{-1}\left(\frac{0.399}{1.00}\right)$$

$$\theta_a = 23.5°$$

Bandwidth Length

As a light pulse moves along a fiber it spreads out, or disperses (see Figure 27–24). Pulses containing information must be spaced to allow for this dispersal. The information-carrying capacity of a fiber may be rated in terms of **pulse dispersion** in units of nanoseconds per kilometer. If a 3.0-km fiber has a pulse dispersal of 2 ns/km, the pulses must be spaced at least 6.0 ns apart.

Another way of expressing carrying capacity is to use **bandwidth length** in units of megahertz · kilometers. If the signal is too high, the pulses will overlap. The upper frequency of the light signal is limited by the signal broadening. Notice that bandwidth length has units of distance per time, the reciprocal of pulse dispersion.

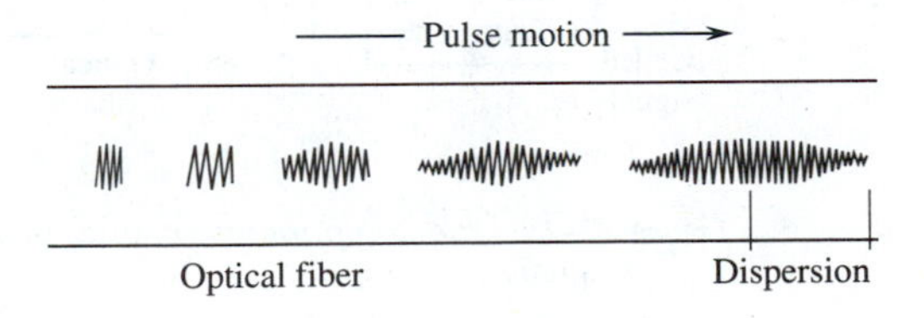

Figure 27–24: *An optical pulse spreads out as it moves along the fiber.*

Light Loss

The intensity of light is reduced by absorption of the glass fiber. Light intensity losses are expressed in decibels per kilometer (db/km). If I/I_0 is the ratio of the light intensity at the source divided by the intensity at the end of a 1-km fiber, then

$$\frac{\text{db}}{\text{km}} = 10 \log \left(\frac{I}{I_0}\right)$$

80% transmission/km = a loss of 1 db/km

10% transmission/km = a loss of 10 db/km

1% transmission/km = a loss of 20 db/km

Fiber Optical Systems

A wide bandwidth (low dispersion) and low intensity losses are what we want in an optical fiber. These can be achieved by cutting down on the lateral part of the zigzag motion to get the smallest possible actual path along the fiber. Low aperture numbers and narrow-diameter fiber cores are favored. For long distances, fiber cores with diameters as small as 6 μm are often used.

Optical fibers can be bundled to transmit images. Endoscopes used in medicine to examine internal organs use this kind of bundled optical fiber system.

Coded information used in telecommunications can be sent along a single fiber. Often, an auxiliary fiber is strung as a redundant system in case the active fiber fails. For long-distance communications, repeater circuits are added to the system at about 20-km intervals.

Figure 27–25 shows a schematic of a light-wave communication system. An electrical signal is coded by the transmitter (A). A laser or a light-emitting diode (LED) acts as a light source (B). The light is transmitted along an optical fiber (C). At the receiving end of the fiber, a photodetector device such as an avalanche photodiode receives the coded light signal and converts it into electrical pulses (D). Finally, the coded signal is amplified and is either translated or retransmitted.

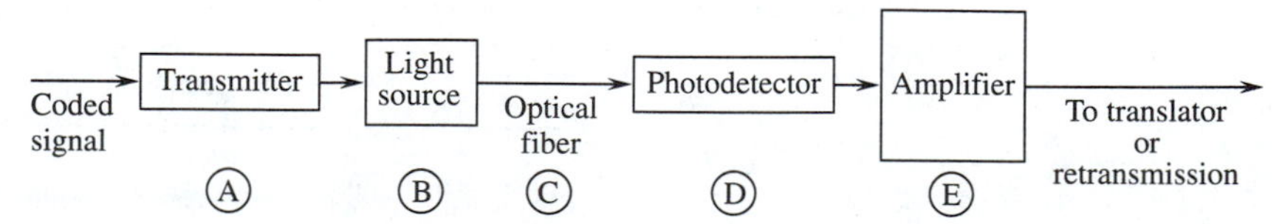

Figure 27–25: *The components of an optical fiber communication system for coded signals.*

Magnifiers

Hold your hand in front of your face and study the lines on your fingers, the ones that form fingerprints. Move your hand away from and then toward your face. If your fingers get too close to your eyes, the lines blur. If they are too far away it is again difficult to see detail. If you have normal vision, you will find that the lines are clearest at a distance of about 25 cm in front of your eyes. This distance is called the **near point.** When you use a magnifier the sharpest image is also formed when the image is 25 cm in front of the eye.

So far we have talked about *lateral* magnification: the ratio of image height to object length. For instruments that use magnifiers, **angular magnification** is used. Look at Figure 27–26. The apparent size of the image formed on the back of the eyes depends on the angle subtended by the image. The ratio of the angle subtended by an image (θ') divided by the angle subtended by the object (θ) is the angular magnification.

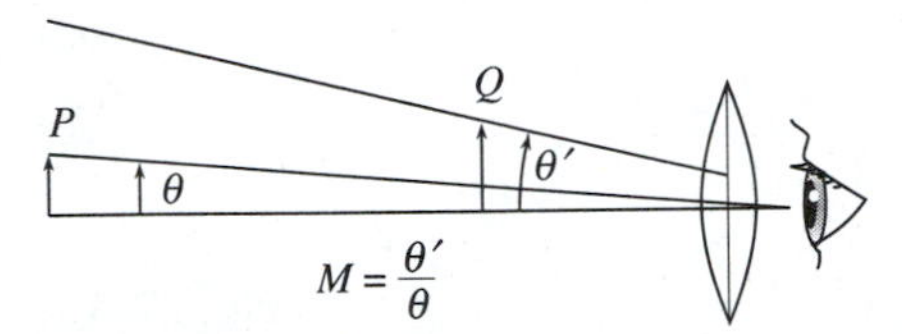

Figure 27–26: *Angular magnification. θ' is the angle subtended by the image; θ, the angle subtended by the object.*

$$M = \frac{\theta'}{\theta} \qquad \textbf{(Eq. 27–10)}$$

A magnifier is usually used with the image at 25 cm. Without the magnifier, the object would be held at the near point for examination (see the triangle formed in Figure 27–27a):

$$\tan\theta = \frac{P}{25 \text{ cm}}$$

or, for small angles:

$$\theta = \frac{P}{25 \text{ cm}}$$

From the triangle in Figure 27–27b we see that with the magnifier:

$$\tan\theta' = \frac{P}{p}$$

or, for small angles:

$$\theta' = \frac{P}{p}$$

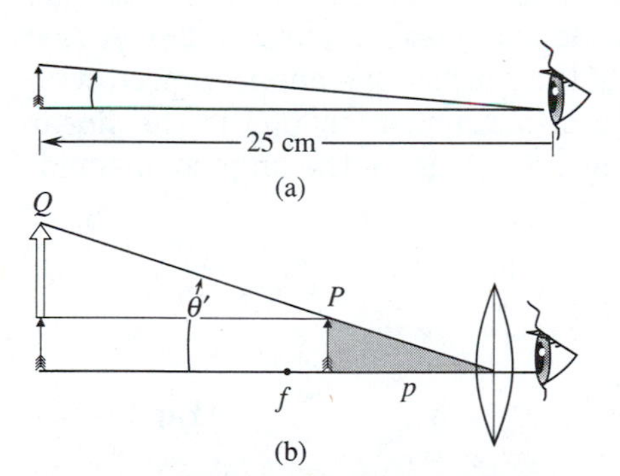

Figure 27–27: (a) *The object is held at the near point. The object angle (θ) is P/25 cm.* (b) *The object is near the lens to give a clear virtual image at the near point. The image angle (θ') is P/p.*

We can express p in terms of the focal length by using the thin lens formula. The image is virtual, so it has a negative image distance.

$$\frac{1}{p} + \frac{1}{-25 \text{ cm}} = \frac{1}{f}$$

or

$$p = \frac{25f}{25 + f}$$

Substituting this in the equation for the image angle we get the angular magnification.

$$M = \frac{\theta'}{\theta} = \frac{P(25 \text{ cm} + f)/25f}{P/25 \text{ cm}}$$

$$M = \frac{25}{f} + 1 \qquad \textbf{(Eq. 27–11)}$$

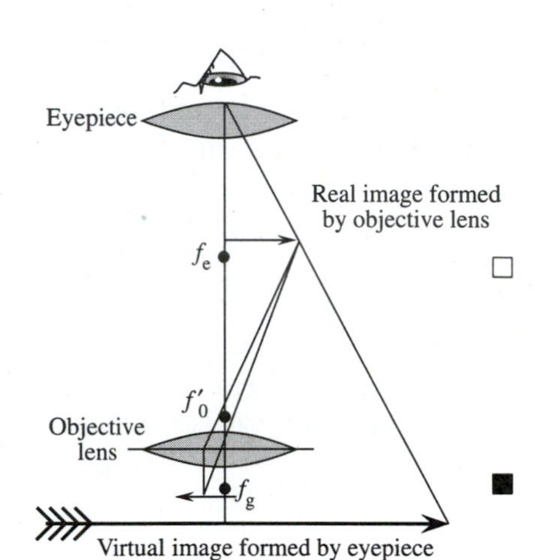

Figure 27–28: A microscope. The objective lens forms a real image that is viewed with a magnifier (eyepiece).

□ **EXAMPLE PROBLEM 27–10: THE EYEPIECE**

The eyepiece on a microscope is a magnifier used to examine a real image formed by the microscope objective lens. If the focal length of the eyepiece is 2.7 cm, what is the magnification of the eyepiece?

■ ***SOLUTION***

$$M = \frac{25 \text{ cm}}{2.7 \text{ cm}} + 1$$

$$M = 10 \text{ X}$$

Microscopes

A microscope is a compound magnifier. The object is just outside the focal length of the objective lens (f_1) (see Figure 27–28). A real image is formed in the barrel of the microscope and is viewed with a magnifying eyepiece. The total magnification (M) is the linear magnification of the objective lens (m_1) times the angular magnification of the eyepiece (m_2).

$$m_1 = \frac{-q}{f_1} \qquad m_2 = \frac{25}{f_2}$$

$$M = \frac{-q}{f_1} \times \frac{25}{f_2} \qquad \textbf{(Eq. 27–12)}$$

Astronomical Telescopes

For a microscope, the objective lens is short. In a telescope, the focal length of the objective lens is very long compared with the focal length of the eyepiece.

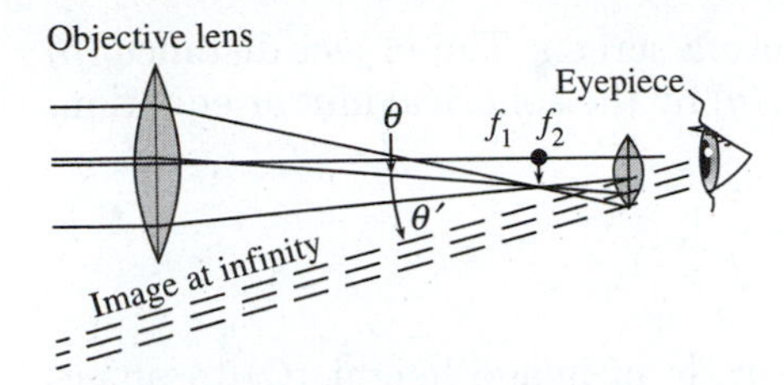

Figure 27–29: The astronomical telescope. The objective lens, with a long focal point, forms a real image on the focal plane. The image formed by the objective lens is viewed with an eyepiece.

The real image formed by the objective lens of a distant object is on the focal point (f_1). For photography, the real image is also on the focal plane of the eyepiece (see Figure 27–29). The effective object distance is $p = f_1 + f_2$, or $p = f_1$ since $f_1 \gg f_2$. The virtual image is at the focal plane of the eyepiece, $q = -f_2$. The magnification is:

$$M = \frac{-f_1}{f_2} \qquad \textbf{(Eq. 27–13)}$$

☐ **EXAMPLE PROBLEM 27–11: THE TELESCOPE**

An astronomical telescope has a 200-cm focal length. Estimate the magnification with a 15-mm eyepiece.

■ ***SOLUTION***

$$M = \frac{-200 \text{ cm}}{1.5 \text{ cm}} = 133 \text{ X}$$

SUMMARY

Geometric optics is the study of the path of light rays through an optical system. When a light ray reflects from a surface the angle of incidence equals the angle of reflection. This principle is known as the law of reflection.

$$\theta_i = \theta_r$$

A spherical mirror is formed from a section of a reflective sphere. Light waves from a distant object form an image at the focal point. The focal point is numerically equal to one-half the radius of curvature of the mirror. The focal length is the distance from the surface of the mirror to the focal point. The focal length of a convex mirror is negative; for a concave mirror it is positive.

$$\text{concave mirror: } f = +\left(\frac{R}{2}\right)$$

$$\text{convex mirror: } f = -\left(\frac{R}{2}\right)$$

Image distances measured behind the mirror are negative; in front of the mirror, they are positive. A negative image length indicates that the image is a virtual image; light does not actually pass through the point at which the image is formed. A positive image distance indicates that light does pass through the image point to form a real image that can be projected onto a screen. The object distance (p) is related to the image distance (q) by the spherical mirror equation.

$$\frac{1}{p} + \frac{1}{q} = \frac{1}{f}$$

Lateral magnification is the ratio of image height (Q) to object height (P). If the magnification is negative, the image is inverted, or upside down. Images that are not inverted are said to be erect.

$$M = \frac{Q}{P} = -\left(\frac{q}{p}\right)$$

Image distances on the side of the lens opposite an object are positive. The image distance and object distance of a thin lens are related by the thin lens formula. It has the same form as the spherical mirror equation.

$$\frac{1}{p} + \frac{1}{q} = \frac{1}{f}$$

$$M = \frac{Q}{P} = -\left(\frac{q}{p}\right)$$

The ratio of incident light intensity to the intensity of light reflected from a surface is called the coefficient of reflection. The coefficient has a value of 1.00 for a perfect reflector and a value of 0.00 for a perfect absorber.

If a light ray inside a material with a high index of refraction strikes a surface in contact with a material of lower index at an angle greater than the critical angle (θ_c), it will undergo internal reflection. The critical angle is related to the indices of refraction by:

$$\sin \theta_c = \frac{n_1}{n_2}$$

Optical light fibers transmit light signals by internal reflection. Factors that affect the performance of an optical fiber include numerical aperture, light absorption, and pulse dispersion.

The magnification of magnifiers is measured in terms of angular magnification. It is the ratio of the angle subtended by an object (θ) at the near point divided by the angle subtended by the image (θ') at the same point.

$$M = \frac{\theta'}{\theta} = \frac{25 \text{ cm}}{f} + 1$$

A microscope uses an objective lens with a short focal length (f_1) to form a real image magnified by an eyepiece of focal length f_2. The overall magnification is:

$$M = \frac{-q\ 25\ \text{cm}}{f_1 f_2}$$

An astronomical telescope uses an objective lens with a long focal length (f_1) to form an image on the focal plane. The eyepiece acts as a magnifier to view the image. If D is the diameter of the objective lens and d is the diameter of the eyepiece, the overall magnification is:

$$M = -\left(\frac{f_1}{f_2}\right) = \frac{D}{d}$$

KEY TERMS

If you can explain the following terms to a friend or classmate, you understand their meaning. If you cannot explain the terms, you should reread the sections in which they are discussed.

acceptance angle
angle of deviation
angle of incidence
angle of reflection
angular magnification
apparent depth
bandwidth length
capture angle
chief ray
coefficient of reflection
critical angle
focal length
focal point
geometric optics
image distance
internal reflection
law of reflection
magnification
near point
numerical aperture
object distance
optical axis
optical center
optical fibers
primary focal point
pulse dispersion
real image
secondary focal point
thin lens
virtual image

EXERCISES

Section 27.1:

1. If you approach a plane mirror at 2.0 mph, how fast does your image approach the mirror? How fast does your image approach you?

2. Explain the following terms in your own words.

   A. Virtual image
   B. Real image
   C. Angle of incidence
   D. Surface normal
   E. Object distance
   F. Image distance
   G. Geometric optics
   H. Coefficient of reflection
   I. Critical angle

3. A good pair of binoculars uses right-angle prisms to move the image in line with the eyepiece (see Figure 27–30). Why is this better than using mirrors?

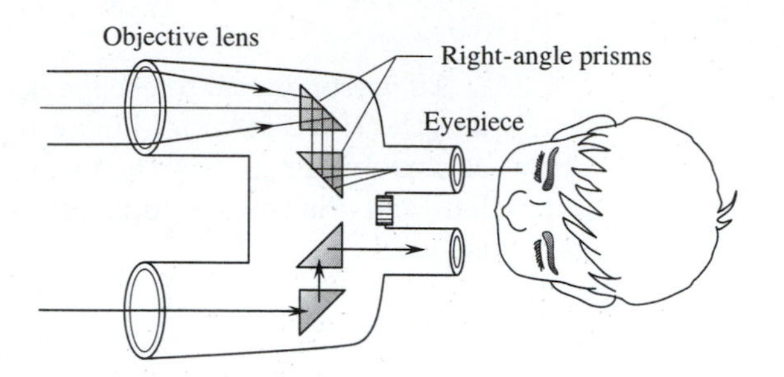

Figure 27–30: *Diagram for Exercise 3. Right-angle prisms are used in binoculars.*

Section 27.2:

4. Explain the following terms in your own words.

   A. Optical axis **B.** Optical center **C.** Focal length
   D. Concave mirror **E.** Convex mirror **F.** Magnification

5. Where must the object be placed in front of a concave mirror to create a real image? A virtual image?

6. An object is placed at the focal point of a concave mirror. Where is the image?

Figure 27–31: *Diagram for Exercise 8. A partly submerged pencil looks bent.*

Section 27.3:

7. What is an angle of deviation?

8. Partly submerge a pencil in a glass of water. Look down at an angle at the pencil. It will look bent like the one in Figure 27–31. Explain why.

Section 27.4:

9. Can a single concave lens form a real image? Explain.

10. Explain the following terms in your own words.

    A. Chief ray
    B. Secondary focal point
    C. Thin lens

11. Which of the following optical devices *can be* converging?

    A. Concave mirror
    B. Concave lens

C. Convex mirror
D. Convex lens

12. Which of the following devices *cannot* form a real image?

    A. Plane mirror B. Concave mirror C. Convex mirror
    D. Concave lens E. Convex lens

13. In a particular situation, the magnification for a *single* thin lens is $M = -2.1$.

    A. Is the image larger or smaller than the object?
    B. Which is closer to the lens: the object or the image?
    C. Is the image virtual or real?
    D. Is the image inverted or erect?

Section 27.5:

14. In your own words explain how the following factors affect the performance of an optical fiber:

    A. Numerical aperture
    B. Pulse dispersion
    C. Absorption

15. A small numerical aperture is preferred for long-distance communications using optical fibers. Should the difference of optical index of refraction between the cladding and the core be large or small for a low numerical aperture?

16. Explain the following terms in your own words.

    A. Angular magnification
    B. Near point
    C. Objective lens

17. Why do objective lenses have a *short* focal point for a microscope and a *long* focal point for a telescope?

PROBLEMS

Section 27.1:

1. A point source of light is 2.00 m from a plane mirror (see Figure 27–32a). The mirror is rotated 3° (Figure 27–32b). The reflection of the light beam is reflected on a screen 2.00 m from the mirror. Find the vertical deflection (y) of the light beam.

2. The system shown in Figure 27–32 is called an optical lever. If the angles are expressed in radians show that the angle of rotation of the mirror (θ) is: $\theta = y/2R$.

3. Two plane mirrors are at right angles (see Figure 27–33). At what angle does the light ray strike the first mirror in order to be reflected through a total angle of 180° by the two mirrors? (Hint: The path is symmetric.)

4. The intensity of the light beam in Figure 27–33 is reduced to what percentage of its original intensity after the two reflections? Assume a coefficient of reflection of 0.85.

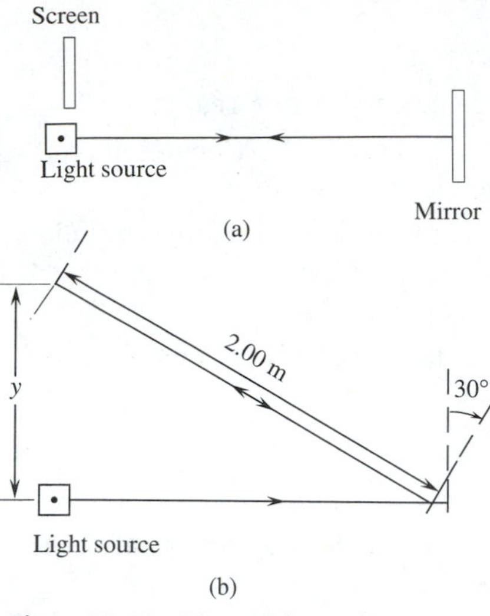

Figure 27–32: Diagram for Problem 1. A light beam is reflected by a rotating mirror.

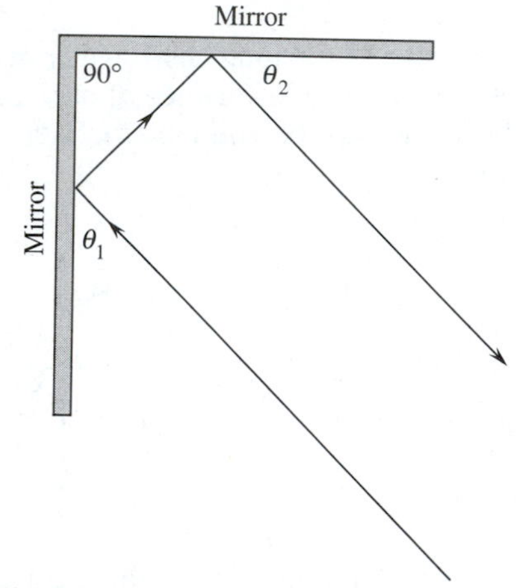

Figure 27–33: Diagram for Problem 3. Mirrors at 90° reflect a light beam back in the direction from which it came.

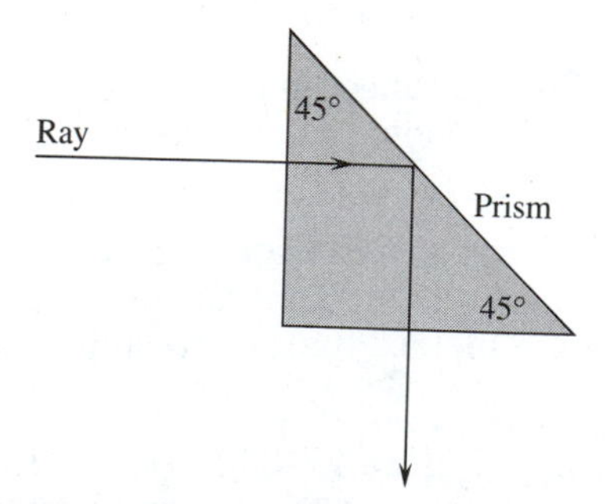

Figure 27–34: Diagram for Problem 5.

5. Prisms are used in binoculars to divert the direction of a light ray by 90° using internal reflection (see Figure 27–34). What is the minimum index of refraction of the prism's glass?

6. What is the critical angle of internal reflection for diamond in air?

Section 27.2:

7. Find the missing quantities for the following spherical mirrors.

a. $p = 14.0$ in, $q = ?$, $f = +3.0$ in, $R = ?$, $M = ?$

b. $p = 6.2$ cm, $q = 6.2$ cm, $f = ?$, $R = ?$, $M = ?$

c. $p = 8.0$ cm, $q = ?$, $f = ?$, $R = -14.0$ cm, $M = ?$

d. $p = ?$, $q = -5.0$ in, $f = ?$, $R = ?$, $M = +3.0$

e. $p = ?$, $q = -6.0$, $f = ?$, $R = 12.0$ in, $M = ?$

8. A handy homeowner decides to make a satellite dish out of chicken wire and wood. A spherical section rather than parabolic shape is chosen for easy construction of the reflector. If the antenna is designed with a radius of curvature of 20 ft, where will the incoming signal focus? (This is why parabolic dishes are preferred.)

9. An inverted stainless steel hubcap forms a concave mirror with a radius of curvature of 38.0 cm. If it is used as a reflector, what are the image distance and the magnification for an object 42.0 cm from the mirror?

10. The cover of an egg cooker is a spherical section of polished aluminum with a radius of curvature of 11.0 in, forming a convex mirror. What are the image distance and the magnification of the nose of a sleepy cook when it is 15.0 in from the mirror?

Section 27.3:

11. A hemispherical Jell-O mold has a cherry embedded at an apparent depth of 3.0 in. The index of refraction of Jell-O is about 1.5. What is the actual depth?

12. Find the angle of deviation for the light ray passing through the prism shown in Figure 27–35.

13. A ray of light strikes a glass plate with parallel sides and an index of refraction of 1.56. The incident angle is 37.0° (see Figure 27–36). What is:

A. the angle of refraction at the first surface?
B. the angle of incidence at the second surface?

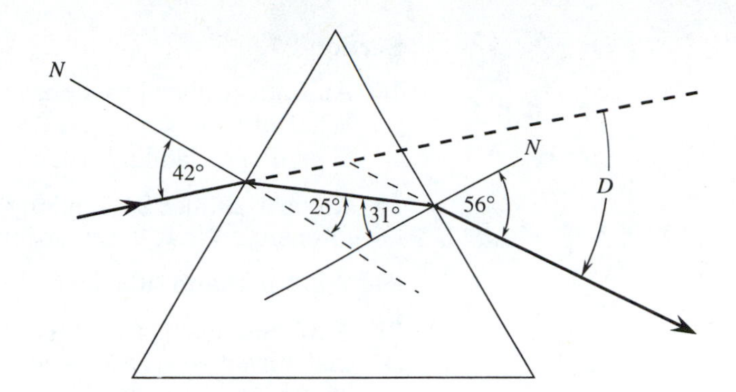

Figure 27–35: Diagram for Problem 12. Find the angle of deviation (D).

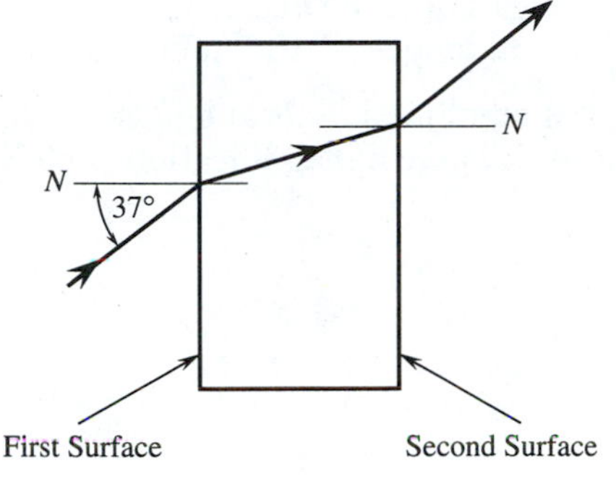

Figure 27–36: Diagram for Problem 13.

C. the angle of refraction at the second surface?
D. the net angle of deviation of a light ray passing through the parallel plate?

14. A 6.0-in layer of transparent oil with an index of refraction of 1.26 lies on top of an 8.0-in layer of water in a metal cylinder. (n[water] = 1.33.) What is the

A. apparent thickness of the oil layer?
B. apparent thickness of the water layer?
C. combined apparent thickness of the two fluids?

Section 27.4:

15. Find the missing data for the following thin lenses.

a.	**b.**	**c.**
p = 24.0 cm	p = 10.0 in	p = 6.0 cm
q = 24.0 cm	q = ?	q = ?
f = ?	f = +5.0 in	f = ?
M = ?	M = ?	M = +2.5

d.	**e.**	**f.**
p = 2000 in	p = 10.0 cm	p = ?
q = ?	q = ?	q = 36.0 in
f = 5.3 in	f = −12.0 cm	f = ?
M = ?	M = ?	M = +0.33

16. An object is 9.0 cm from a convex lens with a 10.0-cm focal point. What is:

A. the image distance?
B. the magnification?
C. the nature of the image?

17. A physics student finds that a cumulus cloud seen through the laboratory window forms a real image on a screen 12.6 cm from the lens the student is using in the laboratory. The student places a light bulb 25.0 cm from the same lens. What are the location and nature of the light bulb's image?

Section 27.5:

18. An optical fiber has a core with an index of refraction of 1.56. What are the numerical aperture and the capture angle for the fiber if the cladding has an index of refraction of 1.48?

19. An optical fiber 2.3 km long transmits 22% of the incident light intensity. What is the absorption in decibels per kilometer?

20. What is the focal length of a 5.6 X single-lens magnifier?

21. A microscope has an objective lens with a focal length of 9.0 mm that forms a real image at 18.0 cm from the object. What focal length eyepiece should be used to give an overall magnification of 100 X for the microscope?

22. What will be the overall magnification of the microscope in Problem 21 if an eyepiece with a focal length of 16.7 mm is used?

23. An astronomical telescope has an objective focal length of 90.0 in. What focal length eyepiece will give a magnification of 60 X?

Chapter 28

PHYSICAL OPTICS

You might think you are looking at a poor photo, but it is actually a photo showing smog in Los Angeles. In this chapter, you will see how light and airborne particles create some of the colors of our environment. (Photograph by Tom Jagoe, copyright © 1986 by Los Angeles Daily News. *Used with permission.*)

OBJECTIVES

In this chapter you will learn:

- how light is scattered by small particles
- that light is polarized by different processes
- how light can undergo constructive and destructive interference when reflected by thin films
- that light waves form fringes as when they pass through closely spaced slits
- that the clarity of an image is limited by the size of the openings through which light waves pass
- how the motion of light-emitting sources can be found using the Doppler effect
- how holograms are formed by light interference
- that the atomic structure of crystals can be inferred from X-ray diffraction patterns

Time Line for Chapter 28

1604 Johannes Kepler shows that light intensity falls off as an inverse square of distance.

1665 Francesco Grimaldi's experiments with diffraction and his wave theory of light are published after his death.

1808 Etienne-Louis Malus finds that reflected light is polarized.

1812 Sir David Brewster discovers the Brewster angle relating the index of refraction with polarization.

1912 Max von Laue finds that X-ray diffraction patterns are created by crystals.

1913 Albert Einstein explains the photoelectric effect as a collision of a photon with electrons on the surface of the metal.

1913 Niels Bohr produces the Bohr model of the atom.

1947 Dennis Gabor develops holography.

1947 Willis Lamb's experiments suggest that the wave theory of light is incorrect.

In Chapter 27, we traced the path of light rays through optical systems. By looking at an assortment of triangles occurring in the light paths we were able to find some equations to predict the behavior of images. This we called geometric optics.

In this chapter, we will look at the wave nature of light, called **physical optics.** We will look at some of the following things.

The way a small particle scatters light depends on wavelength. That is why the sky is blue and the sunset red. Light is a transverse wave. Polarization reduces or eliminates the vibration of a light wave along one transverse axis. That is why polarized sunglasses reduce glare. Absorption is wavelength-selective. It also depends on the way atoms are arranged relative to their neighbors. A hot tungsten filament emits a full spectrum of wavelengths. A low-pressure sodium street lamp emits only selected wavelengths of light, giving it a characteristic yellow color. A variety of devices can split a light beam into parts that are out of phase. Constructive or destructive interference will occur. Interference can be a nuisance. It limits the sharpness of the image of an optical instrument. On the other hand, interference can be very useful. It is used to measure the thickness of thin films and to form holograms.

28.1 SCATTERING

Figure 28–1 shows a piece of wood floating on water. In 28-1a, small water ripples strike the board. Because the ripples are small compared with the dimensions of the board, they are reflected in a

Elephants and the Theory of Light

Early theories of light described light as a particle. Some theories thought light was a particle emitted by light sources. The particles entered the eye, making vision possible. Some early ideas were similar to radar or sonar. According to this theory, light was a particle emitted by the eye that rebounded off objects and then reentered the eye.

This view was changed, or at least should have changed, in the early 1600s. Francesco Grimaldi discovered light made the same kind of diffraction and interference patterns that water waves make. Over a period of several years, he performed diffraction experiments and developed a wave theory of light. At his death in 1665, a book was published containing his wave theories and accounts of his supporting diffraction experiments. Unfortunately, few people paid too much attention to his work.

Isaac Newton developed geometric optics and was able to explain reflection and refraction using a particle theory of light. The development of optics continued under the assumption of particle until 1801, when Thomas Young rediscovered light diffraction. Throughout the nineteenth century, and going into the twentieth, additional experiments confirmed light was a wave. New discoveries, such as the Doppler shift of starlight and polarization, were consistent with the wave theory.

Then things began to get confusing.

Light falling on the clean surface of a metal emits electrons. This is called the photoelectric effect. The wave theory couldn't explain this effect. In 1913, Albert Einstein showed that the photoelectric effect could be explained as a collision of a light particle, or photon, with electrons on the surface of the metal. The photon behaved like a bowling ball striking another solid object. To add to the confusion, the energy of the photon is proportional to the frequency of the light, a wave property.

In 1947, Willis Lamb was studying hydrogen. He discovered an apparent error between experiment and the wave theory of light. This led Richard Feynman and others to develop a theory of the interaction of light and electrons. The theory totally explained Lamb's discovery. Central to this theory is the assumption that light is a particle.

There's a "once upon a time" story about an Asian ruler who lived in a country that had no elephants. One day he was given an elephant as a present. He brought his three wisest sages, blindfolded, to the courtyard where the elephant was standing. The wise men were told to describe the elephant. One by one, the blindfolded sages approached the elephant. The first sage bumped into the side of the elephant. He said, "An elephant is like a wall." The second found the tail. He announced, "An elephant is like a rope." The last one found the trunk and stated, "An elephant is like a snake." None were wrong. None were totally correct.

Is light a particle or a wave? Choose your elephant.

preferred direction. The angle of incidence equals the angle of reflection. In 28-1b, the wavelengths of the waves are about the same size as the object. They bounce off the board, traveling in many directions. The law of reflection no longer applies. This effect is called **scattering**. In Figure 28-1c, the wavelengths are large compared with the size of the piece of wood. The board oscillates up and down, carried by the wave motion of the water. In this case, there is neither scattering nor reflection. Apparently, scattering depends on wavelength.

There are many small particles in the atmosphere about the size of visible light. Wind carries small bits of dust and pollen into the atmosphere. Industrial plants create smoke particles. Droplets of ocean spray evaporate, leaving salt suspended in the air. Many tons

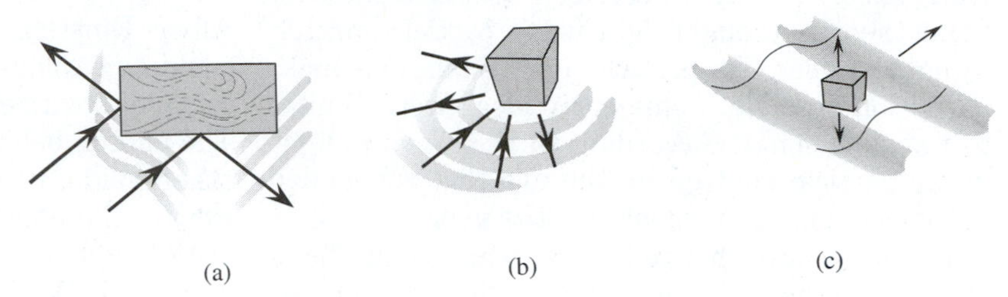

Figure 28–1: (a) Water waves, small compared with the size of a board, are reflected. (b) A floating object scatters ripples with a wavelength comparable to the size of the object. (c) The waves are larger than the floating object; the object bobs up and down on the waves.

of micrometeorites, dust from space, settle on the earth each year. All of these particles can scatter light. Light scattering is a bit different from the scattering of light waves. Particles many times smaller than light wavelengths will cause scattering.

Experimental studies of light scattering have shown that the intensity of scattered light is inversely proportional to the fourth power of a wavelength for particles that are small compared with the length of the wave. This type of scattering is called **Rayleigh scattering.** If I_0 is the intensity of the incident beam of light and I_s is the intensity of the scattered light, then the scattering intensity for light with a wavelength λ is:

$$I_s = \frac{I_0}{\lambda^4} \quad \textbf{(Eq. 28–1)}$$

Figure 28–2: Shorter light waves are more easily scattered by particles in the atmosphere than longer waves. This makes the sky look blue.

□ **EXAMPLE PROBLEM 28–1: THE BLUE SKY**

Sunlight passing through the earth's atmosphere has a full spectrum of visible wavelengths. The blue-violet end of the spectrum has a wavelength near 4.0×10^{-7} m. The red end of the spectrum has a wavelength of about 7.0×10^{-7} m. When you look up into the sky what is the ratio of intensity of scattered wavelengths of the blue end of the spectrum compared with the red end? See Figure 28–2.

■ ***SOLUTION***

The ratio of intensities of scattered light (R) is:

$$R = \frac{I_s(\text{blue})}{I_s(\text{red})} = \frac{I_0\,[\lambda(\text{red})]^4}{I_0\,[\lambda(\text{blue})]^4}$$

$$R = \left(\frac{7.0 \times 10^{-7}\,\text{m}}{4.0 \times 10^{-7}\,\text{m}}\right)^4 = 1.75^4$$

$$R = 9.4$$

The intensity of the blue light is nearly ten times as large as the longer red light. This gives the sky its blue color.

28.2 POLARIZATION

Figure 28–3 shows transverse waves. If the amplitude along one transverse axis is reduced to zero, the wave is polarized. If it is only

Once in a Blue Moon

Did you ever wonder why it gets so hazy in the summer? Haze has a lot to do with light scattering. The air is full of small particles known as aerosols. These particles can be classified into three sizes compared with the wavelength of visible light.

Some particles are much larger than light wavelengths. They scatter white light through small angles. The efficiency of the scattering is not very great. This kind of scattering accounts for the whitish ring you sometimes see around the sun.

Aerosols that are about the same size as visible wavelengths scatter white light in all directions. These are the most efficient light-scattering aerosols. A large percentage of the light hitting them will be scattered in all directions.

(Photograph by Tom Jagoe, copyright © 1986 by Los Angeles Daily News. *Used with permission.)*

Most aerosols are much smaller than light wavelengths. These particles undergo Rayleigh scattering. Short blue wavelengths are scattered with greater intensity than long red or yellow wavelengths. This gives the characteristic blue color to the sky. The intensity of the scattering from these very small particles is not great, but there are many of them in the atmosphere.

For these smallest aerosols, the intensity of the scattering is proportional to the square of the volume. If the diameter of such a particle is doubled, the volume increases by a factor of 8. The square of the volume is 64. The intensity increases by a factor of 64.

At temperatures well below the dew point, water molecules can cling to the surface of small particles. As the relative humidity increases, the water-molecule coating on the particles gets larger. As the diameter increases, the ability to scatter light increases. The particle size approaches the size of light wavelengths and scatters white light. We see this as haze. Higher humidity causes larger numbers of particles to grow to the most efficient scattering size.

There is a particle size just a little larger than one wavelength of light that scatters red light more efficiently than blue light. Under some special conditions created by forest fires and volcanic eruptions a large number of this size aerosol can be created. In these situations people have reported seeing a green sun and a blue moon.

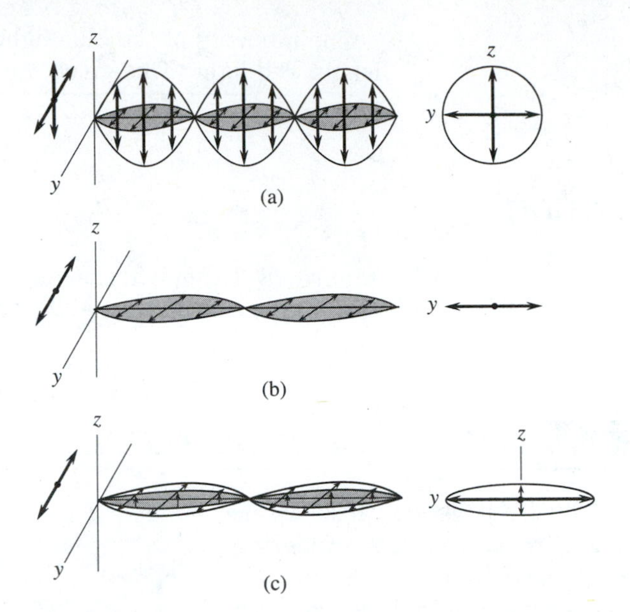

Figure 28–3: *(a) Unpolarized transverse waves. (b) Polarized transverse waves. (c) Partially polarized waves.*

partly reduced, it is partially polarized. Several processes can cause polarization of light. Here are a few of them.

Selective Absorption

If you attempt to send waves along a rope threaded through the slats of a chair, the waves will get through only if their amplitude is parallel to the slats (see Figure 28–4).

Some materials, such as tourmaline crystals and commercially made polarizing films, behave like the slats in a chair. Long, parallel molecules in the crystal let light through parallel to the molecule chains, but absorb the light amplitude at right angles to the chains (see Figure 28–5).

We can use two sheets of polarizing film. The first sheet acts as

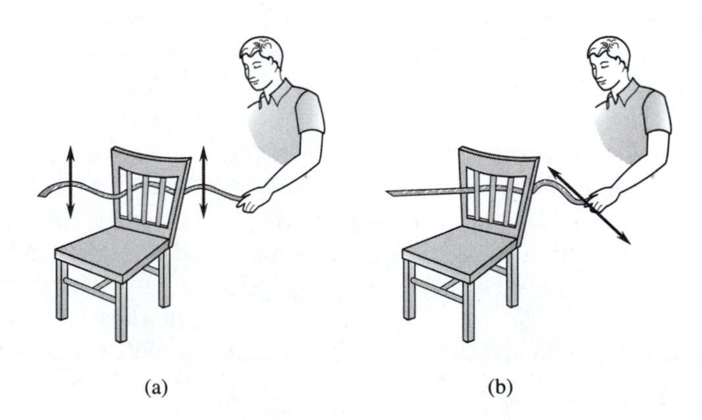

Figure 28–4: *Vertical waves on a rope can pass through the vertical slats of the chair. Horizontal waves are absorbed.*

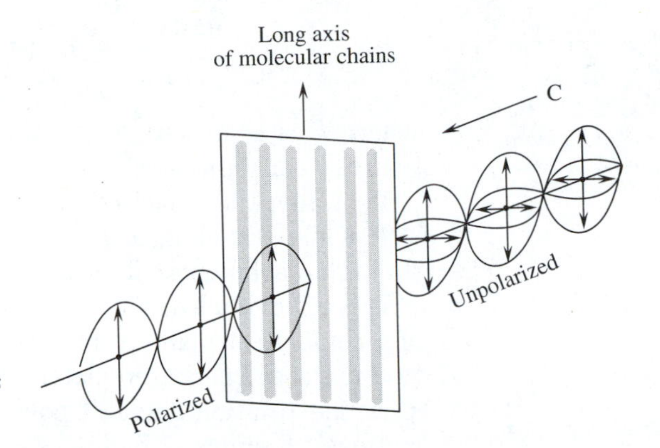

Figure 28–5: The intensity of light is absorbed at right angles to the long axis of molecules in a polarizing film.

a **polarizer.** It polarizes the incoming light. The second sheet is the **analyzer.** If it is at right angles to the first sheet, no polarized light will get through it. (See Figure 28–6a). If it is oriented at an angle θ, only the component of amplitude parallel to the analyzer can pass through it. A_0 is the amplitude of the light incident to the analyzer. From Figure 28–6b, we see that the amplitude transmitted through the analyzer (A) is:

$$A = A_0 \cos \theta$$

Since intensity is proportional to the square of the amplitude, the intensity of the transmitted light is proportional to the cosine squared. This is called the **law of Malus.**

$$I = I_0 \cos^2 \theta \qquad \textbf{(Eq. 28–2)}$$

Some transparent materials, including some plastics, are **optically active** (see Figure 28–7). The molecules making up such a substance have a helical or screwlike symmetry. As plane polarized light passes through the material it rotates. The amount of rotation is propor-

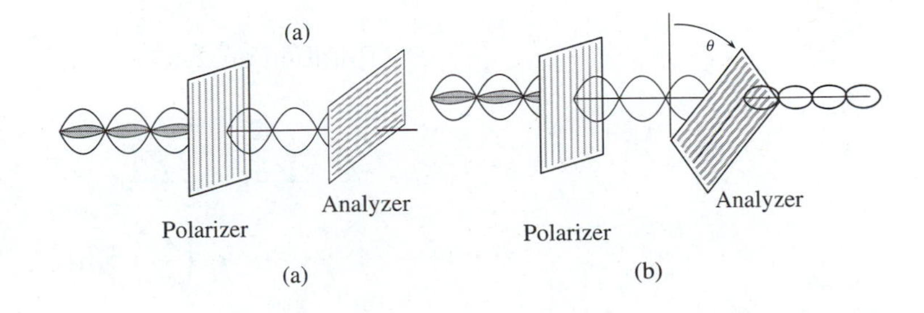

Figure 28–6: (a) Light cannot pass through two polarized films oriented at right angles to each other. (b) The intensity of polarized light passing through the analyzer is reduced for angles between 0° and 90°.

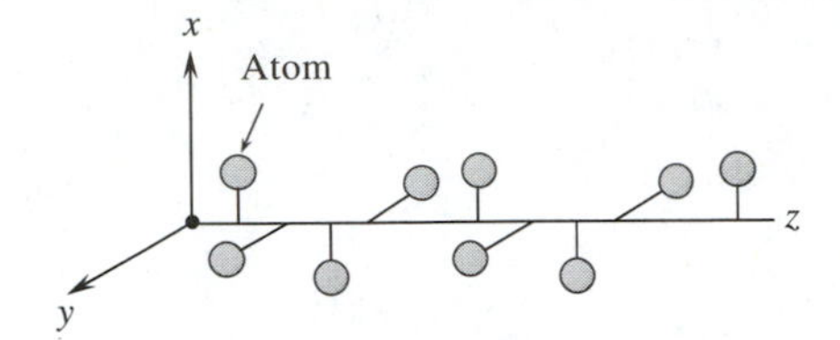

Figure 28–7: In optically active materials, atoms have a spiral symmetry.

tional to the thickness of the material and inversely proportional to wavelength.

$$\alpha = \frac{k\,L}{\lambda} \qquad \textbf{(Eq. 28–3)}$$

where L is thickness, λ the wavelength of light, and k a proportionality constant.

Typically, the polarized wave will make one complete rotation of 360° while traveling a distance of a little less than a centimeter through a solid optically active material.

This effect is used in strain analysis of mechanical systems. Models of a component made from optically active material are placed between crossed polarized plates. When the model is stressed, strains alter the rotation angle of polarized light at different sites in the material. Regions of high stress can be identified.

□ **EXAMPLE PROBLEM 28–2: LAW OF MALUS**

A sheet of optically active material is placed between a polarizer and an analyzer 90° to each other. Polarized light with a wavelength of 6.2×10^{-7} m is rotated through an angle of 37° as it goes through the optically active sheet.

A. What is the relative intensity of the light transmitted through the analyzer?
B. Through what angle will light with a wavelength of 4.6×10^{-7} m be rotated?
C. What is the relative intensity of the transmitted light for the wavelength in part B?

■ ***SOLUTION***

A. Relative intensity:

$$\frac{I}{I_0} = \cos^2\theta = \cos^2 37°$$

$$\frac{I}{I_0} = 0.64$$

B. Rotation angle:

$$\frac{\alpha_2}{\alpha_1} = \frac{k\,L/\lambda_2}{k\,L/\lambda_1}$$

$$\alpha_2 = \left(\frac{\lambda_1}{\lambda_2}\right) \times \alpha_1$$

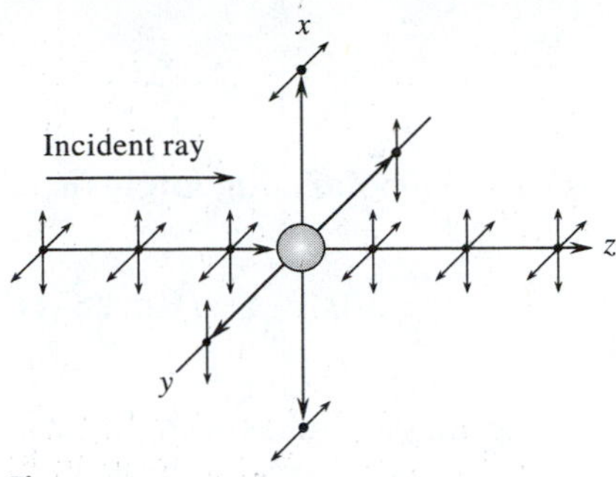

Figure 28–8: Rayleigh scattering polarizes scattered light 90° from the incident ray.

$$\alpha_2 = \left(\frac{6.2 \times 10^{-7}\ \text{m}}{4.6 \times 10^{-7}\ \text{m}}\right) \times 37°$$

$$\alpha_2 = 50°$$

C. Relative intensity:

$$\frac{I}{I_0} = \cos^2 50°$$

$$\frac{I}{I_0} = 0.41$$

Rayleigh Scattering

Light scattered by small particles is polarized in directions at right angles to the incident ray (see Figure 28–8). You can see this effect by looking at the sky through polarized glasses. The glasses act as analyzers. Rotate the glasses. The light intensity will vary, indicating that the scattered light is partially polarized.

Reflection

If you have a pair of polarizing sunglasses or a sheet of polarizing film, look at a number of shiny surfaces with and without the film. Try shiny desktops, water, windshields, polished floors—anything with a bit of glare on the surface. You will notice that the glare either disappears or is reduced by the polarizing material. It acts as an analyzer. Reflected light tends to be polarized.

Reflected light is at its maximum polarization if its incident angle is at the **Brewster angle.** The Brewster angle occurs when the reflected ray and the refracted ray make an angle of 90° (see Figure 28–9).

According to Snell's law:

$$n_1 \sin \theta_1 = n_2 \sin \theta_2$$

At the Brewster angle:

$$\theta_2 = 90° - \theta_1$$

or

$$\sin \theta_2 = \sin (90° - \theta_1) = \cos \theta_1$$

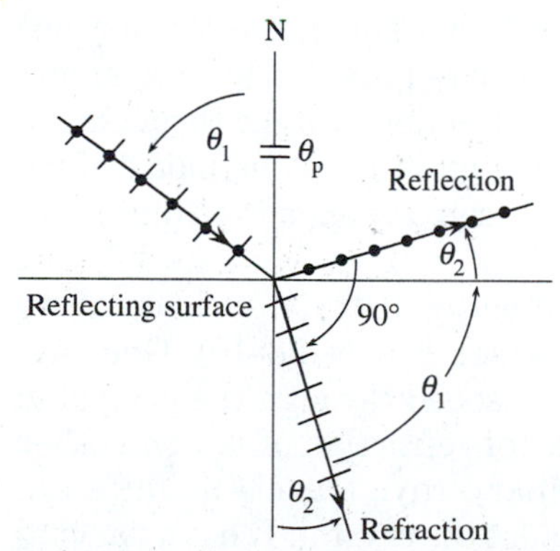

Figure 28–9: Reflected light is completely polarized when the angle of reflection and the angle of refraction add up to 90°.

We can substitute this into Snell's law.

$$\frac{\sin \theta_1}{\cos \theta_1} = \tan \theta_1 = \frac{n_2}{n_1}$$

We will call the incident angle in this case the **polarizing angle** (θ_p).

$$\tan \theta_p = \frac{n_2}{n_1} \qquad \textbf{(Eq. 28–4)}$$

For angles different from the polarizing angle, reflected light will be partially polarized.

□ **EXAMPLE PROBLEM 28–3: POLARIZATION BY REFLECTION**

At what angle of incidence will light reflected from a pond be totally polarized?

■ *SOLUTION*

$$n(\text{water}) = 1.33$$

$$n(\text{air}) = 1.00$$

$$\tan \theta_p = \frac{1.33}{1.00}$$

$$\theta_p = 53.1°$$

Double Refraction

Glass is an amorphous, isotropic material. It is amorphous because the atoms that make up glass are randomly situated. It is isotropic because its mechanical, optical, and electrical properties are the same in all directions in the glass. The speed of light passing through glass is the same no matter what direction it travels through the material.

Calcite, quartz, and many other crystalline substances are neither amorphous nor isotropic. Atoms are situated in regular arrays, and the optical properties depend on the direction of light through the crystal. Many of these substances have more than one index of refraction. The speed of light through the crystal depends on direction. We call this effect **double refraction.**

Ordinary light hitting a double-refracting material such as calcite is split into two plane polarized rays (see Figure 28–10). One ray, called the **ordinary ray,** has the same speed through the crystal in all directions. It has a constant index of refraction of n_o. The other polarized ray, called the **extraordinary ray,** travels at different speeds in different directions throughout the crystal. It has a varying index of refraction, n_e.

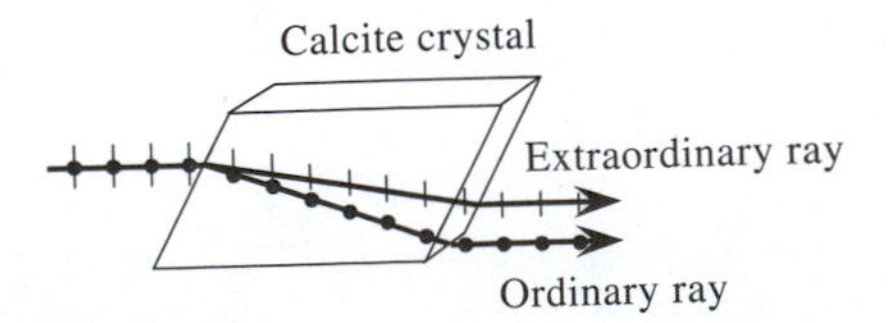

Figure 28–10: *A calcite crystal splits unpolarized light into two polarized rays.*

Table 28–1: Indices of refraction for selected double-refracting minerals, for 20°C and a wavelength of 589 nm.

MINERAL	CHEMICAL FORMULA	n_o	n_e	$(n_o - n_e)$
Beryl	$(Be_3Al_3[SiO_3]_6)$	1.525	1.479	0.046
Calcite	$CaCO_3$	1.658	1.486	0.172
Cinnabar	HgS	2.854	3.201	−0.347
Dolomite	$(CaCO_3 \cdot MgCO_3)$	1.682	1.503	0.179
Greenockite	CdS	2.506	2.529	−0.023
Quartz	SiO_2	1.544	1.553	−0.009
Soda nitre	$NaNO_3$	1.587	1.336	0.251

There is one direction through the crystal in which both rays travel at the same speed. This is called the **optical axis.** Along the optical axis $n_e = n_o$. The maximum difference between n_e and n_o is at right angles to the optical axis. n_e is listed for this direction in tables of indices for double-refracting materials. Table 28–1 lists values for some materials.

28.3 INTERFERENCE

Light is a wave. If a light beam is split into two parts and they are brought back together, it has undergone **interference.** We can predict whether the interference will be constructive or destructive by keeping track of the phase.

Figure 28–11 shows two waves traveling parallel. One wave passes through a piece of glass with an index of refraction of n. The other wave moves through a vacuum. They both travel the same physical distance, but they get out of phase because the wave moving through the glass has a shorter wavelength.

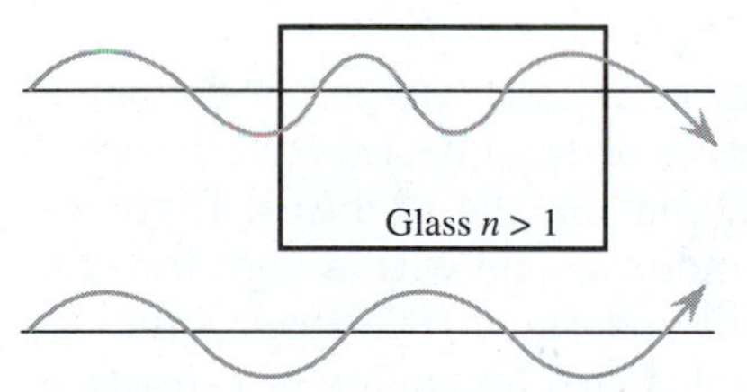

Figure 28–11: The optical path of the ray traveling through the piece of glass is longer. It oscillates through an extra half wavelength compared with the ray traveling the same distance through air.

The wavelength in a vacuum is:

$$\lambda = c/f$$

The wavelength in glass is:

$$\lambda' = \frac{v}{f} = \frac{c}{nf}$$

or

$$n\lambda' = \frac{c}{f}$$

Combining the two equations we get:

$$\lambda = n\lambda'$$

The phase shift depends on the index of refraction. More wavelengths are needed to make up the physical distance in the glass. The **optical path** keeps track of the phase shift.

$$\textit{optical path} = \textit{index of refraction} \times \textit{distance traveled}$$
$$op = n\,d$$

The optical path of light traveling through air for 2.0 cm is:

$$op = 1.00 \times 2.0 \text{ cm} = 2.0 \text{ cm}$$

The optical path of light through 2.0 cm of water is:

$$op = 1.33 \times 2.0 \text{ cm} = 2.66 \text{ cm}$$

Usually, we are interested in optical paths that are only a few wavelengths long.

Look at Figure 28–12. Light is incident on a layer of material only a few wavelengths thick (d). It might be a soap bubble, a layer of oil on water, or a coated lens. Part of the light beam is reflected by the first surface. The other is reflected by the second surface. The difference in the optical paths (Δ) is $[n \times (\text{AFD})] - (\text{AE})$, where n is the index of refraction of the thin film.

After reflection at F, the light beam looks as though it is coming from point B. The optical path $n \times (\text{DC})$ is the same as optical path AE. The total difference in the optical paths is then $n \times (\text{BC})$.

From the right triangle in the diagram we see:

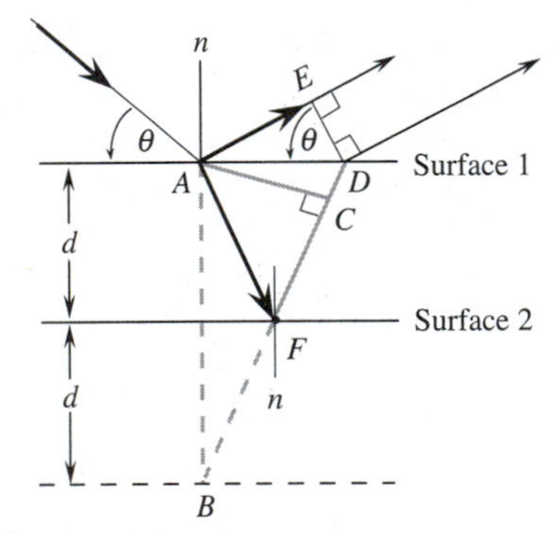

Figure 28–12: The ray reflected from the second surface undergoes the extra optical path n × (BC).

$$\Delta = n \times (\text{BC}) = n\,(2\,d \cos \theta)$$

If the path difference is an integer number of waves, the combined waves will undergo total constructive or total destructive interference. The kind of interference depends on the surfaces. Light reflected from a surface with a low index to one with a high index of refraction (**external reflection**) will undergo a 180° phase shift. An **internal reflection** (from a high index to a low index of refraction) does not involve a phase shift.

A soap film has an external reflection on the first surface and an internal reflection on the second surface. There is a phase shift at the first surface, but not at the second. If the optical path difference is m wavelengths where $m = 1, 2, 3 \ldots$, then the two reflections will be a half wavelength out of phase. In this case, the equation for dark fringes, **destructive interference,** is:

$$m\,\lambda = 2\,n\,d \cos \theta \qquad \textbf{(Eq. 28–5A)}$$

For a coated lens or oil on water the second reflection also is an external reflection. In this case, two reflected waves come together in phase. The equation gives us bright fringes, or **constructive interference.**

$$m\,\lambda = 2\,n\,d\cos\theta \qquad \textbf{(Eq. 28–5B)}$$

☐ **EXAMPLE PROBLEM 28–4: THE COATED LENS**

High-quality camera lenses are often coated to prevent reflection. A lens has an optical index of refraction of 1.72 and a coating with an optical index of refraction of 1.31. For near-normal angles ($\cos\theta = 1.00$) find the minimum thickness the coating should have to prevent reflection for wavelengths of 5.3×10^{-7} m.

■ ***SOLUTION***

This is a case of two external reflections. We want $m = \frac{1}{2}$ in Equation 28–5B.

$$\frac{1}{2}\lambda = 2\,n\,d \times (1.00)$$

$$d = \frac{\lambda}{4\,n}$$

$$d = \frac{5.3 \times 10^{-7}\ \text{m}}{4\,(1.31)}$$

$$d = 1.0 \times 10^{-7}\ \text{m, or } 0.1\ \mu\text{m}$$

28.4 DIFFRACTION

A knife can be used to make two closely spaced slits in a piece of thin cardboard. The two slits split a beam of light into two parts that will interfere (see Figure 28–13). The distance between the slits is d. The two shaded triangles are proportional. The common angle $\theta = \tan^{-1}(S/L)$. A light ray from the bottom slit must travel an extra distance Δ to reach point A.

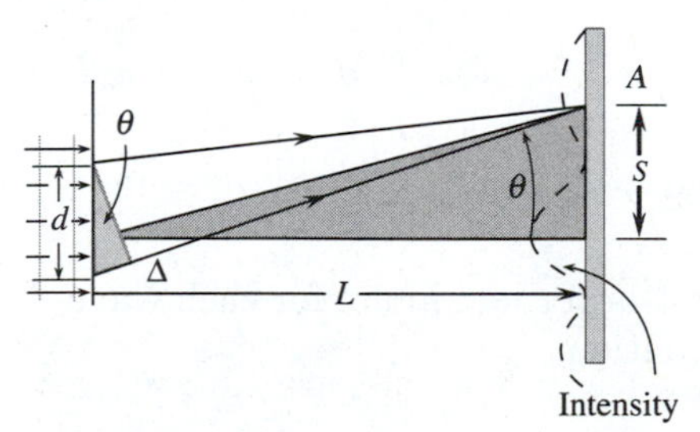

Figure 28–13: *The ray from the bottom slit travels an extra distance (Δ). $\Delta = d \sin\theta$.*

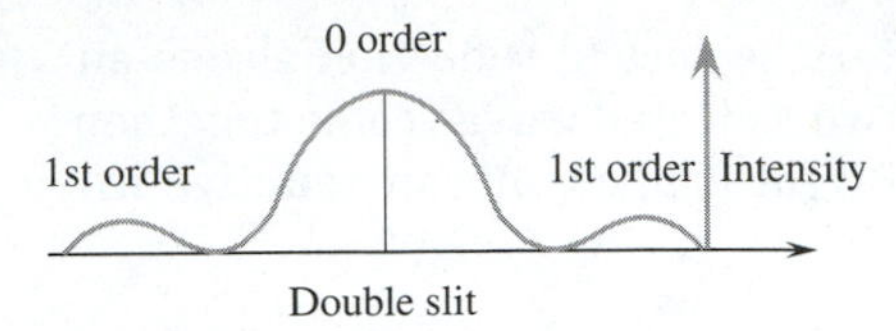

Figure 28–14: The intensity of a double-slit pattern. The first-order maxima occur for N = 1, where N is the number of wavelengths.

$$\Delta = d \sin \theta$$

If Δ is an integer number of wavelengths (N), the two rays will interfere constructively at point A.

$$N\lambda = d \sin \theta \qquad \textbf{(Eq. 28–6)}$$

where $N = 0, 1, 2, 3 \ldots.$

Interference from two or more slits is called **diffraction.** Figure 28–14 shows an intensity plot of the two-slit diffraction pattern. The central maximum is called the **zero-order maximum** ($N = 0$). The two maxima next to the center one correspond to $N = 1$. They are the **first-order maxima.**

If we have several evenly spaced slits, the sharpness of the maxima increases. Look at Figure 28–15. For the first maximum, the path difference for each added slit is one additional wavelength. The added slits increase the intensity of the maximum. Angles a little larger or smaller than the maximum condition cause greater destructive interference.

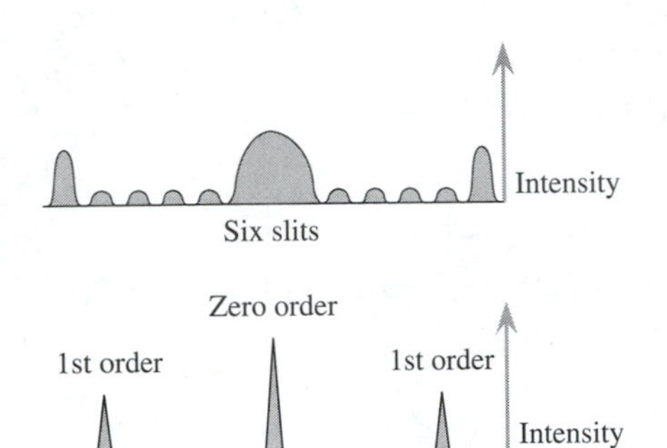

Figure 28–15: As more slits are added to a diffraction grating, the peaks become narrower.

Transmission diffraction gratings can be made by forming many tiny parallel scratches on a transparent plate of glass or plastic. Typically, a diffraction grating will have hundreds of evenly spaced grooves per centimeter. Gratings are usually rated by a **grating constant:** the number of grooves per unit length. The spacing between adjacent grooves can be found by taking the reciprocal of the grating constant.

□ **EXAMPLE PROBLEM 28–5: DIFFRACTION GRATING**

A grating has 1100 lines/cm. What is the spacing between the first-order maxima of wavelengths $\lambda_1 = 6.20 \times 10^{-7}$ m and $\lambda_2 = 5.40 \times 10^{-7}$ m on a screen 2.00 m from the diffraction grating? See Figure 28–16.

■ ***SOLUTION***

First find the distance (d) between adjacent grooves.

$$d = \frac{1}{1100 \text{ line/cm}}$$

$$d = 0.000909 \text{ cm} = 9.09 \times 10^{-6} \text{ m}$$

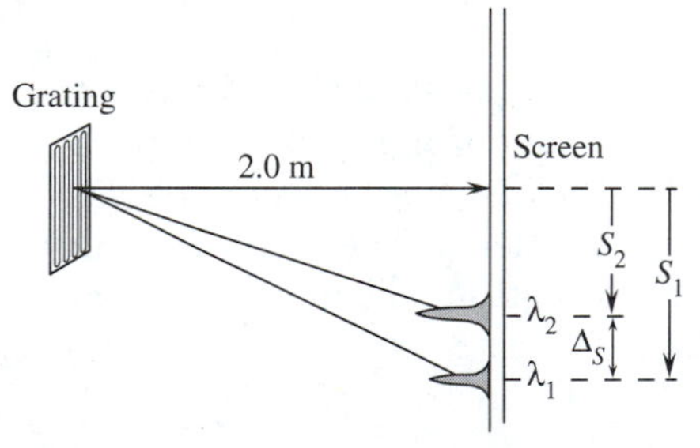

Figure 28–16: Diffraction grating. S is the spacing between diffraction lines on the screen for two different wavelengths of light.

We can find the angle for the first-order maximum for each wavelength by using the diffraction equation.

$$N = 1$$

$$\lambda = d \sin \theta$$

or

$$\theta_1 = \sin^{-1}\left(\frac{\lambda}{d}\right)$$

$$\theta_1 = \sin^{-1}\left(\frac{6.20 \times 10^{-7}\ \text{m}}{9.09 \times 10^{-6}\ \text{m}}\right)$$

$$\theta_1 = 3.91^\circ$$

$$\theta_2 = \sin^{-1}\left(\frac{5.40 \times 10^{-7}\ \text{m}}{9.09 \times 10^{-6}\ \text{m}}\right)$$

$$\theta_2 = 3.41^\circ$$

Use Figure 28–13 as a guide. The large triangle has sides L and S where S is the distance from the central maximum.

$$S = L \tan \theta$$

$$S_1 = 200\ \text{cm}\ (\tan 3.91^\circ)$$

$$S_1 = 13.7\ \text{cm}$$

$$S_2 = 200\ \text{cm}\ (\tan 3.41^\circ)$$

$$S_2 = 11.9\ \text{cm}$$

The spacing between the first-order maxima for the two wavelengths is:

$$\Delta S = S_1 - S_2$$

$$\Delta S = 13.7\ \text{cm} - 11.9\ \text{cm}$$

$$\Delta S = 1.8\ \text{cm}$$

28.5 RESOLUTION

The light passing through a single slit can interfere with itself (see Figure 28–17). The optical path of the light ray traveling to point A from the bottom of the slit is longer than the ray coming from the top of the slit. If the path difference (Δ) is an integer number (M) of wavelengths, constructive interference can occur. Take the example that $M = 1$. By pairing rays, we see that for each ray from the top half of the slit there is a ray from the bottom half that is 180° out of phase. For a slit width of b the ray from the center will be exactly 180° out of phase with the ray from the top edge. A ray from $\frac{3}{4}\,b$ is 180° out of phase with the ray from $\frac{1}{4}\,b$.

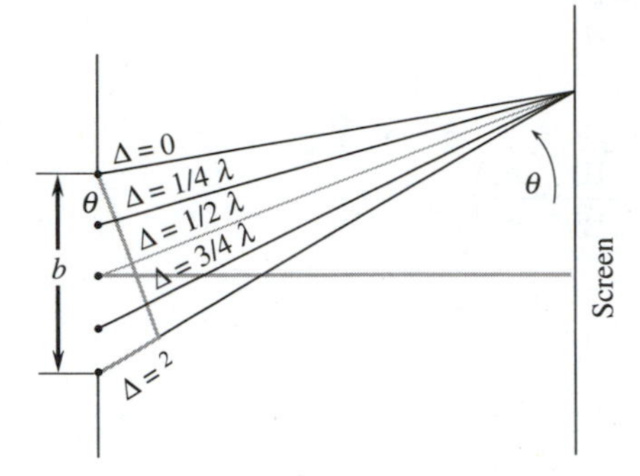

Figure 28–17: *Single-slit interference. Waves from the bottom half of the slit interfere with waves from the top half.*

$$M\,\lambda = b \sin \theta \qquad \textbf{(Eq. 28–7)}$$

For small angles, $\sin \theta = \theta$. Solving for θ, we get:

$$\theta = \frac{M\,\lambda}{b}$$

Notice that the smaller the opening b is, the wider the angle θ will be. A camera, a telescope, or a microscope has a pattern similar to this. The objective lens acts as a circular slit. Figure 28–18 shows the image of two point sources of light seen through different-sized lenses. As the size of the lens is reduced, a situation is reached where we can no longer determine whether there is one or two sources of light.

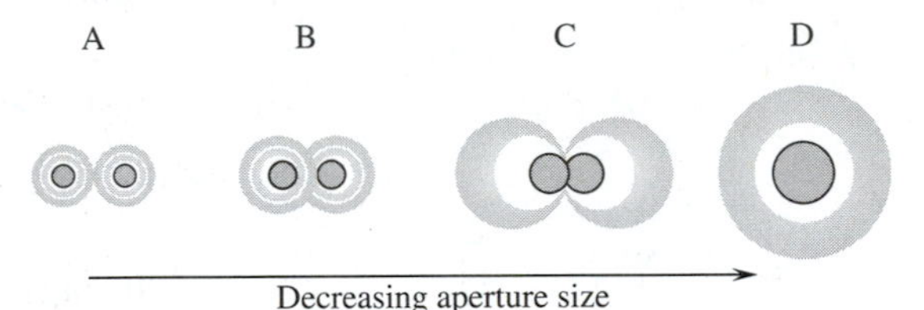

Figure 28–18: *As the size of a circular opening decreases its single-slit pattern broadens. If the opening is too small, the images of two objects are seen as one (D). At the minimum angle of resolution two sources are barely separated (C). For larger openings the two sources are clearly resolved (A, B).*

The ability of an optical instrument to separate two images is called **resolution.** Resolution causes images to blur; it limits the amount of detail we can see. Resolution is usually stated by the **minimum angle of resolution** (θ_R). This is the smallest angle at which we can resolve two point sources into two separate images. For a **circular aperture** such as a lens, the angle of resolution (R) in radians is:

$$R = 1.22\left(\frac{\lambda}{b}\right) \qquad \textbf{(Eq. 28–8)}$$

☐ **EXAMPLE PROBLEM 28–6: THE RESOLUTION OF THE EYE**

The pupil of the eye has an average diameter of 6.0 mm.

A. What is the minimum angle of resolution of light with a wavelength of 5.60×10^{-7} m?

B. What would be the diameter of the objective lens of a camera that had the same sharpness of image as the eye for infrared radiation with a wavelength of 7.0×10^{-6} m?

■ *SOLUTION*

A. The minimum angle of resolution of the eye:

$$R = 1.22\left(\frac{\lambda}{b}\right)$$

$$R = 1.22\left(\frac{5.60 \times 10^{-7}\ \text{m}}{6.0 \times 10^{-3}\ \text{m}}\right)$$

$$R = 1.14 \times 10^{-4}\ \text{rad}$$

B. Camera objective lens for infrared:

Using the angle of resolution from part A, solve Equation 28–8 for b.

$$b = \frac{1.22\ \lambda}{R}$$

$$b = \frac{1.22 \times 7.0 \times 10^{-6}\ \text{m}}{1.14 \times 10^{-4}\ \text{rad}}$$

$$b = 0.075\ \text{m, or } 7.5\ \text{cm}$$

28.6 SPECTRA

Hot solids and liquids emit a black-body spectrum: a continuous wide range of wavelengths of which visible light is only a small portion. Gases are different. Only a few discrete wavelengths are emitted, producing a **discrete spectrum** for a particular gas. Let us look at the structure of an atom to see why this is true.

An atom is composed of a positively charged center, or nucleus, and negatively charged electrons. The electrons are much less massive than the nucleus. They orbit around the nucleus much like planets orbit around the sun.

Electrons have waves associated with them. In most orbits, the wave interferes with itself. Orbits in which the wave interferes with itself are unstable. Electrons stay only in orbits where a standing wave is produced. For stable orbits, the circumference of the orbit is an integer multiple of the electron's wavelength (see Figure 28–19).

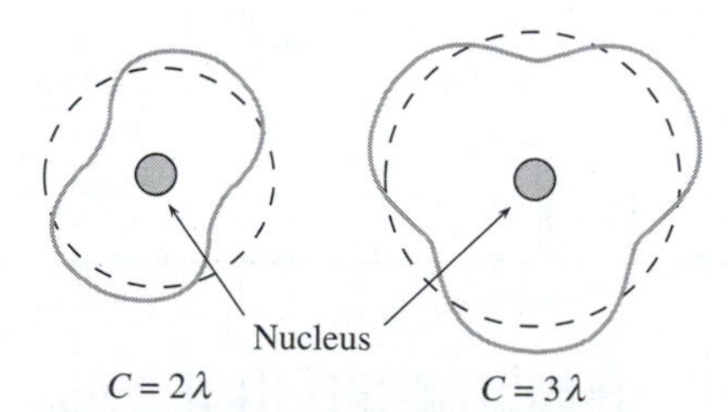

Figure 28–19: *Allowed orbits for an electron occur only when the circumference of the orbit (C) is an integral multiple of the electron's wave.*

Each electron orbit has a certain energy associated with it. No two electrons can have exactly the same energy. This means that they cannot have exactly the same orbit; their waves would interfere. Orbits closest to the nucleus have less energy than orbits farther from it. The electrons cluster in these lower-energy orbits closest to the nucleus, with no two electrons having exactly the same orbit.

An electron orbiting the nucleus in an orbit of energy E_1 can move to a higher stable orbit of energy E_2 if it can gain exactly the energy difference of the two orbits ($\Delta E = E_2 - E_1$). This can happen in two ways. Either the electron gains energy in a collision with another atom, something we would expect in a hot material, or it gains energy by absorbing a photon with the energy $E_2 - E_1$. Recall that the energy of a photon is hf (Planck's constant $= 6.626 \times 10^{-34}$ J · s).

$$E_2 - E_1 = hf$$

Once the electron gets to the higher-energy orbit it tries to return to its low-energy orbit. Usually, it stays in the higher-energy position for a very small fraction of a second. It must give off exactly the energy difference between the two orbits to change position. Usually the electron emits a photon to reduce its energy.

$$hf = E_1 - E_2$$

Notice that the frequency (or wavelength) emitted by the electron just matches the energy difference of the orbits. Since there are only a few stable orbits, only a few frequencies or wavelengths will be emitted. Different elements have different energies for the stable electron orbits. This results in a spatial spectrum for each kind of atom. Figure 28–20 shows the visible light spectra for some common

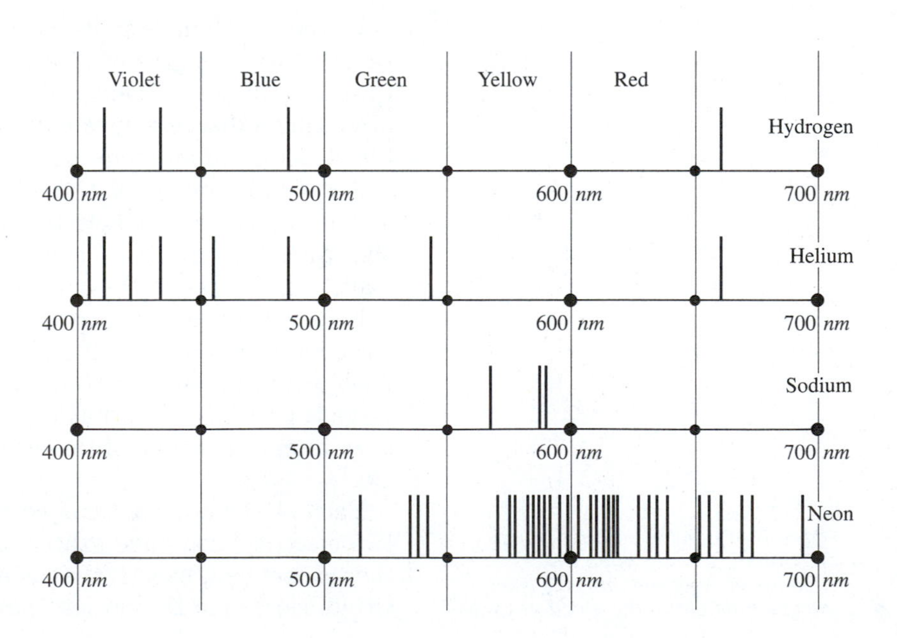

Figure 28–20: *The visible light spectrum for four light sources. Wavelengths are given in nanometers (10^{-9} m). Yellow sodium light is seen in the flame of a burning match. Most of the wavelengths of neon are in the orange and red range.*

gases. We can look at these spectra as optical fingerprints. Astronomers can identify the composition of stars many billions of miles away by looking at the light emitted by them.

A gas consists mostly of empty space where there is room for electrons to move into higher-energy orbits with large radii. In solids and liquids, the situation is more complicated. Atoms are closely packed. When electrons go to higher orbits there is an interaction between atoms. A different kind of analysis is needed to predict the continuous light spectrum of solids and liquids.

28.7 THE DOPPLER EFFECT

You are familiar with the Doppler effect for sound (refer to Chapter 25). The Doppler effect is a property of waves. It occurs for electromagnetic waves as well as for mechanical waves such as sound. As a source approaches an observer, the wave's frequency increases. As a source recedes from an observer, the wave's frequency decreases.

The Doppler effect for sound and the Doppler effect for light differ somewhat. First, how we keep time affects the frequency we observe. Second, motion is relative. In determining the Doppler effect for light, we do not distinguish between the motion of the observer and the motion of the light source. We look only at the speed of approach (u). u will be positive if the observer and the light source are getting closer together. u will be negative if the observer and light source are moving farther apart.

Let f be the frequency of the light source if there is no relative motion, and let f' be the Doppler-shifted frequency. c is the speed of light. We get the following relativistic equation.

$$f' = f\left[\frac{1 + (u/c)}{1 - (u/c)}\right]^{1/2} \qquad \textbf{(Eq. 28–9)}$$

If the speed of approach is much smaller than the speed of light ($u << c$), then the equation can be approximated by:

$$f' = \mathrm{f}\left(1 + \frac{u}{c}\right)$$

The Doppler shift of light has been a very important tool for astronomers. The motion of nearby stars moving toward or away from the sun can be calculated using the Doppler shift. Doppler-shift measurements have shown that all the galaxies surrounding our Milky Way galaxy are moving away from us and away from each other. This observation has led to a model of an expanding universe.

The Doppler effect can be used to measure the temperature of distant stars. As the temperature of a gas rises, the average speed of the atoms making up the gas also increases. The motion of the atoms is random. Some atoms are approaching the observer when light is emitted. Other atoms are moving away. As a result the characteristic wavelengths of light like those shown in Figure 28–20 get smeared out. The lines seen in the figure broaden. The higher the temperature, the wider the characteristic wavelength pattern gets. Astronomers can use this effect to measure the temperature of distant light sources.

□ **EXAMPLE PROBLEM 28–7: A DISTANT GALAXY**

Light received from a distant galaxy is shifted from a frequency of 5.88×10^{14} Hz to 4.21×10^{14} Hz. How fast is the galaxy moving away from the sun?

■ ***SOLUTION***

Solve Equation 28–9 for u.

$$u = c\left[\frac{(f'/f)^2 - 1}{(f'/f)^2 + 1}\right]$$

The ratio $(f'/f)^2$

$$= \left(\frac{4.21 \times 10^{14}\ \text{Hz}}{5.88 \times 10^{14}\ \text{Hz}}\right)^2 = 0.513$$

$$u = 3.00 \times 10^8 \left(\frac{0.513 - 1}{0.513 + 1}\right)$$

$$u = -9.66 \times 10^7\ \text{m/s}$$

The minus sign in the velocity of approach indicates that the galaxy is moving away from the sun.

28.8 APPLICATIONS

Holograms

A black-and-white photograph records the intensity of the light forming an image. Photons break up molecules of silver chloride and

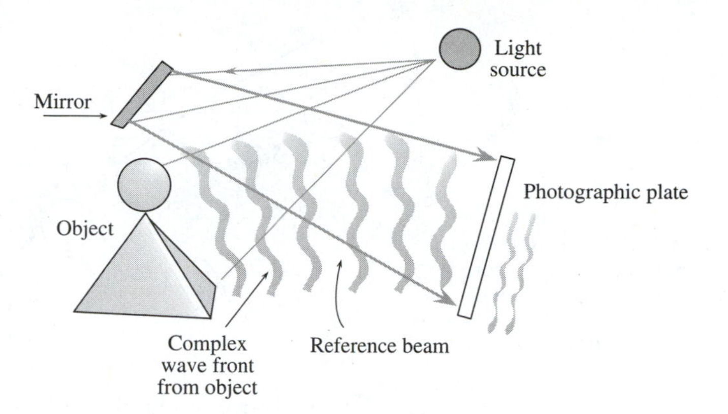

Figure 28–21: *A hologram is formed by interference patterns between a reference beam and a complex wave front reflected from an object. The interference pattern, not an image, is recorded on the photographic plate.*

silver bromide suspended in a gelatin coating. Silver is deposited on the film.

Color photographs record an additional bit of information: the frequency of the light. Three separate coatings are placed on the film. Each layer is coupled to a particular dye in the developer. The overlay of dyed deposits produces the full-range spectrum.

Holograms record a different kind of information. Instead of recording the image's intensity and frequency, an interference pattern of waves is recorded. When you look at a hologram you see the complex wave front complete with phase differences and interferences that were caused by the original object.

Some method of recording the complex phase shifts from the object onto a recording plane is needed. This is done by using a reference beam. The light from the source is split into two beams by a mirror (see Figure 28–21). One beam bounces off from the object, setting up a complex beam. The other beam is a reference beam. It is reflected from the mirror directly to the recording surface. The two wave fronts—the reference beam and the waves scattered by the object—interact where they converge at the recording surface. What is recorded on the photographic plate is not an image, but a record of wave interference.

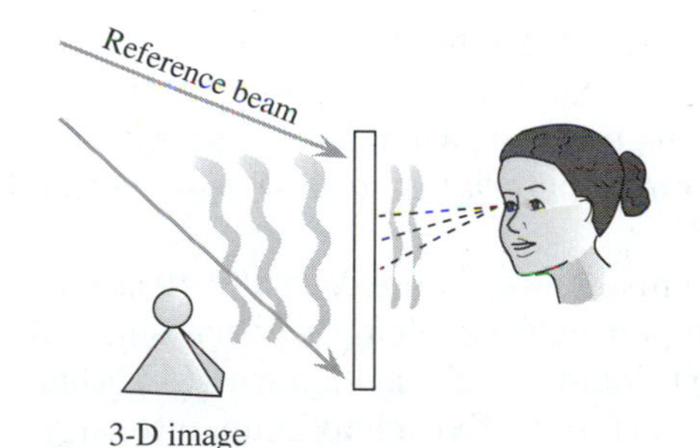

Figure 28–22: *When the reference beam shines on the developed hologram, the original wave front from light scattered from the object is recreated.*

When the photographic plate is developed, it contains a complex interference pattern. It is then illuminated with the primary beam. The original wave front from the source is reconstructed. When we view the wave fronts by looking through the hologram, we see the object, complete with spatial relationships (see Figure 28–22).

X-ray Diffraction

In crystals, atoms arrange themselves in neat patterns (see Figure 28–23). Looking at the crystals at different angles we can find several different layers of atoms with regular spacing between the layers. These layers can reflect waves to cause interference patterns.

In a solid, the distance between atoms is on the order of 10^{-10} m.

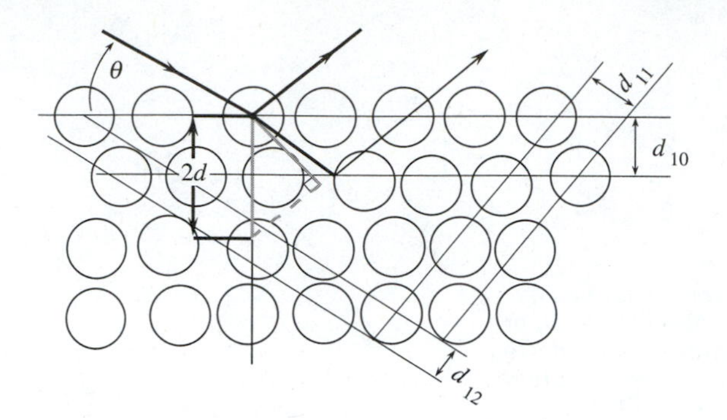

Figure 28–23: *Atoms in a crystal form several differently spaced layers for various directions through the array. Constructive interference occurs when X rays are reflected by atom layers.*

This distance is about a thousand times shorter than visible wavelengths. Fortunately, X rays have wavelengths comparable to these distances. X rays of known wavelengths can be bounced off the atomic layers to find their spacing. By looking at the spacing of several different layer distances, the geometry of the atomic packing can be determined. This process is called **X-ray diffraction** even though it is really an interference pattern of reflected rays. It is a major tool for finding the crystalline structure of a material.

Look again at Figure 28–23. For X-ray diffraction, the angle of incidence is taken from the plane rather than from the plane normal. The path difference is:

$$\Delta = 2\,d \sin\theta$$

An index of refraction has little effect for this process. As the X-ray photons move through the material, they either bounce off from an atom or pass through the empty space between atoms. For X rays, the index of refraction of a material is so close to 1.00 that it can be ignored.

X rays bouncing off from atoms behave like external reflections. All the reflected waves undergo a half-wavelength phase shift. If the path difference is a multiple number of wavelengths (N), then the waves will constructively interfere to give a maximum intensity.

$$N\lambda = 2\,d \sin\theta \qquad \textbf{(Eq. 28–10)}$$

Because the distances between atoms are very small, people who work with X-ray diffraction often use a unit called an **angstrom** (Å).

$$1\ \text{Å} = 1.00 \times 10^{-10}\ \text{m}$$

☐ **EXAMPLE PROBLEM 28–8: X-RAY DIFFRACTION**

One of the spacings between atomic layers of potassium bromide is 3.30 Å. An X-ray beam with a wavelength of 1.92 Å is used to examine the crystal. At what angles will the first- and second-order maximum reflections occur?

■ *SOLUTION*

First-order ($N = 1$):

$$\sin \theta_1 = \frac{N \lambda}{2d}$$

$$\sin \theta_1 = \frac{1 \times 1.92 \text{ Å}}{2 \times 3.30 \text{ Å}}$$

$$\sin \theta_1 = 0.2909$$

$$\theta_1 = 16.9°$$

Second-order ($N = 2$):

$$\sin \theta_2 = \frac{2 \times 1.92 \text{ Å}}{2 \times 3.30 \text{ Å}}$$

$$\theta_2 = 35.6°$$

SUMMARY

Physical optics is the study of the effects of the wave nature of light. The scattering of light by objects that are small compared with the wavelength of light is called Rayleigh scattering. The intensity of Rayleigh-scattered light is inversely proportional to the fourth power of the wavelength.

$$I_s = \frac{I_o}{\lambda^4}$$

Light may be plane polarized by four processes: selective absorption, Rayleigh scattering, reflection, and double refraction.

Polarization by selective absorption can occur in materials that have long, thin parallel molecules that absorb light amplitudes at right angles to the long axis of the molecules. If light that is already polarized passes through such a material at an angle, the intensity will be reduced according to the law of Malus.

$$I = I_o \cos^2 \theta$$

Some materials have optical activity. The axis of polarized light is rotated as it moves through the material. The mount of rotation (α) depends on the thickness (L) of the material and the wavelength (λ) of the light.

$$\alpha = k\frac{L}{\lambda}$$

where k is a proportionality constant.

Polarization by scattering. Rayleigh scattering polarizes light scattered at an angle of 90° relative to the incident beam.

Polarization by reflection. Reflected light will be completely plane polarized if it is reflected at the angle of polarization (θ_p). The angle of polarization depends on the indices of refraction of the materials at the reflecting interface. For surfaces in air, $n_1 = 1.00$.

$$\tan \theta_p = \frac{n_2}{n_1}$$

Polarization by double refraction. Some crystalline materials split light into two polarized rays. One ray, the ordinary ray, has the same index of refraction in all directions through the crystal. The other ray, the extraordinary ray, has a variation of the index of refraction with direction throughout the crystal. The two polarized rays tend to travel along different paths through the crystal.

Light waves reflected by two surfaces such as the two sides of a thin film may form **interference fringes.** If the film has an index of refraction of n and a thickness of d, then the maximum or minimum intensities will occur for the condition:

$$m\lambda = 2\,n\,d\cos\theta$$

where $m = 1, 2, 3 \ldots.$

If both surfaces produce external reflections, as on a lens coating, the equation for interference fringes gives the location of maximum intensities of the fringes. If one surface produces an external reflection and the other an internal reflection, as on a soap bubble film, then the equation gives the location of minimum intensities.

Multiple slits can cause an interference pattern called diffraction, where light waves passing through the slits undergo constructive interference patterns. If d is the spacing of the waves through the slits, then

$$N\lambda = d\sin\theta$$

where $N = 0, 1, 2, 3 \ldots$

When light passes through a single opening it can interfere with itself. This decreases the sharpness of an image. The images of two point sources are barely separated at the minimum angle of resolution (θ_R). For a circular opening of diameter b, θ_R is:

$$\theta_R = 1.22\frac{\lambda}{b}$$

Solids and liquids tend to emit a full-range spectrum of electromagnetic radiation. Gases emit discrete spectra corresponding to the energy differences of electron orbits in an atom (E_1 and E_2). The energy of the emitted photon is:

$$hf = E_1 - E_2$$

Holograms record the interference pattern between a wave front produced by light bouncing off an object and a reference beam. When the developed hologram is illuminated with the reference beam the original wave front is reproduced.

X ray diffraction is reflective interference of waves bounced off atomic layers in a crystal. If d is the spacing of atomic layers, and θ is measured relative to the atomic plane instead of its normal, then constructive interference will occur for:

$$N\lambda = 2d \sin\theta$$

where $N = 1, 2, 3 \ldots$

A unit of distance often used in X-ray work is the angstrom (Å).

$$1 \text{ Å} = 1 \times 10^{-10} \text{ m}$$

KEY TERMS

If you can explain the following terms to a friend or classmate, you understand their meaning. If you cannot explain the terms, you should reread the sections in which they are discussed.

analyzer
angstrom (Å)
Brewster angle
circular aperture
diffraction
discrete spectrum
double refraction
external reflection
extraordinary ray
first-order maximum
grating constant
interference
internal reflection
law of Malus
minimum angle of resolution
optical axis
optically active
optical path
ordinary ray
physical optics
polarizer
polarizing angle
Rayleigh scattering
resolution
scattering
transmission diffraction gratings
X-ray diffraction
zero-order maximum

EXERCISES

Section 28.1:

1. Example Problem 28–1 shows why the sky is blue. Explain how it also explains why the sunset is red.
2. A photographer wants to make a black-and-white picture of a mountain view on a hazy day. He has a choice of films and filters to use to best penetrate the haze.
   A. Which kind of film should he use: ordinary (sensitive to blue light only), orthochromatic (sensitive to all but red light), or panchromatic (sensitive to light of all colors)?
   B. Which filter will give the clearest picture: yellow or blue?
3. Explain the following terms in your own words.
   A. Scattering
   B. Physical optics

Section 28.2:

4. Explain the following terms in your own words.
   A. Optical activity
   B. Double refraction
   C. The law of Malus
   D. Polarizing angle
   E. Extraordinary ray
   F. Selective absorption
5. A red light ray and a blue light ray travel the same path through an optically active material. Both rays are polarized. Which ray will be rotated through the largest angle?
6. Photogray glasses have silver halides in solution with glass. Ultraviolet light dissociates the molecules in a reversible process that produces silver deposits in the glass. The deposits scatter visible light. What advantage do polarizing sunglasses have over photograys?
7. Look at Figure 28–24. Light is reflected from two glass surfaces. The angle of incidence is the same in both cases. In Figure 28–24a, the glass surfaces are parallel; the image of the source can be seen by looking at the second mirror. In Figure 28–24b the second piece of glass has been rotated 90° around the vertical

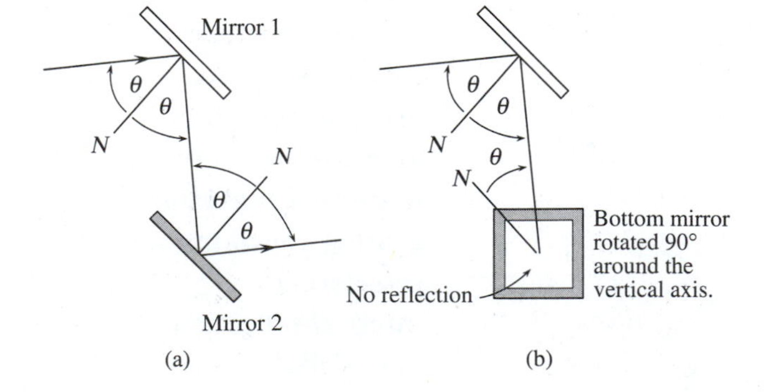

Figure 28–24: *(a) When the mirrors are parallel, the image of a light source is reflected from the second mirror. (b) The bottom mirror is rotated 90° around the vertical axis while maintaining the same angle of incidence as in (a). Light reflected from mirror 2 can no longer be seen.*

axis. An observer looking at the second mirror can no longer see an image of the source. What happened to the light ray reflected off the second surface? What must be true about the angle of incidence of the light?

Section 28.3:

8. Explain the difference between an internal and an external reflection.

9. What is an optical path? Why is it important in interference?

10. Optical flats are perfectly smooth plates of glass with a uniform thickness. Figure 28–25a shows a sheet of thin foil used to separate the ends of two flats to form a thin air wedge. When it is illuminated with light with a wavelength of 5.90×10^{-7} m, the interference fringe pattern shown in Figure 28–25b results in air when the film is viewed from directly above. The dark fringes are minimums. Estimate the thickness of the foil. (The index of refraction of the thin film of air is 1.00.)

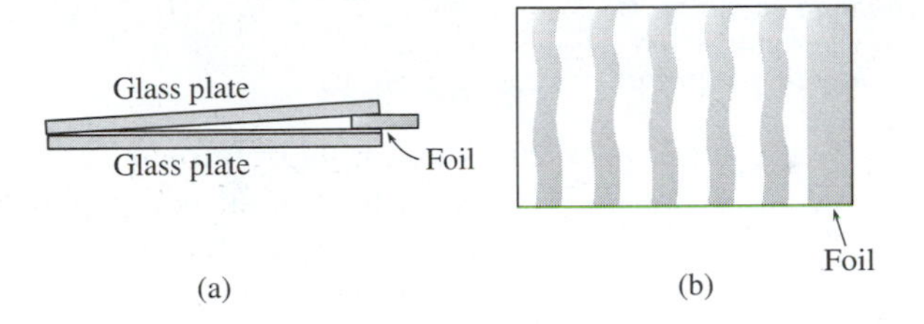

Figure 28–25: *(a) A thin foil separates two perfectly smooth pieces of glass. (b) An interference pattern results.*

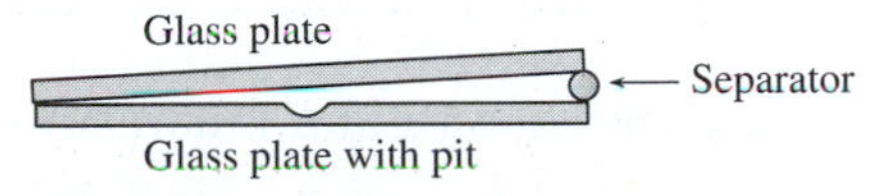

Figure 28–26: *Diagram for Exercise 11. The bottom plate has a pit that will affect the interference patten.*

11. Look at Figure 28–26. The bottom flat has been replaced by a surface containing a pit. Sketch the resulting fringe pattern. (This technique is used to test the smoothness of telescope objective lenses.)

12. Why does Equation 28–5A give the location of maximum intensity fringes for a thin layer of oil on water and minimum intensities for a soap bubble?

13. Using soap bubble solution, blow a bubble and catch it on the ring. Watch the top of the bubble. What do you see that makes it possible to predict when the bubble is about to pop?

Section 28.4:

14. For diffraction, which wavelengths have the greatest angle of dispersion: red wavelengths or blue wavelengths? Compare this result with dispersion by refraction using a prism.

15. Outline an experimental procedure using a diffraction grating to find an unknown wavelength of light.

16. Diffraction grating A has a larger grating constant than diffraction grating B.

A. Which diffraction grating has a larger spacing between lines?
B. Which grating will produce the greater dispersion?

Section 28.5:

17. Two identical black-and-white pictures are taken with the same camera. One picture is taken with infrared-sensitive film. The other is taken using a blue filter with ordinary film. Which picture will have the sharper image? Explain why.

18. Radio telescopes with an effective diameter of many meters have a much poorer resolution than an amateur astronomer's 6.0-in optical telescope. Why?

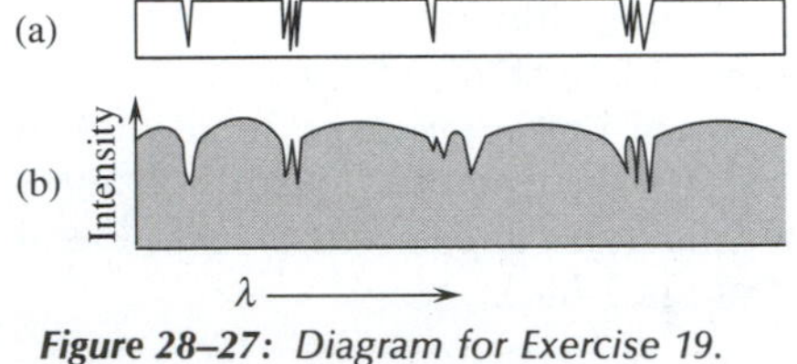

Figure 28–27: Diagram for Exercise 19. The intensity of a full-range spectrum is reduced at selected wavelengths when light passes through a low-temperature gas cloud.

Section 28.6:

19. When a full spectrum of light passes through a cool gas cloud, only selected wavelengths are reduced in intensity. What causes this? See Figure 28–27. (The effect is known as an absorption spectrum.)

Section 28.8:

20. What kind of information is recorded on a photographic film by:
    A. a black-and-white photograph?
    B. a color photograph?
    C. a hologram?

21. Compare the diffraction angle (θ) for X-ray diffraction for the following situations.
    A. The distance between atomic planes in crystal A is narrower than in crystal B.
    B. Compare a second-order pattern with a first-order pattern.
    C. The X-ray wavelength of source A is longer than the wavelength of source B.

PROBLEMS

Section 28.1:

1. Find the ratio of intensity for Rayleigh-scattered wavelengths of 5.0×10^{-6} m (infrared) with wavelengths of 5.0×10^{-8} m (ultraviolet).

2. The ratio of intensities of two scattered wavelengths is 9.30. The shorter wavelength is 4.20×10^{-7} m. What is the intensity of the other?

Section 28.2:

3. A glass has an index of refraction of 1.71 (see Figure 28–28). What is its polarizing angle in:

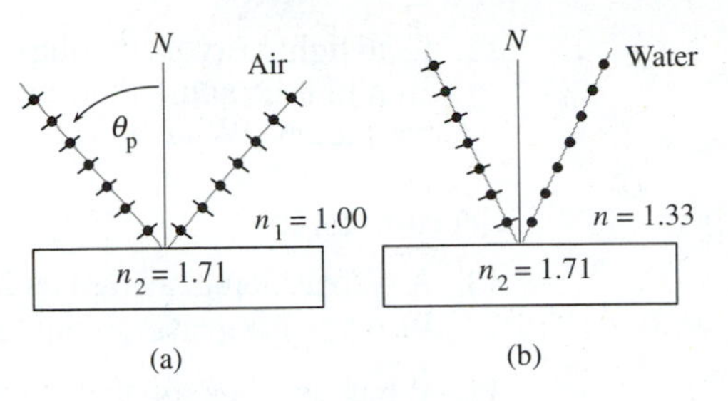

Figure 28–28: Diagram for Problem 3. The polarization angle depends on the index of refraction of the two materials at the interface.

A. air ($n = 1.00$)?
B. water ($n = 1.33$)?

4. Fluorite has a polarizing angle of 55° in air. What is its index of refraction?

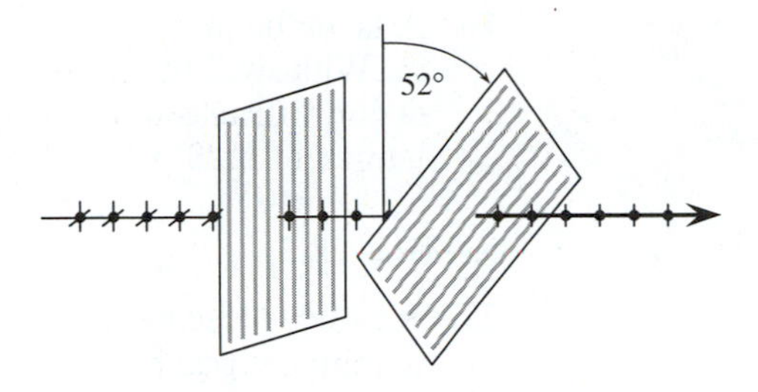

Figure 28–29: Diagram for Problem 5. Polarizing sheets are oriented 52° relative to each other.

5. A polarizer is oriented 52° with respect to an analyzer (see Figure 28–29). What fraction of light striking the analyzer gets through?

6. An analyzer and a polarizer are oriented 80° with respect to each other. What fraction of the light striking the analyzer is *absorbed?* Assume no reflection.

7. A 1.00 cm thick piece of optically active plastic will rotate red light through an angle of 7.20 rad. How thick a piece of this material will cause a rotation of 90°?

8. A 5.30×10^{-7} m wavelength of light is rotated through an angle of 98° as it passes through a sheet of optically active material. Through what angle will a wavelength of 4.27×10^{-7} m be rotated in a sheet with a doubled thickness?

Section 28.3:

9. What is the optical path of light passing through 1.40 in of water?

10. A thin uniform coating of a transparent plastic with an index of refraction of 1.28 is evaporated onto a glass optical flat. The coated surface appears to be red ($\lambda = 6.6 \times 10^{-7}$ m) under white light. What are the three *smallest* thicknesses the coating could have?

11. An air wedge is made like that shown in Figure 28–25. If the foil is 1.60 μm thick, how many fringes will be formed under light with a wavelength of 6.20×10^{-7} m?

12. Blue light travels through a thin soap film 6.7 wavelengths thick. In air, that same distance is only 4.2 wavelengths for the same frequency. What is the index of refraction of the film?

Section 28.4:

13. A diffraction grating has lines with a separation of 0.000125 cm. What is its grating constant?

14. What is the separation of grooves on a diffraction grating with a grating constant of 3200 lines/in?

15. A diffraction grating has a constant of 980 lines/cm. At what angles will the first- and second-order maxima of a 6.24×10^{-7} m wavelength ray occur?

16. A laser beam passes through a pair of slits separated by 0.016 in. What will be the separation in inches of the first- and second-order diffraction pattern on a wall 17.8 ft away? Assume a wavelength of 6.20×10^{-7} m. (Do not forget unit conversions.)

Section 28.5:

17. A radio telescope has an effective diameter of 1.2 km. It is used to map neutral hydrogen clouds emitting a wavelength of 21 cm. What is the minimum angle of resolution for this device?

18. The Hale optical telescope has a diameter $20\bar{0}$ in (5.10 m). What is its minimum angle of resolution for:

    A. light with a wavelength of 4.5×10^{-7} m?
    B. light with a wavelength of 6.8×10^{-7} m?

19. A microscope has a minimum angle of resolution of 1.40×10^{-4} rad for light with a wavelength of 5.2×10^{-7} m. What is the diameter of the objective lens?

Section 28.6:

20. What are the frequency and wavelength of light emitted by an electron when it falls from an orbit with an energy of 2.53×10^{-18} J to an orbit with an energy of 2.11×10^{-18} J? (See Figure 28–30.)

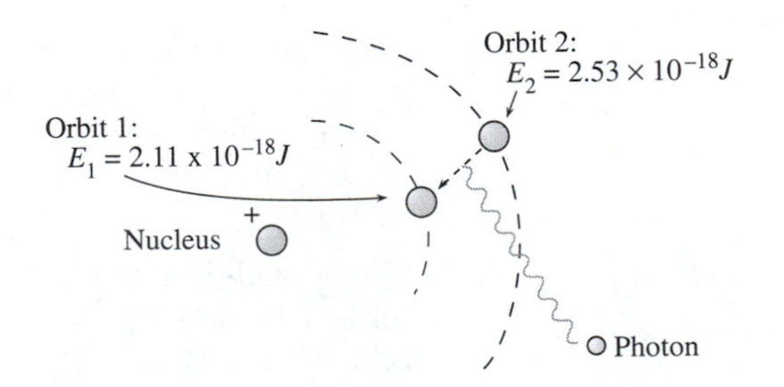

Figure 28–30: *Diagram for Problem 20. An electron emits a photon while changing orbits.*

21. A gas surrounding the sun absorbs light with a wavelength of 5.69×10^{-7} m. This corresponds to an electron energy change of how much? Is the electron gaining or losing energy?

Section 28.7:

22. For the small velocity approximation ($u << c$), the Doppler-shift frequency is: $f' = f(1 + u/c)$. Show that the Doppler-shifted wavelength (λ') is: $\lambda' = \lambda(1 - u/c)$. (Note: $1/(1 + x) = 1 - x$ for $x << 1$.)

23. A star approaches earth with a speed of 5.7×10^5 m/s. What is the Doppler-shifted wavelength for light that emits a wavelength of 623.8 nm at rest? Use the small velocity approximation from Problem 22.

24. A 641.2-nm Doppler-shifted wavelength is received from a galaxy receding from the sun with a speed equal to half the speed of light ($u/c = 0.500$). What would be the wavelength of the light if it were not Doppler-shifted?

Section 28.8:

25. The range of visible light is from 4.0×10^{-7} m to 7.0×10^{-7} m. What is this range in angstroms?

26. What is the spacing of atomic layers if 2.3-Å X rays are diffracted at a reflection angle of 57°? (Assume $N = 1$.)

27. The distance between atomic layers of a crystal is 2.88 Å. At what angles will the first- and second-order reflections occur using 1.20-Å X rays?

Chapter 29

ELECTROSTATICS

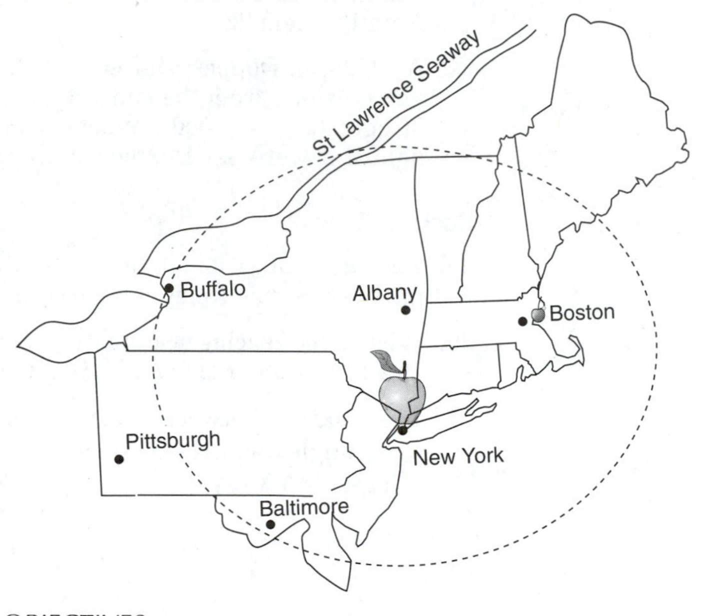

If a large apple placed in New York City represents the nucleus of a hydrogen atom, the first electron orbit could be a basketball traveling hundreds of miles away. With so much empty space in an atom, how can anything be solid? As you learn about electrostatic forces, you will see how empty space can feel hard.

OBJECTIVES

In this chapter you will learn:

- about the nature of electrostatic forces acting on positive and negative charges
- how to calculate the electrostatic force between two charges (Coulomb's law)
- how to calculate electric fields from electrostatic forces
- how electric field lines are associated with electric charges
- how electric flux is related to electric fields (Gauss's law)
- to calculate potential energy per unit charge around a static charge (electrical potential)
- to find the capacitance of charged objects of different shapes
- to describe the motion of a charged particle in a uniform field

Time Line for Chapter 29

1751 Benjamin Franklin explains electricity as a single "fluid" and introduces the concept of positive and negative electricity.

1761 Benjamin Franklin shows that lightning is electricity by flying a kite.

1767 Joseph Priestley suggests that electrical forces obey an inverse square law.

1785 Charles-Augustin de Coulomb experimentally proves the inverse square law of electrical forces

1786 Abraham Bennet invents the electroscope.

1788 Alessandro Volta introduces the idea of electrical capacity and electrical potential, which he calls electric tension.

An electrically charged automobile body is dipped in a solution of water and paint. Particles of paint are attracted to the charged metal to create a uniform prime coat over the entire irregular surface. A piece of chalk is drawn across the smooth surface of a blackboard. A trail of chalk dust is held in place by electrostatic attraction. A student walks across the nylon carpeting in the student union on a dry day. She is shocked to discover she creates sparks when she reaches for a doorknob. A television repair person carefully grounds the anode of a picture tube to the chassis with a light metal chain before removing it from the set. Socks stick to underwear as they are taken from a clothes dryer unless some commercial antistatic substance has been used.

These are examples of **electrostatic effects:** the effects of electricity at rest. In this chapter we will look at electrostatic forces and the work such forces can do.

29.1 ELECTRICAL CHARGE

Electrostatic effects have been known for a long time. In ancient Greece, the decorative material amber was known as electron. The philosophers of the day recorded that bits of fabric and pieces of straw were attracted to amber. In the 1600s, William Gilbert noticed that many materials could be made to behave like amber, or electrified, by friction.

Place a sheet of paper on a flat surface such as a table or a desktop. Briskly rub a flexible plastic ruler over the paper. Th, ruler will be electrified and will attract small bits of paper. The ruler is not the only thing that becomes charged with electricity. If you try to pick up the sheet of paper it will stick to the tabletop. This experiment works best on a day with low humidity.

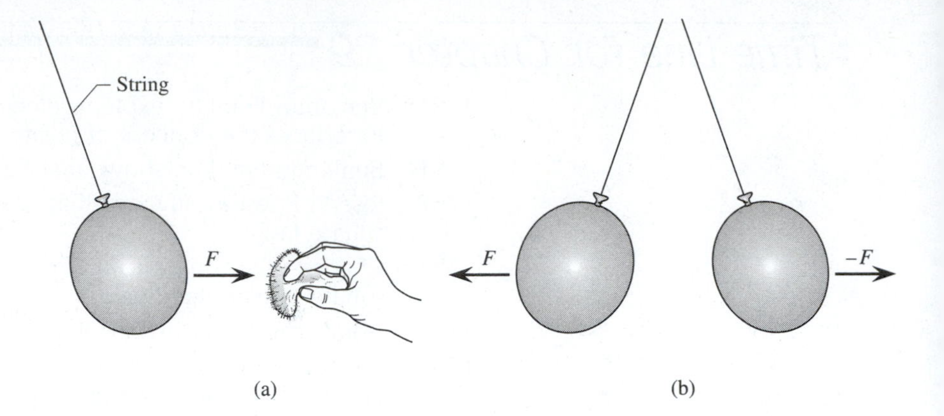

Figure 29–1: (a) *A piece of fur is rubbed against a balloon. When the fur and the balloon are separated, the balloon is attracted to the fur.* (b) *Two balloons that have been rubbed with fur repel each other.*

Early experimenters found that there were two kinds of electrical charges. If a rubber balloon is stroked by fur, it will be electrified. When it is hung by a thread, it will be repelled by another balloon that has also been charged by the fur (see Figure 29–1). If the fur is brought close to either balloon, there will be an attraction between balloon and fur.

A glass rod rubbed with silk will attract a charged rubber balloon, but two similarly charged glass rods will repel each other (see Figure 29–2). The kind of charge a glass rod retains when rubbed with silk is called a positive charge. The kind of charge the rubber retains when stroked with fur is a negative charge.

The two negatively charged balloons repelled each other. The two positively charged glass rods also repelled each other; while a positively charged glass rod and a negatively charged balloon were attracted to each other. In general we can say:

Like charges ([+ *and* +] *or* [− *and* −]) *repel. Unlike charges* (+ *and* −) *attract.*

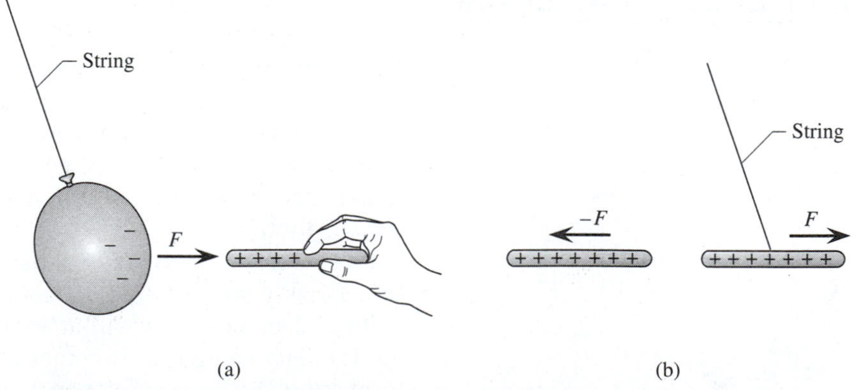

Figure 29–2: (a) *A positively charged glass rod attracts a negatively charged balloon.* (b) *Two positively charged glass rods undergo mutual repulsion.*

We can explain electrification by the idea that when two things are rubbed together "something" is moved from one object to another. One kind of charge is a surplus of this something and the other charge is the lack of this something. Today we are fairly certain that the "something" is electrons. The electron charge is negative. The nucleus of an atom has positively charged particles called protons. Something that is neutral—has neither a positive nor a negative charge—contains equal numbers of positive protons and negative electrons. If we remove some electrons from an object such as a rubber balloon, more positive charges than negative charges will be left on the balloon. We can define electrical charge in terms of a surplus or deficiency of electrons.

A neutral *object has equal numbers of protons and electrons. A* positively charged *object has a deficiency of electrons. A* negatively charged *particle has a surplus of electrons.*

An electroscope is sometimes used to demonstrate the repulsion of electrostatic forces. An **electroscope** consists of a rod with a knob at one end and metal foil leaves at the other end. The assembly is held in a nonconducting cylinder with glass sides for viewing (see Figure 29–3).

If there are no charges on the electroscope, or in the neighborhood of the device, the metal foil leaves are closed (Figure 29–3a). When a negatively charged rubber rod is placed near the knob, electrons flow away from the knob to the foil leaves (Figure 29–3b). The more electrons that flow to the leaves, the more the leaves will repel each other. The angle of separation of the leaves (θ) increases. The overall charge of the electroscope is still neutral at this point. However, the charge distribution is not uniform. If the rod touches the knob, electrons will flow from the rod onto the positively charged knob (Figure 29–3c). If the rod is removed, the electroscope retains a surplus of electrons, a net negative charge (Figure 29–3d).

Figures 29–3e and 29–3f show how to put a positive charge on the electroscope with a negatively charged rod. If you place your finger on the knob as the rod is placed next to it, your body acts as an alternate path for the electrons to escape from the knob. If you remove your finger and then the rod, the electroscope is left with a deficiency of electrons, a net positive charge.

The charge on a single electron is very small and difficult to measure. For most applications, a unit larger than one electronic charge is needed. A larger unit of charge, the coulomb (C), is defined in terms of a collection of electrons.

One coulomb (C) *of electrical charge equals the charge of* 6.24×10^{18} *electrons.*

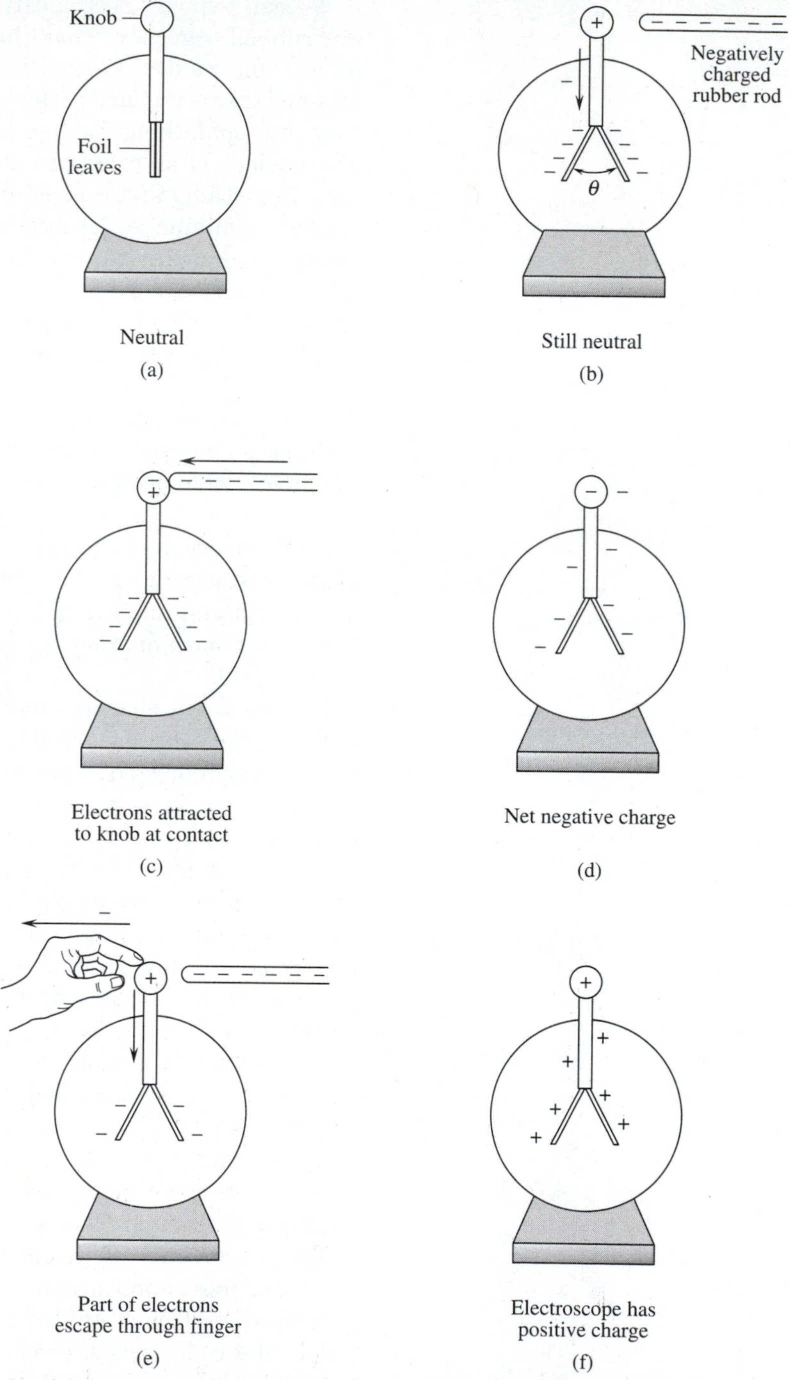

Figure 29–3: *Two ways to charge an electroscope.* (a–d) *A negatively charged rod creates a net negative charge.* (e, f) *A finger provides an alternate path for electrons, creating a net positive charge.*

EMPTY SPACE CAN BE SOLID

Atoms consist mostly of empty space.

Let us look at an up-scale model of an atom. Assume the nucleus of a hydrogen atom is a large apple located in the middle of New York City's theater district. The first orbit of the electron passes south of Baltimore, MD, goes nearly as far west as Pittsburg, PA, slips through Buffalo, NY, crosses the Canadian border at the St. Lawrence Seaway, and swings out over the Atlantic Ocean along the coast north of Boston.

The electron is in orbit through neighboring states, and the nucleus sits in Times Square. There is nothing at all between these two particles according to the Bohr model of the atom. If matter is mostly a void, why does it seem so solid?

Atoms consist of a concentration of positive charge surrounded by clouds of negative electrons. The number of positive charges equals the number of negative charges. The net charge of an atom is neutral.

In a gas, the atoms are widely spaced. The average space between atoms is much larger than the radii of the atoms. The atoms interact as neutral particles.

When atoms are packed close together, the clouds of orbiting electrons are closer together than the positively charged centers. The Coulomb repulsion of the electron clouds is much greater than the attractive force of electrons to the more-distant protons in the nucleus. The atoms cannot penetrate each other.

When your fist strikes a wooden tabletop, what you feel is the electrostatic repulsion of the atoms in your hand by the atoms making up the table. In our world, electrostatic forces make empty space hard.

If one coulomb consists of 6.24×10^{18} electrons, then the charge of one electron is $1.00\ \text{C}/6.24 \times 10^{18}$ electrons.

The electronic charge (e) *is* 1.60×10^{-19} C.

All electrons have the same charge. It is the smallest quantity of electrical charge encountered in electrical devices. Larger charges are made up of multiples of electron charges, just as dollars are multiples of pennies. This is another way of saying that electrical charge is **quantized**, or divided into small but measurable increments.

☐ **EXAMPLE PROBLEM 29–1: A COLLECTION OF ELECTRONS**

A dust particle is found to have a charge of -1.92×10^{-17} C.

A. Does the dust particle have a surplus or a deficiency of electrons?

B. How many electrons are involved with the charge?

■ *SOLUTION*

A. The charge is negative (−). There is a surplus of electrons on the dust particle.

B. We can divide the total charge (q) by the charge of one electron (e) to find the number of electrons.

$$\textit{number of electrons } (N) = \frac{q}{e}$$

$$N = \frac{1.92 \times 10^{-17}\text{ C}}{1.60 \times 10^{-19}\text{ C/electron}}$$

$$N = 120 \text{ electrons}$$

29.2 ELECTROSTATIC FORCES

The electrostatic force between two point charges behaves a lot like gravitational forces. The magnitude of the force obeys an inverse square law. The force also depends on the product of the magnitude of the charges. Like gravitational forces, the two particles exert equal and opposite forces on each other.

The inverse square law of electrostatic forces between small particles is called **Coulomb's law.** The magnitude of the Coulomb force (F_c) between two charged particles, q_1 and q_2, is:

$$F_c = k\frac{q_1 q_2}{R^2} \qquad \textbf{(Eq. 29–1)}$$

The proportionality constant $k = 9.0 \times 10^9 \text{ N} \cdot \text{m}^2/\text{C}^2$.

Be careful. The coulomb force, like any other force, is a vector quantity. Coulomb forces have to be added as vectors. Draw a diagram to decide in what direction the forces act, then label the positive and negative directions.

Do not confuse the + and − charge signs with the + and − signs of direction.

The signs of the charges only tell us whether we have an attractive or a repulsive force. They do not tell us in what direction the force acts.

□ **EXAMPLE PROBLEM 29–2: TWO POINT CHARGES**

A small plastic sphere has a charge of $+1.26 \times 10^{-6}$ C. A particle of fly ash (smoke particle) with a mass of 0.023 g and an electrical

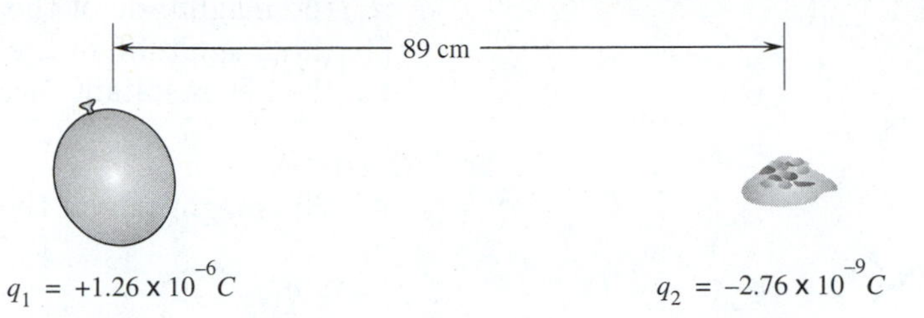

Figure 29–4: Diagram for Example Problem 29–2.

charge of -2.76×10^{-9} C is 89 cm from the plastic sphere (see Figure 29–4).

A. Is the fly ash attracted or repelled by the sphere?
B. What force does the fly ash particle exert on the sphere?
C. What force does the sphere exert on the smoke particle?
D. What is the acceleration of the piece of fly ash?

■ *SOLUTION*

A. The charges have opposite signs. The force is attractive.
B. According to the diagram, the force acting on the sphere is in the positive direction. The magnitude is found by Coulomb's law. Once we have drawn the diagram to find the direction of force we do not need to use the + and − signs of charge to find the size of the force.

$$F_c = +k\frac{q_1 q_2}{R^2}$$

$$F_c = \frac{(+9.0 \times 10^9\ \text{N} \cdot \text{m}^2/\text{C}^2) \times (1.26 \times 10^{-6}\ \text{C}) \times (2.76 \times 10^{-9}\ \text{C})}{(0.89\ \text{m})^2}$$

$$F_c = +4.0 \times 10^{-5}\text{N (force acting on the sphere)}$$

C. The coulomb forces acting on the particle and the sphere make a force couple: equal but opposite forces. The sphere exerts a force on the fly ash in the negative direction.

$$F_c = -4.0 \times 10^{-5}\ \text{N} \quad \text{(force acting on the fly ash)}$$

D.

$$a = \frac{\mathbf{F}}{m} = \frac{-4.0 \times 10^{-5}\ \text{N}}{2.3 \times 10^{-5}\ \text{kg}}$$

$$a = -1.7\ \text{m/s}^2$$

□ **EXAMPLE PROBLEM 29–3: COULOMB FORCES WITH COMPONENTS**

Figure 29–5 shows three electrostatic charges: $q_1 = +1.5 \times 10^{-6}$ C, $q_2 = -2.6 \times 10^{-6}$ C, and $q_3 = 4.0 \times 10^{-7}$ C. Find:

A. the magnitude of the force that q_1 exerts on q_3.
B. the magnitude of the force that q_2 exerts on q_3.
C. the net resultant force acting on q_3.

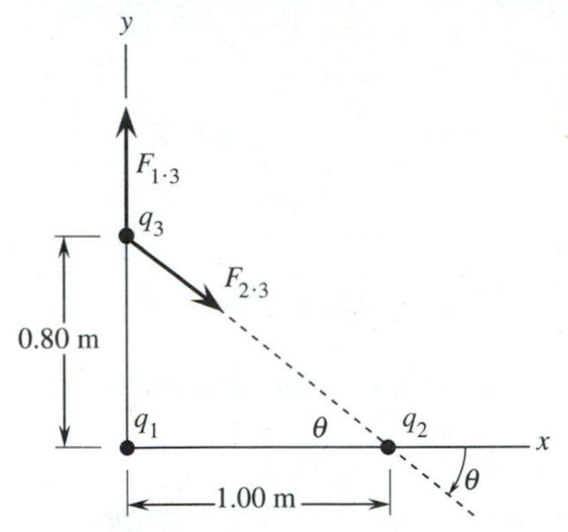

Figure 29–5: Diagram for Example Problem 29–3. Two electrical point charges exert forces on a third charge.

■ *SOLUTION*

A. The magnitude of the force that q_1 exerts on q_3:

$$F_{1\cdot3} = k\frac{q_1q_2}{R^2}$$

$$F_{1\cdot3} = \frac{(9.0 \times 10^9\ \text{N}\cdot\text{m}^2/\text{C}^2) \times (1.5 \times 10^{-6}\ \text{C}) \times (4.0 \times 10^{-7}\ \text{C})}{(0.80\ \text{m})^2}$$

$$F_{1\cdot3} = 8.44 \times 10^{-3}\ \text{N}$$

B. The magnitude of the force that q_2 exerts on q_3:

We can use the Pythagorean theorem to get the distance between the two charges.

$$R^2 = x^2 + y^2$$

$$R^2 = (1.0\ \text{m})^2 + (0.80\ \text{m})^2 = 1.64\ \text{m}^2$$

$$F_{2\cdot3} = \frac{(9.0 \times 10^9\ \text{N}\cdot\text{m}^2/\text{C}^2) \times (2.6 \times 10^{-6}\ \text{C}) \times (4.0 \times 10^{-7}\ \text{C})}{1.64\ \text{m}^2}$$

$$F_{2\cdot3} = 5.7 \times 10^{-3}\ \text{N}$$

C. Now we know the magnitude of the forces; the rest of the problem is simply vector addition. We need to know the x and y components of the force that q_2 exerts on q_3 ($\mathbf{F}_{2\cdot3}$). The angle θ can be found by using the tangent function.

$$\theta = \tan^{-1}(0.80\ \text{m}/1.00\ \text{m}) = 39°$$

The azimuth angle of the direction of the force $\mathbf{F}_{2\cdot3}$ is 360° − 38.7° = 321°.

$$F_{2\cdot3}(x) = \mathbf{F}_{2\cdot3}\cos 321° = 5.71 \times 10^{-3}\ \text{N}\ (0.7809)$$

$$F_{2\cdot3}(x) = 4.5 \times 10^{-3}\ \text{N}$$

$$F_{2\cdot3}(y) = F_{2\cdot3}\sin 321° = 5.71 \times 10^{-3}\ \text{N}\ (-0.6247)$$

$$F_{2\cdot3}(y) = -3.6 \times 10^{-3}\ \text{N}$$

The force q_1 exerts on q_3 is upward along the y axis.

$$F_{1\cdot3}(x) = 0.000$$

$$F_{1\cdot3}(y) = +8.4 \times 10^{-3}\ \text{N}$$

We can now find the resultant vector (R).

$$R(x) = \mathbf{F}_{1\text{-}3}(x) + \mathbf{F}_{2\text{-}3}(x) = 0.00\ \text{N} + (+4.5 \times 10^{-3}\ \text{N})$$

$$R(x) = +4.5 \times 10^{-3}\ \text{N}$$

$$R(y) = \mathbf{F}_{1\text{-}3}(y) + \mathbf{F}_{2\text{-}3}(y) = (+8.44 \times 10^{-3}\ \text{N}) + (-3.57 \times 10^{-3}\ \text{N})$$

$$R(y) = +4.9 \times 10^{-3}\ \text{N}$$

$$R = [R(x)^2 + R(y)^2]^{1/2} = [(4.5 \times 10^{-3}\ \text{N})^2 + (4.9 \times 10^{-3}\ \text{N})^2]^{1/2}$$

$$R = 6.6 \times 10^{-3}\ \text{N}$$

$$\tan \theta_R = \frac{R(y)}{R(x)} = \frac{4.9 \times 10^{-3}\ \text{N}}{4.5 \times 10^{-3}\ \text{N}}$$

$$\theta_R = 48°$$

Finally! To two significant figures the resultant is:

$$R = 6.6 \times 10^{-3}\ \text{N}\ \underline{\angle 48°}$$

29.3 ELECTRIC FIELDS

Charged metal plates in a cathode ray tube (CRT) use coulomb forces to deflect a beam of electrons on their trip to the phosphorescent screen. The force the electrons exert on the plates has little effect.

A high-voltage transmission line exerts coulomb forces on charged particles surrounding it. The coulomb forces the particles exert on the line are, of course, equal and opposite in direction, but they have little effect on the behavior of the much more massive lines.

Electrostatic plates in a smokestack remove charged particles from a furnace exhaust to reduce pollutants from the atmosphere.

There are several situations in which a coulomb force is exerted by a charge Q on a smaller charge q. We are interested in the force on the small charge, but not the one on the large charge.

One way of handling this kind of problem is to introduce the idea of an electric field. Coulomb forces depend on position. We map the forces acting on a small test charge in various positions surrounding the charge Q. We will always use a positive charge to do the mapping. Figure 29–6 shows a force map on a small positive test charge for isolated positive and negative point charges.

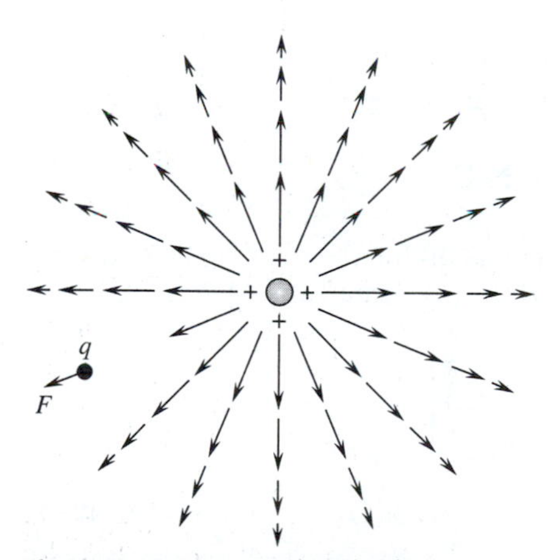

Figure 29–6: We can map the forces exerted on a positive test charge (q). The forces are plotted for several positions around a positive point charge. The map is a vector field.

If we use a smaller test charge, the vectors will be proportionally smaller. If we use a larger test charge, the vectors will be larger. In either case, they will point in the same direction as shown in the map. The map is called a **vector field.**

We can standardize the map by plotting, not the force, but the force per unit charge (**F**/q). This is called the **electric field (E).** To find the coulomb force that would be exerted on a charge at some position in the map, simply multiply the electric field by the charge.

$$\mathbf{F}(\text{coulomb}) = q\,E \qquad \textbf{(Eq. 29–2)}$$

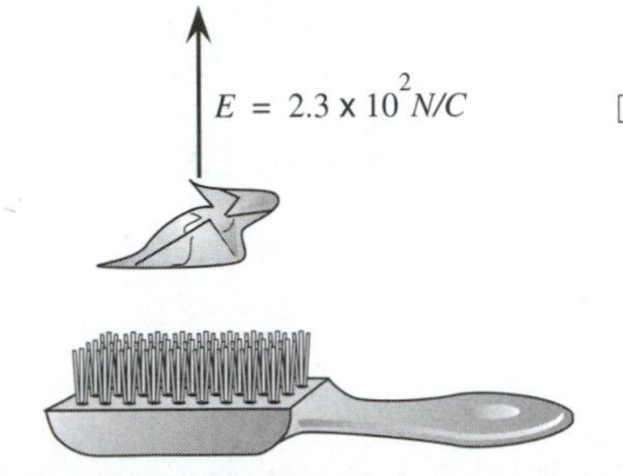

Figure 29–7: *Find the force on a charged object near a hairbrush.*

EXAMPLE PROBLEM 29–4: ELECTRIC FIELD

At a particular point in a field map around an electrically charged nylon hairbrush the electric field has a magnitude of 2.3×10^2 N/C (see Figure 29–7).

A. What is the size of the coulomb force on a small piece of paper at this location with a charge of 1.2×10^{-6} C?

B. At another location, the 1.2×10^{-6} C charge experiences a coulomb force of 4.2×10^{-4} N. What is the magnitude of the electric field at this point?

SOLUTION

A.

$$\mathbf{F} = q\,E$$

$$\mathbf{F} = (1.2 \times 10^{-6}\ \text{C}) \times (2.3 \times 10^{2}\ \text{N/C})$$

$$\mathbf{F} = 2.8 \times 10^{-4}\ \text{N}$$

B.

$$E = \frac{\mathbf{F}}{q}$$

$$E = \frac{4.2 \times 10^{-4}\ \text{N}}{1.2 \times 10^{-6}\ \text{C}}$$

$$E = 3.5 \times 10^{2}\ \text{N/C}$$

We can get a mathematical expression for the electric field around a small charged particle (Q) by dividing Equation 29–1 by q.

$$E = \frac{\mathbf{F}}{q} = k\left[\frac{(Q\,q)/R^2}{q}\right]$$

$$E = \frac{k\,Q}{R^2} \qquad \textbf{(Eq. 29–3)}$$

EXAMPLE PROBLEM 29–5: ELECTRIC FIELD OF A POINT CHARGE

Find the strength of the electric field 1.2 m from a point charge of 5.7×10^{-4} C.

■ *SOLUTION*

$$E = k\frac{Q}{R^2}$$

$$E = (9.0 \times 10^9\ \text{N} \cdot \text{m}^2/\text{C}^2) \times \frac{5.7 \times 10^{-4}\ \text{C}}{(1.2\ \text{m})^2}$$

$$E = 3.6 \times 10^6\ \text{N/C}$$

29.4 ELECTRIC FIELD LINES

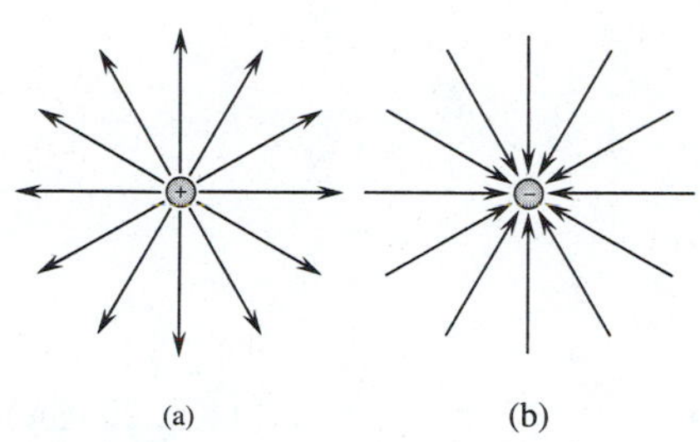

Figure 29–8: *Electric field lines act radially on* (a) *an isolated positive point charge and* (b) *an isolated negative point charge.*

If we space the electric force vectors close together in a map like that shown in Figure 29–6, they will overlap. This can make a rather messy diagram. **Electric field lines** are usually used instead. Figure 29–8 shows the electric field lines drawn for two point charges. The arrows point away from a positive charge, which would repel a positive test charge. The arrows point inward for a negative charge because the negative charge would attract a positive test charge.

Since electric field is force per unit charge, it is a vector in the same direction as the coulomb force. If there is more than one charge, the net field at a point can be found by vector addition of the electric fields arising from the various charges.

$$E(\text{net}) = E_1 + E_2 + E_3 \ldots E_n$$

Figure 29–9 shows the field lines around two sets of charges: a pair of positive charges of the same size, and two charges of the same size but different sign. The force acting on a charge placed on a field line will be tangent to the field line. The force will be in the same direction as the line if it is positive, in the opposite direction if it is negative.

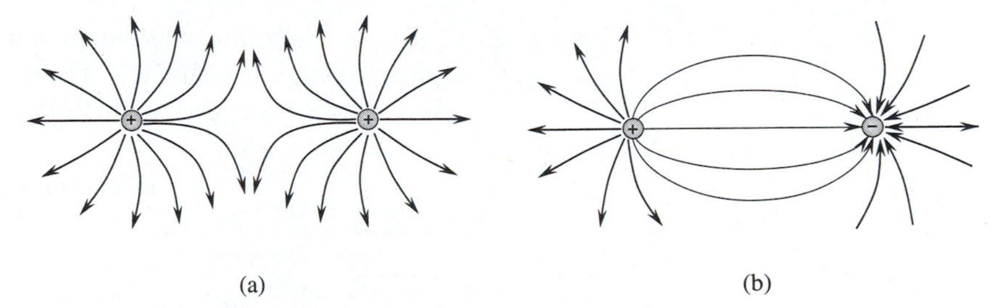

Figure 29–9: *Electric field lines near pairs of point charges.* (a) *Two like charges of the same magnitude.* (b) *Two unlike charges of the same size.*

29.5 ELECTRICAL FLUX

In many ways, an electric field behaves like photons emitted from a light bulb. Light illumination from a point source obeys an inverse

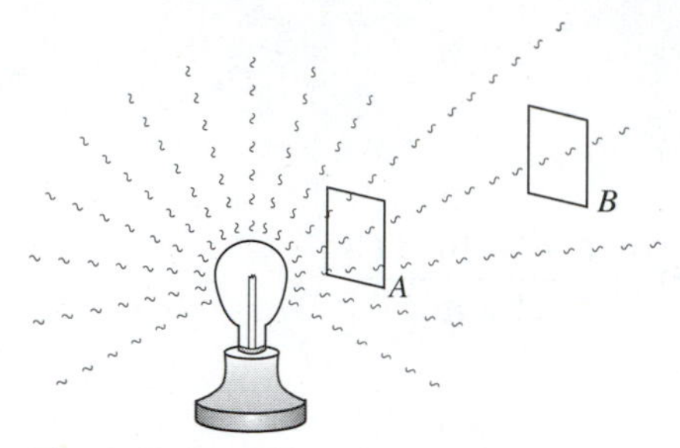

Figure 29–10: More photons pass through a surface near a light bulb (A) than through a surface farther away (B).

law like the electric field from a point source. If we hold a surface A close to a light bulb, more photons strike it than if we hold it farther away (see Figure 29–10). The number of photons striking the surface is called **luminous flux.** The illumination is found by dividing the luminous flux by the area it strikes.

Fewer electric field lines strike a surface far away from a point source than one held closer to the charge. Let us extend this analogy. We can call the number of lines passing through an area **electric flux** (ϕ_e). The electric field would act like illumination. If we can count the number of lines passing through the surface (electrical flux), the electric field can be calculated by dividing the electric flux by the area. We have to be careful to use the perpendicular area (see Figure 29–11).

$$E = \frac{\phi_e}{A}$$

or

$$\phi_e = E\,A \qquad \textbf{(Eq. 29–4A)}$$

It would be reasonable to assume that a charge is proportional to the number of flux lines it produces. We will introduce a proportionality constant (ϵ_0). We can mathematically determine the numerical value of ϵ_0.

$$q = \epsilon_0 \phi_e$$

or

$$\phi_e = \frac{Q}{\epsilon_0} \qquad \textbf{(Eq. 29–4B)}$$

If we make a *closed* surface around a charge, all of the flux will pass through the surface. There cannot be any gaps, or some of the flux might escape without hitting the surface.

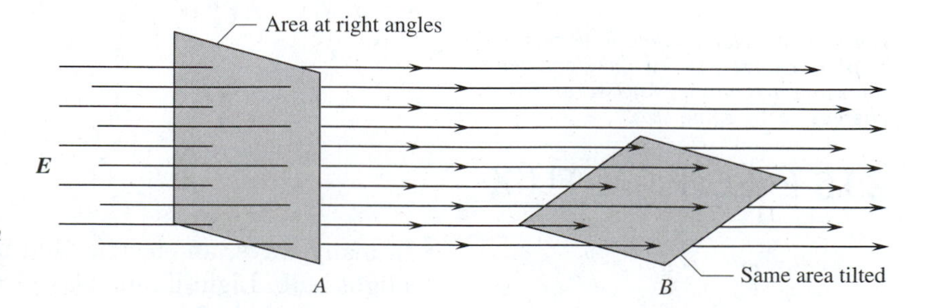

Figure 29–11: The angle between an area and the electric field is important when we calculate electric flux.

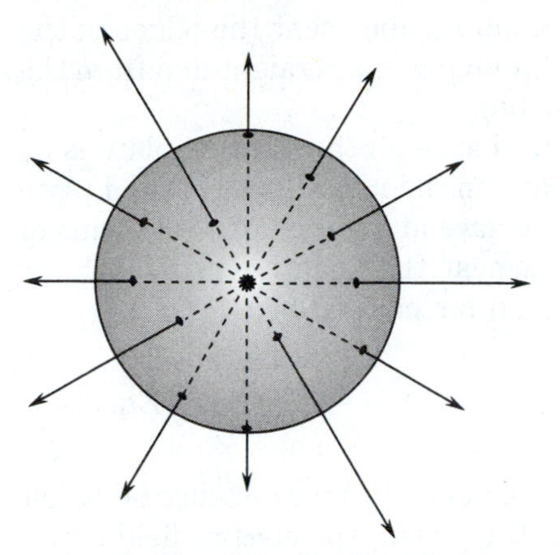

Figure 29–12: All the flux lines from a small source pass perpendicular to a sphere that has the charge at its center.

Look at Figure 29–12. If we have spherical surface with a point charge (Q) in the center, the flux will pass through the surface at right angles, and all of the flux lines will hit the surface. Both Equations 29–4A and 29–4B are satisfied.

$$\phi_e = E\,A = \frac{Q}{\epsilon_0}$$

The area of the sphere is $A = 4\pi\,R^2$. We can substitute this into the equation and solve for the electric field.

$$E = \frac{1}{4\pi\epsilon_0}\,\frac{Q}{R^2}$$

Compare this with Equation 29–3. The only difference is the way the proportionality constant is represented. By equating the two forms of the constant we can find the numerical value of ϵ_0.

$$k = \frac{1}{4\pi\epsilon_0} = 9.0 \times 10^9\ \mathrm{N \cdot m^2/C^2}$$

or

$$\epsilon_0 = 8.85 \times 10^{-12}\ \mathrm{C^2/(N \cdot m^2)}$$

ϵ_0 is called the **permittivity of (free) space.** By free space, we mean a vacuum.

The idea that the total flux from some source passes through a closed surface surrounding it is known as **Gauss's law.** It is handy for solving some problems with gravitational fields and illumination as well as examining the nature of electrostatic fields.

Look at Figure 29–13. Two parallel plates have the same area. They also have equal and opposite charges (q). You might expect this if the plates were connected to the plus and minus plates of a battery. The electric field lines begin at positive charges and end at

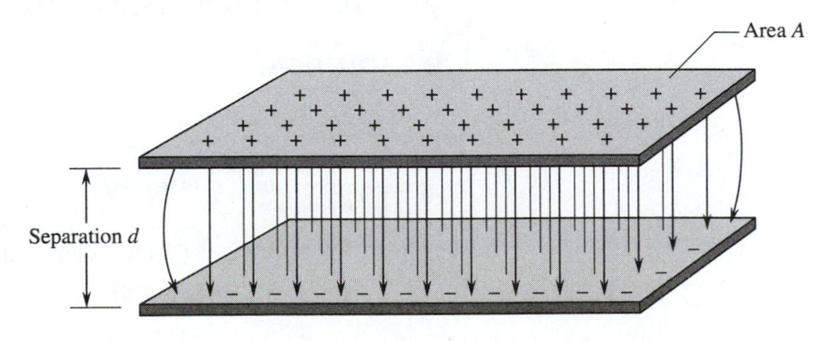

Figure 29–13: The electric field between parallel plates is uniform. The field lines are all parallel except for a negligible number near the plate edge.

negative charges. Except for a small number near the edges of the plates, the field lines are parallel, moving in a straight line from the positive plate to the negative plate.

The total flux generated by the charge on the positive plate is $\phi_e = q/\epsilon_0$. All the field lines terminate on the negatively charged plate of area A where they strike the surface at right angles. In terms of the electric field, the electric flux near the negative plate is $\phi_e = E\,A$. This gives us the electric field for parallel plates.

$$E = \frac{q}{\epsilon_0 A} \qquad \textbf{(Eq. 29–5)}$$

Notice that the electric field does not depend on the distance between the plates. Everywhere between the plates, the electric field is the same. It is constant. This is a **uniform field.**

☐ **EXAMPLE PROBLEM 29–6: ELECTRIC FLUX**

A charge of 3.8×10^{-7} C sits inside the cube shown in Figure 29–14. What total electric flux passes through the cube?

■ ***SOLUTION***

All of the electrical flux generated by the charge will pass through the cube because it is a closed surface.

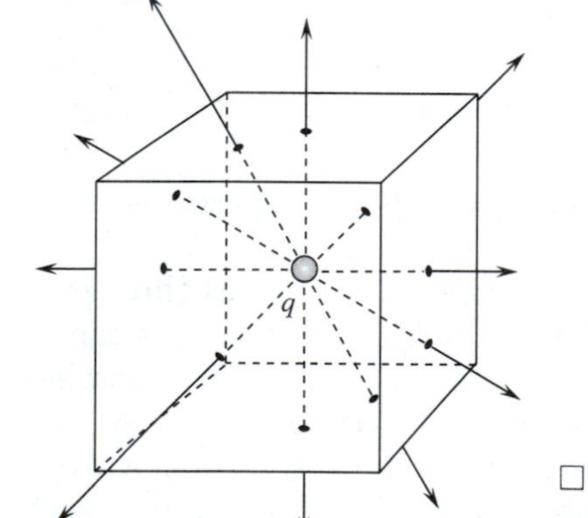

Figure 29–14: *Diagram for Example Problem 29–6.*

$$\phi_e = \frac{Q}{\epsilon_0}$$

$$\phi_e = \frac{3.8 \times 10^{-7}\ \text{C}}{8.85 \times 10^{-12}\ \text{C}^2/(\text{N} \cdot \text{m}^2)}$$

$$\phi_e = 4.3 \times 10^4\ \text{N} \cdot \text{m}^2/\text{C}$$

☐ **EXAMPLE PROBLEM 29–7: PARALLEL PLATES**

An electric field of 2.8×10^6 N/C exists between two parallel plates with an area of 0.023 m^2.

A. What is the magnitude of the charge on one of the plates?

B. What is the total electric flux?

■ ***SOLUTION:***

A.

$$E = \frac{q}{\epsilon_0 A} \quad \text{or} \quad q = \epsilon_0 A\,E$$

$$q = 8.85 \times 10^{-12}\ \text{C}^2/(\text{N} \cdot \text{m}^2) \times 0.023\ \text{m}^2 \times (2.8 \times 10^6\ \text{N/C})$$

$$q = 5.7 \times 10^{-7}\ \text{C}$$

B.

$$\phi_e = E\,A$$

$$\phi_e = 2.8 \times 10^6\ \text{N/C} \times 0.023\ \text{m}^2$$

$$\phi_e = 6.4 \times 10^4\ \text{N} \cdot \text{m}^2/\text{C}$$

29.6 ELECTRICAL POTENTIAL

Work must be done to bring two electrically charged items close together.

work = **force × parallel displacement**

$$\mathbf{W} = \mathbf{F} \cdot \mathbf{D}$$

If we bring a positive charge (q) close to another positive charge, we must do work against repulsive forces. The work goes into increasing potential energy. If we released the charges, they would move away from each other, gaining kinetic energy. If we move a positive charge (q) closer to a negative charge, the charge pulls us along, doing work on us. The charge is using potential energy to do work. Its potential energy is decreased. Let us express the electrical force in terms of field.

$$PE = q\,\mathbf{E} \cdot \Delta\mathbf{R}$$

Instead of potential energy we can consider potential energy per unit charge. This is called **electrical potential.** It has units of **volts (V).**

$$V = \frac{PE}{q} = \mathbf{E} \cdot \Delta\mathbf{R}$$

$$\mathbf{1.00\ V = 1.00\ J/C} \qquad \textbf{(Eq. 29–6)}$$

Figure 29–15 shows a pair of parallel plates separated by a distance (d). A positive charge is moved from the negative plate to the positive plate along two different paths. One path is parallel to a field line. The potential difference for parallel plates is:

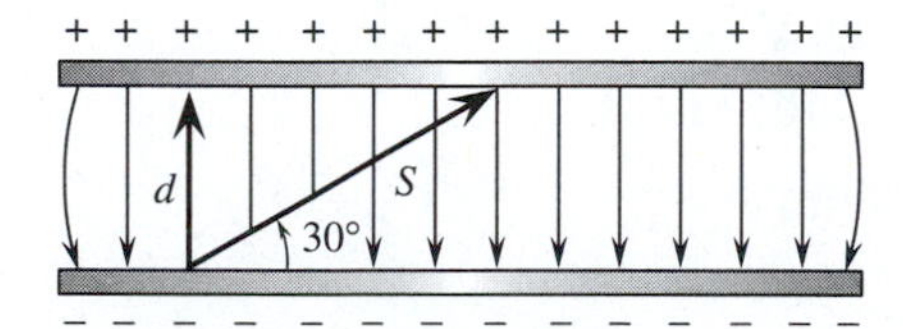

Figure 29–15: The same amount of work is needed to move a charge along paths S and d.

$$\Delta V = \mathbf{E} \cdot \mathbf{d} = \frac{q\,d}{\epsilon_0\,A} \qquad \textbf{(Eq. 29–7)}$$

The other path is along line S, making an angle of 30° with the electric field lines. Only the component parallel to the electrical force on the

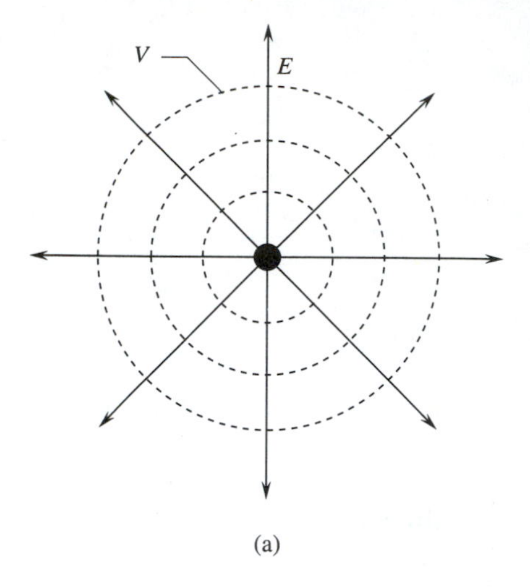

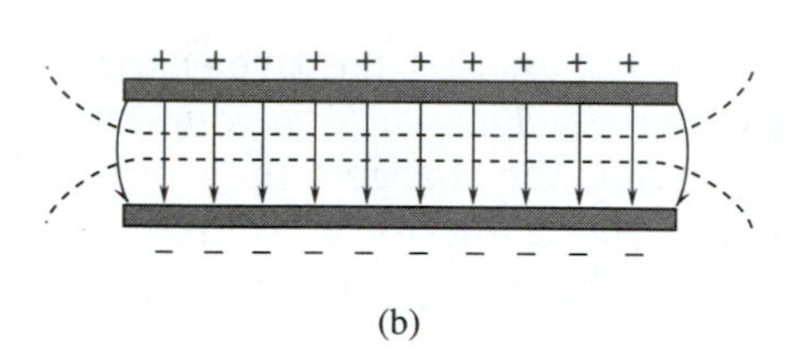

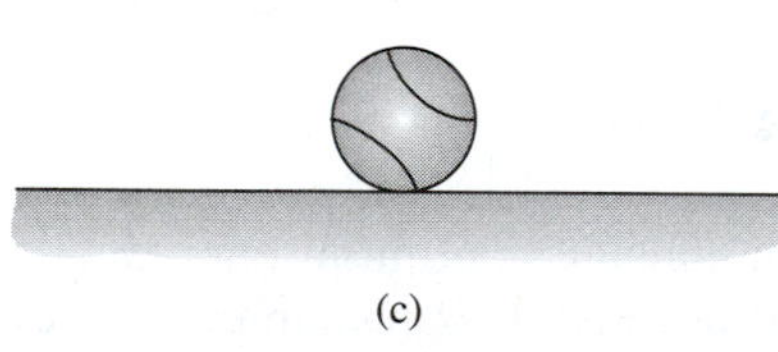

Figure 29–16: Equipotential surfaces. (a) *Electrical equipotential surfaces form concentric spheres around a point charge.* (b) *The electrical equipotential surfaces between parallel plates.* (c) *A level gym floor is a gravitational equipotential surface.*

moving charge is involved with work. That is the part parallel to the field. The parallel component is $S \cos 30° = d$. The same amount of work per unit charge is done along the two paths.

Lines drawn perpendicular to the electric field lines are **equipotential lines** (see Figure 29–16). A charged place anywhere on an equipotential line will have the same electrical potential or the same voltage as it would have anywhere else on the line. It is like a floor. A level floor is a gravitational equipotential surface. A basketball placed at rest on the floor will not start to move along the floor; an electrical charge placed at rest on an electrical equipotential line will not move along it.

EXAMPLE PROBLEM 29–8: PARALLEL-PLATE POTENTIAL

The potential difference across a pair of parallel plates separated by 0.130 mm is 6.20 V (see Figure 29–17).

A. What is the electric field between the plates?

B. An electron ($e = 1.6 \times 10^{-19}$ C) is initially at rest near the negative plate. What is its electrical potential energy (PE)?

C. What is the electrical potential (V) of the electron halfway between the two plates?

D. What is the kinetic energy of the electron halfway between the plates?

E. As the electron reaches the positive plate, what are its electrical potential (V), potential energy (PE), and kinetic energy (KE)?

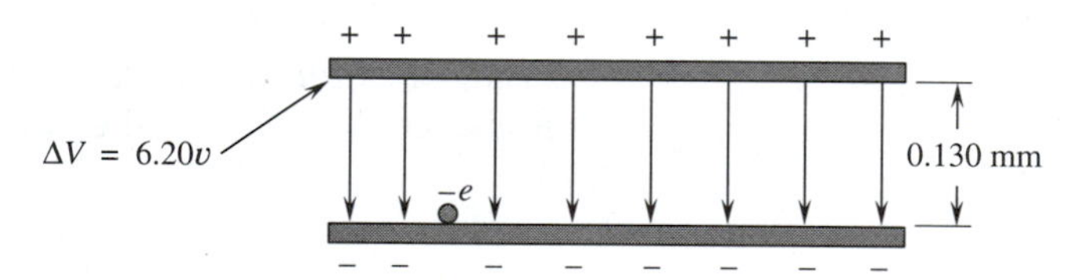

Figure 29–17: Diagram for Example Problem 29–8.

■ *SOLUTION*

A.

$$E = \frac{\Delta V}{d} = \frac{6.20 \text{ V}}{1.30 \times 10^{-4} \text{ m}}$$

$$E = 4.77 \times 10^{4} \text{ V/m}$$

B.

$$PE = q\,\Delta V = 1.6 \times 10^{-19} \text{ C} \times 6.20 \text{ V}$$

$$PE = 9.9 \times 10^{-19} \text{ J}$$

C.

$$V = E\,d = (4.77 \times 10^{-4}\ \text{V/m}) \times (6.5 \times 10^{-5}\ \text{m})$$

$$V = 3.1\ \text{V}$$

D. The gain in kinetic energy equals the loss of potential energy.

$$KE = q\,(V_1 - V_2) = (1.6 \times 10^{-19}\ \text{C})(6.2\ \text{V} - 3.1\ \text{V})$$

$$KE = 5.0 \times 10^{-19}\ \text{J}$$

E. The potential energy has been completely converted to kinetic energy.

$$V = 0$$

$$PE = 0$$

$$KE = 9.9 \times 10^{-19}\ \text{J}$$

Calculating voltage is more involved when we do not have a uniform field. Since Coulomb's law is an inverse square law, we can find the electrical potential in the same way that we found the gravitational potential energy in Section 12.6. This will give us:

$$V = \frac{1}{4\,\pi\epsilon_e}\left[Q\left(\frac{1}{R_2} - \frac{1}{R_1}\right)\right] \qquad \textbf{(Eq. 29–8)}$$

Sometimes **absolute potential** is used rather than potential difference. For a point charge, we can calculate the work needed to bring a charge in from infinity to some point R. Since $1/\infty = 0$, the term in parentheses in Equation 29–8 will always be zero.

$$V(\text{abs}) = \frac{1}{4\pi\epsilon_0}\,\frac{Q}{R} \qquad \textbf{(Eq. 29–9)}$$

If the electrical potential is expressed in absolute potential, then the potential difference between two locations is simply the difference between the two absolute potentials.

$$V = V_2(\text{abs}) - V_1(\text{abs})$$

There is another reason for using absolute potentials. The absolute potentials are additive. If we want the electrical potential at some location, we can add the absolute potentials of all the charged particles that affect our location. It is much easier than adding electric

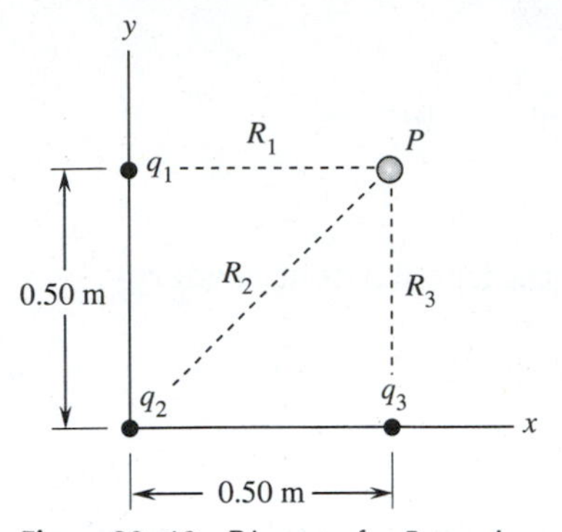

Figure 29–18: *Diagram for Example Problem 29–9.*

fields. Electric fields are vectors, but electrical potential is a scalar quantity. We do not have to keep track of direction, but we do need to know if an electrical potential is positive or negative.

The electrical potential around a positive charge will be positive. The electrical potential around a negative charge will be negative.

$$V(\text{net}) = V_1 + V_2 + V_3 \ldots V_n$$

☐ **EXAMPLE PROBLEM 29–9: ELECTRICAL POTENTIAL**

Find the electrical potential at point P in Figure 29–18.

$$q_1 = 8.3 \times 10^{-7}\ \text{C};\ q_2 = 1.2 \times 10^{-6}\ \text{C};\ q_3 = -2.0 \times 10^{-6}\ \text{C}.$$

■ *SOLUTION*

From the diagram: $R_2 = [(0.50\ \text{m})^2 + [(0.50\ \text{m})^2]^{1/2} = 0.71\ \text{m}$

$$V(\text{net}) = \text{V}_1 + \text{V}_2 + \text{V}_3$$

$$V(\text{net}) = \frac{1}{4\pi\epsilon_0}\left(\frac{q_1}{R_1} + \frac{q_2}{R_2} + \frac{q_3}{R_3}\right)$$

$$V(\text{net}) = 9 \times 10^9\ \text{N} \cdot \text{m}^2/\text{C}^2 \left(\frac{8.3 \times 10^{-7}\ \text{C}}{0.50\ \text{m}} + \frac{1.2 \times 10^{-6}\ \text{C}}{0.71\ \text{m}} - \frac{2.0 \times 10^{-6}\ \text{C}}{0.50\ \text{m}}\right)$$

$$V(\text{net}) = 5.8 \times 10^3\ \text{V}$$

29.7 CAPACITANCE

All sorts of things can collect an electrostatic charge. Metal plates, rain droplets, the human body, and coaxial cables are a few of them. The shape and size of an object determine how much work is needed to build up a charge of a particular size.

Look at Figure 29–19. An electron moves toward a small charged metal plate. Work must be done against coulomb forces exerted by the electrons already on the plate. The same amount of charge on a larger plate is more widely spaced. The coulomb forces on an approaching electron are smaller. Less work is needed to add the additional electron to the collection. This means that the larger plate has an electrical potential that is smaller for the same charge. It has a greater capacity to collect electrons. **Capacitance** (C) measures the ratio of charge to electrical potential of an object.

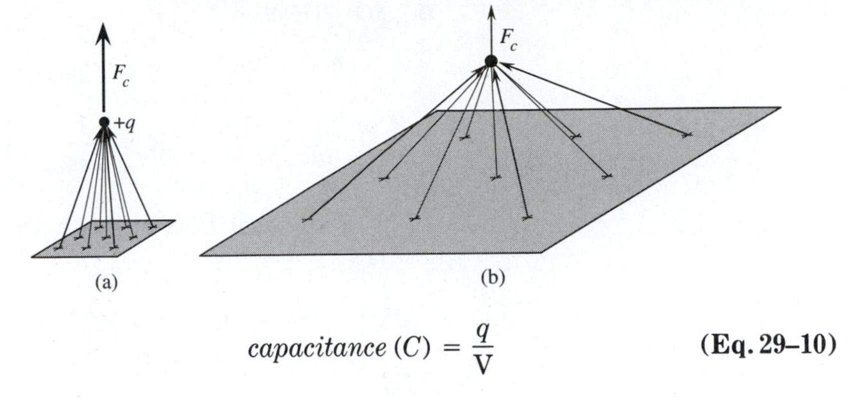

Figure 29–19: Plates a and b have the same total charge. More work is needed to overcome coulomb forces when charges are concentrated in a small area than when the same charge is spread over a large area.

$$\textit{capacitance}\ (C) = \frac{q}{V} \qquad \textbf{(Eq. 29–10)}$$

The unit of capacitance is the **farad** (F).

$$1\ \text{F} = 1\ \text{C/V}$$

The farad is a rather large unit. Most electrical devices have capacitances in microfarads (μF), nanofarads (nF), and picofarads (pF).

If we express electrical potential in terms of dimensions and charge, it is simple to find capacitance. The potential around a metal sphere is the same as that for a point charge.

$$V(\text{sphere}) = \frac{1}{4\pi\epsilon_0}\left(\frac{q}{R}\right)$$

We can solve the equation for the ratio q/V to get the capacitance of a sphere.

$$C = \frac{q}{V} = 4\pi\epsilon_0 R$$

The electrical potential of a pair of parallel plates is:

$$V(\text{plates}) = \frac{q\,d}{\epsilon_0 A}$$

The capacitance of two parallel plates is then:

$$C(\text{plates}) = \frac{q}{V} = \frac{\epsilon_0 A}{d}$$

☐ **EXAMPLE PROBLEM 29–10: A 1-F CAPACITOR**

What is the radius of a sphere with an electrical capacitance of 1 F?

■ *SOLUTION*

$$C = 4\pi\epsilon_0 R$$

$$R = \frac{1}{4\pi\epsilon_0} \times 1.00\ \text{F}$$

Notice that $1/(4\pi\epsilon_0)$ is the proportionality constant for Coulomb's law.

$$R = 9 \times 10^9\ \text{N} \cdot \text{m}^2/\text{C}^2 \times 1\ \text{C/V}$$

$$1\ \text{V} = 1\ \text{N} \cdot \text{m/C}$$

$$R = 9 \times 10^9 \frac{\text{N} \cdot \text{m}^2}{\text{C}^2} \times \frac{1\ \text{C}}{\text{N} \cdot \text{m/C}}$$

$$R = 9 \times 10^9\ \text{m} = 9 \times 10^6\ \text{km}$$

Compare this with the earth's radius, 1.28×10^4 km. It would be difficult to fit a spherical 1-F capacitor in your back yard.

29.8 APPLICATIONS

Coaxial Cables

Coaxial cable is often used to hook up stereophonic equipment, cable television, and other electrical devices where stray radiation could affect performance. Figure 29–20 shows the structure of a coaxial

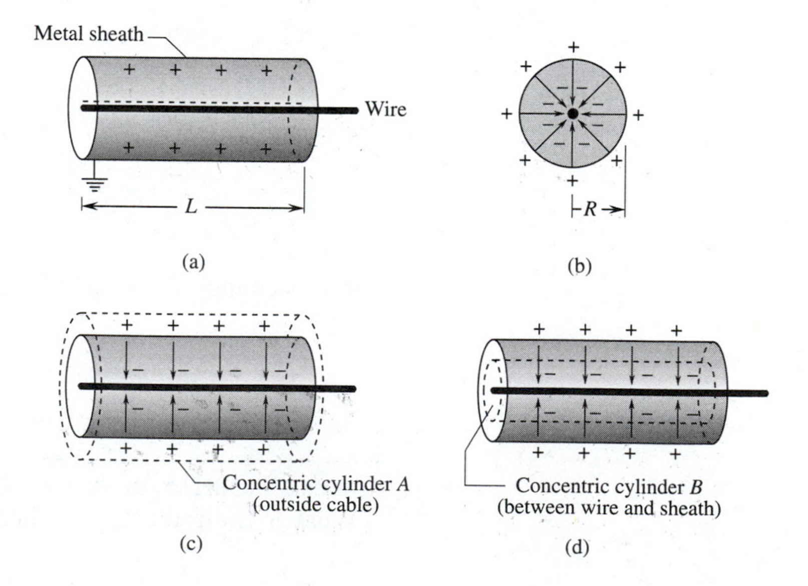

Figure 29–20: *(a) (Side view) A coaxial cable of length L. (b) (End view) The electric field acts along the radii. (c) The net charge enclosed in a concentric cylinder drawn outside the cable is zero. All field lines terminate inside the cylinder. (d) All field lines pass through a cylinder drawn inside the sheath.*

cable. A central wire carries a signal. An outside sheath forms a concentric cylinder around the central wire. It is usually grounded so that the charge on the sheath matches the charge in the central wire (a). Electric field lines act radially beginning at the positively charged sheath and ending at the central wire (b).

If we draw a concentric cylinder (d) between the conducting wire and the sheath, the flux will pass through it at right angles. We can use Gauss's law. For a wire with length L, the area of the cylinder is $A = 2\pi R L$. If $-q$ is the charge on the wire, then

$$E\,A(\text{cylinder}) = \frac{-q}{\epsilon_0}$$

$$E = \frac{-q}{2\pi R\,L\,\epsilon_0}$$

If we draw a cylinder outside the cable, no electric flux passes through the surface (C). The net charge inside this cylinder is zero. The negative charge of the wire cancels out the positive charge of the sheath. The electric field is totally contained inside the cable. It cannot cause interference with surrounding equipment.

$$E = 0$$

The Cathode Ray Tube

Cathode ray tubes (CRT) are used for displays on television sets, oscilloscopes, and computer monitors. A heated filament sends a narrow beam of electrons toward the positively charged screen (see Figure 29–21). On the way to the screen the beam of electrons passes through two sets of charged parallel plates. One set of plates deflects the beam up and down. The other set deflects the beam left and right. By varying the voltage on the plates, the beam can be made to strike any point on the surface of the screen, where it causes a phosphorescent glow.

Figure 29–22 shows two plates that might be used to cause a vertical deflection. As an electron travels the length (L) of the plates, a uniform field (E) will cause a uniform acceleration. Once the electron passes through the plates, there will be no further change of velocity.

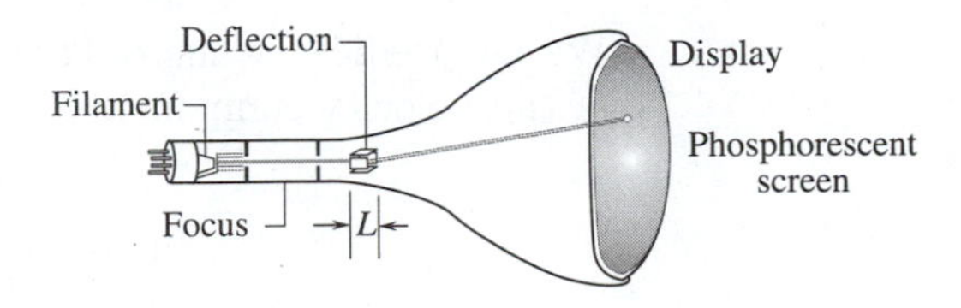

Figure 29–21: *A cathode ray tube. Electrons are emitted by the filament. The beam is focused before it passes through two sets of parallel plates. The plates deflect the beam. One set controls vertical deflection; the other, horizontal deflection. When the electrons strike the coated screen, a fluorescent glow occurs.*

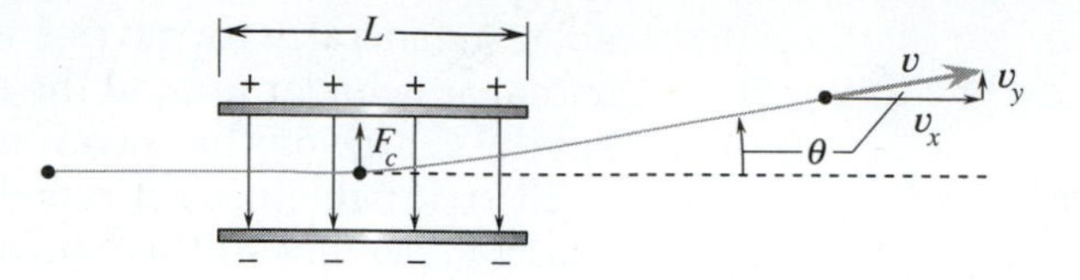

Figure 29–22: Two charged plates deflect a beam of electrons by coulomb force (F_c).

We can use Newton's second law of motion to find the acceleration.

$$F = m\, a_y = e\, E$$

or

$$a_y = \frac{e\, E}{m}$$

If we know the separation of the plates (d), we can write the field in terms of the electrical potential between the plates.

$$E = \frac{V}{d}$$

or

$$a_y = \frac{e\, V}{m\, d}$$

If v_x is the horizontal speed of the electron, we can find the amount of time spent between the plates.

$$\Delta t = \frac{L}{v_x}$$

The vertical component of the electron's velocity as it leaves the parallel plates is:

$$v_y = a_y\, \Delta t$$

$$v_y = \frac{e\, V\, L}{m\, d\, v_x}$$

We can predict the angle of travel of the electron by using the tangent of the velocity components.

$$\tan \theta = \frac{v_y}{v_x}$$

$$\tan\theta = \frac{e\,V\,L}{m\,d\,v_x^{\,2}}$$

EXAMPLE PROBLEM 29–11: THE CRT BEAM

The vertical-deflection plates in a CRT are 12 mm long and have a separation of 8.0 mm (see Figure 29–23). What voltage across the plates will produce an electron beam deflection of 10° if the electrons have an initial horizontal velocity of 6.0×10^6 m/s? ($m_e = 9.11 \times 10^{-31}$ kg, and $e = 1.60 \times 10^{-19}$ C.)

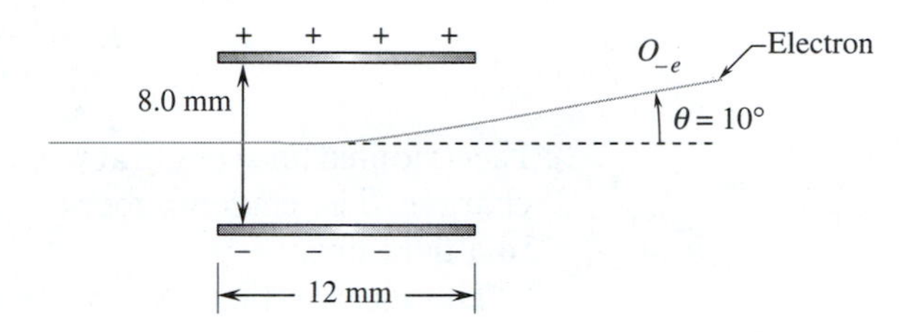

Figure 29–23: Diagram for Example Problem 29–12.

SOLUTION

$$V = \frac{m\,d\,v_x^{\,2}}{e\,L}\tan\theta$$

$$V = \frac{(9.11 \times 10^{-31}\ \text{kg}) \times (8.0 \times 10^{-3}\ \text{m}) \times (6.0 \times 10^{6}\ \text{m/s})^2\,(\tan 10°)}{(1.60 \times 10^{-19}\ \text{C}) \times (1.2 \times 10^{-2}\ \text{m})}$$

$$V = 24\ \text{V}$$

SUMMARY

Electrostatics is the study of electrical charges at rest. A neutrally charged object has equal numbers of electrons and protons. A positively charged object has a deficiency of electrons. A negatively charged object has a surplus of electrons.

The measure of electrical charge is the coulomb (C). One coulomb is the charge of a collection of 6.24×10^{24} electrons. The charge of an electron (e) is 1.6×10^{-19} C. Unlike charges attract; like charges repel.

The magnitude of the Coulomb force ($\mathbf{F}_c$) acting between two charges (q_1 and q_2) separated by a distance R, known as Coulomb's law, is:

$$\mathrm{F}_c = k\,\frac{q_1 q_2}{R^2}$$

where $k = 9.0 \times 10^9\ \text{N} \cdot \text{m}^2/\text{C}^2$.

The electric field (E) is the coulomb force per unit electrical charge.

$$E = \frac{F_c}{q}$$

For a point charge Q, the magnitude of the electric field surrounding it is:

$$E = k\left(\frac{Q}{R^2}\right)$$

The electric field between charged parallel plates is:

$$E = \frac{q}{\epsilon_0 A}$$

Electric field lines begin at positive charges and terminate at negative charges. The coulomb force acting on a charged particle is tangent to a field line.

The electric flux (ϕ_e) is the number of lines passing through an area at right angles (A_{perp}).

$$\phi_e = E\,A_{perp}$$

The total number of electric field lines (total electric flux) associated with a charge (q),

$$\phi_e = \frac{q}{\epsilon_0}$$

where $\epsilon_0 = 8.85 \times 10^{-12}\ \mathrm{C^2/N \cdot m^2}$, is called the permittivity of free space.

If a closed surface is drawn around a net charge, all the electric field lines associated with the net charge will pass through the surface. This is known as Gauss's law.

Electrical potential difference (V) is work done against a coulomb force per unit charge.

$$\Delta V = \frac{\mathbf{F}}{q} \cdot \mathbf{\Delta R} = \mathbf{E} \cdot \mathbf{\Delta R}$$

Electrical potential is measured in units of volts (V):

$$1\ \mathrm{V} = 1\ \mathrm{J/C}$$

No work is required to move a charged particle along an equipotential surface. Equipotential surfaces are at right angles to electric field lines.

The potential difference of a pair of parallel plates of area (A) and separation (d) is:

$$\Delta V = \frac{q\,d}{\epsilon_0 A}$$

The potential difference between two points (R_1 and R_2) near a charge Q is:

$$\Delta V = \frac{Q}{4\pi\epsilon_0}\left(\frac{1}{R_2} - \frac{1}{R_1}\right)$$

The absolute potential of a point charge (q) is the electrical potential relative to infinity. It is the work per unit charge required to bring a positive test charge in from infinity to distance R.

$$V(\text{abs}) = \frac{1}{4\pi\epsilon_0}\frac{Q}{R}$$

The net potential at some location is the sum of the absolute potentials relative to surrounding charges.

$$V(\text{net}) = V_1 + V_2 + V_3 \ldots V_n$$

Capacitance (C) is a measure of the amount of charge an object can hold per volt. Capacitance depends on the size and shape of the object.

$$C = \frac{q}{V}$$

The unit of capacitance is the farad (F):

$$1\text{ F} = 1\text{ C/V}$$

The capacitance of a sphere is:

$$C = 4\pi\,\epsilon_0\,R$$

A parallel-plate capacitor has a capacitance of:

$$C = \frac{\epsilon_0 A}{d}$$

KEY TERMS

If you can explain the following terms to a friend or classmate, you understand their meaning. If you cannot explain the terms, you should reread the sections in which they are discussed.

absolute potential
capacitance
coulomb (C)
Coulomb's law
electrical potential
electric field
electric flux
electronic charge (*e*)
electroscope
electrostatic effect
equipotential lines
farad
Gauss's law
luminous flux
negatively charged
neutral
permittivity of space
positively charged
uniform field
vector field
volts (V)

EXERCISES

Section 29.1:

To perform the experiments in these exercises you will need a flexible plastic ruler. Clear plastic ones that are pliable enough to bend into a complete circle work best. These experiments may not work well if the humidity is high. Moist air contains ions that will react with electrically charged objects.

1. ***A.*** Rub a flexible plastic ruler against a glass windowpane and then let go of the ruler.
   B. Hold the ruler against the glass without rubbing. Again, release it.
   C. Explain what happens in experiments A and B.
2. Place some small bits of paper—about 1.0 cm diameter—on a desktop. Rub the ruler several times on a magazine or pad of paper. Hold one end of the ruler near the bits of paper. Explain what happens.
3. Hang two balloons close together (see Figure 29–24). Charge the plastic ruler on a pad of paper or magazine.
   A. Bring the ruler near one of the balloons. Why is the balloon attracted to the ruler?
   B. Touch the balloon with the ruler. Recharge the ruler and bring it near the balloon again. Why is the balloon repelled by the ruler?
   C. Explain the behavior of the second balloon during this experiment.

Figure 29–24: *Diagram for Exercise 3.*

Section 29.2:

4. The distance between two charged particles is doubled. Assume no other changes. The coulomb force acting on the particles will change by a factor of:
   A. 4 ***B.*** 2 ***C.*** 1/2 ***D.*** 1/4

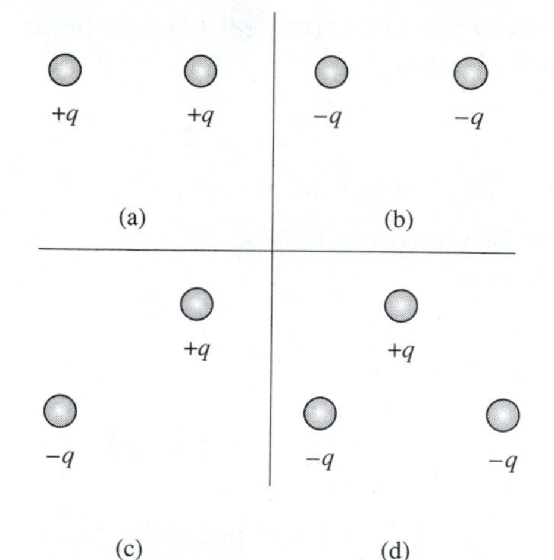

Figure 29–25: Diagram for Exercise 7. Copy the pattern of charges. Draw in the vectors to show the action of forces on each charge. (Do not forget to pay attention to the signs of the charges.)

5. A coulomb force acts between two charged particles, q_1 and q_2. If the charge of q_2 is doubled and nothing else is changed, the electrostatic force acting between the charges will change by a factor of:
   A. 4 ***B.*** 2 ***C.*** 1/2 ***D.*** 1/4

6. A coulomb force acts between charges q_1 and q_2. If charge q_2 is doubled and the distance between the charges is also doubled, then the coulomb force acting between the two charges will change by a factor of:
   A. 4 ***B.*** 2 ***C.*** 1/2 ***D.*** 1/4

7. For the sets of point charges shown in Figure 29–25 draw vectors to show the direction of coulomb forces acting on the point charges.

Sections 29.3, 29.4:

8. Explain the difference between electric field (E) and coulomb force ($\mathbf{F}_c$).

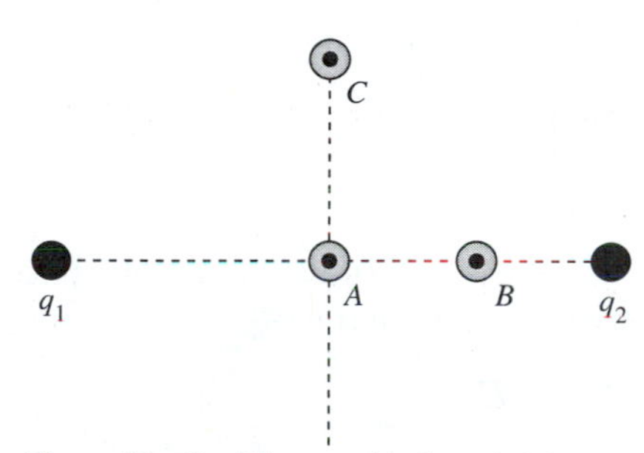

Figure 29–26: Diagram for Exercise 9. Does E = 0 at A, B, or C?

9. Figure 29–26 shows two charged particles (q_1 and q_2).
   A. Is it possible for the field to be equal to zero ($E = 0$) at point A, halfway between the charges? If it is possible, what must be true concerning the two charges?
   B. Is it possible for the field to be equal to zero at point B? If it is possible, explain under what conditions.
   C. Can $E = 0$ at point C? If it is possible, explain the conditions necessary.

10. A particle has a charge of $Q = 3.5 \times 10^{-6}$ C. A test charge of $q_a = 4 \times 10^{-9}$ C is 1 m away. At another position, also 1 m away from Q, is a second smaller test charge of $q_b = 2 \times 10^{-9}$ C (see Figure 29–27).
    A. Is the electrostatic force caused by Q that is acting on q_a larger, smaller, or the same as that acting on q_b?
    B. Is the electric field caused by Q that is acting on q_a larger, smaller, or the same as that acting on q_b?

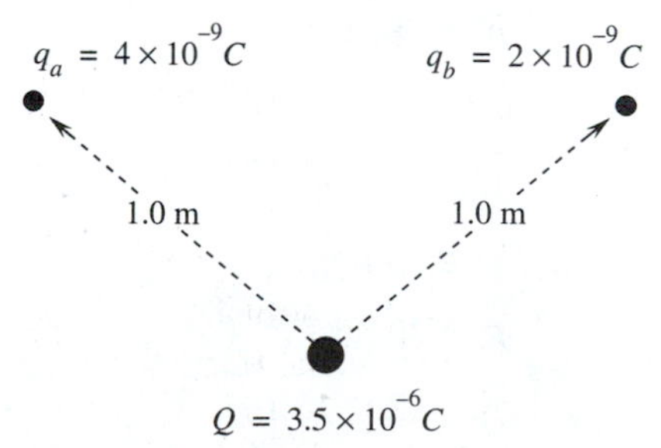

Figure 29–27: Diagram for Exercise 10.

Section 29.5:

11. Explain the following terms in your own words.
    A. Electric flux
    B. Permittivity of free space
    C. Gauss's law

12. Copy Figure 29–28 and draw in the electric field lines using the idea that electric field lines begin at positive charges and end at negative charges.

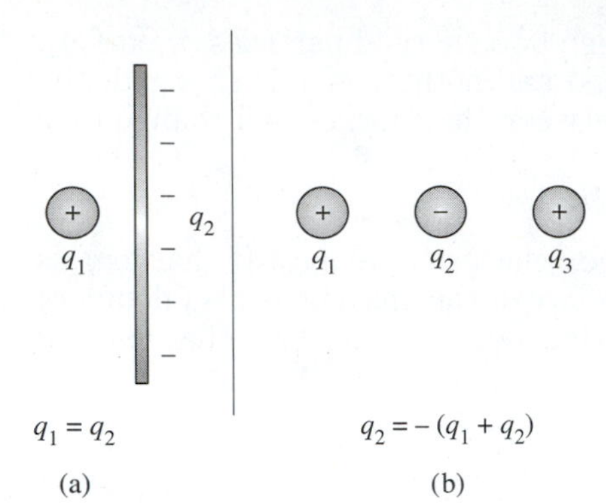

Figure 29–28: *Diagram for Exercise 12. Draw the electric field lines for the two sets of charges.*

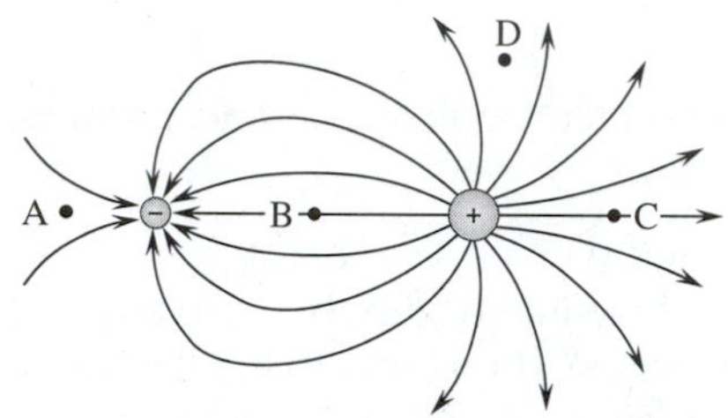

Figure 29–29: *Diagram for Exercise 13. The positive charge is larger than the negative one. Identify the location of the largest field strength.*

13. In Figure 29–29, which location has the strongest electric field: A, B, C, or D? Explain your choice.

Section 29.6:

14. How is electrical potential related to potential energy?

15. Explain the following terms in your own words.
- ***A.*** Electrical potential difference
- ***B.*** Absolute potential
- ***C.*** Equipotential surface, or equipotential line
- ***D.*** Uniform field

16. How much work is required to move an electron around a circular orbit of constant radius R from a point charge Q?

17. What can you say about the absolute electrical potential on the surface of a sphere with an isolated point charge at its center? What can we call this surface?

Section 29.7:

18. Explain the term capacitance in your own words.

19. Two parallel plate capacitors have the same separation (d), but C_2 has twice the plate area of C_1. Capacitor C_2 is different from capacitor C_1 by a factor of:
A. 4 ***B.*** 2 ***C.*** 1 ***D.*** 1/2 ***E.*** 1/4

20. The separation (d) of a parallel capacitor is doubled. The capacitance changes by a factor of:
A. 4 ***B.*** 2 ***C.*** 1 ***D.*** 1/2 ***E.*** 1/4

21. The separation and area of a parallel plate capacitor are both doubled. The capacitance changes by a factor of:

A. 4 ***B.*** 2 ***C.*** 1 ***D.*** 1/2 ***E.*** 1/4

PROBLEMS

Section 29.2:

1. Find the missing quantities for the following point charges.

a. $\mathbf{F} = ?$
$R = 0.45$ m
$q_1 = 2.0 \times 10^{-7}$ C
$q_2 = 3.0 \times 10^{-6}$ C

b. $\mathbf{F} = ?$
$R = 87$ cm
$q_1 = 1.7 \times 10^{-5}$ C
$q_2 = 2.3 \times 10^{-12}$ C

c. $\mathbf{F} = 0.035$ N
$R = ?$
$q_1 = 3.0 \times 10^{-6}$ C
$q_2 = 4.0 \times 10^{-7}$ C

d. $\mathbf{F} = 0.025$ N
$R = 0.30$ m
$q_1 = 2.0 \times 10^{-7}$ C
$q_2 = ?$

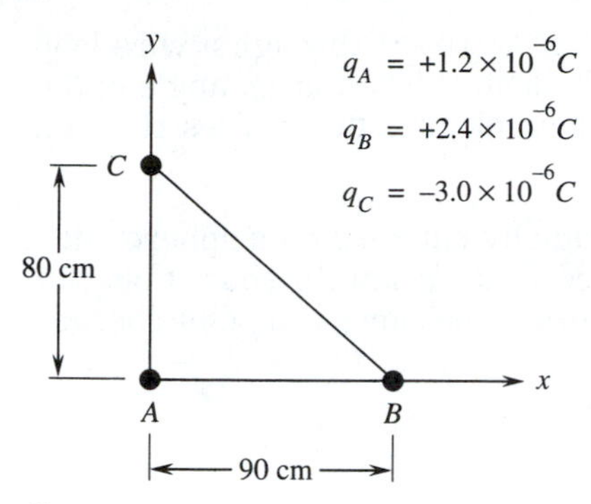

Figure 29–30: Diagram for Problem 4. Find the net forces on charge C.

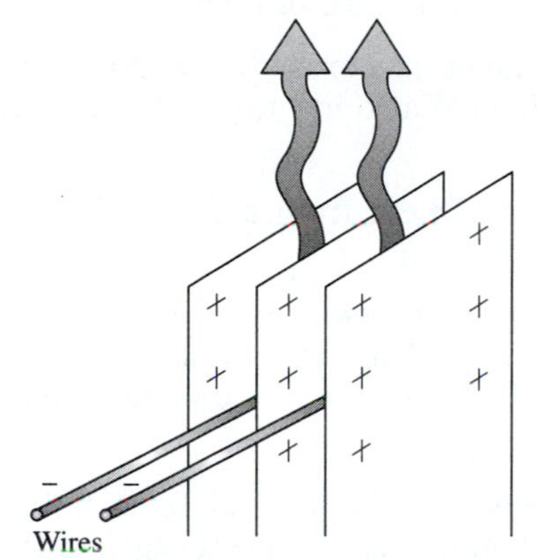

Figure 29–31: Diagram for Problem 6. Smoke travels upward between positively charged plates. Negatively charged wires transfer surplus electrons onto smoke particles as they pass by. The charged particles are attracted to the plates.

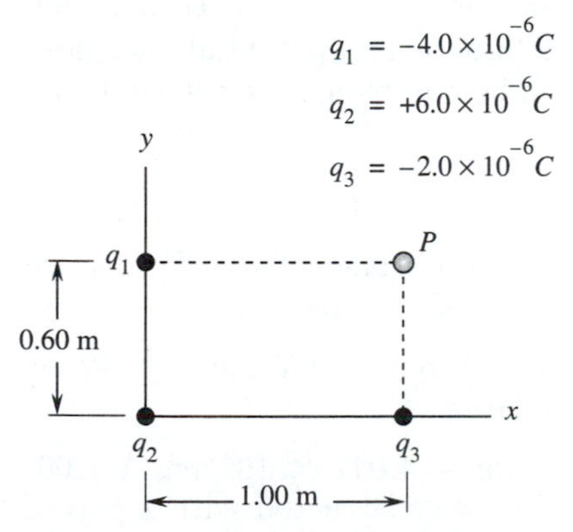

Figure 29–32: Diagram for Problem 9. Find the electric field strength at point P.

2. Particle A has a charge of $+2.3 \times 10^{-6}$ C. It is 1.00 m from particle B, which has a charge of -3.4×10^{-6} C. Particle C is midway between particles A and B. $q_C = +5.4 \times 10^{-8}$ C.
   A. Draw a diagram showing the direction of forces on charge q_c.
   B. Calculate the magnitude and direction of the net coulomb force acting on particle C.
3. For the situation in Problem 2, find the magnitude and direction of the net coulomb force on particle A.

** 4. Find the net force acting on point charge C in Figure 29–30. (Vector addition.)

Sections 29.3, 29.4:

5. When a -1.2×10^{-8} C charge is 1.20 m from charge Q, it experiences a repulsive coulomb force of 0.024 N.
   A. What is the strength of the electric field at this point?
   B. What is the sign (+ or −) of charge Q?
   C. What are the magnitude and direction of the force that would act on a $+5.6 \times 10^{-7}$ C charge at this same location?
6. Figure 29–31 shows a section of an electrostatic precipitator used in some smokestacks. A negatively charged wire transfers electrons to smoke particles. The charged particles are then attracted to the plates by electrostatic forces.
   A. If the electric field surrounding the plates is 10^4 N/C, what force acts on a smoke particle with a charge of 10^{-12} C?
   B. If a particle has a mass of 0.02 mg, what will be its acceleration toward the plate?
7. What is the strength of the electric field 89 cm from a point charge of 7.3×10^{-5} C?

* 8. What are the strength and direction of the electric field midway between a $+2.3 \times 10^{-6}$ C charge and a -1.7×10^{-6} C charge if the two charges are separated by 1.20 m?

** 9. Find the strength and direction of the electric field at point P in Figure 29–32.

Section 29.5:

10. What is the total flux coming from a 1.00-C charge?
11. A uniform field of 2.4×10^3 N/C passes at right angles through a rectangular surface with an area of 0.26 m². How much electric flux passes through the rectangle?
12. What is the electric field between a pair of square parallel plates carrying a charge of 2.0×10^{-7} C if the length of one side is 15.0 cm?

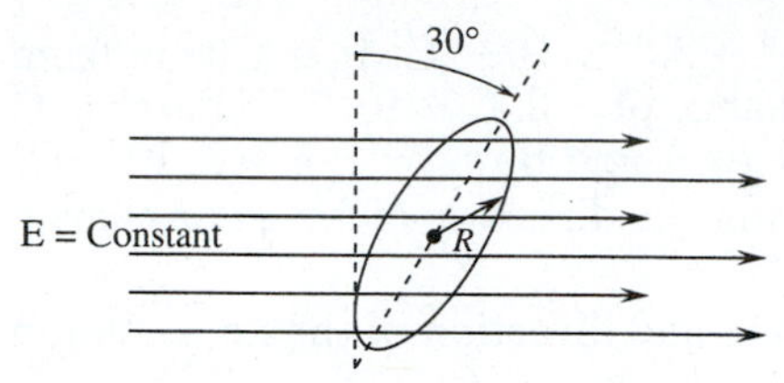

Figure 29–33: *Diagram for Problem 13. A uniform electric field passes through a circular loop tilted at 30° to the field.*

***13.** A uniform field of 6.8×10^3 N/C passes through a wire loop with a radius of 12.0 cm. The loop is tilted at an angle of 30° (see Figure 29–33). How much electric flux passes through the loop?

****14.** The electric field lines act radially outward on a sphere with a positive charge ($+Q$). Use Gauss's law to show that the electric field outside the sphere is the same as a point charge.

$$E = \frac{1}{4\pi \epsilon_0} \frac{Q}{R^2}$$

Section 29.6:

15. What is the potential energy change in moving a 2.30-C charge through an electrical potential difference of 50.0 V?

16. Electric field (E) has units of newtons per coulomb. Do a units analysis to show that electric field also has units of volts per meter.

17. What is the absolute potential 78 cm from a 2.6×10^{-7} C charge?

18. What is the potential difference between two parallel plates with equal but opposite charges with a magnitude of 2.3×10^{-9} C if they have an area of 0.19 m^2 and a 0.66-mm separation?

19. The potential difference between two parallel plates is 6.2 V. If the plate separation is 0.50 mm, what is the strength of the electric field between the plates?

***20.** For the plates in Problem 19, an electron is initially at rest near the negative plate. (Mass of electron = 9.1×10^{-31} kg.)

A. What are the initial kinetic and potential energies of the electron?

B. When the electron has moved half the distance between the plates what will be its kinetic and potential energies?

C. How fast will the electron be traveling just before it hits the electric plate?

Section 29.7:

21. A capacitor requires a potential difference of 12.0 V to store a charge of 3.0×10^{-8} C. What is its capacitance?

22. A 0.22-μF capacitor is connected to a 12.0-V battery. What charge will be stored on its plates?

23. One acre = 1/640 mi^2, or 1 acre = 4.047×10^3 m^2. A 1.00-F parallel plate capacitor is to be constructed with a gap of 0.10 mm between the plates. Find the plate area in acres. In square miles.

24. A 3.6-pF parallel plate capacitor with a 0.30-mm plate separation is connected to a 24.0-V battery.
 A. What is the charge on the capacitor?
 B. What is the area of the plates?
 C. What is the strength of the electric field between the plates?

Section 29.8:

25. The charge on a wire is often given a linear charge density (λ), or charge per unit length ($\lambda = q/L$). What is the linear charge density on a coaxial cable if the electric field between the wire and sheath is 6.3×10^3 N/C and the cable has a radius of 0.35 cm?

26. A beam of electrons passes horizontally between parallel plates separated by 4.6 mm. If the potential difference of the plates is 100 V, what vertical acceleration will the electrons have?

27. The separation of two parallel plates is 0.50 cm. Their length is 1.20 cm. The potential difference is 37.0 V. A beam of electrons with a horizontal speed of 2.0×10^7 m/s passes between the plates. Calculate:
 A. the time of transit.
 B. the vertical acceleration.
 C. the final vertical velocity.
 D. the deflection angle of the beam.

Chapter 30

RESISTANCE AND CAPACITANCE

When you see an electrical storm, most of the lightning stays between charged points in the clouds; about one out of five lightning strokes reaches the ground. As you read more about electricity in this chapter, you will be able to relate your knowledge to the natural environment as well as to controlled environments. (Photograph courtesy of NASA.)

OBJECTIVES

In this chapter you will learn:

- how electrons move in a metal conductor
- to find the current density, conductivity, and electric field in a conductor
- how the flow of electrons is related to electrical current
- that moving electrons collide with atoms within a conductor, creating heat
- to calculate the relationship between electric current and potential difference across a resistor (Ohm's law)
- that resistivity of a conductor increases with temperature
- how capacitors store energy
- how the capacitance of a capacitor can be increased by a dielectric

Time Line for Chapter 30

1820	Hans Oersted invents the ammeter.
1827	Georg Simon Ohm finds that the current through a metal is equal to the ratio of voltage to resistance.
1833	Heinrich Lenz finds that the resistance of a metal increases as temperature increases.
1843	Charles Wheatstone invents the Wheatstone bridge.
1880	Thomas Alva Edison's first electrical power plant goes into operation.

Electrical energy comes in many forms. Electrons streaming from a power plug pass through a heating element on an electric stove. Much of the electrical energy is converted into heat. Electrical energy is converted into thermal energy. From another power plug, electrons flow through the motor in an automatic clothes washer. Electrical energy is converted to mechanical energy. Chemical energy stored in a lead-acid battery in an automobile is converted to moving electrons. The electrical flow turns over the engine and ignites fuel in the cylinder. A capacitor stores electrostatic energy, which can be rapidly released. There are many applications for capacitors.

A high-voltage defibrillation unit uses a high-voltage capacitor to restart a patient's heart rhythm after a heart attack. A battery-operated capacitor unit with a much lower voltage is implanted in the patient to pace the beat of the heart. Capacitors are used in the ignition systems of a variety of burners and gasoline motors to create a spark. Lighting systems and electrical timing circuits use the time required to store energy in a capacitor as an interval timer.

In this chapter we will look at how two kinds of electrical devices—resistors and capacitors—relate to energy. We will also look at electrical charge in motion.

30.1 THE FLOW OF ELECTRONS

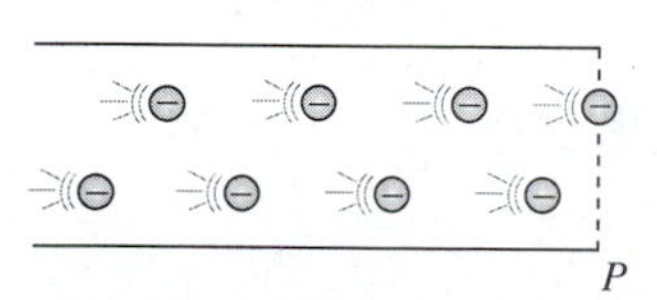

***Figure 30–1:** The magnitude of electrical current (I) is the amount of charge passing a point P in an interval of time.*

Look at Figure 30–1. If we have a piece of copper wire or a beam of electrons in a CRT, we can count the number of electrons passing some point during an interval of time. The total charge (ΔQ) passing the reference point divided by the time interval (Δt) is the **electrical current** (I). Usually charge is measured in coulombs, and time in seconds.

$$I = \frac{\Delta Q}{\Delta t} \quad \textbf{(Eq. 30–1)}$$

The unit of electrical current is an **ampere** (A).

$$1 \text{ ampere (A)} = 1 \text{ C/s}$$

Early experimenters with electricity did not have the atomic models of matter we work with today. They had to guess whether electrical current was carried by positive or negative charges. It was decided that electrical current is carried by moving positive charges. We now know that current is carried through metal wires by moving negatively charged electrons. We need to distinguish between **electron current** and electrical current.

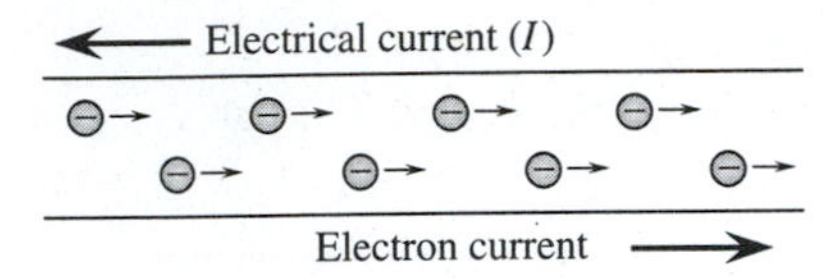

Figure 30–2: The electrical current (I) *flows in the opposite direction of the electron current.*

Electrons move in the opposite direction of electrical current.

Figure 30–2 shows the motion of electrical current and electron current through a wire. In some situations, electrical current is carried by positive charges. In semiconductors, part of the electrical current is made up of moving positively charged **holes.** In a positive ion beam the charge carriers are all positive. The current is in the same direction as the positive carriers.

Figure 30–3 shows one moving electron. It might be an electron moving through a metal wire or a beam of electrons in a television picture tube. We can use Equation 30–1 to find the electrical current.

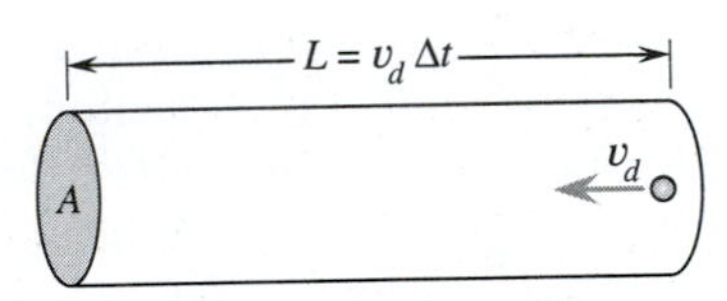

Figure 30–3: During an interval of time (Δt)*, an electron will travel a maximum distance* L *at an average drift velocity* ($\mathbf{v}_d$) *in the wire. All the electrons passing through area A are initially in a cylinder of length* L.

The total charge (ΔQ) passing the end of the cylinder during a time Δt depends on the average speed of the charges. This is called the **drift velocity** (v_d) of the electron. The last electron to pass through the surface A at the end of the time interval travels a distance $L = v_d \Delta t$ from the far end of the cylinder.

All of the electrons that pass through the counting surface are in the cylindrical volume when we start the timing clock. Multiplying the total number of electrons (N) by the electronic charge (e) will give us the charge passing through the area A.

$$\Delta Q = N e$$

It is usually easier to find the **charge density** than the total charge itself. Charge density is the amount of charge per unit volume.

$$n = \frac{\textit{number of electrons}}{\textit{volume}} = \frac{N}{V}$$

The volume is the cross-sectional area of the cylinder times its length.

In terms of charge density, the number of electrons in the cylinder is:

$$N = n\,V = n\,(A \times L)$$
$$N = n\,A\,(v_d\,\Delta t)$$

We now know the charge passing through the surface.

$$Q = N\,e = n\,e\,v_d\,A\,\Delta t$$

Dividing by Δt will give us the electrical current in terms of drift velocity.

$$J = \frac{\Delta Q}{\Delta t} = n\,e\,v_d\,A \qquad \textbf{(Eq. 30–2)}$$

For some engineering applications, current per unit cross section is useful. This is called the **current density** (J).

$$J = \frac{I}{A} \qquad \textbf{(Eq. 30–3)}$$

☐ **EXAMPLE PROBLEM 30–1: THE ELECTRON BEAM**

A beam of electrons in a CRT has a cross-sectional area of 1.2 mm² and carries a current of 48 milliamperes (mA). If the average speed of electrons is 4.3×10^6 m/s, find:

A. the current density (J) of the beam.

B. the density of electrons (n) in the beam.

■ ***SOLUTION***

A. Current density:

$$48 \text{ mA} = 48 \times 10^{-3}\text{ A} = 4.8 \times 10^{-2}\text{ A}$$
$$1.2 \text{ mm}^2 = 1.2 \times 10^{-6}\text{ m}^2$$

$$J = \frac{I}{A} = \frac{48 \times 10^{-3}\text{ A}}{1.2 \times 10^{-6}\text{ m}^2}$$
$$J = 4.0 \times 10^4\ A/\text{m}^2$$

B. Density of electrons:

$$J = \frac{I}{A} = n\,e\,v_o, \quad \text{or} \quad n = \frac{J}{e\,v_d}$$

$$n = \frac{4.0 \times 10^4\ A/\text{m}^2}{(1.6 \times 10^{-19}\text{ C})(4.3 \times 10^6\text{ m/s})}$$
$$n = 5.8 \times 10^{16}\text{ electrons/m}^3$$

How Lightning Strikes

Thunderstorms contain very strong updrafts inside immense clouds. Ice, snow, and air swirl about in the cloud. Friction builds up regions of positive and negative charges in the thunderhead. Very strong electric fields develop, along with very high electrical potentials. Voltage differences of 10 million to 100 million volts may occur.

Usually, air is a poor conductor of electricity. The high electric fields can ionize the air. Outer electrons of the atoms in the air can be pulled loose. This leaves charged particles that can act as conduits for electrical current.

(Photograph courtesy of NASA.)

Most of the lightning travels from one part of the cloud to another. The charged areas of the cloud are usually closer together than the cloud and the ground. The closer together the charged areas are, the stronger will be the electric field between them for the same potential difference ($E = -V/d$).

Negative charges tend to build up on the bottom of the cloud. This induces a positive charge on the ground. If conditions are right, the air between the ground and the cloud can be ionized, allowing a lightning stroke to reach the ground. One out of five lightning strokes strikes the ground.

Typically, a charge of only 20–25 C is transferred by a lightning stroke. Because the potential difference is so large, the energy in a lightning stroke can be 200 million to 2 billion J.

□ **EXAMPLE PROBLEM 30–2: DRIFT VELOCITY IN A COPPER WIRE**

Number 12 gauge copper wire is commonly used for wiring kitchen areas of houses and public buildings. The density of electrons in copper is 8.5×10^{28} electrons/m^3, and the cross-sectional area is 0.062 cm^2. Use these data to estimate the average time required for an electron to travel 2.0 m in a number 12 wire carrying a current of 15 A.

■ ***SOLUTION***

The time will be the distance divided by drift velocity. The drift velocity is:

$$v_d = \frac{I}{n\,e\,A}$$

$$v_d = \frac{15\ \text{A}}{(8.5 \times 10^{28}\ e/\text{m}^3) \times (1.6 \times 10^{-19}\ \text{C}/e) \times (6.2 \times 10^{-6}\ \text{m}^3)}$$

$$v_d = 1.77 \times 10^{-4}\ \text{m/s}$$

The time required to travel 2.0 m is:

$$t = \frac{s}{\mathbf{v}_d}$$

$$t = \frac{2.0\ \text{m}}{1.77 \times 10^{-4}\ \text{m/s}}$$

$$t = 1.13 \times 10^{4}\ \text{s} \times \frac{1\ \text{h}}{3600\ \text{s}}$$

$$t = 3.7\ \text{h}$$

Because electron densities are very large in conductors, small drift velocities create large currents.

30.2 OHM'S LAW

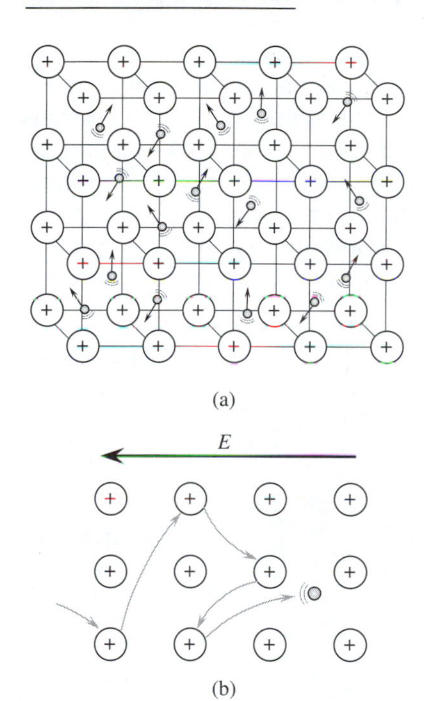

Figure 30–4: (a) *Free electron model of metals. Electrons flow freely in a grid of anions. The free electrons are not chemically bonded to any particular atom.* (b) *Under the influence of an electric field* (E)*, electrons are accelerated. This acceleration gives them curved paths with a net component of motion opposite the field.*

We can visualize a conductor as having two parts. One part is a grid of fixed positive ions. The other is a cloud of electrons wandering freely in the space between the ions. This description is called the **free electron model.** Electrons are free to move about in the conductor. Figure 30–4a shows the random motion of electrons in a metal rod. The electrons will always have some random motion. This is part of the thermal energy contained in the metal. The electrons bounce around in straight-line paths, bumping into the much more massive anions from time to time. After the collision the electrons travel in another random direction (see Figure 30–4a).

We can exert an electric field on the material by connecting the rod to a voltage source such as a battery, creating a potential difference between the ends of the rod. For materials that obey the free electron model, a uniform field is created between the ends of the rod (see Figure 30–5). The electric field (E) is simply:

$$E = \frac{V}{L}$$

The randomly moving electrons begin to be accelerated in the direction opposite the electric field. Their paths look somewhat like the ones shown in Figure 30–4b. Their motion is no longer totally random. They drift toward one end of the rod under the influence of a coulomb force caused by the electric field (E). We can use Newton's second law of motion to estimate the acceleration.

$$\text{F} = e\,E = m\,\mathbf{a}$$

$$a = \frac{e\,E}{m}$$

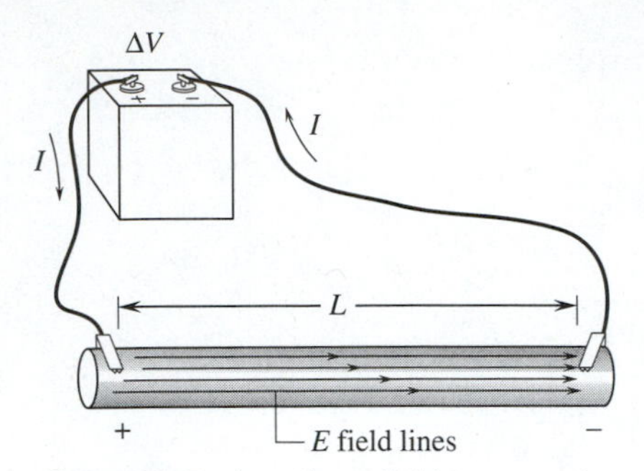

Figure 30–5: A uniform field is created in a conducting rod by a potential difference across its ends.

Let us assume that the electrons lose all their motion caused by the electric field when they collide with the grid of anions. Also, let us say that the average time between collisions, the **mean free time,** is τ. The drift velocity, the velocity of electrons caused by the electric field, is:

$$v_d = a\tau$$

$$v_d = \left(\frac{eE}{m}\right)\tau$$

We can combine Equations 30–2 and 30–3 and substitute this expression for v_d.

$$J = n e v_d = \left[\left(\frac{n e^2}{m}\right)\tau\right]E$$

The current density is directly proportional to the electric field. The quantity in brackets depends on the property of the material. Both the density of electrons and the mean free time depend on the way atoms are stacked. The spacing and symmetry of the anions will vary from material to material. We can lump these things together in a material constant called **conductivity** (σ). The greater the conductivity, the more easily electrons will flow through the material.

$$J = \sigma E \qquad \textbf{(Eq. 30–4)}$$

We can change the equation around.

$$E = \frac{J}{\sigma}, \text{ or } E = \rho J \qquad \textbf{(Eq. 30–5)}$$

where $\rho = 1/\sigma$.

ρ is called **resistivity.** It is the reciprocal of conductivity. A material with high resistivity has a high resistance to electron flow. Table 30–1 shows some typical values of resistivity.

Recall that $J = I/A$ and $E = V/L$ for a conductor. We can rewrite Equation 30–5.

$$E = \frac{V}{L} = \rho\frac{I}{A}$$

Table 30–1: Resistivity for selected materials at room temperature.

MATERIAL	RESISTIVITY (ρ) ($\Omega \cdot m$)	THERMAL COEFFICIENT OF RESISTIVITY (α) (C^{-1})
Metals		
Aluminum	2.8×10^{-8}	3.9×10^{-3}
Brass	7.0×10^{-8}	2.0×10^{-3}
Copper	1.7×10^{-8}	3.9×10^{-3}
Iron	10×10^{-8}	5.0×10^{-3}
Mercury	96×10^{-8}	0.9×10^{-3}
Steel	$\sim 11 \times 10^{-8}$	$\sim 4 \times 10^{-3}$
Tungsten	5.6×10^{-8}	4.5×10^{-3}
Nonmetals		
Bakelite	1×10^{9}	—
Carbon	3.5×10^{5}	-0.5×10^{-3}
Germanium	0.46	-48×10^{-3}
Glass	1×10^{12}	—
Paraffin	1×10^{15}	—
Silicon	64×10^{2}	-75×10^{-3}
Water		
distilled	5×10^{3}	—
salt	4	—

or

$$V = \left(\frac{\rho L}{A}\right) I$$

The part of the expression in parentheses depends on the material and its dimensions. For any specific conductor, we can combine and call the property **resistance** (R).

$$R = \frac{\rho L}{A} \qquad \textbf{(Eq. 30–6)}$$

and

$$V = R\,I \qquad \textbf{(Eq. 30–7)}$$

For many electrical devices the electrical potential difference across the object is directly proportional to the current. The value of the resistance is the proportionality constant, or the slope of the

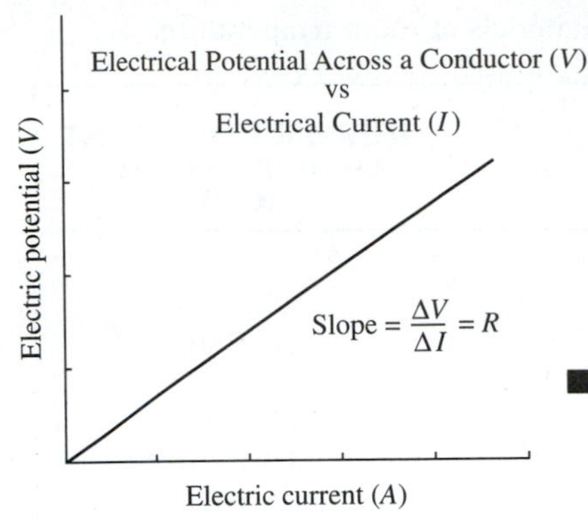

Figure 30–6: Electrical potential across a conductor (V) versus electrical current (I). The slope of a V vs. I graph is resistance (R).

plot of voltage versus current (see Figure 30–6). The equation of direct proportion between voltage and current is called **Ohm's law.**

EXAMPLE PROBLEM 30–3: RESISTIVITY

A transmission line uses number 00 gauge aluminum alloy wire with a cross section of 67.4 mm². If $22\bar{0}0$ m of this wire has a resistance of 1.00 ohm (Ω), calculate its resistivity.

■ *SOLUTION*

$$R = \frac{\rho L}{A}, \quad \text{or} \quad \rho = \frac{R\,A}{L}$$

$$\rho = 1.00\Omega \left(\frac{67.4 \times 10^{-6}\ \text{m}^2}{2.20 \times 10^3\ \text{m}}\right)$$

$$\rho = 3.06 \times 10^{-8}\ \Omega \cdot \text{m}$$

□ **EXAMPLE PROBLEM 30–4: OHM'S LAW**

The current through an electrical resistance is 1.20 A when the potential difference across the resistor is 12.0 V.

A. What is the resistance?

B. What potential difference across the resistor is needed for a current of 2.10 A?

■ *SOLUTION*

A. Resistance:

$$R = \frac{V}{I}$$

$$R = \frac{12.0\ \text{V}}{1.20\ \text{A}}$$

$$R = 1\bar{0}\ \Omega$$

B. Electrical potential:

$$V = R\,I$$

$$V = 1\bar{0}\ \Omega \times 2.10\ \text{A}$$

$$V = 21.0\ \text{V}$$

□ **EXAMPLE PROBLEM 30–5: CONDUCTIVITY**

A 0.150 mm diameter wire is 21.0 m long. When there is a potential difference of 12.0 V across the wire a current of 50.5 mA flows through the wire.

A. What is the current density?
B. What is the electric field strength in the wire?
C. What is the conductivity of the wire?
D. What is the resistivity?

■ *SOLUTION*

A. Current Density (J):

$$J = \frac{I}{A} = \frac{I}{\pi R^2}$$

$$J = \frac{50.5 \times 10^{-3}\ \text{A}}{3.14 \times (0.0750 \times 10^{-3}\ \text{m})^2}$$

$$J = 2.85 \times 10^6\ \text{A/m}^2$$

B. Electric field (E)

$$E = \frac{V}{L}$$

$$E = \frac{12.0\ \text{V}}{21.0\ \text{m}}$$

$$E = 0.571\ \text{V/m}$$

C. Conductivity:

$$\sigma = \frac{J}{E}$$

$$\sigma = \frac{2.85 \times 10^6\ \text{A/m}^2}{0.571\ \text{V/m}}$$

$$\sigma = 4.99 \times 10^6\ (\Omega \cdot \text{m})^{-1}$$

D. Resistivity:

$$\rho = \frac{1}{\sigma}$$

$$\rho = \frac{1}{4.99 \times 10^6\ (\Omega \cdot \text{m})^{-1}}$$

$$\rho = 2.00 \times 10^{-7}\ \Omega \cdot \text{m}$$

30.3 POWER, HEAT, AND RESISTANCE

Electrical potential is energy per unit charge. When an electrical charge (ΔQ) falls through a voltage difference (ΔV), work is done on the charge.

$$work = Q\,V \quad \textbf{(Eq. 30–8)}$$

We saw in Chapter 29 that the change in potential energy is converted to kinetic energy when there is an electrical charge between two charged plates. The situation is a bit different for a conductor. As electrons move through a metal wire, they repeatedly bump into cations. This limits the speed of the electrons to an average drift velocity (see Figure 30–4). The electron's kinetic energy is transferred to the more massive matrix of anions making up the mechanical structure of the material. The kinetic energy shows up as vibration of the atoms. We interpret this as heat.

In a resistor the work done by moving charges generates heat.

$$work = \Delta Q\,\Delta V = heat$$

We can express this in terms of work per unit time, or power.

$$P = \frac{work}{time} = \left(\frac{\Delta Q}{\Delta t}\right)\Delta V = I\,\Delta V = \frac{heat\ dissipated}{time}$$

$$P = I\,\Delta V \quad \textbf{(Eq. 30–9)}$$

Equation 30–9 is true for any electrical device. If we measure the potential difference across the device and multiply it by the effective current traveling through it, we have the electrical power of the device. The power may show up as heat for the case of resistance, as mechanical power in an electric motor, or as chemical energy change in a battery. For the special case of an electric resistor we can invoke Ohm's law ($V = R\,I$).

$$P = I\,(R\,I) = R\,I^2 \quad \textbf{(Eq. 30–10A)}$$

Alternately we can use Ohm's law to eliminate current.

$$P = \left(\frac{\Delta V}{R}\right)\Delta V = \frac{\Delta V^2}{R} \quad \textbf{(Eq. 30–10B)}$$

☐ **EXAMPLE PROBLEM 30–6: THE COFFEEPOT**

A heating coil in an electric coffeepot heats $5\bar{0}0\ cm^3$ of water from 20.0°C to 100.0°C (see Figure 30–7). How long will it take to heat the water if the coil operates at $11\bar{0}$ V and 5.10 A? Let us unrealistically ignore the heat required to heat the pot and surrounding air.

Figure 30–7: *Electrical resistive heating is used to make a pot of coffee.*

■ *SOLUTION*

The heat required is:

$$heat = c\,m\,\Delta T = 1.00\,\frac{\text{cal}}{\text{g}\cdot\text{c}} \times 500\text{ g} \times (100.0°\text{C} - 20.0°\text{C}) \times 4.184\text{ J/cal}$$

$$heat = 1.67 \times 10^5\text{ J}$$

The energy dissipated by resistive heating by the coil is:

$$work = Q\,\Delta V = I\,\Delta V\,\Delta t, \text{ or } \quad \Delta t = \frac{work}{I\,\Delta V}$$

$$\Delta t = \frac{1.67 \times 10^5\text{ J}}{5.10\text{ A} \times 11\bar{0}\text{ V}}$$

$$\Delta t = 2.985 \times 10^2\text{ s, or } 5.0\text{ min}$$

□ **EXAMPLE PROBLEM 30–7: THE ELECTRIC MOTOR**

A DC motor operates at 12.0 V drawing. How much electrical power is used in watts and in horsepower when the motor draws 11.5 A?

■ *SOLUTION*

$$P = I\,\Delta V$$

$$P = 11.5\text{ A} \times 12.0\text{ V}$$

$$P = 138\text{ W}$$

$$P = 138\ \text{J/s} \times 0.7376\ \text{ft} \cdot \text{lb/J} \times \frac{1\ \text{hp}}{550\ \text{ft} \cdot \text{lb/s}}$$

$$P = 0.185\ \text{hp}$$

☐ **EXAMPLE PROBLEM 30–8: THE TRANSMISSION LINE**

A 152-km transmission line operates at $12\bar{0}$ kV carrying a current of 21.5 A. The line is constructed from an aluminum alloy with a resistance of 0.42 Ω/km.

A. What power is delivered by the transmission line?
B. How much heat per second in watts is dissipated by the line?
C. What is the voltage difference between the ends of the line?
D. By delivering the same power at twice the voltage the resistive heat losses would be reduced by how much?

■ ***SOLUTION***

A. Power:

$$P = I\,V = 21.5\ \text{A} \times 12\bar{0}\ \text{kV}$$

$$P = 2.58 \times 10^3\ \text{kW, or } 2.58\ \text{MW}$$

B. Resistive heat losses:

$$heat = R\,I^2 = 0.42\ \Omega/\text{km} \times 152\ \text{km} \times (21.5\ \text{A})^2$$

$$heat = 3.0 \times 10^4\ \text{W} = 3\bar{0}\ \text{kW}$$

C. Potential difference:

$$V = R\,I$$

$$V = 0.42\ \Omega/\text{km} \times 152\ \text{km} \times 21.5\ \text{A}$$

$$V = 1.4 \times 10^3\ \text{V (less than a 1\% drop of potential)}$$

D. Doubling voltage:

$$I = \frac{P}{2V} = \frac{2.58 \times 10^6\ \text{W}}{2 \times 1.20 \times 10^5\ \text{V}} = 10.8\ \text{A}$$

$$heat = R\,I^2 = 0.42\ \Omega/\text{km} \times 152\ \text{km} \times (10.8\ \text{A})^2$$

$$heat = 7.4 \times 10^3\ \text{W}$$

Heat losses are dropped to 0.25 of the original value.

30.4 NONLINEAR DEVICES

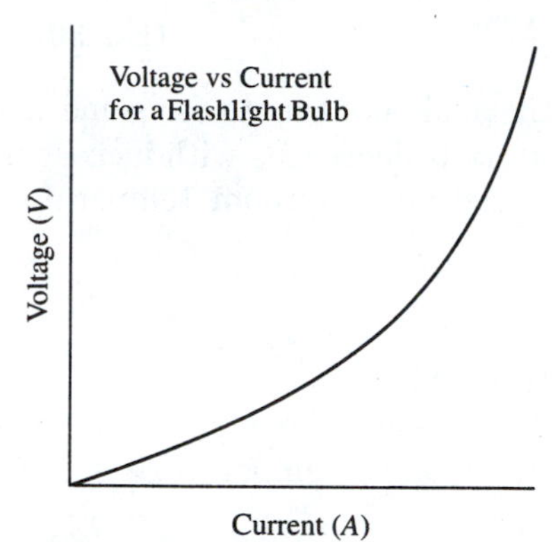

Figure 30–8: *Electrical potential (V) versus current for a flashlight bulb. The slope (ΔV/ΔI) increases as more current is drawn through the filament.*

Electric toasters and incandescent lamps do not obey Ohm's law. Figure 30–8 shows a typical plot of voltage versus current for a 6.0-V light bulb found in a flashlight. As the voltage increases, the rate of increase of current decreases. The slope of the V versus I graph is the resistance of the filament. The slope increases as the voltage increases.

Perhaps the resistance changes with temperature. If we plot slope (R) against temperature, we get Figure 30–9. The resistance is directly proportional to the temperature. We can write the equation of the straight line.

$$y = y_0 + (m\,x)$$

or

$$R = R_0 + \left(\frac{\Delta R}{\Delta T}\right) \Delta T$$

$$R = R_0 \left[1 + \left(\frac{1}{R_0}\frac{\Delta R}{\Delta T}\right) \Delta T\right]$$

The slope and R_0 are both constants. We can replace them with one constant (α).

$$\alpha = \frac{1}{R_0}\frac{\Delta R}{\Delta T}$$

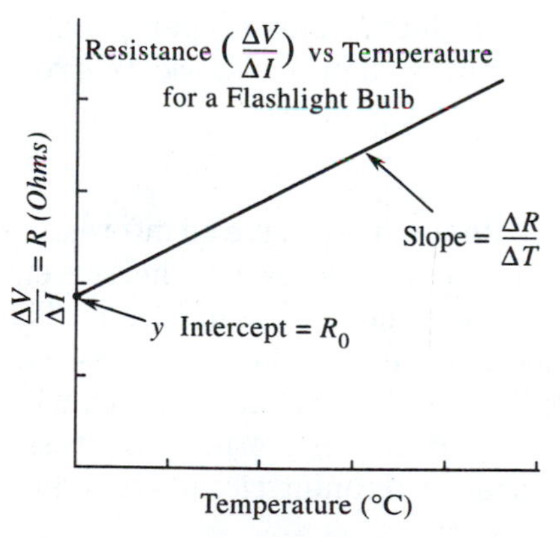

Figure 30–9: *Resistance (ΔV/ΔI) versus temperature for a flashlight bulb. The resistance of a conductor increases linearly with temperature.*

and

$$R = R_0\,[1 + (\alpha \Delta T)] \qquad \textbf{(Eq. 30–11)}$$

Equation 30–11 gives the temperature variation for a specific resistor. We can change this to a more general form by expressing resistance is terms of resistivity ($R = [\rho L]/A$). Length and area cancel out everywhere, even in the constant α. This gives us an equation in terms of material properties. The constant α is called the **thermal coefficient of resistivity.** It has the same value whether we express it in terms of resistance or resistivity.

$$\alpha = \frac{1}{\rho_0}\frac{\Delta \rho}{\Delta T} \qquad \textbf{(Eq. 30–12)}$$

and

$$\rho = \rho_0[1 + (\alpha\ \Delta T)] \qquad \textbf{(Eq. 30–13)}$$

Table 30–1 lists values of the thermal coefficients for some materials. The resistivity of some materials decreases with increasing temperature. Usually ρ_0 is the resistivity at room temperature (20°C).

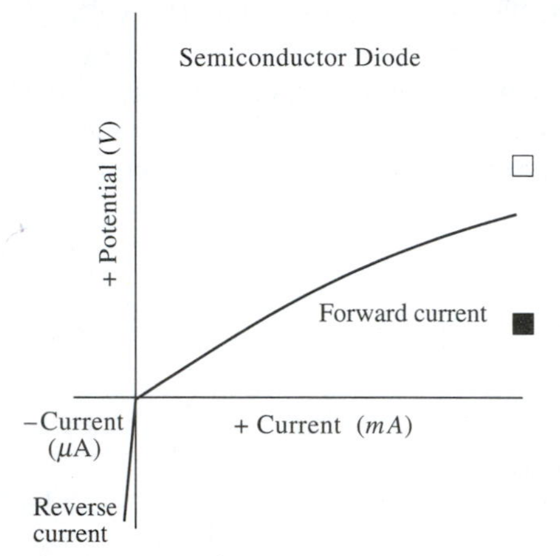

Figure 30–10: The reverse current in a semiconductor diode is very small. Notice that forward current is in units of milliamperes, while the reverse current is in microamperes.

EXAMPLE PROBLEM 30–9: THE HEATED COIL

A copper coil in a television set has a resistance of 22.7 Ω at room temperature. What is its resistance when its temperature is 38°C?

SOLUTION

$$R = R_0[1 + (\alpha\ \Delta T)]$$

$$R = 22.7\ \Omega[1 + (3.9 \times 10^{-3}\ {}^\circ\text{C}^{-1}) \times (38^\circ\text{C} - 20^\circ\text{C})]$$

$$R = 24.3\ \Omega$$

Here are a few other devices that do not follow Ohm's law. We will look at their operation in more detail in Chapter 35.

Semiconductor Diodes

Figure 30–10 shows a plot of voltage versus current across a semiconductor diode. In the "forward" direction, the current increases rapidly with voltage, reaching a few milliamperes. If the voltage is reversed in an attempt to force current in the "reverse" direction, only a tiny microampere current is produced. This property makes diodes useful for checking valves to limit the flow of electrical current in one direction.

Tunnel Diodes

Tunnel diodes are also called **Esaki diodes.** They have an interesting behavior. As the voltage across the device increases, the current reaches a maximum and then decreases before rising again. The slope is negative between the peak and valley of the current versus voltage curve (see Figure 30–11). This region is called the **negative resistance region.** A tunnel diode operating in the negative resistance region can be placed in certain kinds of tuning circuits to reduce their overall resistance.

Transistors

Figure 30–12 shows the current versus voltage curves for a typical transistor. At first the current drawn from the transistor increases

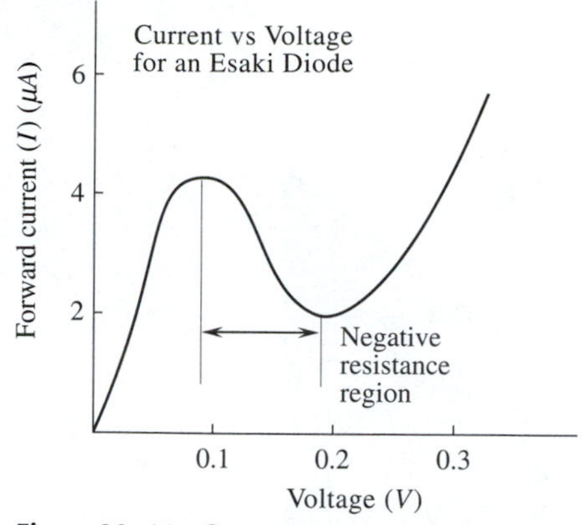

Figure 30–11: *Current versus voltage. A tunnel diode has a negative resistance range of operation.*

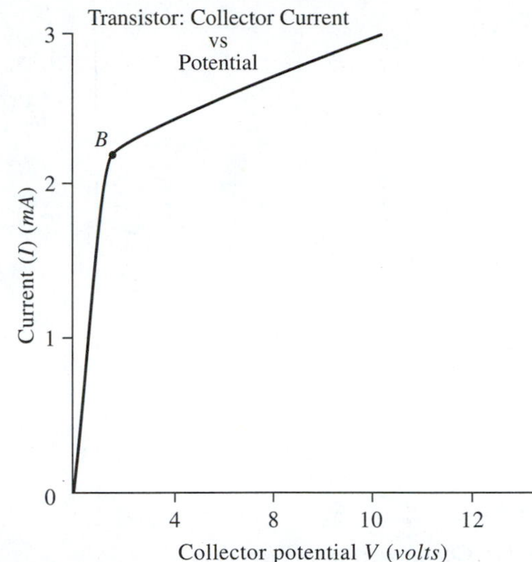

Figure 30–12: *Current versus potential. Current through the collector of a transistor is nonlinear.*

rapidly as the voltage increases. Beyond point B, the current increases much more slowly with increased voltage.

30.5 CAPACITORS AND ENERGY

While resistors dissipate energy through heat loss, capacitors store energy. The charge stored in a capacitor is directly proportional to the voltage ($Q = C\,V$). Figure 30–13 shows the direct proportional relationship. The slope of the line is the capacitance. The shaded area under the line is the product of charge and current. Area is the product of charge and electrical potential. Since voltage is energy per unit charge, the area represents stored energy (PE). The area is that of a triangle (area = ½ *base* × *height*). For a total charge Q stored at a voltage V_0 the area under the line (energy stored) is:

$$area = \left(\frac{1}{2}Q\right)V = PE$$

We can use the relationship $C = Q/V$ to express the energy in terms of capacitance.

$$PE = \left(\frac{1}{2}Q\right)V = \left(\frac{1}{2}C\right)V^2 = \frac{1}{2}\left(\frac{Q^2}{C}\right) \qquad \textbf{(Eq. 30–14)}$$

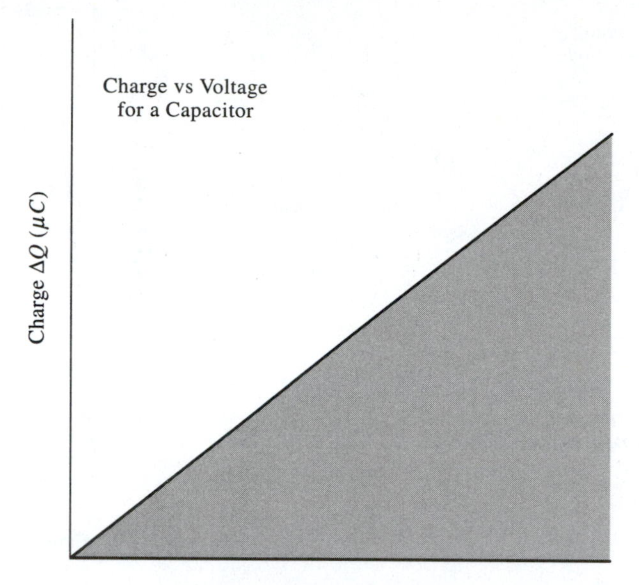

Figure 30–13: Charge versus voltage for a capacitor. The shaded area represents the energy stored in the capacitor, plotted in Figure 30–14.

☐ **EXAMPLE PROBLEM 30–10: ENERGY STORED IN A CAPACITOR**

A 1.26 μF capacitor is charged to 53.0 V. How much charge and energy are stored in the capacitor?

■ ***SOLUTION***

Charge:

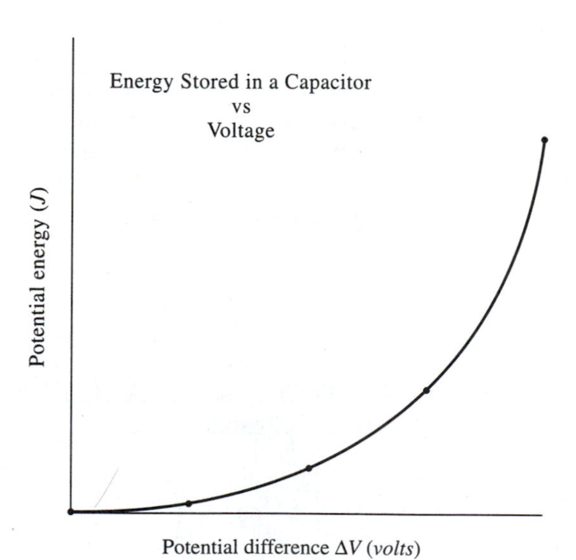

Figure 30–14: Energy stored versus voltage. The energy stored in a capacitor is directly proportional to the square of the voltage across its plates.

$$Q = C\,V$$
$$Q = 1.26\ \mu F \times 53.0\ V$$
$$Q = 66.8\ \mu C$$

Energy: We can use any version of Equation 30–14.

$$PE = \left(\frac{1}{2}\right) Q\,V = \frac{1}{2}\left(\frac{Q^2}{C}\right) = \frac{1}{2}(C\,V^2)$$

$$PE = \frac{1}{2}(66.8\ \Omega\ C)(53.0\ V) = \frac{1}{2}\left(\frac{66.8\ \mu C}{1.26\ \mu F}\right)^2 = \frac{1}{2}[(1.26\ \mu F)(53.0\ V)^2]$$

$$PE = 1.77 \times 10^{-3}\ J$$

30.6 DIELECTRICS

Electrical materials can be divided into three categories: conductors, such as metals; semiconductors, such as germanium; and dielectrics, such as glass and polymers. The electrical behavior of these materials depends mostly on chemical bonding.

Metals have a particular bond in which one or two electrons from each atom are free to move through the material. Electrical current flows through these materials easily (see Figure 30–15a).

Semiconductors have most of their electrons chemically bonded to individual atoms. These **valance electrons** cannot move through the semiconducting material. However, the chemical bonds are only moderately strong. At room temperature, a small percentage of the electrons have enough thermal energy to break loose and roam through the material. The conductivity of semiconductors is highly temperature-dependent (see Figure 30–15b).

Dielectrics, or **electrical insulators,** have very strong chemical bonds. At room temperature very few electrons have enough chemical energy to break free and move through the material. Most covalent materials are dielectrics (see Figure 30–15c).

Molecules in dielectric materials tend to have asymmetric charge distributions. Although the net charge is zero, the positive and negative charges are not evenly distributed. One end of the molecule is net negative and the other end net positive. When the molecule is placed in an electric field, a torque acts on the molecule, causing it to line up with the electric field (see Figure 30–16).

If we place a voltage across a parallel plate capacitor (C_0) by hooking it up to a battery, a maximum charge (Q_0) will be stored on the plates of the capacitor.

$$Q_0 = C_0\,V$$

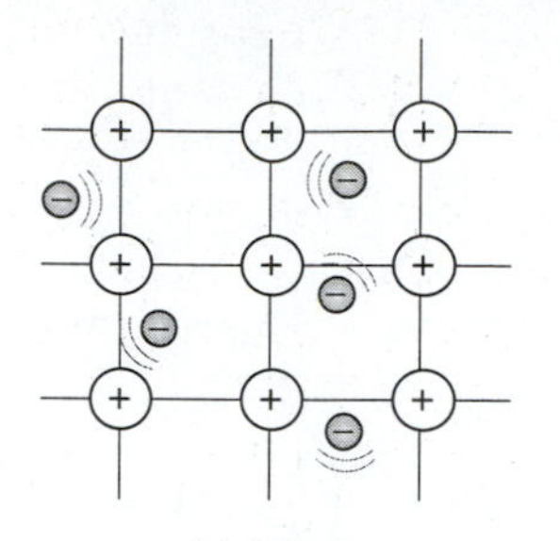

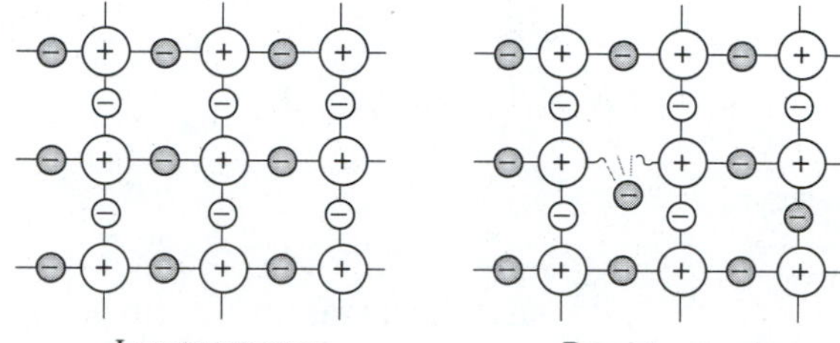

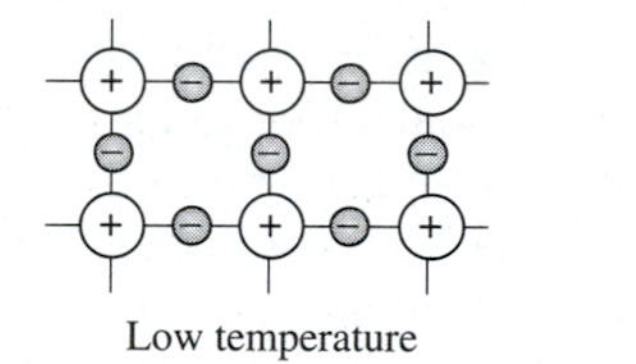

Figure 30–15: (a) *In metals, electrons move freely between a lattice of positive ions.* (b) *In semiconductors, electrons are all chemically bonded to specific atoms at low temperatures, but at room temperature some of the electrons have enough thermal energy to break their chemical bond and move through the lattice.* (c) *Dielectrics have very strong chemical bonds. At room temperature, electrons do not have enough energy to break free.*

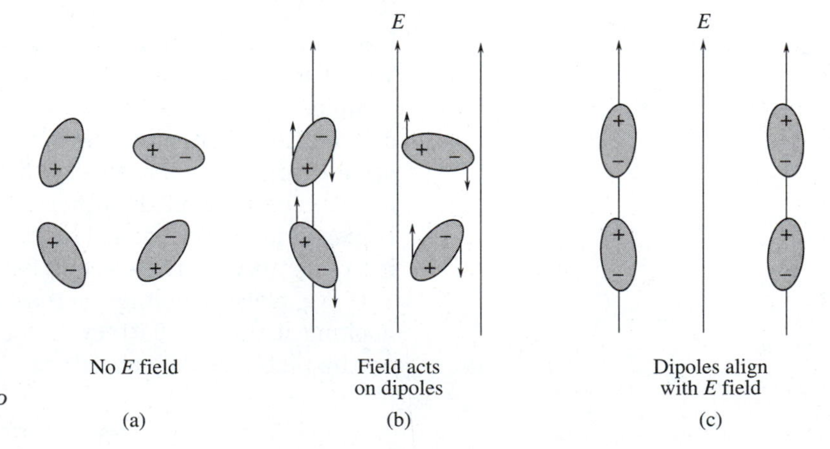

Figure 30–16: (a) *Dipoles are randomly oriented in a material in the absence of an external electric field.* (b) *An electric field creates force couples on the dipoles. The resulting torques rotate the dipoles.* (c) *Dipoles are aligned parallel to the electric field.*

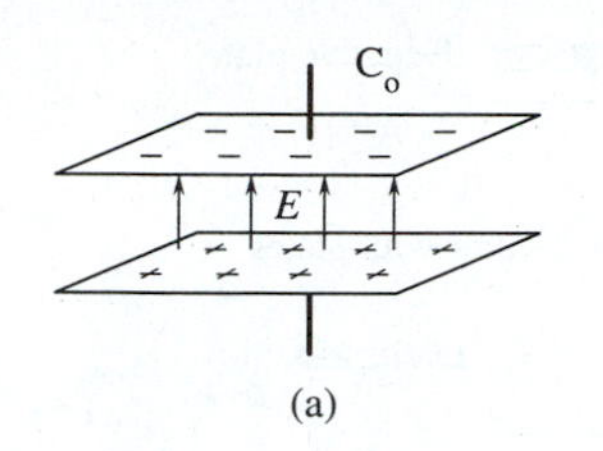

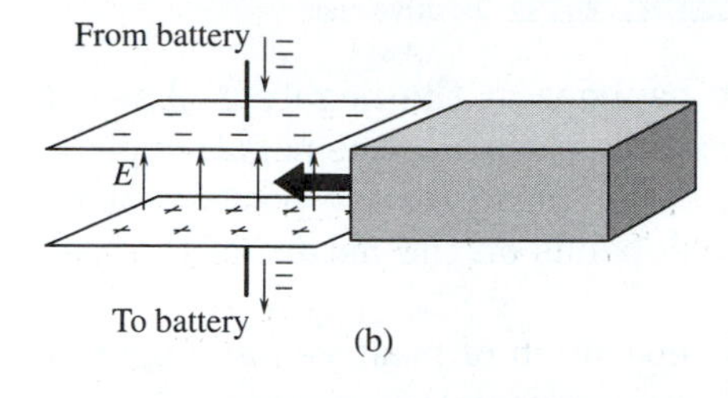

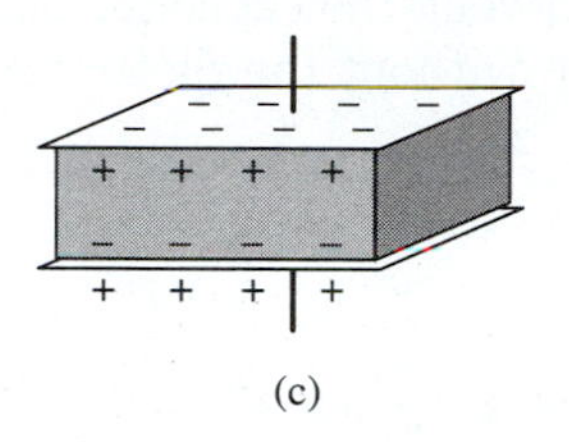

Figure 30–17: (a) *An electric field exists in the gap between a capacitor's plates.* (b) *As a dielectric is inserted between the plates, electrons are drawn from the battery to the negative plate. An equal number of electrons leave the positive plate.* (c) *A surface charge has been induced on the dielectric. The plates of the capacitor have a greater charge than before for the same voltage.*

We can now fill the space between the plates with a piece of dielectric such as plastic. Additional electrons move from the battery to the plates of the capacitor to give it a greater charge (Q) for the same electrical potential. We have increased the capacitance to a new value (C) (see Figure 30–17).

$$C = \frac{Q}{V}, \text{ or } Q = C\,V$$

If we use paraffin instead of plastic, a different number of additional electrons will be attracted to the capacitor. If we use machine oil, we will get another number of electrons on the plate and yet another capacitance. The capacitance of the device depends upon the kind of dielectric material filling the space between the plates. We can define a material constant called the **dielectric constant** (k) by taking the ratio of the capacitance with and without the dielectric material. Table 30–2 lists the dielectric constant for a few selected materials.

$$k = \frac{C}{C_0} \qquad \textbf{(Eq. 30–15)}$$

Let us see why the capacitance increases. The asymmetric molecules in a dielectric are called **electric dipoles.** In the absence of an outside electric field the dipoles are randomly oriented. When a material containing electric dipoles is placed between the plates of a capacitor the dipoles tend to line up with the electric field between the plates (see Figure 30–18). In the bulk of the material, the plus and minus ends of the dipoles are close together. They cancel. On the surface of the material, negative ends of the dipoles near the positive plate do not cancel out. This leaves an effective charge near

Table 30–2: Dielectric constants (k) for selected materials.

MATERIAL	DIELECTRIC CONSTANT
Air	1.0006
Bakelite	5.0
Ethyl alcohol	26
Metals	1.000000
Paper	3.7
Paraffin	2.1
Polyethylene	2.2
Porcelain	5.7
Silicon oil	2.5
Vacuum	1.000000

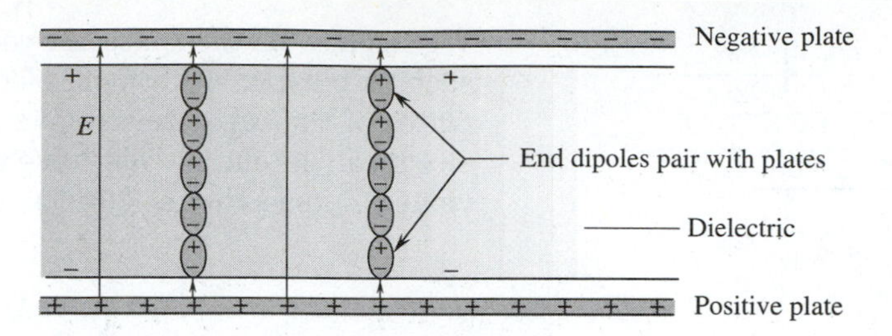

Figure 30–18: Dipoles line up with the electric field of the capacitor. + and − ends of the dipoles are close together; they cancel. The end dipoles do not have another dipole to pair with and induce a charge on the surface of the dielectric.

the positive plate. The reverse happens at the negative plate. In effect, some positive charges are drawn closer to the negative plate, and some negative charges are drawn closer to the positive plate. The size of this effective charge depends on the nature of the molecule.

The effective distance (d') of separation of positive and negative charges decreases when the volume between the capacitor plates is filled with a dielectric. The capacitance without the dielectric is:

$$C_0 = \frac{\epsilon_0 A}{d}$$

With the dielectric the capacitance is larger since $d' < d$.

$$C = \frac{\epsilon_0 A}{d'} = k\,C_0 = \frac{k\epsilon_0 A}{d}$$

We see that the dielectric constant results from the electric dipole nature of dielectric molecules.

□ **EXAMPLE PROBLEM 30–11: A TUBULAR CAPACITOR**

A common way of making electrical capacitors is to roll a long length of metal foil into a cylinder, with a dielectric sandwiched between the layers (see Figure 30–19). A particular tubular capacitor has a capacitance of 2.4×10^{-9} F when polyethylene is used to separate the plates. What would be the capacitance if paper were used instead?

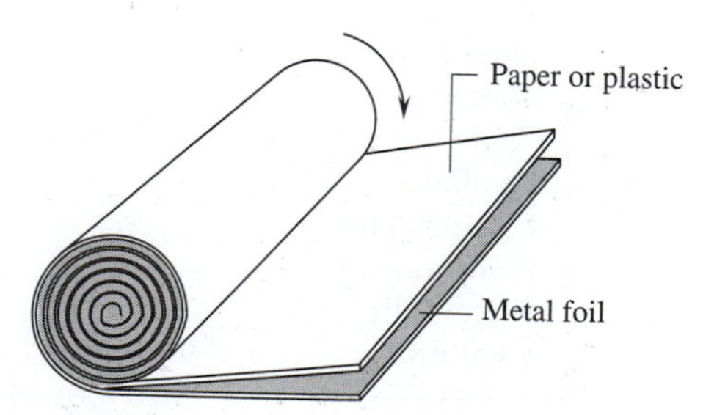

Figure 30–19: The structure of a tubular capacitor.

■ *SOLUTION*

Let C_e and k_e be the capacitance and dielectric constant, respectively, for polyethylene. Let C_p and k_p be the corresponding values for paper. Let C_0 be the capacitance if there were no dielectric material.

$$C_e = k_e C_0 \quad \text{and} \quad C_p = k_p C_0$$

or

$$C_0 = \frac{C_e}{k_e} = \frac{C_p}{k_p}$$

Solving for C_p we get:

$$C_p = k_p \left(\frac{C_e}{k_e}\right)$$

$$C_p = 3.7 \left(\frac{2.4 \times 10^{-9}\ \text{F}}{2.2}\right)$$

$$C_p = 4.0 \times 10^{-9}\ \text{F, or } 4.0\ \text{nF}$$

SUMMARY

Electrical current (I) is the charge per unit time passing a point. It is measured in amperes (A).

$$I = \frac{\Delta Q}{\Delta t}$$

$$1\ \text{A} = 1\ \text{C/s}$$

Electrons move in the opposite direction of electrical current. The average speed of electrons caused by an electric field is called drift velocity ($\mathbf{v}_d$). The electric current through a conductor of cross-sectional area (A) depends on the drift velocity and the electron density (n).

$$I = n\, e\, v_d\, A$$

Current density (J) is electrical current per unit cross section.

$$J = \frac{I}{A}$$

For a conducting wire the electric field is uniform along the length (L) of the wire.

$$E = \frac{V}{L}$$

Current density is directly proportional to the electric field. The proportionality constant is called conductivity (σ).

$$J = \sigma E$$

Resistivity (ρ) is the reciprocal of conductivity.

$$\rho = \frac{1}{\sigma}$$

The potential difference across a conductor is directly proportional to the current passing through it. The proportionality constant is called electrical resistance (R). The relationship is called Ohm's law.

$$V = R\,I$$

Not all electrical devices obey Ohm's law. The resistance of a conductor depends upon its resistivity, length, and cross-sectional area.

$$R = \frac{\rho L}{A}$$

Resistivity and resistance are temperature-dependent. If ρ_0 is the resistivity of a material at room temperature (T_0), then its resistivity at some other temperature (T) is:

$$\rho = \rho_0\{1 + [\alpha(T - T_0)]\}$$

where α is the thermal coefficient of resistivity.

Resistors dissipate energy in the form of heat. The dissipative power of a resistor with a potential difference of V across it is:

$$P = I\,V = R\,I^2 = \frac{V^2}{R} = \frac{\textit{heat loss}}{s}$$

Dielectrics (electric insulators) contain molecules with asymmetric charge distributions. They form electric dipoles. The dipoles rotate to align themselves with an electric field. When placed between the plates of a capacitor, the capacitance is increased. The ratio of capacitance with the dielectric material (C) to the capacitance without it (C_0) is called the dielectric constant. It is a material constant.

$$k = \frac{C}{C_0}$$

KEY TERMS

If you can explain the following terms to a friend or classmate, you understand their meaning. If you cannot explain the terms, you should reread the sections in which they are discussed.

ampere
charge density
conductivity
current density
dielectric constant
dielectrics
drift velocity
electrical current (*I*)
electrical insulators
electric dipoles
electron current
Esaki diodes
free electron model
holes
metals
negative resistance region
Ohm's law
mean free time
resistance
resistivity
semiconductors
thermal coefficient of resistivity
valance electrons

EXERCISES

Section 30.1:

1. Explain the difference between electron current and electrical current.
2. Electrical current flows from positive potential to negative potential. Why then are fuses put on the *negative* side of a power source to protect it?
3. Explain the following terms in your own words.
   A. Drift velocity
   B. Electron density
   C. Current density
   D. Ampere (A)
4. Copper wire A carries the same electrical current as copper wire B. Wire A has twice the diameter of wire B. In all other respects the wires are the same. Complete each of the following statements by adding one of the following: *L*, *S*, *I*, or *N*. (*L*)arger, (*S*)maller, (*I*)dentical, or (*N*)ot possible to determine because of insufficient data.
   A. Compared with wire B, the current density in A is . . .
   B. Compared with wire B, the drift velocity of electrons in wire A is . . .
   C. Compared with wire B, the amount of electrical charge flowing through wire A in 2.0 s is . . .
   D. Compared with wire B, the electron density in A is . . .

Section 30.2:

5. Explain the following terms in your own words.
   A. Free electron model
   B. Conductivity
   C. Mean free time
   D. Resistivity
   E. Resistance
   F. Ohm's law
6. Which of the graphs in Figure 30–20 correspond to devices obeying Ohm's law?
7. Find the resistance of the conductor shown in Figure 30–21.
8. Figure 30–22 shows a plot of electrical potential as a function of length along a wire. Find the strength of the electric field in the wire.

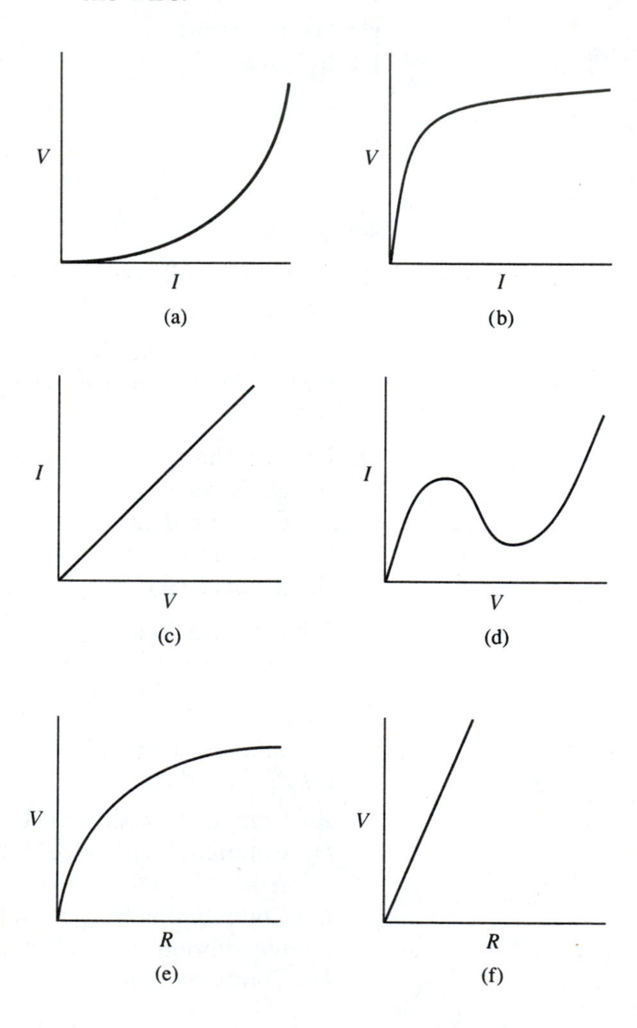

Figure 30–20: *Diagram for Exercise 6. Voltage versus current through a conductor. Notice that not all of these plots are voltage versus current.*

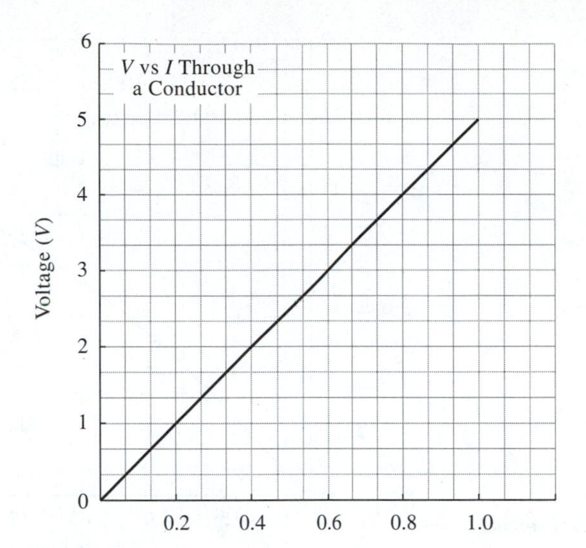

Figure 30–21: Diagram for Exercise 7. Find the resistance.

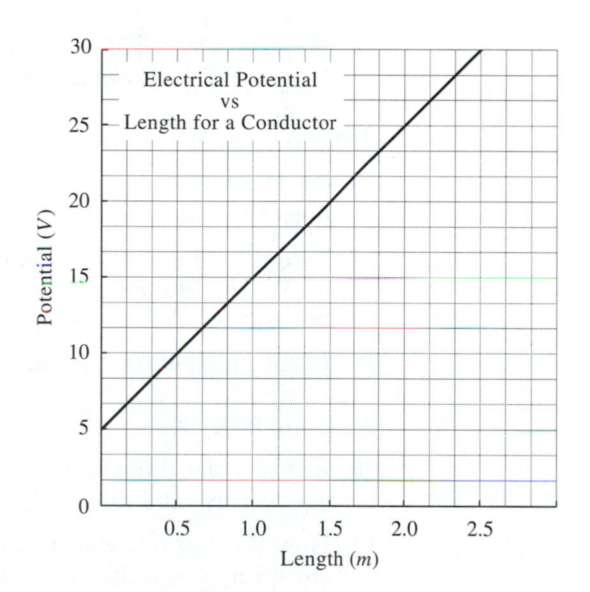

Figure 30–22: Diagram for Exercise 8. Electrical potential versus length for a conductor. Find the electric field.

9. Figure 30–23 shows a graph of voltage versus length. What can you say about the electric field in this device?

Section 30.3:

10. Why are high-voltage transmission lines used to move electrical power over long distances rather than high-current lines?

11. The windings on a 220-V electric motor have a smaller diameter than the windings on a 110-V motor with the same power rating. Why?

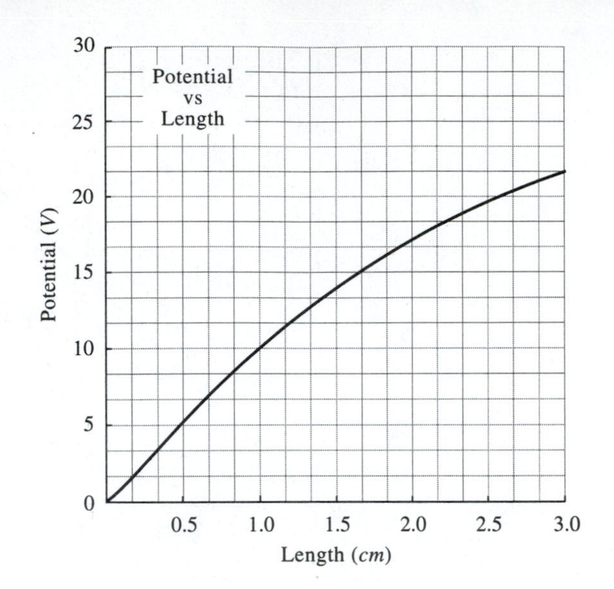

Figure 30–23: Diagram for Exercise 9. Potential versus length.

12. Explain why a 220-V electric motor might burn out if operated from a 110-V power source.

Section 30.4:

13. As an electric toaster with metal heating elements heats up, does the toaster draw more or less current? Why?
14. Two incandescent light bulbs operate at 110 V. One is rated at 50 W; the other at 75 W. Which one has the higher electrical resistance? Which draws the larger current?
15. Light bulbs often burn out when they are turned on. Why?

Sections 30.5, 30.6:

16. Explain the differences that distinguish among metals, semiconductors, and dielectrics.
17. How do the following factors taken individually affect energy stored in a capacitor connected to a battery?
    A. The plate area is increased.
    B. The plate separation is increased.
    C. The electrical potential across the plates is increased.
    D. The empty space between the plates is filled with a dielectric material with $k > 1.0$.

PROBLEMS

Section 30.1:

1. A current of 0.18 A flows through a wire. How many coulombs of charge pass through one end of the wire in 3.1 s? How many electrons?

2. The density of electrons in a metal wire is 5.2×10^{28} electrons/m^3. The wire carries a current of 0.87 A. If its cross section is 1.3×10^{-7} m^2, find:
   A. the current density (J).
   B. the drift velocity of electrons.

3. In a metal wire, the drift velocity is 2.24×10^{-4} m/s. On the average, how long will it take an electron to travel 1.00 m in this particular wire?

4. The drift velocity in a 1.00 mm diameter wire is 7.33×10^{-5} m/s for a current of 1.82 A. What is the density of free electrons in this material?

*5. An electron beam in a CRT tube has a diameter of 2.0 mm and carries a current of 73 mA. The electrons in the beam have a drift velocity of 5.1×10^6 m/s.
   A. What is the *current* density?
   B. What is the *volume* density of electrons?
   C. If the beam is a cylinder 0.28 m long, how many electrons are in the beam at any instant?

Section 30.2:

6. A 3.2 m long wire has a potential difference of 2.1 V across its ends.
   A. How large is the electric field along the wire?
   B. What is the potential difference between the wire's center and one end?

7. For the following conductors find the missing quantities.

   A. $I = 3.0$ A, $V = 110$ V, $R = ?$

   B. $I = 42.0$ mA, $V = ?$, $R = 35.7\ \Omega$

   C. $I = ?$, $V = 12.0$ V, $R = 48\ \Omega$

8. Find the missing quantities.

   A. $J = ?$, $E = 4.0 \times 10^2$ V/m, $\sigma = ?$, $\rho = 5.6 \times 10^{-8}\ \Omega \cdot \text{m}$

   B. $J = 3.3 \times 10^{10}$ A/m^2, $E = 7.2 \times 10^3$ V/m, $\sigma = ?$, $\rho = ?$

   C. $J = 6.2 \times 10^{10}$ A/m^2, $E = ?$, $\sigma = 3.3 \times 10^7\ (\Omega \cdot \text{m})^{-1}$, $\rho = ?$

9. There is a potential difference of 24.2 V across the ends of a 3.24 m long metal wire with a cross section of 2.10×10^{-5} m^2. If the current passing through it is 1.70 A, find the:

A. resistance (R).
B. current density (J).
C. electric field (E).
D. conductivity (σ).
E. resistivity (ρ)

10. Tungsten has a resistivity of $5.6 \times 10^{-8}\ \Omega \cdot \text{m}$. What is the resistance of a filament made from a 11.5-cm length of tungsten with a 0.080-mm diameter?

Section 30.3:

11. A $10\bar{0}0$-W hair dryer operates at $11\bar{0}$ V. What current does it draw?

12. How much heat is dissipated by a $2\bar{0}$-Ω dropping resistor when 0.133 A flows through it?

13. A resistor in an electrical circuit is rated at 1/4 W. The voltage drop across the resistor is 2.6 V.
A. What maximum current can the resistor draw without exceeding its power rating?
B. If the resistance is 120 Ω, how much heat does it dissipate per second with a potential difference of 2.6 V across it?

***14.** A $11\bar{0}$-V DC electric motor draws 8.7 A of current while delivering 523 W of mechanical power.
A. How much electrical energy does the motor consume in 3.0 s?
B. How much mechanical work does the motor perform in 3.0 s?
C. How much heat in joules does it dissipate in 3.0 s?
D. What is the motor's efficiency?

****15.** A cable elevator raises an effective load of 780 kg through a distance of $2\bar{0}$ m in 9.2 s (see Figure 30–24). The overall efficiency of the system is 35%, including a 220-V DC motor. How much current does the motor draw to operate the elevator?

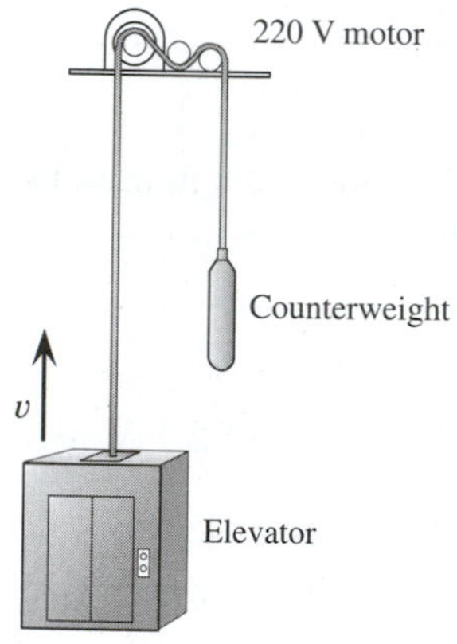

Figure 30–24: *Diagram for Problem 15. A 220-V motor raises an elevator.*

Section 30.4:

16. Figure 30–25 shows a plot of current versus voltage for a flashlight bulb. Use tangent lines to find the resistance of the filament at room temperature (point A) and for a voltage of 4.0 V (point B).

17. A resistor has a resistance of 100 Ω at 20.0°C and a resistance of 97.5 Ω at 70°C. What is the thermal coefficient of resistivity for the material? From Table 30–1 determine the material from which the resistor is made.

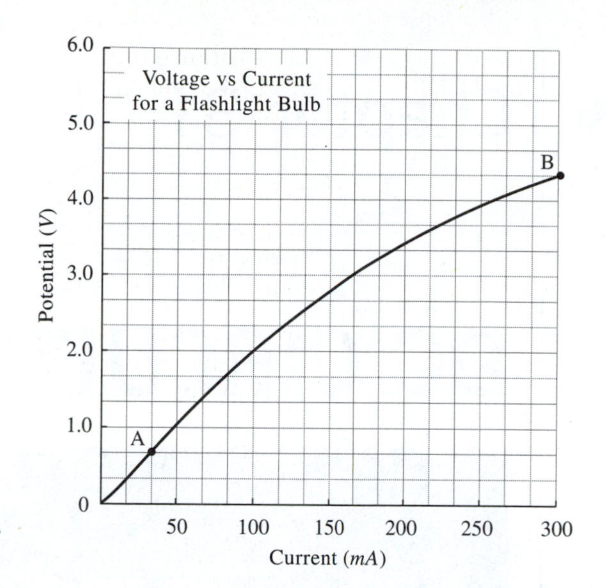

Figure 30–25: *Diagram for Problem 16. Voltage versus current for a flashlight bulb. Use tangent lines to find resistance at two different voltages.*

18. A tungsten filament has a resistance of 4.50 Ω at 20°C. What is its resistance at 750°C?

Section 30.5:

19. A 3.2×10^{-7} F capacitor is connected to a 24.0 V DC voltage source. How much energy is stored in the capacitor?

20. A 3.0-μ F capacitor stores 8.0×10^{-6} J of energy. What is the charge on its plates?

Section 30.6:

21. When a dielectric material is placed in the air gap between the plates of a capacitor, its capacitance increases from 1.24n F to 3.10n F. What is the dielectric constant (k) of the material?

***22.** A parallel-plate capacitor is fully charged across the terminals of a battery to voltage V_0. The capacitor is then disconnected from the power source. A dielectric material is inserted between its plates. The voltage across the plates falls to a new value (V). The charge on the plates remains unchanged. Use Equations 30–15 and 29–10 to show that for this situation $V = V_0/k$.

Chapter 31

DC CIRCUITS

This woman might look a little startled, but she's in no danger. She is experiencing the same kind of electrical charge that you can build up by walking across a rug. As you read about DC circuits, you will encounter some serious considerations for working with any power source. Don't overlook them! (Photograph courtesy of the Ontario Science Centre, Toronto, Ontario Canada.)

OBJECTIVES

In this chapter you will learn:

- how energy and charge are conserved in an electrical circuit
- that voltage gains equal voltage losses around a closed loop in an electrical circuit
- to calculate equivalent capacitance for series and parallel circuits
- to calculate equivalent resistance for series and parallel circuits
- to solve DC networks using Kirchhoff's laws
- how resistance inside a power source affects terminal voltage
- how different kinds of electric cells and batteries are built
- to find unknown resistances using a Wheatstone bridge
- to find unknown potentials using a potentiometer

Time Line for Chapter 31

1771 Henry Cavendish develops the mathematics for a single-fluid theory of electricity.
1800 Alessandro Volta produces the first electric battery.
1833 Michael Faraday states laws of electrolysis.
1880 Jacques-Arsenes d'Arsonval invents an improved galvanometer.

There are a couple of ways we can send electrical energy through a conductor. We can send charge carriers in a steady, one-way path through the conductor. This method is called **direct current (DC).** As work is done on the electrons their potential energy and kinetic energy change. The electrons can also create heat when they bump against the atomic lattice of the material. The flow of electrons and their associated energy are much like the flow of water in plumbing.

Another way of moving electrical energy from one place to another is by **alternating current (AC).** The electrons move back and forth in pretty much the same position in the conductor. The energy is carried by wave motion. This is similar to energy transfer by compressional sound waves.

In this chapter we will discuss steady DC currents. AC circuits will be covered in Chapter 34 after we have learned a little bit about magnetism and magnetic devices.

31.1 ENERGY AND CHARGE CONSERVATION

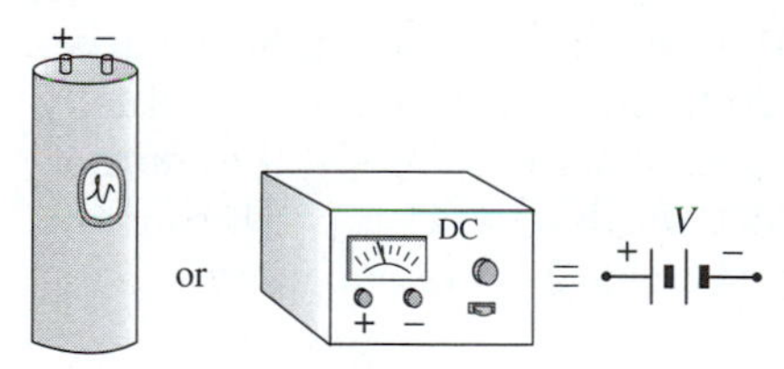

Figure 31–1: DC power sources, such as batteries or generators, are represented in a circuit diagram by a symbol representing the plates.

To set up a DC circuit we first need a constant supply of electrons and a source of energy. We can use a battery, an AC-to-DC converter, or a DC generator. An energy source in an electrical circuit is represented by the symbol shown in Figure 31–1. It looks something like the plates of a wet-cell battery.

Next we need a conduit for the electrons to flow through. Figure 31–2 has added lines to represent hookup wires to the circuit diagram.

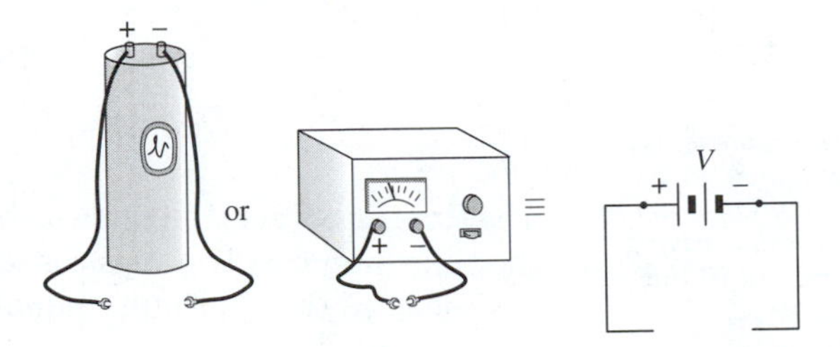

Figure 31–2: Connecting wires are represented by lines.

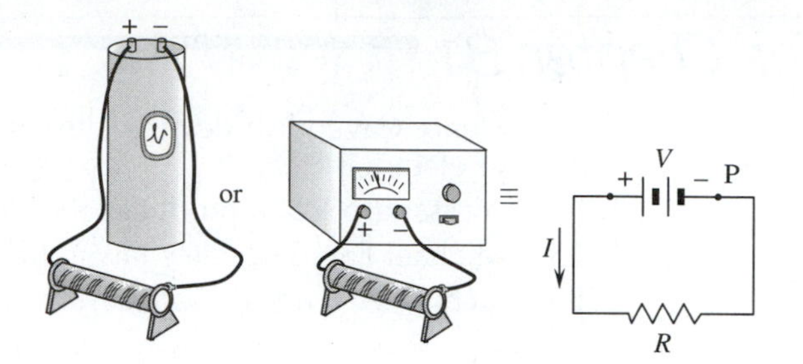

Figure 31–3: Resistors are represented in a circuit diagram by a jagged line.

When the electrons reach the end of a wire they have no place to go. They will pile up and repel any additional electrons trying to reach the end of the wire. We need to close the loop in order to have a constant flow. We will add a resistor to the circuit (Figure 31–3). We now have an energized loop, or circuit, through which the electrons can travel. They gain energy in the power supply and lose energy in the resistor in the form of heat. The electric current flows from the positive terminal of the power supply to the negative terminal through the circuit. Electrons, of course, flow in the opposite direction. Two things must be true concerning this simple circuit. First, there must be a conservation of charge.

The same number of electrons enter and leave a device in an electrical circuit.

The number of electrons entering the resistor equals the number leaving the resistor. This is true for capacitors as well as for resistors. We can replace the resistor with a capacitor, represented by a schematic of the plates of a parallel-plate capacitor (see Figure 31–4). A capacitor is initially uncharged. As electrons pile up on the negative plate, they repel an equal number from the positive plate. Eventually coulomb repulsive forces on the negative plate will resist the addition of approaching electrons, and electron flow will stop. During the time that electrons were moving, the charges entering and leaving the capacitor were the same.

There is a conservation of charge for any instant of time. We can express the charge in terms of current flow for a short period of time (Δt). For any device:

$$Q(\text{in}) = Q(\text{out})$$

$$I(\text{in})\,\Delta t = I(\text{out})\,\Delta t$$

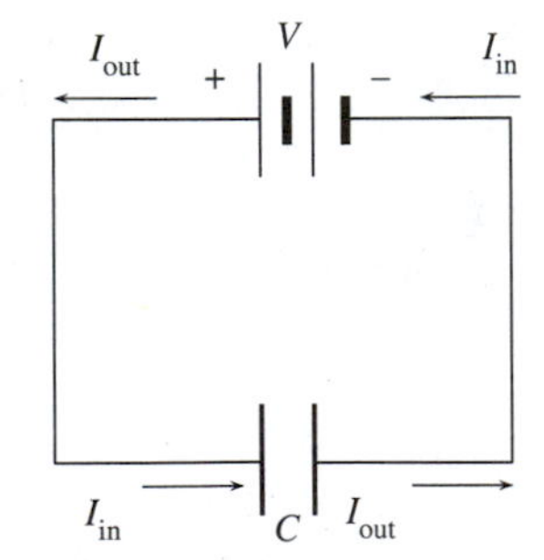

Figure 31–4: Capacitors are represented by a schematic of parallel plates. As a capacitor charges up, the current entering the capacitor equals the current leaving it. The current flows in and out of the power source are also equal.

The time interval is the same on both sides of the equation. Canceling the time interval gives us a conservation of current relationship that is similar to the continuity equation for fluids (see Equation 17–1).

The current going in and coming out of any device is the same.

$$\boldsymbol{I}(\textbf{in}) = \boldsymbol{I}(\textbf{out})$$

The second thing that is true about this circuit is that the energy of the electrons is conserved.

In any closed circuit loop, energy gains equal energy losses.

This is just another statement of the conservation of energy law. Electrons gain energy in the power supply. This energy either shows up as stored potential energy in a capacitor or heat losses in a resistor. Energy is conserved at all times. We can write the energy conservation in terms of power.

$$\frac{\textit{energy gain}}{\textit{time}} - \frac{\textit{energy loss}}{\textit{time}} = 0$$

$$\textit{power gain} - \textit{power loss} = 0$$

Let us look at the circuit in Figure 31–3. We will start at point P and move counterclockwise around the loop. The battery is a power source. As we move from the negative to the positive plate, the electric current undergoes a power increase ($I\ V$). We will assume that power losses in the connecting wire are too small to consider. The next power change is across the resistor. The power lost to heat is $R\ I^2$. Following the connecting wire, we return to point P, ready to go through the battery again. The power gains and losses in the circuit are:

$$I\ V(\text{gain}) - R\ I^2(\text{loss}) = 0$$

Cancel out an I.

$$V(\text{gain}) - R\ I(\text{loss}) = 0$$

$$V(\text{gain}) = R\ I(\text{loss})$$

Remember Ohm's law. The $R\ I$(loss) is numerically equal to the electrical potential across the resistor.

$$\boldsymbol{V}(\textbf{gain}) = \boldsymbol{V}(\textbf{loss})$$

We can also write the conservation of energy in a circuit loop in terms of electrical potential. The potential difference across a resistor will be the IR drop. For a capacitor we can express its voltage drop as $\Delta v = Q/C$. In many applications it is easier to use conservation of electrical potential.

31.2 EQUIVALENT CAPACITANCE

Figure 31–5 shows two ways to hook up three capacitors. In Figure 31–5a the positive side of each capacitor is connected to the positive sides of the other capacitors. All of the negative sides are hooked together. Electrons coming from the battery have a choice at point N. Any particular charge can go through any of the three capacitors. The capacitors are hooked up in a **parallel connection.**

In Figure 31–5b the positive side of each capacitor is connected to the negative plate of the next capacitor. Electrons coming from the battery have no choice. There is only one path for them to travel. This kind of hookup is called a **series connection.**

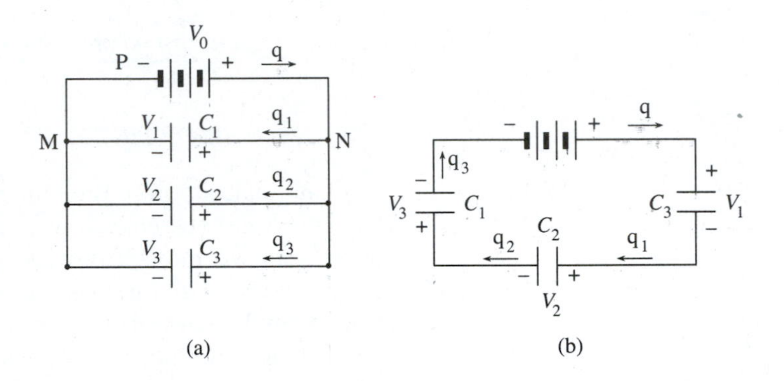

Figure 31–5: *Two ways to hook up three capacitors. (a) Parallel connection. All positive plates are connected to the positive side of the battery. There are several current paths. Each capacitor has the same potential drop across it. (b) Series connection. The positive side of each capacitor is connected to the negative side of the next capacitor. There is only one current path. The sum of the potential drops across the capacitors equals the potential gain of the battery.*

Let us look at the parallel circuit in Figure 31–5a. Point N is called a **node.** Electrons have a choice of paths at a node. We can treat it as a "device"—charge in equals charge out. The charge from the battery (q) enters the node. It leaves in any one of three paths. We can write the conservation of charge equation:

$$q = q_1 + q_2 + q_3$$

We can check the conservation of electrical potential for the three separate circuit loops formed by the capacitors. Start at point P. Charge gains potential (V) as it goes through the battery. At the node, we choose the path through capacitor C_1. The potential drop is V_1. We pass through node M and are back at point P. V(gain) = V_1(loss). We can repeat the process, choosing to go through a different capacitor at node N. We get:

$$V = V_1; \quad V = V_2; \quad \text{and } V = V_3$$

The potential drop across each capacitor equals the potential gain of the battery.

The charge on a capacitor is CV. Substitute this for charge in the charge conservation equation.

$$C\,V = (C_1\,V_1) + (C_2\,V_2) + (C_3\,V_3)$$

The potentials are all the same.

$$C\,V = (C_1\,V) + (C_2\,V) + (C_3\,V)$$

Cancel potential from the equation.

$$C = C_1 + C_2 + C_3 \qquad \textbf{(Eq. 31–1)}$$

If we replace the three capacitors with a single capacitor with a capacitance (C) calculated by Equation 31–1, the same charge will be drawn from the battery and the same amount of energy will be stored on the capacitor plates. C is the equivalent capacitance of the parallel circuit.

Next let us look at the series circuit. Since there is only one path for charge to travel, the charge on each capacitor is the same. The electrons leaving the positive plate of capacitor C_1 show up on the negative plate of C_2 and so on. Let q be the charge leaving the battery.

$$q = q_1 = q_2 = q_3$$

We can add up the potential gains and losses in the single loop.

$$V = V_1 + V_2 + V_3$$

We will express potential in terms of capacitance.

$$\frac{q}{C} = \frac{q_1}{C_1} + \frac{q_2}{C_2} + \frac{q_3}{C_3}$$

Charge cancels out of the equation.

$$\frac{1}{C} = \frac{1}{C_1} + \frac{1}{C_2} + \frac{1}{C_3} \qquad \textbf{(Eq. 31–2)}$$

C is the equivalent capacitor of the three series capacitors. If the three capacitors were replaced by a single capacitor with the value calculated by Equation 31–2, the same charge would be drawn from the battery and an equivalent amount of energy would be stored on its plates.

A Shocking Experience

Your body can act as a capacitor. When you walk across a polyester rug on a dry day, electrical charges can build up on the surface of your body. The salty perspiration on your skin acts as a conductor. The skin itself has a fairly high electrical resistance. Most of the charge remains on the body surface, building up a high electrical potential. When you reach for a metal doorknob or touch someone with a different electrical potential, sparks can jump from the area of contact. This static electricity is a bit startling, but it is no major concern.

Electrical current from an AC power source is something else. Your muscles are triggered by electrical impulses. Again, the resistance of your skin helps protect you. Much of the current stays near the surface of the body. Typically, your skin has a resistance of $10^5\ \Omega$. If you grab a pair of 120-V AC wires, a current of 1.2 mA will travel between the wires through your body. Most of this current will move along your skin. A small portion of it will pass through the interior of your body.

The most dangerous part of the current is that small portion that passes through your heart. If a current as low as 10 mA passes through your heart it could cause your heart to undergo uncontrolled contractions. Your heart would be unable to pump blood, and you would die. This random contraction is called ventricular fibrillation.

Here is a list of things that can happen when an electrical source is in contact with the skin. If the source is interior to the skin, as might occur with medical equipment, currents above 10 mA may be fatal.

(Photograph courtesy of the Ontario Science Centre, Toronto, Ontario Canada.)

Current	Effect
0.5 mA	Painful sensations occur; muscles may contract.
16 mA	The victim is unable to let go.
5–25 mA	Muscle damage occurs from extreme contraction.
25–100 mA	Respiratory paralysis. Victim may be revived by artificial respiration.
1–3 A	Ventricular fibrillation. Prompt cardiac resuscitation may revive the victim.
above 3 A	Fatal. The heart tissue is destroyed. Attempts to revive the victim are useless.

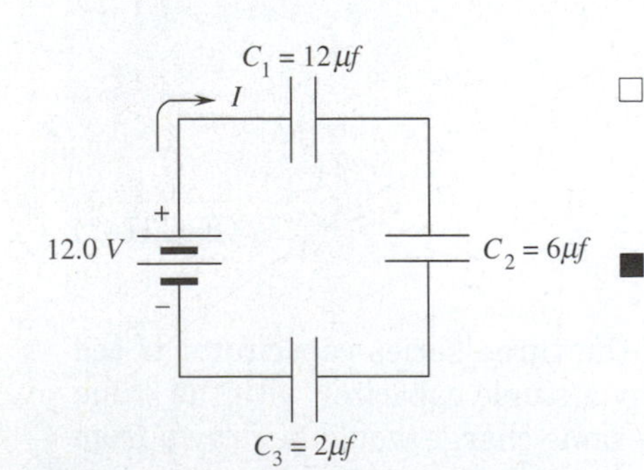

Figure 31–6: *Three capacitors are connected in series.*

☐ **EXAMPLE PROBLEM 31–1: CAPACITORS IN SERIES**

Find the equivalent capacitance of the capacitors shown in Figure 31–6.

■ ***SOLUTION***

$$\frac{1}{C} = \frac{1}{C_1} + \frac{1}{C_2} + \frac{1}{C_3}$$

$$\frac{1}{C} = \frac{1}{12\ \mu\text{F}} + \frac{1}{6.0\ \mu\text{F}} + \frac{1}{2.0\ \mu\text{F}}$$

$$\frac{1}{C} = (0.0833 + 0.1666 + 0.5000\ \mu F)^{-1} = (0.75\ \mu F)^{-1}$$

$$C = \frac{1}{0.75}\ \mu F$$

$$C = 1.3\ \mu F$$

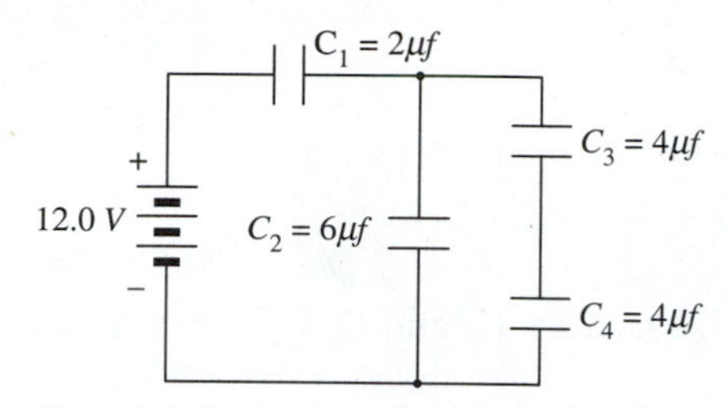

Figure 31–7: *A network of capacitors in combinations of series and parallel connections.*

Often a circuit has a mixture of series and parallel connections. Figure 31–7 is an example. Here is a general strategy for finding the equivalent circuit.

1. Look for the simplest series or parallel connections.
2. Replace that combination with its equivalent capacitance.
3. Redraw the circuit with the new equivalent capacitance.
4. Repeat steps 1, 2, and 3 until you have one equivalent capacitance.

□ **EXAMPLE PROBLEM 31–2: A COMPLEX CAPACITANCE CIRCUIT**

Find the potential drop across capacitor C_1 in Figure 31–7.

■ ***SOLUTION***

All charges leaving the battery must pass through C_1. We can find the charge by calculating the equivalent circuit. We can then find the potential using the capacitance equation. The farthest combination from the source is C_4 and C_3. They are in series.

$$\frac{1}{C_{3,4}} = \frac{1}{C_3} + \frac{1}{C_4}$$

$$\frac{1}{C_{3,4}} = \frac{1}{4.0\ \mu F} + \frac{1}{4.0\ \mu F}$$

$$C_{3,4} = 2.0\ \mu F$$

Next, redraw the diagram.

In Figure 31–8, C_2 and $C_{3,4}$ are in parallel.

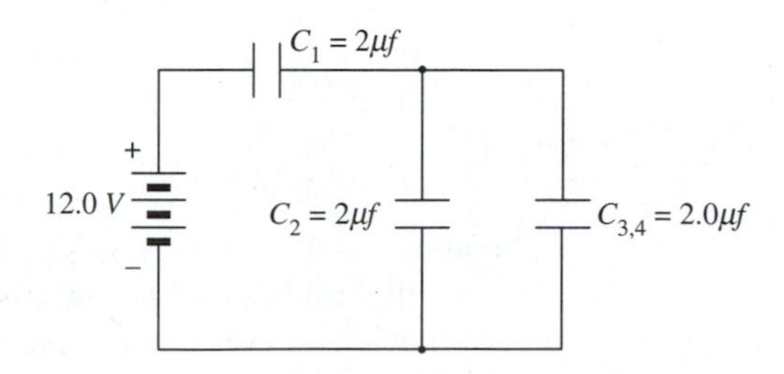

Figure 31–8: *Capacitors C_3 and C_4 have been replaced with their equivalent capacitance.*

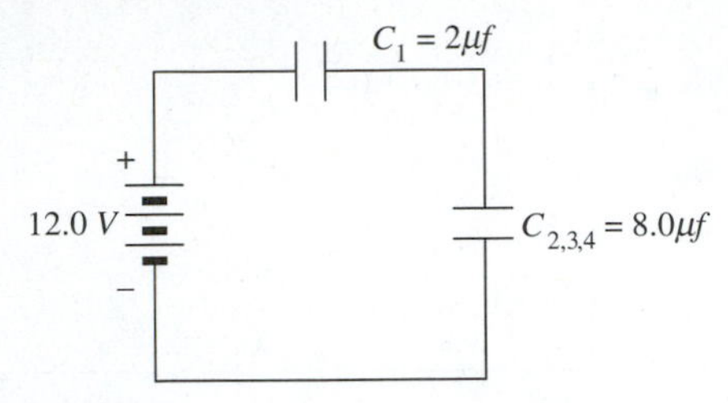

Figure 31–9: *Capacitors C_4, C_3, and C_2 have been replaced by their equivalent capacitance.*

$$C_{2,3,4} = C_2 + C_3$$

$$C_{2,3,4} = 6.0\ \mu\text{F} + 2.0\ \mu\text{F}$$

$$C_{2,3,4} = 8.0\ \mu\text{F}$$

Again, redraw the diagram. Capacitors C_1 and $C_{2,3,4}$ are in series (Figure 31–9).

$$\frac{1}{C} = \frac{1}{C_1} + \frac{1}{C_{2,3,4}}$$

$$\frac{1}{C} = \frac{1}{2.0\ \mu\text{F}} + \frac{1}{8.0\ \mu\text{F}}$$

$$C = 1.6\ \mu\text{F}$$

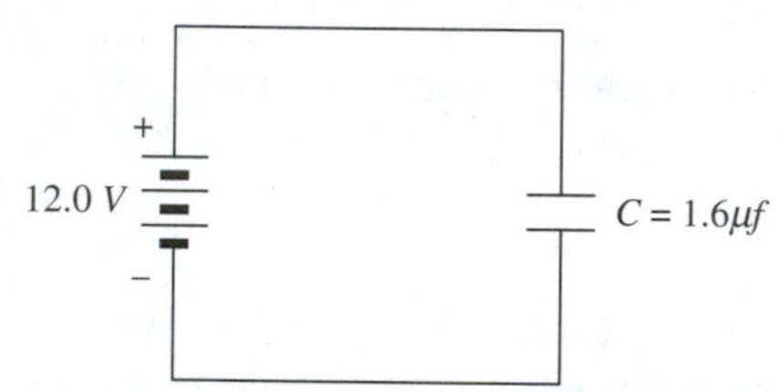

Figure 31–10: *The complete network of capacitors can be represented by a single equivalent capacitance, which is used to find the charge drawn from the battery.*

Here is our equivalent circuit (Figure 31–10). The charge drawn from the battery is:

$$Q = C\,V = 1.6\ \mu\text{F} \times 12.0\ \text{V}$$

$$Q = 19.2\ \mu\text{C}$$

Since C_1 is in series with the rest of the capacitors (look at Figure 31–9), Q is the charge on its plates.

$$V_1 = \frac{Q}{C_1} = \frac{19.2\ \mu\text{C}}{2.0\ \mu\text{F}}$$

$$V_1 = 9.6\ \text{V}$$

If we wanted the voltage across the other capacitors, we could work backward with the help of the equivalent circuits we have drawn and conservation rules. For instance, the potential drop across C_2 and the equivalent capacitance $C_{3,4}$ is 2.4 V from the conservation of potential around a loop.

31.3 EQUIVALENT RESISTANCE

The analysis of resistors is similar to the analysis of capacitors, but the parallel and series equations will be a bit different. We will use conservation of current in place of conservation of charge.

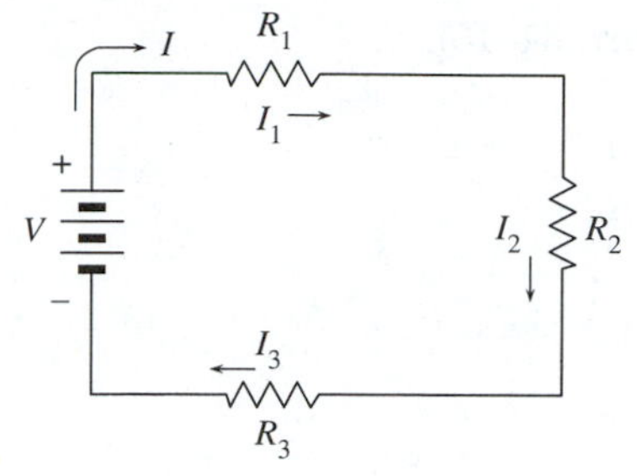

Figure 31–11: Three resistors are connected in series. There is a single current path. The sum of potential drops across the three resistors equals the potential gain of the power source.

Figure 31–11 shows three resistors in series. Since there is only one current path, they all carry the same current.

$$I = I_1 = I_2 = I_3$$

The only potential gain is at the DC power supply. This must equal the sum of the potential losses across the three resistors.

$$V = V_1 + V_2 + V_3$$

$$I\,R = (I\,R_1) + (I\,R_2) + (I\,R_3)$$

$$R = R_1 + R_2 + R_3 \qquad \textbf{(Eq. 31–3)}$$

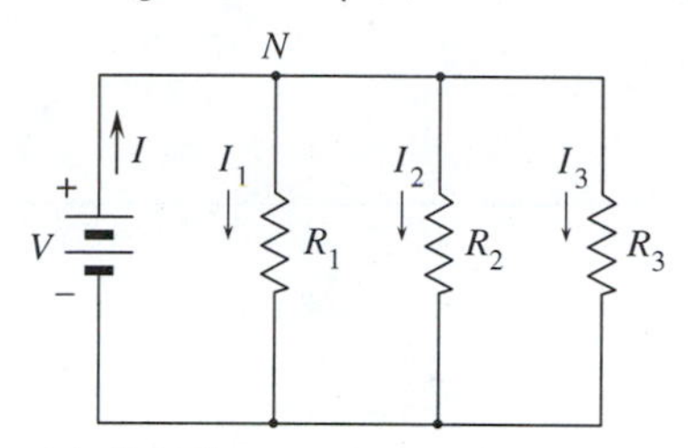

Figure 31–12: Three resistors are connected in parallel. The current drawn from the power source is the sum of the currents across the resistors. The potential drop across each resistor equals the potential gain of the power source.

Figure 31–12 shows three resistors in parallel. The current from the power supply splits into three separate currents at node N. The current leaving the node passes through the three resistors. Each resistor has the same potential drop across it.

$$I = I_1 + I_2 + I_3$$

or

$$\frac{V}{R} = \frac{V}{R_1} + \frac{V}{R_2} + \frac{V}{R_3}$$

$$\frac{1}{R} = \frac{1}{R_1} + \frac{1}{R_2} + \frac{1}{R_3} \qquad \textbf{(Eq. 31–4)}$$

For complicated mixtures of parallel and series resistors we can use the same technique outlined in Section 31.2 for capacitors.

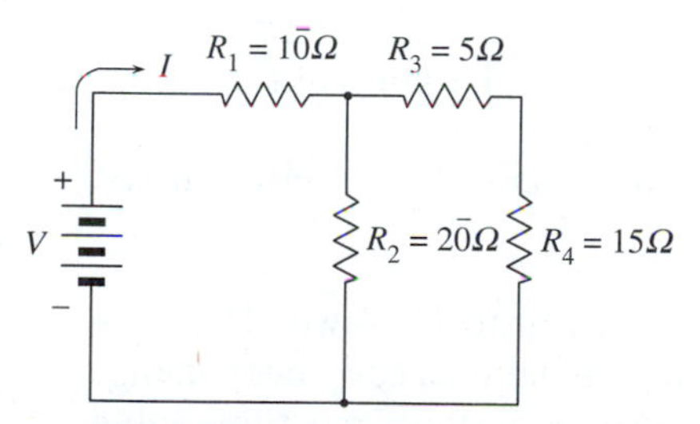

Figure 31–13: A network of resistors is connected to a voltage source.

□ **EXAMPLE PROBLEM 31–3: A NETWORK OF RESISTORS**

Find the equivalent resistance of the network of resistors in Figure 31–13.

■ ***SOLUTION***

R_3 and R_4 are in series (Figure 31–14).

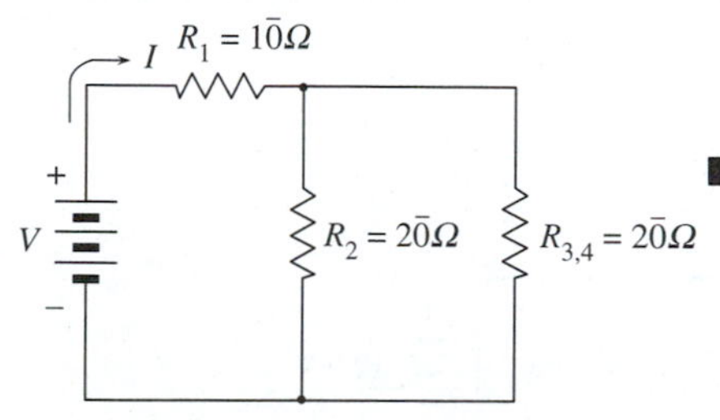

Figure 31–14: Resistors R_3 and R_4 have been replaced by their equivalent resistance.

$$R_{3,4} = R_3 + R_4$$

$$R_{3,4} = 5\Omega + 15\Omega$$

$$R_{3,4} = 20\Omega$$

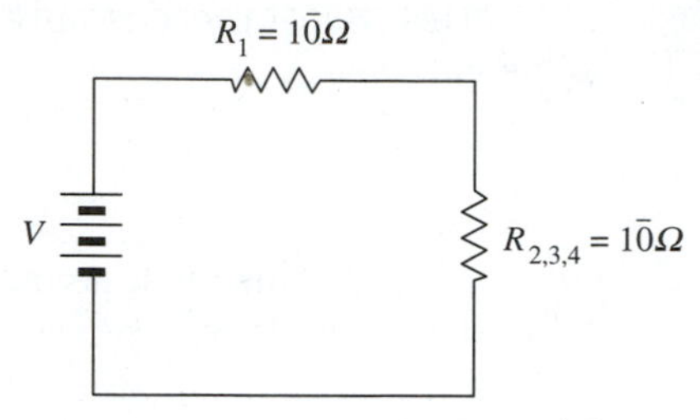

Figure 31–15: Resistors R_2, R_3, and R_4 have been replaced by their equivalent resistance.

$R_{3,4}$ and R_2 are in parallel (Figure 31–15).

$$\frac{1}{R_{2,3,4}} = \frac{1}{R_{3,4}} + \frac{1}{R_2}$$

$$\frac{1}{R_{2,3,4}} = \frac{1}{2\bar{0}\Omega} + \frac{1}{2\bar{0}\Omega}$$

$$R_{2,3,4} = 1\bar{0}\Omega$$

R_1 and $R_{2,3,4}$ are in series.

$$R = R_1 + R_{2,3,4}$$

$$R = 1\bar{0}\Omega + 1\bar{0}\Omega$$

$$R = 2\bar{0}\Omega$$

31.4 DC NETWORKS

When there is more than one power source in a circuit, equivalent resistance Equations 31–3 and 31–4 do not work. We need a different strategy. Figure 31–16 shows a simple circuit with two power sources. We can find the currents flowing through each device by applying two simple rules.

1. For each node in the circuit, the sum of the currents entering equals the sum of the currents leaving.
2. The sum of all of the potential gains equals the potential losses around any closed circuit loop.

These rules are often referred to as **Kirchhoff's laws.** They are a restatement of the conservation laws we have already been using. We can use these ideas to create linear equations and then solve them simultaneously. We need as many equations as we have unknowns. The idea is simple, but it can result in a bit of tedium while grinding out the results.

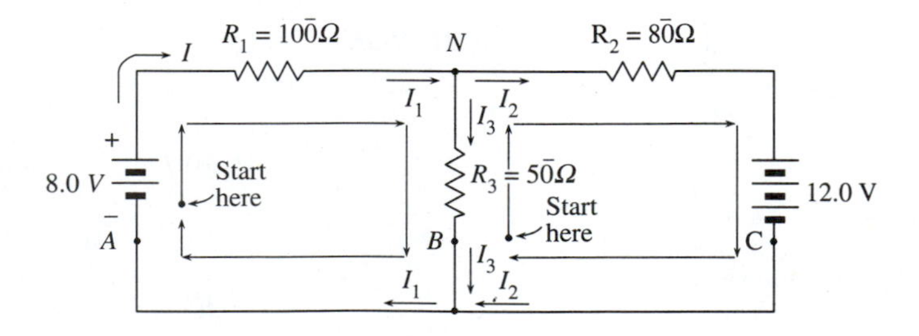

Figure 31–16: A simple circuit with two power sources.

☐ EXAMPLE PROBLEM 31–4: KIRCHHOFF'S LAW

Find the currents through resistors R_1, R_2, and R_3 in Figure 31–16.

■ *SOLUTION*

Let I_1 be the current through R_1. Let I_2 be the current through R_2. Let I_3 be the current through R_3.

We need three equations because we have three currents.

A. Nodal equation:

At node N assume I_1 enters the node. I_2 and I_3 leave the node. If we have chosen the wrong direction of current flow, the magnitude of the calculated current will be unaffected. We will just get a negative current in out answer. The currents entering and leaving the node are shown in the diagram.

$$I_1 = I_2 + I_3 \qquad \textbf{(Eq. 31–5)}$$

B. Loop equations:

Let us write an equation for loop AB. Start at point A. We will go clockwise around the loop. We gain potential as we go from − to + in the battery. The potential drop across R_1 is $R_1 \times I_1$. The potential drop across R_3 is $R_3 \times I_3$. For convenience we will write the equation without units.

$$gain = loss$$
$$8.0 = 10\bar{0}\, I_1 + 50\, I_3 \qquad \textbf{(Eq. 31–6)}$$

Next choose another loop. We will use loop BC. Start at B and go around clockwise. As we trace around the loop we pass through R_3 in the direction opposite to current I_3. We are going from low potential to high potential. We see a potential gain of $R_3 \times I_3$. The potential drop across R_2 is $R_2 \times I_2$. As we travel through the battery, we go from the positive terminal to the negative terminal. This is a drop of potential.

$$gain = loss$$
$$5\bar{0}\, I_3 = 8\bar{0}\, I_2 + 12.0$$

or

$$-12.0 = -5\bar{0}\, I_3 + 8\bar{0}\, I_2 \qquad \textbf{(Eq. 31–7)}$$

We now have three equations. We can eliminate two of the unknowns. First eliminate I_1 by substituting the nodal equation (Equation 31–5) into loop Equation 31–6.

$$8.0 = [10\bar{0}(I_3 + I_2)] + 5\bar{0}\, I_3$$

or

$$8.0 = 15\bar{0}\, I_3 + 10\bar{0}\, I_2 \qquad \textbf{(Eq. 31–8)}$$

Multiply Equation 31–7 by three and add it to Equation 31–8.

$$3(-12.0) = 3(-5\bar{0}\, I_3 + 8\bar{0}\, I_2)$$

$$\begin{array}{rcl} -36.0 &=& -15\bar{0}\, I_3 + 24\bar{0}\, I_2 \\ +\quad 8.0 &=& +15\bar{0}\, I_3 + 10\bar{0}\, I_2 \\ \hline -28.0 &=& 0 \quad + 340\, I_2 \end{array}$$

$$I_2 = \frac{-28.0}{34\bar{0}} = -0.0824 \text{ A, or } -82.4 \text{ mA}$$

Apparently, the current is flowing in the opposite direction to the arrow in the diagram since we have a negative current in our answer. We can now substitute this value into Equation 31–7 or 31–8 to find I_3.

$$-12.0 = -5\bar{0}\, I_3 + 8\bar{0}\,(-0.0824)$$

$$I_3 = 0.108 \text{ A}$$

or

$$8.0 = 15\bar{0}\, I_3 + 10\bar{0}\,(-0.0824)$$

$$I_3 = 0.108 \text{ A}$$

The nodal equation can be used to find I_1.

$$I_1 = 0.108 \text{ A} - 0.0824 \text{ A}$$

$$I_1 = 0.026 \text{ A}$$

In summary:

$$I_1 = 0.026 \text{ A}$$

$$I_2 = -0.082 \text{ A}$$

$$I_3 = 0.108 \text{ A}$$

31.5 TERMINAL VOLTAGE

The voltage across the terminals of an ignition dry cell is checked before it is hooked up to a circuit. The voltage is 1.55 V. The cell is then connected to a circuit. The voltage across its terminals, measured as current is drawn from the cell, is called the **terminal voltage.** When a 0.50-A current is drawn from the cell, the terminal voltage falls to 1.45 V. If the current is increased to 1.000 A, the terminal voltage drops to 1.35 V. Further increases of the current load reduce the voltage across the terminals.

The cell is replaced with an AC-to-DC converter. The terminal voltage is set at 1.55 V before the circuit is hooked to the power supply. Again the terminal voltage is reduced as current is drawn from the power source. In this case the change of terminal voltage with load current is not as drastic as with the dry cell.

We can explain the behavior of terminal voltage by assuming the power supply has some **internal resistance.** Electrical current must pass through this internal resistance before it reaches the terminals of the power source (see Figure 31–17).

The maximum electrical potential of a dry cell occurs when no current is drawn. It is determined by the chemical potential difference of the materials that make up the cell. This is called the **electromotive force (EMF).** The EMF equals the terminal voltage when no current is drawn from the cell. For an AC-to-DC converter or other power supply we can use the same idea. The EMF is the terminal voltage when no current is being drawn from the power supply.

When current is drawn from the power source, an Ir potential drop occurs across the internal resistance. The potential across the terminals (V_t) is the difference between the EMF and the Ir drop across the internal resistance.

$$V_t = EMF - (I\,r) \qquad \textbf{(Eq. 31–9)}$$

We can find the internal resistance of a power supply experimentally. Simply monitor the terminal voltage as more and more

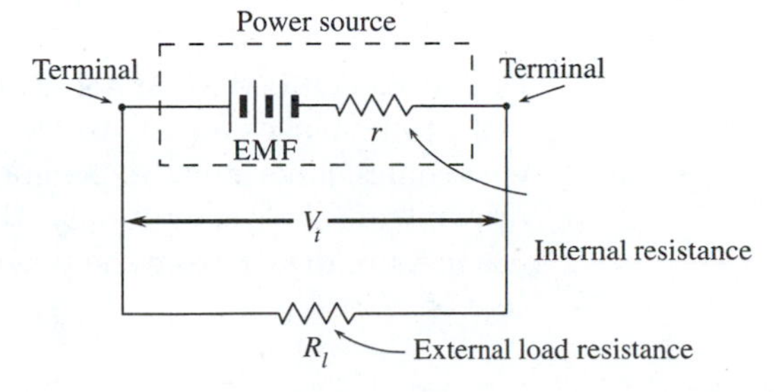

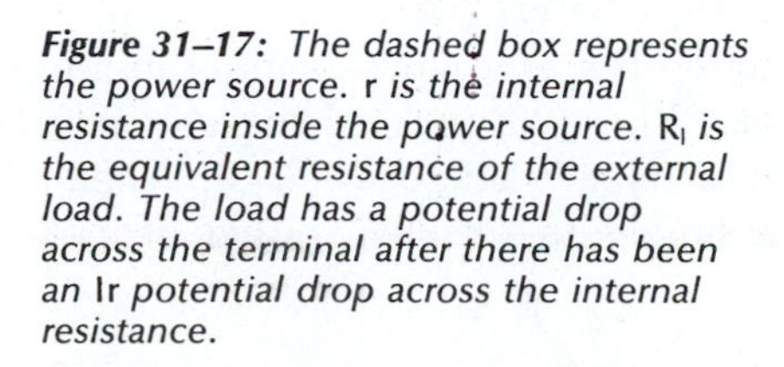
Figure 31–17: *The dashed box represents the power source. r is the internal resistance inside the power source. R_l is the equivalent resistance of the external load. The load has a potential drop across the terminal after there has been an Ir potential drop across the internal resistance.*

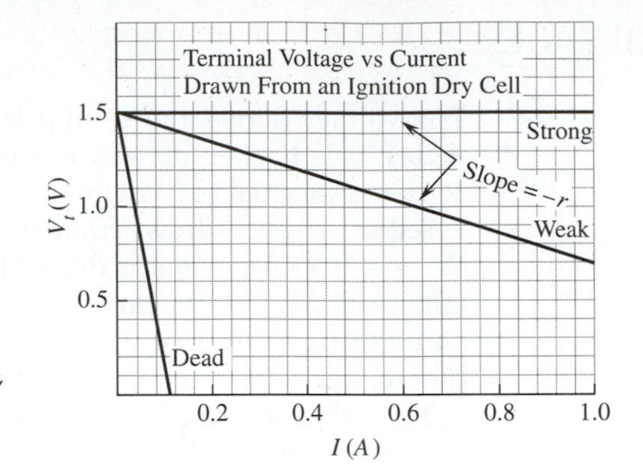

Figure 31–18: *A plot of terminal voltage versus current drawn from an ignition dry cell. As a dry cell ages, the internal resistance increases.*

current is drawn from the power source. The slope of a plot of terminal voltage versus current will be $-r$. The y intercept is the EMF of the source. Figure 31–18 shows plots for three ignition dry cells. As the dry cell ages, chemical reaction rates slow down. This increases the internal resistance.

☐ **EXAMPLE PROBLEM 31–5: INTERNAL RESISTANCE**

From Figure 31–18 find the internal resistance and EMF of the "weak" cell.

■ ***SOLUTION***

The EMF is the y intercept.

$$EMF = 1.50 \text{ V}$$

The resistance can be obtained from the slope.

$$slope = -r = \frac{\Delta V_t}{\Delta I}$$

$$-r = \frac{(0.75 - 1.50) \text{ V}}{(1.0 - 0.0) \text{ A}}$$

$$r = 0.75 \ \Omega$$

☐ **EXAMPLE PROBLEM 31–6: AN AC-TO-DC CONVERTER**

A student uses an AC-to-DC converter as a power supply in an electronics laboratory experiment. When 200 mA of current are drawn from the power supply, the terminal voltage is 5.82 V. When 500 mA are drawn from the power supply its terminal voltage drops to 5.55 V.

A. Find the internal resistance of the power source.
B. What is the terminal voltage when no current is drawn from the converter?

■ *SOLUTION*

A. Internal resistance:

$$-r = \frac{V_t}{I}$$

$$r = \frac{-(5.55 - 5.82)\text{ V}}{(0.50 - 0.20)\text{ A}}$$

$$r = 0.90\ \Omega$$

B. EMF

$$V_t = EMF - (r\,I)$$

or

$$EMF = V_t + (r\,I)$$

$$EMF = 5.55\text{ V} + 0.90\ \Omega \times 0.50\text{ A}$$

$$EMF = 6.00\text{ V}$$

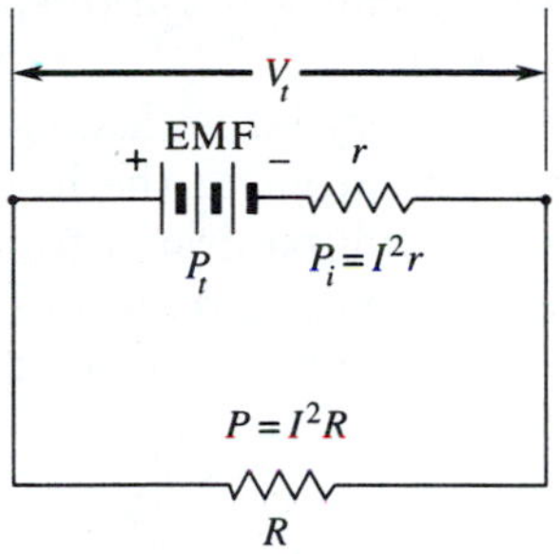

Figure 31–19: The total power P_t supplied by an EMF is divided between an internal loss (P_i) and the power delivered to the external circuit (P).

Internal resistance can waste useful energy. Figure 31–19 shows a power source connected to a load resistance (R). The power source might be an electric generator, signal generator battery, AC-to-DC converter, or an amplifier. The load resistance might be the equivalent resistance of a complex circuit or something simple such as a speaker. The potential gain (EMF) from the power supply is lost at the internal resistance as well as at the load resistance.

$$EMF = (I\,r) + (I\,R)$$

Let us solve this for current.

$$I = \frac{EMF}{R + r} \qquad \textbf{(Eq. 31–10)}$$

The power gains and losses are:

$$I\,(EMF) = (I^2\,R) + (I^2\,r)$$

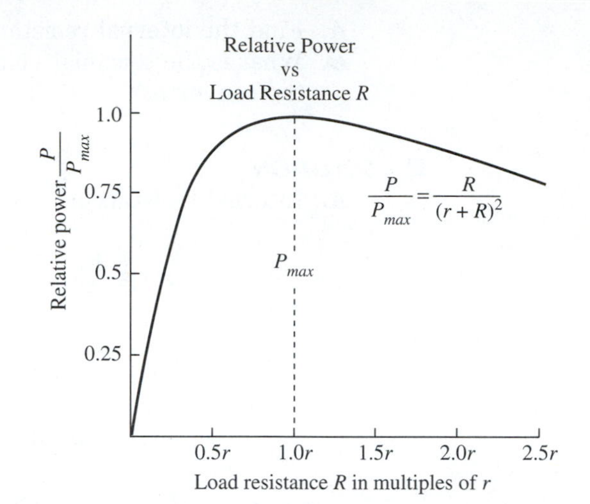

Figure 31–20: A plot of relative power output from a power source versus load resistance (R). R *is expressed in terms of multiples of the internal resistance* r.

The power delivered to the load resistance is $P = I^2 R$. The power lost to the internal resistance is $I^2 r$. If r is small, most of the power will go to the load resistance (R).

Often the EMF has a fixed value. If R is very large, less current will be drawn from the source. The power output to the load will be small. If R is very small, more current will be drawn from the *EMF*, but much of the power is lost to the internal resistance. We can find the value of R that will give the maximum power to the load resistance by using Equation 31–10.

$$P = I^2 R = \frac{(EMF)^2 R}{(R + r)^2}$$

Both the EMF and r are constants. We can plot power delivered (P) against the load resistance R. (See Figure 31–20.) The maximum power is delivered when the load resistance is equal to the internal resistance. That is why it is important to match speaker impedences (effective resistance) with stereo systems and to match the impedence of antenna lead-ins with television sets.

31.6 APPLICATIONS

Batteries and Electric Cells

An electric cell can be made by sticking a piece of copper wire and a steel paper clip into a lemon. A potential difference shows up between the two dissimilar metals. This is not a very good battery, but it shows the three essential parts needed to make an electric cell.

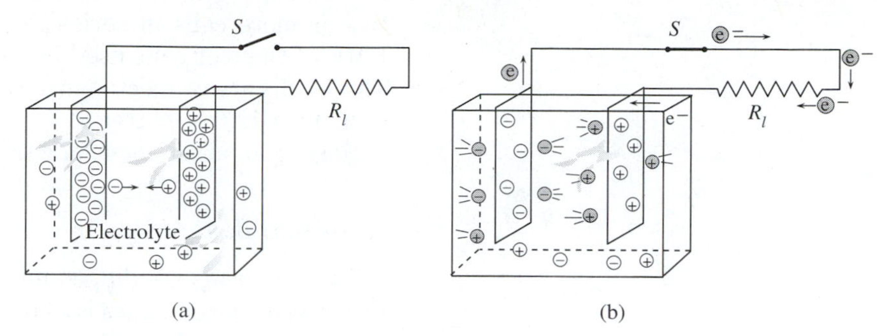

Figure 31–21: (a) *When switch S is open, ions pile up on the plates. Coulomb forces prevent additional ions from approaching the plates.* (b) *When the external circuit is closed, electrons can leave the negative plate, travel through the external circuit, and reach the other plate to neutralize the positive ions. Coulomb forces are reduced, and the chemical reaction can proceed.*

We need two materials with different chemical potentials. These two materials are the **electrodes** for the battery. The electrodes are immersed in an **electrolyte,** a liquid or paste that can be easily ionized.

A chemical action occurs in which positive ions pile up on one electrode and electrons on the other. Coulomb forces shortly stop the reactions. Additional electrons are repelled by the negatively charged electrode, and anions are repelled from the positive electrode (see figure 31–21).

If we supply an external conducting path between the electrodes, electrons can leave the negative electrode and travel outside the cell to reach the positive electrode and neutralize the anions. The chemical reaction can then continue until the external circuit is broken.

The EMF developed between the electrodes of a cell depends on the materials from which they are made. Typically, the EMF of a single cell is from 1 to 2 V. Higher voltages are obtained by combining

Ben Franklin Was a Lucky Fellow

Benjamin Franklin was many things in his life. He was a printer, a publisher, an inventor, a statesman, and a diplomat.

He was also a scientist. His greatest impact was in the area of electricity. European physicists thought electricity was made up of two different kinds of fluids corresponding to two different charges. Franklin advanced the idea that electricity consisted of a single fluid. A number of scientists such as Franz Aepinus and Henry Cavendish developed the one-fluid theory. This theory predicted many of the things the later atomic theory found. (In Franklin's time the atomic theory of matter had not been developed.)

Benjamin Franklin's most famous experiment was done with a kite. He suspected that lightning had some connection with electricity. During a thunderstorm in 1752 he flew a kite. Lightning struck nearby, and a spark jumped from a key on the kite string to a ring on his finger. The spark showed that lightning was a large spark of electricity.

News of his experiment spread rapidly. A year later, George Richmann recreated Ben Franklin's experiment. Lightning hit not nearby, but directly on his kite. He was killed.

two or more cells in series to make a **battery.** For example, the EMF of the wet cells used in a car battery have an EMF of 2.0 V. Three cells are connected in series to make a 6.0-V battery; six cells to make a 12.0-V battery.

Here is an assortment of commonly used batteries and cells.

Lead-Acid Batteries

Two materials are dipped into an acid solution. The lead-acid battery used in automobiles is a common example of this sort of battery. These are **secondary,** or rechargeable, **batteries.** Alternate plates of lead (Pb) and lead oxide (PbO_2) are immersed in a water-sulfuric acid solution (see Figure 31–22).

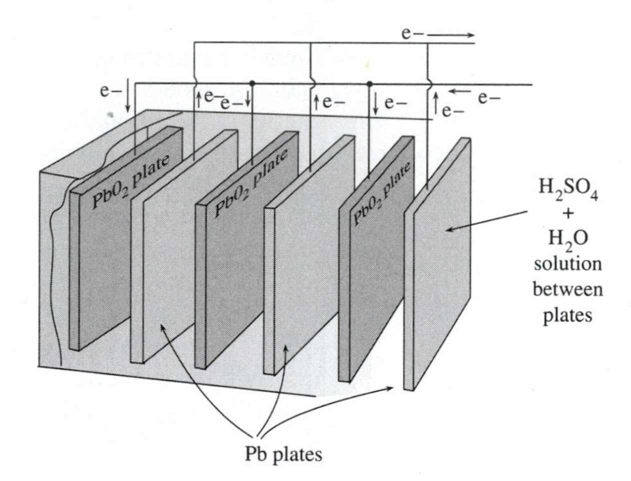

***Figure 31–22:** The simplified structure of a lead-acid storage battery. A solution of sulfuric acid and water is circulated between banks of plates made of lead oxide (PbO_2) and lead (Pb).*

When the battery delivers current, the lead on the negative plate is oxidized. Pb^{2+} ions combine with SO_4^{2-} from the sulfuric acid to form lead sulfate. The lead sulfate precipitates onto the lead plate. Two free electrons result from this reaction. At the positive plate, lead oxide is reduced to Pb^{2+}, which also forms lead sulfate. The oxygen ions removed from the lead go into solution in combination with the hydrogen from the sulfuric acid to increase the water content of the solution. Two electrons must enter this reaction for it to continue.

The reaction at the Pb plate releases two electrons. The reaction at the PbO_2 plate captures two electrons. Here's the overall chemical reaction.

$$Pb + SO_4^{2-} \rightarrow PbSO_4 + 2\,e^- \qquad \text{(Pb plate)}$$

$$PbO_2 + 4H^+ + SO_4^{2-} + 2\,e^- \rightarrow PbSO_4 + 2H_2O \qquad (PbO_2\ \text{plate})$$

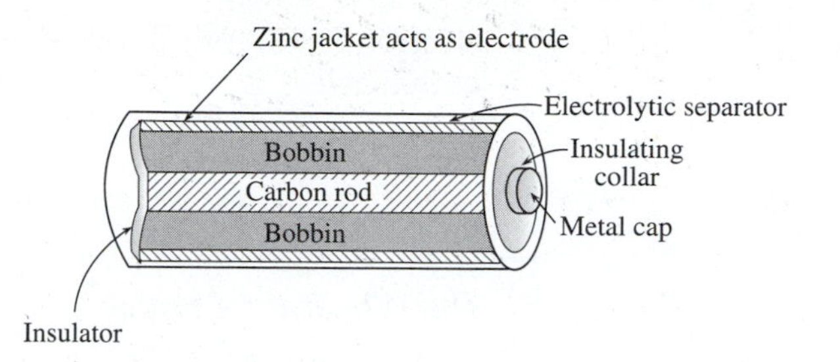

Figure 31–23: A simplified structure of a primary flashlight battery. A carbon post is surrounded by a paste called a bobbin. The second electrode is the zinc jacket.

The net effect is to combine sulfuric acid with lead and lead oxide to form lead sulfate and water, while moving electrons through an external circuit and creating an EMF of 2.0 V. The chemical process can be reversed by passing electrical current through the battery in the reverse direction.

Carbon-Zinc Cells

An ordinary flashlight battery or ignition cell is a carbon-zinc cell. This is a **primary cell.** Once the chemical reactants are used up the battery is spent. It cannot be recharged.

A positive electrode consists of a carbon rod surrounded by a salt paste, called a **bobbin,** shaped into a cylinder. The bobbin is a mixture of magnesium dioxide, acetylene black particles, ammonium chloride, and zinc chloride. The magnesium oxide is the chemically reacting material of the cathode. The bobbin is separated from the zinc case by a thin layer of gel-like electrolyte. The zinc case acts as the negative electrode. Typically, these cells produce an EMF of 1.5 V (see Figure 31–23).

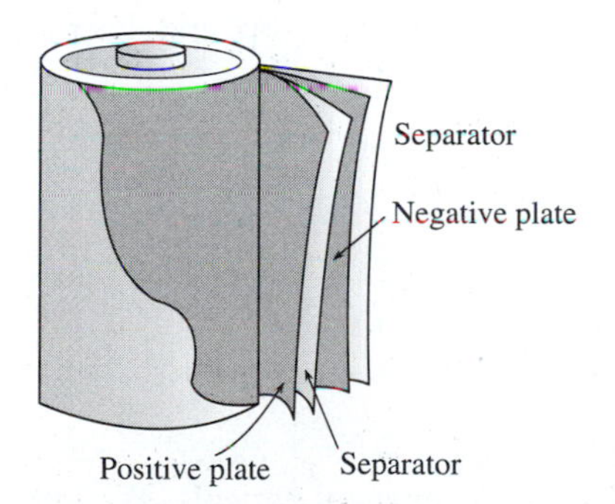

Figure 31–24: Many rechargeable cells have a jelly-roll structure.

Alkaline Cells

Instead of using an acid or a salt paste, such as aqueous ammonium chloride as an electrolyte, an alkaline base is used in alkaline cells. Potassium hydroxide is most popular. In cheaper cells, sodium hydroxide is used. A rechargeable nickel-cadmium cell is this sort of cell. Rechargeable batteries often have the two metal plates rolled in a cylinder like a jelly roll (see Figure 31–24).

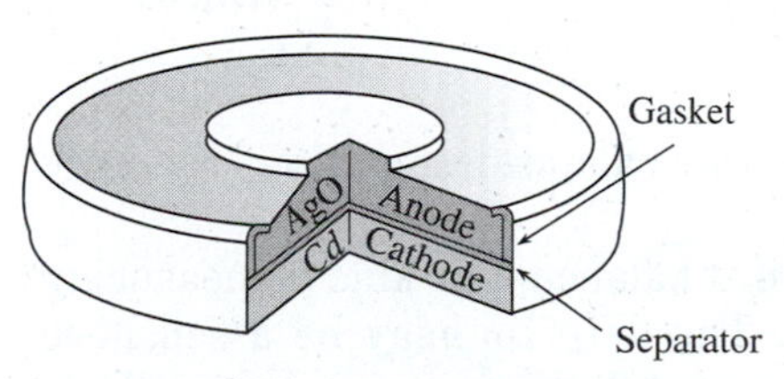

Figure 31–25: A button cell like those used in pocket calculators and watches.

Button Cells

For relatively small cells, such as those used in pocket calculators and watches, the cost of materials make up a small portion of the production cost. More expensive materials can be used to increase the storage of energy per unit volume. More efficient electrode combinations such as silver oxide-zinc (AgO-Zn), copper oxide-zinc (CuO-Zn), and mercury oxide-zinc (HgO-Zn) are combined with an alkaline electrolyte. Figure 31–25 shows the typical structure of a small button cell.

Special DC Circuits

The Wheatstone Bridge

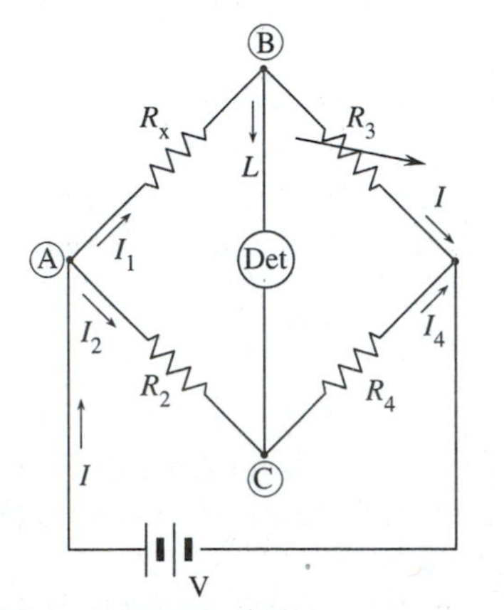

Figure 31–26: *The Wheatstone bridge. If the potential drop across R_x is the same as the drop across R_2, no current will flow through the bridge connecting nodes A and B.*

Figure 31–26 shows a circuit called a **Wheatstone bridge** often used to measure an unknown resistance (R_x). It has several applications as part of many more complex measuring instruments.

R_3 is a variable resistor. This is indicated by the arrow through the resistor symbol. Current from the power source takes two possible paths at node A. I_1 passes through the unknown resistor to node B. At node B, some of the current will flow through the detector (DET) if there is a potential difference between the bridge connecting nodes B and C. If B and C have the same potential, no current will flow through the detector. B and C will have the same potential only if the potential drop across R_x is the same as that across R_2. We say the bridge is **balanced** for this condition. We can vary R_3 to get a balance. When the bridge is balanced:

$$I_1 = I_3 \quad \text{and} \quad I_2 = I_4$$

We can look at the potential drops across the top and bottom branches of the circuit.

$$I_1 R_x = I_2 R_2$$

and

$$I_1 R_3 = I_2 R_4$$

When we divide one equation by the other, current cancels. We get:

$$R_x = R_3 \left(\frac{R_2}{R_4}\right) \qquad \textbf{(Eq. 31–11)}$$

Equation 31–11 is true only for the balanced condition, the only situation in which $I_1 = I_3$.

Since the detector will read zero at balance, this kind of measuring circuit is called a **null detector.** The detector may be a sensitive galvanometer used to measure current flow, or it may be a voltage detector such as a voltmeter or an oscilloscope.

☐ EXAMPLE PROBLEM 31–7: A HEATED COIL

The thermal coefficient of resistivity is found for a coil by heating it from 20°C to 100°C while monitoring with a Wheatstone bridge. At room temperature, the coil has a resistance of 22.4 Ω. At 100°C the setting of the bridge is:

$$R_3 = 14.2\ \Omega; \qquad R_2 = 50\ \Omega; \qquad R_4 = 30\ \Omega$$

Find the thermal coefficient of resistivity for the coil.

■ *SOLUTION*

First find the resistance of the coil at 100°C.

$$R_x = R_3\left(\frac{R_2}{R_4}\right)$$

$$R_x = 14.2\ \Omega\left(\frac{50\ \Omega}{30\ \Omega}\right)$$

$$R_x = 23.7\ \Omega$$

The change of resistance is $\Delta R = (23.7 - 22.4)\ \Omega = 1.3\ \Omega$.

$$\alpha = \frac{1}{R}\frac{\Delta R}{\Delta T}$$

$$\alpha = \frac{1.3\ \Omega}{22.4\ \Omega\,(100°\mathrm{C} - 20°\mathrm{C})}$$

$$\alpha = (7.3 \times 10^{-4}\ °\mathrm{C})^{-1}$$

The Potential Divider

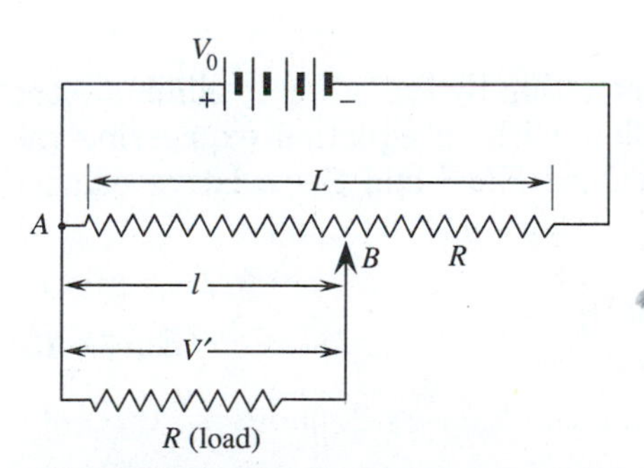

Figure 31–27: *A potential divider. If the load resistance, R(load), is large compared with the resistance of the slide wire (R), the potential drop across the load V′ is proportional to the length of wire between A and B.*

Figure 31–27 shows a potential divider. A DC voltage source is connected across a wire resistor R which has a sliding contact (B). The sliding contact is indicated by the total length of the wire L. Its resistance is $R = \rho L/A$. The load resistance R_1 is parallel to a fraction of the total wire length (l).

If the load resistance is large compared with the slide wire, the current is approximately constant along its length. The potential drop along the entire length of the wire is:

$$V_0 = \left(\frac{\rho L}{A}\right) I$$

The potential along length l is:

$$V' \cong \left(\frac{\rho l}{A}\right) I$$

One equation can be divided into the other to express the voltage across the load in terms of the length of the wire.

$$V' \cong V_0 \left(\frac{l}{L}\right) \quad \textbf{(Eq. 31–12)}$$

Potentiometer

A potential divider can be used to make a null detector to measure electrical potential. The load resistance is replaced by a potential source such as a battery or a thermocouple. Notice in Figure 31–28 that the positive sides of the batteries are connected to the positive side of the power source.

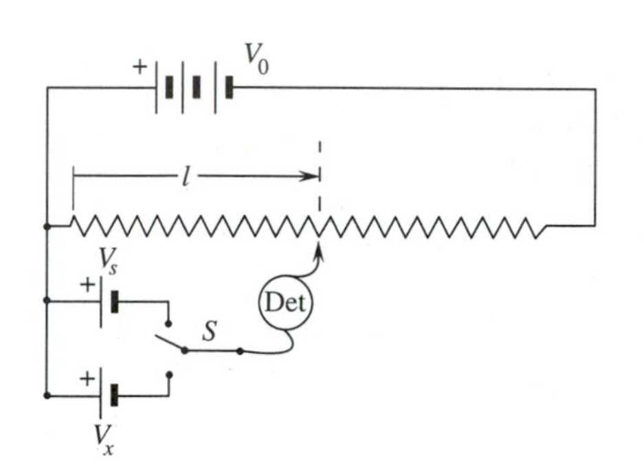

Figure 31–28: *a potentiometer uses a potential divider. The ratio of voltage to length is found by using the standard cell (V_s). The unknown EMF is obtained by finding a new balance length (l_x) for it.*

First a known, or **standard cell** (V_s), is connected to the circuit by the double-pole switch (S). The slide is moved until a null is indicated on the detector. This will occur when the EMF across the battery matches the potential drop across the slide wire length l_s.

$$V_s = \frac{V_0 \, l_s}{L}$$

The switch is changed to connect the unknown potential source. A new balance is found.

$$V_x = \frac{V_0 \, l_x}{L}$$

We can divide one balance equation by the other to eliminate the working voltage V_0. We are left with an equation expressing the unknown in terms of the standard EMF and the relative balance lengths of the slide wire.

$$V_x = \frac{V_s \, l_x}{l_s} \quad \textbf{(Eq. 31–13)}$$

☐ **EXAMPLE PROBLEM 31–8: THE POTENTIAL DIVIDER**

An unknown battery potential is compared with a standard cell that has an EMF of 1.0192 V. The standard balances for a slide wire length of 23.4 cm. The unknown balances at 71.4 cm. What is the EMF of the unknown battery?

■ *SOLUTION*

$$V_x = \frac{V_s\, l_x}{l_s}$$

$$V_x = \frac{1.0192 \text{ V} \times 71.4 \text{ cm}}{23.4 \text{ cm}}$$

$$V_x = 3.11 \text{ V}$$

SUMMARY

The sum of the currents going into a device in an electrical circuit equals the sum of the currents coming out. This is a statement of conservation of charge.

In any closed-circuit loop, the sum of potential gains equals the sum of potential losses. This is a statement of conservation of energy. Using these two principles we can find equivalent capacitance or resistance. Add the capacitance of capacitors connected in parallel to find the equivalent capacitance.

$$C = C_1 + C_2 + C_3 + \ldots C_n$$

Add the reciprocal capacitance of capacitors in series to find the equivalent capacitance.

$$\frac{1}{C} = \frac{1}{C_1} + \frac{1}{C_2} + \frac{1}{C_3} + \ldots \frac{1}{C_n}$$

Add the reciprocal resistance of resistors in parallel to find equivalent resistance.

$$\frac{1}{R} = \frac{1}{R_1} + \frac{1}{R_2} + \frac{1}{R_3} + \ldots \frac{1}{R_n}$$

Add the resistance of resistors in series to find equivalent resistance.

$$R = R_1 + R_2 + R_3 + \ldots R_n$$

The following rules are followed in analyzing an electrical network.

1. Nodal equations. The sum of current going into a node equals the currents coming out.

$$\Sigma I(\text{in}) = \Sigma I(\text{out})$$

2. Loop equations. The sum of potential gains equals the sum of potential losses around any closed loop.

$$\Sigma V(\text{gain}) = \Sigma V(\text{loss})$$

3. Write as many independent equations for the network as there are unknowns, using rules 1 and 2. Solve the equations simultaneously.

Electric power sources contain internal resistance (r). The voltage across the terminals of the source is called the terminal voltage (V_t). The value of the terminal voltage when no current (I) is drawn is called the electromotive force (EMF).

$$V_t = EMF - r\,I$$

An electric power source delivers maximum power to a load when the resistance of the load matches the internal resistance of the source.

Electric cells have two poles or plates made of materials with different chemical potentials. The poles are immersed in an electrolyte, an ionized fluid. A chemical reaction occurs that depends on the transport of electrons between the two poles through an external circuit.

A simple but important DC circuit is the Wheatstone bridge. It is a null-detecting circuit used to measure an unknown resistance (R_x). At balance:

$$R_x = R_3\left(\frac{R_2}{R_4}\right)$$

Another important DC circuit is the potentiometer. It compares the EMF of a standard cell (V_s) against an unknown potential (V_x) using a potential divider circuit. If l_s is the balance length for the standard cell and l_x is the balance length for the unknown, then:

$$V_x = V_s\left(\frac{l_x}{l_s}\right)$$

KEY TERMS

If you can explain the following terms to a friend or classmate, you understand their meaning. If you cannot explain the terms, you should reread the sections in which they are discussed.

alternating current (AC)
balanced
battery
bobbin
direct current (DC)
electrodes
electrolytes
electromotive force (EMF)
internal resistance
Kirchhoff's laws
node
null detector
parallel connection
primary cell
secondary batteries
series connection
standard cell
terminal voltage
Wheatstone bridge

EXERCISES

Section 31.1:

1. Explain the difference between alternating current and direct current.
2. Figure 31–29 shows two electric light bulbs. One light bulb is rated at 25 W; the other at 100 W. (Useful equations: $P = I\,V = I^2\,R = V^2/R$)
   - **A.** When hooked up separately to a 110-V power source, how much power does the source supply for each bulb?
   - **B.** Which bulb draws the larger current?
   - **C.** Which bulb has the larger resistance?
   - **D.** The bulbs are now hooked in series to the 110-V source. Across which bulb will the potential drop be larger?
   - **E.** Which bulb will dissipate more power in series?

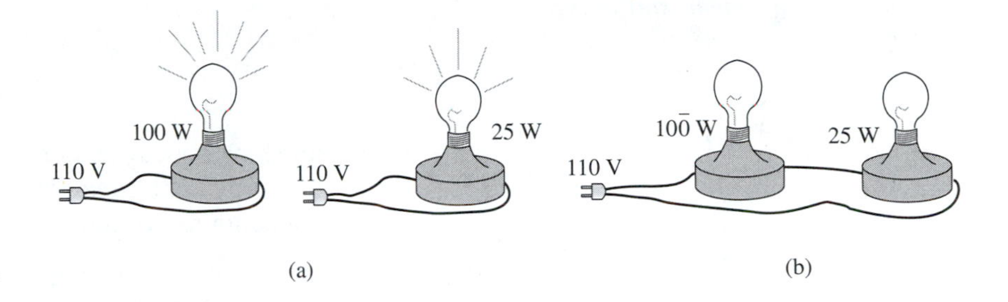

Figure 31–29: *Diagram for Exercise 2.*

3. A calculator draws a current of 0.00012 A from its 2.5-V battery for a period of 30 s.
   - **A.** How much power is drawn from the battery?
   - **B.** How much charge passes through the computer chip in 30 s?
   - **C.** How much energy is supplied by the battery in 30 s?
4. Find the missing potential drops and missing currents in Figure 31–30.

Section 31.2:

5. In which kind of circuit—series or parallel—is:
   - **A.** current (or charge) the same for each device in the circuit?
   - **B.** the potential drop across each device the same?
   - **C.** current (or charge) additive?
   - **D.** the potential drops additive?
   - **E.** power loss additive?
   - **F.** energy loss additive?
6. Capacitor A has half the capacitance of capacitor B. When the two capacitors are connected in *parallel* to a battery:
   - **A.** is the voltage across A larger, smaller, or the same as across B?

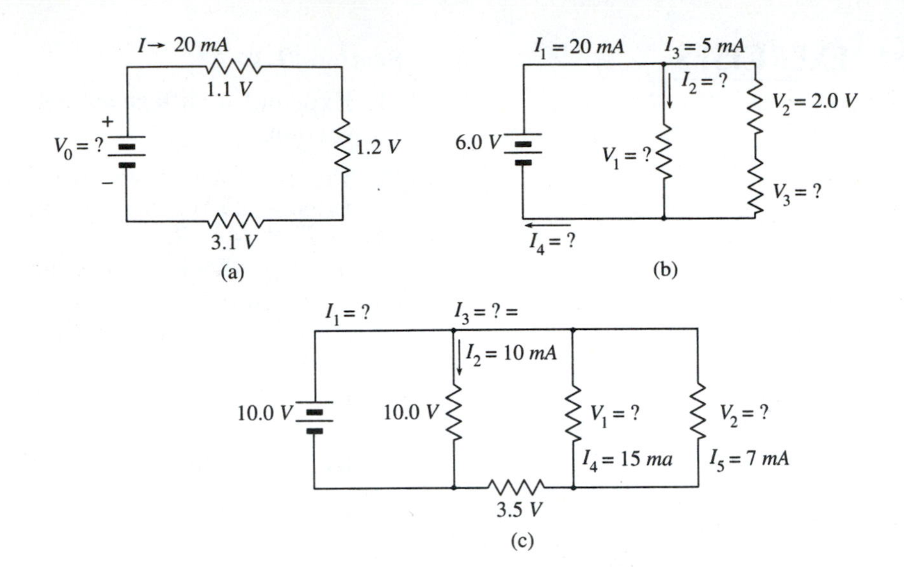

Figure 31–30: *Diagrams for Exercise 4. Find the missing currents and voltage drops.*

B. is the charge on A larger, smaller, or the same as on B?
C. is the equivalent capacitance of the circuit larger, smaller, or the same as capacitance B?

7. Capacitor A has half the capacitance of capacitor B. When the two capacitors are connected in *series* to a battery:
A. is the charge on A larger, smaller, or the same as on B?
B. is the potential drop across A larger, smaller, or the same as the potential drop across B?
C. is the equivalent capacitance of A larger, smaller, or the same as the capacitance of B?
D. is the equivalent capacitance of A larger, smaller, or the same as the capacitance of A?

8. How should two identical capacitors be hooked up—in series or parallel—to create an equivalent capacitance with twice the value of one of the capacitors?

9. How should two identical capacitors be hooked up—in series or parallel—to create an equivalent capacitance with half the value of either capacitor?

Section 31.3:

10. Resistor A has half the resistance of resistor B. They are connected in *parallel* to a battery.
A. Is the voltage drop across A larger, smaller, or the same as across B?
B. Is the current through A larger, smaller, or the same as through B?

C. Is the equivalent resistance larger, smaller, or the same as through resistor B?
D. Is the equivalent resistance larger, smaller, or the same as through resistor A?

11. Resistor A has half the resistance of resistor B. They are connected in *series* to a battery.
   A. Is the voltage drop across A larger, smaller, or the same as across B?
   B. Is the current through A larger, smaller, or the same as through B?
   C. Is the equivalent resistance larger, smaller, or the same as through resistor B?
12. How should two identical resistors be hooked up—in series or parallel—to create an equivalent resistance that is twice that of either resistor?
13. How should two identical resistors be hooked up—in series or parallel—to create an equivalent resistance that is half that of either resistor?
14. Compare your answers to Exercises 12 and 13 with the answers to Exercises 8 and 9.

Section 31.4:

15. Explain the following terms.
   A. Kirchhoff's laws
   B. Node
   C. Circuit loop
   D. Loop current

Section 31.5:

16. Explain the difference between an EMF and terminal voltage.
17. When a car is started with the headlights on, the lights dim. Why?
18. An old flashlight battery has an increased internal resistance, but the EMF does not change. Why is the light dim when old batteries are used?
19. A flat-wire, 75-Ω antenna lead-in can be connected directly to the terminals at the back of a television set. When a coaxial cable is connected, it must pass through a coupler that alters its impedence. Why?

Section 31.6:

20. What is the difference between a primary and a secondary cell?
21. Why does the chemical reaction in a cell not continue at a rapid rate when the external circuit is removed from the electrodes?

22. Why are bridge circuits and potentiometers called "null detectors"?

23. Why are Equations 31–11 and 31–13 true only for the condition of balance?

24. Does a potentiometer measure EMF or terminal voltage when hooked up directly to an unloaded battery? Explain.

PROBLEMS

Section 31.2:

1. Find the equivalent capacitance of the following sets of capacitors connected in *series*.

A.	B.	C.
$C_1 = 3.0\ \mu F$	$C_1 = 9.0\ \mu F$	$C_1 = 2.0\ \mu F$
$C_2 = 6.0\ \mu F$	$C_2 = 12.0\ \mu F$	$C_2 = 2.0\ \mu F$
	$C_3 = 6.0\ \mu F$	$C_3 = 2.0\ \mu F$

2. Find the equivalent capacitance of the sets of capacitors in problem 1 when they are connected in *parallel*.

3. A 3.60-μF capacitor and a 6.20-μF capacitor are connected in *series* to a 12.0-V battery. Find the charge on each capacitor and the voltage drop across each capacitor. (Hint: Find the equivalent capacitance first.)

4. A 7.2-μF capacitor and a 4.2-μF capacitor are connected in *parallel* to a 9.0-V battery. Find the voltage across each capacitor and the charge stored in each capacitor.

5. Find the equivalent capacitance for the combinations of capacitors shown in Figure 31–31.

** 6. A 4.0-nF capacitor is charged to 10.0 V. It is then disconnected and placed across the plates of an uncharged 2.0-nF capacitor (see Figure 31–32). Some charge moves from one capacitor to another. Find the final charge on the plates of the two capacitors. (Hint: This is equivalent to a parallel connection. Charge is conserved.)

Section 31.3:

7. Find the equivalent resistance of the following sets of resistors connected in *series*.

A.	B.	C.
$R_1 = 13\bar{0}\ \Omega$	$R_1 = 24\bar{0}\ \Omega$	$R_1 = 8\bar{0}\ \Omega$
$R_2 = 39\bar{0}\ \Omega$	$R_2 = 40\bar{0}\ \Omega$	$R_2 = 8\bar{0}\ \Omega$
	$R_3 = 56\bar{0}\ \Omega$	$R_3 = 8\bar{0}\ \Omega$

8. Find the equivalent resistance of the sets of resistors in Problem 7 when they are connected in *parallel*.

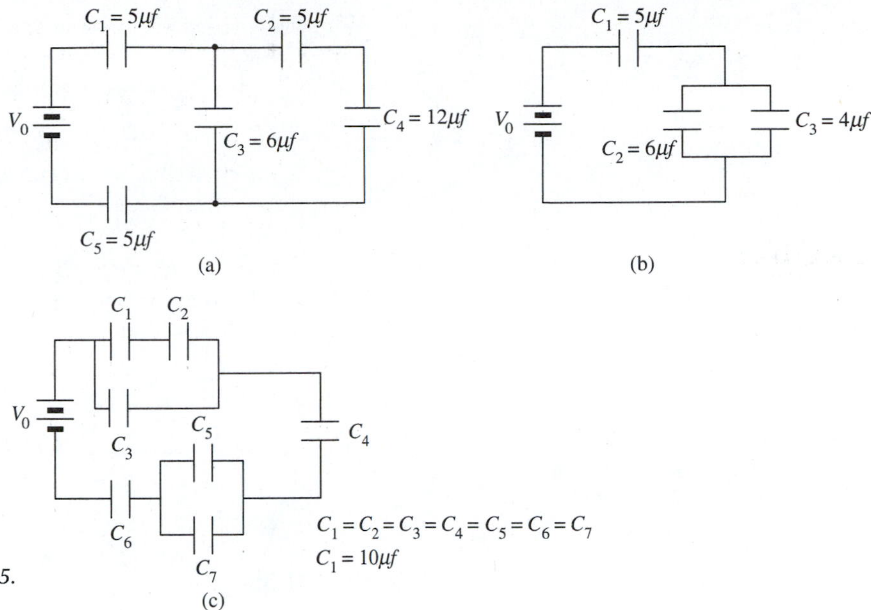

Figure 31–31: *Diagrams for Problem 5. Calculate equivalent capacitance.*

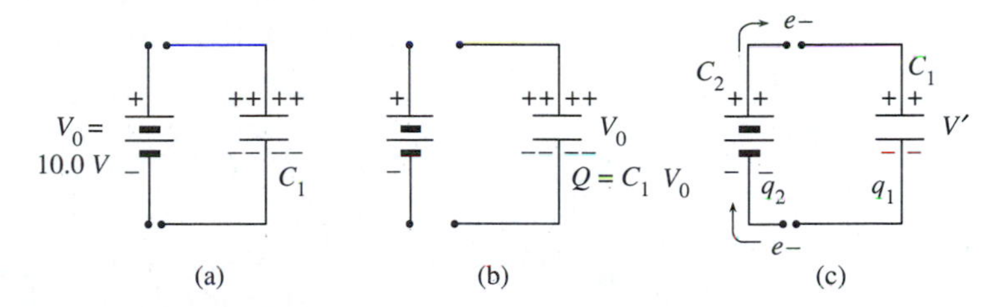

Figure 31–32: *Diagram for Problem 6.* (a) *Capacitor C_1 is charged up,* (b) *disconnected, and* (c) *then connected to an uncharged capacitor, C_2.*

9. A 228-Ω resistor and a 372-Ω resistor are connected in series to a 6.0-V battery. What is the current through each resistor? What is the potential drop across each resistor?

10. A 25$\bar{0}$-Ω resistor and a 175-Ω resistor are connected to a 12.0-V battery. What is the current through each resistor and the voltage drop across each resistor?

11. Find the equivalent resistance of the combinations of resistors shown in Figure 31–33.

Section 31.4:

12. Use Kirchhoff's laws to find current through the 50-Ω resistor in Figure 31–34.

13. Find the current and voltage for the 75-Ω resistor in Figure 31–35.

14. Use Kirchhoff's laws to find the potential drop across the 5$\bar{0}$-Ω resistors in the circuits shown in Figure 31–36.

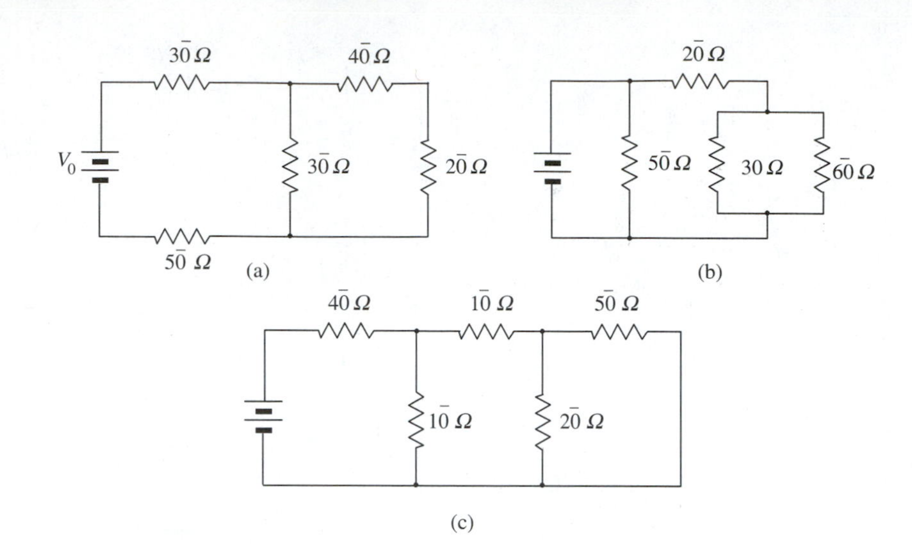

Figure 31–33: Diagram for Problem 11. Find the equivalent resistance.

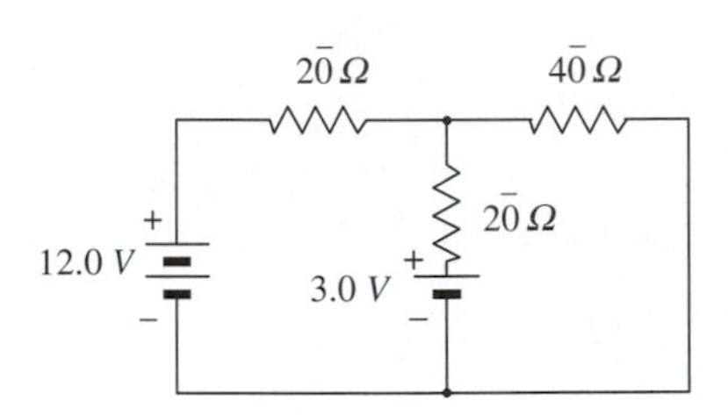

Figure 31–34: Diagram for Problem 12.

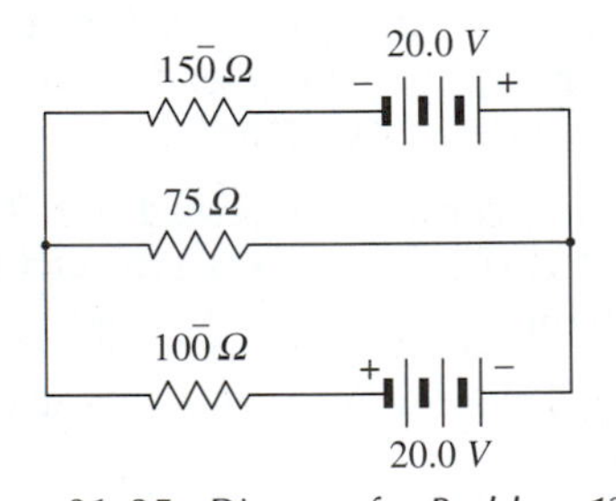

Figure 31–35: Diagram for Problem 13.

Section 31.5:

15. A piece of copper wire is placed across the terminals of a 1.5-V dry cell to form a short circuit. A current of 32 A flows through the wire. What is the internal resistance of the cell?

16. A 9.0-V battery with an internal resistance of 0.10 Ω is connected to a circuit with an equivalent resistance of 17 Ω. What current will be drawn from the battery and what will be its terminal voltage?

17. A 12.0-V automobile battery has an internal resistance of 0.03 Ω. What is its terminal voltage when 90.0 A of current are drawn to start the car?

18. If the internal resistance of a power source is r, show that the maximum power output is: $P(\text{max}) = EMF^2/4r$.

Section 31.6:

19. Find the unknown resistances for the bridge circuits shown in Figure 31–37.

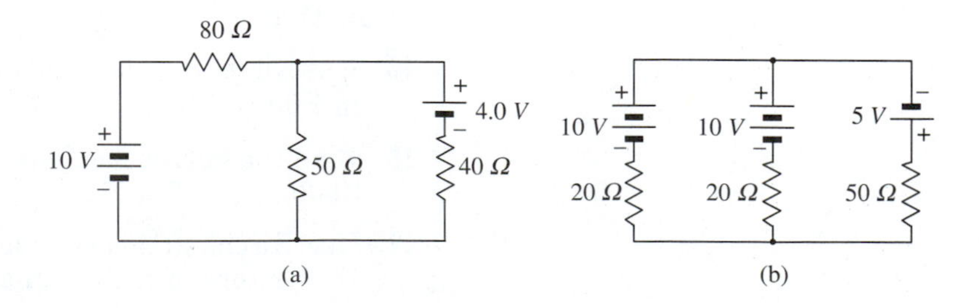

Figure 31–36: Diagrams for Problem 14.

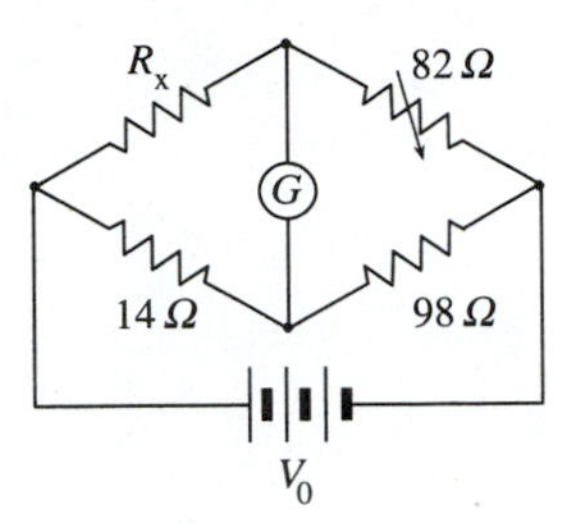

Figure 31–37: Diagram for Problem 19. The bridge is in balance; find the unknown resistance.

20. One form of the Wheatstone bridge replaces resistors R_2 and R_4 with a slide wire similar to that used in a potentiometer (see Figure 31–26). At balance $R_3 = 56\ \Omega$; $l_2 = 56.2$ cm; and $l_4 = 43.8$ cm. What is the unknown resistance?

21. A potentiometer uses a meter-long slide wire. A standard cell with an EMF of 1.0186 V balances at a length of 27.6 cm. An unknown cell balances at 41.7 cm.

A. What is the potential drop across the full length of the slide wire?

B. What is the EMF of the unknown cell?

****22.** The idea of series resistance can be used to increase the sensitivity of a slide-wire potentiometer. If an added resistance (R_a) is placed in series with the slide wire (R_w), only a fraction of the potential drop will be across the wire. In Figure 31–38 assume $R_a = 9.000\ R_w$. The standard cell is balanced across both resistances ($R_a + R_w$). The unknown is balanced only across the slide wire (R_w).

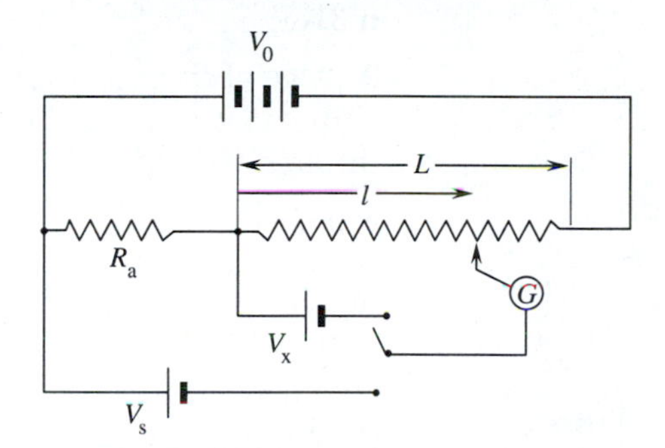

Figure 31–38: Diagram for Problem 22. The sensitivity of a potentiometer can be increased by using a series resistance (R_a) with the slide wire.

A. Show that at balance:

$$V_x = V_s\left(\frac{l_x}{l_s + 9L}\right)$$

B. Find V_x if $V_s = 1.0186$ V, $l_x = 23.4$ cm, $l_s = 32.0$ cm, and L = 100.0 cm.

Chapter 32

MAGNETISM

This atomic clock passes a stream of atoms through a magnetic field to make extremely small time measurements. It operates on a frequency that is billionths of a second long. Read about magnetism in this chapter to learn how scientists make such small measurements using this clock. (Photograph courtesy of Hewlett-Packard Company.)

OBJECTIVES

In this chapter you will learn how:

- magnetic fields exert forces on moving charges
- to calculate the motion of a charged particle moving through a uniform magnetic field
- to calculate the force on a current-carrying wire
- to find the torque on a current-carrying coil in an external magnetic field
- to determine the magnetic moment of a current-carrying coil
- to find the amount of magnetic flux through a coil
- magnetic fields are created by moving charges and electrical currents
- to calculate the magnetic fields produced by flat coils, solenoids, and toroids
- solenoids can be used in a variety of electromechanical devices
- magnetic fields are involved with electrical power generation

Time Line for Chapter 32

1050 B.C.	The Duke of Chou (China) creates the first magnetic compass.
1070	The Chinese use the compass for navigation.
1600	William Gilbert publishes work concerning magnetism. He suggests that the world is a giant magnet.
1750	John Michell explains magnetic induction and discovers the inverse square law for magnetism.
1751	Benjamin Franklin shows that electricity will magnetize and demagnetize an iron needle.
1819	Hans Oersted finds that magnets are deflected by electrical current.
1823	Andre-Marie Ampere develops a theory relating electrical current and magnetism.
1846	Wilhelm Weber develops a relationship between magnetic force and electrical current.
1864	James Clerk Maxwell creates a set of mathematical equations connecting electricity and magnetism.
1929	M. Matuyama finds evidence that the earth's magnetic field reverses poles periodically.

Coulomb forces arise from electrically charged particles. Coulomb forces interact with charged particles whether they are at rest or in motion. Magnetic forces are a bit different. They are caused by *moving* charges and interact only on *moving* charges. Magnetic forces will not affect a charged particle at rest. Even in permanent magnets, magnetic forces are caused by orbiting and spinning electrons, which are certainly moving electric charges.

In this chapter we will explore the creation of magnetic fields and their interaction with moving charges.

32.1 MAGNETIC FIELDS

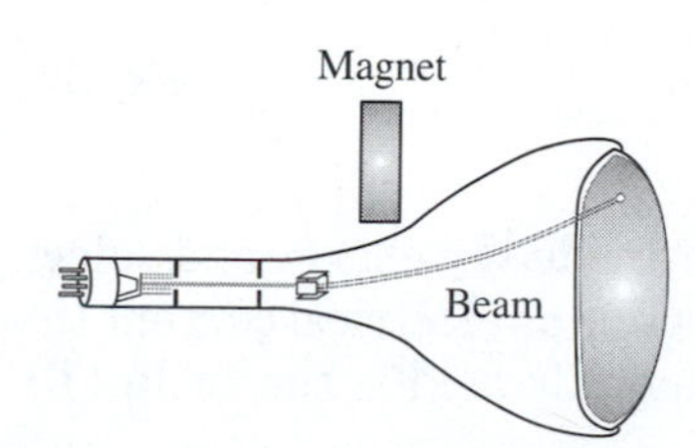

Figure 32–1: *A bar magnet deflects a beam of electrons in a CRT.*

If we hold a permanent magnet near a CRT, we will see that the beam of electrons is deflected by the magnetic field set up by the magnet. As the beam enters the magnetic field, it travels a curved path (see Figure 32–1). According to Newton's laws of dynamics, a force must be acting on the electrons in the beam.

We can improve the experiment. By passing the beam through the pole of a laboratory magnet we can produce a region of constant, uniform magnetic field. We discover a few things about the uniform magnetic force on moving charges.

1. The force is a centripetal force. It acts at right angles to the electron's velocity.
2. The force is at right angles, not only to the charge's velocity, but also to the magnetic field.
3. The magnitude of the force is directly proportional to the strength of the magnetic field. If we increase the strength of the magnetic field the beam is deflected through an arc with a shorter radius of curvature.
4. The size of the magnetic force acting on a charge is directly proportional to the size of the charge. Doubly ionized particles have twice the force acting on them as singly ionized particles.
5. The size of the magnetic force acting on a charge is directly proportional to the component of velocity at right angles to the magnetic field. If a charged particle moves *parallel* to the magnetic field there is *no* force acting on it.

We can combine all this information into one vector equation. We will represent the magnetic field by the symbol **B**.

$$\mathbf{F} = q\,(\mathbf{v} \times \mathbf{B}) \qquad \textbf{(Eq. 32–1)}$$

$\mathbf{v} \times \mathbf{B}$ is the vector cross product. Its magnitude is $\mathbf{v}\,\mathbf{B} \sin \theta$ where θ is the angle between the particle velocity and the magnetic field. The strength of the magnetic field is:

$$|\mathbf{F}| = q\,\mathrm{v}\,\mathrm{B} \sin \theta \qquad \textbf{(Eq. 32–2)}$$

Figure 32–2 shows how to use the right hand to sort out the directions. Curl the fingers to form the arc made by angle θ as it rotates *from* **v** *to* **B.** The thumb will point in the direction of the magnetic force. This trick is called the **right-hand rule.**

Earlier we used Coulomb's law to define electric field as the coulomb force per unit charge ($E = \mathrm{F}_c/q$). We can do something similar here. We can use the magnetic force equation to define a magnetic field, by solving equation 32–2 for **B.**

$$|\mathbf{B}| = \frac{\mathrm{F}}{q\,\mathrm{v} \sin \theta} \qquad \textbf{(Eq. 32–3)}$$

The magnetic field has units of force divided by charge and velocity. Charge times velocity ($q \times s/t$) is the same as electrical current times distance ($q/t \times s$). The SI unit of magnetic field is the **tesla (T).**

$$1 \text{ tesla (T)} = 1 \text{ N/C} \cdot \text{m/s} = 1 \text{ N/A} \cdot \text{m}$$

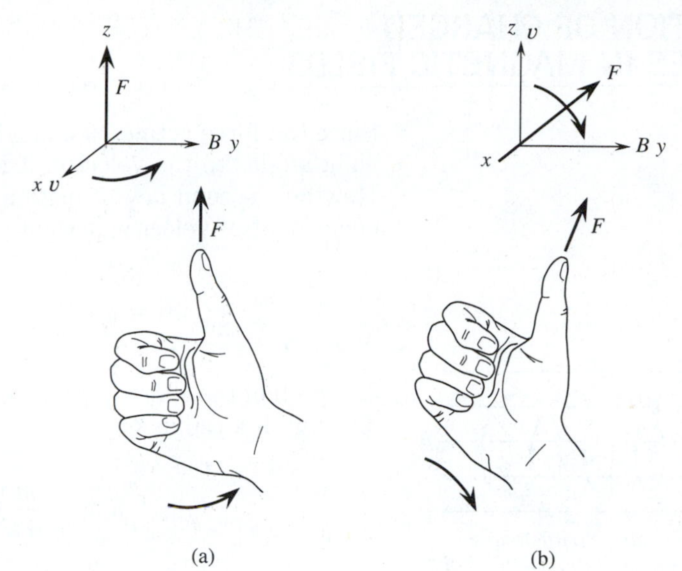

Figure 32–2: The right-hand rule. Curl the fingers to match the rotation of **v** *into* **B.** *The thumb points in the direction of the magnetic force.*

A tesla is a fairly large unit. Quite often an older, smaller unit called the **gauss (G)** is used.

$$1 \text{ tesla (T)} = 10^4 \text{ gauss (G)}$$

☐ **EXAMPLE PROBLEM 32–1: MAGNETIC FIELD**

A beam of electrons ($e = 1.6 \times 10^{-19}$ C) passing through a magnetic field at right angles with a speed of 5.7×10^6 m/s experiences a force of 2.1×10^{-14} N on each electron (see Figure 32–3). What is the magnetic field strength in tesla and in gauss units?

■ ***SOLUTION***

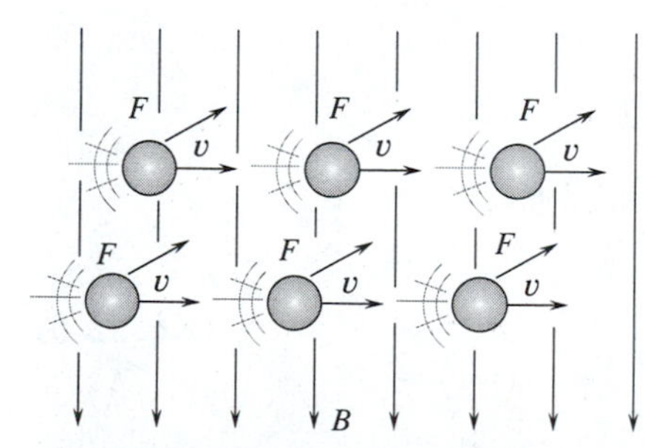

Figure 32–3: Electrons move perpendicular to a magnetic field. (Notice: The direction of magnetic force on a negative charge is opposite to that on a positive charge.)

$$|\mathbf{B}| = \frac{F}{q \text{ v} \sin \theta}$$

$$|\mathbf{B}| = \frac{2.1 \times 10^{-14} \text{ N}}{1.6 \times 10^{-19} \text{ C} \times 5.7 \times 10^6 \text{ m/s} \times \sin 90°}$$

$$|\mathbf{B}| = 0.023 \text{ T}$$

$$|\mathbf{B}| = 0.023 \text{ T} \times 10^4 \text{ G/T}$$

$$|\mathbf{B}| = 230 \text{ G}$$

32.2 MOTION OF CHARGED PARTICLES IN MAGNETIC FIELDS

Since the force acting on a moving charge in a uniform field acts at right angles to the velocity, it is a centripetal force. We can write Newton's second law of motion with centripetal acceleration. The component of velocity at right angles to the field is v sin θ.

$$|\mathbf{F}| = q\,\mathrm{v}\,\mathrm{B}\sin\theta = m\frac{\mathrm{v}^2}{R}$$

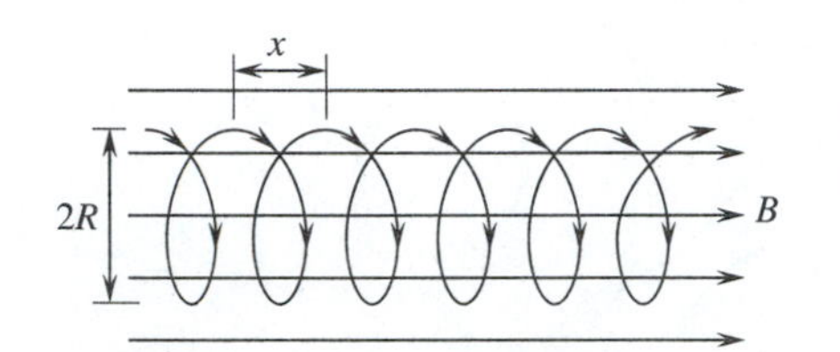

Figure 32–4: A charged particle in a uniform magnetic field moves in a helical pattern.

The path of the charged particle will be a helix or spiral (see Figure 32–4) with a radius of:

$$R = \frac{m\,\mathrm{v}}{q\,\mathrm{B}\sin\theta} \qquad \textbf{(Eq. 32–4)}$$

The angular speed of the particle is $\mathrm{v}/R = \omega$ rad/s. We can find the period (T) for one complete rotation by dividing two radians by the angular speed.

$$T = \frac{2\pi}{\omega} = \frac{2\pi}{\mathrm{v}/R}$$

Solve Equation 32–4 for the ratio v/R and substitute this into the expression for period.

$$T = \frac{2\pi\,m}{q\,\mathrm{B}\sin\theta} \qquad \textbf{(Eq. 32–5)}$$

The distance between adjacent coils (X) of the helix depends on the period and the component of speed parallel to the **B** field (the magnetic field).

$$X = (\mathrm{v}\cos\theta)\,T$$

$$X = \frac{2\,\pi\,m\,\mathrm{v}\cos\theta}{q\,\mathrm{B}\sin\theta}$$

$$X = \frac{2\pi\,m\,\mathrm{v}}{q\,\mathrm{B}\tan\theta} \qquad \textbf{(Eq. 32–6)}$$

☐ **EXAMPLE PROBLEM 32–2: ORBIT OF PARTICLE IN A UNIFORM B FIELD**

A proton travels at right angles to a uniform magnetic field at a speed of 6.8×10^6 m/s (see Figure 32–5). If the strength of the field is 0.92 T, calculate the radius and frequency of its orbit. (Proton mass = 1.67×10^{-27} kg; proton charge = $+1.6 \times 10^{-19}$ C.)

SOLUTION

The angle is 90°. The separation of the coils is zero. The proton will be trapped in a circular orbit in the magnetic field.

A. Radius of orbit:

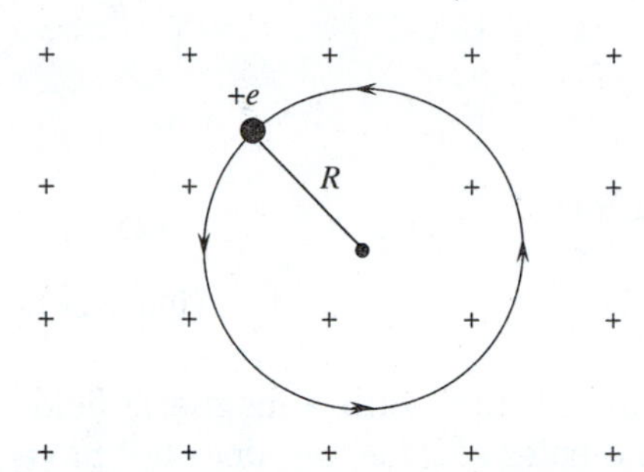

Figure 32–5: *If the motion is at right angles to the magnetic field, the path is a circular orbit. (+'s indicate a* **B** *field into the page.)*

$$R = \frac{m\,v}{q\,B \sin\theta}$$

$$R = \frac{1.67 \times 10^{-27}\,\text{kg} \times 6.8 \times 10^{6}\,\text{m/s}}{1.6 \times 10^{-19}\,\text{C} \times 0.92\,\text{T} \times 1.000}$$

$$R = 0.074\text{ m, or } 7.4\text{ cm}$$

B. Frequency:

The frequency is the reciprocal of the period ($f = 1/T$). Find the period and convert it to frequency.

$$T = \frac{2\pi\,m}{q\,B \sin\theta}$$

$$T = \frac{2\pi \times (1.67 \times 10^{-27}\,\text{kg})}{(1.6 \times 10^{-19}\,\text{C}) \times 0.92\,\text{T} \times 1.000}$$

$$T = 7.13 \times 10^{-8}\,\text{s}$$

$$f = \frac{1}{T} = \frac{1}{7.13 \times 10^{-8}\,\text{s}}$$

$$f = 1.4 \times 10^{7}\,\text{Hz}$$

32.3 FORCES ON A CURRENT-CARRYING WIRE

The current density (J) is directly proportional to the drift velocity (v_d) of electrons moving through a wire.

$$J = n\,e\,v_d$$

For a wire that has a length l and cross-sectional area A we can rewrite the equation in terms of electrical current (I) and the total number of conduction electrons (N) in the length of wire. Remember, n is the charge density, N/V(olume).

$$\left|\frac{I}{A}\right| = \frac{N\,e\,v_d}{A\,l}$$

or

$$N e v_d = I l$$

$N e$ is simply the total charge of the conduction electrons in the length of wire.

$$q \mathbf{v}_d = I \mathbf{l}$$

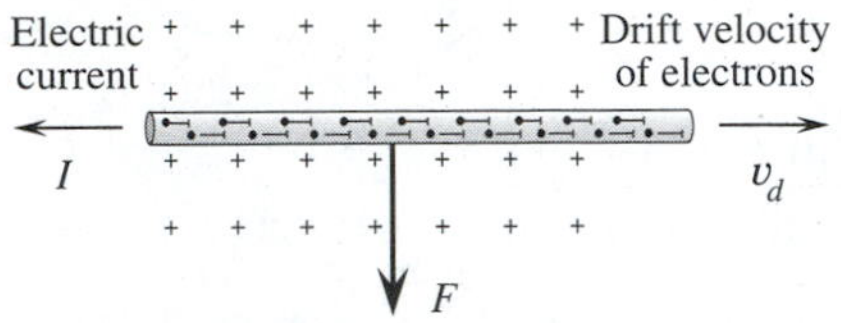

Electric current I — Drift velocity of electrons v_d — F

Figure 32–6: A conductor carrying an electric current through a magnetic field will have a force acting on it. The direction of the force can be found by using the right-hand rule; rotate I *into* **B**. *(+'s indicate a* **B** *field into the page.)*

We can substitute this into Equation 32–1.

$$\mathbf{F} = I(\mathbf{l} \times \mathbf{B}) \qquad \textbf{(Eq. 32–7)}$$

Any wire carrying a current can interact with a magnetic field. A copper wire passing between the poles of a magnet does not have a magnetic force acting on it (see Figure 32–6). If we pass a current through this wire, the magnetic field will react with the moving electrons. The wire will be pushed at right angles to both the magnetic field and the direction of current.

☐ **EXAMPLE PROBLEM 32–3: A WIRE IN A B FIELD**

An aluminum wire passes through the poles of a laboratory magnet making an angle of 67° with respect to a 0.096-T magnetic field. The effective length of wire in the **B** field is 8.9 cm.

A. What magnetic force acts on the wire when there is no electrical current through the wire?

B. What is the magnitude of the force on the wire when it carries a current of 2.4 A?

■ ***SOLUTION***

A. **F** = 0. The electrons in the wire do not have a drift velocity.

B.

$$|\mathbf{F}| = I\, l\, B \sin\theta$$

$$|\mathbf{F}| = 2.4\ \text{A} \times 0.089\ \text{m} \times 0.096\ \text{T} \times \sin 67°$$

$$|\mathbf{F}| = 1.9 \times 10^{-2}\ \text{N}$$

32.4 MAGNETIC TORQUES

Figure 32–7 shows a rectangular coil in a uniform magnetic field. The coil has a width of w and a length l. The coil is tilted at an angle with respect to the magnetic field lines. We will use the normal to

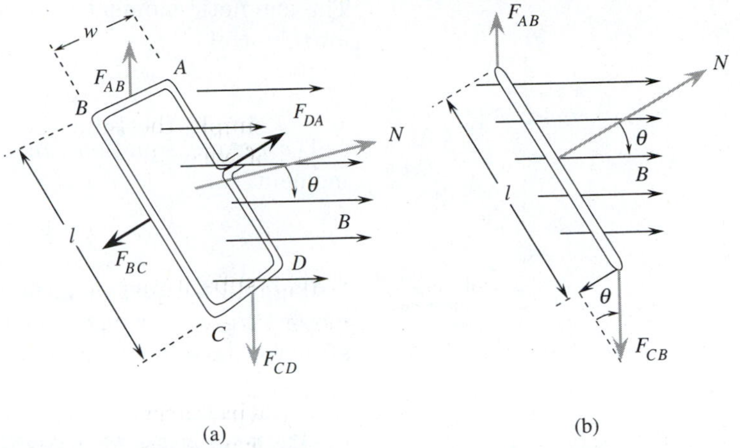

Figure 32–7: *The forces on a coil.* (a) *Perspective view. Forces* $\mathbf{F}_{BC}$ *and* $\mathbf{F}_{DA}$ *are concurrent. Forces* $\mathbf{F}_{AB}$ *and* $\mathbf{F}_{CD}$ *are nonconcurrent.* (b) *Side view. The angle* (θ) *between the coil normal* (N) *and the* **B** *field is the same as the angle needed to find the components of* $\mathbf{F}_{AB}$ *and* $\mathbf{F}_{CD}$ *causing rotation.*

the face of the coil to measure the angle of tilt. As current passes through the coil, magnetic forces act on each of the four sides of the coil.

The forces acting on sides DA and BC are equal and opposite concurrent forces. They add to a net force of zero.

The forces acting on sides AB and CD are nonconcurrent forces. This can be seen better in part B of the diagram. The net torque acting on the coil is:

$$\tau = l\,\mathrm{F}\sin\theta$$

The wire sections AB and CD are at right angles to the **B** field. The force acting on each end is $I\,w\,\mathbf{B}$.

$$\tau = l\,(I\,w\,\mathrm{B})\sin\theta$$

$$\tau = I\,A\,\mathrm{B}\sin\theta$$

where $A = l\,w$, the area of the coil.

If there is more than one turn on the coil, the force on each end will be larger. For two turns, the effective current through each end of the coil is doubled. For N turns, the torque will be N times as large.

$$\tau = N\,I\,A\,\mathrm{B}\sin\theta \qquad \textbf{(Eq. 32–8)}$$

Equation 32–8 applies not only to a rectangular coil, but to a coil of any shape. We do not need to know the length and width of the coil, just its area (A), to calculate the torque. The product $N\,I\,A$ is called the **magnetic moment** (μ) of the coil. For most situations, the number of turns and the area of the coil are fixed quantities.

The magnetic moment is directly proportional to the current going into the coil.

$$\mu = N I A \qquad \textbf{(Eq. 32–9)}$$

The torque equation can be rewritten in terms of the magnetic moment.

$$|\tau| = \mu \beta \sin \theta \qquad \textbf{(Eq. 32–10)}$$

Magnetic torques on a coil make it possible to operate an electric motor. We will examine the operation of DC motors in Chapter 33 after we have looked at some other magnetic effects.

☐ **EXAMPLE PROBLEM 32–4: MAXIMUM TORQUE**

A $2\bar{0}$-turn circular coil with a radius of 12 cm is in a uniform magnetic field of 0.150 T. A current of 3.2 A flows through the coil.

A. What is the magnetic moment of the coil?
B. What is the maximum torque acting on the coil?
C. What is the torque on the coil when the maximum number of magnetic field lines pass through the face of the coil?

■ *SOLUTION*

A. Magnetic moment:

$$\mu = N I A = N I (\pi R^2)$$

$$\mu = 2\bar{0} \text{ turns} \times 3.2 \text{ A} \times \pi \times (0.12 \text{ m})^2$$

$$\mu = 2.9 \text{ A} \cdot \text{m}^2$$

B. Maximum torque:
The maximum torque occurs when the tilt is 90° (see Figure 32–8b).

$$\tau = \mu \text{ B} \sin 90°$$

$$\tau = 2.9 \text{ A} \cdot \text{m}^2 \times 0.150 \text{ T} \times 1.000$$

$$\tau = 0.44 \text{ A} \cdot \text{m}^2 \times \text{T} = 0.44 \text{ A} \cdot \text{m}^2 \cdot \text{N/A} \cdot \text{m}$$

$$\tau = 0.44 \text{ N} \cdot \text{m}$$

C. Maximum magnetic lines through coil:
Look at Figure 32–8b. The maximum number of magnetic field lines pass through the coil when the coil normal is aligned parallel to the magnetic field. The torque when the maximum number of field lines pass through the coil is zero.

$$\tau = A \text{ B} \sin 0° = 0$$

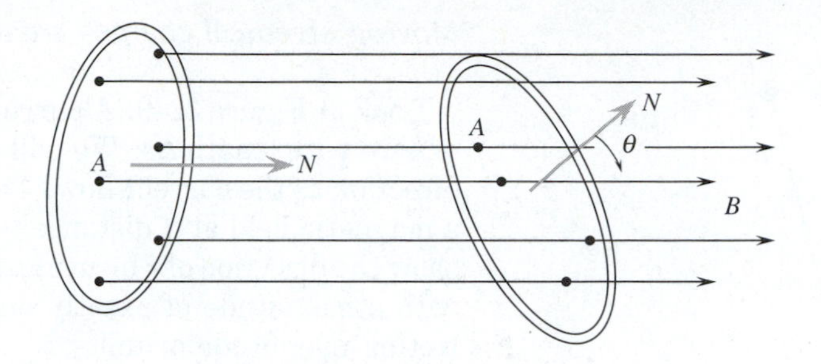

Figure 32–8: The number of magnetic field lines that pass through a coil depends on the orientation of the coil. The maximum flux occurs when the coil normal (N) is parallel to the magnetic field lines.

Parts B and C of Example Problem 32–4 should convince us that a randomly tilted coil will rotate in a magnetic field to align itself parallel to the field with the maximum possible number of magnetic lines passing through the coil.

It will be helpful to define a term called **magnetic flux** (ϕ_m). It is similar to electric flux. If we represent the orientation of a coil by its normal, we see from Figure 32–8 that the flux is maximum at 0° and minimum at 90°. A larger area means more flux through the coil; the density of magnetic field lines is proportional to the strength of the field. This is the scalar product of magnetic field and cross-sectional area. The normal to the surface is the area "direction."

$$\phi_m = \mathbf{B} \cdot \mathbf{A} = B\,A \cos\theta \qquad \textbf{(Eq. 32–11)}$$

In SI units magnetic flux is measured in **webers** (Wb).

$$1 \text{ weber (Wb)} = 1 \text{ T} \cdot \text{m}^2 = 1 \text{ N} \cdot \text{m/A}$$

We can say that the torque acting on a coil is in a direction to maximize the magnetic flux through the coil.

See Example Problem 32–5 in the next section to see how magnetic flux is calculated.

32.5 SOURCES OF MAGNETIC FIELDS

In the discussion of electrostatics, we discovered that coulomb forces act on electrically charged particles. These coulomb forces and their corresponding electric fields originated from the electric charges themselves.

We should not be surprised to see the same pattern in magnetism. Magnetic forces act on moving charges. These magnetic forces and corresponding magnetic fields originate from the moving charges themselves.

Moving electrical charges create magnetic fields.

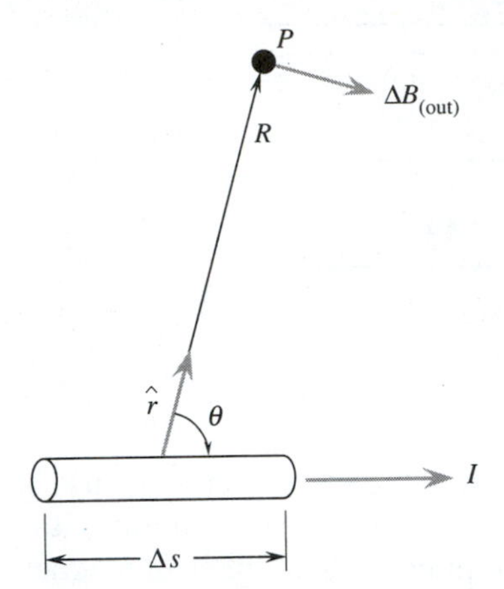

*Figure 32–9: Current passing through a small segment of wire (Δs) makes a contribution (**ΔB**) to the magnetic field a distance r away. r̂ is a unit vector. **ΔB** is at right angles to both Δs and r̂.*

Look at Figure 32–9. A current (I) flows through a short segment of wire with length Δs. We will choose Δs to be a vector in the same direction as the current flow. The moving charges in the wire create a magnetic field at a distance (r) from the wire at point P. We can show the direction of r by using a unit vector $\hat{r}$. $\hat{r}$ is a unitless quantity with a magnitude of exactly one. It indicates direction without affecting magnitude or units.

The direction of the field at point P can be found by using a small magnetic compass. *The direction of the **B** field is at right angles to both the wire segment and the vector* $\hat{r}$. In the diagram, $\mathbf{\Delta B}$ is pointing out of the page.

By testing the action of the produced magnetic field on another current-carrying wire we can find the strength of the field at point P. *The strength of the magnetic field is inversely proportional to distance.*

The strength of the magnetic field can be written in an inverse square law similar to the coulomb force equation. This relationship is called the **Biot-Savart law** (pronounced *be-oh sah-var* law).

$$|\mathbf{\Delta B}| = k_{\mathrm{m}} \frac{I}{r^2} \Delta s \qquad \textbf{(Eq. 32–12)}$$

where k_{m} is a proportionality constant. $k_{\mathrm{m}} = 10^{-7}$ Wb/A. $\mathbf{\Delta B}$ is at right angles to the wire segment (Δs) and the displacement (Δr). If you rotate Δs into $\hat{r}$, the right-hand rule shown in Section 32.1 will give you the field direction.

The value of the proportionality constant k_{m} is usually written in terms of another physical constant found in electromagnetic wave theory.

$$k_{\mathrm{m}} = \frac{\mu_0}{4\pi} = 10^{-7} \frac{\mathrm{Wb}}{\mathrm{A \cdot m}}$$

or

$$\mu_0 = 4\pi \times 10^{-7} \frac{\mathrm{Wb}}{\mathrm{A \cdot m}}$$

μ_0 is called the **permeability of free space.** By free space we mean empty space, a vacuum.

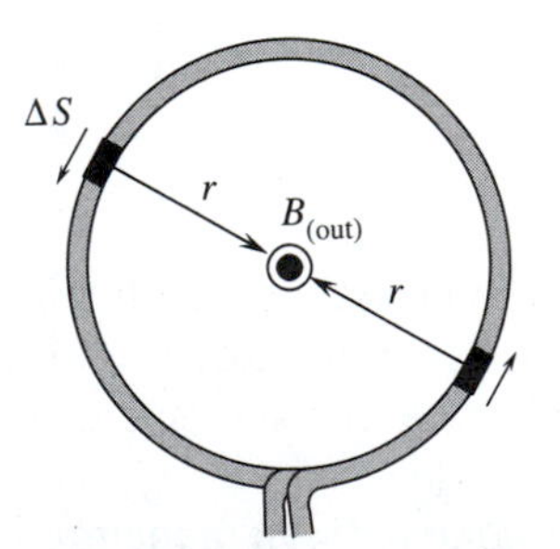

Figure 32–10: The magnetic field at the center of a coil is the sum of the contributions of the wire segments surrounding the center.

Look at Figure 32–10. We can use the Biot-Savart law to find the strength of the electric field in the center of the coil. The coil has N turns. The effective current through a small segment of the coil (Δs) is $N\,I$. r is the radius of the coil.

The contribution to the field at the coil's center by each segment of the coil is:

$$|\Delta \mathbf{B}| = \frac{\mu_0}{4\pi} \frac{N\,I\,\Delta s}{r^2}$$

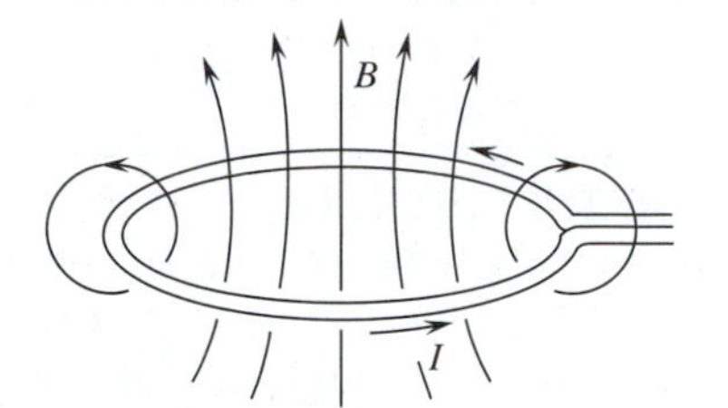

Figure 32–11: A cross section of the **B** field for a flat coil.

We can add up the contributions to the field made by each line segment. The total length of the segments is $\Sigma\Delta s = 2\pi r$, all a distance r from the center.

$$|\mathbf{B}| = \Sigma \Delta \mathrm{B}_i = \frac{\mu_0}{4\pi} \frac{N\,I\,\Delta s}{r^2}$$

$$|\mathbf{B}| = \frac{\mu_0}{4\pi} \frac{N\,I\,(2\pi\, r)}{r^2}$$

$$|\mathbf{B}| = \frac{\mu_0\, N\, I}{2\, r} \qquad \textbf{(Eq. 32–13)}$$

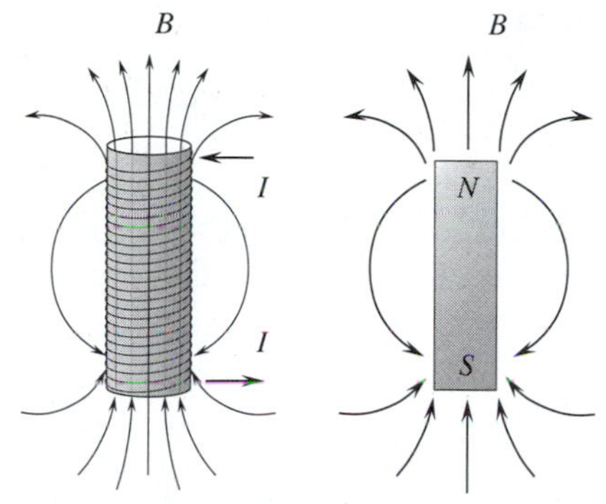

Figure 32–12: The magnetic field outside a solenoid is similar to the field around a bar magnet. Notice that the **B** field lines of a permanent magnet begin at the north (N) pole and terminate at the south (S) pole.

Figure 32–11 shows the magnetic field for the coil. The maximum strength is at the coil center. The field falls off sharply with distance from the coil.

A **solenoid** is a long coil. It is formed by wrapping windings around a straight cylinder. The magnetic field looks similar to the field of a flat coil, but the field passing inside the solenoid is nearly uniform except at the ends. We can treat the field as a uniform field for a solenoid that has a length (l) much larger than its radius. If n is the number of turns per unit length (N/l), the field inside a solenoid is:

$$|\mathbf{B}| = \frac{\mu_0\, N\, I}{l} = \mu_0\, n\, I \qquad \textbf{(Eq. 32–14)}$$

Solenoids are very useful. The magnetic field outside the solenoid looks very much like the **B** field of a bar magnet (see Figure 32–12). Inside, away from the ends of the coil, a uniform field is produced.

We can bend a long solenoid into a closed circle (see Figure 32–13). This doughnut-shaped coil is called a **toroid.** It has the interesting property that the magnetic field is all trapped inside the ring.

For a toroid with a radius r, the length of this "bent" solenoid is $l = 2\pi\, r$. We can substitute this into Equation 32–14 to find the strength of the **B** field inside the toroid.

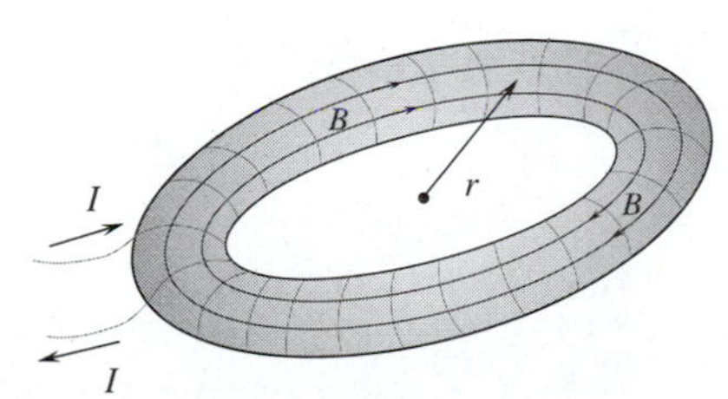

Figure 32–13: The magnetic field of a toroid is completely trapped inside the toroid.

$$|\mathbf{B}| = \frac{\mu_0\, N\, I}{2\pi\, R} \qquad \textbf{(Eq. 32–15)}$$

Large toroids are used in plasma physics to confine clouds of high-energy ions. The magnetic field causes the charged particles to spiral around the doughnut without touching the wall. If the ions touched the wall of the toroid they would lose energy. The magnetic field acts as a giant magnetic Thermos bottle.

Smaller toroids are used to measure the magnetic properties of materials. The core of the doughnut is filled with the test material. The strength of the resulting magnetic field for a given current through the windings tells us something about the material's magnetic behavior.

EXAMPLE PROBLEM 32–5: THE COIL AND THE SOLENOID

A flat coil with a radius of 2.3 cm is inside a 24 cm long solenoid with 1200 turns. The coil is tilted at 45° relative to the axis of the solenoid (see Figure 32–14). When the current through the solenoid is 2.6 A, how much magnetic flux passes through the flat coil?

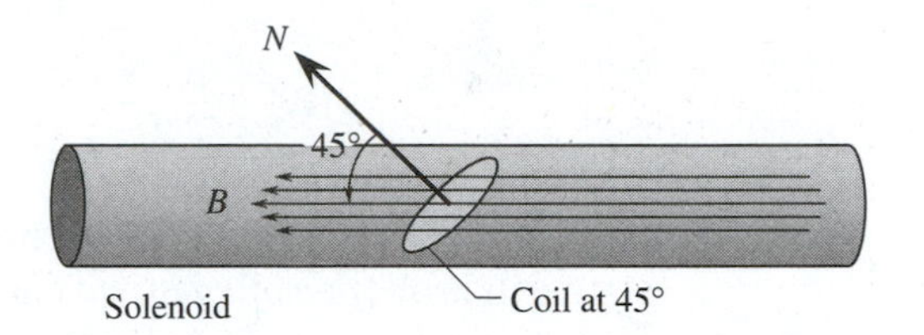

Figure 32–14: *A flat coil is inside a long solenoid.*

SOLUTION

We can break the problem into two parts. First, find the strength of the magnetic field created in the solenoid. Second, calculate the magnetic flux through the coil.

1. **B** field in the solenoid:

$$|\mathbf{B}| = \frac{\mu_0 N I}{l}$$

$$|\mathbf{B}| = (4\pi \times 10^{-7}\ \mathrm{Wb/A \cdot m}) \times \frac{1200\ \text{turns} \times 2.6\ \mathrm{A}}{0.24\ \mathrm{m}}$$

$$|\mathbf{B}| = 0.016\ \mathrm{Wb/m^2} = 0.016\ \mathrm{T}$$

2. Magnetic flux through the coil:
The area of the coil is:

$$A = \pi R^2 = 3.1416 \times (0.023\ \mathrm{m})^2 = 1.66 \times 10^{-3}\ \mathrm{m^2}$$

The magnetic flux is then:

$$\Phi_m = B\,A \cos\theta$$

$$\Phi_m = 0.016\ \mathrm{T} \times (1.66 \times 10^{-3}\ \mathrm{m^2}) \times \cos 45°$$

$$\Phi_m = 1.9 \times 10^{-5}\ \mathrm{Wb}$$

32.6 APPLICATIONS

Magnetic Devices

Many devices use the magnetic effects we have covered in this chapter. Here are a few of them.

Solenoids in Controls

Solenoids are used in a variety of controls. Figure 32–15 shows how a circuit breaker works. A steel rod is held in place by a compressional spring. When the current through the solenoid increases to a critical value, the **B** field in the coil becomes large enough to pull the steel rod away from the contact. This breaks the circuit b. In practice, additional features hold the circuit open for a period of time.

Figure 32–16 shows one scheme for using a solenoid to close a circuit. Circuit b is closed when the magnetic field in the solenoid is large enough to overcome the resisting force of the tension spring.

Figure 32–17 shows one way a solenoid can be used to open and

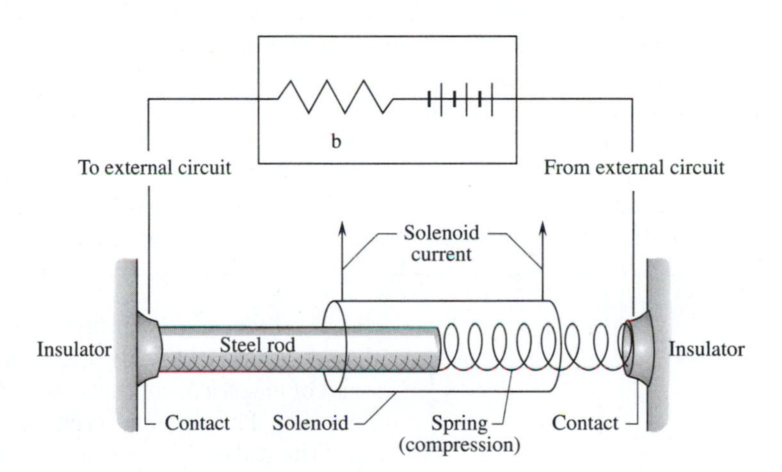

Figure 32–15: *A simplified version of a circuit breaker, or relay. Current in the solenoid creates a magnetic field that acts on the steel bar to pull it back. In a circuit breaker, the solenoid is in series with the current from the external circuit.*

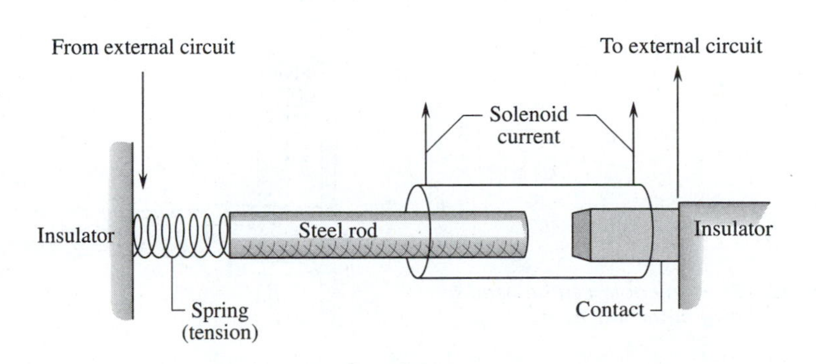

Figure 32–16: *Another kind of relay. The current through the solenoid causes the circuit to close rather than open.*

Atomic Flips in Time

(Photograph courtesy of Hewlett-Packard Company.)

A spinning electron has a magnetic moment with a north pole and a south pole. Spinning protons and neutrons also have magnetic moments with north and south poles.

If we place two bar magnets side by side, we will discover that like poles repel. If two north poles are beside each other, a force will keep the two north poles separated. If we place a north pole beside a south pole, the two poles will attract each other.

In an atom, it takes more energy for an electron to align its north pole with the nuclear pole than to have a north-to-south alignment.

Electrons tend to pair up in an atom. Two electrons on the same orbit have opposite magnetic directions; one has north up and the other has north down. If there is an odd number of electrons, the unpaired electron will usually have its magnetic pole in the opposite direction from the nuclear pole. Radiation with just the correct amount of energy can cause an electron to flip so that its north pole is up in the same direction as the nuclear magnet. Radiant energy is proportional to frequency. This gives us a very precise frequency we can use to calibrate time.

An atomic clock uses a stream of cesium atoms. The atoms pass through a magnetic field. Atoms with north poles aligned will travel at a different angle than the other atoms. We can separate the beams so that we have a beam of only atoms with one orientation. This beam passes through a radiation chamber. The atoms are then separated by another magnet, so we can count the ones that were flipped. The frequency of the radiation is adjusted to get the largest fraction of atoms to flip. This frequency turns out to be 9,192,631,770 Hz. A second can be subdivided into small units of 1/9,192,631,770 Hz. The speed of light is 3.0×10^8 m/s. In one of these small time units, light could travel only 3.3 cm.

---

close a valve. You will find such devices in automatic washing machines. Electrical current is converted to mechanical activity. Instead of electrical contacts, a spring-loaded valve is connected to the steel bar. The solenoid creates a strong enough force on the bar to open the valve. When the current through the solenoid stops, the compression spring closes the valve.

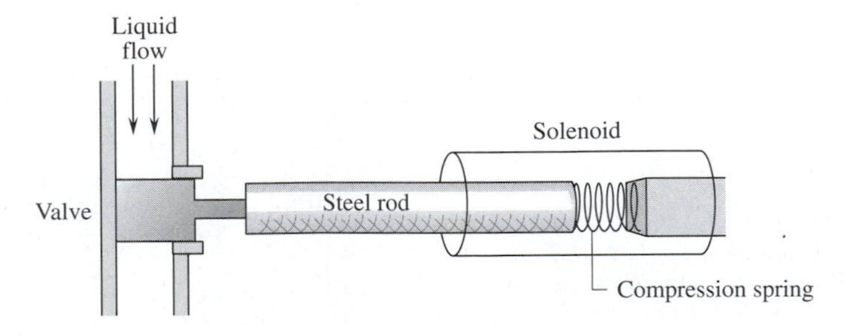

Figure 32–17: *A solenoid can be used to activate a flow-control valve.*

Electric Meters

Galvanometers, ammeters, and voltmeters all operate on the same principle. A flat coil is placed between the poles of a permanent magnet. The magnetic moment of the coil is proportional to the current passing through it. The poles of the magnet are curved to keep the **B** field constant as the coil turns, so that the current (I) is the only variable in the torque equation. We do not need to worry about a sine factor (see Figure 32–18).

$$|\tau| = \mu_0 N I A \mathrm{B}$$

or

$$|\tau| = k I$$

where $k = \mu_0 N A \mathrm{B}$.

A coil spring is attached to the rotating coil. It creates a countertorque proportional to the angle of rotation.

$$\tau_{(\text{spring})} = C \theta$$

where C is a proportionality constant. The system maintains equilibrium.

$$k I = C \theta$$

or

$$I = \frac{C \theta}{k}$$

The current is directly proportional to the angle of rotation. A pointer is attached to the moving coil. It indicates the appropriate reading on a scale.

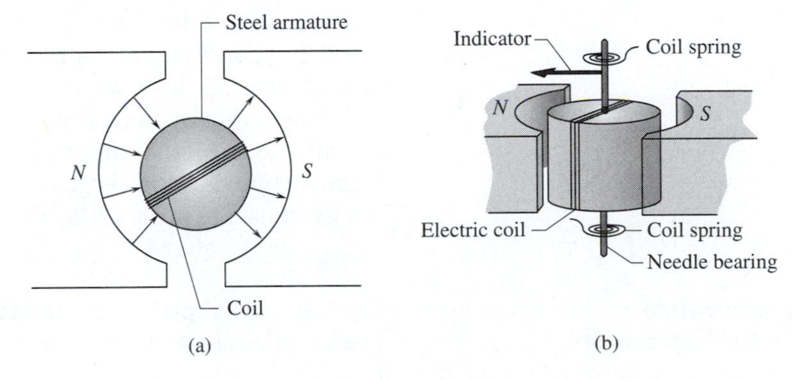

Figure 32–18: *An electric current indicator used in a galvanometer.* (a) *Top view. A cylinder of steel is used as an armature so that the* **B** *field is the same for a wide range of coil orientations.* (b) *Perspective view. The magnetic torque on the electric coil works against two coil springs.*

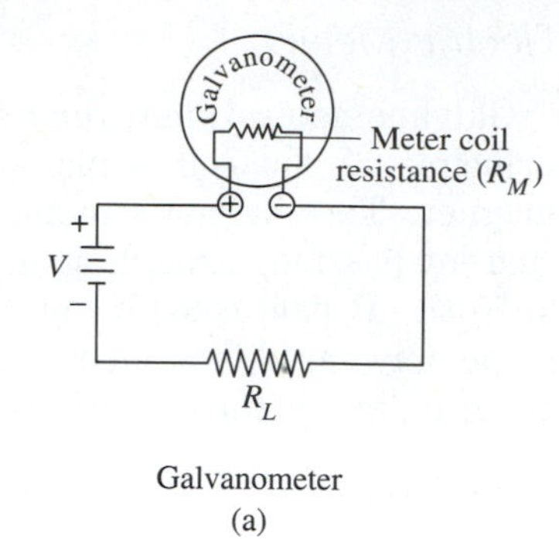

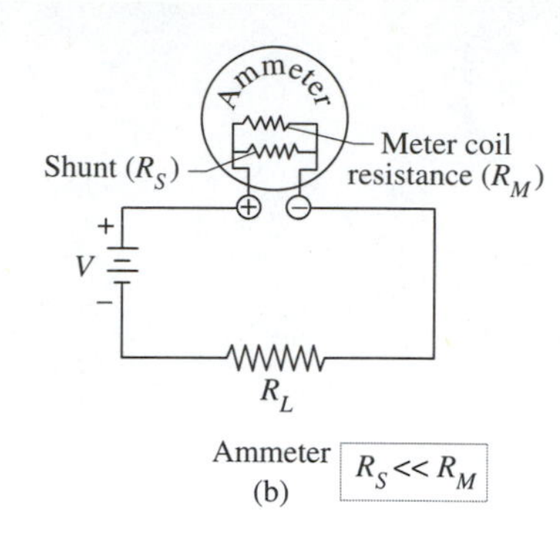

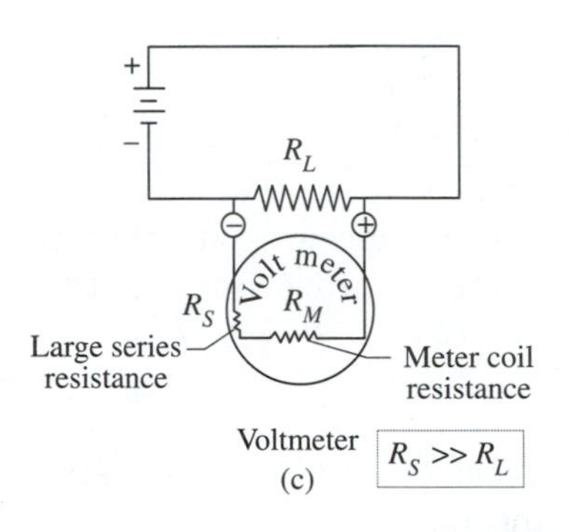

Figure 32–19: *Three ways to use a galvanometer movement. (a) Hook the device in series with the circuit as a galvanometer to measure very small currents. (b) A low resistance shunt is parallel to the galvanometer. An ammeter is created. (c) A very large resistance is in series with the galvanometer. This creates a voltmeter.*

Figure 32–19 shows three ways this device can be hooked up. If the current is fed directly through the movement, the device is a **galvanometer.** Galvanometers typically measure currents in the range of microamperes.

We can measure larger currents through a circuit with a device called an **ammeter.** Ammeters have a **shunt,** a very low resistance in parallel with the coil. The shunt has a resistance of a small fraction of an ohm. Most of the current passes through the shunt. A small fixed fraction of the current passes through the higher resistance of the coil to produce the operating torque.

If we want to measure potential differences, we use a very high resistance in series with the coil. The meter is then connected in parallel to the resistance between the two potential points we choose to measure. As long as we measure potential differences across resistances that are small compared with the meter resistance, only a small fraction of the current will pass through the coil. Since voltage is proportional to the current by Ohm's law, the scale on the meter can be marked out in volts. Such a device is called a **voltmeter.**

Magnetohydro-dynamic Generation

Nuclear power plants or traditional fossil-fuel electric generating plants generate electricity in the same way. Some source of heat produces steam. The steam turns turbines. The turbines work electric generators to produce electricity. This process loses a lot of energy through thermal and mechanical limits in efficiency.

Magnetohydrodynamic (MHD) generators use the heat released in burning chemical fuels in a different way. The fuel is burned at very high temperatures, so hot that the by-products are ionized to form a plasma. This ionized gas is passed through a tunnel with metal plates on opposite sides. The other two sides are electrically insulating materials (see Figure 32–20).

A uniform magnetic field is at right angles to the flow of the ionized gas. In accordance with Equation 32–1 positive ions are diverted to one metal plate, the negative ions to the other. A giant capacitor is created. The potential between the plates is several kilovolts. The plates discharge through an external load resistance, neutralizing the ions striking the plates. A continuous flow of ions maintains the plate charge.

In addition to high temperatures, MHD generators require a rapid flow of the ionized gas. Burners designed in a fashion similar to jet engines are used to create these two requirements.

Because the exhaust from MHD units is still hot, the exhaust can be used to heat boilers to supply steam for conventional steam turbines. MHD units are likely to be used for auxiliary power in fossil-fuel electric generating plants.

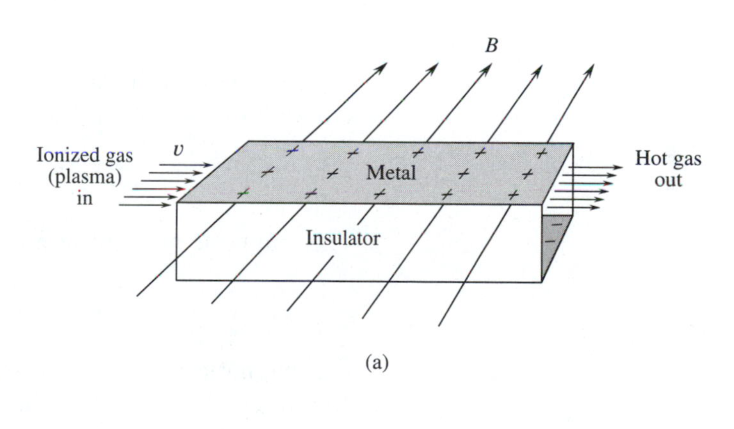

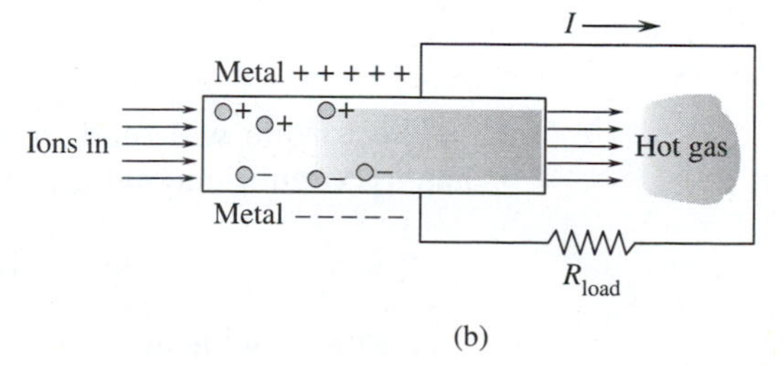

Figure 32–20: *Magnetohydrodynamic generation (MHD).* (a) *Perspective view. A high-speed plasma enters a chamber with metal plates on two opposite sides and a perpendicular magnetic field.* (b) *Side view. A flow of electrons through an external circuit neutralizes the ions driven to the metal plates by magnetic forces.*

SUMMARY

When an electric charge moves through a magnetic field a force acts on it. The magnitude of the force is:

$$|\mathbf{F}| = q \, \mathrm{v} \, \mathrm{B} \sin \theta$$

where θ is the angle between the charged particle's velocity and the magnetic field. The direction of the force is at right angles to both the charged particle and the magnetic field. If **v** is rotated into **B**, the direction of the field is along the rotational axis.

The magnetic field strength is measured in tesla (T) or gauss (G).

$$1 \text{ tesla (T)} = 1 \text{ N/A} \cdot \text{m} = 10^4 \text{ G}$$

The magnetic force acting on a moving charge in a uniform magnetic field is a centripetal force. The orbit of the charge will be a spiral or circle with the following radius (R), period (T), and spacing of coils (X):

$$R = \frac{m \, \mathrm{v}}{q \, \mathrm{v} \sin \theta}$$

$$T = \frac{2\pi \, m}{q \, \mathrm{B} \sin \theta}$$

$$X = \frac{2\pi \, m \, \mathrm{v}}{q \, \mathrm{B} \tan \theta}$$

An electric current is a moving charge. The magnitude of the magnetic force acting on the current (I) in a length of wire (l) is:

$$|\mathbf{F}| = I \, l \, \mathrm{B} \sin \theta$$

The magnetic moment of a coil of any two-dimensional shape is:

$$\mu = N I A$$

In a magnetic field a coil will have a torque acting on it proportional to the magnetic moment of the coil.

$$|\tau| = \mu \mathrm{B} \sin \theta$$

The torque will tend to rotate a coil toward an orientation that has maximum magnetic flux (Φ_m) through the coil of area A.

$$\Phi_m = \mathrm{B} \, A \cos \theta$$

Magnetic fields are created by moving electric charges. The contribution (**ΔB**) toward the magnetic field a distance r from a small segment of wire (Δs) carrying a current I is called the Biot-Savart law:

$$|\Delta \mathbf{B}| = \frac{k_m I \Delta s}{r^2}$$

The net field at some point can be found by adding up vectorially the contributions ($\Delta \mathbf{B}$) for each small segment of a wire. The proportionality constant (k_m) in the Biot-Savart law is related to the permeability of free space (μ_0).

$$k_m = 10^{-7} \frac{\text{Wb}}{\text{A} \cdot \text{m}} = \frac{\mu_0}{4\pi} \frac{\text{Wb}}{\text{A} \cdot \text{m}}$$

and

$$\mu_0 = 4\pi \times 10^{-7} \frac{\text{Wb}}{\text{A} \cdot \text{m}}$$

The strengths of the magnetic field at the center of some special coils are listed below.

Flat coil of radius R and N turns:

$$|\mathbf{B}| = \frac{\mu_0 N I}{2R}$$

Solenoid of length l and N turns:

$$\text{B} = \frac{\mu_0 N I}{l} = \mu_0 n I$$

Toroid of radius r and N turns:

$$\text{B} = \frac{\mu_0 N I}{2\pi r}$$

KEY TERMS

If you can explain the following terms to a friend or classmate, you understand their meaning. If you cannot explain the terms, you should reread the sections in which they are discussed.

ammeter
Biot-Savart law
galvanometer
gauss (G)
magnetic moment
magnetohydrodynamic generators (MHD)
right-hand rule
shunt
solenoid
tesla (T)
toroid
voltmeter
webers (Wb)

EXERCISES

Section 32.1:

1. Figure 32–21 shows some moving charges in magnetic fields. Find the direction of the magnetic force acting on the particle in each case.

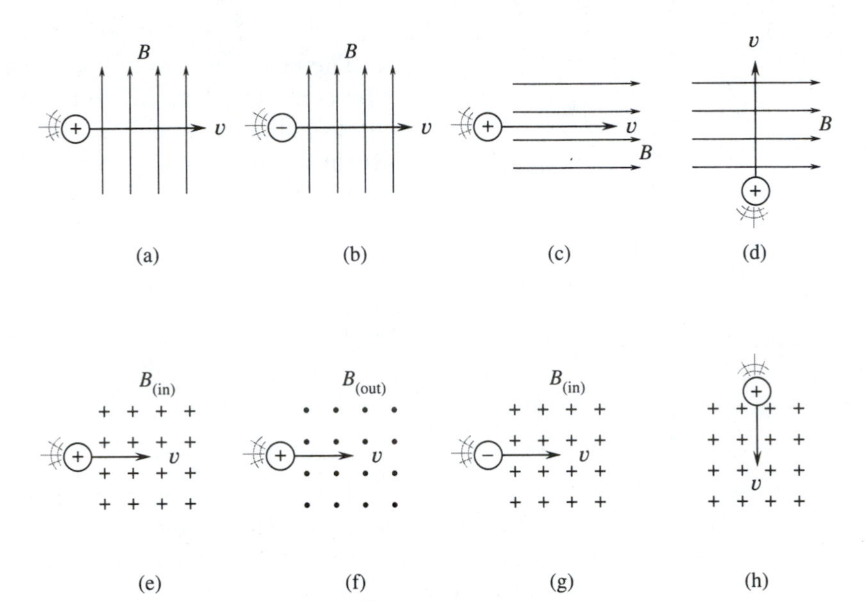

Figure 32–21: Diagram for Exercise 1. Magnetic forces on charged particles. (+'s indicate a **B** *field into the page. ●'s indicate a* **B** *field out of the page.)*

2. Under what conditions will an electric charge moving through a magnetic field *NOT* experience a magnetic force acting on it?

3. A velocity selector is shown in Figure 32–22. A uniform magnetic field and the electric field created by a pair of charged plates balance out, resulting in no net deflection of a beam of charged particles for a particular speed. The magnetic force and electric force are equal. ($F_m = F_e$, or $q\,v\,B = qE$. The speed of undeflected particles is $v = E/B$. Particles traveling at other speeds will not pass through the second slit.)

Electric plate 1
B field
Slit Slit
Ions of velocity v only
Electric plate 2

Figure 32–22: Diagram for Exercise 3. A velocity selector.

   A. Which plate (1 or 2) should have a positive charge for the electrostatic and magnetic forces on a positive ion beam to balance out?

   B. If the beam contains negative charges, should the charge on the plates be the reverse of your answer to part A?

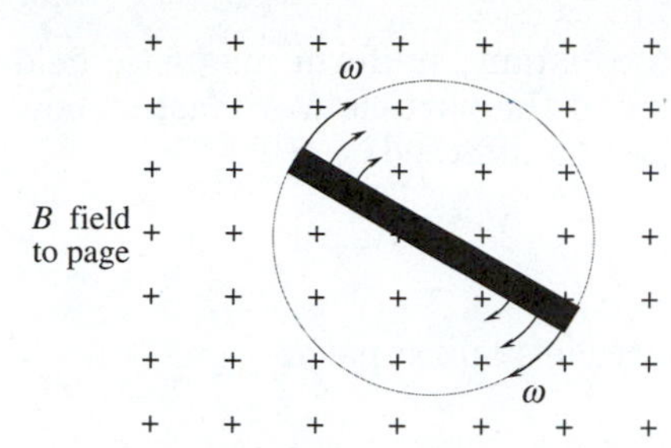

Figure 32–23: *Diagram for Exercise 4. A simplified illustration of a homopolar generator. (+'s indicate a **B** field into the page.)*

4. Figure 32–23 shows a metal bar rotating at right angles to a uniform magnetic field. Use Equation 32–1 to show why a potential difference develops between the ends of the rod and its center. This idea is used in a homopolar electric generator to produce DC current.

Section 32.2:

5. A beam of electrons and a beam of protons both pass at right angles to the same uniform magnetic field (see Figure 32–24). The speeds of the particles in the two beams are the same. The magnetic field is into the page in each case.
   A. Which beam will have the shorter radius of curvature?
   B. Which will have the longer period of orbit?
   C. Which diagram in Figure 32–24 best depicts the orbits of the two ion beams?

6. A charged particle enters at right angles to a uniform magnetic field. If the strength of the field (**B**) is increased, how will the following change?
   A. The radius of the particle's orbit?
   B. The period of the particle.
   C. The frequency of the particle.
   D. The speed of the particle.

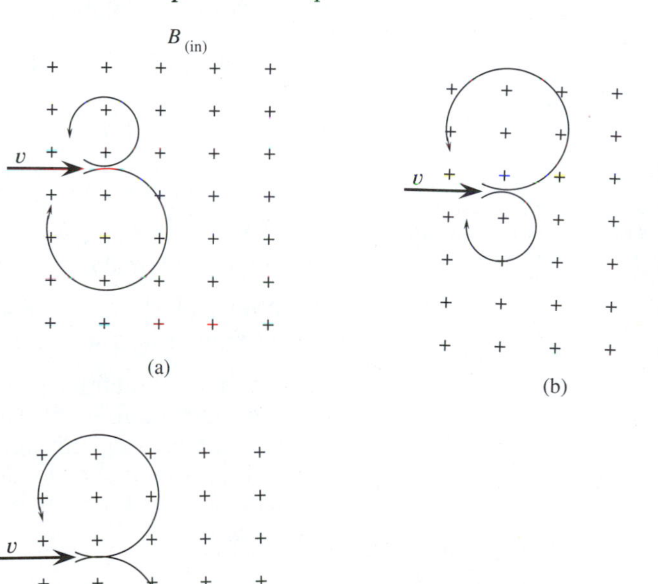

Figure 32–24: *Diagram for Exercise 5. Relative sizes of orbits are not exact. (+'s indicate a **B** field into the page.)*

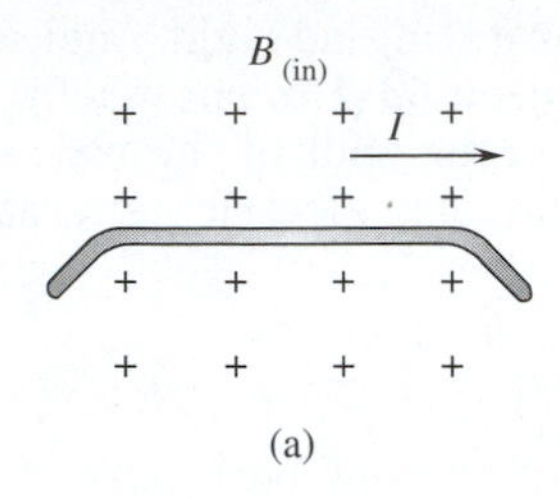

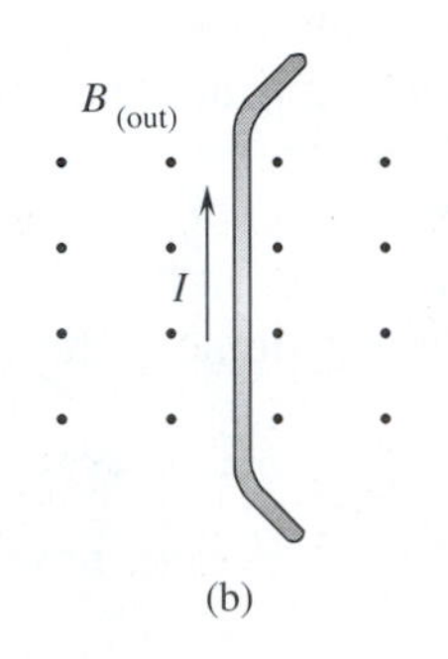

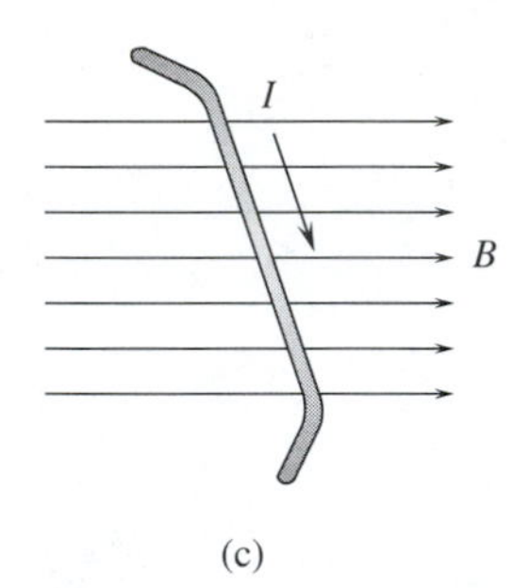

Figure 32–25: *Diagram for Exercise 9. Magnetic forces on wires. (+'s indicate a* **B** *field into the page; ●'s indicate a* **B** *field out of the page.)*

7. A charged particle enters a constant, uniform magnetic field at an angle of 88°. If the speed of the particle were faster, how would the following quantities be affected?
   A. The radius of orbit.
   B. The period of orbit.
   C. The frequency.
   D. Coil separation in the particle's spiral path.

Section 32.3:

8. How does the size of the force on a current-carrying wire change with an increase of the following factors?
   A. Current through the wire.
   B. The length of the wire in the field.
   C. The strength of the perpendicular component of the magnetic field.
   D. The angle between the wire and the magnetic field as angles increase from 0° to 90°.

9. Indicate the direction of the magnetic force on the wires shown in Figure 32–25.

Section 32.4:

10. Explain the following terms:

    A. Magnetic flux *B*. Magnetic moment *C*. Weber
    D. Tesla *E*. Permeability of free space

11. Coil A has 10 turns and a radius twice that of coil B. Coil B has 20 turns. The same current flows through the two coils. Which coil has the larger magnetic moment? Justify your answer.

12. Do a units analysis to show that 1 Wb = 1 N · m/A.

*13. Which of the two coils in Exercise 11 will have the larger maximum magnetic flux passing through it? Explain.

*14. Two rectangular coils are identical except for shape. They have the same number of turns, the same area, and the same electric current passing through them. They are in identical **B** fields. Coil A is long and slender. Coil B is nearly square. Which one will have the larger maximum torque acting on it? Explain your answer.

Section 32.5:

15. Explain the following terms:
    A. Biot-Savart law
    B. Solenoid
    C. Toroid

16. Two flat coils are identical except coil A has twice the radius of coil B. The same current flows through the coils. Compared with coil B, the magnetic field produced in coil A's center will be different by a factor of:

a. 4 **b.** 2 **c.** 1 **d.** ½ **e.** ¼

17. Flat coil A has twice as many turns as coil B. Coil A also has twice the diameter of coil B. For identical conditions, the magnetic field at the center of coil A will be:
A. larger than coil B.
B. the same as coil B.
C. smaller than coil B.

18. How does the magnetic field inside a solenoid change for the following factors?
A. Current through the coil increases.
B. Total number of turns increases with the length constant.
C. Length increases with the number of turns constant.
D. Length increases with the number of turns per unit length constant.

*19. A toroid is formed by wrapping 23 m of copper wire around a doughnut-shaped core. What length of wire is needed to build another toroid that will create the same **B** field for the same current? Its radius (r) is twice as large as that of the first toroid. (Hint: The length of wire is proportional to the number of turns.)

Section 32.6:

20. Make a list of electrical and electromechanical devices that use a solenoid.

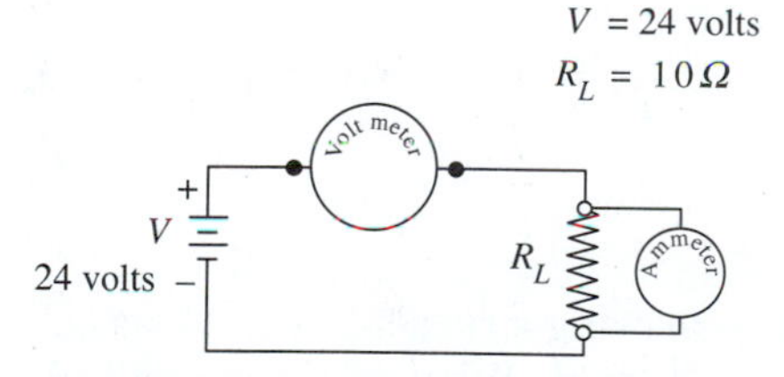

Figure 32–26: *Diagram for Exercise 23. The ammeter and voltmeter hookups are* WRONG.

21. An ammeter connected in series with a very low resistance may give a poor reading because its shunt resistance may become a significant factor in the circuit. In this circumstance, would the meter read too high or too low? Explain.

22. A voltmeter connected in parallel across a very large resistance may give a poor reading. Explain why. Would the reading be too high or too low?

23. In Figure 32–26 the ammeter and the voltmeter are both improperly connected. The voltmeter is in series with the circuit, and the ammeter is in parallel. What is the scale reading on the voltmeter likely to be? On the ammeter?

PROBLEMS

Section 32.1:

1. Here are some different sets of data for charged particles moving in uniform magnetic fields. Find the missing quantity in each set.

A. $q = +2.7 \times 10^{-14}$ C	**B.** $q = -1.6 \times 10^{-19}$ C
$\mathbf{v} = 3.2 \times 10^{2}$ m/s	$\mathbf{v} = 1.7 \times 10^{7}$ m/s
$\mathbf{B} = 0.067$ T	$\mathbf{B} = ?$
$\theta = 73°$	$\theta = 90°$
$\mathbf{F} = ?$	$\mathbf{F} = 3.3 \times 10^{-13}$ N
C. $q = +1.6 \times 10^{-19}$ C	**D.** $q = ?$
$\mathbf{v} = 2.3 \times 10^{6}$ m/s	$\mathbf{v} = 7.2 \times 10^{5}$ m/s
$\mathbf{B} = 0.23$ T	$\mathbf{B} = 0.092$ T
$\theta = 0°$	$\theta = 90°$
$\mathbf{F} = ?$	$\mathbf{F} = 1.3 \times 10^{-13}$ N

2. An electron travels at a speed of 6.3×10^{6} m/s at right angles to a uniform magnetic field. If the magnetic force exerted on the electron is 1.2×10^{-13} N, what is the strength of the magnetic field?

3. A magnetic field strength is 271 G. What is its strength in tesla?

Section 32.2:

4. A proton that is traveling at 5.4×10^{5} m/s enters a 0.130-T uniform magnetic field at right angles to the field. (The mass of a proton is 1.67×10^{-27} kg.)
   A. What is the radius of the proton's orbit?
   B. How much time is needed to make one complete orbit?
   C. What is the frequency of the orbit?

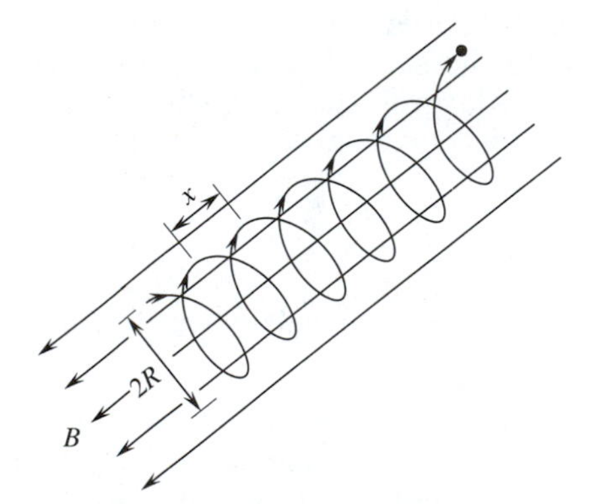

Figure 32–27: Diagram for Problem 5. An electron travels through a uniform magnetic field.

5. An electron travels through a 0.0012-T magnetic field at a speed of 1.5×10^{7} m/s at an angle of 87°. (The mass of an electron is 9.1×10^{-31} kg.) See Figure 32–27.
   A. What is the radius of the orbit?
   B. What is the period of the orbit?
   C. What is the separation of the coils in the helical path of the orbit?

Section 32.3:

6. A 23-cm length of copper wire carrying a current of 1.3 A passes at right angles through a 0.23-T field. What is the magnitude of the force acting on the wire?

7. A magnetic force of 0.12 N acts on a 12 cm long piece of aluminum wire passing at an angle of 55° to a 0.26-T field. How much current is flowing through the wire?

Section 32.4:

8. A rectangular coil has 45 turns, a length of 14 cm, and a width of 9.2 cm. A 2.1-A current flows through the coil.
   A. What is the coil's magnetic moment?
   B. What maximum torque could a 0.31-T field exert on the coil?

C. What maximum magnetic flux passes through the coil in the 0.31-T magnetic field?

9. The rotor of a small starter motor has a magnetic moment of 250 A · m^2. What maximum torque is produced when the **B** field from the field coils is 0.80 T?

10. In a 1200-G field, the maximum flux through a particular coil is 0.036 Wb. What is the cross-sectional area of the coil?

Section 32.5:

11. A flat coil with 125 turns has a diameter of 32 cm. What is the magnetic field strength when a current of 2.7 A passes through the coil?

12. A 2.0-cm section of wire carries a current of 1.2 A. Estimate its contribution to the magnetic field strength 10 cm from the wire and at right angles to the wire.

13. A solenoid with a radius of 0.90 cm and a length of 32 cm is made with 1300 turns of wire. When 1.7 A of current pass through the solenoid:

A. what is the strength of the magnetic field in the center of the solenoid?

B. how much magnetic flux passes through the center of the solenoid?

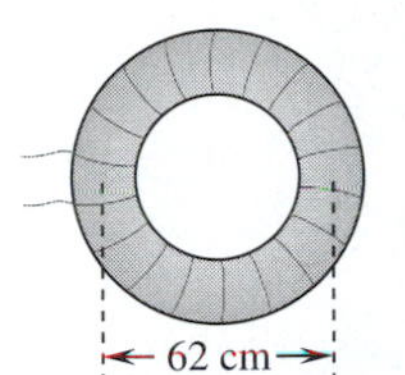

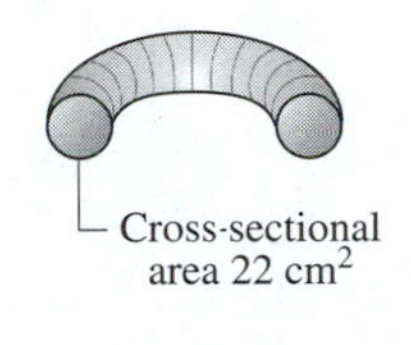

Figure 32–28: *Diagram for Problem 14. Calculate the magnetic flux inside the toroid.*

14. A 62 cm diameter toroid has 120 turns. The cross-sectional area of the doughnut is 22 cm^2 (see Figure 32–28). Calculate the amount of magnetic flux trapped in the toroid when a current of 6.2 A flows through the windings.

Chapter 33

INDUCTANCE

This is a close-up photo of tape recording equipment. Magnetic induction plays an important role in how this machinery records sound signals on tape. (Photograph reprinted from Communications Technology *by Barden and Hacker, copyright © 1990 by Delmar Publishers Inc. Used with permission.)*

OBJECTIVES

In this chapter you will learn how:

- a changing magnetic field in a coil can produce a voltage (Faraday's law)
- a current-carrying wire can induce a back-EMF (self-inductance)
- two coils can induce back-EMFs in each other (mutual inductance)
- energy can be stored in a coil's magnetic field
- different kinds of materials affect the magnetic field in a current-carrying coil
- to compare the properties of diamagnetic, paramagnetic, and ferromagnetic materials
- to calculate input and output voltages of a transformer
- DC field shunt motors and DC series motors work

Time Line for Chapter 33

1830	Joseph Henry discovers the principle of the dynamo, but fails to publish his result.
1831	Independent of Henry, Michael Faraday discovers that electricity can be induced by a magnetic field.
1845	Michael Faraday discovers paramagnetism and diamagnetism.
1890	J. Alfred Ewing discovers magnetic hysteresis.
1907	Pierre Weiss develops the domain theory of ferromagnetism.
1921	Arthur Compton relates ferromagnetism with electron spin.

We have looked at electrical charges at rest. They are the source of electric fields. Charges can pile up in capacitors and store energy in the resulting E field.

We have looked at moving electrical charges, or electrical current. Moving charges create and react with magnetic fields. Moving charges bump into the atomic lattice that makes up the structure of a conductor. In this process, energy is dissipated in the form of heat.

In this chapter we will take things one step further. We will look at the effect of changing electrical current. When electrical charges are accelerated, they resist the change. The resistance to change of motion is not simply an inertial effect caused by the electron mass, but a reaction to changing magnetic fields. This kind of reaction is called **magnetic induction.**

33.1 FARADAY'S LAW

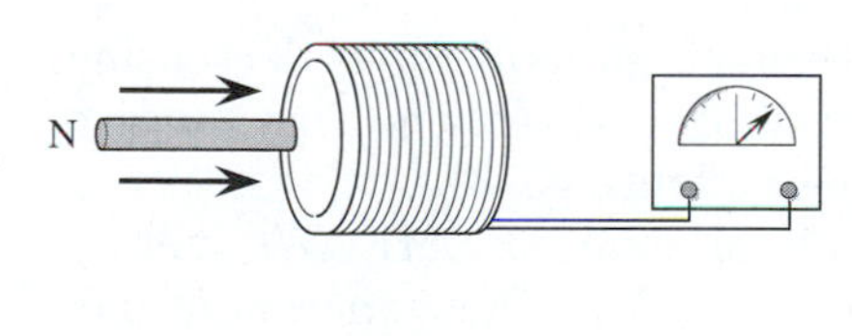

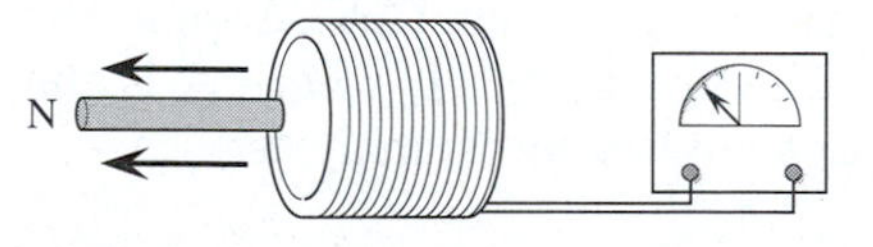

Figure 33–1: A permanent magnet is inserted into a coil. As the magnet moves into the coil, a galvanometer shows that a current is flowing. When the magnet is pulled out, the current flows in the opposite direction.

We can perform a simple experiment, shown in Figure 33–1. A coil is connected to a galvanometer. There is no battery or other voltage source in the circuit to cause an electrical current to move through the circuit. The galvanometer reads zero.

If we insert a bar magnet into the coil, a current will briefly flow as long as the magnet is moving. When the magnet is stationary, there is no current.

When we pull the magnet out of the coil, current again flows, but in the opposite direction as before (Figure 33–1, bottom).

We can perform the experiment several times with slight variations. We can change the ends of the magnet. This reverses the direction of current flow. We can insert and remove the bar magnet at different speeds. We will find that the current is stronger for faster insertions and removals of the magnet. We can change coils. The current will be stronger for a coil with many windings than for a coil with only a few windings (see Figure 33–2).

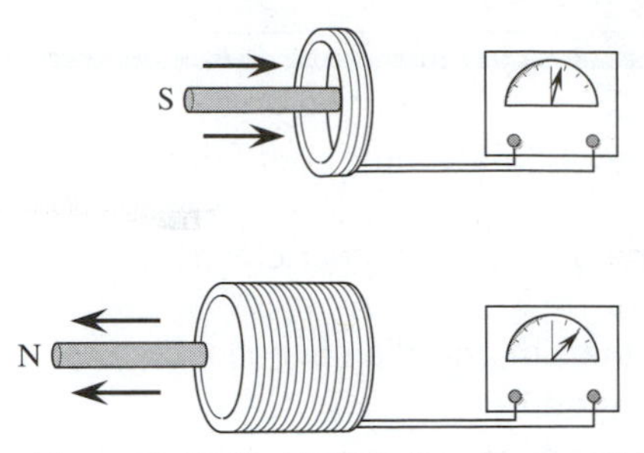

Figure 33–2: The EMF induced in a coil is proportional to the number of turns in the coil. It also depends on the direction of the field and the direction of motion.

A coil is a long conductor wrapped in a series of loops. A current moving through it implies two things. The first is that there must be a voltage difference between the two ends of the conductor. The second is that the current must be creating a magnetic field of its own. Let us look at the second item first.

As we insert the bar magnet into the coil, the magnetic flux going through the coil increases. At the same time, current flows through the coil, creating its own magnetic field. This type of current is called an **induced current.** We will note the direction of the induced current and apply the right-hand rule. We discover that the magnetic field caused by the induced current is in a direction *opposite* to the increasing magnetic field from the bar magnet. The current flows in such a direction as to create a magnetic field resisting the change in magnetic flux through the coil. This observation is called Lenz's law. See Figure 33–3.

Since the current through the conducting wire in the coil is proportional to the voltage difference between the ends of the wire, we can talk about the **induced EMF** rather than the induced current.

We see induced EMFs in all kinds of coils: toroids, solenoids, flat coils, or shapes of your own invention. Such a device is called an **inductor.** An inductor is any device that creates an induced EMF when the magnetic flux changes in it.

Lenz's law: *The induced EMF (or induced current) in an inductor is in a direction to resist the change of the magnetic flux through the conductor.*

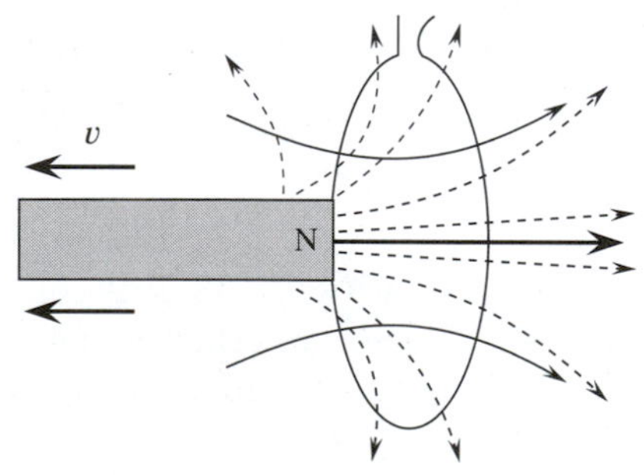

Figure 33–3: The magnetic field lines of the permanent magnetic (dashed lines) are enforced by the induced field of the coil (solid lines) as the magnet is withdrawn. An EMF is created in a direction to resist the change of magnetic field strength.

If the magnetic field passing through an inductor is increasing, a magnetic field in the opposite direction is created by a coil. This tends to reduce the net magnetic field.

If the magnetic field passing through an inductor is decreasing, the induced magnetic field is in the same direction. This tends to increase the net magnetic field. See Figure 33–3.

Now let us look more closely at the induced EMF. We will use the symbol V_L to indicate the induced EMF. Our experiment indicates that V_L is proportional to the change of flux ($\Delta\phi_m/\Delta t$) and the number of turns in the coil (N). The flux is $\mathbf{B} \cdot A$, where A is the cross-sectional area of the coil. Let us do a units analysis.

$$\frac{\Delta\phi_m}{\Delta t} = \frac{\text{T} \cdot \text{m}^2}{\text{s}} = \frac{\text{N}}{\text{C} \cdot \text{m/s}}\left(\frac{\text{m}^2}{\text{s}}\right) = \frac{\text{N} \cdot \text{m}}{\text{C}}$$

The time rate of magnetic flux has units of work (or energy) per coulomb. This is a unit of voltage. We can write a simple equation for V_L called **Faraday's law.**

$$V_L = -N\left(\frac{\Delta\phi}{\Delta t}\right) \qquad \textbf{(Eq. 33–1)}$$

The minus sign occurs because of Lenz's law. The induced EMF opposes the changing magnet flux.

☐ **EXAMPLE PROBLEM 33–1: FARADAY'S LAW**

A 310-turn coil with a cross-sectional area of 1.9×10^{-3} m^2 is connected to a galvanometer. The magnetic field through the coils changes from 0 to 0.12 T in a period 0.20 s at a uniform rate.

A. What maximum magnetic flux passes through the coil?

B. Calculate the induced EMF (V_L).

■ ***SOLUTION***

A. Maximum magnetic flux:

$$\phi_m(\text{max}) = B(\text{max})\,A$$

$$\phi_m(\text{max}) = 0.12 \text{ Wb/m}^2 \times 1.9 \times 10^{-3} \text{ m}^2$$

$$\phi_m(\text{max}) = 2.3 \times 10^{-4} \text{ Wb}$$

B. Induced EMF:

$$V_L = -N\left(\frac{\Delta\phi}{\Delta t}\right)$$

$$V_L = -310 \text{ turns} \times \frac{(2.3 \times 10^{-4} - 0.0)\text{ Wb}}{(0.20 - 0.0)\text{ s}}$$

$$V_L = -0.36 \text{ V}$$

33.2 SELF-INDUCTANCE

An induced EMF is produced in an inductor for any change of magnetic flux through it. The source of the magnetic flux is not important. It does not have to be from some external source.

Coils, toroids, solenoids, and any other shaped inductor creates a magnetic field when current passes through it. If the current changes, the magnetic field will also change. The inductor can react to changes in its own magnetic field. This is called **self-inductance.**

Table 33–1 shows the magnetic flux for the three kinds of inductors we looked at in Chapter 32 (see Figure 33–4). Notice that the only

Table 33–1: Magnetic field strength and self-inductance for three kinds of inductors.

SHAPE	B (MAGNETIC FIELD STRENGTH)	L (SELF-INDUCTANCE)
Flat coil (Figure 33–4a)	$\sim\mu_0 \dfrac{N}{2R} i$	$\sim \dfrac{\mu_0 N^2 \pi R}{2}$
Long solenoid (Figure 33–4b)	$\mu_0 \left(\dfrac{N}{l}\right) i$	$\mu_0 \left(\dfrac{N^2}{l}\right) A$
Toroid (Figure 33–4c)	$\mu_0 \left(\dfrac{N}{2\pi r}\right) i$	$\dfrac{\mu_0 N^2 A}{2\pi r}$

variable in the magnetic flux is the current in each case. When we insert these expressions into Faraday's law we find that the induced EMF is proportional to the change of current flowing through an inductor. The proportionality constant depends on the shape, size, and number of turns in the inductor. We will denote the proportionality constant by the symbol L and call it **inductance.**

$$V_L = -L\left(\frac{\Delta I}{\Delta t}\right) \quad \textbf{(Eq. 33–2)}$$

The minus sign indicates a **back-EMF.** It is in the reverse direction to the power source in a circuit. We will find that Equation 33–2 is very important for AC circuits.

We can find the units for inductance by solving Equation 33–2 for L.

$$L = \frac{-V_L}{\Delta I/\Delta t} = \frac{\text{V}}{\text{A/s}}$$

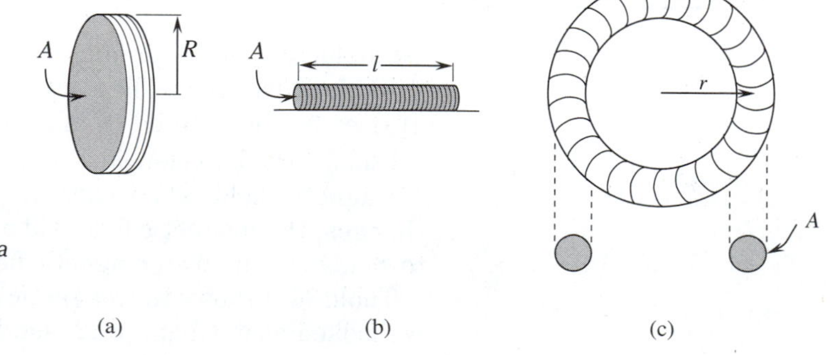

Figure 33–4: *See Table 33–1.* (a) *A flat coil with radius* R *and cross-sectional area* A. (b) *A long solenoid with length* l *and cross-sectional area* A. (c) *A toroid with radius* r *and cross-sectional area* A.

The unit of inductance is called a **henry (H).**

$$1 \text{ henry (H)} = 1 \text{ volt} \cdot \text{second/ampere} = 1 \text{ V} \cdot \text{s/A}$$

☐ **EXAMPLE PROBLEM 33–2: SELF-INDUCTANCE OF A TOROID**

A 645-turn toroid has a cross-sectional area of 12.7 cm^2 and a radius of 18.2 cm. Calculate its self-inductance, L.

■ ***SOLUTION***

From Table 33–1 we see that:

$$L = \frac{\mu_0 N^2 A}{2\pi r}$$

$$L = \frac{4\pi \times 10^{-7} \frac{N}{A^2} \times (645)^2 \times 1.27 \times 10^{-3} \text{ m}^2}{2\pi \times 0.182 \text{ m}}$$

$$L = 5.81 \times 10^{-4} \text{ H}$$

$$L = 0.581 \text{ mH}$$

We can see from Example Problem 33–2 that the henry is a fairly large unit. Most inductors have self-inductances on the order of a millihenry or less.

☐ **EXAMPLE PROBLEM 33–3: BACK-EMF ACROSS A SOLENOID**

A long solenoid has a self-inductance of 4.30 mH. If the electrical current through the device linearly increases from 0 to 2.32 A in 50 ms, what is the back-EMF induced in the solenoid?

■ ***SOLUTION***

$$V_L = -L\left(\frac{\Delta I}{\Delta t}\right)$$

$$V_L = -(4.30 \times 10^{-3} \text{ H}) \times \frac{(2.32 - 0.00) \text{ A}}{(0.050 - 0.00) \text{ s}}$$

$$V_L = -0.20 \text{ V}$$

33.3 MUTUAL INDUCTANCE

Figure 33–5 shows two coils, one inside the other. They share magnetic flux. They react with one another. The current change in one

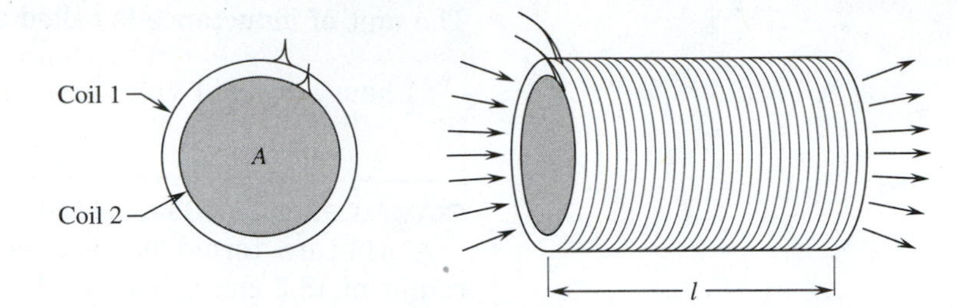

Figure 33–5: *Two concentric coils share the same magnetic flux.*

coil creates a changing magnetic flux, inducing a current flow in the other coil.

The EMF induced in the first coil (V_1) is caused by the change of the magnetic flux in the second coil ($\Delta\phi_2/\Delta t$).

$$V_1 = -N_1\left(\frac{\Delta\phi_2}{\Delta t}\right) \qquad \textbf{(Eq. 33–3A)}$$

The EMF induced in the second coil (V_2) is caused by the change of magnetic flux in the first coil ($\phi_1/\Delta t$).

$$V_2 = -N_2\left(\frac{\Delta\phi_1}{\Delta t}\right) \qquad \textbf{(Eq. 33–3B)}$$

If the two coils are nested solenoids, then the two changes of magnetic flux are:

$$\frac{\Delta\phi_1}{\Delta t} = \frac{\mu_0 N_1 A_1 \Delta I_1}{l_1 \Delta t}$$

and

$$\frac{\Delta\phi_2}{\Delta t} = \frac{\mu_0 N_2 A_2 \Delta I_2}{l_2 \Delta t}$$

We can substitute these expressions for flux into Equations 33–3A and 33–3B.

The EMF induced in coil 1 of cross-sectional area A_1 is caused by the flux generated by the second coil.

$$V_1 = -\frac{\mu_0 N_1 N_2 A_1 \Delta I_2}{l_2 \Delta t}$$

The corresponding relationship for the second coil is:

$$V_2 = -\frac{\mu_0 N_1 N_2 A_2 \Delta I_1}{l_1 \Delta t}$$

Usually the two coils fit together snugly. For all practical purposes the areas and lengths of the coils are the same.

$$l_1 = l_2 = l, \text{ and } A_1 = A_2 = A$$

This gives us:

$$V_1 = -\frac{(\mu_0 N_1 N_2 A)\,\Delta I_2}{l\,\Delta t}$$

and

$$V_2 = -\frac{(\mu_0 N_1 N_2 A)\,\Delta I_1}{l\,\Delta t}$$

Notice that the terms in parentheses are the same and depend only on the geometry of the coils. They can be combined into one symbol (M). This constant is called the **mutual inductance** of the two coils. This gives us a much simpler expression.

$$V_1 = -M\left(\frac{\Delta I_2}{\Delta t}\right) \quad \textbf{(Eq. 33–4A)}$$

$$V_2 = -M\left(\frac{\Delta I_1}{\Delta t}\right) \quad \textbf{(Eq. 33–4B)}$$

The main idea here is that the EMF induced in one coil is proportional to the change of current of the other coil. The two coils are linked by a common magnetic flux with which they interact. We can use this idea to connect two separated circuits by magnetic flux.

The two coils can have different shapes. They do not have to be one inside the other. All that is necessary for mutual inductance to occur is that they share a common changing magnetic flux (see Figure 33–6).

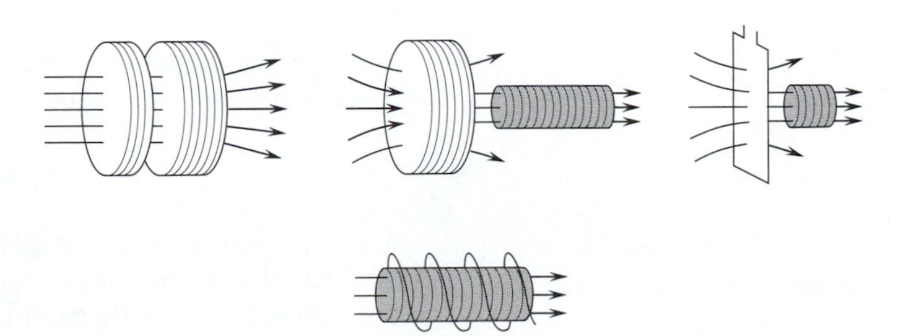

Figure 33–6: *Mutual inductance can occur between a variety of pairs of coils as long as some of the flux is shared.*

☐ **EXAMPLE PROBLEM 33–4: MUTUAL INDUCTANCE**

Two nesting coils have a mutual inductance of 0.016 H. If the current in one coil changes from 0.00 to 2.90 A in 0.030 s, what EMF is induced in the other coil?

■ ***SOLUTION***

Use Equation 33–4.

$$V_2 = -M\left(\frac{\Delta I_1}{\Delta t}\right)$$

$$V_2 = -0.016\ \text{H} \times \frac{(2.90 - 0.00)\ \text{A}}{(0.030 - 0.000)\ \text{s}}$$

$$V_2 - -1.5\ \text{V}$$

33.4 ENERGY STORED IN INDUCTORS

We found that a capacitor stores energy in its electric field. Let us look at the storage of energy in an inductor to see if there is a similar process.

We will start with the equation for self inductance. A back-EMF is produced by the changing current. This negative voltage drop indicates that energy is being taken out of the circuit.

$$-V_L = L\left(\frac{\Delta I}{\Delta t}\right)$$

During some period of time (Δt) an amount of charge (ΔQ) passes through the inductor. The change of stored energy (ΔU) is then:

$$\Delta U = -V_L\,\Delta Q = L\left(\frac{\Delta I}{\Delta t}\right)\Delta Q = L\,\Delta I\left(\frac{\Delta Q}{\Delta t}\right)$$

$\Delta Q/\Delta t$ is the current (I) passing through the inductor. We can substitute this into the equation.

$$\Delta U = L\,\Delta I\,I$$

When we rearrange the equation we see that the change of stored energy relative to current change ($\Delta U/\Delta I$) is directly proportional to the current flowing through the inductor.

$$\frac{\Delta U}{\Delta I} = L\,I \qquad \textbf{(Eq. 33–5)}$$

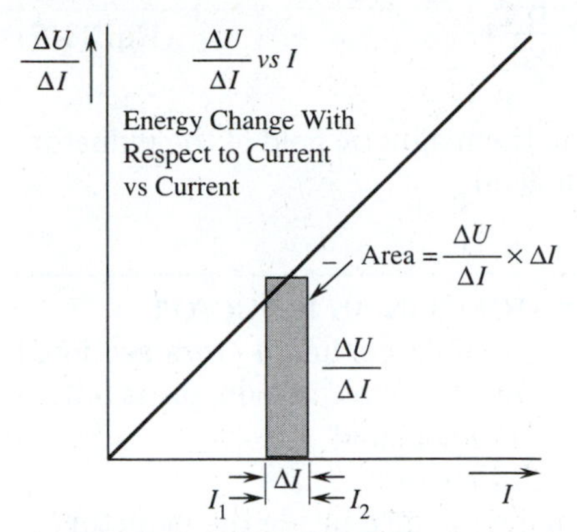

Figure 33–7: *A plot of change of energy with change of current. (ΔU/ΔI) is plotted against current through an inductor (I). The area of the shaded rectangle is the energy stored as the current increases from I_1 to I_2 through a current change of ΔI.*

A plot of Equation 33–5 is shown in Figure 33–7. Since it is a direct proportion, the plot is a straight line. The shaded area has a width of ΔI and a height of $\Delta U/\Delta I$. Its area is $(\Delta U/\Delta I) \times \Delta I$, or ΔU. This is the energy stored for an increase of current ΔI.

The total energy stored is the shaded area under the curve shown in Figure 33–8. The area is a triangle with a base of I and a height of $(\Delta U/\Delta I)$.

$$U = \frac{(\Delta U/\Delta I) \times I}{2}$$

We can substitute the right-hand side of Equation 33–5 into this expression to eliminate $\Delta U/\Delta I$.

$$U = \frac{L\,I^2}{2} \qquad \textbf{(Eq. 33–6)}$$

For a solenoid

$$L = \frac{\mu_0 N^2 A}{l}$$

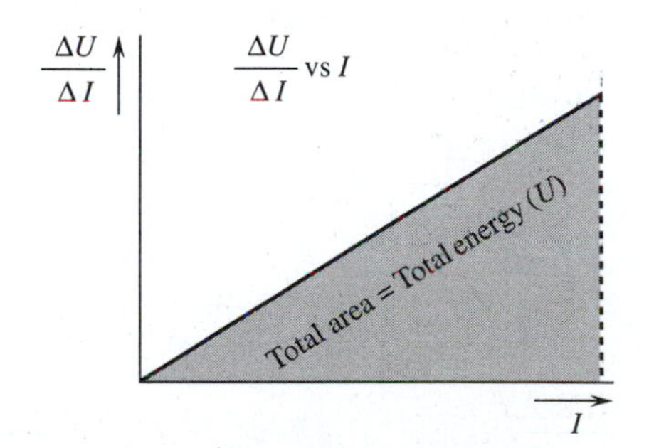

Figure 33–8: *The shaded triangular area is the total energy stored in an inductor as the current increases from zero.*

We can express the current passing through the solenoid in terms of the magnetic field it creates.

$$B = \mu_0 \left(\frac{N\,I}{l}\right), \text{ or } I = \frac{B\,l}{\mu_0 N}$$

Substituting the expressions for self-inductance and current into Equation 33–6, we get:

$$U = \left(\frac{\mu_0 N^2 A}{2\,l}\right)\left(\frac{B\,l}{\mu_0 N}\right)^2$$

This reduces to:

$$U = \frac{l\,A\,B^2}{2\,\mu_0}$$

A l is the volume of the solenoid. We can find the energy density (u) by dividing by volume.

$$u = \frac{U}{A\,l} = \frac{B^2}{2\mu_0} \qquad \textbf{(Eq. 33–7)}$$

The energy density depends on the magnetic field of the inductor. Energy is stored in its magnetic field.

☐ **EXAMPLE PROBLEM 33–5: ENERGY STORED IN AN INDUCTOR**

A 2300-turn solenoid has a length of 37 cm and a cross-sectional area of 17.2 cm^2. The current passing through the inductor is 5.2 A.

A. What is the self-inductance of the solenoid?
B. How much energy is stored in the coil?
C. What is the strength of the magnetic field inside the inductor?
D. How much energy is stored in the inductor per unit volume?

■ ***SOLUTION***

A. Self-inductance (L):

$$L = \frac{\mu_0 N^2 A}{l}$$

$$L = 4\pi \times 10^{-7}\ \text{N/A}^2 \times \frac{(2300^2 \times 1.72 \times 10^{-3}\ \text{m}^2)}{0.37\ \text{m}}$$

$$L = 0.031\ \text{H}$$

B. Stored energy (U):

$$U = \frac{L I^2}{2}$$

$$U = \frac{(0.031\ \text{H}) \times (5.2\ \text{A})^2}{2}$$

$$U = 0.42\ \text{J}$$

C. Magnetic field (**B**):

$$|\mathbf{B}| = \frac{\mu_0 N I}{l}$$

$$|\mathbf{B}| = (4\pi \times 10^{-7}\ \text{N/A}^2) \times \frac{2300 \times 5.2\ \text{A}}{0.37\ \text{m}}$$

$$|\mathbf{B}| = 4.1 \times 10^{-2}\ \text{T}$$

D. Magnetic energy density (u):

$$u = \frac{B^2}{2\mu_0}$$

$$u = \frac{(4.06 \times 10^{-2}\ \mathrm{T})^2}{2 \times 4\pi \times 10^{-7}\ \mathrm{N/A^2}}$$

$$u = 660\ \mathrm{J/m^3}$$

33.5 MAGNETIC MATERIALS

So far we have assumed that materials have no effect on magnetic fields. We can check the effect of materials by placing a core of material in a toroid. We choose a toroid because the entire **B** field is trapped inside it (see Figure 33–9).

The strength of the **B** field with no core we will denote by $\mathbf{B}_0$. This is the same magnetic field we would calculate using the equations in Chapter 32. The strength of the magnetic field in the toroid with a material inside we will denote by **B** with no subscript.

Tape Recorders

Recording tape has a thin coating that contains tiny ferromagnetic crystals of iron oxide. The crystals are in a soft bonding material that holds them on a plastic tape.

While recording, the tape passes through a recording head. The head is a tiny electromagnet. The magnet is a split ring of iron with wire windings around the iron core. A current through the wire creates a strong magnetic field that is almost completely enclosed within the ring. The exception is the split in the ring. Here the magnetic field spills out between the two cut faces making a north and south pole like a horseshoe magnet.

As the tape passes between these poles, the crystals on the tape are magnetized. If the field is strong, when one part of the tape passes through, the crystals will be strongly magnetized. If the field is weak, the crystals will be only weakly magnetized. A permanent record of the changing magnetic field of the electromagnet is recorded on the tape. Of course, the varying current through the electromagnet is an electrically amplified sound signal.

(Photograph reprinted from Communications Technology *by Barden and Hacker, copyright © 1990 by Delmar Publishers Inc. Used with permission.)*

A playback head reverses the process. The tape passes through another set of poles on an electromagnet. The varying pattern of the magnetic field of the iron oxide induces a current in the playback head. This passes through an amplifier to the speaker.

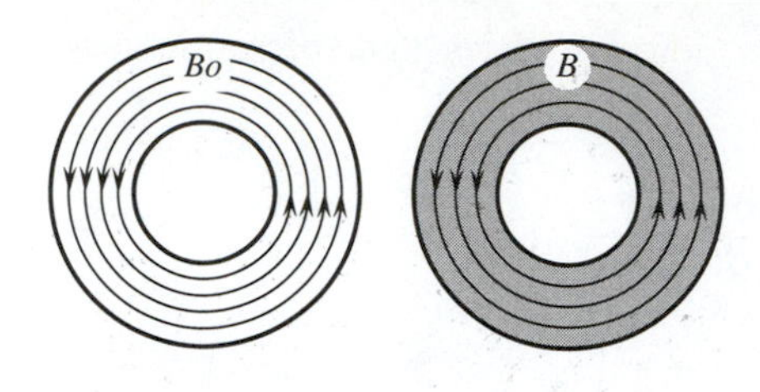

Figure 33–9: The magnetic field strength of a toroid changes from $\mathbf{B}_0$ to $\mathbf{B}$ when a material is placed in the core.

We will measure the magnetic field inside the toroid with a variety of materials and with a wide range of currents through its windings.

We find that there *is* a material effect. The magnetic field in the toroid with a core is directly proportional to the magnetic field without a core. We can write this as:

$$\mathbf{B} = k_m \mathbf{B}_0 \qquad \textbf{(Eq. 33–8)}$$

The proportionality constant (k_m) is called **relative permeability.** It is a unitless quantity since we are comparing the relative strength of two magnetic fields.

We find that materials fall into three obvious groups on the basis of relative permeability.

1. **Diamagnetic materials.** The largest group of materials has a relative permeability that is very slightly less than one. This group is called diamagnetic materials.

$$k_m < 1$$

2. **Paramagnetic materials.** Another group of materials has a relative permeability slightly larger than one. This group is called paramagnetic materials.

$$k_m > 1$$

3. **Ferromagnetic materials.** The third group, called ferromagnetic materials, has a relative permeability that is many times larger than one.

$$k_m >> 1$$

It would be convenient to break down the permeability into two parts. Part of the field we observe in the toroid is caused by the field created by the current through its windings. When nothing is in the toroid, $k_m = 1$, since $\mathbf{B} = \mathbf{B}_0$ for this situation. The other part of the observed field is dependent on the material in the core. It adds or subtracts from the existing field of the toroid windings. We will represent this with a material constant called **magnetic susceptibility** (χ_m). We can write the relative permeability in these two parts.

$$k_m = 1 + \chi_m \qquad \textbf{(Eq. 33–9)}$$

- For diamagnetic materials, χ_m is a small negative number.
- For paramagnetic materials, χ_m is a small positive number.
- For ferromagnetic materials, χ_m is much larger than one.

Table 33–2 shows the relative permeabilities and magnetic susceptibilities for a few materials.

Table 33–2: Magnetic properties of selected materials.

MATERIAL	RELATIVE PERMEABILITY ($k_m = 1 + \chi_m$)	MAGNETIC SUSCEPTIBILITY (χ_m)
Diamagnetic*		
Calcium carbonate	0.99996	-3.8×10^{-5}
Copper	0.999994	-5.5×10^{-6}
Graphite	0.999994	-6.0×10^{-6}
Silver	0.99998	-2.0×10^{-5}
Sulfur	0.99998	-1.5×10^{-5}
Water	0.99999	-1.3×10^{-5}
Paramagnetic		
Aluminum	1.00002	$+1.6 \times 10^{-5}$
Calcium	1.00004	$+4.0 \times 10^{-5}$
Copper chloride ($CuCl_2$)	1.00103	$+1.03 \times 10^{-3}$
Iron oxide	1.0072	$+7.2 \times 10^{-3}$
Oxygen	1.0035	$+3.5 \times 10^{-3}$
Uranium	1.00041	$+4.1 \times 10^{-4}$
Ferromagnetic		
Cobalt	250	250
Iron	5000	5000
Nickel	600	600
Permalloy	1×10^5	1×10^5
Superpermalloy	1×10^6	1×10^6

* Nearly all organic compounds are diamagnetic.

33.6 MOLECULES AND MAGNETIC MATERIALS

Diamagnetism

Diamagnetism is an example of magnetic induction. Look at Figure 33–10. We can view an orbiting electron as a coil with a cross-sectional area equivalent to the area of its orbit. If f is the frequency of the orbiting electron, its effective current is $e\,f$.

When we want to measure a magnetic material we bring it to the poles of a magnet. If we use a toroid or other electromagnet we place the sample in the device and then turn it on. In each case, for a period of time there is an increasing magnetic field passing through our sample material.

Inside our material, an induced back-EMF is created for each

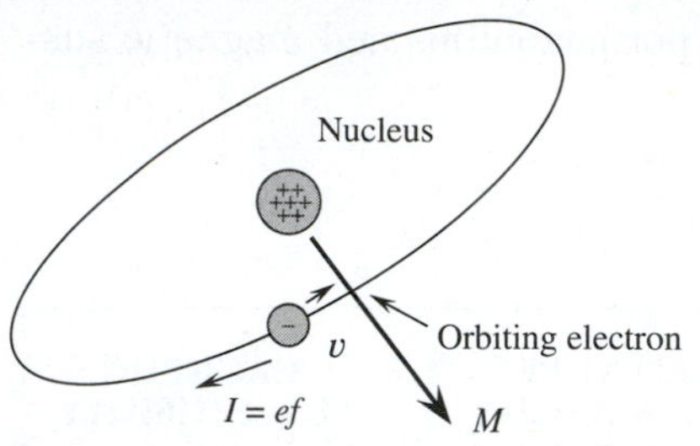

Figure 33–10: *An orbiting electron has an effective current of I = e f. It behaves like a coil with a magnetic moment.*

orbiting electron according to Faraday's law, as the magnetic flux through its orbit changes. We can write the equation for the back-EMF by using either Faraday's law or the self-inductance relationship.

$$V = -N\left(\frac{\Delta\phi_m}{\Delta t}\right) = -L\left(\frac{\Delta I}{\Delta t}\right)$$

Since $I = ef$, then:

$$-N\left(\frac{\Delta\phi_m}{\Delta t}\right) = -L\left(\frac{e\,\Delta f}{t}\right)$$

We can rewrite this as:

$$f = K\,\Delta\phi_m$$

where we have combined all of the constants into one symbol (K).

The effect of the changing field is to alter the frequency of the electron in orbit. According to Lenz's law, the frequency change creates an induced magnetic field in a direction to resist the change of total flux through the electron orbit. The electron slows down proportionately to the change of the magnetic flux. The net **B** field is slightly reduced. Typically, the field reduction is on the order of one part in 10 thousand. This can be seen in Table 33–2.

In this treatment we have ignored the orbital magnetic moment that existed before the material was put into an external field. Certainly the orbiting electrons have magnetic moments whether they are in an external magnetic field or not. However, the electron orbits are randomly oriented and tend to cancel out. Only the induced part of the magnetic moment is consistently against the external field.

Paramagnetism

Paramagnetism is caused by a different effect. We can view an electron as a spinning sphere with its charge distributed throughout the volume of the sphere. The electron itself behaves like a coil. The cross-sectional area of the electron is very small compared with its orbital area. As a result, the magnetic flux through the electron is very small compared with its orbit, so that magnetic induction has little effect.

The spinning charge of the electron does have a magnetic moment. Each electron behaves like a very tiny permanent magnet.

According to quantum mechanics, all electrons have the same size spin. The only choice they have is to spin clockwise or counterclockwise relative to their orbital motion around the nucleus of the atom.

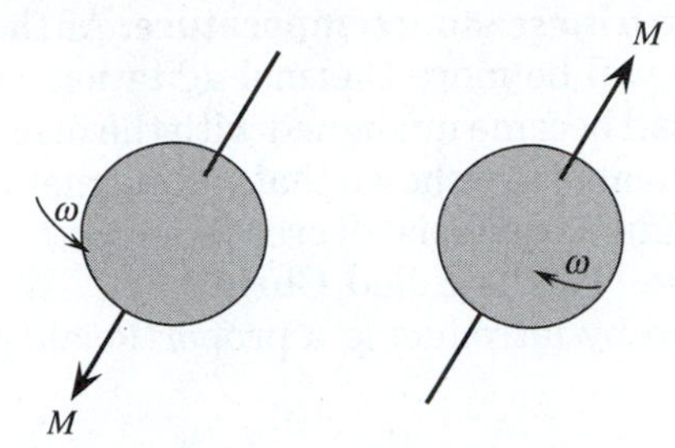

Figure 33–11: The spinning negative charge distribution of an electron creates a magnetic moment. All electrons have the same spin rate. Clockwise and counterclockwise spins are possible.

A counterclockwise spin creates a magnetic field that is oppositely directed to one created by a clockwise spin (see Figure 33–11).

In individual molecules and in solid state materials, the electrons are usually arranged so that they can pair up. Each pair has a clockwise- and a counterclockwise-spinning electron, resulting in a net magnetic field of zero. Each pair cancels out.

There are exceptions to the rule; some atoms and molecules have one or more unpaired electrons. A few of these are shown in Table 33–3.

In a paramagnetic material the permanent magnetic moments associated with molecules are randomly oriented. In the absence of an external field there is no net magnetic moment.

Table 33–3: Some elements with unpaired electrons.

ATOMIC NUMBER	22	23	24	25	26	27	28	29	30
	ELEMENT								
	Ti	V	Cr	Mn	Fe	Co	Ni	Cu	Zn
Number of unpaired electrons	2	3	5	5	4	3	2	0	0
Magnetic property at 20°C*	P	P	P	P	F	F	F	D	D

* P, paramagnetic; F, ferromagnetic; D, diamagnetic.

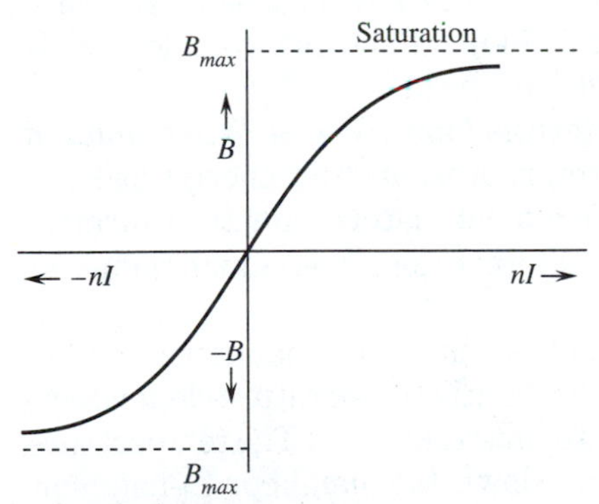

Figure 33–12: A plot of **B** *vs. nI for a paramagnetic material.* $\mathbf{B}_{max}$ *occurs when all unpaired electrons have lined up with the magnetic field.*

When a magnetic flux passes through the material, the magnetic fields of the individual atoms tend to line up with the external field and enhance it (see Figure 33–11). The alignment is not perfect. Thermal kinetic energy of the molecules works against it.

This model predicts two additional things about the paramagnetism we observe.

1. **There is saturation.** When the magnetic field is strong enough to align all the magnetic moments in the material, an additional increase of the external magnetic field will not cause a further increase in the total field.

   Figure 33–12 shows a plot of **B** vs. $n\,I$. For the toroid experiment, **B** is the total magnetic field found inside the toroid, and $n\,I$ is the number of turns per unit length times the current through the toroid winding. As we increase I, we increase the external field acting on the sample material. We see there is a point at which an increase in current will have no additional effect on **B**.

2. **Paramagnetism decreases with increasing temperature.** As the temperature increases, there will be more thermal agitation. A greater number of molecules will become unaligned with the magnetic field. Theory and experiments have shown that the magnetic susceptibility (χ_m) of a paramagnetic material decreases inversely with the absolute temperature. This is called **Curie's law.** We can write this in equation form by introducing a proportionality constant (C).

$$\chi_m = \frac{C}{T} \qquad \textbf{(Eq. 33–10)}$$

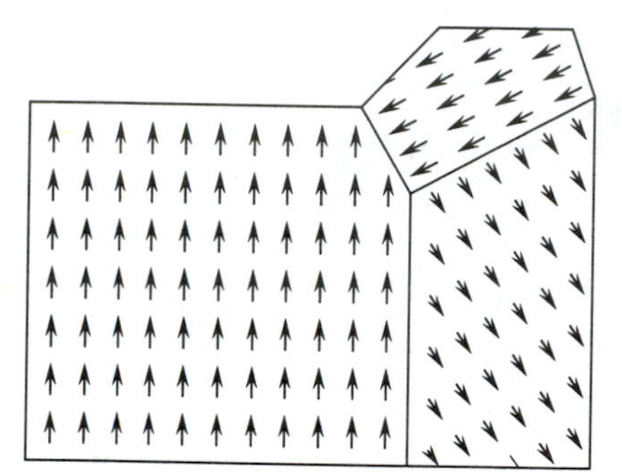

Figure 33–13: Unpaired electrons in a ferromagnetic material line up with neighbors to create magnetic domains, blocks of atoms with parallel spins.

Ferromagnetism is closely related to paramagnetism. If the temperature is low enough and the magnetic moments of individual atoms and molecules are strong enough, the atoms and molecules begin to clump together. Each lines up with its neighbor to form blocks of molecules grouped together with parallel magnetic moments. These blocks are called **magnetic domains** (see Figure 33–13).

There are usually a few billion or more atoms in each domain. Many randomly oriented domains make up a sample of ferromagnetic material.

When an external magnetic field is applied, the domains orient in the direction of the applied field. Once the field is removed, the domains tend to maintain their parallel orientation, to create a permanent magnet.

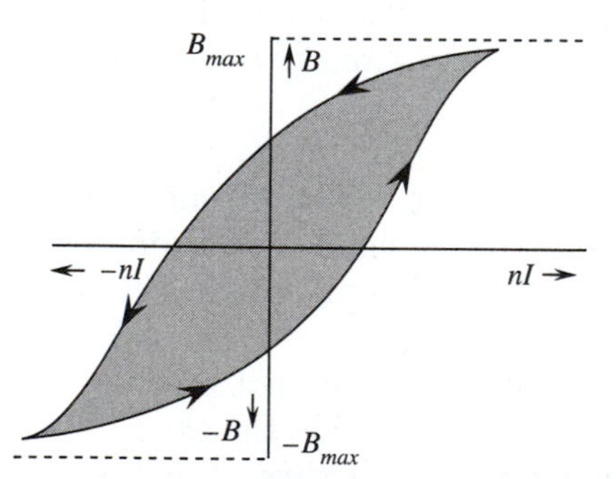

Figure 33–14: A plot of **B** *vs.* nI *for a ferromagnetic material. Saturation occurs at* $\mathbf{B}_{max}$ *when all domains have oriented parallel to the magnetic field. The enclosed shaded area is the work done in one cyclic change of* nI*. This is called hysteresis.*

Figure 33–14 shows a plot of **B** vs. $n\ I$. Notice a reverse current in the toroid is needed to reduce the value of **B** to zero. As with paramagnetism, there is saturation. Once the domains have all oriented to the field, no further increase in $n\ I$ will cause an increase in **B**. The plot is called a **hysteresis loop.** It shows a complete cycle of reversing the current through the toroid.

The area enclosed by the hysteresis loop ($\mathbf{B} \times n\ I$) has units of energy per unit volume. The area represents the energy lost per unit volume through one cycle. An ideal material for transformers would have a small area in the hysteresis loop and a high value for **B** when $n\ I$ is zero.

The strong interaction of adjacent atoms in ferromagnetism resists thermal agitation. The effect is only slightly temperature-dependent. There is, however, an upper limit to this resistance. The temperature can become high enough to break down the neighbor-to-neighbor interaction. At this temperature, the material changes from a ferromagnetic material into a paramagnetic material. This transition temperature is called the **Curie temperature.** Table 33–4 lists the Curie temperature for a few ferromagnetic materials.

Table 33–4: Curie temperature for selected ferromagnetic materials.

	ELEMENT					
	COBALT (Co)	DYSPROSIUM (Dy)	GADOLINIUM (Gd)	IRON (Fe)	NICKEL (Ni)	TERBIUM (Tb)
Curie temperature (K)	1404	85	289	1043	631	230

33.7 APPLICATIONS

Transformers

Mutual inductance can be used to change the voltage in an AC circuit where the current is continually changing. Figure 33–15 shows two solenoids formed by wrapping wire around the same form. The center of the form may be empty to form an air-core transformer or it may contain a ferromagnetic material like that shown in Figure 33–15.

The EMF inducted in coil 1 by the flux from coil 2 is similar to that shown in Figure 33–5.

$$V_1 = -N_1\left(\frac{\Delta\phi_2}{\Delta t}\right)$$

The EMF inducted in the second coil by the magnetic flux from the first coil is:

$$V_2 = -N_2\left(\frac{\Delta\phi_1}{\Delta t}\right)$$

If the two coils share the same flux, we can drop the subscript on $\Delta\phi$. When we write the ratio of voltages flux cancels.

$$\frac{V_2}{V_1} = \frac{-N_2\,\Delta\phi/\Delta t}{-N_1\,\Delta\phi/\Delta t}$$

or

$$V_2 = \left(\frac{N_2}{N_1}\right)V_1 \qquad \textbf{(Eq. 33–11)}$$

The voltage across the two coils depends on the ratio of turns. In a transformer, the input coil, the one connected to the power source, is called the **primary,** and the output coil is called the **secondary.**

Usually, the two windings are wrapped one over the other on an

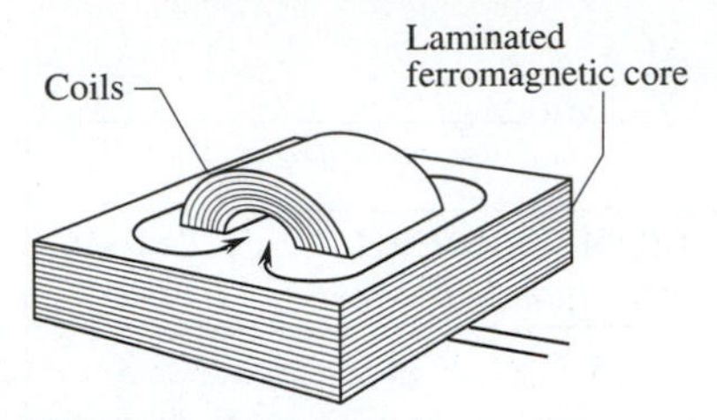

Figure 33–15: A transformer. Two coils are wrapped around a common ferromagnetic core.

insulated form (see Figure 33–15). A ferromagnetic material is often used. This keeps nearly all the magnetic flux within the core to reduce power losses. The varying field can cause electrical currents to flow through a metal conductor at right angles to the field lines. For this reason, the core is constructed of laminated pieces that are electrically insulated from each other.

□ **EXAMPLE PROBLEM 33–6: THE TRANSFORMER**

Some people for whom money is not a concern prefer the performance of old-fashioned vacuum tubes to solid-state circuits in high-quality audio equipment. A transformer connected to a 110-VAC source has a primary with 454 turns. One secondary winding supplies 6.3 V to heat the vacuum tube filaments. Another secondary winding supplies 240 V to other parts of the circuit. Find the number of turns in each of the two secondaries.

■ ***SOLUTION***

A. Filament voltage:

$$N_2 = \frac{N_1 V_2}{V_1}$$

$$N_2 = 454 \text{ turns} \times \frac{6.3 \text{ V}}{110 \text{ V}}$$

$$N_2 = 26 \text{ turns}$$

B. Circuit voltage:

$$N_2 = 454 \text{ turns} \times \frac{240 \text{ V}}{110 \text{ V}}$$

$$N_2 = 908 \text{ turns}$$

Electrical DC Motors

Figure 33–16 shows a very simple direct current motor. A current-carrying loop is in a constant magnetic field. In Chapter 32, we found that a time-varying torque acts on the loop. It is described by the equation

$$\tau = N\,I\,A\,\mathbf{B}\sin\theta$$

The angle θ changes with time. It can be expressed in terms of the rotational rate of the loop.

$$\theta = \omega t$$

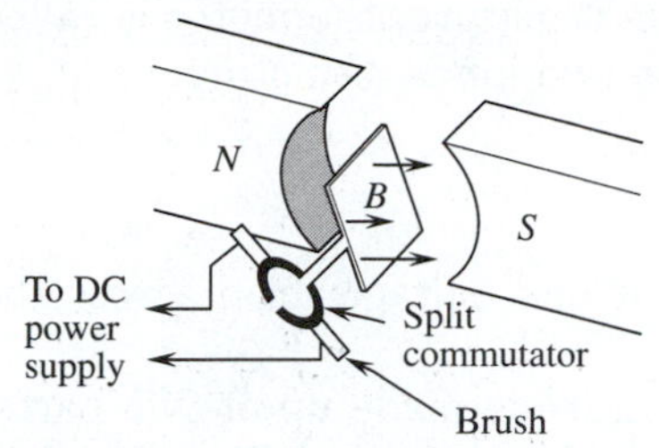

Figure 33–16: A single current loop in a magnetic field demonstrates the idea behind the operation of a DC motor. The magnetic field exerts a torque on a current-carrying coil. The split commutator reverses the current flow through the loop as the torque reaches zero.

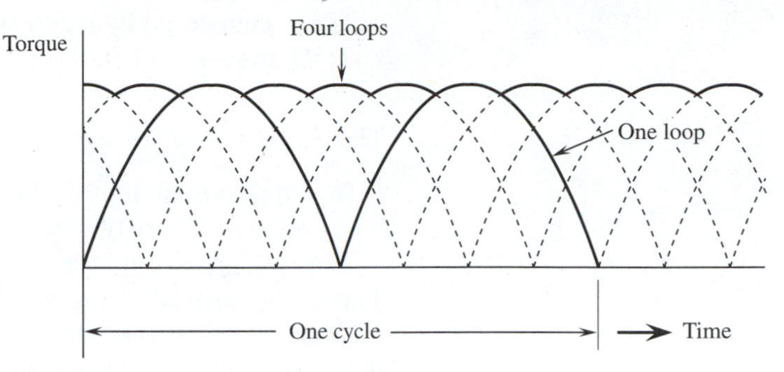

Figure 33–17: Torque versus time. The use of four loops and additional sections in the commutator increases the efficiency of the motor.

Figure 33–17 shows a plot of torque versus time. As the loop passes through the position of zero torque, its momentum continues its motion, and the current is reversed by a **split commutator. Brushes,** or sliding contacts, ride on the commutator. When the split passes under the brushes the current reverses.

The efficiency of the machine can be improved by increasing the number of coils and the number of sections in the commutator. Each loop has current passing through it only near the maximum torque orientation. Figure 33–17 also shows the torque versus time for a four-loop motor.

The torque depends on the strength of the magnetic field **B** as well as on the current through the loop. We can increase the power output of the motor by replacing the permanent magnets with electromagnets with ferromagnetic cores. The coils that produce the magnetic field are called **field coils.** The current-carrying loops are called the **armature.** Figure 33–18 shows a typical armature.

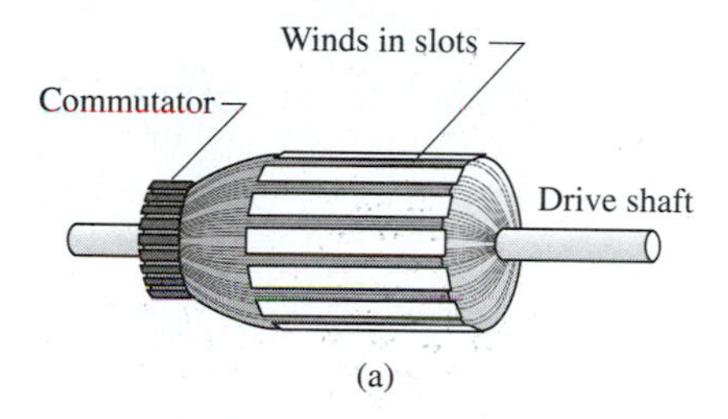

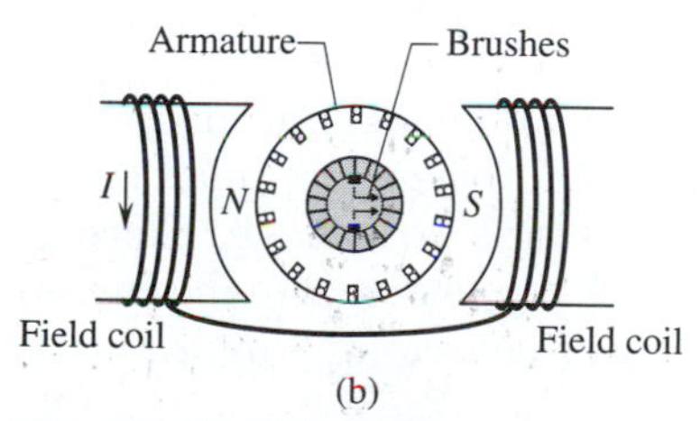

Figure 33–18: The structure of a DC motor. (a) *The armature has several coils embedded in slots of the soft iron core.* (b) *The brushes are inside the commutator. The* **B** *field is created by electromagnets (field coils).*

Both the current through the armature and the current through the electric magnets are supplied by the same current source. There are a couple of popular ways of connecting the armature and field coils (see Figure 33–19). If the field coils are connected in parallel to the armature the motor is called a **field shunt motor.** If the field

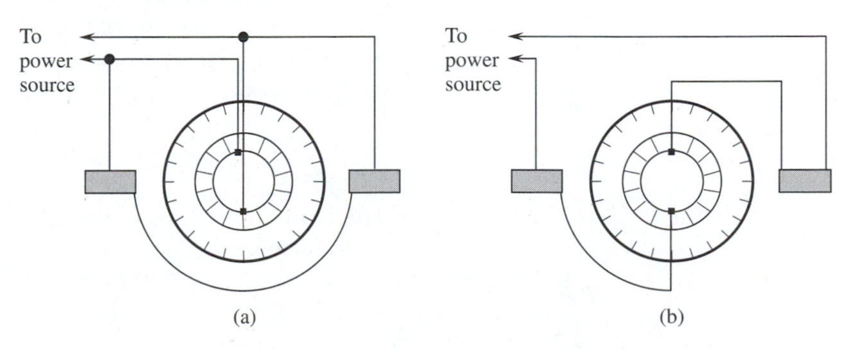

Figure 33–19: Two ways of hooking up field coils. (a) *Field shunt motor. The field coils are connected in parallel to the armature.* (b) *Series motor. The field coils are in series with the armature.*

coils are connected in series with the armature, the motor is called a **series motor.** Let us look at these two kinds of motors.

The Field Shunt Motor

In a field shunt motor the current and voltage drop across the field coils (V_f) is fairly constant.

The voltage drop across the armature is made up of two parts. There is a resistive drop across the windings ($V_R = I R$) and a back-EMF induced in the coil ($V_1 = N\Delta\phi_m/\Delta t$). We have dropped the negative sign here because we are talking about a potential drop rather than a potential gain.

Since the field coil and the armature are in parallel they must have the same total potential drop.

$$\left(N\frac{\Delta\phi_m}{\Delta t}\right) + (I R) = V_f = \text{constant}$$

How rapidly the magnetic flux changes through the armature coils depends on its angular speed.

$$\frac{\Delta\phi_m}{\Delta t} = K\omega$$

where K is a proportionality constant.

$$V_L + V_R = (N K\omega) + (I R) = \text{constant}$$

When the angular speed is large, much of the potential drop across the armature is caused by the back-EMF. When ω is small most of the potential drop is across the armature's resistance. Since the potential drop across a fixed resistance is proportional to the current, a large current is drawn from the power source for slow rotational speeds of the motor.

When a field shunt motor starts up, the current drawn through the armature is large. The starting current is dependent on the resistance of the armature. As the motor speeds up, the current drawn from the power source decreases until the motor reaches a speed to match the load.

The power output for a single-loop armature would be:

$$\textit{power} = \tau\omega = N I \omega \mathrm{B} A \sin \omega t$$

or

$$\textit{power}(\max) = N I \omega \mathrm{B} A$$

The variables in the power equation are angular speed and current through the armature. A large current results in large resistive power losses and low efficiency. Larger speeds with lower resistive power losses would increase the efficiency of the machine.

Notice that $N \omega B A \sin \omega t = N \Delta\phi_m/\Delta t = V_L$.

The power output depends on the time average, or effective EMF, of the armature times the current passing through it.

$$power = I_a V_L$$

The Series Motor

In the series motor the field coils are connected in series with the armature. The voltage drops across the armature and field coils add up to the potential of the power source (V_0).

$$V_L + (I R_A) + (I R_F) = V_0$$

The same current passes through the field coil and the armature. The back-EMF (V_L) is dependent upon the angular speed of the motor in a similar fashion as the shunt field motor.

The **B** field is proportional to the current through the field coils. At low speeds, the current through the field coils is large, since V_L is small. As the motor speeds up, V_L increases, reducing the current through both the field coils and the armature. There is an advantage to this idea. The starting torque of series motors is large. This makes them useful for hoists and traction motors.

☐ **EXAMPLE PROBLEM 33–7: THE FIELD SHUNT MOTOR**

A field shunt motor operates from a line voltge $V_0 = 120.0$ V and draws 40.0 A of DC current at full load. The field coils have a resistance of 60.0 Ω (R_f) and the armature has a resistance of 0.30 Ω.

A. What is the current through the armature?

B. How many volts of EMF are produced in the armature?

C. How much power is produced by the armature?

■ ***SOLUTION***

A. Armature current:

The current through the armature and the current through the field coils must add up to the total current from the line since we have a parallel connection.

$$40.0\text{ A} = I_a + I_f$$

or

$$I_a = 40.0\text{ A} - \left(\frac{V}{R_f}\right)$$

$$I_a = 40.0\text{ A} - \left(\frac{120.0\text{ V}}{60.0\ \Omega}\right)$$

$$I_a = 38.0\text{ A}$$

B. Back-EMF:
The total voltage drop across the armature is the sum of the back-EMF and the resistive loss in the armature.

$$V_0 = V_L + V_r$$

or

$$V_L = V_0 - V_r = V_0 - (I_a R_a)$$

$$V_L = 120.0\text{ V} - (38.0\text{ A} \times 0.30\ \Omega)$$

$$V_L = 108.6\text{ V}$$

C. Armature power:

$$P = V_L I_a = 108.6\text{ V} \times 38.0\text{ A}$$

$$P = 4.13\text{ kW}$$

Frictional losses and electrical losses at the commutator will reduce the actual power output of the motor.

☐ **EXAMPLE PROBLEM 33–8: THE FIELD SHUNT MOTOR AGAIN**

If the motor in Example Problem 33–7 is operating at 1800 rpm estimate the maximum torque exerted by the armature.

■ ***SOLUTION***

$$P = \tau(\text{max})\,\omega$$

or

$$\tau(\text{max}) = \frac{P}{\omega}$$

$$\tau(\text{max}) = \frac{4.13 \times 10^3\text{ W}}{1800\text{ rev/min} \times 1\text{ min/60 s} \times 2\pi\text{ rad/rev}}$$

$$\tau(\text{max}) = 22\text{ N}\cdot\text{m}$$

SUMMARY

When the magnetic field passing through an inductor changes, an EMF is induced across it. This is known as Faraday's law. The induced potential is in a direction to oppose the *change* of the magnetic flux (Lenz's law). These two ideas can be placed in one equation for the induced EMF in an inductor. For a coil with N turns:

$$V_L = -N\left(\frac{\Delta\phi_m}{\Delta t}\right)$$

An inductor will react with its own magnetic flux. This is called self-inductance (L). The magnetic flux is proportional to the current through the inductor. The potential across the inductor can be expressed in terms of current rather than flux. The equation for self-inductance is:

$$V_L = -L\left(\frac{\Delta I}{\Delta t}\right)$$

The unit of inductance is called a henry (H).

$$1 \text{ henry (H)} = 1\,\frac{\text{volt} \cdot \text{second}}{\text{ampere}}$$

If two coils share the same magnetic flux, the EMF induced in one coil depends upon the change of current in the other coil. This is called mutual inductance (M). If the flux through the two coils is the same then the mutual inductance is:

$$V_1 = -M\left(\frac{\Delta I_2}{\Delta t}\right)$$

$$V_2 = -M\,\frac{\Delta I_1}{\Delta t}$$

Inductors store energy in magnetic fields. The energy (U) stored in an inductor with a self inductance of L is:

$$U = \frac{L\,I^2}{2}$$

The energy density (u), or energy per unit volume, can be expressed in terms of the strength of the inductor's magnetic field.

$$u = \frac{\mathbf{B}^2}{2\,\mu_0}$$

When a material is placed inside an inductor the magnetic field strength is altered. The ratio of the magnetic field strength of a toroid with the material in the core (**B**) and without the material ($\mathbf{B}_0$) is called the relative permeability (k_m).

$$\frac{\mathbf{B}}{\mathbf{B}_0} = k_m$$

The magnetic field strength can be broken down into two parts—the part created by the current passing through the windings of the coil and the material's contribution.

$$\mathbf{B} = \mathbf{B}_0 (1 + \chi_m)$$

The constant χ_m is called magnetic susceptibility. It describes the magnetic behavior of the material.

Materials can be sorted into three major categories.

Diamagnetic materials: $\chi_m < 0$

Paramagnetic materials: $1 > \chi_m > 0$

Ferromagnetic materials: $\chi_m \gg 0$

A transformer is a pair of coils used to change potential using a varying current. The ratio of EMFs across the two coils is proportional to the ratio of turns.

$$V_2 = \left(\frac{N_2}{N_1}\right) V_1$$

In a DC motor, electromagnets, or field coils, produce a magnetic field that exerts a torque on a set of rotating coils called an armature. Reversible DC current is supplied to the armature through a split commutator. The power output of a DC motor is proportional to the rotational speed of the armature. For a single-loop armature with the current I_a flowing through the armature the power is:

$$P = \tau\omega = N\, I_a\, \omega \mathbf{B}\, A \sin \omega t$$

The power is also proportional to the EMF generated across the armature.

$$P = I_a V_L$$

In a DC motor, the field coils usually are connected either in parallel to the armature (field shunt motor) or in series with the armature (series motor). In the field shunt motor, the magnetic field of the field coils is fairly constant. In the series motor, the starting magnetic field is quite high, but decreases as the motor speeds up. Series motors have higher start-up torques than field shunt motors.

KEY TERMS

If you can explain the following terms to a friend or classmate, you understand their meaning. If you cannot explain the terms, you should reread the sections in which they are discussed.

armature
back-EMF
brushes
Curie's law
Curie temperature
diamagnetic materials
Faraday's law
ferromagnetic materials
field coils
field shunt motor
hysteresis loop
induced current
induced EMF
inductance
inductor
Lenz's law
magnetic domains
magnetic induction
magnetic susceptibility
mutual inductance
paramagnetic materials
primary
relative permeability
secondary
self-inductance
split commutator

EXERCISES

Section 33.1:

1. Figure 33–20a shows a plot of the magnetic field through a coil versus time. Which one of the plots in Figure 33–20b best depicts the behavior of the EMF induced in the coil?
2. A permanent magnet is inserted into a solenoid while the potential across the coil is monitored. How is the induced EMF affected by the following factors?
   A. The speed of insertion is increased.
   B. The magnet is pulled out of the coil rather than pushed in.
   C. The number of coils in the solenoid is increased.

Section 33.2:

3. How would the following factors change the self-inductance (L) of a solenoid? In each case, assume all other factors remain unchanged.
   A. An increase in cross-sectional area.
   B. An increase in length, with the total number of turns the same.
   C. An increase in the number of turns, with the same length.
   D. An increase of current through the coil.

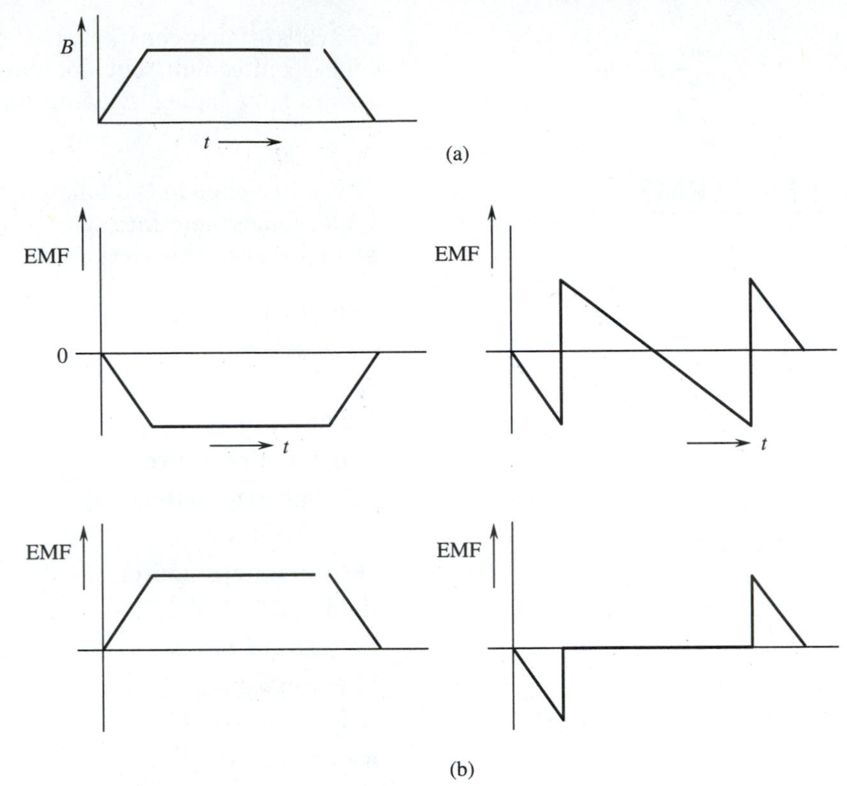

Figure 33–20: Diagram for Exercise 1. Which plot in (b) *best corresponds to the plot of magnetic field versus time?*

4. Figure 33–21a shows a coil connected in series with an open switch and a battery through the open switch. When the switch is closed, a voltage will briefly be induced across the coil as the current builds up. Which end of the coil will be positive, end A or end B? Explain.
5. Figure 33–21b shows a coil connected in series with a battery and a closed switch. When the switch is turned from pole 1 to pole 2, current will continue to flow through the coil for a brief period of time as its magnetic field collapses. Does the current flow from A to B or from B to A? Explain.

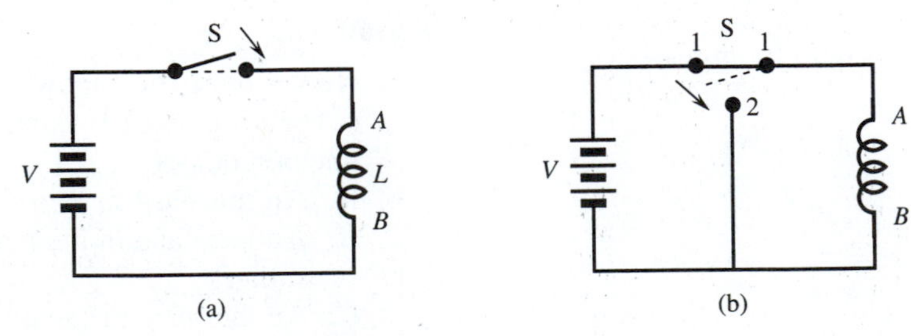

Figure 33–21: Diagram for Exercises 4 and 5. When the switch is opened or closed an EMF occurs across the coil.

Section 33.3:

6. How is mutual inductance different from self-inductance?

*7. A student in a physics laboratory connects a coil to a battery. At the same instant, another student at the adjacent table notices that an EMF is induced in the coil setup for another experiment even though it is not connected to a power supply (see Figure 33–22). Explain what has happened.

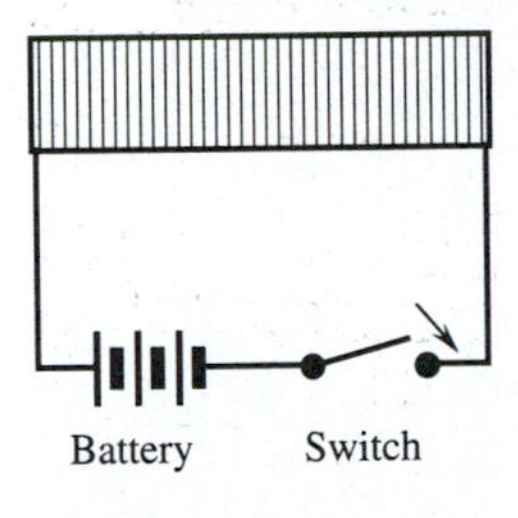

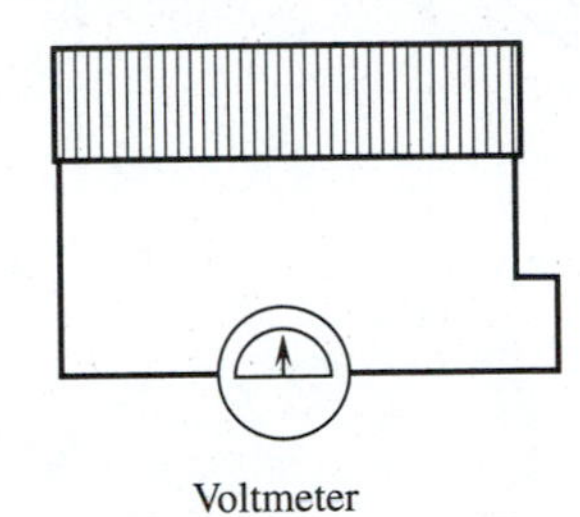

Figure 33–22: *Diagram for Exercise 7.*

Section 33.4:

8. Solenoid A and solenoid B have the same number of turns per unit length (n). Solenoid A has three times the volume of B. The same DC current flows through the two solenoids.
   A. Compare the energy density of the two solenoids.
   B. Compare the total energy stored in the magnetic fields of the two coils.

*9. See Figure 33–23. A toroid is connected to a battery through switch S1. Switch S2 is closed, shorting out the toroid at the same time S1 is opened to disconnect the battery.
   A. Under normal circumstances, what happens to the energy stored in the toroid?
   B. What would happen if the windings of the toroid were made from a superconducting material?

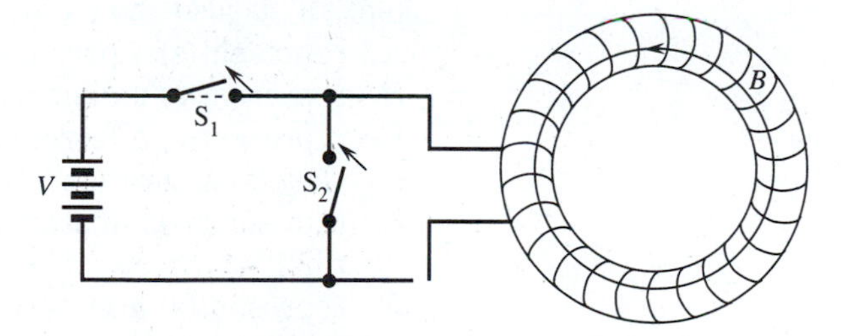

Figure 33–23: *Diagram for Exercise 9. What happens to the magnetic field when the circuit is shorted out?*

Sections 33.5, 33.6:

10. A toroid has n turns per unit length. The magnetic field strength of the toroid without a core is $\mathbf{B}_0$; with a core material it is $\mathbf{B}$. Figure 33–24 shows some plots of $\mathbf{B}$ vs. $n\ I$ for the

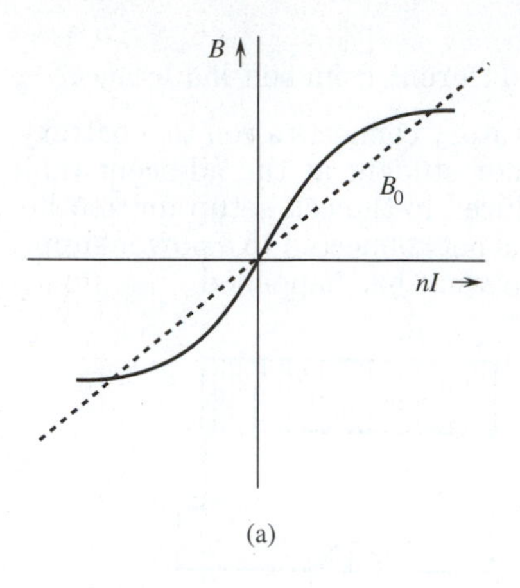

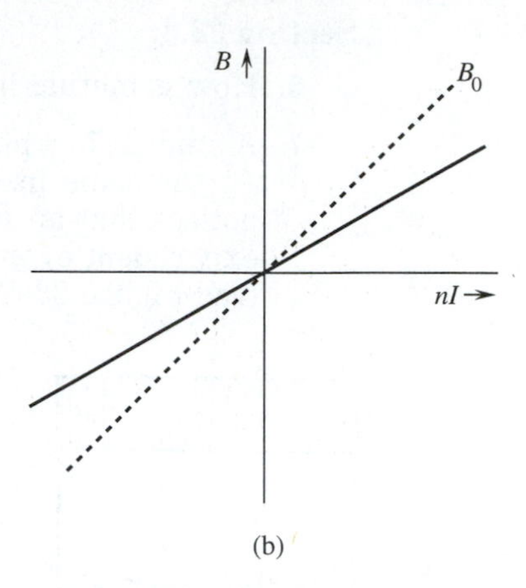

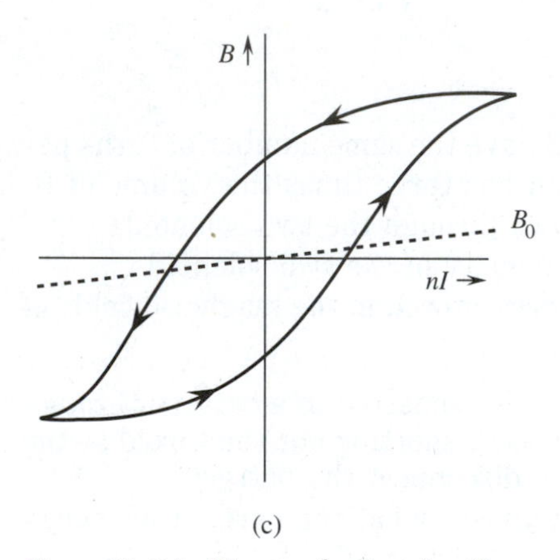

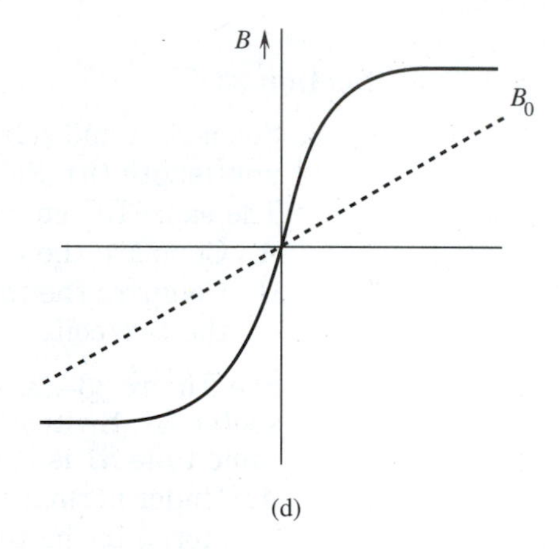

Figure 33–24: *Diagram for Exercise 10.*

toroid, where I is the current through its windings. In each plot, $\mathbf{B}_0$ is also shown for comparison. Which plot(s):

A. represent(s) a paramagnetic material?
B. represent(s) a diamagnetic material?
C. represent(s) a ferromagnetic material?
D. show(s) saturation?
E. represent(s) a material with a negative magnetic susceptibility?
F. represent(s) a material with a positive magnetic susceptibility?
G. exhibit(s) hysteresis?

11. How are paramagnetic materials related to ferromagnetic materials?

12. What is the Curie temperature?

13. A paramagnetic material has a susceptibility of $+2.4 \times 10^{-5}$ at a temperature of 120 K. Estimate its susceptibility if its temperature is doubled. Its Curie temperature is below 120 K.

Section 33.7:

14. A step-down transformer is used to reduce voltage. Which coil should have the larger number of turns, the primary or the secondary?

**15. Equation 33–11 assumes that the same maximum flux passes through the two coils making up the transformer. Show that for an air-core transformer with different cross-sectional areas for the coils Equation 33–11 should be modified to read:

$$V_2 = \frac{N_2 A_1 V_1}{N_1 A_2}$$

16. Explain what each of the following does in a DC motor.

A. Split commutator
B. Field coils
C. Armature

17. Compare a field shunt motor with a series motor.

18. In an overload circuit in a home workshop the switch breaker may be activated when a tablesaw is first turned on. Why?

19. The windings of the field coils in a series motor are made with much heavier wire than the field coils in a field shunt motor. Why?

PROBLEMS

Section 33.1:

1. A solenoid has 23 turns. Calculate the EMF produced across the coil when the magnetic flux changes from 0.00 to 0.78 Wb in 0.085 s.

2. Figure 33–25 shows the variation of flux through a 200-turn inductor as a function of time. Estimate the induced EMF for regions A-B, B-C, C-D, and D-E.

Section 33.2:

3. A 1240-turn solenoid is 24 cm long and has a cross-sectional area of 12.8 cm^2. Calculate the self-inductance (L) of the solenoid.

4. An $82\bar{0}$-turn toroid has a radius of 10.0 cm and a cross-sectional area of 21.0 cm^2.
A. Calculate the self-inductance (L) of the device.

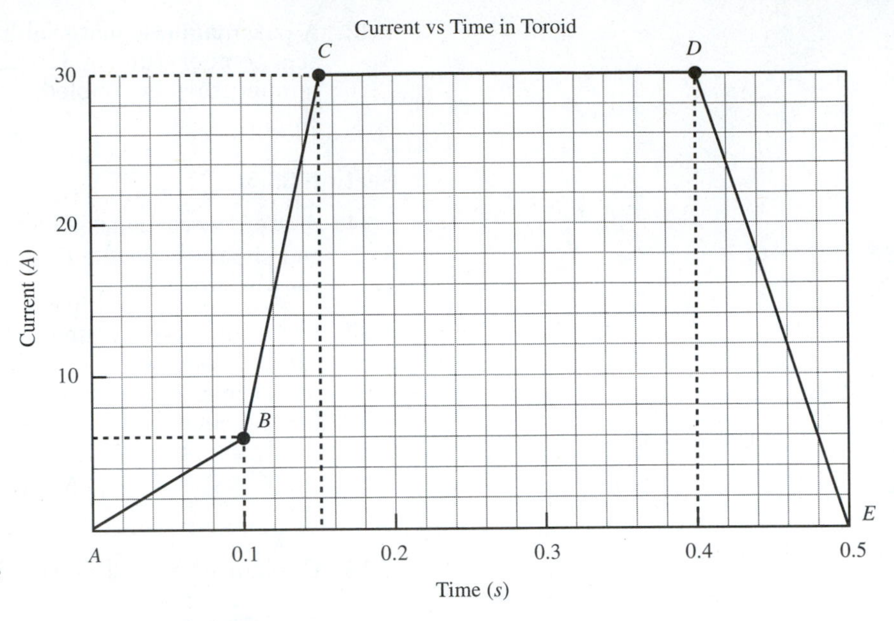

Figure 33–25: *Diagram for Problem 2. Current versus time in a toroid. Estimate the induced EMFs.*

B. What magnetic flux flows through the toroid when a current of 2.8 A passes through the toroid?

C. What EMF is induced when the current varies from 0.00 to 5.20 A in 0.010 s?

5. An inductor has a self-inductance of 12.3 mH. What rate of current change is needed to induce an EMF of 1.00 V?

Section 33.3:

6. Two coils are wrapped around a common hollow cylindrical frame with a cross-sectional area of 11.0 cm^2 and a length of 7.0 cm. One coil has 45 turns; the other $9\overline{0}$ turns. Calculate the mutual inductance (M).

7. Two coils have a mutual inductance of 32 mH. If the rate of change of current in one coil is 474 A/s, what is the EMF induced in the other coil?

Section 33.4:

8. An inductor has a self-inductance of 12.5 mH. How much energy is stored in its magnetic field when a current of 4.7 A flows through the coil?

9. A toroid has a magnetic field of 1.3 T. The cross-sectional area is 12.0 cm^2, and the radius is 9.3 cm.

A. What is the energy density of the magnetic field?

B. What total energy is stored in the toroid?

Sections 33.5, 33.6:

10. A toroid has a magnetic field of 71.00 mT when it has an empty core ($\mathbf{B}_0$) and a current of 4.00 A passes through its windings. When the core is filled with material X, the field is 70.97 mT for the same current.

A. What is the relative permeability (k_m)?

B. What is the magnetic susceptibility (χ_m) of material X?

C. Is material X paramagnetic, diamagnetic, or ferromagnetic?

11. An air-core solenoid has a magnetic field of 0.023 T. Estimate its magnetic field when it has a core of iron. Use Table 33–2.

12. A paramagnetic material has a magnetic susceptibility of $+6.4 \times 10^{-4}$ at $20\bar{0}$ K. What is χ_m at $2\bar{0}$°C?

Section 33.7:

13. A transformer is to be used to convert $11\bar{0}$ VAC to 12.6 VAC. What should be the ratio of turns on the transformer assuming 100% efficiency?

14. The primary of a transformer has 32 turns; the secondary, 272 turns.

A. If the primary voltage is $11\bar{0}$ VAC, what voltage is induced across the secondary?

B. If the voltage across the primary is 40.0 VDC, what voltage is induced in the secondary?

15. A series DC motor has an armature resistance of 0.70 Ω and a series field coil resistance of 0.40 Ω. It draws 12.0 A current.

A. Estimate the EMF across the armature.

B. What average power does the armature produce?

C. What is the motor's efficiency?

16. A 335-W (¼ hp) DC motor operates at a speed of $24\bar{0}0$ rpm. Estimate the maximum torque produced by the motor.

***17.** A field shunt motor draws 35.0 A from a $4\bar{0}$-VDC source at full load. The field coils have a resistance of $2\bar{0}$ Ω; the armature, a resistance of 0.40 Ω.

A. How much current passes through the field coils?

B. How much current passes through the armature?

C. How large an EMF is produced in the armature?

D. What is the average power output of the armature?

E. Estimate the efficiency of the motor using the answer to part C of this problem. (This will be a high estimate. We have overlooked bearing friction and voltage drops at the commutator caused by contact resistance.)

Chapter 34

AC CIRCUITS

How do FM radio stations remain so static-free, even during electrical storms? This chapter discusses how AC current waves behave in different settings. You also will learn the difference between AM and FM frequencies. (Photograph by Ellen Senisi.)

OBJECTIVES

In this chapter you will learn:

- how AC voltages are generated
- to find root mean square voltages and currents
- to calculate inductive reactance
- to find the impedance of an RL circuit, an inductor and resistance in series
- how the phase angle between current and voltage affects electric power (power factor)
- to calculate capacitive reactance
- to find the impedance of a series RLC circuit
- to calculate the resonance frequency of a series RLC circuit

Time Line for Chapter 34

1867 Zenobe Gramme builds the first practical AC generator.
1884 Nikola Tesla invents the electric alternator.
1904 John Fleming files for a patent for the first diode, a device to convert alternating current into direct current.
1915 Manson Benedicks discovers that a solid-state diode can be made from germanium.
1923 Charles Proteus Steinmetz, the architect of the vector model of alternating current, dies.

Direct current occurs when small charged particles drift in a particular direction under the influence of a steady electric field. Energy is transferred in the particles' linear kinetic energy.

Alternating current occurs when charged particles slosh back and forth under the influence of an alternating electric field. In this case, energy is transferred by a wave motion that is similar to the motion of compressional sound waves.

In this chapter, we will look at the way alternate current waves are generated, how to calculate the power associated with these waves, and how they behave in some kinds of electrical devices.

34.1 AC GENERATION

Figure 34–1 shows a simple AC generator. An external mechanical power source rotates a wire coil in a magnetic field. The external power source might be a gasoline engine in an automobile or a steam or water turbine for commercial generation. The ends of the rotating loop are connected to **slip rings** on which brushes ride to make a sliding electrical contact.

The EMF induced across the ends of the coil depends on the number of loops in the coil (N), the cross-sectional area (A), the magnetic field strength ($\mathbf{B}$), and how rapidly the coil is spinning. Faraday's law applies here.

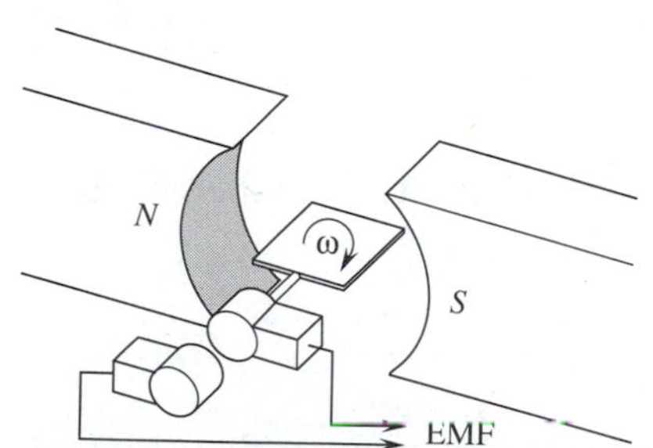

Figure 34–1: *A simple generator. An external torque rotates a coil in a magnetic field. A time-varying EMF is created.*

$$EMF = -N\left(\frac{\Delta\phi}{\Delta t}\right)$$

The flux passing through the coil is $\phi = \mathbf{B}\,A \sin\theta$.
The angle θ at any given time depends on the angular speed (ω).

$$\theta = \omega\, t$$

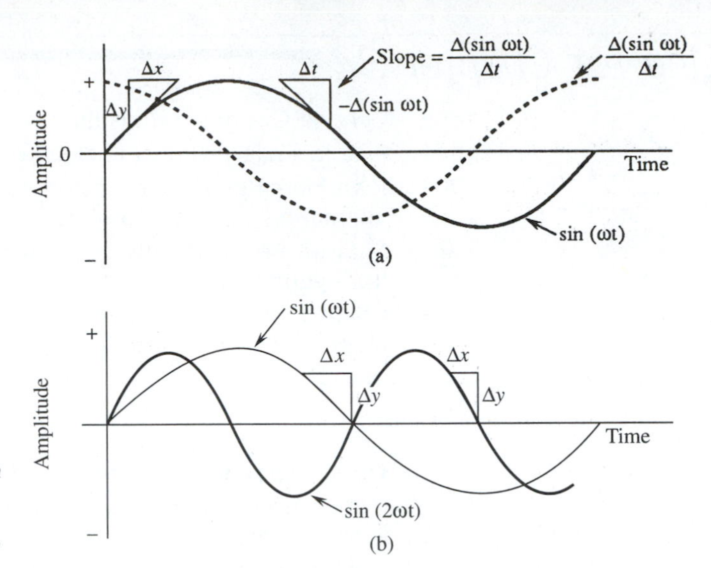

Figure 34–2: (a) *If the slope of a sine wave (solid line) is plotted against time, a cosine wave (dashed line) is produced.* (b) *As the frequency* ($\omega = 2\pi f$) *increases, the slope of the sine wave also increases.*

or

$$EMF = -(N\,\mathbf{B}\,A)\frac{\Delta \sin(\omega\, t)}{\Delta t}$$

Figure 34–2a shows a plot of sin ωt versus time. The slope of this curve is Δ (sin ωt)/Δt. The dashed line is a plot of the slope. It is another sine wave shifted by 90°. In other words, it is a cosine wave.

Figure 34–2b shows two sine waves. One is twice the frequency of the other. Notice for the higher frequency the slope is generally steeper. A careful analysis will show us that the slope is directly proportional to the rotational speed (ω).

In general we can write:

$$\frac{\Delta(\sin \omega t)}{\Delta t} = \omega\,(\cos \omega t) = \omega\,[\sin(\omega t + 90°)] \qquad \textbf{(Eq. 34–1)}$$

Equation 34–1 will be useful in a number of situations in this chapter.

We can now calculate the EMF generated.

$$EMF = -(N\,\mathbf{B}\,A\,\omega)\cos(\omega t) \qquad \textbf{(Eq. 34–2)}$$

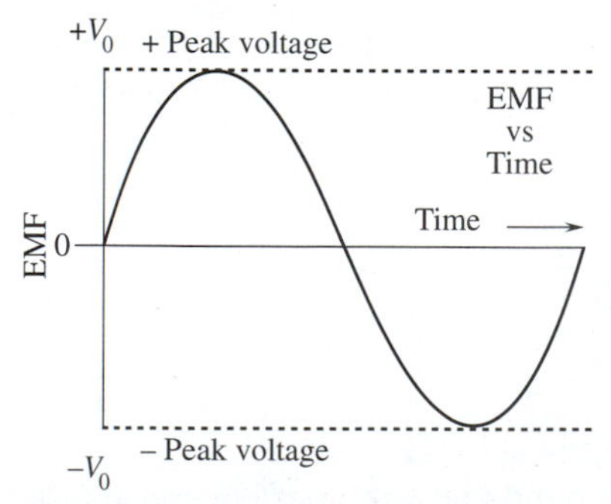

Figure 34–3: EMF versus time. A generator produces a sinusoidal voltage with a peak voltage (V_0).

The time-varying EMF will cause a varying current to flow through the rotating coil. The size of the current will depend on the load resistance and on the load of the coil. The first expression in parentheses is the **peak voltage** (V_0). Figure 34–3 shows the voltage generated during one cycle of an AC generator. Neglecting the 90° shift, the generator output of voltage and current may be written as:

$$V = V_0 \sin(\omega t)$$
$$I = \frac{V}{R} = I_0 \sin(\omega t) \qquad \textbf{(Eq. 34–3)}$$

□ **EXAMPLE PROBLEM 34–1: THE AC GENERATOR**

A 35-turn coil in a small AC generator has a cross-sectional area of 120 cm². If the coil rotates at a frequency of 60.0 Hz in a 0.134-T field, calculate the peak EMF induced.

■ ***SOLUTION***

The angular speed is related to the frequency by:

$$\omega = 2\pi f$$

$$V_0 = N\mathbf{B}A\,2\pi f$$

$$V_0 = 35 \times 0.134\ \text{T} \times (1.20 \times 10^{-2}\ \text{m}^2) \times 2\pi \times 60.0\ \text{Hz}$$

$$V_0 = 21.2\ \text{V}$$

34.2 POWER IN A RESISTOR

The current through a resistor is directly proportional to the electrical potential across it. This is in agreement with Ohm's law. The current and voltage vary in step with each other. As the potential increases across the resistor the current also increases. When the potential reverses, the direction of current flow reverses with it. When the potential is maximum, the current is maximum (see Figure 34–4).

The power transferred through a pure resistance is the product

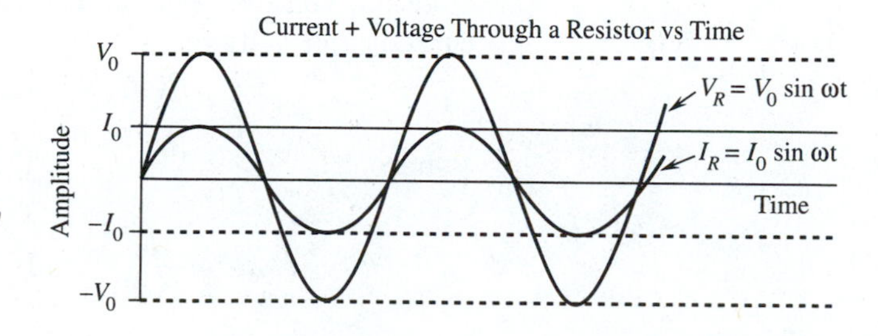

Figure 34–4: Current and voltage through a resistor versus time. The AC current and voltage through a resistor are in phase. They peak at the same time.

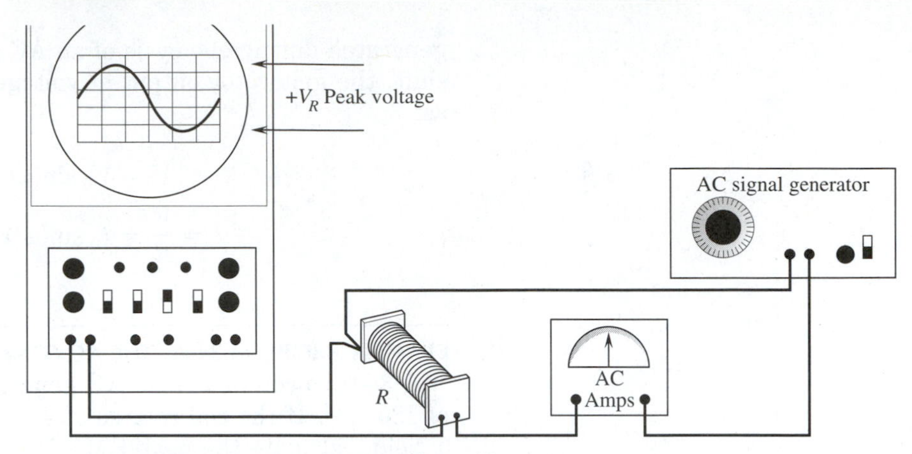

Figure 34–5: Diagram for Example Problem 34–2.

of current and voltage. By **pure resistance** we mean that there is no capacitance or inductance present in the circuit.

$$P = I\,V$$

We can use Equation 34–3 to show the time dependence.

$$P = I_0\,V_0\,(\sin \omega t)^2$$

This expresses the power in terms of peak voltage and peak current to give the **instantaneous power.** For most applications we are more interested in the time average of power. When $\sin \omega t = 1$, the instantaneous power is simply $I_0\,V_0$. At two other times $\sin \omega t = 0$ during a half cycle. At these instants no power is being transferred. At other times the power transmission falls between these extremes. The **effective power,** or time average, is half the peak power (see Figure 34–4).

$$P(\text{eff}) = \frac{I_0 V_0}{2} \qquad \textbf{(Eq. 34–4)}$$

We can write the power in terms of an effective current and an effective voltage. This is also called the **root mean square (rms)** current and voltage.

$$P(\text{eff}) = I_{\text{RMS}}\,V_{\text{RMS}}$$

where

$$I_{\text{RMS}} = \frac{I_0}{(\sqrt{2})} \quad \text{and} \quad V_{\text{RMS}} = \frac{V_0}{(\sqrt{2})}$$

Since $\sqrt{2} = 1.414$

$$I_{RMS} = 0.7071\, I_0 \qquad V_{RMS} = 0.7071\, V_0 \qquad \textbf{(Eq. 34–5)}$$

These are the time average values that AC voltmeters and ammeters record. The voltages of power outlets are usually given in terms of RMS values. For instance, the outlets in a home are mostly 110-VAC sources. This is RMS voltage. A hair dryer may indicate it draws 0.90 A. Again this is usually an RMS value. The power drain can be easily calculated by multiplying the two values together: (0.90 A × 110 V = 990 W).

EXAMPLE PROBLEM 34–2: AC POWER IN A RESISTOR

A technician measures the voltage drop across a $2\bar{0}$-Ω resistor using an oscilloscope. According to the oscilloscope the peak voltage drop across the resistor is 8.5 V (see Fig. 34–5). The current through the resistor is found by using an AC ammeter. The current is 0.300 A. The technician is puzzled to discover the RI^2 power dissipated by the resistor is 1.80 W while the power drain calculated by $P = I\,V$ is 2.55 W. Why are the two answers different?

SOLUTION

The ammeter reads RMS current. The power calculated by the equation $P = R\,I^2$ is correct. It gives effective power.

$$P = R\,(I_{RMS})^2$$

$$P = 2\bar{0}\,\Omega\,(0.300\text{ A})^2$$

$$P = 1.8\text{ W}$$

The voltage reading from the oscilloscope must be converted to RMS voltage before calculating power.

$$V_{RMS} = 0.7071\, V_0$$

$$V_{RMS} = 0.7071 \times 8.5\text{ V}$$

$$V_{RMS} = 6.0\text{ V}$$

$$P = V_{RMS}\, I_{RMS}$$

$$P = 6.0\text{ V} \times 0.300\text{ A}$$

$$P = 1.8\text{ W}$$

The two calculations agree when the correct quantities are used.

EXAMPLE PROBLEM 34–3: HOUSE VOLTAGE

Most of the power outlets in a house are rated as 110 VAC. This is an RMS value. What is the peak voltage?

SOLUTION

$$V_{RMS} = 0.7071\, V_0$$

or

$$V_0 = 1.414 \times V_{RMS}$$

$$V_0 = 155 \text{ V}$$

34.3 AC CURRENT AND INDUCTANCE

The potential drop across an inductor depends not on the current (I), but on the change of current ($\Delta I/\Delta t$) inducing a back-EMF.

$$V_L = \frac{L\,\Delta I}{\Delta t}$$

We have left out the minus sign since we are dealing with a voltage drop. The current from an AC source is $I = I_0 \sin \omega t$. According to Equation 34–1 the voltage drop across the inductor is:

$$V_L = \omega L\, I_0 \sin(\omega t + 90°)$$

The potential drop across the inductor is proportional to the current, but shifted by 90°. We can lump together the proportionality factor ωL. We will represent it by the symbol X_L and call it **inductive reactance.**

$$X_L = \omega L \qquad \textbf{(Eq. 34–6)}$$

In AC circuits inductive reactance behaves very much like resistance. It is the proportionality constant between voltage and current. However, there are a couple of differences. Resistance does not depend on frequency in the normal operating range, while reactance increases in direct proportion to the frequency of the AC wave. In a resistor, the voltage and current are *in phase.* The current and voltage reach a peak value at the same instance. In an inductor, the

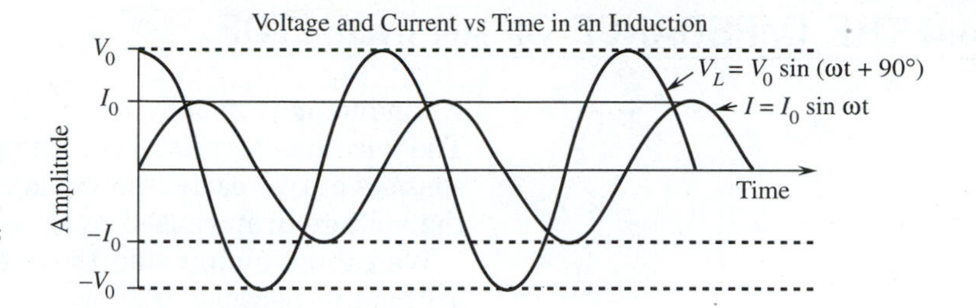

Figure 34–6: Voltage and current versus time in an inductor. The AC voltage leads the current in a pure inductor. Voltage is out of phase with the current by +90°.

voltage peaks and the current peaks are a quarter-cycle apart (see Figure 34–6).

The magnitude of the induced voltage can be written in a form that looks like Ohm's law.

$$|V_L| = X_L I \qquad \text{(Eq. 34–7)}$$

□ **EXAMPLE PROBLEM 34–4: INDUCTIVE REACTANCE**

An inductor has an inductance of 0.074 H.

A. What is its inductive reactance for a frequency of 60 Hz? (This is the usual frequency of power generation in North America.)

B. What is its inductive reactance for 50 Hz? (This is the frequency of AC power used in Europe and Australia.)

■ ***SOLUTION***

A. X_L at $6\bar{0}$ Hz:

$$X_L = \omega L$$
$$\text{where } \omega = 2\pi f$$

$$X_L = 2\pi \text{ rad/cycle} \times 6\bar{0} \text{ cycle/s} \times 0.074\ \Omega \cdot \text{s}$$
$$X_L = 28\ \Omega$$

B. X_L at $5\bar{0}$ Hz:

Since inductive reactance is directly proportional to frequency we can solve it this way:

$$X_L(50) = \frac{5\bar{0}\text{ Hz}}{6\bar{0}\text{ Hz}} X_L(60)$$
$$X_L(50) = \frac{5 \times 28\ \Omega}{6}$$
$$X_L(50) = 23\ \Omega$$

34.4 THE IMPEDANCE OF AN INDUCTOR

An inductor is usually a long piece of wire looped around a form. The wire has a resistance. The potential drop across the inductor consists of two parts: the voltage drop caused by induction (V_L) and the voltage drop caused by resistance (V_R).

We cannot simply add these two voltages together using scalar arithmetic because the two voltages are not in phase. We must superpose the two waves (see Figure 34–7). The superposition is a wave that is out of phase with both V_R and V_L.

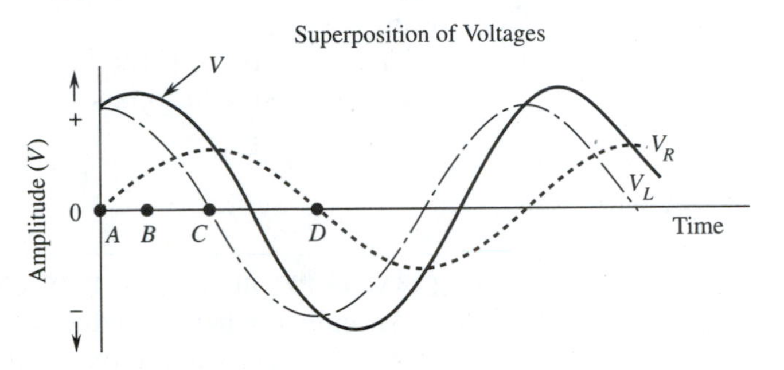

Figure 34–7: Superposition of voltages: the combined voltage of a circuit containing a resistor and an inductor connected in series. V is found by the superposition of the two voltage waves through the devices. Curve (—) is the sum of the voltages A and B.

We can visualize the superposition as a vector addition of the two voltages acting at an angle of 90° (see Figure 34–8). We will use V_R as a reference from which to measure the **phase angle** (θ). Because the current is in phase with the resistance, the phase angle measures the angle between the resulting voltage and the current. The vector equation is:

$$\mathbf{V}_R + \mathbf{V}_L = \mathbf{V}$$

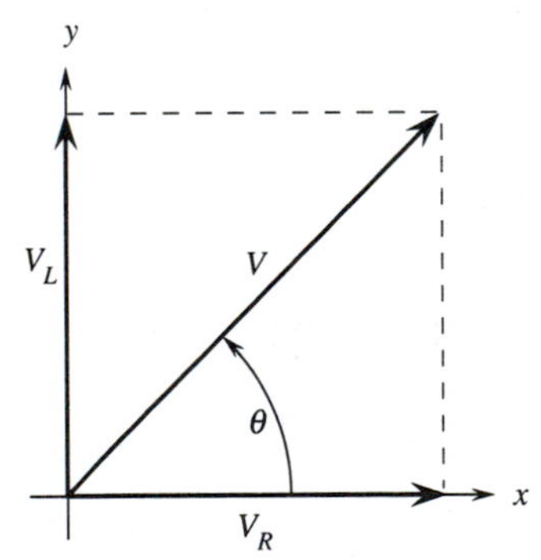

Figure 34–8: $\mathbf{V}_L$ *and* $\mathbf{V}_R$ *are the vector components of the resultant voltage* (**V**). *θ is the phase angle between* **V** *and the current through the circuit.*

In polar form it looks like this:

$$V_R \angle 0° + V_L \angle 90° = V \angle \theta \quad \textbf{(Eq. 34–8)}$$

Visualize a rotation of the vector diagram shown in Figure 34–8. The perpendicular projections of the spinning vectors create the wave patterns seen in Figure 34–7. The orientation of the vectors in Figure 34–9 correspond to the positions A, B, C, and D in Figure 34–7.

Let us write Equation 34–8 in terms of current. Current is always in phase with the resistor so we will attach phase angles to the proportionality constants.

The proportionality constant for the resulting voltage is called **impedance.** We will use the symbol Z to represent it.

$$(R \angle 0° \, I) + (X_L \angle 90° \, I) = Z \angle \theta \, I$$

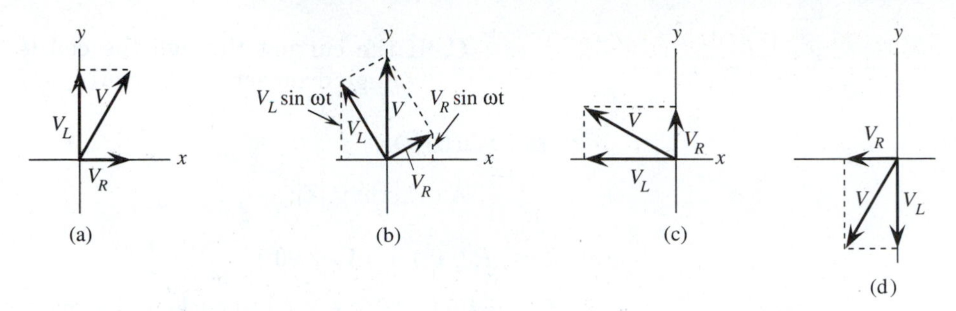

Figure 34–9: *The sine waves in Figure 34–7 can be produced by rotating the voltage vector diagram. The vertical projections are the corresponding amplitudes to positions A, B, C, and D in Figure 34–7.*

In each term, I is the same: $I = I_0(\sin \omega t)$. We can cancel it. We get an equation that tells us how to calculate impedance.

$$Z \angle\theta = (R \angle 0°) + (X_L \angle 90°) \qquad \textbf{(Eq. 34–9)}$$

Notice that R and X_L are the vector components of Z. We can use the same methods of vector addition we used in force problems to find magnitude and angle.

Once we have the impedance for an inductor we can use an Ohm's law type of equation to calculate the voltage across the device for any current flowing through it.

$$V = (Z \angle\theta)\, I \qquad \textbf{(Eq. 34–10)}$$

☐ **EXAMPLE PROBLEM 34–5: IMPEDANCE IN A COIL**

A coil has an inductance of 34 mH and a resistance of 54.5 Ω. An AC current with a frequency of 160 Hz passes through the coil. See Figure 34–10.

A. Find the magnitude of the coil's impedance.

B. What is the phase angle between current and voltage through the coil?

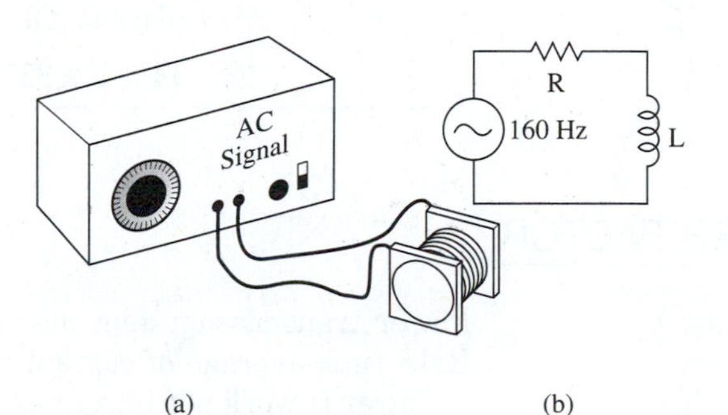

Figure 34–10: *Diagram for Example Problem 34–5.*

C. If the current through the coil is 0.23 A, what will be the voltage drop across the coil?

■ *SOLUTION*

A. Finding |Z|:

$$Z = (R \angle 0°) + (X_L \angle 90°)$$

$$Z = (54.5\ \Omega \angle 0°) + 2\ \pi \text{ rad/cycle} \times 160 \text{ cycle/s} \times (0.034\ \Omega \cdot \text{s} \angle 90°)$$

$$Z = (54.5\ \Omega \angle 0°) + (34.2\ \Omega \angle 90°)$$

This is the answer in component form. We can convert it into polar form by using the Pythagorean theorem. Remember this is nothing more than vector addition.

$$|Z| = [(54.5\ \Omega)^2 + (34.2\ \Omega)^2]^{1/2}$$

$$|Z| = 64.3\ \Omega$$

B. The phase angle:

Since phase angles are measured from R, R behaves as an x axis.

$$\tan \theta = \frac{y}{x} = \frac{X_L}{R}$$

$$\tan \theta = \frac{34.2\ \Omega}{54.5\ \Omega}$$

$$\theta = 32°$$

C. Voltage:

The results of parts A and B give us:

$$Z = 64.3\ \Omega \angle 32°$$

$$V = Z\ I$$

$$V = (64.3\ \Omega \angle 32°) \times 0.23 \text{ A}$$

$$V = 14.8 \text{ V} \angle 32°$$

34.5 THE POWER FACTOR

Power transmission depends on the phase angle as well as on the RMS time average of current and voltage.

Power is work per unit time. When we calculate mechanical work we understand that only the component of force in the direction of displacement does work. We found that the component of force in the direction of displacement (Δs) is:

$$work = F\ \Delta s \cos\theta$$

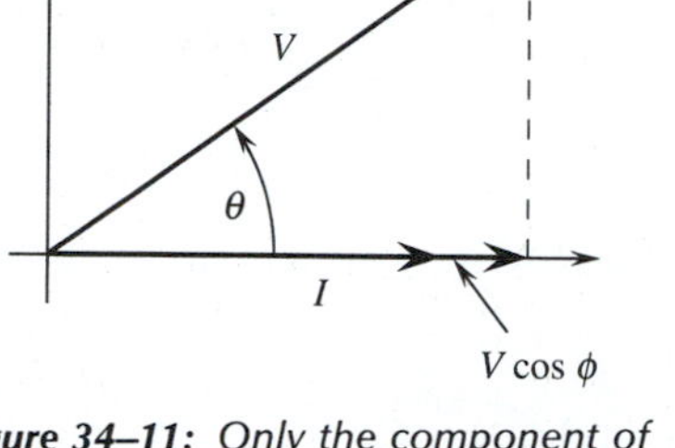

Figure 34–11: *Only the component of voltage parallel to the current does work. Cos θ is the power factor.*

where θ is the angle between force and displacement.

Then the power to move an object at a velocity $v = s/t$ is:

$$power = \frac{work}{time} F\ v \cos\theta$$

We have the same problem with electrical power. Only the vector component of the voltage in the direction of the current flow affects power transmission (see Figure 34–11). For an electrical circuit that has a phase angle between current and voltage the power transmission is:

$$\text{power} = V\,I \cos\theta \qquad \textbf{(Eq. 34–11)}$$

θ is the same phase angle found in the circuit's impedance. The term cos θ is called the **power factor.**

□ **EXAMPLE PROBLEM 34–6: POWER**

Assume that the voltages and current in Example Problem 34–5 are peak values. Calculate the power transmission.

■ ***SOLUTION***

The power depends on RMS values as well as on the power factor. Peak values need to be reduced to RMS values.

$$P = 0.7071\ \text{V} \times 0.7071\ I \cos\theta$$

or

$$P = \frac{V\,I \cos\theta}{2}$$

$$P = \frac{14.8\ \text{V} \times 0.23\ \text{A} \cos 32°}{2}$$

$$P = 1.4\ \text{W}$$

□ **EXAMPLE PROBLEM 34–7: MORE POWER**

A watt meter indicates that the AC power drop through a circuit is 45.6 W. An AC voltmeter across the circuit measures a potential drop of 35.0 V. An AC ammeter measures a current of 1.64 A. Calculate the power factor and phase angle for the circuit.

■ *SOLUTION*

The meters read RMS values. We do not need a factor of one-half in the power equation.

$$\cos\theta = \frac{P}{I\,V}$$

$$\cos\theta = \frac{45.6\ \text{W}}{1.64\ \text{A} \times 35.0\ \text{V}}$$

$$\cos\theta = 0.794\ \text{(power factor)}$$

$$\theta = \cos^{-1}(0.794)$$

$$\theta = 37.4^\circ\ \text{(phase angle)}$$

34.6 AC CURRENT AND CAPACITANCE

The potential drop across a capacitor does not depend on current. It depends on the charge (Q) on the plates and capacitance (C).

$$V_c = \frac{Q}{C}$$

The charge is related to the AC current through a circuit by:

$$I = \frac{\Delta Q}{\Delta t}$$

We found in Equation 34–1 that for a sine wave:

$$\frac{\Delta(\sin\omega t)}{\Delta t} = \omega\,[\sin(\omega t + 90^\circ)]$$

We can work this idea backward. Since the current is a sine wave, the charge on the capacitor must also vary as a sine wave. We have:

$$\frac{\Delta Q}{\Delta t} = I_0(\sin\omega t)$$

Compare this with Equation 34–1. When we find the slope of the sine wave, there is a shift forward of 90°. The charge must lag behind by 90° to give us no phase shift for current.

$$Q \propto I_0 [\sin(\omega t - 90°)]$$

The factor ω must also be taken into account. If Q is inversely proportional to ω, then ω will cancel out when we take the slope. This gives us the expression for Q.

$$Q = \frac{I_0}{\omega} [\sin(\omega t - 90°)] \qquad \textbf{(Eq. 34–12)}$$

If we use Equation 34–1 to find the slope ($\Delta Q/\Delta t$) we get the correct expression for current ($I = I_0 \sin \omega t$). Figure 34–12 is a plot of current and charge versus time for a capacitor.

The voltage across the capacitor is:

$$V_C = \frac{Q}{C} = \left(\frac{1}{\omega C}\right) [I_0 \sin(\omega t - 90°)]$$

or

$$V_C = \left(\frac{1}{\omega C}\right) I \angle -90°$$

Again we have an Ohm's law-like equation. The part of the equation in parentheses is **capacitive reactance** (X_C). The voltage in a capacitor will trail the current and voltage drop of a resistor by a quarter cycle.

$$|X_C| = \frac{1}{\omega C} \qquad \textbf{(Eq. 34–13A)}$$

and

$$V_C = |X_C| I \angle -90° \qquad \textbf{(Eq. 34–13B)}$$

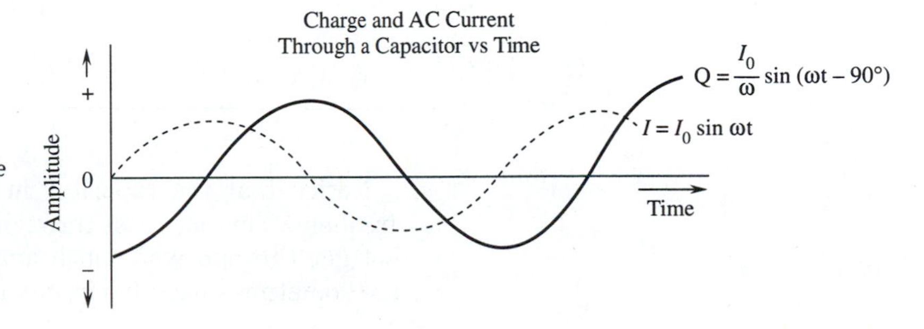

Figure 34–12: Charge and AC current through a capacitor versus time. In a pure capacitor, voltage lags behind current by −90°. The dotted line is the current flowing through the capacitor. The solid line is the voltage drop across the capacitor.

☐ **EXAMPLE PROBLEM 34–8: CAPACITIVE REACTANCE**

A dielectric capacitor has a capacitance of 5.60 μF. Find the magnitude of the capacitive reactance when the AC current passing through it has a frequency of:

A. 60.0 Hz.
B. 500.0 Hz.
C. 10.0 kHz.

■ ***SOLUTION***

A. X_C at 60 Hz:

$$X_c = \frac{1}{\omega C} = \frac{1}{2\pi f C}$$

$$X_C = \frac{1}{2\pi \text{ rad/cycle} \times 6\bar{0} \text{ cycle/s} \times 5.6 \times 10^{-6} \text{ C/V}}$$

$$X_C = 470 \text{ V/A}$$

$$X_C = 470\ \Omega$$

B. At 500 Hz:

Since the capacitive reactance is inversely proportional to frequency we can solve it this way.

$$X_C(2) = \frac{f(1)\, X_C(1)}{f(2)}$$

$$X_C(2) = \frac{60.0 \text{ Hz} \times 474\ \Omega}{500.0 \text{ Hz}}$$

$$X_C(2) = 57\ \Omega$$

C. At 10 kHz:

$$X_C(2) = \frac{60.0 \text{ Hz} \times 470\ \Omega}{1.00 \times 10^4 \text{ Hz}}$$

$$X_C(2) = 2.8\ \Omega$$

Notice that the capacitor in Example Problem 34–8 lets high-frequency currents pass through it easily, while low frequencies cannot get through with much amplitude. For this reason capacitors are sometimes used in circuits as **high-pass filters.**

34.7 APPLICATION: ELECTRICAL RESONANCE

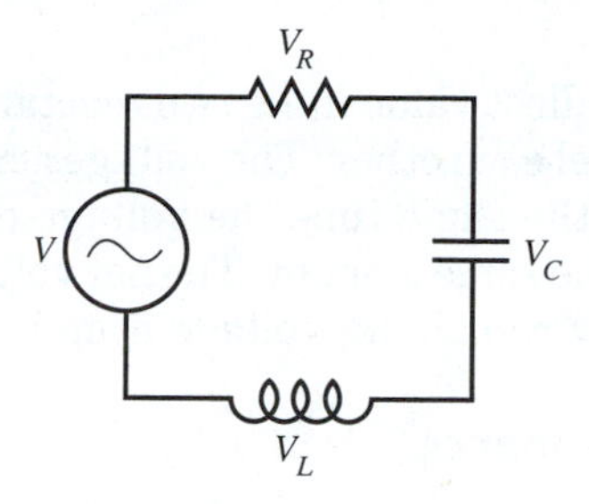

Figure 34–13: An LRC series circuit.

Let us hook up a resistance, a capacitor, and an inductor in series, an LRC circuit. We will complete the LRC circuit with an AC power supply (see Figure 34–13).

In a series circuit the voltages are additive and the current is the same through each device. Let $V = V_0$ (sin ωt) be the voltage of the power source with the corresponding current $I = I_0$ (sin ωt).

$$V = V_{\mathrm{R}} + V_{\mathrm{C}} + V_{\mathrm{L}}$$

or

$$V = [R\, I_0 (\sin \omega t)] + [X_{\mathrm{C}}\, I_0 \sin (\omega t - 90°)] + [X_{\mathrm{L}}\, I_0 \sin (\omega t + 90°)]$$

or

$$V = [\mathrm{R} + (X_{\mathrm{L}} \angle +90°) + (X_{\mathrm{C}} \angle -90°)]\, I$$

We can use vector addition to sum the terms in brackets in the last equation to get the total impedance of the circuit (Z). This reduces everything down to a very simple expression.

$$V = (Z \angle \theta)\, I$$

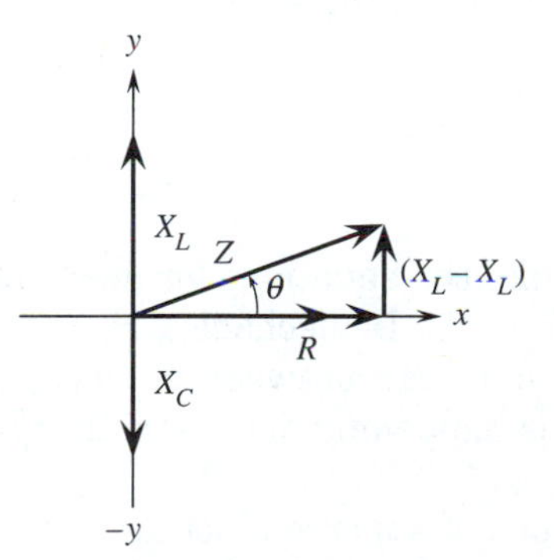

Figure 34–14: The vector diagram for an LRC circuit. The two reactances are the vertical components. R is the horizontal component. The impedance (Z) is the resultant.

Let us draw a vector diagram similar to the one we used in Section 34.4.

We will draw R along the x axis. The vertical components will be X_{L} along the plus y axis and X_{C} along the minus y axis. This corresponds to phase angles of $+90°$ and $-90°$, respectively (see Figure 34–14).

We can combine the vertical components.

$$y = (X_{\mathrm{L}} - X_{\mathrm{C}})$$

and

$$x = R$$

We will use the Pythagorean theorem to find the magnitude of the impedance.

$$Z = [R^2 + (X_{\mathrm{L}} - X_{\mathrm{C}})^2]^{1/2}$$

The phase angle is:

$$\tan \theta = \frac{X_L - X_C}{R}$$

The impedance would be at its smallest value if the two reactances were the same size. They would cancel each other. The voltage across the capacitor would be negative at the same time the voltage drop across the inductor was positive by the same amount. The net voltage drop would add up to zero. The only remaining voltage drop in the circuit would be across the resistor.

This effect is called **electrical resonance.**

$$X_L = X_C \qquad \textbf{(Eq. 34–14)}$$

Let us write the reactances according to their defining expressions.

$$\omega L = \frac{1}{\omega C}$$

Solve for ω.

$$\omega = \left(\frac{1}{L\,C}\right)^{1/2}$$

Since $\omega = 2\pi f$, we can express this in frequency.

$$f_0 = \frac{1}{2\pi}\left(\frac{1}{L\,C}\right)^{1/2} \qquad \textbf{(Eq. 34–15)}$$

This equation looks remarkably like the equations for mechanical resonant systems described in Chapter 23. In mechanical resonance, mechanical energy oscillates back and forth between potential and kinetic energy. Something similar is happening here—another kind of energy transfer.

Assume we can construct a coil out of superconducting material. Its electrical resistance would be zero. We will hook the coil across a charged capacitor with a switch.

If we close the switch shown in Figure 34–15, current will start to flow away from the capacitor. Its electric field will be reduced, losing energy, but the change of current through the coil will produce a magnetic field. The energy is transferred from electric field energy to magnetic field energy.

The capacitor cannot discharge forever. Eventually, the magnetic field starts to collapse. It induces a back-EMF that reverses current flow. The capacitor is recharged. The cycle starts over again.

The frequency of the cycle depends upon the size of the inductor

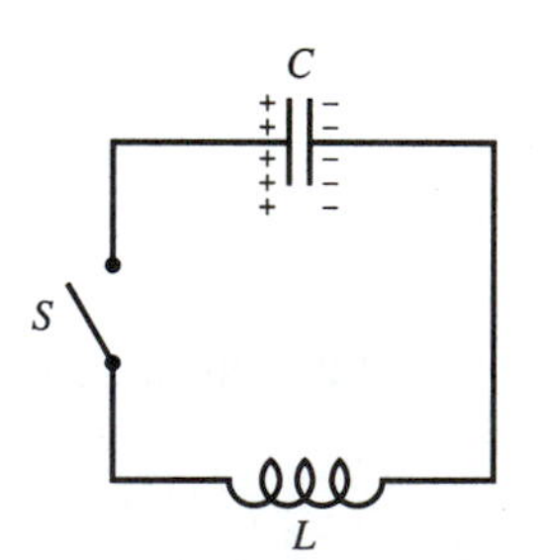

Figure 34–15: *A charged capacitor (C) is connected to an inductor. S is an open switch.*

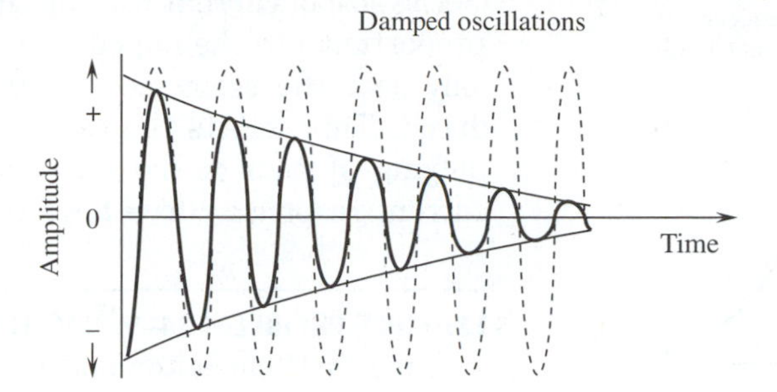

Figure 34–16: *The dashed line is a plot of current in a resonant circuit versus time. The solid line occurs when there is resistance in the circuit. Energy is lost to heat in each cycle.*

and capacitor. Large capacitors and large inductors slow up the cycle.

Real coils have some resistance. Part of the energy is lost in each cycle. Figure 34–16 shows a plot of oscillating current versus time for a resonating circuit.

Many kinds of electrical equipment use resonating circuits. When you tune your radio to the local FM station or flip channels on the television, you are changing the capacitance of a resonating circuit. You are matching the frequency of an oscillating circuit to the carrier frequency of a broadcast station.

No Static at All

When thunderstorms are hovering nearby, AM radio receives a lot of static. At the same time, FM radio is barely affected.

The difference in response to lightning has a lot to do with the way signals are transmitted by FM and AM broadcasts.

AM means amplitude modulation. The signal is imposed on a carrier wave at some fixed frequency. You tune in the carrier wave frequency on your AM radio. The imposed signal is located in variations of the amplitude of the carrier wave. Electronic devices separate the signal from the carrier wave and amplify these variations in amplitude.

FM means frequency modulation. The signal shows up in the carrier wave as small variations in frequency rather than changes of amplitude. Since variations in amplitude are not important, the amplitude is clipped in an FM receiver. This is to say the amplitude above some preset level is discarded. After clipping, the amplitude has flat plateaus where there used to be peaks.

(Photograph by Ellen Senisi.)

Lightning causes an interference that alters the amplitude of a radio signal. The noise you hear is an alteration of amplitude. Since an FM receiver clips these changes in amplitude, you do not hear any static.

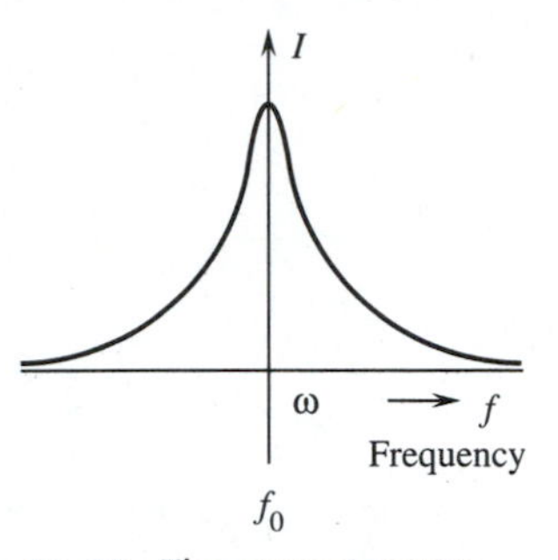

Figure 34–17: *The current versus frequency through a resonating RLC circuit is plotted against frequency.* f_0 *is the resonant frequency.*

The size of current passing through a resonant circuit is inversely proportional to the impedance ($I = V/Z$). The current will be large only near the resonance frequency. Other frequencies will be reduced. This selects the carrier frequencies to be fed into the amplifying system of the television or radio. Figure 34–17 shows a current response curve for a resonating circuit.

□ **EXAMPLE PROBLEM 34–9: IMPEDANCE IN AN RLC CIRCUIT**

A 0.23 μF capacitor and a 2.5-mH coil with an 86-Ω resistance are connected in series. Find the size and phase angle of the circuit impedance at a frequency of 12 kHz.

■ ***SOLUTION***

$$|X_L| = 2\pi f L = 2\pi \times (1.2 \times 10^4 \text{ Hz}) \times 0.0025 \text{ H}$$

$$|X_L| = 190 \ \Omega$$

$$|X_C| = \frac{1}{2\pi f C} = \frac{1}{2\pi \times (1.2 \times 10^4 \text{ Hz}) \times (2.3 \times 10^{-7} \text{ F})}$$

$$|X_C| = 58 \ \Omega$$

$$R = 86 \ \Omega$$

$$|Z| = [R^2 + (X_L - X_C)^2]^{1/2}$$

$$|Z| = [(86 \ \Omega)^2 + (190 - 58 \ \Omega)^2]^{1/2}$$

$$|Z| = 160 \ \Omega$$

$$\tan\theta = \frac{(X_L - X_C)}{R}$$

$$\theta = \tan^{-1}\left(\frac{132 \ \Omega}{86 \ \Omega}\right)$$

$$\theta = 57°$$

□ **EXAMPLE PROBLEM 34–10: RESONANCE**

Find the resonance frequency for the circuit in Example Problem 34–9.

■ ***SOLUTION***

$$f_0 = \frac{1}{2\pi}\left(\frac{1}{LC}\right)^{1/2}$$

$$f_0 = \frac{1}{2\pi}\left[\frac{1}{0.0025 \text{ H} \times (0.23 \times 10^{-6} \text{ F})}\right]^{1/2}$$

$$f_0 = 6.6 \text{ kHz}$$

SUMMARY

Energy is transferred through electric circuits by wave motion. An AC generator induces a sinusoidal voltage and current. The induced EMF is proportional to the rotor's angular speed (ω). The induced EMF of an AC generator is

$$EMF = -(N\,B\,A\,\omega)\cos(\omega t)$$

The power source for a circuit can be viewed in phase voltage and current waves with amplitudes of V_0 and I_0.

$$V = V_0 \sin(\omega t)$$

$$I = I_0 \sin(\omega t)$$

The power passing through a circuit depends on the phase angle (θ) between the voltage and the current. Since current and voltage vary with time, a time average value called the RMS value needs to be used to find the effective power.

$$P_{\text{eff}} = I_{\text{RMS}}\,V_{\text{RMS}}\cos\theta$$

where

$$I_{\text{RMS}} = \frac{I_0}{(2)^{1/2}} \text{ and } V_{\text{RMS}} = \frac{V_0}{(2)^{1/2}}$$

In a pure resistance the voltage and current are in phase. The phase angle is zero ($\theta = 0$).

In a pure inductance the voltage leads the current by a quarter cycle. The phase angle is $+90°$ ($\theta = +90°$).

In a pure capacitor, one with no resistance, the voltage lags behind the current by a quarter cycle. The phase angle is $-90°$ ($\theta = -90°$).

Equations having the form of Ohm's law can be written for voltage and current through different kinds of devices.

$$V_{\text{R}} = (R\,\angle 0°)\,I$$

$$V_{\text{L}} = (|X_{\text{L}}|\,\angle +90°)\,I$$

$$V_{\text{C}} = (|X_{\text{C}}|\,\angle -90°)\,I$$

$|X_{\text{L}}| = \omega L$ is called inductive reactance.

$|X_{\text{C}}| = \dfrac{1}{\omega C}$ is called capacitive reactance.

The combination of resistance and reactance in a circuit is called impedance (Z). It is found by the vector addition of resistance and reactance with resistance along the horizontal axis and reactances along the vertical axis.

For a series circuit, impedance is:

$$Z = [R^2 + (X_L - X_C)^2]^{1/2}$$

The phase angle is:

$$\theta = \tan^{-1}\left[\frac{(X_L - X_C)}{R}\right]$$

The impedance with its phase angle is the proportionality constant between voltage and current across a circuit containing a variety of electrical devices.

$$V = (Z \angle\theta)\, I$$

Resonance can occur between a capacitor and an inductor. Energy oscillates back and forth between the electric field of the capacitor and the magnetic field of the inductor. The resonance frequency is:

$$f_0 = \frac{1}{2\pi}\left(\frac{1}{L\,C}\right)^{1/2}$$

KEY TERMS

If you can explain the following terms to a friend or classmate, you understand their meaning. If you cannot explain the terms, you should reread the sections in which they are discussed.

capacitive reactance
effective time
electrical resonance
high-pass filters
impedance
inductive reactance
instantaneous power
peak voltage
phase angle
power factor
pure resistance
slip rings

EXERCISES

Section 34.1:

1. Why are slip rings used in an AC generator rather than a split commutator?
2. A simple generator of the type shown in Section 34.1 is used to generate electricity from the torque created by a wind turbine. Explain how the EMF will be affected by gusts of wind.

Section 34.2:

3. Explain the difference between peak voltage and RMS voltage.

4. If uncorrected peak voltage and uncorrected peak current are used to calculate power, the power calculation would be too large by what factor?

Section 34.3:

5. The frequency of the AC current through an inductor is doubled. By what factor will its inductive reactance change?
6. At the instant the voltage through a pure inductor is maximum what can you say about the value of the current?
7. At the instant the voltage through a pure resistor is at maximum what can you say about the size of the current?

Section 34.4:

8. In your own words explain the meaning of "phase angle."
9. Between what extreme values can the impedance of a real coil have as frequency changes from zero to a very high frequency? A real coil has resistance (R) as well as an impedance.
10. Why can't we use scalar addition to add resistance and reactance to calculate impedance?

Section 34.5:

11. A coil and a resistor are connected in series with a power source (see Figure 34–18). The voltage delivered by the power supply is 15 VAC. The voltage drop across the resistor is 5.0 VAC, and the voltage drop across the coil is 12.0 VAC. The scalar sum of the voltage drops is larger than the voltage of the power supply. Explain why this is possible.

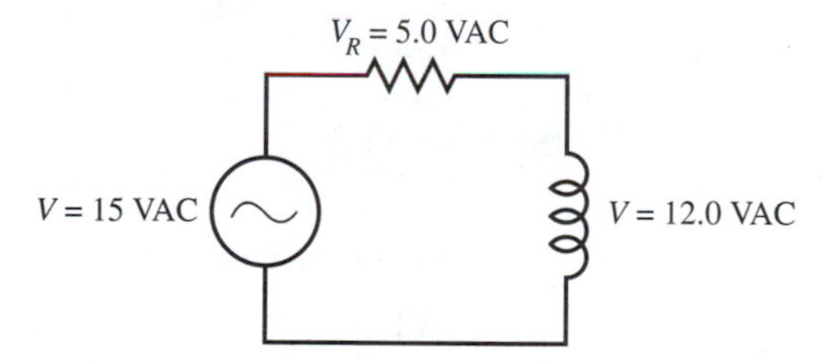

Figure 34–18: *Diagram for Exercise 11. The scalar sum of the voltage drops is larger than the voltage of the source.*

12. An ammeter reads an AC current of 1.6 A through a circuit at the same time a voltmeter connected across the coil is 20 VAC. A watt meter reads a power of 32 W. What can you say about the phase angle between voltage and current for this circuit?

Section 34.6:

13. Does voltage lead or lag behind current in a capacitor?

14. If the frequency through a capacitor is doubled, how will the capacitive reactance change?

15. Which kind of capacitor would give a larger capacitive reactance: a medium-sized capacitor with no dielectric between its plates, or an oil-filled capacitor of the same size?

Section 34.7:

16. A variable capacitor is used to tune a radio. When we turn to a station of higher frequency are we increasing the capacitance of the tuning circuit or decreasing it?

17. Explain what is meant by electrical resonance in your own words.

PROBLEMS

Section 34.1:

1. An AC generator has a coil with 34 turns and a cross-sectional area of 34.2 cm^2. What is the peak EMF produced when the coil turns in a 0.26-T field at a frequency of $6\bar{0}$ Hz?

2. At what frequency should the coil in the generator in Problem 1 turn to create a peak EMF of $20\bar{0}$ V?

Section 34.2:

3. Calculate the power dissipated by a $2\bar{0}$-Ω resistor when the AC current through it has a peak current of 3.40 A.

4. An voltmeter reads $23\bar{0}$ VAC. What is the peak voltage?

5. A power transmission line operates at 765 kVAC. What is the potential difference between peak voltages of a conducting cable that are separated by one-half cycle? The voltage rating of the line is RMS voltage.

6. The RMS current through a transmission line is $12\bar{0}$ A. If the average peak potential of the line is 226 kV, how many megawatts does it transmit?

Section 34.3:

7. Find the inductive reactance of a 42-mH inductor at an angular speed of 314 rad/s.

8. What is the inductive reactance of a 23-mH inductor at a frequency of $6\bar{0}$ Hz?

9. A $6\bar{0}$-cycle current of 1.2 A passes through a pure inductance of 0.141 H. What is the voltage drop across the inductor?

Section 34.4:

10. Find the magnitude of the impedance for the following combinations of resistance and inductive reactance connected in series. f is the frequency of the AC current and voltage.

A. $R = 67$
$L = 33$ mH
$f = 6\bar{0}$ Hz

B. $R = 42\bar{0}$
$L = 47.0$ mH
$f = 10\bar{0}0$ Hz

C. $R = 2.3$
$L = 82$ mH
$f = 12$ Hz

11. Find the phase angle for the impedances for the data in Problem 10.

***12.** A solenoid has an impedance of 128 Ω at a phase angle of 28.0° for a 60.0-Hz AC current. (Hint: Think in terms of vectors. Figure 34–8 should help.)
A. What is the resistance of the coil?
B. What is the inductive reactance?
C. What is the inductance of the solenoid?

Section 34.5:

13. An AC current of 0.67 A passes through a circuit. The voltage across the circuit is 110 VAC at a phase angle of 58°. Assume RMS values. Calculate the power dissipated by the circuit.

14. An AC current of 234 mA passes through a pure inductor at a peak voltage of 50.0 VAC. Show that no power is dissipated by the inductor.

***15.** A toroid has an impedance of 78 Ω ∠23° at $6\bar{0}$ Hz. If the current through the toroid is 8.4 A:
A. What is the voltage across the leads to the toroid?
B. How much power is consumed by the device?

Section 34.6:

16. What is the capacitive reactance of a 0.32 μF capacitor at $10\bar{0}0$ Hz?

17. A capacitor has a capacitive reactance of 2.65×10^3 Ω for a $6\bar{0}$-Hz AC current. What is the value of the capacitance?

18. A capacitor has a capacitive reactance of 234 Ω at $10\bar{0}0$ Hz. What will be its reactance at $40\bar{0}0$ Hz?

Section 34.7:

19. A 53-μF capacitor is in series with a 0.230-H solenoid whose coils have a resistance of 63 Ω. What is the magnitude of the impedance for a $6\bar{0}$-Hz AC current?

20. Find the resonant frequency of a 2.1-nF capacitor in series with a 1.2-mH coil.

21. What inductance should be connected to a 4.5-nF capacitor to produce a resonance of 23 kHz?

****22.** A capacitor, a resistor, and a coil are connected in series. The capacitive reactance is 45 Ω, the inductive reactance is 93 Ω, and the resistance is 36 Ω. An AC current with a peak value of 1.30 A flows through the circuit.

A. How large is the series impedance?

B. What is the phase angle?

C. What is the peak voltage?

D. Calculate the electrical power consumption.

Chapter 35

SOLID-STATE ELECTRONICS

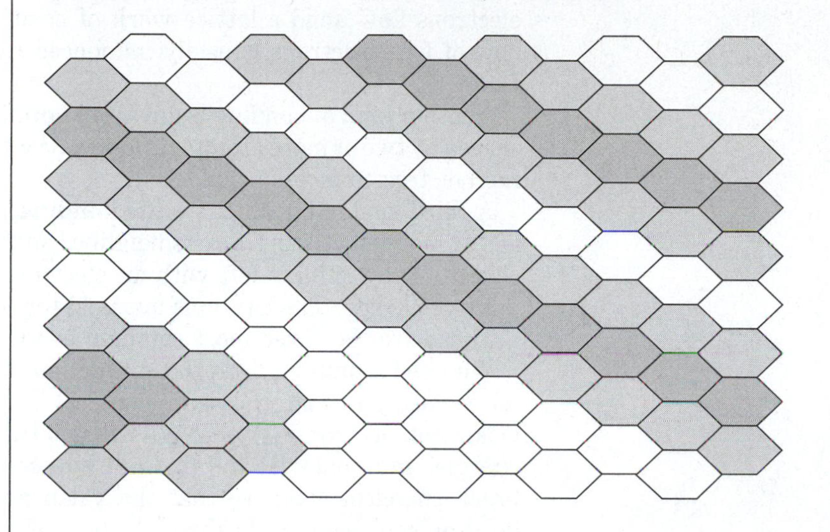

When you watch a color television, you are really watching tiny, phosphorescent dots of red, blue, and green. Read this chapter to see how your mind helps you see the full spectrum on a television screen.

OBJECTIVES

In this chapter you will learn:

- to express energy in units of electron volts
- how electrons gain and lose energy in an atomic orbit
- how electron energy levels are formed in solids (band theory)
- some different ways electrons can escape from a metal
- how electrons break chemical bonds and participate in conduction in a semiconductor
- how "holes" in chemical bonds assist in electrical conduction
- to explain luminescence using the band theory
- how adding impurities to a pure semiconductor affects conductivity (doping)
- how temperature alters the conductivity of a semiconductor
- to explain the operation of some semiconducting devices

Time Line for Chapter 35

1911 Heike Onnes discovers superconductivity in mercury.
1929 Paul Dirac introduces the idea of a semiconductor hole.
1948 William Shockley, Walter Houser Brattain, and John Bardeen together discover the transistor effect.
1957 Leo Esall discovers the tunneling effect in semiconducting diodes.
1986 The first high-temperature superconductors are developed.

The electronic behavior of solids (solid-state electronics) depends on the way atoms arrange themselves in a material and on the way they bond together. One kind of bonding is the **metallic bond.** Free electrons flow amid a lattice work of positively charged ions. The flow of free electrons is easily influenced by electric and magnetic fields.

Another kind of bonding is **covalent bonding.** Outer electrons are shared by two or more atoms. As long as the bond holds, the electrons are not free to move around.

A third kind of bonding is **ionic bonding.** One kind of atom steals one or more electrons from its neighbor and develops a net negative charge. The victim is left with an electron deficiency. It has a net positive charge. The two ions are held together by coulomb forces.

These are the three most common bonds in solid material.

We need a unifying idea to tie together the behavior of electrons for a variety of materials and chemical bonds. We can find one by looking at electron energies. We will find the distribution of electron energies in a material and find out how much energy is needed to break chemical bonds so that the valance electrons involved with bonding can roam freely through the material.

The study of electron energies in different materials will lead us to the band theory and several interesting electronic properties of solids.

35.1 THE ELECTRON VOLT

The electronic mass and electronic charge are very small. The kinetic energy of an electron traveling a million meters per second is:

$$KE = \frac{1}{2} m v^2 = \frac{1}{2} \times 9.11 \times 10^{-31}\,\text{kg} \times (10^6\,\text{m/s})^2$$

$$KE = 4.56 \times 10^{-19}\,\text{J}$$

The potential energy of an electron with a thousand volts of electrical potential is:

$$PE = e\,V = 1.6 \times 10^{-19}\,\text{C} \times 1000\,\text{V}$$
$$PE = 1.6 \times 10^{-16}\,\text{J}$$

The joule is an inconveniently large quantity of energy when we discuss electrons. It would be better if we defined a smaller unit of energy.

The energy an electron gains or loses when its potential changes by 1 V is numerically equal to the electronic charge. We can use this as the unit of energy. It is called the **electron volt (eV).**

If the electron has an increase of electrical potential of 1 V, its energy increases by 1.0 eV. If its electrical potential increases by 50 V, then the increase in energy is 50 eV. When an electron is accelerated from zero by a potential drop of 1500 V in a CRT, its kinetic energy gain is 1500 eV.

When a charge of 1 C has a 1.00-V increase in electrical potential, its energy increases by 1.0 J. The corresponding increase of 1.0 V for a single electron gives us an energy of 1.0 eV. If we take a ratio we will get a conversion factor.

$$\frac{1\,\text{eV}}{1\,\text{J}} = \frac{1.602 \times 10^{-19}\,\text{C} \times 1.000\,\text{V}}{1.000\,\text{C} \times 1.000\,\text{V}}$$
$$1\,\text{eV} = 1.602 \times 10^{-19}\,\text{J}$$

The electron volt *is the amount of energy a single electronic charge gains or loses when its potential changes by 1 V.*

☐ **EXAMPLE PROBLEM 35–1: THE SPEEDING ELECTRON**

Express the kinetic energy of an electron traveling at a million meters per second in electron volts.

■ ***SOLUTION***

We have already calculated the energy in terms of joules. All we need is a units conversion.

$$KE(\text{eV}) = \text{KE(J)} \times \frac{1}{1.602 \times 10^{-19}\,\text{J/eV}}$$
$$KE(\text{eV}) = \frac{4.56 \times 10^{-19}\,\text{J}}{1.602 \times 10^{-19}\,\text{J/eV}}$$
$$KE(\text{eV}) = 2.85\,\text{eV}$$

EXAMPLE PROBLEM 35–2: A DUST PARTICLE

A dust particle has an electrostatic charge of 2.24×10^{-18} C. How large is its change of kinetic energy in electron volts when it falls through a potential difference of 7.5 V?

SOLUTION

The change of kinetic energy is equal to the change of potential energy according to the conservation of energy law. In metric units $|KE| = |\Delta PE| = q\,|\Delta V|$. To convert to electron volts divide by the electronic charge (e).

$$KE = \frac{q}{e} V$$

$$KE = \frac{2.24 \times 10^{-18}\ \text{C} \times 7.5\ \text{V}}{1.602 \times 10^{-19}\ \text{C}}$$

$$KE = 105\ \text{eV}$$

35.2 A SIMPLE ATOMIC MODEL

Before we look at the electrons in an assembly of atoms we had better look at the electrons orbiting a single atom. Here is a very simplified model of how they behave.

Figure 35–1 shows the orbit of electrons near an atom's nucleus. Electrons have a wave nature. Their orbit is a standing wave. An integral number of wavelengths must fit into the orbit; otherwise, the electron destructively interferes with its own wave pattern. Orbits that accommodate a standing wave are called **stable orbits.**

Only a few standing wave orbits exist near the nucleus. Each orbit has a different energy. The moving electron has kinetic energy. Its average distance from the positively charged nucleus determines its electrostatic potential energy. Since electrostatic forces are attractive, the potential energy is negative.

This means that the closest orbits have the lowest energies. Electrons will try to congregate at these low energy levels. If two electrons try to occupy the same orbit (same energy), they encounter a problem. In order to keep their standing waves from interfering with each other they must occupy the same position at the same time. This possibility is excluded. Two electrons in an atom cannot have exactly the same energy or orbit. This rule is called Pauli's exclusion principle.

Pauli's exclusion principle: *No two electrons in a system can have exactly the same energy.*

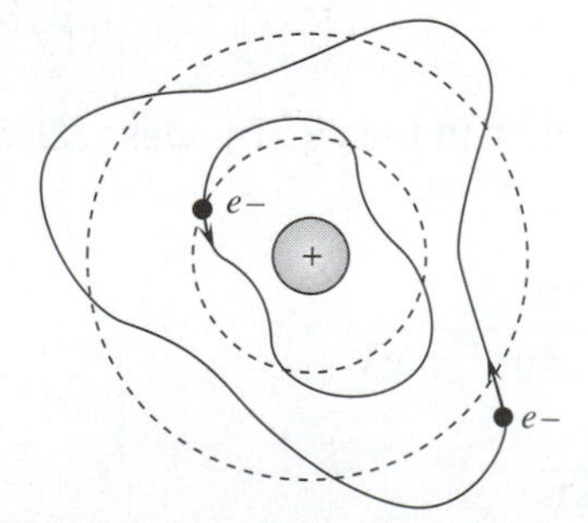

Figure 35–1: *Orbiting electrons form standing waves.*

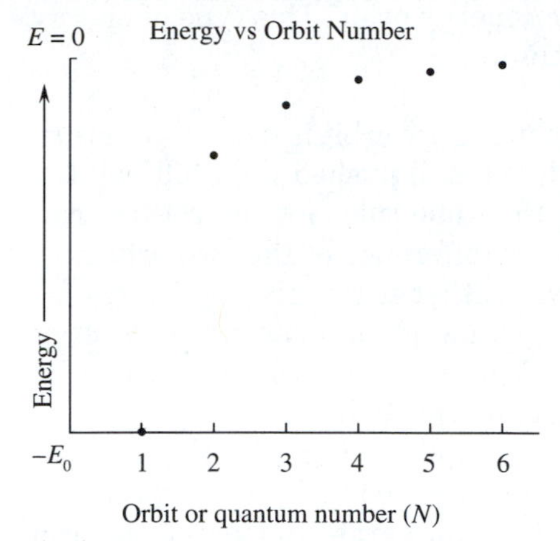

Figure 35–2: Energy versus orbit number. The orbit number, or quantum number (N), indicates the number of standing waves. Energies increase with larger orbits.

Each electron orbiting an atom must have a different energy or orbit. Each orbit must be a standing wave.

We can number the possible orbits using $n = 1$ for the innermost orbit, $n = 2$ for the next orbit, and so forth. Figure 35–2 is a simplified plot of energy versus the orbit number.

Notice that the net energy of the electrons is negative. An electron must gain energy to go to a higher numbered orbit farther away from the nucleus.

In order to get to a higher orbit, the electron must absorb exactly the amount of energy equal to the difference of the two orbits' energy. Remember, it cannot exist in an orbit that does not give it a standing wave. In order to go from orbit $n = 1$ to orbit $n = 2$, it has to gain an energy increase of exactly $\Delta E = E_1 - E_2$. We say the energy is **quantized.** It has to come in a certain size package: no more and no less than the required amount.

There are two ways an electron can gain the energy to make a quantum jump to a higher energy orbit (see Figure 35–3).

1. The electron can gain thermal energy. If the atom it orbits collides with another atom, the electron can absorb the required energy from the collision. This type of energy gain is called **thermal activation.**
2. The electron can gain radiant energy. If a photon strikes it with exactly the right energy, the electron will absorb the photon

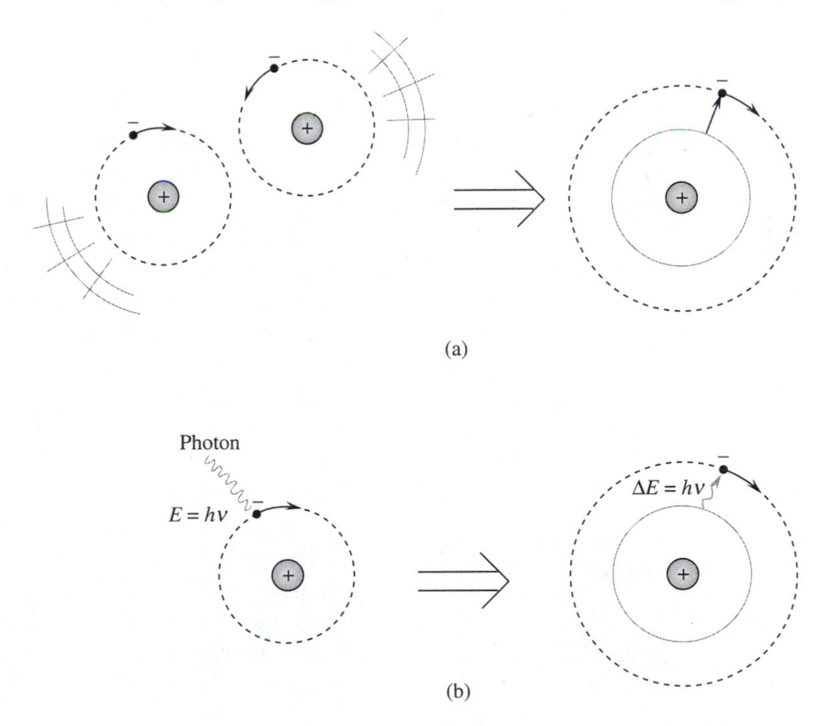

Figure 35–3: There are two ways an electron can gain energy to move to a higher orbit. (a) Thermal activation. Atoms collide. Kinetic energy is transferred to the electron during the collision. (b) Photon activation. An electron absorbs a photon with exactly the energy difference of the two orbits.

energy and move to the higher energy orbit. This type of energy gain is called **photon activation.**

Once an electron gets to a higher energy orbit it will try to return to a lower energy state, much like a ball pushed up a hill will try to roll back down. It must obey the same rules for the return trip. It must give up exactly the energy difference of the two orbits.

It can lose energy in two ways. Either it can give up energy by emitting a photon with just the right energy or it can give up energy in a collision with another atom. These are the same processes by which the electron gained energy, but in reverse.

A more detailed model is more complicated. Each orbit number can have suborbits because differently shaped elliptical orbits with slightly different energies can fit an integral number of electron wavelengths. We will deal with a more complicated model because we already have the ideas we need.

The ideas we want to use are:

1. No two electrons in a system can have exactly the same energy because their waves would interfere.
2. Electrons need to absorb or give up a certain size package of energy to change energy levels.
3. Electrons cannot exist very long between stable orbits.

35.3 THE BAND THEORY

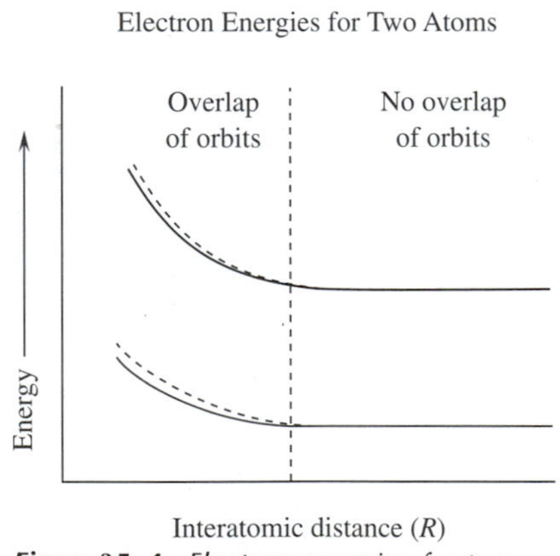

Electron Energies for Two Atoms

Overlap of orbits

No overlap of orbits

Energy

Interatomic distance (R)

***Figure 35–4:** Electron energies for two atoms. When two atoms are brought close together their orbits overlap. The two separate atoms become a single system. The orbital energies split into two closely spaced energies to obey Pauli's exclusion principle.*

The explanation of the behavior of electrons changing orbits is called the **band theory.** When we bring two atoms close together the wave patterns of the electron orbits begin to interact. This is shown in Figure 35–4. The system becomes the two combined atoms rather than two individual atoms. No two electrons can have exactly the same energy level in the combined system. The individual energy levels begin to split into sets of closely spaced energy levels.

If we bring together ten atoms or a thousand atoms or several trillion atoms, the same thing will happen. The system becomes the combination of atoms. Each individual energy level splits into closely spaced **bands** of energy levels (see Figure 35–5). The number of energy levels within a band is proportional to the number of atoms in the system of interacting atoms. If there are a thousand atoms in the collection, then there will be a few thousand closely spaced levels in each band. If there are a trillion atoms in a small crystal then each band will have a few trillion energy levels. Exactly how many energy levels occur depends on the way the atoms stack up in a crystal and on how many suborbits each atomic orbit contains.

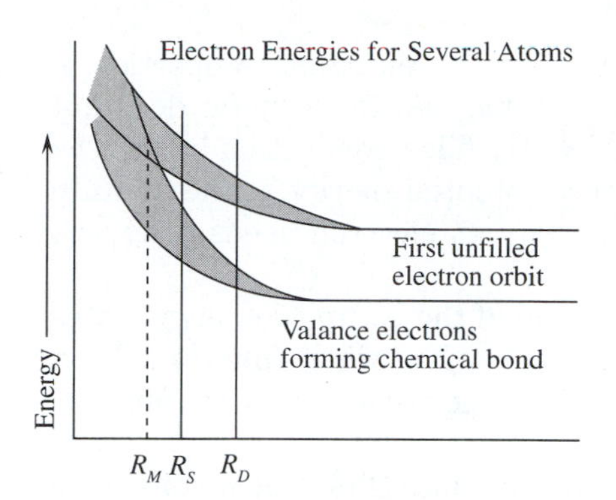

Figure 35–5: *Electron energies for several atoms. When the many atoms that make up a crystal are brought together each overlapping electron orbit broadens into an energy band. The electrical properties of the material depend on the interatomic spacing. Metals (R_M): The band filled with valance electrons overlaps the empty band above it. A partially filled conduction band is created. Semiconductors (R_S): There is a small gap between bands. Dielectrics (R_D): There is a large gap between bands.*

We can sort materials into three categories according to their bands. Look at Figure 35–6. In many materials, there is a band that is completely filled with electrons with an "empty" band above it. The electrons in the filled band are involved with chemical bonds. This filled band is called the **valance band.**

If an electron can get into the empty band above it, the electron can move around from atom to atom by jumping from one closely spaced energy level in the band to another. (The spacing of energy levels within a band is on the order of Planck's constant, or 10^{-34} J). This empty band is called the **conduction band.**

The energy space between the two bands is the **forbidden gap.** Electrons cannot exist at these energy levels. It is like the energy gaps between stable orbits in a single atom. In order for an electron to get into the conduction band it must absorb enough energy to jump the gap.

Materials are categorized into three groups according to the band theory: dielectrics, metals, and semiconductors.

1. **Dielectrics.** In a dielectric, the valance band is completely filled and the conduction band is empty. The forbidden gap is large. At normal temperatures, electrons have a very small chance of absorbing enough thermal energy to jump the gap. Visible light photons do not supply enough energy to electrons to allow them to surmount the gap. The conductivities of these materials are very small.
2. **Metals.** Metals have a partially filled conduction band. The electrons in this band can freely move about in the material.
3. **Semiconductors.** Semiconductors are very similar to dielectrics except the forbidden gap is smaller. At room temperature, a moderate number of electrons gain enough thermal energy to overcome the gap and move into the conduction band. Some semiconductors are designed with a gap energy equivalent to the photon energy of visible light. These semiconductors can emit or absorb visible light.

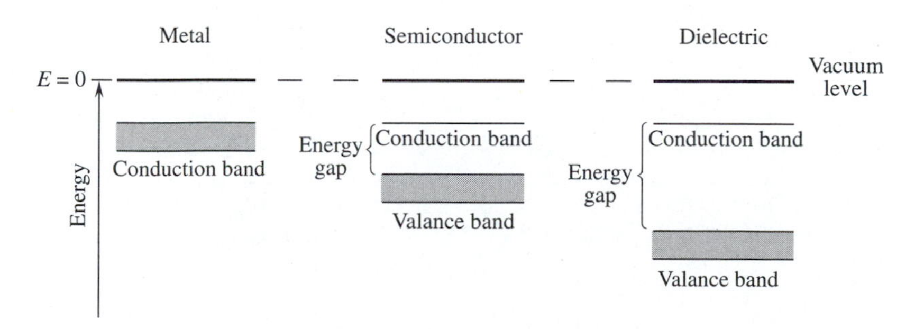

Figure 35–6: *A simplified illustration of energy bands. A metal has free electrons that can move in the conduction band. A semiconductor has all its outer electrons involved with chemical bonds in the valance band. Electrons must gain energy to break free and move into the conduction band. A dielectric is much like a semiconductor except the energy gap between the valance band and the conduction band is larger.*

35.4 ELECTRICAL PROPERTIES OF METALS

The valance band has little effect on the electrical properties of metals. In Figure 35–6 the energy diagram for a metal does not show the valance band, just the partly filled conduction band. The diagram includes energies up to zero potential energy, or the **vacuum level.** The vacuum level is the energy an electron needs to escape from the metal.

The energy difference from the top of the occupied energy states in the band to the vacuum level is called the **work function.** If an electron absorbs energy equal to, or larger than, the work function it can escape from the metal.

The work function is a material constant. Different metals have different work functions. The work functions for some metals are shown in Table 35–1.

If an electron near the surface of a metal is given more energy than the work function, it can escape from the metal. There are three ways this can happen.

1. **Thermionic emission.** If the metal is heated to a high temperature, some of the electrons will gain enough thermal kinetic energy to overcome the work function and escape. This is the process that occurs in the filament of a cathode ray tube.
2. **Secondary electron emission.** A high-energy electron called a primary electron can strike the surface of a metal and knock loose additional secondary electrons in a collision process. The primary electron collides with several electrons, giving each enough energy to overcome the work function, much as a bowling ball striking bowling pins knocks them down.
3. **Photoelectron emission.** Photons can collide with individual electrons. When the surface of a metal is irradiated with photons of

Table 35–1: Work function for selected metals.

METAL	WORK FUNCTION (eV)
Aluminum (Al)	4.08
Barium (Ba)	2.48
Calcium (Ca)	2.71
Cesium (Ce)	1.9
Magnesium (Mg)	3.68
Nickel (Ni)	5.01
Potassium (K)	2.24
Sodium (Na)	2.28

energy hf, all of the photon's energy can be absorbed by a single electron in a collision-like process. Part of the radiant energy goes into overcoming the work function (W). The surplus energy shows up as the kinetic energy of the photoemitted electron.

$$hf = KE + W \qquad \textbf{(Eq. 35–1)}$$

☐ **EXAMPLE PROBLEM 35–3: PHOTOEMISSION**

Ultraviolet radiation with a wavelength of 357 nm strikes the surface of barium. Calculate the kinetic energy of the emitted photoelectrons. The work function of barium is 2.48 eV.

■ ***SOLUTION***

First find the radiant energy.

$$\lambda f = c$$

then

$$E = hf = \frac{hc}{\lambda}$$

$$E = \frac{6.626 \times 10^{-34}\ \text{J} \cdot s \times 3.00 \times 10^{8}\ \text{m/s}}{3.57 \times 10^{-7}\ \text{m}}$$

$$E = 5.57 \times 10^{-19}\ \text{J}$$

In terms of electron volts this energy is:

$$V = \frac{E}{e} = \frac{5.57 \times 10^{-19}\ \text{J}}{1.60 \times 10^{-19}\ \text{C}} = 3.48\ \text{eV}$$

The kinetic energy is the difference between the radiant energy and the work function.

$$KE = V - W$$

$$KE = 3.48\ \text{eV} - 2.48\ \text{eV}$$

$$KE = 1.00\ \text{eV}$$

A **photomultiplier tube** uses both the photoelectric effect and secondary emission (see Figure 35–7). An evacuated tube has a window. Light passing through the window strikes a material that has a small work function. Photoemitted electrons are accelerated

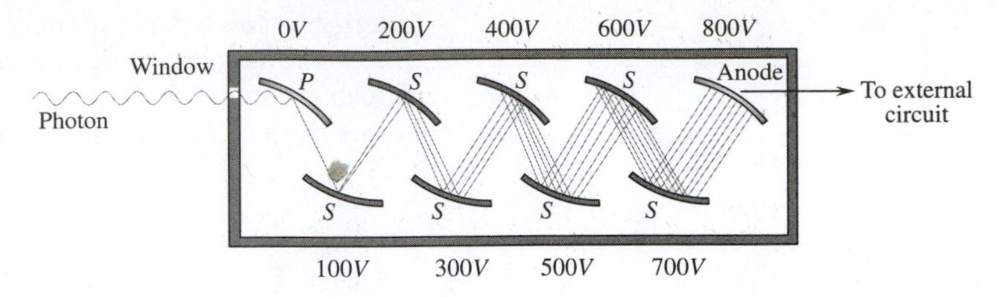

Figure 35–7: A photomultiplier. Photons enter a window at the end of an evacuated tube and strike a photoelectric material (P). Photoelectric electrons are ejected. These electrons successively gain energy and strike secondary electron emitters (S). Additional electrons are ejected with each collision.

as they fall through a high voltage. Then they strike a parabolic surface causing secondary emission. The pulse is further multiplied by additional accelerations to more secondary emitting surfaces until an expanded pulse of electrons reaches the cathode.

35.5 INTRINSIC SEMICONDUCTORS: CONDUCTIVITY

Semiconductors and dielectrics have a similar band structure. At very low temperatures, the valance band is filled and the conduction band is empty. The size of the energy gap determines the electrical properties.

The semiconductor has a smaller energy gap. At room temperature, electrons gain enough thermal energy to jump the gap and enter into the conduction band. This means they gain enough energy to break loose from a chemical bond and roam about in the crystal lattice. They become free electrons subject to electric fields, much like an electron in a metal (see Figure 35–8).

Because dielectrics have a larger energy gap, only a few electrons gain enough thermal energy to break free at room temperature. They are poor electrical conductors.

Conductivity is proportional to the number of free electrons. This is the number of electrons jumping the forbidden gap in the band model. This number can be calculated statistically.

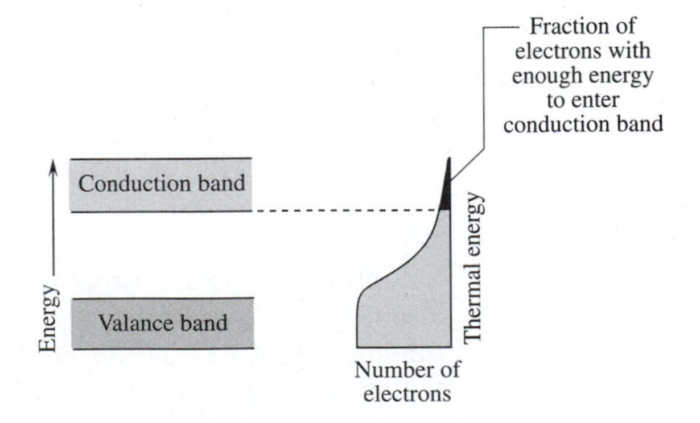

Figure 35–8: Only a small fraction of the electrons gain enough thermal energy to jump the gap between the valance band and the conduction band in a semiconductor.

Table 35–2: Semiconductor energy gap for selected materials, at room temperature.

MATERIAL	ENERGY GAP (eV)
Covalent	
Germanium (Ge)	0.67
Silicon (Si)	1.11
Ionic	
Aluminum Arsenide (AlAs)	2.2
Cadmium Sulfide (CdS)	2.42
Gallium Phosophide (GaP)	2.24
Indium Arsenide (InAs)	0.36
Magnesium Stannide (Mg_2Sn)	0.21
Lead Selenide (PbSe)	0.26

Let n_n be the number of electrons per cubic meter (electrons/m^3) entering the conduction band. The number of electrons jumping the gap will depend on the thermal energy (kT). (Remember, k is Boltzmann's constant, k = 1.381×10^{-23} J/K, or 8.61×10^{-5} eV/K.)

The forbidden energy band width is called the **Fermi energy** (E_f). The Fermi energy level has some interesting properties.

1. At any temperature above absolute zero the statistical probability of the Fermi energy level being filled is one-half.
2. For most practical purposes, E_f is half the energy of the forbidden gap an electron must jump.
3. When two materials are placed in contact, electrons flow from one material to the other until the Fermi energy levels of the two materials are the same.

Table 35–2 shows the forbidden gap energies for some pure, or **intrinsic, semiconductors.** Impurities can alter the electrical properties. We will look at other ways of activating electrons in the next section.

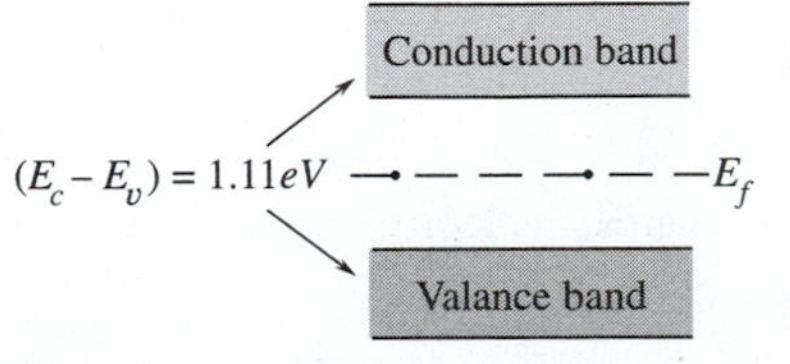

Figure 35–9: *Diagram for Example Problem 35–4. Energy gap for silicon.*

□ **EXAMPLE PROBLEM 35–4: FERMI ENERGY**

How far is the Fermi energy level above the top of the valance band in silicon? See Figure 35–9.

■ ***SOLUTION***

The Fermi energy for an intrinsic material is half the energy gap obtained from Table 35–2.

$$E_f = \frac{(E_c - E_v)}{2}$$

$$E_f = \frac{1.1 \text{ eV}}{2}$$

$$E_f = 0.55 \text{ eV}$$

The conductivity of conduction electrons (σ_n) depends on the density of electrons in the conduction band (n_n), the electronic charge (e), and the mobility of the electrons (μ_n). The **mobility** is a measure of the charge carrier's ability to move through the lattice under the influence of an electric field.

$$\sigma_n = n_n \, e \, \mu_n$$

The density of electrons in the conduction band is exponentially dependent on temperature.

$$n_n = 2.52 \times 10^{25} \, e^{-E_f/kT} \qquad \textbf{(Eq. 35–2)}$$

☐ **EXAMPLE PROBLEM 35–5: CONDUCTIVITY OF GERMANIUM**

Germanium has a mobility of 0.36 $m^2/V \cdot s$ for conduction electrons. Calculate the following quantities for a temperature of 300 K.

A. Height of the Fermi energy level above the valance band.
B. Density of conduction electrons.
C. Conductivity of conduction electrons.

■ ***SOLUTION***

A.

$$E_f = \frac{(E_c - E_v)}{2}$$

$$E_f = \frac{0.67 \text{ eV}}{2}$$

$$E_f = 0.335 \text{ eV}$$

B.

$$n_n = 2.52 \times 10^{25} \, e^{-E_f/kT}$$

$$n_n = 2.52 \times 10^{25} \, e^{-0.335 \text{ eV}/(8.61 \times 10^{-5} \text{eV/K} \times 300 \text{ K})}$$

$$n_n = 2.52 \times 10^{25} \, e^{-12.97}$$

$$n_n = 5.87 \times 10^{19} \text{ electrons/m}^3$$

C.

$$\sigma_n = n_n\, e\, \mu_n$$

$$\sigma_n = (5.87 \times 10^{19}\ \text{electrons/m}^3) \times (1.6 \times 10^{-19}\ \text{C}) \times 0.36\ \text{m}^2/\text{V} \cdot \text{s}$$

$$\sigma_n = 3.38\ (\Omega \cdot \text{m})^{-1}$$

So far we have ignored what is happening in the valance band. Electrical current can flow here once we have removed some of the bonding electrons (see Figure 35–10). When an electron is removed from a chemical bond and allowed to roam about the lattice, an electron from a neighboring atom can easily move into the hole left behind. We still have a hole, but it has moved. An electron from another atom can fill it. If an electric field is present, the hole will move in the direction of the field, much as a positive charge carrier does.

The electrons are playing a game of musical chairs with the hole. Mathematically it is much easier to describe the behavior of the hole than of the several electrons that keep occupying the hole and moving it along.

The **hole,** or absence of an electron in the valance band, behaves much like a positive electron. It is called a **positive charge carrier.** It has an electronic charge of $+e$.

In an intrinsic semiconductor there are exactly as many holes as there are conduction electrons ($n_p = n_n$).

The process that moves a hole through the valance band is slower than the process that moves a free electron through the conduction band. That is, holes have a smaller mobility than electrons (μ_p). Table 35–3 lists mobilities for electrons and holes for some materials.

The conductivity of holes in the valance band looks much like the conductivity of free electrons.

$$\sigma_p = n_p(e^+)\mu_p$$

The total conductivity of the semiconductor is the combined conductivity of conduction electrons and holes.

$$\sigma = [n_p(e^+)\,\mu_p] + [n_n(e^-)\,\mu_n] \qquad \textbf{(Eq. 35–3)}$$

Figure 35–10: *A neighboring electron replaces the missing electron in the incomplete chemical bond. The "hole" moves in the opposite direction to the electrons.*

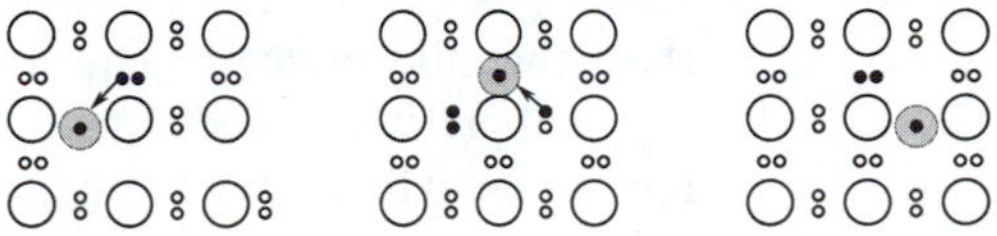

Table 35–3: Mobilities of selected materials, at 300 K.

MATERIAL	MOBILITY OF ELECTRONS ($m^2/V \cdot s$)	MOBILITY OF HOLES ($m^2/V \cdot s$)
Aluminum Arsenide (AlAs)	0.12	0.042
Cadmium Sulfide (CdS)	0.040	—*
Gallium Phosophide (GaP)	0.02	0.012
Germanium (Ge)	0.36	0.23
Indium Arsenide (InAs)	3.26	0.026
Magnesium Germanide (Mg_2Ge)	0.040	0.007
Lead Selenide (PbSe)	0.15	0.15
Silicon (Si)	0.19	0.043

* Not available

□ **EXAMPLE PROBLEM 35–6: SEMICONDUCTOR CONDUCTIVITY**

Find the total conductivity of germanium at 300 K.

■ ***SOLUTION***

We already have the conductivity of conduction electrons from Example Problem 35–5. We need to add the conductivity of holes to this quantity.

$$n_p = n_n$$

$$n_p = 5.87 \times 10^{19} \text{ electrons/m}^3$$

$$\sigma_p = n_p e^+ \mu_p$$

$$\sigma_p = (5.87 \times 10^{19} \text{ e/m}^3) \times (1.6 \times 10^{-19} \text{ C}) \times (0.23 \text{ m}^2/\text{V} \cdot \text{s})$$

$$\sigma_p = 2.16 \, (\Omega \cdot \text{m})^{-1}$$

$$\sigma = \sigma_p + \sigma_n$$

$$\sigma = (2.16 + 3.38) \, (\Omega \cdot \text{m})^{-1}$$

$$\sigma = 5.5 \text{ A} \, (\Omega \cdot \text{m})^{-1}$$

35.6 LUMINESCENCE

Electrons can be activated in a semiconductor by processes other than thermal energy. The kinetic energy from a beam of cathode rays can activate electrons in a material. This process occurs on the face of a cathode ray tube.

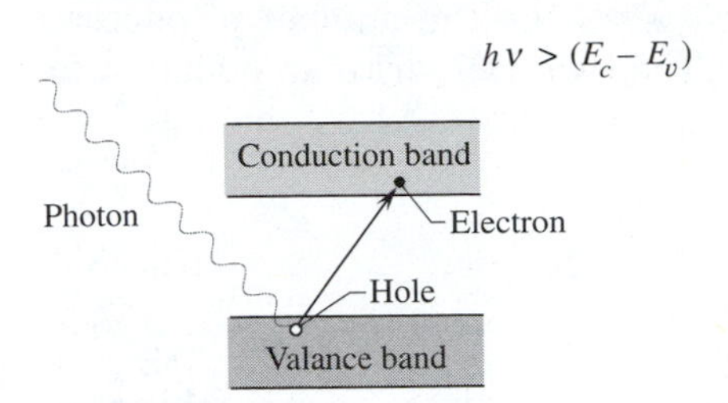

Figure 35–11: A valance electron can absorb a photon to create a free electron-hole pair.

Photoelectric current can be created by radiant energy. If the radiant energy is large enough, an electron can absorb the energy, enabling it to jump the forbidden gap and enter into the conduction band. Some semiconductors have forbidden gap energies in the range of visible light. When light falls on the surface of the material, the conductivity of the material increases in direct proportion to the light intensity (see Figure 35–11).

An important process that occurs in semiconductors is **electron-hole recombination.** We have seen that when an electron is activated, it breaks free from its chemical bond and roams through the lattice of atoms making up the semiconductor. Eventually the electron will encounter a hole in the lattice, an incomplete chemical bond with a missing electron. The free electron can go into the chemical bond by giving up energy. It can reduce its energy by emitting radiant energy or by transferring mechanical energy to the lattice to increase the lattice's thermal vibration.

The net result is that the free electron leaves the conduction band to complete a chemical bond. A free electron disappears from the conduction band and a hole disappears from the valance band. The electron and the hole have recombined.

For thermally activated electrons, there is a balance at any temperature between the number of electrons that are being activated each second and the number of electrons that are recombining with holes. This is much like the water vapor in a sealed bottle of water. The number of water molecules evaporating from the water is matched by the number of water molecules striking the water's surface and sticking to it (see Figure 35–12).

Things can be a bit different for other kinds of electron activation. When cathode rays or light rays activate electrons, more electrons are added to the conduction band than would normally be activated by thermal energy. When the cathode ray or light source is turned off, these added electrons will eventually recombine. The number of electrons in the conduction band is reduced to the thermal equilibrium level.

It takes time for the added free electrons to find a hole for recombination. The amount of time needed for an electron to find a hole involves probability. Let the total number of electrons plus

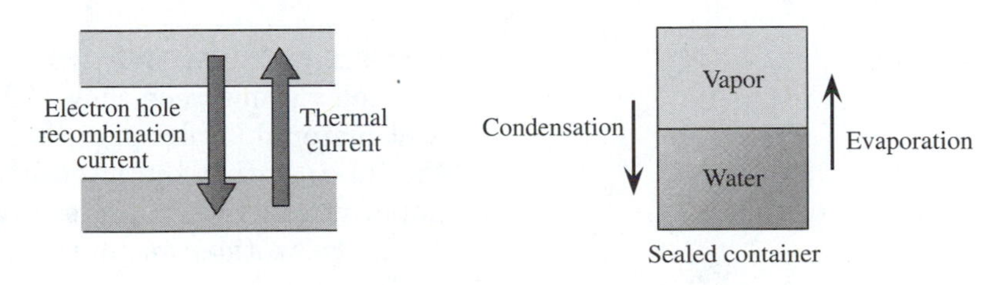

Figure 35–12: There is a thermodynamic balance between thermally activated electrons and electron-hole recombination. This is much like the evaporation and condensation balance of vapor over water in a sealed container.

holes that recombine be N, and let N_0 be the number of photoactivated electrons at time zero. Then we can write an equation for recombination of nonthermal free electrons.

$$N = N_0 e^{-t/\tau} \qquad \textbf{(Eq. 35–4)}$$

The constant τ is called the **relaxation time,** or **recombination time.** If τ is large it will take longer for the electrons to find a suitable recombination hole (see Figure 35–13).

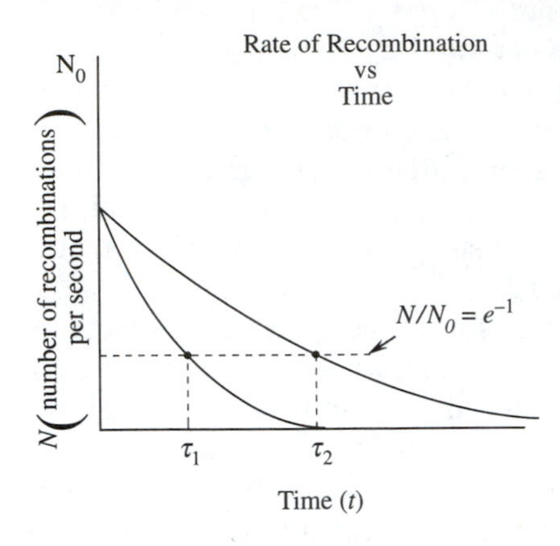

Figure 35–13: The rate of recombination versus time for photoelectrons is an exponential decay curve. The relaxation time, or recombination time (τ), is the time needed for the rate to fall to 1/e, or 0.37, of its initial value.

Often Equation 35–4 is written as a logarithmic equation.

$$\ln\left(\frac{N_0}{N}\right) = \frac{t}{\tau} \qquad \textbf{(Eq. 35–5)}$$

If the free electrons emit radiation as they recombine, the intensity of the radiation will be proportional to the number of recombinations per unit time. If I_0 is the intensity of the radiation emitted at the instant the light source is turned off, then

$$\ln\left(\frac{I_0}{I}\right) = \frac{t}{\tau} \qquad \textbf{(Eq. 35–6)}$$

The light emitted by this process is called **luminescence.**

If the relaxation time is very short ($\tau < 10^{-5}$ s), the process is referred to as **fluorescence.** Fluorescent lights have short recombination times. Ultraviolet (UV) light activates electrons in the coating inside the tube. The electrons rapidly recombine, emitting a longer visible wavelength of light.

Luminescence with longer relaxation times is called **phosphorescence.** The light emission may persist for only a small fraction of a second or for several minutes or hours, depending on the recombination time.

☐ **EXAMPLE PROBLEM 34–7: THE TELEVISION PICTURE**

A television screen shows 30 pictures per second. A phosphorescent material (a phosphor) is chosen as a coating inside the face of the tube, so that the intensity of a picture dot falls to 15% of its initial intensity when it is reactivated by the next sweep. What should be the relaxation time of the phosphor?

■ *SOLUTION*

$$\ln\left(\frac{I_0}{I}\right) = \frac{t}{\tau}$$

$$\tau = \frac{t}{\ln(I_0/I)}$$

$$\tau = \frac{(1/30)\ \text{s}}{\ln(1.00/0.15)}$$

$$\tau = 0.018\ \text{s}$$

Color Television and the Eye

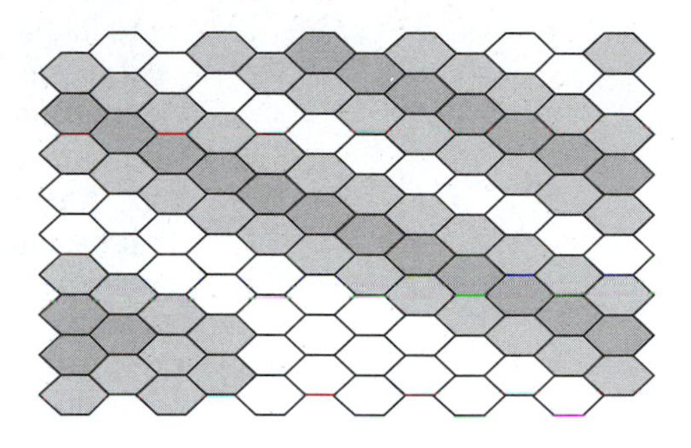

The inside of the face of a television screen is covered with a phosphor. A beam of electrons strikes the phosphor, causing it to glow. The beam passes horizontally across the screen. As the beam moves, its intensity varies, giving the shades of light and darkness that you see. Once the beam completes its path across the screen, it returns to the other side and begins another trace a little lower on the screen. The phosphor must have a relaxation time long enough for the image to persist in our minds, but fast enough for the beam to have a greatly reduced intensity by the time it returns to trace out another image.

Color television has groups of colored phosphors. Three colors are used—red, blue, and green. Separate beams activate the different phosphorescent dots. The activation of these different dots allows the full spectrum of the rainbow to be sensed by your eye. Actually, the television tricks you. Here's how.

The lining of your eye has two kinds of light-sensitive cells. One type is cone-shaped and is sensitive to color. These cones can be divided into three groups. Each group is sensitive to a different area of the spectrum. One type of cone best detects blue. Another has a sensitivity that peaks at green, and the last is most sensitive to red light. These are the same colors found on the television screen. A red cone is activated by long wavelengths such as red, orange, and yellow; it senses red best. A green cone responds to green and yellow and weakly to orange or blue. The blue cones respond to short wavelengths—blue and violet.

If yellow light strikes the lining of your eye, both green and red cones will be activated by different amounts. The wavelength of yellow is in the range of overlap of the two types of cones. A signal goes to your brain and you see yellow.

A television set does not emit yellow light, but you can see yellow on the screen anyway. Two separate colored beams, one red and one green, hit the red and green cones. The red light activates the red cones. Green light triggers the green cones. If the relative intensities of the red and green beams have the correct ratio of intensities, the same signal goes to the brain that occurred when yellow light struck the two kinds of cones. Voila! You see yellow!

35.7 EXTRINSIC SEMICONDUCTORS

In intrinsic semiconductors, the number of free electrons in the conduction band is the same as the number of holes in the valance band. We can add extra electrons or extra holes by a process called **doping.** This will give us a material with more of one kind of charge carrier. Doped semiconductors are called **extrinsic semiconductors.**

Silicon and germanium are the best naturally occurring semiconductors. They are both from period IV of the periodic table. We can form a lattice with a small amount of some other element.

We can choose atoms from period V, such atoms as arsenic (As), antimony (Sb), or bismuth (Bi). When these period-V atoms go into the lattice, four of these electrons complete chemical bonds with the surrounding period-IV atoms (see Figure 35–14). The fifth electron is left without a chemical bond with a neighbor. It is loosely connected to its parent atom. Only a small amount of energy is needed to break it loose. This creates a free electron in the conduction band without creating a corresponding hole in the valance band. The period-V atoms are called **donors**; they donate electrons.

A semiconductor with donor states has more free electrons than it has holes. It is called an **N-type** semiconductor because it has more negative charge carriers than positive charge carriers.

Notice in Figure 35–14 that the Fermi energy level is halfway between the donor states and the conduction band. If E_d is the energy of the donor states and E_c is the energy of the bottom of the conduction band, then the number of activated electrons at temperature T is:

$$n = n_0 e^{-\Delta E_f/kT}$$

where

$$E_f = (E_c - E_d)^{1/2}$$

The equation is the same for intrinsic semiconductors. Only the size of the energy gap is smaller.

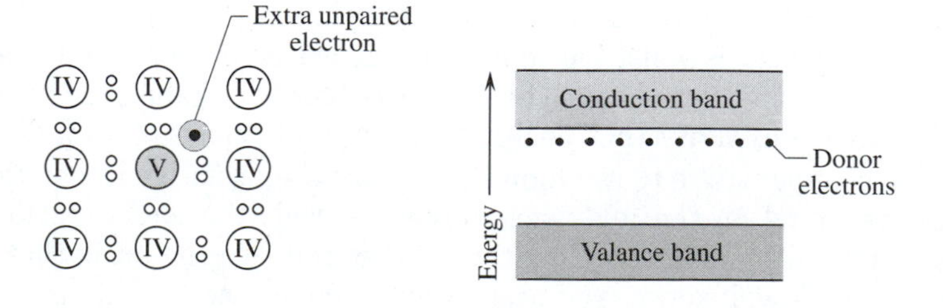

Figure 35–14: Donor states. A period-V impurity in a period-IV lattice has an extra electron that cannot enter into a covalent bond with neighboring atoms. It easily breaks free and enters the conduction band.

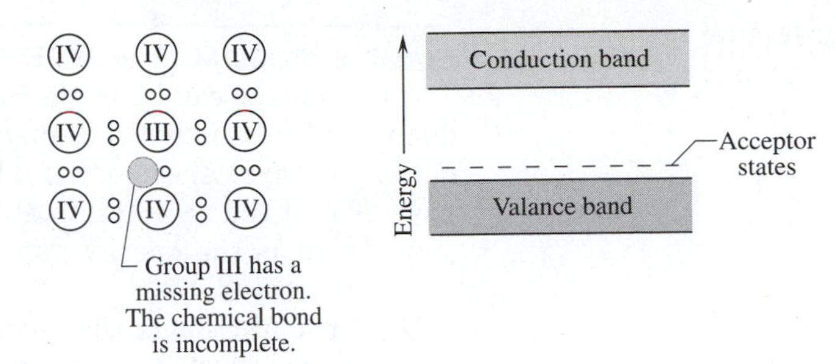

Figure 35–15: Acceptor states. A group-III impurity lacks one electron to complete covalent bonding with its four nearest neighbors. An electron from a nearby atom can complete the bond. Holes are injected into the valance band without creating free electrons.

We can make a **P-type** semiconductor, one with more holes than free electrons, by adding impurities from group III of the periodic table. Gallium (Ga), indium (In), and thallium (Tl) are the most common ones used.

Look at Figure 35–15. The group-III impurity atom lacks one electron to complete a chemical bond with its neighboring group-IV atoms. This creates an acceptor state. The site will accept an electron from a neighbor to complete the bond. The acceptor has an energy a bit higher than the rest of the bonding electrons in the valance band. Once an electron enters the acceptor state, a hole is produced in the valance band without creating a free electron in the conduction band. We have a material with more positive carriers than negative carriers.

Notice that a P-type material has the Fermi energy level between the valance band (E_v) and the acceptor band (E_a).

☐ **EXAMPLE PROBLEM 35–8: DONOR STATES**

Silicon is doped with a group-V element, creating donor states 0.03 eV below the conduction band. What percentage of the donor states are activated at 30$\bar{0}$ K?

■ ***SOLUTION***

$$E_f = \frac{0.15 \text{ eV}}{2}$$

$$\% = \frac{N}{N_0} \times 100 = 100\, e^{-\Delta E_f/kT}$$

$$\% = 100 \times e^{-[0.15 \text{ eV}/(8.61 \times 10^{-5}\text{eV/K} \times 300 \text{ K})]}$$

$$\% = 100 \times e^{-5.81}$$

$$\% = 0.30\%$$

EXAMPLE PROBLEM 35–9: ACCEPTOR STATES

A thermistor can be made from a semiconductor so that its conductivity (or resistivity) changes rapidly through the design range of temperature. Assume that a P-type semiconductor is designed to have 20% of the acceptor states filled at $30\bar{0}$ K.

A. What is the energy gap between the valance band and the acceptor states?

B. What fraction of the donor states will be filled at 350 K?

C. The thermistor has a conductivity of 0.96 $(\Omega \cdot \text{m})^{-1}$ at 300 K. What is its conductivity at 350 K?

SOLUTION

A.

$$\frac{N}{N_0} = 0.20 = e^{-\Delta E_f/kT}$$

Take the natural log of the equation and solve for E_f.

$$\ln(0.20) = -\Delta E_f/kT$$

$$\Delta E_f = -[\ln(0.20) \times kT]$$

$$\Delta E_f = -[-1.61 \times (8.61 \times 10^{-5}\ \text{eV/K}) \times (30\bar{0}\ \text{K})]$$

$$\Delta E_f = 0.0415\ \text{eV}$$

Twice ΔE_f gives us the energy gap.

$$\Delta E = E_a - E_v = 2\,\Delta E_f$$

$$E = 0.083\ \text{eV}$$

B.

$$\frac{N}{N_0} = e^{-\Delta E_f/kT}$$

$$\frac{N}{N_0} = e^{-[0.0415\ \text{eV}/(8.61 \times 10^{-5}\text{eV/K} \times 35\bar{0}\ \text{K})]}$$

$$\frac{N}{N_0} = 0.252$$

C. The conductivity is directly proportional to activated holes.

$$\frac{\sigma_2}{\sigma_1} = \frac{N_2}{N_1} = \frac{(N_2/N_0)}{(N_1/N_0)}$$

$$\sigma_2 = 0.96\ (\Omega \cdot \text{m})^{-1} \times \frac{0.252}{0.200}$$

$$\sigma_2 = 1.21\ (\Omega \cdot \text{m})^{-1}$$

35.8 SEMICONDUCTOR DEVICES

Diodes

Look at Figure 35–16. A P-type semiconductor and an N-type semiconductor are fused together, forming a **diode.** Free electrons in the conduction band of the N-type material will diffuse across the junction of the two materials. Holes in the P-type material will diffuse in the opposite direction in the valance band.

The conduction band in the N-type material will be partially depleted of electrons. This will show up as a plus charge at the junction. Coulomb attraction will maintain the diffused electron in the P-type material near the junction. A similar process occurs with holes in the valance band.

An electrical potential builds up across the junction, called a **P-N junction.** The assembly is called a **semiconductor diode.**

An electric field has built up across the P-N junction.

Heat activates electrons, moving them from the valance band into the conduction band. Thermally activated electrons from the P region move across the junction into the N region under the influence of the field (see Figure 35–17). The moving electrons are called the **thermal current.** The thermal current is proportional to the number

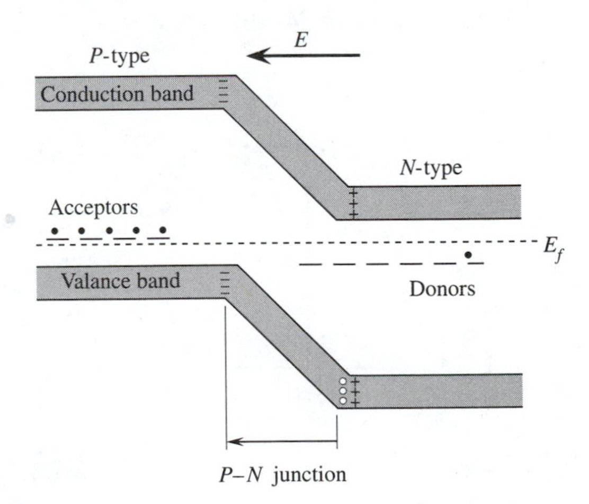

Figure 35–16: *At a P-N junction, holes and electrons diffuse across the junction to create an electric field in a semiconductor.*

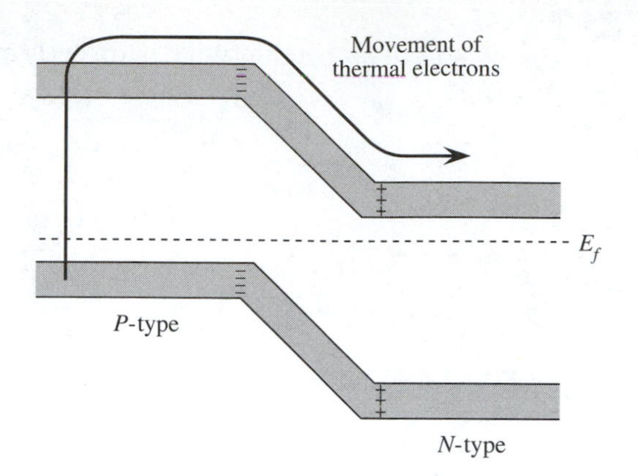

Figure 35–17: Thermally activated electrons can move easily across the junction.

of thermally activated electrons (N_1). If n is a proportionality constant then:

$$N_1 = n\, e^{-\Delta E_f/kT}$$

In the meantime, another electron current moves in the opposite direction. The N region has many more electrons in the conduction band. Some of these have enough energy to overcome the potential barrier. They enter the P region and recombine with holes in the valance band. This group is called the **recombination current** (see Figure 35–18). The recombination current is proportional to the number of recombining electrons (N_2).

If there is no external voltage across the diode caused by a battery or other source, the recombination current and the thermal current balance to give a net current of zero. That is, $N_1 = N_2$.

Let us put a reverse bias on the diode. We will hook up a battery so that the energy of the P-N junction is increased by $e\Delta V$ (see Figure 34–19).

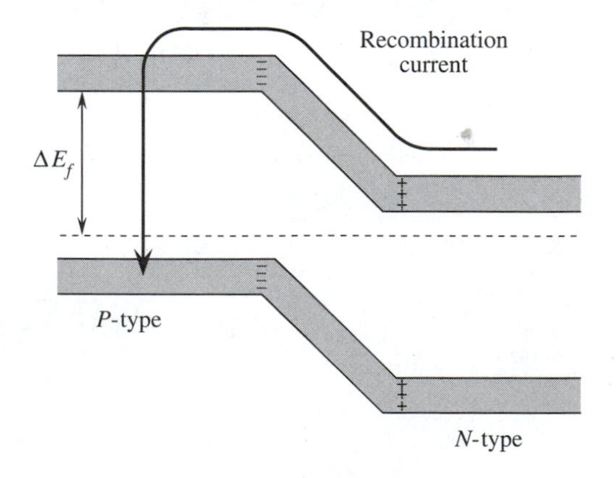

Figure 35–18: A number of recombination electrons from the electron-rich N-type material can overcome the barrier.

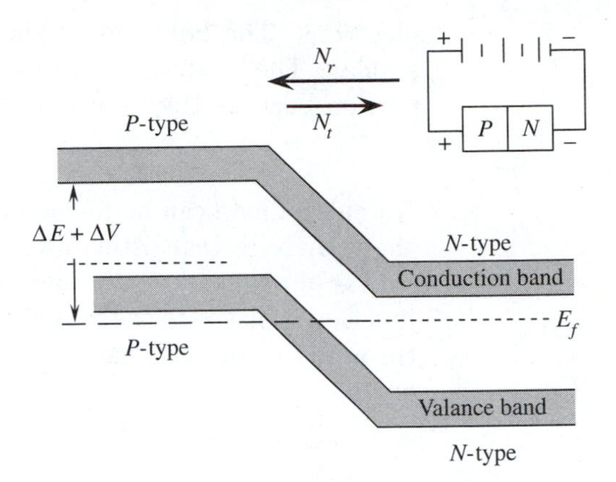

Figure 35–19: A reverse bias increases the potential barrier at the junction. The number of recombination electrons crossing the potential barrier is decreased. The flow of thermal electrons is unaffected. A small net reverse current results.

The thermal current from the P side of the junction is not affected. Electrons move from the P region with the same ease as before. The number of thermal electrons is still $N_1 = n\,e^{-\Delta E_f/kT}$.

The situation is different for the recombination electrons. They must overcome a larger potential barrier. Fewer electrons will have enough energy to get over the larger barrier. N_2 is reduced to:

$$N_2 = n\,e^{-(\Delta E_f + eV)/kT}$$

The net flow of electrons is no longer zero. It is proportional to the difference of the thermal current and the recombination current.

If we forward bias the diode, we get the situation shown in Figure 35–20. Again, the thermal current is unaffected. The recombination electrons have a smaller potential barrier to overcome. A larger number of electrons have enough energy to overcome the barrier.

$$N_2 = n\,e^{-(\Delta E_f - eV)/kT}$$

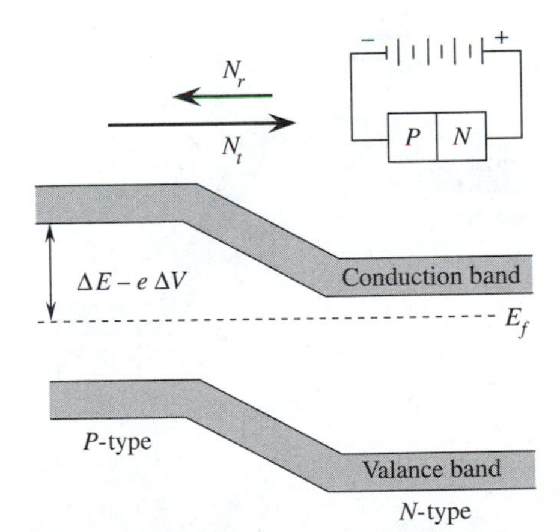

Figure 35–20: A forward bias reduces the barrier. A much larger number of recombination electrons can surmount the potential barrier. A large net forward current results.

The net current is the difference between the thermal and recombination currents.

$$I(\text{net}) = N_2 - N_1 = (n\,e^{-\Delta E_f/kT})(e^{e\Delta V/kT} - 1)$$

The first term in parentheses is independent on the external voltage. We will denote this as I_0. The net current is:

$$I(\text{net}) = I_0\,(e^{e\Delta V/kT} - 1) \qquad \textbf{(Eq. 35–7)}$$

Figure 35–21 is a plot of Equation 35–7. Notice the difference in scale between the forward and reverse bias. Current in the reverse direction is very small.

In our analysis of the diode we concentrated on the behavior of electrons. The behavior of the holes in the valance band is very similar. The main difference is that they move in the opposite direction because they are positive charges.

Transistors

A double diode can be formed to create a **transistor.** Figure 35–22 shows an NPN transistor. Two N regions are separated by a P region.

One junction is usually biased in the forward direction. The other junction is biased in the reverse direction. Electrons can easily move through the forward bias and then fall down the reverse bias, gaining energy.

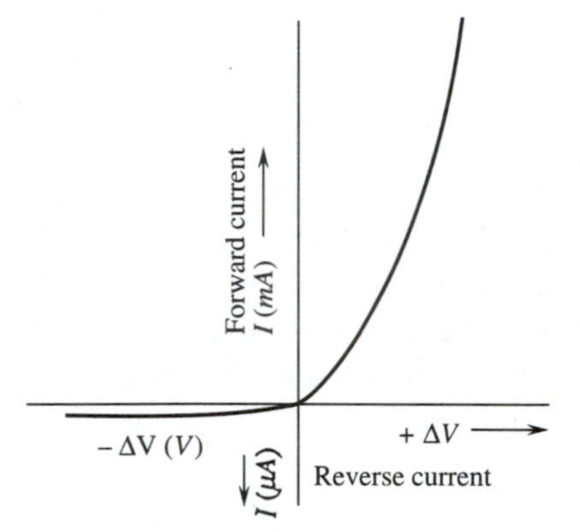

Figure 35–21: *Current versus voltage for a P-N junction. The forward current increases exponentially with increased electrical potential. The reverse current is very small.*

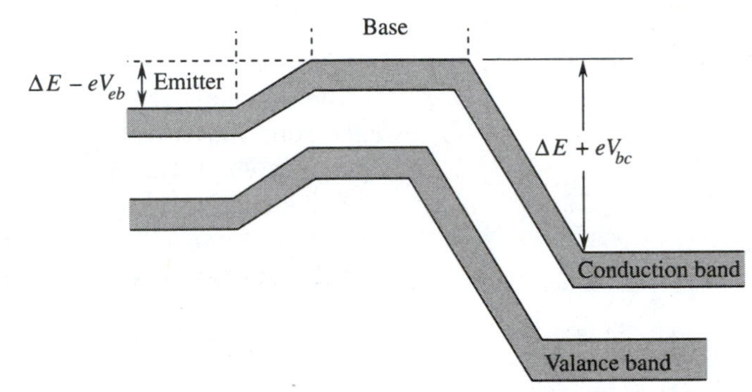

Figure 35–22: *A transistor has two P-N junctions. Potential V_{eb} controls the recombination electron current passing from the emitter into the base. Potential V_{bc} increases the energy of electrons flowing through the base into the collector.*

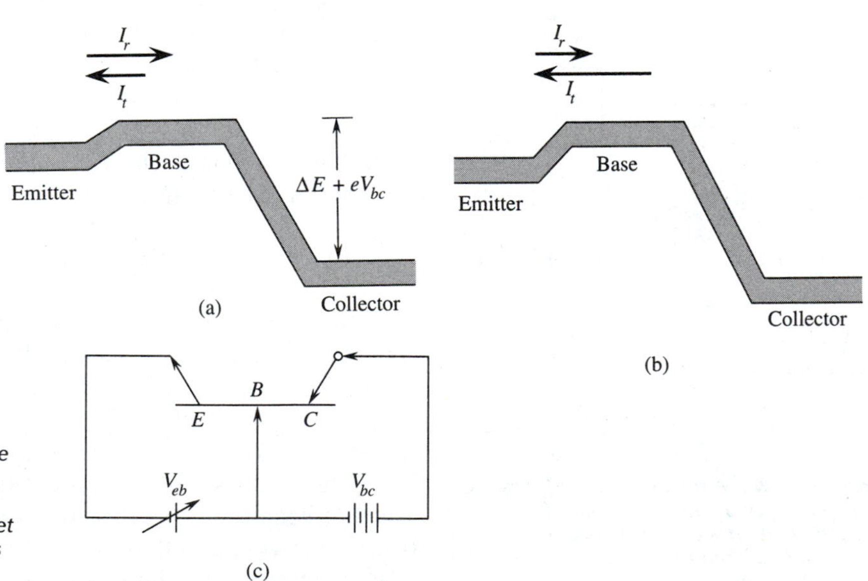

Figure 35–23: (a) *A small potential difference between the emitter and the base creates a large net current.* (b) *A large potential difference between the emitter and the base creates a small net current.* (c) *The electrical circuit shows how an NPN transistor is hooked up.*

The three parts of the transistor are:

1. The *emitter* acts as the source of recombination electrons.
2. The *base* is the central region. A varying voltage signal to be amplified is introduced into the base.
3. The *collector* collects the thermal current passing through the base from the emitter.

Figure 35–23 shows what happens if we vary the voltage of the base. The percentage change of the emitter-base junction voltage is large. This causes large changes in the recombination current from the emitter. There is only a small percentage change in the base-collector junction voltage. The energy gain of electrons passing from the base to the collector stays fairly constant.

SUMMARY

The electronic behavior of solids depends on the way atoms arrange themselves in a material. There are three major types of bonding between atoms: metallic bonding, covalent bonding, and ionic bonding.

The energy of individual electrons is small. A convenient unit for electron energies is the electron volt (eV). This is the amount of energy a single electronic charge gains when it falls through a potential difference of 1 V.

$$1 \text{ eV} = 1.602 \times 10^{-19} \text{ J}$$

No two electrons in a system can have exactly the same energy. This rule is called the Pauli exclusion principle.

No two electrons in an atom can have exactly the same orbit (that is, the same energy).

No two electrons with overlapping orbits in a solid can have exactly the same energy. This causes overlapping orbits to split into bands.

Electrons involved in chemical bonding are called valance electrons. Their energy band is called the valance band. Electrons that have broken free of their chemical bonds and roam freely through the crystal are called conduction electrons. They have energies in the conduction band. The energy gap between the conduction band and the valance band is called the forbidden energy gap. Electrons cannot exist with energies in this range.

- Dielectrics have a large forbidden gap.
- Semiconductors have a small forbidden gap.
- Metals have a partially filled conduction band.

Metals

The work function (W) of a metal is the minimum energy needed to remove an electron from the surface of the metal. Electrons can escape from a metal surface by three major processes.

- Thermionic emission: Electrons in a hot metal gain enough energy to overcome the work function.
- Secondary emission: An ion strikes the surface of the metal. Electrons gain enough energy from the collision to be ejected.
- Photoelectric effect: A single electron completely absorbs the radiant energy of a photon striking the surface of the metal. It is ejected with kinetic energy (KE).

$$hf = KE + W$$

Semiconductors

A semiconductor composed of a pure substance is called an intrinsic semiconductor. At room temperature some of the electrons in the valance band gain enough energy to break their chemical bonds and become free electrons in the conduction band. These free electrons can participate in electrical current. The holes left in the valance band behave like positive charges. They can move around in the crystal. They also are involved in electrical current.

Free electrons move more easily than holes. A measure of the ability of a hole or an electron to move through a crystal under the influence of an electric field is called mobility (μ).

The density of free electrons in the conduction band is n_n. The density of holes in the valance band is n_p. In an intrinsic semiconductor $n_p = n_n$. The density of thermally activated electrons is given by:

$$n_n = 2.52 \times 10^{25}\, e^{-\Delta E_f/kT}$$

where ΔE_f is half the forbidden gap energy, k is Boltzmann's constant, and T is the absolute temperature.

The conductivity of a semiconductor is the combined conductivity of holes and free electrons.

$$\sigma = (n_n\, e\, \mu_n) + (n_p\, e\, \mu_p)$$

Electrons and holes are continually recombining in a semiconductor. In the absence of an external field there is a balance between the number of thermally activated electrons and the number of electron-hole recombinations.

Electrons can be activated by nonthermal processes. Cathode rays or photons can also give valance electrons enough energy to move into the conduction band. When the source is turned off, electrons activated by the process recombine with holes. If the energy gap is in the range of visible light, light will be emitted during recombination. This is called luminescence. Light emission will occur for a period of time after the source is turned off. For an initial light intensity of I_0:

$$\ln\left(\frac{I_0}{I}\right) = \frac{t}{\tau}$$

τ is called the relaxation time, or the recombination time.

Covalent semiconductors (group-IV elements) can be easily doped with impurities.

If group-V impurities are added, donor states are created near the conduction band. These produce extra electrons in the conduction band and electrons become the majority charge carrier in the material. This is an N-type semiconductor.

If group-III elements are used for doping, acceptor states are produced just above the valance band. This creates added holes in the valance band. Holes become the majority charge carrier in the material. This is a P-type semiconductor.

When a P-type and an N-type material are fused together an electric field is produced across the junction of the two materials. This is called a P-N junction or diode. Electrical current moves easily in the forward direction of a diode. When it is biased in the reverse direction, the current is very small.

A transistor has two P-N junctions. The central region is called the base. It controls current flow by voltage fluctuations. One side of the transistor is a source of electrons. It is called the emitter. The other side is called the collector. As electrons move from the base to the collector they gain energy.

KEY TERMS

If you can explain the following terms to a friend or classmate, you understand their meaning. If you cannot explain the terms, you should reread the sections in which they are discussed.

bands
band theory
conduction band
covalent bond
dielectrics
diode
donors
doping
electron-hole recombination
electron volt (eV)
extrinsic semiconductors
Fermi energy
fluorescence
forbidden gap
hole
intrinsic semiconductors
ionic bond
luminescence
metallic bond
metals
mobility
N-type
Pauli's exclusion principle
phosphorescence

photoelectric current
photoelectron emission
photomultiplier tube
photon activation
P-N junction
positive charge carrier
P-type
quantized
recombination current
recombination time
relaxation time
secondary electron emission
semiconductor diode
semiconductors
stable orbits
thermal activation
thermal current
thermionic emission
transistor
vacuum level
valance band
work function

EXERCISES

Section 35.1:

1. A doubly ionized helium atom (alpha particle) and an electron both fall through a potential difference of 10 V. Do they undergo the same-size energy change? Explain.

Section 35.2:

2. Explain the following terms in your own words.
   A. Stable orbit
   B. Pauli's exclusion principle
   C. Quantized energy
3. In what ways can an electron in an atom gain energy to change stable orbits?
4. In what ways can an electron orbiting an atom lose energy in order to move to a lower stable orbit?

Section 35.3:

5. Explain the following terms in your own words.
   A. Valance band
   B. Conduction band
   C. Forbidden gap
6. How are the energy bands different between semiconductors and dielectrics? Between semiconductors and metals?

Section 35.4:

7. Explain the following terms in your own words.
   A. Vacuum level
   B. Work function
   C. Thermionic emission

D. Secondary electron emission
E. Photoelectric emission

8. The radiant energy of a photon is absorbed by a single electron in the photoelectric effect. What happens to this energy?

Section 35.5:

9. Explain the following terms in your own words.
A. Fermi energy
B. Hole
C. Mobility
D. Charge carrier
E. Intrinsic semiconductor

10. In intrinsic silicon or germanium will electric current be carried mostly by electrons or mostly by holes? (Hint: Look at Table 35–1.)

11. Why are there equal numbers of holes and conduction electrons in intrinsic silicon and germanium?

Section 35.6:

12. Explain the following terms in your own words.
A. Electron-hole recombination
B. Luminescence
C. Phosphorescence
D. Fluorescence
E. Relaxation time
F. Photoelectric current

13. Glass and many dielectric plastics are transparent. Use the band theory to explain why a transparent material is likely to be a dielectric.

Section 35.7:

14. Explain the following terms in your own words.
A. Extrinsic semiconductor
B. Donor
C. Acceptor
D. P-type semiconductor
E. N-type semiconductor

15. In intrinsic semiconductors, the number of holes and the number of conduction electrons are the same. How does doping a semiconductor change this situation?

Section 35.8:

16. Would the current through a diode in a cold car radio in the winter be the same as in the hot summer with the same external voltage across the diode?

17. Identify and explain the function of each of the following parts of a transistor:
    A. Emitter
    B. Base
    C. Collector

PROBLEMS

Section 35.1:

1. Convert the following energies to electron volts.
   A. 2.3×10^{-18} J
   B. 1.0 J
   C. 7.8×10^{-19} J

2. Convert the following energies to joules.
   A. 2.07 eV
   B. 7.2 MeV
   C. 0.022 eV

3. A small water droplet has an electronic charge of 1.76×10^{-17} C. The droplet falls through an electrical potential of 1200 V.
   A. How much energy does the droplet gain in joules?
   B. How much energy does it gain in electron volts?

Section 35.4:

4. What maximum wavelength can a photon have that will produce photoelectrons from calcium?

5. Ultraviolet light with a wavelength of 248 nm strikes the surface of magnesium.
   A. What is the radiant energy in electron volts?
   B. What is the maximum kinetic energy of a photoelectric electron ejected from the magnesium?

6. Calculate the density of conduction electrons in Indium Arsenide (InAs) at room temperature.

7. Calculate the density of conduction electrons and the density of holes in intrinsic silicon at 300 K.

8. Use your answer to Problem 7 to find the conductivity of intrinsic silicon at room temperature.

*9. What fraction of the total electric current in Gallium Phosophide (GaP) is carried by holes at room temperature or 300 K?

Section 35.6:

10. A phosphorescent material has a time constant of 2.7 s. What is its relative intensity 5.0 s after illumination ceases?

11. A phosphor's intensity drops to 5% in 0.10 s. What is its recombination time?

Section 35.7:

12. What percentage of acceptor states are activated at room temperature (300 K) in a doped silicon semiconductor in which the acceptors lie 0.008 eV above the valance band?

13. At room temperature (330 K), 80% of the acceptor states in a doped semiconductor are filled. What is the energy gap between the valance band and the acceptor states?

**14. A P-type semiconductor has acceptor states 0.022 eV above the valance band in germanium.

A. Calculate the density of acceptor states that are filled at room temperature (300 K). (Use Equation 35–2 with E_f for acceptor states.)

B. Calculate the density of intrinsic hole-conduction electron pairs.

C. Calculate the total conductivity of the semiconductor.

D. What fraction of the current is carried by acceptor-activated holes?

Chapter 36

NUCLEAR PHYSICS

This is Nova, the world's most powerful laser. The photo in the upper corner shows a pinpoint-sized star created when the laser hit a tiny fuel capsule and created a fusion reaction. In the fifty trillionths of a second that it lasted, the fusion burst gave off ten trillion neutrons. Scientists estimate that one ounce of fuel in a fusion reactor could power a four-person household for about fifty years. (Photographs courtesy of the Lawrence Livermore National Laboratory.)

OBJECTIVES

In this chapter you will learn:

- that the nucleus of an atom is composed of positively charged particles and particles that have no electrical charge
- how mass equivalent energy holds the nucleus of an atom together
- to calculate half-lives and decay constants for radioactive isotopes
- how to distinguish among alpha, beta, and gamma radiation
- why a neutrino, a massless particle, is necessary for conservation of angular momentum in beta decay
- how a nuclear fission reactor works
- how nuclear fusion can release energy
- how radioactive isotopes are used
- how radiation dosage is determined

Time Line for Chapter 36

1898 Marie and Pierre Curie discover radioactivity.
1910 Joseph J. Thomson discovers isotopes.
1923 Louis de Broglie introduces the concept of the particle-wave duality of matter.
1926 Erwin Schrodinger writes his first paper on quantum mechanics.
1932 James Chadwick discovers the neutron.
1938 Otto Hahn is the first to split the uranium atom.
1939 Frederic and Irene Curie demonstrate that a chain reaction can occur with the fission of uranium.
1942 Enrico Fermi directs the first controlled nuclear chain reaction.
1952 Edward Teller directs the development of a thermonuclear bomb exploded on a Pacific Island on November 6.

A nuclear power plant generates many megawatts of electrical power.

Radioactive steel is used to examine the wear of piston rings. Radioactive tracers are also used to find the biological concentration of phosphorus in lake trout.

Astronomers use nuclear physics to explain the source of energy generated inside stars. The birth, evolution, and death of a star are guided by nuclear reactions and gravity.

The federal government continues to search for a safe way to dispose of high-level radioactive wastes from nuclear power plants.

In Chapter 19 and again in Chapter 35 we looked at the Bohr model of an atom. Electrons orbited about a much more massive positively charged nucleus. In this chapter, we will take a closer look at the atom's center. The atom's nucleus is a source of mass equivalent energy. Unstable nuclei supply us with radioactive materials. On one hand, these radioactive materials have many beneficial industrial and medical applications. On the other hand, radioactivity can create serious biological damage.

36.1 NUCLEAR STRUCTURE

A **nucleon** is a particle in the nucleus of an atom. We will look at two kinds of nucleons: the proton and the neutron.

The **proton** is a particle with a positive charge equal in size to the electronic charge. A neutral atom has no net charge. The number of negative electrons orbiting the nucleus is equal to the number of protons. Indirectly, the protons determine the chemical behavior

In Search of Atoms

The idea of atoms is very old. In 430 B.C. Democritus expanded Leucippus's idea of atoms. Matter was made of atoms. The shape, size, and motion of these atoms accounted for the physical properties of matter.

In 390 B.C. Plato explained matter in terms of an old theory of elements instead of atoms. Matter was made of four elements: earth, water, air, and fire. The Chinese had a similar idea. They added a fifth element, woods, to the list. A few years later Aristotle adapted the elementary theory of matter with a few modifications.

Because Aristotle was viewed as the ultimate authority of science for many centuries, his idea persisted. There were a few weak suggestions of atoms over the years, but it was not until the nineteenth century that the atomic theory forced itself into the scientific arena.

In 1803, John Dalton, a chemist, noted that chemical elements combined in simple proportions. This indicated, he suggested, an atomic theory of matter. In less than 10 years, some chemists were suggesting that chemical bonds and electric forces were the same. Atoms had electrical charges.

By the turn of the century, Hendrich Lorentz was suggesting that vibrating charges in atoms cause visible light. By 1897, electrons were discovered in cathode rays. People began to develop models of atoms made of positive and negative particles.

In 1905 Einstein showed that Brownian movement was caused by molecules bumping into small smoke particles. This was the first direct indication that atoms actually existed. In 1912, Marx von Laue irradiated crystals with X rays. He found the kind of diffraction patterns one would expect if the crystals were made of evenly spaced particles. This was additional strong evidence that atoms are real.

A year later, in 1913, Neils Bohr was able to develop a model of the atom consisting of negative electrons orbiting around a positive nucleus. This model neatly explained the wavelengths of light emitted by atoms at the beginning of the periodic table.

During the twentieth century details of the atom have been highly developed. Atomic models are used to explain the magnetic, mechanical, optical, and electrical properties of matter.

---

of an atom. Because the number of protons in the nucleus determines the chemical behavior of the atom, this number is called the atomic number (Z).

The atomic number (Z) *is equal to the number of protons in an atom.*

- Hydrogen has a single proton in its nucleus; its atomic number is $Z = 1$.
- Helium has two protons in its nucleus; its atomic number is $Z = 2$.
- Carbon has six protons in its nucleus; its atomic number is $Z = 6$.

With the exception of the very simplest atom, hydrogen, all nuclei have neutrons. **Neutrons** are particles with no charge. If an atom has too many or too few neutrons in its nucleus, the nucleus will be unstable and will spontaneously alter its structure. Neutrons seem to have something to do with holding the nucleus together.

The number of neutrons in the nucleus is called the neutron number (N).

Protons and neutrons are much more massive than electrons. They make up most of the mass of the atom. The sum of the atomic number and the neutron number gives us a good estimate of the mass of the atom.

The atomic mass number (A) *is the sum of the atomic number and the neutron number.*

$$A = Z + N \qquad \textbf{(Eq. 36–1)}$$

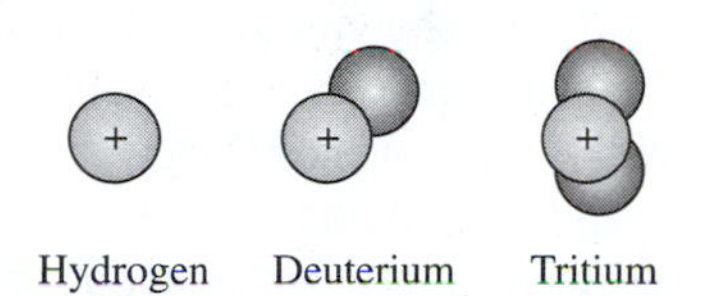

Figure 36–1: *Three isotopes of hydrogen. Normal hydrogen has only a proton in the nucleus. Deuterium has an added neutron; tritium has two neutrons. All three isotopes are chemically identical. They vary only in mass and stability.*

Atoms that have the same number of protons, but different numbers of neutrons, are called **isotopes.** An element may have two or more isotopes. Look at Figure 36–1. There are three known isotopes of hydrogen. The most common form of hydrogen consists of a single proton with an electron orbiting it. A rare form of hydrogen has an added neutron in the nucleus. This kind of hydrogen is called deuterium. Another isotope of hydrogen, known as tritium, has two neutrons in the nucleus.

Since each element may have several isotopes, we must have some notation to uniquely show the nuclear structure. To the chemical symbol of an atom we can add the atomic number as a subscript. We can also place the atomic mass number above it as a superscript. We can then use Equation 36–1 to find the neutron number.

The atoms shown in Figure 36–1 are: hydrogen, ${}^{1}_{1}H$; deuterium, ${}^{2}_{1}H$; and tritium, ${}^{3}_{1}H$.

□ **EXAMPLE PROBLEM 36–1: CARBON 14**

Carbon 14 (${}^{14}_{6}C$) is used by archaeologists to estimate the age of materials that were once living, such as pieces of wood or bone. For carbon 14 find:

A. the number of nucleons.
B. the number of protons.
C. the number of neutrons.

■ ***SOLUTION***

A. $A = 14$. The atomic mass number equals the total number of nucleons.
B. $Z = 6$. The number of protons equals the atomic number.
C. $N = A - Z = 14 - 6$
$N = 8$

36.2 BINDING ENERGY

An attractive coulomb force acts between the positive protons in the nucleus and the orbiting electrons. The coulomb force holds the electrons in orbit.

Within the nucleus, the coulomb force is repulsive between closely packed protons. Attractive forces must exist inside the nucleus of an atom that are much stronger than coulomb forces. These attractive forces have little effect outside of the nucleus. They must fall off much faster with distance than the inverse square law of electrostatic forces.

The forces that bond nucleons together are called **strong forces.** Figure 36–2a shows how the net force changes with distance from the center of the nucleus for a positively charged particle such as a proton. Figure 36–2b shows the forces on a neutron. Since the neutron has no net charge, coulomb forces do not occur. Strong attractive forces hold nucleons together. Repulsive coulomb forces tend to push like charges apart. A proton located outside the nucleus of radius R_n will be repelled from the nucleus by coulomb forces. Inside R_n the same proton is attracted to the nucleus by strong forces.

The net force inside R_0 does work on the nucleons when the nucleus is formed. This work appears as binding energy associated with nucleons. Because strong forces are large inside the nucleus the amount of work done is very large.

The binding energy shows up as mass equivalent energy. The mass of the nucleus is less than the sum of the masses of protons and neutrons that form the nucleus. Here is an example.

The mass of a proton is 1835.7 times the mass of an electron ($m_p = 1835.7\ m_e$). The mass of a neutron is 1838.6 times the mass of an electron ($m_n = 1838.6\ m_e$). The mass of the neutron is a little bit more than that of the proton.

Let us combine two protons and two neutrons to form a helium nucleus ($^4_2$He). We can express mass in terms of multiples of electron masses.

$$\text{neutron masses} = 2 \times 1838.6\ m_e = 3677.2\ m_e$$

$$\text{proton masses} = 2 \times 1835.7\ m_e = 3671.4\ m_e$$

$$\text{total of masses} = 7348.6\ m_e$$

$$\text{The mass of a } ^4_2\text{He nucleus} = 7294\ m_e$$

$$\text{The mass difference} \approx 55\ m_e$$

There is a mass difference equivalent to the mass of 55 electrons. The particles in the nucleus have a smaller collective mass than their

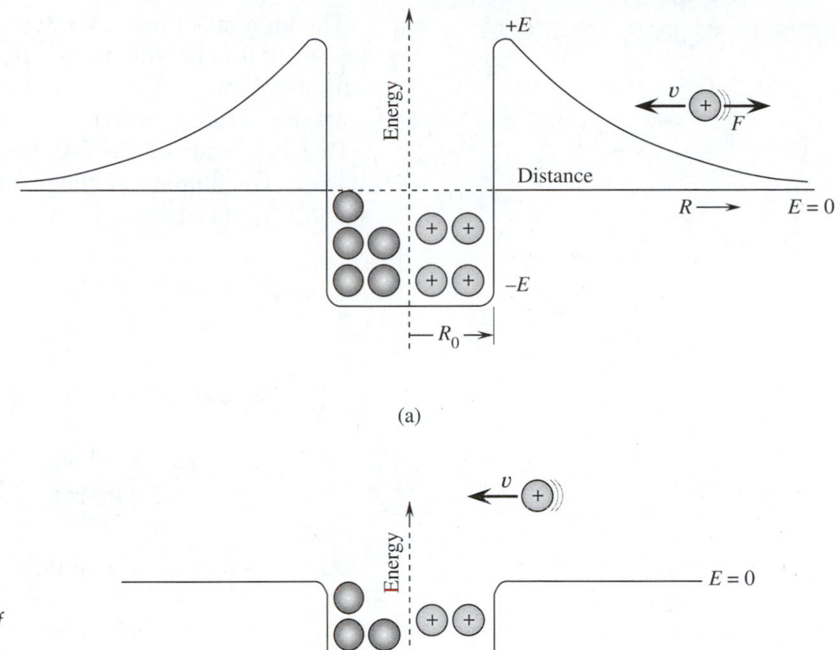

Figure 36–2: *The potential energy surrounding a nucleus. Strong forces bind the nucleons together.* (a) *A proton approaching the nucleus is repelled by a coulomb force. The proton needs a lot of kinetic energy to enter the nucleus.* (b) *A neutron does not encounter a potential energy barrier as it comes near the nucleus. The neutron can easily penetrate the nucleus.*

individual masses. This mass is the mass equivalent energy used to bond the nucleons together.

Assume that we have 7348.6 g of equal parts of neutrons and protons. This is a little over 16 lb of material. Let us also assume that we have a way to *fuse* pairs of neutrons together with pairs of protons to make helium. The result would be 7294 g of helium. Fifty-five grams of matter have been converted to energy. This is a reduction in energy of the particles. It is released in terms of heat and radiant energy.

The amount of energy released is:

$$E = m\,c^2 = 0.055 \text{ kg} \times (3.0 \times 10^8 \text{ m/s})^2$$
$$E = 5.0 \times 10^{15} \text{ J}$$

We can write this in terms of energy released for each kilogram of "fuel."

$$\frac{E}{M} = \frac{5.0 \times 10^{15} \text{ J}}{7.35 \text{ kg}} = 6.8 \times 10^{14} \text{ J/kg}$$

The amount of energy released by burning 1 kg of natural gas is 5.6×10^7 J. The amount of energy released by nuclear reactions is tens of millions of times larger than the energy released by chemical reactions. Here is another comparison.

One kilogram of TNT (trinitrotoluene) releases 1.5×10^7 J of energy. The number of kilograms of TNT needed to release the same amount of energy as 55 g of mass equivalent energy is:

$$M = \frac{6.8 \times 10^{14}\text{ J}}{1.5 \times 10^7\text{ J/kg}} = 4.5 \times 10^7\text{ kg of TNT}$$

TNT has a density of 1620 kg/m^3. The volume of TNT is:

$$V = \frac{M}{d} = \frac{4.5 \times 10^7\text{ kg}}{1620\text{ kg/m}^3} = 2.8 \times 10^4\text{ m}^3\text{ of TNT}$$

Assume a baseball stadium with a square playing area of 400 ft on a side. This is close to 120 m on a side. The area is 1.4×10^4 m^2. If we cover the whole area with dynamite, the height of the pile will be:

$$h = \frac{V}{A} = \frac{2.8 \times 10^4\text{ m}^3}{1.4 \times 10^4\text{ m}^2} = 2.0\text{ m, or 6.5 ft}$$

TNT is used as a standard for nuclear weapons. A 1-megaton nuclear bomb releases the same energy as 1 megaton of dynamite. This is equivalent to two billion pounds of dynamite. We would have to stack the dynamite to a height of 40 m in our baseball field to get the same chemical energy that is available in a one-megaton nuclear explosion.

Let us look at the amount of average binding energy needed for each nucleon in the atom.

For $^4_2$He we found the binding energy was equivalent to 55 electron masses. An electron has a mass of 9.11×10^{-31} kg. The binding energy is:

$$E = m\,c^2 = 55 \times (9.11 \times 10^{-31}\text{ kg})\,(3.0 \times 10^8\text{ m/s})^2$$

$$E = 4.5 \times 10^{-12}\text{ J}$$

or

$$E = \frac{4.5 \times 10^{-12}\text{ J}}{1.6 \times 10^{-19}\text{ J/eV}} = 28.1 \times 10^6\text{ eV} = 28\text{ MeV}$$

There are four nucleons. The average binding energy per nucleon is about 7 MeV. Figure 36–3 shows a plot of average binding energy per nucleon versus the atomic mass number.

The peak of the chart has the highest stability. This corresponds with isotopes that have atomic mass numbers between 50 and 70. The strongest bond per nucleon occurs in $^{56}_{26}$Fe. Elements beyond iron have less and less binding energy per nucleon as the atomic mass number increases.

Light atoms with atomic mass numbers below that of iron can increase the binding energy per nucleons by fusing together with the release of mass equivalent energy. This is called **nuclear fusion.** Hydrogen can undergo fusion to create helium with the release of energy. Three helium atoms can fuse to create carbon with the release of energy.

Atoms at the right of the curve shown in Figure 36–3 can release energy by breaking up into smaller pieces with atomic mass numbers

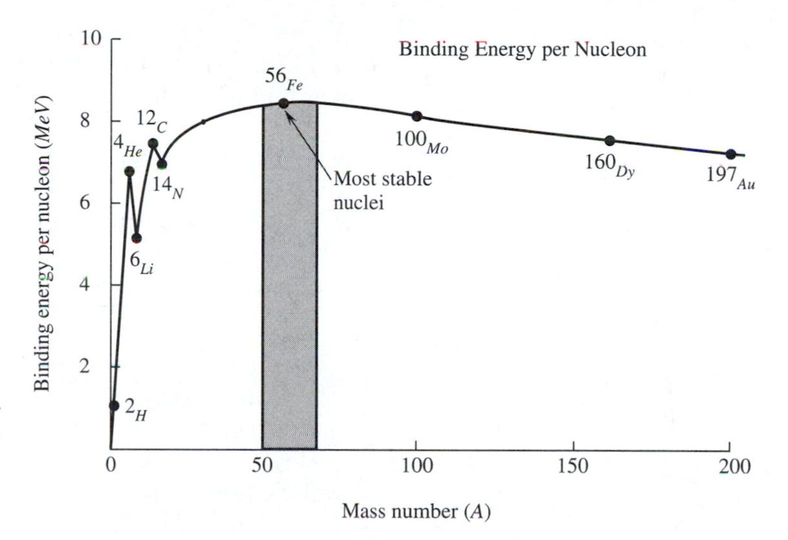

Figure 36–3: A plot of binding energy per nucleon versus atomic mass number. The slope of the curve shows the amount of energy gained or lost when one more nucleon is added to the nucleus.

near the peak of the binding energy curve. The fragmentation of heavy atoms is called **fission.**

☐ **EXAMPLE PROBLEM 36–2: A FISSION OF $^{235}_{92}$U**

In one possible fission a uranium 235 nucleus fragments into two carbon 15 nuclei, a nucleus of mercury 203, and there are a couple of free neutrons left over. The reaction is:

$$^{235}_{92}\text{U} \rightarrow \, ^{203}_{80}\text{Hg} + 2(^{15}_{6}\text{C}) + 2\text{ n}$$

The masses of the nuclei are:

$$^{235}_{92}U = 235.043925 \text{ u}$$

$$^{203}_{80}Hg = 202.972864 \text{ u}$$

$$^{15}_{6}C = 15.010599 \text{ u}$$

$$^{1}_{0}n = 1.008665 \text{ u}$$

A. Is there a conservation of charge in this reaction?
B. Find a conversion between atomic mass units (U) and MeV units of energy.
C. How much energy is released with this reaction?

■ *SOLUTION*

A. Charge conservation:

We need to count protons. Each has a +1 electronic charge. If we add up the total of protons in the products of the fission, we should find 92 protons, the number we started with.

$$U(+92) = [Hg(+80)] \times [C(+6) \times 2] \times n(0) = +92$$

B. $1 \text{ u} = 1.66 \times 10^{-27} \text{ kg}$

The mass equivalent energy is:

$$E = \frac{m\,c^2}{e \times 10^6 \text{ eV/MeV}} = \frac{1.66 \times 10^{-27} \times (3.00 \times 10^8 \text{ m/s})^2}{1.602 \times 10^{-19} \text{ J/eV} \times 106}$$

$$E = 931 \text{ MeV}$$

$$1 \text{ u} = 931 \text{ MeV}$$

C. Energy released:
The total masses of the products of the fission are:

$$\begin{aligned} 2 \times {}^{15}_{6}C &= 2\,(15.010599 \text{ u}) = 30.021198 \text{ u} \\ {}^{203}_{80}Hg &= 202.972864 \text{ u} \\ 2 \times {}^{1}_{0}n &= 2\,(1.008665 \text{ u}) = 2.01733 \text{ u} \\ \text{Total} &= 235.011392 \text{ u} \end{aligned}$$

This is 0.03253 U less than the original mass of the uranium nucleus. The energy released is:

$$E = 0.03253 \text{ u} \times 931 \text{ MeV/u} = 30.3 \text{ MeV}$$

36.3 RADIOACTIVITY

The electrons outside the nucleus have orbits with quantized energies. The nucleons behave in a similar way. Inside the nucleus, the protons and neutrons have discrete energy levels (see Figure 36–4). In some cases, particles or electromagnetic waves are emitted from a nucleus in order to reduce the energies of the remaining nuclei. This process is called **radioactivity.**

When a nucleus reduces its energy by emitting or absorbing a particle to become a new kind of atom, or emitting radiant energy from the nucleus, we say the atom has **decayed.**

Radioactivity is a probability process. If we examined one particular atom, we could not predict when it would decay. If we had many atoms, we could predict on the average how many would decay in a particular time.

Radioactivity is much like the life expectancy tables kept by insurance companies. It is not easy to predict the life expectancy of a single person, but one can predict how many people will die between the ages of 35 and 40 in a population of 100,000 to a very good accuracy. Insurance companies make their living by being able to predict this kind of information.

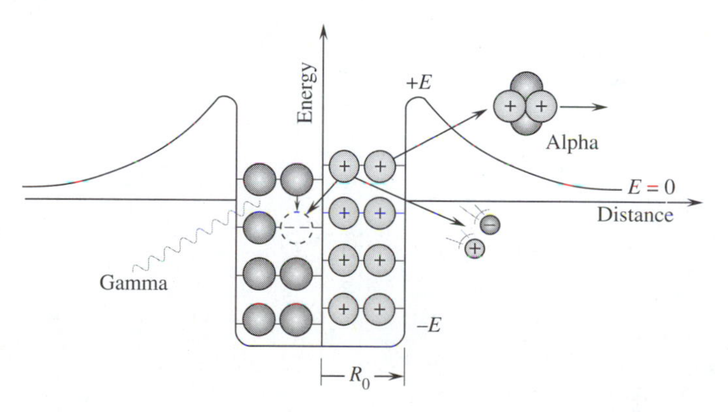

__Figure 36–4:__ Energy levels inside a nucleus. Coulomb forces make the proton energy levels slightly higher than neutron energy levels. A neutron can fall into the empty level by emitting a gamma ray (gamma decay). A proton can fall into the energy level by emitting a positron (beta decay). The top level of nucleons can reduce their potential energy by escaping from the nucleus together as an alpha particle (alpha decay).

In a population of people, the number of deaths or "decays" is proportional to the population. Ten times as many people will die in a year from a population of 100,000 as from a population of 10,000. In a population of radioactive nuclei the same is true. The number of decays that occur during a time (ΔT) is proportional to the number of atoms in the population.

Let ΔN equal the number of decays during time ΔT, and N equal the total number of atoms in the population. The rate of decay is proportional to the number of unaltered atoms left in the population.

The unchanged atom is called the **parent atom.** The new kind of atom created by radioactive decay is called the **daughter atom.**

$$-\left(\frac{\Delta N/\Delta t}{N}\right) = \text{constant} = \lambda \qquad \textbf{(Eq. 36–2)}$$

The proportionality constant (λ) is called the **decay constant.** The minus sign indicates that the change (ΔN) is decreasing the number of parent atoms.

As atoms decay, fewer atoms are left in the parent population. This means that the rate of decay decreases proportionately. This proportional decrease produces the exponential decay curve seen in Figure 36–5. N_0 represents the initial number of parent atoms. The number of parent atoms left at any time (N) is then:

$$N = N_0 e^{-\lambda t} \qquad \textbf{(Eq. 36–3)}$$

From Equation 36–2 we see that the activity or rate of decay (A) is proportional to the number of parent atoms left.

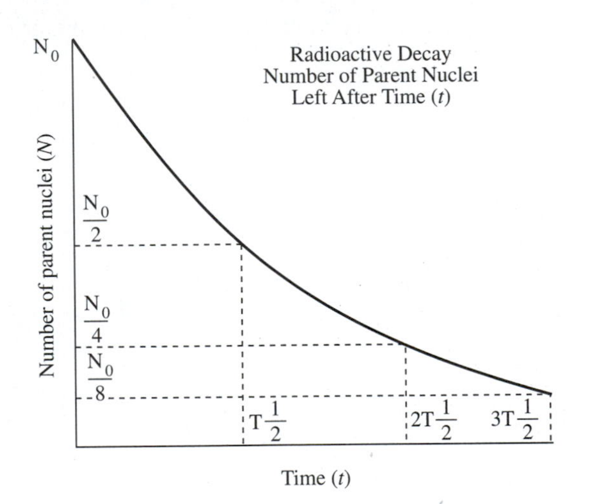

Figure 36–5: Radioactive decay. The number of parent nuclei versus time. The number of remaining parent atoms decreases exponentially with time.

$$A = -\left(\frac{\Delta N}{\Delta t}\right) = \lambda N \qquad \textbf{(Eq. 36–4)}$$

Substitute Equation 36–3 into 36–2. We get:

$$A = \lambda N_0 e^{-\lambda t} \qquad \textbf{(Eq. 36–5)}$$

Two units are used most often to measure activity. For large activity rates, the **curie (Ci)** is used. This is a rather large unit for practical use. Millicuries and microcuries are more often encountered.

$$1 \text{ curie (Ci)} = 3.7 \times 10^{10} \text{ decays/s}$$

The **becquerel** (Bq) is the SI unit of activity.

$$1 \text{ becquerel (Bq)} = 1 \text{ decay/s}$$

Look at Figure 36–5 again. Notice that we can find a time interval in which half the atoms have decayed. This is called a **half-life** ($T_{1/2}$). It does not matter whether we have billions of atoms or just a few hundred. Half of them will decay in the same time interval. After a second half-life, half the remaining parent atoms will have decayed. This leaves us with one-quarter the number we started with. After another half-life we will have only one-eighth the original number.

The half-life is related to the decay constant. Let us rewrite Equation 36–3 for one half-life.

$$\frac{N}{N_0} = \frac{1}{2} = e^{-\lambda T_{1/2}}$$

Take the natural logarithm of the equation.

$$\ln(0.50) = -\lambda T_{1/2}$$

or

$$\lambda T_{1/2} = 0.6931 \qquad \textbf{(Eq. 36–6)}$$

☐ **EXAMPLE PROBLEM 36–3: URANIUM**

A sample of $^{230}_{92}U$ has a half-life of 20.6 days and an activity of 48 Bq.

A. What is the decay constant?
B. How many parent atoms are in the sample?
C. How long will it take for the activity to fall to 20 Bq?

■ ***SOLUTION***

A. Decay constant:
First convert the half-life from days to seconds.

$$20.6 \text{ d} \times 24 \text{ h/d} \times 3600 \text{ s/h} = 1.78 \times 10^6 \text{ s}$$

$$\lambda = \frac{0.6931}{T_{1/2}} = \frac{0.6931}{1.78 \times 10^6 \text{ s}}$$

$$\lambda = 3.89 \times 10^{-7} (\text{s})^{-1}$$

B. Parent atoms:

$$A = \lambda N, \quad \text{or} \quad N = \frac{A}{\lambda}$$

$$N = \frac{48 \text{ Bq}}{3.89 \times 10^{-7} \text{ (s)}^{-1}}$$

$$N = 1.2 \times 10^{8} \text{ atoms of } ^{230}_{92}\text{U}$$

C. Decreased activity:

$$A = A_0 e^{-\lambda t}$$

Take the natural logarithm of the equation.

$$\ln\left(\frac{A}{A_0}\right) = -\lambda t$$

$$t = -\frac{\ln(A/A_0)}{\lambda} = -\left[\frac{\ln(20 \text{ Bq}/48 \text{ Bq})}{3.89 \times 10^{-7}}\right] = 2.25 \times 10^{6} \text{ s}$$

$$t = 2.25 \times 10^{6} \text{ s} \times \frac{1}{3600 \text{ d}} \times \frac{1}{24 \text{ h}} = 26 \text{ d}$$

Alpha Emission

Many of the atoms with high atomic mass numbers emit helium nuclei known as **alpha particles.** This process is known as **alpha emission** (α).

Two neutrons and two protons tend to group together inside the nucleus to form alpha particles. In the more massive atoms the alpha particles have positive energies inside the nucleus. The combination of the strong forces and coulomb forces hold the alpha particles inside a potential barrier.

Quantum mechanics predicts that an alpha particle may occur outside this barrier. Visualize a narrow tunnel passing through the potential barrier. As the alpha particle bounces back and forth inside the nucleus, it strikes the sides of the potential barrier (see Figure 36–6). Eventually the alpha particle will strike the tunnel instead of the barrier. The alpha particle will pass through the tunnel. Once

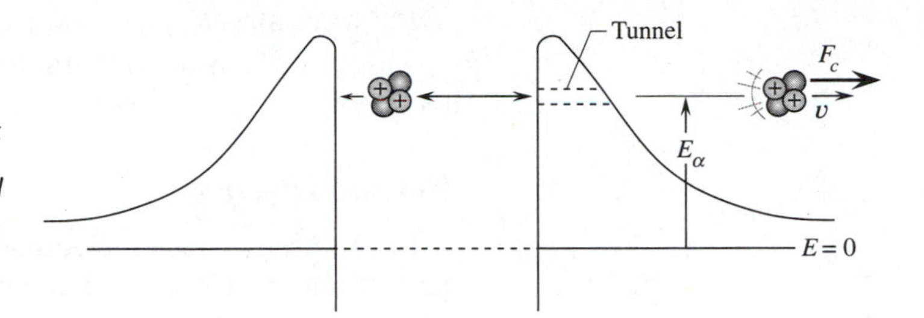

Figure 36–6: Alpha particles bounce back and forth inside the coulomb potential. There is a probability that the particle will tunnel through the potential barrier. The final kinetic energy of the alpha particle (E_α) depends on the initial energy level inside the nucleus.

it is outside, coulomb forces push the alpha particle away from the nucleus, giving it a kinetic energy.

Alpha particles with higher energies will be more likely to escape the nucleus of an atom. At higher energies the "tunnel" through the potential barrier is shorter.

See Table 36–1. The second column lists the kinetic energy of alpha particles emitted by some radioactive atoms.

The third column lists the half-life of the atom. A long half-life indicates a low probability of an alpha particle tunneling out of the nucleus. If the probability of escape is large, the alpha particle will quickly leave the nucleus. It will have a short half-life.

Table 36–1: Selected alpha decay energies in order of decreasing half-life.

ISOTOPE	KINETIC ENERGY (MeV)	$T_{1/2}$	s^{-1}
$^{232}_{90}Th$	4.01	1.4×10^{10} y	1.6×10^{-18}
$^{238}_{92}U$	4.19	4.5×10^{9} y	4.9×10^{-18}
$^{244}_{94}Pu$	4.66	8×10^{7} y	2.7×10^{-16}
$^{242}_{94}Pu$	4.98	3.8×10^{5} y	5.8×10^{-14}
$^{238}_{94}Pu$	5.50	88 y	2.5×10^{-10}
$^{235}_{89}Ac$	5.83	10.0 d	8.0×10^{-7}
$^{220}_{86}Rn$	6.29	56 s	1.2×10^{-2}
$^{222}_{89}Ac$	7.01	5 s	0.14
$^{219}_{87}Fr$	7.30	20 ms	35
$^{216}_{86}Rn$	8.05	45 μs	1.5×10^{4}
$^{212}_{84}Po$	8.78	0.3 μs	2.3×10^{6}

Beta Decay

Sometimes nuclei make adjustments by changing the ratio of neutrons to protons in the nucleus without changing the atomic mass number. This process is called **beta decay** (β). Either a proton transforms into a neutron or a neutron transforms into a proton. There are three ways in which beta decay occurs: electron emission, positron emission, and electron capture.

Electron Emission

In the process called **electron emission,** a neutron decays into a proton and an electron. The electron is ejected from the nucleus. The atomic number Z increases by one. The neutron number $(A - Z)$ is reduced by one.

$$n \rightarrow p + e^-$$

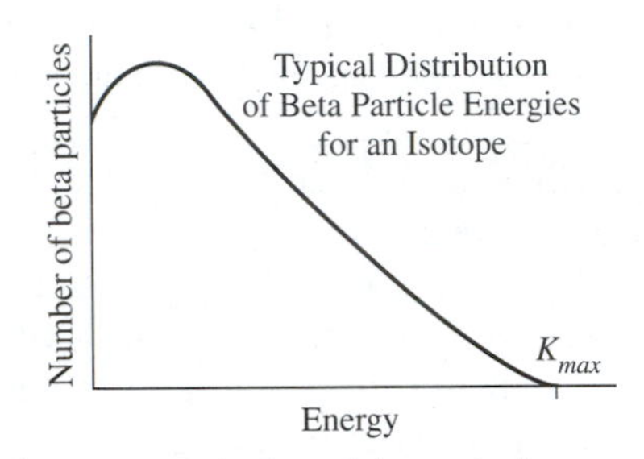

Figure 36–7: A plot of the typical distribution of beta particle energies of electrons ejected from an isotope. The electron can have zero energy because of attractive coulomb forces between it and the nucleus. The energy K_{max} is shared between the electron and a small particle called a neutrino (ν).

The energy of alpha particles is the same for all particles ejected from the same kind of nucleus. This is not true for beta particles. The kinetic energy can vary from zero to some maximum value. At first sight, it appears that energy is not being conserved (see Figure 36–7). The maximum kinetic energy varies from one kind of nucleus to another.

The beta particle spins just as all other electrons do. It has an angular momentum. The angular momentum of the electron is added to that of the nucleus after the beta emission. This gives us the final total angular momentum of the system. This final momentum is *less* than the angular momentum before the beta decay. Some angular momentum is unaccounted.

Conservation of energy and angular momentum can be restored by assuming that another small particle is ejected from the nucleus along with the electron. This particle is called an **antineutrino** ($\bar{\nu}$). It has precisely the spin direction and angular momentum necessary to cancel out the angular momentum of the beta particle. A neutrino has no electrical charge. Its mass is too small to measure.

Our corrected reaction for beta decay is:

$$n \rightarrow p + e^- + \bar{\nu} + KE$$

Positron Emission

The second process of beta decay is **positron emission.** If there is a surplus of protons in the nucleus, a proton can be converted into a neutron by emitting a positive electron called a **positron** (e^+). The positron is a mirror image of an electron. It is called **antimatter.** If a positron and an electron collide, the two particles are destroyed. Their combined mass is replaced by radiant energy. A **neutrino** (ν) is ejected with the positron. It is the mirror image of a neutrino. The reaction is:

$$p \rightarrow n + e^+ + \nu$$

Electron Capture

The third process of beta decay is **electron capture.** Again, the nucleus changes a proton into a neutron. In atoms with a high atomic number (Z) the innermost electrons orbit close to the nucleus because of a very strong coulomb attraction between them and the many protons in the nucleus. These innermost electrons are called **K electrons.** In some cases, the nucleus absorbs an electron to combine with a proton, creating a neutron. This process is called **electron capture.** If a K electron is involved, this process is called K capture.

Gamma Radiation

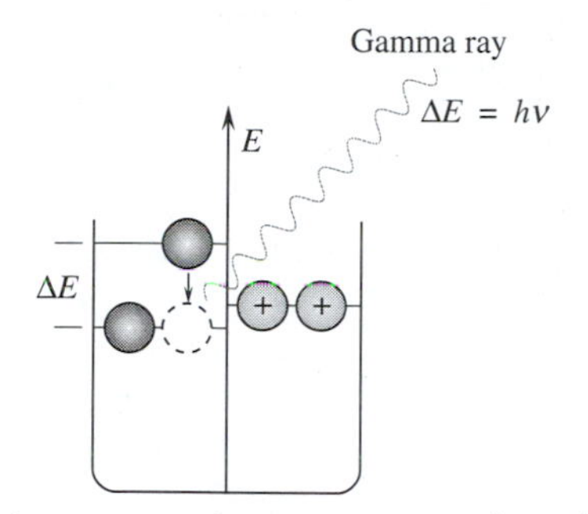

Figure 36–8: Gamma rays are released when the nucleons adjust energy states inside the nucleus.

After a nucleus has undergone emission of an alpha or a beta particle, it may make some internal energy adjustments of its nuclei by emitting radiant energy. This is a process similar to the emission of light waves from electrons orbiting an atom. When an electron falls to an orbit with lower energy, it must give up the energy difference of the energy levels. Usually this discrete quantum of energy is emitted as electromagnetic radiation. The nucleons inside the nucleus can also change energy levels by emitting radiation (see Figure 36–8). Electromagnetic radiation that originates from inside the nucleus is called **gamma radiation** (γ).

Heavy radioactive elements tend to form radioactive chains. The parent atom decays into a daughter atom that is also radioactive. The daughter atom decays into yet another radioactive atom. Figure 36–9 shows the radioactive chain for U 238. This is only one radioactive chain.

Uranium is found in many kinds of granite. Notice that one of the elements created in the U 238 chain is radon (Rm). Radon is a gas rather than a solid. It can seep out of granite rock. In some areas of the United States this is a problem. In unventilated areas such as basements this radioactive gas can build up in the air.

36.4 NUCLEAR ENERGY

While discussing binding energy we noticed that energy could be obtained from the nucleus by two processes. Either we could fuse light atoms together (nuclear fusion) or we could fragment very heavy atoms (nuclear fission). Either process created surplus heat.

Let us look at these two processes in a little more detail.

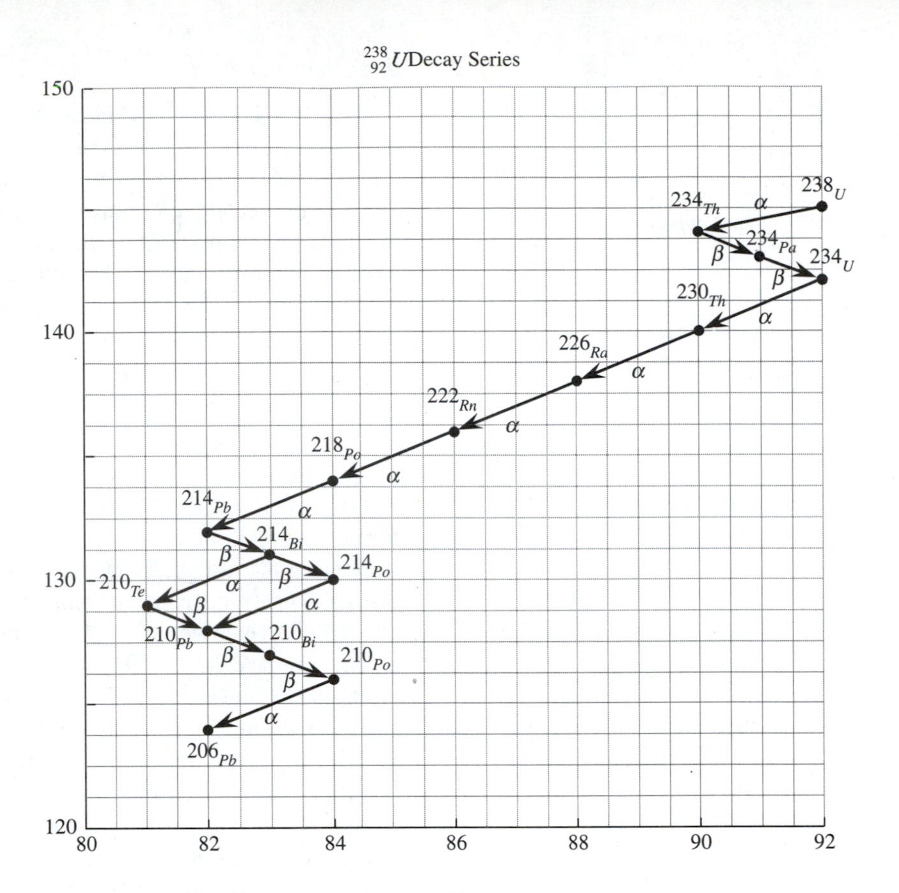

Figure 36–9: *U 238 decays through a series of alpha and beta emissions into stable lead, Pb 206.*

Nuclear Fission

Some heavy atoms will fragment after they have absorbed a neutron. Such atoms are said to be **fissionable.**

$^{235}_{92}U$ is a fissionable atom. When it absorbs a neutron it fissions into atoms lower in the periodic table. For lighter atoms, the ratio of neutrons to protons is usually smaller. Some free neutrons are released from the U 235 along with the fragments. These added neutrons can be absorbed by other fissionable atoms. This releases more neutrons and more energy. Energy and neutrons are created in much the same way as a chain letter generates more letters. The process that yields extra neutrons to generate more fissions is called a **chain reaction** (see Figure 36–10).

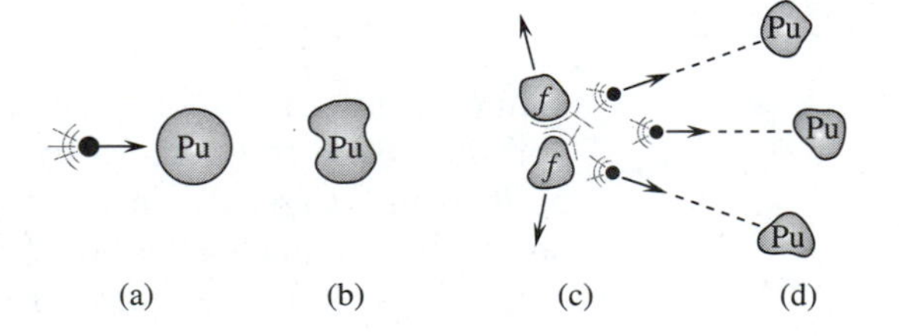

Figure 36–10: *A chain reaction.* (a) *A neutron is absorbed by a plutonium nucleus.* (b) *The nucleus becomes unstable.* (c) *The nucleus breaks apart. Fragments* (f) *and three free neutrons are produced.* (d) *Three plutonium atoms absorb the neutrons released by the fission. They in turn are now unstable and* (c) *and* (d) *are repeated.*

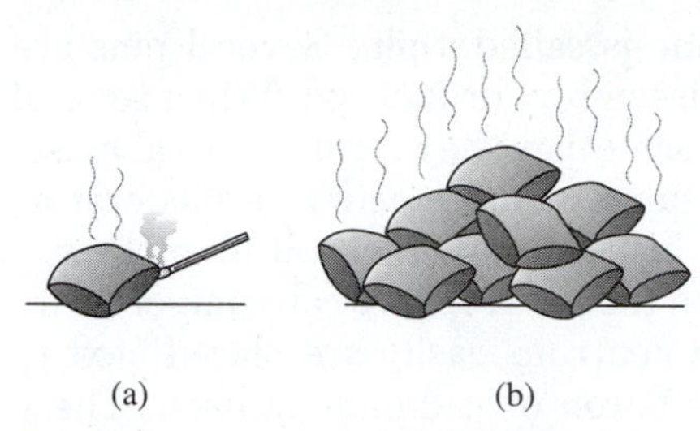

Figure 36–11: *(a) It is difficult to light one piece of charcoal. Most of the heat is lost. The temperature of the charcoal is not maintained above the ignition temperature. (b) A pile of charcoal burns easily. There is a balance of heat reflected between inner surfaces and heat escaping from the outer surfaces.*

It is not possible to start a charcoal fire with a single piece of charcoal. Too much heat is radiated away from the coal. Its temperature will not stay above the ignition temperature (see Figure 36–11).

To start a fire we stack a few pieces of charcoal in a pile. Much of the heat created by burning charcoal at the bottom of the pile is reflected back into the charcoal pile. The center of the pile maintains a hot enough temperature to continue burning.

A nuclear chain reaction behaves in a similar way. In a small mass of fissionable material, many of the free neutrons will escape through the surface without causing fissions. The chain reaction will not be maintained. If we increase the size of the mass, more of the free neutrons will encounter fissionable nuclei before they escape from the mass. As we increase, the mass will eventually reach a minimum mass in which the chain reaction can be maintained. This is called the **critical mass.**

If we form a critical mass of a fissionable material with no way to control the flux of neutrons, we have created a nuclear bomb. An uncontrolled chain reaction will quickly generate enough thermal energy to cause a very rapid expansion. In other words, the material explodes. An early form of the fission bomb consisted of two hemispheres of fissionable material separated by enough distance to prevent an explosion. The bomb was detonated by driving the two hemispheres together. A critical mass was created. Modern fission bombs are shaped into a complete sphere. The density of fissionable material in the sphere is too low to sustain an explosion. The bomb is detonated by setting off explosive charges planted all around the sphere. The sphere is compressed, creating the necessary critical mass to set off the explosion. If only a few charges go off accidentally, the sphere is not compressed enough to cause an uncontrolled chain reaction.

Bombs are easy to make. A more detailed technology is needed to control the neutron flux so we can have a controlled chain reaction. This will allow us to obtain the energy created by fission at a slower and more useful rate.

If we can slow down the flux of neutrons, we can slow down the chain reaction. We can slow down the neutrons by having them bump into atoms that will not absorb them. The neutrons will bounce off the atoms and transfer some of their kinetic energy to them.

The material that slows down the neutrons in a nuclear reactor is called a **moderator.** Water works as a moderator in commercial reactors.

Pellet Pin Subassembly

Figure 36–12: *Fuel assembly. Fuel pellets are encased in zirconium alloy rods to form pins. The pins are bundled to form a subassembly, or fuel rod.*

We will spread out the fissionable fuel so we can place the moderator in the midst of the critical mass. In a commercial reactor, the fuel is in the form of pellets about the size of a pencil eraser (see Figure 36–12). Several of the pellets are placed in hollow tubes made of a zirconium alloy. A filled tube is called a pin. Several pins are bundled together to form a subassembly, or fuel rod. When several subassemblies are placed near each other they form a critical mass. Water circulating between the subassemblies acts as a moderator, slowing down the neutrons passing from one fuel rod to another.

Now that we have slowed down the neutrons we can control them. Rods of a material that absorbs neutrons easily are placed next to the fuel subassemblies. Usually boron or cadmium is used. These rods are called **control rods.**

When the control rods are fully inserted into the fuel subassembly, they absorb many neutrons. There are not enough neutrons to sustain a chain reaction.

As we remove control rods from the assembly, fewer neutrons are absorbed. A slow-acting chain reaction begins. We say the reactor has "gone critical." If the number of fissions per second passes beyond a desired rate, the rods can be partially reinserted. The reactor becomes subcritical. That is to say, there are no longer enough free neutrons to sustain a chain reaction. The rate of fission slows down. The control rods can be alternately moved in and out of the fuel assembly to maintain an average fission rate.

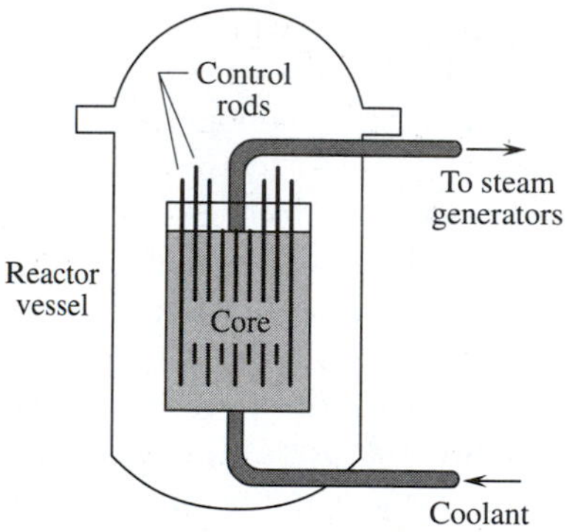

The energy obtained from the nuclear reactor is in the form of heat. Radiant gamma energy is partly absorbed within the reactor. The moderator picks up kinetic energy from neutrons. The kinetic energy of fission fragments is transferred to other parts of the system by collision. A large amount of thermal energy is created.

Water has a double duty in this system. It acts as a **coolant** as well as a moderator.

The core of the reactor is made up of the fuel assembly, moderator, and control rods. The core is placed inside a stainless steel pressurized **reactor vessel** (see Figure 36–13). It acts like a pressure cooker. The water flowing through the reactor vessel boils at a much higher temperature than water at atmospheric pressure. Typically, pressures of 1000 psi or larger allow the reactor to operate at temperatures between 540°F and 600°F.

Figure 36–13: *The reactor vessel acts like a giant pressure cooker. The core inside the reactor vessel contains the fuel assembly, moderator, and control rods. The coolant flows through the reactor vessel.*

The superheated water is converted to steam. The steam drives a turbine that generates electricity. Waste steam is cooled in condensers and recycled. Essentially, a nuclear power reactor is a high-tech way to heat water.

Figure 36–14 shows two types of fission reactors used in the United States. The boiling water reactor (BWR) is an older type. It uses a direct cycle between the reactor vessel and the turbines. It operates with a vessel pressure of 1000 psi and a temperature of 545°F. The pressurized water reactor (PWR) is the design used on more-recent reactors. A heat exchanger creates the steam to operate the turbines. Water passing through the reactor does not pass through the generating equipment. A PWR system can operate at 2250 psi and 600°F.

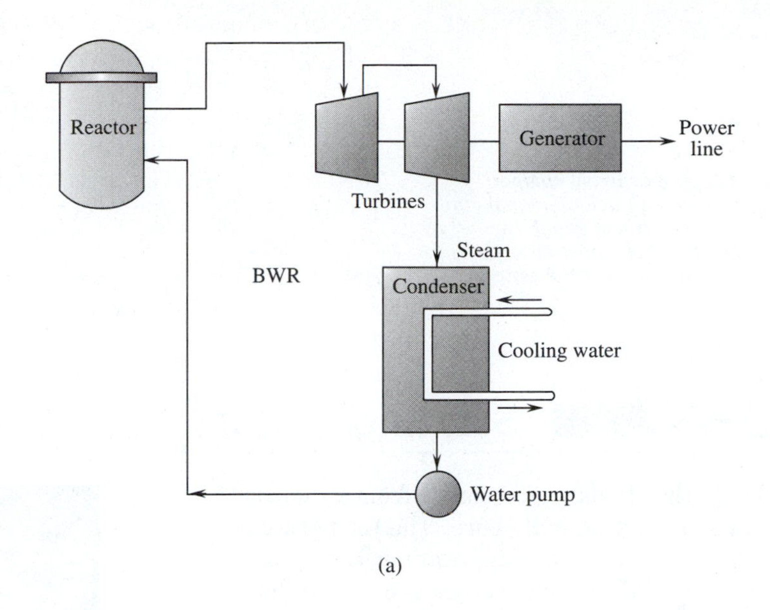

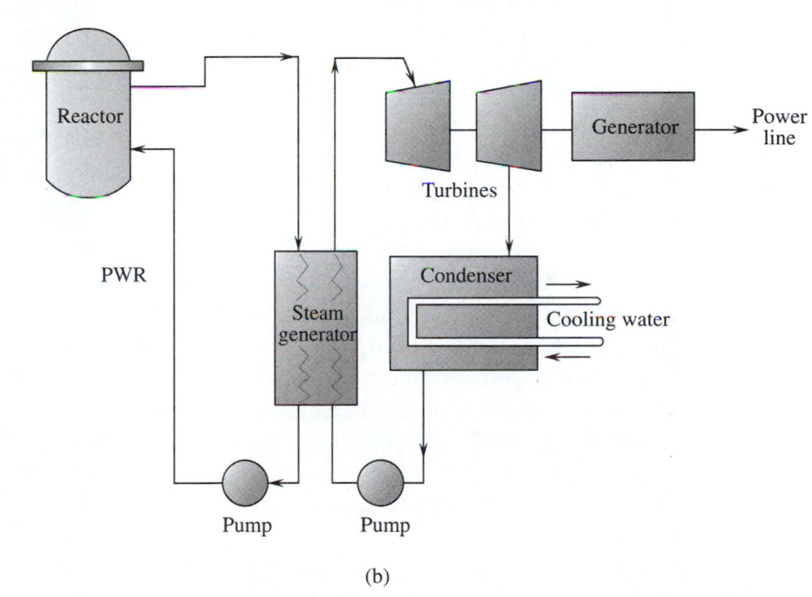

Figure 36–14: (a) *The boiling water reactor (BWR). Coolant in the form of steam passes through steam turbines to generate electricity. The spent steam is cooled in a condenser. It is then pumped back through the reactor.* (b) *The pressurized water reactor (PWR). Heat is transferred from the coolant to steam in a steam generator. A higher pressure and temperature can be maintained in the reactor vessel.*

Nuclear Fusion

In nuclear fusion, light nuclei are bonded together to make more massive nuclei. The problem is getting the two nuclei close enough so that the strong forces overcome the coulomb repulsion between the two positively charged nuclei (see Figure 36–15). The two particles need a very large kinetic energy to overcome the coulomb potential. The necessary kinetic energy is found in a gas at very

Figure 36–15: *As positively charged particles approach a nucleus, equal and opposite coulomb forces repel the particles. The particles must have enough kinetic energy to get over the potential barrier* E_{max}.

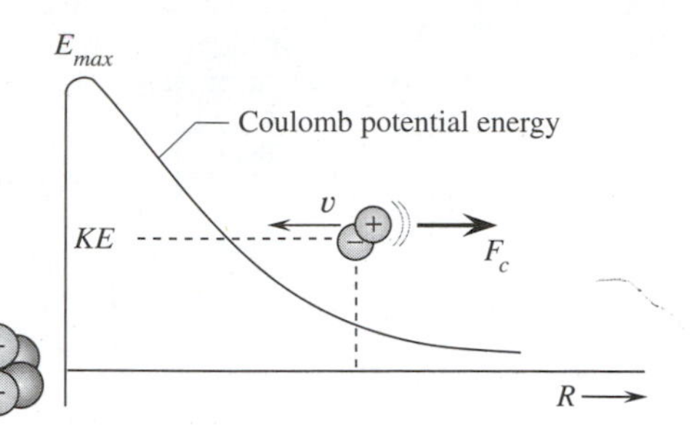

Starting a Star

The energy that fuels a star comes from an uncontrolled fusion reaction in its core. The temperature at the center of a medium-sized star such as our sun is 15,000,000 K. This temperature is certainly high enough to maintain the nuclear fusion of hydrogen into helium. The energy generated by the reaction maintains the temperature.

How does a star start up? Fusion is a thermal nuclear reaction. How does a new star get a temperature hot enough to trigger hydrogen fusion?

There are regions of space between stars that contain immense clouds of hydrogen gas. Eddy currents can develop in these clouds. These swirls of gas create regions of higher than normal density. The denser regions have a slightly larger gravitational attraction than the surrounding gas.

Gas from a spherical region surrounding the eddy begins to drift inward under the gravitational pull of the center. As the mass of the center increases, the gravitational force near the center increases. The gas molecules are accelerated toward the center at an increasing rate.

As the core increases in size and mass, incoming atoms of hydrogen have a larger and larger speed as they fall through an increasing gravitational potential energy. According to the kinetic theory of gases this creates larger temperatures.

The temperature the core of a star attains is proportional to its mass. Atoms falling through a greater gravitational potential will gain more kinetic energy, or temperature. If the kinetic energy of a core of hydrogen atoms is high enough, the atoms can get close enough together to undergo nuclear fusion.

(Photograph courtesy of the Lawrence Livermore National Laboratory.)

Whether the core of an imploding gas ball will get hot enough to trigger nuclear fusion depends on the mass of the ball. A would-be star that does not attain a hot enough temperature to trigger fusion is called a brown dwarf. A brown dwarf cools down very slowly over a period of billions of years, radiating infrared radiation rather than visible light.

We have a ball of gas in our solar system that just missed being massive enough to be a star. Jupiter, unlike the earth, is a large gas ball. Measurements show that this planet is radiating more energy in infrared wavelengths than it is receiving in radiant energy from the sun. Jupiter is a brown dwarf.

high temperatures. Because fusion occurs only at high temperatures, it is called a **thermonuclear reaction.**

In the center of a star temperatures are high enough for thermonuclear reactions to occur. Most stars are made up of hydrogen and helium. Our own star, the sun, has a core temperature estimated at 15 million degrees Kelvin. One type of thermonuclear reaction thought to create solar energy is the **proton-proton reaction.** In this process, four hydrogen atoms are converted into helium.

Two hydrogen nuclei fuse to form deuterium.

$$^{1}_{1}H + ^{1}_{1}H \rightarrow ^{2}_{1}H + e^{+} + \tilde{\nu} + KE$$

The deuterium combines with another hydrogen nucleus.

$$^{2}_{1}H + ^{1}_{1}H \rightarrow ^{3}_{2}He + \gamma + KE$$

Two things can happen now. The most probable reaction is:

$$^{3}_{1}He + ^{1}_{1}H \rightarrow ^{4}_{2}He + e^{+} + \nu + KE$$

The other possible reaction is:

$$^{3}_{2}He + ^{3}_{2}He \rightarrow ^{4}_{2}He + 2(^{1}_{1}H) + KE$$

The end result of these reactions is to convert four protons into a helium nucleus with the release of 25 MeV of energy. In the center of a star, a continuous uncontrolled thermonuclear reaction is in progress.

The high temperatures necessary to create fusion pose a major technical problem in building a controlled fusion reactor. The density of an ionized gas is an additional factor. The center of the sun has a high density. This helps the fusion reaction. At the lower operating densities found in a fusion reactor a much higher temperature is needed for fusion to occur. Temperatures on the order of 10^8 K have been found necessary. Two major approaches to solve these problems are being investigated: magnetic confinement and inertial confinement.

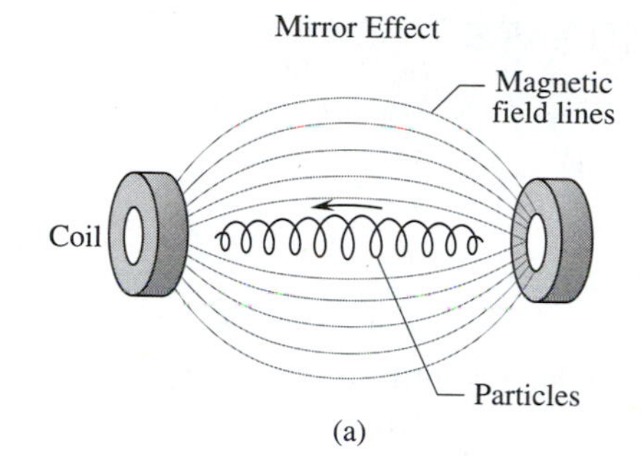

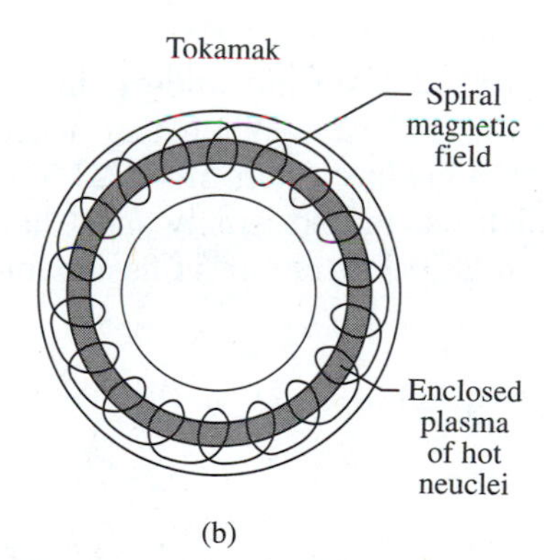

Figure 36–16: *Two methods of magnetic confinement of charged particles.* (a) *Mirror confinement. Charged particles spiral back and forth between the pinched ends of a nonuniform magnetic field.* (b) *The tokamak. A spiral magnetic field causes the charged particles to race around in a circle inside a doughnut-shaped container.*

1. *Magnetic confinement.* One solution for maintaining high temperatures in a stream of charged particles is to put them in a magnetic field. The charged particles spiral around the center of the magnetic field. They are trapped in the magnetic field. This acts as a vacuum bottle. In the absence of the magnetic field, the charged particles would touch the sides of their container and lose energy. This would reduce the temperature of the ion gas, making ignition impossible. Figure 36–16 shows two confinement schemes.

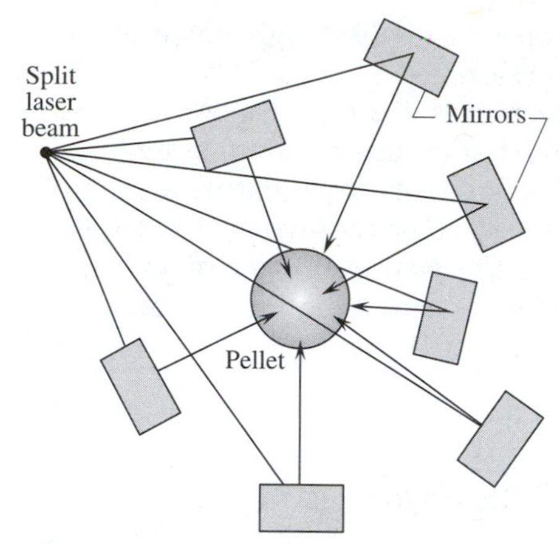

Figure 36–17: Inertial confinement. A pellet containing the fusion fuel is blasted on all sides by an intense laser beam. The pellet is compressed, increasing the fuel's density and temperature.

2. *Inertial confinement.* Deuterium and tritium are held in a small pellet, less than a millimeter in diameter. Powerful laser beams strike the pellet on all sides at the same time. The pressure and energy from the laser beam drive the pellet into a much smaller volume. The density and temperature are driven upward to the ignition temperature (see Figure 36–17).

In both methods most of the energy derived from the fusion reaction occurs in the kinetic energy of neutrons given off in a deuterium-deuterium reaction or a deuterium-tritium reaction. Neutrons do not react strongly with many materials. This makes it difficult to strip them of their energy. One technique is to surround the reactor with lithium. Lithium reacts with neutrons to create tritium and helium, from which the energy can be obtained.

$$^{6}_{3}\text{Li} + \text{n} \rightarrow {}^{4}_{2}\text{He} + {}^{3}_{1}\text{H} + KE$$

36.5 RADIOACTIVE ISOTOPES

More than 99% of naturally occurring uranium is $^{238}_{92}\text{U}$ rather than fissionable $^{235}_{92}\text{U}$. A diffusion process is used to "enrich" the ratio of fissionable to nonfissionable uranium. The process takes advantage of the RMS speeds of the two isotopes. The heavier isotope will have a lower RMS speed. After enrichment there is still a great deal of nonfissionable uranium in the fuel.

Neutrons can be absorbed by $^{238}_{92}\text{U}$. Rather than undergoing fission, the nucleus is transformed by beta decay. When the neutron is absorbed uranium 239 is formed. This has a half-life of 23.5 min. It decays into neptunium 239, which has an extremely short half-life. It quickly decays into plutonium 239. Plutonium is fissionable and has a half-life of 24,000 years.

$$^{238}_{92}\text{U} + \text{n} \rightarrow {}^{239}_{92}\text{U} \rightarrow {}^{239}_{93}\text{Np} + \text{e}^- + \tilde{\nu}$$

$$^{239}_{93}\text{Np} \rightarrow {}^{239}_{94}\text{Pu} + \text{e}^- + \tilde{\nu}$$

Plutonium can be chemically separated from the spent fuel and purified. It is the primary fissionable material used in nuclear weapons.

The process that creates plutonium is called **neutron activation.** The absorption of a neutron creates a new isotope that is usually radioactive.

Neutron activation also affects other parts of the reactor. The core and the reactor vessel eventually become highly radioactive. Repairs and decommissioning of the reactor become very expensive.

On the other hand, many useful radioactive isotopes have been developed for use in medicine, metallurgy, and a variety of industrial applications.

Table 36–2 shows some of the medical applications. The artificially produced isotopes are chemically the same as the stable atoms found in organic compounds. Different parts of the body use chemicals and elements selectively.

Radioisotopes have been used for biological research, diagnosis, and therapy. Here are a few examples:

Biological Research

Carbon 14 has been used in biological research to trace the steps of the metabolism of fructuse sugar ($C_5H_{12}O_6$). Carbon 14 replaces one of the carbon 12 atoms in the sugar. The radioactive carbon can be traced through the chemicals formed by the sugar.

Diagnosis

The uptake of radioactive iodine 131 by the thyroid gland can be measured by monitoring radiation activity in the thyroid. This can be used to diagnose abnormal behavior of the gland.

Fluorescein dye is strongly absorbed by brain tumors. The dye can be tagged with iodine 131 to locate tumors. Other radioisotopes can be used to find tumors elsewhere in the body. Phosphorus 32 can be used to locate breast tumors and gallium 72 to find bone tumors.

Table 36–2: Some medical uses of radioactive isotopes.

ISOTOPE	HALF-LIFE	RADIATION	USE
Arsenic 74	18 days	β^+	locate brain tumors
Chromium 51	28 days	γ	blood tracer
Cobalt 60	5.3 years	β^- and γ	therapeutic radiation
Iodine 131	8.0 days	β^- and γ	measure cardiac output, liver activity, and fat metabolism; locate thyroid cancer
Iron 59	45 days	β^- and γ	determine use of iron in the digestive tract
Phosphorus 32	14 days	β^-	trace brain tumors

Therapy

Cobalt 60 is a gamma emitter. Used as an external radiation source the gamma rays can be focused on tumors that are not operable. With sufficient concentration the α radiation destroys the cells in the tumor.

Some tumors absorb phosphorus selectively. If large doses of radioactive phosphorus are administered, much of it will concentrate in a tumor and destroy it.

36.6 BIOLOGICAL EFFECTS OF RADIOACTIVITY

High-energy particles and radiation lose energy by ionization. When an *alpha particle* passes through a material, it knocks electrons loose from atoms in its path. This creates free electrons and positive ions in the material. It is like rolling a bowling ball down a hallway covered with Ping-Pong balls. The energy of the much more massive bowling ball is distributed among many of the lighter balls before it comes to rest. An alpha particle leaves a dense track of ion pairs in its wake (see Figure 36–18a). Because it reacts so strongly with matter, an alpha particle does not penetrate far into a material before its energy is used up in making ion pairs. A sheet of paper will normally stop an alpha particle.

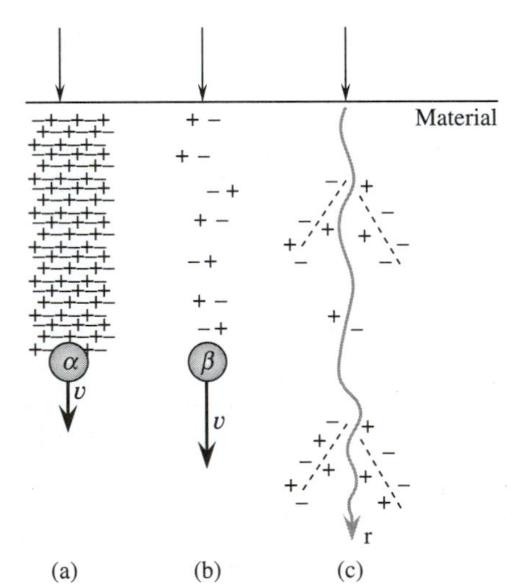

Figure 36–18: *Ionization tracks.* (a) *Slow massive alpha particles leave a dense trail of ion pairs.* (b) *Electrons have less charge and mass. They move faster than alpha particles. Their trail has fewer ion pairs per unit length.* (c) *X rays and gamma rays may transfer enough kinetic energy to ions to create short ionized recoil trails. Fewer ion pairs are produced per length by electromagnetic radiation than by particles.*

A *beta particle* has much less mass and only half the electrical charge of an alpha particle. The change of momentum of the particle as it encounters an electron in the material depends on two things: the force between the particle and the electron, and the period of time in which the particle and electron interact. This forms the impulse ($\mathbf{F}\,\Delta t$). Because the beta particle has less mass it must move faster to get the same kinetic energy as an alpha particle. The amount of interaction time (Δt) is less than for an alpha particle with the same kinetic energy. Because the electric charge of a beta particle is less than that of an alpha particle the interacting force is also less. Beta particles leave a less dense ionization trail than alpha particles (see Figure 36–18b). Beta particles can penetrate a bit farther than alpha particles. A piece of aluminum foil will normally stop beta particles.

Gamma rays and X rays are both high-energy photons. A gamma ray and an X ray can have the same frequency. Their only difference is the source. Gamma rays are radiation from the nucleus of an atom. An X ray is produced when the inner electron of an atom is ejected from its orbit. High-energy photons can also cause ionization. Photons move at the speed of light and have no charge. Conditions for ionization are less probable than for producing a charged particle such as an electron (see Figure 36–18c). The photon's frequency decreases with each ion pair it produces ($E = hf$). If it travels through a large thickness of material, eventually the photon's frequency becomes too low to cause ionization as the photon reaches ultraviolet frequencies. It begins to transfer energy to whole atoms and is totally absorbed as heat.

The number of ionizations per unit length caused by photons depends on the density of electrons in the photon's path. Atoms with high atomic numbers, such as lead, have more orbiting electrons for each atom than light atoms, such as carbon. More ionizations per unit length are produced in lead than in carbon. If we pack atoms closer together, there will be more electrons in the photon's path per unit length. This also reduces the photon's ability to penetrate.

X-ray photography depends on this density factor. The photons can pass more easily through less dense areas, such as flesh, than through more dense bone or enamel.

Ionization of atoms in a biological cell is a matter of concern. If certain key molecules in a cell are destroyed, the cell may die. If the DNA material in cell genes is fragmented or altered cancer may result.

The amount of ionization caused by radioactivity is more important in the medical field than the number of decays per second.

Here are some different ways of defining radioactive dosage.

A. **Roentgen (R).** The Roentgen is measured by ionization of air. One roentgen delivers 8.4×10^{-3} J of energy for each kilogram of air at standard conditions.

B. **Rad.** One rad equals the absorption of 0.01 J/kg in any material. This measures the ionization occurring in tissue. Rad stands for *r*adiation *a*bsorbed *d*ose.

C. **Rem.** Different kinds of radiation cause different amounts of biological damage. Radiation that causes closely spaced ion pairs is more likely to cause damage. Cells have less of a chance for repair if several ionizations occur in the same cell. The rem is obtained by multiplying the rad by a biological damage factor (see Table 36–3).

$$\text{rem} = \text{rad} \times \text{biological damage factor}$$

Table 36–3: Biological damage factors.

TYPE OF EXPOSURE	DAMAGE FACTOR
X rays, gamma rays, electrons	1
Neutrons, protons, alpha particles	10
High-speed heavy nuclei	20

SUMMARY

The nucleus of an atom contains nucleons: neutrons and protons. The atomic mass number (A) is equal to the number of protons plus the number of neutrons in the nucleus. The neutron number is N. The number of protons is called the atomic number (Z).

$$A = N + Z$$

Isotopes are atoms that have the same number of protons, but a different number of neutrons.

The atomic mass number (A) is given as a leading superscript; the atomic number (Z) is a subscript in the notation for an isotope. An isotope of element X is:

$$^{A}_{Z}\mathrm{X}$$

Short-range forces called strong forces hold the nucleons together. The binding energy of nucleons can be found from the mass difference between the individual masses of the nucleons and the mass of the composite nucleus. The binding energy of iron is the largest. Nuclei lighter than iron will release energy when they are fused together (nuclear fusion). Heavy elements can release energy if they are broken into smaller nuclei (nuclear fission).

Radioactive isotopes decay exponentially. Let λ equal the decay constant. If N is the number of undecayed atoms and N_0 is the initial number of atoms, then the number of undecayed atoms left after time t is:

$$N = N_0 e^{-\lambda t}$$

The number of decays per unit time is called activity (A). Two units of activity are the curie and the becquerel.

$$1 \text{ curie (Ci)} = 3.7 \times 10^{10} \text{ decays/s}$$

$$1 \text{ becquerel (Bq)} = 1 \text{ decay/s}$$

The activity is related to the number of undecayed nuclei by:

$$A = \lambda N$$

The half-life ($T_{1/2}$) of a radioactive atom is the amount of time required for half of the atoms to decay. Half-life is related to the decay constant.

$$T_{1/2} = \frac{0.693}{\lambda}$$

Alpha emission is a $^{4}_{2}$He nucleus emitted from the nucleus of an atom. Beta decay occurs when a nucleus changes its value by one atomic number by emitting a positron or an electron, or when it captures one of the electrons orbiting near the nucleus (K capture). Gamma radiation occurs when highly energetic photons are emitted from the nucleus of an atom.

In a nuclear reactor, the fission of U 235 releases more free neutrons than it absorbs. The additional neutrons fission more U 235. This is called a chain reaction. The chain reaction is controlled by slowing down neutrons with a moderator and then absorbing surplus neutrons in control rods. A coolant carries usable thermal energy away from the reactor.

Very high temperatures are needed for nuclear fusion. This process is known as a thermonuclear reaction.

Useful radioactive isotopes can be created by neutron activation. Stable isotopes absorb neutrons and become radioactive isotopes. Radioactivity causes biological damage through ionization. The biological dose is usually expressed in rems, a measure of ionization times a biological damage factor.

KEY TERMS

If you can explain the following terms to a friend or classmate, you understand their meaning. If you cannot explain the terms, you should reread the sections in which they are discussed.

alpha emission
alpha particles
antimatter
antineutrino
atomic mass number
atomic number
becquerel (Bq)
beta decay
chain reaction
control rods
coolant
critical mass
curie (Ci)
daughter atoms
decay constant
decayed
electron capture
electron emission
fission
fissionable
gamma radiation
half-life
isotopes
K electrons
moderator
neutrino
neutron activation
neutron number
neutrons
nuclear fusion
nucleon
parent atom
positron
positron emission
proton
proton-proton reaction
radioactivity
strong forces
thermonuclear reaction

EXERCISES

Section 36.1:

1. What is the total number of nucleons in each of the following atoms?

   A. ${}^{23}_{11}Na$ *B.* ${}^{24}_{11}Na$ *C.* ${}^{96}_{42}Mo$ *D.* ${}^{185}_{76}Os$ *E.* ${}^{205}_{82}Pb$

2. Which two atoms in Exercise 1 are related isotopes?
3. What is the number of neutrons in each of the atoms listed in Exercise 1?

Section 36.2:

4. Explain the difference between nuclear fusion and nuclear fission.
5. Exothermic reactions give off energy or heat; endothermic processes absorb energy or heat. Characterize the following reactions as exothermic or endothermic.
   A. Deuterium and tritium are fused to form helium.
   B. Phosphorus and sulfur nuclei are fused to form gallium. (Gallium has a higher atomic number and higher atomic mass number than iron.)
   C. U 235 absorbs a neutron and splits into two fragments plus four free neutrons.

Section 36.3:

6. Compare decay constants with half-lives. Does the decay constant increase or decrease as the half-life becomes longer?
7. Phosphorus 32 has a half-life of 14 days. Strontium 90 has a half-life of 25 years. If we had an equal number of parent atoms of each kind of isotope, which one would have the greater initial activity or would the activities be the same?
8. For each of the following modes of radioactive decay does the atomic number of the atom increase, decrease, or remain unchanged?

   A. Alpha decay *B.* Positron emission *C.* Gamma emission *D.* K capture *E.* Electron emission

9. For each of the following modes of radioactive decay does the neutron number increase, decrease, or remain unchanged?

   A. Alpha decay *B.* Electron emission *C.* K capture *D.* Positron emission *E.* Gamma emission

10. For each of the following modes of radioactive decay does the atomic mass number increase, decrease, or remain unchanged?

    A. Alpha decay *B.* Electron emission *C.* K capture *D.* Gamma decay *E.* Positron emission

11. Assume that when a volcanic rock is formed the daughter products of U 238 are not present in measurable quantities. Use Figure 36–9 to suggest a scheme for dating very old rock. This process is called radiometric dating.

Section 36.4:

12. What function do the following have in a fission reactor?

    A. Moderator ***B.*** Coolant ***C.*** Control rods

13. Plutonium 239 is a by-product of nuclear reactors. It is used extensively in nuclear weapons. Why is it easier to extract plutonium from spent fuel rods than to extract U 235 from uranium ore?

Sections 36.5, 36.6:

14. Explain why an alpha particle creates more ions per unit length of its path than a beta particle.
15. Would you expect a neutron to cause more or fewer ions per unit length of its path than an alpha particle? Explain.
16. Iodine 131 decays into xenon. Fill in the missing particle(s) in the reaction.

$$^{131}_{53}\text{I} \rightarrow {}^{131}_{54}\text{Xe} + ?$$

17. Cobalt 60 undergoes gamma decay. If the emitted photons have an energy of 1.33 MeV, estimate the number of hydrogen atoms they could ionize. Assume 13.6 eV are needed for each ionization.

PROBLEMS

Section 36.2:

1. A hydrogen atom has a mass of 1.007825 U and a neutron has a mass of 1.008665 U. Use this information to find the binding energy per nucleon for the following neutrally charged atoms.

    A. $^{42}_{20}\text{Ca} = 41.958622$ u ***B.*** $^{49}_{20}\text{Ca} = 48.955677$ u
    C. $^{6}_{3}\text{Li} = 6.015123$ u ***D.*** $^{239}_{94}\text{Pu} = 239.052158$ u
    E. $^{56}_{26}\text{Fe} = 55.934939$ u

2. How much energy is released when the following fission reaction occurs? Use the masses of hydrogen and neutrons given in Problem 1.

$$^{171}_{68}\text{Er} = 170.038041 \text{ u}$$

$$^{56}_{24}\text{Cr} = 55.940671 \text{ u}$$

$$^{235}_{92}\text{U} = 235.045563 \text{ u}$$

$$^{235}_{92}\text{U} + \text{n} \rightarrow {}^{56}_{24}\text{Cr} + {}^{171}_{68}\text{Er} + 9\,\text{n}$$

Section 36.3:

3. Iodine 131 has a half-life of 8.08 days. If a sample has an initial activity of 32 μCi, how long will it take for its activity to fall to 4.0 μCi?

4. Cesium 131 has a half-life of 9.69 days. Calculate its decay constant.

5. Radon 220 has a decay constant of 0.14 $(s)^{-1}$. What is its half-life?

6. Rubidium 86 has a half-life of 18.8 days. If a sample of this isotope has an activity of 1220 Bq, how many atoms of rubidium 86 are present in the sample?

7. Gold 198 has a half-life of 2.70 days. A sample has an initial activity of 2.4 μCi. How long will it take for the activity to fall to 1.7 μCi?

Sections 36.4–36.6

8. Cosmic radiation continually forms radioactive carbon 14 in the atmosphere, creating radioactive carbon dioxide (CO_2). The carbon dioxide enters the food chain through plant photosynthesis. The ratio of carbon 14 to carbon 12 remains constant as long as an organism lives. When the organism dies, the ratio begins to decrease. The half-life of carbon 14 is 5730 years. At an ancient Australian campsite a sample of carbon has an activity of 0.38 Bq. If freshly charred wood with the same amount of carbon has an activity of 0.88 Bq, how old is the sample of carbon from the campsite?

9. A patient is given a dose of 2.3 millirads.
   A. If the dose is in X rays, what is the dosage in rems?
   B. If the dosage is in beta emitters, what is the dosage in rems?

*10. Isotope A has a half-life of 34.0 days. B, the daughter atom, is stable. A sample that was initially all isotope A is found to be composed of 1.53 g of A and 28.47 g of B. How much time has elapsed since the sample was initially formed? (A and B have the same number of nucleons. The number of atoms (N) is proportional to mass.)

**11. A patient is injected with 5.0 ml of blood labeled with chromium 51. The activity of the injection is $100\bar{0}$ Bq. Five-milliliter blood samples are drawn and tested at intervals. The activity of the blood samples stabilizes at 1.4 Bq. How many total liters of blood are in the patient's body?

Appendix A

MATHEMATICS SUMMARY

A.1 BASIC ALGEBRA

Algebra is a branch of mathematics that compares the size of one number with another. Before we look at the rules for algebra, we need to define a few ideas.

Kinds of Numerals

A numeral is a symbol used to represent a number.

1. Some numerals have fixed values. *Examples:* 3, 27, π, 10.3, 1.2×10^{-6}. This kind of numeral is called a **constant.**
2. Other numerals may represent more than one value. These numerals are called **variables.** Variables are usually represented by letters. *Examples:* x, y, θ, T, a.

Kinds of Operations

An operation tells us how to combine numbers. For every operation there is an inverse operation. **Inverse operations** "undo" operations. Here are a few operations with their inverse.

OPERATION	INVERSE OPERATION	EXAMPLES OF OPERATION	OPERATION WITH INVERSE
+	−	$4 + 2 = 6$	$(4 + 2) - 2 = 4$
−	+	$4 - 2 = 2$	$(4 - 2) + 2 = 4$
×	/	$4 \times 2 = 8$	$(4 \times 2)/2 = 4$
/	×	$4/2 = 2$	$(4/2) \times 2 = 4$
y^n	$y^{1/n}$	$4^2 = 16$	$(4^2)^{1/2} = 4$
$y^{1/n}$	y^n	$(4)^{1/2} = 2$	$[(4)^{1/2}]^2 = 4$

Kinds of Relationships

The size of one number can be related to the size of another number in many ways. Here are the three most commonly used **relationships.**

RELATIONSHIP	EXAMPLE
Equal (=)	$9 = 3^2$
Smaller than (<)	$9 < 10$
Larger than (>)	$9 > 8$

Kinds of Identities

For an operation and its inverse there exists a number that will not change the value of the original number. Such a number is called the **identity** of the operation.

Additive identity:	0	$6 + 0 = 6$	$6 - 0 = 6$
Multiplicative identity:	1	$6 \times 1 = 6$	$\frac{6}{1} = 6$

Rules of Algebra

The three basic rules of algebra are simple.

1. If the same number is added to equal numbers, the sums are equal.

   If $a = b$, then $a + c = b + c$ or $a - c = b - c$.

   If $9 = 3^2$, then $9 + 2 = 3^2 + 2$ or $9 - 2 = 3^2 - 2$.

2. If two equal numbers are multiplied by the same number, the results are equal.

   If $a = b$, then $a \times c = b \times c$ or $\frac{a}{c} = \frac{b}{c}$.

   If $9 = 3^2$, then $9 \times 2 = 3^2 \times 2$ or $\frac{9}{2} = \frac{3^2}{2}$.

3. If two equal numbers are raised to the same power, then the results are equal.

   If $a = b$, then $a^n = b^n$ or $a^{1/n} = \mathrm{b}^{1/n}$.

   If $9 = 3^2$, then $9^{1/2} = (3^2)^{1/2}$ or $9^{1/2} = (3^2)^{1/2}$.

Basic algebra is a game in which inverse operations are used to place all the constants on one side of the equation. This leaves the variable alone on the other side of the equation so we can "read" the answer. Below are a few example problems.

The rules of algebra require us to do identical operations on both sides of the equation.

□ EXAMPLE PROBLEM 1: BASIC ALGEBRA

What value of d makes the equation below true?

$$2d + 3 = 7$$

■ *SOLUTION*

First let us operate on the term that does not contain the variable. The inverse of adding 3 is subtracting 3.

$$2d + 3 - 3 = 7 - 3$$

$$2d + 0 = 4$$

The additive identity is 0.

$$2d = 4$$

The inverse of multiplying by 2 is dividing by 2.

$$\frac{2d}{2} = \frac{4}{2}$$

$$1d = 2$$

The multiplicative identity is 1.

$$d = 2 \qquad \text{Answer}$$

We can check our answer by substituting 2 for d in the original equation.

$$2(2) + 3 = 7$$

$$7 = 7 \qquad \text{Check}$$

□ EXAMPLE PROBLEM 2: BASIC ALGEBRA

What values of t make the equation below true?

$$2t^2 - 2 = 30$$

■ *SOLUTION*

Again we will eliminate the term without the variable first. The inverse of subtracting 2 is adding 2.

$$2t^2 - 2 + 2 = 30 + 2$$

$$2t^2 + 0 = 32$$

The additive identity is 0.

$$2t^2 = 32$$

The inverse of multiplying by 2 is dividing by 2.

$$\frac{2t^2}{2} = \frac{32}{2}$$

$$1t^2 = 16$$

The multiplicative identity is 1.

$$t^2 = 16$$

Taking the square root of a number is the inverse of squaring the number.

$$(t^2)^{1/2} = (16)^{1/2}$$

The square of -4 and the square of $+4$ are both $+16$. We have two solutions.

$$t = +4 \text{ and } -4 \qquad \text{Answer}$$

Substitute the solutions into the original equation to check the solutions.

$$2(+4)^2 - 2 = 30$$

$$2(16) - 2 = 30$$

$$30 = 30 \qquad \text{Check}$$

and

$$2(-4)^2 - 2 = 30$$

$$2(16) - 2 = 30$$

$$30 = 30 \qquad \text{Check}$$

☐ **EXAMPLE PROBLEM 3: BASIC ALGEBRA**

Solve the equation below for R.

$$I = \frac{C}{R^2}$$

■ *SOLUTION*

Often we need to solve an equation for one variable in terms of the other variables. In this example, the variable we want is in the denominator of a ratio. The first step of the solution is to get R out of the denominator.

The inverse of dividing by R^2 is multiplying by R^2.

$$I \times R^2 = \frac{C}{R^2} \times R^2$$

$$I \times R^2 = C \times 1$$

The multiplicative identity is 1.

$$I \times R^2 = C$$

Dividing by I is the inverse of multiplying by I.

$$\frac{I \times R^2}{I} = \frac{C}{I}$$

$$1 \times R^2 = \frac{C}{I}$$

The multiplicative identity is 1.

$$R^2 = \frac{C}{I}$$

Taking the square root is the inverse operation of squaring a number.

$$(R^2)^{1/2} = \left(\frac{C}{I}\right)^{1/2}$$

$$R = \pm\left(\frac{C}{I}\right)^{1/2} \qquad \text{Answer}$$

□ **EXERCISES:** Solve the equations for the indicated variables.

a. $2.3\,y = 4.6$ for y

b. $3t - 4 = 16$ for t

c. $8.2 = 2\,v - 4.6$ for v

d. $16t^2 - 3.8 = 0$ for t

e. $F = m\,a$ for a

f. $v - v_o = a\,t$ for t

g. $2\,a\,s = v^2 - v_o^2$ for s

h. $2\,a\,s = v^2 - v_o^2$ for v

■ *SOLUTIONS*

a. $y = 2$ **b.** $t = 6\frac{2}{3}$ **c.** $v = 6.4$ **d.** $t = +1.54$

e. $a = \frac{F}{m}$ **f.** $t = \frac{v - v_o}{a}$ **g.** $s = \frac{(v^2 - v_o^2)}{2a}$

h. $v = (2\,a\,s + v_o^2)^{1/2}$

A.2 THE QUADRATIC EQUATION

An equation that contains a squared variable is a **quadratic equation.** Here are some examples.

1. $3t^2 = 7t$
2. $4x^2 = 12x - 9$
3. $v^2 = 81$

Let us write each equation with the right-hand side equal to zero. This is the **standard form** of writing a quadratic equation.

1. $3t^2 - 7t = 0$
2. $4x^2 - 12x + 9 = 0$
3. $v^2 - 81 = 0$

In the standard form a general equation is written as:

$(a\,x^2) + (b\,x) + c = 0$ **Quadratic Equation**

where a, b, and c are constants. b or c may be equal to zero.

In each case we can rewrite the quadratic equation as the product of two factors.

1. $(3t - 7)\,t = 0$
2. $(2x - 3)(2x - 3) = 0$
3. $(v - 9)(v + 9) = 0$

We can find the values of the variables that satisfy these equations by introducing one more idea. **If the product of two numbers is zero, then at least one of the numbers must be zero.** In symbols this is:

If a × b = *0, then* a = *0 or* b = *0.*

Let us apply this idea to the three equations.

1. $(3t - 7)\,t = 0$

then

$$(3t - 7) = 0$$

which gives us

$$t = \frac{7}{3}$$

or

$$t = 0$$

Two values of t satisfy the equation. We can check this by substituting the values back into the equation in the standard form.

$$3t^2 - 7t = 0$$

$$3\left(\frac{7}{3}\right)^2 - 7\left(\frac{7}{3}\right) = 0 \quad \text{or } 3(0) - 7(0) = 0$$

$$\frac{49}{3} - \frac{49}{3} = 0, \text{ or } 0 = 0$$

The solutions to the other two equations are:

2.

$$(2x - 3)(2x - 3) = 0$$

then

$$2x - 3 = 0, \text{ or } x = \frac{3}{2}$$

3.

$$(v - 9)(v + 9) = 0$$

then

$$v - 9 = 0$$

which gives us

$$v = +9$$

or

$$v + 9 = 0$$

which gives us

$$v = -9$$

The general form of the quadratic equation can be factored. For

$$(a\,x^2) + (b\,x) + c = 0$$

$$\left\{x + \frac{[b - (b^2 - 4\,a\,c)^{1/2}]}{2a}\right\}\left\{x + \frac{[b + (b^2 - 4\,a\,c)^{1/2}]}{2a}\right\}$$

This gives us the two solutions of:

$$x = \frac{[-b \pm (b^2 - 4a\,c)^{1/2}]}{2a}$$ **General Solution to the Quadratic Equation**

☐ **EXAMPLE PROBLEM 4: USING THE QUADRATIC EQUATION**

Find the values of t that satisfy the equation:

$$-16 = 2t - 3t^2$$

■ ***SOLUTION***

First rewrite the equation in the standard form.

$$3t^2 - 2t - 16 = 0$$

Compare the coefficients with the general standard equation:

$$(a\,x^2) + (b\,x) + c = 0$$

The variable is t with constants $a = 3$, $b = -2$, and $c = -16$.

$$t = \frac{-b \pm (b^2 - 4\,a\,c)^{1/2}}{2a}$$

$$t = \frac{-(-2) \pm [(-2)^2 - 4(3)(-16)]^{1/2}}{2(3)}$$

$$t = \frac{2 \pm (4 + 192)^{1/2}}{6} = \frac{2 \pm 14}{6}$$

$$t = +2.67, \text{ or } -2.0$$

☐ **EXERCISES:** Solve equations a–d by setting each factor equal to zero.

a. $(x - 3)(2x + 4) = 0$ **b.** $(2v - 0.5)(0.7v + 6) = 0$

c. $(3t - 8)\,t = 0$ **d.** $(5p + 8)(7p - 1) = 0$

Use the solution to the general quadratic equation to solve the following equations.

e. $3x^2 + 4x - 5 = 0$ **f.** $t^2 - 5t + 2 = 0$

g. $v^2 = 3v - 2$ **h.** $8 - 2y^2 = 8y$

■ ***SOLUTIONS***

a. $x = +3$ or -2 **b.** $v = 0.25$ or -8.57 **c.** $t = 2.67$ or 0

d. $p = 0.14$ or -1.6 **e.** $x = 0.79$ or -2.12 **f.** $t = 4.56$ or 0.44

g. $v = 2$ or 1 **h.** $y = 0.83$ or -4.83

A.3 BASIC TRIGONOMETRY OF RIGHT TRIANGLES

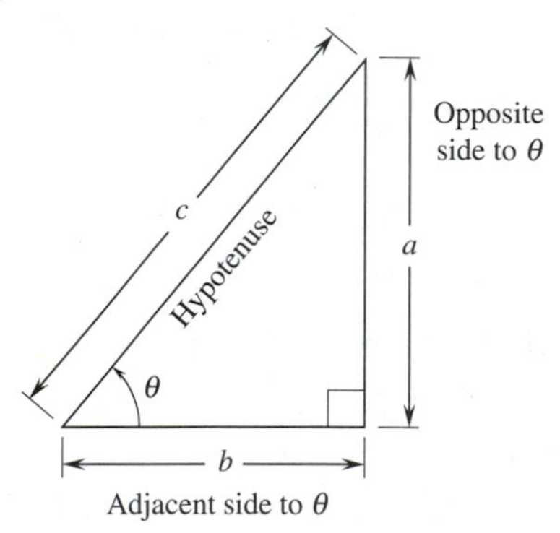

Figure A–1

The sum of the interior angles in a triangle is 180°. If one angle is 90°, the sum of the other two angles must also add up to 90°. The two smaller angles are called **complementary angles.**

Many problems in this book can be solved using the ratio of the sides of a **right triangle.** A right triangle is a triangle that contains a 90° angle. There are three commonly used ratios: sine, cosine, and tangent. These are found on your scientific calculator.

Look at Figure A–1. The leg of the triangle opposite the angle θ has a length a. The leg of the triangle adjacent to angle θ has a length b. The long side of the triangle opposite the 90° angle is called the **hypotenuse.** It has a length c.

All right triangles with an angle θ are proportional to all other right triangles with the same angle θ. Corresponding sides will be proportional.

The three basic ratios of sides of right triangles are:

$$\text{sine of } \theta = \frac{\text{opposite side}}{\text{hypotenuse}} \quad \text{or } \sin\theta = \frac{a}{c} \quad \text{or } a = c \sin\theta$$

$$\text{cosine of } \theta = \frac{\text{adjacent side}}{\text{hypotenuse}} \quad \text{or } \cos\theta = \frac{b}{c} \quad \text{or } b = c \cos\theta$$

$$\text{tangent of } \theta = \frac{\text{opposite side}}{\text{adjacent side}} \quad \text{or } \tan\theta = \frac{a}{b} \quad \text{or } a = b \tan\theta$$

Pythagorean Theorem

For a right triangle, the sum of the squares of the legs of the triangle equals the square of the hypotenuse (see Figure A–1). In symbols this is:

$$c^2 = a^2 + b^2$$

NOTE: This equation is true only for *right* triangles.

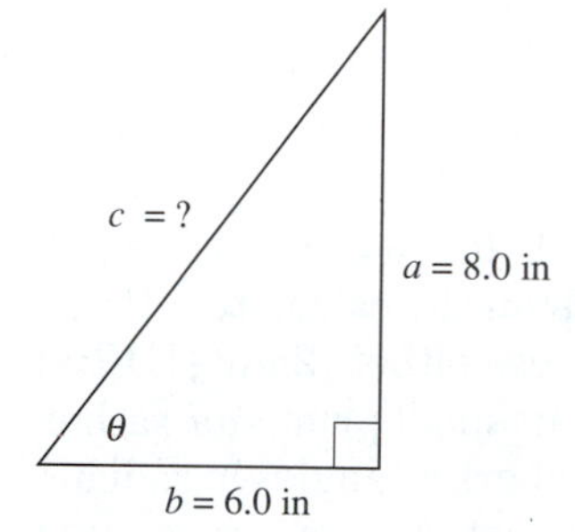

Figure A–2

☐ **EXAMPLE PROBLEM 5: FINDING TRIGONOMETRICAL RATIOS**

Find the sine, cosine, and tangent of angle θ in Figure A–2.

■ ***SOLUTION***

First find the length of the hypotenuse.

$$c = (a^2 + b^2)^{1/2}$$

$$c = [(8.0\text{ in})^2 + (6.0\text{ in})^2]^{1/2} = (100\text{ in}^2)^{1/2}$$

$$c = 10\text{ in}$$

$$\sin\theta = \frac{a}{c} = \frac{8.0\text{ in}}{10\text{ in}} \quad \text{or } \sin\theta = 0.80$$

$$\cos\theta = \frac{b}{c} = \frac{6.0\text{ in}}{10\text{ in}} \quad \text{or } \cos\theta = 0.60$$

$$\tan\theta = \frac{a}{b} = \frac{8.0\text{ in}}{6.0\text{ in}} \quad \text{or } \tan\theta = 1.3$$

☐ **EXAMPLE PROBLEM 6: FINDING TRIGONOMETRIC FUNCTIONS FROM ANGLES**

Find the cosine, sine, and tangent for an angle of:

A. 27°.
B. 27 rad.
C. 27 grad.

■ ***SOLUTION***

A. Trigonometric functions in degrees:

When you turn on your calculator, it records angles in degrees. Simply enter the angle and press the desired trigonometric function to find the value.

Enter: 27

Press: [cos]

Display: [0.8910065] cos 27° = 0.8910065

Enter: 27

Press: [sin]

Display: [0.4539904] sin 27° = 0.4539904

Enter: 27

Press: [tan]

Display: [0.5095254] tan 27° = 0.5095254

B. Trigonometric functions in radians:

Look for a key labeled [DRG]. This stands for degree, radian, or grad. It may be one of the second functions on the calculator. Press the [DRG] key. If it is a second function, press either [2ndF] [DRG] or [INV] [DRG]. The display should show in small print *rad* rather than *deg*. The calculator now interprets numbers as angles in radians rather than degrees. Repeat the steps in part A. Your calculator should give you values like this.

$$\cos(27 \text{ rad}) = -0.2921388$$
$$\sin(27 \text{ rad}) = 0.9563759$$
$$\tan(27 \text{ rad}) = -3.2737037$$

C. Trigonometric functions in grads:

Press the [DRG] key one more time. The display should change from *rad* to *grad*. Proceed in the same way as in part A to find the functions. Your calculator will give you values like this.

$$\cos(27 \text{ grad}) = 0.9114032$$
$$\sin(27 \text{ grad}) = 0.4115143$$
$$\tan(27 \text{ grad}) = 0.4515173$$

Inverse Trigonometric Functions

Once we have a trigonometric ratio, we can find the angle associated with it. This is the inverse process of looking up a trigonometric function given the angle. The angle is called the **inverse function.** If we have the tangent ratio, we can find the angle associated with the function by looking up the inverse tangent. In symbols this is:

$$\theta = \tan^{-1}(a/b) \quad \text{or} \quad \theta = \text{inv}\tan(a/b)$$

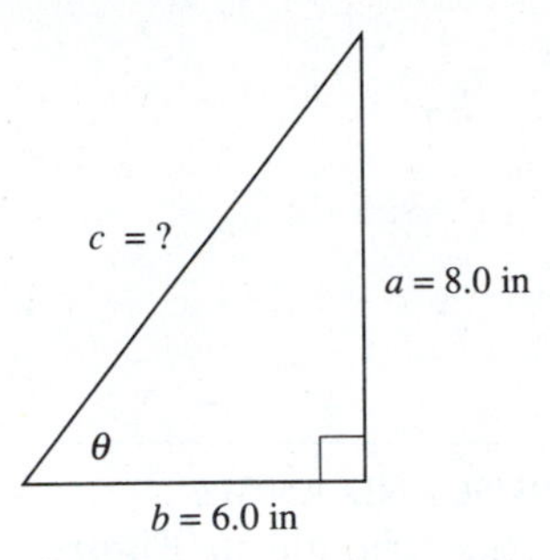

Figure A–3

EXAMPLE PROBLEM 7: FINDING ANGLES

Find the two acute angles in Figure A–3.

SOLUTION

Use the sine ratio calculated in Example Problem 5. If your calculator has a [2ndF] key, do it this way.

Enter: 0.80

Press: [2ndF]

Press: [sin]

Display: [53.13 . . .] $\theta = \sin^{-1}(0.80) = 53.13°$

If your calculator has an [INV] key, press it instead of the [2ndF] key.

Enter: 0.80

Press: [INV]

Press: [sin]

Display: [53.13 . . .] $\theta = \text{inv sin}(0.80) = 53.13°$

Since the two acute angles are complementary angles, we have:

$$53.13° + \phi = 90°$$
$$\phi = 36.87°$$

If we are given one side of a *right* triangle and either an angle or another side, we can calculate all the sides and angles of the triangle using basic trigonometry. Here are two examples.

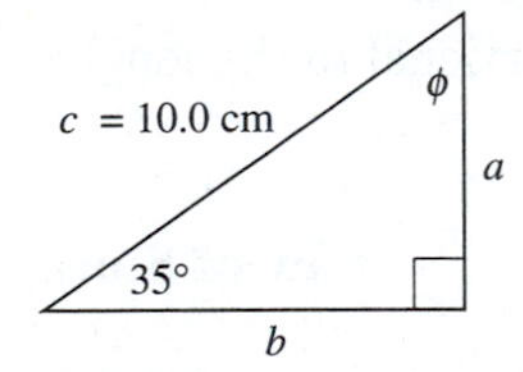

Figure A–4

EXAMPLE PROBLEM 8: FINDING THE SIDES OF A TRIANGLE

Find the lengths of the legs of the triangle shown in Figure A–4.

SOLUTION

$$\sin 35° = \frac{a}{c}$$
$$a = c \times \sin 35°$$
$$a = 10.0 \text{ cm} \times (0.5736)$$
$$a = 5.74 \text{ cm}$$

$$\cos 35° = \frac{b}{c}$$

$$b = c \times \cos 35°$$

$$b = 10.0 \text{ cm} \times (0.8192)$$

$$b = 8.19 \text{ cm}$$

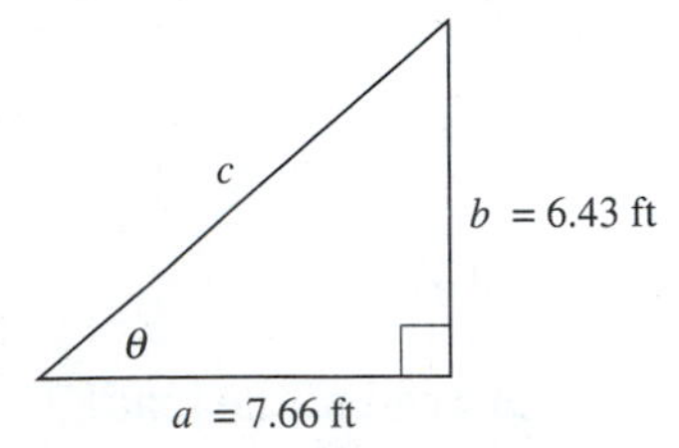

Figure A–5

☐ **EXAMPLE PROBLEM 9: FINDING THE HYPOTENUSE AND ANGLE**

Find the hypotenuse (c) and angle θ for the triangle in Figure A–5.

■ ***SOLUTION***

$$c = (a^2 + b^2)^{1/2}$$

$$c = [(7.66 \text{ ft})^2 + (6.43 \text{ ft})^2]^{1/2}$$

$$c = 10.0 \text{ ft}$$

$$\theta = \tan^{-1}\left(\frac{a}{b}\right)$$

$$\tan\theta = \frac{6.43 \text{ ft}}{7.66 \text{ ft}} = 0.839$$

$$\theta = \tan^{-1}(0.839)$$

$$\theta = 40°$$

A.4 TRIGONOMETRIC EQUATIONS

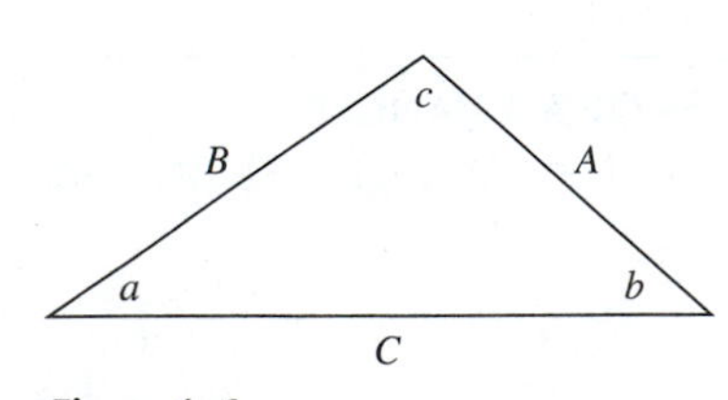

Figure A–6

Here are some relationships that can be used with triangles that are *not* right triangles. They refer to Figure A–6.

The ratios of the sines of a triangle are proportional to the length of the opposite sides.

$$\frac{\sin a}{A} = \frac{\sin b}{B} = \frac{\sin c}{C}$$ **Law of Sines**

The law of cosines is also sometimes useful.

Law of Cosines

$$A^2 = B^2 + C^2 - 2BC \cos a$$

$$B^2 = A^2 + C^2 - 2AC \cos b$$

$$C^2 = A^2 + B^2 - 2AB \cos c$$

The following approximation can be made for small angles (less than 5°).

$$\sin\theta = \tan\theta$$
$$\cos\theta = 1$$

If the angle is expressed in radians, then the small angle approximation is:

$$\sin\theta = \tan\theta = \theta \text{ (radians)}$$
$$\cos\theta = 1$$

A.5 PERIMETERS, AREAS, AND VOLUMES OF GEOMETRIC FIGURES

Plane Figures

GEOMETRIC FIGURE	PERIMETER (P)	AREA (A)
Rectangle	$P = 2(a + b)$	$A = a\,b$ 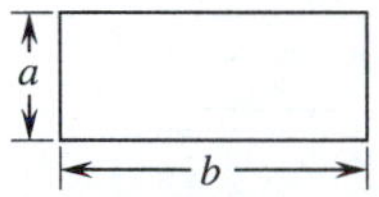
Parallelogram	$P = 2(a + b)$	$A = b\,h$ 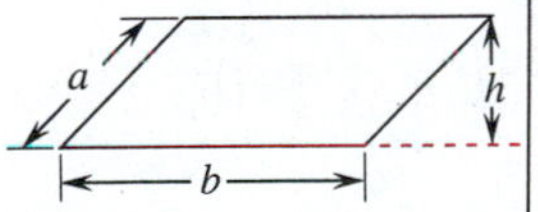
Trapezoid	$P = a + b + c + d$	$A = \frac{(a + b)h}{2}$ 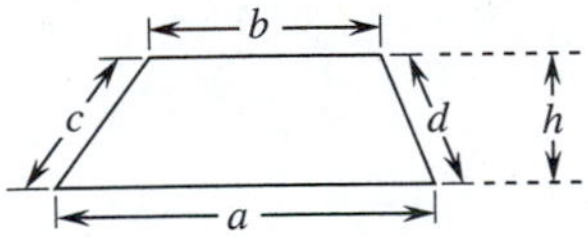
Triangle	$P = a + b + c$	$A = \frac{b\,h}{2}$ 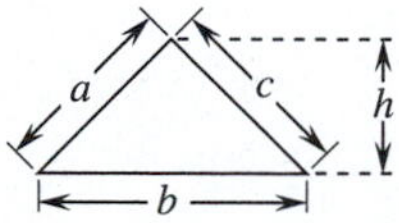
Circle	$P = \pi d$, or $P = 2\pi r$	$A = \frac{\pi d^2}{4}$, or $A = \pi r^2$ 

Solid Figures

GEOMETRIC FIGURE	VOLUME (V)	LATERAL SURFACE (LS)
Circular cylinder	$V = \pi r^2 h$	$LS = 2 \pi r h$ 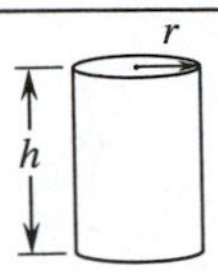
Pyramid	$V = \frac{B h}{3}$	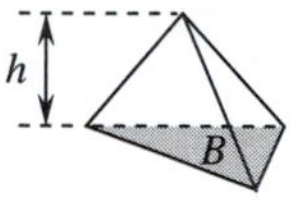
Cone	$V = \frac{\pi r^2 h}{3}$	$LS = \pi r s$ 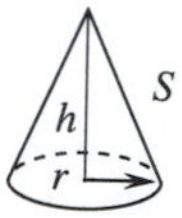
Sphere	$V = \frac{4 \pi r^3}{3}$	$LS = 4 \pi r^2$ 
Any solid figure with a constant cross-sectional area	V = cross-sectional area × height, or $V = V = B h$	LS = perimeter × height, or $LS = P h$ 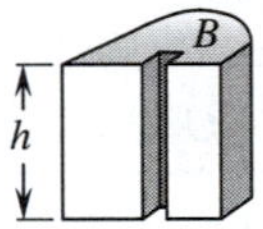

Appendix B

CHAPTER 1

Exercises

3. Here are some possibilities:

$$9 + 9 + \frac{9}{9} + \frac{9}{9}, \quad \frac{99}{9} + 9 - 9 + 9$$

5. $10^n \times 10^m = 10^{n+m}$
   a. 6×10^{15} **b.** 4.9×10^{11} **c.** 4.4×10^{17} **d.** 4.2×10^{-11}

7. $\log 100 = 2$ $\log 10 = 1$ $\log 1 = 0$
   $\log 0.1 = -1$ $\log 0.01 = -2$ $\log 0.001 = -3$
   $\log 0.0001 = -4$ $\log 0.00001 = -5$ $\therefore \log 0 = -\infty$

Problems

1. 317 in^2 **3.** *Volume = cross section × height*, or $V = (CS) \times h$

5. **a.** 3.2×10^{-3} **b.** 1.3×10^2 **c.** 7.6×10^5
   d. 6.5×10^{-5} **e.** 2.3×10^7 **f.** 1.4×10^1
   g. 2.70×10^{-2} **h.** 7.6×10^{-1}

7. **A.** b, c, d, f
   B. $23 = 2.3 \times 10^1$
   $11.6 \times 10^{-4} = 1.16 \times 10^{-3}$
   $0.034 = 3.4 \times 10^{-2}$
   $0.93 \times 10^{14} = 9.3 \times 10^{13}$

9. **a.** 9266 **b.** 0.4424 **c.** 0.0929 **d.** 0.6286

CHAPTER 2

Exercises

1. **a.** 6.4×10^{-5} s **b.** 4.5×10^{-12} m **c.** 2.3×10^4 m
   d. 3.4×10^4 tons **e.** 3.7×10^7 W **f.** 5.82×10^{-1} cm
   g. 1.4×10^{-3} m **h.** 7.4×10^{-3} W

3. **a.** 12.8 m **b.** 0.432 m **c.** 2.58×10^3 m
   d. 1.71×10^1 m **e.** 1.43 m

5. **a.** 64.8 ft^2 **b.** 4.61×10^3 in^2 **c.** 2.21 m^2
   d. 3.19 in^2 **e.** 1.8×10^4 cm^2 **f.** 1.20×10^{-6} km^2

7. Yes, we are using a smaller unit (43.2 m = 142 ft).

9. (1) 8.2 (2) 9.1 cm (3) 9.9 cm (4) 10.0 cm
   (5) 10.1 cm (6) 11.0 cm (7) 12.2 cm (8) 13.3 cm

11. **a.** 3.20×10^2 lb **b.** 5.60×10^{-4} mm **c.** 2.300×10^4 mg
    d. 7.30×10^3 lb **e.** 8.030×10^3 ft **f.** 9×10^1 ton
    g. 7.30×10^{-2} m **h.** 5.300×10^2 kg

13. **a.** 8.9 m^2 **b.** 14.4 lb · ft **c.** 6.76 g/cm^3 **d.** $3\bar{0}$ mph
    e. 129.0 cm **f.** 32.0 ft

Problems

1. 30 AU **3. A.** 76 cm **B.** 86 cm Actual sizes are 75 cm and 85 cm.

5. Yes, the width is 5.1 cm. **7.** 8.2 yd^3

9. **a.** 18.9 acre · ft **b.** (8.23×10^5) ft^3 **11.** 72 kg

CHAPTER 3

Exercises

1. Yes. **3.** As liquid is added to the beaker the weight increases, but the density of the liquid remains constant. **5.** 4

7. **A.** Same magnitude, but opposite directions.
   B. Same direction, but different magnitudes.
   C. Same magnitude, but different directions.
   D. Vectors are the same.

9. 85 lb **11. A.** B, E, F **B.** C, D, E **C.** E **D.** A

13. $V_x = +31.5$ m $V_y = +37.9$ m

Problems

1. **B.** 0.0327 $lb/in^2/R$ **C.** 22.9 lb/in^2

3. $W = 22.5\ lb/ft^3 + 5.0\ lb$ **5. A.** $f = \frac{c}{\lambda}$, or $f \propto \lambda^{-1}$
   B. $3.0 \times 10^8\ m/s = c$

7. $P = k\ R^{3/2}$, where k is a proportionality constant.

9. **A.** 405 mi **B.** 387 mi $\angle 193°$

11.

	V_x	V_y
a.	46.9 mph	110.5 mph
b.	−20.9 km	11.1
c.	−54.2 lb	−50.5 lb
d.	0	−460 m/s^2

13. **a.** 55 ft $\angle 50°$ **b.** 56 m $\angle 165°$ **c.** 23 m/s $\angle 180°$ **d.** 34 mph $\angle 69°$
    e. 265 ft/s $\angle 208°$

15. 175 mph $\angle 224°$ **17.** 27 ft **19. a.** 1.8×10^3 ft · lb **b.** 36.5 N · m
    c. 52.8 m^2/s **d.** 0 ft · lb

CHAPTER 4

Exercises

1. Yes. A car may move around a curve at constant speed, but its direction is changing.

3. Yes. Acceleration has direction. Speed does not have direction. A ball thrown upward will increase in speed after it has reached the highest point in its flight.

5. B and D. Equation 4–3 is true only for constant acceleration, indicated by a straight line in a velocity versus time graph.

7. No. Speed has no direction associated with it. If there is motion, there will be a nonzero average speed.

9. **A.** A, E **B.** D, E, H **C.** C **D.** A, C, E, H

11. $\frac{S_A}{S_B} = 4$

13. 16 ft:64 ft:144 ft, or a ratio of $1{:}4{:}9 = 1{:}2^2{:}3^2$ **15.** **B** and **E**

17. No. The acceleration is downward in both cases. As the ball moves upward the acceleration is in the opposite direction from the velocity, causing the ball to slow down. As the ball falls, velocity and acceleration are in the same direction, causing the ball to speed up.

19. As water falls its speed increases. The volume of water passing some point per second should remain constant. This produces a constant flow rate. The water stream narrows to keep the flow rate constant. As the speed increases gaps occur to maintain the constant volume flow rate.

21. *A* is in the air the longest time; *D* the shortest. The time of flight is determined solely by the vertical component in the absence of friction.

23. **A.** As the angle of elevation increases, the time of flight also increases.
    B. On a flat surface the maximum range occurs at an angle of 45° if air friction is negligible. The angle of maximum range will vary around 45° depending on factors such as firing uphill or downhill or from an elevated platform.
    C. As the angle of elevation increases up to 90°, the maximum height increases.

Problems

1. $\bar{v} = 10.0$ m/s

3. Average speed = 22.2 ft/s; average velocity = 0 ft/s

5. **A.** 14.0 mi **B.** 14 mph

7. **A.** Average speed = 12 mph **B.** Average velocity = 4 mph toward B

9. 15.5 ft/s^2 **11.** 28 m/s **13.** 9.47×10^3 ft/s^2 **15.** 10.0 m/s

17. **A.** -17.5 ft/s^2 **B.** 186 ft

19. 6.0×10^3 ft/s^2

21. **A.** 2.1 s **B.** $6\bar{0}$ ft/s **23.** 2.3 s **25.** 6.7 m/s **27.** A **29.** 0.8 s

31. 250 ft **33.** 156 m

CHAPTER 5

Exercises

1. $S = R\theta$, *or* $\theta = S/R$ will work only for θ in radians. For example: S = circumference of a circle $= C = 2\pi R$. $\theta = 2\pi R/R = 2\pi$. If an angle of 360° were used, we would get the wrong numerical result: $S = C = 360° R$ for the circumference.

3. All parts of the hammer will have the same angular velocity because it is a rigid body. The center of gravity is near the head of the hammer.

The hammer rotates about this point. The end of the hammer must travel through a circle with a larger radius than the head with the same angular speed. Its tangent speed will be larger ($v_t = \omega R$).

5. Smaller, 1 rpm = 0.105 rad/s.

7. The turbine is slowing down. The acceleration is in the opposite direction from the initial velocity.

9. Yes. + and − are merely used to indicate clockwise or counterclockwise directions of spin.

11. The three transform equations. The radius is involved with each of these relationships.

Problems

1. **a.** 156° **b.** 23.6 rad **c.** 3.42 rev **d.** 234 rad

3. **a.** 6.80 in **b.** 35.4 cm **c.** 0.250 m **d.** 0.699 ft **5.** 807 rev

7. **A.** 6.0 rad/s **B.** 24 rad/s **C.** 6.87 rad/s **9.** 1 rad/s = 9.55 rpm

11. **A.** 1880 rad/s **B.** 1440 m/s **13.** 8570 rpm **15.** 44 rad/s^2

17. −6.63 rad/s^2 **19.** −21.2 rad/s

21. **A.** −3.68 rpm/s, or −0.39 rad/s^2 **B.** 13.8 rev

23. **A.** 21.4 rad **B.** 8.65 rad/s

CHAPTER 6

Exercises

3. **A.** No, the path is not a straight line on the vertical plane.
   B. Yes, we have constant speed and a straight line.
   C. No, the rock's speed is changing.
   D. No, the earth moves through a curved path.
   E. Yes, the girder moves in a straight line at a constant speed.

5. **A.** F_{AB}, F_{BA}, F_{EA}, and F_{AE} **B.** $F_{TE} = F_{RB} + F_{RA} + F_{RB}$ **7.** 0

9. **A.** All forces are internal forces except the frictional force of the ground acting on the wheel (F_{12}).
   B. F_{12}, F_{21}, F_{34}, and F_{43} are internal forces. F_{65} is the external force acting on the engine. The reacting force F_{56} drives the clutch.
   C. F_{21} acts on the piston.

11. **A.** 5.00 m/s^2 **B.** 1.00 m/s^2 **C.** 9.8 m/s^2 **D.** 5.00×10^{-3} m/s^2
    E. 1.00×10^3 m/s^2 **F.** 0.225 ft/s^2 **G.** 7.3 ft/s^2

13. The frictional force between the seat and the passenger is too small to slow the passenger as rapidly as the car. As the car comes to rest, the passenger continues forward until hitting an obstruction.

15. **A.** 0.0196 N **B.** 19.6 N **C.** 64 lb

17. **A.** The 4.6-lb pail of paint is larger. **B.** The chain is slightly heavier. **C.** The 100-lb box of nails is heavier.

19. No. The spring scale measures the weight of an object hung on the spring. If it is used on the moon, it will read too low. The balance compares masses instead of weights. It will read the same on the moon and on the earth.

21. B **23.** C **25.** The center of mass is closer to the more massive steel head of the hammer than to the center of the lighter handle.

27. The force is created by a reaction to the mass of air expelled from the balloon.

Problems

1. 800 lb **3.** A or B **5.** 47 lb

7. A. 19.6 N **B.** 0 (constant speed) **C.** 100 N **9.** 280 lb

11. A. cars c and d **B.** car e **13.** 340 lb **15.** 89 m/s

17. A. 89 N **B.** 3.0 lb **C.** 530 N **D.** 2.0×10^4 lb **E.** 0.36 N

19. A. 72.2 lb **B.** 12.2 ft/s^2 **21.** 0.59 cm **23.** $4\bar{0}0$ g

25. A. 3.2 ft from the camper to the center of the canoe **B.** 18.2 ft **C.** The center of gravity of the canoe remains unchanged. Only internal forces act between the canoe and camper (neglecting friction between the canoe and water). **D.** 8.6 ft **27.** 75 N **29.** 7.00×10^3 N

31. A. It remains in its initial position. **B.** $6\bar{0}$ g **C.** $+6000$ g · m/s and -6000 g · m/s **D.** 0

CHAPTER 7

Exercises

1. A. $\mathbf{g}$ = 11.8 m/s^2 **B.** $\mathbf{g}$ = 11.8 m/s^2 **C.** $\mathbf{g}$ = 11.8 m/s^2

3. Yes, with an acceleration of more than 4.0 ft/s^2.

5. The float moves toward the front or rear of the tank pulling the cable, which indicates the liquid level. **7.** d. $\mathbf{W}$ cos 5°

9. A. Yes **B.** Yes **C.** No **D.** No **11.** C

Problems

1. A. 32 lb **B.** 36 lb **C.** 28 lb

3. A. 590 N **B.** 390 N **C.** 490 N

5. 31.2 ton **7.** 1.6 m/s^2

9. A. 16 ft/s^2, or 4.9 m/s^2 **B.** 26 ft/s^2, or 7.8 m/s^2 **C.** 26 ft/s^2, or 7.8 m/s^2 **D.** No

11. 2.4×10^4 N **13.** 1.6×10^3 lb **15.** $12\bar{0}$ lb **17.** 285 N

19. a = 0.47 m/s^2 T = 10.3 N

CHAPTER 8

Exercises

1. No. $F_1 > F_2$

3. The rotating tire responds to the higher coefficient of static friction. A skidding or spinning tire interacts with the road surface with the lower sliding coefficient. At lower speeds the force exerted by the engine can match the higher static frictional force.

5. No. The car that is moving into the wind has a larger velocity relative to the moving air than the car moving with the wind. Since air drag is proportional to the square of the air speed, the car with the headwind has the larger air friction.

7. At higher elevations, the air density is less than near the ground. The terminal velocity will be greater at a higher altitude. The terminal velocity can be reduced by increasing the drag coefficient by taking a spread-eagle position.

9. The table tennis ball would be slower than the golf ball. They have essentially the same size and shape and thus the same air resistance. This gives them the same upward force, but they have different weights acting downward. The ball with the larger weight will have the larger terminal speed.

11. In general, the stiffness of a spring is inversely proportional to its length. Each coil can stretch by a certain amount with a fixed tension. If we increase the number of coils by tying the two springs together, we will have a larger Δx for the same tension (F). Since $k = -F/\Delta x$, k is smaller.

13. 8.0 ft/s^2 and 3.6 ft/s^2

15. Infinite. Equation 8–7 is true only *outside* a mass with spherical symmetry.

Problems

1. **A.** 7.1 N **B.** 6.1 N **C.** 2.4 N **D.** 0.0 N

3. **a.** 14.0 lb **b.** 22.0 lb **c.** 14.5 lb **5.** 87 N

9. $W_p < f_r$. The blocks remain stationary.

11.

R	V_t
0.40 cm	14.3 m/s
0.10 cm	7.1 m/s
0.01 cm	2.2 m/s

13. 190 lb **15.** 48 lb **17.** 15 N/mm, or 15×10^3 N/m **19.** 0.69 m

21. 83 N/m **23.** 6.67×10^{-11} N **25.** 3.8 m/s^2 **27.** 2.83 R_e

CHAPTER 9

Exercises

1. Yes. When the car is traveling at a constant speed, the net force must be zero.

3. The two forces must be equal in magnitude and opposite in direction. Also, they must be concurrent forces. Otherwise there would be a net torque.

5. A jet or prop engine supplies a forward thrust. Air drag resists forward motion. Weight acts downward and is balanced by the lift of the wings.

7. No. There is no lever arm.

9. A larger torque can be created with the larger radius wheel.

11. The axle acts as a fulcrum. Weight placed behind the wheels creates a torque that must be balanced by a countertorque at the trailer hitch. As a result, the car or truck pulling the trailer experiences an upward force on the rear end of the vehicle and the hitch is under tension rather than compression.

13. 87

15. Yes. If the rate of spin is constant, the sum of the external forces and the sum of the torques can both be zero.

17. The center of gravity is below the radius of rotation of the toy's base. The center of gravity creates a countertorque when the Weeble is tilted over wide angles.

19. *Pipe wrench:* The torque exerted by forces on the handle is balanced by a countertorque exerted on the jaws of the wrench by the pipe. The ratio of lever arms allows a small force on the handle to create a large torque.
Nutcracker: The torque exerted on the handle is balanced by a resisting countertorque supplied by the nut. Because the nut is closer to the hinge of the cracker it must exert a larger force than is exerted on the handle.
Bumper jack: Again, a long lever arm makes it possible to use a small force at one end of the device to overcome a larger resisting force. In this case the force that is closer to the fulcrum is part of the weight of the car.
Screwdriver: The screwdriver acts like two concentric axles. The handle has a wider radius, which allows a small force to transmit a fairly large torque to the screw.

21. There needs to be a vertical component in the rope's tension acting upward to balance the downward weight of the wet sheet.

23. This is an example of vector equilibrium. Guys or posts are needed to balance horizontal components of force.

Problems

1. **a.** $F_1 = 64$ N; $F_2 = 77$ N **c.** $F_1 = F_2 = 200$ N
b. $F_1 = 50$ N; $F_2 = 0.0$ N **d.** $F_1 = 165$ N; $F_2 = 56.6$ N

3. $M = 0.30$ kg 5. 12.2 lb

7. **a.** 26 N · m **b.** −40 N · m **c.** 2.6 ft · lb **d.** −6.7 ft · lb

9. 1.6 N · m

11. $X_{cg} = 6.0$ cm; $Y_{cg} = 5.7$ cm **13.** 12 cm

15. 37 lb = Sam; 23 lb = Wilma

17.

	T	F_x	F_y
a.	220 lb	191 lb	10 lb
b.	120 lb	120 lb	120 lb
c.	150 lb	75 lb	−9.9 lb

19. $F_2 = 246$ lb
$F_3 = 172$ lb
Torque = 0

21. −79 ft · lb, tilts **23.** 120 lb **25.** 52 lb

27. $T = 690$ lb; $C = 870$ lb

CHAPTER 10

Exercises

1. **A.** Centripetal **B.** Both **C.** Both **D.** Both
   E. Centripetal **F.** Neither

3. **A.** 6 ft/s^2 **B.** 48 ft/s^2 **C.** 16 ft/s^2

5. The adhesive force between the mud and the tire is too small to supply the necessary centripetal force to keep the mud traveling in a circular path. The mud flies off tangent to the tire.

7. A fast-spinning cylinder pushes inward on the clothes, holding them in a circular path. Holes in the spinning cylinder allow water to escape the spinning path.

9. Gravity acts into the center of the circle at the top and outward at the bottom. Halfway up, gravitational acceleration is opposite velocity, slowing down the bob. On the way down, gravity helps increase the bob's speed.

11. This allows a wide range of nonskidding speeds.

13. Gravitational force rather than rocket fuel was used to change the direction of motion.

15. 85.2 h

Problems

1. 9.7 ft/s^2 **3.** 12 ft/s^2, 3.0 ft/s^2 **5.** 22 m/s^2 **7.** 2200 lb

9. **A.** 2.5 N **B.** 6.5 N **11.** 39 mph **13.** 67 rad/s **15.** 16 m/s

17. 360 m **21.** **A.** 1.90×10^{27} kg **B.** 318 times as massive

CHAPTER 11

Exercises

1. Not necessarily. An impulse is the product of force and the length of time the force acts. A small force of long duration may give a larger impulse than a large force of short duration.

3. D
   A. Velocity increases, momentum changes.
   B. Direction changes, momentum changes.

C. Velocity increases, momentum changes.
D. Velocity constant, momentum constant.

5. As the car moves up (+ momentum), the counterweight moves downward (− momentum). The winch needs to supply only the difference between the two momentums.

7. 1. At separation the backward momentum of the expended stage is matched with an equal-sized momentum forward by the remaining stages.
   2. Once the used stage is separated, the remaining mass is smaller. It will accelerate more rapidly than it would with the expended stage attached.

9. C   11. Northeast

13. No. The one with the mass near the rim has a larger moment of inertia. It will have a larger angular momentum.

15. The shoes create the maximum torque at the rim.

17. The tail rotor exerts a countertorque for stability. Without the tail rotor the body of the helicopter would rotate in the direction opposite the main rotor to maintain conservation of angular momentum.

19. The linear momentum of the piston is continually changing. However, over one complete cycle the average linear momentum is zero. In the unlikely case that there is no friction there would be conservation of momentum over a complete cycle. If there is friction, forces on the piston will slow it down. These same frictional forces transmit a retarding torque to the crankshaft.

21. Momentum is the product of mass and velocity. Since the water delivered by the ram is greater than the water entering the device it must have a smaller mass in order to maintain conservation of momentum.

23. 45°

Problems

1. **A.** 1.2 N · s   **B.** 1.2 kg · m/s   **C.** 3.0 m/s

3. 120 m/s   **5. A.** 9.9×10^3 slug · ft/s   **B.** 1.5×10^3 lb

7. 0.40 lb   **9.** 4.4 ft/s   **11.** 21 mph, east

13. 15 m/s eastward   **15.** 0.125 kg · m^2 **17.** 5.0-lb force produces a torque of 150 lb · ft.   **19.** 1.65 rad/s   **21.** −10 rad/s

23. 4.1 rad/s

CHAPTER 12

Exercises

1. Yes. An object must be displaced for work to occur. A force acting on a stationary object does no work.

3. None. Work is done against kinetic friction.

5. No. It will be proportional to a function of distance.

7. Yes. If the object does not rotate no work is done.

9. The power is converted into light and radiant heat.

11. A kilowatt-hour has units of power × time = energy.

13. No. The speed is squared. The square of any real quantity is positive. Kinetic energy can be added to or subtracted from a system, but cannot be less than zero.

15. **A.** 6.75×10^5 J **B.** 2.52×10^6 J **C.** 3.0×10^3 J **D.** There is less relative kinetic energy to transfer in the collision. **17.** 6.3×10^7 J

19. Yes, because potential energy is proportional to position, which may be negative.

21. Kinetic energy is converted to heat.

23. Frictional force will heat up and wear down the clutch lining from relative motion.

25. Most of the kinetic energy is converted to potential energy as the rock rises. Air friction may cause part of the kinetic energy to be converted to heat.

Problems

1. **A.** 276 ft · lb **B.** 239 ft · lb **C.** 0.00 ft · lb **D.** 205 J **E.** 205 J **F.** 1.33×10^3 J

3. 350 J **5.** **A.** 73.8 ft · lb **B.** 369 J **C.** 170 ft · lb

7. 2.7×10^6, or 3.6×10^6 **9.** 0.83 J

11. **A.** 12×10^3 ft · lb **B.** 2.0×10^4 J **C.** 3.13×10^5 ft · lb

13. **A.** 276 ft · lb **B.** 158 ft · lb **C.** 176 ft · lb **D.** 60.0 N · m **E.** 812 N · m **15.** 4.0×10^2 J

17. **A.** 11 hp **B.** 9.4×10^2 ft · lb/s **C.** 1.23×10^3 W **D.** 170 ft · lb/s **E.** 41 kW

19. 1.3×10^3 ft · lb or 1800 J **21.** 12 hp **23.** $2\bar{0}$ W

25. **A.** 2.4×10^{-11} J **B.** 740 J **C.** 89 ft · lb **D.** 272 ft · lb **E.** 9.6×10^3 J, or 7.1×10^3 ft · lb **F.** 9.8×10^5 ft · lb

27. 30.0 ft/s **29.** 0.14 J **31.** 2.29×10^9 J; 11% **33.** 384 ft · lb

35. **A.** 2.81×10^4 ft · lb **B.** 2.56×10^4 ft · lb **C.** 2.5×10^3 ft · lb

37. 620 ft · lb/s **39.** $1\bar{0}$ HP

41. **A.** 14.4% **B.** 75 lb **C.** 6.34×10^5 ft · lb **D.** 22 mi

CHAPTER 13

Exercises

1. **A.** $I_a = 2\ I_B$ **B.** $KE(A) = 1/2\ KE(B)$

3. The linear speed is directly proportional to the angular speed of the wheels. The proportion of energy is constant.

5. 99.1%

7. Use $m = I/R^2$ and $v = \omega R$. 9. $PE = 1/2\ (C\omega^2)$

11. **A.** A mass hung from a spring or rubber band
 B. A falling object or a wheel rolling down an incline
 C. A mass sliding down an incline at constant speed or a meteoroid striking the earth's atmosphere

13. **A.** A **B.** D **C.** B **D.** C

15. The total kinetic energy is $(1/2)mv^2 + (1/2)I\ (v/R)^2$. On the incline the object rolls on the dowel with a small radius. The rotational part of the kinetic energy is large because we divide by a small R in the energy equation. When the large-radius wheels touch the floor, the object rotates about the axle with the larger radius wheel. The second term of the energy equation is smaller because of the larger R. Some of the rotational kinetic energy is converted to linear kinetic energy.

17. The momentum of one car before the collision is mv, and, of the other car, $-mv$, for a total momentum of zero. After collision both will have a velocity of zero. Initially each car has a kinetic energy of $(1/2)mv^2$ and a final kinetic energy of zero. This is the same result that would occur if each car ran into an immovable object such as a concrete wall separately.

19. The lighter car undergoes a greater change of velocity during the collision. This means a larger acceleration. The forces between the car and its passengers would be larger.

Problems

1. **A.** 240 ft · lb **B.** 460 ft · lb **C.** 594 ft · lb

3. **A.** 1.46×10^3 kg · m **B.** 9.1×10^3 J **5.** 0.173 slug **7.** 5.6 ft/s

9. **A.** 7.64×10^5 ft · lb **B.** 113 ft/s **11. A.** 9.69×10^3 J
 B. 0.547 **13.** 1.9 m/s

15. **A.** 3.1 ft · lb **B.** 3.1 ft · lb **C.** 6.2 ft · lb

17. 76 ft · lb/s **19.** about 4%, assuming disk-shaped wheels **21.** 73 ft/s

23. **A.** -29 ft/s **B.** 7.2×10^5 ft · lb

25. 56-g object has a velocity of 1.4 m/s.
 78-g object has a velocity of 7.4 m/s.

CHAPTER 14

Exercises

1. No. An efficiency larger than one means more work is gotten out of the machine than is put into it. This violates the law of conservation of energy.

3. Distance or velocity

5. No. The IMA assumes no friction in a machine. It is the largest MA the machine can produce. AMA is an experimental value with friction present.

7. Machine screws go into prethreaded holes. A wood screw is a combination screw and wedge used to draw the thread into the wood fibers.
9. The efficiency increases. Work done against friction decreases since the normal force decreases and the length of the incline also decreases; the force acts against friction for a shorter distance.
11. The trailer axle acts as a lever arm. Loads placed on the back of the trailer will tend to lift the front end of the trailer. This reduces the normal force on the back wheels of the car, reducing control. The trailer hitch will be under tension rather than compression, making a failure more likely.
13. Lifting with the back requires a larger lever arm, and the upper body becomes part of the load. The legs have more-massive muscles, and the upper body has a negligible lever arm.

15. 0.62 **17.** Decreasing **19.** Torque

Problems

1. 0.842 **3.** 2.2×10^4 J

5. A. 5.4 **B.** 3.1 **C.** 0.57 **7. A.** 0.405 **B.** 0.455 **9.** 2.0

11. 1.7×10^3 N **13.** 12 **15.** $IMA = 4.8$; $AMA = 3.3$

17. A. 35.9 ft · lb **B.** 27 **19. A.** 29 ft **B.** 5.0 **C.** 71%

21. A. 12.4 **B.** 0.47 **23. A.** 2.3 **B.** 0.43

25. 84:32, or 2.63:1 **27. A.** 5000 rpm **B.** 1.3 N · m

CHAPTER 15

Exercises

1. **A** and **D** are material constants; they depend only on the nature of the materials. **B** and **C** depend on dimensions; they are not material constants.
3. The number of pounds is directly proportional to the volume.
   A. Larger **B.** Smaller **C.** Smaller **5.** 920 kg
7. Of or pertaining to tension
9. Under identical compressions and tensions the two materials will have the same change of length, but material A will undergo a larger change of cross section.
11. Aluminum. Aluminum has a smaller shear modulus.
13. θ is directly proportional to the length (L) of the rod.
    θ is inversely proportional to the radius of the rod.
    θ is directly proportional to ϕ for larger strain moduli (G).
    θ decreases for smaller ϕ, or larger G will decrease.

15. A. D **B.** C **C.** A **D.** B **E.** C **F.** A

Problems

1. 62.5 lb/ft^3 **3.** 4.23×10^{-3} **5.** \$53.82 **7. A.** 4.37×10^3 psi
B. 1.51×10^{-4} **C.** 0.022 in **D.** 7.9×10^{-6} in

11. Thermal stresses caused by the much larger thermal coefficient of expansion of aluminum loosen the contact. Loose contacts tend to heat up when electricity flows, causing a fire hazard.

13. The nails contract more than the surrounding wood. Friction between the nail and wood is reduced, releasing the nail.

15. Polymers, liquids, metals, ceramics

17. The kinetic energy is effectively zero.

19. The surrounding air pressure may be different.

Problems

1. **A.** 357°C **B.** −33°C **C.** 17.8°C **D.** 38°C **3.** 9.4°C

5. 0.31 mm **7.** 0.155 in **9.** 0.28 ft **11.** 2.5 in^2 **13.** 0.00 cm, both expand by the same amount **15.** 123 in^2 **17.** 54 gal

19. $\Delta V = (V + \Delta V) - V = (l + \Delta l)(w + \Delta w)(h + \Delta h) - lwh$. Expand the product and drop small terms. Divide by $V = lwh$. $\Delta V/V = \Delta l/l + \Delta w/w + \Delta h/h = 3\alpha T$. **21.** **A.** 690 R **B.** 428 R **C.** 510 R

23. **A.** 800 K **B.** 93 K **C.** 293 K **25.** 105 ft^3 **27.** 0.45 atm

CHAPTER 19

Exercises

3. **A.** 21 **B.** 10 **C.** 21 **5.** **A.** 2.18 moles **B.** 96 g **C.** 1.31×10^{24} molecules

7. **A.** The pressure increases because temperature increases.
   B. The pressure increases because the number of molecules increases.

9. **A.** The kinetic energies are the same.
   B. The less massive carbon monoxide would have a larger average velocity.

11. **A.** Smaller speed **B.** Decreased **C.** Temperature decreases with expansion.

13. Absolute temperature and mass

15. See Figure 19–9. The velocity distribution curve has an exponential tail. A small fraction of the molecules have speeds of more than 10 times the rms speed.

Problems

1. **A.** 28.01 u **B.** 17.03 u **C.** 16.03 u
   D. 120.9 u **E.** 64.04 u **F.** 34.08 u

3. 2.82×10^{24} atoms

5. 3.39×10^{-25} kg

7. 9.0×10^{-4} g/cm^3

9. **A.** 950 l **B.** $8\bar{0}0$ in^3 **C.** 7.16×10^{-3} m^3

11. **A.** 44.09 u **B.** 57.4 moles **C.** 3.46×10^{25} molecules **13.** 385 l

15. 286 ft^3 17. a. 534 m/s b. 489 m/s c. 472 m/s d. 1270 m/s 19. 5.66×10^{-24} ft · lb/R 21. 22.5 K

CHAPTER 20

Exercises

3. Yes. The kinetic energy gained by the falling water is converted to heat when the water strikes the rocks.

5. Iron and copper

7. Gain: ice and water. Lose: lead and cup

9. The steam will change to water in contact with the skin, releasing the latent heat of vaporization. There is much more thermal energy in steam at 100°C than in water at the same temperature.

11. The melting ice must gain thermal energy from the brine solution to change state. This reduces the temperature of the ice-brine solution.

13. Water on your skin evaporates. Some of the heat needed to evaporate the water comes from your body.

15. The wind moves saturated air away from the clothes and replaces it with air that has a lower relative humidity.

17. One scheme would be to place the juice in a partial vacuum. This would enhance boiling at a low temperature.

Problems

1. 29.6 Btu

3. A. $12\bar{0}$ kcal B. 1.81 Btu C. 4.89 kcal D. 163 therm

5. 0.39 therm 7. 1.86×10^6 J 9. 48 Btu/°F 11. A. 308 Btu/°F B. 270 min 13. 0.1 Btu/(lb · °F) 15. 2.2°C 17. 7.4°C

19. 5.19×10^4 cal 21. 574 cal/g 23. 9.9×10^3 Btu 25. 250°F

27. A. 70.2 lb/in^2 B. T > 257°F

CHAPTER 21

Exercises

1. A. 800 cal B. 173 Btu C. 1300 cal 3. 0% 5. 80%

7. The pressure decreases the volume of the gas. The gas does negative work. This is another way of saying that work is done on the gas [$P(-\Delta V) = -W$].

9. The drop in pressure and large increase in volume of the gas occur rapidly. There is little opportunity for the gas to gain energy from its surroundings to do this work. This is an adiabatic process. Internal energy of the gas is used for the work of expansion, reducing the gas's temperature below its sublimation temperature.

13. It would involve the 100% conversion of heat into work. This violates the second law of thermodynamics.

15. The heat lost by the ice cream is absorbed by the walls of the ice cream container and the ice-brine solution surrounding it. These things undergo

an increase of entropy. For any process, the net entropy of the *universe* increases. There can be a decrease of entropy in some component parts. If we add up *all* of the increases and decreases involved in the process, the net change is always an increase.

17. 0 K **19. A.** *COP* is infinite. **B.** Heat is transferred from a hot reservoir to a cold reservoir without work being done on the system. This violates the second law of thermodynamics.

21. Substitute $Q_2 = Q_3 + Q_4 - Q_1$ into $COP = Q_2/Q_1$.

23. The two-cycle engine does not require a complicated valve system and it has a more frequent power stroke than a four-cycle engine operating at the same revolutions per minute.

Problems

1. 187 Btu **3.** 2.52×10^4 cal **5.** 0.76 **7.** 85% **9.** 6.9×10^5 J

11. 150 ft · lb **13.** 1.1 Btu **15.** 0.880 Btu/R **17.** 119 cal/K

19. 0.474 **21.** 2.61 **23. A.** 101 Btu **B.** 2.05 Btu

CHAPTER 22

Exercises

1. The rug has a lower thermal conductivity than the floor. Heat is conducted away at a slower rate by the rug.

3. A. Wall A, 5°F/in; wall B, −5°F/in **B.** Wall A, heat travels to the outside. Wall B, heat travels to the inside. **C.** No. The magnitudes of the gradients, the areas, and the coefficient of conductivity are all the same.

5. 8×10^4 Btu/h **7.** Infinitely thick

9.

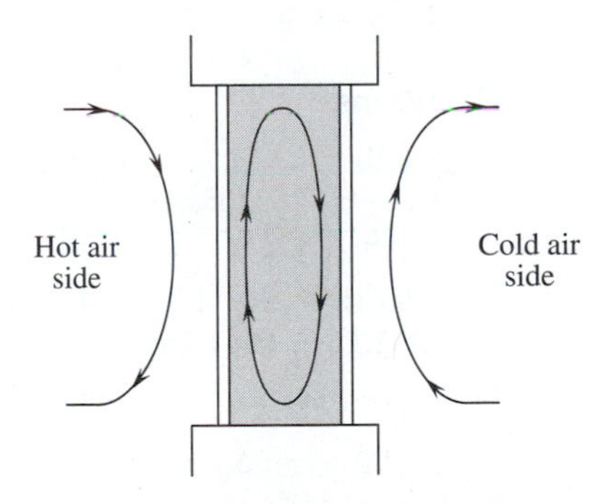

11. The spun glass captures small pockets of air, which is a good insulator.

13. The book absorbs radiant heat from its surroundings at the same rate that it radiates heat.

15. Some radiant heat reaches the room. Most of the convective heat goes up the chimney, drawing warm air from the room. A fireplace without a glass screen produces a net heat loss.

17. The temperature difference between the collector and the air temperature is smaller on a hot day. There is a smaller net radiant heat loss, leading to higher efficiency.

19. **A.** More radiant energy gets in—a heating process.
B. More long-wavelength radiant heat is radiated back to the earth from the carbon dioxide gas molecules—a heating process.
C. The dust particles will scatter some of the incoming short-wave radiation back into space—a cooling process.

Problems

1. 10°F/in **3.** 5.5×10^6 Btu/h **5.** 68°F **7. A.** 12.1 (ft^2 · h)/Btu **B.** 0.0826 Btu/(ft^2 · °F · h) **9.** 7.2×10^5 Btu **11.** 83 kcal
13. A. 2.5×10^{-3} kcal/(m^2 · s · °C) **B.** 2.9×10^5 kcal **C.** 1.21×10^6 kJ **D.** 337 kW · h **15.** 4.5×10^{26} J/s
17. 1.1×10^{-5} m **19.** At 67°C, $\frac{17 \text{ cal}}{(\text{s} \cdot \text{ft}^2)}$; at 89°C, $\frac{22 \text{ cal}}{(\text{s} \cdot \text{ft}^2)}$

CHAPTER 23

Exercises

3. A. Increase **B.** No change **C.** Decrease **5.** Displacement = 0, $a = 0,\ v = v_{max}$

7. It does not affect the frequency. Although the restoring force is larger for a larger mass, the inertia of the bob increases as well.

9. Lengthened

11. A.

$$f = \frac{1}{2\pi}\left(\frac{CAY}{mL}\right)^{1/2}$$

B. It will increase the frequency.

13. A. Decrease **B.** Decrease **C.** No change
D. Decrease **E.** No change

Problems

1. 8.6 lb **3.** 11 lb **5. A.** 2.1 in **B.** −330 in/s **C.** -5.2×10^4 in/s^2 **7. A.** 0.625 rev/s **B.** −12.8 ft/s **C.** −50.1 ft/s^2

9. 0.95 Hz **11.** 4.9 s **13.** 68 Hz
15. A. 1.75 **B.** 1.75 **C.** 3.7 m/s

17. 16 J **19.** $\Delta y = L - h \quad h/L = \cos\theta = 1 - \theta^2/2 \quad \theta = \Delta x/L$
$\therefore PE = mg\Delta y = mg\,\frac{\Delta x^2}{2L}$

CHAPTER 24

Exercises

1. Radiant heat, water waves, AC electricity, and sound waves are examples of energy transfer by waves.

3. A. 5.5 cm **B.** 40 cm **C.** 48 Hz **5.** A

7. Sound waves have a greater decrease of intensity with distance. The energy of the sound is spread over a three-dimensional space, where intensity falls off as the inverse square of distance ($I = I_0/R^2$). Ripples are constrained to the two-dimensional surface of the water, where the intensity falls off as a simple inverse proportion ($I = I_0/R$).

9. The propellers are spinning at slightly different frequencies. Alternate reinforcement and cancellation of the sound waves occur with a frequency equal to the difference of the two propeller frequencies.
11. Resonance occurs for light frequencies of $f = (nc)/(2L)$, where c is the speed of light and n is an integer.
13. Sound travels faster in the earth's rock than in the air. The long-wavelength sound waves in the earth we feel as an earthquake.
15. **A.** Increases in temperature decrease the air density. This increases the speed of sound.
    B. Water molecules are less massive than oxygen or nitrogen molecules. Higher humidity decreases the air density. This increases the speed of sound.

Problems

1. **A.** 1.3×10^9 Hz **B.** 3.78 ft
   C. 3.94×10^4 m/s **D.** 1.9×10^5 ft/s

3. 2.8 cm/s **5.** 0.44 in **7.** 0.11 W/m^2 **9.** 2.1×10^8 Hz

11. $\lambda_0 = 39.0$ in; $f_0 = 342$ Hz
    $\lambda_1 = 19.5$ in; $f_1 = 684$ Hz

13. 6.2×10^{-7} m, or 620 nm

15. **A.** 1.2×10^4 ft/s **B.** 1.9×10^4 ft/s **C.** 7.1×10^3 ft/s

17. 1.5×10^5 N/m^2

CHAPTER 25

Exercises

1. **A.** Increases **B.** Increases **C.** Decreases

3. The speed of sound is fastest in material A, which has the largest Young's modulus. Sound will be slowest in material C.

5. $\frac{2}{3}L$ $\frac{1}{3}L$ $\frac{3}{4}\lambda = L$, or $\lambda = \frac{4}{3}L$ $\frac{1}{2}\lambda = \frac{1}{3}L$, or $\lambda = \frac{2}{3}L$

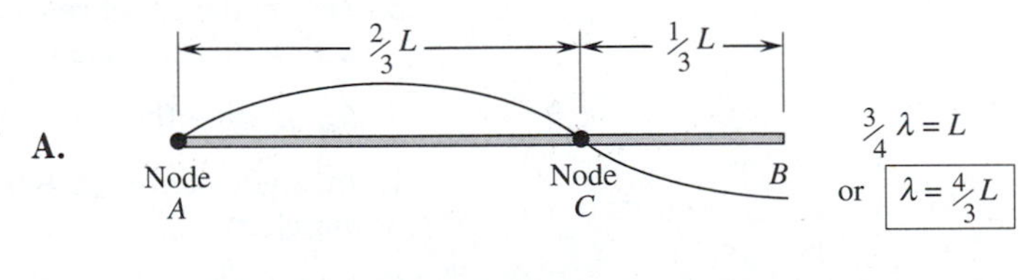

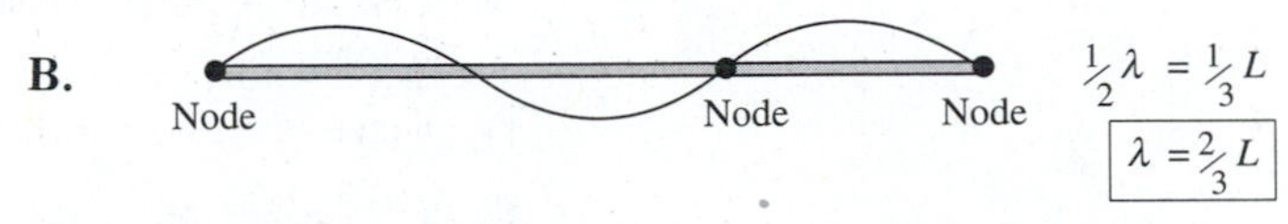

7. 4.00 in **9.** The speed of sound decreases with decreasing temperature ($f = v/\lambda$). If v decreases, then λ must also decrease to maintain the same frequency. The instrument should be shortened.

11. 2.0 b, or 20 db **13.** 2 **17.** No. Although there is a change in wavelength, there is a compensating change of the speed of sound.

Problems

1. 305 K **3.** $-231°$ C = 42 K **5.** 1.7×10^4 ft/s **7.** 0.76 **9.** 2.00 m **11.** 700 Hz **13. A.** 13.0 ft **B.** 1130 ft/s **15.** 84 dB **17.** 63 N/m^2 **19.** 8.74×10^4 Hz **21. A.** 352 Hz **B.** 348 Hz **C.** 307 Hz **D.** 300 Hz **23. B.** 10.6 m/s **25.** From 2.04×10^6 to 2.25×10^6 Hz

CHAPTER 26

Exercises

1. A, C, F **3.** The speed of light in water is slower than in air. $v = c/n = (3 \times 10^8 \text{ m/s})/1.33 = 2.56 \times 10^8$ m/s. Our eyes detect different frequencies of visible light through a color code. Since the color remains yellow in water, the frequency must be the same in water and air. The wavelength of yellow light must be shorter in water than in air.

5. Away from the normal. According to Snell's law the product of the index of refraction and the sine of the angle is a constant. In this case $n_2 < n_1$. Therefore, sin θ_2 must be larger than sin θ_1 and $\theta_2 > \theta_1$.

7. $E = hf$ and $f = c/\lambda$. Therefore, $E = hc/\lambda$. The energy of a photon is inversely proportional to wavelength.

9. 2π sr

11. The luminous flux (lumen) rating would be numerically larger by a factor of 4π than the illumination (lux) rating.

Problems

1. 31° **3.** 1.45 **5.** 0.3° **7.** $n = c/v$. Substitute $v = f\lambda'$ and $c = f\lambda$ and solve for λ'. **9.** 7.2×10^{16} photons/s **11.** 2.19×10^5 lm **13.** 74 ft · c

CHAPTER 27

Exercises

1. A. The image approaches the mirror at 2.0 mph.
B. The image approaches you at 4.0 mph.

3. The coefficient of reflection is 100%. This is better than can be obtained with a polished aluminized or silvered mirror.

5. A. Outside the focal point **B.** Inside the focal point

7. An angle of deviation is the angle between the incident ray and its final direction.

9. Yes. If the object is outside the focal length, a real, inverted image is formed.

11. A, D

13. A. Larger; $|M| > 1$. **B.** Object; $|M| = q/p$, then $q > p$. **C.** Real; only real images are inverted by a single lens. **D.** Inverted; the ratio of Q/P is negative.

15. Low. Only rays striking the interface at a small angle undergo internal reflection.

17. A microscope objective focuses a cone of light from a small source just outside the focal length of the lens. If the focal length is small, the lens is closer to the source and can catch a wider cone of light, allowing a brighter image than a longer focal length could provide.
    A telescope gathers light from a source far away. If the focal length is long, the image from the objective is formed near the focal point in a narrow beam. The narrower the beam the more easily light can pass through a small-diameter eyepiece.

Problems

1. 21 cm **3.** 45° **5.** 1.42

7. **a.** $q = 3.8$ in; $f = -6.0$ in; $M = -0.27$
b. $f = 3.1$ cm; $R = 6.2$ cm; $M = -1.0$
c. $f = -7.0$ cm; $q = -3.7$ cm; $M = -0.47$
d. $p = 1.67$ in; $f = 2.5$ in; $R = 5.0$ in
e. $f = 24.0$ in; $p = 4.8$ in; $M = 1.25$

9. 35.6 in **11.** 4.5 in

13. There is a lateral displacement of the light ray, but there is no change of the ray's angular direction after it passes through the plate. **A.** 22.7° **B.** 22.7° **C.** 37.0° **D.** 0.0°

15. **a.** $f = 12.0$ cm; $M = -1.0$ **b.** $q = \bar{1}0$ in; $M = -1.0$
c. $f = \bar{1}0$ in; $q = -15$ in **d.** $q = 5.3$ in; $M = 12.65 \times 10^{-3}$
e. $q = -5.45$ cm; $M = 0.545$ **f.** $f = -54$ in; $p = 110$ in

17. 25.4 cm, real, inverted image **19.** 4.7 dB/km **21.** -5.0 cm

23. 1.5 in

CHAPTER 28

Exercises

1. At sunset, sunlight has a much longer path through the earth's atmosphere than at midday. Much of the shorter wavelengths of light are scattered by small particles in the air. As you look at the sun you see the longer wavelength (red and orange) that has not been scattered.

5. Blue light

7. θ must be at the polarizing angle, θ_p. Light reflected from the first mirror is completely linearly polarized. When the polarized light strikes the second mirror the lateral component of the light wave is mostly reflected. A 90° rotation causes the polarized light to strike the second mirror as a vertical component and it is absorbed when it strikes the mirror.

9. The optical path is the product of the index of refraction and the distance traveled (op $= nd$). The speed of light varies inversely with the index of refraction for a fixed frequency. The number of wavelengths enclosed in the path is proportional to the optical path.

11.

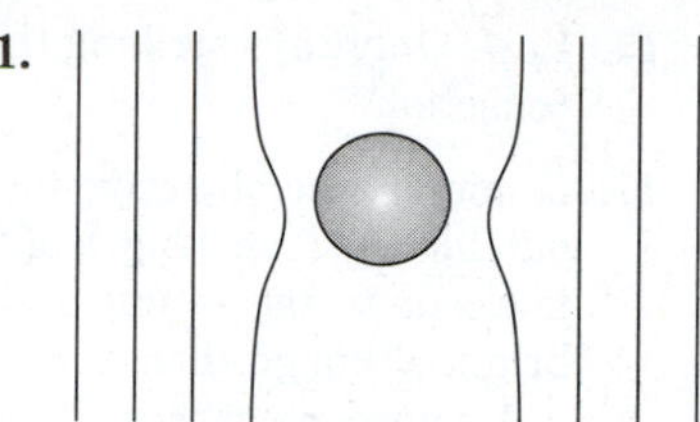

13. Colored rings will float down from the top of the bubble as gravity pulls material to the bottom. When the top of the bubble reaches a thickness of /4 of violet light, constructive interference will no longer occur. The top of the bubble will no longer reflect light of any color.

15. **A.** The spacing between the lines on the grating is the reciprocal of the grating constant. The grating with the smaller grating constant (B) has the wider spacing.
 B. Grating A will give the wider dispersion.

17. Optical telescopes work with wavelengths on the order of 10^7 m. Radio telescopes operate at wavelengths from 10^{-1} m to 10^{-2} m. The objective lens of a radio telescope needs to be 10^5 times larger to have the same resolution as the optical telescope.

19. Outer electrons in the atoms are excited only by frequencies of the light spectrum that just match the energy difference between allowed orbits ($\Delta E = hf$). These electrons move to outer orbits for a brief period. They then fall back to their original lower orbits emitting the frequencies corresponding to the same frequencies the electrons absorbed. This radiation is emitted in all directions. Only part of the radiation is in the original direction of the full spectrum.

21. **A.** Larger angle for A **B.** Larger angle for second order **C.** Larger angle for longer wavelength

Problems

1. 1.0×10^{-8} **3. A.** 60° **B.** $\frac{I}{I_0} = 0.38$ **5.** 0.38 **7.** 120°

9. 1.86 in **11.** 5 **13.** $80\bar{0}$ lines **15.** 3.5° **17.** 2.1×10^{-4} rad

19. 4.5 mm **21.** 3.49×10^{-14} J; gain **23.** $4\bar{0}00$ to $7\bar{0}00$ Å

27. $\theta_1 = 24.6°$; $\theta_2 = 56.4°$

CHAPTER 29

Exercises

1. Rubbing transfers electrons from one object to another. Opposite electrical charges hold the two objects together.

3. **A.** An opposite charge is induced on the balloon on the side nearest the ruler.
 B. Charge is transferred when the ruler and balloon touch by conduction. The balloon and ruler now have the same charge.
 C. The second balloon is first attracted to the first by induction. After they touch they have the same charge and are repelled.

5. B

7. (a) (b)

(c) (d)

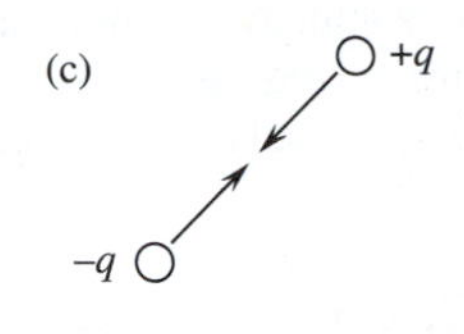

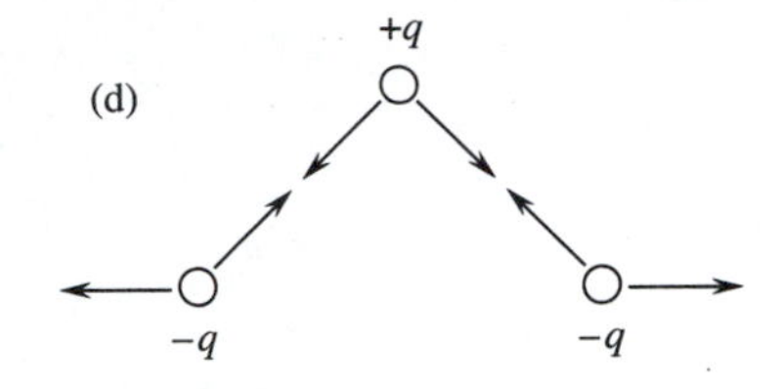

9. **A.** Yes, if the two charges are like charges of the same magnitude.
B. Yes, if for like charges of magnitude with a ratio of $q_1/q_2 = (r_{1b}/r_{b2})^2$.
C. No. The vertical components of field will not cancel for like charges. The horizontal components will not cancel for unlike charges.

13. **B.** The field lines are denser than the other choices.

17. The absolute potential will be the same everywhere on the surface of the sphere. It is an equipotential surface. **19. B** **21. C**

Problems

1. **a.** 2.7×10^{-2} N **b.** 4.6×10^{-7} N **c.** 1.8 m **d.** 1.2×10^{-6} C

3. 6.6×10^{-2} N

5. **A.** -2.0×10^{6} N/C **B.** Negative **C.** 1.1 N

7. 8.3×10^{5} N/C **9.** 3.0×10^{4} N/C $\angle$ 275° **11.** 620 (N · m²)/C

13. 270 (N · m²)/C **15.** 120 J **17.** 3.0×10^{3} V **19.** 1.2×10^{4} V/m

21. 2.5 nF **23.** 2.8×10^{3} acres, 4.4 mi² **25.** 1.2 nC/m

27. **A.** 6.0×10^{-10} s
B. 1.3×10^{15} m/s²
C. 7.8×10^{5} m/s **D.** 2.2°

CHAPTER 30

Exercises

1. Electron current is the actual flow of electrons through a conductor. The electrical current used in circuit calculations flows in the direction opposite the electron flow. Electrical current assumes a positive charge while electrons have a negative charge.

7. 5.0 Ω

9. E is not constant. The field is stronger at the position 0.5 cm than at 3.5 cm. This can be seen from the slopes of the graph at these points.

11. A 220-V motor requires only half the current of a 110-V motor to produce the same power. Thinner wires with a higher resistance can be used.

13. The toaster draws less current. The resistance of the heating element increases with temperature.

15. The filament is initially cool. The surge of electrical current causes rapid heating and expansion, resulting in thermal strains that can cause a weak point in the filament to fail.

17. **A.** More energy can be stored with increased plate area.
B. Less energy is stored as the plates are separated.
C. More energy is stored as the voltage across the plates increases.
D. More energy is stored by placing a material with a higher dielectric constant between the plates.

Problems

1. 0.56 C, 3.5×10^{18} electrons **3.** 1.24 h **5. A.** 2.3×10^{3} A/m^2 **B.** 2.8×10^{15} electrons/m^3 **C.** 2.5×10^{9} electrons

7. **A.** 37 Ω **B.** 1.5 V **C.** 0.250 A

9. **A.** 14.2 Ω **B.** 8.10×10^{4} A/m^2 **C.** 7.47 V/m **D.** 1.08×10^{4} $(\Omega \cdot m)^{-1}$ **E.** $9.21 \times 10^{-5}\ \Omega \cdot m$ **11.** 9.1 A **13. A.** 96 mA **B.** 56 mW **15.** 212 A **17.** -5.0×10^{-4} (°C)$^{-1}$, carbon **19.** 9.2×10^{-5} J **21.** 2.5

CHAPTER 31

Exercises

1. In DC current, electrons flow steadily in one direction. In AC current, electrons oscillate around a fixed position and energy is transferred by wave motion of the electrons.

3. **A.** 0.30 mW **B.** 2.2×10^{16} electrons **C.** 9.0×10^{-3} J

5. **A.** Series **B.** Parallel **C.** Parallel **D.** Series **E.** Both **F.** Both

7. **A.** The two capacitors have the same charge. **B.** A has a larger potential drop. **C.** Smaller for series connection **D.** Smaller

9. Connect in series. **11. A.** Smaller **B.** The same current passes through the resistors. **C.** Larger **13.** Connect in parallel.

17. A car battery has a small voltage, usually 12 V. Because the power supply has a low voltage, a large current must be drawn from the battery to turn over the motor. The terminal voltage decreases substantially as the large current passes through the battery's internal resistance.

19. Matched input impedance (effective resistance) will transfer the maximum antenna signal to the set's amplification circuits (see Figure 31–22).

21. Positively charged ions (cations) pile up on the positive pole of the battery. These cations repel other approaching cations. Only when electrons are supplied to the positive pole through an external circuit can the chemical reaction proceed.

23. If the Wheatstone bridge is not balanced then $I_1 \neq I_3$ and $I_2 \neq I_4$. Current will no longer drop out of the loop equations when we take a ratio. The potential drops are no longer equal on the two segments of the branches ($I_1 R_x \neq I_2 R_2$).

Problems

1. **A.** 2.0 μF **B.** 2.8 μF **C.** 0.67 μF

3. $Q_1 = Q_2 = 27.4\ \mu C$. $V_1 = 7.61\ V$; $V_2 = 4.42\ V$

5. **a.** 2.0 μF **b.** 3.3 μF **c.** 3.2 μF

7. **A.** $52\bar{0}\ \Omega$ **B.** $120\bar{0}\ \Omega$ **C.** $24\bar{0}\ \Omega$

9. $I = 1.0 \times 10^{-2}\ A$; $V_1 = 2.3\ V$; $V_2 = 3.7\ V$

11. **a.** 100 Ω **b.** 22 Ω **c.** 46 Ω

13. 146 mA, 11.0 V **15.** 0.047 Ω **17.** 9.3 V **19.** 12 Ω

21. **A.** 3.69 V **B.** 1.54 V

CHAPTER 32

Exercises

1. **a.** Out of page **b.** Into page
   c. No force **d.** Into page
   e. Up, to top of page **f.** Down, to bottom of page
   g. Down, to bottom of page **h.** To the right

3. **A.** The magnetic force is upward. The *E* field should point down, and plate 1 should have a plus charge.
   B. No. The magnetic force will be downward on a negative charge. Negative charges will be repelled by plate 2, which is also negative.

5. **A.** The electrons. The radius is proportional to mass. **B.** The protons **C.** Positive ions will spin counterclockwise; negative ions, clockwise. The clockwise orbits will be smaller. Choice B best fits the condition.

7. **A.** Increase **B.** No change **C.** No change **D.** Increase

9. **a.** Up, to top of page **b.** To the right **c.** Out of page

11. Coil A has a larger magnetic moment. Although it has only half as many turns as coil B, it has four times the area of B.

13. They have the same maximum flux since the cross-sectional areas are the same.

15. $1\ Wb = 1\ T \cdot m^2$; substitute $1T = 1\ N/(m \cdot A)$

17. C **19.** 46 m

Problems

1. **A.** $5.5 \times 10^{-13}\ N$ **B.** 0.12 T **C.** 0 N **D.** $2.0 \times 10^{-18}\ C$

3. 27.1 mT **5. A.** 7.1 cm **B.** 30 ns **C.** 2.4 cm **7.** 4.7 A

9. $2\bar{0}0\ N \cdot m$ **11.** 1.3 mT, or 13 G **13. A.** 8.7 mT, or 87 G
   B. $2.2 \times 10^{-6}\ Wb$

CHAPTER 33

Exercises

1. Lower right 3. **A.** L increases **B.** L decreases **C.** L increases **D.** No change

5. From A to B. Current will flow in a direction to resist the decrease of current and magnetic flux.

7. This is an example of mutual induction. The magnetic flux building up in the first coil is partly passing through the second coil.

9. **A.** Current would flow through the shorted circuit. Magnetic field energy would be converted into heat through the resistance of the toroid's windings.
   B. Current would flow indefinitely since there is no resistance.

11. Paramagnetic materials have atoms with permanent magnetic moments that react with an external magnetic field independently of neighboring atoms. Ferromagnetic materials also have permanent magnetic moments. The magnetic moments are strong enough to cause neighboring atoms to align with each other to form magnetic domains. At high temperatures the coupling between neighboring atoms in a ferromagnetic material breaks down, and the material becomes paramagnetic. The transition temperature is called the Curie temperature.

13. 1.2×10^{-5}

17. In the field shunt motor the field coils and armature are connected in parallel. The current through the field coils is constant at any motor speed. As the motor speeds up the current through the armature decreases. In the series motor the field coils and the armature are connected in series. At low speeds the current through the field coils is large, causing a large starting torque. As the motor speeds up the current through the field coils decreases.

19. The field coils of a field shunt motor carry a constant current, while the field windings of a series motor have a varying current. The current is particularly large when the motor speed is low. Heavier wire is needed to accommodate these larger low-speed currents.

Problems

1. 210 V 3. $\bar{1}0$ H 5. 81.2 A/s 7. 15 V 9. **A.** 6.7×10^5 J/m^3 **B.** 75 J 11. 120 T 13. 8.73:1 15. **A.** $11\bar{0}$ V **B.** 1.3 kW **C.** 0.90 17. **A.** 2.0 A **B.** 33.0 A **C.** 27 V **D.** 890 W **E.** 0.64

CHAPTER 34

Exercises

1. The split commutator is used to invert current of the rotating coil in the DC generator. If the current were not inverted it would move in the reverse direction during half of the cycle. In the AC generator we want the current to reverse so continuous sliding contacts are used.

3. The peak voltage is the maximum magnitude of the instantaneous volt-

age during a cycle. The rms voltage is the root mean square average voltage over a half cycle.

5. 2

7. The magnitude of the current in the resistor is also maximum. Current and voltage are in phase in a resistor.

9. For very low angular speeds X_L near zero, or $|Z| = R$. For high frequencies X_L is much larger than R, or $|Z| = |X_L|$.

11. The voltage across the resistor is out of phase with the voltage across the inductor. Vector addition must be used.

13. Voltage lags current in a capacitor.

15. The smaller air-filled capacitor would have a larger capacitive reactance.

Problems

1. 11 V **3.** 120 W **5.** 2.16 MV **7.** 13 Ω **9.** 64 V

11. A. $1\bar{0}^\circ$ **B.** 35° **C.** $7\bar{0}^\circ$ **13.** 39 W **15. A.** 660 V **B.** 5.1 kW

17. 1.0 μF **19.** 73 Ω **21.** 11 mH

CHAPTER 35

Exercises

1. The doubly ionized alpha particle will have twice the energy gain as the electron ($E = q\,V$).

3. 1. When atoms collide some of the energy can be absorbed by the electron. 2. The electron may absorb a photon.

11. The holes are created when electrons jump from the valance band to the conduction band.

13. Transparency indicates that the energy gap is larger than the energy of visible light photons. Otherwise electrons could jump the energy gap by absorbing photons. This energy range is larger than would be expected for conductors or semiconductors.

15. Acceptor states allow electrons with energies less than the energy gap to leave the valance band. This creates a surplus of holes. Donor states inject electrons into the conduction band without creating holes in the valance band. This gives a surplus of electrons.

17. The emitter is a charge carrier source. The base controls the number of electrons leaving the emitter. The collector increases the energy of electrons passing through the base by boosting their electric potential.

Problems

1. A. 14 eV **B.** 6.3×10^{18} eV **C.** 4.9 eV

3. A. 2.11×10^{-14} J **B.** 1.32×10^{5} eV

5. A. 5.01 eV **B.** 1.33 eV

7. 1.0×10^{16} electrons/m^3 **9.** 1/3 **11.** 3.3 ms **13.** 0.0091 eV

CHAPTER 36

Exercises

1. A. 23 B. 24 C. 96 D. 185 E. 205
3. A. 12 B. 13 C. 54 D. 109 E. 123
5. A. Exothermic B. Endothermic C. Exothermic
7. The phosphorus would have a larger initial activity than strontium for the same number of parent atoms. After a long period of time most of the parent phosphorus atoms would have been transformed. The phosphorus activity would then fall below the slower decaying strontium.
9. A. Decrease B. Decrease C. Increase D. Increase E. No change
11. The relative portions of daughter products compared with the number of parent atoms would indicate age. Another scheme is to compare the density of helium atoms in the rock created by alpha decay with the density of parent atoms.
13. Plutonium is chemically different from uranium, so a chemical means can be used for separation. Some physical means such as gas diffusion is needed to separate U 235 and U 238, which are chemically the same.
15. Far fewer. The neutron has no charge. Ions are formed mostly by direct collision between neutrons and electrons, while charged particles can ionize with the aid of coulomb forces.
17. 9.04×10^3 ions

Problems

1. A. 21.2 MeV/nucleon B. 8.56 MeV/nucleon C. 5.33 MeV/nucleon D. 7.56 MeV/nucleon E. 8.79 MeV/nucleon
3. 24 days 5. 5.0 s 7. 1.3 days 9. A. 2.3 mrem B. 23 mrem
11. 3.57 l

Index

NOTE: Entries with page numbers beginning with the letter A are references to Appendix A. Italicized entries indicate material other than text material on the referenced page.

Mathematical Relationships

Quadratic Equation

The solution to the quadratic equation in the form:

$$a X^2 + b X + c = 0$$

is:

$$X = \frac{-b \pm (b^2 - 4 a c)^{1/2}}{2a}$$

Basic Trigonometry

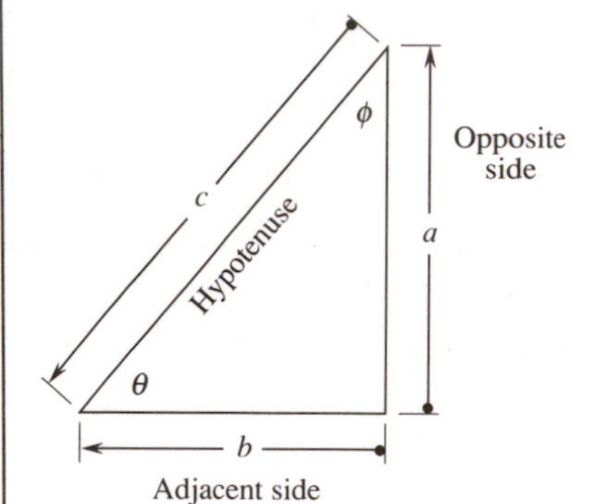

The sides of a right triangle.

$$\sin \theta = \frac{a}{c} = \cos \phi \qquad a^2 + b^2 = c^2$$

$$\cos \theta = \frac{b}{c} = \sin \phi \qquad \theta + \phi = 90°$$

$$\tan \theta = \frac{a}{b} \tan \phi = \frac{b}{a} \qquad \sin \theta = \cos (90° - \theta)$$

Small Angle Approximation

$\tan \theta = \sin \theta$, and $\sin \theta = \theta$, if the angle is in radians.

Rules of Exponents

$$X^n \cdot X^m = X^{n+m} \qquad \frac{X^n}{X^m} = X^{n-m}$$

$$X^{-n} = \frac{1}{X^n} \qquad X^0 = 1$$

$$(X^n)^m = X^{nm} \qquad \log X^n = n \log X$$